Lecture Notes on Data Engineering and Communications Technologies

Volume 15

Series editor

Fatos Xhafa, Technical University of Catalonia, Barcelona, Spain
e-mail: fatos@cs.upc.edu

The aim of the book series is to present cutting edge engineering approaches to data technologies and communications. It will publish latest advances on the engineering task of building and deploying distributed, scalable and reliable data infrastructures and communication systems.

The series will have a prominent applied focus on data technologies and communications with aim to promote the bridging from fundamental research on data science and networking to data engineering and communications that lead to industry products, business knowledge and standardisation.

More information about this series at http://www.springer.com/series/15362

S. Smys · Robert Bestak
Joy Iong-Zong Chen · Ivan Kotuliak
Editors

International Conference on Computer Networks and Communication Technologies

ICCNCT 2018

Editors
S. Smys
Department of CSE
RVS Technical Campus
Coimbatore, Tamil Nadu, India

Dr. Joy Iong-Zong Chen
Department of Electrical Engineering
Dayeh University
Taiwan, Taiwan

Robert Bestak
Department of Telecommunication Engineering
Czech Technical University in Prague
Czechia, Czech Republic

Ivan Kotuliak
Faculty of Informatics and Information Technology
Slovak University of Technology in Bratislava
Bratislava, Slovakia

ISSN 2367-4512 ISSN 2367-4520 (electronic)
Lecture Notes on Data Engineering and Communications Technologies
ISBN 978-981-10-8680-9 ISBN 978-981-10-8681-6 (eBook)
https://doi.org/10.1007/978-981-10-8681-6

Library of Congress Control Number: 2018948583

This Springer imprint is published by the registered company Springer Nature Singapore Pte Ltd.
The registered company address is: 152 Beach Road, #21-01/04 Gateway East, Singapore 189721, Singapore

I dedicate this to all the participants of the conference ICCNCT 2018

Preface

International Conference on Computer Networks and Inventive Communication Technologies (ICCNCT 2018) provided a forum for academic scientists, leading engineers, industry researchers, and scholar students to exchange and share their experiences and research results. ICCNCT 2018 aims to cover the recent advancements and trends in the area of computer networks and advanced communication technologies to facilitate knowledge sharing and networking interactions on emerging trends and new challenges.

ICCNCT 2018 tends to collect the latest research results and applications on computer networks and communication technologies. It includes a selection of 96 papers from 548 papers submitted to the conference from universities and industries all over the world. All of the accepted papers were subjected to strict peer-reviewing by 2–4 expert referees. The papers have been selected for this volume because of quality and relevance to the conference.

We would like to express our sincere appreciation to all authors for their contributions to this book. We would like to extend our thanks to all the referees for their constructive comments on all papers; especially, we would like to thank the organizing committee for their hard work. Finally, we would like to thank the Springer Publications for producing this volume.

Coimbatore, India — Dr. S. Smys
Czechia, Czech Republic — Dr. Robert Bestak
Taiwan, Taiwan — Dr. Joy Iong-Zong Chen
Bratislava, Slovakia — Dr. Ivan Kotuliak

Acknowledgements

We thank all the participants of ICCNCT 2018 and the respected session chairs for their useful suggestions.

We also thank all the conference committee members for their support.

Contents

About the Editors

Dr. S. Smys received his bachelor of engineering degree in electronics and communication engineering from Periyar University, India, in 2002 and his master of engineering degree in digital communication and networking from Anna University, India, in 2004. He received his Ph.D. degree on virtual structure constructions in wireless networks from Karunya University, Coimbatore, India, in 2012. He has 13 years of teaching experience and 8 years of research experience. He has written more than 50 publications and organized 14 international conferences. He is Associate Editor of the Journal of Computers and Electrical Engineering (published by Elsevier).

Prof. Robert Bestak obtained his Ph.D. degree in computer science from ENST Paris, France, in 2003 and M.Sc. degree in telecommunications from Czech Technical University (CTU) in Prague in 1999. Since 2004, he has been Assistant Professor in the Department of Telecommunication Engineering, Faculty of Electrical Engineering, CTU. He is the Czech representative in the IFIP TC6 Working Group. He has served in the steering and technical program committees of numerous IEEE/IFIP international conferences (networking, WMNC, NGMAST, etc.), and he is a member of the editorial board of several international journals (Electronic Commerce Research Journal, etc.). He has participated in several national, EU, and third-party research projects (FP7-ROCKET, FP7-TROPIC, etc.). His research interests include 5G networks, spectrum management, and big data in mobile networks.

Prof. Joy Iong-Zong Chen is a Chinese communications engineer and researcher and is Full-time Professor in the Department of Electrical Engineering, Dayeh University, Taiwan. He obtained his Ph.D. degree from the National Defense University, Chung Cheng Institute of Technology, Taiwan, in 2002. He has written over 70 research publications and is a member of the Institute of Electrical and Electronics Engineers and the Institute of Electronics, Information and Communications Engineers. His achievements include the development of new patents for cellular communication systems. He was the recipient of the 2006 International Association of Engineers Best Paper Award.

Dr. Ing. Ivan Kotuliak, Ph.D. is currently associated with Slovenská technická univerzita v Bratislave, Bratislava, Slovakia, as Vice-Dean for International Relations and the Faculty of Informatics and Information Technologies. He has written 57 research publications. He has over 17 years of research experience in computer engineering. His areas of interest are computer engineering, communication engineering, wireless communications, computer networks, wireless ad hoc networks, and wireless LAN.

A Novel on Biometric Parameter's Fusion on Drowsiness Detection Using Machine Learning

V. B. Hemadri, Padmavati Gundgurti, G. Dharani Chowdary and Korla Deepika

Abstract The operator driving vehicle in night has become a major problem nowadays. The largest number of accidents in the world is due to drowsiness. To overcome this problem, we have developed a machine for a longer period which detects drowsiness and alert the operator. Early detection of fatigued state has become important to develop a detection system. According to the previous work, we found a lot of issues in detecting drowsiness when wearing spectacles and in dark and light condition. In our research paper, we have overcome these issues to detect drowsiness based on the fusion of visual parameters like face detection, eye detection, and yawning in all conditions and wearing spectacles.

Keywords Face detection · Eye detection · Mouth detection · SVM function

1 Introduction

In recent days, increase in the number of accidents is becoming the major threat for the society. The person with drowsiness may lead his life and others live in danger; most of the accidents nowadays are due to drowsiness.

Machines are developed for the safety of the driver based on the recent computer vision and new technologies. Intelligent systems in the vehicle will be totally automatic with the use of intelligent control systems in the present system. The main aim

V. B. Hemadri
SDMCET, Dharwad 580002, Karnataka, India
e-mail: vidya_gouri@yahoo.com

P. Gundgurti · G. Dharani Chowdary · K. Deepika (✉)
BVRITH, Hyderabad 500090, Telagana, India
e-mail: korladeepika@gmail.com

P. Gundgurti
e-mail: padmavati.eg@gmail.com

G. Dharani Chowdary
e-mail: dharani1497@gmail.com

S. Smys et al. (eds.), *International Conference on Computer Networks and Communication Technologies*, Lecture Notes on Data Engineering and Communications Technologies 15, https://doi.org/10.1007/978-981-10-8681-6_1

is to improve the driver or any person safety; these systems will not only automate the detection of traffic signal and lanes but also for measuring the operators' behavior. The major behavior of the operator is to detect the driver drowsiness.

Recent reports say that a leading cause of fatal or injury-causing accidents is due to person with a fatigue level which is due to the working for a longer period and boring environment which often causes lack of concentration in an operator and leads to accidents. Other than fatigue, it might be due to states like happy, depression, short temper, and disturbance which lead to accident.

India is the highest for road accidents around 800 due to drowsiness according to recent report of Team-BHP.com. According to this, driver should take break and sleep for 7 h or need to stop the vehicle, take break, and continue the journey.

2 Literature Survey

In [1], the authors proposed a very good system where it is of low cost using sensors. The parameters used for detecting drowsiness are PERCLOS, eye closure duration, blink, face position, and fixed eye gaze. The machine is fully automatic and detected for real time. It works only without wearing spectacles but fails wearing spectacles.

In [2], the authors proposed an algorithm where it fails to detect because of the use of low infrared web camera. The driver is the main operator where the light illumination should be proper to extract the features of the operator and send to the database to match the frames in dark intensity and no clear frames are formed.

In [3], the authors implemented the identification of fatigue on head position and geometrical features of mouth. Test was conducted based on the example of 50 video frames and experiments that head movement contributes about 8% and yawning contributes about 49%. The result fails to identify for very dark condition.

In [4], the authors introduced the dependable system for operator's drowsiness detection. The analyzed data is required from real traffic. The information are pre-processed according to assumptions about driver's behavior, and it is sent to the frequency state by means of orthogonal transform. The data is recorded by the bus system operator. Features are extracted from the operator.

In [5], the author presents the author reviews of the different procedures to resolve driver drowsiness. The author concluded that by the implementation of hybrid drowsiness detection system, it combines non-intrusive physiological measures with other system to get accurate and detect the drowsiness of the driver. They also use ECG and other physical measures for the detection of fatigue of driver.

In [6], the author reviews the different procedures to resolve driver drowsiness. The author concluded that by the implementation of hybrid drowsiness detection system, it combines non-intrusive physiological measures with other system to get accurate and detect the drowsiness of the driver. They also use ECG and other physical measures for the detection of fatigue of driver.

In [7], the author presents a novel Non-intrusive Intelligent Driver Assistance and Safety System (Ni-DASS) for assessing within motors. In this challenge to reduce accidents by focusing on creating Advanced Driver Assistance Systems (ADAS) which is able to identify motor, operator, and surrounding conditions to use the information, it uses an onboard CCD camera to observe the driver's face. A template matching approach is used to evaluate the driver's eye-gaze pattern with a set of eye-sign frames of the operator looking at different sides rather than concentrating on driving. The results indicate that the proposed technique could be useful in situations where it fails in low-resolution estimates of the driver's drowsiness level.

In [8], the authors presented a brain system which uses electroencephalogram signals for the identification of fatigue. The system controls the human muscle and movements of the human communication with the other person. The system is used to detect the activities of the brain. The result is calculated by placing the electrodes on the scalp which contains small metal. The main intention is to find the drowsiness of the driver but fails because every time the driver cannot place the electrodes on the body; it irritates and it creates irritation and rashes.

In reference to the [9], the authors implemented a non-disturbing model system for actual time monitor a driver's drowsiness. The parameters include eyelid movement, gaze, and face orientation. The main apparatus of the considered system consists of a hardware system for actual time acquirement of video frames of the operator and algorithms and their software implementations for real-time eye tracking, eyelid movement parameters computation, face pose discrimination, and gaze evaluation. It was tested under several conditions for the illumination of different genders and ages. It fails for gaze movement and position of the head.

In reference to the [10], the authors say that drowsiness is detected with experimental lab setup of eye movement and yawning of the lower jaw is selected for test cases where the upper portion of mouth is fixed; it is not going to affect. The results fail to detect in very dark condition and more lightening.

3 Proposed Method

Our research paper focuses on the design and improvement of drowsiness detection system based on the eye detection, head movement, and face detection on human fatigues under different situations. The proposed algorithm to detect the fatigue is described below:

Procedure fatigue_detection

Step 1 Record the webcam online.
Step 2 Process the video clip into frame by frame.
Step 3 Detect face using HaarCascade classifier.
Step 4 Detect eye using HaarCascade classifier.
Step 5 Detect mouth using HaarCascade classifier.
Step 6 Extract the features of eye and mouth parameters like head position and yawning.

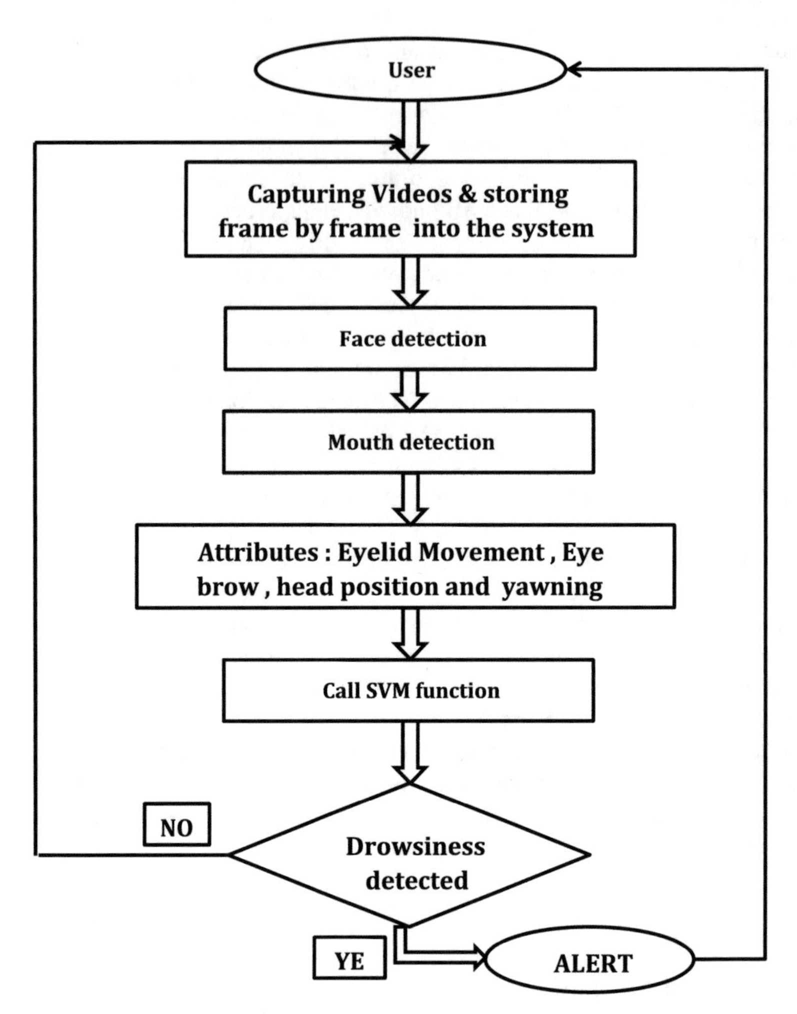

Fig. 1 System flow diagram

Step 7 Support vector machine is called to find whether drowsy or not.
End of procedure

In the first step, the videos are captured from webcam online completely and then stored frame by frame. Face is detected using Viola–Jones classifier, then using HaarCascade classifier eye, mouth is located with a rectangle drawn around them. Then, the eye opening and closing, and mouth opening and closing are calculated for each frame and counted. The mouth opening and closing and head positions are given as input for machine learning technique called machine and Harr classifier to identify the level of fatigue. The drowsiness detection is shown in Fig. 1.

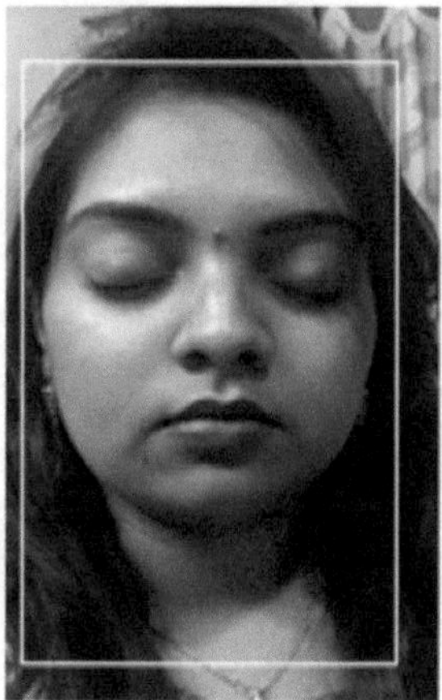

Fig. 2 Detected faces marked with rectangle box

Implementation The four major modules are involved in the implementation of the proposed system as follows:

a. Online data collection,
b. Face detection,
c. Eye detection, and
d. Mouth detection.

4 Online Data Collection

The major step in the implementation of drowsiness detection system is online data collection from the web camera. The camera is placed in front of the driver by which the video is recorded continuously.

5 Face Detection

After the videos being framed, each frame is sent for the face detection process. The standard HaarCascade classifier of file *haarcascade_frontalface* is used for frontal face detection. The detected face region is marked with a rectangle box as shown in Fig. 2.

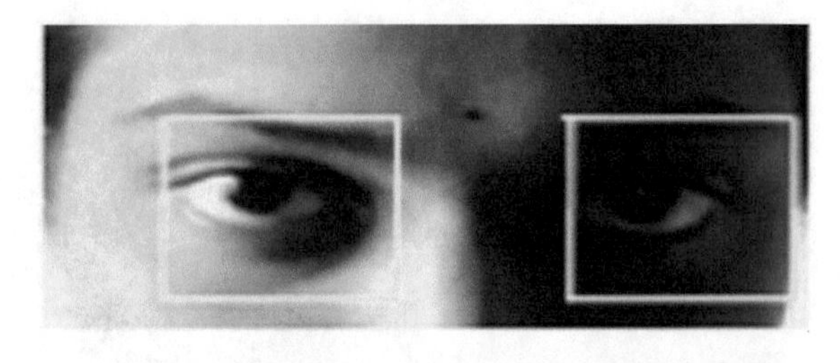

Fig. 3 Detected eyes marked with rectangle box

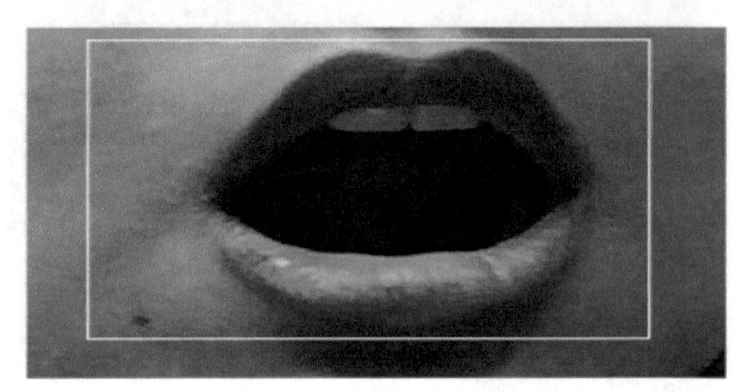

Fig. 4 Detected mouth marked with rectangle box

6 Eye Detection

After successful online data collection, the face and eyes are detected and matched with stored data, and alarm sounds loud, if 50% eyes close. We use HaarCascade classifier to locate the region of eye. Eye is detected and the rectangle box is drawn for the eye region of the face as shown in Fig. 3.

7 Mouth Detection

After successfully detecting the face and locating the eye in each frame, the frame is processed for locating mouth region. As the region of interest, mouth is detected as shown in Fig. 4. Yawning is 100% if the voice message is given.

8 Experimental Results

Almost 80 videos were collected of different attributes like with glasses, different hairstyle conditions, with beard and without beard, head changing positions in different environmental circumstances, etc.

9 Conclusion

Our research paper describes how the accidents are going on due to drowsiness and what are the major steps to be taken throughout India, because ours is the largest in total number of road accidents. We developed a system using an algorithm projected and developed which can capture the online video which detects the fatigue. The parameters of face detection, eye detection, and yawning are used to check the fatigue level. The process is done under different circumstances like hairstyle, wearing glass or not wearing glasses, with beard or without, etc. The algorithm works fine till now and gives 95% result. If drowsiness detected, an alarm that sounds loudly is provided.

Acknowledgements This project was carried out under Research Promotion Scheme grant from All India Council for Technical Education (AICTE), project Ref. No: 8023/RID/RPS-114(Pvt)/2011-12. Authors wish to thank AICTE, New Delhi.
Consent of all participants was taken.

References

1. Reddy, B.P.G., Vudasreenivasarao, Mohan, B.V., Srinivas, P.C.: A low cost real time embedded control system design using infrared signal processing with application to vehicle accident prevention. Comput. Inf. Syst. Dev. Inf. **3**(4) (2012)
2. Khan, M.F., Aadil, F.: Efficient car alarming system for fatigue detection during driving. Int. J. Innov. Manag. Technol. **3**(4) (2012)
3. Gundgurti, P., Patil, B., Hemadri, V.B., Kulkarni, U.P.: Experimental study on assessment on impact of biometric parameters on drowsiness based on yawning & head movement using support vector machine. IJCSMR **2**(5) (2013). ISSN: 2278-733X
4. Haupt, D., Honzik, P., Raso, P., Hyncica, O.: Steering wheel motion analysis for detection of the driver's drowsiness. Math. Models Methods Mod. Sci. ISBN: 978-1-61804-055-8
5. You, C.-W., Lane, N.D., Chen, F., Wang, R., Chen, Z., Bao Martha Montes-de-Oca, T.J., Cheng, Y., Lin, M., Torresani, L., Andrew, T.: CarSafe App: alerting drowsy and distracted drivers using dual cameras on smartphones. In: Campbell MobiSys'13, 25–28 June 2013, Taipei, Taiwan
6. Sahayadhas, A., Sundaraj, K., Murugappan AI-Rehab Research Group, Universiti Malaysia Perlis (UniMAP).: Detecting driver drowsiness based on sensors: a review. *Malaysia; Sensors* 16937–16953
7. Hafizah, N., Zaid, M., Maguid, M.A., Soliman, A.H.: Eye Gesture Analysis with Head Movement for Advanced Driver Assistance Systems. World Academy of Science, Engineering and Technology (2012)
8. Rajendra Kumar, G.., Raju, S.V.P., Santhosh Kumar, D.: Classification of Eeg signals for drowsiness detection in brain and computer interface. GESJ Comput. Sci. Telecommun. **4**(36) (2012). ISSN: 1512-1232
9. Ji, Q., Yang, X.: Real-time eye, gaze, and face pose tracking for monitoring driver vigilance. Real-Time Imaging **8**, 357–3177 (2002)
10. Vidyagouri, B.H., Kulkarni, U.P.: Detection of drowsiness using fusion of yawning and eyelid movements. In: ICAC3 2013, CCIS 361, pp. 583–594. Springer (2013)
11. Patil, B., Gundgurti, P., Hemadri, V.B., Kulkarni, U.P.: Experimental study on assessment on impact of biometric parameters on drowsiness detection. In: Proceedings of International Conference of Computer Science and Information Engineering (ICCIE), 24 May 2013, pp. 1–6. ISBN: 978-93-5104-130-6 and International Journal of Advanced Trends in Computer Science and Engineering (IJATCSE), vol. 2, no. 3, pp. 01–06 (2013). ISSN 2278-3091

V. B. Hemadri working as an Assistant Professor in SDMCET, Dharwad, Karnataka, 580002. Her research area includes machine learning and SVM detection approaches.

Padmavati Gundgurti studying BE in BVRITH, Telagana, Hyderabad, 500090. Her research area includes machine learning and SVM detection approaches.

G. Dharani Chowdary studying BE in BVRITH, Telagana, Hyderabad, 500090. Her research area includes machine learning and SVM detection approaches.

Korla Deepika studying BE in BVRITH, Telagana, Hyderabad, 500090. Her research area includes machine learning and SVM detection approaches.

An IoT-Based Smart Classroom

Chinju Paul, Amal Ganesh and C. Sunitha

Abstract Rapid development of automation technology makes people life very simple and easy. In today's world, all of them depend only on automatic systems over manual system. An IoT-based smart classroom system mainly deals with the automation of electronic appliances in a classroom based on the Internet of Things (IoT) protocol called MQTT. The system architecture is comprised of several wireless nodes, a middleware, and user interface. All the wireless nodes communicate over dedicated or existing network with the middleware. This communication is based on Message Queue Telemetry Transport (MQTT) connectivity protocol which is designed for Internet of things. The MQTT protocol uses publish/subscribe-based messaging on the top of TCP/IP protocol. Through the user interface section, a user can interact with middleware of the system. The interaction is done by recognizing the command of the user through his speech. Mainly, secret commands are used for the interaction with the middleware. Raspberry Pi is the backbone of the system. It operates as a middleware, in the system architecture. Wireless nodes used in this system are called as Node MCU, and this Node MCU is placed in each classroom. After the execution of user interface section, the control is forwarded to the middleware which is placed in the staffroom. Finally, based on the user secret commands, the automation of appliances of each classroom will be done through Node MCU resulting in classroom automation.

Keywords Message queue telemetry transport (MQTT) · Internet of things (IoT)

C. Paul (✉) · A. Ganesh · C. Sunitha
Department of CSE, Vidya Academy of Science and Technology, P.O. Thalakottukara, Thrissur, Kerala, India
e-mail: chinjupaul123@gmail.com

A. Ganesh
e-mail: amal.ganesh@vidyaacademy.ac.in

C. Sunitha
e-mail: sunitha@vidyaacademy.ac.in

S. Smys et al. (eds.), *International Conference on Computer Networks and Communication Technologies*, Lecture Notes on Data Engineering and Communications Technologies 15, https://doi.org/10.1007/978-981-10-8681-6_2

1 Introduction

Automation is a method of operating a process using electronic devices with a view to reduce human effort. The need for developing an automation system in an office or at home is increasing rapidly. Automation makes not only an efficient system but also results in reducing the energy wastage. IoT is a platform to connect people and things at anytime, anyplace, and with anyone. Automation is the most relevant application of IoT. It monitors the energy consumption and controls the environment in buildings, schools, offices, and museums using different types of sensors and actuators that control lights, temperature, and humidity. In today's world, energy wastage is an important issue. The major reason for this is people are forgetting to turn off the appliances. Especially in educational institutions, also students are busy with their studies and nobody is bothered about the energy wastage. So in order to avoid such conditions and reduce the human interventions, automation systems have a great role. Thus, in order to overcome the energy wastage and save money, a system is introduced here. The system mentioned here is a portable system, and by using this anyone can be controlled from anywhere in the world using IoT protocol. The general architecture for the automation system contains the following components:

- **User interface**: It is used to give order or control the electrical equipments.
- **Mode of transmission**: Either wired or wireless connection is used to transmit the control. The protocols which are used for data transfer and control are Ethernet, MQTT, etc.
- **Central controller**: It is a hardware component which acts as an intermediate between user interface and mode of transmission. Sometimes, all the electrical equipments are connected to this central controller also, e.g., Raspberry Pi, Node MCU, etc.
- **Electronic devices**: All the electrical equipments are compatible with the structure.

So, this paper mainly discusses how to automate electrical appliances in a classroom using an architecture mentioned below. Here, we apply an IoT-based protocol called Message Queuing Telemetry Transport (MQTT) and to incorporate a speech recognition service from Amazon (ALEXA) to control various electrical appliances in classroom such as fan, light, etc. This work is complete on its own in remotely and automatically switching on and off of any electrical appliances in a classroom. The system aims to automate IoT and to control various electrical appliances such as fan, light, etc., through speech and saves energy and money.

2 Related Works

In paper [1], the authors point out an efficient implementation using IoT for monitoring and automation system. It aims at controlling electrical equipments in home via smartphone using Wi-Fi as a communication protocol and Arduino Uno as a central controller. The user here will move directly with a portable system through a web-based interface over the web, whereas home appliances are remotely controlled through a website. This system also provides a full smart environment condition and monitored by various sensors for providing necessary data to automatic detection and resolution of any problem in the devices. The paper [2] consists of variety of sensors in the system architecture. Using Wi-Fi module, the Intel Galileo connects to the Internet and after this connection it will start reading the parameters of sensors. Then, set the threshold levels for each sensor. Data from the sensor are sent to the web server and stored in the cloud. These data can be analyzed at anytime from anywhere. If the sensor parameters are greater than the threshold level, then the respective alarm will be raised and the required action will be performed for the control of parameters. This model monitors the temperature, gas leakage, and motion in the house. The temperature and the motion detection are stored in a cloud. When the temperature exceeds the threshold level, the cooler will automatically turn on and when the temperature comes to control it will turn off. If there is a leakage of gas in the house, then alarm is raised giving the alert sound. The required lights are turned on/off automatically by detecting the light outside the house. The user can also monitor the electric appliances through the Internet. By simply typing the IP address of the web server, the lights or any electrical appliances in the home are turned off remotely. In paper [3], a home automation architecture based on a remote password operated appliances is mentioned. The system reads the data from Bluetooth module, initializes the LCD and UART protocol, and displays the status of the electrical loads on LCD. The system mainly uses two graphical user interfaces. The status of the appliances can be known by using this interface. Any changes in the status of the appliances will give an immediate intimation by showing it in GUI. The window GUI will act as a server to transmit any data to and from the smartphone. If there is any failure, then connection can be reestablished using USB cable. The user can monitor and control the devices from any remote location at any time using IoT. In paper [4], the authors have proposed a protocol standard for smart homes called Home Automation Device Protocol (HADP). This system aims for the capacity of home automation devices across different platforms. The IFTTT (IF This Then That) service used here to define a set of device communication protocols and actions are combined to generate and manage interactions through a central node. The system demands less power consumption, and bandwidth requirements are done using the minimum data packets to trigger an action on a home automation device.

3 System Overview

The system architecture (Fig. 1) is comprised of several wireless nodes, a middleware, and user interface. All the wireless nodes communicate over dedicated or existing network with the middleware. This communication is based on Message Queue Telemetry Transport (MQTT) connectivity protocol which is designed for Internet of things. The MQTT protocol uses publish/subscribe-based messaging on the top

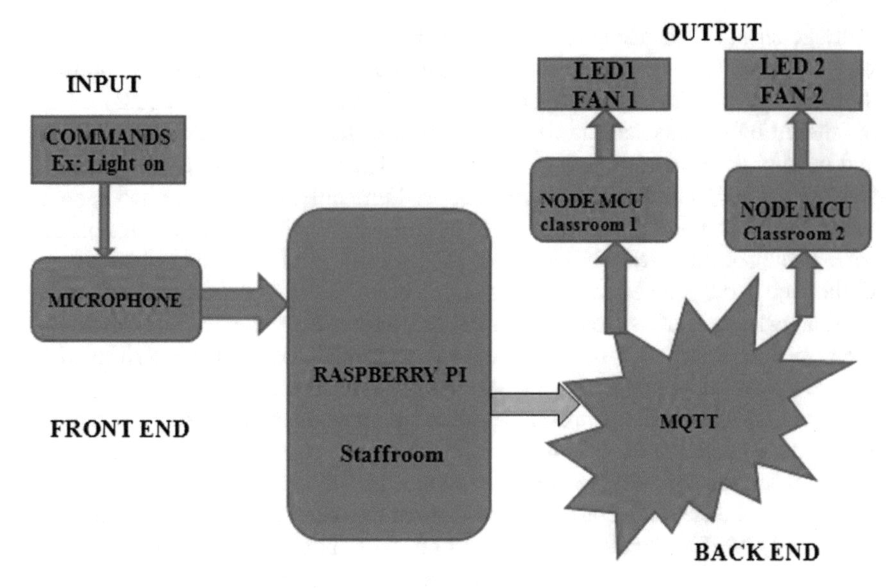

Fig. 1 Architecture

of TCP/IP protocol. Through the user interface section, a user can interact with middleware of the system. The interaction is done by recognizing the command of the user through his speech. Mainly secret commands are used for the interaction with the middleware. Raspberry Pi is the backbone of the system. It operates as a middleware in the system architecture. Wireless nodes used here are called as Node MCU and this Node MCU is placed in classroom. After the execution of user interface section, the control is forward to the middleware which is placed in the staffroom or office room and finally based on the user secret command, the automation of appliances of each classroom room will be done through Node MCU.

3.1 MQTT Architecture

MQTT stands for message queuing telemetry transport. The major features of MQTT protocol are low bandwidth and high latency. It is a most suitable protocol for machine-to-machine (M2M) communication. Basically, there are three components in MQTT: publisher, subscriber, and broker. The process of receiving and publishing the data is very much secure and accurate. Whenever the user wants to check or go through any data, it sends the request to broker and upon receiving the request it sends to the publisher, and it responds to the requests and sends the data that is requested by the subscriber and hence publishes the data; in overall process, the communication is secure and up to the topic of interest. MQTT broker acts like a filter allowing only those data which are requested, thereby saving the flow of ambiguous data.

3.2 Implementation

The system mainly contains two modules. The module 1 is speech recognition and Raspberry Pi. In this phase, there are two stages. The stage 1 is called IFTTT account creation, and stage 2 is called working of speech recognition system and Raspberry Pi placed at staffroom. For the working of module 1, we require a microphone, USB sound dongle card, Raspberry Pi, power bank or power supply to Raspberry Pi, high-speed Wi-Fi for Internet, etc. At first, user turns on the Wi-Fi and gives a power supply to Raspberry Pi. Then, user connects the USB dongle adapter to the Raspberry Pi to receive the speech from microphone. Before that, user first creates an IFTTT account and creates the triggers to turn on/off the appliances through MQTT server called Adafruit IO. Then, user just gives commands to the microphone, and it converts the speech into text messages by Alexa voice service provided by Amazon. While using IFTTT service, if a command to turn on a device is given (IFT), then it (TT) forwards to Node MCU which is placed in the corresponding classroom through Adafruit IO MQTT server. The module 2 of the system is the Node MCU connected with electrical appliances. This phase also contains two stages: stage 1 is called Adafruit IO (MQTT server) account creation, and stage 2 is called working of Node MCU connected with electrical appliances through relay which is placed at each classroom. Before receiving the output from the Adafruit MQTT server, user first creates an Adafruit IO account and adds all the electrical appliances in the dashboard. After the IFT (if this) section of the IFTTT service in the Raspberry Pi, the "THEN THAT(TT)" section of IFTTT is forward to Node MCU through Adafruit IO MQTT server, and corresponding device will turn on/off. So for this module 2, we require Node MCU, relay, high-speed Wi-Fi for Internet and electrical appliances like fan, bulb, projector, etc., and give power supply to Node MCU and relay.

4 Results

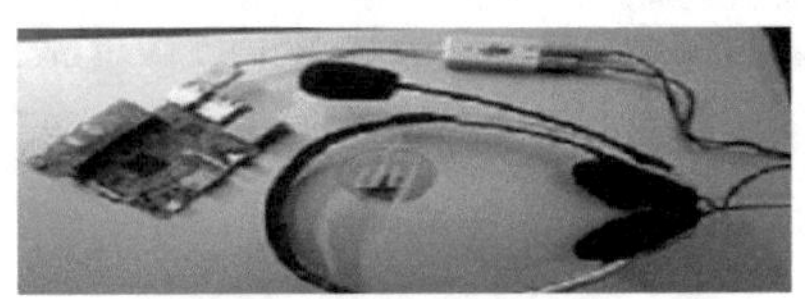

Image of Module 1

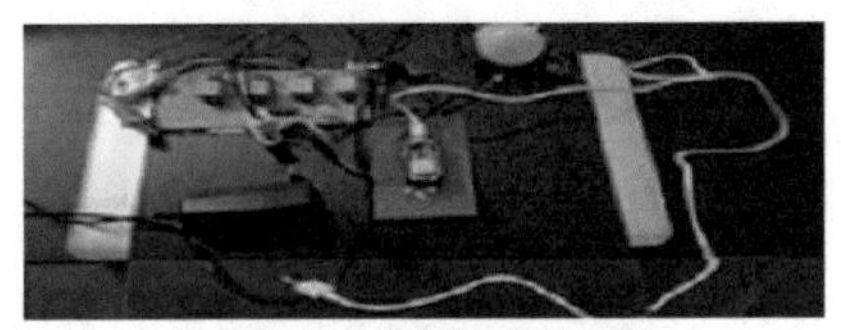

Image of Module 2

5 Conclusion

Thus, the system can automate the electrical appliances in a classroom using IoT from anywhere and control various electrical appliances such as fan, light, projector, etc. through speech. Through this system, we can save energy and money. The architecture which specified here reduces all the limitations of existing system architectures and capable to give a high-speed multitasking and can control the electrical equipments from anywhere at any time, and a friendly interface for all type of users. This IoT-based automation architecture is a challenge to other architectures because of its low-cost capacity compared to others.

References

1. Gondaliya, T.: A survey on an efficient IOT based smart home. Int. J. Rev. Electron. Commun. Eng. **4**(1) (2016)
2. Vinay sagar, K.N., Kusuma, S.M.: Home automation using internet of things. Int. Res. J. Eng. Technol. (IRJET) **02**(03) (2015). e-ISSN: 2395-0056
3. Pawar, P.N., Ramachandran, S., Singh, N.P., Wagh, V.V.: A survey on internet of things based home automation system. Int. J. Innovative Res. Comput. Commun. Eng. **4**(1) (2016)
4. Gonnot, T., Yi, W.-J., Monsef, E., Saniie, J.: Home automation device protocol (HADP): A protocol standard for unified device interactions. Adv. Internet Things **5**, 27–38 (2015)

Chinju Paul studying M.Tech CSE in Vidya Academy of Science and Technology, P.O. Thalakottukara, Thrissur, Kerela, India. Her research area includes Automation and IoT.

Amal Ganesh working as an Assistant Professor in the Department of CSE at Vidya Academy of Science and Technology, P.O. Thalakottukara, Thrissur, Kerala, India. Her research area includes Automation and IoT.

C. Sunitha as an Associate Professor Department of CSE in Vidya Academy of Science and Technology, P.O. Thalakottukara, Thrissur, Kerela, India. Her research area includes Automation and IoT.

Internet and Web Applications: Digital Repository

R. Nayana and A. Bharathi Malakreddy

Abstract Digital repository is a collection of online resources, in order to make web-based repository accessible, we are developing a repository which describes information resources about all the seminar and projects of the students in an educational institutes available on the users on the browser. In this repository it is sectioned into all the departments in an institution, stores the details of the student's seminars and project reports; hence it would be easier to retrieve the information available on the college website. Basically students and professors do not have time to go to the library or to their respective department to check out every citation. We need a system which deals with both the information overload problem, and difficulties in obtaining copies of the published works. We are incorporating few features in this project to meet the needs using eclipse IDE and tomcat server.

Keywords Digital repository · Internet · Web applications

1 Introduction

In today's technological generation, Internet plays a cost-effective source of medium. The business requires to store and make the data available at a faster rate in effective means of medium, also to process the data and make it presentable to the users. Similarly, Internet has been of great importance to education institute in various ways. Through this source of medium we can develop web application which acts as a front end with usage of internet to communicate with end users, likewise various application has been developed in an educational institute such as college websites and many more. Using these technologies we can develop a digital repository for educational institution.

R. Nayana (✉) · A. Bharathi Malakreddy
B.M.S. Institute of Technology and Management, Bengaluru, Karnataka, India
e-mail: nayanar2g@gmail.com

S. Smys et al. (eds.), *International Conference on Computer Networks and Communication Technologies*, Lecture Notes on Data Engineering and Communications Technologies 15, https://doi.org/10.1007/978-981-10-8681-6_3

The university provide certain services to members for the management of materials which are digitalized created by the institution and its students is viewed as institutional repository.

Few objectives of an institutional repository are to offer open access to research output by self-archiving it, to globally develop visibility for an institution's search, and to preserve institutional digital assets, such as a student's seminar and project technical reports.

The remainder of the paper is structured as follows Sect. 2 presents the literature survey, Sect. 3 describes the benefits of having digital repository, Sect. 4 shows the design styled for development of the application, Sect. 5 deals with the implementation of the application, Sect. 6 shows the results and discussion, Sect. 7 summarize the whole idea behind building a repository.

2 Literature Survey

Stephen Abrams, John Kratz in their work discusses how data sharing made easy at their university, California [1]. Stephen Abrams, Andrea Goethal have shown in their study about collaborative collection of web archiving [2].

Applications based on web are usually combination of server-side scripts and client side script to handle storage and retrieve information, and to present information to users respectively. The web application allow users to create, share and store the information or data regardless of location or device. Web application is developed in languages supported by browser such javaScript and HTML as languages like these rely on the browser to render the program executable [3].

One of widely used open source web server container is Apache Tomcat. A java servlet program provides the server capabilities and it could respond to all kinds of request for the application hosted on the web servers [4, 5].

3 Traditional Versus Proposed System

3.1 Traditional Overview of a Repository

A oldschool repository consists of the following objectives:

- Storage and preservation of items physically such as technical seminar and project reports is emphasized.
- High-level cataloging indexing by using text rather than one of detail, e.g., student name, department year.
- Physical proximity of related materials, e.g., students report from particular year are browsed.

- Every information has to preserved in on elocation physically in a wearhouse, e.g.; students/faculties must travel to a particular location where all the reports are placed to learn what is there and make use of it.

3.2 *Proposed Overview of Repository*

A digital institutional repository differs from the traditional repository by following ways:

- Digitized materials can be accessed wherever they may be located, by eliminating the need to have or preserve a physical item.
- Individual words or parameter can be used to search file or catlog.
- At any place, any time user can have access to their file electronically provide with the Internet.

4 Design

System design is the solution to for the creation of new system. It provides the understanding and procedural details necessary for implementing the system. Three forms of design are, a data design, an architectural design, and a procedural design. The data design transforms the information domain model created during the analysis into the data structured that will be required to implement the software (Fig. 1).

Users register on to the site in order to upload their document on to database or to search for any document. Only a valid user will be able to access the information. To search a particular report stored in the cloud we need to give any keyword related to the file you are searching in the search field. The entire file with matched keyword is displayed on the page and we can check for the details of the students who owns

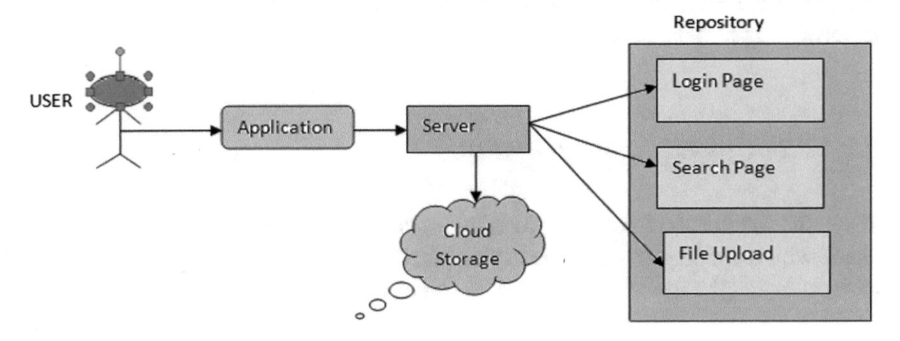

Fig. 1 System design

the file and one can download the file form the page. A registered student can upload their report onto the cloud.

During the process of uploading the program is designed to check for two validations. First, Checks for title match. If the file has the same title to the file which was uploaded by a student in the past it rejects the file and the file is not uploaded. Second, Checks for content match, when the file is submitted for uploading program it takes the abstract of the file and check if the keywords of the file matches with previously uploaded file, if the abstract matches the file is rejected and file is not uploaded onto the cloud. Also we have given a tab when clicked opens up free online plagiarism checker that check the plagiarism of the file which is ready to upload to the site.

5 Implementation

5.1 Providing User Credentials for Login

The purpose of providing credentials to system is to ensure that only authorized users gain access to facilities or data. Entering a password into a website or using a key to unlock a door both represent single-tier authentication systems, since a single factor is required to pass through the security.

Once the user is given the credential can login into the website. Unauthorized user will not be able login into the site.

5.2 Search Implementation

In this module, when a user enters a keyword related to his file on the search tab the search engine search for the corresponding file and give the resultant output to the user. Search engines useful in an academic setting for finding and accessing paper in academic journals, repositories.

5.3 File Upload Implementation

When the file is submitted for uploading program takes the abstract of the file and check with if the keywords of the file matches with previously uploaded file, if the abstract match the file is rejected and file is not uploaded onto the cloud.

6 Results and Discussion

After evaluating the proposed solution we found the following results. The web-based application shows the following results as per the requirements stated in the proposed solution (Figs. 2 and 3).

In Fig. 4a, when the user trys to upload new file, the file gets uploaded if there is no similar file in the database as shown in Fig. 4b, when the user uploads similar as perviously uploaded file the paper gets rejected as shown in Fig. 4c.

Fig. 2 Search tab to the search the existing file with a keyword

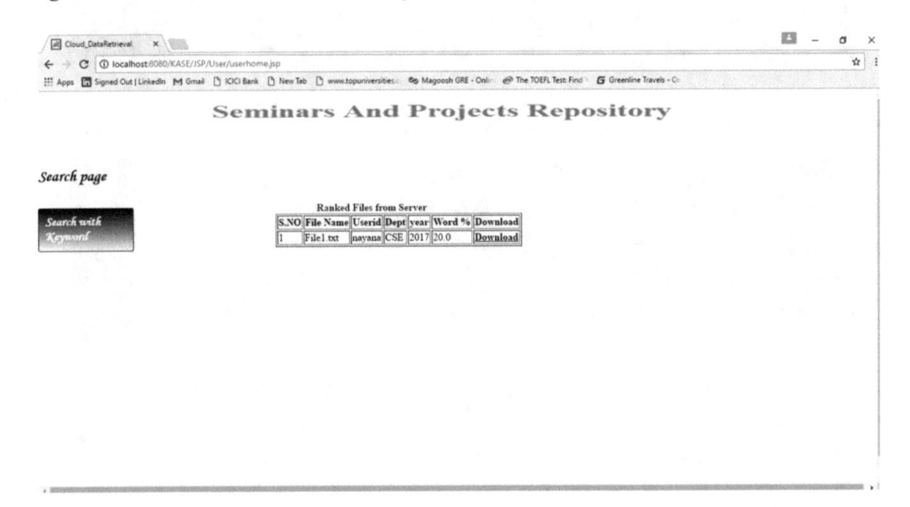

Fig. 3 Result of the searched file

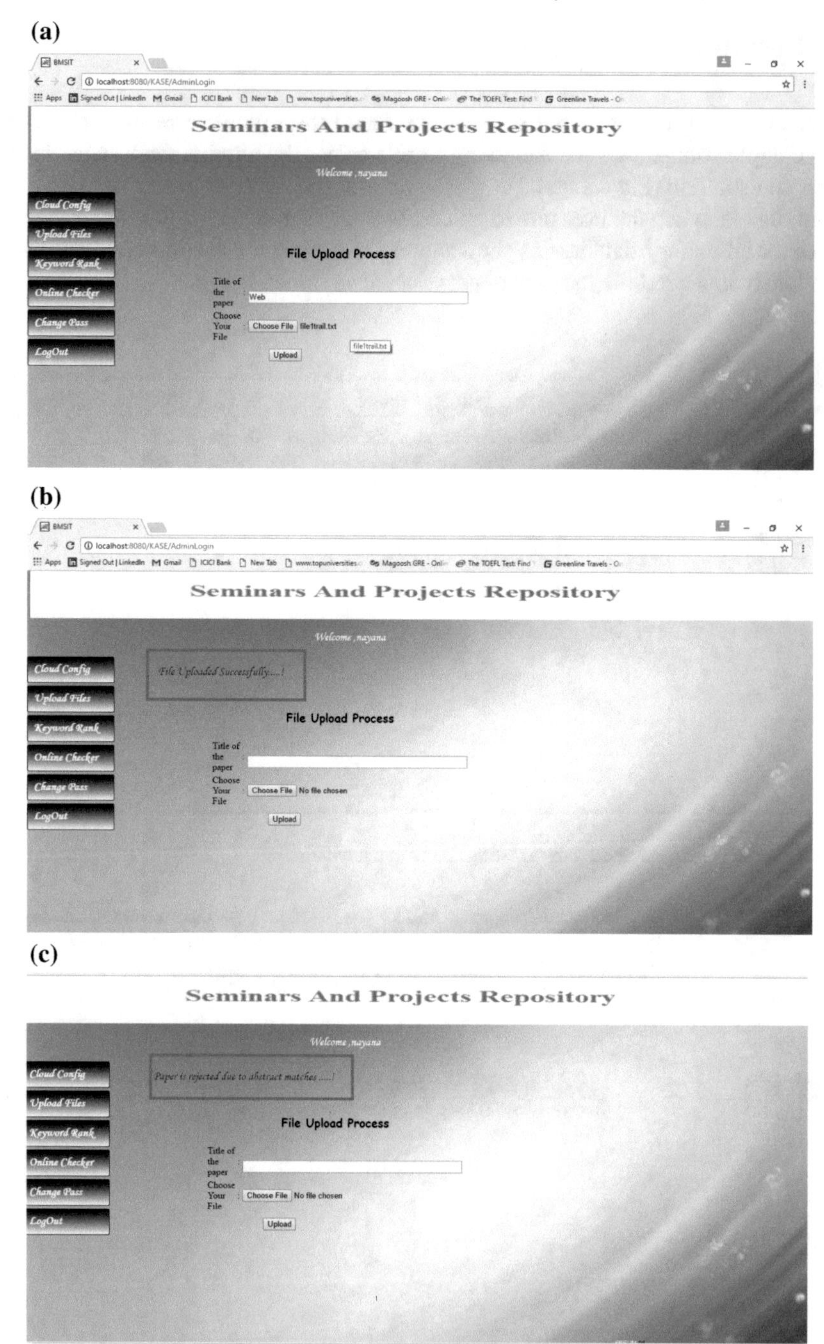

Fig. 4 **a** Upload new file. **b** File uploaded successfully. **c** Upload fails

7 Conclusion

In the proposed repository system, we can maximize the visibility and impact for better accessibility of the required information, repositories provide an easy way to make your work available, as a result of being openly accessible. Also provides an easy approach to search the data, support faculty and students through open access to and archiving pre- and post information and other details. Repositories are managed easily, where all the data is preserved for long-term access as data presented is on the backend database. It will also ensure ongoing maintenance and back-ups.

References

1. Abrams, S., Katz, J., Simms, S., Strong, M., Dash, W.P.: Data sharing made easy at the university of california. In: 11th International Digital Curation Conference, Amsterdam, Netherlands
2. Abrams, S., Goethals, A., Klein, M., Lack, R.: Cobweb: a collaborative collection development platform for web archiving (2016)
3. Sebesta, R.W.: Programming the World Wide Wed. Pearson
4. Schildt, H.: Java—The Complete Reference, 7th edn. Tata McGraw Hill (2007)
5. Daniel Liang, Y.: Introduction to JAVA Programming, 6th edn. Pearson Education (2007)

R. Nayana studying Master degree in B.M.S. Institute of Technology and Management, Bangalore. Her research field includes Digital repository, internet and web applications.

A. Bharathi Malakreddy working as a Professor and PG Cordinator in B.M.S. Institute of Technology and Management, Bangalore. Her research field includes Digital repository, internet and web applications.

Mining Frequent Itemsets Using Proposed Top-Down Approach Based on Linear Prefix Tree (TD-LP-Growth)

M. Sinthuja, N. Puviarasan and P. Aruna

Abstract Plenty of algorithms are available for datamining. LP-Growth occupies an important place in data mining. LP-Growth algorithm constricts data required for mining frequent itemsets in LP-tree and recursively builds LP-tree to mine entire frequent itemsets. In this study, an algorithm of top-down linear prefix tree (TD-LP-Growth) is proposed for mining frequent itemsets. The proposed TD-LP-Growth algorithm searches LP-tree from top to down order which is opposite to the old LP-Growth algorithm. TD-LP-Growth does not generate conditional pattern base and conditional LP-tree. Thus, it improves the performance of proposed TD-LP-Growth algorithm. In this paper, the benchmark databases considered are Online shopping dataset 1, Chess and Mushroom. While using online shopping dataset, the frequent purchaser of the dataset is visualized using Google map in geographical method. From the experimental results, it is concluded that the proposed TD-LP-Growth algorithm consumes lower runtime and memory space during the process of mining. Thus, the proposed TD-LP-Growth algorithm outperforms LP-Growth algorithm in mining frequent itemsets.

Keywords Association rule mining · Bottom up approach · Data mining
Frequent itemset mining · Linear tree · Minimum support · Pruning
Top-down approach

M. Sinthuja (✉) · P. Aruna
Department of Computer Science and Engineering, Annamalai University,
Chidambaram, Tamil Nadu, India
e-mail: sinthujamuthu@gmail.com

P. Aruna
e-mail: arunapuvi@yahoo.co.in

N. Puviarasan
Department of Computer and Information Science, Annamalai University,
Chidambaram, Tamil Nadu, India
e-mail: npuvi2410@yahoo.in

S. Smys et al. (eds.), *International Conference on Computer Networks and Communication Technologies*, Lecture Notes on Data Engineering and Communications Technologies 15, https://doi.org/10.1007/978-981-10-8681-6_4

1 Introduction

Association rule mining plays a significant part in the domain of data mining [1–5]. The famous algorithm of Apriori is introduced to mine frequent itemsets [6]. However, this approach faces the obstacle of exploring a cart load of itemsets. To overcome this issue [7] an algorithm based on tree structure was introduced known as frequent pattern tree (FP-tree). This algorithm adopts an approach of bottom up to search FP-tree. This algorithm needs to build conditional pattern base and conditional FP-tree for each itemset in its scan for lengthy itemsets. Thus, the consumption of time and memory is more in mining frequent patterns [8, 9]. The IFP-growth is introduced to mine frequent itemsets [10].

The LP-Growth algorithm which adopts linear structure for mining frequent itemsets [11]. The merit of this algorithm is that it uses minimum usage of pointers. The conduct of this algorithm is improved than FP-Growth algorithm by its special structure [12]. But it still it has a gray area in mining frequent itemsets—usage of bottom up approach which needs to construct conditional pattern base and conditional LP-tree for each itemsets.

2 Proposed Algorithm

2.1 *Top-Down Approach in Mining Frequent Itemsets Using Linear Prefix Tree (TD-LP-Growth)*

In this study, the proposed algorithm of TD-LP-Growth is proposed. This algorithm adopts top-down approach for mining frequent itemsets. The advantage of the proposed TD-LP-Growth algorithm is: it does not establish conditional pattern base and conditional LP-tree. In TD-LP-Growth algorithm, the construction of tree is performed on two scans of transactional database. In the first scan of transactional database count is calculated for all itemsets in the transactional database. In the second scan, each transactional item is interpolated into the LP-tree as a node. A rule of tree construction: if the different transaction contains same few items at the top they share their upper path in the LP-tree. The items of the transactional database are sorted in lexicographical order. The structure of the proposed TD-LP-Growth is defined as follows.

Each node in the tree is specified as an item. An item *l_node* is assigned to a node labeled by an item "*l*". Each node is combined with an attribute count. Count of each node is denoted as "v". It means the frequency of items occurred in the database. For each item *l*, all *l-nodes* are linked by *item-link* using pointer and it is denotes as *ptr.*

Header table is managed to store the information of items. Header table consist of three attributes namely Item name, Count and Item-link. Header table is denoted as *H*(*l_node*, v, *l-link*). An entry of an item in the header table *H*(*l_node*, v, *l-link*)

Table 1 Transaction database

TID	Sorted items
1	Disk, mobile
2	Cable, disk, laptop, mobile
3	Disk, laptop, mobile
4	Cable, driver, laptop

Table 2 Prioritize the items

Item name	Count
Cable	2
Disk	3
Laptop	3
Mobile	3

records the name of the item specified by *l_node*, overall count of the item specified by v and the head of the *l-link* for item *l* specified by *l-link*.

Example 1: Transactional Database displayed in Table 1 is taken as example. The transactions are categorized in lexicographical format and pruning the items based on minimum support which are infrequent itemsets. Thus, the item "Driver" is pruned. After sorting the items, the database is displayed in Table 2.

An item in the transactional database is infused into LP-tree as shown in Fig. 1. Let us take the first transaction items of the database purchase by user {Disk, Mobile}. As the early construction of LP-tree is straight forward, the transactions are added in the LP-tree linking to the node "root" following the item "Disk and Mobile". It is similar to the insertion of string in a tie. As the parent of LPN1 is "root" update the header node to P_{root}. Here, the attribute of the node are updated. As soon as entering the item "disk and Mobile" the count of the item are updated to "1" and item-link to "null" as there is no link to the same node. Tagging the flag value to "false" as branch node is not created. In the meantime branch node list are created as $P_{1,1}$ and $P_{2,1}$. Likewise insert remaining items as shown in Fig. 1.

The process of top-down approach is described as follows: Initially, an entry of item "Cable" is present in the header table H. To create sub-header-table walk through the path of the item "Cable" from LP-tree. It is found that Cable_node appears only on the first level of the LP-tree. Thus, item "Cable" is explored as frequent. Subsequently for item "Disk" in H, by following the item-link walk through the path of Disk_node in the LP-tree to construct sub_header_table denoted as SH_Disk. At the time of walk up, link up all scanned nodes of the same item by an item-link and aggregate the count for such nodes. Two paths are found that start with Disk_node: Root-Disk and Root-Cable-Disk. By walking up these paths, gather the count of the Cable_node to 1. It is because the path Root-Cable-Disk occurs only once in the database. Create an entry Cable in the sub_header_table SH_Disk. The count of entry Cable is 1 sine the count of Cable is actually the count of itemset {Cable, Disk}. Since the minimum support threshold value is 2, itemset {Cable, Disk} is infrequent. This completes the mining of itemsets with Disk as the last item. Now, SH_Disk can be erased from

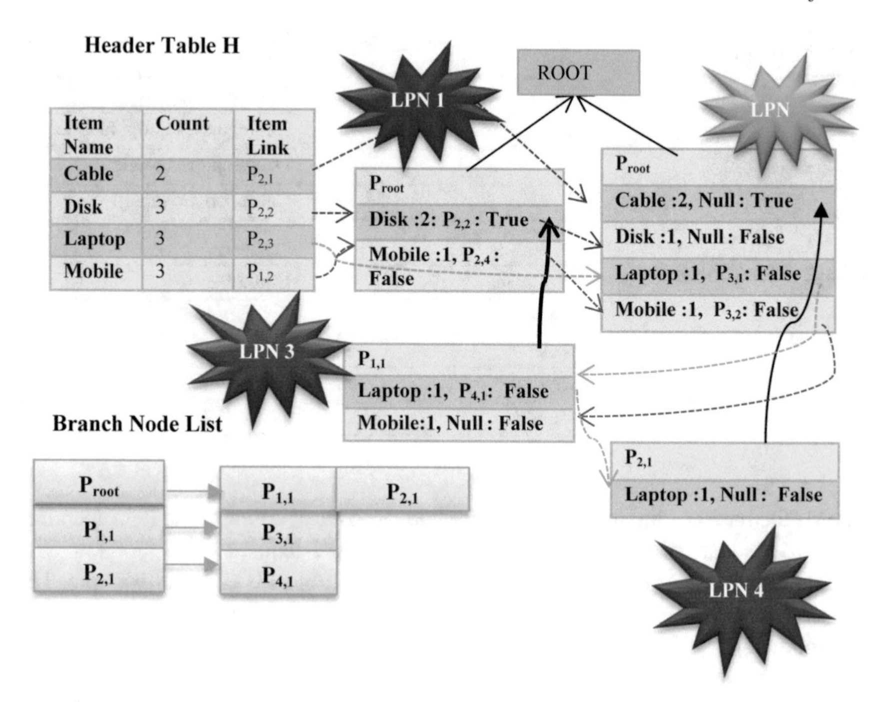

Fig. 1 Construction of LP-tree

memory. In case if {Cable, Disk} is frequent continue to construct sub_header_table SH_ Cable, Disk and mine LP-tree recursively. For item Laptop in H, walk up the path starting from Laptop_node in LP-tree. Construct sub_header_table denoted as SH_Laptop.

In this case three paths are found that starts with Laptop_node: Root-Cable-Disk-Laptop, Root-Disk-Laptop and Root-Cable-Laptop. For the path Root-Cable-Disk-Laptop the count of the Cable and Disk is induced in SH_Laptop to 1. In the next path Root-Disk-Laptop, while inserting the item disk the count of disk is updated to 2. While inserting the item Cable in SH_Laptop for the third path the count of Cable in incremented to 2. As it satisfy minimum support {Cable, Laptop: 2} {Disk, Laptop: 2} are frequent. Likewise find sub_header_table for item Laptop as shown in Fig. 2. Overall whenever there is new entry *I* in H_x mine all frequent itemsets that ends with *Ix*. The frequent itemsets explored are {Disk, Mobile: 3} {Laptop, Mobile: 2} {Cable, Laptop: 2} {Disk, Laptop: 2}{Disk, Laptop, Mobile: 2}{Cable: 2} {Disk: 3}.

Dissimilar to FP-growth algorithm, TD-LP-Growth algorithm follows a top-bottom method of processing. As a result, any change made in the higher levels would not have effect at the lower levels. A node at lower level is processed after its predecessor nodes. A Conditional pattern can be achieved by passing over the current node links with modification in counts. In such scenario, the count is updated without

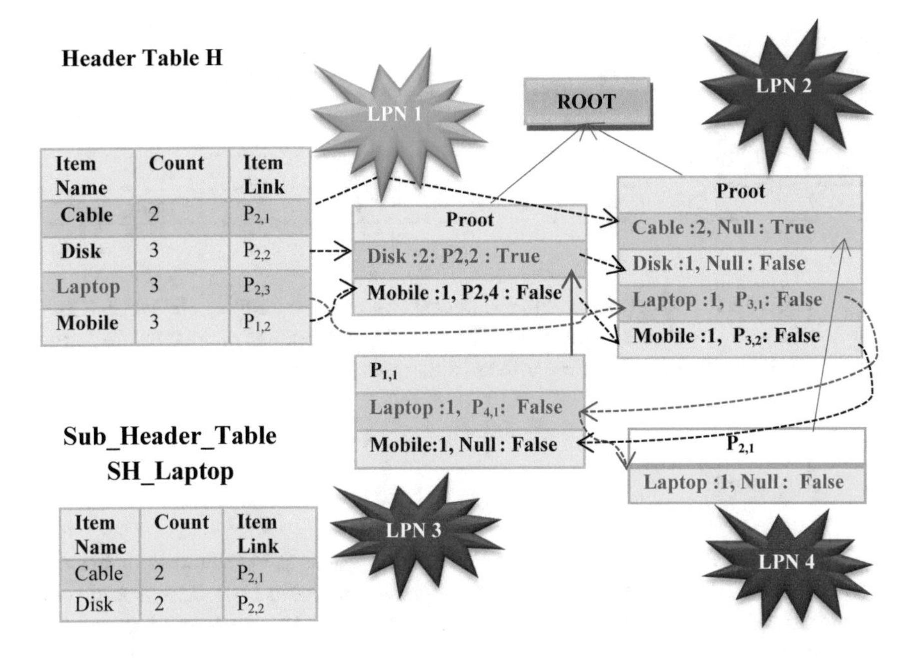

Fig. 2 Sub_header_table SH_Laptop

recreation of paths. FP-Growth follows a bottom-top approach and it requires a conditional pattern and sub-tree to be created for each pattern found. Whereas in the case of proposed TD-LP-Growth algorithm, generation of conditional pattern and sub-tree is not required. Thus, the latter is meritorious than the FP algorithm.

3 Performance Evaluation

In this section, the proposed TD-LP-Growth is compared with LP-Growth and FP-Growth algorithms which are a tree based algorithm. Experimental environment is explained first and the databases in the first subsection and then the results of runtime and memory usage.

3.1 Experimental Environment

The experiments are conducted in Intel® corei3™ CPU with 2.13 GHz, and 2 GB of RAM computer. The algorithms have been implemented in Java.

Table 3 Characteristics of databases

Datasets	File size (kb)	# of transactions	# of distinct items	Type
Online shopping dataset 1	100	725	114	Sparse
Chess	335	3196	75	Dense
Connect	558	67,557	116	Dense

The analysis is performed on three databases of online shopping dataset 1, Chess and Mushroom which are widely used for evaluating frequent pattern mining algorithms. The significant attributes of these databases are shown in Table 3.

3.2 Runtime

In this subsection, the proposed TD-LP-Growth algorithm is related with existing algorithms of LP-Growth and FP-Growth algorithms. From Figs. 3, 4, 5 and 6 the latitudinal axis shows various minimum support threshold values and longitudinal axis shows runtime in milliseconds. The chart portrayed in Fig. 3 shows that TD-LP-Growth is twice faster than LP-Growth algorithm for Online Shopping dataset 1. Here, FP-growth algorithm needs to construct FP-tree from the dataset which takes more time in building. Comparatively, the proposed TD-LP-Growth algorithms outperforms other algorithms as it uses top-down approach which avoids the generation of conditional pattern base and conditional LP-tree in mining frequent itemsets. The location of frequent user of the Online Shopping dataset 1 is visualized in Google map as shown in Fig. 4. As illustrated in Fig. 5, mining times of TD-LP-Growth algorithm are 2252 ms when the minimum support threshold value is 2000 for the dataset chess. LP-Growth consumes 3444 ms and FP-Growth algorithm consumes 6888 ms. As the existing algorithm has to generate conditional patterns consumption of runtime is more. While decreasing the minimum support FP-growth algorithm and other algorithms cannot mine itemsets normally, so the consumption of runtime is more that system could not bear. Thus, it is clearly shown that the runtime performance of TD-LP-Growth is better than other algorithms. Figure 6 portrays the runtime of all algorithms for the dataset Mushroom.

From these result, TD-LP-Growth algorithm outperforms other algorithm at higher minimum support threshold values. In case of lower minimum support threshold values, the conduct of other algorithms is poor when compared to the proposed algorithm. Since LP-Growth algorithm is the best algorithm that outperforms all other algorithm, this is because the generation of conditional pattern base and sub-tree are avoided. On the whole the productivity of the proposed TD-LP-Growth algorithm is best at most among all datasets. It is considered as the best among all the algorithms.

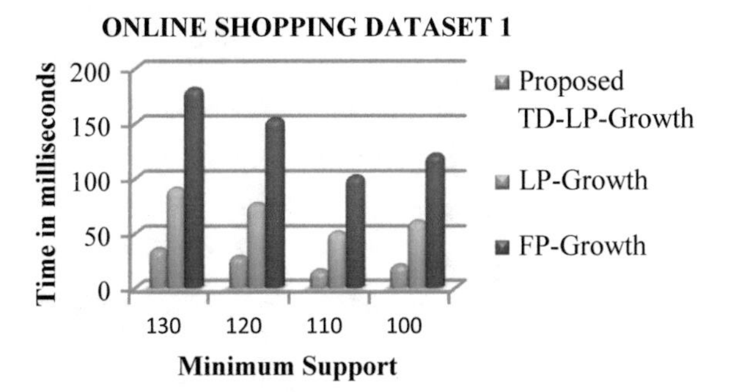

Fig. 3 Runtime of FIs using online shopping dataset 1

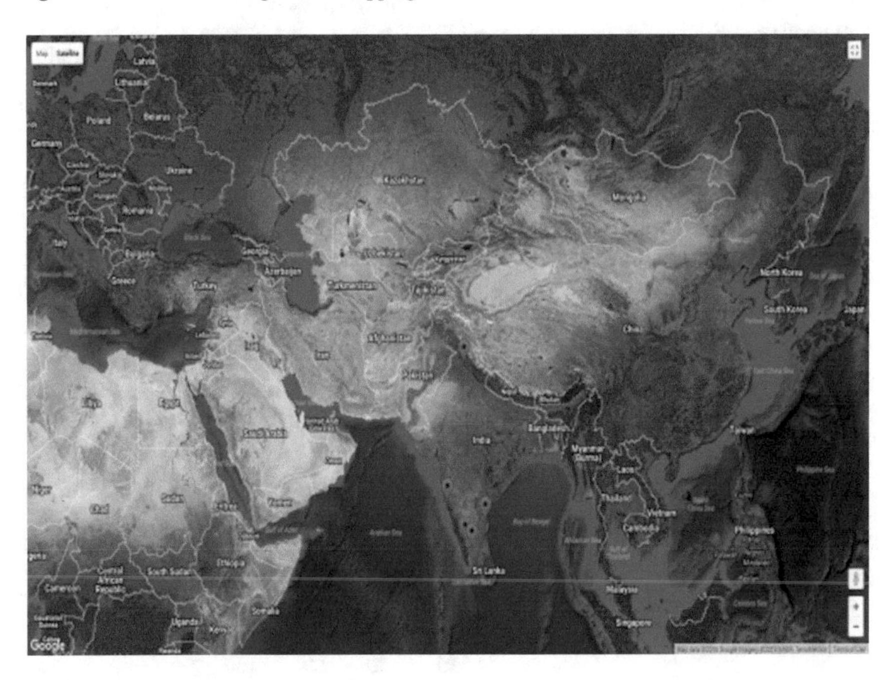

Fig. 4 Location of frequent purchaser using Google map for online shopping dataset 1

3.3 Memory Usage

The experimental results of consumption of memory are shown in Figs. 7, 8 and 9. The memory consumption of proposed TD-LP-Growth, LP-Growth and FP-Growth algorithms for mining frequent itemsets are plotted for varying minimum support threshold values of 100, 110, 120, 130 for the sparse Online Shopping dataset 1. From the chart it is come to know that LP-Growth and FP-Growth algorithms have to do a considerable amount of work to deal with an exploration of conditional pattern base. Thus memory utilization to store itemset is more. Meanwhile, the proposed TD-LP-

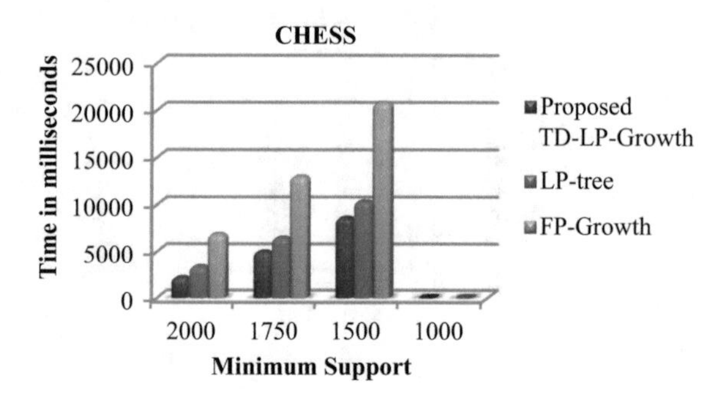

Fig. 5 Runtime of FIs using chess dataset

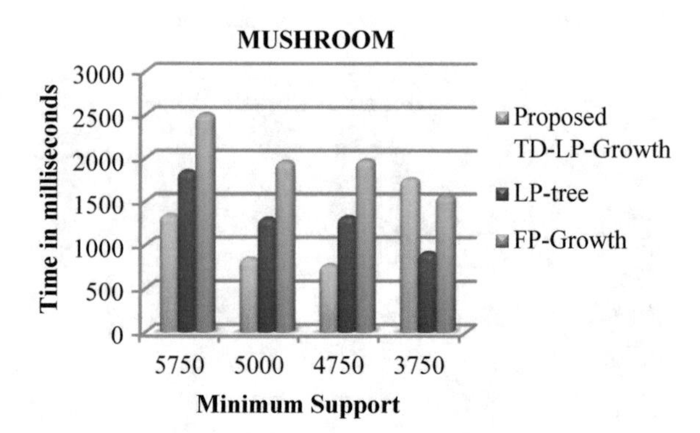

Fig. 6 Runtime of FIs using mushroom dataset

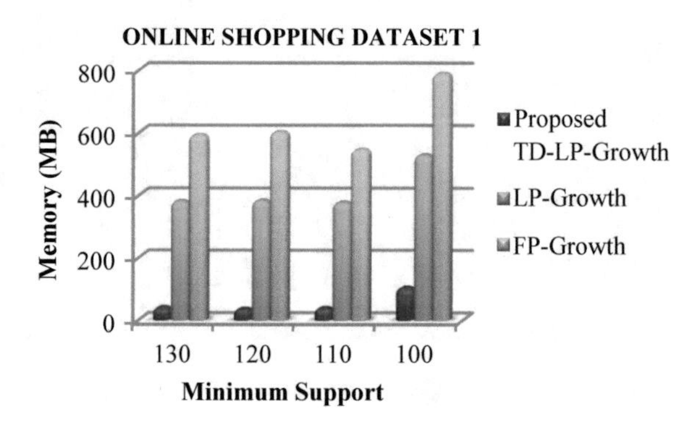

Fig. 7 Memory consumption of MFIs using online shopping dataset 1

Growth algorithm requires lower consumption of memory. Because this algorithm uses sub_header_table to generate frequent itemsets which avoids conditional pattern base.

Figure 8 portrays the consumption of memory of all algorithms for the dataset Chess. Typically, the proposed TD-LP-Growth algorithm makes better use of memory, enabling it to mine larger and more complicated databases. TD-LP-Growth algorithm tends to work well and is quicker for higher minimum support thresholds. While decreasing the minimum support for the dataset Chess the FP-growth algorithm and other algorithms cannot mine itemsets normally, so the consumption of

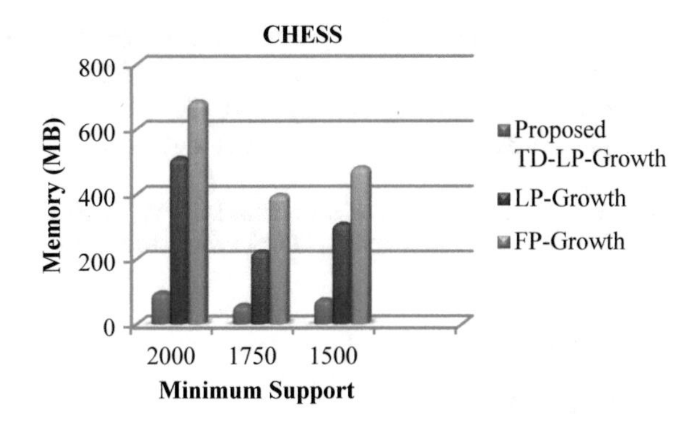

Fig. 8 Memory consumption of MFIs using chess dataset

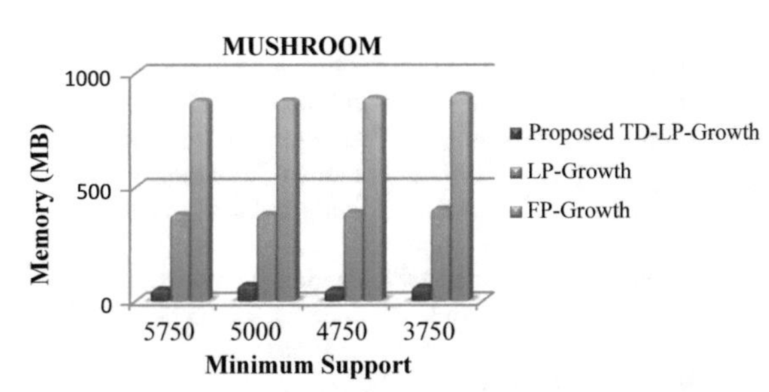

Fig. 9 Memory consumption of MFIs using mushroom dataset

memory is more that system could not bear. Thus, it is not shown in the chart. Since, the proposed TD-LP-Growth algorithm is efficient than other algorithms. Figure 9 depicts the consumption of memory of all algorithms for the dataset Mushroom. Specifically, when the minimum support was both higher and lower the algorithm FP-Growth is less effective for all datasets.

When FP-Growth is compared with LP-Growth memory required to store frequent itemset is higher. Because it has to construct bushy tree and has to generate conditional pattern base to mine frequent itemsets. So, its effect is not available for dataset Mushroom and Connect. Furthermore, TD-LP-Growth is relatively consistent and stable in consumption of memory independent of the minimum support in contrast to all other algorithms.

4 Conclusion

In this study, the proposed algorithm of TD-LP-Growth is applied for mining frequent itemsets. As the proposed algorithm adopts top-down approach the efficiency of the mining is improved. The major benefit of the TD-LP-Growth algorithm is: it does not establish conditional pattern base and conditional LP-tree. Whereas the existing FP-Growth and LP-Growth algorithm need to generate conditional pattern base and sub-tree. The generation of frequent itemset requires huge amount of runtime and memory. From the experimental results it is observed that the proposed TD-LP-Growth algorithm is efficient than the existing algorithms of LP-Growth and FP-Growth. By geographical method, the frequent purchaser present in the online shopping dataset is visualized using Google map.

Acknowledgements We gratefully acknowledge the funding agency, the University Grant Commission (UGC) of the Government of India, for providing financial support for doing this research work.

References

1. Agarwal, R., Imielinski, T., Swami, A.: Mining association rules between sets of items in large databases. ACM-SIGMOD, 207–216 (1993)
2. Agrawal, R., Srikanth, R.: Fast algorithm for mining association rules in large databases. In: Proceedings of the 1994 International Conference VLDB, Santiago, Chile, pp. 487–499 (1994)
3. Wang, K., He, Y., Han, J.: Mining frequent pattern using support constraints. VLDB, 43–52 (2000)
4. Sinthuja, M., Puviarasan, N., Aruna, P.: Evaluating the performance of association rule mining algorithms. World Appl. Sci. J. **35**(1), 43–53 (2017). ISSN: 1818-4952
5. Mannila, H., Toivonen, H.: Level wise search and borders of theories in knowledge disco, very. Data Min. Knowl. Disc. **1**(3), 241–258 (1997)
6. Agarwal, R., Srikanth, R.: Mining Sequential Patterns. ICDE (1995)
7. Sinthuja, M., Puviarasan, N., Aruna, P.: Experimental evaluation of apriori and equivalence class clustering and bottom up lattice traversal (ECLAT) algorithms. Pak. J. Biotechnol. **13**(2), 77–82 (2016)
8. Lin, K.-C., Liao, I.-E., Chen, Z.-S.: An improved frequent pattern growth method for mining association rules. Expert Syst. Appl. **38**, 5154–5161 (2011)
9. Sinthuja, M., Puviarasan, N., Aruna, P.: Comparative analysis of association rule mining algorithms in mining frequent patterns. Int. J. Adv. Res. Comput. Sci. **8**(5) (2017)
10. Sinthuja, M., Puviarasan, N., Aruna, P.: Research of improved FP-growth (IFP) algorithm in association rules mining. Int. J. Eng. Sci. Invention, 24–31 (2018)
11. Pyun, G., Yun, U., Ryu, K.H.: Efficient frequent pattern mining based on linear prefix tree. Knowl. Based Syst. **55**, 125–139 (2014)
12. Tanbeer, S.K., Ahmed, C.F., Jeong, B.S., Lee, Y.: Efficient single pass frequent pattern mining using a prefix-tree. Inf. Sci. **179**(5), 559–583 (2008)

M. Sinthuja is a Research scholar in the Department of Computer Science and Engineering at Annamalai University, Chidambaram, Tamil Nadu, India. Her research area includes Association Rule Mining and data mining.

N. Puviarasan working as a professor in the Department of Computer and Information Science, Annamalai University, Chidambaram, Tamil Nadu, India. Her research area includes Association Rule Mining and data mining.

P. Aruna working as a professor in the Department of Computer Science and Engineering at Annamalai University, Chidambaram, Tamil Nadu, India. Her research area includes Association Rule Mining and data mining.

Target Localization Algorithm in a Three-Dimensional Wireless Sensor Networks

Sangeeta Kumari and Govind P. Gupta

Abstract In a Wireless Sensor Networks, localization techniques are needed to identify the exact position of an event. In this paper, we proposed Fuzzy Inference System (*FIS*) based target localization algorithm in a three-dimensional wireless sensor networks. In the proposed scheme, distance between anchor and target nodes are calculated by adding correction factor with hop size of an anchor node. The concept of fuzzy logic-based edge weight calculation is introduced to improve the localization accuracy. Simulation result shows that proposed scheme achieves less localization error and better accuracy as compared with existing localization technique.

Keywords Fuzzy inference system · Wireless sensor networks
Target localization · Correction factor

1 Introduction

Wireless sensor networks (*WSNs*) generally refers to a distributed network, which is composed of numerous number of sensor nodes, allocated over a monitored region to observe the environmental conditions. In *WSNs*, sensor nodes can be mounted with one or more sensors for sensing the different physical phenomenon. These sensor nodes can sense, process, and transmit observed data to the Base station through multi-hop communication [1, 2]. In many real-time applications, location information about sensor nodes is required to identify the exact position of an event. Thus node localization is a very challenging research issues in this domain.

WSNs localization techniques are classified into two methods, range-based and range-free. In Range-based method, position of each sensor node can be determine

S. Kumari (✉) · G. P. Gupta
Department of Information Technology, National Institute of Technology,
Raipur 492010, Chhattisgarh, India
e-mail: sangeetak2606@gmail.com

G. P. Gupta
e-mail: gpgupta.it@nitrr.ac.in

S. Smys et al. (eds.), *International Conference on Computer Networks and Communication Technologies*, Lecture Notes on Data Engineering and Communications Technologies 15, https://doi.org/10.1007/978-981-10-8681-6_5

through various approaches such as Time of Arrival (*ToA*), Received Signal Strength Indicator (*RSSI*) and Angle of Arrival (*AoA*) from localized anchor nodes in a network [3, 4]. Conversely, Range-free method is introduced to reduce the additional requirements of hardware cost, energy consumption, computational overhead, etc. of the range-based method. Existing Range-free techniques are Distance vector-hop (*DV-Hop*), Centroid and Approximate Point in Triangulation (*APIT*) [5–7]. In range-free based scheme, with help of anchor nodes, location of the target nodes are estimated. Here, anchor node knows its location and target node does not know its location. Range-free method is most often used to calculate the position of the target nodes based on radio connectivity information and sensing capabilities among neighbouring nodes. Here target nodes are those nodes whose location needs to be estimated.

Most of the existing localization algorithm proposed for *2D* plane based *WSNs*, but in real applications of *WSNs*, sensor nodes are generally deployed in *a 3D* plane such as building, forest, mountain, and under water monitoring in sea. Thus, node localization for *3D* plane is paramount requirement in the application of *WSNs*. In this paper, we proposed Fuzzy Inference System (*FIS*) based node localization technique in three-dimensional *WSNs* [8, 9]. In this scheme, correction factor and estimated hop size of an anchor node is used to modify the distance between the anchor and target anodes. To enhance the localized node proportion, fuzzy logic-based edge weight calculation is proposed. Performance evaluation of the proposed method is analyzed in terms of the percentage of anchor nodes, number of sensor nodes and communication range.

The remaining work is structured as follows: Sect. 2 presents a brief discussion of the related work on *3D* Wireless Sensor Networks. In Sect. 3, a brief overview of fuzzy inference system is presented. In Sect. 4, fuzzy inference system based target localization scheme is presented. Simulation results have been discussed in Sect. 5. Section 6 finally, conclude the proposed work.

2 Related Work

In this section, a brief overview of the related work on localization techniques in Wireless Sensor Networks is discussed.

Zhang et al. [10] proposed a hybrid localization method by combining range-based *RSSI* and range-free hop value to achieve higher accuracy in *3D* node localization. Chaurasiya et al. [11] have explored multi-dimensional scaling (*MDS*) based localization scheme in *3D* to estimate the position of nodes and convert the local coordinates to the global coordinates by using Helmert transformation. In [12] multi-hop range-free localization technique with approximate shortest path is proposed to reduce the effect of network anisotropy and enhance the localization accuracy. Tomic et al. [13] have addressed both distance- and angle-based localization problems in cooperative and non-cooperative three-dimensional wireless sensor networks with known and unknown transmit power based on distance and angle measurements model. Xu

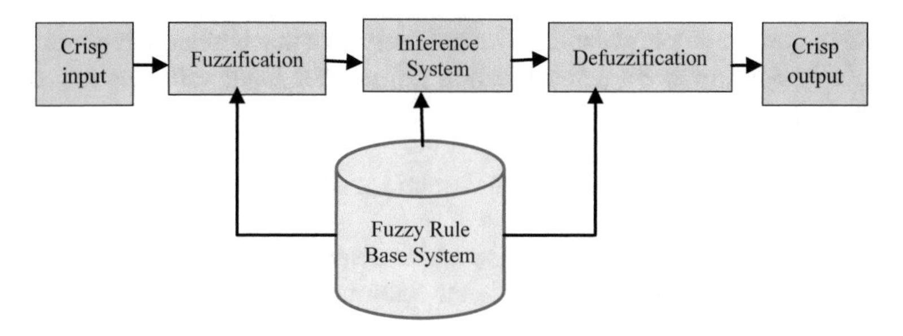

Fig. 1 Block diagram of fuzzy inference system (*FIS*)

et al. [14] introduced degree of co-planarity and quasi-newton technique based on traditional.

DV-hop to enhanced the position of nodes, it shows better coverage and accuracy as compared to *DV-hop*.

Yun et al. [15] used soft computing optimization technique to solve the problem of the range-free based localization method. In first scheme, fuzzy logic-based approach [16] is applied to model the edge weight and enhanced it by using genetic algorithm-based optimization. In second scheme, application of neural network is used to calculate the position of the sensor nodes. Chen et al. [17] used particle swarm optimization (*PSO*)-based optimization for localization and compared it with the traditional *DV-hop* method. In [18] fuzzy logic system is applied to modelled the edge weight between nodes and enhanced it with *H-best PSO* (*HPSO*) and biogeography based optimization (*BBO*) to reduce the localization errors. Sharma et al. [19] proposed application of invasive weed optimization (*IWO*) bacterial foraging optimization (*BFO*) techniques to modelled the edge weight to reduce the localization errors.

3 Overview of Fuzzy Inference System

In this section, we present a brief discussion on Fuzzy Inference System (*FIS*). *FIS* is one of the most pragmatic approach to imitate human ideas in a realistic manner [17, 18]. It is a fuzzy logic-based system which consists of a fuzzification, inference system, defuzzification and fuzzy rule base system as shown in Fig. 1. It maps an input data to an output data using *IF-THEN* rules.

Fuzzy rules are written as:

R_n: if I_1 is f_{n1} and I_2 is f_{n2} and $\ldots I_m$ is f_{nm} then O_n is $Z_{n.}$

where R_n is the *n*th rule, I_m is the *n*th input values, O_n is the *n*th output variable of the fuzzy logic system, f_{nm} is the membership function of input variable and Z_n is the membership function of output variable which represent the degree of truth in reasoning.

In this paper, we employ Mamdani Inference System that implies maximum aggregation and centre of sums defuzzification process for input variables that can be described as:

$$\mu(O) = \max(\min(I_m,\ Z_i)),$$

where max is aggregation method that combine output of all fuzzy rule in a single set and min is implication method to scale the output fuzzy set in Mamdani Inference System.

4 Fuzzy Inference System Based Target Localization Scheme

This section presents FIS-based target localization scheme to calculate the exact position of the target nodes. The working of the proposed FIS-based location algorithm consists of three main parts such as calculation of distance between anchor node target node, edge weight calculation using *FIS* and calculation of target location. These steps are described as follows.

4.1 Calculation of Distance Between Anchor Node Target Node

For this calculation, following five steps are used which is described as follows:

Step 1: Each anchor node broadcasts a packet containing its location, node_id and hop value (initial value is 0) in a multi-hop communication throughout the network. Each neighbouring node (anchor or target node) stores the location of an anchor node in its routing table and updates the hop-size by hop-size +1. The actual distance between anchor nodes is calculated using Euclidean Distance formulae as follows:

$$\text{Actual\_Dist}_{as} = \sqrt{(x_a - x_s)^2 + (y_a - y_s)^2}, \tag{1}$$

where (x_a, y_a) and (x_s, y_s) are the coordinates of an anchor nodes a and s respectively. After this step, each anchor node knows its hop distance from another anchor node (i.e., hopValue_{as}) and also its actual location coordinate. Each anchor node calculates its distance and hop value from other anchor nodes in a network. The estimated hop-size of a hop is calculated as follows:

$$\text{Esti\_HopSize}_a = \frac{\sum_{a \neq s} \sqrt{(x_a - x_s)^2 + (y_a - y_s)^2}}{\sum_{a \neq s} \text{hopValue}_{as}}, \tag{2}$$

where hopValue_{as} represents the hop value between the anchor nodes a and s $(a \neq s)$.

Step 2: After calculation of actual distance and estimation of hop-size, each anchor node calculates its estimated distance (Esti_Dist_{as}) using Eq. (3) with respect to other anchor node.

$$\text{Esti\_Dist}_{as} = \text{Esti\_HopSize}_a \times \text{hopValue}_{as} \quad (3)$$

Step 3: After calculation of estimated distance between each pair of anchor node, each anchor node calculates distance error using Eq. (4).

$$\text{Dist\_Error}_{as} = |\text{Actual\_Dist}_{as} - \text{Esti\_Dist}_{as}| \quad (4)$$

Step 4: In this step, correction factor (cF_a) for each anchor node a is calculated using Eq. (5). Correction factor (cF_a) is used to correct the distance between an anchor node and its target node for which location is being calculated.

$$cF_a = \frac{\sum_{a \neq s} \text{Dist\_Error}_{as}}{\sum_{a \neq s} \text{hopValue}_{as}} \quad (5)$$

Step 5: In this step, modified distance between an anchor and target node is calculated using Eq. (6). Modified distance between an anchor and target node is obtained by adding correction factor to the estimated hop size of an anchor node and multiplied it by hop value between them.

$$\text{Modi\_Dist}_{at} = \left(\text{Esti\_HopSize}_a + cF_a\right) \times \text{hopValue}_{at}, \quad (6)$$

where Modi_Dist_{at} denoted as the modified distance between anchor-target pair and cF_a is the correction factor of an anchor node.

4.2 *Edge Weight Calculation Using FIS*

Table 1 illustrates the fuzzy rule base system to reduce the nonlinearity between modified and actual distance and computes the edge weight between each anchor and target node. If anchor node is closer to the target node, it must have a maximum weight. Conversely, if anchor node is far away from the given target node, it must have low weight.

The input data x from the modified distance of each anchor–target pair mapped into the five fuzzy membership functions, i.e., VLess, Less, Moderate, Max and VMax. It takes input in the interval [0, $\text{Modi\_dist}_{\max}$], where $\text{Modi\_dist}_{\max}$ is the maximum modified distance as shown in Fig. 2.

Figure 3 illustrate the output data y from the edge weight of each anchor-target pair that composed of five fuzzy membership functions i.e., VLess, Less, Moderate, Max and VMax. It takes an output in the interval [0, $u_{\max}$], where $u_{\max}$ is the maximum edge weight.

Table 1 Fuzzy rule base system

S. No.	Input	Output
1	If Modi_Dist is VLess	Then weight is VLess
2	If Modi_Dist is less	Then weight is less
3	If Modi_Dist is moderate	Then weight is moderate
4	If Modi_Dist is max	Then Weight is max
5	If Modi_Dist is VMax	Then Weight is VMax

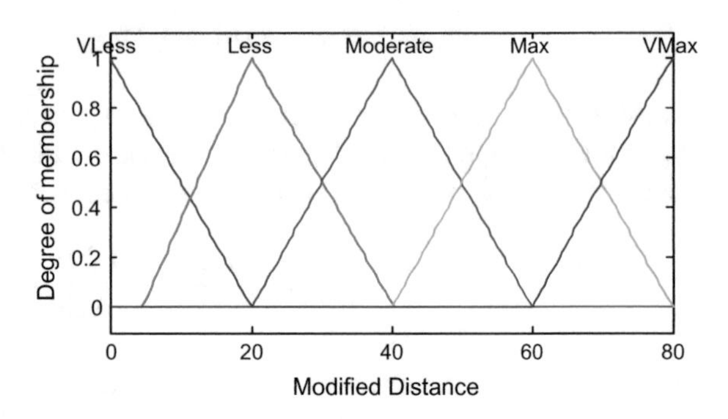

Fig. 2 Fuzzy membership function of modified distance

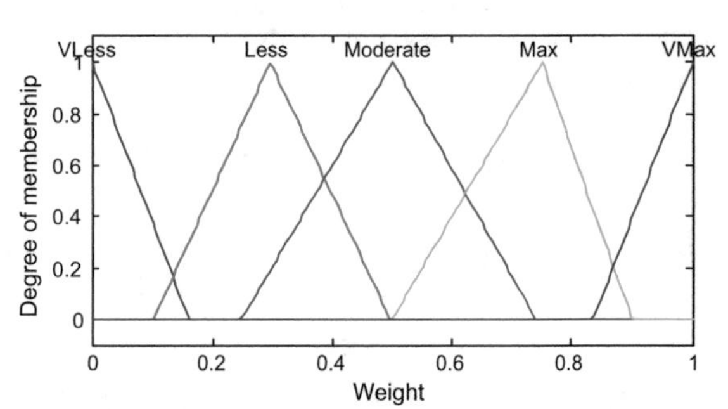

Fig. 3 Fuzzy membership function of weights

4.3 *Calculation of Target Location Coordinate*

After evaluation of edge weight using *FIS*, the weighted centroid method is used to determine the position of the target nodes. In *3D* plane, each anchor node broadcasts their location information towards the other nodes and the estimated position of the target nodes is calculated by the combination of at most four anchor nodes by using Eq. (7). Let us the location of closet nodes are (x_1, y_1, z_1), (x_2, y_2, z_2), (x_3, y_3, z_3) *and* (x_4, y_4, z_4). Then location coordinates of target node (x_t, y_t, z_t) respectively as follows:

$$(x_t, y_t, z_t) = \left[\frac{(u_1 x_1) + \cdots (u_4 x_4)}{\sum_{i=1}^{4} u_i}, \frac{(u_1 y_1) + \cdots (u_4 y_4)}{\sum_{i=1}^{4} u_i}, \frac{(u_1 z_1) + \cdots (u_4 z_4)}{\sum_{i=1}^{4} u_i} \right], \tag{7}$$

where u_1 is the edge weight between target 1 and anchor 1, u_2 is the edge weight between target 1 and anchor 2, u_3 is the edge weight between target 1 and anchor 3, u_4 is the edge weight between target 1 and anchor 4.

5 Simulation and Result Analysis

In this section, we analyze the performance evaluation of the proposed work with the existing localization algorithm *SR-WLS* [13] using MATLAB R2009. Sensor nodes are randomly scattered in three dimensional areas of $200 \times 200 \times 200\ \text{m}^3$. The proposed result was evaluated by varying different parameters such as percentage of anchor nodes, number of sensor nodes, and communication range.

Figure 4 shows the effect of the percentage of anchor nodes over localization error. In this experiment, number of anchor nodes are varied from 10 to 50% of the sensor nodes and observed the localization error. It can be observed from Fig. 4 that with increase of percentage of anchor nodes, i.e., 10–50, the average localization error keeps decreasing. This is due to fact that if number of anchor nodes is more, modified distance between anchor nodes and target node has less error which results in better localization accuracy, i.e., less localization error.

To investigate the impact of node density on average localization error, the number of sensor nodes varied from 50 to 200 as shown in Fig. 5. As the number of sensor nodes increases, the average localization error decreases.

Figure 6 shows the effect of communication range over localization error. In this experiment, communication range varied from 15 to 45 m and the number of sensor nodes and anchor nodes were fixed 200 and 30 respectively. It can be observed from Fig. 6 that localization error for the proposed scheme is significantly less than the existing scheme.

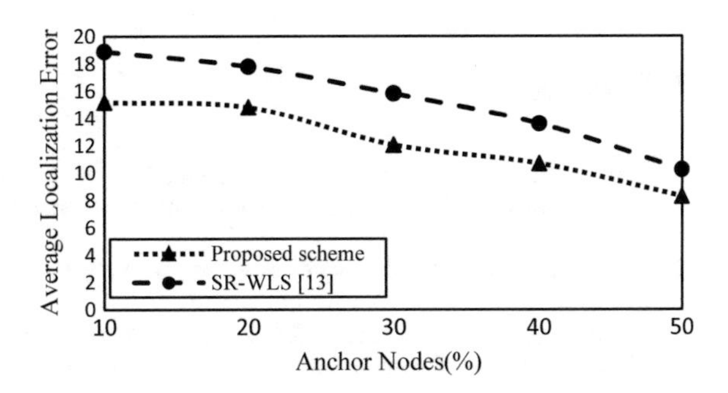

Fig. 4 Average localization error versus number of anchor nodes (%)

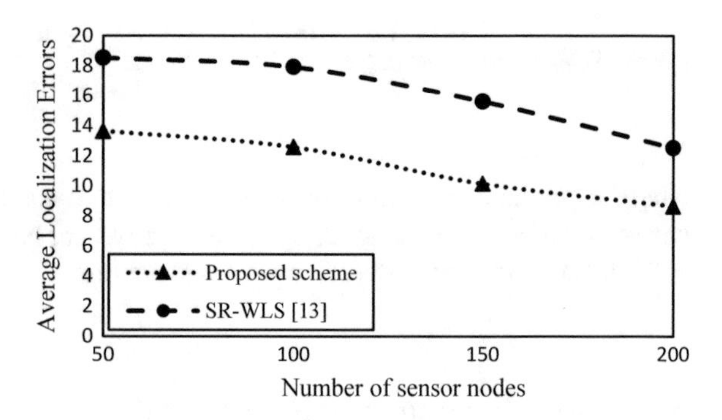

Fig. 5 Average localization error versus number of sensor nodes

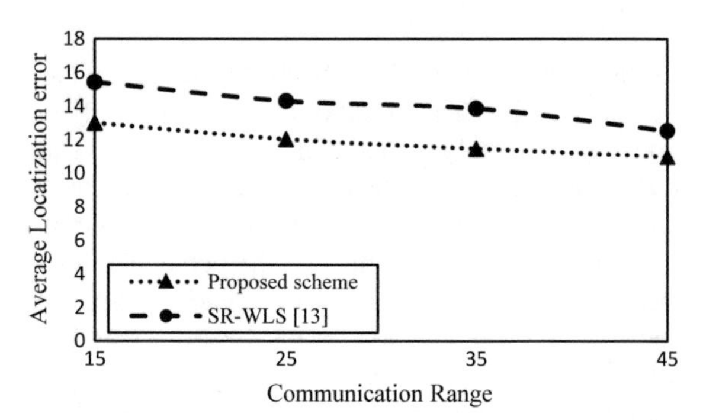

Fig. 6 Average localization error with communication range

6 Conclusion

This paper proposed a *FIS*-based target localization technique for three-dimensional *WSNs*. In proposed scheme, correction factor and estimated hop-size of an anchor node is used to modify the distance between anchor and target nodes. Fuzzy based edge weight calculation scheme is also introduced to enhance the localization accuracy. Our simulation results show that proposed scheme achieves less localization errors and higher accuracy by varying different parameters such as percentage of anchor nodes, number of sensor nodes, and communication range.

References

1. Patwari, N., Ash, J.N., Kyperountas, S., Hero, A.O., Moses, R.L., Correal, N.S.: Locating the nodes: cooperative localization in wireless sensor networks. IEEE Signal Process. Mag. **22**(4), 54–69 (2005)
2. Pottie, G.J., Kaiser, W.J.: Wireless integrated network sensors. Commun. ACM **43**(5), 51–58 (2000)
3. Cai, S., Pan, H., Gao, Z., Yao, N., Sun, Z.: Research of localization algorithm based on weighted Voronoi diagrams for wireless sensor network. EURASIP J. Wireless Commun. Netw. **50**(1), 1–10 (2014)

4. Niculescu, D., Nath, B.: Ad hoc positioning system (APS) using AOA. In: Twenty-Second Annual Joint Conference of the IEEE Computer and Communications, vol. 3, pp. 1734–1743 (2003)
5. Bachrach, J., Christopher, T.: Localization in sensor networks. In: Handbook of Sensor Networks: Algorithms and Architectures, vol. 1, pp. 277–289 (2005)
6. Bulusu, N., Heidemann, J., Estrin, D.: GPS-less low-cost outdoor localization for very small devices. IEEE Pers. Commun. **7**(5), 28–34 (2000)
7. Niculescu, D., Nath, B.: Ad hoc positioning system (APS). Glob. Telecommun. Conf. **5**, 2926–2931 (2001)
8. Niculescu, D., Nath, B.: DV based positioning in ad hoc networks. Telecommun. Syst. **22**(1), 267–280 (2003)
9. He, T., Huang, C., Blum, B.M., Stankovic, J.A., Abdelzaher, T.: Range-free localization schemes for large scale sensor networks. In: Proceedings of the 9th Annual International Conference on Mobile Computing and Networking, ACM, pp. 81–95 (2003)
10. Zhang, B., Fan, J., Dai, G., Luan, T.H.: A hybrid localization approach in 3D wireless sensor network. Int. J. Distrib. Sens. Netw. **11**(10), 1–10 (2015)
11. Chaurasiya, V.K., Jain, N., Nandi, G.C.: A novel distance estimation approach for 3D localization in wireless sensor network using multi-dimensional scaling. Inf. Fusion **15**, 5–18 (2014)
12. Lee, S., Park, C., Lee, M.J., Kim, S.: Multihop range-free localization with approximate shortest path in anisotropic wireless sensor networks. EURASIP J. Wireless Commun. Netw. **14**(1), 1–12 (2014)
13. Tomic, S., Beko, M., Dinis, R.: 3-D target localization in wireless sensor networks using RSS and AoA measurements. IEEE Trans. Veh. Technol. **66**(4), 3197–3210 (2017)
14. Xu, Y., Yi, Z., Gu, J.: An improved 3D localization algorithm for the wireless sensor network. Int. J. Distrib. Sensor Netw. **11**(6), 315714 (2015)
15. Yun, S., Lee, J., Chung, W., Kim, E., Kim, S.: A soft computing approach to localization in wireless sensor networks. Expert Syst. Appl. **36**(4), 7552–7561 (2009)
16. Zadeh, L.A.: Fuzzy logic = computing with words. IEEE Trans. Fuzzy Syst. **4**(2), 103–111 (1996)
17. Chen, X., Zhang, B.: 3D DV–hop localisation scheme based on particle swarm optimisation in wireless sensor networks. Int. J. Sens. Netw. **16**(2), 100–105 (2014)
18. Kumar, A., Khosla, A., Saini, J.S., Sidhu, S.S.: Range-free 3D node localization in anisotropic wireless sensor networks. Appl. Soft Comput. **34**, 438–448 (2015)
19. Sharma, G., Kumar, A.: Fuzzy logic based 3D localization in wireless sensor networks using invasive weed and bacterial foraging optimization. Telecommun. Syst. 1–14 (2017)

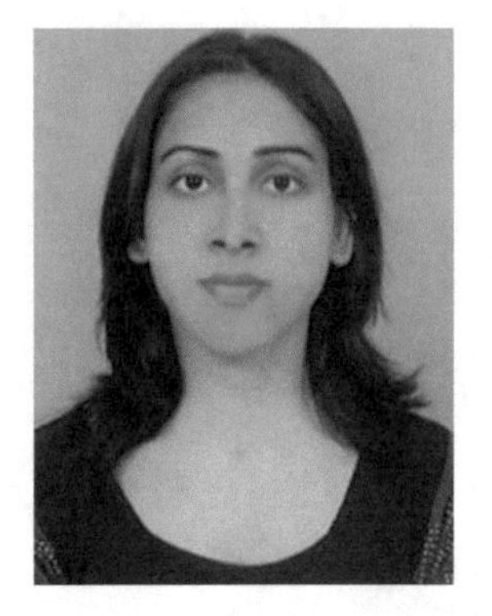

Sangeeta Kumari received her M.Tech. degree from National Institute of Technical Teachers' Training and Research, Shamla Hills, Bhopal, India, in 2015. She is currently pursing Ph.D. in department of Information Technology, National Institute of Technology, Raipur, India. Her current research interests include wireless sensor networks, and Sensor Cloud.

Govind P. Gupta received his Ph.D. degree from Indian Institute of Technology, Roorkee, India, in 2014. He is currently an Assistant Professor in the Department of Information Technology at National Institute of Technology, Raipur, India. His current research interests include efficient protocol design for Wireless Sensor Networks and Internet of Things, mobile agents, Network Security and Software-defined Networking. He is a professional member of the IEEE and ACM.

Green Energy Efficiency in Cellular Communication on LTE-A-Based Systems

V. Manickamuthu, B. V. Namrutha Sridhar and I. Chandra

Abstract Green energy efficiency technology provides environment-friendly approach towards mobile communication. In order to achieve this energy efficiency, we considered renewable energy resources and efficient use of eNB (BS) in our work. We designed an optimal eNB ON/OFF operation and renewable resources like solar and wind energy to reach the effective energy and power consumption. The eNB's active/idle state is identified by measuring the responder frequency. We proposed a unique algorithm to carry out the switching between eNB by considering number of users as load. The potency of our work is endorsed by simulation results and to control the pollution by mainly reducing the CO_2 in order to reduce the global warming.

Keywords BS/eNB · Renewable resource · Responder frequency
Energy efficiency

1 Introduction

Greening is not solely a contemporary approach, but is an emerging urgency to maintain our economic and environment sustainability. As a part of the international efforts for conserving energy and reducing the carbon footprint, moving to an energy-efficient mobile framework is of high precedence to the communication industry. In recent years, the green energy-efficient communication has acknowledged more considerations due to overwhelming growth of mobile Internet access demand. The

---

V. Manickamuthu · B. V. Namrutha Sridhar · I. Chandra (✉)
Department of Electronics and Communication Engineering, R.M.D. Engineering College, Chennai, India
e-mail: chandra.rajaguru@gmail.com

V. Manickamuthu
e-mail: manickamuthu97@gmail.com

B. V. Namrutha Sridhar
e-mail: namruthas97@gmail.com

S. Smys et al. (eds.), *International Conference on Computer Networks and Communication Technologies*, Lecture Notes on Data Engineering and Communications Technologies 15, https://doi.org/10.1007/978-981-10-8681-6_6

cellular wireless systems are currently changing over to LTE. In order to meet all the mobile broadband applications, the next generation will suffer from bursting growth in traffic which results in increased use of power for maintaining the eNB in the active state [1, 2]. To overcome this, energy saving is one of the key factor for mobile operators. By deploying the energy-efficient eNB, operators can reduce the carbon footprint that is released into the environment from their network [3]. Basically, we depend on fossil fuels which lead to pollution, climate change [4, 5]. Instead of fossil fuels, using the renewable energy resources leads to power consumption by improving the energy efficiency of base station key components, such as power amplifiers and air conditioners, is of greater importance [6, 7]. In the life cycle of a base station (eNB), power consumption is ascendant in the deployment and operational phase than in the productive phase. Therefore, ease of deployment and low-power operation are detracting for an energy-efficient mobile infrastructure. Independent of the actual energy source that is used for powering the access network achieving the highest possible energy efficiency is very substantial. This applies to conventional, grid-powered network elements in larger cities as well as even more developing countries without reliable grid-based energy [8, 9]. Productive energy management is thus a key requirement for favorable and profitable operation of mobile communication networks. In order to achieve this, here we show a heuristic approach to handle the traffic, energy consumed by eNB and the process of switching the eNB based on the traffic experienced by the eNB.

2 System Model for Energy-Efficient Communication

The conceivable model for the future LTE-A standard with energy-efficient communication is shown in Figs. 1 and 2. The system components include main towers, sub-towers and UE's. The main towers are in active mode always independent of the traffic. The sub-towers are positioned in between the main towers. The sub-towers will be activated only based on the requirement if the number of users starts to increase beyond limit.

Those sub-towers are kept in idle state and are shown in Fig. 1 where ($n > N$). Each tower has the limit to UE's. If the number of users exceeds the assigned limit, the main tower instructs the sub-towers to become active by considering the direction of demand and is shown in Fig. 2.

With this system model we can realize an energy-efficient communication. Along with this, the power conservation is achieved by using renewable energy resources such as gird-based energy for maintaining power amplifiers, air conditioners, lighting to the towers and is of greater importance [10, 11]. The first thing is effective use of eNB to achieve energy-efficient communication. The second thing is conserving power and reducing carbon footprint by using renewable resources instead of fossil fuels.

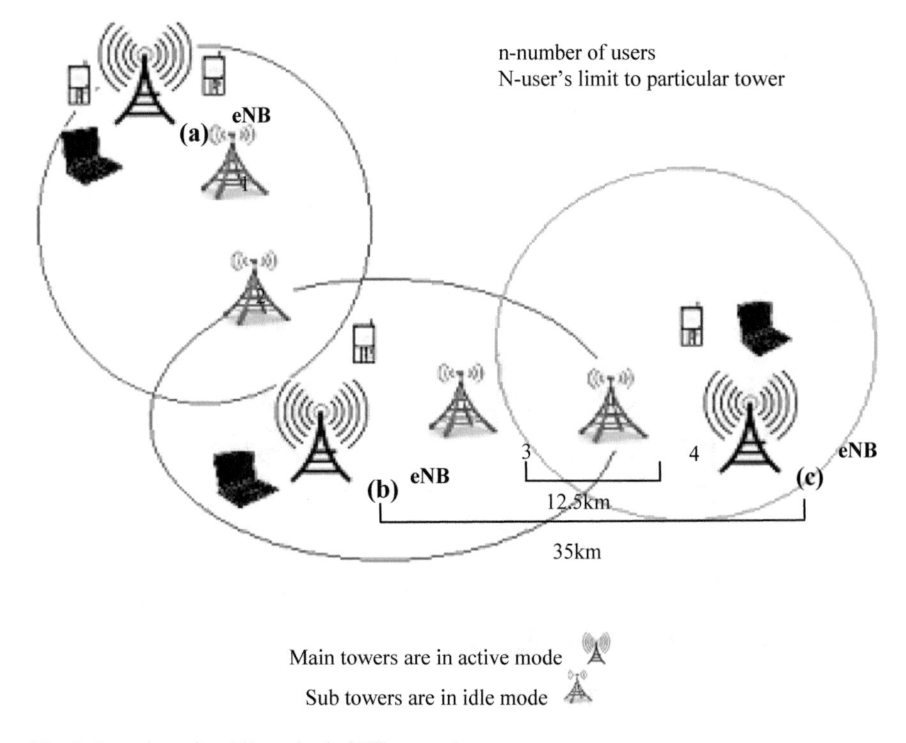

Fig. 1 Location of mobile nodes in LTE network

2.1 LTE-A

The guidelines which forced the further advancement of LTE (Long-Term Evolution) towards LTE-Advanced (LTE-A) are to provide higher data rate in worth efficient way and it is also credited to as 4G [12, 13]. It focuses on higher data rate UL 1.5 Gbps, increased spectral efficiency maximum of 16 bps/Hz in R10 and growth in the number of active subscribers in a simultaneous way [14].

2.2 Related Work

In the existing system, all the mobile towers are kept ON in a particular locality irrespective of the number of users. This results in high power consumption. A typical mobile network may consume approximately 40–50 MW, expelling the power consumed by user's handsets. The service providers use diesel to power their network when direct electricity connections are not readily accessible. As a result, a polluted environment is organized and as whole of about 1% is utilized by the mobile network

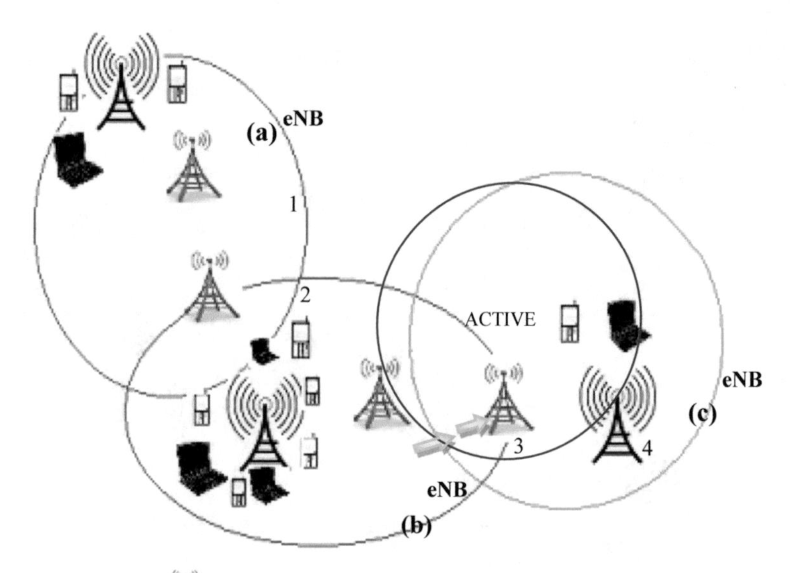

The Main tower **b** exceeds the user limit and transfers the control to the nearby Sub tower 3 by considering the direction of demand

Fig. 2 Power consumption in LTE network

itself from the total generated power. In addition to this, the lighting and cooling units are always functioning, thereby significantly proliferating the power consumption rate day by day.

3 Proposed Method

Proper base station management can lead to significant and efficient power and energy consumption in Long Term Evolution-Advanced (LTE-A) systems. In the proposed algorithm, optimum number of towers (BS) is kept in active state. As the load, i.e., number of user's increases, corresponding responder frequency also slowly increases from zero. The responder frequency is the one which is responsible for the switching between the BS. Depending on the angle in which the users are increasing the responder frequency accordingly activates the neighboring BS. The number of users can be determined by using the formula

$$\text{Number of users} = \frac{\text{Total power} - 87\% \text{ of power}}{3\text{ W}} \tag{1}$$

The average power consumption (APC) is measured by the ratio [11] and given by

$$\text{APC} = \text{Power consumption}/\text{Area} \quad (2)$$

These 3 W of power denotes the power required for the signal to reach the cellular device from the BS. Generally in the BS 80% of the power is utilized by the local users and the rest 20% is left for mobile users. If the BS load exceeds predetermined value say 80%, then the responder frequency is sent to the nearby efficient tower to take up the remaining load by means of wireless sensors networks. In addition to this, we have also proposed the green renewable energy resources like solar and wind energy which generates the voltage for air conditioning, fan cooling and lighting. Moreover with the help of ambient analyzer maximum power saving can be achieved by means of localized power controller where the lighting and cooling units turned on as per the requirement. The simulations of this method are implemented with Visual Basic (VB) 6.0; while in the conventional scheme it was done with MATLAB.

3.1 Power Abatement Model

In the overall block diagram shown in Fig. 3 the voltage and current sensing circuit senses the power from the power amplifier and feeds it to the monitoring system. When the number of users increases the main server will generate the responder frequency which is sent to the corresponding tower. The responder frequency which is received from the main server will activate the corresponding tower's base station components from idle state to charging state. When the responder frequency is received it is also displayed in the monitoring unit. The temperature and light sensor senses the temperature and illumination level of environment and then through localized power control, corresponding relays for cooling and lighting units are operated. If the power amplifier output is found to be zero then there will be no users present in that particular instant. Similarly once the number of user's level exceeding more than 80% then relay to activate the neighboring tower is energized.

3.2 Power Consumption Design Flow

The algorithm proposed includes monitoring the number of users (n) along with the surrounding environmental parameters like temperature, humidity and light. If the temperature is less than the determined temperature (say "X") then air conditioning is switched ON. Similarly if humidity is less than the estimated/required value (say "Y"), then fan cooling is needed and if light is less than the required figure (say "Z"), then lighting is switched ON. The voltages required for all the above three (i.e.) AC,

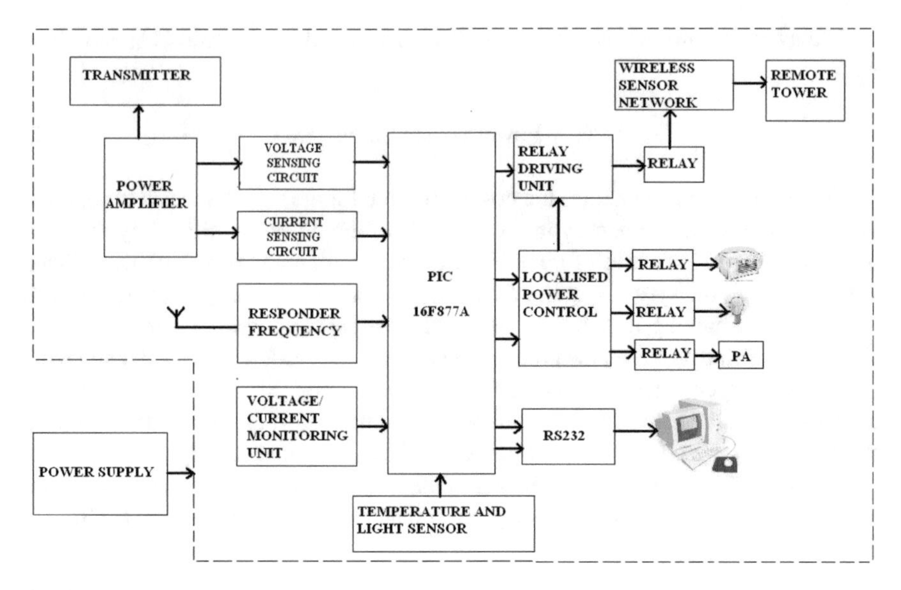

Fig. 3 Overall block diagram of abatement of power consumption in mobile base stations

fan, lighting is provided by using green renewable resources like solar and wind power. On the other side if the number of users (n) (i.e.) loads is greater (Fig. 4).

On the other hand side if the number of users (i.e.) the load is greater than a particular amount (here we considered it as 700) than a particular amount (N) (here we have considered it as 700), the responder frequency is generated accordingly which turns ON the nearby eNB. By adopting this algorithm, we can achieve effective power consumption.

4 Simulation Scenario

This section studies the proposed network level power minimization by providing the simulation results. In this simulations are performed by using Visual Basic (VB) version 6.0. VB lets you to take full advantage of window graphical environment to built powerful application quickly. The new OLE (Object Linking Embedding) control allows in place editing facility.

We simulate a simple LTE-A network using VB 6.0. The network comprises of a main tower (eNB) along with the surrounding BS as in Fig. 1. The assumption is that all the main towers are always in active state and only the sub-towers will be activated depending on the load.

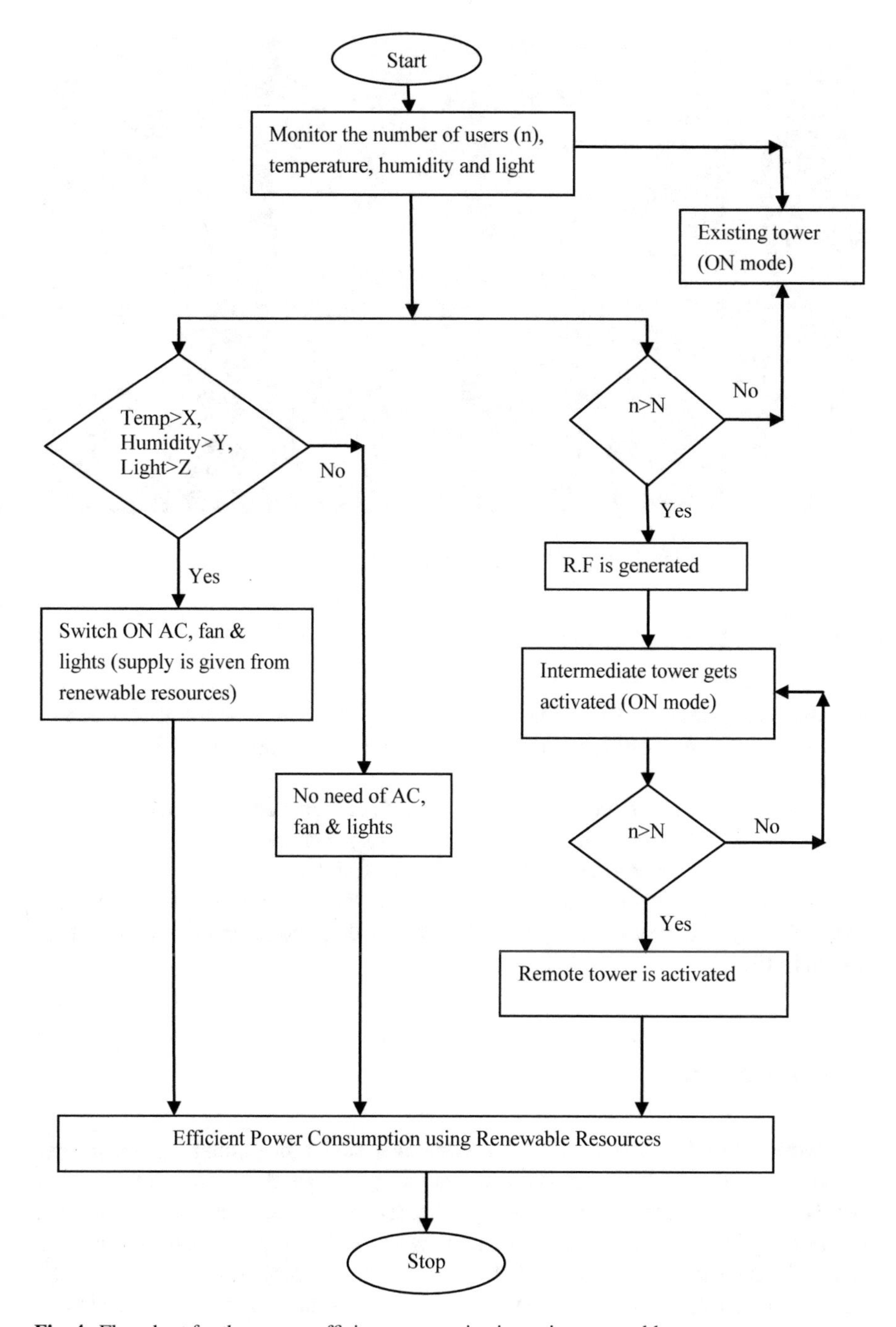

Fig. 4 Flowchart for the energy-efficient communication using renewable resources

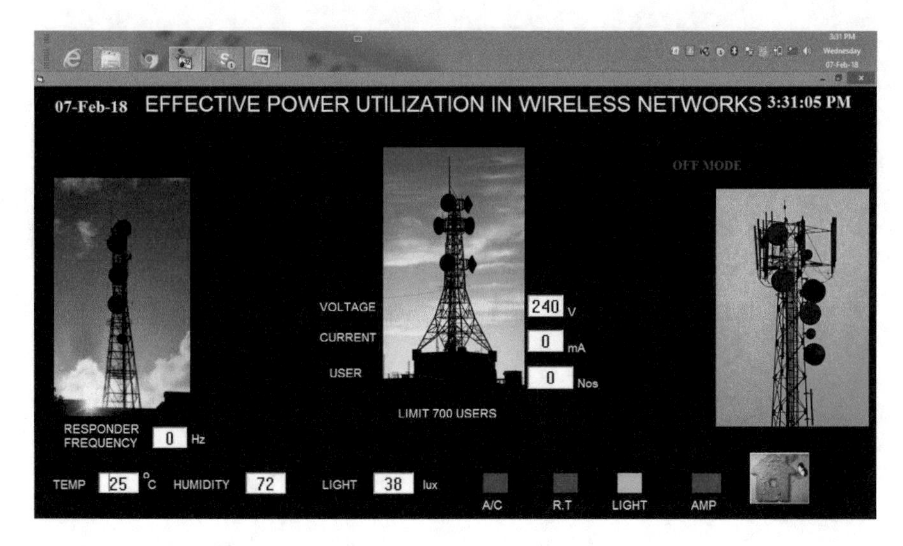

Fig. 5 Main tower in active mode and sub-towers are in idle mode

4.1 Experimental Results

Initially when number of users is within the BS limit, the remote tower is in idle state as there is no signal transmission because the power amplifier is in OFF state as shown in Fig. 5.

When the load starts increasing, the responder frequency is the responder frequency is sent from one BS to the neighboring sub-tower and the amplifier is turned ON as shown in Fig. 6. As the number of users (load) increases further, the command is sent to the remote tower to turn ON. By sensing the surrounding temperature and light intensity and depending upon the condition they are turned ON and OFF as shown in Fig. 7.

5 Conclusion

This work focused on energy-efficient communication for the future LTE-A standard. Here we proposed an LTE-A model which supports energy-efficient communication. We also worked for conserving power by using renewable resources such as solar and wind energy instead of fossil fuels and thereby worked for reducing carbon footprint and made a way to control pollution. We proposed an algorithm to execute our model. Simulation results validated that our designs can considerably reduce pollution and realize an energy-efficient communication.

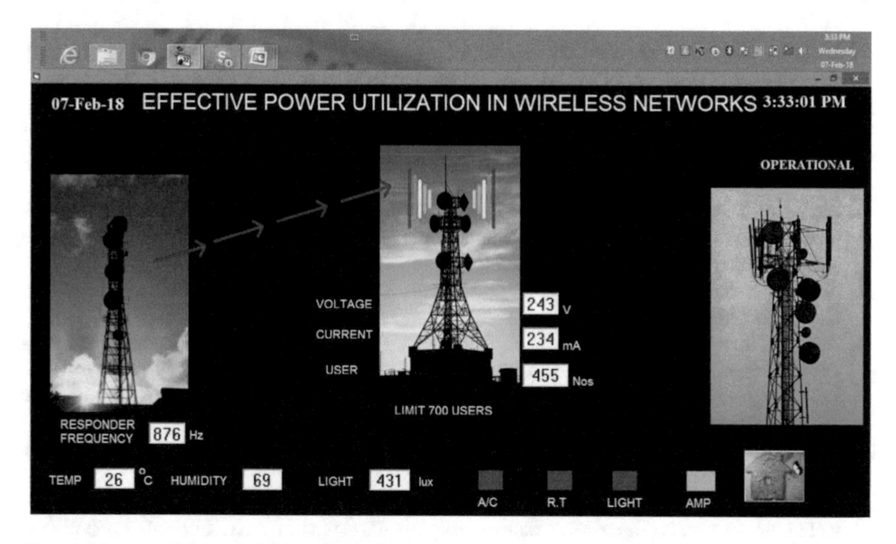

Fig. 6 The condition $n > N$ is achieved and main tower sends responder frequency to ON the sub-tower

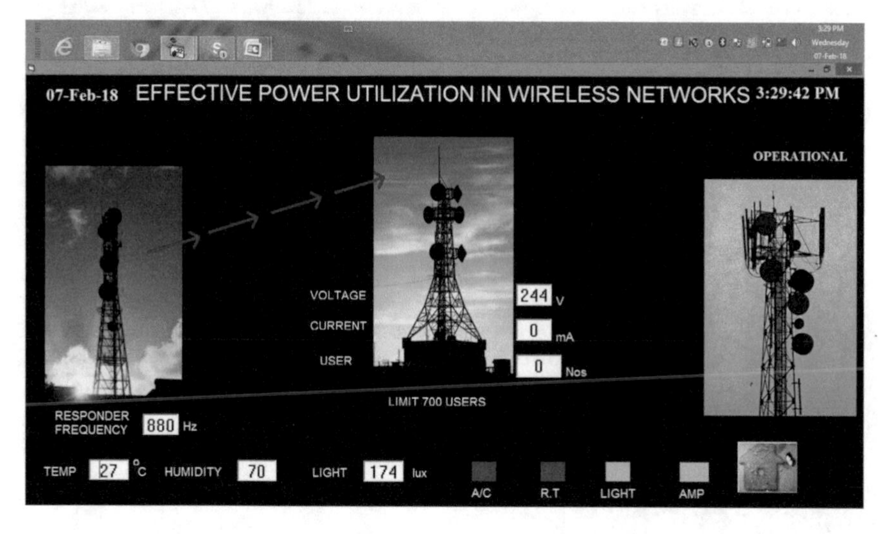

Fig. 7 Based on the surrounding condition, light is switched ON

References

1. Ma, R., Xia, N., Chen, H.-H., Chiu, C.-Y., Yang, C.-S.: Mode selection, radio resource allocation and power coordination in D2D communications. IEEE Wireless Commun. (2016)
2. Gao, C., Zhang, W., Tang, J.: Building elastic hybrid green wireless networks. IEEE Internet Things J. **4**(6) (2017)
3. Yaacoub, E.: RESCUE: renewable energy small cells for utility enhancement in green LTE HetNets. IEEE Syst. J. **11**(4) (2017)

4. Deruyck, M., Renga, D., Meo, M., Martens, L., Joseph, W.: Accounting for the varying supply of solar energy when designing wireless access networks. IEEE Trans. Green Commun. Netw. (2017)
5. Xie, R., Li, Z., Huang, T., Liu, Y.: Energy-efficient joint content caching and small base station activation mechanism design in heterogeneous cellular networks. China Commun. **14**, 70–83 (2017)
6. Shakir, M.Z., Qaraqe, K.A., Tabassaum, H., Alouini, M.-S., Serpedin, E., Imran, M.A.: Green heterogeneous small-cell networks: toward reducing the CO_2 emissions of mobile communications industry using uplink power adaptation. IEEE Commun. Mag. (2013)
7. Han, T., Ansari, N.: On greening cellular networks via multicell cooperation. IEEE Wirel. Commun. (2013)
8. Andreev, S., Pyattaev, A., Johnsson, K., Galinina, O., Koucheryavy, Y.: Cellular traffic offloading onto network-assisted device-to-device connections. IEEE Commun. Mag. (2014)
9. Davaslioglu, K., Ayanoglu, E.: Quantifying potential energy efficiency gain in green cellular wireless networks. IEEE Commun. Surv. Tutorials **16**(4) (2014)
10. Che, Y.L., Duan, L., Zhang, R.: Dynamic base station operation in large-scale green cellular networks. IEEE J. Sel. Areas Commun. **34**(12) (2016)
11. Wu, J., Zhang, Y., Zukerman, M., Yung, E.K.-N.: Energy-efficient base-station sleep mode techniques in green cellular networks—a survey. IEEE Commun. Surv. Tutorials **17**(2) (2015)
12. Lee, H., Vahid, S., Moessner, K.: A survey of radio resource management for spectrum aggregation in LTE-advanced. IEEE Commun. Surv. Tutorials **16**(2) (2014)
13. Coskun, C.C., Ayanoglu, E.: Energy-efficient base station deployment in heterogeneous networks. IEEE Wireless Commun. Lett. **3**(6) (2014)
14. Chung, W.-C., Chang, C.-J., Teng, H.-Y.: A Green Radio Resource Allocation Scheme for LTE-A Downlink Systems with CoMP Transmission. Springer (5 Jan 2014)
15. Chandra, I., Helenprabha, K.: Efficient resource allocation with QoS guarantee on LTE-A downlink network systems. J. Netw. **9**(9) (2014)

V. Manickamuthu was born in Karaikudi, Tamil Nadu, India on 4th March, 1997. She received the B.E. degree in Electronics and Communication Engineering from R.M.D. Engineering College affiliated to Anna University. Her areas of interest include Energy-efficiency in next-generation mobile networks, wireless communication and sensor networks.

B. V. Namrutha Sridhar was born in Chennai, India on July 1997. She received the B.E. degree in Electronics and Communication Engineering from R.M.D. Engineering College affiliated to Anna University. Her areas of interest include Energy-efficiency in next-generation mobile networks, load balancing and wireless communication and sensor networks.

Dr. I. Chandra B.E., M.Tech., Ph.D. is Associate Professor in the Department of Electronics and Communication Engineering at R.M.D. Engineering College where she has been a faculty member since October 2005. She completed her B.E. Degree in Electronics and Communication Engineering in the year 1996 at Thiagarajar College of Engineering affiliated to Madurai Kamaraj University and M.Tech. Degree in Applied Electronics under Dr. M.G.R. University in the year 2007. She has completed Ph.D. in the area of Wireless Communication under St. Peter's University. She has 17.6 years of teaching experience to UG classes and has guided many B.E. projects. She has published four papers in National/Inter-National journals and National/Inter-National Conferences. She has organized AICTE, IETE Sponsored workshops and Anna University approved FDTP. She has delivered guest lectures on FDP. She has attended many Workshops, FDP and Seminars. She has received the fund of Rs. 60,000/- from AICTE for conducting NATIONAL LEVEL SEMINAR by 2009. Her Areas of Interest include Electromagnetic Fields and Wireless Communication. She is a life time member of ISTE and also in various Associations.

Spam Detection Using Machine Learning in R

K. R. Vidya Kumari and C. R. Kavitha

Abstract Social Network (SN) is an online platform broadly used as communication tool by millions of users in order to build social relationships with others for knowledge point of view, career purposes and many more. Social Networks such as Twitter, Facebook, and LinkedIn have become the most leading tools on the web. Spam, floods the Internet with many copies of the same message and it can be manifest in numerous ways, it includes bulk messages, malicious links, fake friends, fraudulent reviews and personally identifiable information. The aim of this paper is to classify the tweets into spam and non-spam using Machine Learning and which will give the best results.

Keywords Social network · Hashtag · Spam detection · R · Machine learning

1 Introduction

Data mining is the process of sorting through a huge amount of data set to identify patterns or examining the existing large databases. Also, to generate new information or to solve problems through data analysis and to predict future trends.

Social Networks have become trendier in the entire world. The purpose of the Social networks are sharing one's personal thoughts, views, and Social activities. In Social Network a small information is spread widely. So, many chose Social Network for sharing information. Social Networks such as Twitter, Facebook, and Snapchat has been a part of our daily lives. In Parallel, the spam started to spread broadly all around Social networks in various forms. People share their thoughts or activities in the form of post in Facebook or tweets in Twitter to their set of friends or followers.

K. R. Vidya Kumari (✉) · C. R. Kavitha
Department of Computer Science & Engineering, Amrita School of Engineering,
Amrita Vishwa Vidyapeetham, Bengaluru, India
e-mail: krvidyakumari29@gmail.com

C. R. Kavitha
e-mail: cr_kavitha@blr.amrita.edu

S. Smys et al. (eds.), *International Conference on Computer Networks and Communication Technologies*, Lecture Notes on Data Engineering and Communications Technologies 15, https://doi.org/10.1007/978-981-10-8681-6_7

A feature is available in social network that helps to share the post of one social network to another. Again and again, the same post would be shared in various social network so that the information can reach millions of people. Altogether, makes high similarity among different Social Network for a Single User [1].

Spammers are one who are misusing these Social Network platforms. They spread spam by spreading rumors, misinformation, hoax news, uninvited messages. Spammer subscribes to dissimilar mailing list to advertise any commercial purpose to send spam messages regularly. Due to these activities, it disturbs the Non-Spammers that is genuine Users and it decreases the opinions of Social Networks platforms.

The target of the Spammers/Malicious users are databases of Social Networks with larger number of users. Spam is not easy to detect and it takes many different forms. Day by day many people are facing spam on Internets such as email spam and newsgroup. Electronic message system which sends spontaneous junk messages can also be defined as Spam. Some Spammers post URLs which will contain sensitive data or it may be phishing websites which steal user's data. So, it is observed that the spams are spreading widely with the growth of Social Networks [2].

In this work, we have chosen Twitter as a Social Network platform. On March 21, 2006 Twitter was launched and has 330 million monthly active users who share information in the form of tweets. Twitter uses a chirping bird as its logo and hence the name Twitter. Anyone can read and send message (i.e.) tweets which are 140 characters long. User's tweets may contain opinions, photos, news, videos, messages, and links. There are standard terminology in the twitter which are Tweets, Retweets, Followers and Followings, Hashtags, Lists, Mentions and Direct Messages. By default, who all are following a specific user, can read all the tweets tweeted by that user [2].

Spam profiles can be easily identified by if the user post consists of popular hashtags (#) with unrelated information and posting other user's tweet as their own. And the user follows a huge number of users in a short span of time.

2 Related Work

The social networks have important distinct features from the security perspective and it usually regulates to whom the personal information can be viewed. Usually, User's friends know personal information of that user, the shared information may be public or private. If only the user's network can view the information then the information or message is private.

Social Network graph is a Directed graph. Based on the social graph model as shown in Fig. 1 is Twitter Social Graph [3]. It represents the followers and following of the certain number of the random users collected from the public timeline of Twitter.

Social Network sites do not have a mechanism to provide secure system confirmation for identity, therefore the user spread and hidden in another person's account. For the popularity, the users disclose their personal details by accepting unknown

Fig. 1 Twitter social graph

request. Most of the Users, even they do not know in real may click a link or URLs sent by them. Many of the spammers use this technique for advertising a website [4].

Spam is a common problem in online media and can be found in the different form as emails, microblogging, websites, comments, videos, and reviews to name a few. Mainly Spammer chooses a platform-specific techniques to avoid spam detection systems such as Spam on Email, Web, Comment, and Product Review. Moreover, nowadays it became difficult to segregate between spammers and legitimate users because they post legitimate content and may also have many followers. The Spammer concentrate on trending hashtag and post spam tweet by using trending hashtag. Similarly, the credibility of tweets on trending topics is estimated based on the tweet content, topic, user profile, and propagation-based features [5].

Spam accounts are targeting and abusing social network platform to influence normal user's involvement. According to the Rules given by the Twitter, there are some unusual activities that are considered as what spammers actually do that includes:

- Propagating malicious links;
- creates misleading or false messages;
- Trading followers; and
- Aggressive friending.

Spam can be spread to non-spammer users through the following functions:

- Home timeline: it shows all original and reposted messages from those being followed by the users of that profile;
- Search timeline: display result of a query or a word is given in the search box;
- Hashtags: a word prefixed with the symbol #, and it marks messages with topics;
- Profile bio: Spam account target normal users would click and view the URLs appears on their bios; and
- Personal messages: a private chat between two users [6].

Every day, a huge amount of user's data being handled by sites of Social Networks and it faces vulnerable to security and privacy issues. Various attacks on the Social Network are revealed, they are:

- Social bots—User's personal data are collected by bogus profiles.
- Sybil (fake) attack—The main intention is to harm the status of non-spammer users by obtaining various fake identities and act like a legitimate user in the system [7].
- Spammers—In Social Network it sends spam messages.
- Phishing—It acquires non-spammer's sensitive data by portraying as an ethical third party.
- Viruses—In Social Networks it spread malicious information [8].
- Clone and identity theft attacks—To trick a non-spammer user's friend in the network by creating a profile which already exists on various social networks by Spammer. The cloned identities steal victim's information, which is sent by the spammer as a friend request to the victim [2].

The Twitter's legitimate profiles divided into two groups: Graph and Content-based features and a few are defined:

- Content-based features—Trending topics, Repetitive Tweets, Links and Replies, and Mentions these are the four important feature to find spam user profile.
- User Based Techniques are Followers count, Following count, Mentions count, Reply count, Date of creation and Time of tweet.
- Hybrid Based Techniques—If the Spammer manipulates the attributes then user based techniques are not effective. Similarly, content-based techniques have the limitation if the spam contains a few words in the link or URLs. So hybrid uses both content and user to find or identify the spam.
- Relation Based Techniques—In the user-based technique, the spammer is detected after the spam has been sent to a valid user. So, there is an unavoidable postponement between the creation of spam account and detection. Therefore the relation based is difficult to manipulate.

Two types of relation features are:

Distance—It is the length of the shortest path between users.
Connectivity—The connectivity resembles the relationship strength [9].

Our work classifies based on features from the extracted spam and non-spam tweets. The values are found from tweets and the dataset is applied in the Machine Learning Algorithm to find the best accuracy of the algorithms.

3 Methodology

3.1 R-Studio

R-GUI is an open-source and free integrated development environment (IDE) for R, a programming language for statistical computing and graphics.

R has an OAuth Package for authentication of Twitter API's. To run Twitter API, a Twitter application have to be created. Creation of App developer is done using following steps:

- The user who wants to create application has to register with Twitter. It provides a consumer key and secret key to the user which have to be used for authentication of user's application.
- To initiate the Authentication process these keys are used. Twitter does the Verification of User identity and issues verifier that is PIN 3. The PIN is used for application by the user.
- Access Token and Access Secret are requested by using the PIN information in next step, exclusive to the user from Twitter API.
- By using GetUserAccessKeySecret, to get information of Token and secret key for future use [10].
- And a few R packages have to be loaded.

3.2 Dataset

This paper uses R-studio to extract tweets from Twitter API. From Twitter API, tweets are extracted based on one of trending hashtag. In this work, a trending hashtag is "#ChennaiRains". Based on the hashtag the tweets were collected from October 2017 to December 2017.

Northeast Monsoon gave heavy rains with very heavy showers over the capital of Tamil Nadu. During this period, three digit rains lashed Chennai that led to water logging and traffic chaos at many places. In Twitter, the Chennai Rains hashtag ranked 1 or 2 for a few weeks. And there was a raising fear of a repeat of the 2015 rain havoc. Therefore, the extracted tweets include both spam and normal tweets.

The tweets were cleaned by removing the duplication of tweets.

3.3 Features

In this work, we have chosen eight features as shown in Table 1. The features are based on Account-Based and Tweet-Based Features. The values were extracted from tweets in R through Twitter's Public Streaming API. These features are referred to

Table 1 Extracted features

Feature category	Feature name	Description
Account-based features	account_age	The age of an account
	no_tweets	The number of user posted tweets
	no_follower	The number of followers
	no_following	The number of followings
Tweet content-based features	no_retweets	The number of times this tweet has been retweeted
	no_hashtag	The number of hastags in this tweet
	no_url	The number of URLs contained in this tweet
	no_usermention	The number of times this tweet being mentioned

as Lightweight Statically features for detecting spam. It requires some computation work. During the training and testing a model the processes it helped to minimize the calculation complexity, due to this it makes to detect spam in real life with the huge number of training and testing data.

3.4 Experiment Environment

Hardware Specification

Processor	Processor Intel(R) Core(TM) i7-3537U CPU@ 2.00 GHz
RAM	8.00 GB
System type	64-bit Operating System, x64-based processor

Software Specification

R-GUI (64 bits) or R-Studio

3.5 Machine Learning Algorithm

Support Vector Machine—Support vector machines (SVM) [11] are models for supervised learning. SVM is mostly used in the classification problem. With given data, it will create hyperplane to classify the objects correctly. More formally, it is used in classification or regression by constructing a hyperplane or number of hyperplanes in a high dimension space or detecting outliers. Possibly, a best separation or classification is obtained by the largest distance of the hyperplane to any class which has the immediate training data point. In common, the larger the margin the lower the generalization error of the classifier.

Naïve Bayes—Naive Bayes is a conditional probability model, given a problem instance to be classified, represented by a vector using n features (independent variables), it assigns to this instance probability [12].

Naive Bayes classifier is a simple probabilistic classifier based on applying Bayes' theorem with strong independence assumptions between the features. Naïve Bayes classification can be two class (binary) or a multi-class classification algorithm. In a simple term, Naive Bayes classifier considers that the presence of a specific feature in a class is unrelated to any other feature's presence. Naïve Bayes is used when the dimensionality of input is high. The conditional probability and class probability are stored in files to train Naïve Bayes. It comes in supervised learning. The Naive Bayesian classifier is based on the well-known Bayes theorem:

$$P(c|m) = \frac{P(m|c).P(c)}{P(m)} \tag{1}$$

The conditional probability of $P(c|m)$ is also known as the posterior probability for c, as opposed to its prior probability $P(c)$. Where $P(m|c)$—Likelihood and $P(m)$—Predictor Prior Probability.

4 Result and Analysis

The analysis in each tweet is done if the twitter id is pointing to exact tweet, otherwise by using a screen name or twitter profile name. By reading first 20 to 50 tweets, it helps to conclude that the tweets are spam or not.

Based on the tweets, it is found that if the tweets are tweeted from the legitimate accounts then they are non-spam tweets. If it is a spam profile then most of the tweeted tweets are spam tweets.

In Fig. 2, we have analyzed the features of the dataset.

Most of the case the non-spammer users have the large count of the followers. Comparatively the spammer user have a small count of the followers as shown in Fig. 2a, where X and Y represent the number of users and number of Followers.

For the Spammers, the count of following is larger than the non-spammers following counts as shown in Fig. 2b where X and Y axis represents number of users and Number of following.

In the case of User Mentions of tweets, the non-spammer's tweet has the high count of mentions than spammer's as shown in Fig. 2c Where X and Y axis depicts number of users and Number of User Mentions.

The spammer mostly targets on hashtags. In Fig. 2d, it shows that the spammer uses more number of hashtags than legitimate user. The X and Y axis represents number of users and Number of Hashtags.

From Fig. 2e, the Spammers Account age is less than Non-spammers Accounts. The X and Y axis represents Number of days and Number of Users.

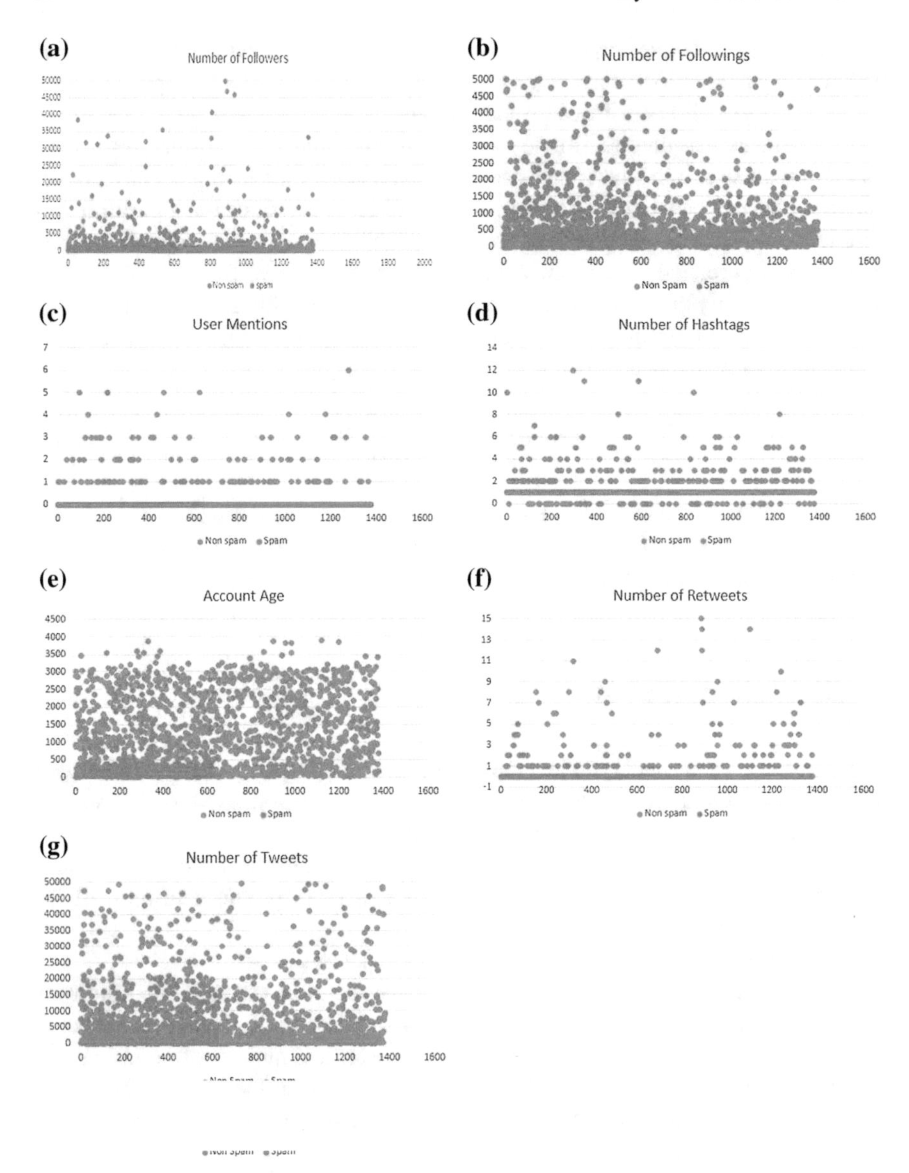

Fig. 2 Anaysis of each feature, **a** number of followers, **b** number of followings, **c** usermentions, **d** number of hashtags, **e** account age, **f** number of retweets and **g** number of tweets

Number of Retweets for the Spam account is less than Non-spammer account shown in Fig. 2f.

Number of Tweets shown in Fig. 2g, *X* and *Y* axis represents Number of Tweets and Number of Users.

The dataset is split into training and testing, where for training 70 and 30% for the testing.

When the dataset is applied in SVM it is trained, then tested and got 76% accuracy. And the same dataset is applied in Naïve Bayes where it is also trained and tested and got 92%. From this comparative study, we can conclude that Naïve Bayes is better than SVM for this dataset. The result may vary based on the feature selection.

5 Conclusion and Future Enhancement

In this work, we have extracted 15,000 tweets from which we have found user profile on the twitter. From this extracted lightweight features, it is feasible to distinguish non-spam tweets and spam tweets from the labeled dataset. Based on the features also the result varies.

Discrete dataset have been used in the work. For this dataset, the Naïve Bayes gives the best accuracy than SVM. In Continuous dataset, the classifier's ability will be better to detect spam, which can be taken as future scope.

References

1. Xu, H., Sun, W., Javaid, A.: Efficient spam detection across online social networks. In: 2016 IEEE International Conference on Big Data Analysis (ICBDA). IEEE (2016)
2. Verma, M., Sanjeev, S.: Techniques to detect spammers in twitter-a survey. Int. J. Comput. Appl. **85**(10) (2014)
3. Wang, A.H.: Don't follow me: spam detection in Twitter. In: Proceedings of the 2010 International Conference on Security and Cryptography (SECRYPT). IEEE (2010)
4. Eshraqi, N., Mehrdad, J., Moattar, M.H.: Spam detection in social networks: a review. In: 2015 International Congress on Technology, Communication and Knowledge (ICTCK). IEEE (2015)
5. Sedhai, S., Sun, A.: An analysis of 14 Million tweets on hashtag-oriented spamming. J. Assoc. Inf. Sci. Technol. **68**(7), 1638–1651 (2017)
6. Liu, C., Wang, G.: Analysis and detection of spam accounts in social networks. In: 2016 2nd IEEE International Conference on Computer and Communications (ICCC). IEEE (2016)
7. Suresh, A., Babu, V., Jithin, J., Manne, J., Abirami, K.: Sybil defense mechanisms: a survey. Int. J. Appl. Eng. Res. **10**, 24693–24700 (2015)
8. Gregory Paul, T.G., Gireesh, K.T.: A framework for dynamic malware analysis based on behavior artifacts. Adv. Intell. Syst. Comput. **515**, 551–559 (2017)
9. Kaur, P., Singhal, A., Kaur, J.: Spam detection on Twitter: a survey. In: 2016 3rd International Conference on Computing for Sustainable Global Development (INDIACom). IEEE (2016)
10. Rani, V.V., Sandhya Rani, K.: Twitter streaming and analysis through R. Ind. J. Sci. Technol. **9**(45) (2016). https://doi.org/10.17485/ijst/2016/v9i45/97914
11. Pathan, S., Goudar, R.H.: Detection of spam messages in social networks based on SVM. Int. J. Comput. Appl. **145**(10) (2016)
12. Pandey, S., Supriya, M., Shrivastava, A.: Data classification using machine learning approach. In: The International Symposium on Intelligent Systems Technologies and Applications. Springer, Cham (2017)

Ms. K. R. Vidya Kumari Currently pursuing M.Tech., CSE in Amrita School of Engineering, Bengaluru, Amrita Vishwa Vidyapeetham, India. Received B.Tech. in CSE from the Anna University, Chennai in 2014. Research interests are Data Mining, Machine Learning and Social Network Mining.

Ms. C. R. Kavitha Currently serves as Asst. Professor in Amrita School of Engineering, Bengaluru, Amrita Vishwa Vidyapeetham, India. Pursuing Ph.D. in the area Cloud Security in VTU, received M.Tech. in CSE from VTU in 2010. Research interests are Cloud Computing and Security, Cryptography and Network Security, Data Mining.

QOS-Based Technique for Dynamic Resource Allocation in Cloud Services

Jeevan Jala and Kolasani Ramchand H. Rao

Abstract Cloud computing is a growing trend which provides us less cost with many types of facilities like resources of maximum, elasticity, dynamic resource provisions, and self-service providing to the user on an Internet. Cloud offers a huge and vast platform in allocating and provision of resources to the user with less computing speed and makes it very easy to use. Resource allocation and provision is done based on the quality of service in an effective manner, but it lack with certain cases of dynamic resource provision, many of the works done are based on resource management and utilization with high cost and speed. But our proposed method wills effectively management in allocation the resources with less time dynamically based on the available of the resource and manage it very effectively, effectiveness is based on low cost and high utilization of resource. Our aim is to allocate the resource in peak time and control it very effectively in managing the resource. The proposed algorithm will analyze and allocate the resource in based on various policies and is compared with the others.

Keywords Service · Provision · Allocation · Computing and dynamic

1 Introduction

Cloud Computing has been a prominent and virtualization in providing an on-demand sharing of information and resource rather than in handling personal unit of information for handling Internet per pay use of application in driving the system, which provides a reliability in servicing the customer with QOS. The most and major platform in cloud computing is to allocate resource very easily with less amount of time along with cost efficiently. The model of cloud along *i*th the services are classified

J. Jala (✉)
Department of Computer Science, Acharya Nagarjuna University, Guntur 522006, India
e-mail: jeevanjala@gmail.com

K. R. H. Rao
Department of Computer Science, ASN Degree College, Tenali, India

S. Smys et al. (eds.), *International Conference on Computer Networks and Communication Technologies*, Lecture Notes on Data Engineering and Communications Technologies 15, https://doi.org/10.1007/978-981-10-8681-6_8

based on the following services a community cloud, hybrid cloud, private cloud, and public cloud. The services of public are been accessible to each and every one, there is no restriction in accessibility of service. In case Private cloud certain security feature are proposed, where as in hybrid cloud it is a combination of public and private. Community cloud is a type of cloud which is managed by a third party manager, it is PaaS and IaaS. Which manages the user CPU, memory and power processing with the help of user software remotely; it also provides various types of software's to the user.

Virtualization process means which may not be real but functionally real, which are handled by numerous servers at a time bound. Various virtualization techniques are available which will run on different types of operation system which will be executed on a single physical machine. Problems related to on-demand features are the main interference issues which are to be executed in maintain the workload for long-term resource reservation which may decrease the performance related to QOS provision and also maintain the necessary for managing the resource for which allocation has to be done necessary for effective management of resource based on the consumer request for effective service and provision.

QOS requirement is the basic feature of cloud in maintain the resource, managing the resources has to be done based on the agreement or negotiation between the provider of service and user. A plan of agreement in allocation of the resource has to be done for effective communication, without knowing the SLA, there wont be any quarantee for QOS service.

2 Related Work

Kaur (2014) proposed a literature survey of different computing techniques on resource provision and allocation in cloud management. All the methods are compared based on the plan and QOS metric and verified with basic efficiency in managing and allocation of resource.

Magoul (2015) proposed a pricing of resource and policy of allocation of cloud computing. The allocation of resource is done based on the pricing of the customer budget which has to be satisfied regardless of predicted budget. This allocation can only manage limit number of resource based on QOS.

Janaki (2015) devised an algorithm name advanced reservation, which uses a counter in providing the resource to the user request, but this method does not support the policy of multi-agent request which are done at the same time. This method is implementation using effective QOS policies in an effective manner in allocation of resource.

Garg (2015) proposed a provider SAas method for allocation of resource based on provision, this method only minimizes the cost parameter in a metric but violation is done on SLA and QOS service parameter. Y. Zhao devised a dynamic algorithm for resource management based on metrics, the proposed techniques wont follow the QOS metric in allocation of resource like time, cost and reliability.

3 Architecture Diagram

In Fig. 1 shows the architecture used in providing QOS in allocation of resource provision, initially a request which is made by the user will be applied with a legacy process in providing the service based on the agreement and the level of legacy. Based on the availability in resource, the manager of the resource will handle the user based on the conditions of policy. Based on the analyzed process, it will direct the link with a datacenter. Then the algorithm is added with features of QOS metric which may be heterogeneous and homogenous based on the reliability, capacity of computing, cost and time, etc., the allocated resource is checked for the capability Akhani (2011). Then a process of management will go through a pool of resource, a provision of resource check is done in either way, at first, availability of resource check is done based on QOS, if the resource is not available, then the manage will submit QOS resource based on SLA violation, otherwise it then will submit directly to a scheduler. Then the scheduler will added the resource request in a resource allocation plan. By the process of QOS and plan of resource allocation resource is allocated to the request user. It also allocates the resource based on the resource reservation plan, by its time of peak in allocation of resource.

In recent days, resource provision and allocation efficiently is a complex task and issues occurred in cloud computing. Knowing the details of resource available and allocation process for increasing the efficiency and flexibility in cloud based on the request done by the user. The user do not require any program of software to be installed or requirement of hardware in using the resource of the cloud over the Internet in accessing the applications and for developing the application which are done on host over the global net. A plan of allocation of resource is prepared as a flow and is compared with various parameters of resource allocation. In Fig. 2 illustrates a plan of allocation the resource (Wu et al. 2011). The basic parameters of VM which possess cost, load, type, speed, etc., basic utility of profit, time of response, bid of market auction, satisfaction of application and certain policies related to security, parametric conditions and response time of GSLA, QOS, throughput, turnover time, and the parameters of application are database of large scale, gossip of real-time, connectivity related to peer to peer, expert of resource management, and knowledge-based expert.

The request of the user related to resource is done based on the lease condition. The lease of resource is done in the form of an Advance reservation mode. Initially reservation of resource is done in advance and allocation is done based on the time of peak and available of resource time. BE is done next, i.e., all the resource requests are in a queue area based on provision allocation will be done soon or immediately.

This is done based on the resource availability. At first, the user will submit a request, which is done based on the availability or discard based on sensitivity of deadline. DS will make a flexible assessment either on the provision of resource or at the time of constraints. RA-ALT is a provision state which provides or assigns the resource on the availability of user and a time slice is provided to the resource.

Dynamic provision of resource is done in two different ways.

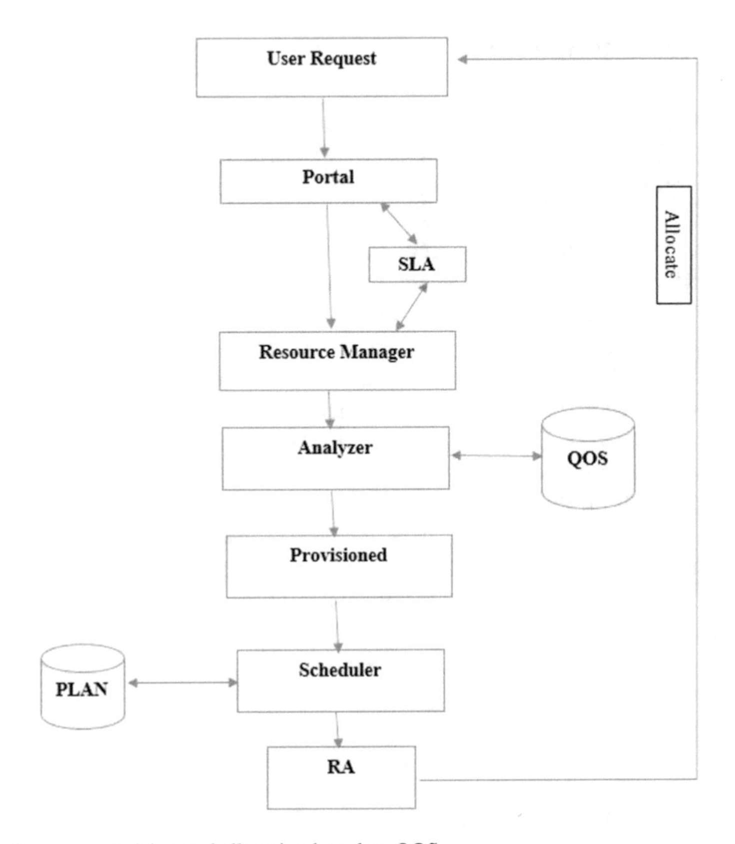

Fig. 1 Resource provision and allocation based on QOS

1. Over-provision on on-demand
2. Under provision for long-term reservation.

In case of on-demand resource provision a resource is allocated urgently, where as in a reservation process on long term the resources are assigned and used later. To overcome the interference problems, we require a process of effective management of resources. These two ways may affect the process of under and over-provision of resource which may increase in time waste and cost increase. Our work is based on development of effective load balance method which effectively manages the resource provision in cloud computing and also verified with the metric of QOS. The metric of QOS contain both heterogeneous and homogenous in nature, the metric parameter in QOS are bandwidth of network, utilization or resource, cost, time for execution, confidence level of customer and serviceability.

The region of QOS satisfaction done by the customer is based on the SLA form and is shown below in the table form, which shows the workload, workload analysis,

RESOURCE ALLOCATION BASED PLAN

PARAMETERS	AGENT	POLICY	NEGOTIATION	LOYALITY	MARKET	GSLA	STATISTICAL
VM	v	×	v	×	×	×	×
UTILITY	×	v	v	×	×	×	×
AUCTION	×	×	×	×	v	×	×
POLICY	×	v	×	v	v	×	v
GSLA	×	×	×	×	×	v	×
APPLICATION	×	×	×	v	×	×	v
GOSSIP	v	v	×	v	×	×	×

Fig. 2 Plan of allocation of resources

Table 1 QOS-based workload identification, analysis, pattern, and requirement

Workload identification	Analysis	Pattern identification	Requirement of quality of service
Website	Web service interface and API	Website = web service interface and API	Reliability
Online transition	Cloud deployment	Online transition = cloud deployment	Testing time
E-commerce	Storage base system	E-commerce = storage base system	Network bandwidth
Financial service	Instant service management	Financial service = instant service management	Computing capacity

fixing the workload, etc., we need to find out the requirement of the workload accordingly with the requirement of QOS based on resource provision made by the user in an effective and efficient way (Table 1).

4 Results of Experiments

Figure 3 shows the comparison with the series of lowest to the highest based on plan of allocation of resource. We have chosen various types of plan from different researchers based on the papers of existing on resource allocation and provision in

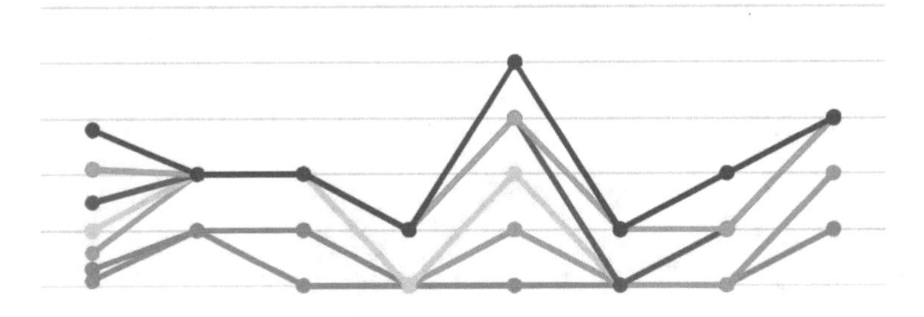

Fig. 3 Plan through highest and lowest series

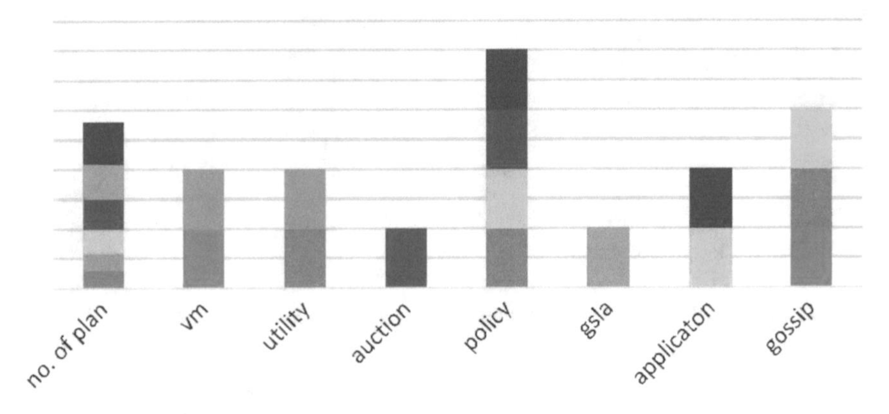

Fig. 4 Policy based resource utilization efficiency

cloud computing policies. We concentrated on GSLA-based allocation of resource technique. Majority of researcher will follow this plan of policy. So resource provision and allocation based policy technique will work efficiently based on accuracy and policy driven and is used more too.

Figure 4 shows the plan of implementation related to allocation of resource, and are then compared with the plan of allocation with other resources with certain parameter and a process of computation is done using resource parameter allocation based on the plan schedule and also compared with all the resource based policy of allocation is done.

Use case diagram

The figure below shows the resource allocation and provision in detail. The diagram of use case first has a login page, then registration is validate, based on QOS-based SLA resource provision availability. The manager of the resource will check for the availability and will give detailed information of the resources which are allocated and the resources which are freely available to user. The sequence diagram will show a resource based plan which is allocated.

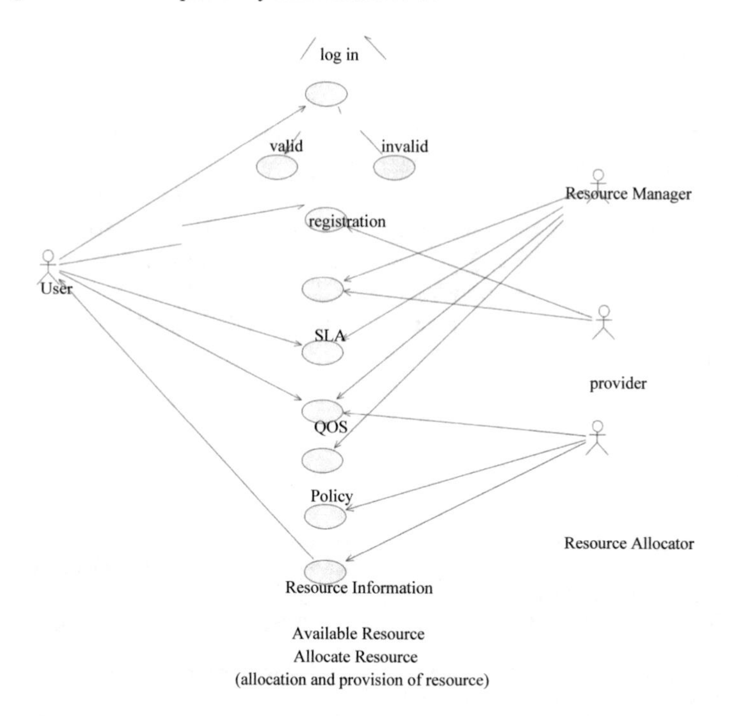

Available Resource
Allocate Resource
(allocation and provision of resource)

Sequence diagram:

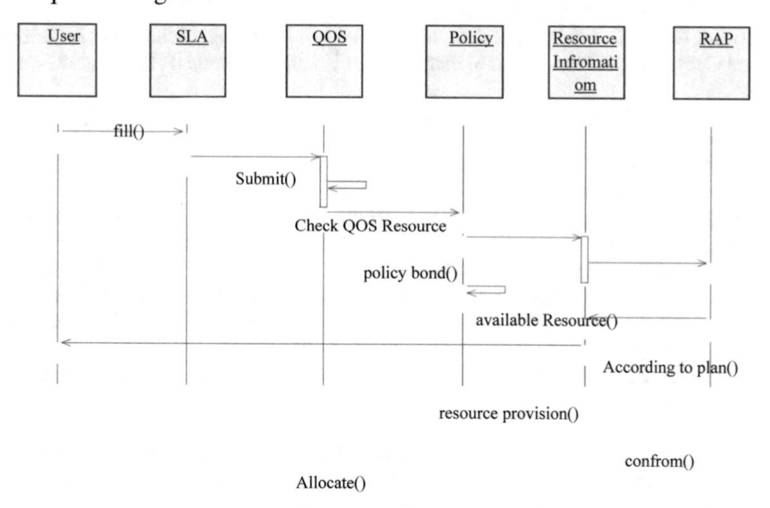

resource provision()

confrom()

Allocate()

(Pan of Allocation of Resource)

5 Conclusion

In our work we propose a resource provision and allocation based on QOS, a detailed plan of flow is being explained. In cloud technology, resource provision is down on demand which should be based on QOS. As from the previous work done by the researcher is based on resource allocation done using policy-based technique. We first concentrated only effective QOS-based resource provision method without violating the plan of allotment and policy of SLA. The proposed flow and method is compared using lab tool. This paper mainly discuses on QOS-based allotment of resource and provision efficiently.

References

1. Akhani, J.: Negotiation for Resource Allocation in IaaS Cloud (2011)
2. Aron, R., & Chana, I.: Formal QOS Policy Based Grid Resource Provisioning Framework, pp. 249–264 (2012). https://doi.org/10.1007/s10723-012-9202-y
3. Bharti, K., Kaur, K.: A Survey of Resource Allocation Techniques in Cloud Computing, vol. 2, pp. 31–35 (2014)
4. Duran-Limon, H.: A QOS and Energy-Aware Load Balancing and Resource Allocation Framework for IaaS Cloud Providers, pp. 410–415 (2016)
5. Ghumman, N. S., Kaur, R.: Dynamic Combination of Improved Max-Min and Ant Colony Algorithm for Load Balancing in Cloud System (2015)
6. Masdari, M., Salehi, F., Jalali, M., Bidaki, M.: A Survey of PSO-Based Scheduling Algorithms in Cloud Computing. J. Netw. Syst. Manag. **25** (2016). https://doi.org/10.1007/s10922-016-9385-9
7. Nayak, N.R.: QOS & GSLA Based Resource Provision Technique in Cloud Computing, vol. 3, no. 24, pp. 382–386 (2017)
8. Nayak, N.R., Brintha, R., Bhuvaneshwari, S.: Intelligent & automated vm Green Optimization, vol. 5, no. 6, pp. 289–294 (2015)
9. Raicu, I., Zhao, Y., Dumitrescu, C., Foster, I., Wilde, M.: Dynamic resource provisioning in grid environments (n.d.)
10. Singh, S., Chana, I.: Q-aware: quality of service based cloud resource provisioning. Comput. Electr. Eng. **47**, 138–160 (2015a). https://doi.org/10.1016/j.compeleceng.2015.02.003
11. Singh, S., Chana, I.: QRSF: QOS-aware resource scheduling framework in cloud computing, 241–292 (2015b). https://doi.org/10.1007/s11227-014-1295-6
12. Teng, F.: Mathematics, A., Paris, E.C., Mathematics, A., Paris, E.C.: Resource pricing and equilibrium allocation policy in cloud computing, (cit) (2010). https://doi.org/10.1109/cit.2010.70
13. Wu, L., Garg, S.K., Buyya, R.: SLA-based resource allocation for software as a service provider (SaaS) in cloud computing environments (2011). https://doi.org/10.1109/ccgrid.2011.51
14. Magoul: Policy of resource allocation in cloud resource managment, IJET, vol. 6, (2015)
15. Janaki: Reservation for resource allocation based on multi-agent policy in cloud IJEST, vol. 7, (2015)
16. Garg: Metric measure for quality for resource metric allocation in cloud managment IJATIT, vol. 7, (2015)
17. Kaur, survey of Efficiency in resourc managment in cloud, IJEST journal vol. 4, (2014)

Mr. Jeevan Jala acting as Research Scholar in Department of Computer Science, Acharya Nagarjuna University, Guntur, India 522006. His research field includes Cloud computing and QOS services.

Dr. Kolasani Ramchand H. Rao working as Professor, Department of Computer Science ASN Degree College Tenali. His research area includes Cloud computing and QOS services.

Performance Analysis of Energy-Efficient MANETs-Using Modified AODV (M-AODV)

H. K. Sampada and K. R. Shobha

Abstract Mobile Ad hoc Networks (MANETs) have no centralized infrastructure; they are dynamic in nature. A node can act as a host or a router and can freely join or leave the network any time. The challenges in creating MANETs are mainly because of their limited physical security and energy constraints. As most of the nodes are battery operated their energy gets drained rapidly. This paper presents a modification of the existing AODV routing protocol called M-AODV. AODV uses the HELLO packets to discover and maintain the neighbors; we have added two new parameters in HELLO packet that is Trust and the Willingness (based on the remaining battery of the node). The intermediate nodes for routing packets are selected based on these two parameters. NS2 simulator is used for analyzing the performance of the protocol in terms of throughput, PDR, Packet drop, and routing overhead. All the performance parameters for the proposed algorithm were compared with the regular AODV and the results for M-AODV compared to AODV are better.

Keywords MANETs · M-AODV · PDR · RREQ · RREP · RRER · NS2

1 Introduction

The growth of smart electronic gadgets since 1990s like laptops PDAs, Wi-Fi networks have made MANETs a popular research topic. MANETs are [1] are infrastructure less wireless network. That is they do not require a base station to communicate and can be deployed very fast in any of the remote places. They do not have a central control, hence the nodes move freely in any direction wirelessly. This leads to unique routing and communication challenges in MANETS. MANETs [2] are networks in

H. K. Sampada (✉)
Department of Electronics and Communication Engineering, Atria IT, Bengaluru, India
e-mail: sampada.hk@atria.edu

K. R. Shobha
Department of Telecommunication Engineering, MSRIT, Bengaluru, India
e-mail: shobha_shankar@msrit.edu

S. Smys et al. (eds.), *International Conference on Computer Networks and Communication Technologies*, Lecture Notes on Data Engineering and Communications Technologies 15, https://doi.org/10.1007/978-981-10-8681-6_9

which the node which is a peer can communicate with other peer only if they reach each other's range. MANETs are also self-configurable and self-healing networks. Each node should perform three duties: a host or a router or an intermediate node. MANETs have their applications [3], in almost each and every field. The energy of MANET nodes gets drained very quickly as the node should act as a router and also an intermediate node. The efficient energy usage of each node becomes a critical requirement to keep the network alive and to get good QoS (Quality of Service) parameters of the network. The routing protocols play a major role to keep the power consumption level as low as possible in MANETs. There are many routing protocols introduced for MANET communication over the years which are broadly divided as the proactive or table driven routing protocols, reactive or on-demand routing protocols and the hybrid which is a mixture of both the routing protocols.

AODV is a reactive routing protocol. Here the route discovery is on demand, it works in two phases route discovery phase and route maintenance phase. During route discovery the source establishes a connection with the destination. For doing this the source sends the HELLO packets to its reachable nodes. Nodes which can hear HELLO are in each other's communication range will respond to the HELLO. Then the source transmits a route request (RREQ) packet to its nearest neighbors for computing route to the destination node. If one of the neighbors is the destination it will reply back with a Route Reply (RREP), else the RREQ will be propagated to the next set of neighbors. This process continues until the destination node responds with a RREP [4, 5]. Then the actual data transfer happens between the sender node and the receiver node. In the route maintenance phase all the intermediate nodes receiving the RREP will update their routing table for any topology changes. In case of any link breakage a Route Reply with Error (RRER) will be sent back to the source node so that it initiates the route discovery process again. One of the major causes for link failure in AODV is due to limited battery of the nodes. Hence energy efficiency becomes a major concern in MANETs. There are many methods to achieve energy efficiency in MANETs. Like load balancing in which the routing load is balanced between the nodes so that they are not drained. Another method is power saving which deals with energy loss during idle mode where the node is just a part of the network and not participating in communication. In multipath routing the energy of the entire link is calculated before actual communication. In maximum lifetime method a node with less power is never considered as the intermediate node for routing packets [6, 7]. Instead a node with more power is considered for communication.

In this paper we have proposed an energy-efficient routing protocol algorithm based on the existing AODV routing protocol called M-AODV. The main objective of the paper is to increase the energy efficiency, network lifetime, throughput, PDR and decrease the packet drop and routing overhead. Performance of the proposed algorithm is tested for varying node density using NS2 [8–10] simulator.

This paper is divided in different sections; Sect. 2 explains the related work on various energy aware routing protocols in MANETs. Section 3 explains proposed work with the M-AODV algorithm details. Section 4 explains the experimental results of the proposed system and comparisons. Section 5 concludes the paper.

2 Related Work

Many types of research were done on link failure and energy efficiency of the network to come up with a reliable and energy-efficient routing algorithms. Most of the related work are based on the static networks or wired networks where the topologies of the nodes do not vary often and the mobility is either not given or very less mobility is given to the nodes.

Shivashankar et al. [11] proposes a new power aware routing protocol called Efficient Power Aware Routing (EPAR) that increases the network lifetime of MANET. EPAR identifies the capacity of the node by residual energy and also by the expected energy required to forward the data packets over the link. Using the minimum-maximum formulation EPAR selects the most suitable link among many links using the residual packet transmission capacity. The paper evaluates network protocols EPAR, Minimum Transmission Power Routing (MTPR) protocol, and Dynamic Source Routing (DSR) protocol for various network scales. The scheme decreases energy consumption by 20%, decreases mean delay for high loads achieving a good packet delivery ratio.

Patel et al. [12] proposes a work which is a new variation of Ad hoc on Demand Distance Vector (AODV) routing protocol, which deals with making the network more adaptive to the topology changes and also makes the network more energy efficient. The authors have achieved this by calculating the available energy of the nodes and forwarding packets along the best and energy-efficient path, making the network more adaptive to the network lifetime. Performance evaluation for throughput, network lifetime packet delivery, and delay is done using NS2.34/QualNet simulators.

Hannon et al. [13] the paper has adopted two different approaches in the algorithm to enhance the energy for MANETs. It uses threshold energy level which is calculated and updated in the request packet available energy level is calculated and updated in the reply packet. Based on RREQ and RREP the source nodes and the intermediate nodes select the path for data transmission between source and destination. The performance of the proposed protocol is checked with the help of NS-2 simulator for various scenarios using different parameters along with a very brief survey on various energy aware algorithms. The paper is concluded saying that the proposed protocol enhances the node energy, hence the network energy is enhanced, and also the throughput and lifetime are improved to a certain extent.

Reena et al. [14] this paper proposes an Energy-Efficient Ad Hoc Distance Vector (EE-AODV) protocol in this the modification to the existing AODV routing protocol is done. EE-AODV has modified the RREQ and RREP handling process to make the mobile devices more efficient in terms of energy usage. The algorithm considers some level of energy as the minimum energy level which should be available in the node to be used as an intermediate node (IN) taking part in future communication. When the node energy gets drained below the specified level, the node will not be reflected as an intermediary node. Simulation results show that lifetime of network is increased in EE-AODV as compared to AODV. The algorithm works only for the static networks like the sensor networks and not for the dynamic networks like the mobile ad hoc networks.

Abdelkabir et al. [15] has used the modified OLSR protocol called the power aware OLSR as the routing protocol to manage the available energy better. Authors have used, modification to Multi Path Route (MPR) selection mechanism to make the routing decision and increase the network lifetime. To achieve this the willingness parameter whose value ranges from 0 to 7 is used. A node with willingness 0 is never chosen for communication and the node with willingness 7 is always chosen for communication as a Multi Path Relay (MPR) node. Simulations were carried out using NS3 simulator in different mobile scenarios and the results show a significant improvement of 20% gain in lifetime of the network and also observed that power aware OLSR shows a little improvement in terms of packet delivery parameter.

Paraskevas et al. [16] has suggested a new MPR selection method that allows network to be alive for a longer time, by taking the value of the residual energy of the nodes in the MPR node selection criteria, using a new variable that specifies whether a node is willing to participate in the communication process or not based on its remaining energy and this variable is defined as the willingness variable. The simulations are carried out for different mobility and varying network size. The authors have compared modified OLSR, in terms of packet loss ratio, MPR count and delay, with the standard OLSR. Simulations show that the modified OLSR is able to enhance the network performance in delay calculations and packet loss ratio is enhanced compared to the precedent work without significant loss in terms of end-to-end delay using NS3 simulator.

Yu et al. [17] proposes a new energy aware routing protocol which is the enhancement of the Dynamic Source Routing (DSR) protocol. Here the load balancing approach is used where the packets are not always routed using the shortest path between the source and the destination but the remaining energy of the node and the transmission power are calculated as a function called joint function (JF) variable and the JF value is stored as a part of RREQ packet. In the available routes the route with the maximum JF is taken as the optimal route for sending data. The joint function value is the sum of the remaining energy and the required transmission energy which decides on the path to be chosen. Simulations are carried out for two approaches shortest path function (SPR) without the energy model and compared with the joint function routing (JFR). The results are better for the proposed model.

Anjana et al. [18], proposed a very simple and an efficient technique which is the modification to the AODV route discovery process. In regular AODV the source node initiates the route discovery with a RREQ the intermediate nodes will pass the same RREQ packet to the destination node, the destination node will send the reply back using RREP packet and if link is failed the node which cannot reach the destination will send the RRER error packet to the source and the source node will start the entire route discovery process again. This will reduce the energy efficiency of the entire network. To avoid this the author has proposed a new algorithm in which a recvReverse() packet is sent back to the previous node instead of the source node who will initiate the route discovery with a fresh RREQ packet and will ensure that at least the link which was established before will not break and thus the network performance is enhanced.

3 Proposed Work

Most of works which are carried out till now are done in a static topological environment. In the proposed work an attempt is made do the simulations in the dynamic topological environment. This paper mainly deals with the efficient energy usage of the nodes in dynamic topology mobile ad hoc networks along with the performance analysis of the various QoS parameters of the network. The proposed algorithm is modified AODV (M-AODV) with the following steps.

The algorithm depicts the operation of the Modified-AODV protocol, the process of route discovery and storage since the nodes are prone to dynamic changes (physically) and the routes are stored as and when a new (better) route is discovered, which are used for communication between nodes, thus enhancing the energy efficiency. M-AODV reduces energy consumption as continuous broadcast for route change update is not required and thus it is able to overcome the disadvantages of AODV. The algorithm works for any node acting as a source node or the intermediate node in the network. In the route initiation procedure the node having data to transmit (source node) starts transmitting and receiving the HELLO packets used to discover the neighbors. A node is a neighbor if the distance between the two nodes is less than or equal to 250 m. In our M-AODV algorithm the HELLO packet format is modified with two new fields Willingness and trust, willingness ranges from 0 to 7 and trust ranges from 0 to 1. Based on the number of experimental iterations the threshold value for our experiments for willingness and threshold are set to willingness being less than or equal to 3 and trust value less than equal to 0.5. Table 1 shows the simulation configuration.

Table 1 Simulation configuration

Parameter	Value
Channel	Wireless channel
Dimension of topology	6400 * 6400 (m)
Number of nodes	20, 30, 50
Network simulation tool	NS2
Simulation time	80 s
Wireless protocol	802.11b
Initial energy	50 J
Speed	0.5 (m/s)
Mobility model	Random way point

Algorithm 1: Modified AODV (M-AODV)

```
    N = {N1, N2, N3...........Nn} set of N nodes in the
network
        for i = 1 to N do
              Discovery of neighbour nodes using
          HELLO packets
              Check for the [willingness] and [Trust]
              Update data received from neighbours
        if
              willingness ≥ 3 and Trust ≥ 0.5
              accept the neighbours and send RREQ
        else
              discard the neighbour
              check for new neighbours
        end if
        end for
```

Route discovery process: The source that wants to communicate with the destination listens to the HELLO packets from the neighbors checks for willingness and trust value. If the willingness and trust are equal to or more than the threshold value the node is accepted as the neighbor and sent a RREQ packet to find the route to the destination else the node is discarded and the node discovery process is again started.

Data communication process: After RREQ is sent the reply from the destination as RREP is got and the actual data transfer is initiated. Hence the modified-AODV routing protocol is better than AODV protocol as the RREQ are not sent to all the neighboring nodes but only to the limited nodes whose willingness and trust are above threshold value. The routing overhead is decreased and other QoS parameters are increased. The simulation experiments are carried out for 20, 30, 50 nodes and found that the algorithm performs better than AODV routing protocol.

3.1 Formulae Used

When a node receives a HELLO packets from its neighbors, the willingness and trust values from each node are calculated by the source node to select that particular node for further communication.

$$\text{Willingness} = \left\{ \frac{\text{Ei}}{\text{Emax}} * 7 \right\} \tag{1}$$

Ei = remaining energy at any time and Emax = initial energy assigned to a node.

The value ranges from 0 to 7 with 0 being the minimum value and 7 being the maximum value. Trust of a node defines its efficiency in routing the packet sent from the previous node to the next hop node which is calculated as follows:

$$\text{Trust} = \text{Total No of packets forwarded/Total No of packets dropped} \tag{2}$$

Minimum value of trust is 0 and maximum is 1. For any node first trust is checked which should be more than 0.5 and then willingness is checked which should be more than 3, only then the node is selected as an intermediate node for further communication.

4 Results and Discussion

The simulations are done in NS2 simulator creating a dynamic environment where the nodes are moving randomly and the QoS parameters like throughput, packet delivery ratio, packet drop, control overhead and routing overhead are calculated using M-AODV and compared with the simulation results of AODV. The results show that M-AODV routing protocol performance is better compared to AODV routing protocol.

Figure 1 shows the simulation results captured at a particular time of simulations. Figure 2 shows the simulation graphs for throughput comparison of AODV and M-AODV. Hence proposed system gives better results compared to AODV. Figure 3 shows the graph for PDR comparison between AODV and M-AODV. PDR initially is better for AODV than M-AODV but the PDR of the proposed system is better later. This is because the data transmission from source to the destination starts only after the route discovery process. Also since the battery usage of the nodes is better in proposed algorithm the PDR is improved. Figure 4 shows packet drop comparison between AODV and M-AODV and it shows that the packet drop for M-AODV is 0 for a scenario of 50 nodes ideally as there are no malicious nodes in the system, also the packet drop for AODV and M-AODV for 20 nodes scenario is almost same, but for AODV with 30 nodes the packet drop is more than the M-AODV with 30 nodes. Hence in each case our proposed algorithm performs better. Figure 5 is the graph of number of control packets used versus total number of nodes for AODV and M-AODV. Number of control packets used for AODV with 50 nodes is least and for M-AODV with 50 nodes is moderate. This may be probably because of the dynamic scenario chosen for the simulations. Figure 6 shows the routing overhead comparison between AODV and M-AODV with respect to time, graph shows that performance of M-AODV for 20, 30, and 50 nodes is better than the performance of AODV with 20, 30, and 50 nodes.

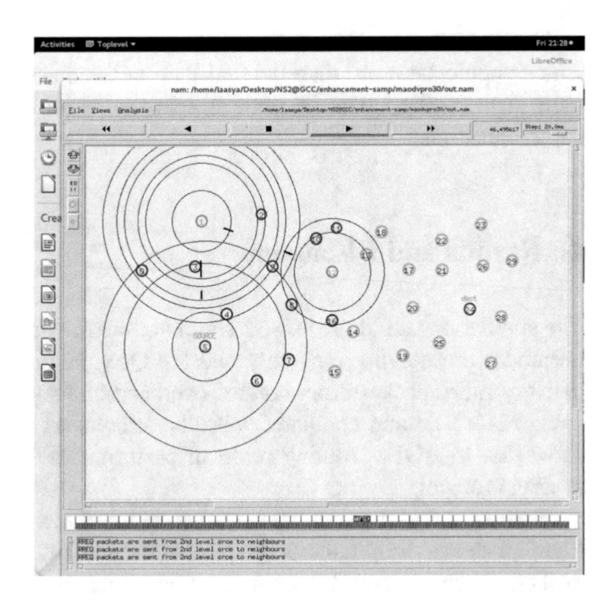

Fig. 1 Simulation scenario

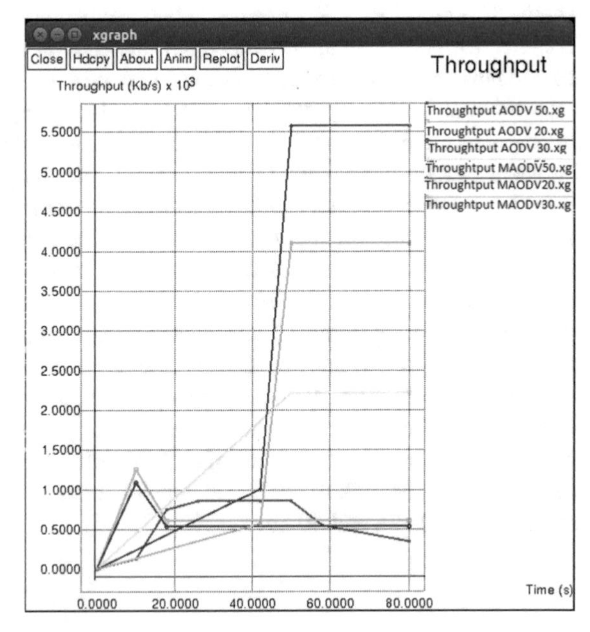

Fig. 2 Throughput comparison

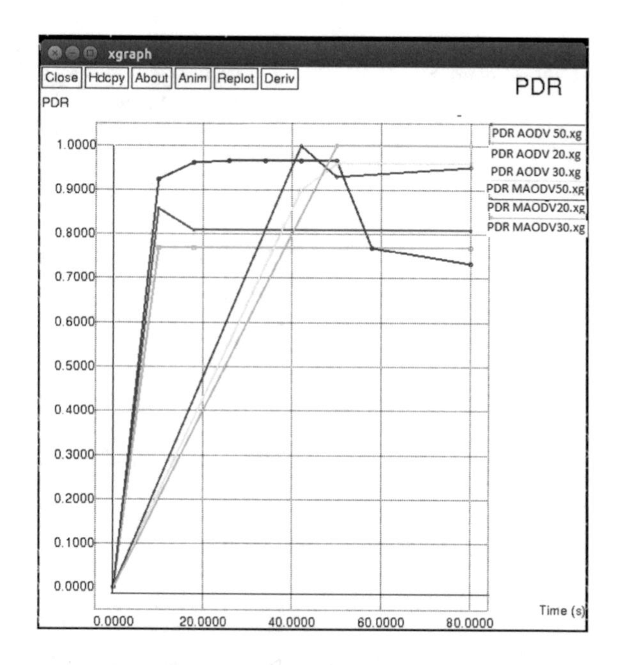

Fig. 3 Packet delivery ratio comparison

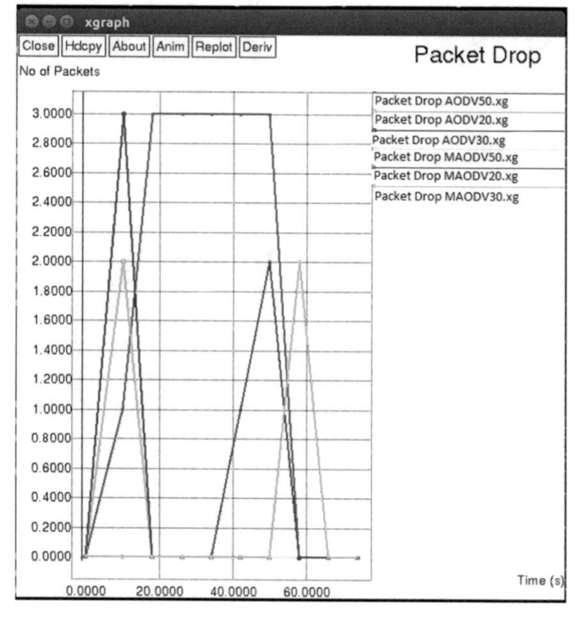

Fig. 4 Packet drop comparison

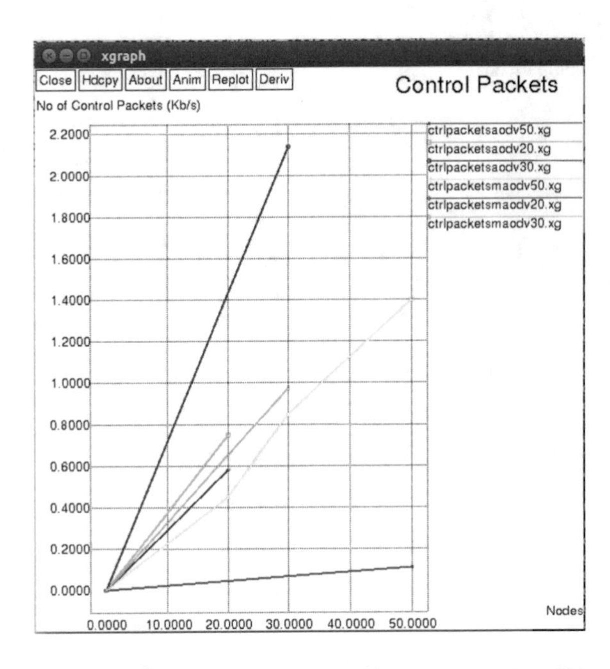

Fig. 5 Control packet versus no. of nodes

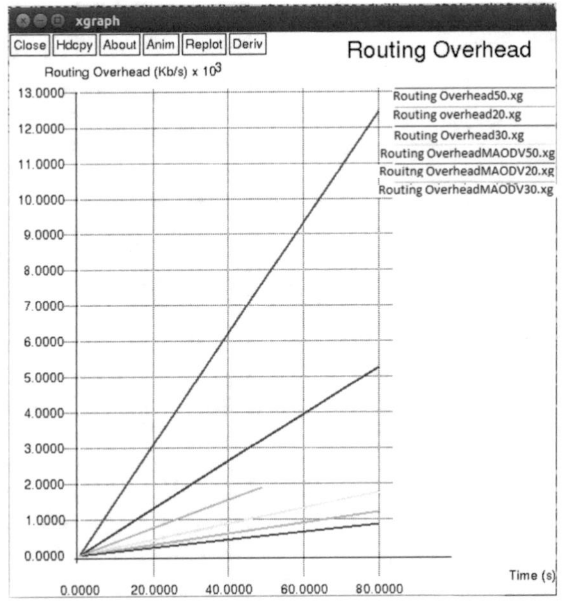

Fig. 6 Routing overhead versus time

5 Conclusions

Energy efficiency in any network is a major issue and is of more concern in MANETs as the nodes are mobile and each node should behave as both a host and a router, hence energy of the nodes is a major concern.

In this paper we have demonstrated an effective and energy-efficient proactive routing protocol called M-AODV which is a modification to the existing AODV protocol to increase network lifetime and performance of the network. M-AODV algorithm integrates two salient parameters which are: trust and residual energy as a part of HELLO packet. This allows the network lifetime to be preserved for longer time, by involving the residual energy and trust as the selection criteria for the intermediate nodes. The algorithm was evaluated for the modified AODV under a range of different scenarios of varying traffic load. The results were plotted for modified AODV (M-AODV) and standard AODV in terms of throughput, PDR, packet drop, routing overhead and control overhead. The comparison of the graphs showed that the performance of M-ADOV is better than that of AODV.

References

1. Jubin, J., Tornow, J.: The DARPA packet radio network protocols. Proc. J. Electr. Eng. Comput. Sci. **5**(1), 21–32 (IEEE) (1987)
2. Perkins, C.: Ad Hoc Networking. Addison-Wesley Longman Publishing Co., Inc., Boston, MA, USA (2001)
3. Perkins, C.E., Bhagwat, P.: Highly dynamic destination sequenced distance vector routing (DSDV) for mobile computer. In: Proceedings of the SIGCOMM '94 Conference on Communications Architectures-Protocols and Applications, pp. 234–244 (1994)
4. Shenbagapriya, R., Kumar, N.: Survey on proactive routing protocols in MANETS. In: IEEE 14th International Conference on Science Engineering and Management (ICSEMR), pp. 1–7. IEEE Press, India (2014)
5. Patel, D. N., Patel, S. B., Kothadiya, H. R., Jethwa, P. D., Jhaveri, R. H.: A survey on reactive routing protocola in MANET. In: International Conference on Information Communication and Embedded Systems (IICICES), pp. 1–6. IEEE Press, India (2014)
6. Liu, Y., Guo, L., Ma, H., Jiang, T.: Energy efficient on–demand multipath routing protocol for multi-hop ad hoc networks. In: 10th International Symposium on Spread Spectrum Techniques and Applications, pp. 25–28. IEEE Press, Italy (2008)
7. De Rango, F., Fotino, M., Marano, S.: EE-OLSR: energy Efficient olsr routing protocol for mobile ad-hoc networks. In: Military Communications Conference, 2008 (MILCOM), pp. 1–7. IEEE Press, USA (2008)
8. Sivakumar, D., Suseela, B., Varadharajan, R.: A survey of routing algorithms in MANET. In: International Conference on Science and Management, pp. 625–640. IEEE Press, India (2012)
9. Perkins, C.E., Royer, E.M.: Ad-hoc on-demand distance vector routing. In: Second IEEE Workshop on Mobile Computing Systems and Applications, pp. 90–100. IEEE Press, USA (1999)
10. The Network Simulator—ns—2. https://www.isi.edu/nsnam/ns
11. Shivashankar, S., Suresh, H.N., Golla, V., Jayanthi, G.: Designing energy efficient routing protocol with power consumption optimization in MANET. Trans. Emer. Top. Comput. IEEE **2**(2), 192–197 (2014)

12. Patil, A.P., Chandan, B.V., Aparna, S., Greeshma, R., Akshatha, H.P.: An improved energy efficient routing protocol for MANETs. In: Eleventh International Conference on Wireless and Optical Communications Networks (WOCN), pp. 978–1–4799–3083–8. IEEE Press, India (2014)
13. Hannan, X., Ibrahim, D.M., Christianson, B.: Energy consumption in mobile ad hoc networks. In: IEEE WCNC'14 Track 3 Mobile and Wireless Networks, pp. 978–1–4799–3083–8/14. IEEE Press, Turkey (2014)
14. Reena, S., Shilpa, G.: EE-AODV energy efficient AODV routing protocol by optimizing route selection process. Int. J. Res. Comput. Commun. Technol. **3**(1), 2278–5841 (2014)
15. Abdelkabir, S., Abbadi, J.E., Habbani, A.: Increasing network lifetime by energy-efficient routing scheme for OLSR protocol. In: International Conference on Information Informatics and Communication Systems. 978-1-4673-8743-9/16/$31.00. IEEE Press, UAE (2016)
16. Paraskevas, E., Manousakis, K., Das, S., Baras, J.S.: Multi-metric energy efficient routing in mobile ad-hoc networks. In: IEEE Military Communications Conference. IEEE Press, Baltimore USA (2014)
17. Yu, W.: Study on energy conservation in MANET. J. Netw. **5**(6), 708–714 (Academy Publisher) (2010)
18. Anjana, T., Kaur, I.: Performance evaluation of energy efficient for MANET using AODV routing protocol. In: 3rd EEE International Conference on Computational Intelligence and Communication Technology IEEE-CICT. IEEE Press, India (2017)

H. K. Sampada completed her M.-Tech. in Digital Communication and Networking in the year 2004 from SJC Institute of Technology (SJCIT), Chikballapur, under Visvesvaraya Technological University (VTU). She is currently persuing her Ph.D. under VTU, in the department of TCE, MSRIT, Bengaluru.

She is currently working as Assistant Professor in the department of ECE, Atria IT, Bengaluru. Her research interests are in security issues in Mobile Ad-hoc Networks, VANETs, IoT and Data Science. She has presented her research papers in several international conferences and Journals.

Dr. K. R. Shobha received her M.E. degree in Digital Communication Engg from Bengaluru University, Karnataka, India and Ph.D. from Visveswaraya Technological University. She is currently working as an Associate Professor in the department of Telecommunication Engineering, M.S. Ramaiah Institute of Technology, Bengaluru. Her research areas include Mobile Adhoc Networks, IoT and Cloud Computing. Se has more than 25 Papers publications to her credit.

She is a Senior IEEE Member serving as Secretary of IEEE Sensor Council, Bengaluru Section. She is also an active member of IEEE Communication Society and Women in Engineering under IEEE Bengaluru Section.

A Methodology for Meme Virality Prediction

E. S. Smitha, S. Sendhilkumar and G. S. Mahalakshmi

Abstract In current era, online social networking performs an important role in content sharing. Information diffusion takes place very rapidly through social network. There are so many positive and negative impacts for information spreading. An Internet meme is a unit of information, which is copied and spread very fast in internet, may be with slight variations. This movement will spread from person to person via social networks. In this paper, we put forward a method for predicting the virality of meme in future. The data collected from Twitter and machine learning algorithms are used to predict the virality.

Keywords Diffusion · Information · Internet · Meme · Social network
Twitter · Viral

1 Introduction

The term memes have become a new social phenomenon. The main aim of Internet meme is to convey some social thoughts or ideas which can be in the form of voice, text and image. Word or text memes are easily generated and spread very fast. One of the main sources of text meme is Twitter. In the book Selfish Gene [1] Dawkins coined the word memes similar to gene. Memes are a unit of information or cultural propagation for individuals to pass ideas and views to each other. Memes have the

---

E. S. Smitha (✉) · S. Sendhilkumar
Department of Information Science and Technology, College Engineering Guindy, Anna University, Chennai 600025, India
e-mail: smithaengoor@gmail.com

S. Sendhilkumar
e-mail: ssk_pdy@yahoo.com

G. S. Mahalakshmi
Department of Computer Science, College Engineering Guindy, Anna University, Chennai 600025, India
e-mail: gsmaha@annauniv.edu

S. Smys et al. (eds.), *International Conference on Computer Networks and Communication Technologies*, Lecture Notes on Data Engineering and Communications Technologies 15, https://doi.org/10.1007/978-981-10-8681-6_10

ability to replicate, mutate and respond to certain situations, which make memes to resemblance with genes. A meme will spread out from brain to mind with or without change or modification.

The meme is a social phenomenon; it will transmit content or idea rapidly from one person to another through online networking. Memes can be seen in different forms like photo or video, image, word and audio. As like the name suggests, the photo memes and video memes are formed from photos or from videos. Image memes are common, they have some image part and text part, the text part (caption) is usually seen on the top or bottom of the image. Word memes utilize the use of well-liked phrases and words in which the people can add their own comments. Twitter hashtags are good example of word memes.

The Twitter is a popular social networking platform. Using as platform messages called tweets can be sent and received using Twitter. A tweet has a maximum length of 140 (now it is extended) characters, containing regular text as well as special characters. The markers are the hashtag sign ("#") and the at the rate sign ("@"). These can be used to tag tweets by topic or to reply to other specific users. A huge amount of tweets are sent every day in social networks. The amount of data that can be gathered from this social platform is thus very large and complex, but important. The main work proposed in this paper is predicting the virality of memes. The virality prediction has many applications in the field of marketing, trend analysis and psychological study.

Section 2 discussed about some papers related to this topic. Section 3 gives an idea about the proposed prediction model. The next section is about result and discussion. The last section discuss about conclusion and future work.

2 Related Works

Memes have become an important way to transmit information in social networkis. In this work tweet are considered as memes. The social interactions on Twitter is used to study the sparse hidden network and its usage by considering social interactions between the friends and followers [2] in the underlying interactions. Java et al. [3] noticed that community structure can be identified by social network analysis. They had also detected different types of user intentions on Twitter. Jansen and others [4] have treated the Twitter as a method for advertising and analysing the sentiments in web.

In their paper Leskovec et al. [5] proposed that new topics and trends can be discovered by monitoring social media memes. From the text and metadata features in tweets Joshi et al. [6] use a linear regression method to predict popularity for movies. Mishne and Glance [7] use a correlation technique for analysing the sentiments in blog posts with movie box-office scores. But using the correlations, it is observed that positive sentiments are too low and not enough to use for predictive purposes. In [8] Jamali and Rangwala have studied the previous popularity of online content is correlated with its upcoming popularity. So many works are related to community and sentiment analysis. The Weng et al. [9] have found the community structure has high influence in the prediction of meme virality.

Another approach to determining if a tweet is likely to go viral is by analysing and modelling the contents of the tweets. There is a wealth of information in this area of research, as there are many potential aspects to investigate. Bandari et al. studied news articles that were shared on Twitter and found that the source of the article, as well as what entities the article contained, played an important role in predicting virality [10]. Tweets containing articles from famous or popular sources and that write about well known organizations, people or events. Tweets of that category will be mostly retweeted. The users with a so may number of followers are to be retweeted. A strong correlation between the followers of a user and the likelihood of tweets have found [11] by Suh et al., that will lead virality. They also found the use of hashtags and links also increase the likelihood of a tweet going viral. But, they could not determine why certain popular hashtags produced many retweets while others did not.

3 Methodology

The following framework (Fig. 1) shows the prediction model. The main processes involved in meme prediction are as follows: Word meme collection, Feature extraction, Model training (Regression/classification) and predictions.

This paper focus on the approach of predicting the virality of word memes from its content and some important attributes. The problem of virality prediction is very wide, but this work is restricted to focus on the virality of hashtags by predicting the virality of the tweets containing a particular hashtag.

The whole work is implemented using Python code. The prediction of virality of a tweet depends on the number of retweets it obtained is the main discussion overhere. Some features related to the author of the tweet, such as the author's number of followers, friends, appearances on lists, statuses and favourites as well as the age

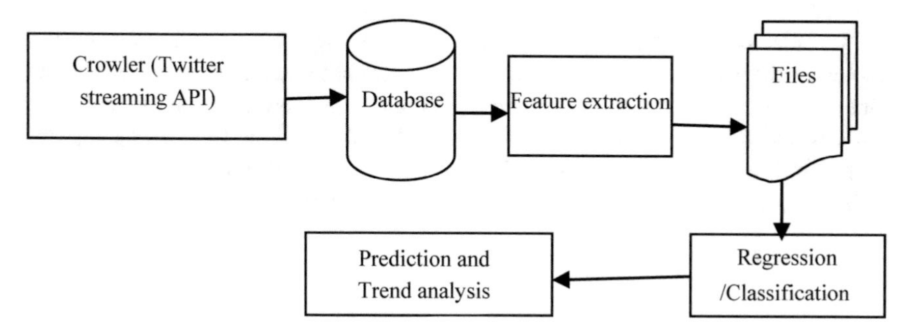

Fig. 1 Framework for virality prediction

of the account in days are taken as features for prediction. Features related to the content of the tweet itself like the number of hashtags, media, user mentions, and URLs contained as well as the length of the tweet are also included.

3.1 Data Collection and Storage

In this work the tweets which are spreading with or without modification are considered as memes. The first component is a crawler which aims at retrieving random English tweets using the Twitter streaming API, which gives developers access to Twitter's global stream of tweet data. To receive a representative subset of tweets, the streaming API is queried for a list of English stop words, which are commonly contained in tweets.

Each tweet is returned as a JSON document and stored in a MongoDB instance. Further, each tweet is analysed for hashtags. Based on the hashtags contained, an inverted index is created which contains the IDs of all tweets containing a particular hashtag. Three days after the data acquisition, the number of retweets of each tweet in the database is updated by performing an additional call to the Twitter API for each tweet. This time frame between the data retrieval and the retweet count update determines the scope of predictions. The virality predictions made by this model will be valid for this specific time interval. After collecting the raw data, the data needs to be cleaned before extracting features from it. Outliers with a very high retweet count and tweets which do not contain any hashtag are removed. Counters lower than 0 are set to 0 and tweets longer than 140 characters are set to 140.

3.2 Feature Selection

As features for our predictor we chose some features related to the author of the tweet, such as the author's number of followers, friends, appearances on lists, statuses and favourites as well as the age of the account in days.

The selected features are mostly from metadata, which means that our current implementation does not analyse the text of the tweet to predict its virality. The raw data stored in the database is not altered during this process and store the extracted features and the number of retweets in a HDF5 file. Mainly 12 features selected for prediction, they are followers, friends, statuses, listed, favourites, account age, verified, hash tags, media, user mentions, urls, tweet length.

3.3 Data Analysis: Model Training Using Regression and Classification

In this step, the trained model to predict the virality of hashtags is used. To discover potentially viral hashtags, iterate over all hashtags in the inverted index and predict the virality of the tweets containing that hashtag. The virality of the hashtags is then simply calculated by adding the virality of the tweets containing that hashtag. In a final step, the hashtags are sorted in decreasing order by their predicted virality. This approach enables the creation of a top-K list of viral hashtags. If the predicted virality for that hashtag exceeds a certain threshold, the hashtag is then assumed to be viral.

3.3.1 Regression

For the prediction of virality, the first approach is a linear regression model. The idea of this model is to treat the number of retweets as the dependent variable and all the other features as independent variables. To optimize the weights in such a way that the root mean squared error between the actual values and the predicted values is minimized. The algorithm used is a generalized linear model (GLM) called Bayesian Ridge.

Equation (1) gives a basic representation of a GLM. $w0$ represents the intercept term, w_1 to w_n are the values of the coefficients, and $y \in 0; +\alpha$.

$$\hat{y}(w, x) = w_0 + w_1 x_1 + w_2 x_2 + \cdots + w_n x_n \tag{1}$$

3.3.2 Classification

The second approach is to treat the problem as a classification problem. Here, the dependent variable Y is the not number of the retweets, but a class label. The classes are viral and non-viral indicated by 1 or 0, where the boundary between those classes is defined by a threshold applied to the retweet count. This approach should make the model less sensitive to noise or some fake viral retweets.

For classification, logistic regression was chosen. Equation 2 gives a basic representation of a logistic regression

$$\hat{y}(w, x) = \frac{\left[\exp(w_0 + w_1 x_1 + w_2 x_2 + \cdots w_n x_n)\right]}{\left[1 + \exp(w_0 + w_1 x_1 + w_2 x_2 + \cdots w_n x_n)\right]} \tag{2}$$

Logistic regression is used to measure the probability of binary response based on one or more independent variables or features, which is also called predictors. Logistic regression is a simple classification algorithm for learning to make decisions on discrete values such as high or low (0 for failure, 1 for success) data.

3.4 *Virality Prediction*

In a next step, a trained model is used to predict the virality of hashtags. To be able to discover potentially viral hashtags, iterate over all hashtags in the inverted index and predict the virality of the tweets containing that hashtag. The virality of the hashtags is then simply computed by taking the sum of the virality of the tweets containing that hashtag. In a final step, the hashtags are sorted in decreasing order by their predicted virality. This approach enables the creation of a top-K list of viral hashtags. If the predicted virality for that hashtag exceeds a certain threshold, the hashtag is then assumed to be viral.

4 Result and Discussion

4.1 *Data Set*

The data set collected using a twitter streaming API. More than 70,000 tweets had collected for continuous three days. After loading the features and retweet count from the HDFS file, dataset is balanced in order to increase the weight of viral tweets. The balanced dataset is shuffled and divided into two parts. The first 80% used as the training set and the rest 20% used as the testing set. The data will be experimented with different threshold values and got the best results with threshold equal to 55,000 (Fig. 2; Table 1).

4.2 *Regression*

The first part of Fig. 3 illustrates the predicted retweet count (*Y*-axis) according to the expected retweet count, the second part shows the difference between the prediction and expectation.

From Fig. 3 we can observe that the prediction error for viral tweets is quite low. However, lots of false positive are generated by our model, i.e. high retweet counts

```
966 @sweetmanjuan Thank you bunches....I too love it
967 RT @SincerelyTumblr: rt if these pancakes look better than u https://t.co/MizRJyZBWl
968 RT @TimeaFanclub: #sexy #beautiful #anal #submissive #escort Timea Antala. Booking at her webp
969 WAIT A MIN I FORGOT RUN TODAY
970 RT @UraSwallow: Outdoor blowjob and thick semen swallow! #Blowjob #CumSwallow #SuckThatCock #T
971 RT @TheSoneSource: Sunny, by jinho_seo
https://t.co/CiHQtDxVXu
https://t.co/zww87oQRHi https://t.co/KPLuiJOpSy
972 RT @vleks_: @hubertgriffe Fuck u then https://t.co/ip6ce0CS3h
973 RT @wxSpinner89: Put my "spin" on the forecast today. #FidgetSpinner https://t.co/om8plSFuMh
```

Fig. 2 Sample data extracted using Twitter API

Table 1 Mean, standard deviation and correlations between each feature and the retweet count

Features	Mean	Std. deviation	Correlation
Followers	576220.62	4365041.66	0.194274
Friends	8327.40	35412.00	0.033895
Statuses	43056.14	97635.02	−0.016041
Listed	3514.62	34121.9	0.159713
Favourites	3800.77	11700.45	−0.021614
Account age	1156.66	777.73	0.06927
Verified	0.11	0.31	0.077538
Hashtags	1.80	1.37	−0.052247
Media	0.30	0.46	0.071278
User mentions	0.95	1.00	0.028637
URLs	0.45	0.54	−0.034712
Tweet length	113.53	28.27	0.060207

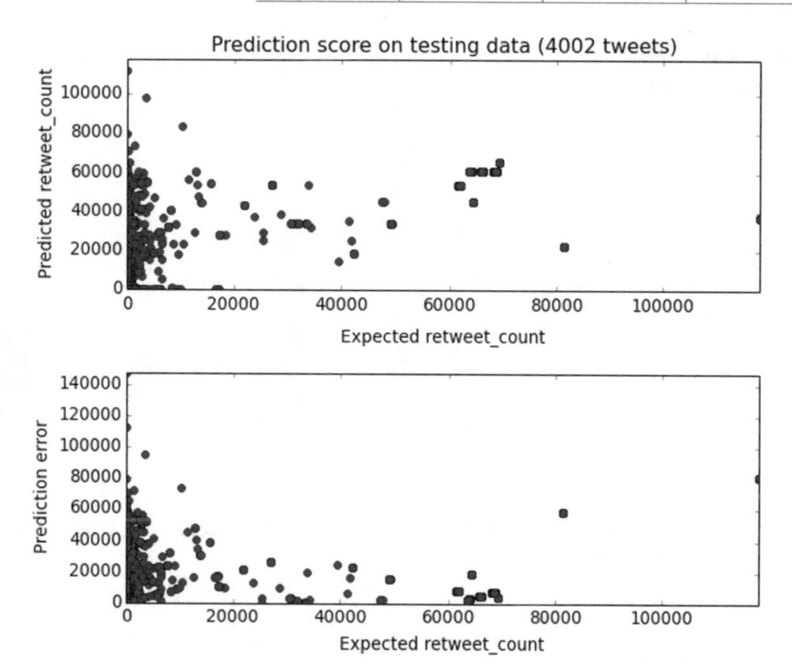

Fig. 3 Linear regression predictions on testing dataset

are predicted when the actual retweet count is low. The most likely explanation for this is that the relationship between some of the features and the retweet count simply are not linear.

The weights of the coefficients in Fig. 4 illustrate the relative importance of the features. A significant positive or negative value indicates that this feature is an

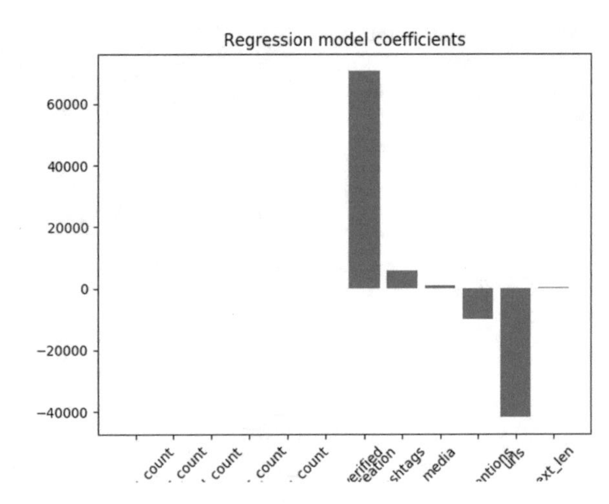

Fig. 4 Linear regression model coefficients

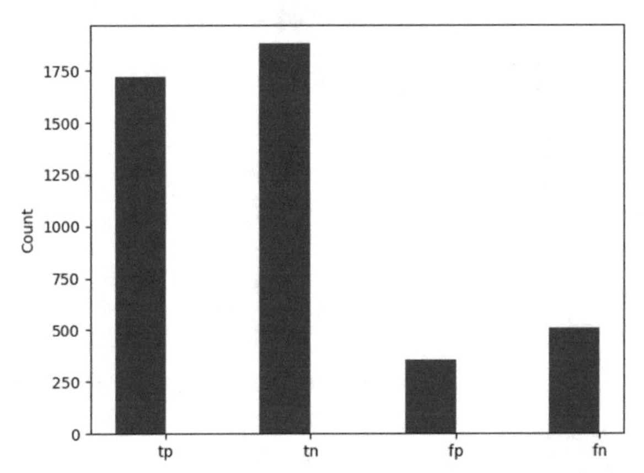

Fig. 5 Logistic regression predictions on testing dataset

important factor to calculate the retweet count. It can be seen that many features have very low regression coefficients.

4.3 Classification

Figure 5 visualizes the number of true positives tp (classifier correctly predicts virality), true negatives tn (classifier correctly predicts non-virality), false positives fp (classifier incorrectly predicts virality) as well as false negatives fn (classifier incorrectly predicts non-virality) for a threshold of 50,000. It can be seen that the classifier correctly predicts both virality scores of 1 and 0 with a similar rate (true positives and true negatives). The error rate of false positives and false negatives is also similar.

Table 2 Top-10 predicted viral hash tags for the dataset

No.	Regression	Classification
1	World mental health day	World mental health day
2	The flash	The flash
3	Hiphop awards	Hiphop awards
4	The last jedit	MAGA
5	MAGA	The Last Jedit
6	ProjectXATL	Messi
7	Messi	ProjectXATL
8	Harvey Weinstein	USMNT
9	USMNT	Harvey Weinstein
10	Now playing	Now playing

Table 3 ROC performance of the classifier

Threshold	TP rate	FP rate
25,000	89.05	6.71
50,000	90.94	10.00
55,000	96.60	5.60
65,000	96.33	5.93
68,000	96.21	4.25

The performance of the classifier for different values of thresholds in Table 3 shows a jump in the true positive rate. By changing the threshold value to 55,000, the true positive rate saturates and a reduction in the false positive rate can be noticed. However, this reduction is not very steep. This indicates that for this data the modelling function becomes much more discriminative at a threshold of 55,000.

Table 2 shows the top-10 predicted viral hash tags for the first week October 2017.

Predicting the virality of hashtags, which are not popular is a difficult problem. Because popular hashtags at the time will be more often featured in the inverted index and therefore more tweets will be taken into account for already popular hashtags The regression model does allow the prediction of high virality for non popular hashtags if the contained tweets have high retweet count predictions, for example if the tweets are created by very influential authors. The classification model only distinguishes between two classes as viral and non-viral. Therefore, one extension to the classification approach would be to classify the virality of a tweet into more than two classes. The virality could be divided into different classes of a certain number of retweets. This would potentially allow for a more precise and fine-grained virality prediction when using the classification approach.

5 Conclusion and Future Work

The work done in this paper shows the feasibility of predicting the virality of tweets using a feature-based approach. This work integrates the ideas with concepts from information retrieval to create a novel approach to predicting the virality of hashtags. The result of the work shows some promise but further research into other fields such as social media analysis will be needed to obtain more accurate and reliable results. Trend analysis can be applicable in the areas like marketing, politics, psychological study and so on.

References

1. Dawkins, R.: The Selfish Gene. Oxford University Press (1989)
2. Huberman, B. A. H., Romero, D. M., Wu, F.: Social networks that matter: Twitter under the microscope. First Monday **14**(1) (2009)
3. Java, A., Song, X., Finin, T., Tseng, B.: Why we twitter: understanding microblogging usage and communities. In: Proceedings of the 9th WebKDD and 1st SNA-KDD 2007 Workshop on Web Mining and Social Network Analysis, pp. 56–65 (2007)
4. Jansen, B., Zhang, M., Sobel, K., Chowdury, A.: Twitter power: Tweets as electronic word of mouth. J. Am. Soc. Inf. Sci. Technol. (2009)
5. Leskovec, J., Backstrom, L., Kleinberg, J.: Meme-tracking and the dynamics of the news cycle. In: Proceedings of 15th ACM International Conference on Knowledge Discovery and Data Mining, Paris, France, June (2009)
6. Joshi, M., Das, D., Gimpel, K., Smith, N.A.: Movie Reviews and Revenues: An Experiment in Text Regression NAACL-HLT (2010)
7. Mishne, G., Glance, N.: Predicting movie sales from blogger sentiment. In: AAAI 2006 Spring Symposium on Computational Approaches to Analysing Weblogs (2006)
8. Jamali, S., Rangwala, H.: Digging digg: comment mining, popularity prediction, and social network analysis. In: Proceedings of International Conference on Web Information Systems and Mining (WISM), pp. 32–38 (2009)
9. Weng, L., Menczer, F., Ahn, Y.-Y.: Virality prediction and community structure in social networks. Sci. Rep. **3**(2522) (2013)
10. Bandari, R., Asur, S., Huberman, B. A.: The pulse of news in social media: forecasting popularity. CoRR, abs/1202.0332 (2012)
11. Suh, B., Hong, L., Pirolli, P., Chi, E. H.: Want to be retweeted? Large scale analytics on factors impacting retweet in twitter network. In: Second International Conference on Social Computing (SocialCom), 2010 IEEE, pp. 177–184 (Aug 2010)

E. S. Smitha received the Master of Technology in Computer Science and Engineering degree from Anna University in 2009. Presently she is perusing Full time research in College of Engineering Guindy, Anna University, Chennai. Her doctoral research is on 'Meme Mining in Social Networks'.

S. Sendhilkumar received his Ph.D. during 2009 in Web Search Personalisation. Presently he is an Assistant Professor (Senior Grade) in Department of Information Science and Technology, Anna University, Chennai. He has authored numerous research articles in Reputed Journals and International Conferences. His research interests include Data Mining, Social Networks, Text Mining and Big Data Analytics.

G. S. Mahalakshmi received her Masters in CSE from Anna University, Chennai. She received the Ph.D. during 2009 in the field of Artificial Intelligence. She has authored numerous research articles in Reputed Journals and International Conferences. Presently she is an Assistant Professor (Senior Grade) in Department of Computer Science and Engineering, Anna University, Chennai. Her research interests include Machine Learning, Social Networks, Text Mining and Big Data Analytics.

Localization of Nodes with Ocean Current Mobility Model in Underwater Acoustic Sensor Networks

C. Kayalvizhi, R. Bhairavi and Gnanou Florence Sudha

Abstract Underwater Acoustic Sensor Networks (UASN) have different applications such as oil platform monitoring, earthquake, tsunami forewarning, water pollution tracking, etc. In UASN, localization is the fundamental step to make the sensed data meaningful and the localization is performed after random deployment of nodes. It is useful in different ways such as node tracking, target detection, etc. Existing works do not consider the ocean current drift during node localization. In this paper, node localization in UASN is achieved by incorporating the ocean current drift for the sensor nodes. As underwater environment is usually complex, the Meandering Current Mobility (MCM) model is considered as the ocean current model and the unlocalized node position is estimated using Dive and Rise Localization method with Distance Vector Hop algorithm. Aquasim software is the simulation tool used to implement the UASN model. Simulation results indicate that the proposed localization method incorporating the ocean drift model outperforms the existing localization scheme in terms of packet delivery ratio, delay, localization ratio, and coverage.

Keywords Underwater acoustic sensor networks · Localization
Meandering current mobility model · Distance vector hop algorithm

1 Introduction

Localization is the process of determining the physical coordinates of a group of sensors in a sensor network. The Global Positioning System (GPS) used for localiza-

C. Kayalvizhi (✉) · R. Bhairavi · G. F. Sudha
Department of Electronics and Communication Engineering, Pondicherry Engineering College, Pillaichavadi 605014, Puducherry, India
e-mail: kayal0804@pec.edu

R. Bhairavi
e-mail: bhairavi@pec.edu

G. F. Sudha
e-mail: gfsudha@pec.edu

S. Smys et al. (eds.), *International Conference on Computer Networks and Communication Technologies*, Lecture Notes on Data Engineering and Communications Technologies 15, https://doi.org/10.1007/978-981-10-8681-6_11

tion in the terrestrial environment does not work underwater. Hence the localization of underwater sensor node is a challenge. Several works related to localization of Underwater Acoustic Sensor Networks (UASN) are available in literature. In [1], a Three-Dimensional Underwater Localization (3DUL) using only three surface buoys for localization is proposed. 3DUL uses two steps, which are Ranging and Projection and dynamic trilateration phase. In the first step, the sensor node at unknown location determines the distance to neighboring anchors. In the second step, the algorithm uses these pairwise distances and depth information to project the anchors onto its horizontal level and forms a virtual geometric structure.

Estimation and prediction based localization techniques are discussed in [2, 3]. DV Hop localization algorithm is discussed in [4]. Localized and Precise Boundary Detection in 3-D Wireless Sensor Network (WSN) [5] uses Unit Ball Fitting (UBF) algorithm and Isolated Fragment Filtering (IFF) algorithm. The UBF algorithm first finds a set of boundary nodes which is followed by a refinement IFF algorithm. This removes the isolated nodes that are misinterpreted as boundary nodes by UBF. Anchor based and anchor free localization techniques are discussed and it is compared with different techniques in [6].

The time Synchronization free Distributed Localization method, [7] uses DNR (Dive and Rise) buoys. All underwater sensors including DNR buoys are connected to each other through acoustic link. DNR buoys have GPS receiver and they are attached with the cable which has different length. Node mobility is not considered in this work.

The Single Anchor Node Based Localization proposed in [8] with Single Anchor node Support uses five steps: range computation, transformation from Geographic coordinates to Cartesian coordinates, coordinate computation, transformation from Cartesian coordinates to Geographic coordinates and anchor node position update. However, this is not suitable for large areas. Localization for Double-head maritime Sensor Networks (LDSN) [9] scheme consists of three steps: Self-Moored node Localization (SML), Underwater Sensor Localization (USD) and Floating-node Localization Algorithm (FLA). Localization ratio is high for LDSN in comparison to 3DUL method and localization error is high for 3DUL.

Dual-Hydrophone Localization (DHL) Method is used to localize the node. DHL uses dual-hydrophone nodes to convert the localization problem into half-plane intersection issues. Each node is attached to a surface buoy by a cable, which can transmit data between the node and the surface buoy. To locate a single unknown node more number of dual hydrophone are required [10]. In [11], Localization scheme free from time synchronization concept is proposed for node localization. Dive and Rise (DNR) localization technique is used to estimate the position of nodes. Speed of sound under water is assumed as a constant. Here, unlocalized nodes are assumed to be static and beacon node is used to find the position of unknown nodes. However in real time, speed of sound in water depends on temperature, pressure, etc., due to various oceanographic forces.

In order to overcome this drawback, in this paper the drift of sensor nodes due to ocean current drift is considered and is incorporated in DNR localization technique. Beacon node is known as DNR beacon node or anchor node. In this work, the

Meandering Current Mobility (MCM) drift model is considered as the ocean current mobility model and implemented for the DNR method. To estimate the position of unlocalized node, DNR localization with Distance Vector (DV) Hop algorithm [12] is used. Unlocalized nodes are restricted to move within the area of 50 m range. It is inferred from the results that this algorithm improves the packet delivery ratio, delay, localization ratio and coverage.

The rest of the paper is organized as follows. Section 2 describes the localization of UASN nodes by considering the ocean drift model. The performance evaluation is discussed in Sect. 3. Finally, Sect. 4 presents the conclusion.

2 Localization of UASN Nodes with Ocean Drift Model

In this section, the localization of underwater sensor node using the proposed method is discussed. The drift model and the DV hop algorithm incorporating the drift model for finding the location of the sensor nodes are described.

2.1 *Drift Model*

Meandering current mobility model (MCM) [13] is used for modeling the drift of sensor nodes with respect to underwater environmental conditions like ocean current, tides and oceanographic forces. A sensor at (x, y, z) shifts its position by considering horizontal velocity is the same as that of the surrounding flow.

In oceanography, the absence of vertical movements is a design feature of drifters, where the sensors hang at a fixed (small) depth. Nodes are initially deployed in small area and move according to their mobility model which is given by Eq. (1)

$$\varphi(x, y, t) = -\tanh\frac{[y - B(t)\sin(K(x - ct))]}{\sqrt{1 + K^2B^2(t)\cos^2(K(x - ct))}} \tag{1}$$

$$B(t) = A + \varepsilon\cos(\omega t),$$

where A is average meander width, c is phase speed, K is no. of meanders, ω is the frequency and ε is the amplitude. The current field is assumed to be a superposition of the tidal and the residual current field. The tidal field is assumed to be a spatially uniform oscillating current in one direction, and the residual current field is assumed to be an infinite sequence of clockwise and anticlockwise rotating eddies. The dimensionless velocity field in the kinematical model can be approximated as

$$\begin{aligned} V_x &= K_1 * \lambda * v * \sin(K_2 * x) * \cos(K_3 * y) + K_1 * \lambda * \cos(2K_1 t) + K_4 \\ V_y &= -\lambda * v * \cos(K_2 * x) \sin(K_3 * y) + K_5 \end{aligned} \tag{2}$$

where V_x is the speed in x-axis, V_y is the speed in y-axis, K_1, K_2, K_3, K_4, v and λ are variables which are closely related to environmental factors, such as tides, etc. These parameters will change in different environments. Using Eqs. (1) and (2), the drift model is implemented. After implementing the MCM model, the nodes which are randomly deployed at different depths start to move due to oceanographic forces. Hence the node position changes to (x', y', z'). To estimate the position of unlocalized node, DNR with DV Hop algorithm is used. The DNR beacon node will move up and down to broadcast its position to the unlocalized sensor node. Position of unlocalized node can be determined by using DV Hop algorithm which is used to find the distance. Estimated distance is used in multilateration technique to find the position of unlocalized node.

2.1.1 DV Hop Algorithm

DV Hop [12] is a range-free, distributed hop-by-hop localization scheme that gives an approximate location of all nodes in a network. DV Hop localization consists of three phases, determining minimum hop counts of every node, determining average hop distance and determining coordinates of an unlocalized node.

(a) *Determining minimum hop counts of every node*

In this step, each beacon node conveys its location information to its neighbor nodes, so that all the nodes in the network get this information by broadcasting a small packet. The packet contains $\langle x_i', y_i', h_i \rangle$ information where, x_i', y_i' are the coordinates of anchor i and h_i represents the hop count. The initial value of h_i is 0. Each node maintains its hop count table containing $\langle i, x_i', y_i', h_i \rangle$ for each anchor i. When the packet is received by any node, it checks its own table and if the value of h_i stored in its table is less than h_i value received by it, then it ignores that received value; otherwise it increments h_i value by 1 and stores the new value of h_i for anchor i in its table. After saving this value, it forwards the packet with updated value of h_i to all its neighbors. So each node gets minimum hop count from every anchor node and has updated hop count table.

(b) *Determining average hop distance*

In this phase, each anchor (A_i) estimates average distance per hop (average hop distance$_i$) using Eq. (3).

$$\text{average hop distance}_i = \frac{\sum_{i=0;j=0}^{n} \sqrt{(x_i' - x_j')^2 + (y_i' - y_j')^2}}{h_i}, \tag{3}$$

where n is the total number of anchors in the network, j denotes all other anchors and h_i is the number of hops between anchor i and anchor j, $\left(x_i', y_i'\right)$ and $\left(x_i', y_i'\right)$ represents coordinates of anchors i and j respectively. After computing average hop distance, each anchor A_i broadcasts it in the network. Then, each node whose location is unknown computes its distance from the anchor A_i using Eq. 4.

$$d_i = \text{average hop distance}_i * h_i \tag{4}$$

Thus the anchor node is used to find the distance by using Eqs. (3) and (4).

(c) *Determining coordinate of an unknown node*

In this phase, each unlocalized node uses multilateration method to determine its position coordinates. Unlocalized node requires the coordinates of all anchor nodes and their distance from each anchor to perform multilateration. The multilateration method works as follows: Let $\left(x_n', y_n'\right)$ be the coordinates of unknown node N and $\left(x_i', y_i'\right)$ be the coordinates for anchor A_i and let there be total m anchors. Hence,

$$\begin{gathered}(x_n' - x_1')^2 + (y_n' - y_1')^2 = d_1^2 \\ (x_n' - x_2')^2 + (y_n' - y_2')^2 = d_2^2 \\ \vdots \\ \vdots \\ (x_n' - x_m')^2 + (y_n' - y_m')^2 = d_m^2\end{gathered} \tag{5}$$

Subtracting all equations one by one from the last equation in (5), the resultant Eq. (6) is formed as

$$\begin{gathered}x_1'^2 - x_m'^2 + y_1'^2 - y_m'^2 - d_1^2 - d_m^2 = 2 * x_n' * \left(x_1' - x_m'\right) + 2 * y_n' * \left(y_1' - y_n'\right) \\ x_2'^2 - x_m'^2 + y_2'^2 - y_m'^2 - d_2^2 - d_m^2 = 2 * x_n' * \left(x_1' - x_m'\right) + 2 * y_n' * \left(y_1' - y_n'\right) \\ \vdots \\ \vdots \\ x_{m-1}'^2 - x_m'^2 + y_{m-1}'^2 - x_m'^2 - d_{m-1}^2 - d_m^2 = 2 * x_n' * \left(x_{m-1}' - x_m'\right) + 2 * y_n' * \left(y_1' - y_n'\right)\end{gathered} \tag{6}$$

Equation (6) can be written in matrix form as

$$AX_n' = B \tag{7}$$

$$A = 2 * \begin{bmatrix} x'_1 - x'_m \ y'_1 - y'_m \\ x'_2 - x'_m \ y'_2 - y'_m \\ \vdots \\ \vdots \\ x'_{m-1} - x'_m \ y'_{m-1} - y'_m \end{bmatrix}, x'_n = \begin{bmatrix} x'_n \\ y'_n \end{bmatrix}$$

$$B = \begin{bmatrix} x'^2_1 - x'^2_m + y'^2_1 - y'^2_m - d^2_1 - d^2_m \\ x'^2_2 - x'^2_m + y'^2_2 - y'^2_m - d^2_2 - d^2_m \\ \vdots \\ \vdots \\ x'^2_{m-1} - x'^2_m + y'^2_{m-1} - y'^2_m - d^2_{m-1} - d^2_m \end{bmatrix}$$

Equation (7) can be written as

$$x' = (A^{\mathrm{T}}A)^{-1}A^{\mathrm{T}}B \tag{8}$$

The coordinates of the nodes are determined by solving Eq. (8) using the Least square method. The advantage of DV Hop algorithm is that, it is cost-effective and can be used to find the location of all nodes even if the node has less than three neighboring anchors as compared to the other range-free algorithms. Estimating the depth from pressure, based on the Saunders and Fofonoff method the results deviate by 0.08 m at 5000 decibars and 0.44 m at 10000 decibars. The depth z' for unlocalized node in meters can be calculated by using the gravity variation (g), specific volume (V) and geopotential anamoly (Del).

The gravity variation with latitude Φ in degree is calculated by:

$$g = 9.880318 * (1 + 5.2788^{-3} \sin^2 \varnothing + 2.36^{-5} \sin^4 \varnothing) \tag{9}$$

The depth (z) in meters can be calculated from the gravity variation (g), the specific volume (V) and geopotential anamoly (D) in J/Kg.

$$V = 9.72659 * P - 2.2512^{-5} * P^2 + 2.279^{-10} * P^3 - 1.82^{-15} * P^4 \tag{10}$$

where $\varnothing$—30 degree and P—10000 decibars

By substituting Eqs. (9) and (10) in Eq. (11)

$$z' = \frac{V}{g} + \frac{D}{9.8} \tag{11}$$

the coordinates (x', y', z') of unlocalized node is determined.

3 Performance Evaluation

Performance of the proposed localization method incorporating ocean drift model and existing DV Hop localization was evaluated through simulations in Aquasim 1.0. This section gives the simulation details and the results obtained.

(a) *Aquasim simulation parameters and results*

The proposed localization technique is evaluated using Aquasim 1.0. The total number of deployed sensor nodes are 57 nodes (7 DNR beacons and 50 unlocalized nodes) and the communication range of nodes are 200 m. These sensor nodes are deployed randomly with different depth in 1000 m × 650 m monitored space. Unlocalized nodes are restricted to move within 50 m range to avoid harsh environmental conditions. Aquasim simulation parameters are listed in Table 1.

(b) *Performance metrics*

The performance evaluation metrics for localization of nodes are localization ratio, packet delivery ratio, delay and coverage. The existing localization technique (with static unlocalized nodes) is compared with the proposed localization scheme with MCM (Meandering Current Mobility Model).

Localization ratio: The localization ratio can be defined as the number of localized sensor nodes to the ordinary sensor node. The localization ratio can be computed using L_Ratio = N_L/N_O, where N_L is the number of localized ordinary nodes and N_O is the total number of ordinary nodes.

The performance of proposed localization method incorporating ocean drift model is compared with the existing DV Hop in terms of localization ratio. From Fig. 1, it can be inferred that when the total number of sensor node is 25, the proposed localization scheme achieves localization ratio of 100%. Localization ratio decreases

Table 1 Aquasim simulation parameters

Simulator	Aquasim on NS2
Number of nodes	57
Interface type	Phy/Underwater Phy
MAC type	802.11
Queue type	Queue/Drop Tail/Pri Queue
Antenna type	Omni-antenna
Propagation type	Underwater propagation
Transport agent	UDP
Application agent	CBR
Meandering current mobility model settings	Mean: $A - 1.2, c - 0.2,$ $K - \frac{2\pi}{7.5}, \varepsilon - 5, \omega - 0.4$
Ocean current parameter settings	$K_1, K_2 - \pi, K_1 - 2\pi$, K_4, $K_5 - 1, \upsilon - 1, \lambda$ $-0.5 - 3.0$

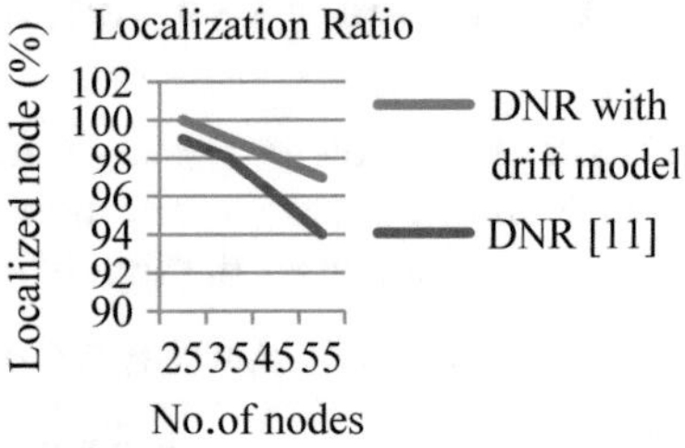

Fig. 1 Localization ratio

Localization Ratio
Localized node (%)
102 100 98 96 94 92 90
25354555
No.of nodes
DNR with drift model
DNR [11]

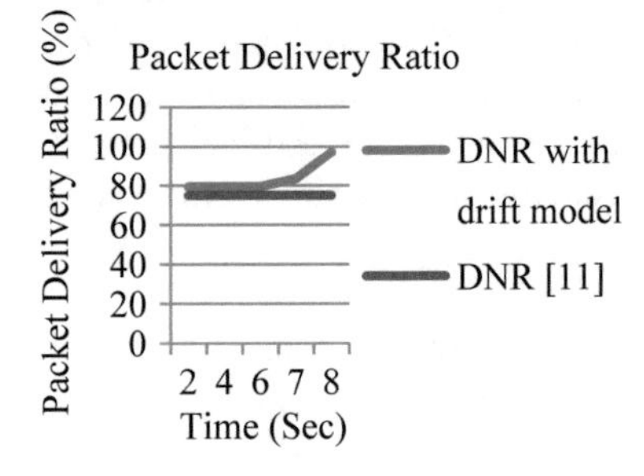

Fig. 2 Packet delivery ratio

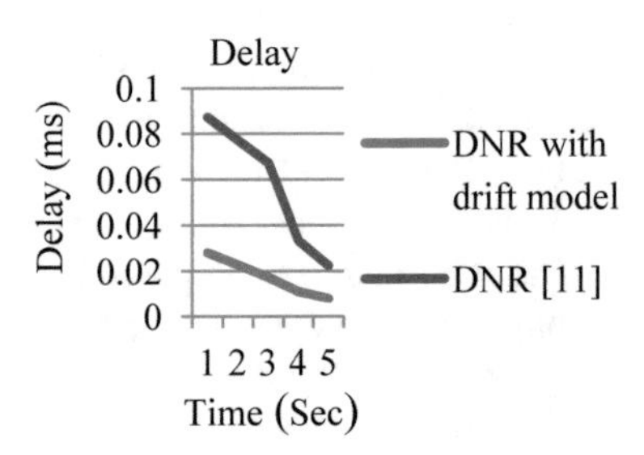

Fig. 3 Delay

as number of nodes increases. It can be observed that DNR with node having drift still maintains a better localization ratio of 97% even when the total number of sensor node is increased to 55 nodes.

Packet delivery ratio: Packet delivery ratio is defined as the ratio of data packets received by the destination node to those generated by the sources. Beacon node broadcasts its position coordinates to its neighbor nodes. Based on the number of packets received successfully, packet delivery ratio is calculated. Figure 2 shows the packet delivery ratio. For the proposed method, maximum packet delivery ratio obtained is 97.02% while it is 74.04% for the existing method. Thus, Packet delivery ratio is high for the proposed method compared to the existing method.

Delay: Delay refers to the time taken for a packet to be transmitted across a network from DNR beacon node to nearest node. From Fig. 3, it can be observed that the proposed localization scheme has lesser delay when compared with the existing DV Hop algorithm.

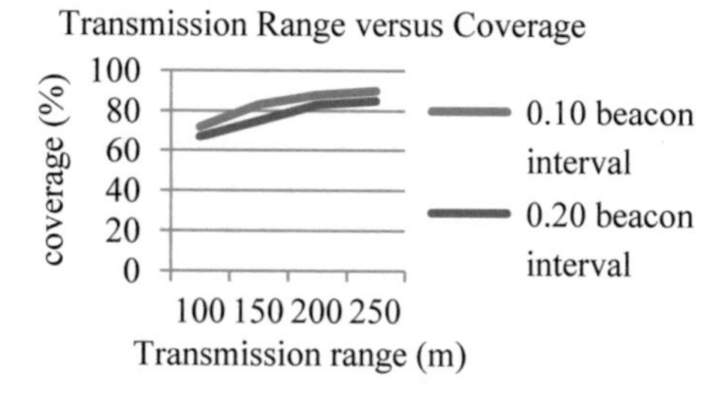

Fig. 4 Transmission range versus Coverage

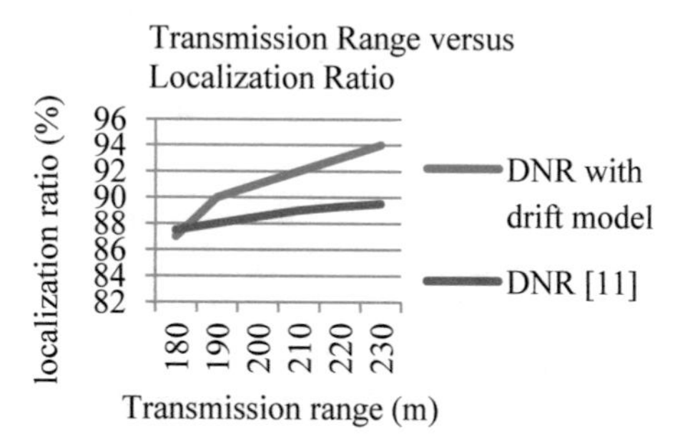

Fig. 5 Transmission range versus Localization ratio

Coverage: Coverage can be defined as how well or to how much extent each point of a deployed network is under the vigilance of a sensor node. The transmission range for sensor nodes varies from 150 to 250 m. Let R be the transmission range of a mobile beacon. The distance traveled by a mobile beacon before sending the next message is denoted by D_b. Then, the coverage for a mobile beacon is given by:

$$\text{coverage} = \frac{\pi}{12}(4R + D_b)(2R - D_b)^2, \tag{12}$$

where D_b denotes beacon interval and R denotes the transmission range.

The performance of proposed localization method incorporating ocean drift model is next tested by varying beacon interval as 0.1 and 0.2 s. Initially, the transmission range is taken as 150 m and it is gradually increased regularly with a fixed interval of 50 up to 250 m. And, the beacon intervals are set to the 0.1 and 0.2 s. Most of the sensor nodes are capable of receiving two or more messages if the beacon interval is kept small with higher transmission range.

Figure 4 describes transmission range versus coverage. It is clearly observed that coverage increases with an increase in transmission range and the coverage can be achieved up to 88% at higher transmission range keeping the beacon interval at its lowest end, of 0.1 s. Beacon interval is set as 0.10 and 0.20 s with transmission range of 250 m. Coverage percentage increases for 0.10 s beacon interval as compared with 0.20 s.

Figure 5 shows transmission range versus localization ratio for the proposed DNR with MCM ocean drift model. As the transmission range increases localization ratio also increases for the proposed method.

4 Conclusion

This paper aims at analyzing the existing localization technique and to improve the localization ratio. Meandering current mobility (MCM) model is designed as the ocean drift model. In this paper, DNR with DV Hop algorithm localization approach is proposed to estimate the coordinates of unlocalized nodes. DV Hop localization consists of three phases: determining minimum hop counts of every node, determining average hop distance and determining coordinates of an unlocalized node. Simulation results validate that the proposed method shows considerable improvement in the localization performance evaluated in terms of localization ratio, packet delivery ratio, delay, and coverage.

References

1. Talha Isik, M., Akan, O. B.: A three dimensional localization algorithm for underwater acoustic sensor networks. IEEE Trans. Wirel. Commun. **8**, 4457–4463 (2009)
2. Erol-Kantarci, M., Mouftah, H. T., Oktug, S.: Localization techniques for underwater acoustic sensor networks. IEEE Commun. Mag. 152–158 (2010)
3. Erol-Kantarci, M., Mouftah, H.T., Oktug, S.: A survey of architectures and localization techniques for underwater acoustic sensor networks. IEEE Commun. Surv. Tutorials **13**(3), 487–507 (2011)
4. Khurana, M., Payal, A.: Analysis of DV-hop localization algorithm in wireless sensor networks. In: Proceedings of the 5th National Conference; INDIACom, pp. 1–4 (2011)
5. Zhou, H., Xia, S., Jin, M., Wu, H.: Localized and precise boundary detection in 3-D wireless sensor networks. IEEE/ACM Trans. Netw. 1–14 (2014)
6. Anil, C. B., Mathew, S.: A survey and comparison study of localization in underwater sensor networks. Int. J. Comput. Sci. Mob. Comput. **3**(12), 23–29 (2014)
7. Kundu, S., Sadhukhan, P.: Design and implementation of a time synchronization-free distributed localization scheme for underwater acoustic sensor network. IEEE Appl. Innovations Mob. Comput. (AIMoC), 74–80 (2015)
8. Das, A. P., Thampi, S. M.: Single Anchor Node Based Localization in Mobile Underwater Wireless Sensor Networks. Springer International Publishing Switzerland, pp. 757–770 (2015)
9. Luo, H., Kaishun, W., Gong, Y.-J., Ni, L.M.: Localization for drifting restricted floating ocean sensor networks. IEEE Trans. Veh. Technol. **65**, 9968–9981 (2016)
10. Sanfeng, Z., Naigao, J., Lei, W., Xueshu, Z., Shuailing, Y., Ming, Z.: A novel dual-hydrophone localization method in underwater sensor networks. In: IEEE/OES China Ocean Acoustics Symposium (2016)
11. Beniwal, M., Singh, R. P., Sangwan, A.: A localization scheme for underwater sensor networks without time synchronization. Springer Wirel. Pers. Commun. **88**, 537–552 (2016)
12. Kaur, A., Gupta, G. P., Kumar, P.: A survey of recent developments in DV-hop localization techniques for wireless sensor network. J. Telecommun. Electron. Comput. Eng. **9**(2), 69–71 (2017)
13. Tsai, P.-H., Tsai, R.-G., Wang, S.-S.: Hybrid localization approach for underwater sensor networks. Hindawi J. Sens. 1–13 (2017)

C. Kayalvizhi completed her B.E. in Electronics & Communication Engineering from Kongu Engineering College, Erode affiliated to Anna University in 2016. Obtained M.Tech in Electronics & Communication Engineering from Pondicherry Engineering College, Puducherry affiliated to Pondicherry University in 2018. Her areas of interests include Wireless Sensor Networks and Security.

R. Bhairavi received her B.Tech in Electronics and Communication Engineering in 2014 from Sri Manakula Vinayagar Engineering College and M.Tech in Wireless Communication in 2016 from Rajiv Gandhi College of Engineering and Technology, Puducherry, India. She is currently doing her research in the area of Underwater Acoustic Sensor Network in Pondicherry Engineering College, Puducherry. Her other areas of interests include Wireless Communication and Wireless Sensor Network.

Dr. Gnanou Florence Sudha , born in 1971 at Pondicherry. Completed B.Tech in Electronics & Communication Engineering from Pondicherry Engineering College affiliated to Pondicherry University in 1992. Obtained M.Tech in Electronics & Communication Engineering from Pondicherry Engineering College in 1994. Completed Ph.D. in Electronics & Communication Engineering in 2005. She joined the Department of Electronics & Communication Engineering, Pondicherry Engineering College as Lecturer in 1995. She worked as Senior Lecturer in the same institute from 2000 to 2005. Currently, she is working as a Professor in the same department in Pondicherry Engineering College. She has published more than 29 journals and conference papers, 23 years of experience and her field of research interests are in Signal Processing and Communication.

Air Quality Prediction Data-Model Formulation for Urban Areas

Kavita Ahuja and N. N. Jani

Abstract A continuous time model called Air Quality Data-Model was developed to assist air quality measurement in urban areas. This model can be utilized in several large areas to estimate the Air Quality parameters for measurement of pollinations in urban area because of urbanization. The model can used to predict the air quality for upcoming time of interval for the city. This work is pertaining to Surat Urban area which includes southern part of Tapti River and Arabian Sea to its western coast. The predictive model developed for the air quality monitoring based on training model. This predictive model will present the forecasting of the air quality measurement of the upcoming years.

Keywords Air quality · AQI prediction model

1 Introduction

Air quality is measurement of purity and level of contamination in the air. The quality of air is measured using Air quality index which is for the purpose of measuring the air contamination and whether the environment is healthy, or cause to health hazard. The Air quality might be affected by various factors. The gross study of Air quality measurement and its prediction is very a lot essential to understand the growth of deterioration in the quality of air. It also helps to understand the pattern of possible pollution in air and respective measures can be taken for. The present study is pertaining to the Air quality and its causes based on the databases available for the Surat city. It is important to understand that the Surat city is highly developing city situated in the south region of Gujarat state. The city is at the bank of Tapti River and

Kavita Ahuja (✉)
Vimal Tormal Poddar BCA College, Surat, India
e-mail: prof.ahujakavita@gmail.com

N. N. Jani
SKPIMCS-MCA, Gandhinagar, India
e-mail: drnnjanicsd@gmail.com

S. Smys et al. (eds.), *International Conference on Computer Networks and Communication Technologies*, Lecture Notes on Data Engineering and Communications Technologies 15, https://doi.org/10.1007/978-981-10-8681-6_12

close to the bay of Cambay. It is necessary to understand certain terminologies and basics related to the air quality. It is also essential to understand the large datasets and perform the data analytics to explore the pattern for air quality. The aim of this research paper is formulation strategies for the air quality prediction model for urban areas. The formulation of model is based on database available of the air quality and steps for processing on it.

1.1 Background

A. **About AQI**

AQI can be considering as a standard having range from 0 to 500. The measurement of air pollution rank depends on AQI level, if its high that leads to increase the pollution level and health concern [1]. For example, an AQI value of 40 to be at good air quality level with little possible to affect public health, while an AQI value over 300 represents unsafe air quality.

B. **Air Pollutant**

Major Air pollutants elements are consist of six important components. It includes Carbon Dioxide (CO_2), Nitrogen Dioxide (NO_2), Methane (CH_4) Carbon Monoxide (CO), Ozone (O_3), and Particulate Matter (PM) [2]. To maintain the efficiency of air quality the parameters temperature, humidity, dew of climate are also consider for the study.

2 Related Work

Khedo et al. [3] has undertaken the issue of air pollution monitoring for the island Mauritius. They proposed the design of Wireless Air Pollution Monitoring System (WAPMS) which used Recursive Converging Quartiles (RCQ) data aggregation algorithms for data smoothing by that way the rate of data passed to sink is minimize and power management of the network can be achieved. Devarakonda et al. [4] have undertaken the issue and challenges of designing and implementing a vehicular based mobile approach for air quality monitoring. For that two models has been undertaken one that is deployed on public transportation as on bus for same route every day and another is in personal sensing device for multiple route. The experiments are purely based on considering only carbon monoxide (CO). Prasad et al. [5] have undertaken the issue of air quality Monitoring system in Hyderabad city. Appropriate gas Sensors is calibrated to generate gas data. These pre-calibrated sensors are then interfaced with the wireless sensor nodes forming multi-hop mesh network. The experiments carried out in different physical conditions for making reliable source of real-time fine-grain pollution data. Brienza et al. [6] have proposed the air quality monitoring system in urban area. The developed tool has been tasted in three different urban areas. The proposed approach can be more efficient if it consider huge amount of data for the monitoring with implementation in the whole urban city. Houyoux et al. [7] has proposed the air model based on season for some period of time in years. They have consider two pollutants NO and CO. Elbir et al. [8] has proposed the meteorological model to predict the air pollutant SO_2 emission in the air. The model was design with the considerations of all inferences. Yu et al. [9] proposed to evaluate the self-determining signals and prediction of air quality and at last the author checked it's efficiency also. Rosario et al. [10] work on to calculate air pollutant with the use of model. The work was experimented in the city Catania. The statistical analysis has been applied to find the relationship between the parameters. Al-Haija et al. [11] has proposed the implementation of wireless sensor network to calculate the air quality with respect to pollutants. Nuria et al. [12] have constructed cost-effective sensor environment to measure and easy to predict and monitoring of air quality. The undertaken work targeted to define the model and its strategies to develop the model which used to predict the air quality measurement for the future.

3 Research Methodology

Data collected using three sensors placed at three different locations in the city are recorded and obtained at regular intervals. Three sensors at different locations record the Environmental variables data at regular intervals. These intervals are not uniform. Data stored are filtered and Data of Interest are obtained. The data collected for these three filters are of three years 2014, 2015, and 2016 with month extracted are further stored into seven groups based on the time intervals.

As shown in Table 1, total seven slots are arranged for the purpose of biphergation of database for Monthly analysis. It can be observed that T2 and T7 are peak hours where the maximum vehicle traffic is available.

Table 1 Data reading time intervals

Time interval	Duration	Description
T1	12–5 pm	Afternoon hours
T2	5–8 pm	Evening-peak hours
T3	8–10 pm	Late evening-moderate time
T4	10–12 pm	Night hours
T5	12–6 am	Early morning hours
T6	6–9 am	Morning moderate hours
T7	9–12 am	Morning peak hours

3.1 Designing the Research Model

The model is developed based on supervised learning. It is important to layout the research model design and flow of work which leads to the ultimate goal to meet the objectives of the research. The research design flow is steps of procedures that leads to the development of model and includes various procedures and tools that deal with the data which yield the desired results. The results are then need to be verified and outcomes are analyzed.

As shown in Fig. 1. Prediction model is established based on (a) For each sensor, inter dependency and correlation among the parameters are observed, (b) The pattern of inter dependency is analyzed for individual Sensors and for separate time slots, (c) Correlation and inter dependency pattern is compared among the pair of respective sensors for (a) time slots (b) For each month between 2 years (viz. 2014 and 2015, 2015 and 2016). (c) Seasons (winter, summer and monsoon). (d) Pattern of dependency is formed and it is used as learning and training pattern sets, (e) The prediction model is framed using the inferences as mentioned in step (a) to (d), (f) Independent variables include, rain (Other than in Monsoon), cyclone, wind are also affecting the pattern marginally, (g) Independent variables and affecting parameters also includes increase in vehicles, factories, constructions, (h) The pattern of progressive growth in parameter values are considered and used for designing Prediction model.

The model is developed using certain tools which is based on the algorithm described in following section. The flow of proposed research work is designed based on certain assumptions.

- The AQI is based on nine parameters and it varies from location to location.
- Sensors are arranged at three different locations of Surat city. These three sensors termed as S1, S2, and S3 are measuring parameters and record them into the

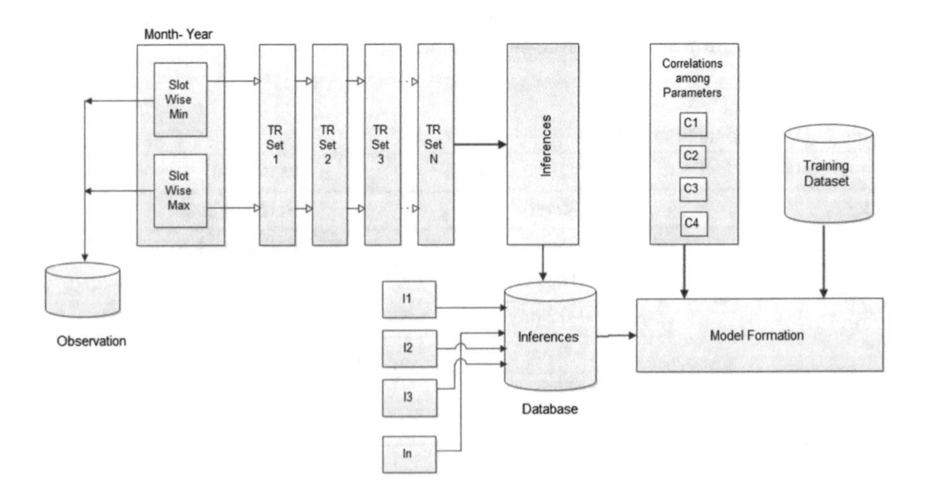

Fig. 1. Training model for AQI prediction model development

terminal linked with the sensors at a fix intervals for round clock. The data is stored in the linked terminals. These three sensors are located at different geographical area of Surat city. S1: This sensor is located at region of Katargam, which is near Tapti River front and having residential area nearby. S2: This sensor is located at Pandesara region which is highly dense industrial and residential area. S3: This sensor is located at region of RingRoad, which is having dense commercial places and having heavy traffic throughout the day. All these three locations have different geographical locations and properties.

- Temperature, Dew, Humidity are captured by all sensors at intervals of 12 min. Parameters including CO, CO_2, and CH_4 are captured at intervals of 16 min and NO_2, PM2.5 and PM10 are captured at intervals of 22 min by the sensors.
- It is important to synchronize the readings of all parameters for every sensor at regular time intervals. The nearest reading algorithm is used to synchronize the readings of all parameters. Per hour readings obtained are 5 for each sensor. Each sensor capture, 120 readings per day for every nine parameters. Together, 360 readings for nine parameters are captured and recorded in database.
- Database for the year 2015 and 2016 are used to train the algorithm. The algorithm is supervised self learning algorithm, which is depends on predictor or dependent variables. The methodologies depend on set of these variables as taken input to the algorithm.

3.2 *Algorithm Using R-Code*

The design of model is based on the algorithm, which is generated in R-code that representation of flow and working of proposed model.

```
#Import Train Datasets
#Verification of Variables
X-T<- input_train_variables
Y-T<- output_train_variables
X-Test <- input_test_data
X<-cbind(X-T, Y-T)
# Model Training using training sets
Linearity<-lm(Y-T ~ ., data=X)
Summary(Linearity)
#prediction of output
Priction<- predict(Linearity, X-Test)
```

The alternative approach to present algorithm using Python:

```
#import Library
#load Train and Test datasets
#identification of variable
X-T<- input_train_variables
```

```
Y-T<- output_train_variables
X-Test <- input_test_data
# linear regression
Linearity = linear_model_LinearRegression()
#Model Training using the training sets
Linearity.fit(X-T, Y-T)
Linearity_score(X-T, Y-T)
#Equation coefficient and Intercept
print('coefficient: \n', Linearity.coef_)
print('Intercept: \n',Linearity.intercept)
# Predict Output
predicted = Linearity.predict(x_test)
```

As shown in Fig. 2, the Data-Model Generation is to be implementing considering the S2 sensor as the pivot sensor value. There is a difference among the parameter values as far as the readings of obtained values of different sensors are concerned. The main objective of designing the Training Data-Model is based on following steps. The whole methodology of research is divided into three steps and their sub-steps:

(a) Observations obtained by all three sensors are required to divide into total seven time segments. These time slots are based on Early night hours, Mid-night hours, Early morning hours, day time (peak hours), Mid-Afternoon (off Peak Hours), Early evening hours and Evening Hours (Peak Hours).

   a. Average Observations are recorded for all three sensors for all parameters using the database of 2015 and 2016.
   b. Considering the Sensor-S2 as pivot observations, Sensor-S1 and Sensor-S3 observations are related for each parameters and a fix pattern is analyzed slot-wise for each parameters.
   c. Correlation among the parameters is obtained for each slot.
   d. Slot-wise patterns are obtained which are further analyzed and Month-Wise pattern are obtained for individual slots for each Parameter.
   e. The pattern and correlation among the parameters are obtained which is slot-wise and month-wise.
   f. For generation of all above patterns and correlation among various parameters, database of 2015 and 2016 is used. This database is used to train the pattern generation algorithm and generating Data-Model.

(b) Database of year 2013 and 2014 are used to test the generated data-model and algorithm. The margin of error and accuracy of the range of parameter predicted values along with the correlation among the parameters are analyzed.
(c) The developed Data-model is used to test for known datasets of 2013 and 2014. The predictions are generated month-wise for the year 2013 and 2014. Obtained pattern and predicted observations are verified and compared with the actual known values for the year 2013 and 2014. The accuracy level of observations are measured and documented. Further, the observations and predictions are

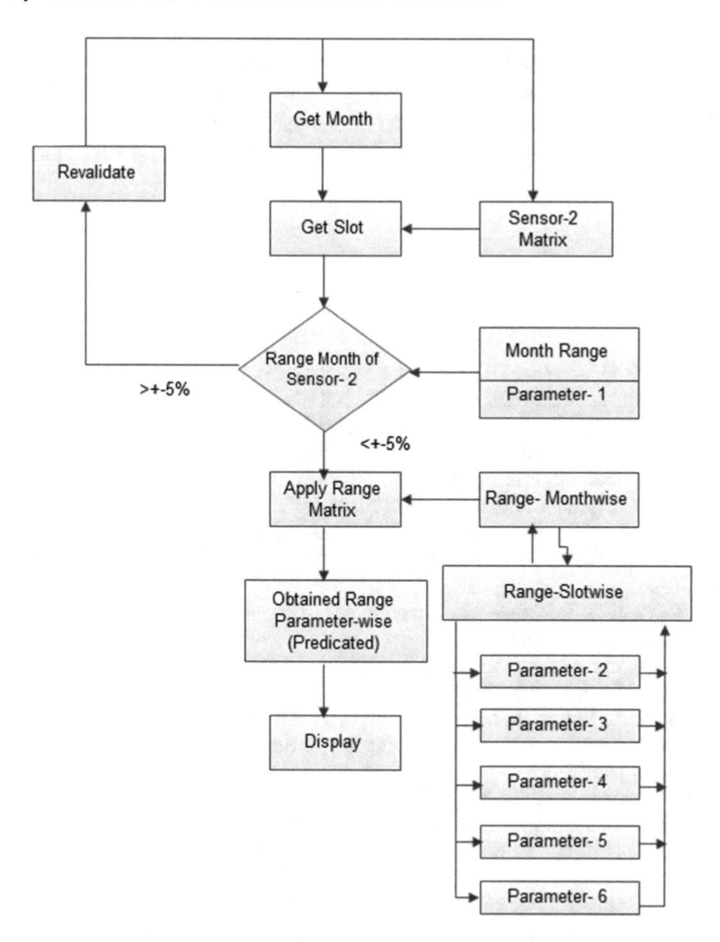

Fig. 2. AQI prediction data-model

generated for the year 2017 month-wise and they are compared with actual observations obtained for each month of the year.

It is important to note that the predicted model generate predicted observations for the Sensor-S2. Using the pattern algorithm, the results are obtained for the Sensor-S1 and Sensor-S3.

4 Conclusion and Future Enhancement

The algorithm is developed to prepare and represents the flow of the work. The air quality monitoring database is based on existing known data of the year 2013 and 2014 can be checked and further we can forecast the air quality level in the future.

We can predict the air quality level for the year 2016 and 2017 and so on. The future work will enhanced by the implementation of the database and predicting the range of each parameter for the all the months of the year with respectively each slot, and will check the accuracy level of predicting the air quality by our formulated proposed model.

References

1. Shooter, D., Brimblecombe, P.: Air quality indexing. Int. J. Environ. Pollut. **36** (2009). ISSN: 0957-4352. https://doi.org/10.1504/ijep.2009.021834
2. Raipure, S.: Calculating pollution in metropolitan cities using wireless sensor network. Int. J. Adv. Res. Comput. Sci. Manag. Stud. **2**(12) (2014). ISSN: 2321-7782 (Online)
3. Khedo, K.K., Perseedoss, R., Mungur, A.: A wireless sensor network air pollution monitoring system. arXiv preprint arXiv: 1005.1737 (2010)
4. Devarakonda, S., et al.: Real-time air quality monitoring through mobile sensing in metropolitan areas. In: Proceedings of the 2nd ACM SIGKDD International Workshop on Urban Computing. ACM (2013)
5. Prasad, R. V., et al.: Real time wireless air pollution monitoring system. ICTACT J. Commun. Technol. **2.2**, 370–375 (2011)
6. Brienza, S., et al.: A low-cost sensing system for cooperative air quality monitoring in urban areas. Sensors **15.6**, 12242–12259 (2015)
7. Houyoux, M. R., Vukovich, J. M., Coats Jr., C. J., Wheeler, N. J. M., Kasibhatla, P. S.: Emission Inventory Development and Processing for the Seasonal Model for Regional Air Quality (SMRAQ) Project (2000)
8. Elbir, T.: Comparison of model predictions with the data of an urban air quality monitoring network in Izmir, Turkey. Atmos. Environ. **37**(15), 2149–2157 (2003)
9. Yu D., Rao, Q., Cai, S., Yu, S.: Blind Detection Algorithm Based on WSN Air Quality Monitoring System. IEEE (2016)
10. Rosario, L., Pietro, M., Francesco, S.P.: Comparative analyses of urban air quality monitoring systems: passive sampling and continuous monitoring stations. Energy Procedia **101**, 321–328 (71st Conference of the Italian Thermal Machines Engineering Association, Elsevier) (2016)
11. Al-Haija, Q.A., Al-Qadeeb, H., Al-Lwaimi, A.: Case study: monitoring of AIR quality in King Faisal University using a microcontroller and WSN. Procedia Comput. Sci. **21**, 517–521 (The 4th International Conference on Emerging Ubiquitous Systems and Pervasive Networks (EUSPN-2013), Elsevier) (2013)
12. Nuria, C., Dauge, F.R., Schneider, P., Vogt, M., Lerner, U., Fishbain, B., Broday, D., Bartonova, A.: Can commercial low-cost sensor platforms contribute to air quality monitoring and exposure estimates? Environ. Int. xxx–xxx (Elsevier) (2016)

Kavita Ahuja working as an Assistant Professor in Vimal Tormal Poddar BCA College, Surat, India. Her research area includes AQI Prediction Model and wireless communication systems.

Dr. N. N. Jani working as Director in SKPIMCS-MCA, Gandhinagar, India. Her research area includes AQI Prediction Model and wireless communication systems.

Secure Data Deduplication System with Tag Consistency in Cloud Data Storage

Pramod Gorakh Patil, Aditya Rajesh Dixit, Aman Sharma, Prashant Rajendra Mahale and Mayur Pundlik Jadhav

Abstract Cloud computing technique is most generally used technique nowadays, in that clouding up is completed over giant communication network. And provides massive space for storing in all sectors like administrative unit, private enterprises, etc. Additionally stores personal information on cloud, but the foremost necessary drawback in cloud is that giant quantity of space for storing is needed and conjointly duplicate copies of knowledge is store on cloud. There is a unit of several techniques that is employed for preventing duplicate copies of continuation information. The necessary technique is data deduplication, Data Deduplication is specializes data compression techniques for removing duplicate copies of repeated data and has been widely utilized in cloud storage to reduce the quantity of space for storing. To safeguard the confidentiality of sensitive data on cloud, the confluent encryption technique is utilized to cipher the data before storing it on cloud. In projected system we have used Shamir Secrete sharing algorithm.

Keywords Cloud · Deduplication · Network · Security

P. G. Patil (✉)
Sandip Institute of Technology and Research Centre (SITRC), Mahiravni, Nashik, India
e-mail: pgpatil11@gmail.com

A. R. Dixit · A. Sharma · P. R. Mahale · M. P. Jadhav
Department of Computer Engineering, Sandip Institute of Technology and Research Centre (SITRC), Mahiravni, Nashik, India
e-mail: adityadxt1@gmail.com

A. Sharma
e-mail: sharmaaman896@gmail.com

P. R. Mahale
e-mail: prashantmahale18@gmail.com

M. P. Jadhav
e-mail: mayurjadhav6896@gmail.com

S. Smys et al. (eds.), *International Conference on Computer Networks and Communication Technologies*, Lecture Notes on Data Engineering and Communications Technologies 15, https://doi.org/10.1007/978-981-10-8681-6_13

1 Introduction

Storage efficiency and information measure efficiency is not used properly due to duplicate information on public cloud. This is often main focus of development. To avoid duplication, information on public cloud is compared to the uploaded information. However main drawback is encrypted information on cloud cannot compared the main task to match encrypted [1] information. The deduplication [2] check need to be the user specific, it got to check the deduplication at cloud storage. Level information is shared among multiple users thus there got to be some constraint that prohibits the deduplication check for unauthorized user. The software will be generating convergent keys for encryption and maintaining tag consistency. It will also deduplicate the files in the given cloud using Block Deduplication. It will also handle the dynamic [3] ownership management for the cloud service. The software will encrypt the files using [4] converging keys alongside deduplication [5]. It will increase the speed of service and reduce the bandwidth use. Application of the Software: Cloud Service Providers. And methodologies used to solve the problem and efficiency issues are: Convergent encryption [6], Symmetric encryption [7], Shamir secret sharing algorithm and Proof of ownership, using all the above together will increase the tag consistency [8] in the current available systems and will also increase the [9] efficiency. After applying this software to the cloud service the duplicated files will be deleted and the files will be encrypted [10] with convergent keys according to hast value of file. The files will have consistence tags to increasing throughput. This will also help in levitating the speed of cloud service.

2 Goals and Objectives

- To protect data confidentiality
- To increase the security of data over the cloud using Secret Sharing
- Reduce risk of data stealing
- To deduplicate the files over cloud data storage
- To protect data integrity using tag consistency
  - Input: An insecure Cloud with duplicate and non-encrypted files
  - Output: Encrypted deduplicated files in cloud

3 Project Schedule

3.1 Task Network

See Fig. 1.

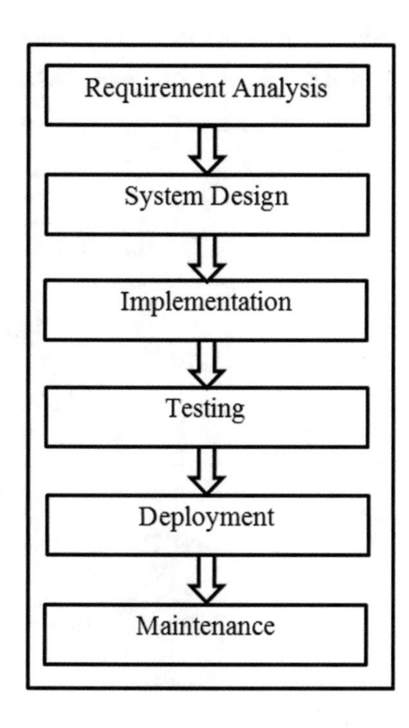

Fig. 1 Task network

3.2 Project Task Set

Major Tasks in the Project stages are:

Task 1: Search for the Problem in Current Systems
Task 2: Define the requirement for Project
Task 3: Do Literature Survey on Project Area
Task 4: Search for the Proper Solution
Task 5: Apply the Solution in algorithmic form
Task 6: Do the Coding and create Software based on algorithm prepared with Documentation
Task 7: Develop the software project

4 Architecture Diagram

See Fig. 2.

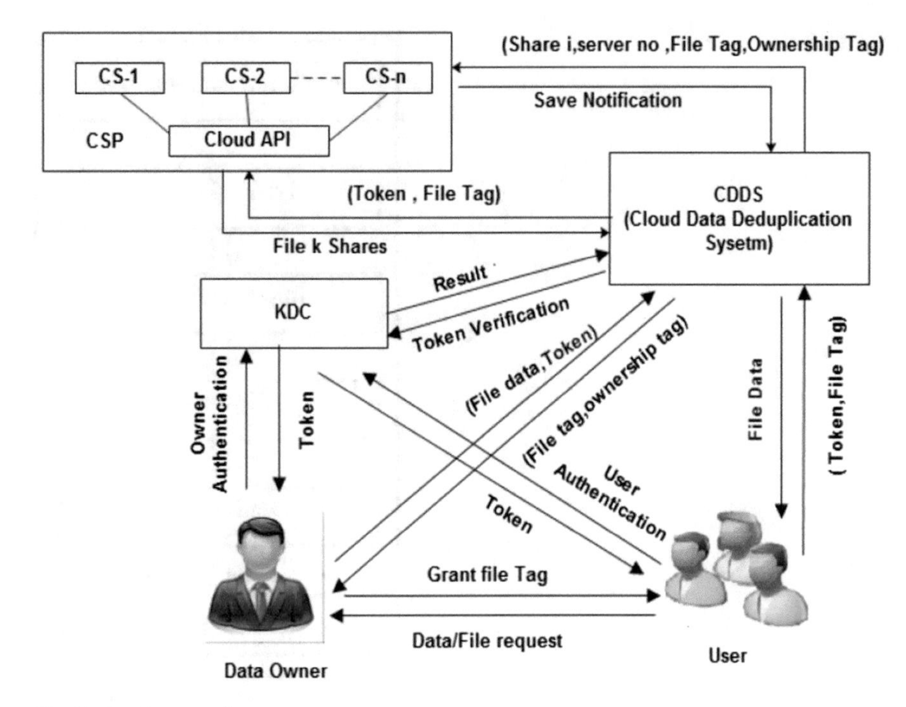

Fig. 2 Architecture diagram

5 Hardware and Software Resources Required

See Tables 1 and 2.

Table 1 Hardware requirements

S. No.	Parameter	Requirement
1	CPU pentium	IV (above)
2	RAM	2 GB
3	Hard disk	180 GB (min)
4	I/p devices	Monitor, keyboard, mouse

Table 2 Software requirements

S. No.	Parameter	Requirement
1	Operating system	UBUNTU/windows
2	IDE	Eclipse
3	Programming language	JAVA
4	Cloud	Amazon Ec2/Ec3
5	File transfer	Win Scp
6	SSH	Putty

6 Applications

The software will be primarily applied by cloud service providers who are providing cloud services to the big number of users. E.g. Amazon cloud, Drop-Box, Google Drive.

7 Conclusion

By doing this paper we come to a conclusion that the Cloud Deduplication can be done more efficiently by using Deduplication by hash values. The tag consistency is also and major issue in cloud which was solved in this. By using Shamir Secret Sharing algorithm alongside of MD5 algorithm for hash value calculation we can increase the security and confidentiality of data over the cloud efficiently. By applying this project Cloud data can be more secured.

Acknowledgements It gives us immense pleasure in presenting the preliminary project report on "Secure data Deduplication System with Tag Consistency in Cloud data Storage", we would wish to take this chance to convey my grateful feelings to my internal guide professor. P. G. Patil for providing us with all the guidance and help I required. I am very grateful to them for their kind support. Their invaluable suggestions were extremely helpful. We are additionally thankful to professor. Dr. A. D. Potgantwar, Head of Computer Engineering Department, SITRC for his indispensable support, suggestions.

References

1. Patil, P.G., Verma, V.K.: A recent survey on different symmetric key based cryptographic algorithms. IJCAT (2016)
2. Quinlan, S., Dorward, S.: Quinlan, S., Dorward, S.: Venti: a new approach to archival storage. In: Proceedings of USENIX Conference on File and Storage Technologies, FAST, January (2002)
3. Halevi, S., Harnik, D., Pinkas, B., Shulman-Peleg, A.: Proofs of ownership in remote storage systems. In: Chen, Y., Danezis, G., Shmatikov, V. (eds) ACM Conference on Computer and Communications Security, p. 491500. ACM (2011)
4. Li, J., Chen, X., Li, M., Li, J., Lee, P., Lou, W.: Secure deduplication with efficient and reliable convergent key management. IEEE Trans. Parallel Distrib. Syst. (2013)
5. Ng, C., Lee, P.: Revdedup: a reverse deduplication storage system optimized for reads to latest backups. In: Proceedings of APSYS (2013)
6. Bellare, M., Keelveedhi, S., Ristenpart, T.: Dupless: Serveraided encryption for deduplicated storage. In: USENIX Security Symposium (2013)
7. Bellare, M., Keelveedhi, S., Ristenpart, T.: Message-locked encryption and secure deduplication. In: EUROCRYPT, pp. 296 312 (2013)
8. Li, J., Li, Y.K., Chen, X., Lee, P.P.C., Lou, W.: A Hybrid Cloud Approach for Secure Authorized Deduplication
9. Stanek, J., Sorniotti, A., Androulaki, E., Kencl, L.: A secure data deduplication scheme for cloud storage. Technical Report (2013)

10. Yuan, J., Yu, S.: Secure and constant cost public cloud storage auditing with deduplication. IACR Cryptol. (e-Print Archive) **2013**, 149 (2013)

Pramod Gorakh Patil working as an Assistant Professor in Sandip Institute of Technology and Research Centre (SITRC). Mahiravni, Nashik. His research area includes network security and cloud computing.

Aditya Rajesh Dixit studying Engineering academics in Department of Computer Engineering Sandip Institute of Technology and Research Centre (SITRC). Mahiravni, Nashik. His research area includes network security and cloud computing.

Aman Sharma studying Engineering academics in Department of Computer Engineering Sandip Institute of Technology and Research Centre (SITRC). Mahiravni, Nashik. His research area includes network security and cloud computing.

Prashant Rajendra Mahale studying Engineering academics in Department of Computer Engineering Sandip Institute of Technology and Research Centre (SITRC). Mahiravni, Nashik. His research area includes network security and cloud computing.

Mayur Pundlik Jadhav studying Engineering academics in Department of Computer Engineering Sandip Institute of Technology and Research Centre (SITRC). Mahiravni, Nashik. His research area includes network security and cloud computing.

High-Resolution Weather Prediction Using Modified Neural Network Approach Over the Districts of Karnataka State

L. Naveen and H. S. Mohan

Abstract Forecasting of future rainfall from previous years data samples has always challenging and major area to focus. There are various factors are applied to anticipate the rainfall such as Mean sea-level, temperature, pressure, wind speed, humidity, etc. We have inaugurated a strategy for predicting the average ground rainfall over the districts of Karnataka state from the past rainfall data applying modified ANN approach without conceiving the rainfall parameters, but considering the average rainfall rates of the previous years and primarily focus on optimization techniques to reduce the error rate during training process. The proposed approach predicts the average rainfall of next consequent year, on inputting anyone year's rainfall data of any districts taken into account. The suggested technique is implemented in MATLAB and the results are tested.

Keywords Social spider optimization (SSO) · Artificial neural network (ANN) Back-propagation algorithm (BPA) · Drag temperature model (DTM)

1 Introduction

Weather condition is state of atmosphere at afforded time in charge of weather variables like temperature, pressure, rainfall, etc. In modern weather forecasting a combo of computer models, observation, and knowledge of trends and patterns are introduced. Weather forecasts entrenched on temperature and precipitations are vital to agriculture and industry sector [1]. Temperature forecasts are completed by collecting quantitative information regarding the present state of the atmosphere. The key advantage of the neural network methodology is that it will fairly approximate an outsized category of functions. In climatic conditions, the climate predictions

L. Naveen · H. S. Mohan (✉)
ISE, SJBIT, Bengaluru, India
e-mail: mohan_kit@yahoo.com

L. Naveen
e-mail: naveen_lingaraju@yahoo.com

S. Smys et al. (eds.), *International Conference on Computer Networks and Communication Technologies*, Lecture Notes on Data Engineering and Communications Technologies 15, https://doi.org/10.1007/978-981-10-8681-6_14

are unit applied to warn regarding natural disasters that area unit evoked by abrupt amendment.

Forecasting the speed of wind and solar radiation are a vital research area under weather prediction since they are essential origins of energy. In combining large-scale alternative energy and because of the high changeableness and finite prognostication power of wind speed, current power systems planning ways face challenge [2]. The radiation could be an important criterion for alternative energy analysis however is international organization accessible for many of the sites because of non availableness of radiation evaluating instrumentality at the meteoric stations. Therefore, for a location applying many environmental condition variables, it's required to predict radiation [3]. Rainfall being the most vital factors is an end product of a number of complex processes in the atmosphere [4]. Prognostication downfall is troublesome atmospheric method because of the irregular downfall characteristics series that is dependent of area and time [5].

For money time-series foretelling, the neural network will apply for optimum of the prediction aspects, radial basis perform neural network have exploit a lot [6, 7]. Formerly, prediction was completed manually by analyze gift climatic conditions, atmospheric pressure and sky condition. However these days weather prediction is completed mistreatment models supported data processing, soft computing approaches [8]. In [9] inaugurated the advance and determination of out of doors air temperature mistreatment neural network based mostly identification algorithms implement nonheritable information from four European cities. ANNs area unit advanced strategies appropriate in creation of difficult systems like wind speed [10]. The AI-based strategies outline the link between input and output information from statistic of the past by a non-statistical approach like artificial neural network, formal logic, fuzzy-inference and data processing [11]. Different methods applied for weather Prediction applying Neural Network has been discussed below.

2 Literature Survey

Advertisement and Radial Basis operate (RBF) neural network model to predict the facility output of a turbine has been projected by Shetty et al. [12] and performance has enlarged by hybrid PSO-FCM (Particle Swarm optimization primarily based Fuzzy C Means) bunch formula. ELM (Extreme Learning Machine) formula and PSO (Particle Swarm Optimization) has been applied to boost neural network model.

Nemesio et al. [13] have introduced a procedure to retrieve soil wet (SM) on a routine. Neural network uses this technique to catch the applied math relationship of reference SM information set and input. Neural Network is trained by reference SM dataset in European Centre For Medium-Range Weather Forecasts model predictions.

James et al. [14] have presented chaotic oscillatory-based neural networks, to calculate the evolution of wind in airport vicinity. This method has better achievement while compared with traditional multilayered perceptron model neural network.

Bhaskar et al. [15] have developed a two-stage forecasting method to conclude wind power output up to 30 look-ahead hours. In stage-I, Adaptive Wavelet Neural Network (AWNN) model has improved approximation and rapid training ability as compared to the Feed Forward Neural Network (FFNN). In stage-II, the forecasted wind speed was used to forecast wind power output through nonlinear input–output mapping using FFNN.

Choury et al. [16] have presented a method for prediction of the thermosphere density by forecasting exospheric temperature. The impact of artificial neural network predictions has been quantified and an order of magnitude smaller orbit errors were found when compared with orbits propagated using the thermosphere model DTM2009.

Sideratos et al. [17] have proposed a methodology for probabilistic wind power forecasting and forecast the prediction uncertainties due to the defect of the numerical weather predictions (NWP), the weather stability and the deterministic forecasting model.

Quan et al. [18] have proposed a short term load and wind power forecasting using NN-based Prediction Intervals to beaten the failure of point forecasts to handle ambiguity. With the enlarged searching capacity, Particle Swarm Optimization (PSO) was then used to clarify the new problem and train the NN models using mutation operator.

Quan et al. [19] have proposed an ensemble short-term wind power forecasting model that was based on advanced forecasting techniques in the modern literature. In parallel this model uses 52 sub-models, 5 sub-models of Neural Network and Gaussian Process respectively. This model forecasts the future wind power for 48 h using historical data.

3 Problem Definition

Weather prediction has been revolutionized due to novel technological improvements such as the invention of the telegraph, electronic computer, and remote sensing. To predict the weather, originally statistical methods had utilized on building linear models and time series analysis of evaluated variables. However, due to inaccuracy meteorological data, data are affected by different kinds of perturbations and uncertainty. To resolve this trouble, more sophisticated stochastic models were improved. However, for local topography, such models have limitations such as the finite observation set for amount (a serious issue given the extremely broad number of variables in these methods) relatively finite spatial resolution possible over such a wide area (typically the whole earth) and the impossibility of accounting. However, found insufficient solution in this area and motivated me to do research.

4 Proposed High-Resolution Weather Prediction Method Using Modified Neural Network Approach

To raise the being Weather prediction models and to improve a new way of weather prediction, we have suggested an effective high resolution Weather Prediction model to take full advantage of big observation data. The forecasting model takes the average ground rainfall data from previous twenty years. With more accuracy, the weather forecasting is created and applying the Modified Artificial Neural Network technique. Districts around Karnataka state, the proposed forecasting model is aligned to predict the future weather feasibility. The database applied and comprises the average rainfall data of every month from 1993 to 2013 years. The proposed neural network can be generalized to predict the future average rainfall details of every month through learning established on the trained data. The block diagram of proposed rainfall detection model is given in Fig. 1.

Here, the training will be caused in the basis of, for any given (tth) year data of a district as input; the next consequent ($t+1$th) data must be gotten as output. The

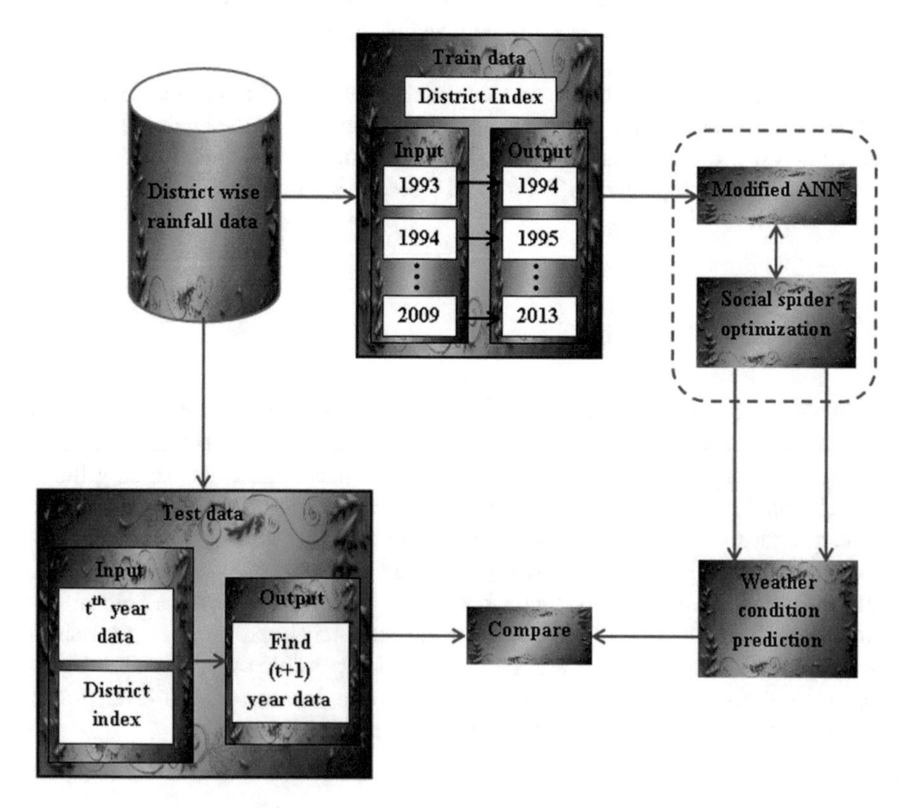

Fig. 1 Proposed rainfall detection model

network can be generalized to predict the next consequent year for any afforded input year rainfall data by this way.

a. **District-Wise Yearly Weather Prediction Using Modified ANN**

The proposed technique utilizes a modified Artificial Neural Network (ANN) approach for predicting the yearly average ground rainfall. ANN is inspired by biological neuron and human cognitive capability to relate dissimilar elements together. However, there are few drawbacks in the traditional neural networks, particularly, the greater computational burden, proneness to over fitting, and the empirical nature of model improvement.

Random selection of number of hidden nodes may induce the trouble of over fitting and under fitting is the main trouble. Among neural network training stage the over fitting trouble will found. During training, Over fitting is a condition, in which error value is small or large, when novel data is induced. The network has trained with samples, but not learned for neither abnormal nor novel situations. Efficient ways of data preprocessing, selection of architecture and training for the network plays vital role in getting proper result ANN.

Hence, to develop the training performance, we have inaugurated the Social Spider optimization (SSO) algorithm for optimal number of hidden layers and its nodes for the artificial neural network. The advantage of proposed method is that, it is not approximately computing number of hidden nodes but established on accuracy of outcomes. In suggested method number of hidden nodes are not predefined but generated among the training time. During training, the selection of optimal network architecture is done along with the back-propagation training algorithm, where the BP algorithm tends to minimize the prediction errors and based on which the network will be modified using SSO to produce optimal network structure. Thus the objective function of SSO algorithm is the maximum accuracy of the network (i.e., minimum prediction error). The Structure of Social Spider Optimization in Artificial Neural Network is given in Fig. 2.

4.1 Layer Neuron Optimization Using Social Spider Optimization Technique

For determining its optimal network architecture, the suggested modified ANN approach applies the SSO algorithm. Cuevas et al. [18] have been developed the SSO algorithm is dependent on the simulation of the biological cooperative behavior of social spiders. Heavier individuals rule the lighter ones in nature. This behavior is duplicated to the algorithm and the spider's weight is representing to the solution decision. In the colony, the spiders are partitioned by gender and everyone have an alternate conduct. Gender adjust on the colony is ordinarily around 70% of females. Let us conceive the spider population 'X' is composed of 'G' number of agents, which are separated into two sub-groups of male 'T' and female 'D'.

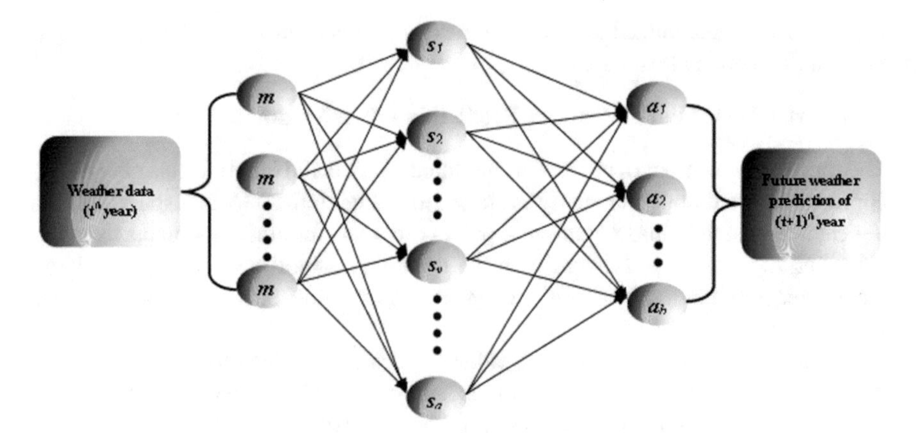

Fig. 2 Structure of social spider optimization in artificial neural network

Initialization

The phase factor values are started as the original solution spider. Let the phase factor values are represented as X_s which is shown in below,

$$[X_s = \{x_1, x_2, \ldots, x_r\}], \tag{6}$$

where X_s indicates an original solution and rth measure is denoted as length of solution. Here, the primary solution is the number of hidden layers and its corresponding neurons.

Generate Male-Female Population

The number of female spider is selected randomly among 65–70%. Now the input population 'G' is characterized by the quantity of female G_d and male G_t spiders in the entire population. The characterization of male and female spiders are afforded as,

$$G_d = \text{floor}[0.9 - I * (0.25) * G] \tag{7}$$

$$G_t = G - G_d, \tag{8}$$

where I denotes a random number among [0, 1] and floor(·) matches a real number to the nearest integer number. The resultant female and male population is afforded as,

$$\left(d = \left\{d_1, d_2, \ldots, d_{G_d}\right\}\right) \tag{9}$$

$$\left(t = \left\{t_1, t_2, \ldots, t_{G_t}\right\}\right) \tag{10}$$

Evaluation of Fitness Function

Now, by using below equation fitness value for each iteration solution can be calculated,

$$\text{Fit} = \text{Max}(\text{Accuracy}) \tag{11}$$

To determine the fitness for the all spiders and the spiders with best fitness alone survive. The best fitness solution denotes that the solution (i.e., the number of hidden nodes selected) affords mores prediction accuracy of the ANN.

Spider Weight Evaluation

The weight of each spider is computed. The spider weight can be determined by means of the following equation.

$$P_s = \frac{\text{Fit}_s - L_{\text{fit}}}{H_{\text{fit}} - L_{\text{fit}}}, \tag{12}$$

where Fitness_i is the fitness value received by the evaluation of the sth spider location and the values worst_t and best_t are calculated and applying the below expression.

$$H_{\text{fit}} = \text{Fit}_s^{(\text{max})}; \quad L_{\text{fit}} = \text{Fit}_s^{(\text{min})} \tag{13}$$

Modeling of the Vibrations

The data is transmitted through the communal web in the form of vibration. Moreover, the communication is produce in three ways, that is vibration emitted by closest heavier member, closest female and best spider.

The vibration emitted by the r_Hth spider (closest heavier member) on the 'sth' spider is modeled by the following equation:

$$\text{Vi}_{s,r_H} = P_{r_H} \text{e}^{-\left(\frac{\text{ED}_{s,r_H}}{A_G}\right)}, \tag{14}$$

where, ED_{s,r_H} is the Euclidian distance among the spiders 's' and 'r_H' and A_G is the gain adaptive factor. Similarly, the vibration emitted by the r_dth spider (closest female member) and r_bth spider (best spider in the web) on the 'sth' spider is modeled by the following Eqs. (15) and (16) respectively:

$$\text{Vi}_{s,r_d} = P_{r_d} \text{e}^{-\left(\frac{\text{ED}_{s,r_d}}{A_G}\right)} \tag{15}$$

$$\text{Vi}_{s,r_b} = P_{r_b} \text{e}^{-\left(\frac{\text{ED}_{s,r_b}}{A_G}\right)} \tag{16}$$

Female Cooperative Operator (Repulsion/Attraction)

Female spiders demonstrate an attraction or repulsion over others regardless of gender. For every female spider, the selection is modeled as a stochastic decision by means of generating a random number within the range 0 and 1. If the random number is less than threshold Th, an attraction motion is developed or else a repulsion motion is created.

Hence the operator is represented as follows:

$$d_s^{\text{itr}+1} = \begin{cases} d_s^{\text{itr}} + \delta * \rho * \left(\text{Vi}_{s,r_H}\right)\left(H_{(x)} - d_s^{\text{itr}}\right) + \gamma * \left(\text{Vi}_{s,r_b}\right)\left(b_{(x)} - d_s^{\text{itr}}\right) + \lambda * (I - 0.5), & \text{if prob} = \text{Th} \\ d_s^{\text{itr}} - \delta * \left(\text{Vi}_{s,r_H}\right)\left(H_{(x)} - d_s^{\text{itr}}\right) - \gamma * \left(\text{Vi}_{s,r_b}\right)\left(b_{(x)} - d_s^{\text{itr}}\right) - \lambda * (I - 0.5), & \text{if prob} = 1 - \text{Th} \end{cases} \tag{17}$$

In the above equation γ, δ and λ are the arbitrary values among [0, 1] whereas $d_s^{\text{itr}+1}$ denotes the next iteration. The individual $H_{(x)}$ and $b_{(x)}$ denote the coordinate of closest heavier member and the position of best individual of the entire population, G. Also, ρ represents the gain adaptive factor.

Male Cooperative Operator (Dominance)

Male individuals, as the predominant individuals, with median weight inside the male populace are viewed. The movement of dominant and non-dominant male individuals can be established as follows.

$$t_s^{\text{itr}+1} = \begin{cases} t_s^{\text{itr}} + \delta * \rho * \left(\text{Vi}_{s,r_d}\right)\left(d_{(x)} - t_s^t\right) + \rho(I - 0.5), & \text{if } P_{G_d+y} > P_{\text{med}} \\ t_s^{itr} + \delta * \rho\left(\frac{\sum_{y=1}^{G_t} t_y^{\text{itr}} . P_{G_d+y}}{\sum_{y=1}^{G_t} P_{G_d+y}} - t_s^{\text{itr}}\right), & \text{if } P_{G_d+y} \le P_{\text{med}} \end{cases}, \tag{18}$$

where, the individual $d_{(x)}$ indicates the closest female member to the male individual 'i' while $\left(\frac{\sum_{y=1}^{G_t} t_y^{\text{itr}} . P_{G_d+y}}{\sum_{y=1}^{G_t} P_{G_d+y}} - t_s^{\text{itr}}\right)$ indicate the weighted average of the male populace G_t and P_{med} is the median weight. Also, the weight of yth male individual greater than the median weight ($P_{G_d+y} > P_{\text{med}}$) are conceived as the dominant and those with weight lesser than the median weight ($P_{G_d+y} \le P_{\text{med}}$) are the non-dominant male individuals.

Mating Procedure After moving, the male and female individuals, mating procedure is done. Mating in a social-spider state is done by dominant males and the female individuals. The mating process is done among the female individual and the dominant males for those within particular radius. The mating radius is computed as,

$$M_{\text{Rad}} = \frac{\sum_{n=1}^{N}\left(m_n^U - m_n^L\right)}{2N}, \tag{19}$$

where, m_n^U and m_n^L are the upper and lower bounds of the dimension N. Among the mating, novel candidate spiders were generated established on Roulette Wheel Selection method. For each spiders then fitness will be estimated. Established on the fitness value, best solutions will be chosen from the novel candidate solution and the population members. Once mating procedure is fulfilled, the new spiders are created. Now, the new spiders are again fitness determined, and the process takes place continuously till the maximum iteration or till the desired accuracy is attained. Once, the SSO algorithm ends, the optimal network architecture is found.

4.1.1 Back-Propagation of Error in the Optimal Network Structure

The optimal network structure derived from the SSO optimization approach, is then applied for the further training procedure applying Back-Propagation (BP) learning algorithm. For finding the minimum error function, the gradient descent learning algorithm is applied and it is employed with BP to modify the weight of neurons. The learning rate was enforced in BP algorithm to resolve how directly the gradient descent should be utilized to the weight matrix and threshold values. The gradient is multiplied by the learning rate and then added to the weight matrix or threshold value. This will slowly enlarge the weights to values that will develop a lower error. The second parameter is as momentum, which specifies, to what degree, the weight changes from the previous iteration should be utilized to the current iteration. This helps back-propagation algorithm to come out of local minimal. The anticipated Adaptive Artificial neural network presumes the values of learning rate and momentum value as 0.001 and $1.0000e^{-03}$ respectively. Moreover, the number of training iterations is taken as 1000 and the error convergence rate to be less than 0.001.

Let $\{m_u(t)\}$ be the tth year input rainfall data, where $1 \leq u \leq p$ and $\{a_w(t+1)\}$ be the output variable determining the $(t+1)$th year rainfall data. The generalized model of the neural network can be afforded as A_{OL} for output of the entire network and S_{HL} for output at the hidden layer. Let W be the weights of the interconnection links and B be the constant value which acts as bias.

4.1.2 Back-Propagation Algorithm

The steps admitted in the Back-Propagation Algorithm are shown below.

1. Initialization
2. Feed Forward Propagation
3. Back-Propagation of Errors
4. Terminate

If Error is reduced then the condition is stopped.

After the end up of training procedure, the network will be able of developing rainfall data for any next consequent year.

5 Results and Discussions

This section contains result and discussion about the proposed Weather forecasting approach using modified ANN approach from the past rainfall data even without considering the rainfall parameters. The proposed SSO based algorithm is implemented using MATLAB software.

For evaluating the performance of the proposed prediction approach, average ground rainfall over the districts of Karnataka were gathered from the year 1993 to 2013. The experimental results for the proposed Weather forecasting were analyzed in terms of error factors with existing methods like Gray Wolf Optimization (GWO)-ANN, Social Spider Optimization (SSO)-ANN and default ANN.

Rainfall Data samples used for my research is collected from Karnataka State Natural Disaster Monitoring Centre through Department of Agriculture, Bangalore. Below results are evaluated by implementing the algorithm using MATLAB software.

5.1 Performance Analysis

The proposed modified ANN (SSO) approach for predicting the average ground rainfall is shown with various existing methods. Below analysis is done by dividing the data samples of 80% for training and 20% for testing. The performance parameters incorporated are Root Mean Square Error (RMSE), Mean Absolute Percentage Error (MAPE) values are tabulated in Table 1.

From Table 1, it is clear that our proposed SSO-ANN approach minimizes the Root Mean Square Deviation (RMSE), Mean Absolute Percentage Error (MAPE) while comparing with other existing GWO-ANN and default ANN approaches. To prove the efficiency of our proposed approach the ground rainfall was predicted for different districts of Karnataka such as Bangalore, Belgaum and Hassan has been discussed below.

Rainfall Prediction, Error Rate and Actual Rainfall for Bangalore
Figure 3 shows the Rainfall prediction for Bangalore 2011. Here, Bangalore 2010 rainfall data was given as input and get the rainfall prediction for 2011 as output by using different existing approaches such as GWO-ANN, ANN and it is compared with our proposed SSO-ANN approach. Actual indicates the actual rainfall in Bangalore 2011. It shows that our proposed SSO-ANN approach predicted the rainfall rate correctly while comparing with other existing GWO-ANN, ANN approaches.

Table 1 RMSE and MAPE values of proposed and existing methods

	RMSE	MAPE
SSO-ANN	55.58544	13.43822
GWO-ANN	62.65511	12.93093
ANN	69.7444	20.1869

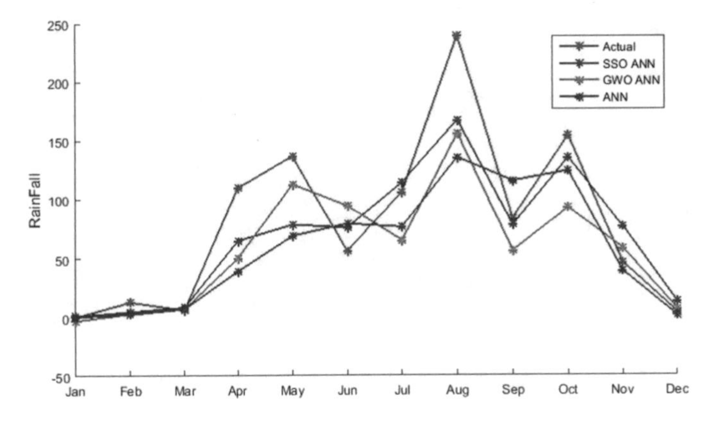

Fig. 3 Rainfall prediction in Bangalore 2011

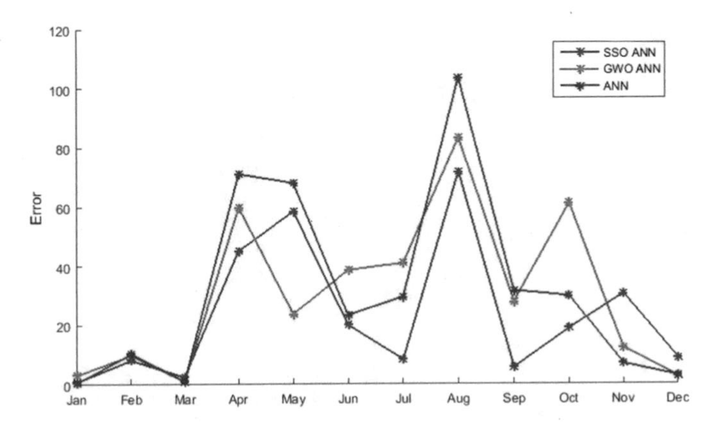

Fig. 4 Error rate for rainfall prediction Bangalore 2011

Figure 4 shows the error rate for rainfall prediction in Bangalore 2011. It clearly shows that our proposed SSO-ANN approach minimizes the error rate while comparing with other existing GWO-ANN, ANN approaches.

Figure 5 shows the Rainfall prediction for Bangalore 2013. Data of Bangalore 2012 was given as input and get the rainfall prediction for 2013 as output by using different existing approaches such as GWO-ANN, ANN and it is compared with our proposed SSO-ANN approach. Actual indicates the actual rainfall in Bangalore 2013. It shows that our proposed SSO-ANN approach predicted the rainfall rate correctly while comparing with other existing GWO-ANN, ANN approaches.

Figure 6 shows the error rate for rainfall prediction in Bangalore 2013. It clearly shows that our proposed SSO-ANN approach minimizes the error rate while comparing with other existing GWO-ANN, ANN approaches.

Figure 7 shows the actual rain fall rate in Bangalore from 1992 to 2012.

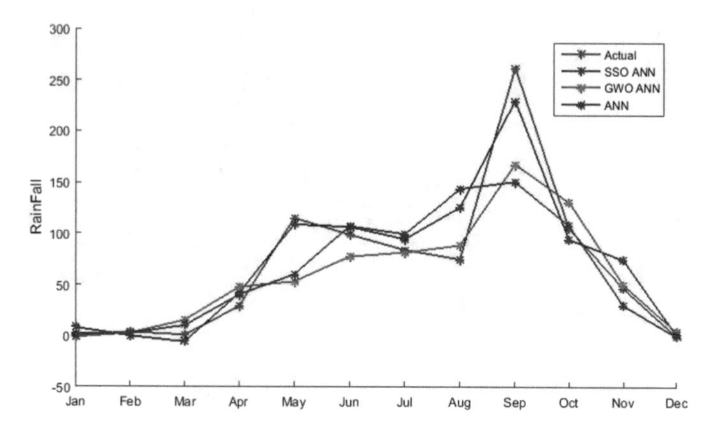

Fig. 5 Rainfall prediction in Bangalore 2013

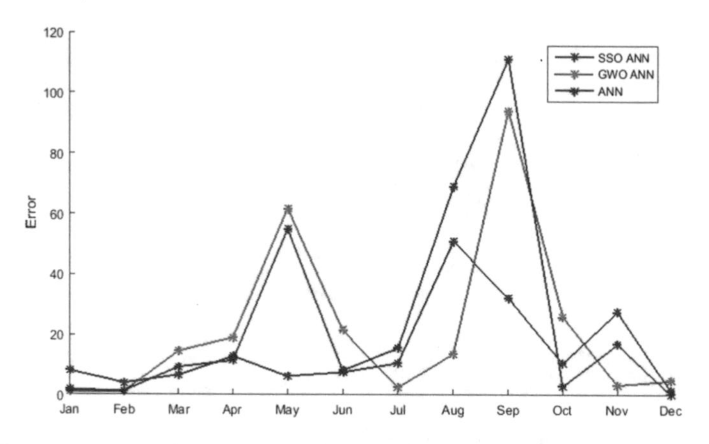

Fig. 6 Error rate for rainfall prediction Bangalore 2013

Rainfall Prediction, Error Rate and Actual Rainfall for Belgaum
Figure 8 shows the Rainfall prediction for Belgaum 2011. Rainfall data of Belgaum 2010 was given as input and get the rainfall prediction for 2011 as output by using different existing approaches such as GWO-ANN, ANN and it is compared with our proposed SSO-ANN approach. Actual indicates the actual rainfall in Belgaum 2011. It shows that our proposed SSO-ANN approach predicted the rainfall rate correctly while comparing with other existing GWO-ANN, ANN approaches.

Figure 9 shows the error rate for rainfall prediction in Belgaum 2011. It clearly shows that our proposed SSO-ANN approach minimizes the error rate considerably while comparing with other existing GWO-ANN, ANN approaches.

Figure 10 shows the Rainfall prediction for Belgaum 2013. Rainfall data of Belgaum 2012 was given as input and get the rainfall prediction for 2013 as output by using different existing approaches such as GWO-ANN, ANN and it is compared with our proposed SSO-ANN approach. Actual indicates the actual rainfall in Bel-

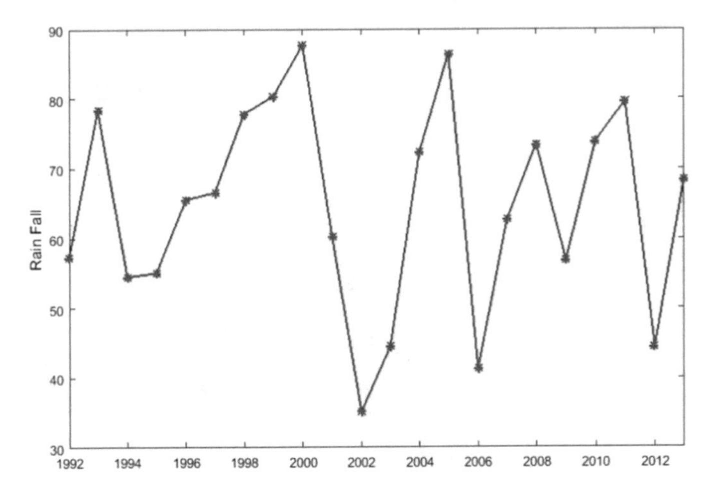

Fig. 7 Actual rainfall in Bangalore 1992–2012

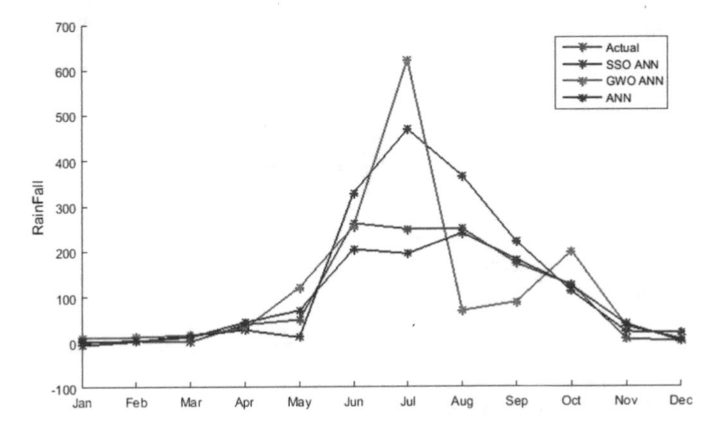

Fig. 8 Rainfall prediction in Bangalore 2011

gaum 2013. It shows that our proposed SSO-ANN approach predicted the rainfall rate correctly while comparing with other existing GWO-ANN, ANN approaches.

Figure 11 shows the error rate for rainfall prediction in Belgaum 2013. It clearly shows that our proposed SSO-ANN approach minimizes the error rate while comparing with other existing GWO-ANN, ANN approaches.

Figure 12 shows the actual rain fall rate in Bangalore from 1992 to 2012.

Rainfall Prediction, Error Rate and Actual Rainfall for Hassan

Figure 13 shows the Rainfall prediction for Hassan 2011. Rainfall data of Hassan 2010 was given as input and get the rainfall prediction for 2011 as output by using different existing approaches such as GWO-ANN, ANN and it is compared with our proposed SSO-ANN approach. Actual indicates the actual rainfall in Hassan 2011.

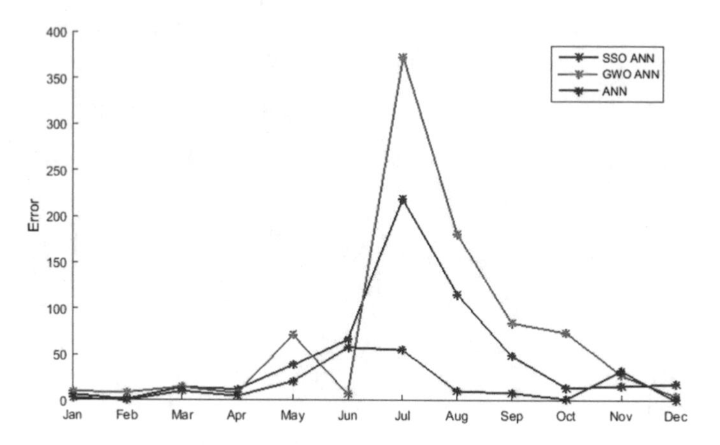

Fig. 9 Error rate for rainfall prediction Belgaum 2011

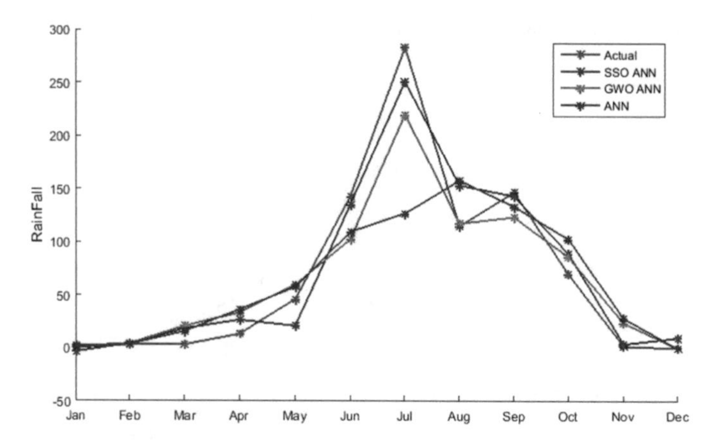

Fig. 10 Rainfall prediction in Belgaum 2013

It shows that our proposed SSO-ANN approach predicted the rainfall rate correctly while comparing with other existing GWO-ANN, ANN approaches.

Figure 14 shows the error rate for rainfall prediction in Hassan 2011. It clearly shows that our proposed SSO-ANN approach minimizes the error rate while comparing with other existing GWO-ANN, ANN approaches.

Figure 15 shows the Rainfall prediction for Hassan 2013. Rainfall data of Hassan 2012 was given as input and get the rainfall prediction for 2013 as output by using different existing approaches such as GWO-ANN, ANN and it is compared with our proposed SSO-ANN approach. Actual indicates the actual rainfall in Hassan 2013. It shows that our proposed SSO-ANN approach predicted the rainfall rate correctly while comparing with other existing GWO-ANN, ANN approaches.

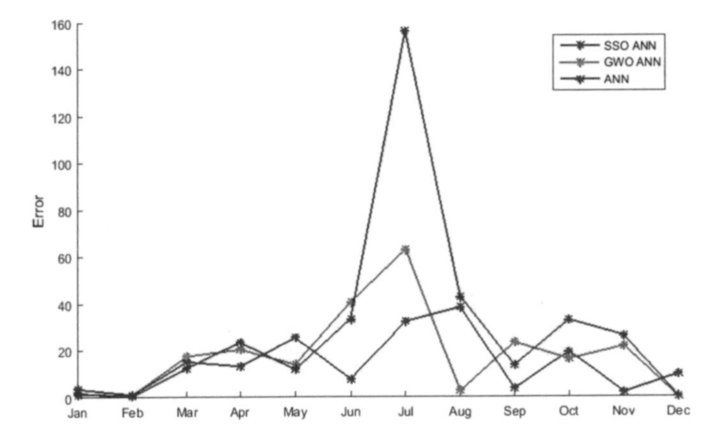

Fig. 11 Error rate for rainfall prediction Belgaum 2013

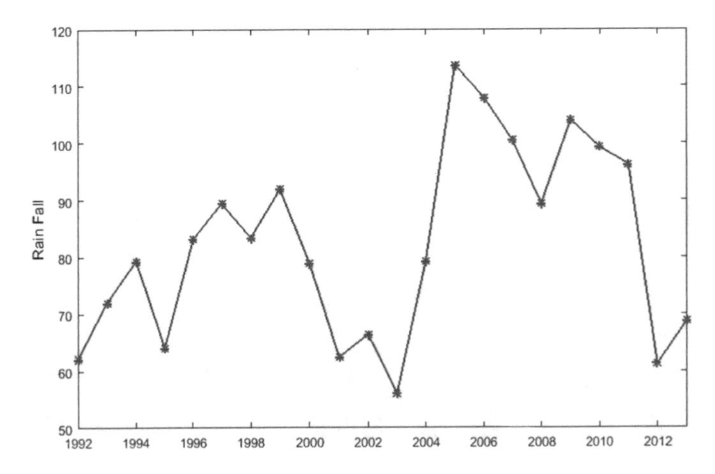

Fig. 12 Actual rainfall in Belgaum 1992–2012

Figure 16 shows the error rate for rainfall prediction in Hassan 2013. It clearly shows that our proposed SSO-ANN approach minimizes the error rate while comparing with other existing GWO-ANN, ANN approaches.

Figure 17 shows the actual rain fall rate in Bangalore from 1992 to 2012.

From the above graphs it clearly shows that our proposed SSO-ANN approach predicted the rainfall rate correctly while comparing with other existing ANN and GWO-ANN approach. And it also minimizes the error rate while comparing with other existing methods.

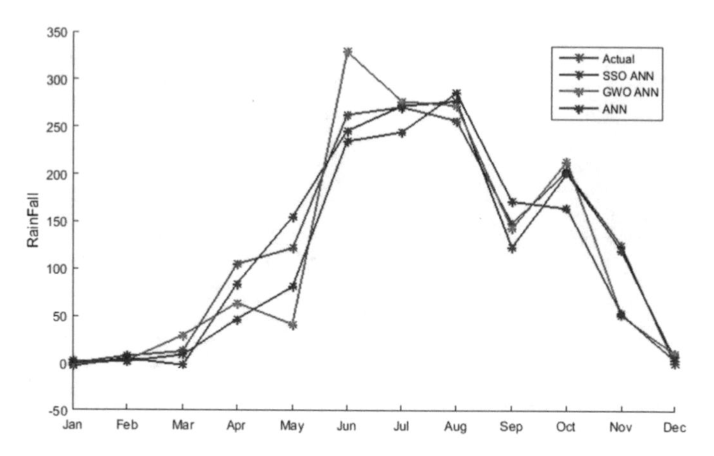

Fig. 13 Rainfall prediction in Hassan 2011

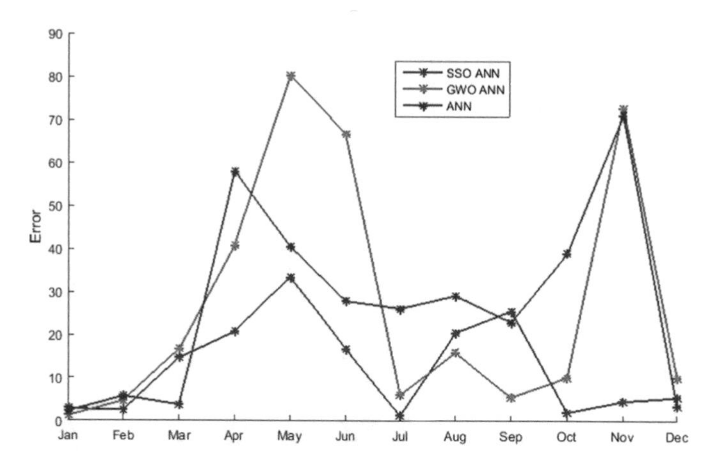

Fig. 14 Error rate for rainfall prediction Hassan 2011

6 Conclusion

We have developed an efficient approach for forecasting the average ground rainfall automatically using Adaptive ANN. The proposed Adaptive ANN uses Social Spider optimization algorithm for finding the optimal layer neuron sets in the hidden unit. The performance of the proposed AANN is compared with GWO-ANN and ANN methods in terms of RMSE and MAPE error measures. The result outcome shows that error rate is considerably less for the proposed prediction method when compared to the other existing methods. This shows that the proposed prediction approach performs accurately when compared to the other existing methods.

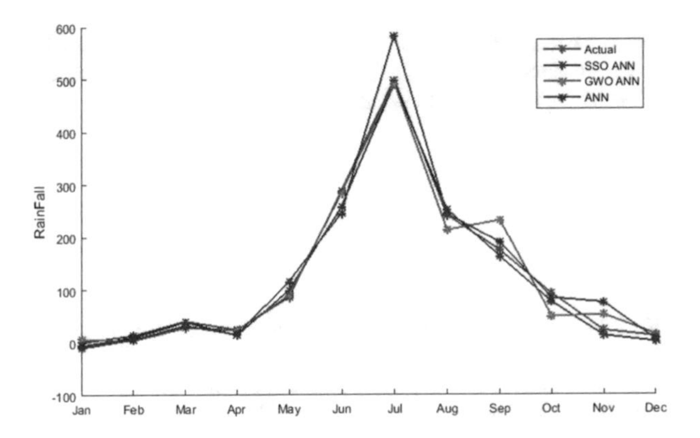

Fig. 15 Rainfall prediction in Hassan 2013

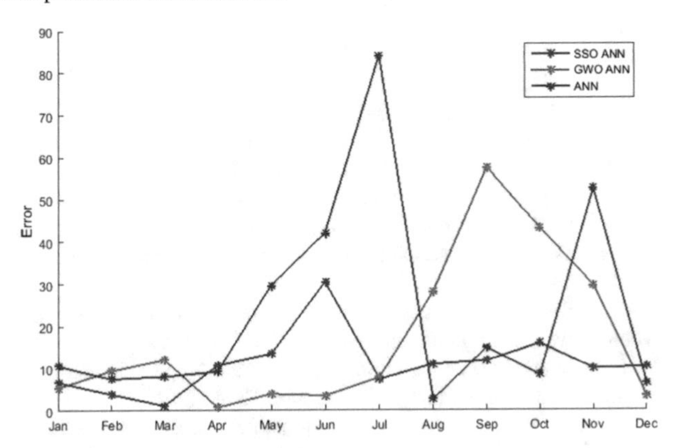

Fig. 16 Error rate for rainfall prediction Hassan 2013

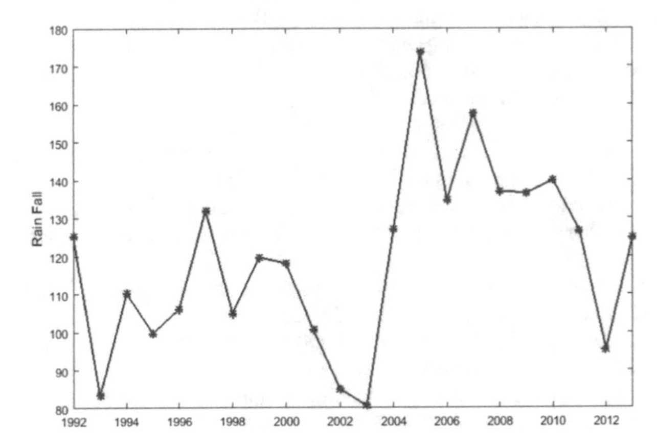

Fig. 17 Actual rainfall in Hassan 1992–2012

Acknowledgements We would like to thank Ms. Shalini Deepak, Agriculture Officer, Bangalore, India, Karnataka State Natural Disaster Monitoring Centre, Bangalore, Dr. K. C. Gouda, Senior Scientist, CSIR-CMMACS, Bangalore, India and Linyi Top Network Pvt. Ltd, Shandong province, Linyi, China.

References

1. Abhishek, K., Singh, M.P., Ghosh, S., Anand, A.: Weather forecasting model using artificial neural network. Procedia Technol. **4**, 311–318 (2012)
2. Moustris, K.P., Zafirakis, D., Alamo, D.H., Nebot Medina, R.J., Kaldellis, J.K.: 24-h ahead wind speed prediction for the optimum operation of hybrid power stations with the use of artificial neural networks. In: Perspectives on Atmospheric Sciences, pp. 409–414 (2016)
3. Yadav, A.K., Chandel, S.S.: Solar radiation prediction using artificial neural network techniques: a review. Renew. Sustain. Energy Rev. **33**, 772–781 (2014)
4. Venkata Ramana, R., Krishna, B., Kumar, S.R., Pandey, N.G.: Monthly rainfall prediction using wavelet neural network analysis. Water Resour. Manage. **27**, 3697–3711 (2013)
5. Goyal, M.K.: Monthly rainfall prediction using wavelet regression and neural network: an analysis of 1901–2002 data, Assam, India. Theor. Appl. Climatol. **118**(1–2), 25–34 (2014)
6. Nekoukar, V., Hamidi, M.T.H.: A local linear radial basis function neural network for financial time-series forecasting. Springer Sci. **23**, 352–356 (2010)
7. Akpinar, M., Fatih Adak, M., Yumusak, N.: Time series forecasting using artificial bee colony based neural networks. In: International Conference on Computer Science and Engineering (UBMK), pp. 554–558
8. Ghosh, S., Nag, A., Biswas, D., Singh, J.P.: Weather data mining using artificial neural network. In: IEEE Recent Advances in Intelligent Computational Systems, pp 192–195 (2011)
9. Papantoniou, S., Kolokotsa, D.: Prediction of outdoor air temperature using neural networks; application in 4 European cities. Energy Build. **114**, 72–79 (2016)
10. Noorollahi, Y., Jokar, M.A., Kalhor, A.: Using artificial neural networks for temporal and spatial wind speed forecasting in Iran. Energy Convers. Manag. **115**, 17–25 (2016)
11. Elattar, E.E.: Prediction of wind power based on evolutionary optimised local general regression neural network. IET Gener. Transm. Distrib. **8**(5) (2014)
12. Shetty, R.P., Srinivasa Pai, P., Sathyabhama, A., Adarsh Rai, A.: Optimized radial basis function neural network model for wind power prediction. In: Second International Conference on Cognitive Computing and Information Processing (2016)
13. Rodríguez-Fernández, N.J., Aires, F., Richaume, P., Kerr, Y.H., Kolassa, J., Cabot, F., Jiménez, C., Mahmoodi, A., Drusch, M.: Soil moisture retrieval using neural networks: application to SMOS. IEEE Trans. Geosci. Remote Sens. **53**(11) (2015)
14. Liu, J.N.K., Kwong, K.M., Chan, P.W.: Chaotic oscillatory-based neural network for wind shear and turbulence forecast with LiDAR data. IEEE Trans. Syst. Man Cybern. Part C (Appl. Rev.) **42**(6) (2012)
15. Bhaskar, K., Singh, S.N.: AWNN-assisted wind power forecasting using feed-forward neural network. IEEE Trans. Sustain. Energy **3**(2) (2012)
16. Choury, A., Bruinsma, S., Schaeffer, P.: Neural networks to predict exosphere temperature corrections. Space Weather **11**(10) (2013)
17. Sideratos, G., Hatziargyriou, N.D.: Probabilistic wind power forecasting using radial basis function neural networks. IEEE Trans. Power Syst. **27**(4) (2012)
18. Quan, H., Srinivasan, D., Khosravi, A.: Short-term load and wind power forecasting using neural network-based prediction intervals. IEEE Trans. Neural Netw. Learn. Syst. **25**(2) (2014)
19. Lee, D., Baldick, R.: Short-term wind power ensemble prediction based on Gaussian processes and neural networks. IEEE Trans. Smart Grid **5**(1) (2014)

L. Naveen (Naveen Lingaraju) obtained his Bachelor's degree in Information Science & Engineering from Visvesvaraya Technological University (VTU), Belgaum, Karnataka, India. Then he obtained his Master's degree in Computer Science & Engineering from VTU. Currently, he is pursuing his Ph.D. in VTU. His research interests are in Neural Networks, Data mining and Networking.

Dr. H. S. Mohan (Mohan Hosaagrahara Savalegowda) received his Bachelor's degree in computer Science and Engineering from Malnad college of Engineering, Hassan during the year 1999, M.Tech in computer Science and Engineering from Jawaharlal Nehru National College of Engineering, Shimoga during the year 2004 and Ph.D. in Computer Science & Engineering from Dr. MGR University Chennai. He is working as a Professor and Head in the Department of Information Science and Engineering at SJB Institute of Technology, Bengaluru-60. He is having total 18 years of teaching experience. His area of interests are Networks Security, Image processing, Data Structures, Computer Graphics, finite automata and formal languages, Compiler Design. He has obtained a best teacher award for his teaching during the year 2008 at SJBIT Bengaluru-60. He has published and presented papers in journals, international and national level conferences.

6LowPan—Performance Analysis on Low Power Networks

Nikshepa, Vasudeva Pai and Udaya Kumar K. Shenoy

Abstract The modern world had been dominant to the automation applications. We can see that the smart applications are being leveraged significantly. The deployment of Smart watch, Smart Cars, Smart Homes are the common examples that can be referred to automation. The core of the automation technology lies in the domain of Wireless Sensor Networks referred as WSN and the Internet of Things, i.e., IoT. The prominent part or the concept in the technology is the sensors. The common name for this sensors are nodes in common network terms. The major consideration in any of the sensor networks or the IoT environment is the energy availability at each nodes. Therefore the minimal energy consumption by the nodes of the network or the energy consumption factors are highly observed. The nodes are also supposed to perform immediate regrouping of the network whenever the nodes die. In this paper, the behaviour of the low power protocol called 6LowPan abbreviated as IPv6 over Low-Power Wireless Personal Area Networks is analysed under various topological scenarios. Cooja Contiki simulator is the tool used in for the simulation purposes of the 6LowPan.

Keywords Internet of things · 6LowPan · RPL

Nikshepa (✉)
NMAM Institute of Technology, NITTE, Karkala, India
e-mail: nikshepa8594kumar@gmail.com

V. Pai
Department of Information Science and Engineering, NMAM Institute of Technology, NITTE, Karkala, India
e-mail: paivasudeva@nitte.edu.in

U. K. K. Shenoy
Department of Computer Science and Engineering, NMAM Institute of Technology, NITTE, Karkala, India
e-mail: ukshenoy@nitte.edu.in

S. Smys et al. (eds.), *International Conference on Computer Networks and Communication Technologies*, Lecture Notes on Data Engineering and Communications Technologies 15, https://doi.org/10.1007/978-981-10-8681-6_15

1 Introduction

1.1 IPv6 Over Low Power Wireless Personal Area Networks (6LowPan)

The 6LowPan works on a solution to empower remote IPv6 correspondence over the institutionalized IEEE 802.15.4 low-radio for gadgets with restricted space, energy and storage, for example, sensor devices [1, 2]. The major difference with the IoT protocol layer is the presence of a new adaption layer making its presence between the data link and the network layer. As a delineation, 1280 bytes of length was the value for the maximum transmission rate, and the payload of 127 bytes. From the bottom to top perspective, the 6LowPan possess the features of the two layers between them and is independent of the lower stack layers.

The 6LowPan convention layering is depicted in Fig 1. The first layer is the conventional Physical (PHY) layer with the specifications of 250 kbits/s throughput. The frequency range of this layer is the standard ISM band with the range of 2400–2483.5 MHz.

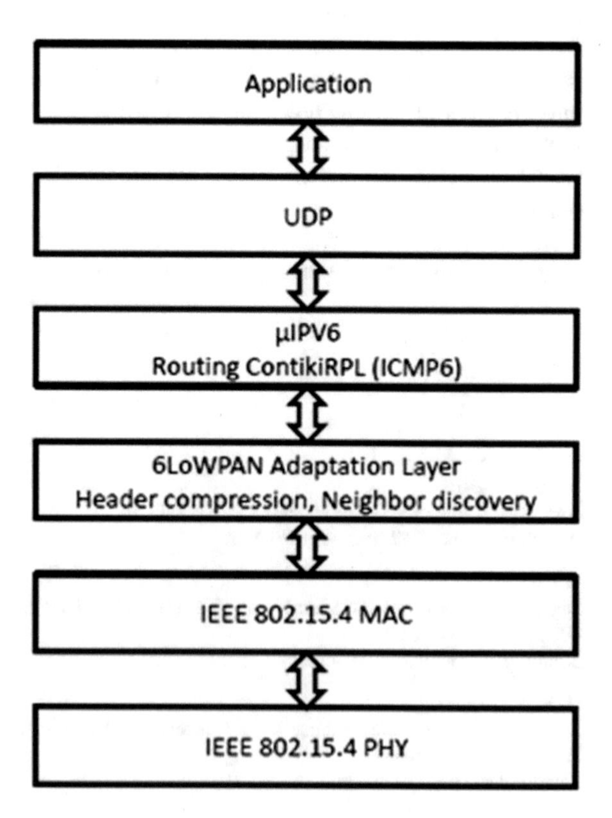

Fig. 1 6LowPan protocol stack

The Medium Access Control (MAC) layer gives the devices that empower the channel usage to perform data transfer for example, in case of multiple channels requiring to access the line of channels, then the MAC layer performs the job of allocating the same without any collision between the data communication through the CSMA/CD mechanism.

The presence of adaption layer really speaks to the primary distinction in contrast with customary OSI (Open Systems Interconnection) convention stack [3]. The functions associated with the adaption layer include compression and decompression of headers, fragmentation and reassembly, address auto-design, and supports the application specific assist.

The task of the network layer is mainly focused on the packet routing. This is where the RPL protocol is introduced that eases the routing the low power packets across destination and hence the layer is additionally in charge of steering. Two mechanisms of the routing process is observed across the RPL routing:

1. The packet transfer within the 6LowPan network environment.
2. The packet transfer outside the 6LowPan network environment.

The transport layer is backed by the task enhancing the smooth flow of packet transmission across the ends. Concisely the Transmission Control Protocol, i.e., TCP and the User datagram Protocol (UDP) rules and is processed along this layer. In case of 6LowPan the UDP protocol has a dominance due to the energy economical and the connection-less features. The lightweight behaviour of the datagram protocol makes it an easy choice for the use in the low power networks.

The application layer processes the transfer and the reception IPv6 packets allowing users to actually work on with the applications deployed. The topmost layer of the protocol stack requires suitable alterations as to work suitably with the low power networking requirements [4].

1.2 RPL

The RPL is one of a foundation conventions, it is a separation vector and a source directing convention that is composed over a few network layer systems including the physical layer and the Mac layer belonging to the standard IEEE 802.15.4 standard. It targets accumulation systems (WSNs) which involve up to a great many switches (nodes), where the larger part such switches have extremely compelled and restricted assets. RPL has adopted three major modes of communication namely P2P, P2MP and MP2P. P2P is a one-to-one way of data transfer where only two devices are in the part of communication. Similarly as the name suggest P2MP is a point to multipoint mechanisms where the transmission is a multicast communication mechanism. The last mode MP2P is a reverse of the P2MP model. The RPL has a two to three components namely DODAG, RPL control messages and the Objective functions (OF).

1.3 Cooja Contiki Simulator

A recreated Contiki Mote in Cooja is a genuine gathered and executing Contiki framework. The framework is controlled and broke down by Cooja. This is performed by ordering Contiki for the local stage as a common library, and stacking the library into Java utilizing Java Native Interfaces (JNI). A few distinctive Contiki libraries can be aggregated and stacked in the same Cooja reproduction, speaking to various types of sensor nodes (heterogeneous systems). Cooja controls and breaks down a Contiki framework by means of a couple of capacities. For example, the test system illuminates the Contiki framework to deal with an occasion, or gets the whole Contiki framework memory for examination. This approach gives the test system full control of reproduced frameworks. Sadly, utilizing JNI additionally makes them pester symptoms. The hugest is the reliance on outside apparatuses, for example, compilers and linkers and their run-time contentions. Cooja was initially created for Cygwin/Windows and Linux stage, however has later been ported to MacOS.

2 Related Works

The work carried out so far has a lots of contribution from the research papers surveyed during the period. A lot of research and experimentation has been carried out in the domain of IoT, their protocols and similar others stuffs. This chapter details about the references used in accommodating significant amount of knowledge for the project in IoT protocols.

Dragomir et al. [5] explained the various IoT communication protocols along with different security considerations for the different stack layers. It also stated the security mechanisms that can be implemented across different layers for better robustness and stability.

In a survey carried out by Sonavane in [6] the authors were keen into specifying the concerns related to IoT and the support provided by different IoT protocols for security requirements. They also went on to explain the security methods to be incorporated into layers.

Granjal et al. in [7] performed a survey on existing IoT protocols and also stated the different insecurities associated with the different protocols and also proposed the protection mechanism to be implemented across these protocols for their secure functionality.

The survey on RPL presented by Aljarrah [8] has a detailed information on the working of the protocol. The paper describes the resource constraint in the IoT nodes and inclusion of DODAG models and Objective function for efficient routing of the data among nodes in a low power and Lossy networks.

Kamble et al. in [9] describes the different types of attacks across the RPL networks. The different types of attacks include the direct attacks, indirect attacks, Traffic attacks and Topological attack. It also details the counter measures for these attacks. They have described the DODAG formation in a RPL network and their maintenance.

Pongle et al. in [10] has explained the different types of attacks on the RPL and the 6LowPan topology. The major attacks on RPL networks Selective forwarding, Sybil attack, Sinkhole attack, Hello forwarding attacks, Wormhole attacks etc. The common attacks on 6LowPan networks are Fragmentation Attack, Authentication Attack, Confidentiality Attack and Security threats from Internet side.

Chen et al. in [11] quantitatively compared the performance of IoT protocols, namely MQTT (Message Queuing Telemetry Transport), CoAP (Constrained Application Protocol), DDS (Data Distribution Service) and a custom UDP-based protocol in a medical setting. The performance of the protocols was evaluated using a network emulator, allowing us to emulate a low bandwidth, high system latency, and high packet loss wireless access network.

Al-Sarawi et al. analysed the different protocols in various types of IoT networks like Low Power Wide Area Network (LPWAN) and Short Range Network. Various new IoT protocols were introduced like Sigfox, BLE and Z-Wave etc. [12].

Gardasevic et al. in [13] experimented on the 6LowPan using the motes and analysed the different behaviour of the protocol in different scenarios. The report results in terms of average round-trip-time, packet loss rate and throughput for different payload sizes and number of hops, both for unicast and multicast traffics.

Cody-Kenny et al. in [14] evaluated the performance of the 6LowPan on different types of motes. The common parameters were evaluated and also specified the advantages of the 6LowPan.

3 Objectives of the Proposed Work

The aim of the proposed work is to analyse the performance of the IoT adaption layer protocol IPv6 over Low Power Wireless Personal Area Networks (6LowPan)

1. Determining the behaviour of the static 6LowPan nodes in different networking topologies.
2. Determining the total network lifetime under simulation environment.
3. Analysing the different evaluation parameters in each topology.

4 Experimental Setup

The 6LowPan protocol was subjected to work in different environments. Each simulation was performed under four types of the networking topologies namely dumbbell, butterfly, cubic and the hierarchical topology. During each of this case study the number of nodes were constant with ten nodes (source) and one sink used. All the nodes were in static with no mobility.

Let us go through the introduction of all these topologies in brief before looking at the results of the simulation.

4.1 Dumbbell Topology

The name of this topology is termed after the way the nodes are placed in the network like the dumbbell. In the topological structure there are two ends where the both the ends are connected by the link. Thus forming the dumbbell structure. In each of these ends there may be any structure of nodes. In case of our simulation the original shape is placed but in the place of the link a sink node is placed to analyse the working of the 6LowPan protocol (Fig 2).

4.2 Butterfly Topology

The butterfly topology is defined after the packets or the routes in this topology takes the path forming that of the structure of the butterfly. The placement of the nodes

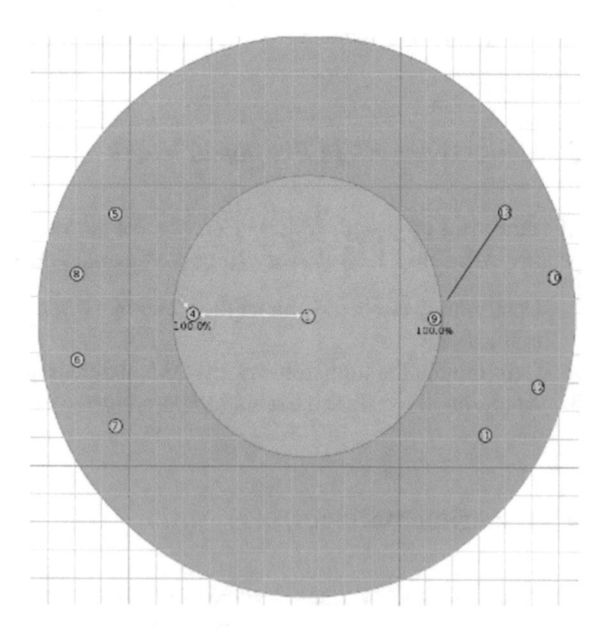

Fig. 2 Dumbbell topology

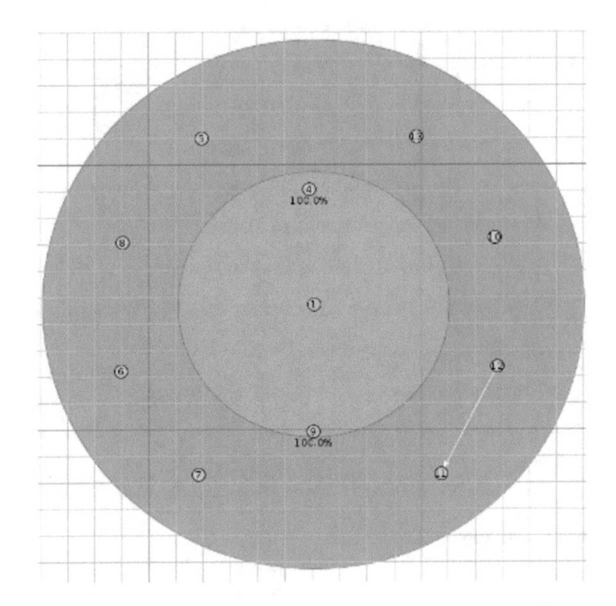

Fig. 3 Butterfly topology

are also placed in the butterfly structure. In our simulation the topology is designed similarly and in the centre of the entire network a single sink node have been placed for analysis purposes (Fig. 3).

4.3 Hierarchical Topology

The hierarchical topology of networks will have a level/layered structure. Where within each layer there will be certain number of nodes with or without similarities. Every layer an upper and lower layers with few number nodes. The simulation follows a similar structure with a sink node at the topmost layer followed by other sink/source nodes at following lower layers (Fig. 4).

4.4 Cubic Topology

The cubic topology involves the node placements such a way that all the nodes are deployed along the positions forming a vertices of cubic structure. In the simulation the centre of the topology has a sink node. The packet transmission along this topology has normal routing mechanism of the RPL protocol (Fig. 5).

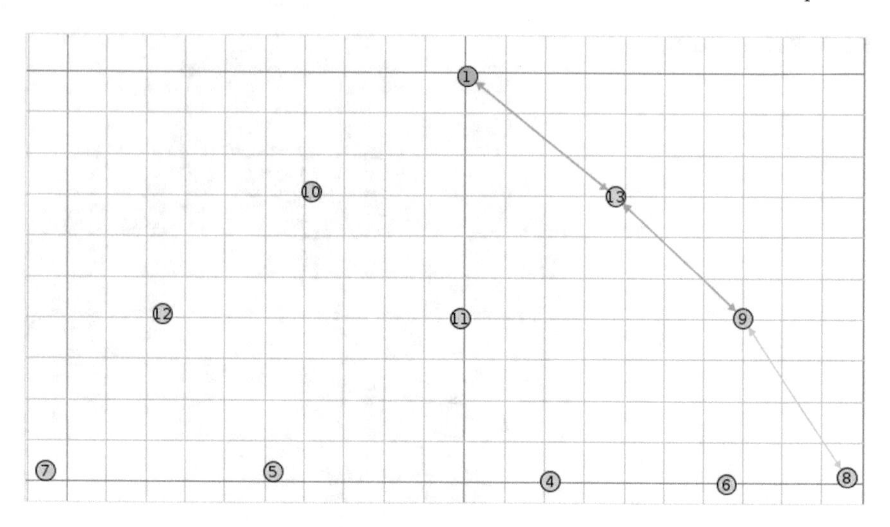

Fig. 4 Hierarchical topology

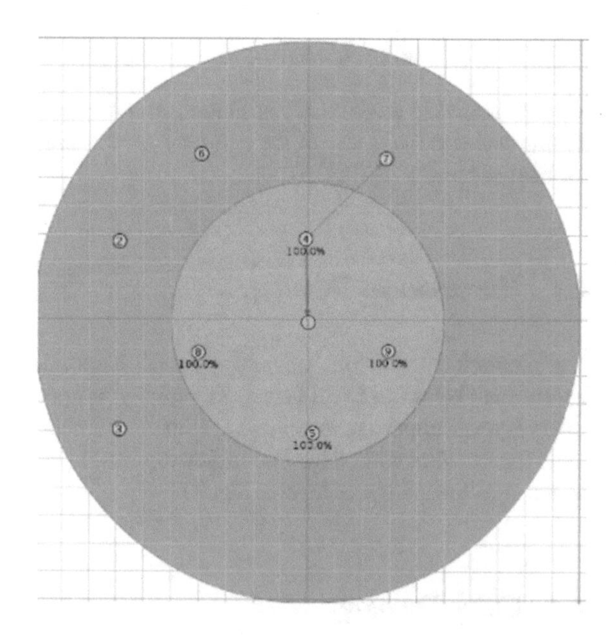

Fig. 5 Cubical topology

5 Result and Analysis

The metrics used for the performance analysis of 6LowPan are Energy Consumption, Total network simulation lifetime and other packet related parameters.

5.1 Simulation Parameters

Table 1 shows the network parameters used for the simulation.

The following results and the observations are carried out during the simulation of the 6LowPan protocol analysis.

5.2 Total Simulation Lifetime

From Table 2 we can see the variation in the lifetime across different topologies. The dumbbell has a highest lifetime among the mentioned topology due to the division into sub networks across the sink node. While along the remaining topology since all the nodes equally participate in the routing of the packets, the total energy consumption is little high and hence they have a reduced lifetime.

5.3 Total Energy Consumption

The energy consumption values across all the topologies are almost the same but the hierarchical one has the highest energy usage level. As shown in the figure in the previous section the distribution of the network is wide compared to other topologies,

Table 1 Simulation parameters

Parameters	Value
Simulator	Cooja Contiki
Routing protocol	6LowPan with RPL
Traffic generated	UDP
Number of sender/source nodes	10
Number of sinks	1
Topologies used	Dumbbell Butterfly Hierarchical Cubical

Table 2 Lifetime value

Network topology	Total lifetime (min)
Dumbbell	195
Butterfly	180
Hierarchial	150
Cubical	165

Table 3 Energy values

Network topology	Average energy consumption (mW)
Dumbbell	0.896
Butterfly	0.882
Hierarchial	0.954
Cubical	0.838

Table 4 RT metric values

Network topology	Average RT metric
Dumbbell	490.33
Butterfly	537.80
Hierarchial	645.13
Cubical	433.525

Table 5 Latency values

Network topology	Average packet interval (s)
Dumbbell	53
Butterfly	54
Hierarchial	48
Cubical	28

the energy for routing across all the nodes and or all the nodes requiring the send the packets, energy consumption is higher in this topology (Table 3).

5.4 *RT Metric*

The RT Metric is the ratio of the packet reception and transmission across the network. This means the packets arrived and left across all the nodes. From the values obtained from the simulation, the table shows that the hierarchical topology has the maximum value. The reason could be that the large range of the network distribution requires the large number of packets to be transmitted for routing and also for communication. These packets also involve the data and the control packets (Table 4).

5.5 *Average Packet Interval*

This parameter includes the values associated with all the nodes in the respective network topology. The packet interval is the time required to transmit a packet from one node to another. This parameter allows the user to analyse the link capacity with respective to the packet sizes of the data transmitted (Table 5).

The cubical topology has a least interval or the latency value as all the nodes are nearer to the sink compared to other networks.

6 Conclusion

The proposed work involved the performance analysis of the 6LowPan IoT adaption layer protocol. 6LowPan protocol was studied for its performance are analysed for different topological scenarios. The protocol was tested for energy consumption and packet related information like RT Metric, etc., under each topology. Topology changes effected the performance with the protocol varying differently in most of the scenarios. However in the future work on variation in the nodes behaviour on adding mobility should be evaluated. Also the phenomenon of dynamic node property changes can be introduced to determine the changes among the node/network performance.

References

1. Bhat, S., Pai, V., Kallapur, P.V.: Energy efficient clustering routing protocol based on LEACH for WSN. Int. J. Comput. Appl. **120**(13) (2015)
2. Pavan, R., Pai, V., Kallapur, P.V.: Energy efficient clustered routing protocols of LEACH. Int. J. Res. Appl. Sci. Eng. Technol. **3**(V) (2015)
3. Suchitra, C., Vandana, C.P.: Internet of things and security issues. IJCSMC **5**(1) (2016)
4. Roman, R., Najera, P., Lopez, J.: Securing the internet of things. IEEE Comput. **44**, 51–58 (2011)
5. Dragmoir, D., Gheaorge, L., Costea, S., Radovici, A.: A survey on secure communication protocols for IoT systems. In: International Workshop on Secure Internet of Things (2016)
6. Sonavane, S.S., Deshmukh, S.: Security protocols for internet of things: a survey. In: IEEE Conference (2017)
7. Granjal, J., Moteiro, E., Silva, J.S.: Security for the internet of things: a survey of existing protocols and open research issues. IEEE Commun. Surv. Tutorials **17**(3) (2015)
8. Aljarrah, E., Yassein, M.B., Aljawarneh, S.: Routing protocol of low-power and lossy network: survey and open issues. In: IEEE Conference (2016)
9. Kamble, A., Malemath, V.S., Patil, D.: Security attacks and secure routing protocols in RPL-based internet of things: survey. In: International Conference on Emerging Trends & Innovation in ICT (ICEI) (2017)
10. Pongle, P., Chavan, G.: A survey: attacks on RPL and 6LoWPAN in IoT. In: International Conference on Pervasive Computing (ICPC) (2015)
11. Chen, Y., Kunz, T.: Performance Evaluation of IoT Protocols Under a Constrained Wireless Access Network (2016)
12. Al-Sarawi, S., Anbar, M., Alieyan, K., Alzubaidi, M.: Internet of things (IoT) communication protocols: review. In: 8th International Conference on Information Technology (ICIT) (2017)
13. Gardasevic, G., Mijovic, S., Stakjic, A., Buratti, C.: On the performance of 6LoWPAN through experimentation. In: IEEE Conference (2015)
14. Cody-Kenny, B., Guerin, D., Ennis, D., Carbaj, R.S., Huggard, M., Goldrick, C.M.: Performance Evaluation of the 6LoWPAN Protocol on MICAz and TelosB Motes. School of Computer Science and Statistics, Trinity College Dublin, Dublin 2, Ireland

Nikshepa received the Bachelor's degree and Master's degree M.Tech. from Visvesvaraya Technological University (VTU), Belagavi, in 2016 and 2018 in Computer Science and Engineering and Computer Network Engineering respectively. His research interests include Internet of Things, Network Security, Wireless Sensor networks.

Vasudeva Pai received the B.E. and M.Tech. degrees from Visvesvaraya Technological University, Belagavi, India, in 2005 and 2010, respectively, both in Computer Science Engineering. He is currently working as Assistant Professor in NMAM Institute of Technology, NITTE, India. His research interests include Wireless Sensor Networks, mobility management, Cryptography and Network Security, SDN, and IoT.

Udaya Kumar K. Shenoy recevied the bachelor's degree (1992) in computer science, master's degrees (2000) in computer application from Mangalore University and Ph.D. (2009) from National Institute of Technology Karnataka, Surathkal. He is currently working as Professor in NMAM Institute of Technology, NITTE, India. His research interests include Wireless Networks, Optimization, Multimedia communication and Network Security.

IMU-Based Indoor Navigation System for GPS-Restricted Areas

R. M. Bommi, V. Monika, R. Narmadha, K. Bhuvaneswari and L. Aswini

Abstract Indoor positioning system is a challenge for most military operations because conventional technologies like GPS do not work in indoor conditions. Also at this point in time there is no proven technology that is globally accepted that provides indoor navigation and positioning. In this project a indoor positioning system based on a combination of technologies and features is proposed that can provide the best possible solution to this requirement. This project uses a combination of 3D accelerometer to provide information about the movement of the person. A 3D 360° Digital magnetometer which provides information about the direction of movement. A Gyroscope which provide information about the person's orientation and finally a barometric pressure transducer that uses atmospheric pressure as reference to calculate the height or elevation of the individual. Thus using these multiple inputs and a slew of processing and calculations can provide information about the movement and location of a person in 3D space additionally this information can be transmitted using wireless communication technologies to keep track of the respective position on soldiers inside a building where GPS and other conventional technologies do not work. If every soldier is provided with a device like this then a military operation can be coordinated easily from a remote location by knowing the respective location of all the soldiers. The location of various soldiers can be plotted on a software like LabVIEW on a remote computer terminal in a control room away from the danger zone.

R. M. Bommi · V. Monika · K. Bhuvaneswari · L. Aswini (✉)
Department of ECE, Jeppiaar Maamallan Engineering College, Sriperumpudur, India
e-mail: aswinilogan1998@gmail.com

R. M. Bommi
e-mail: rmbommi@gmail.com

V. Monika
e-mail: monika.saravan@gmail.com

K. Bhuvaneswari
e-mail: kbhuvaneswari48@gmail.com

R. Narmadha
Department of EEE, Jeppiaar Maamallan Engineering College, Sriperumpudur, India
e-mail: rnarmadhaeee@gmail.com

S. Smys et al. (eds.), *International Conference on Computer Networks and Communication Technologies*, Lecture Notes on Data Engineering and Communications Technologies 15, https://doi.org/10.1007/978-981-10-8681-6_16

Keywords GPS · Gyroscope · Barometer · Magnetometer · LabVIEW

1 Introduction

In the modern age, we rely heavily on GPS to find our way around while driving, walking or biking. However, there's a big problem! The satellite signals of GPS are using a frequency that cannot move through solid objects easily. In addition, a GPS device requires a series of satellites to successfully position the device. When trying to use a GPS device indoors, there is no free line-of-sight to the satellites. This means a wide variety of physical objects stand in line between the satellites and device needed for an accurate position. What's amazing is that Indoor Positioning System IPS are more accurate than GPS, something that is needed when for example soldiers or fire fighters going on a rescue mission. We can get a really accurate indoor position by combining a range of sensors. The sensors that are relevant for indoor positioning are the gyroscope, the compass, the altimeter and the accelerometer. By combining all the data from these sensors and transmitting them wirelessly, a process called sensor fusion—our algorithm is able to calculate an accurate and robust position indoors.

2 Literature Review

In this paper the distribution transformer outage detection system based on radio mobile communication system is proposed [1]. The protection and maintenance of a power system is availed by detecting and locating faults on power transmission lines [2]. Grounding short circuit fault are produced by high voltage overhead lines due to various reasons [3]. Error models for this gyro and accelerometer are presented with a study of their perturbation in trajectory prediction. A full inertial [4]. Although GPS measurement are the essential information for currently developed land vehicle [5]. The high cost of inertial parameters is the main hindrance for their existence in precision navigation systems [6]. The demand for fall safe navigation systems which can be used on large autonomous land vehicles such as those found in container terminals, agriculture and in mines [7]. The current industry standard and most widely used algorithm for this purpose is the Extended Kalmah Filter (EKF) [8]. In this paper, the algorithm are further developed and illustrated [9]. This article provides a comprehensive introduction into the field of robotic mapping, with a focus on indoor mapping [10].

3 Objective of the Project

The main objective of our project is to overcome the drawbacks produced by the current navigation system GPS. In our project, the position of a person can be determined with the help of a combination of four sensors and not with the help of satellite. As we know GPS signals tend to become weak inside a building, using this drawback we have found a fusion of sensors that helps to determine the current position by calculating the parameters such as motion, direction of the motion, pressure, and direction. With the help of these four parameters we will be able to find the current position of a person by wirelessly transmitting these values to a receiver and then displaying the output in the form of a three dimensional graph. Such graphs are used in applications like high security applications and defence to keep track of moving objects or people. Previous approaches of such applications were based on technologies like RFID which are limited practically useless in most applications. Now, we use a module called IMU (Inertial Measurement Unit) that consists of these four sensors such as the gyroscope, the magnetometer, the accelerometer and the barometer and with the help of these sensors a person's current position can be easily determined.

4 Existing System

As we all know, the current existing system for tracking a human being is GPS. The **Global Positioning System** (**GPS**) is a navigation system that provides time and location information in all weather conditions, anywhere on or near the Earth where there is an unhindered line of sight to four or more GPS satellites. GPS system can be used to get location which includes details like longitude and latitude along with the timestamp details. To track the movement of a person we have to use Google maps for mapping the location sent by the mobile phone which is GPS enabled. The General Packet Radio Service is a server which communicates towards GPS location that is drawn from mobile phone.

5 Proposed System

5.1 System Overview

The GPS system works in two stages: the tracking phase and the mapping phase. In the tracking phase, the mobile application developed in android using the mobile phone GPS receiver, draws the GPS location, after computing the exact location it transfer a GPRS packet which includes all the location details. The application then sends the packet to the server which data is stored in a database. The next phase is the mapping phase, in which the data is drawn from the database and is displayed on the Google map on the application that has been developed for the tracking system.

5.2 IMU

The proposed system overcomes the various disadvantages produced by this current navigation system GPS, we have found an alternate method of determining a person's location. This alternate method is called as "**The Indoor Navigation System**" with the help of Sensors. As we already know that the present GPS system gives us the necessary coordinates of a person's location so that we might track his location, but its biggest disadvantage is that it makes it difficult when the person goes indoors or if he enters a place which has a very poor signal connectivity. Using this as our concept, we have developed module called as Inertial Measurement Unit (IMU).

This module consists of four sensors such as the Gyroscope, the Magnetometer, the Accelerometer and the Barometer and these four sensors are used to measure the orientation, the direction, the motion and the pressure respectively. The values generated by these four sensors are combined together with the help of a microcontroller and are transmitted wirelessly to a receiver that is supposed to be a control base in case of a military rescue operation.

After these values have been transmitted to the receiver, it is then a child's play to find out the exact location of a person inside any building. It does not matter even if the building has any number of floors. This IMU module will clearly transmit the necessary details to the receiver. Once the receiver has obtained these values, he will be able to see in which direction the person has moved with the help of a Lab view. In this lab view, all the necessary reading such as the direction of the movement, the position in a discrete manner, the altitude and the orientation will be displayed in the form of a three-dimensional graphical representation.

6 Circuit Description

A. **Data Flow Programming**

In LabVIEW programming language is used, which is a dataflow programming language which is also referred to as G. The structure of a graphical block diagram (the LV-source code) determines the execution on which the programmer draws wires for connecting different function-nodes. The built-in scheduler automatically accomplishes the multi-threading and multi-processing hardware, which multiplexes multiple OS threads over the nodes (Fig. 1).

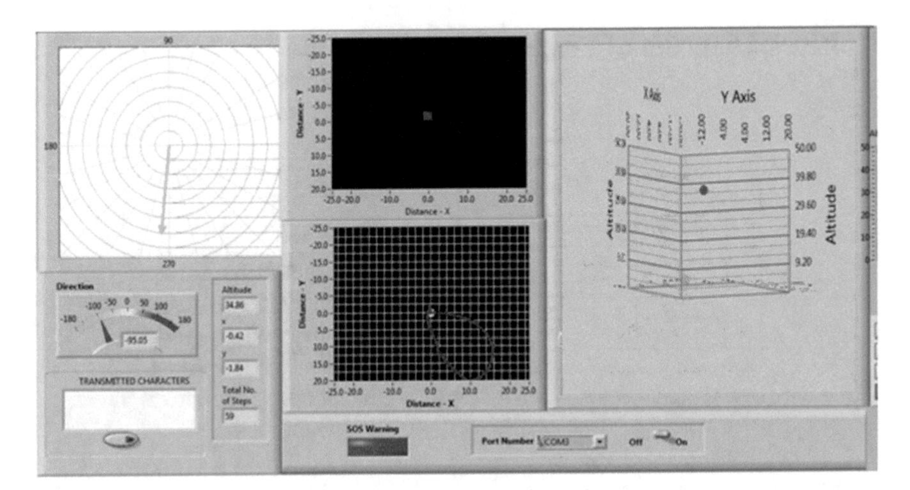

Fig. 1 Results

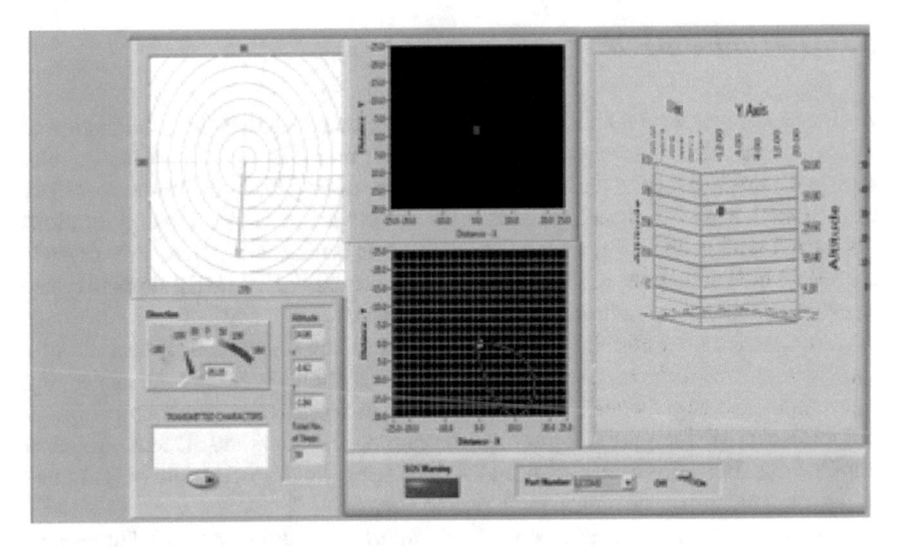

Fig. 2 LabVIEW output

B. Data Flow Programming

In LabVIEW programming language is used, which is a dataflow programming which is also referred to as G. The structure of a graphical block diagram determines the execution on which the programmer draws wires for connecting different function-nodes. The built-in scheduler automatically accomplishes the multi-threading and multi-processing hardware, which multiplexes multiple OS threads over the nodes (Fig. 2).

7 Block Diagram of Proposed System

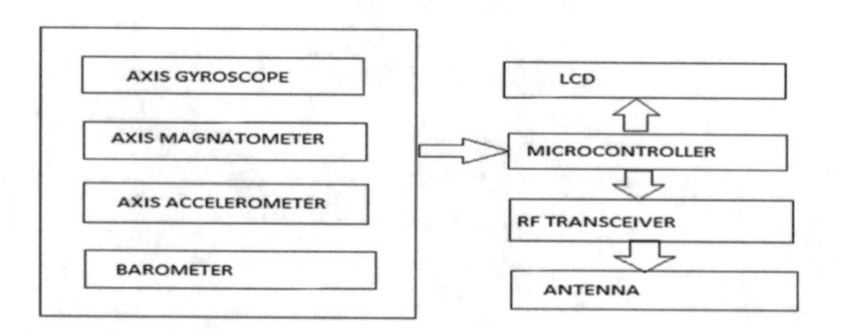

8 Hardware Requirements

8.1 IMU Module

An Inertial Measurement Unit is an electronic device which uses the combination of gyroscope, accelerometer and magnetometer that measures and reports an angular rate, body specific source, and magnetic field surrounding the body. When the object is integrated, it goes from an idle velocity to any velocity; to respond to the vibrations associated with such movement the accelerometer is designed. Micro strip crystals are used which undergoes stress when vibrations occurs, and from that stress to create a reading on any acceleration a voltage is generated.

A gyroscope uses Earth's gravity to help determine orientation. It consists of a rotor which is a freely rotating disk mounted on a spinning axis in the center of a more stable and larger wheel. "The ITG-3200 is the world's first single-chip, digital-output, 3-axis MEMS gyro IC optimized for gaming, 3D mice, and 3D remote control applications. Magnetometers are used for measuring the magnetic field of Earth and to detect magnetic anomalies in geophysical surveys.

Low-field magnetic sensing with a digital interface is designed with a **Honeywell HMC3883L** which is a multi-chip, surface-mount module for applications such as magnetometer and low-cost compassing. This RF transceiver plays an important role in transmitting and receiving the data between the soldier and the control base. The combined values of the sensors are transmitted wirelessly to the control base and any instruction from the control base can also be received by the soldier from the control base. The RF module that has been used in this module is a TARANG module. TARANG modules receive the data from host through the serial port. Packetization is the next step before transmitting the data on air. The packet size can be configured by user to desired value from 0 to 90 with "ATSPK" command (Refer to Serial Interfacing Commands [S]).

Table 1 Technical specification

Microcontroller	ATmega328
Operating voltage	5 V
Input voltage	7–9 V
Digital I/O pins	14 (of which 6 provide PWM output)
Analog input pins	8 (of which 4 are broken out onto pins)
DC current per I/O pin	40 Ma
Flash memory	32 kB (of which 2 kB used by boot loader)
SRAM	2 kB
EEPROM	1 kB
Clock speed	16 MHz

Transmit buffer buffers the serial data. The serial data is stored in the Transmit Buffer. The data is packetized and sent at any timeout or when maximum packet size is received. TARANG module extracts the contents from the data packet recovered from air and pushes out to serial port is pushed according to the serial parameters configured. The strength of the signal packet received can be viewed through RSSI (Receive Signal Strength Indicator) parameter through "ATPRS" command.

9 Software Requirements

The Arduino Mini is a small microcontroller board intentional for use on breadboards and when space is at a premium. To enable all components to be on the top of the board The revision 05 has a new package for the ATmega328.

Technical Specifications
See Table 1.

10 Results and Discussions

This three-dimensional view also shows in which floor or at what height the soldier is present with the help of the barometer which measure the pressure thereby determining the height or altitude the soldier is present. In addition to all this we will also be able to count the number of steps that the soldier has walked which will be useful in determining the turns or the floors that he have climbed inside a building.

11 Conclusion

The IMU module, which have been mentioned in the proposed system produces an output that is an integration of all the four values provided by the four sensors and transmits it wirelessly to the receiver. According to the current navigation system that is present now, determining a person's position under the sky is easy with the help of a GPS satellite. The main advantage of our project is that it can be used to for determining the position as well as trace the person's whereabouts inside as wells outside building. By the use of Indoor navigation system users can navigate without any inconvenience. Thus our solutions combine internal measurements to estimate user's position. The proposed system has the advantages of high reliability, high safety and user friendly and it can also be used inside any type of building with any number of stores since it utilizes the sensors in such a way.

References

1. Lu, H., Yao, L.: Design and implement of distribution transformer outage detection system. National Science Council of the Republic of China, NSC 94-2213-E-027-055.2007 (2007)
2. Ferreira, K.J.: Fault Location for Power Transmission Systems Using Magnetic Field Sensing Coils. ECE Department of Worcester Polytechnic Institute, Apr 2007
3. Dong, A., Geng, X., Yang, Y., Su, Y., Li, M.: Overhead line fault section positioning system based on wireless sensor network. PRZEGLĄD ELEKTROTECHNICZNY (2013). ISSN 0033-2097
4. Rogne, R.H., Johansen, T.A.: Observer and IMU based detection and isolation if faults in position reference systems and gyrocompasses with dual redundancy in dynamic positioning
5. Asiminoaei, L., Blaabjerg, F., Hansen, S.: Detection is key—harmonic detection methods for active power filter application. IEEE Ind. Appl. Mag. **133**(4), 22–33 (2007)
6. Bhattacharya, A., Chakraborty, C., Bhattacharya, S.: Shunt compensation. IEEE Ind. Electron. Mag. **3**, 38–49 (2009)
7. Cirrincione, M., Pucci, M., Vitale, G.: A single-phase DG generation unit with shunt active power filter capability by adaptive neutral filtering. IEEE Trans. Ind. Electron. **55**(5), 2093–2110 (2008)
8. Costa-Castello, R., Grino, R., Fossas, E.: Odd-harmonic digital repetitive control of a single-phase current active filter. IEEE Trans. Power Electron. **19**(5), 2093–2110 (2008)
9. Fujita, H.A., Kagi, H.: Voltage-regulation performance of a shunt active filtering trend for installation on a power distribution system. IEEE Trans. Power Electron. **22**(4), 1046–1053 (2007)
10. Fujita, H.: A single-phase active filter using an H-bridge PWM converter with a sampling frequency quadruple of the switching frequency. IEEE Trans. Power Electron. **24**(4), 934–941 (2009)

R. M. Bommi obtained B.E. in Electronics and Communication Engineering from SRM Easwari Engineering college, Madras University in the year 2003, M.Tech. in VLSI Design from Sathyabama University in 2009. She is currently doing Ph.D. in the specialization of VLSI design in Sathyabama University. Furthermore, she is working as an Assistant Professor in the department of Electronics and Communication Engineering in Jeppiaar Maamallan Engineering College. She has published around 10 papers in various national and international journals and conferences.

V. Monika is Assistant Professor in the Department of Electronics and Communication Engineering, Jeppiaar Maamallan Engineering College, Chennai. She obtained her B.E. in ECE from PSNA college of Engineering, M.E. in Embedded Systems from Sathyabama University and M.B.A. in Project Management from Alagappa University. Her areas of interest include Digital Electronics, Embedded Systems and IoT. She has published around six papers in various national and international conferences and Journals.

R. Narmadha obtained B.E. in Electrical and Electronics Engineering from Pallavan Engineering college, Madras University in the year 2002. M.E. in Power System in Annamalai University in 2006. She is working as an Assistant professor in the department of Electrical and Electronics Engineering in Jeppiaar Maamallan Engineering College. She has produced around 3 papers in various national and international journals and conferences.

K. Bhuvaneswari is pursuing B.E. in Electronics and Communication Engineering from Jeppiaar Maamallan Engineering College, Chennai. Areas of interest are VLSI and COMPUTER NETWORKS. Currently undergoing a training in CCNA in NIIT, Chennai.

L. Aswini is pursuing B.E. in Electronics and Communication Engineering from Jeppiaar Maamallan Engineering College, Chennai. Area of interest is COMPUTER NETWORKS. Interested in Arduino Projects.

An Intelligent Approach to Demand Forecasting

Nimai Chand Das Adhikari, Nishanth Domakonda, Chinmaya Chandan, Gaurav Gupta, Rajat Garg, S. Teja, Lalit Das and Ashutosh Misra

Abstract Demand Forecasting, undeniably, is the single most important component of any organizations Supply Chain. It determines the estimated demand for the future and sets the level of preparedness that is required on the supply side to match the demand. It goes without saying that if an organization does not get its forecasting accurate to a reasonable level, the whole supply chain gets affected. Understandably, Over/Under-forecasting has deteriorating impact on any organizations Supply Chain and thereby on P and L. Having ascertained the importance of Demand Forecasting, it is only fair to discuss about the forecasting techniques which are used to predict the future values of demand. The input that goes in and the modelling engine which it goes through are equally important in generating the correct forecasts and determining the Forecast Accuracy. Here, we present a very unique model that not only pre-processes the input data, but also ensembles the output of two parallel advanced forecasting engines which uses state-of-the-art Machine Learning algorithms and Time-Series algorithms to generate future forecasts. Our technique uses data-driven statistical techniques to clean the data of any potential errors or outliers and impute missing values if any. Once the forecast is generated, it is post processed with Seasonality and Trend corrections, if required. Since the final forecast is the result of statistically pre-validated ensemble of multiple models, the forecasts are stable and accuracy variation is very minimal across periods and forecast horizons. Hence it is better at estimating the future demand than the conventional techniques.

Keywords Supply chain · Demand forecasting · Time series · Machine learning Ensemble · Optimization · Cognitive analysis · Outliers

---

C. Chandan · G. Gupta · R. Garg · S. Teja · L. Das
SS Supply Chain Solutions Pvt. Ltd., Bengaluru, India

A. Misra
Enterprise Information Management, Philips Lighting, Bengaluru, India

N. C. D. Adhikari (✉)
AIG, Bengaluru, India
e-mail: dasadhikari.nimaichand@gmail.com

N. Domakonda
Novartis, Bengaluru, India

S. Smys et al. (eds.), *International Conference on Computer Networks and Communication Technologies*, Lecture Notes on Data Engineering and Communications Technologies 15, https://doi.org/10.1007/978-981-10-8681-6_17

1 Introduction

Demand Forecasting is the key activity which more or less controls all other activities of Supply Chain Management. It is the key driving factor in planning and decision making in Supply Chain Management as well as Enterprise level. All major firms truly depend on the efficiency of demand forecasting to take major decisions such as capacity building, resource allocation, expansion and forward or backward integration, etc. Forecasting is a prediction or an estimation of an actual value in a future time period. It is usual that there might be forecast error as actual results differ from the projected value, long time horizon has chance of more error. This is important to measure the forecast error for adapting corrective action plan. The development of advanced technology enables the Supply Chain Management stakeholders to share real time data and information across the network which helps dual benefit to inventory and customer service. This underlying result of the process is accuracy of forecast which firmly ensures the successful and sustainable business operations. This is called Collaborative Planning, Forecast, and Replenishment. Sales and Operations Planning (S&OP) is an integration process used in business organization to ensure efficient coordination among cross functional units to align company strategy with Supply Chain planning. Globalization, new opportunities and Supply Chain Management differentiation compel the organization even struggle further as the integration process among suppliers, market, and stakeholders become complex and data intensive. Various forecasting models are used to predict what the demands on the system will be in the future so that appropriate designs and operating plans can be devised. Ironically, the basic premise behind generating forecasts is that they are mostly wrong. Forecasts must include analysis of their potential errors. The longer the forecast horizon, the less accurate the forecast will generally be. Benefits can often be obtained from transforming a forecasted quantity into a known quantity. The information systems play a key role mainly in operative planning and management. It is the requirement of almost all the companies to start run their business without having relevant information from ERP (Enterprise Resource Planning) systems. In consequence of more efficient information use through ERP, the partial planning methods as MRP II (Manufacturing Resource Planning), Sales and Operations Planning (S&OP) or APS (Advanced Planning and Scheduling) are developed. To maximize the effect of accessible methods for any internal company processes, the company must be built on an objective and evaluated demand forecasts. The choice of optimum forecasting procedures and following use of obtained forecasts may become a competitive advantage. Together with other modern methods it accelerates other company processes, reduces the costs and increases the value for the customer. The demand forecast determines the volume of products, place and time horizon in which they will be needed. In relation with the demand forecast it is necessary to deal not only with the quantitative aspect of the needs (the volume demanded by customers) but also their qualitative aspect (the type of customer's needs). The accurate demand forecast is thus important for the production and distribution management but also for, e.g., areas of marketing (distribution of sales forces, communication, promotion and planning of new products), finance (current need of money, budgets and calculations), investment designs (production facilities, workshops and warehouses), research and development (innovations) and human resources (structure and labour force volume planning, training). Demand planning represents a set of methodologies and information technologies for the use of demand forecasts in the process of planning. The aim is to accelerate the flow of raw materials, materials and services beginning with the suppliers through transforming to products in the company and to their distribution to their final consumers. The demand planning process is done to help the business understand its profit potential. Indirectly it sets the stage for capacity, financing, and stakeholder confidence. The implementation of the demand planning enables to determine the closest possible forecast to the planning horizon and decide the volume of production, stock and sources capacity distribution among particular products to maximize the profits of the whole company. The key requirement for efficient company management is sharing the mutual forecast. However, the research carried in production companies showed individual departments of the company in some cases draw up forecasts on their own and thus they base their planning on different figures. This provokes conflicts among the resulting activities of in-company plans. This kind of situation happens in case when the company prefers to approve the financial plan which does not correspond with the updated forecast results. The forecasting should always be the process which is essential and determining for other company processes, including financial planning. The financial plan often represents the main motivation source for the company managers as this reflects the requirements of the company top management and company's strategic goals.

A variety of modelling techniques are available for producing forecasts. Based on data patterns, forecasting horizon, data availability and business requirements the choice of technique differs [1]. Through appropriate combining of the forecasting techniques it is possible to estimate quantitative influences of the identified factors and set the demand forecast [2]. The most frequently used statistical forecasting method is the time-series technique. It uses historical data sequenced by time and projects future demand by the same time sequence [3]. POS data is rich in information for building forecast models. Building a good forecasting model with POS data is demonstrated in many case studies. It is important to realize that demand planning used in practice is not a mere creation of a perfect system to carry out the demand and sales forecast. The objective of forecasting is to predict demand while the aim of demand planners is to shape the demand and produce a resource requirement plan [4]. Without the right forecast planning system integration, it is not possible to use efficiently the information provided in the forecasts.

2 Existing Procedure

2.1 Demand Planning Process

The objective of Demand planning is to make reliable demand plans as base for the S&OP process. Game-playing, the intentional manipulation of the forecasting process to gain personal, group, or corporate advantage, is clearly not allowed. Examples of such games:

- Sandbagging: underestimating sales in order to set expectations lower than the demand actually anticipated
- Hedging: overestimating sales in order to secure additional product or production capacity
- Enforcing: maintaining a higher forecast than the anticipated sales, in order to keep forecasts in line with the organizations sales or financial goals
- Spinning: manipulating forecasts to obtain the most favourable reaction from individuals or departments in the organization

Demand plans should reflect what we really think what we will sell in the future.

2.2 Statistical Forecast

The ambition of the business in using the statistical forecast is to minimize the sales planner work load and, in the meantime, give a good forecast that will benefit the organization. Statistical forecasting might not be suitable for every product, but it is proven successful for products with a fairly stable sales history and a large number of lighting products fit this description. Statistical forecasting significantly reduces the number of products that will have to be touched by every individual sales planner every week/month.

- Data pre-processing: Outlier alerts are used as input for the stat forecast manager to do history cleaning. All individual alerts will be listed, which means that if a certain product has more than one outlier, all these events will be specified. A value is considered an outlier if:

  (1) Value is higher than (average + 3*standard_deviation) which is the upper limit or the
  (2) Value lower than (average - 3*standard_deviation)

- Algorithms: Based on the chosen statistical model, the existing business model will calculate a forecast. Available statistical forecast models are listed as:

  (1) *Moving Average* The moving average model is used to exclude irregularities in the history pattern. The model calculates the average of all historical values.

All values are weighted equally. It does not take leading zero values into account.

(2) *SES Model* The SES model will also calculate an average, but instead of weighting all historical values equally, the SES model will give more weight to the more recent history. Within the SES model the smoothing factor (alpha) will determine how quickly the forecast reacts to a change in the history pattern. The higher the alpha factor, the more weight will be assigned to the most recent history, the quicker the forecast reacts to the recent history.

(3) *Croston Model* The Croston model is used for products with intermittent demand. It consists of exponential smoothing calculation, together with an average interval between demands. These two inputs will then be used in a form of a constant model to predict future demand. In case no zero values exist in the history pattern, Croston model will have exactly the same result as the SES model (in case of an equal alpha factor).

(4) *Seasonal Linear Regression* Seasonal linear regression will automatically determine trend values and seasonal indices. The system will first try to find a seasonal pattern. In case the system cannot find a seasonal pattern, it will calculate a forecast using linear regression. In case it can find a seasonal pattern, it will first correct the historical values using calculated seasonal indices. A linear regression will calculate a forecast using these corrected historical values. Last step is to apply the seasonal indices to the calculated linear regression.

(5) *Double Exponential Smoothing* Double exponential smoothing uses the same technique as the single exponential smoothing but will also apply a trend. As in the single exponential smoothing, an alpha factor will determine the weight to determine how quickly the forecast reacts to a change in recent history. The beta will determine how quickly the forecast picks up a trend in history.

- Forecast Corrections:

(1) A manually maintained seasonality can be applied to SES and moving average models. The system will first de-seasonalize the history using seasonality factors as put in by the regional planning manager. It will then calculate a forecast using SES or moving average. The result of the forecast will be re-seasonalized again using the seasonality factors.

(2) A workday correction will be applied to all stat forecasts that do not use the manual seasonality or statistical forecasts that are using SLR.

(3) A manually maintained trend can be applied to SES and moving average models. After calculating a forecast using SES or moving average, it will apply the trend values to the result.

3 Demand Forecasting Procedure

The following set of procedures describes the demand forecasting process, quality control and exceptions management process for forecasts.

3.1 Extraction of Corrected Baseline History (CBH) with Links

(1) The Data in the real world has the life-cycle which varies a lot. The correct linking to the new product introduction, sister SKU and de-cleaning SKU is very important in the forecasting methodology.
(2) The Data which the model forecasts is CBH (Corrected Baseline History), which has the correct linking with the sister SKU or correction of the sales taking out the effect of the external driving factors like promotions, stockouts, etc. If the corrections of the data is not done or maintained, then the model will correct the data as required using the outliers correction and missing value imputation.

3.2 Validation

The first step of the validation should be done by checking for the availability of the lowest Characteristics Vector Code (CVC) elements. For each Market, comparison of the sum for last three months with that of the Month-wise Sum for the data downloaded in the previous month should be done. This is done to have a validation on the actual which is downloaded is matching the figures that is going into the model for forecast.

3.3 Data Shaping

The Data should be aggregated or dis-aggregated to the level in which the forecast should be generated. This helps in making the generalization capability of the forecasting to a better level. The cost function for the time-series forecasting in this domain differs for different organization.

Some common cost functions used are

1. Accuracy $= 1 - \frac{\sum_{i=1}^{N} \left|\sum_{j=1}^{3} (y_p - y_a)\right|}{\sum_{j=1}^{3} y_p}$
2. Accuracy $= 1 - \frac{\sum_{i=1}^{N} \left|\sum (y_p - y_a)\right|}{\sum_{j=1}^{3} y_a}$

3. $\text{Accuracy} = 1 - \frac{\sum_{i=1}^{N} |\sum (y_p - y_a)|}{\sum_{j=1}^{3} y_a}$,

where N is the total SKU's present for a market and y_p and y_a are the predicted forecast for a month and actual for that month respectively. j varies for the forecasted done for the consecutive months. The first equation is to evaluate the accuracy in the three months of forecast against the actuals and the last two for evaluating the accuracy for one month.

3.4 Data Quality Assurance

(1) Once the CBH data with Links is prepared as per markets requirement, the CBH data is now required to be analysed and the quality of the data is to be ensured before it is used for forecasting demand. The demand forecasting for each market is done from different stand points, as per market needs. Some of the examples are sales region wise, SKU wise, Channel level and material level.
(2) It is mandate to validate the data sets used for the purpose of generating Baseline Statistics Forecast. If some garbage data is put into the model the result will be garbage in itself.
(3) The Data complete description is made which makes it important and easy to validate the data which input and output from the model.
(4) The Data needs to be validated by the market demand planners to find if the forecast follows the trend, pattern and level along with the actual. In case of any discrepancy, the forecast will be modified with the intelligence of the market.

3.5 Models Used for Generating the Forecast

Once the Data Quality is assured then the data is uploaded in the forecasting model prepared on R and Python.

(1) *Outliers Detection, Data Treatment and Data Pre-processing* The most important part for any data analysis is the data processing. If there are ambiguities in the data, the forecast will also be ambiguous and un-useful. Outlier detection and removal or correction of the data are a part of the model [5–7]. Each Market requires the forecast at different level.
(2) *Model Selection* The forecast done for each SKU, is done using the time-series and machine learning algorithms. Then each of the SKU is matched with the results given by time-series, machine learning and a model with the ensemble of the outputs of time-series and machine learning.

    (1) Time-Series Algorithms: The actual data is fed into the model consisting of the time-series algorithms. For different SKU's different algorithms are

selected based on the *mse* in the validation or testing data. Before passing the data into the model, it is pre-processed [8].

- Data Pre-processing: The actual sales do have great impact with the *promotions*, *stockouts* and *dumping*, etc. These behave as the outlier in the system and need to be removed or treated. The values which are more than 40% or less than 10% for a particular month in a range of 12-month are replaced with the mean of the actual sales data or if the sales for a particular month is above the *particular limit Inter Quartile Range* is treated and imputed with the accepted sales for that month. *Particular limit* varies SKU to SKU. Also, the sparse SKU, i.e., the actual data containing more number of zeros are either passed into *croston model* [9, 10] else the zero sales data in a month is replaced with mean of the data.
- Algorithms: The algorithms which are used in the time-series algorithms are Weighted Moving Average, Simple moving average, Autoregressive integrated moving average (ARIMA) and Holts Winter [11–15]. For Weighted Moving Average 0.5, 0.3 and 0.2 weights are given starting from the most recent observation to the 3rd last observation whereas equal weights are assign to the last n (no of dependent variables) observations to find out the new forecast. On the basis of residuals, we are finding out the value of n. To find out the order of the Arima model we have used Arima optimization function ML and CSS and then using residuals RMSE and MSE. To find out the value of α, β and γ of Holt's Winter, the optimization functions L-BFGS-B and Nelder-Mead and for finding the residuals RMSE and MSE is used. In case the model gives 0 in the middle of the forecast [16] value is replaced with the mean of the whole vector. In the validation set, the algorithm which gives the minimum *mse* is selected for the future forecast [17].

(2) Regression-Based Algorithms: There is a bit of difference in the data that is fed into the models containing machine learning regression algorithms. This regression-based model contains the following steps to be done to take care of the forecasting of sales data.

- Data Processing: The sales of a particular period (day, week, month or year) is matched with the actual data and the values or the sales which contribute to more than 40% or less than 10% of the total sales for 7 days (if the data is weekly), 53 weeks (data is weekly) or 12 months (data is monthly) is treated as the outlier and is either brought down/up to the mean/median based on maximizing the cost function of the organization. This basically means that each SKU will have a different set of the outlier treatment. The missing value is nothing but to find out whether the "sales zero" in a particular data is actually the case or not. If the zero sales are not possible for the SKU, then the missing value is imputed with mean of the actual data. For getting the seasonal pattern of the data, the data is transformed into the lags of the months for finding a relationship between the current actual sales. The problem of forecasting

is a regression problem. The regressors are the lags of the actual data. Here to find a seasonality pattern in the data, each yearly data is taken as a lag. The two-important constraint are taken into the consideration:
  - *Time Constraint* It takes a lot of time to find the correct hyper-parameter settings.
  - *Seasonality Pattern* It was found that if the lag of 12 for monthly data and 53 for weekly data was taken into the consideration, the accuracy [cost function] increased.
- Algorithms: There is a lot of constraints for the forecasting using machine learning.
  - The hyper-parameters for each SKU for different algorithms will be different all the time. This takes a lot of time to optimized to the required parameters settings.
  - Seasonality capturing is very less when used machine learning algorithms.

  The algorithms that are used are Support Vector Regression [18–20], Decision Trees Regression [21], Linear Regression, Ridge Regression [22] and Random Forest Regression [23]. Random Forest Regression is used for the selection of important features out of the lag training set [24]. Other feature selection techniques can be used which is the future implementation and research for our work. As the dataset is a time-series dataset, the sequence of the data or the training examples is important. The training and testing sets are divided into the 70.30% using the holdout method. For the optimizing the hyper-parameters, Grid Search and Random search methods are used. Bayesian Search and other evolutionary optimization search methods are the future work. The best algorithm which results in the minimum mse in the validation set is selected for the future prediction of the sku. Out of all the algorithms the one which gives the minimum mse is matched again with the ensemble of the algorithms. Ensemble is done with the averaging of the error in the validation set.

$$\text{model} = \left[\text{algorithm}_1, \text{algorithm}_2, \ldots, \text{algorithm}_n\right]$$
$$\text{error}_i = \left|y_p - y_a\right|,$$

  where y_p is the predicted value and y_a is actual. Also, the weights assigned to each model/algorithm [algoritm$_i$] is given by

$$\text{weight}_i = \frac{\frac{1}{\text{error}_i}}{\sum_{i=1}^{n} \frac{1}{\text{error}_i}},$$

  where $i \in (1, n)$, n varies in the model set.
- Future Forecast Generation: Once the forecast is done, then the forecast results are validated and different methodology are taken into the account

to see which all process can be taken into the account for improving the accuracy. The comparison is required to be done at the level at which the forecast is generated. Identify the Models [algorithm with hyper-parameter settings] producing SKU wise best results and separate the SKUs as per the models giving best results consistently in the validation phase. Once the Model for specific variable is identified, initiate the run to forecast for future 18 months. Seasonality factor is applied to the model if required. This depends on the pattern in which the result is produced.

Algorithm for the Proposed Model

- *Input the Data*
- *for sku 1:N*, where N = total number of SKU for a market
  - *Data Pre-process* Outlier treatment, Missing value treatment
  - *Dataset Creation* Convert the Actual data to 12-lag dataset
  - *Train, Test Subset* Holdout method is used and time sequence for the data is maintained [As the data is time-series]
  - *Feature Selection* Transform the dataset to the most correlated dataset
  - *Training the Algorithms* Pass the Data through different algorithms
  - *Algorithm Selection* Based on the MSE in the validation/test period the model which results in the minimum error is selected [25, 26]
  - *Future Prediction* Generate the future forecast

(3) Ensemble of the Algorithms: The results/forecast generated by the time-series and regression-based algorithms can either be over-forecasting or under-forecasting. To achieve higher accuracy 100%, the forecast should be very much near to the actual sales or the error in the forecasting should minimize. Here, the term that defines the error is *deviation*.

$$\text{deviation} = \sum_{j=1}^{k} \left(y_{\text{actual}_j} - y_{\text{predicted}_j} \right),$$

where k is the time period for the calculation of the deviation. Thus, if the deviation is minimum, the generalization capability of the model increases. Here, we used averaging of the time-series and regression-based model results to give an ensemble forecast which brings the deviation near to zero or minimum of the deviation of time-series output and regression-based model output. Hence, this more enhances the generalization capability of the model. The weights that is calculated for the two kinds of models is by calculating the deviation from the validation period of the data.

$$\text{model} = [\text{model}_{\text{time-series}}, model_{\text{machine-learning}}]$$

$$\text{weight}_i = \frac{\frac{1}{\text{deviation}_i}}{\sum_{i=1}^{k} \frac{1}{\text{deviation}_i}}$$

here, k varies in the model dataset. The model that is selected in case of the ensemble is through the *greedy selection*. Here the algorithms that gives better results consistently is selected to use in the ensemble of the results [27, 28].

(4) Market Intelligence or Cognitive Approach: After the results are generated, the forecasts are plotted against the history actual and level, trend and seasonality pattern of the SKU is observed. If the forecast for the SKU doesn't follow any pattern with the history sales, it is changed or tweaked a bit. The following are some of the market intelligent inputs that are implemented in the model:

- *Zero Forecast* If the actual sales for a SKU is zero for the consecutive 2 months, its converted into zero forecast for the forecasting periods. This case will not be valid for the SKU having sparse data.
- *Trend Imbalance* If the trend of the forecast doesn't follow the trend of the history of the sales, the trend is corrected. If the mean of the initial forecast value is 4 times or 0.25 time of the mean of the last 6 months then the forecast value is replaced with the last 3 months weighted moving average.
- *Level Correction* There will be some SKU's which are going to have a change in the level due to some promotional impacts and other external conditions. This information which is externally present is input to the model and the level is shifted for the future forecast for that many number of periods which the market Demand Planner have informed.
- *Seasonality Correction* Some of the months will have lower peaks or depths in the forecast data than expected. This can be externally corrected before the forecast is submitted.
- *PIPO* Phase-In and Phase-Out SKU's are clustered and the forecast is changed in that manner. The Phase-In products are basically the NPI or New Products Introduced and will have a higher or lower sale which are difficult to predict. In case of those SKU's having only one data point, are either used Naive Forecast or done the same result as the forecasting output. Mostly Market Demand Planner's intelligence helps a lot to take care of this.

4 Results

The model comprising of the methods discussed above is used to generate the forecast for the different markets of Philips and the results are compared with the tool which is mostly used in the forecasting for different organization, Incumbent Business Model(I-B-M). For generating the forecasts two different platforms are used R (Version 1.0.136) and Python (Version 2.7). The results are generated for different markets and for different consecutive months to check the validity and the stability of the models. Two different terminology which are used are "$t - 1$" and "$t - 5$"

forecast accuracy. "$t - 1$" is the accuracy generated after two months of the current month the forecast is generated and "$t - 5$" is after 5 months of the current month.

Dataset Description: This data set contains the information of 7 markets. Due to confidentiality, actual names of the markets are masked with *market*– "*no*". In this section, first the comparison of the I-B-M is done with the developed model with different statistical measures followed by the result improvement comparison of *ensemble method* over the *Regression based model* and *time-series methods*.

4.1 Forecast Comparison for the Month April-2017 Using the Actual Data till February-2017

Market-name	Total SKU's	Model FACC (%)	I-B-M (%)
Market-1	3256	53	43
Market-2	982	36	28
Market-3	1806	29	31
Market-4	923	64	62
Market-5	1078	30	51.1
Market-6	2620	48	48
Market-7	2301	49	50

Total number of SKU's that has been forecasted is 12966. Using the model developed by us, the overall average accuracy that has been achieved is 44.14% and the overall accuracy that has been achieved by Incumbent Business Model is 44.72%. In checking the market-wise accuracy comparison, our model is equivalent to Incumbent Business Model. This comparison is the overall comparison of the total SKU's for each market in the month of April-2017.

4.2 Forecast Comparison for the Month May-2017 Using the Actual Data till March-2017

Total number of SKU's that has been forecasted is 16422. Using the model developed by us, the overall average accuracy that has been achieved is 52% and the overall accuracy that has been achieved by Incumbent Business Model is 49%. In checking the market-wise accuracy comparison, our model is better than Incumbent Business Model in five markets out of seven markets. This comparison is the overall comparison of the total SKU's for each market in the month of May-2017.

Market-name	Total SKU's	Model FACC (%)	I-B-M (%)
Market-1	2999	54	45
Market-2	1164	49	49
Market-3	2007	49	39
Market-4	1446	56	58
Market-5	3265	58	56
Market-6	3026	47	43
Market-7	2515	51	53

4.3 *Forecast Comparison for the Month June-2017 Using the Actual Data till April-2017*

Market-name	Total SKU's	Model FACC (%)	I-B-M (%)
Market-1	3480	55	38
Market-2	1391	39	33
Market-3	2685	48	41
Market-4	480	58	68
Market-5	3262	58	59
Market-6	3028	49	50
Market-7	2549	54	56

Total number of SKU's that has been forecasted is 16875. Using the model developed by us, the overall average accuracy that has been achieved is 51.57% and the overall accuracy that has been achieved by Incumbent Business Model is 49.28%. This comparison is the overall comparison of the total SKU's for each market in the month of June-2017. If we check the above two table, we can find that our model is consistently beating Incumbent Business Model by 3%.

If we take the above three results, our model is stable and giving good results. In case of the April-2017, the results of our model is comparable to that of the Incumbent Business Model.

Below are the results for the comparison of the individual models for different markets.

4.4 *Forecast Comparison for Different Algorithms in the Model for Different Markets*

Market-1:

Month	Time-series (%)	Regression models (%)	Ensemble (%)
April'17	66	63	66
May'17	68	64	69
June'17	65	63	66

Market-5:

Month	Time-series (%)	Regression models (%)	Ensemble (%)
April'17	60	58	62
May'17	62	59	64
June'17	60	59	62

Market-3:

Month	Time-series (%)	Regression models (%)	Ensemble (%)
April'17	53	48	53
May'17	45	46	49
June'17	49	47	51

Market-7:

Month	Time-series (%)	Regression models (%)	Ensemble (%)
April'17	49	53	54
May'17	41	38	42
June'17	55	50	55

The above table tells about the comparison of the different algorithms that we are using in out model. We can see that the ensemble of the results produces better results than the individual algorithms in consecutive three months of April'17, May'17 and June'17. The weights for the ensemble is found out from the previous months and the average weight is assigned for the final month.

5 Conclusion

The model which we have presented has all the state-of-art statistical methods used in the demand forecasting fields. In the results above, we see that ensemble of the results of time-series model and regression-based model gives a better result due to the fact of nullifying the over-forecasting and under-forecasting and bringing the forecast values near to the actual. These results are far better than considering individual algorithms used in the two models.

Accurate forecast is very important for the demand planning team. The data used in this research and building the model is using the sales-in data for different markets. The important factor to be considered is the stability of the model and removing the *game-playing*. Two open-source platforms are used to build the model. Time-series model is developed in R-Studio and Regression-Based model using the data mining algorithms developed in Python. After the results are generated, Ensemble of results is validated and generated using the Microsoft Excel. In the future work, different techniques will be considered and researched. Time-Series and Machine Learning to be built in one platform and check how the minimization of mse produces the forecast.

Acknowledgements The complete list of the team which has worked in this project: Rajiv Ranjan*, Arashdeep Singh*, Shaivya Datt*, Aamir Ahmed Khan*, Mitav Kulshrestha*, Pawan Kulkarni*, Anubhav Rustogi*, Aditya Iskande*, Mandar Shirsavakar*, Rabiya Gill*, Heine van der Lende†, Srinivas Deshpande†, Srijan A†.
*Advanced Analytics Team, SS Supply Chain Solutions Pvt. Ltd., Bangalore.
†Enterprise Information Management, Philips Lighting, Bangalore.

References

1. Vlckova, V., Patak, M.: Role of demand planning in business process management. In: The 6th International Scientific Conference Business and Management 2010, pp. 1119–1126 (2010)
2. Gregory Daniel Noble.: Application of Modern Principles to Demand Forecasting for Electronics, Domestic Appliances and Accessories (2009)
3. Wei, M., Liu, Y.: Key Factors and Key Obstacles in Global Supply Chain Management: A Study in Demand Planning Process (2013)
4. Vlckova, V., Patak, M.: Barriers of demand planning implementation. Econ. Manag. **16**, 1000–1005 (2011)
5. Barnett, V., Lewis, T., et al.: Outliers in Statistical Data, vol. 3. Wiley, New York (1994)
6. Fox, A.J.: Outliers in time series. J. R. Stat. Soc. Series B (Methodological), 350–363 (1972)
7. Watson, S.M., Tight, M., Clark, S., Redfern, E.: Detection of Outliers in Time Series (1991)
8. Du, K.-L., Swamy, M.N.S.: Fundamentals of machine learning. In: Neural Networks and Statistical Learning, pp. 15–65. Springer (2014)
9. Ghobbar, A.A., Friend, C.H.: Evaluation of forecasting methods for intermittent parts demand in the field of aviation: a predictive model. Comput. Oper. Res. **30**(14), 2097–2114 (2003)
10. Shenstone, L., Hyndman, R.J.: Stochastic models underlying croston's method for intermittent demand forecasting. J. Forecast. **24**(6), 389–402 (2005)
11. Granger, C.W.J., Joyeux, R.: An introduction to longmemory time series models and fractional differencing. J. Time Ser. Anal. **1**(1), 15–29 (1980)

12. Hibon, M., Makridakis, S.: Arma Models and the Box—Jenkins Methodology (1997)
13. Valipour, M., Banihabib, M.E., Behbahani, S.M.R.: Comparison of the arma, arima, and the autoregressive artificial neural network models in forecasting the monthly inflow of dez dam reservoir. J. Hydrol. **476**, 433–441 (2013)
14. Wagner, N., Michalewicz, Z., Schellenberg, S., Chiriac, C., Mohais, A.: Intelligent techniques for forecasting multiple time series in real-world systems. Int. J. Intell. Comput. Cybern. **4**(3), 284–310 (2011)
15. Peter Zhang, G.: Time series forecasting using a hybrid arima and neural network model. Neurocomputing **50**, 159–175 (2003)
16. Tersine, R.J., Tersine, R.J.: Principles of Inventory and Materials Management (1994)
17. Filzmoser, P., Liebmann, B., Varmuza, K.: Repeated double cross validation. J. Chemom. **23**(4), 160–171 (2009)
18. He, W., Wang, Z., Jiang, H.: Model optimizing and feature selecting for support vector regression in time series forecasting. Neurocomputing **72**(1), 600–611 (2008)
19. Lu, C.-J., Lee, T.-S., Chiu, C.-C.: Financial time series forecasting using independent component analysis and support vector regression. Decis. Support Syst. **47**(2), 115–125 (2009)
20. Muller, K.-R., Smola, A.J., Rätsch, G., Schölkopf, B., Kohlmorgen, J., Vapnik, V.: Predicting time series with support vector machines. In International Conference on Artificial Neural Networks, pp. 999–1004. Springer (1997)
21. Lai, R.K., Fan, C.-Y., Huang, W.-H., Chang,P.-C.: Evolving and clustering fuzzy decision tree for financial time series data forecasting. Expert Syst. Appl. **36**(2), 3761–3773 (2009)
22. Golub, G.H., Heath, M., Wahba, G.: Generalized cross validation as a method for choosing a good ridge parameter. Technometrics, **21**(2), 215–223 (1979)
23. Rasmussen, C.E., Williams, C.K.: Gaussian Processes for Machine Learning, vol. 1. MIT Press Cambridge (2006)
24. Menze, B.H., Michael Kelm, B., Masuch, R., Himmelreich, U., Bachert, P., Petrich, W., Hamprecht, F.A.: A comparison of random forest and its gini importance with standard chemometric methods for the feature selection and classification of spectral data. BMC Bioinf. **10**(1), 213 (2009)
25. Adhikari, N. C. D.: Prevention of heart problem using artificial intelligence. Int. J. Artif. Intell. Appl. (IJAIA) **9**(2), (2018)
26. Adhikari, N. C. D, Alka, A., Garg, R. Hpps: Heart problem prediction system using machine learning
27. Opitz, D.W., Maclin, R.: Popular ensemble methods: an empirical study. J. Artif. Intell. Res. (JAIR) **11**, 169–198 (1999)
28. Adhikari, N. C. D., Garg, R., Datt, S., Das, L., Deshpande, S., & Misra, A. (2017, December). Ensemble methodology for demand forecasting. In International Conference on Intelligent Sustainable Systems (ICISS), (pp. 846–851). IEEE (2017)

Nimai Chand Das Adhikari received his Master's in Machine Learning and Computing from Indian Institute of Space Science and Technology, Thiruvananthapuram in the year 2016 and did his Bachelor's in Electrical Engineering from College of Engineering and Technology in the year 2011. He is currently working as a Data Scientist for AIG. He is a vivid researcher and his research interest areas include computer vision, health care and deep learning.

Nishanth Domakonda received his Bachelor's and Master's from National Institute of Technology, Rourkela. He is working as a Data Scientist in Novartis Hyderabad and was a member of the Data Science Team of Philips in Bangalore. His interest areas are Time Series, Machine Learning using real time data, Health Care etc.

Chinmaya Chandan is working as a consultant in SS Supply Chain Solutions Pvt. Ltd. and has done his MBA from NITIE Mumbai.

Gaurav Gupta working as Advanced Analytics Team, SS Supply Chain Solutions Pvt. Ltd., Bangalore. His research area includes Supply Chain and cognitive analysis.

Rajat Garg received his Bachelor's in Biotechnology Engineering from National Institute of Technology, Jalandhar and is currently working as Data a Scientist in Philips Lighting (SS Supply Chain Solutions Pvt. Ltd.). His interest areas include Machine Learning, Computer Vision and Data Analysis.

S. Teja working as Advanced Analytics Team, SS Supply Chain Solutions Pvt. Ltd., Bangalore. His research area includes Supply Chain and cognitive analysis.

Lalit Das is the CEO of SS Supply Chain Solutions Pvt. Ltd. His research area includes Supply Chain and Cognitive Analysis.

Dr. Ashutosh Misra is the director of Enterprise Information Management, Philips Lighting for the Supply Chain Division in India. He has done his post doctorate from IIT Kanpur.

Reliable Digital Twin for Connected Footballer

S. Balachandar and R. Chinnaiyan

Abstract Digital twin will gain new insight in sports domain. It will not only help to connect and monitor sportsperson virtually in a lab but also help to simulate actions, movements in the ground and play shots which recommend coach and trainer to redefine the game strategy on the fly or even half time or break during the match.

Keywords IoT · Digital twin · RFID tags · Wearables · Mobile application
Simulation · Monitoring · Connected things · Gateway · Augmented reality
Edge network · 3D modeling · Big data · Analytics

1 Introduction

"Connected [1] Sportsman" concept is becoming popular now to measure the effectiveness of individual player's health, sports performance, predicted [2, 3] shots. It will help the clubs or coach to plan the game well during the game, assess the player's profile on the fly and recommend a similar player who can replace based on their health conditions and playing conditions (e.g. Shot style, penalty corner time). The feature of digital twin will level up the game strategy by virtually connecting the things [4] (e.g., Wearables, RFID Tags, Smart Watches, Cameras, body temperature, pulse and humidity sensors) and simulate the values in real-time and recommend the possible outcomes of that sportsperson (e.g., likely they play this shot or not, recommend a right place for him to place the goal, avoid penalties based on his position, pulse, blood pressure).

S. Balachandar (✉)
Shell India Market Private Limited, Bengaluru, India
e-mail: aaathibala@gmail.com

R. Chinnaiyan
Dept.of Information Science and Engineering, CMR Institute of Technology, No.132, AECS
Layout IT Park Road, Bengaluru 560037, India
e-mail: vijayachinns@gmail.com

S. Smys et al. (eds.), *International Conference on Computer Networks and Communication Technologies*, Lecture Notes on Data Engineering and Communications Technologies 15, https://doi.org/10.1007/978-981-10-8681-6_18

2 Background

IoT (Internet of Things) [5] plays an important role on connecting things with different networks (e.g., WIFI, LORA, Bluetooth). To control these things in real-time with the help of gateway and edge server and defined business rules. The connected sportsperson (e.g. footballer) devices can be monitored and data can be analyzed on the fly. Building prescriptive analytics by simulating physical things virtually through 3D modeling which adds lot of values and it is not possible without a digital twin. If we need to simulate outside the data without devices doesn't give reliable results.

3 What Is Digital Twin

Digital twin [6] technology is expected to increase IoT deployments in the future, given its ability to add value for end-customers. Digital twin [7–9] refers to a digital replica of physical assets, processes, and systems that can be used for various purposes. The digital representation provides both the elements and the dynamics of how an Internet of Things device operates and lives throughout its life cycle. In the context of "Connected Sportsperson (Footballer)" digital twin plays a vital context of monitoring physical things seamlessly and measures reliable readings from different devices. The coach or club constantly assess things virtually and test with different values through 3D modeling and simulate it and play via Augmented Reality devices with pre-built simulation programs. Quite simply, a digital twin is a virtual model of a process, product or service. This pairing of the virtual and physical worlds allows analysis of data and monitoring of systems to head off problems before they even occur, avoid downtime, develop new opportunities and even plan for the future by using simulations.

4 Literature Survey

E. H. Glaessgen and D. Stargel in 2012 presented the Digital Twin paradigm for future NASA and US Air Force vehicles. S. Boschert, R. Rosen—Mechatronic Futures, 2016 presented Digital twin the simulation aspect. A. Schwirtz, D. Hahn, A. Huber, A. Neubert in 2004 presented Biomechanical power analysis in Nordic and alpine skiing. Chinnaiyan and Somasundaram [10–15] presented reliability evaluation of software systems. The authors evaluated the reliability of software systems with novel methods and ensured that the reliability models provides better reliability results. The reliability models presented by the authors can be applied for this problem and optimized reliability results can be obtained. As per these literature review, following are key findings depicted in Fig. 1 which help us to solve this problem for sports domain.

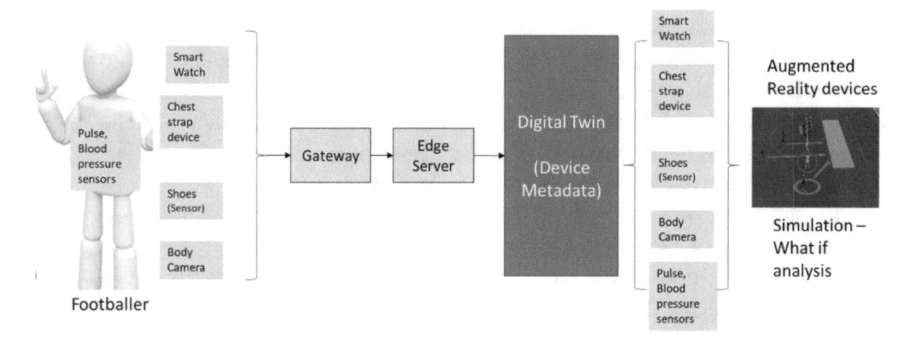

Fig. 1 Connected footballer using digital twin model

5 How Relevant to This Problem

In the recent era of connected stadium or player, it is difficult to simulate or replicate physical devices, sensors and simulate on the fly Digital twin is the solution which collects data from these devices will help to replicate these devices virtually with the help of 3D Modeling tools and simulate the values and analyze the behavior with different scenarios and test these values. The output can be visually experienced through Augmented and virtual reality device and test the shots or movements. Here the output will not only help the coach or club to assist the player and formulate right gaming strategy for a match.

a. Coach can monitor player's condition and they can simulate it with other environmental variables or correlation factor to determine the player's comfort level, they also can adjust and notify the player to move the geo position accordingly during break or half time.
b. During penalty shoot, coach can simulate based on the player's geo position value, they can simulate the probable geo position and they can recommend the player to shoot the penalty corner with right positions based on their health and distance to the goal post and his preferred kick styles (e.g. head or foot).

6 High-Level Approaches

a. Wearables (e.g., Smart Watches, Fit bands) and other physical sensors (e.g., Body Temperature, Pressure, pulse) attached in their chest strap device data will communicated through WIFI or Bluetooth network and it will be attached to IoT gateway.
b. The RFID tag of Shoes and shoe's position will constantly communicates over WIFI and it will send the events to nearest gateway box.

c. The edge server collects and stores various data from these things and data will be stored in unified format (e.g., JSON), the protocol translation and message format will be done at gateway level with the help of protocol translation software.
d. Data from edge will be continuously ingested into digital twin storage which not only keeps the device metadata [e.g., Sensor value ranges, Threshold ranges, Mac Address, device property and IDs (e.g., Mac Address)].
e. 3D modeling software will be used to simulate these sensor values and experimented with various model algorithms (e.g., Monto Carlo Simulation, etc.).
f. The output will be tested with different ranges to analyze "What if "scenarios for the player. It will help the coach to predict the worst case, best case and good case situation.
g. Coach and club members can analyze these values through AR (Augmented Reality) or VR (Virtual Reality) devices.
h. Reliable readings will be observed at different layers of the above solution.
i. Mobile application (e.g. Android, iOS App) can help the player or coach to visualize the analytics and dashboards during or after the match. The historical data will also help to learn the model constantly and improve the player's movement accuracy level.

7 Key Benefits of This Solution

a. Helps the player, sports club and coaches to formulate a right gaming strategy.
b. Assess player's health and plan the gaming schedule and effectively utilize other player based on their similarities or playing style with these players.
c. Monitor the IoT devices health and replicate the same data and measure the reliability of device events.

8 Role of Big Data and Analytics in Digital Twin and its Key Components

8.1 Big Data

The digital twin's storage needs to keep both structured and unstructured data coming from IoT gateway. Hadoop or large file systems (e.g. GFS or Amazon S3 or Azure Data lake) can be used to keep these large data set and processed with large compute engines like Spark. The streamed data coming from cameras, sensors will be ingested through Kafka/Flink tools which not only process the data but also compute and replicate in parallel to digital twin layer.

8.2 *Metadata*

Device metadata will be ingested from IoT Gateway and it will help the digital twin to build these devices virtually with the help of 3D model and drawing tools (e.g., CAD—Computer Aided Design).

8.3 *Analytics*

This capability is used to produce a digital profile of both the historical and current behavior of the IoT devices with the resulting profile used to: Detect device behavior that is less than optimal ad Predict future behavior of the device. Simulation might be used to find the optimal operating configuration for the wearables or devices from the player, study its failure modes or evaluate the device behavior under extreme operating conditions. Simulation model to perform what if analysis on one or more control strategies applied to the player's devices when it is in the simulated state. The physical data never been manipulated for this simulation.

8.4 *Visualization*

Once an accurate model of the sportsperson's device is available in the digital domain, its status can be considered as a proxy for the device's status. If a suitable visualization is available the status can be displayed on a virtual version of the physical asset. For example, if the device is an shoe it might be possible to build a 3D visualization of the shoe with necessary sensor objects and then zoom into the shoe's sensor to view the simulated speed and position of the shoe or view a visualization of the actual sportsperson shoe.

9 Conclusion

The above solution is specific to footballer it needs to be customized based on other sports devices or sports type. The digital twin solution can be complemented if there is an IoT platform, 3D modeling tool, AR/VR tools already exists.

References

1. World Urbanization Prospects: The 2014 Revision. Department of Economic and Social Affairs, United Nations (2014) [Online]. Available: http://esa.un.org/unpd/wup/Publications/Files/WU

P2014-Highlights.pdf
2. Su, K., Li, J., Fu, H.: Smart city and the applications. In: 2011 International Conference on Electronics, Communications and Control (ICECC). IEEE, pp. 1028–1031 (2011)
3. Al-Hader, M., Rodzi, A., Sharif, A.R., Ahmad, N.: Smart city components architecture. In: International Conference on Computational Intelligence, Modelling and Simulation, CSSim'09. IEEE, pp. 93–97 (2009)
4. Cocchia, A.: Smart and digital city: a systematic literature review. In: Smart City. Springer, pp. 13–43 (2014)
5. Machine-to-machine communications (m2m); impact of smart city activity on IoT environment. European Telecommunications Standards Institute (ETSI) (2015)
6. https://en.wikipedia.org/wiki/Digital_twin
7. https://www.researchgate.net/publication/312646475_C2PS_A_Digital_Twin_Architecture_Reference_Model_for_the_Cloud-based_Cyber-Physical_Systems
8. forbes.com/sites/bernardmarr/2017/03/06/what-is-digital-twin-technology-and-why-is-it-so-important/
9. http://vantiq.com/wp-content/uploads/2017/12/VANTIQ-Digital-Twin-Architecture-1.pdf
10. Chinnaiyan, R., Somasundaram, S.: Evaluating the reliability of component based software systems. Int. J. Qual. Reliab. Manag. **27**(1), 78–88 (2010)
11. Chinnaiyan, R., Somasundaram, S.: An experimental study on reliability estimation of GNU compiler components—a review. Int. J. Comput. Appl. **25**(3), 13–16 (2011)
12. Chinnaiyan, R., Somasundaram, S.: Reliability of object oriented software systems using communication variables—a review. Int. J. Softw. Eng. **2**(2), 87–96 (2009)
13. Chinnaiyan, R., Somasundaram, S.: Reliability assessment of component based software systems using test suite—a review. J. Comput. Appl. **1**(4) (2008)
14. Chinnaiyan, R., Somasundaram, S.: Reliability of component based software with similar software components—a review. J. Softw. Eng. **5**(2) (2010) (i-Manager)
15. Chinnaiyan, R., Somasundaram, S.: Monte Carlo simulation for reliability assessment of component based software systems. J. Softw. Eng. (2010) (i-Manager)
16. Lee, J., Bagheri, B., Kao, H.-A.: A cyber-physical systems architecture for industry 4.0-based manufacturing systems. Manuf. Lett. **3**, 18–23 (2015)
17. Jia, D., Lu, K., Wang, J., Zhang, X., Shen, X.: A Survey on Platoon Based Vehicular Cyber-Physical Systems (2015)
18. Perera, C., Zaslavsky, A., Christen, P., Georgakopoulos, D.: Sensing as a service model for smart cities supported by internet of things. Trans. Emerg. Telecommun. Technol. **25**(1), 81–93 (2014)

S. Balachandar is working as a, Application Delivery Lead - Technology Shell India Private Limited, Bengaluru, Karnataka, India. He is having 17 years of industrial experience in the areas of Big Data and Cloud Computing. Now he is doing his Ph.D. in Computer Applications in VTU. His research interest includes Security, Reliability, IOT, Machine Learning, Big Data, Cloud Computing and Data Science.

Dr. R. Chinnaiyan is working as Professor in the Department of Computer Applications, New Horizon College of Engineering, Marathalli, Bengaluru, Karnataka, India. He is having 17+ years of teaching experience. He is a life member of ISTE and CSI of India. He completed his research in Anna University-Chennai at Coimbatore Institute of Technology, Coimbatore in 2012. His research interest includes Object Oriented Analysis and Design, Qos and Software Reliability.

Centralized Reliability and Security Management of Data in Internet of Things (IoT) with Rule Builder

S. Balachandar and R. Chinnaiyan

Abstract The number of connected things or devices will be 125 billion by 2030 as per IHS market survey. It indirectly translates that large number of users will be accessing these devices for collecting, controlling, monitoring and building edge analytics. The device management policy and data policy on accessing data from different resources needs proper governance and control. The data security will help these users to assure regulatory compliance, security, privacy and protection. This paper is a centralized data security management across different layers of Internet of Things along with rule builder. Mainstream of this study focusses on compliance standard, identity management, data management and policy engine and audit reports. The specific regulatory compliance, security standards or firmware or hardware security and vulnerabilities at IoT layer has not been captured in this paper.

Keywords LoraWan · NFC (near field communication) · Zig Bee
Identity management · Data audit · Metadata management · Business rules
Privacy · MQTT (message queuing telemetry transport) · Audit engine · HTTP
IoT message broker · REST

1 Introduction

Internet of things (IoT) is ringing the bell on every corner and it is disrupting across different verticals. It is nothing but a network of physical objects [1] or "things" [2]. The term "Connected" and "Smart" closely linking the usage of "things" or "devices" and building the business application which take the advantage of Internet of Things

S. Balachandar (✉)
Shell India Market Private Limited, Bengaluru, India
e-mail: aaathibala@gmail.com

R. Chinnaiyan
Dept.of Information Science and Engineering, CMR Institute of Technology, No.132, AECS
Layout IT Park Road, Bengaluru 560037, India
e-mail: vijayachinns@gmail.com

S. Smys et al. (eds.), *International Conference on Computer Networks and Communication Technologies*, Lecture Notes on Data Engineering and Communications Technologies 15, https://doi.org/10.1007/978-981-10-8681-6_19

or “Connected X” to drive the business value. In 1990 the first IoT-based toaster [3] device created by “John Romkey” that could be turned off and on over internet. He proved it during October’89 INTEROP conference. It is using TCP/IP network to control the toaster. 1999 was the big year for IoT and MIT. “Kevin Ashton” coined the term “Internet of Things” [4]. The evolution of IoT started from RFID-based communication to all the way to “LTE Advanced” now. Many different protocols and networks are used today such as IEEE 802.15.14, Bluetooth, WIFI, Zig Bee, Z-wave and NFC and LoraWan (Low Power Wide area network). The protocol across varies from device to device and device to gateway.

The data security at IoT platform will help on accessing and controlling the data from different edge or gateway or devices or actuators. It mandates the IT driven approach to adhere data security policies for an organization.

2 Data Security in the IoT Edge or Hub

Data security in IoT Edge is managing the overall data availability, usability, integrity and security of data across different IoT layers. It is driven by set of policies and procedures defined by a governing body or council who manages the IoT platform (Fig. 1).

2.1 *Device Identity Management*

Each device identifier is either user specified or system generated device key or 128-bit UUID and that is being tracked by device identity registry. Nowadays Edge platform assigns a universally unique identifier (UUID) to a gateway and each connected device. Every UUID is a 128-bit value.

Gateway ID—e91f6aaa-2ff0-4890-9235-04f8ee8335c0

Smart Plug—446ae5a3-afc5-4f42-94c1-ca07c692f4dc

Fig. 1 Data security functions at IoT platform

2.2 User Authentication and Authorization

User management is generally associated with authentication [5–8] service supported by IoT Edge server or cloud IoT platform. It can be linked with one of the authentication service

- MQTT or HTTP over server authentication
- Open LDAP
- Windows Active Directory
- X.509 Certificates
- Kerberos

Data Policies determine what an authenticated user can do. Authorization will allow the users to invoke a MQTT client call or access a REST API or listen set of messages from message broker (e.g., Kafka or AWS IoT Message broker).

2.3 Data Policy and Rules Management

Policy manager or policy controller defines the user policy or role based policies. It will enforce the necessary business rule based on the user identity. Example: a device (e.g., Temperature Sensor) data can be accessed through MQTT subscription service for user. Here the business rule is Temperature Sensor data with defined data format for allowed to access (e.g., Sensor ID, Location, Timestamp, value) with defined action (read or write).

2.4 Encryption and Masking

Data needs to be encrypted and masked based on the access roles assigned to the user. The user's individual information should not be visible to other users (e.g., a wearable which track's one user's pulse, walking steps, calorie index, vital signs). Privacy related data such as location (geo-location) or personal info such as age, sex, etc.,) this information might be masked and it will not be visible to other users when the platform wants to share data to an application.

2.5 Audit Engine

Audit engine is keep tracks the system changes and what changes performed by different user. Access to audit engine can be defined at data policy level (e.g., System activity reports, User activity reports) for a specific user who can be audit administrator or audit user. Audit log can be configured for different time frames to visualize the user security, policy logs.

3 Literature Survey

Internet of things is helping to measure the physical characteristics of the surrounding things such as sensors, cameras, engine, pump, signal lights and street lamps and fire alarms and wearables like fit band. Security management on different IoT layer needs a comprehensive framework to track, control and access the necessary data from gateway or sensors or edge server or cloud server.

(a) Xiruo Liu, Meiyuan Zhao, Sugang Li, Feixiong Zhang and Wade Trappe researched on June 2017 "**A Security Framework for the Internet of Things in future internet architecture**" [9]. It explains the potential threats for the IoT Systems, it mentions about Mobility first infrastructure which mentions about "Global Name Resolution Service" which provides a clean separation of the identifiers and the dynamic network address locators and supports on the fly name binding to network address for dynamic mobility. That research also covered how current challenges that IoT is facing such as salacity, mobility and context retrieval, interoperability and security and privacy.

(b) Wei Zhou, June 2014 mentioned in his research paper about "**Security/privacy of wearable fitness tracking** IoT **devices**" [10]. He explains about the privacy and vulnerabilities of fit bit integration with social media feeds and how sensitive health data might get exchanged and how to use fit lock as an example to show vulnerabilities. It also helped to learn the "clear text login information" and "clear text HTTP processing" lead to security and privacy related vulnerabilities.

(c) Qi Jing, Athanasios V. Vasilakos, Jiafu WanEmail author, Jingwei Lu, Dechao Qiu, [11] 17 June 2014 explained "**Security of the Internet of Things**: **perspectives and challenges**" IoT contains three different layers application, transportation and perception layer, explained the security problems in each layer and explains the importance of cross layer heterogeneous integration issues and security issues at different layer of IoT.

(d) George Corser, Ph.D., Assistant Professor of Computer Science and Information Systems [12]. Saginaw Valley State University, May 2017. "**Internet of Things (IoT) Security Best Practices**" It explains about three key areas "securing devices", "securing network" and "secure the overall IoT System". In the "Securing Network", it is evident that following best practices are also followed in Sect. 2 (e.g., Use Strong Authentication, Use Strong Encryption and Protocol, authorization and role based access).

(e) Reliability Management of Data in IoT
The Reliability model for IoT systems represents a clear picture of the data functional interdependencies providing a means to trade-off in IoT data design

alternatives and to identify areas for IoT design improvement of databases. The reliability models for IoT are also helpful in:

i. Identifying of critical items and single points of failure of data
ii. Allocating reliability goals to portions of the design of database and IoT
iii. Providing a framework for comparing estimated reliability of IoT methods
iv. Trading-off alternative fault tolerance approaches for IoT approaches.

Reliability of systems is presented by Chinnaiyan and Somasundaram [13–18]. The authors evaluated the reliability of systems with novel methods and ensured that the reliability models provide optimized reliability results. The same can be applied to the data management of IoT devices and for providing reliable and optimized results.

4 How This Is Relevant to This Problem

The device level security, network level security is independently handled by different products or tools. The data exchange from central data layer (e.g., gateway server or edge server or IoT Hub) requires a sophisticated security control to simplify the data security with central authentication and policy and audit. It will not only solve the issues of controlling central data policy but also achieves the consistent control and visibility of who is accessing what data through audit engine.

5 High Level Approaches

Below figure illustrates how central data security management to be done at IoT Layers (Fig. 2).

- Device will be registered along with its client ID or source IP address given at IoT Gateway. The respective MAC Address or UUID or device id will be registered in the device identity management module.
- Users at different IoT Layers (e.g., Gateway or Edge or IoT Hub) will try to access their data or API or Invoke a subscription call to respective message broker.
- User will be authenticated based on the authentication mechanism provisioned during the registration (e.g., what gateway allowed to access using a MQTT based subscription or direct REST API from gateway). Users will be authenticated using token based messages or certificates or LDAP access or Active Directory authentication model.
- Data policy engine will allow the users to create the policy based on the business rules of the IoT data access model (e.g. PUT or GET) the policy will be created for different user roles or device level or gateway level or edge server levels.
- Cross policy configuration between two different gateways or two different edge servers is also taken care with a separate policy map.
- Encryption and masking will be configured during the business rule creation.

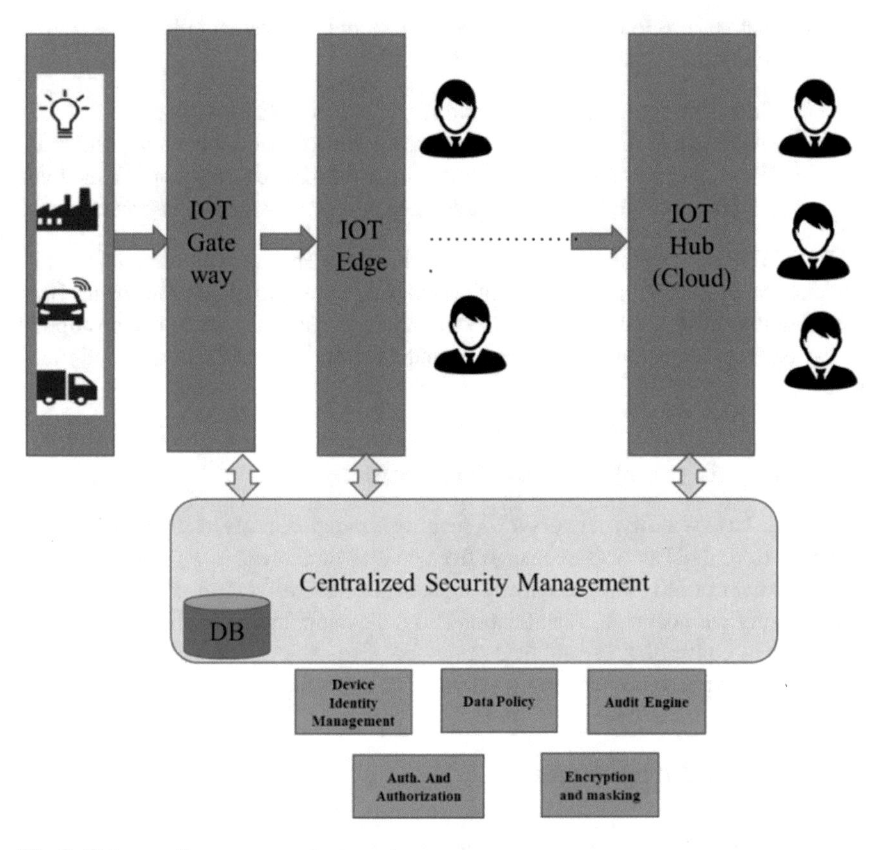

Fig. 2 Data security management approach

- Audit engine helps users to create audit reports for their transactions or audit admin can generate the complete system audit logs from the database (Audit DB).
- The audit DB can be a lightweight DB and it can be deployed using PostgreSQL or MySQL.

Business Rules

Rule name	Details	Access	Mask	Encrypt
Motor1	http://9.9.9.8:6889/res/fulcrum/device/gcc77f91-288f-427d-9d5d-49ff1693e3d9	READ	Loc_id	Gateway_id

Policy Details

Policy name	Rule name	Action	Filter data
Gateway access	Motor1	Allow	–
	Camera1	Allow	C2052 (camera ID)
	Humidity_Sensor2	Deny	

User Policy Mapping

User name	Policy name	Action	Active
User1	Gateway access	Allow	A
User1	Edge1 access	Deny	N
User1	Edge2 access	Allow	A

Sample message from** IoT **gateway: http://1.1.1.1:9999/res/gw/


```
{
"fields": {},
"attributes": {
"TEST": "LG",
"runtime_energy_preset": "00",
"host_address": "8.8.1.1",
"site_description": "BalaLab",
"City_State": "BLR_KA",
"post_code": "560102",
"image_url": "/tmp/image/gn.jpg",
"Address": "",
"Gateway_Manufacturer": "XX",
"Gateway_Model": "304"
},
"schema_version": 2,
"remote": false,
"last_mod_date": 84611785093,
"created_date": 3233795988,
"id": "9212ed19-f27d-32d4-8276-0f9d837feb42",
"name": "BalaLab",
"description": "IOT GW",
"vendor_id": 9999,
"firmware_revision": "3.0",
}
```

6 Conclusion

The above problem statement is not generic to all the IoT devices, some of the devices may not communicate through gateway and it might not be accessible through REST APIs however the above approach will solve the following key issues in IoT Security today.

- Policy-based data access with encryption and masking
- User level access to heterogenous IoT gateway or IoT Edge Server
- Portable security rules and authentication for different authentication services (e.g., Windows Active Directory to LDAP or HTTP/REST based)
- User interface design is not limited to above model and it can be extended different user levels or functionality

It doesn't immediately comply to any InfoSec certification standard (CEH—Certified Ethical Hacking or CISM: Certified Information Security Manager or Security+ or CISSP: Certified Information Systems Security Professional or GSEC: SANS GIAC Security Essentials).

References

1. https://www.sciencedirect.com/science/article/pii/S2352864817300214#s0310
2. https://blog.lftechnology.com/internet-of-things-iot-overview-a38eabbd0c99
3. Toast of the IoT: The 1990 Interop Internet Toaster. http://ieeexplore.ieee.org/document/7786805/
4. That 'Internet of Things' Thing published in RFID Journal article number 4986. http://www.rfidjournal.com/articles/view?4986
5. https://www.ericsson.com/assets/local/publications/white-papers/wp-iot-security-february-2017.pdf
6. https://dzone.com/articles/understanding-iot-security-part-3-of-3-iot-security
7. https://www.pubnub.com/static/papers/IoT_Security_Whitepaper_Final.pdf
8. https://docs.aws.amazon.com/iot/latest/developerguide/iot-security-identity.html
9. Liu, X., Zhao, M., Li, S., Zhang, F., Trappe, W.: Researched on June 2017 "A Security Framework for the Internet of Things in Future Internet Architecture"
10. Zhou, W.: June 2014 mentioned in his research paper about "Security/privacy of wearable fitness tracking IoT devices"
11. Jing, Q., Vasilakos, A.V., Wan, J., Lu, J., Qiu, D.: Security of the Internet of Things: Perspectives and Challenges
12. Internet of Things (IoT) Security Best Practices, May 2017 George Corser, Ph.D., Assistant Professor of Computer Science and Information Systems. Saginaw Valley State University. Co-authors: Glenn A. Fink, Ph.D., Cyber Security Researcher, Secure Cyber Systems Group, Pacific Northwest National Laboratory, Mohammed Aledhari, Doctoral Associate, Center for High Performance Computing and Big Data, Western Michigan University, Jared Bielby, Independent Consultant; Co-chair, International Center for Information Ethics, Rajesh Nighot, Independent Consultant, Sukanya Mandal Nagender Aneja, M. Engg. in Computer Technology and Applications, Universiti Brunei Darussalam Chris Hrivnak, Management Consultant, Marketing, Lucian Cristache, Lucomm Technologies
13. Chinnaiyan, R., Somasundaram, S.: Reliability assessment of component based software systems using test suite—a review. J. Comput. Appl. **1**(4) (2008)

14. Chinnaiyan, R., Somasundaram, S.: An experimental study on reliability estimation of GNU compiler components—a review. Int. J. Comput. Appl. (0975–8887) **25**(3) (2011)
15. Chinnaiyan, R., Somasundaram, S.: Reliability of component based software with similar software components—a review. J. Softw. Eng. **5**(2) (2010) (i-Manager)
16. Chinnaiyan, R., Somasundaram, S.: Monte Carlo simulation for reliability assessment of component based software systems. J. Softw. Eng. (2010) (i-Manager)
17. Chinnaiyan, R., Somasundaram, S.: Evaluating the reliability of component based software systems. Int. J. Qual. Reliab. Manag. **27**(1), 78–88 (2010)
18. Chinnaiyan, R., Somasundaram, S.: Reliability of object oriented software systems using communication variables—a review. Int. J. Softw. Eng. **2**(2), 87–96 (2009)

S. Balachandar is working as a Application Delivery Lead - Technology, Shell India Private Limited, Bengaluru, Karnataka, India. He is having 17 years of industrial experience in the areas of Big Data and Cloud Computing. Now he is doing his Ph.D. in Computer Applications in VTU. His research interest includes Security, Reliability, IOT, Machine Learning, Big Data, Cloud Computing nd Data Science.

Dr. R. Chinnaiyan is working as Professor in the Department of Computer Applications, New Horizon College of Engineering, Marathalli, Bengaluru, Karnataka, India. He is having 17+ years of teaching experience. He is a life member of ISTE and CSI of India. He completed his research in Anna University-Chennai at Coimbatore Institute of Technology, Coimbatore in 2012. His research interest includes Object Oriented Analysis and Design, Qos and Software Reliability.

User Location-Based Adaptive Resource Allocation for ICI Mitigation in MIMO-OFDMA

Suneeta V. Budihal, Beena Kumari and V. S. Saroja

Abstract The paper proposes a user location-based dynamic radio resource allocation in MIMO-OFDMA systems to mitigate Inter-Cell Interference (ICI) using evidence theory. In order to utilize the limited radio resources like subcarrier and its associated power efficiently, it is needed to develop a scheme. In cellular communication, for better utilization of resources, if the radio resources are repeatedly used in the neighboring cells, it leads to ICI. In order to address this issue, a two-step resource allocation technique is devised. In the first step, the evidence of the user in a cell is calculated in order to categorize the users as Cell Edge Users (CEU) and Cell Center Users (CCU) and in the second step, the dynamic allocation of radio resources is carried out. The criterion for resource allocation is the frequency of occurrence of a user evidence in a cell. It increases the datarate and efficiency of the system.

Keywords Radio resource management · Evidence theory · ICI · Subcarrier

1 Introduction

Future-generation wireless communication systems sustain wireless Internet access that demands high datarate and tangled designs. Communication across wideband channels with high datarate are crucially limited by ICI due to the time-dispersive or frequency-selective essence of channels. In multiuser systems like cellular systems, ICI is experienced by user for which numerous copies of transmitted signal are created by surrounding objects (such as building, cars, etc.). To counter ICI, techniques such as multicarrier modulation, including Orthogonal Frequency-Division Multiplexing (OFDM) are among the conceivable arrangements that are recommended. OFDM

S. V. Budihal (✉) · B. Kumari · V. S. Saroja
School of ECE, KLETU, Hubli 580031, India
e-mail: suneeta_vb@bvb.edu
URL: http://www.bvb.edu

V. S. Saroja
e-mail: sarojavs@bvb.edu

S. Smys et al. (eds.), *International Conference on Computer Networks and Communication Technologies*, Lecture Notes on Data Engineering and Communications Technologies 15, https://doi.org/10.1007/978-981-10-8681-6_20

separates a broadband channel into limited subcarrier such that the response of a channel on a specific subcarrier appears to be flat. Adding a guard band or Cyclic Prefix (CP), whose length is equal to the dispersion time of the channel, and to transmitted symbols, builds each of the subcarriers parallel, independent, and additive white Gaussian noise channels. This arrangement permits the received signal to be ISI free.

Development of telecommunication from 1990s to the present generation techniques in telecommunication is discussed in [1]. It mainly focuses on implementation, application, advantages, disadvantages, and cause of failure to meet the quality of service. In paper [2], the authors deal with characteristic features, architecture, RF and Hardware considerations, constraints, and trial results of initial implementation fourth-generation MIMO-OFDMA broadband wireless communication. The authors in reference [3] discuss various subcarrier allocations, power allocations, budgets, constraints, and complexities of algorithms described by various authors. To defeat restrictions of existing methodologies talked about, it investigates the analytic approach and algorithmic advancement of subcarrier, rate, power allocation, and scheduling problems for multiuser MIMO-OFDMA systems. It also shows that greedy water-filling algorithm is the best approach that a number of users and subcarriers are with linear complexity, assuming that the random fading channel is continuous in the distribution function.

In paper [4], the authors use co-operative game theory to study cooperation among the node pairs found on the Nash bargaining solution. Using Hungarian method and user negotiation scheme, multiple users' partner selection systems are designed. In the first step, user pairs are made in which coalition of two co-operating nodes is formed. Then, using Hungarian method, the coalition assignment problem is resolved. All pairs adopt fair strategy of co-operative gain distribution for each user bargains with respective partners and all nodes are satisfied instantly. Then, co-operative game theory is applied to formulate allocation of subcarriers with guaranteed fairness and increase in secrecy rate, without increment in total subcarriers and transmission power.

In [5], solution for joint optimization problem is worked out using combined scheme of MIMO-OFDMA for wireless broadband communication. In power allocation, an effective water-filling algorithms are used. Subcarrier allocation is carried out in two steps. Initially, every user is assigned with minimum amount of subcarriers. In the second step, subcarrier reallocation is carried out iteratively for each user to satisfy user datarate constraints using cost function. The user datarate is directly affected by the cost function resulting in reduced overall throughput of the system. In [6], MIMO-OFDMA resource allocation is formulated using Modified Multidimensional Genetic (MMDG) algorithm. To demonstrate the effectiveness of algorithm, four different operation scenarios are studied to meet the system requirements. Equal number of subcarriers and users are considered in first case. In the second case, the amount of subcarriers is double the amount of users. The third case is vice versa of the second case with limited resources where time–frequency sharing scheme is designed to provide minimum datarate to each user. Based on the results obtained from three cases, MMDG for multiusers is designed in case four.

The authors in [7] consider both sum capacity and user datarate for designing algorithm for resource allocation. In the first step, each user is allocated with one subcarrier to encounter minimum datarate requirements. Then, subcarriers are reallocated to each to meet datarate requirements. When every user is satisfied, the power allocation is carried out by effective water-filling algorithm. In [8], first the subcarrier allocation algorithms assign the subcarrier to the users. The second step concentrates on subcarrier allocation based on the multiuser water-filling theorem. This magnifies the total amount of bit rate under the restriction of a maximum transmission power per user. In [9], joint optimization of resources in MIMO-OFDMA is achieved using iterative method of subcarrier and the power allocation to meet adaptive user datarate requirements. An efficient Lagrange dual method is used for resource allocation algorithm to decrease the transmit power and keep the secrecy datarate of users.

The authors in [10, 11] propose Basic Particle Swarm optimization (BPSO) and Genetic Algorithm (GA) for subcarrier and bit allocation. BPSO is simple but in the case of complicated and multi-peaks issues, local optima occur. The rate of convergence of GA is slow as well as solution is delicate to selection of initial population. In [12], a combination of BPSO and GA is proposed which has better performance. The paper [13] deals with survey on swarm intelligence and artificial bee colony scenario which can be applied to resolve any real-world problem. In [14], a joint power and subcarrier algorithm is proposed using water-filling algorithm. Subcarriers are allocated in the first step with adaptive rate constraint. Then, the power of each subcarrier is obtained by water-filling algorithm, and the user datarate is calculated. The remaining resources are distributed according to required user datarate.

In [15], a low-complexity suboptimal algorithm is proposed, to distribute power and subcarrier separately. In this type of algorithm, a subcarrier allocation is executed by presuming uniform power distribution, and then best subcarrier efficiency function is assigned to the user. Then, following subcarrier allocation, they utilize the finest and linear water filling for allocation of power. It was shown that the algorithm is higher level to the conventional power allocation that reduces the complications from exponential to linear for the number of subcarriers. A limited amount of resources, i.e., power and bandwidth, are available at the transmitter. Rate adaption achieves higher goodput by combating channel variability, which is common to all wireless communication system due to factors such as fading mobility and multiuser interference due to number of user increases system also scalable. In some of the paper, they use complex algorithm for cellular wireless system which is hard to understand and implement. The basic problem of resource allocation is the scarcity of available spectrum, the extensive servicing area, as well as large number of users.

The above algorithms and methodologies adopted improved the performance and capacity of the network. But the improvement in parameters of edge users is average as compared to the center users. The following are the contribution of paper. In this paper, the authors

1. Proposed an algorithm to dynamically allocate the radio resources to edge and center users in LTE downlink system.

In this paper, two-step subcarrier and power allocation are designed to achieve maximum datarate for each user based on location. Section 2 deals with the proposed system model. The simulation results are discussed in Sect. 3. The conclusion is discussed in Sect. 4.

2 System Model

In this paper, evidence of the user is taken as a parameter to find the user location in a cell for further categorization as discussed in [16]. The clustering of users into center and edge is done on the basis that fewer users are considered as edge users. This helps in management of resources. Consider a cell with K total users and S subcarriers in an area of ASq. Mts. Let P_t be the total power of the antenna in watts. Let m and n be the number of transmitters and receivers in the cell. Datarate of a user is expressed as

$$P_i = B_i \log 10\left(1 + \frac{\dot{P}_t}{S * t_i}\right) * m \tag{1}$$

where B_i is the user bandwidth and t_i is the symbol rate. SINR is the ratio of signal to interference plus noise ratio.

2.1 Frequency Planning

The total subcarriers are classified into four sectors. The center users of all the cells are allocated with same frequency band c. The remaining three sectors E1, E2, and E3 are allocated to edge users in such a way that no ICI occurs. Figure 1 shows the seven cell structures with predefined frequency bands for center and edge users. The marginal value between center and edge user is average of summation of distance between the user and the base station of the cell and the distance between the user and the base station of neighboring cell. With the marginal value, the user is termed as edge or center.

$$\frac{\sum_{i=1}^{n} \sum_{i=1}^{2} U_{in} - U_{jn}}{N} \tag{2}$$

where U_{in} is the distance between the user and base station of the respective cell and U_{jn} is the distance between the user and base station of the neighboring cell.

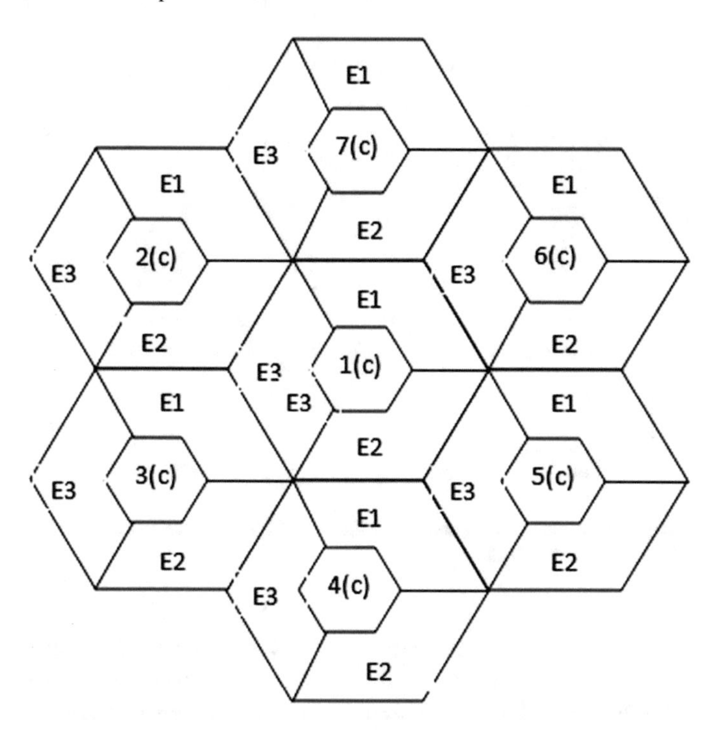

Fig. 1 Frequency spectrum partitioning as center and edge spectrum in seven cell layouts to avoid ICI for MIMO-OFDMA system

Consider two cells that use the same frequency bands and each cell having few User Terminals (UTs) that use the same frequency bands. The UTs who are at the center tend to communicate with higher power as they are nearer to the base station. The UTs which are away from the center and nearing edges communicate with lesser power as the transmitted signal undergoes degradation due to path loss and fading effects. In Fig. 2, A1 is a Cell Edge User (CEU) in cell A and it is allocated with certain resources which are similar to resources allocated to B1, which is a CEU in cell B. Both the cell controllers are not aware of the frequency bands allocated to one another. Unfortunately, the frequency band allocated to both CEUs is same, and hence, the transmitted signals of both the cells tend to interfere with each other, thereby causing the degradation in the received signal. This degradation leads to an overall decrease in the throughput of the sector.

2.2 *Dynamic Subcarrier Allocation*

The allocation of subcarriers is brought out in two steps. Initial allotment of subcarriers is brought out in the first step. Each user is allocated with minimum two number of subcarriers to achieve marginal datarate and SINR. In the second step, based on

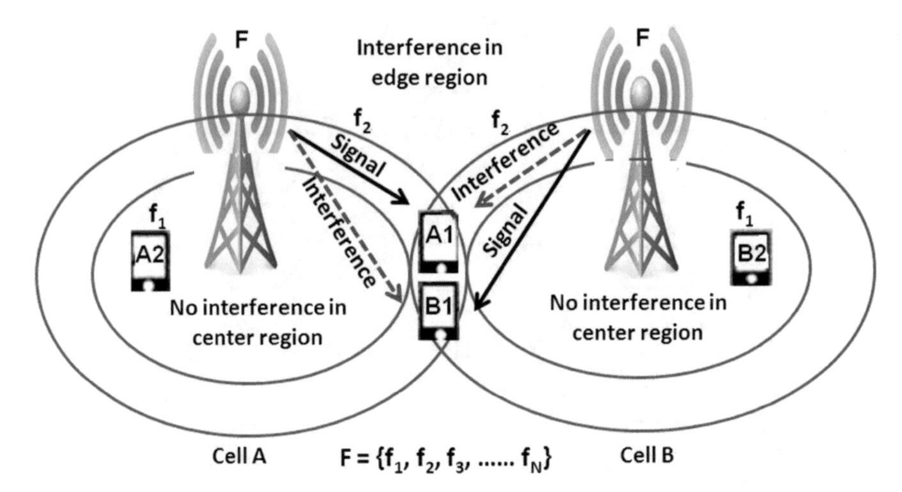

Fig. 2 A scenario depicting ICI experienced by the UTs A1 and B1 sharing the common spectrum f_2 at edge of the cell A and cell B

user frequency of evidence and the percentage of center and edge user of total number of users, available subcarriers are divided into center and edge bandwidth. High datarate and minimum SINR are achieved by the above scheme. In initial allocation of resources, each user is given minimum two subcarriers:

$$s_i = \frac{e \cdot f_i}{\sum_{i=1}^{N} (e_i \cdot f_i)} \cdot s \tag{3}$$

This expression gives the final subcarriers to be allocated to each user with reference to the evidence.

2.3 The Proposed Algorithm

The radio resources are allocated as per the discussed algorithm effectively for the CEU and CCU.

1. Start
2. Read the user data and the respective evidences.
3. Calculate P_t and S given in Eqs. 3 and 4.
4. Calculate power of each P_i using the formula analyzed in Eq. 5.
5. Each user is provided with s_i of subcarriers.
6. Calculate initial D_i and $SINR_i$.
7. Calculate s_e of each user using Eq. 2.
8. Calculate D_f and $SINR_f$ after final allotment of resources.
9. If there is any change in data, the process is repeated.

2.4 Power Allocation

P_t of the antenna in terms of gain of antennae is expressed as

$$P_t = \frac{10^{\frac{P}{10}}}{100} \quad (4)$$

where P is the total gain of the antenna. Each subcarrier power P_i is expressed as

$$P_i = 10\log\left(\frac{\dot{P}_t - P_L}{S}\right) \quad (5)$$

where P_l represents the channel path loss.

3 Simulation Results

The results are obtained using MATLAB simulation software. The MIMO-OFDMA system is analyzed with the following specifications of cell as shown in Table 1. Figure 3 represents the computed evidences of all users in a cell considering the evidence theory [16]. The larger value of evidence indicates that the user is away from the center and is considered as the CEU else CCU. Figure 4 depicts the plot of frequency of occurrence of individual evidences in a cell. Also the frequency distribution of evidences is used to calculate the margin between center and edge users. The margin value approaches a value of 29.88% of edge users. Initially, each user is allocated with minimum two subcarriers to achieve marginal datarate. The total datarate achieved is 150 Mbps and individual SINR of 0.0008. The initial allocation is depicted in Fig. 5. Figure 6 represents the datarate and SINR achieved after initial allocation of resources. Figure 7 represents final subcarrier and power allocation distributed on percentage of frequency distribution and center and edge user ratio. There is a considerable increase in datarate with increment of 148 Mbps. Figure 8 represents the datarate and SINR achieved after final allocation of resources. Total user datarate achieved is 298 Mbps closer to the practical downlink datarate value of 300 Mbps and optimal individual SINR value is achieved.

Table 1 MIMO-OFDMA system parameters considered for simulation

Parameter	Specification
N	100 per cell
P_t	43 dBm
Cell radius	720 m
Carrier frequency	20 MHz
Sise of antenna (Tx * Rx)	2 × 2

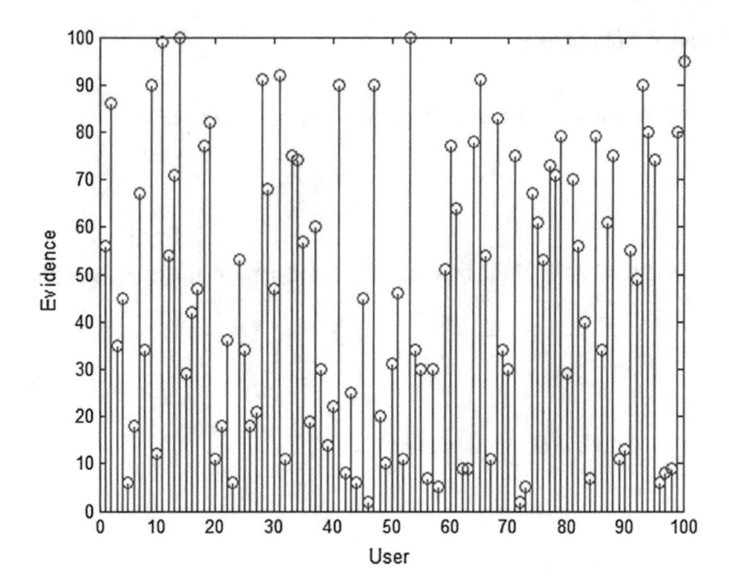

Fig. 3 Graph represents the evidence of each users

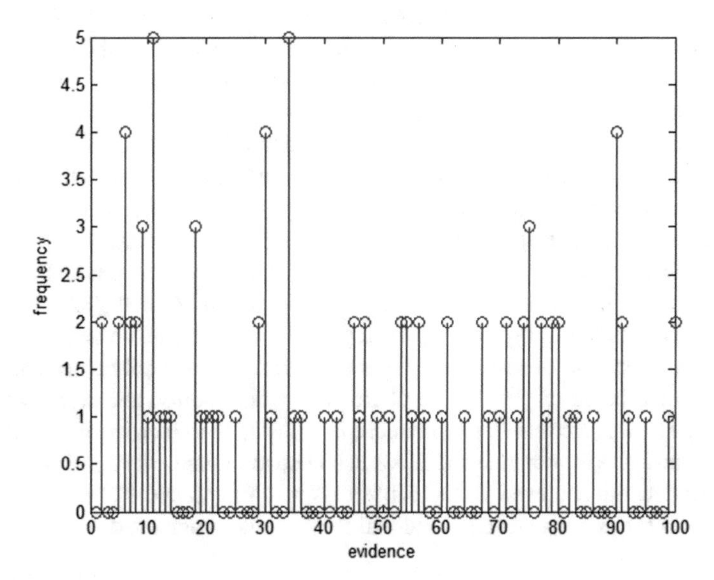

Fig. 4 Graph represents the frequency of each evidence

Figure 9 represents a plot of comparison of evidence and resource allocation versus users. It is clear from the graph that there is less increment in datarate of center users with minimum channel fading and edge users are allocated with enough resources to achieve required datarate with increase in channel fading.

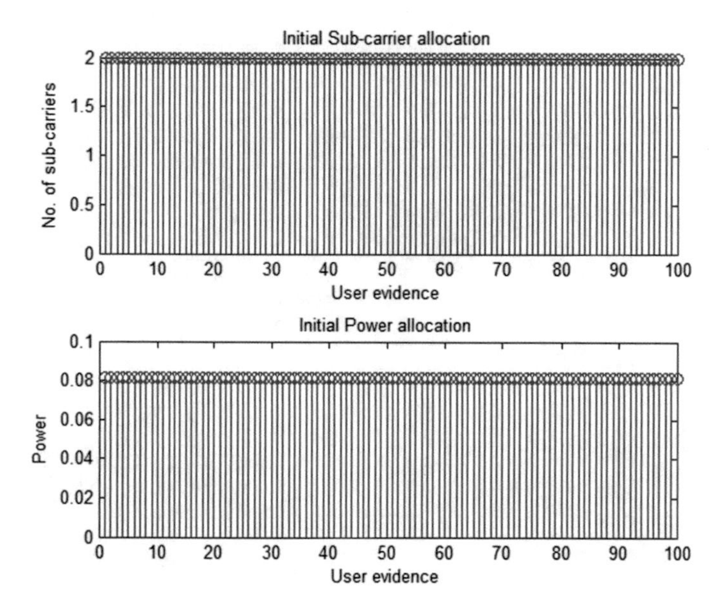

Fig. 5 Graph represents initial allotment of resources

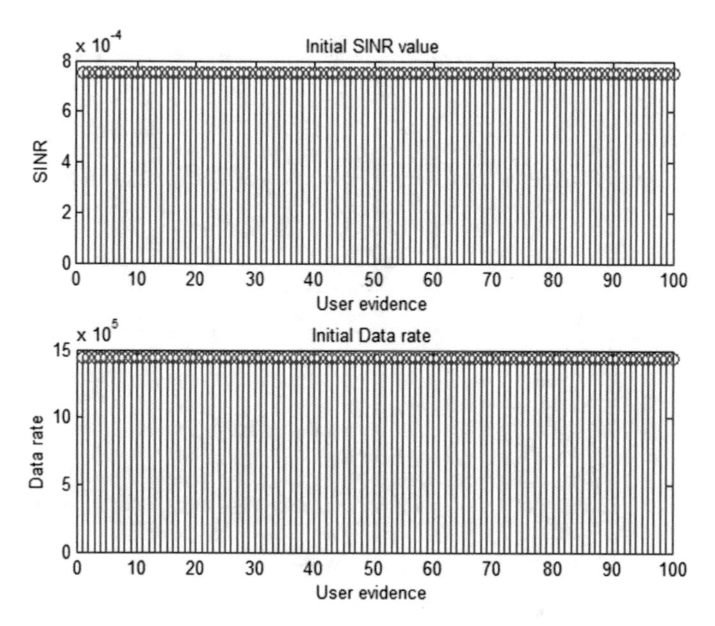

Fig. 6 Graph represents datarate and SINR achieved after initial allotment of resources

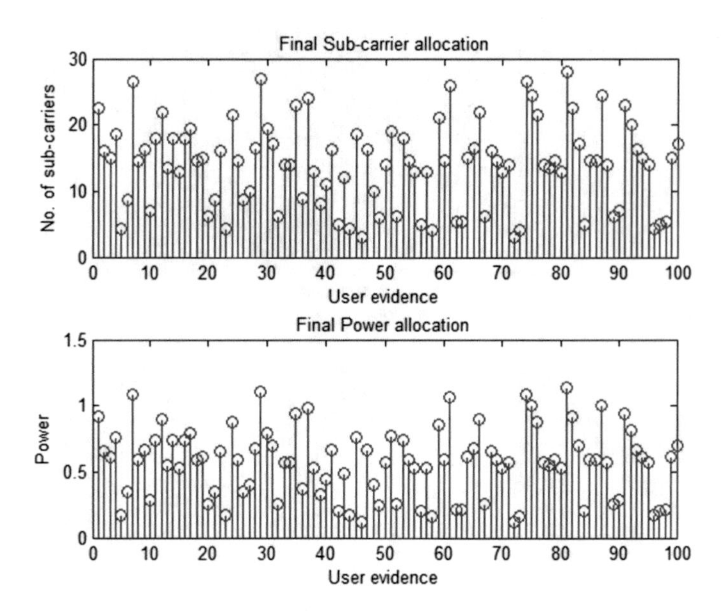

Fig. 7 Graph represents final allotment of resources

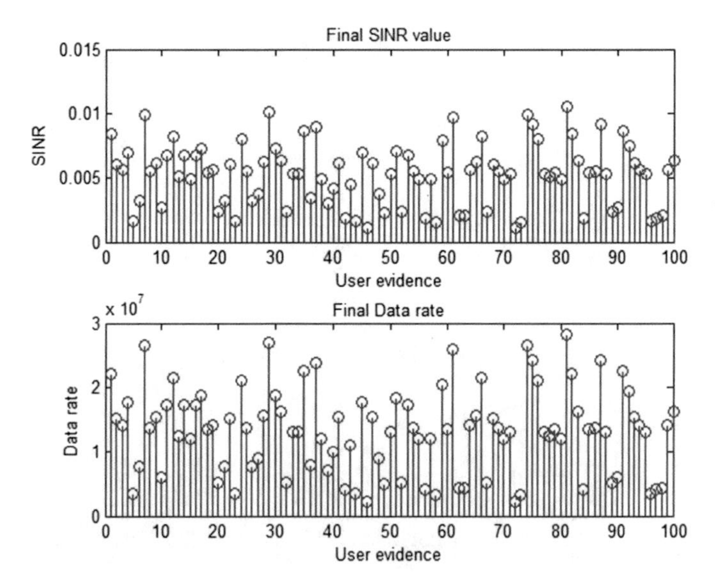

Fig. 8 Graph represents datarate and SINR achieved after final allotment of resources

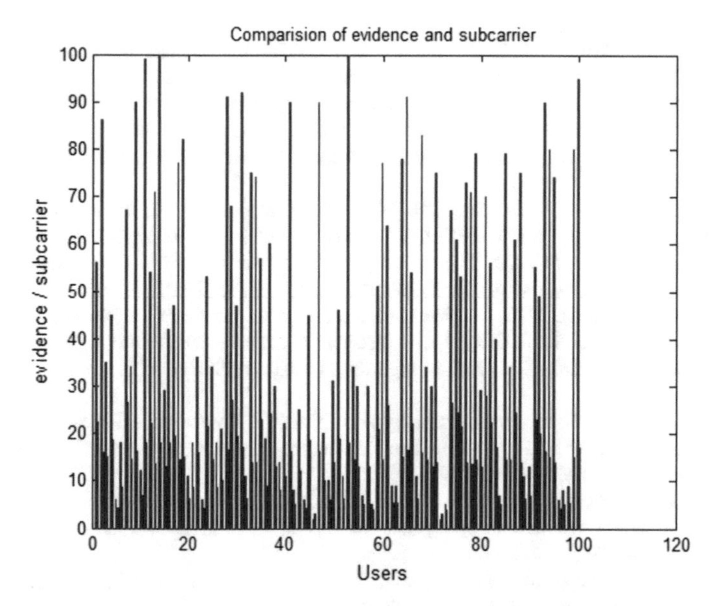

Fig. 9 Graph represents the comparison of evidence and resource allocation of each user

4 Conclusion

The users are classified into center and edge users using the frequency of occurrence of evidences. Based on the population ratio, resources are classified into center and edge which helps in management of resources. After final allocation of resources, maximum datarate is achieved based on user location using statistical analysis. It is an efficient method of resource categorization and allocation that enhances the achievable datarate of users.

References

1. Wang, X., Giannakis, G.B.: Resource allocation for wireless multiuser OFDM networks. IEEE Trans. Inf. Theory **57**(7), 4359–4372 (2011)
2. Rappaport, T.S., Annamalai, A., Buehrer, R.M., Tranter, W.H.: Wireless Communications: Past Events and a Future Perspective, May 2002
3. Zhang, R., Ho, C.K.: MIMO broadcasting for simultaneous wireless information and power transfer. IEEE Trans. Wirel. Commun. **12**(5), 1989–2001 (2011)
4. Huang, S., Jing, A., Tan, J., Xu, J.: Subcarrier allocation and cooperative partner selection based on nash bargaining game for physical layer security in OFDM wireless networks. In: Concurrency and Computation: Practice and Experience, vol. 29, No. 3. Wiley Online Library, Jan 2017
5. Patel, H., Gandhi, S.D., Vyas, D.: A research on spectrum allocation using optimal power in downlink wireless system. Int. Res. J. Eng. Technol. **3**(4), Apr 2016

6. Ibrahim, M.K., AlSabbagh, H.M.: Adaptive OFDMA resource allocation using modified multi-dimension genetic algorithm. J. Electr. Electron. Eng. **12**(1), Dec 2016
7. Zhao, W., Wang, S.: Joint subchannel and power allocation in multiuser OFDM systems with minimal rate constraints. Int. J. Commun. Syst. **27**(1), 1–12 (2014)
8. Jang, J., Lee, K.B.: Transmit power adaptation for multiuser OFDM systems. Int. J. Commun. Syst. **21**(2), 171–178 (2003)
9. Wong, C.Y., Cheng, R.S., Letaief, K.B., Murch, R.D.: Multiuser OFDM with adaptive subcarrier, bit and power allocation. IEEE J. Sel. Areas Commun. **17**(10), 1747–1758 (1999)
10. Jinfeng, J.F.Y.: Adaptive resource allocation in multiuser OFDM system based on particle swarm optimization. J. Jisuanji Yingyong yu Ruanjian **28**(4) (2011)
11. Wang, Y.X., Chen, F.J., We, G.: Resource allocation for multiuser OFDM system based on genetic algorithm. J. South China Univ. Technol. (Natural Science Edition) (2011)
12. Yao, J.K., Li, F.F., Liu, X.Y.: Hybrid algorithm based on PSO and GA. Comput. Eng. Appl. (2014)
13. Rhee, W., Cioffi, J.M.: Increase in capacity of multiuser OFDM system using dynamic subchannel allocation. In: Vehicular Technology Conference Proceedings, 2000. VTC 2000-Spring, Tokyo, vol. 2, pp. 1085–1089, May 2000
14. Sampath, H., Talwar, S., Tellado, J., Erceg, V., Paulraj, A.: A Fourth generation MIMO-OFDM broadband wireless system: design, performance, and field trial results. IEEE Commun. Mag. **40**(9), 143–149 (Sep. 2002)
15. Xiang, Z., Tao, M.: Robust beamforming for wireless information and power transmission. IEEE Wirel. Commun. Lett. **1**(4), 372–375 (2012)
16. Budihal, S.V., Saroja, S., Sneha, V.S.S., Banakar, R.M.: Framework for dynamic resource allocation to avoid intercell interference by evidence theory. In: Springer Book Chapter in Progress in Intelligent Computing Techniques: Theory, Practice, and Applications, pp. 391–402, Sept 2016

Suneeta V. Budihal is from India and received B.E degree from Karnataka University in Electronics and Communication, from Karnataka University, Dharwad in 1997 and M.E. degree in Digital Electronics in 1999. She is pursuing her PhD in cooperative communication. She is a life member of ISTE. Presently she is serving as associate professor at KLE Technological University, Hubli. She has 25 publications in national and international conferences. Her research areas include communications, MIMO, wireless communications etc. Professor may be reached at, suneeta_vb@bvb.edu.

V. S. Saroja received the B. Tech. degree in Electronics and Communication Engineering, and M.E in Digital Electronics from Karnataka University, Dharwad in 1995 and 1997. She received Ph.D degree from Department of Electronics and Communication, the Jawaharlal Nehru Technology University at India in 2013. Presently she is serving as Professor in School of Electronics and Communication at KLE Technological University , Hubli. Her research interests include VLSI Signal Processing which includes Architectural design and BIST Testing. She advises doctoral students at Vishweshwaraya Technological University, Belgaum and Bangalore. She is faculty advisor for E-Baja since 2015, ESVC 2014–2016, HVVC-2015-2016 events. She has authored or co-authored over 25 technical papers in archival journals and refereed international conferences. She is the reviewer of many international conferences. She received best poster award in 2015 International Conference on Transformations in Engineering Education. She is life member of IETE and ISTE.

An Approach to URL Filtering in SDN

K. Archana Janani, V. Vetriselvi, Ranjani Parthasarathi and G. Subrahmanya VRK Rao

Abstract Phishing is considered as a form of the Internet crime. To detect a phishing website, human experts compare the claimed identity of the website with the features of the website along with its content. Every website URL has its own lexical features like length, domain names, etc. The phishing websites may appear to perform the same activities of another website but the content of the two websites will be different. In traditional networks, a proxy server handles the URL requests and determines whether an URL is malicious or not. In this paper, URL filtration is incorporated into an SDN framework as a security application. The proposed system uses deep packet inspection and machine learning techniques at the controller and the rule installation in the switches for efficient URL phishing detection. The phishing system analyzes the lexical and content-based features of the URLs. Based on the categorization, the rules are formed and installed in the switches. The performance of the system is evaluated based on the response time and accuracy in the detection of phishing URLs using a simulation framework.

Keywords Phishing · SDN framework · URL filtration

K. Archana Janani · V. Vetriselvi (✉) · R. Parthasarathi
College of Engineering Guindy, Anna University, Chennai, India
e-mail: vetri@annauniv.edu

K. Archana Janani
e-mail: janani.archana@gmail.com

R. Parthasarathi
e-mail: rp@auist.net

G. Subrahmanya VRK Rao
Cognizant, Chennai, India
e-mail: SubrahmanyaVRK.Rao@cognizant.com

S. Smys et al. (eds.), *International Conference on Computer Networks and Communication Technologies*, Lecture Notes on Data Engineering and Communications Technologies 15, https://doi.org/10.1007/978-981-10-8681-6_21

1 Introduction

Phishing is a web-based attack which allures end users to visit fraudulent websites and give away personal information (e.g., user id, password, etc.). The information is used to perform illegitimate activities. Phishing attacks cost billions of dollars losses to business organizations and end users. The attack jeopardizes the prospects of the e-commerce industries. Therefore, addressing phishing attacks is important. There are two main activities performed by phishers to make an attack successful. They are (i) developing fraudulent websites and (ii) motivating (or urging) users to visit those sites. The phishing attacks mainly happen by tricking the users' with false URLs.

URL attacks can be broadly classified as URL obfuscation with other domains, URL obfuscation with keywords, typosquatting domains or long domains, URL obfuscation with IP address and obfuscation with URL shortened. URL obfuscation with IP address and keywords is a similar kind of attacks in which the domain being phished is a part of the URL, either a path, query or upper-level domain. In attacks by typosquatting domains, the domain in the URL is either misspelled or combined with other targeted brand domains to look like a legitimate URL. URL obfuscation with IP address deals with replacing the domain name of the URL with an IP address and the domain being phished is a part of the URL. Attacks due to the URL shortening services happen to cover the tracks of the real name of the host. These URLs are not meaningful and are used for phishing attacks; they target services like Twitter. In the traditional methods, the phishing websites are detected by the lexical feature analysis of the URLs or webpage layout analysis. To detect all these phishing attacks, supervised classification techniques are used [1].

In this paper, we propose a novel approach of identifying the malicious URLs in the SDN domain. Since SDN controllers [2] are flexible and can be programmed using high-level programming languages, we propose the use of an application running at the controller to provide the defense against phishing attacks. The malicious URLs are found using a machine learning based analysis of the packets at the controllers. The controllers which are responsible for setting up and tearing down flows and paths in the network translate this application's finding into packet forwarding/dropping rules at the switches to prevent phishing attacks from entering the network. We present the details of this SDN-based URL filtration application and an analysis of its performance in this paper.

This paper is organized as follows: Sect. 2 discusses the related work. Section 3 emphasizes the proposed system and the algorithm involved. Section 4 discusses the evaluation metrics to be used for the performance evaluation of the proposed system. Section 5 discusses the conclusion and scope of future work.

2 Related Work

Various solutions have been proposed in the area of URL filtration in the recent past focusing on URL analysis based on content, ranking mechanism, lexical features, etc.

Google's PageRank value can be considered as one of the heuristics to detect whether a page is a legitimate or phished site. Google's PageRank value is robust and updated frequently. Naga Venkata Sunil et al. [3] have proposed a technique using PageRank value as a heuristic and obtained an accuracy of up to 98% in detecting a phished site. As an enhancement of this idea, Feroz and Mangel [4] proposed a system capable of clustering, classifying, categorizing and ranking URLs in real time. URL categorization is done by the Microsoft reputation services (MRS) and put into several bags. Then, by using a novel URL ranking approach, the URLs are ranked and an internal threat scale is derived from them.

Classification of the URLs can be done using predefined blacklist and whitelist of the URLs. Shahriar and Zulkernine [5] attempted to address the URL phishing issues by performing a comprehensive analysis of anti-phishing approaches. They classify related approaches based on employed information sources into five types: whitelisted, blacklisted, hybrid, standalone, and random.

Lee and Kim [6] have proposed WARNINGBIRD, a suspicious URL detection system for Twitter. Because attacker's resources are generally limited and need to be reused, their URL redirect chains usually share the same URLs. Hence, instead of investigating the landing pages of individual URLs in each tweet, which may not be successfully fetched, a method is created to detect correlated URL redirect chains using such frequently shared URLs.

Aburrous and Hossain [7] have assessed phishing sites using fuzzy data mining to detect e-banking phishing websites in an automated manner. Phishing sites can also be detected by the structure and composition of the websites of emails. Basnet et al. [8] employed a few novel input features that can assist in discovering phishing attacks with very limited a prior knowledge about the adversary or the method used to launch a phishing attack.

Analyzing the URL itself to predict whether a website contains undesirable content is another way of achieving phishing detection. Fang et al. [9] proposed a RayScan method to find the webpage image layout. The similarity between the webpage image characteristics is analyzed using distance functions. In addition to it, reverse lookup and domain transformation techniques are also used. Hence, it detects phishing websites before it is widely spread.

The lexical features of the URLs can also be analyzed. Lawrence et al. [10] pursued this approach. In particular, the authors make predictions from the lexical and host-based features of URLs without examining the actual content of web pages. Once automated, the approach attempts to improve on blacklisting while providing a lightweight, low-overhead alternative to systems that download and analyze the full content of web pages. This paper focuses on detecting the phishing websites by combining both the lexical features of the URLs and the content-based features of the websites, and implementing it as an SDN application. Not much work has been done on implementing the URL filtering in SDN. Hence, this paper explores the SDN approach.

3 Proposed System

In the traditional systems, the URLs have been blocked or passed by analyzing whether the URL is malicious or not at a proxy or firewall. In the proposed system, this technique is extended to the SDN environment. The architecture of the proposed system is shown in Fig. 1.

A switch receives all kinds of packets. When an SDN switch receives a packet, it matches the different fields in the packet headers with the rules in its match–action table. If it finds a matching rule, it executes the corresponding action. A packet that does not match any rule is sent to the controller by default. Thus, any new HTTP packet is sent to the controller. The controller extracts the URL requests from the packets using deep packet inspection and uses machine learning techniques to classify the URLs. Based on the analysis, it determines if the URL is malicious or not, and installs appropriate rules at the switches to forward or drop the packets.

The controller analyzes different fields of the received packets using deep packet inspection. For the HTTP packets, the controller opens the packet and extracts the URL from the packet. A modified version of Aho–Corasick algorithm [11] is used to check for the string patterns using regular expression in the packet payload and the required part is extracted. The extracted URL is sent to the phishing detection system. The phishing detection system is separated from the controller to support large networks, where more than one controller may be needed. Any controller can get service from this detection system which simplifies their work and the detection process is executed only once.

In the phishing detection system, the naïve Bayes classifier is trained with the URL dataset. The URL dataset consists of various features of the different URLs. It involves both the lexical- and content-based features of the URLs. The URL is fed into the system. Then, the URL is checked against the blacklist URL database. If the URL is already present in the database, then it is directly classified as phishing without further processing. If it is not present, then the URL is suspected for an attack and appropriately trained classifier is selected. Then, the lexical- and content-based features are extracted.

Lexical features involve the syntax and source-based features of the URL. The syntax features include the syntactical properties of the URL like length, number of domains, the presence of symbols, etc.

The algorithm for syntax-based features extraction is shown in Fig. 2. A rule-based algorithm is used to extract these features. The source-based analysis uses the information from the third-party websites such as Alexa [12]. The Alexa website stores the traffic information of all the websites. Domain knowledge and the certificate are gathered from the WHOIS database. This database provides the latest information

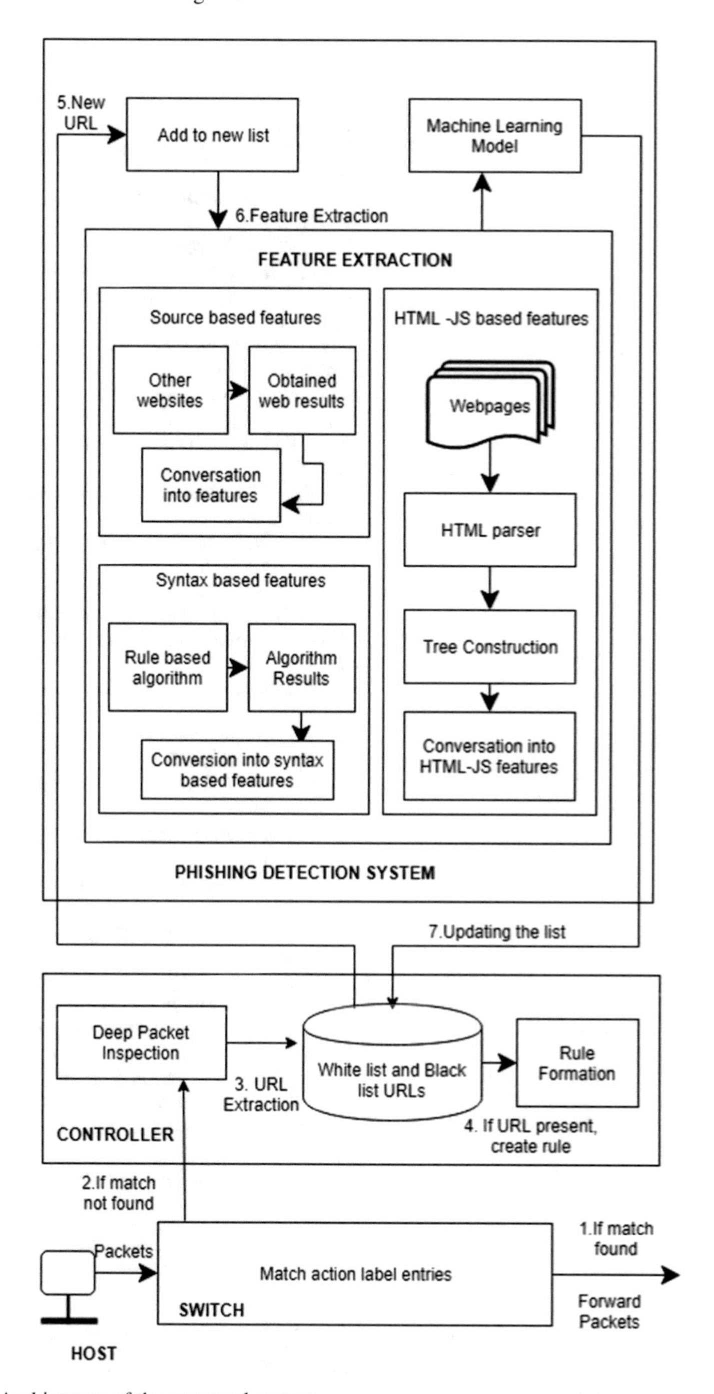

Fig. 1 Architecture of the proposed system

Fig. 2 Algorithm for syntax-based feature extraction

Input: Valid URL
Output: Syntax based features
Algorithm:
Step 1: For the URL, extract the following features.
Step 2: Check for the @ ,(-),// symbols and HTTPS token.
Step 3: Check for the HTTPS token.
Step 4: Check for the length of the URL.
Step 4(a): If length is less than 54, then URL is legitimate.
Step 4(b): If length is between 54 and 75, then URL is suspicious.
Step 4(c): Otherwise it is phishing.
Step 5: Check for the IP address.
Step 6: Check for the number of dots in domain part of U.
Step 6(a): If the number of dots is one, then the URL is legitimate.
Step 6(b): If the number of dots is two, then the URL is suspicious.
Step 6(c): Otherwise it is phishing.
Step 7: Convert the rules into array of features.
Step 8: Write the features in binary form into file.

Fig. 3 Algorithm for source-based feature extraction

Input: Valid URL
Output: Rank and traffic based features
Algorithm:
Step 1: For the URL, extract the following features.
Step 2: Establish a connection to the website.
Step 3: Retrieve the traffic from the Alexa website.
Step 4: Check the DNS record, domain registration length and age of certificate.
Step 5: Check for domain name using Whois client app.
Step 6: Get the website statistics from the Phish Tank website.
Step 7: Convert the rules into array of features.
Step 8: Write the features in binary form to a file.

about the domain age, the expiry dates of the certificates, etc. These features differ for the legitimate and phishing websites. The algorithm for source-based features extraction is shown in Fig. 3.

Content-based features include the HTML–JS features, and the algorithm for content-based feature analysis is shown in Fig. 4. Every webpage has a structure

Input: Valid URL
Output: DOM tree
Algorithm:
Step 1: For the URL, extract the following features.
Step 2: Parse the HTML page using a parser.
Step 3: Find the tags and construct DOM tree.
Step 4: Manipulate the HTML elements, attributes and text.
Step 4(a): If the onmouseover/event is present, then it is phishing.
Step 4(b): If the right click is disabled, the URL is phishing.
Step 4(c): If there is an invisible iframe border present, then the website is phishing.
Step 4(d): If the anchor does not link to any webpage then it is considered phishing.
Step 5: Check the number of redirections in the website.
Step 5(a): If redirection is less than or equal to 1, then it is legitimate.
Step 5(b): If count is greater than 2 and less than 4, then it is suspicious.
Step 5(c): If count is greater than 4, then it is phishing.
Step 6: Check if the URL has a shortener.
Step 7: Convert rules into tag based features.
Step 8: Write the features in binary form to the file.

Fig. 4 Algorithm for HTML–JS-based feature extraction

and content. This structure is analyzed by constructing a DOM tree for a webpage. For constructing the DOM tree, an HTML parser is used. The DOM tree shows the hierarchy of the webpage elements. We traverse the tree and extract the required webpage tags. These features are combined and written into a file. Then, the file is supplied as a test file to the classifier. If the input URL is classified as phishing, then it is inserted into the database and the controller creates a flow rule with respect to the packet fields and URL. The rule is installed in the switch by the controller, and the request packet is dropped. If it is classified as legitimate, then the controller forwards the response to the switch. The switch forwards the packet to the requested host.

The system ensures that there is no duplication in processing the malicious URLs. The processing time is saved by using match–action tables in the switches and the database installed in the controller. This way the workload is distributed among the switches and controller in handling the URL requests. Different switches may have different rules depending upon the requests done by the hosts.

All the hosts connected to a particular switch will have the same restrictions of accessing the URLs. In this way, we separate the users into different groups to access specific domain of URLs in addition to phishing URLs.

In a proxy-based networking, the detection system is one service among many other functionalities. This will increase the delay in identifying the phishing URL. Our architecture takes the advantage of SDN by installing new rules or updating the existing rules in the switches, which helps to identify the phishing URLs at the edge of the customer itself. This might increase the performances of SDN-based networks in terms of latency and network congestion.

4 Performance Evaluation

This system has been studied and evaluated using a simulation environment consisting of Mininet simulator [13], open network operating system (ONOS) SDN controller [14] and SCAPY [15] for traffic generation. The performance of the system is evaluated based on the processing time of packets and detection rate.

(i) Variation of processing time with the number of packets: This metric helps in measuring the latency for the packet processing in the switches and the controller. Packets in batches of 100, 1000, 2000, and 5000 were sent between different hosts in the network. Figure 5 shows the variation of the processing time with the number of packets. The difference between the time taken to send the packets and the reception of the packets has been considered as processing time. This time corresponds to checking the packet contents, checking against the rules, and installing any rules, if needed. As shown in Fig. 5, the time per packet processing decreases. This is because, once rules are installed, the packets with matching rules need not be sent to the controller. In a typical environment, as more packets are received, the number of packets with matching rules is expected to increase.

(ii) Variation of processing time based on attacks: This metric will show how much time is taken for detecting each of these attacks. Figure 6 shows processing time based on the attacks.

From the chart, it can be seen that the attack types III and V takes longer time to detect, i.e., 0.69 s and 0.71 s/packet. This is because these attacks can be detected only after redirecting through multiple sites. Due to this redirection, the time for the attack detection takes a bit more time as compared to others. The attack types I and II deal with the results obtained from external sources. The lowest processing time is taken by the attack type IV, that is, 0.55 s, which mainly covers attacks where IP addresses are used in the URL.

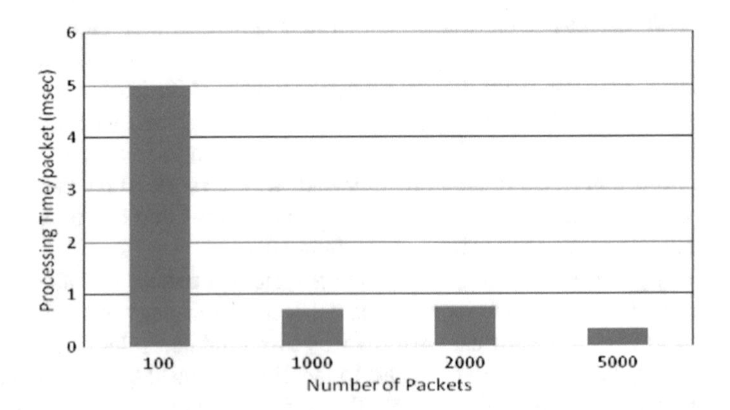

Fig. 5 Variation of processing time with the number of packets

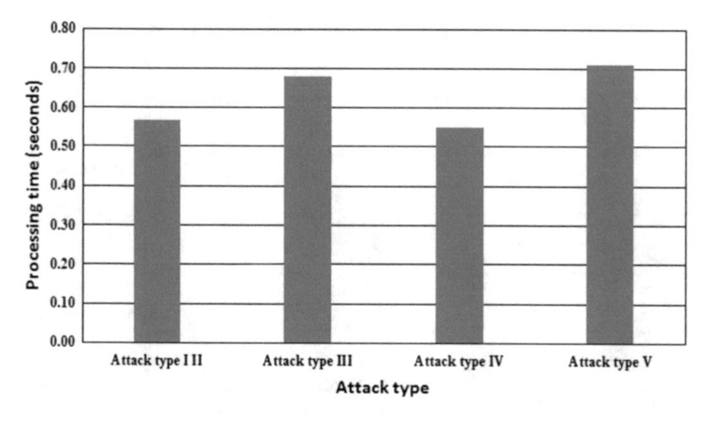

Fig. 6 Variation of processing time based on attacks

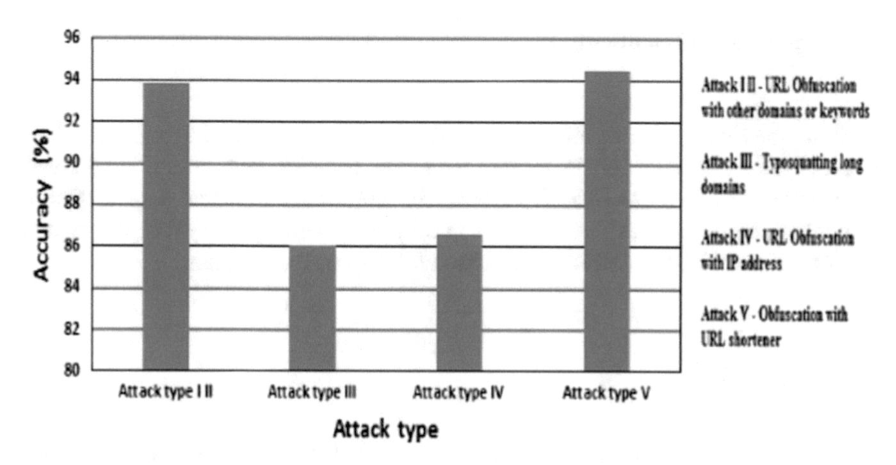

Fig. 7 Accuracy chart

(iii) Accuracy: We next look at the accuracy of detecting the five types of URL attacks encountered in the network.

Figure 7 shows how many attacks are detected correctly. All the packets are considered malicious. Based on the number of packets sent and received, the detection rate can be measured.

From Fig. 7, it can be seen that for attacks type I, type II, and type V, a detection rate of nearly 95% is achieved. The attack types III and IV have about 87% detection rate.

(iv) Precision–Recall measure: The performance of the system can be evaluated based on the precision and recall values for each of the URL attacks. Figure 8 shows the precision and recall values obtained for each URL attack type.

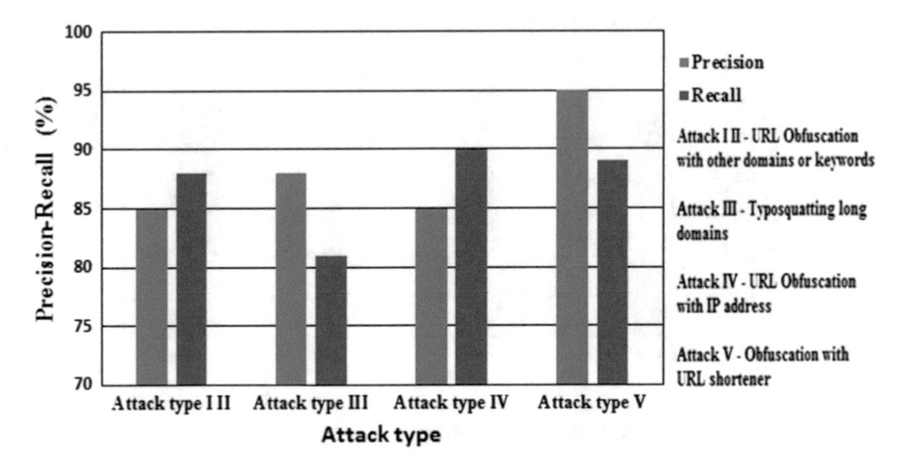

Fig. 8 Precision based on attack type

From the figure, we can infer that the attack types III and IV have lower values of precision and recall. This is because the model has been built based on a limited number of URL features like IP address and long domain names. Attack type V provides a high value as these attacks can be identified by analyzing the syntax features of the URLs and the content-based features of the websites.

5 Conclusion

This paper has proposed a mechanism to perform URL filtration in the SDN environment. In the traditional networks, the URL filtration is mainly performed in the proxy servers. The proxy servers are administered by the admin group. With SDN, the rules in the switches can be dynamically updated by analyzing the URL. The URL analysis includes the study of the lexical features of the URL and content-based features of the corresponding websites. Hence, the workload of the servers can be minimized when the switch becomes capable of handling URL requests from different hosts. The performance evaluation carried out in the limited simulation environment shows reasonable accuracy and latency. Hence, this approach is promising and need to be evaluated in a practical setting.

Further, the rules installed in the switches are dependent on the strength of the model created from the URL analysis. Hence, various machine learning techniques can be used to enhance the model. Approaches like detailed website content analysis using imaging techniques, textual analysis, etc. can be used for further strengthening of the model.

References

1. Marchal, S., François, J., State, R., Engel, T.: Phishstorm: detecting phishing with streaming analytics. IEEE Trans. Netw. Serv. Manag. **11**(4) (2014)
2. Stancu, A.L., Halunga, S., Vulpe, A., Suciu, G., Fratu, O., Popovici, E.C.: A comparison between several software defined networking controllers. In: Telecommunication in Modern Satellite, Cable and Broadcasting Services, Serbia, Oct 2015
3. Naga Venkata Sunil, A., Sardana, A.: A PageRank based detection technique for phishing web sites. In: IEEE Symposium in Computers and Informatics, Malaysia, Mar 2012
4. Feroz, M.N., Mengel, S.: Phishing URL detection using URL ranking. In: IEEE International Congress on Big Data, New York, USA, July 2015
5. Shahriar, H., Zulkernine, M.: Information source-based classification of automatic phishing website detectors. In: IEEE/IPSJ International Symposium on Applications and the Internet, Germany, July 2011
6. Lee, Sangho, Kim, Jog: WarningBird: a near real-time detection system for suspicious URLs in twitter stream. IEEE Trans. Dependable Secure Comput. **10**, 183–195 (2013)
7. Aburrous, M., Hossain M.A.: Modelling intelligent phishing detection system for e-banking using fuzzy data mining. In: International Conference on CyberWorlds, UK, Sept 2009
8. Basnet, R., Mukkamala, S., Andrew H.S.: Detection of phishing attacks: a machine learning approach. In: Soft Computing Applications in Industry, Studfuzz, vol. 226. Springer, Berlin, Heidelberg (2008), pp. 373–383
9. Fang, L., Wang, B., Huang, J., Sun, Y., Wei, Y.: A proactive discovery and filtering solution on phishing websites. In: IEEE International Conference on Big Data, Washington, USA, Oct 2015, pp. 2348–2355
10. Lawrence, K.S., Savage, S., Geoffrey Voelker, M.: Learning to detect malicious URLs. ACM Trans. Intell. Syst. Technol. **2**(30), 3–5 (2011)
11. Aho, A.V., Corasick, M.J.: Efficient string matching: an aid to bibliographic search. Commun. ACM **18**(6), 333–340 (1975). (New York, USA)
12. Alexa: Website Ranking [Online]. Available: http://www.alexa.com/
13. Mininet [Online]. Available: http://mininet.org/
14. Onos [Online]. Available: http://onosproject.org/
15. Scapy [Online]. Available: http://www.secdev.org/projects/scapy/

Ms. K. Archana Janani was a student at the Department of Computer Science and Engineering, at College of Engineering, Guindy, Anna University, Chennai, India. She received her M.E. degree (Software Engineering) from Anna University, in 2017. Her research interests are in Software Defined Networks and Security.

Dr. V. Vetriselvi is an Associate Professor at the Department of Computer Science and Engineering, at College of Engineering, Guindy, Anna University, Chennai, India. She received her B.E. degree (Electronics and Communication) and M.E. degree (Communication Systems) from Madurai Kamraj University in the year 1997 and 1999, and her Ph.D. from Anna University, in 2008. She has been working in Anna University for the past 18 years. Her current research interests are in Computer Networks, Ad hoc networks and Security. She has been actively involved in security related projects sponsored by MiTY and Cognizant. She has many publications in International journals and conferences.

Dr. Ranjani Parthasarathi is a Professor at the Department of Information Science and Technology, at College of Engineering, Guindy, Anna University, Chennai, India. She received her B.E. degree (Electronics and Communication) from University of Madras in 1983, her M.S. degree (Electrical and Computer Engineering) from Illinois Institute of Technology, Chicago, in 1988, and her Ph.D. from IIT Madras, in 1996. From 1983 to 1992, she has held various responsibilities in the computer industry in India, handling software and hardware design and development. Her current research interests are in Computer Architecture, Multi-core computing, adhoc networks and Indian Language Technology solutions. She has been actively involved in both Govt. sponsored and industry sponsored projects in these areas for the past 15 years. She has more than 50 publications in International journals and conferences. She is particularly interested in Indian traditional knowledge systems. She is a member of ACM, and IEEE. She received the IBM faculty award in the year 2008 for research on multi-core architectures.

Dr. G. Subrahmanya VRK Rao (Dr. Rao) is currently associated with Global Technology Office of Cognizant Technology Solutions as *AVP-Technology*. Dr. Rao has over Twenty four years of work experience which include Government, Industry, Academia and Research careers. While Dr. Rao earned multiple International Technology awards from ITU (United Nations), EPA (USA), Computerworld etc., Dr. Rao was also awarded multiple Government/Industry/Public Private Partnership (PPP) Research grants from Intel, NWO (The Netherlands), DEiTY (Govt. of India) etc. A Ph.D. Gold-medalist, Dr. Rao did his Postdoctoral Research work at Telkom Center of Excellence attached to University of Fort Hare, South Africa. Author of the Book entitled "WiMax—A Wireless Technology Revolution" and Editor of the book entitled "Web Services Security and E-Business", Dr. Rao regularly files Patents and Trademarks. Widely Published/Presented his research at International Conference/Journals, Dr. Rao is the Founder and Organizing Chair for 'Cloud for Business Industry and Enterprises (C4BIE)'. He is a Senior Member of IEEE and a Senior Member of ACM, and his Bio is available in multiple editions of "Who's Who in the World". Dr. Rao is an active (White Hat) Certified Ethical Hacker.

Mitigation of DoS in SDN Using Path Randomization

N. A. Bharathi, V. Vetriselvi and Ranjani Parthasarathi

Abstract SDN is a recent blooming architecture which provides greater flexibility for the network professionals. SDN decouples the control logic from the forwarding devices, and the centralized controllers decide the forwarding rules in the network. In spite of the flexibility provided, it is vulnerable to many kinds of attacks. Our focus is on mitigating the denial-of-service attack on flow tables which can result in severe degradation of the network switches. In order to address this issue, we propose a path randomization technique and flow aggregation algorithm. The performance of the system has been evaluated in a simulation environment which has shown a positive result.

Keywords SDN · DoS attack · Path randomization

1 Introduction

Software-Defined Networking (SDN) is an emerging networking architecture which simplifies the network management. In SDN, the control plane is separated from the data plane making the networking device a simple packet forwarding hardware. These devices are controlled by a logically centralized software-based controller through an open interface. In this architecture, the network becomes directly programmable which enables network administrators to create innovative applications that aid in network management, configuration, traffic analysis and traffic engineering [1]. As shown in Fig. 1, the southbound interface is the basic communication protocol which defines the APIs between the data plane elements and the control plane elements.

N. A. Bharathi (✉) · V. Vetriselvi · R. Parthasarathi
College of Engineering Guindy, Anna University, Chennai, India
e-mail: baru.cheers@gmail.com

V. Vetriselvi
e-mail: vetri@annauniv.edu

R. Parthasarathi
e-mail: rp@auist.net

S. Smys et al. (eds.), *International Conference on Computer Networks and Communication Technologies*, Lecture Notes on Data Engineering and Communications Technologies 15, https://doi.org/10.1007/978-981-10-8681-6_22

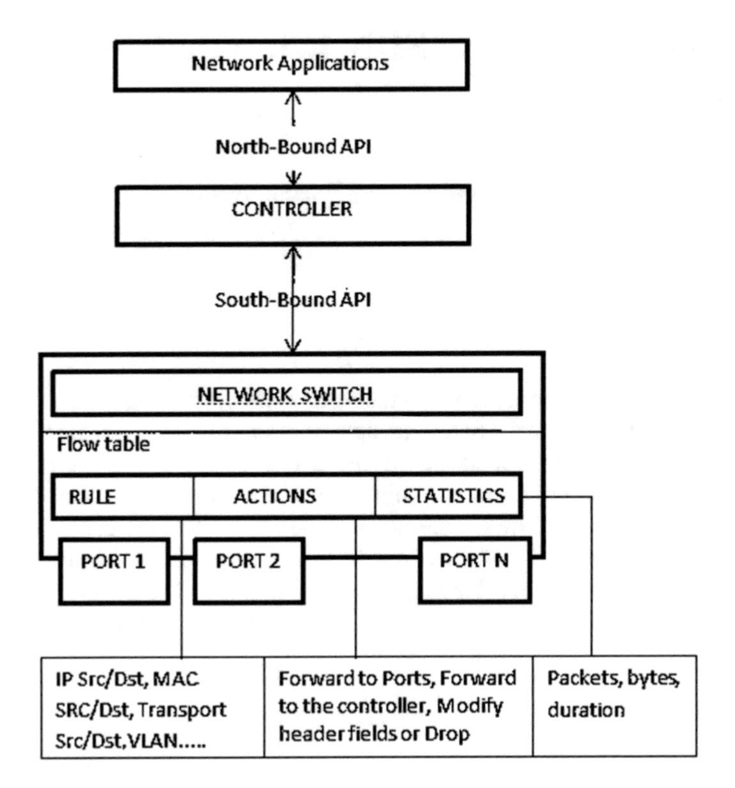

Fig. 1 SDN architecture

OpenFlow is an open-standard southbound protocol used. The northbound interface defines the APIs for developing applications. Forward decisions in the devices are flow-based, instead of destination-based. The flow consists of the matching fields such as port number, MAC address, Ethernet type, IP address and TCP port details. As soon as the packet enters the switch, the packet header fields are extracted and matched with the matching field's portion of the flow entry in flow table [2].

Although SDN provides the above benefits, it leads into new threats and attack planes. A denial-of-service attack in SDN involves enormous computation in the networking resources which makes the switches cripple and unable to forward the packets. In the data plane, switches have the hardware unit called Ternary Content Addressable Memory (TCAM) where the rules are actually placed. TCAMs have limitations of small capacity, high power consumption, high heat generation and high cost. The rules created will be infinite when we consider highly scalable networks. As the switches have limited memory to store the flow rules, it becomes the attack point to the attackers. This attack is one of the considerable threats to SDN as they can easily flood switch TCAM tables, which results in a vital reduction of the overall network performance. In this paper, we focus on the DoS attack on the flow table and mitigate it.

Flow aggregation is considered as one of the solutions to the rule placement problem. Here, we consider flow aggregation in the context of security solution for the DoS attacks occurred in the switch table. There are few works related to the flow aggregation for flooding switch table and the closest work would be SDN-Guard. In SDN-Guard, Dridi et al. [3] focus on the malicious traffic alone and forward that in a path with least-utilized links in terms of bandwidth consumption. But we focus on two things to achieve a better result of reducing the rules in the switch. The first solution is randomizing the paths to be forwarded and then aggregating the rules in network. By choosing the shortest path or path with least-utilized links in the network, the traffic is forwarded in a single path which causes flooding in specific devices in the path. But by randomizing the paths, the rules are distributed in the network and further rule aggregation is performed over the network to mitigate the DoS attack on the switch table, which is better than the existing ones.

This paper is organized as follows: Sect. 2 speaks about the related work which has two subsections: one is DoS attacks and other one is related to flow aggregation. Section 3 explains about the proposed work. The evaluation and results are given in Sect. 4 and finally conclusion in Sect. 5.

2 Related Work

2.1 Denial-of-Service Attacks

Zhang et al. [4] give a review on DoS attacks and some countermeasures in SDN environment. They mainly proposed a queuing-based solution known as Multilayer Fair Queuing (MLFQ) for maintaining queues in the switch in order to avoid DoS attack in the switch. Kandoi et al. [5] focus on the DoS attacks specific to OpenFlow. They discussed two attacks: one related to control plane bandwidth and other related to the flow table. They emulated these two attacks and assessed. They suggested two possible ways, namely, choosing an optimal time-out and flow aggregation to prevent flooding of the flow table.

Shang et al. [6] proposed a protocol-independent framework named flood defender to defend against SDN-based DoS attacks. This framework has four modules: attack detection, table-miss engineering, packet filter and flow rule management to defend DoS attacks and install new rules accordingly.

Kuerban et al. [7] introduced a mitigation strategy known as FlowSec, in which the rate limit is applied to the number of packets sent to the controller. FlowSec uses the floodlight controller module to collect the switch statistics and instructs the switch port to slow down.

2.2 Flow Aggregation

In general, flow aggregation is the process of merging the similar flows which have the same matching criteria and same action. The rule aggregation depends on the matching fields of the packet header. This has to be done by considering the priority of the rules and conflicts if any.

Yoshioka et al. [8] introduce a new routing method with aggregated flows in SDN network. This method reduces number of flow entries by routing with flow aggregation and minimizes the link utilization by selecting routes with low utilization.

Kang et al. [9] addressed different ways to aggregate the rules. The rule space capacity is always initialized for all the switches present in the network. They performed rule aggregation by applying policies in a single switch, over the network perimeter. By doing this, they achieved optimization in rule placement.

All these works give the solution of flow aggregation for the rule placement problem. As mentioned earlier, the SDN-Guard [3] is the work which addresses the flow aggregation as one of the mitigation strategies for DoS attacks on flow table. In this work, they achieved an acceptable performance in the SDN networks during the attack by rerouting the malicious traffic to a least-utilized link using an external Intrusion Detection System (IDS), adjust the flow timeouts and aggregate the flow entries of the malicious traffic.

3 Proposed Work

In this section, we provide a natural and organic solution for the DoS attack on flow tables, using the existing SDN features. The flow aggregation is a suggested solution to reduce the flow entries in the device [10]. We propose a mitigation technique which effectively reduces the number of rules in the switches and avoids the flooding of the flow table. This process includes two steps.

The first step is forwarding the packets using the randomized paths. Normally, packets are forwarded using the shortest path in the SDN network. Choosing the shortest path each time leads to the overuse of particular switches in the network. Using the shortest path is an advantage in legitimate scenario. But in DoS scenario, using the shortest path alone will not be a good option. So we preferred to have randomization in selecting the path in such a way that no route will be overloaded. By randomizing the paths in the network each time, the rules are distributed evenly in all the devices over the path.

The mitigation process shown in algorithm uses reactive forwarding method as the forwarding application in controller to forward the packets reactively as they enter the network. In the reactive forwarding, the rules are not preinstalled for the new flows; they are installed spontaneously as the packet enters the network. The PacketContext contains the incoming packet plus some metadata about the packet. We read the inbound packet, parse it and extract the source and destination device ids, source and destination device ports. In the algorithm, device refers to the switch in the path. The lines 4–8 in the algorithm checks whether the source device id is equal to destination device id; if they are equal, then it checks whether the source device port is equal to the destination device port. If they are not equal, then the rule is installed on the same device. From the lines 9–18 in the algorithm, set of paths from the source to the destination are calculated and if the set is empty the packet context is flooded to all the ports. If there are paths, then the paths are randomized such a way that each time when a new flow enters, a random path is assigned. Then, the rules are installed in all the devices in the random path.

The next step in the mitigation process is flow aggregation. Many authors proposed the flow aggregation in the network to reduce the flow rules. But we use this reduction in flow rules for increasing the performance of the network in case of a DoS attack in the network. We perform flow aggregation under two conditions of routing. In the first method of flow aggregation, the routing is done by choosing the shortest path in the network which is the normal way of routing and in the second method, the paths of the network are chosen randomly.

The method flow aggregation in the algorithm reduces the number of rules with a common destination IP addresses and source IP prefix match. The lines 21–26 fetch the rules from each device; for each port in the device with all common destination IP addresses, the output treatment is found and stored in an array. Then, for each rule in the array, the common source prefix is found using the mask length. The prefix match can be 8, 16 and 24. In order to choose the prefix match, we have done an analysis with the real-time DDoS attacks data using the UCLA datasets [11] and found that the prefix match 8 will give a better aggregation result. For all the rules in the device, prefix match is checked and if the rule matches, it is stored in matched array else stored in unmatched array. After storing in separate arrays, the rules are removed from the table. And for all the rules in the matched and unmatched array, rules are installed in the device. So, the rules in the device which matches once will not be added to the matched array. Thus, the rules get reduced in a device using the algorithm.

Algorithm for Randomized Path Selection and Flow Aggregation
Reactive Forward (Packet Context)

```
 1:  Read the inbound packet from the context
 2:  Parse the packet and obtain the source and the destination device ids.
 3:  If packet source device id is equal to the destination device id then
 4:         If the source port is not equal to the destination port then
 5:            Install the rule in the same device with device id
 6:         End if
 7:  End if
 8:   else
 9:    Get the set of paths that lead from source to destination device
10:    Check if the paths are empty
11:  If there are no paths then
12:    Flood the context to the available ports
```

13: else
14: Randomize the paths
15: Pick the random path from the set of paths
16: For all the devices in the random path
17: Install rule in the device
18: End For
19: End if
Aggregation on a Device (Device)
20: For each Device D in the path do
21: Read all the rules installed in the device.
22: For each port p in the Device "D" do
23: Find rules with output treatment to the port p
24: Store in the array common[].
25: End For
26: End For
27: For each rule r in common array do
28: Find common prefix rule by reducing the mask length
29: If packet matches one of the rules contained in the then
30: Store it in array matched [].
31: Remove the matched rules.
32: else
33: Store it in the array nomatch[].
34: End if
35: End For
36: For each rule r in nomatch array do
37: Install the rule on the device D.
38: End For
39: For each rule r in matched array do
40: Install the rule on the device D.
41: End For

In order to understand the behaviour of the network, we applied the flow aggregation in two cases. The first case is with the shortest path and other one with the randomized path. The number of rules that can be merged in the network is compared between the two cases.

4 Evaluation

4.1 Emulation Environment

This section explains the experimental setup used to emulate the SDN network. To emulate the network topology, we used the Mininet. The ONOS controller is used to control the whole network. ONOS stands for Open Network Operating System.

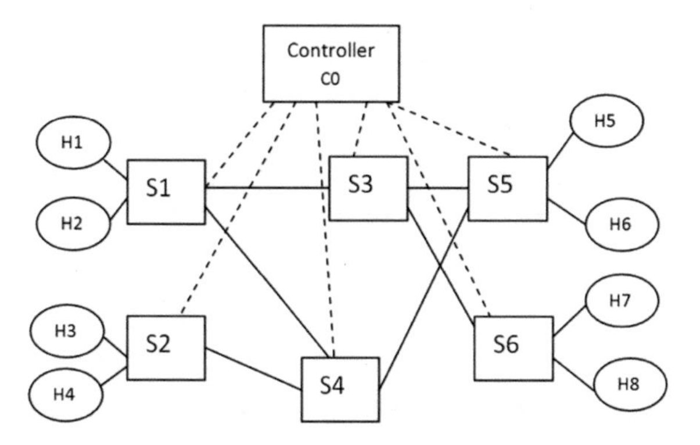

Fig. 2 Emulated topology

OpenFlow is used as the API between the controller and the OpenFlow switches. The topology used for the experiment has six switches and eight hosts connected as shown in Fig 2.

The attack traffic is generated using the tool hping3. Host h1 is considered as the attacker and the host h8 as the victim. Traffic is generated to attack the h8 from h1 with random source IP addresses. Hping3 tool floods with more number of packets with different source IP addresses. By doing this, the switches connected along the path are filled with more number of rules which leads to the switch table flooding DoS attack.

4.2 Result Analysis

In our experiment, we focus on the result analysis using two parameters: First, the number of rules installed in the switch and the other one is the round trip time (RTT). We focus on analysing the results using two cases.

Case1: Normal path selection
The traffic is generated and the forwarding is done choosing the shortest path. The flow aggregation is done in all the switches along the shortest path and the rules are minimized. The number of rules reduced can be viewed in the ONOS GUI, and the RTT for the number of the packets is observed.

Case2: Randomized path selection
Similarly, the traffic is generated and forwarded using a random path and results are noted. As the rules are distributed along the network and the aggregated in all the devices over the paths, the prevention of flooding the switch table is more effective than choosing the shortest path.
The comparative analysis of the number of rules installed in the switch using the shortest path and the randomized paths is shown in Fig 3. From Fig 3, we notice that the numbers of rules installed in the random path are fairly distributed and also the reduction of the rules is better compared to the rules reduced in the shortest path.

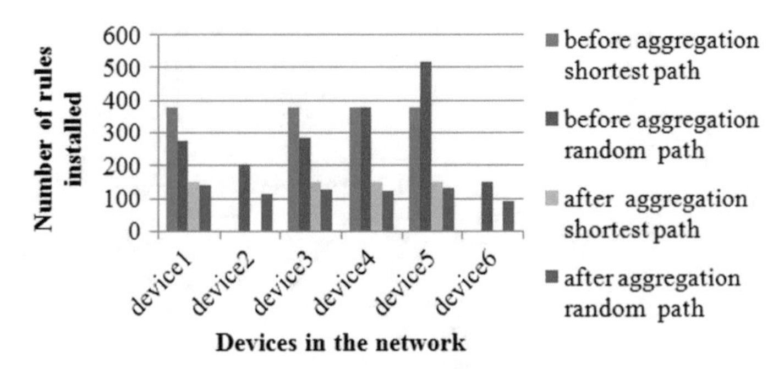

Fig. 3 Rules before and after aggregation in shortest path and random path

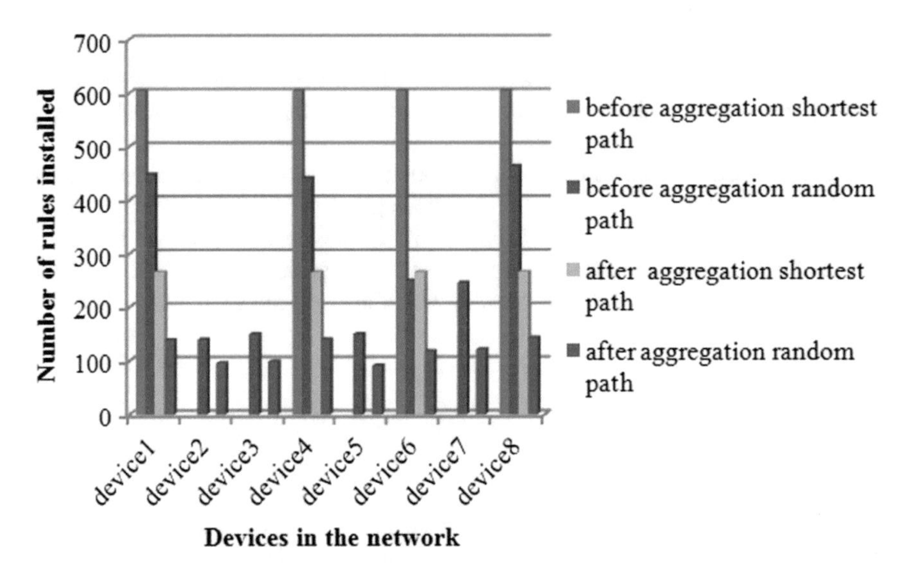

Fig. 4 Rules before and after aggregation in two cases: SDN-guard scenario

Figure 5 shows the average RTT taken between h1 and h8 for the two cases. The latency occurs only for the first few packets of forwarding in the randomized path, and other than that, it shows a similar performance. It shows that randomizing of paths is much suitable for the prevention of the DoS attacks in the switch table.

As the SDN-Guard is the closest work related to our work, we analyse the results with the topology used by them and compare the number of rules reduced. The attack is generated from h1 to h6 using hping3. We use the same topology and scenario of attack and implement both the shortest path and the random path forwarding with aggregation of rules in the network. The results are shown in Fig 4. On an average, the total number of rules aggregated is 58% in our work which is higher than 26% reported in SDN-Guard [3].

The RTT is analysed for both the topologies, and Figs. 5 and 6 show that the first few packets of the network's RTT alone are high in the case of random path. This

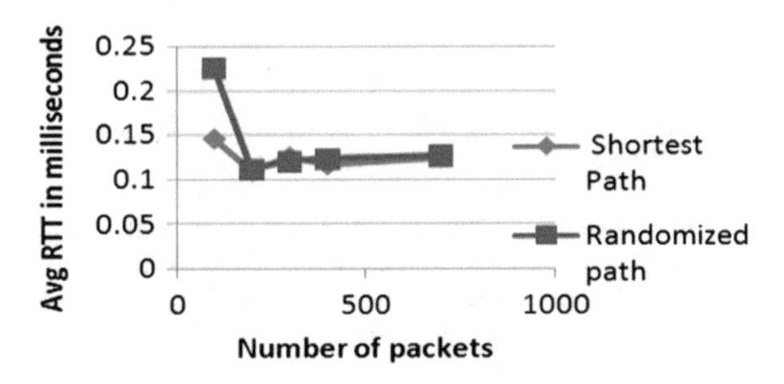

Fig. 5 Average RTT for experiment

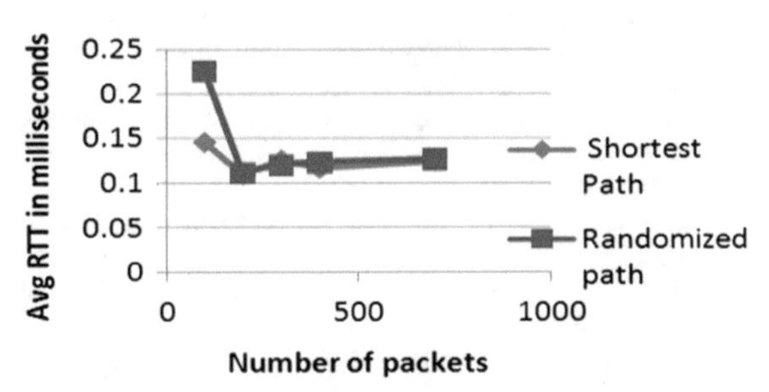

Fig. 6 Average RTT for the SDN-guard

may affect a minimal impact of latency for the first few packets of forwarding for the legitimate flows. Otherwise, it works well in mitigating the DoS attacks on the switch table.

The randomization of paths works good in case of malicious traffic where the rules are distributed all over the network, unlike choosing only one path for forwarding. There is a latency impact only for the first packet which is a small price to pay for mitigating DoS attack.

5 Conclusion

In this paper, we have proposed a solution to mitigate DoS attacks on flow table. Our method of flow aggregation and randomizing the paths effectively protect the switch from flooding. We evaluated the idea for two scenarios: one for shortest path and other using randomized path using Mininet and found that more than 50% rules are reduced. By doing this, the switch memory is saved and prevents the flooding of flow table. To further extend this work, we can think of different flow aggregation techniques to reduce the number of rules.

References

1. Kreutz, D., Ramos, F.M.V., Verissimo, P., Rothenberg, C., Azodol-molky, S., Uhlig, S.: Software-defined networking: a comprehensive survey. Proc. IEEE **130**(01), 14–76 (2014)
2. Open Networking Foundation, 'OpenFlow Switch Specification'. Available from: http://OpenFlowSwitch.org, Version 1.3.4 (Protocol version 0x04), pp. 1–40 (2014)
3. Dridi, L., Faten Zhani, M.: SDN-guard: DoS attacks mitigation in SDN networks. In: 5th IEEE International Conference on Cloud Networking, pp. 213–217 (2016)
4. Zhang, P., Wang, H., Hu, C., Lin, C.: On denial of service attacks in software defined networks, network forensics and surveillance for emerging networks. IEEE Netw., 28–33 (2016)
5. Kandoi, R., Antikainen, M.: Denial-of-service attacks in OpenFlow SDN networks. In: 1st International Workshop on Security for Emerging Distributed Network Technologies (DISSECT). IEEE, pp. 1322–1326 (2015)
6. Shang, G., Zhe, P., Bin, X., Aiqun, H., Kui, R.: FloodDefender: protecting data and control plane resources under SDN-aimed DoS attacks. In: IEEE INFOCOM, IEEE Conference on Computer Communications (2017)
7. Kuerban, M., Tian, Y., Yang, Q., Jia, Y., Huebert, B., Poss, D.: 'FlowSec: DOS attack mitigation strategy on SDN controller. IEEE (2016)
8. Yoshioka, K., Hirata, K., Yamamoto, M.: Routing Method with Flow Entry Aggregation for Software-Defined Networking. IEEE (2017)
9. Kang, N., Liu, Z., Rexford, J., Walker, D.: Optimizing the one big switch abstraction in software-defined networks. IEEE Trans. Comput. **4**(1), 1–10 (2013)
10. UCLA CSD packet traces dataset: https://lasr.cs.ucla.edu/ddos/traces
11. Giroire, F., Moulierac, J., Khoa Phan, T.: Optimizing rule placement in software-defined networks for energy-aware routing. In: Globecom Symposium on Selected Areas in Communications, GC14 SAC Green Communication Systems and Networks. IEEE, pp. 2523–2529 (2014)
12. Huang, H., Guo, S., Li, P., Ye, B., Stojmenovic, I.: Joint optimization of rule placement and traffic engineering for QoS provisioning in software defined network. IEEE Trans. Comput. **3**(1), 1–14 (2015) (Digests 9th Annual Conference on Magnetics Japan, p. 301)

N. A. Bharathi received the M.E. degree in Multimedia technology from College of Engineering, Guindy, India, in 2016. Currently, she is pursuing the Ph.D. degree in the Department of Information Science and Technology at Anna University. Her research interests are software defined networks and network security.

Dr. V. Vetriselvi is an Associate Professor at the Department of Computer Science and Engineering, at College of Engineering, Guindy, Anna University, Chennai, India. She received her B.E. degree (Electronics and Communication) and M.E. degree (Communication Systems) from Madurai Kamraj University in 1997 and 1999, and her Ph.D. from Anna University, in 2008. She has been working in Anna University for the past 18 years. Her current research interests are in Computer Networks, Ad hoc networks and Security. She has been actively involved in security related projects sponsored by MiTY and Cognizant. She has many publications in International journals and conferences.

Dr. Ranjani Parthasarathi is a Professor at the Department of Information Science and Technology, at College of Engineering, Guindy, Anna University, Chennai, India. She received her B.E. degree (Electronics and Communication) from University of Madras in 1983, her M.S. degree (Electrical and Computer Engineering) from Illinois Institute of Technology, Chicago, in 1988, and her Ph.D. from IIT Madras, in 1996. From 1983 to 1992, she has held various responsibilities in the computer industry in India, handling software and hardware design and development. Her current research interests are in Computer Architecture, Multi-core computing, adhoc networks and Indian Language Technology solutions. She has been actively involved in both Govt. sponsored and industry sponsored projects in these areas for the past 15 years. She has more than 50 publications in International journals and conferences. She is particularly interested in Indian traditional knowledge systems. She is a member of ACM, and IEEE. She received the IBM faculty award in the year 2008 for research on multi-core architectures.

Smart Homes Using Alexa and Power Line Communication in IoT

Radhika Dotihal, Ayush Sopori, Anmol Muku, Neeraj Deochake and D. T. Varpe

Abstract This paper deals with smart home which caters for the comfort of citizens. Smart homes are those where household devices/home appliances could be monitored and controlled remotely. This aims at controlling home appliances via smartphone and voice by using Alexa acting as a client. Local network created by home router or cloud is used for monitoring and controlling of the connected devices such as LEDs, fans, motors, dimmers, PWM devices, etc. System consists of devices connected to a central hub called a "gateway" from which the system is controlled with a user interface which is provided by the Android application or through Alexa. The communication between the devices and gateway takes place through Power Line Communication (PLC) and the RF links either through TCP protocol or Message Queue Telemetry Transport (MQTT) protocol.

Keywords ZigBee · Gateway · Alexa · Power line communication (PLC)
Radio frequency links (RF Links)

1 Introduction

The terms "Smart Homes" and "Intelligent Homes" have been used to introduce the concept of networking appliances and devices in home. Home automation includes controlling of lights, appliances, security of home to provide improved comfort, and energy efficiency. Smart home automation uses the smartphone devices as user interface. User can communicate with home automation network either through an Internet gateway or locally created network. The aim is controlling home appliances via smartphones acting as a client for Android application and using Alexa acting as a client for voice interaction. ARM-powered microcontroller acts as a server for IoT-based system with various interfaces mounted on board, where other components and

---

R. Dotihal (✉) · A. Sopori · A. Muku · N. Deochake · D. T. Varpe
Department of Information Technology, Pune Vidhyarthi Griha's College of Engineering and Technology, Pune, India
e-mail: radhikakatraj@gmail.com

S. Smys et al. (eds.), *International Conference on Computer Networks and Communication Technologies*, Lecture Notes on Data Engineering and Communications Technologies 15, https://doi.org/10.1007/978-981-10-8681-6_23

devices are connected. The home appliances are connected through these interfaces mounted on microcontroller and their status is passed to the controller. Accordingly, action is performed by controller and the status of the appliance gets updated which is thereby reflected in the Android application.

2 Internet of Things

The term "Internet of Things" can be seen as an umbrella term for interconnecting the various aspects which includes extension of the Web and Internet into the realm of day-to-day devices. The term also provides a vision that these "things" of the real world will seamlessly integrate into the world of virtualization which will allow access to these "things" anytime from anywhere. It forms an integrated part of the future Internet and can be defined as the global network which has got the capabilities of self-configuration on the basis of the interoperable protocols and the "things" (devices) that have got the identities and virtual personalities are seamlessly capable of integrating themselves into the information network. In the near future, it is expected that the things are going to be more active in the spheres like business, information, and social process which provide them with the capability to interact among themselves and with the physical environment, thereby trigging the appropriate actions without any human intervention. The application of the IoT paradigm to an urban context is of particular interest which corresponds to the term known as smart city. Home automation (also known as domotics) is one of the concepts which fall under the smart city.

Following is the layer of IoT in smart home automation system (Fig. 1).

The hardware layer consists of the devices which are to be controlled like tunable lights, fan, DC motors, fire detection, and many more PWM-based devices. The data link layer consists of IoT gateway and various communication protocols such as HTTPS, TCP/IP, and MQTT. The ARM processor is used in the IoT gateway which communicates to smartphone with/without Internet in the gateway and network layer. The application and presentation layers consist of Android application in which we can control the appliances.

3 Related Work

Seeing the past work, the communication between different appliances and server takes place through Bluetooth technology. IoT devices are driven by short-range communication. Bluetooth technology would perfectly work for shorter distances [1]. With the help of mobile telephony, one can control home appliances from faraway locations. It provides us with the capabilities of voice and data transfer services. Data is transferred by SMS and by using other enhanced data rate services like GPRS and EDGE. One of the telecommunication protocols known as Short Message Service

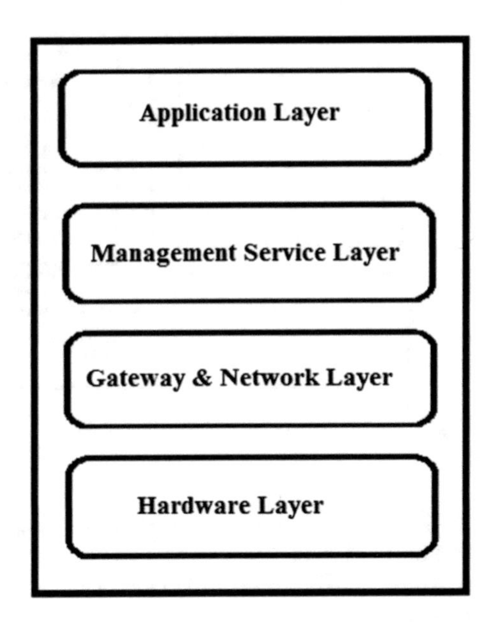

Fig. 1 Layers of IoT

(SMS) is used for sending text messages with limit of 160 characters or less [2]. The drawback of the abovementioned systems is that the Bluetooth works in the real time but the range of the connected devices is not much which becomes the big hurdle in process. In the second scenario, the GMS and SMS services do not offer much control and add to the cost.

4 Proposed Work

The proposed system is based on the concept of distributed home automation system, consisting of devices and gateway. A central gateway handles and controls various devices and can be easily configured to handle more devices. This proposed home automation system is capable enough to switch ON/OFF different appliances, control tunable lights, control light level, control different PWM devices, fire, and smoke detection.

Internet, a means of communication, is easily available and affordable. Android mobile phones and applications are already an important part of life. So, combination of these technologies will make life more simple and easy to live as these devices can be accessed through it.

The application will work in two modes, i.e., Cloud mode and local mode. In local mode, the devices will be connected to the gateway (controller) and the communication between the devices and gateway will take place through power line communication. In local mode, the application will be sending the events to the microcontroller by using the existing wireless network created by the router.

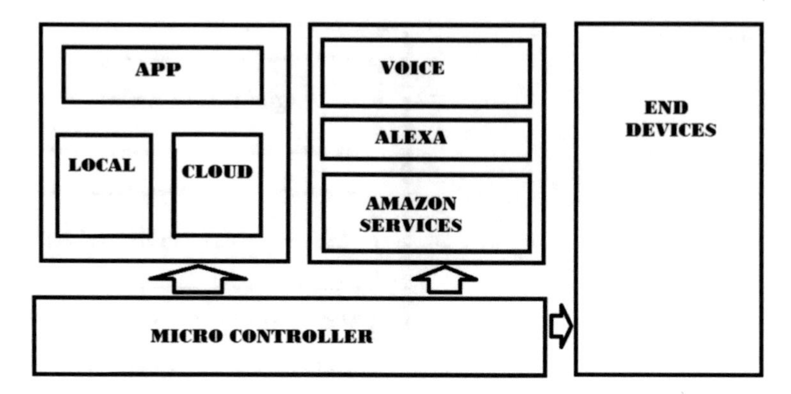

Fig. 2 Architecture of smart home automation

Cloud will improve the system potential to a greater extent by providing the flexibility of accessing the gateway from remote area. End devices and the microcontroller will be connected in the same fashion as it was in local mode. But in cloud mode, the events are sent to the cloud and from there to microcontroller. Popular communication protocols for communication include ZigBee, TCP/IP, and MQTT. The architecture of smart home automation is as follows (Fig. 2).

4.1 Ngrok

Ngrok is a handy tool and services that allows to tunnel request from the wide-open Internet to your local machine when it is behind a firewall. It tracks requests through the tunnel. By default, ngrok forwards HTTP and HTTPS traffic to our local machine. Ngrok exposes the IP address of appliances to microcontroller. An extensive feature of ngrok is its ability to run multiple tunnels simultaneously. Thus, ngrok is a multiplatform tunneling service. It is a reverse proxy software that establishes secure tunnels from a public endpoint such as Internet to a locally running network service. It captures all traffic for detailed inspection and replay.

4.2 Alexa

Alexa proves to be an intelligent personal assistant that was developed by Amazon. Amazon Echo and the Amazon Echo Dot devices make the use of Alexa for voice support. Alexa is available in English and German languages. Alexa interacts through a microphone and speaker. This voice-activated virtual assistant can be used to control smart home gadgets, giving it the ability to turn ON/OFF the appliances, dim the lights, etc. The list of commands that Alexa can understand is known as skills. There is no activation button to press. So we simply have to say a trigger word "Alexa" followed by what we want to happen. Alexa is very natural and responsive.

4.3 ZigBee Devices

ZigBee devices are the global, standard-based wireless solution that can conveniently and affordably control the widest range of devices to improve comfort, security, and convenience. It ensures that the devices in a network stay connected and network performance remains constant even if it is dynamically changing. These devices are the communication paths between the devices. ZigBee devices are used to create personal area network for home automation. It is suitable for Home Area Network (HAN). With the emergence of the ZigBee/IEEE 802.15.4 standard, systems are expected to transition to standard-based approaches, allowing sensors to transfer information in a standardized manner. C2WSNs (and C1WSN, for that matter) that operate outside a building and over a broad geographic area may make use of any number of other standardized radio technologies. The (low data rate) C2WSN market is expected to grow significantly in the near future.

4.4 Power Line Communication

Power line communication is the trending technologies which is unfolding its ways toward the development of the communication channel within the existing electrical network. Based on power line communication technology, research of smart home system has gained more potential. Combined with high-speed power line, data communication system based on low-voltage distribution network forms the backbone of this technology. Smart application of the home network system and power line communication technology put forward new ideas. Based on power line communication technology, development of the intelligent home automation system is in progress. By using the specially designed PLC modems, we are able to transform the data in such a pattern so that it will easily integrate with the waveforms present in the electrical current. Power line technology is also capable of providing the fast and secure data communication overcoming the certain limitations of the wireless technology. As mentioned, transport of information on these electrical lines is performed with a power line communication module (PLC). Each PLC provides an interface with a bandpass filter to avoid interference with existing audio and light systems. The main parameters of PLC communications are the transmission bandwidth (normally expressed in ranges of kHz or MHz) and theoretical bits per second rate (bps) reached during the communication. This network architecture can be used both at train or vehicle levels, providing two separated layers like the architecture adopted for WTB and MVB communication. In this system, a PLCNODE is defined as the transceiver allowing bridging from train to vehicle networks and vice versa (Fig. 3).

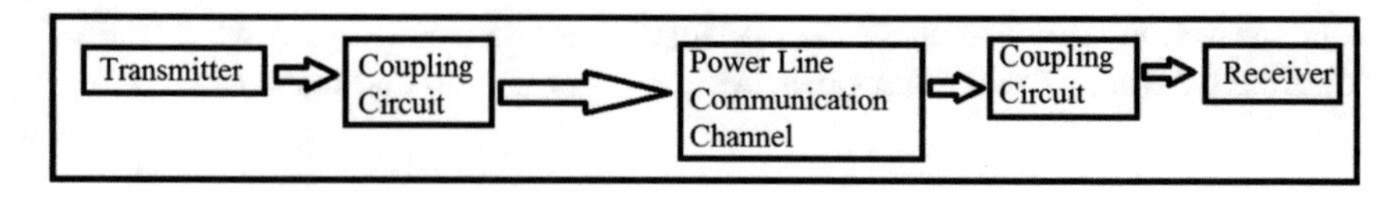

Fig. 3 Power line communication

4.5 Microcontroller

The controller consists of ARM 9 processor with LINUX OS which also includes Wi-Fi module, Ethernet, USB port, and UART port. The controller also consists of PLC and RF module which are linked to the processor with the help of SPI interface. The controller consists of ZigBee module for handling the ZigBee devices. It enables controllability to LED lights from ON/OFF and dimming to cooler temperature adjustment via PLC or ZigBee. The Wi-Fi module is used to establish the connection with the outside world.

5 Mathematical Model

$$S = \{I, O, F_n, S, F\}$$

where

I Set of Inputs = {Voice, Buttons}
O Set of Outputs = {Check the status of device, Control the device}
F_n Set of Functions
S Success
F Failure

6 Conclusion

In this paper, we presented our smart home automation system. This system will ease the problem of people to manage and monitor the home appliances either using application or using Alexa.

In future work, this home automation system can be implemented in schools, small offices, and big malls.

References

1. Shahriyar, R., Hoque, E., Sohan, S.M., Naim, I., Mostafa Akbar, M., Khan, M.K.: Remote controlling of home appliances using mobile telephony. Int. J. Smart Home **2** (2008)
2. Sriskanthan, N., Tan, F., Karande, A.: Bluetooth based home automation system. Microprocess. Microsyst. **26,** 281–289 (2002)
3. Domingue, J., et al. (eds.).: The Future Internet Assembly 2011: Achievements and Technological Promises. Springer (2011)
4. Alkar, Z., Buhur, U.: An internet based wireless home automation system for multifunctional devices. IEEE Trans. Consumer Electron. **51**(4) (2005)
5. Syed Anwaarullah, S., Altaf, V.: RTOS based home automation system using android. Int. J. Adv. Trends Comput. Sci. Eng. **2**. ISSN 2278-3091
6. Zanella, A., Vangelista, L.: Internet of things for smart cities. IEEE Internet Things J. **1** (2014)
7. Baraka, K.: Smart power management system for home appliances and wellness based and wireless sensors and mobile technology. In: 2015 XVIII AISEM Annual Conference. IEEE. 978-1-4799-8591-3/15
8. Mannan, A., Saxena, D.K., Banday, M.: A study on power line communication. Int. J. Sci. Res. Publ. (2014). ISSN 2250-3135
9. Akarte, V., Punse, N., Dhanorkar, A.: Power line communication systems. Int. J. Innov. Res. Electr. Electron. Instrum. Control Eng. (2014). ISSN 2321-5526
10. Unsal, D.B., Yalcinoz, T.: Applications of new power line communication model for smart home grids. Int. J. Comput. Electr. Eng. **7** (2015)
11. Agarwal, A., Singh, R., Gehlot, A., Gupta, G., Choudhary, M.: IoT enabled home automation through nodered and MQTT. Int. J. Control Theor. Appl. (2017). ISSN 0974-5572
12. Stergioou, C., Psannis, K.E., Kim, B.-G., Gupta, B.: Secure Integration of IoT and Cloud Computing. Elsevier 0167-739X (2016)
13. Mazhar Rathore, M., Ahmed, A., Paul, A., Rho, S.: Urban Planning and Building Smart Cities Based on the Internet of Things Using Big Data Analytics. Elsevier, pp. 1389–1286 (2016)
14. Credits to https://developer.amazon.com and Greenvity Communications

Radhika Dotihal is born and brought up in Pune. She has completed her secondary and higher secondary education from Sinhgad Spring Dale School and Maharashtra Vidyalaya in 2012 and 2014 respectively. She has completed her bachelor's degree in Information Technology from Pune Vidhyarthi Griha's College of Engineering and Technology under Savatribai Phule Pune University in 2018. Her interests are Internet of Things and sensor networks.

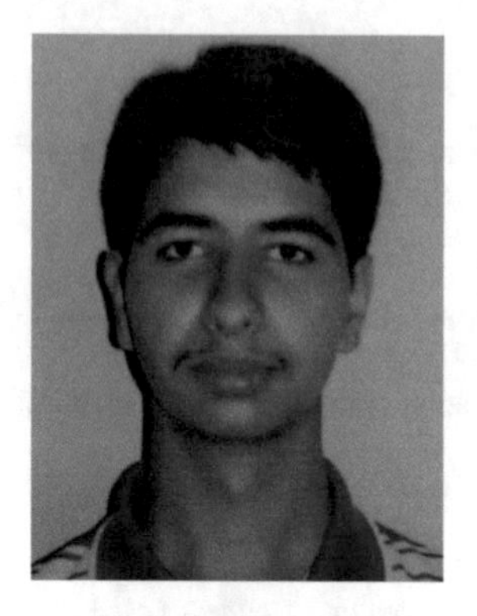

Ayush Sopori native of Jammu and Kashmir, the little paradise of earth, has completed this secondary as well as higher secondary education in 2012 and 2014 respectively form Jammu itself. There after moved to Pune to pursue his Bachelor Degree from Savatribai Phule Pune University. During this course he worked on the projects related to plc and home automation. His interest area comprises of automation and IoT.

Neeraj Deochake native of Ahmednagar, Maharashtra, India completed his secondary as well as higher secondary education in 2012 and 2014 respectively in AshokBhau Firodiya English Medium School and Junior College. Later on he moved to Pune to complete his graduation course from Savitribai Phule Pune University. During his graduation as a graduation project he worked on the Home Automation Systems. His interest area specifies Machine Learning and IOT.

Acquisition and Mining of Agricultural Data Using Ubiquitous Sensors with Internet of Things

M. R. Suma and P. Madhumathy

Abstract In this paper, a modular architecture for Internet of Things (IoT) based on an agricultural setting for data acquisition and mining is proposed, developed, and implemented for validation. This work is an attempt to modularize the general IoT architecture and separate functional layers to make them independent of each other. The purpose of the presented solution is to make development and deployment of IoT systems, easy and cost-effective in agriculture to accelerate the adoption of precision agriculture. The architecture involves four functional layers: things layer (sensors, controllers, and actuators), edge computing layer (microcomputer), gateway layer (Wi-Fi Modem/Broadband), and cloud layer (Database, web application). The local machine-to-machine communication between things layer and the edge computing layer uses a high-level implementation of the message queue transport telemetry protocol.

Keywords Internet of things · Modular architecture · MQTT · Sensors
Raspberry Pi · Web application · Adriano · Edge computing · Ubiquitous sensors
Cloud

1 Introduction

In India, about 70% of population depend upon farming and one-third of the nation's capital comes from it. Even with new concepts like smart farming and precision farming introducing new methods of farming control, adoption has been scarce. Agricultural information technology broadly applies to every aspect of agriculture and has become the most effective means and tools for enhancing agricultural pro-

M. R. Suma (✉)
Dayananda Sagar College of Engineering, Bengaluru, India
e-mail: sumamrvp@gmail.com

P. Madhumathy
Dayananda Sagar Academy of Technology and Management, Bengaluru, India
e-mail: sakthi999@gmail.com

S. Smys et al. (eds.), *International Conference on Computer Networks and Communication Technologies*, Lecture Notes on Data Engineering and Communications Technologies 15, https://doi.org/10.1007/978-981-10-8681-6_24

ductivity [1]. The lack of high agricultural output even with these proposed systems is mainly due to their high cost of installation, maintenance, difficulty in use and discrete acquisition of agricultural data, and its lack of accessibility to those who might have had tremendous use of them. Acquired data over similar parameters can imply useful correlations on how they affect the agricultural output. This would be hard to do with only discrete set of data from a single field. However, if we could have access to a huge data store consisting of data from a large number of fields across a given region that can be mined using simple machine learning and data science algorithms, the opportunities would be open-ended.

This paper proposes a four-layered modular architecture for agricultural data acquisition and mining using an existing set of sensors, in combination with an Atmel ATmega328p microcontroller, called as a sensor node, acts as a data acquisition and transmission center from each sensor location to the Edge Computing Device (ECD)—Raspberry Pi Embedded Linux Board. The communication uses the popular machine-to-machine (M2M) communication standard, MQTT protocol. The edge computing device connects to the Cloud using any available Internet gateway device such as Wi-Fi modem or a broadband over Hypertext Transfer Protocol (HTTP) to accumulate farming data in the cloud server [2]. A web application built using Django web framework makes it accessible to those who want to build control applications based on acquired data.

2 Background

The motivation for the work comes from multiple sources. One of which is the problem in the present agricultural ecosystem across the country. Adoption of new practices and techniques in agricultural technology is scarce. This is a chicken and egg problem. The main barriers have been expensive equipment, complexity in operation and maintenance, and the standards are still under development. An ideal solution would be cheap, easy to setup, maintain, and would last long. Working on a solution made it clear that the modular architecture and the intention to use non-proprietary devices (wherever possible, while reducing the cost) makes it a more fitting solution to spaces beyond just agriculture [3]. Tools like Arduino, Python, MQTT, C++, Raspberry Pi, and 433 MHz RF modules are sufficiently mature technologies with high amount of reliability and attract a large community of developers and makers, who we think can make this work more extensive.

3 Architecture

The current systems of IoT infrastructure employ a direct connection to the Internet using some kind of network interfacing tools like GPRS/GSM modules, which makes these systems unnecessarily costly and complex to develop. In addition, the functions

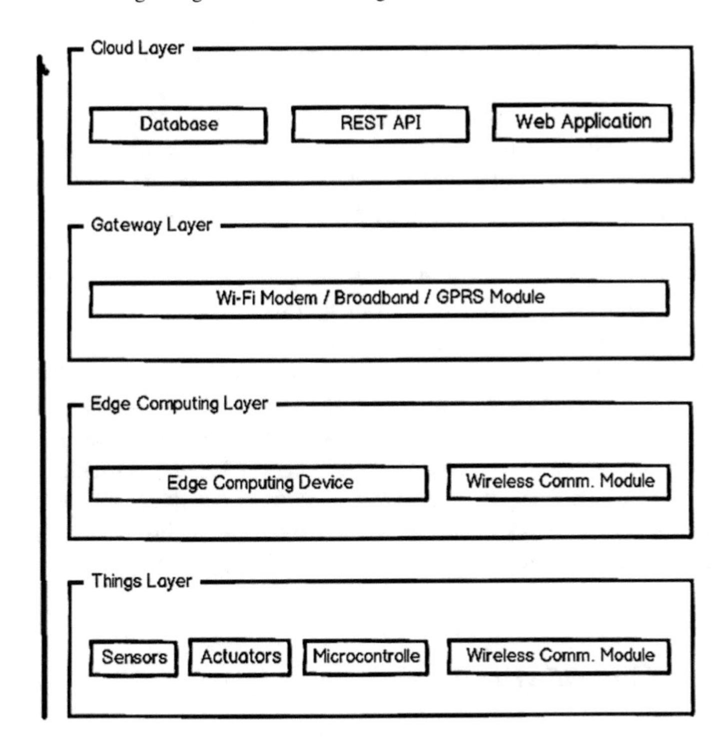

Fig. 1 Proposed architecture

integrate singularly for a particular system, making the switching costs high for the user. The freedom to choose your own stack of technology does not exist. This paper proposes an architecture that attempts to modularize the current IoT infrastructure based on the functions necessary in each context. Figure 1 shows the proposed architecture. It consists of four functional layers—Things layer, edge computing layer, gateway layer, and cloud layer.

In brief, a smart sensor in the "Things Layer" collects necessary data and uses MQTT protocol to transmit data to a message broker located on the edge of the local network called the Edge Computing Device (ECD), which forms the edge computing layer. Then, the ECD uses the hypertext transfer protocol (HTTP) to transmit valuable data from the sensor to a remote server (in the cloud layer) over the Internet using an internet gateway device, which forms the gateway layer [4].

3.1 Things Layer

Internet of Things (IoT) is a network of objects equipped with sensors, actuators, and network-access modules that connect them to the Internet. The objects here in reference are mainly "things" or "tools" we come across every day. Things layer mainly consists of sensors, actuators, and controllers that enable these devices to act on request. A sensor node is a device that consists of a set of sensors, actuators, and a wireless communication module that enables it to transmit and receive data. This layer is the lowest level of abstraction in the layered architecture and consists of a node either independently, like a specific device for monitoring data or in combination with existing consumer devices in the market for gaining insights about these devices. Sensor nodes will have a microcontroller at the heart of all operations that controls the actions like reading data from sensors, acting on requests using actuators, and encoding and decoding data during communication [5]. It is important to note that these nodes are not connected to the Internet by themselves and the wireless communication module used is only to transmit the data to an intermediate computing engine before sending it to the Internet. The intention being that if there are several hundred or thousand nodes deployed across agricultural field or any other place, transmitting these massive amounts of raw data over a network puts tremendous load on network resources. The edge computing layer eases this load by processing most of it locally.

3.2 Edge Computing Layer

Edge computing is a paradigm in which the data processing occurs at the point of origin or at the edge of the local network by an intelligent computing device that is in close geographical proximity to the client. An Edge Computing Device (ECD) acts as a buffer between the cloud and the client [6]. "In some cases, it is much more efficient to process data near its source and send only the data that has value over the network to a remote data center. Instead of continually broadcasting data about the oil level in a car's engine, for example, an automotive sensor might simply send summary data to a remote server on a periodic basis. Rises or falls outside acceptable limits" [7].

3.3 Gateway Layer

Gateway layer consists of a network appliance residing in the client premises or field that connects the ECD to the Internet over standard Internet protocol, e.g., Wi-Fi modem, broadband, etc.

3.4 Cloud Layer

The cloud layer is the topmost layer that collects all the important data transmitted by the ECD over the Internet. This layer consists of a remote database, a web application to access, monitor, and analyze the data. There is sufficient amount of freedom for a developer to design the server system in any language/framework of his choice as long as there is a REST API interface for the ECD to send and receive data over HTTP.

4 Implementation

We chose to work with an agricultural setting for data acquisition and monitoring to test and validate the proposed architecture. We have also made sufficient effort to keep the use of proprietary technology to a minimum to avoid a platform lock-in for the architecture.

The problem statement we worked on demanded that we collected data from multiple points in an agricultural field cyclically during the day and communicated that data to the ECD located at the farmhouse, which also had access to the Internet via a Wi-Fi modem.

For the implementation of the things layer, the choice of the ubiquitous sensors DHT11 and the standard soil moisture sensor was obvious. This lets us to collect three data points—Temperature, humidity, and soil moisture. Soil moisture sensor and DHT11 are as shown in Figs. 2 and 3.

The node was constructed using the sensors mentioned above and Atmel ATmega328p microcontroller built into the Arduino Uno board. The board was programmed using the Arduino IDE with C++ and a library called VirtualWire was used to program parts related to RF communication. 433 MHz radio frequency transmitter and receiver modules are used to design a custom single plug-in transceiver module for the Uno board. The sensor node is powered using a Li–ion battery and forms a stand-alone scalable device.

Fig. 2 DHT11 sensor module

Fig. 3 Soil moisture sensor module

Edge computing device needed to have sufficient computing power to process the data and control processes quickly. It was also necessary that it could be interfaced with any gateway device. Raspberry Pi 3 Model B was an easy choice given its popularity and adoption, which will make it easy for developers to make any developments to the system in future. The Pi was setup using Raspbian Jessie OS, a toned down flavor based on Linux. The process logic was programmed using Python 3.5. "Requests" Python library is used to connect to the web application API over HTTP. ECD also uses a custom-designed RF transceiver module to communicate with the things layer.

The communication between the things layer and the edge computing layer was done using a high-level implementation of Message Queue Transport Telemetry (MQTT) protocol. Andy Stanford-Clark of IBM and Arlen Nipper of Cirrus Link Solutions authored the first version of the protocol in 1999 and is now maintained and improved by IBM [8].

It is a publish/subscribe, very simple, and lightweight electronic messaging protocol, designed for constrained devices and low-bandwidth, high-latency, or unreliable networks. The design principles are to minimize network bandwidth and device resource requirements while also attempting to ensure reliability and some degree of assurance of delivery. MQTT defines three levels of Quality of Service (QoS)—0, 1 and 2. The QoS defines how hard the broker/client will try to ensure that a message is received. Messages may be sent at any QoS level, and purchasers might decide to take topics at any QoS level [9].

Only QoS Level 0 (The broker/client will deliver the message once, with no confirmation) was implemented for validation. The topics used in the implementation are shown in Table 1.

The abovementioned topics are encoded into a packet and sent across the RF link during communication between the ECD and the node. The packet is decoded to get the action to be performed—"pub" for publish, "get" for requesting data, etc., are as shown in Table 1.

Table 1 MQTT protocol packet implementation

Packet format	Function
get/{node_id}/{topic}	Used for requesting data of a particular {topic} from a sensor node with {node_id}
pub/{node_id}/{topic}/{data}	Used for publishing data from a particular "node" with {node_id} to the broker for the requested {topic}
con/{node_id}	Used to connect a sensor node with {node_id} with ECD
dcn/{node_id}	Used to disconnect the sensor node with {node_id} from ECD

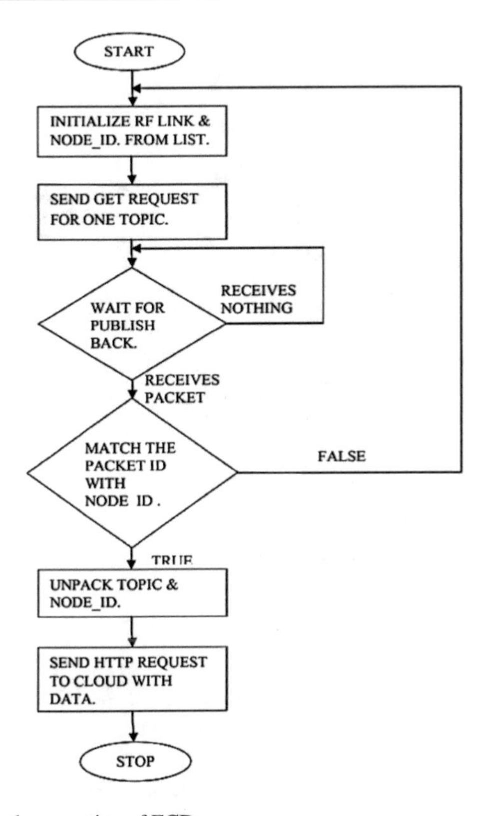

Fig. 4 Flowchart for the operation of ECD

For implementation, a single ECD with three identical sensor nodes were deployed with consecutive ID numbers 1, 2, and 3 as shown in Fig. 4. Simultaneous communication is not possible owing to the limitations of 433 MHz RF link that works in a single frequency with a defined bit rate at a moment. To avoid interference, a simple tokenized communication was set up where the nodes by default are in "receive" mode and wait for a request with their ID numbers. The ECD cyclically sends the request for all the nodes and topics (temp—Temperature, moist—Moisture, humid—Humidity) one after the other as shown in Fig. 5. All the nodes receive

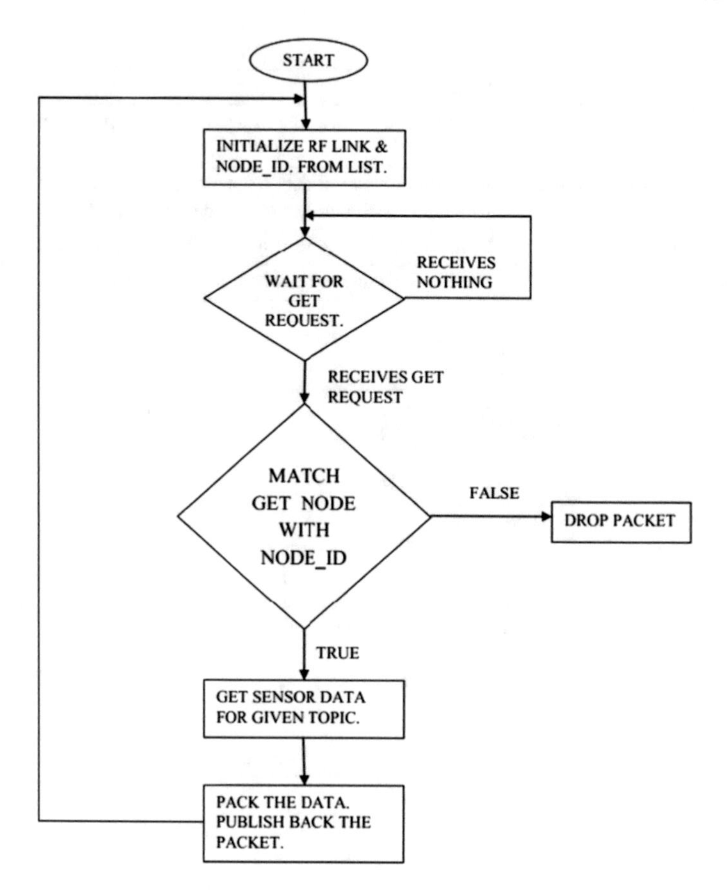

Fig. 5 Flowchart for the operation of things layer

the packets and decode them, but choose to ignore if the ID on the packet does not match. The node whose ID matches will go on to fetch the sensor value for the topic for which the request was received. Once it fetches the data point, it transmits it back to the ECD with "pub" action and the packet construction mentioned above. This goes on for each topic and ID cyclically, and the nodes respond systematically as expected. The range for the RF link between the ECD and the nodes was successfully tested with no packet loss and no loss in synchronization to a maximum distance of 12 ft [9, 10].

Then, the received data on the ECD is fed into a URL that is called using the "requests" Python library to raise an HTTP request to the server deployed at Heroku. The web server is built using the Django web framework written in Python and deployed at Heroku platform-as-a-service.

The web server receives the data and then writes it into the PostgreSQL Database v9.5 integrated with the Django server. The data can be monitored on the project site URL by logging in with an appropriate credential.

5 Results and Discussion

Figure 6 shows the implementation block diagram for the proposed system. For the implementation of the things layer, the choice of the ubiquitous sensors DHT11 and the standard soil moisture sensor was obvious. This lets us to collect three data points—Temperature, humidity, and soil moisture. The node was constructed using the sensors mentioned above and Atmel ATmega328p microcontroller, built into the Arduino Uno board. The board was programmed using the Arduino IDE with C++ and a library called VirtualWire was used to program parts related to RF communication. 433 MHz radio frequency transmitter and receiver modules are used to design a custom single plug-in transceiver module for the Uno board. The sensor node is powered using a Li–ion battery and forms a stand-alone scalable device. Edge computing device needed to have sufficient computing power to process the data and control quickly. It was also necessary that it could be interfaced with any gateway device. Raspberry Pi 3 Model B was an easy choice given its popularity and adoption, which will make it easy for developers to make any developments to the system in future.

The Pi was setup using Raspbian Jessie OS, a toned down flavor based on Linux. The process logic was programmed using Python 3.5. "requests" Python library is used to connect to the web application API over HTTP. ECD also uses a custom-designed RF transceiver module to communicate with the things layer. The communication between the things layer and the edge computing layer was done using a

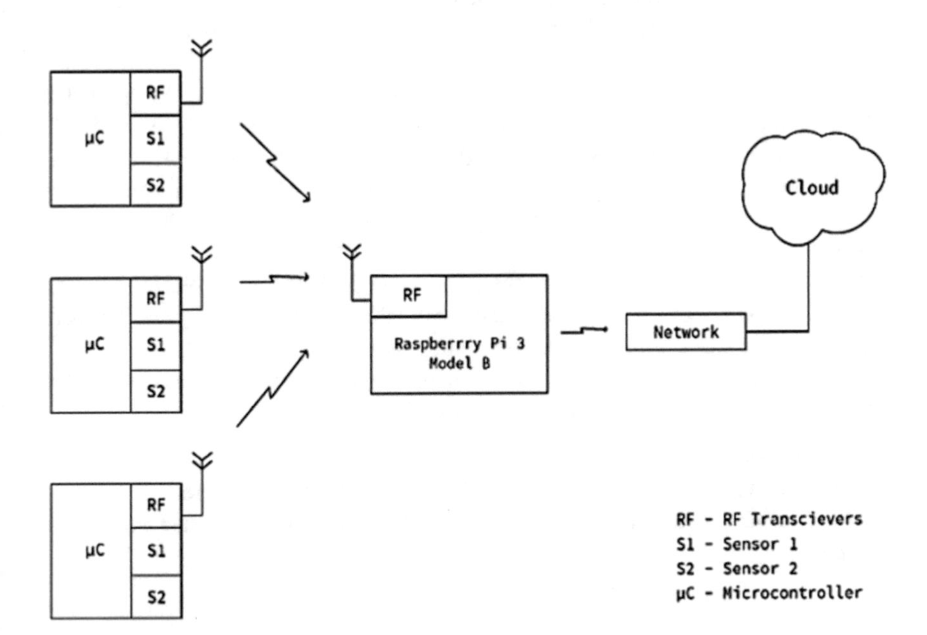

Fig. 6 Proposed block diagram

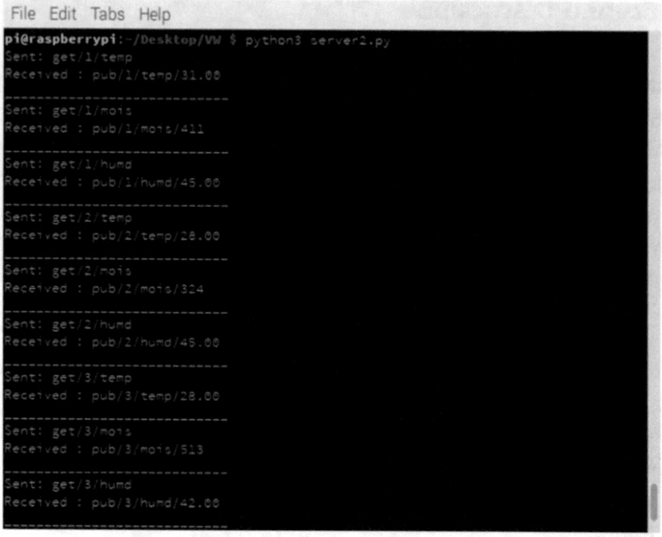

```
File Edit Tabs Help
pi@raspberrypi:~/Desktop/VW $ python3 server2.py
Sent: get/1/temp
Received : pub/1/temp/31.00
----------------------------
Sent: get/1/mois
Received : pub/1/mois/411
----------------------------
Sent: get/1/humd
Received : pub/1/humd/45.00
----------------------------
Sent: get/2/temp
Received : pub/2/temp/28.00
----------------------------
Sent: get/2/mois
Received : pub/2/mois/324
----------------------------
Sent: get/2/humd
Received : pub/2/humd/45.00
----------------------------
Sent: get/3/temp
Received : pub/3/temp/28.00
----------------------------
Sent: get/3/mois
Received : pub/3/mois/513
----------------------------
Sent: get/3/humd
Received : pub/3/humd/42.00
----------------------------
```

Fig. 7 The ECD terminal output of {get} requests and {pub} responses

high-level implementation of Message Queue Transport Telemetry (MQTT) protocol. Andy Stanford-Clark of IBM and Arlen Nipper of Cirrus Link Solutions authored the first version of the protocol in 1999 and is now maintained and improved by IBM.

Figure 7 shows the general operation of the ECD node request–response cycle. The ECD continuously sends {get} requests cyclically over a set of node IDs and registered "topics". Simultaneous communication with multiple nodes is avoided to eliminate interference between RF links on multiple nodes and ECD. The MQTT request and responses can be seen from the above screenshot of the terminal. The ECD is cyclically pinging nodes in list [1, 2, 6] where the numbers are the IDs of the registered nodes. It also picks topics from a list of topics for which sensors are attached on the nodes.

Figure 8 shows the home page of the project site deployed at Heroku PAAS. The home page summarizes the platform status in a table as seen in the figure. The user can log into the site admin page via the login page as shown in Fig. 9.

As the user/farmer logs in, he shows the admin page of the platform from where we can manage the entire project operation such as

1. Creation of new fields,
2. Addition of new nodes and linking them to appropriate fields, and
3. Temperature, humidity, and moisture database models are seen in the DataStacker app.

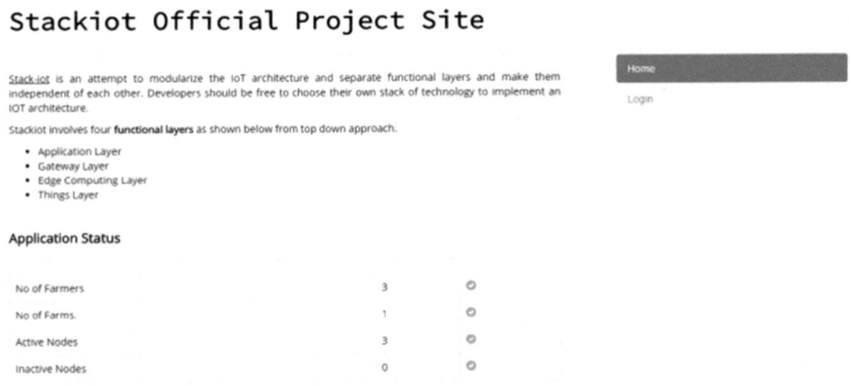

Stackiot Official Project Site

Stack-iot is an attempt to modularize the IoT architecture and separate functional layers and make them independent of each other. Developers should be free to choose their own stack of technology to implement an IOT architecture.

Stackiot involves four **functional layers** as shown below from top down approach.

- Application Layer
- Gateway Layer
- Edge Computing Layer
- Things Layer

Home
Login

Application Status

No of Farmers	3
No of Farms.	1
Active Nodes	3
Inactive Nodes	0

Fig. 8 The home page of the website

Stackiot Official Project Site

Please Enter your credentials to Sign In

Username
admin
Password
••••••••••••
Submit

Home
Login

Fig. 9 The login page

6 Conclusion

The proposed system was successfully developed and deployed into the field using tools like Adriano, Raspberry Pi 3 Model B, Python, Django Web Framework, and Standard 433 MHz RF link. The proposed architecture was successfully validated with inference from the experimental results where the entire modular stack was integrated and tested. Major problems were noticed in communication between the things layer and the edge computing layer where the implementation was done only in a half-duplex communication mode, and a single real-time thread execution of the logic was possible in the things layer. Both of these factors can be improved with hardware that is more capable of such execution. However, even with limitations observed, it is clear that it is possible to choose your own stack of technology for deployment of the architecture. Each layer in itself is independent of the layers around.

References

1. Rajalakshmi, P., Devi Mahalakshmi, S.: IOT based crop-field monitoring and irrigation automation. In: 2016 10th International Conference on Intelligent Systems and Control (ISCO). IEEE (2016)
2. Duan, Y.: Design of intelligent agriculture management information system based on IoT. In: 2011 International Conference on Intelligent Computation Technology and Automation (ICICTA), vol. 1. IEEE (2011)
3. Zhao, J., et al.: The study and application of the IOT technology in agriculture. In: 2010 3rd IEEE International Conference on Computer Science and Information Technology (ICCSIT), vol. 2. IEEE (2010)
4. Rahman, N.A.A., Jambek, A.B.: Wireless sensor node design. In: 2016 3rd International Conference on Electronic Design (ICED). IEEE (2016)
5. Gutiérrez, J., Francisco Villa-Medina, J., Nieto-Garibay, A., Porta-Gándara, M.A.: Automated irrigation system using a wireless sensor network and GPRS module. IEEE Trans. Instrum. Meas. **63**(1) (2014)
6. Grgić, K., Špeh, I., Heđi, I.: A web-based IoT solution for monitoring data using MQTT protocol. In: International Conference on Smart Systems and Technologies (SST). IEEE (2016)
7. War1e, T., Corke, P., Sikka, P., et al.: Transforming agriculture through pervasive wireless sensor networks. IEEE Pervasive Comput. **6**(2), 50–57 (2007)
8. Hu, J., et al.: Design and implementation of wireless sensor and actor network for precision agriculture. In: 2010 IEEE International Conference on Wireless Communications, Networking and Information Security (WCNIS). IEEE (2010)
9. MQTT Version 3.1.1.: May 14 2017 [Online]. Available: http://docs.oasisopen.org/mqtt/mqtt/v3.1.1/os/mqtt-v3.1.1-os.html
10. Ma, X., Valera, A., Tan, H.-X., Tan, C.K.-Y., Thangavel, D.: Performance evaluation of MQTT and CoAP via a common middleware. In: 2014 IEEE Ninth International Conference on Intelligent Sensors, Sensor Networks and Information Processing (ISSNIP)

M. R. Suma Assistant professor, Dayananda Sagar College of Engineering, Bengaluru, Karnataka, India. She completed her engineering from Mysore University in 1990. M.Tech. from VTU in 2014. She has Professional work Experience of over 22 years in the area of Teaching and Industry, Project guidance, Administration and other academic related activities. Authored 5 research papers in International Journal and Conference.

P. Madhumathy Associate Professor at Dayananda Sagar Academy of Technology and Management, Bengaluru, Karnataka, India. She completed her engineering from Anna University in 2006. M.E. from VMU in 2009 and Ph.D. from Anna University in 2016. She has Professional work Experience of over 12 years in the area of Teaching, Project guidance, Research, Training, Administration and other academic related activities. Experience includes 5 years of research in the area of Wireless Communication and Sensor networks. Authored 25 research papers in International Journal and Conference.

Reliable AI-Based Smart Sensors for Managing Irrigation Resources in Agriculture—A Review

R. Divya and R. Chinnaiyan

Abstract The advanced development of the wireless sensor today is termed as the “smart sensors”. The smart sensor in the field of agriculture is one of the new levels of developing factor in artificial intelligence system. The smart sensor in the field of agriculture is a need for country which is having a water scarcity, as well the growing countries that take the irrigation to the next level with the help of technology. This motivates an enormous attempt in the research activities of the irrigation field, standardization, and investment in the field of agriculture for the better management of the water resources. This survey paper aims at reporting an overview of sensors Saturas on miniature SWP technology for irrigation, major applications, and also provides the features of the Saturas, designs, and a case study based on a real implementation is also reported. Trends and probable evolutions are traced based on the agro-sensors for irrigation purpose.

Keywords Smart sensor · Miniature SWP · Saturas

1 Introduction

With the help of smart sensor devices in the agro-based system, it is used to analyze data generated from crops to irrigation commands, through which the farmer gets a real-time alert about the plants, like actual water needs of plants in the agricultural fields. The miniature SWP for irrigation to reduce water consumption and increase fruit production and quality (Stem water potential) sensor is embedded in the trunk of trees, vines, and plants. This is a part of automatic irrigation system, which is broadly

R. Divya (✉)
VTU, Bengaluru, Karnataka, India
e-mail: divya09chinnu@gmail.com

R. Chinnaiyan
Department of Information Science and Engineering, CMR Institute of Technology, Bengaluru, Karnataka, India
e-mail: vijayachinns@gmail.com

S. Smys et al. (eds.), *International Conference on Computer Networks and Communication Technologies*, Lecture Notes on Data Engineering and Communications Technologies 15, https://doi.org/10.1007/978-981-10-8681-6_25

named a Saturas sensor that provides precise information for optimized irrigation to reduce water consumption and increase fruit production and quality. The system is developed with the miniature implanted sensors and wireless transponders for determining the water status of fruits trees easily and inexpensively.

2 Working of Wireless Sensor

2.1 Working of Simple WSN

The working of simple single path WSN, here the wireless sensor network (WSN), is a network formed by a large number of low-power and low-complexity wireless nodes that can sense a variable parameter from the physical environment and transmit the collected data to a sink (or possibly multiple sinks), typically through multiple hops as depicted [4] in Fig. 1.

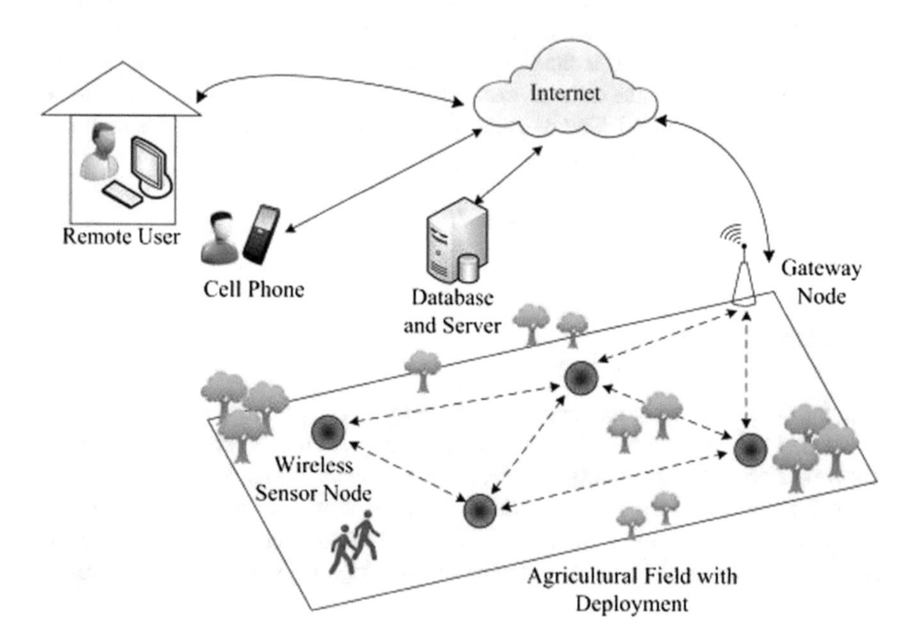

Fig. 1 Wireless sensor network w.r.t. irrigation resources of agricultural field

Fig. 2 Miniature implanted sensor

2.2 *Working of Wireless Smart Sensor in Agriculture*

The wireless sensor is most likely called as the smart sensor in this paper. These smart sensors play a vital role in the communication network of the real-time world. These WSNs usually consists of hundreds to thousands of sensor points that help in detection of the sudden event that occurs in the environment, and these events are sensed by the sensor and transmitted to the user through sink in a single-hop or multiple-hop method. Generally, this particular process of WSN's will apply to the devices; hence, we hereby consider sensor device named Saturas that are great need for the agricultural environment; this sensor device in particular helps to overrule the irrigation process by implementing the sensor devices to the stem of the plant. Just like the working of the other devices that are associated in the sensors filed and are related to implantable and wearable device for the mankind to monitor the health of the individual. This device is also one of the kinds but exclusively made to monitor the plants, to yield the better crop, and helps in the conservation of the water. Hence, the device named Saturas is basically a miniature-implanted sensor device of the plant, which will monitor the plants' stem, through which it can sense the requirement of water for plant only when it is needed. It also rapidly helps to transfer the sensed data and thereby sends the signal to the mobile of the host (farmer) that the plant is in need of the water. Saturas's sensing system comprises miniature-implanted sensors and wireless transponders that can measure stem water potential (SWP), a metric that is widely recognized as one of the most accurate for determining the water status of plants (Fig. 2).

2.3 *Need for Smart Sensors "Saturas" in the Agriculture*

The advance sense of precision irrigation is used to understand the Saturas, where this precision irrigation is a method of allowing to save the water along with the increased crop quality and quantity. Most of the farmer in the world irrigate without any limitation. But to control the water usage, management of the water with low cost, qualitative and quantitative yields, also helps in the groundwater contamination. For example, in wine grapes, smart irrigation can make the difference of the wine quality. So, the precise irrigation controls the precious resources of water for optimal yield. Unless and until we measure, we cannot control it. The need arises here is how can a framer measure the real water status of the crops.

2.4 *Reliability of Wireless Sensors*

A reliability model for sensor systems represents a clear picture of the data functional interdependencies providing a means to trade-off in sensor design alternatives and to identify areas for sensor design improvement of sensors. The reliability models for sensors are also helpful in

(i) Identifying critical items and single point of failure of sensor.
(ii) Allocating reliability goals to portions of the design of sensors.
(iii) Providing a framework for comparing estimated reliability of sensor design methods.
(iv) Trading-off alternative fault tolerance approaches for sensor models and its approaches.

Reliability of systems is presented by Chinnaiyan and Somasundaram [8–16]. The authors evaluated the reliability of systems with novel methods and ensured that the reliability models provide optimized reliability results. The same can be applied to all the sensors and techniques for providing reliable and optimized results.

3 Limitation of the Current Sensor Technology Available in Irrigation

The current technology enables the farmers to know many details about the agriculture details related to the soil moisture or like measuring the thickness of the stem.

(1) soil based: basic water needs only, empirical data.
(2) climate based: high variability, limited accuracy.
(3) plant based: indirect indication, difficult interpretation, high variability, expensive.

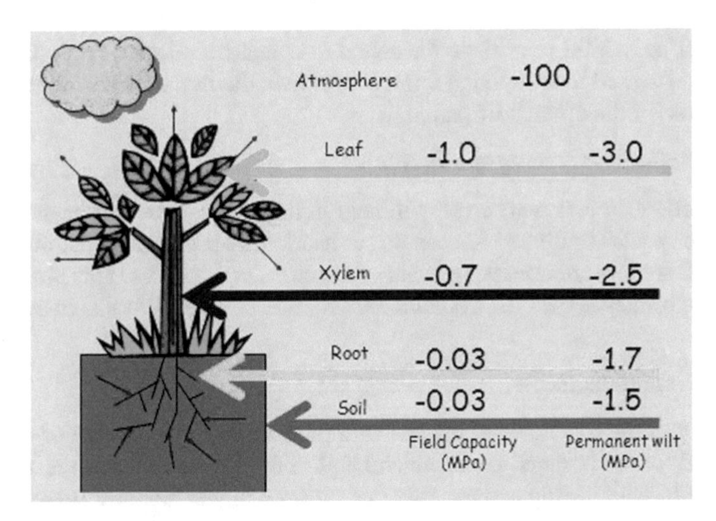

Fig. 3 Water moves upward in tree (stem water potential upward movement from root to shoot)

3.1 Measuring the Water as It Moves Upward in the Tree

Here, we are able to follow the water movements from woods to leave which helps in the identification of the stem water potential (SWP) along the tree. This SWP is one of the scientifically recognized and highly accurate parameters for determining the water status from root to shoot in the tree (Fig. 3). These scientific parameters considered best indication of water status in many crops. But today it can be only being measured by the complex manual device.

3.2 Solution for the Problem Is Saturas

Due to the limitation of the current technology, the farmers will search for the accurate, direct, and cost-effective solutions to determine the real water status for the irrigation as follows:

(A) ***Sensor embedded in tree stem***

Saturas miniature sensors are embedded in the tree stem. The stem is equal to the best location due to integration, it is with the direct contact with water tissue, and also direct water status measurement is possible with this miniature sensor. This gives online, direct, and reliable measurement of the precise status.

(B) ***Continuous and accurate data***

The continuous activity of the sensors helps to retrieve the data associated with the tree that is done through the sensors. The sensors help in sensing the data continuously which will be helpful to retrieve the actual data that is collected from the sensors and then possessed to send to the farmers, and then the farmer is able to receive the instructions for the optimal irrigation.

(C) ***Cost-effective*: *1–2 sensors/hectare***

The accurate data that is received will help in the accurate measurement which in turn will be useful to insert 1–2 sensors per hectare. This is a very cost-effective, as hardly 1–2 sensors are used per hectare of the crops in the farms. This is also one of the main advantages over the available sensors that require 5–9 units to average the results.

(D) ***Low maintenance***

The ability of the Saturas is to irrigate with the help of exact amount of water that is required for the farmers' need, which helps in the irrigation. It also results in the better yields, with quality and quantity that means a higher income. There is always a less risk in losses and saving.

4 Innovation Features of the Technology

4.1 Background Scenario

Today, farmers can only use a manual, labor-intensive procedure for SWP measurement for optimal irrigation. The need for the technology arises as the part of the agricultural system where there are large numbers of countries that face the water scarcity in the agricultural fields. There are around 80% of farmers who irrigate their trees without any scientifically based information. When there is no much knowledge about the scientific factors for growing the crops, the farmer obviously opt for the old and traditional method of the growing of the crop, by watering the trees with the water in a particular instance of the time; this process actually causes water waste, affects the quality and quantity of the fruit, and reduces profitability. The watering of the trees is done even when there is no requirement of the water at all, which in turn leads to a water wastage. Hence, there is a need for the miniature-implanted sensor devices like Saturas.

4.2 Saturas the New Technology

As a part of the automated irrigation system, these sensors provide the accurate information for the optimized irrigation to reduce water consumption and increase the fruit production and quality. It is a decision support system (DSS) based on the miniature stem water potential (SWP) sensors that are basically embedded in the trunks of the trees, veins, and plants. Saturas extremely correct sensors are the mere ones implanted in the stem of the tree which will help to tender automatic SWP measurement united significantly lower to provide easy to use the cost than crop currently on the market related to the crops. The SWP is calculated through a sensor implanted in the tree trunk which provides

- The best location due to integration,
- Direct contact with water tissues, and
- Direct water status measurement.

4.3 Advantages of the Implanted Sensors

Implanting the sensor within the tree trunk has a great advantage with respect to the other parts of the plants like leaves or the branches, soil, and also the problem that arises with these are avoided when implanting the sensors. These sensors usually provide accurate information for optimized irrigation which will allow to reduce water consumption without any stress to the plants and also increases fruit production and fruit quality. SWP is the least spatially variable feature in an orchard or vineyard because stems integrate the water status of the plant from the soil water around and under the plant and up to the atmosphere. The advantages of these sensors with respect to the plants are as follows:

- Continuous plant water status measurement,
- Intuitive and efficient information display,
- Low maintenance,
- Minimal spatial variability, and
- Cost-effectiveness.

4.4 Capabilities of Saturas

According to Saturas, its sensor is the only one that is embedded in tree trunks, providing direct contact with tree/plant water tissues. This enables an accurate measurement of the water status of the plant and eliminates any inaccuracies that are associated with placing sensors in the soil or on leaves and branches. With this more accurate information, farmers can irrigate more precisely and at the right time to help boost the yield and quality of fruit crops. Saturas estimates that this can increase revenues by up to 20% while reducing water use by 10–20%. And the market for irrigated orchards, vineyards, and cotton is worth more than $1 billion a year. It will be able to prove to the farmers that Saturas' precision irrigation technology can give them ongoing data on the actual water status of their crop with

- high efficiency and
- low cost.

4.5 Network-Driven Requirements in Terms of Network Operation and Management

These could be scalability, energy efficiency, capacity, cost, security and privacy, network flexibility, coverage, and spectrum management.

5 The Value Proposition Scenario

5.1 Value Proposition of Saturas Sensors

The technological value is précised as a part of the precision irrigation technology, to which can provide the solutions to farmers to irrigate at the right time and use the precise amount of the water. The benefits of the farmers are as follows:

- Minimizes risks and losses,
- Saves up to 10–20% water,
- Enables more planted area within water allotments,
- Reduces water costs,
- Increases fruit yields and quality higher income by 5–30%,
- Avoids overwatering,
- Produces healthier trees, and
- Lowers groundwater contamination.

5.2 Handling the Saturas Sensors

(1) optimal irrigation model for optimal yield.
(2) advanced decision support system for optimal irrigation.
(3) optimal irrigation management for the optimal yield.

6 Implementation of the Saturas Sensors

The advancing of the product usually helps to determine the way to powerful usage of the miniature sensors. The miniature sensors are nothing but the smart sensors that are made exclusively to fit the tress to sense and measure their water potential and supply water only when the tree is need or requirement of the water. Here, the process of embedding the sensors into the stem starts with simple jewel and under wet condition and then the placing of the sensors in the tree 1, seal it with the common ceiling material and connect the embedded sensors to the communication box attached to the tree 3. The stepwise implementation is as follows:

1. wet condition of the tree and placing sensor in tree: the embedding of the smart sensors into the tree where there is a wet condition of the tree is found, i.e., the water upward moment happening area from the root to shoot, the movement of water happens called as wet area. These sensors are placed in a wet area of the tree called wet condition placing of sensors.
2. ceiling of sensor: once the sensor is placed in the wet area or the wet condition of the tree, the sensor is tightly sealed with the sealers that are sealed under the wet conditions, through which the achievement of the successful placement of the sensor is achieved. So, there is disruption in further time of the plant sensing area.
3. communication box: after the ceiling process, the communication box is fitted to the tree of certain number of the sensors that are allotted for each tree to achieve the communication link between the two devices that are associated here for the proper relying of the data, when should the water be supplied to the tree.

7 Technique Used: Use Case

7.1 The Techniques Used

Currently used tests in the miniature smart sensor is with the help of the alpha test, in the avocado, mango, and peach.

- Poc in laboratory tests plus tree stem.
- Penetration protocol in the tree stem.
- Successful alpha test in trees.

7.2 Use Case: 1

Consider the condition where the alpha test technique in the peach tree is conducted, the peach tree implemented as above Saturas implementations. Here, it is tested to check and get the expected pattern for the precise irrigation with the help of the Saturas Miniature sensors that are embedded into the peach tree. Where the sensors are operated continuously in the peach tree, the data here shows the accurate data retrieved, which is helped with the measurements to know about that the data obtained is whether the expected pattern during the day (Fig. 4). Here, we can note that the stem water potential during the mid-day peak can indicate the stress condition and also the minimum level just before the sun high's when the tree is full of water. The resulting pool which we can get is the contact with the sensor membrane

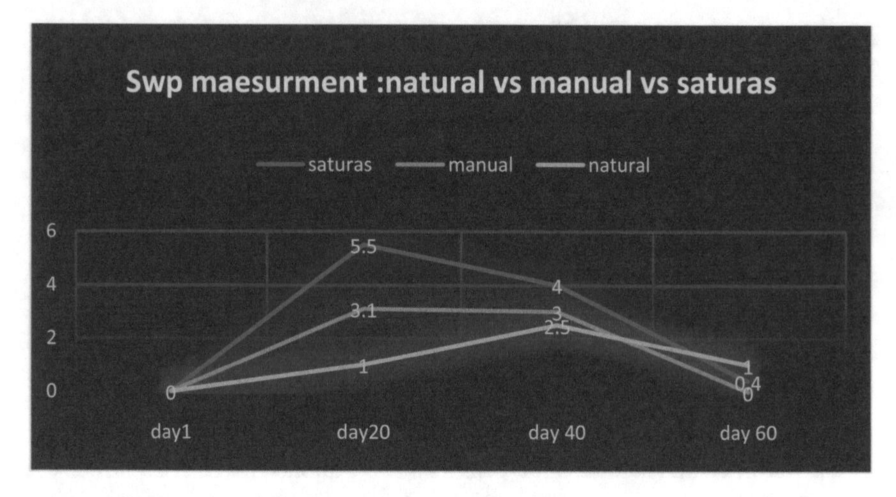

Fig. 4 SWP measurement natural versus manual versus Saturas

and trees water tissues, their compatibility between these measurements, and the stem water potential measured by the manual devices. Hence, the continuous SWP measurements manually are checked with the parameter against the Saturas:

- Shows daily pattern,
- Compatible with manual measurement at set intervals, and
- Reflect climate changes.

8 Conclusion

The smart sensors in the field of agriculture are used in the current generation to obtain a better agriculture environment. In this paper, we are trying to explain about the Saturas the miniature sensors that are implemented in the agricultural fields for the precise irrigation; here, we have given an overview, need, working, implantations, and techniques as well the use case for the better understanding of the miniature technology sensors. As future working, we are intended to work on the overall handling of the Saturas with the 5G technology to embedded to remotely access the data and operate the irrigation process in an automated remote handling method.

References

1. MacGillivray, C.: The Internet of Things Is Poised to Change Everything, Says IDC. Business Wire, 3 Oct 2013
2. McLellan, C.: The Internet of Things and Big Data. ZDNet, 2 Mar 2015

3. The device number comes from Ian King, "5G Networks Will Do Much More Than Stream Better Cat Videos. Bloomberg News, 2 May 2016
4. West, D.M.: How 5G Technology Enables the Health Internet of Things. Center for Technology Innovation at Brookings, July 2016
5. Arrobo, G.: Improving the Throughput and Reliability of Wireless Sensor Networks with Application to Wireless Body Area Networks. University of South Florida (garrobo@mail.usf.edu), Jan 2012
6. Agyapong, P., et al.: Design considerations for a 5G network architecture. Commun. Mag. **52**(11), 65–75 (2014). (IEEE)
7. Ullah, S., Higgins, H., Braem, B., Latre, B., Blondia, C., Moerman, I., Saleem, S., Rahman, Z., Kwak, K.S.: A comprehensive survey of wireless body area networks. J. Med. Syst. (2010)
8. Chinnaiyan, R., Somasundaram, S.: Reliability assessment of component based software systems using test suite—a review. J. Comput. Appl. **1**(4) (2008)
9. Chinnaiyan, R., Somasundaram, S.: An experimental study on reliability estimation of GNU compiler components—a review. Int. J. Comput. Appl. (0975–8887) **25**(3) (2011)
10. Chinnaiyan, R., Somasundaram, S.: Reliability of component based software with similar software components—a review. J. Softw. Eng. (2010). (i-Manager)
11. Chinnaiyan, R., Somasundaram, S.: Monte Carlo simulation for reliability assessment of component based software systems. J. Softw. Eng. (2010). (i-Manager)
12. Chinnaiyan, R., Somasundaram, S.: Evaluating the reliability of component based software systems. Int. J. Qual. Reliab. Manag. **27**(1), 78–88 (2010)
13. Chinnaiyan, R., Somasundaram, S.: Reliability of object oriented software systems using communication variables—a review. Int. J. Softw. Eng. **2**(2), 87–96 (2009)
14. Chinnaiyan, R., Kumar, A.: Estimation of optimal path in wireless sensor networks based on adjacency list. In: 2017 IEEE International Conference on Telecommunication, Power Analysis and Computing Techniques (ICTPACT-2017), 6–8 Apr 2017. IEEE. 978-1-5090-3381-2
15. Divya, R., Chinnaiyan, R., Ilango, V.: Reliability evaluation of wireless sensor networks—a review. Int. J. Eng. Sci. Res. Technol. **5**(12), 598–604 (2016)
16. Divya, R., Chinnaiyan, R.: Reliability evaluation of wireless sensor networks (REWSN—reliability evaluation of wireless sensor networks). In: International Conference on Intelligent Computing and Control Systems (ICICCS). IEEE (2017), pp. 847–852

R. Divya completed her Master Degree in the field of Computer Applications. Now she is doing her Ph.D. in Computer Applications in VTU. Her research interest includes Security, Reliability of Wireless Sensor Networks, Machine Learning, Big Data, Cloud Computing and Data Science.

Dr. R. Chinnaiyan is working as Associate Professor in the Department of Information Science and Engineering, CMR Institute of Technology, Bengaluru, Karnataka, India. He is having 17+ years of teaching experience. He is a life member of ISTE and CSI of India. He completed his research in Anna University-Chennai at Coimbatore Institute of Technology, Coimbatore in 2012. His research interest includes Object Oriented Analysis and Design, Qos and Software Reliability.

Dynamic Data Auditing Using MongoDB in Cloud Platform

P. Akilandeswari, Siddharth D. Bettala, P. Alankritha, H. Srimathi and D. Krithik Sudhan

Abstract Shared computing resources in a pool enable various services that accessed over Internet for storing large company data. It has reduced the cost and infrastructural needs but there is a concern for data safety and integrity. The existing remote data auditing technique integrates the application of algebraic signature properties of a cloud and new architecture, and divides rule table to allow users to carry out data manipulations quickly. However, majority of the auditing techniques are done on static dataset and also incurs a computational overhead when data size increases. In this paper, dynamic data updating which is a pivotal function in data auditing is used. To reduce the overhead and increase the computational efficiency, the proposed system uses MongoDB. This enables the system to be scaled to larger files and also reduces the computation time elapsed in identifying updated data by using JSON format of stored files in the database. Additionally, the security concern and client's overhead are minimized by using RSA signatures to conserve the confidentiality of the data uploaded by the cloud consumer. The auditor quickly and efficiently scans only the updated data block for any viruses or malware, thereby reducing the cost and computational power requirements of the third-party auditor and also improves the overall speed and efficiency, encouraging more people to approach the cloud space with trust.

Keywords Dynamic data auditing · MongoDB · RSA encryption
Cloud security · Data privacy

P. Akilandeswari · S. D. Bettala · P. Alankritha · H. Srimathi · D. Krithik Sudhan (✉)
Department of Computer Science and Engineering, SRM Institute of Science and Technology, Chennai, India
e-mail: krithikapple96@gmail.com

P. Akilandeswari
e-mail: akilandeswari.p@ktr.srmuniv.ac.in

S. D. Bettala
e-mail: siddharth.betala1997@gmail.com

S. Smys et al. (eds.), *International Conference on Computer Networks and Communication Technologies*, Lecture Notes on Data Engineering and Communications Technologies 15, https://doi.org/10.1007/978-981-10-8681-6_26

1 Introduction

Nowadays, organizations have a large amount of data and it gets harder to store everything in local storages. So, cloud storage becomes an ideal solution for this. It can store a large amount of data with ease. The stored data can be accessed from anywhere when required via Internet connection. This storage can be used as an emergency backup. It reduces the cost of storing in local storages. But data integrity and security of the outsourced data continues to be a matter of major concern for data owners.

Data security plays an important role in data stored in cloud datacenter. Data security remotely applied to the stored data by the client as the client does not have direct access to the storage units. This makes the security the most vulnerable element in the cloud architecture. There are few important things related to security, (1) accessing secured data with confidentiality and transferring from and to the cloud user, and (2) checking without data being tampered. Organizations and users trust the cloud service provider on privacy of their data so it is an important factor for cloud storage. But for implementing this, finding the best encryption technique continues to be a major concern.

To implement data security, encryption techniques are used. These techniques help the data to be secure and be accessed only by authentic users. Different cloud providers use different encryption techniques. In the existing system, techniques like searchable encryption and encryption based on fragmentation are used. Here, RSA and AES algorithms are used. It is a simple and most secure encryption technique. Here, the data is encrypted using public key and private key. The date owner gets the public key and the auditor has the private key to decrypt the data when auditing has to be performed. Now handling the data efficiently plays an important role for better storage.

Data handling also plays an important role in cloud storage. This also varies from one provider to another. Here, MongoDB is used for handling data. Earlier auditing techniques can be implemented only on static word files. Here, MongoDB is used to handle all types of data. Here, unique IDs are assigned to each file and are stored. So when data owner wants to access the data, data can be accessed using the IDs and this also applies for auditing when a data is modified and requires auditing. This system is complex but very effective to store and access data easily. The data stored or modified should be audited before it is sent to the cloud for storage, so data auditing is done.

Data auditing is the final part before the data is uploaded to cloud storage. Data auditing is done by a remote auditor. Data auditing is the process of checking the data for malware or any viruses before the data is uploaded to the cloud, as corrupted data can completely destroy the cloud storage. Here, MongoDB is used, so files are uniquely identified with their IDs, so when a file is updated or new file is added, the auditor does not have to download entire file but only the updated file and auditing is done. This saves a lot of time and cost. And finally, the data is uploaded back to cloud.

2 Literature Survey

Amjad Alsirhani et al. [1], **Improving Database Security in Cloud Computing by Fragmentation of Data**, IEEE 2017: This system uses a combination of fragmentation of data and different encryption algorithms like AES, OPE, and HOM for improving database security. Here, the data is secure but delay time is high.

Faurholt-Jepsen [2], **Electronic monitoring in bipolar disorder**, Medicine 2015

Iroju Olaronke et al. [3], **Big Data in Healthcare: Prospects, Challenges and Resolutions**, IEEE 2017: This system uses cloud data with healthcare services, software, platform, and infrastructure to handle data efficiently. It leads to better data storage but resistance to change affects this system.

Delishiya Moral et al. [4], **Improve the Data Retrieval Time and Security Through Fragmentation and Replication in the Cloud**, IEEE 2017: Here, fragmentation and sole replication technique are used for security and faster retrieval time. But implementation of this requires more work.

Ahmed El-Yahyaoui et al. [5], **A Verifiable Fully Homomorphic Encryption Scheme to Secure Big Data in Cloud Computing**, IEEE 2017: It introduces new FHE based on mathematical structure for secure data. It preserves data confidentiality and it is noise free. It suffers from resistance to change and requirement of more work.

Nagesh et al. [6], **Study on Encryption Methods to Secure the Privacy of the Data and Computation on Encrypted Data Present at Cloud**, IEEE 2017: Here, new homomorphic encryption technique (BGV's leveled FHE scheme) is used to secure data. It preserves data confidentiality and it is noise free but suffers from resistance to change.

Neha Pramanick et al. [7], **Searchable Encryption with Pattern Matching for Securing Data on Cloud Server**, IEEE 2017: This paper uses searchable encryption to secure data. It does not require server-level decryption. It can be easily attacked by Chosen Keyword Attack (CKA) and Keyword Guessing Attack (KGA).

Kamalakanta Sethi et al. [8], **A Novel Implementation of Parallel Homomorphic Encryption for Secure Data Storage in Cloud**, IEEE 2017: This paper uses practical and efficient homomorphic cryptosystem with parallel implementation for secure data storage in cloud. It does not require decryption of ciphertext. It requires more work as experimental work was only 80% successful.

3 Architecture Diagram

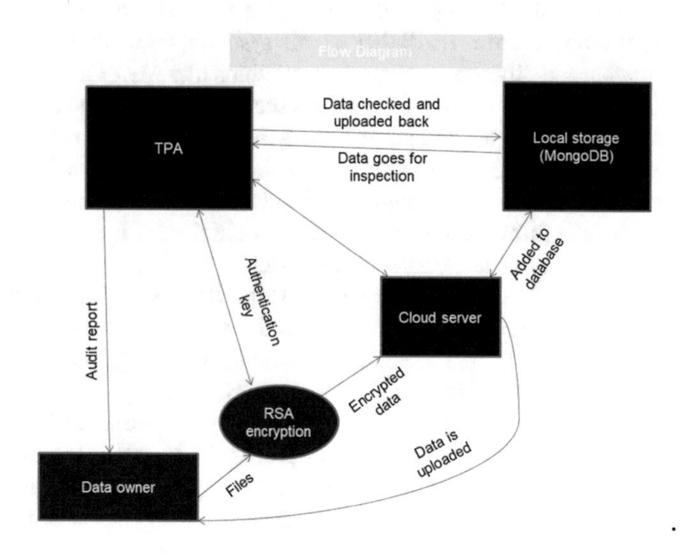

- **TPA**-Third-party auditor

4 Modules

4.1 Data Owner

Data owner is basically an organization or an individual user and uploads the data files to the cloud server. The person or organization can authorize or deny access to certain data that gets the IDs for modifying data and also audit report once the data is audited. Data owner interacts with the front-end HTML page.

Welcome to our system

- When user selects the "user" option, the user is redirected to authentication page.

file:///C:/Users/krithik%20sudhan/Desktop/user.html

Username Enter Username **Password** Enter Password Login Remember me
Cancel Forgot password?

- Once the user is authenticated, the user is redirected to the main page.

- Now the user performs desired functions.
- If a new file is added, the user gets the ID.

Sample ID:

```
"_id":ObjectId("52ffc33cd85242f436000001"),
```

4.2 *RSA Encryption*

When the user uploads the file, the file is encrypted for security. Here, the file is encrypted using public key and private key. These keys are shared between the data owner and auditor. During audit, the data is accessed by the auditor using the private key of the data owner.

Key Generation	
Select p, q	p and q both prime
Calculate n	$n = p \times q$
Select integer d	$gcd(\phi(n), d) = 1;\ 1 < d < \phi(n)$
Calculate e	$e = d^{-1} \bmod \phi(n)$
Public Key	$\mathrm{KU} = \{e, n\}$
Private Key	$\mathrm{KR} = \{d, n\}$

Encryption
Plaintext: $\mathrm{M} < n$
Ciphertext: $\mathrm{C} = \mathrm{M}^e \pmod n$

Decryption
Ciphertext: C
Plaintext: $\mathrm{M} = \mathrm{C}^d \pmod n$

Here, encryption is done after the user uploads the file to the system. The auditor is responsible for decrypting the particular portion of the data in the data file.

4.3 Cloud Server

The encrypted data file reaches the cloud server; the file gets unique IDs and is added to the local database (MongoDB). The cloud server also sends a notification to the user stating whether the data is uploaded successfully or not.

file:///C:/Users/krithik%20sudhan/Desktop/success.html

File successfully uploaded

4.4 Local Storage

This is a MongoDB database. It is a column-based database. It uses IDs to identify the files present in the database. Every time the file is modified, the IDs change as the IDs are assigned dynamically. It is a temporary storage before the file is uploaded to the cloud.

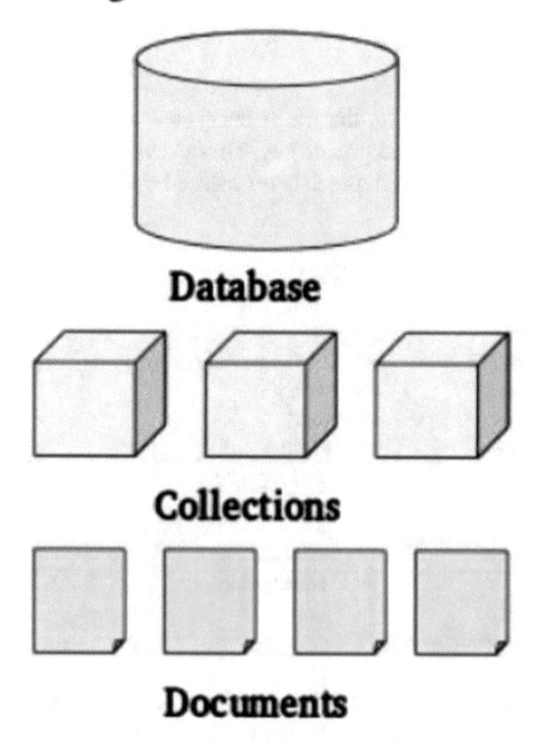

- This database has collections which in turn has documents in it.

4.5 *Third-Party Auditor*

TPA plays the most important role in this process.

Data uploaded in the cloud should not be corrupted as it destroys the entire cloud storage. So when a file is uploaded or file is modified, TPA decrypts the file and checks the file for malware and viruses and when the file is secure, upload the file to the cloud. TPA also sends audit report to the data owner. Auditor also interacts with the system using the same front end.

- Now, the auditor is redirected to authentication page.

- Once the auditor is authenticated, the auditor is redirected to the auditor page.

- The auditor decrypts the user's data to perform auditing.
- Once the auditing is complete, the auditor uploads the data to the cloud and sends audit report to user.

4.6 Audit Report

File successfully uploaded and the file is secure

5 Algorithm

Front end acts as an interface between user and system, so here html is used for composing it. MongoDB is used as temporary database. Authentic auditor audits the data. The algorithm of this process is given as follows:

- Get the user ID and password from user for authentication.
- Once the user is authenticated, the user is provided to choose whether to add or retrieve data.
- **For adding new file**:

1. User selects and uploads the file.
2. File is now encrypted using RSA algorithm.
3. Encrypted file is sent to MongoDB and assigned unique ID.
4. If the file size exceeds 16 MB, then the file is sent to grids.
5. This file is now sent to TPA.
6. TPA decrypts the file.
7. The file is scanned for viruses or malware.
8. If the file is secure, the file is uploaded to cloud.
9. Audit report is sent to the user.

- **Data updation**

1. User retrieves the required file from cloud using ID.
2. User modifies the data.
3. User uploads the file to front end.
4. File is again encrypted using RSA algorithm.
5. Encrypted file is sent to MongoDB and assigned the unique ID.
6. If the file size exceeds 16 MB, then the file is sent to GridFS.
7. This file is now sent to TPA.
8. TPA decrypts the set of available files.
9. TPA compares the old IDs and new ID.
10. When there is a new ID, the file is scanned for viruses or malware.
11. If the file is secure, the file is uploaded to cloud.
12. Audit report is sent to the user.

- END.

Auditing

- Get the user ID and password from the user for authentication.
- Once the auditor is authenticated, the auditor is provided to choose whether to scan available files.

Scan available files.

1. download the file available from server.
2. decrypt the file using RSA algorithm.
3. check the time stamp of the file.
4. if the time stamp is new, then scan the entire file.
5. else scan the part of the file where the timestamp is changed.
6. if the file is free of malware, then upload the file to server.
7. else send it back to user.
8. generate an audit report of the file and send it user.

6 Analysis of Work

The data integrity enforcing capability of a traditional database is usually traded off by the large number of tables, keys, and indices that are created to facilitate simple operations on data. Therefore, in this project, we have chosen to work with MongoDB. MongoDB is an open-source cross-platform document-oriented database application for a specific field, range of certain queries, and searching for regular expression [9]. The database queries can return particular fields of document and contains user-defined JavaScript functions. Column database queries can also be combined in the pattern to return a random sample of results of a given size. The data load balancing and replicated data over multiple machines for storing file can be done by MongoDB. MongoDB also allows server-side Java execution in queries. The following statistical analysis was the driving force behind up-scaling the cloud service using MongoDB [10]. To insert, select, update, and delete, a tuple in MongoDB takes less seconds than MySQL.

The above-stated sentences are very persuading in terms of operating on very large database and also provide many applications on opportunity to be scaled to accommodate a larger audience [11].

The project proposes to use a MongoDB driver, GridFS, which enables the conversion of a file upload (or vice versa) by converting the operations to a necessary JSON-object format. This format aids dynamic data in the cloud storage that is based on MongoDB.

6.1 Security Analysis

The entire process was carried out on metadata that was easily arrived at by using MongoDB. The encryption process on metadata furthered secured the process by not only increasing the time the attackers would take in breaching but also making it very difficult to access the original data uploaded by the user.

The security and trust issues on the TPA for authenticity can further be resolved not only by the fact that the TPA performs audit tasks on metadata but also the usage of encryption keys. These keys can resolve the authenticity issue as only approved third-party auditors will be provided with the generated keys during the audit process [12].

7 Conclusion

The implemented database (MongoDB) helps to work on all types of data. This paper reduces overhead on data owner regarding data integrity and security by the use of RSA and AES encryption. This paper also reduces the time, energy, and cost associated for inspection post an update. The future work can be given as the files can be encrypted with more advanced encryption techniques and can be implemented on larger scale and to implement this for commercial purposes.

References

1. Alsirhani, A., Bodorik, P., Sampalli, S.: Improving database security in cloud computing by fragmentation of data. In: Computer and Applications (ICCA), 2017 International Conference on, pp. 43–49. IEEE, 2017
2. Faurholt-Jepsen, M.: Electronic monitoring in bipolar disorder. Medicine **1**, 14 (2015)
3. Olaronke, I., Oluwaseun, O.: Big data in healthcare: Prospects, challenges and resolutions. In: Future Technologies Conference (FTC), pp. 1152–1157. IEEE, 2016
4. Moral, W. D., Kumar, B. M.: Improve the data retrieval time and security through fragmentation and replication in the cloud. In: Advanced Communication Control and Computing Technologies (ICACCCT), 2016 International Conference on, pp. 539–545. IEEE, 2016
5. El-Yahyaoui, A., El Kettani, M. D. E. C.: A verifiable fully homomorphic encryption scheme to secure big data in cloud computing. In: Wireless Networks and Mobile Communications (WINCOM), 2017 International Conference on, pp. 1–5. IEEE, 2017
6. Nagesh, H. R., Thejaswini, L. Study on encryption methods to secure the privacy of the data and computation on encrypted data present at cloud. In: Big Data Analytics and Computational Intelligence (ICBDAC), 2017 International Conference on, pp. 383–386. IEEE, 2017
7. Pramanick, N., Ali, S. T.: Searchable encryption with pattern matching for securing data on cloud server. In: Computing, Communication and Networking Technologies (ICCCNT), 2017 8th International Conference on, pp. 1–8. IEEE, 2017
8. Sethi, K., Majumdar, A., Bera, P.: A novel implementation of parallel homomorphic encryption for secure data storage in cloud. In Cyber Security and Protection Of Digital Services (Cyber Security), 2017 International Conference on, pp. 1–7. IEEE, 2017

9. https://dzone.com/articles/why-mongodb-is-worth-choosing-find-reasons
10. Eason, G., Nichols, L.: A Comparison of Object-Relational and Relational Databases, Presented to the Faculty of California (Chapter 4)
11. http://ieeexplore.ieee.org/document/7433067/
12. https://www.sciencedirect.com/science/article/pii/S1877050916001411

Mrs. P Akilandeshwari working in the Department of computer science, SRM University, Chennai, India. Her research filed includes network security and cloud computing.

Siddharth D. Bettala working in the Department of computer science, SRM University, Chennai, India. His research filed includes network security and cloud computing.

P. Alankritha working in the Department of computer science, SRM University, Chennai, India. Her research filed includes network security and cloud computing.

H. Srimathi working in the Department of computer science, SRM University, Chennai, India. Her research filed includes network security and cloud computing.

D. Krithik Sudhan working in the Department of computer science, SRM University, Chennai, India. Her research filed includes network security and cloud computing.

Securitization of Smart Home Network Using Dynamic Authentication

Shruthi Sreedharan and N. Rakesh

Abstract Smart home networks today span from technological advances that introduced highly networked devices of high density to applications which are more vulnerable due to its increasingly invasive nature to personal space and data. Such networks are becoming susceptible to multiple types of attacks that are seen today with a high number of third-party framework vendors in increasing risk of malicious attack. In such scenarios, dynamic- or context-aware authentication may provide a reinforced measure that takes into account environmental changes or behavioral changes in the network to enable the administrator in the decision-making process. The following work intends to explore a methodology in the incorporation of context-based processing that attaches the context-based filtering intelligence deep within the IoT network at the data processing center.

Keywords Authentication · Threat intelligence · Intrusion detection · IoT
Smart home

1 Introduction

Dynamic flagging-based authentication methods are explored as an option, as opposed to static authentication, since

- Simple authentication is not sufficient in a dynamic network,
- Biometric methods are hardware-intensive and induce hassle in reconnection,
- Token-based methods are also hardware-intensive and not cost-effective for a solution based on smart home use-case scenario,
- These methods are also limited to location-based context [1, 2], gathering, and

S. Sreedharan (✉) · N. Rakesh
Department of Computer Science and Engineering, Amrita School of Engineering, Amrita Vishwa Vidyapeetham, Bengaluru, India
e-mail: shruthi.sreedharan@gmail.com

N. Rakesh
e-mail: n_rakesh@blr.amrita.edu

S. Smys et al. (eds.), *International Conference on Computer Networks and Communication Technologies*, Lecture Notes on Data Engineering and Communications Technologies 15, https://doi.org/10.1007/978-981-10-8681-6_27

- Predefined security questions during identity checks can be challenging to remember for the users.

A non-static authentication method is proposed in the following paper which applies each time a device user logs into the IoT network of the connected smart home. This will utilize suitable feature selection algorithms to identify specific user profiles that access the network and isolate strange or unfamiliar profiles by flagging it for future denial of service at the database.

The behavioral entities that capture a user's history in a session need to be analyzed and forwarded for the decision step. The entities require to be recognized from the data with the help of multiple metadata [3, 4], or features which may work in isolation or have correlation along with other features.

This filtering would be based on these features [5, 6] such as time difference of logins, device ID, brute force attempts, failed login histories, etc. The software aspect of the system can be viewed as a combination of the following parts, and their respective roles in the system are described further below:

- Log stream generator,
- Smart gateway, and
- Predictive model.

1.1 Log Stream Generator

The log generator script is the program simulating real-time entry of users hitting the firewall. The code constructs log streams per user login and defines each feature for that stream. The data supplied for these features are suitably randomized and adjusted for simulating real-world scenarios [7].

The features consist of username, timestamp, IP source, IP destination, port destination, file name, file size, MAC address, hostname, source domain, and a final feature—"IsFlagged", which denotes whether the user has been flagged malicious (value 1) or not (value 0).

1.2 Smart Gateway

The smart gateway script is the code with the intelligence to flag the users based on certain supplied characteristics, otherwise known as Intrusion Detection Schemes (IDS). Many IDS schemes have been formulated by numerous authors and corporations. The selectivity of the IDS is defined by the application [8], the features or metadata describing the sessions, and the quality of the data itself.

In the smart gateway that is constructed, two IDS are implemented based on commonly accepted network intrusion scenarios:

- Same user via multiple IPs (Fraud logins): Logins having the same user profile but observed to use different source IPs are flagged using this rule.
- Synflood attack: This scenario is a type of Denial-of-Service (DDoS) [9, 10] attack achieved when a malicious actor sends incomplete/ half-opened requests to the server. This utilizes multiple ports in the request, and hence attacker achieves DDoS.

The output csv from the log generator containing the set of log streams is fed to the smart gateway which would be used to implement IDS on it. The filtered users would then be considered in the next run of log generation to flag the incoming suspected user.

1.3 *Predictive Model*

The dataset so produced is then sent as input into a suitable modeling tool such as Rstudio or Weka. Weka tool was chosen in order to perform a comparative study of the classifiers applied to the dataset. The method of training and test split was chosen as 10-fold cross-validation. This is done in order to minimize the effect of bias from any single training set on the model that would be subjected to tests.

2 Software Tests and Predictive Classifier Analysis

The first step of the testing requires the generation of the initial dataset. This is done by executing the log generator script which produces log streams for 50 user instances. The output logfb.csv is stored in the same root folder for access to the gateway script. At this juncture, the "IsFlagged" feature contains all 0 s except for the set of users deliberately flagged 1 (as blacklisted IP users) (Fig. 1).

Datetime	IPSrc	Host	MACad	User	Filename	Filesize	DomainSrc	IPDest	PortDest	IsFlagged
11/11/2017 23:59	62.0.0.0	WIN-9EOZPTSNK	Y1:12:W7:7O:2F:7	6	ff99u4x7g.exe	215040	cardenas.com	123.19.0.8	63	0
11/11/2017 23:59	142.16.0.0	WIN-AOFN6	MC:I3:AL:Z6:TP:P!	14	i6nol6db7.txt	932864	lynch-knapp.com	123.19.0.8	53	0
11/11/2017 23:59	142.16.0.0	WIN-15L1	MC:I3:AL:Z6:TP:P!	9	ijzy6shzv.exe	215040	garcia.org	192.168.1.1	98	0
11/11/2017 23:59	244.192.0.0	WIN-65C7VF9WA	B1:PQ:NH:QZ:VV:	5	8vkv156fi.exe	932864	elliott-nguyen.co	123.19.0.8	73	0
11/11/2017 23:59	123.73.19.0	WIN-AOFN6	2M:S6:6S:TE:R3:E)	14	nn7ubiy58.xls	932864	lopez.info	123.19.0.8	36	0
11/11/2017 23:56	123.73.19.0	WIN-Q5MO0C0C1	K4:9Q:X6:G6:R5:1!	1998	ff99u4x7g.exe	23562	baker.com	123.19.0.8	91	1
11/11/2017 23:53	183.0.0.0	WIN-Q5MO0C0C1	F8:RU:L7:PQ:1I:0C	1998	nn7ubiy58.xls	215040	garcia.org	192.168.1.1	22	1
11/11/2017 23:49	183.0.0.0	WIN-AOFN6	B1:PQ:NH:QZ:VV:	1998	ff99u4x7g.exe	215040	elliott-nguyen.co	123.19.0.8	76	1
11/11/2017 23:46	239.79.8.0	WIN-TJQZXZG	K4:9Q:X6:G6:R5:1!	1998	na043b2fz.exe	215040	elliott-nguyen.co	123.19.0.8	97	1
11/11/2017 23:43	244.192.0.0	WIN-9EOZPTSNK	0G:OD:ZJ:R7:UC:H	1998	ijzy6shzv.exe	215040	hayes.info	192.168.1.1	62	1
11/11/2017 23:43	214.196.224.0	WIN-Y306D04X	OL:EG:E9:11:Q4:PI	27	ijzy6shzv.exe	23562	hayes.info	192.168.1.1	45	0
11/11/2017 23:43	247.0.144.0	WIN-9EOZPTSNK	F8:RU:L7:PQ:1I:0C	21	ijzy6shzv.exe	932864	lopez.info	192.168.1.1	70	0
11/11/2017 23:43	62.0.0.0	WIN-Y306D04X	Y1:12:W7:7O:2F:7	27	ijzy6shzv.exe	23562	lopez.info	192.168.1.1	26	0
11/11/2017 23:43	214.196.224.0	WIN-AOFN6	2M:S6:6S:TE:R3:E)	27	8vkv156fi.exe	215040	garcia.org	192.168.1.1	94	0
11/11/2017 23:42	123.73.19.0	WIN-NX	OL:EG:E9:11:Q4:PI	15	8vkv156fi.exe	215040	garcia.org	192.168.1.1	57	0
11/11/2017 23:42	183.0.0.0	WIN-AOFN6	2M:S6:6S:TE:R3:E)	16	ff99u4x7g.exe	23562	baker.com	192.168.1.1	83	0
11/11/2017 23:42	142.16.0.0	WIN-TJQZXZG	5G:TJ:ST:EY:7O:E0	68	nn7ubiy58.xls	932864	miller.com	192.168.1.1	90	0
11/11/2017 23:42	239.79.8.0	WIN-NX	Y1:12:W7:7O:2F:7	68	nn7ubiy58.xls	215040	nguyen.com	123.19.0.8	40	0
11/11/2017 23:42	142.16.0.0	WIN-S7M8X	2M:S6:6S:TE:R3:E)	20	na043b2fz.exe	932864	miller.com	123.19.0.8	46	0
11/11/2017 23:42	183.0.0.0	WIN-9EOZPTSNK	K4:9Q:X6:G6:R5:1!	12	8vkv156fi.exe	215040	nguyen.com	192.168.1.1	71	0
11/11/2017 23:39	142.16.0.0	WIN-NX	5G:TJ:ST:EY:7O:E0	4995	i6nol6db7.txt	932864	cardenas.com	192.168.1.1	23	1
11/11/2017 23:35	244.192.0.0	WIN-Y306D04X	4Z:26:F1:Z9:SO:KC	4995	aochnowte.xls	932864	lopez.info	192.168.1.1	84	1

Fig. 1 The dataset produced from log generator script indicating feature columns

The logfile is put through the two IDS tests in the gateway script, which results in two separate identification processes. The users shortlisted are stored in a new csv which is then read back as a list in the log generator script.

This would result in a dynamic adjustment in future runs of log generation, which would flag the users incoming through the log stream based on the gateway scanning on historical logs.

2.1 Classifier Performance Analysis

The final component of the dynamic authentication component is the predictive analysis. This is obtained using the LogitBoost [11, 12], model applied to the gateway modified dataset by means of Weka 3.8.1.

The argument for applying a machine learning algorithm to train the model on the dataset is based on the statistics of malicious attacks. An administrator may prevent most attacks with commercial IDS tools. However, zero-day attacks, i.e., unfamiliar intrusions to the system require to be flagged by the system promptly and prevented throughout system lifetime.

LogitBoost has been considered here as a means for predictive analysis of future input log streams. The comparison with other candidate models has been done to check maximum F-measure of each class. This would indicate the model we need considering both precision and recall of the model.

The ratio of correctly classified instances to incorrectly classified instances is significantly higher in the LogitBoost method when compared to other models suitable for binary level nominal feature classification. The ratio measures are given in Table 3.

The dataset is first preprocessed in the first stage of the classification and applied to each classifier in turn, during which the comparative measures are obtained and discussed in the following section (Table 1).

Table 1 Accuracy of the multilayer perceptron model trained on flagged dataset

Model	Classification ratio
Multilayer perceptron	1.38
Multischeme	1.27
Attribute selected classifier	6.14
LogitBoost	5.25

Table 2 Accuracy of the logistic boosting model trained on flagged dataset

Measure	Class Y	Class N
Precision	0.85	0.833
Recall	0.773	0.893
F-measure	0.81	0.862

Table 3 Logistic boosting model confusion matrix of true and false label classes

Prediction	Class a	Class b
Target		
$a = N$	25	3
$b = Y$	5	17

Table 4 Accuracy of the attribute selected classifier model trained on flagged dataset

Measure	Class Y	Class N
Precision	0.941	0.818
Recall	0.727	0.964
F-measure	0.821	0.885

3 Classification Error Analysis and Inferences

The accuracy measures on LogitBoost and attribute selection shown in Tables 2 and 4 draw comparisons on the precision, recall, and F-measure on both models. The data indicate that attribute selected classifier is a better performer when considering F-measure alone. However, the tendency to add more precision to class Y than class N may be indicative of bias introduced in the training of the model (Table 3).

4 Conclusion

The effort on securitization of the smart home network is largely focused on the identification of the features of the dataset and the intrusion detection capabilities. Hence, the future scope of expansion would first include the addition of new and more improvised threat intelligence into the gateway [9]. Better IDS coupled with an unbiased trained predictive model based on the LogitBoost algorithm would allow for more intruder alerts and increase the dynamicity of the system to newer types of attacks. The selection of features is also a crucial method to find incoming intruders [7]. Network traffic can be analyzed but the information that is obtained is limited to IP, port, and domain information. Analyzing the user sessions requires more features that will allow for deeper intrusion detection. A further enhancement would be to add qualitative features such as security-related keywords, number of non-printable characters, the presence of media attachments, and the calculation of entropy of each of these sessions. The addition of such a feature would be ideal for maximum accuracy and gaining greater true positives ratio.

The gateway script when finalized containing an effective classifier model in place can be made to run on a Raspberry Pi 3 terminal. The log generator would then be replaced with the user network—the firewall, router, and the users of the network. The profiles of the users on the network can then be materialized and studied for behavior in comparison to the results of the above-obtained statistics. This would give us a window on a better method to create a seamless security network within a smart home system.

Acknowledgements This research is supported by Amrita School of Engineering, Bangalore.

References

1. Srivastava, J.R., Sudarshan, T.S.B.: Intelligent traffic management with wireless sensor networks. In: ACS International Conference on Computer Systems and Applications (2013)
2. Kim, Y., Yoo, S., Yoo, C.: DAoT: Dynamic and energy-aware authentication for smart home appliances in internet of things. In: IEEE International Conference on Consumer Electronics (2015)
3. Srividya, Ch., Rakesh, N.: Enhancement and performance analysis of epidemic routing protocol for delay tolerant networks. In: International Conference on Inventive Systems and Control, pp. 1–5 (2017)
4. Ashwini, M., Rakesh, N.: Enhancement and performance analysis of LEACH algorithm in IoT. In: International Conference on Inventive Systems and Control (2017)
5. Lakshmi, R.V., Krishnan, D., Parvathy S., Vishnudatha, K., Poroor, J., Dhar, A.: JPermit: usable and secure registration of guest-phones into enterprise VoIP network. In: International Conference on Advances in Computer Engineering (2010)
6. Manmadhan, N., Achuthan, K.: Behavioural analysis for prevention of intranet information leakage. In: International Conference on Advances in Computing, Communications and Informatics (2014)
7. Staudemeyer, R.C., Omlin, C.W.: Extracting salient features for network intrusion detection using machine learning methods. S. Afr. Comput. J. **52** (2014)
8. Santoso, F., Yun, N.: Securing IoT for smart home system. In: IEEE International Symposium on Consumer Electronics (2015)
9. Lee, W., Stolfo, S., Mok, K.: Mining in data-flow environment: experience in network intrusion detection. In: Proceedings of the Fifth ACM SIGKDD International Conference on Knowledge Discovery and Data Mining, pp. 114–124 (1999)
10. Hodo, E., Bellekens, X., et al.: Threat analysis of IoT networks using artificial neural network intrusion detection system. In: International Symposium on Networks, Computers and Communications (2016)
11. Habib, K., Leister, W.: Context-aware authentication for the internet of things. In: The Eleventh International Conference on Autonomic and Autonomous Systems (2015)
12. Friedman, J., Hastie, T., Tibshirani, R.: Additive logistic regression: a statistical view of boosting. In: Annals of Statistics, vol. 028 (2000)

Shruthi Sreedharan pursuing second year in Master of Technology Embedded Systems program at Amrita School of Engineering, Bengaluru. Her areas of Research interests include Internet of Things and Network security.

Dr. N. Rakesh working has Vice chair in Department of Computer Science & Engineering, at Amrita University, Bengaluru Campus. He has almost 14 years of teaching along with research experience. His area of interests includes Computer Networks, Wireless Communication, Wireless sensor Networks, Internet of Things, Wireless Channel Modeling, Mobile Communication and topics related to Networks and Wireless Communication domains. He served has Session chair, Technical Program Chair, International Program Chair for various International Conference in India and abroad. Designated Reviewer for various International Journals and Conferences. Also currently leading "Computer Networks & Internet of Things" Tag.

Design and Performance Evaluation of an Efficient Multiple Access Protocol for Virtual Cellular Networks

B. Sreevidya and M. Rajesh

Abstract Virtual Cellular Network (VCN) is a mobile communication network which is wireless in nature. This communication network includes one Mobile Station (MS) which has the provision of communicating with many (more than one) Base Stations (BSs) at the same time. One of the inherent properties of VCN is that when the mobile station sends a signal to the base station, the base station receives it, thereby reducing the probability of outage that could be caused by shadowing. A mobile station is in continuous and parallel communication with two base stations during the phase of handing over. This communication is achieved by communicating the same data of downlink from the two base stations to the mobile station. When such a downlink data of a single mobile station is shared over multiple base stations, higher bandwidth efficiency is achieved in full virtual cellular network implementation. The VCN also gives the provision that will enable many base stations to receive the signal from one mobile station, thereby setting an environment to improve the reliability of satellite communication. In the proposed scheme, distributed multiple access protocol is used and simulation of this protocol is done using stationary users and mobile users. Every BS is considered to be a separate object which is linked by a simulated backbone which is wired. The simulator used is a time-based simulator where all the base stations are synchronized to one common clock. This paper presents the evaluation of multiple access protocol in combination with link protocol. The proposed work can be a contribution toward the design of multidisciplinary Mobile Multimedia Communications (MMC). The proposed scheme aims to support Asynchronous Transfer Mode, Internet Protocol (IP), and also a support for inexpensive mobile stations.

Keywords Cellular networks · Virtual cellular network · Mobile station
Base station · Multiple access protocol

B. Sreevidya · M. Rajesh (✉)
Department of Computer Science & Engineering, Amrita Scool of Engineering,
Amrita Vishwa Vidyapeetham, Bengaluru, India
e-mail: m_rajesh@blr.amrita.edu

B. Sreevidya
e-mail: b_sreevidya@blr.amrita.edu

S. Smys et al. (eds.), *International Conference on Computer Networks and Communication Technologies*, Lecture Notes on Data Engineering and Communications Technologies 15, https://doi.org/10.1007/978-981-10-8681-6_28

1 Introduction

A virtual cell is defined as the extent of measurement with respect to area in which the strength of the signal has the ability to control a port when no other terminal is transmitting the packet within the virtual circuit network.

Hence, a VCN can be defined as the network in which a single MS (Mobile Station) can communicate with more than one BS (Base Station simultaneously). There a number of disadvantages for the traditional mobile communication which can be reduced by the usage of VCN, e.g., reduced interference (adjacent and co-channel). The VC is mainly used for the indoor mobile communication or for the communication in the femtocell. Here, the coverage area of the virtual cell will be in meters. Size of the cell depends on the RSS (Received Signal Strength) by BS from MS, propagation attenuation, and channel noise.

In VCN, a single mobile station can simultaneously communicate with more than one base station, thus by reducing the outage probability caused by shadowing. A VCN consists of stationary users, mobile users, base stations, FDMA–TDMA hybrid for slot allocation, and a bandwidth of 60 GHz. Multiple access protocols are used for the uplink and downlink. In uplink, we make use of hybrid FDMA–TDMA. In downlink, we make use of CDMA protocol. The TDMA, FDMA, and CDMA are different MAC protocols. We make use of NS2 simulator to simulate our VCN and to calculate our outage probability. In uplink, it provides the site diversity and in downlink, it provides the distributed downlink data.

A virtual cell is defined as an area in which the received signal strength is enough to capture a base station where no other terminal is transmitting the data. Since there is no predetermined boundary for the virtual cell, any mobile can communicate with any base station which resides in the virtual cell of the mobile. A group of base station in a bus topology can contribute to a VCN. As a result, one mobile station can interact with more than one base station.

The main goals of the introduction of the VCN are being noted down:

High bandwidth efficiency

Since there is no frequency reuse factor, the efficiency is increased. The entire bandwidth utilization also improvises the throughput, which indirectly increases the efficiency and performance. In the downlink, the downstream data is being divided into sub-streams which will be allocated to multiple BSs.

Site diversity

In a VCN, the MS communicates with more than one BS simultaneously. Hence, when we consider the uplink, site diversity occurs because it can communicate with more BSs, and different packets can be transferred to different BSs.

Handling the handover phase

In traditional cellular networks, the handover takes place when one MS moves from the coverage area of one BS to the coverage area of another BS. But in case of

VCN, the boundary of virtual cell is determined by the MS and its transmitted signal strength. Hence, there is no predetermined boundary of cells. Here, the scenario of the handover does not come into picture.

High performance

Due to various reasons like high capacity of users and minimized cost BSs, handling of the heterogeneous traffic and the use of MAC for both uplink and downlink the performance can be increased.

Minimized cost ports

In traditional mobile communication, we make use of BSs which handles the functionalities like handover and channel assignment. But in case of VCN, the port does not handle these functionalities since there is no concept of handover in the VCN and channel assignment is done using the slot allocation procedure.

2 Related Works

In general cellular networks, each mobile node is attached with a base station and communication of every mobile node in the area takes place through the base station. In order to restrict the usage of the network and to provide security, registration process is introduced with base station. Each mobile node when it enters into the area of a base station has to register with the base station so as to get access to the channels for communication. So typically a mobile node is attached with single base station. When a mobile node moves from the coverage area of a base station, it has to deregister with the existing base station and register with a new base station. This is a time consuming as well as costly process. Also, during this hand over process, packets need to be re-routed to the base station to which the mobile node is going to register.

In order to avoid this handover process, if a mobile station is allowed to register with multiple base stations simultaneously, handover can be made simple. Wireless network architecture with multiple access ports [1] tries to have multiple access channels for communicating with different base stations in the vicinity.

Donghan et al. [2] attempted to evaluate the issues associated with cell migration of mobile nodes. In the attempt, it is observed that overlapping of base station coverage is a solution to smoothen the hand over process.

Bakker et al. [3, 4] designed a multiple access protocol which enables the mobile node to communicate with various base stations using time slot feature. In this, different timeslots are assigned to communicate to different base stations even though the slots belong to single channel assigned by the current base station to which mobile node is registered.

Bakker et al. [5] further developed a high-frequency bandwidth allocation schemes for wireless network which helps channel assigning process dynamic in nature.

The concept of femtocells [6] further drives the idea of coverage area being associated with mobile stations rather than attaching it with base station. Cluster-based architecture [7, 8] of wireless networks in certain applications further demanded the need of nodes being attached with multiple peers.

3 Proposed System

The detailed description of the virtual cellular network is as given in Fig. 1.

The blocks in the above diagram are described as below:

- Wired nodes (1.1.1): It acts as a gateway between the base stations to transfer the data and thus the bus topology is being created.
- Added (1.1.2): Mobile stations are added to the bus topology.
- Traces BSs under the MS's RSS (1.1.3): The base stations are traced by the mobile station within the received signal strength and time slot is being allocated.
- Communicates (1.1.4): According to the time slots allocated for each base station, they start communicating with the mobile station, and thus the VCN is generated.

For the implementation of the proposed system, ns-2 simulator [9, 10] is used. In VCN, we consider the scenario that one mobile station can simultaneously communicate with more than one base station. Here, MAC layer [11, 12] is modified for the implementation.

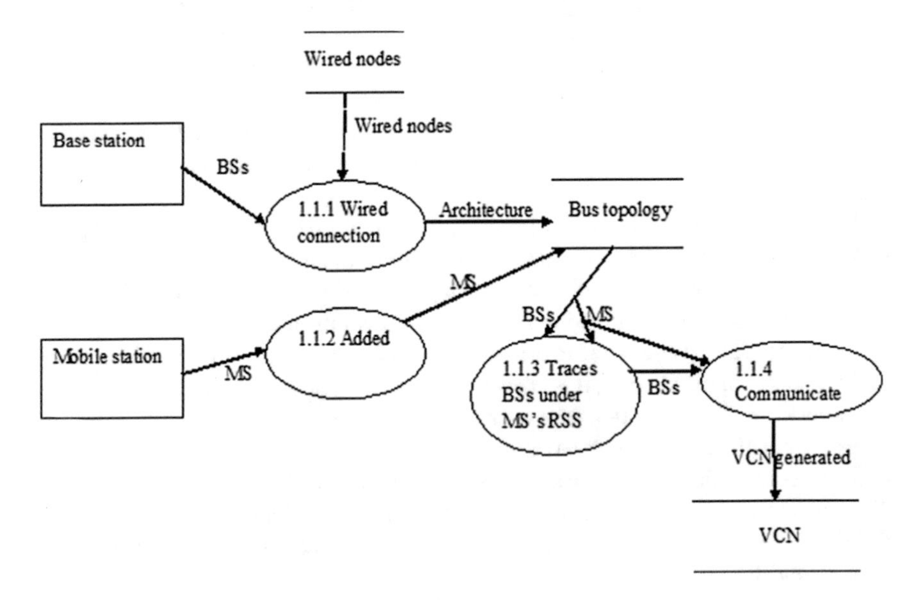

Fig. 1 Schematic representation of virtual cellular network

A virtual cell is defined as the area in which the received signal strength has enough power to capture the port when no other terminal is transmitting the packet within the VCN. In VCN, the main advantage is that there is no predetermined boundary which reduces the interference which is present in the traditional cellular networks. Thus, it increases the throughput and high performance is attained. It also provides the site diversity in the uplink and distributed downlink data. It also avoids handover hotspots. It improves the bandwidth efficiency since the entire bandwidth has been used.

Here, for the proof of concept, simulation is carried out with 3 mobile stations, 22 base stations, and 23 wired nodes which act as the gateway between the base stations. A bus topology is being created, which in turn tells that the base stations are wired. The mobility of the mobile station is being provided.

Now the mobile station traces the base station within its received signal strength. Now the mobile station runs distributed channel allocation algorithm and the time slot is been provided to the base stations. Within that particular time slot mobile station can communicate with that base station. At a time a mobile station can get synchronized to maximum of eight base stations. All base stations are synchronized to a common clock.

The proposed system consists of three modules:

- Architecture,
- Modification of MAC layer, and
- Visualization.

3.1 Architecture

This is the architecture of the proposed system. Here, there are six main components like ports, virtual cell, port network, port server, backbone network, and terminals. Here, the ports are the base stations and terminals are the mobile stations. Port server is required for a VCN. The port server manages the communication between the mobiles and the communication between the VCN and other networks which may be connected through a high-speed backbone network. The ports are interconnected by a port network, and each terminal sends packet using the entire system bandwidth while any nearby ports can pick up the signal (Fig. 2).

Consider the terminal 1. Here a, b, c, and d are the ports which can receive the signals from terminal 1. Hence, this together forms the virtual cell of terminal 1. Consider the terminals 2 and 3. For terminal 2, the ports which form the VCN are e, f, g, h, and for terminal 3 and the ports which form the VCN are g, h, i, and j. Here, the ports g and h are the common ports for both the terminals. So according to the different time slots allocated by the terminals to the ports, overlapping does not occur.

All the attributes of the wireless are being set, such as:

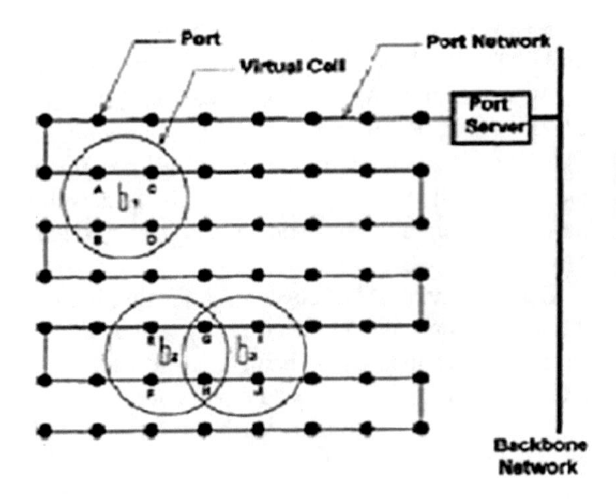

Fig. 2 Architecture of a VCN (only 3 mobiles are transmitting packets simultaneously during a slot)

```
set opt(adhocRouting) DSDV
set opt(mac)          Mac/802_11
```

In the topology, there are 22 base stations, 3 mobile stations, and 21 wired nodes.

```
set opt(nn)            3
set num_wired_nodes 21
set num_bs_nodes      22
```

A new instance of the network simulator ns_ is created. A hierarchical address is being provided to all nodes present. Each domain is divided into clusters, and each cluster is divided into nodes. In our topology, we created the below-given hierarchy:

```
$ns_node-config-addressType hierarchical
AddrParams set domain_num_ 2
lappend cluster_num 21 1
AddrParams set cluster_num_ $cluster_num
lappend eilastlevel 1 1 1 1 1 1 1 1 1 1 1 1 1 1 1 1 1 1 1 1 1 25
AddrParams set nodes_num_ $eilastlevel
```

Here, the entire topology is being divided into two domains. First, domain consists of 21 clusters. Each cluster has only one wired node. Second, domain has only one cluster. In this cluster, there are 25 nodes wherein 22 are BSs and 3 are MSs. Base stations are created and hierarchical address is being set. Random order is being disabled.

```
set BS(0) [$ns_node [lindex $temp 0]]
```

Mobile stations are created and source and destination points are given. Hierarchical address is being set. Each node has to be attached to the BS.

```
set node_($j)[$ns_node[lindex $temp [expr $j+22]]
$node_($j) base-station [AddrParams addr2id[$BS(0) node-addr]]
```

The tcp connection is being provided between the mobile_node(0) and BS(0) at a time slot of 160 s.

3.2 MAC Layer File (802_11.cc) Modification

The 802_11.cc file is supported by the 802_11.h file. The 802_11.h file has preprocessor directives used in the 802_11.cc and the definition. It has three main functions which are used:

- To get RTS length, we make use of this function:
  ```
  inline u_int32_t getRTSlen() {
  return(getPLCPhdrLen() + sizeof(struct rts_frame));
  }
  ```
  Size of rts frame is obtained and added with PLCP header length.
- To get CTS length, we make use of this function:
  ```
  inline u_int32_t getCTSlen() {
  return(getPLCPhdrLen() + sizeof(struct cts_frame));
  }
  ```
  Size of cts frame is obtained and added with PLCP header length.
- To get ACK length, we make use of this function:
  ```
  inline u_int32_t getACKlen() {
  return(getPLCPhdrLen() + sizeof(struct ack_frame));
  }
  ```
  Now, the mac802_11.cc is being modified, it has four functions.
- In the first function (sendRTS(int dst)), the distance of coverage is given as the arguments to the function. The value for attributes like rts_frame and PLCPhdrLen is obtained and the function getRTSlen() is called. It returns the total length. Now the mobile station traces all the base station within its RSS. According to the destination (base stations) identified by the source (mobile station), a packet is assigned as rts packet and send.

3.3 Visualization

For visualization, nam and trace file are used which are part of NS2. The nam file we use here is wireless2-out.nam. The trace file we use here is wireless2-out.tr. The codes to set these nam and trace files are

$$\text{set tracefd}\left[\text{open wireless2-out.tr w}\right]$$
$$\text{set namtrace}\left[\text{open wireless2-out.nam w}\right]$$

The below command writes the traces that is being done at the execution of the program to the trace file instance (trace-all).

$$\text{\$ns\_trace-all \$tracefd}$$

The nam trace file topology is given in terms of x and y coordinates; the bus topology can be visualized within these coordinates.

$$\text{\$ns\_namtrace-all-wireless \$namtrace \$opt}(x)\text{\$opt}(y)$$

Before we run the tcl script, we execute the nam file, so that the visualization can be seen during the compilation of the program.

$$\text{exec nam wireless2-out.nam \&}$$

4 Results and Conclusion

The proposed system of having a Virtual Cellular Network (VCN) is a system concept for serving users in indoor environments. When a number of BSs are deployed in indoor space, interference from other indoor cells may cause significant performance degradation. The proposed system, which forms a virtual cell with several transmitters and performs cooperative transmission, improves the performance of the edge users.

The VCN which is proposed consists of mobile nodes which can communicate with more than one BS. The MAC layer protocol is modified. The time duration is modified so that the RTS and CTS provide a simultaneous communication between one MS to various BSs at different time slots. Hence, the performance of the VCN is high compared to the traditional cellular networks.

Simulation of the Virtual Cellular Networks (VCNs) using MAC layer protocol is mentioned, thereby to improve bandwidth efficiency using the multiple access protocol and in order to maximize the performance gain of cooperative transmission.

References

1. Kim, H.J., Linnartz, J.P.: Virtual cellular network: a new wireless communications architecture with multiple access ports. In: 44th Vehicular Technology Conference, vol. 2, pp. 1055–1059 (1994)
2. Chee, D., Kim, J., Ahn, W.-G., Jo, O., Lee, J.Y., Cho, D.H.: Design and performance evaluation of virtual cellular networks mitigating inter-cell interference in indoor environments. In: International Conference on ICT Convergence (ICTC), pp. 199–203 (2011)
3. Bakker, J.D., Prasad, R.: A multiple access protocol implementation for a virtual cellular network. In: 49th IEEE Vehicular-Technology-Conference, vol. 2, pp. 1707–1711 (2004)
4. Vanishree, K., Rajesh, M., Sudarshan, T.S.B.: Stable route AODV routing protocol for mobile wireless sensor networks. In: 2015 International Conference on Computing and Network Communications (CoCoNet'15), Dec 2015
5. Bakker J.D., Schoute, F.C., Prasad, R.: An air interface for high bandwidth cellular digital communications on microwave frequencies. In: 48th IEEE Vehicular Technology Conference, vol. 1, pp. 132–138 (1998)
6. Chandrasekhar, V., et al.: Femtocell networks: a survey. IEEE Commun. Mag. **46**(9), 59–67 (2008)
7. Sreevidya, B., Rajesh, M.: Enhanced energy optimized cluster based on demand routing protocol for wireless sensor networks. In: 6th International Conference on Advances in Computing, Communications and Informatics (ICACCI), Sept 2017
8. Meena Kousalya, A., Sukanya A.: Clustering algorithms for heterogeneous wireless sensor networks—survey. Int. J. Ad hoc Sens. Ubiquitous Comput. (IJASUC) **2**(3) (2011)
9. en.wikipedia.org/wiki/Mobile_virtual_network_operator
10. NS-2 study material. http://www.sprinkerlink.com. http://www.isi.edu/nsnam/ns/tutorial/index.html
11. Taneja, K., Taneja, H., Kumar, R.: Multi-channel medium access control protocols: review and comparison. J. Inf. Optim. Sci. **39**(1) (2018): 239–247 (2017)
12. AlSkaif, T., Bellalta, B., Zapata, M.G., Ordinas, J.M.B.: Energy efficiency of MAC protocols in low data rate wireless multimedia sensor networks: a comparative study. J. Ad hoc Netw. **56** (2017) (Elsevier Publications)

B. Sreevidya has obtained her B.Tech. degree in Computer Science and Engineering from Visvesvaraya Technological University, Karnataka, India. Later she completed her M.Tech. degree in Computer Science and Engineering from R.V. College of Engineering Bengaluru under Visvesvaraya Technological University, Karnataka, India. Currently she is serving as Assistant Professor in the Department of Computer Science and Engineering, Amrita School of Engineering, Amrita Vihswa Vidyapeetham Bengaluru Campus. She has 12+ years of teaching experience. She has published quite a good number of publications in the domain of Wireless Sensor Networks and Data Mining. Her current research interest is in Secured Data Transmission over Wireless Sensor Networks.

M. Rajesh has obtained his B.Tech. degree in Computer Science and Engineering from Calicut University, Kerala, India. Later he completed his M.Tech. degree in Computer Science and Engineering from R.V. College of Engineering Bangalore under Visvesvaraya Technological University, Karnataka, India. Currently he is serving as Assistant Professor in the department of Computer Science and Engineering, Amrita School of Engineering, Amrita Vihswa Vidyapeetham Bengaluru Campus. He is pursuing his Ph.D. He has 14+ years of teaching experience. He has published quite a good number of publications in the domain of Wireless Sensor Networks and Robotics. His current research interests are in the area of Wireless Sensor Networks, Robotics and Internet of Things.

Healthcare Monitoring System Based on IoT Using AMQP Protocol

C. S. Krishna and T. Sasikala

Abstract Healthcare monitoring system is introduced to reduce the daily visit of patients to the hospital. The patients can collect their vital parameters and send those measured values to the doctor or hospital. Healthcare monitoring system introduces the real-time monitoring of vital parameters of the patients such as collecting all the information from the patients and sending it to the web server. The authorized person can log into his account and see the patients' vital parameters. These vital data can be stored and analyzed for further decision-making and analysis. By identifying the pattern of the vital parameters which is observed, the nature of the disease can be predicted. The system collects patient's body temperature, oxygen saturation percentage, heart rate, blood glucose, and blood pressure using Raspberry Pi board and cloud computing. The communication protocol used in the system is AMQP protocol. It supports reliable communication via message delivery guarantee primitives including at-most-once, at-least-once, and exactly once delivery. The collected data from various patients could not be handled easily by the physician. The main concern of the physician is that he should take the critical decisions about their patient's health from this huge volume of health information. The huge amount of information can be stored in cloud and the data can be analyzed using data mining.

Keywords Raspberry pi board · Temperature sensor · Pulse oximeter Glucometer · AMQP protocol · Internet of things · CloudAMQP Cloud computing

---

C. S. Krishna · T. Sasikala (✉)
Department of Computer Science and Engineering,
Amrita School of Engineering, Amrita Vishwa Vidyapeetham, Bengaluru, India
e-mail: t_sasikala@blr.amrita.edu

C. S. Krishna
e-mail: krishnacs87@gmail.com

S. Smys et al. (eds.), *International Conference on Computer Networks and Communication Technologies*, Lecture Notes on Data Engineering and Communications Technologies 15, https://doi.org/10.1007/978-981-10-8681-6_29

1 Introduction

Internet of Things is the connection between the physical devices or objects through the Internet. These devices can be mobile phones, laptops, tablets, and also sensors. The different sensors can communicate with a gateway implemented using acquisition devices [1]. In recent years, the huge development of IoT makes all physical objects to be interconnected and everyone can control the devices from anywhere. Many of the applications of IoT are smart retail, home automation, smart city, smart environment, automated transportation, and health care. IoT brings an extended development in healthcare system, in the region of patients' care, and communication. The benefits of the Internet of things in healthcare organizations are reduced errors, decreased costs, improved outcomes of treatment, improved disease management, enhanced patient experience, and enhanced management of drugs [2].

Pulse oximeter, glucometer, and temperature sensors are mostly used in homes and in hospitals to measure the oxygen saturation value, glucose level, pulse rate, and body temperature [3]. Pulse oximetry is referred to as SpO_2, which is a noninvasive method of measuring a person's blood oxygen saturation percentage without taking the blood samples. Pulse oximetry allows doctors to assess and monitor a patient's respiratory functions with ease [4]. Manually measuring the vital parameters of every patient at all the time is not possible. It can be solved by implementing the IoT, and these vital readings can be measured and sent through the cloud to display in doctor's laptop or mobile. The communication is done through the AMQP protocol. The health parameters collected from the sensors (temperature, blood glucose level, pulse rate, and SpO_2) use the fully functioning devices (Raspberry Pi). These devices are responsible for establishing the cloud connectivity and transmitting the vital parameters to a cloud storage using an IOT protocol (AMQP). These received values can be mapped and reduced using some mining algorithms [5]. MQTT and AMQP are publish–subscribe-based messaging protocols. These are the application layer protocols for IoT. Both are placed on top of the TCP/IP protocol. For MQTT, publisher can publish the message to the broker, and then the broker will transfer it to the desired subscriber. For AMQP, publisher can publish the messages to the broker; here, the broker consists of an exchange for different queues and the subscriber can subscribe it from there. In the proposed system, both the protocols are compared and observed for their reliability and for the way of publishing the messages in terms of topics.

2 Literature Survey

Perumalsami and Riffath Shehla [6] describe the patient monitoring system—The PMS is based on the latest buses connectivity, using the AMQP protocol as an internal bus. The AMQP is a current efficient protocol. It should be a good approach to use AMQP as an internal bus for reducing the cost. During the treatment, the patient monitor is continuously monitoring the vital values of the patient to transfer the important information. In paper [7], Luzuriaga et al. represent the comparison and observation of AMQP and MQTT protocols with some scenarios. Those scenarios are characterizing both protocol behaviors in terms of jitter, message loss, saturation boundary values, and latency. Also, evaluate MQTT and AMQP, and then compare MQTT and AMQP capacities and capabilities through measurements under an unstable or mobile wireless network tested. In conclusion, they found that during transitions the mean jitter value leads to oscillation between 3 and 6 s, and also found that for high transmission rates in 7 s, peaks have been detected. The disadvantage of the AMQP is that they have noticed that if the publisher's connection suffers any interruption during publishing, then the client persists the messages for a limited time only.

In paper [8], Al-Fuqaha et al. provide an entire view of the IoT with few features which includes enabling technologies, application issues, and protocols. Transpire technologies including big data analytics, fog, and cloud computing are the new technologies with IoT which is observed. The demand for better horizontal integration among the Internet of things services has been presented. At last, they introduced detailed service of use cases, for instance, in that how the different protocols fit together to deliver desired IoT services. In paper [9], Xiong et al. have mentioned the active status certificate to publish and subscribe based on AMQP. Based on observing the advanced message queue protocol, the requirement of publish/subscribe active status certificate has obtained and designed the active status certificate, and the publish and subscribe based on AMQP which supports the ordered service by user certificate authentication mechanism. In paper [10], John and Liu found that when the broker goes down in AMQP protocol, it can recover not only the persistent messages but also the messages published during the breakdown. In the proposed system, it can be overcome by providing the memory location to store the messages in Raspberry Pi. In paper [11], Ani et al. introduce IoT in medical applications for patient monitoring. The system provides patient monitoring for stroke disease. The patient health parameters are collected and sent to the cloud. This methodology motivates to do the current proposed system.

In the proposed system, AMQP protocol has been introduced in healthcare monitoring system with few parameters. These parameters are more user-friendly and comfortable for the patients.

3 System Overview

Figure 1 shows the block diagram for health monitoring system. The proposed system is designed to measure the body temperature, oxygen saturation value, and blood glucose level and pulse rate. The main part of the system design is a server based on cloud AMQP, which hosts RabbitMQ broker. The Wi-Fi router is connected to the Internet. Raspberry Pi board is used to connect the sensors as a gateway. The pulse oximeter, temperature sensor, and glucometer are connected to the Raspberry Pi board. Pulse oximeter measures the patient's oxygen saturation value and pulse

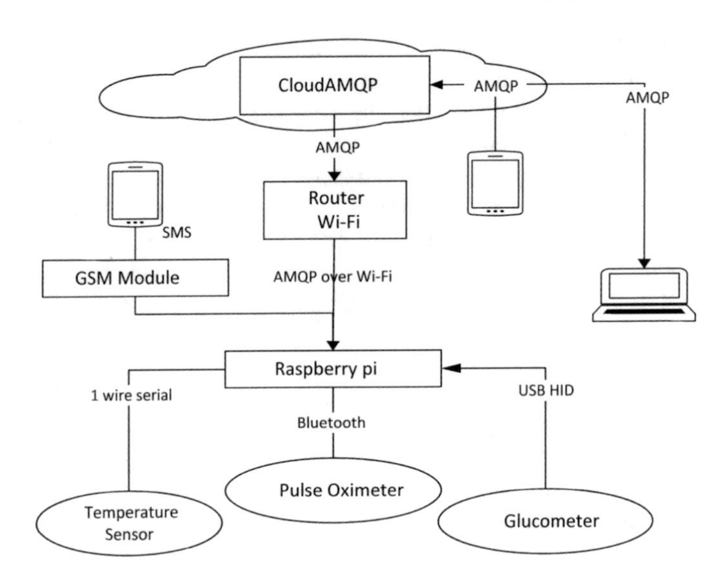

Fig. 1 System design

rate via inbuilt Bluetooth. Temperature sensor measures the temperature via one-wire communication. Glucometer measures the blood glucose level through the USB communication. Raspberry Pi sends the information to the server using AMQP protocol through the inbuilt Wi-Fi connected to the router. The information can be displayed in a web application and an Android application in mobile phone. The GSM module is connected to the Raspberry Pi to send the message to the doctor if any criticality found in the temperature, blood glucose level, pulse rate, and blood saturation value. Figure 2 shows the architecture of the system.

4 Implementation

In Fig. 3, the flowchart represents the continuous flow of the proposed system. Turn ON the Raspberry Pi and all the devices or the sensors to measure the vital values. First, check whether the Bluetooth is paired for the pulse oximeter; if it is connected, then it will give the reading of heart rate and SpO_2. If it is not connected, take the readings of glucometer and temperature. Take the count as 0. Check whether the server is reachable or not. If it is not reachable, store the data in the list and increment the count to 1. Again check whether the server is reachable or not; if it is not, do the same procedure and increment the count to 1. If the server is reachable, publish the data to the CloudAMQP through AMQP protocol and delete all the data from the list. Then, route the data to the queue. If the subscriber does not subscribe the data, it will remain in the queue and wait for the subscriber to subscribe.

Fig. 2 System architecture

4.1 Pulse Oximeter Module

Pulse oximeter module includes a Raspberry Pi 3 with a microprocessor Broadcom BCM 2837 and a pulse oximeter ChoiceMMed MD300C318T. Figure 4 shows the connection diagram of pulse oximeter and Raspberry Pi. The OS used in Raspberry Pi 3 is Ubuntu MATE 15.10 with Kernel 4.1.15. Internet connectivity is established by connecting to a Wi-Fi access point. The library used for AMQP communication to the cloud server is RabbitMQ.

The communication protocol which is used to communicate between the Raspberry Pi 3 and the pulse oximeter is a serial port profile (SPP) over Bluetooth. Libbluetooth is the library used for Bluetooth communication. The heart rate and SpO_2 can be triggered by inserting the finger inside the pulse oximeter and pressing the power button on the pulse oximeter. Raspberry Pi 3 connects to the pulse oximeter through Bluetooth and starts collecting the SpO_2 and the heart rate reading. Once a

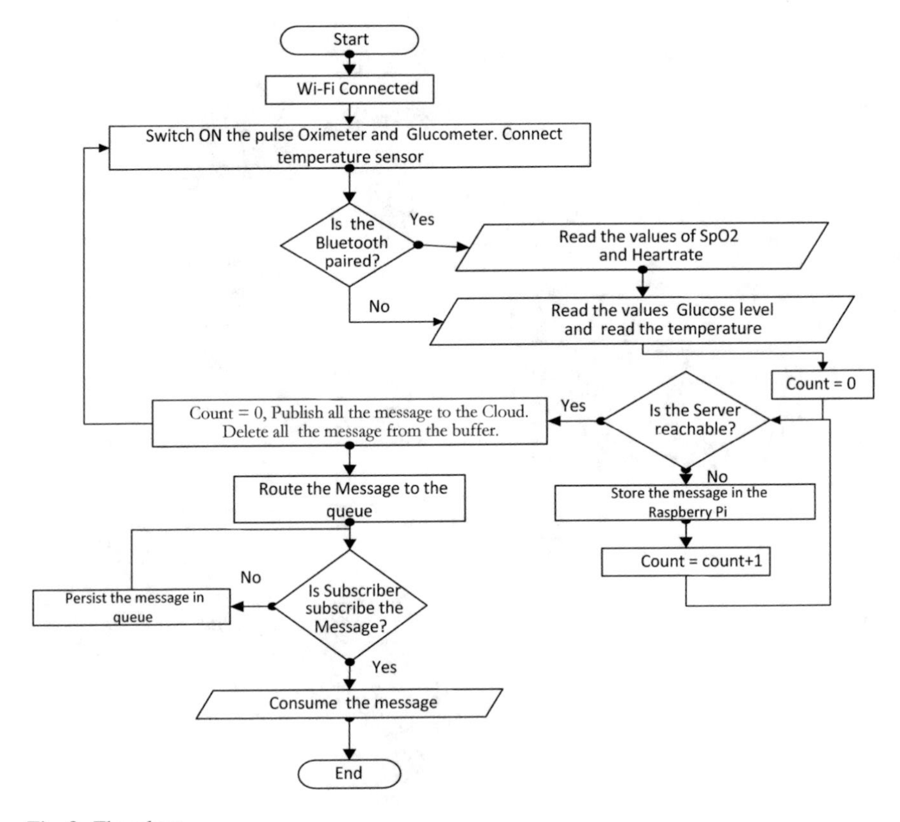

Fig. 3 Flowchart

stable reading is achieved, Raspberry Pi 3 commands the pulse oximeter to power off and sends across the stable SpO_2 and heart rate reading to the CloudAMQP server.

4.2 Temperature Module

Temperature module consists of Raspberry Pi 3 with microprocessor Broadcom BCM 2837 and temperature sensor DS18B20. Figure 5 shows the connection diagram of Raspberry Pi and the temperature sensor. 1-Wire is the communication mechanism used to communicate between Raspberry Pi 3 and temperature sensor. Internet connectivity is established by connecting to a Wi-Fi access point. Eclipse is the development environment used, and the programming is done in C/C++. AMQP is the protocol used to send the temperature value to the client. The previously occurred temperature value is checked with the current value for up to 5–10 s. Whenever a stable value is received, it is considered as a patient's temperature and it is transmitted to the AMQP broker RabbitMQ, which is hosted in AMQP cloud server.

Fig. 4 Pulse oximeter module

4.3 Glucometer Module

Glucometer module consists of Raspberry Pi 3 with microprocessor Broadcom BCM 2837 and *CONTOUR®PLUS ONE* glucometer. Figure 6 shows the connection diagram of Raspberry Pi and the glucometer. The physical connection utilizes the USB HID protocol. Internet connectivity is established by connecting to a Wi-Fi access point. Eclipse is the development environment used and the programming is done in C/C++. The library used for USB communication is USB. AMQP is the protocol used to send the blood glucose value to the client. The RabbitMQ broker exchanges the data with the routing key and sends it to the subscriber. The received data are sent to the cloud server. The meter communicates with the host application using HID interrupt transfers.

4.4 GSM Module

GSM module consists of GSM SIM900A, which is connected to the Raspberry Pi. Figure 7 shows the connection between Raspberry Pi and GSM module. The communication mechanism which is used here is AT command over UART. The

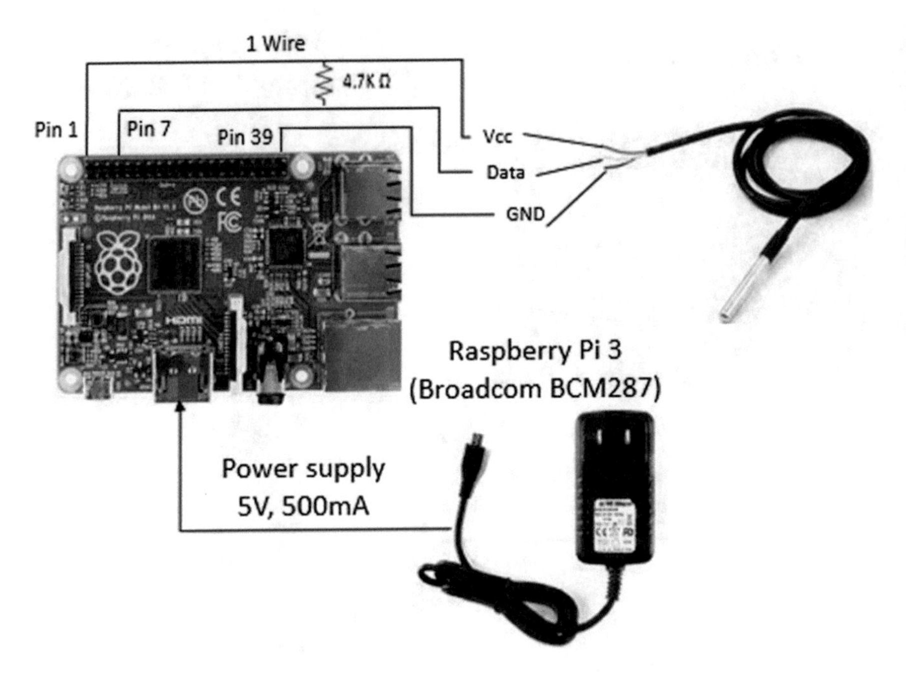

Fig. 5 Temperature module

Fig. 6 Glucometer module

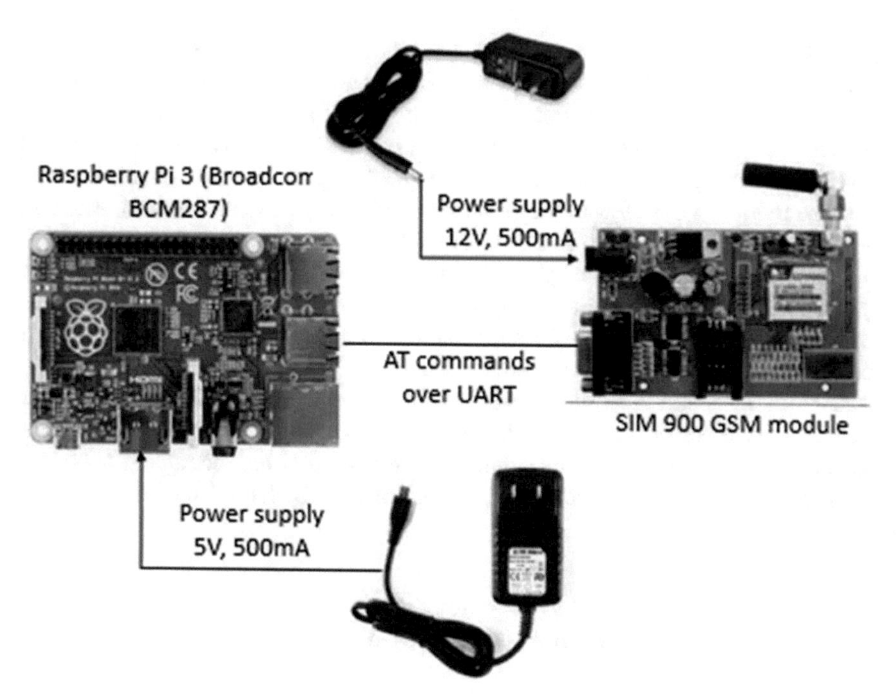

Fig. 7 GSM module

Fig. 8 MQTT clients

data will be checked according to the condition given and the message will be sent to the mobile phone using GSM SIM900A [12].

4.5 Clients

Clients are MQTT Dashboard and MQTT Lens. For Android applications, the MQTT Dashboard is used. Figure 8 shows the symbol of the MQTT Dashboard and MQTT Lens. For web application, the MQTT Lens is used in Chrome browser. These clients provide result of the subscribed messages and can publish the messages to broker.

4.6 Cloud Server

Cloud computing is a recent computing technology that enables the users to use the cloud infrastructures for analyzing the data and applications without having any physical infrastructures themselves. The users have to pay only for the services and resources they need [13]. A cloud server is called as a virtual server or a virtual private server, which is actually a logical server that is built and hosted in the cloud and delivered over the Internet. Various operating systems can be installed on the cloud server and the user can customize the resource configurations like RAM, storage, etc. The following are a few of the cloud servers [2]: IBM Cloud solutions, Amazon web services, Google Cloud, and CloudAMQP.

CloudAMQP is managed by RabbitMQ servers in the cloud-hosted message queues. The cloud server passes the information between processes or the controllers and other system applications. Producer published the messages to a queue through exchanges. When the consumer wants to handle the messages, he can pull the messages from the queue. According to the rule, the queue can buffer, route, and persist the messages. Messages can be sent across platforms, languages, and OS. The highly scalable system can be created by handling the way of message decouple. RabbitMQ is a high-performance message broker, built in Erlang, which implements the AMQP protocol [14].

5 Comparison Between MQTT and AMQP

AMQP and MQTT are the application layer protocols for IoT. Both AMQP and MQTT protocols are publish–subscribe model. MQTT supports the publish–subscribe messaging to topics. In MQTT, each information is published and subscribed with unique topics. For example, the glucometer value is published with the topic name "Glucose". AMQP supports the publish–subscribe messaging, which includes message queue. Here, messages are classified in exchanges and forwarded in different queues with respect to exchanges. Each queue has a unique key to identify. For example, the temperature value is published with a unique key called "Temp".

The message reliability is less for MQTT protocol because theoretically we can say that the message is published and subscribed by the consumer but practically it is not, and if any of the messages are published that does not mean that these messages can be subscribed by the consumer and if any disturbance happens in between the communication, the message could not be reached to subscribe. In AMQP, the message delivery is more reliable. If any of the disturbance happens in between, then the broker will persist the messages in queue and subscribe it once the consumer is available, which is shown in Fig. 11.

6 Result

The health parameters used in the healthcare monitoring system are temperature, heart rate, blood saturation value, and blood glucose level. All these sensors are connected to Raspberry Pi. The patient should touch the temperature sensor and

Fig. 9 Vital values in mobile phone

measure the body temperature. Blood glucose level can be measured by a glucometer. It needs a blood glucose test strip and a lancing device to take the reading of glucose level. First, take the reading and then connect to the Raspberry Pi. So, it will collect the reading from the glucometer.

The finger is inserted into the pulse oximeter and then switch ON the meter. It will measure the blood saturation value and heart rate, and it is connected to the Raspberry Pi via Bluetooth. All these measured values are shown in a mobile app called MQTT Dashboard. We can see the unsubscribed data in the AMQP cloud queue, which is not subscribed by any of the consumers. The information, which is coming from the temperature sensor and pulse oximeter can be displayed in a smartphone and laptop. Here, for displaying the measurements, a software Android application named MQTT Dashboard is used. Figure 9 shows the Android application MQTT Dashboard, which is displaying the collected information from each sensor or device.

In laptop, the Google Chrome application called MQTT Lens is used to display the measurements. Here, the data are subscribed according to the topics. Figure 10 shows the subscribed values of information in MQTT Lens on a web browser.

The data, which is not subscribed by any of the consumer, makes queue in AMQP Cloud. In the Cloud, the data are given to the exchange and bind with the routing keys. In the queue, the data will make the queue if any of the consumers does not subscribe. Figure 11 shows the queued values of information in RabbitMQ management.

If any criticality is found in the measurements, then the GSM module will send the messages to the selected mobile number to inform the doctor or nurse. Figure 12 shows the SMS messages sent to the mobile phone of any authorized person.

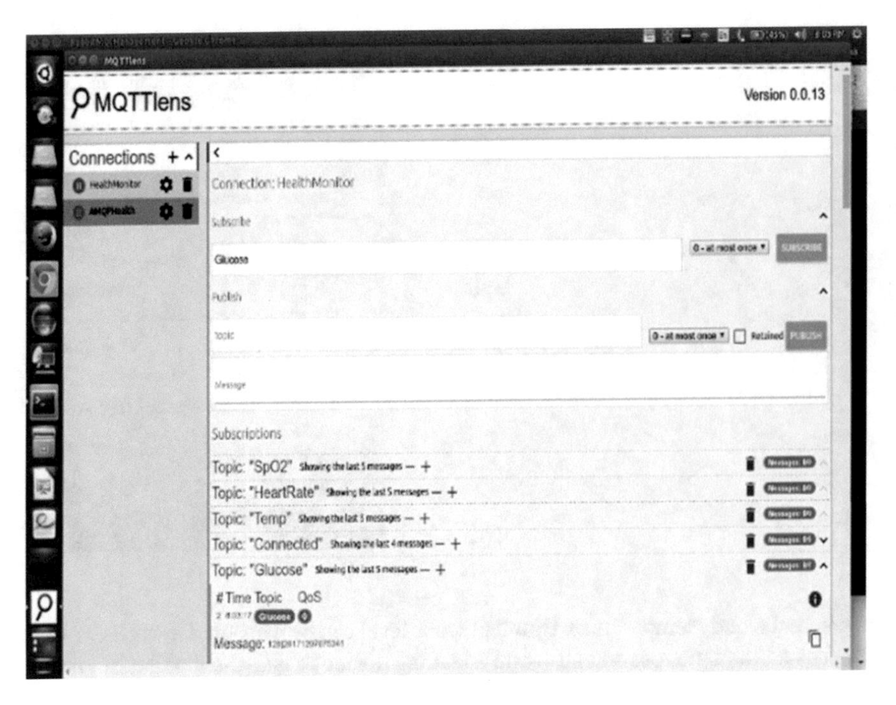

Fig. 10 Vital values in laptop

7 Conclusion and Future Enhancement

The implementation of IoT technology in health care decreases the cost and makes the solution vastly scalable as it supports the trouble-free expansion of nodes for vital collection and processing. The IoT-based system is highly distributed so that the failure of a single node still makes the system functional. The use of IoT application layer protocols, that is, AMQP and MQTT protocols, enables easy integration to third-party applications and devices. The system could be compared using different protocols (AMQP and MQTT) and found that AMQP has more reliable message transferring than MQTT. As per the conclusion, by using AMQP protocol, the healthcare monitoring system gives accurate and secure result. Using data analytics and clustering, this system can analyze and find out the diseases in the future enhancement.

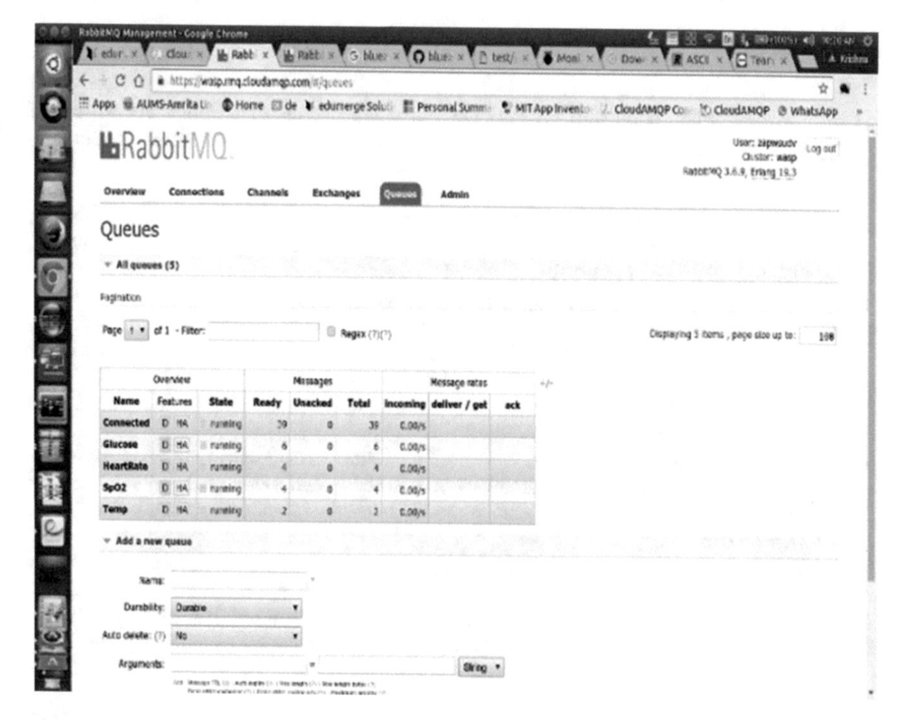

Fig. 11 CloudAMQP queue

Fig. 12 Message in mobile phone

References

1. Binu, P.K., Thomas, K., Nithin, P.V.: Highly secure and efficient architectural model for IoT based health care systems. In: International Conference on Advances in Computing, Communications and Informatics, pp. 487–493. IEEE, Udupi (2017)
2. IoT in healthcare reference available. http://www.readwrite.com/
3. Ray, I., Alangot, B., Nair, S. and Achuthan, K.: Using attribute—based access control for remote healthcare monitoring. In: 2017 Fourth International Conference Software Defined Systems, pp. 137–142, IEEE, Spain (2017)
4. The SpO_2 reference is available. https://www.blog.withings.com/
5. Binu, P.K., Akhil, V., Mohan, V.: Smart secure IoT based child behavior and health monitoring system using Hadoop. In: International Conference on Advances in Computing, Communications and Informatics, pp. 418–423. IEEE, Udupi (2017)
6. Perumalsami, R., Riffath Shehla, M.: AMQP based patient monitoring system. Int. J. Res. Appl. Sci. Eng. Technol. **3**(III): 213–228 (2015)
7. Luzuriaga, J.E., Perez, M., Boronat, P., Cano, J.C., Calafate, C., Manzoni, P.: A comparative evaluation of AMQP and MQTT protocols over unstable and mobile networks publication details. In: 12th Consumer Communications and Networking Conference, pp. 931–936. IEEE, Las Vegas (2015)
8. Al-Fuqaha, A., Guizani, M., Mohammadi, M., Aledhari, M., Ayyash, M.: Internet of things: a survey on enabling technologies, protocols, and applications. In: IEEE Communications Surveys & Tutorials, vol. 17, issue 4, pp. 2347–2376, Fourth Quarter 2015. IEEE (2015)
9. Xiong, X., Fu, J.: Active status certificate publish and subscribe based on AMQP. In: International Conference on Computational and Information Sciences, pp. 725–728. IEEE, China (2011)
10. John, V., Liu, X.: A survey of distributed message broker queue. In: arXiv:1704.00411v1 [cs.DC], April (2017)
11. Ani, R., Krishna, S., Anju, N., Aslam, M.S, Deepa, O.S.: IoT based patient monitoring and diagnostic prediction tool using ensemble classifier. In: International Conference on Advances in Computing, Communications and Informatics, pp. 1588–1593, IEEE, Udupi (2017)
12. GSM/GPRS TTL UART MODEM-SIM900A reference available. http://www.rhydolabz.com/wiki/?p=853
13. Divya, K., Jeyalatha, S.: Key technologies in cloud computing. In: Proceedings of 2012 International of Cloud Computing, Technologies, Applications & Management, pp. 196–199, IEEE, Dubai (2013)
14. CloudAMQP reference is available. http://www.cloudamqp.com

C. S. Krishna completed B.Tech. in Electronics and Communication, Pursuing M.Tech. in Embedded Systems. Amrita School of Engineering, Bengaluru. Her research area includes wireless communication systems.

T. Sasikala currently working as an Assistant Professor (Sr. Gr) in Department of Computer Science, Amrita School of Engineering, Bengaluru. She was completed B.E. in Computer Science Engineering, M.E. in CSE, and Pursuing Ph.D. in area of Data Mining. Her research area includes wireless communication systems and data mining.

Performance Analysis of an Efficient Data-Centric Misbehavior Detection Technique for Vehicular Networks

S. Rakhi and K. R. Shobha

Abstract In current scenario of automotive industry, Vehicular Networks are providing enhanced support to improve driver security and efficiency. But still, security is an essential requirement in vehicular ad hoc networks since VANET packets contain life critical information. A rogue node can transmit fake messages to their neighbors. In this paper an effort is made to explore an environment with simulated Sybil attack and its effects on network performance. We propose an efficient and scalable data centric approach based on the comparison of average flow rate or mobility information exchanged between the vehicles in the network. Our approach does not require any help from the infrastructure during attacker detection. Simulation results show the effectiveness of the proposed system to locate rogue nodes and inform their peers to avoid communication with them, by checking a Boolean flag. So the proposed system will make VANETs more fault tolerant and robust against transmission of fake messages.

Keywords VANETs · WAVE · Security · Data centric · Scalable · Sybil attack

1 Introduction

The development in the areas of wireless communication and internet emerged a new type of network called Vehicular Ad hoc Network (VANET). Currently several efficient routing protocols have been proposed for VANET. Next area for research is security. VANETs have high mobility and occasional network fragmentation. VANETs incorporate safety and security with the help of Vehicle to Vehicle (V2V) and Vehicle to Infrastructure (V2I) communication. Most of the VANET applications nowadays

S. Rakhi (✉)
Department of Electronics & Communication Engineering, Atria IT, Bengaluru, India
e-mail: rakhi.s@atria.edu

K. R. Shobha
Department of Telecommunication Engineering, MSRIT, Bengaluru, India
e-mail: shobha_shankar@msrit.edu

S. Smys et al. (eds.), *International Conference on Computer Networks and Communication Technologies*, Lecture Notes on Data Engineering and Communications Technologies 15, https://doi.org/10.1007/978-981-10-8681-6_30

concentrate more on congestion avoidance, lane change warnings, post-crash warnings etc. [1]. The misbehavior detection needs to be addressed in a broader way in VANETs.

Majority of the data dissemination in VANET is happening basically through cooperative behavior of the vehicles. Messages exchanged among the nodes sometimes carry life critical information. In that type of scenarios any abnormal behavior of nodes lead to dangerous situations. So any misbehavior from any node for that matter in VANETs is a major issue to be addressed. In a normal environment if any one node is acting differently from the other nodes, can be treated as misbehavior. To be more precise misbehavior can be described as, any abnormal behavior of a vehicular node that deviates from the average behavior of other vehicular nodes in VANETs [2]. In VANET the communication is mainly due to the periodic exchange of beacon packets which are exchanged between the nodes over a shared channel. During this communication, it is possible that a particular node can become malicious and send multiple messages by changing the identities. These nodes create an illusion of an unreal event by transmitting duplicate messages using those identities at the same time. Once an attacker is able to launch this kind of attack, it is very easy for them to cause any other related attacks. This paper is divided into six sections; Sect. 2 describes the related work of Sybil attack and their detection methods. Section 3 explains VANET attack model in brief. The overview of the proposed model is explained in Sect. 4. Section 5 explains the experimental results. Section 6 concludes the paper.

2 Related Work

Kerrache et al. [3] describes a trust based approach, where the method takes the advantage of alert messages to carry necessary information to establish trust and select most trustable vehicles as next hop neighbors. The broadcast storm is avoided here. High detection and low packet loss are maintained in this approach. However, the author has not taken care of privacy preservation. The author has formulated another approach [4] where the aim is at finding the shortest and most trusted path to destination considering real traffic information and the distribution of dishonest nodes. The effectiveness in this protocol is packet delivery and end-to-end delay.

Hussain and Oh [5] aim at two conflicting goals privacy and Sybil attack. Privacy is preserved using pseudonymless beaconing and Sybil attack is detected with tokens provided by RSU. Since each time a particular node has to store tokens received from RSU, the memory requirements also increases. Tomandl et al. [6] proposed a dynamic rule based intrusion detection system called REST-Net. This approach has a dynamic detection engine which observes and analyzes the fake data sent in network with plausibility checks to identify attacks. It uses rules or patterns to outline invalid actions of users. However this approach is not feasible in dynamic scenario since each node has to maintain all the rules or patterns.

Ruj et al. [7] have presented a misbehavior detection scheme which detects false alert messages and misbehaving nodes by observing the actions of a particular node before it sends the message and after it sends out the message. The decision is taken based on the steadiness of the previous and the new messages which carry vehicle position information. Since no revocation of nodes, bandwidth is preserved here, but the approach does not evaluate their findings and restricted the concept only in multilane. Feng et al. [8] proposed an event based reputation system, in which each event is given a dynamic reputation and trusted value and are used to avoid the spreading of fake messages. A threshold value is assigned for both reputation and trusted values and compared against the values of events. The approach studies the effect of Sybil attack on packet delivery ratio. However, the system needs the help of RSU and it is time consuming.

Saggi and Kaur [9] have proposed a system which uses neighbor node information to detect the attacker nodes in two phases. In first phase the vehicles have to register with RSU. In the second stage RSU assigns threshold speed limit to each nodes. The information from any attacker node is compared with neighbor nodes. However this approach is purely RSU based and the authors have not explained how RSU is storing the adjacent node information. Bojnord [10] analyzes the effectiveness of a Sybil attack with the help of fuzzy logic technique. It provides three different modules. They are fuzzy parameter extraction, fuzzy decision maker and fuzzy verification. The method uses various fuzzy rules and compares the extracted values and threshold values. The method does not incorporate any cryptography techniques. However the fuzzy logic systems are not well suited for VANETs due to the high mobility.

Gu et al. [11] proposes a Sybil detection method which uses driving pattern of vehicles in urban scenario. The Mahalanobis distance is used to measure resemblance of traffic patterns than the usual Euclidean distance. For each beaconing message driving pattern matrices are created. Then Mahalanobis distance is used to evaluate the nodes driving pattern and detect any unusual patterns if any. Shikha et al. [12] have proposed timestamp based Sybil attack detection. RSU generates timestamp for each vehicle. Each time stamp is digitally signed. If any node contains multiple timestamps from the last RSU, it is treated as attacker. It claims better results than [11] but, since it uses cryptographic techniques the system experiences more delay.

Weerasinghe et al. [13] developed a scheme which is based on position as well as velocity verification for one hop neighbors. The scheme is simulated in various conditions like rural, urban and Manhattan. An automotive grade differential global positioning system is used in the system to verify position of the neighbors. The approach by Shaobe et al. [14] uses Received Signal Strength (RSS) of each node to compute the distance to other nodes. Each node in the network forms a group of neighboring nodes by using similarity in RSS. The group results are periodically broadcasted in the network. The neighboring nodes with comparable RSS values are grouped into a group called suspect group. This approach can create false positives since the similar values are grouped.

Grover et al. [15, 16] have presented a structure for VANET security which is mainly based on machine learning approach. In order to classify different types of attack in VANETs, several features are extracted from different attack cases. Huang et al. [17] have proposed a cheater detection protocol which detects the rogue nodes with the help of radars. The local velocity and distance of a node are measured by the radar which is used to verify the congestion messages sent by selfish nodes. The kinematic wave theory of physics is used in this approach. The wave packets are exchanged between the nodes which are signed. The method is quite effective but the control overhead is more.

3 Vanet Attack Model and Assumptions

The Sybil attack in a VANET network refers to a malicious node taking identities of multiple nodes. In VANETs, nodes broadcasts beacon messages periodically to find new neighbors in a wireless medium. A malicious node can easily assert multiple identities without being detected as malicious. For an attack free environment, network should make sure that each physical node is assured with only one identity. A vehicle is referred as a node. The node which has multiple identities is called malicious or rogue node and the normal nodes are legitimate nodes. The rouge nodes can cause an illusion of traffic bottlenecks. A greedy driver can convince the neighbors about a traffic jam ahead and also convince them to choose an alternate path. Later he can use such free path for his selfish motives. The performance of the network can be impacted by Sybil nodes when they inject false data in the network [18].

The following assumptions are incorporated while the system is designed and simulated. First, we assume that most of the vehicles are travelling independently and most of the vehicles are trusted. Few vehicles are assumed to be greedy in the network. The vehicles are equipped with DSRC standard based radio modules. Even though our detection scheme is not implemented in RSU, an indirect support from base stations is always required for any node in VANET.

4 Proposed Detection Scheme

4.1 Algorithm

The proposed detection scheme is intended to the vehicles which continuously monitor the beacon messages received from the neighbors. Each vehicle creates a beacon message with flow rate, density and a suspicious flag in addition to the normal fields like id, position, and velocity with a message type. Our scheme works under a trust that a legitimate node will always behave in honest manner in carrying out data dissemination. The detection is based purely on data centric approach for faster decision making. This approach deals with analyzing and finding out the discrepancy in the data transmitted among the nodes. This approach can detect both insider and outsider attacks.

Algorithm 1

$V = \{V_1, V_2 \cdots V_n\}$, Set of n vehicles in the network
$S = \{S_1, S_2 \cdots S_m\}$, Set of m Sybil vehicles in the network
for i=1 to n do
    Pseudonym creation
    Find Neighbor by hello packets
    Exchange beacons with node-id,position,$[\text{velocity}]_{av}$,[density],$[\text{flowrate}]_{own}$
    Update data received from neighbors
        Diff= $[\text{flowrate}]_{own}$ - $[\text{flowrate}]_{recv}$
            if Diff > Threshold then
                Malicious node[S_m] detected
                Suspicious Flag=true
                Discard data
                Inform neighbors
            else
                Accept data
                Suspicion Flag=false
            end if
end for

In the proposed scheme pseudonyms are created under the assumption of privacy preservation of the vehicles. As per the DSRC standard the communication range is taken as $R < 300$ m between the neighbors. It means that the vehicle can receive and transmit beacon packets within the radius of 300 m which is calculated by Euclidean distance. Graph theory can be used to model a VANET. The graph $G = (N, E)$ represents a set of N nodes in the Euclidean plane and set of E edges.

$$L_{ij} = \sqrt{(y_i - y_j)^2 - (x_i - x_j)^2} \tag{1}$$

L_{ij} is referred as the Euclidean distance between vehicles i and j. The speed-flow-density is related by a fundamental equation,

$$\text{Flow Rate} = \text{Density} * \text{Mobility} \tag{2}$$

Each node calculates its own parameters of density (k), velocity and flow rate (v). These parameters are transmitted to the neighbors with the beacon messages periodically. In a similar traffic condition, the vehicles which are travelling close by should have same flow rate value. An attacker can transmit a low flow value to create an illusion of traffic congestion ahead. The attacker broadcasts these messages in the network with multiple identities to create an illusion that multiple vehicles ahead are transmitting the same information. Once a node receives beacon with the alert message, it compares the parameter of flow rate in the received and the own calculated value as shown in the Algorithm 1. If the difference value is less than the threshold

Table 1 Simulation parameters

Parameter	Value
Simulation time	200 s
Scenario	Single lane
Road length	1000 m
Speed (max)	40 m/s
Mobility tool	MOVE
Network simulation tool	NS2
Traffic simulation tool	SUMO
Wireless protocol	802.11p
Transmission range	300 m
Number of vehicles	50 and 100

value, the data is accepted and the reporting node is treated as genuine, otherwise an attacker is identified and informed to the peers. The threshold value is calculated as the average value of the flow rates of the neighbor nodes which are communicating with the observing node. A suspicion flag is a Boolean value which is made true if the node is an attacker. A node can store this value in its neighbor table and transmit this information when the beacons are retransmitted at later stage.

4.2 Simulation Setup

In our simulation setup MOVE-SUMO and NS2 are used as vehicular traffic generator and network simulator respectively. The simulated data under various conditions has been used extensively to study the efficiency of the network. In a highway scenario the road is modelled as a straight road. In the simulated scenario each vehicle broadcasts beacon message at a constant time interval. Table 1 shows the detailed configurations of the simulation setup. A simulation scenario using MOVE and SUMO for an ad hoc vehicular network with vehicles moving in a single lane is considered. MOVE is used as mobility generator and SUMO generates vehicular traffic.

5 Results and Perfomance Analysis

In order to evaluate and validate the performance of the network with different parameters we have taken different scenarios. The performance of the network is validated with average packet delivery ratio (PDR) and average mobility or speed of the vehicles. Packet delivery ratio is calculated from the number of packets delivered over number of packets transmitted in the network. Two traffic scenarios and three attack scenarios are considered and simulated here. In the first scenario the total number of vehicles in the network is taken as $N = 100$ with varying number of attackers. The number of attackers is 3, 5 and 10. Same attack model is repeated for $N = 150$.

We have implemented these attackers for varying vehicle mobility in the network. As the mobility increases the probability of the link breakage is more and hence the packet loss is more. So at high speed the PDR drops and the performance of the network reduces gradually. Figure 1 shows the variation of average PDR against average mobility. A Sybil attacker is implemented according to the attack model explained earlier. According to this model, an attacker sends multiple beacons with different identities and all these identities participate concurrently. The graph shows that the average packet delivery ratio reduces more when the number of attackers is more. When the attackers are 3, 5 and 10 the graph shows the variation in average PDR vs average mobility. The PDR reduces when there is more number of attackers since the data is lost due to collision, affecting the network performance.

In order to evaluate the proposed detection method, the simulation results are evaluated using metrics like detection delay, detection rate, false positives rate. Detection delay is defined as the time difference from when the attacker enters the network to the time when the attack gets detected. Detection rate is the ratio of number of

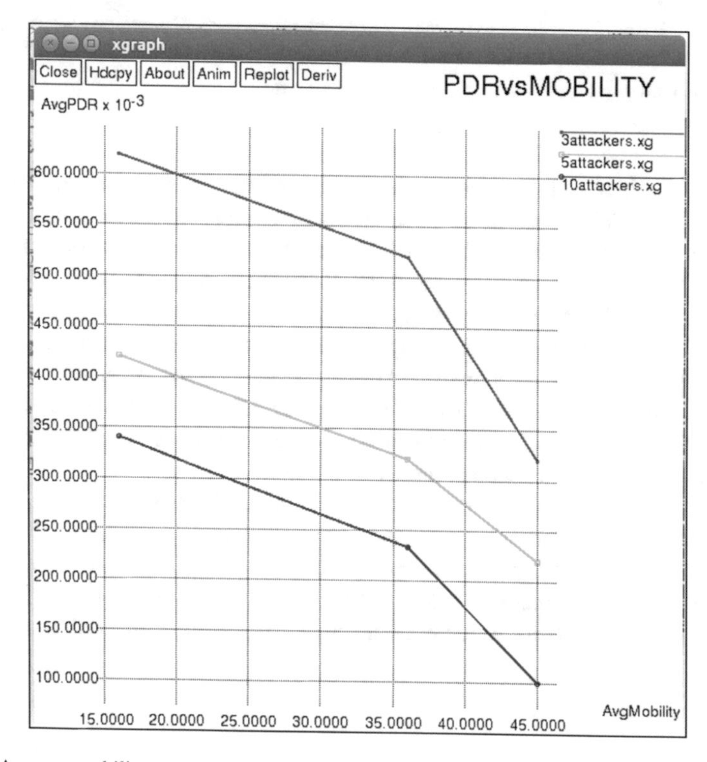

Fig. 1 Average mobility versus average PDR

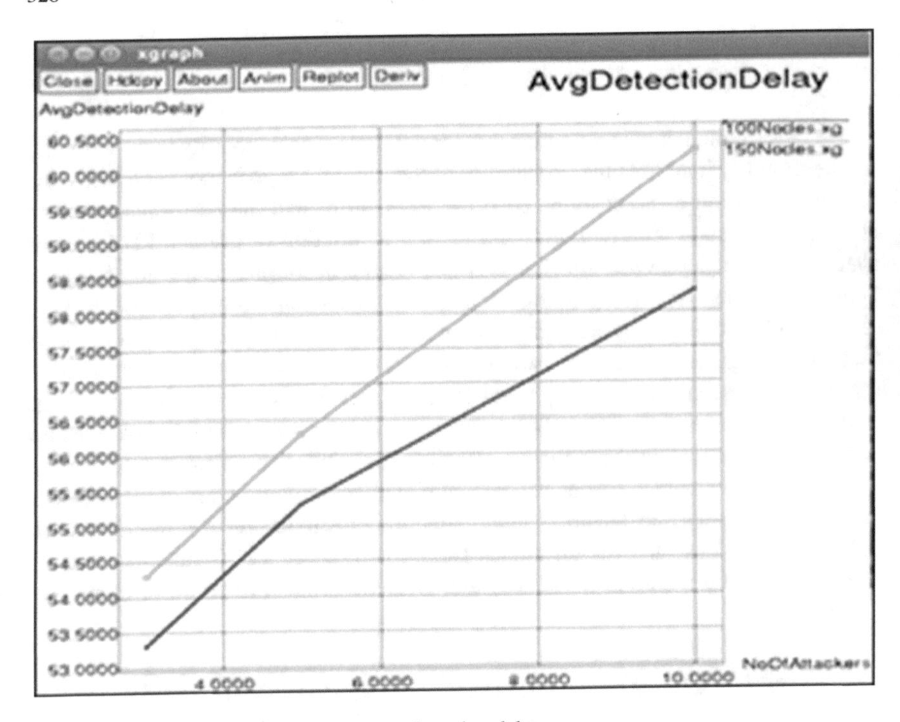

Fig. 2 Number of attackers versus average detection delay

correctly identified Sybil nodes to total number of nodes in the network. We have analyzed the scalability of the proposed work by comparing the detection delay for number of nodes $N = 100$ and 150 for three cases, when the number of attackers are 3, 5, 10. The delay increases as the number of attackers increases as shown in Fig. 2. But it is observed that the delay is not increasing much when the number of nodes is increasing with proposed model. Figure 3 shows that the detection rate decreases as the number of attackers increases. We have evaluated the detection rate for two scenarios, when $N = 100$ and $N = 150$. Since the detection rate increases with the number of nodes in a particular transmission range, it is observed that for $N = 150$ the values are high for the same number of attackers.

False positive rate is the percentage of honest nodes incorrectly detected as malicious nodes. We have measured the false positive rate for our misbehavior detection method for different number of nodes in the network. From Fig. 4 it is observed that as the numbers of attackers are more the amount of false positive rate increases. But if we compare two scenarios with different number of nodes, our system gives

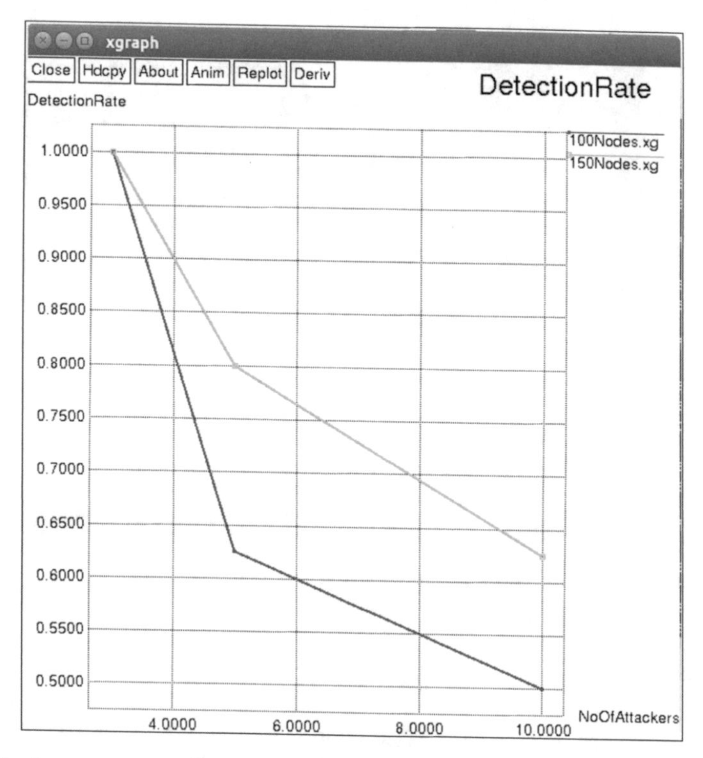

Fig. 3 Number of attackers versus detection rate

reasonable results without much variation in false positive rates. Thus we can conclude that the system shows satisfactory performance and supports scalability of the network.

6 Conclusion

VANET is a preferred target for numerous types of security attacks. If an attacker can launch a Sybil attack in the network any other attacks can be introduced easily. In this paper we have implemented two different scenarios with varying number of nodes and attackers. The main advantage of the proposed detection method is it is purely data centric and scalable. No infrastructure support is required for the detection as we are analyzing only the data which are exchanged periodically. But our approach can give an increase in false positive value even though the node is legitimate because we are considering low flow rate value which is exchanged in the beacon. This issue has to be still addressed.

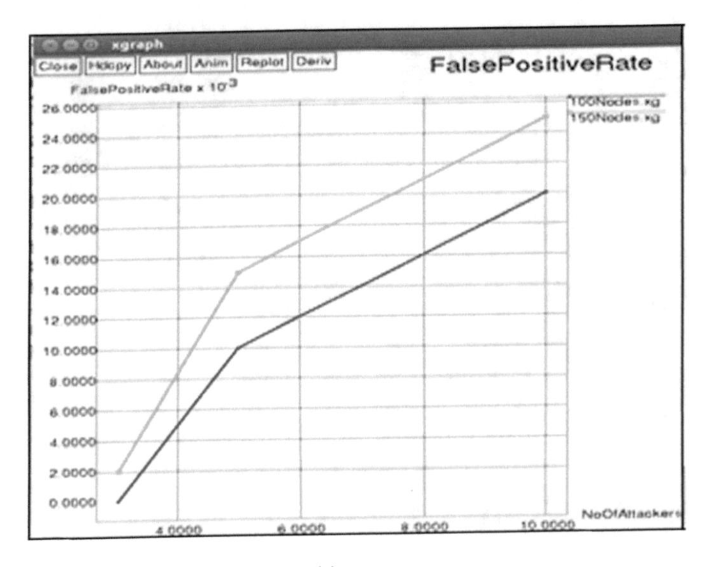

Fig. 4 Number of attackers versus false positive rate

References

1. Rakhi, S., Shobha, K.R.: A comprehensive survey on security issues in VANETS for safe communication. In: Int. J. Emerg. Technol. Comput. Sci. Electron. **14**(2), 446–454 (2015)
2. Security issues in vehicular ad hoc networks. www.intechopen.com
3. Kerrache, C.A., Calafate, C.T., Lagraa, N.: Trustaware opportunistic scheme for VANET safety applications. In: IEEE International Conference on Big Data and IoT, pp. 153–160. IEEE Press, France (2016)
4. Kerrache, C.A., Mellouk, A.H.: A trusted routing protocol for urban vehicular environment. In: 11th IEEE International Conference on Mobile Computing and Networking, pp. 260–267, IEEE Press, Abu Dhabi (2015)
5. Hussain, R., Oh, H.: On secure and privacy aware sybil attack detection in vehicular communication. Int. J. Wirel. Pers. Commun. **77**(4), 2649–2673 (2014)
6. Tomandl, A., Fuchs, K.P., Federrath, H.: REST-Net a dynamic rule based approach for VANETS. In: IEEE International Conference on Mobile Networking, pp. 1–8. Portugal (2014)
7. Ruj, S., Marcos, A., Huang, Z.: On data centric misbehavior detection in VANETS. In: Vehicular Technology Conference. IEEE Press, San Fransisco (2011)
8. Feng, X., Yan, C., Tang, J.: A method for defensing against multisource sybil attacks in VANETS. Peer Peer Netw. Appl. **10**(2), 305–314 (2017)
9. Saggi, M.K., Kaur, R.: Isolation of sybil attack in VANET using neighbor infomation. In: IEEE International Conference on Advance Computing Conference. India (2015)
10. Bojnord, A., Bojnord, H.S.: A secure model for prevention of sybil attack in vehicular adhoc network. Int. J. Comput. Sci. Netw. Secur. **17**(1), 30–34 (2017)
11. Gu, P., Khatoun, R., Begriche, Y., Serhrouchni, A.: Vehicle driving pattern based sybil attack detection. In: 18th IEEE International Coference on High Performance Computing and Communication. IEEE Press, Sydney (2016)
12. Sharma, S., Sharma, S.: A novel mechanism of detection of sybil attack in VANETS using timestamp approach. Int. J. Innovations Eng. Technol. **8**(1), 200–204 (2017)
13. Weerasinghe, H.D., Tackett, R., Fu, H.: Verifying position and velocity in vehicular adhoc networks. Spec. Issue Secur. Commun Netw, **4**(7), 785–791 (2011)

14. Abbas, S., Merabti, M., Llewellyn-Jones, D.: Signal strength based sybil attack detection in wireless adhoc networks. Researchgate, 574–588 (2009)
15. Grover, J., Prajapati, N.K., Laxmi, V., Gaur, M.S.: Machine learning approach for multiple misbehavior detection in VANET. In: International Conference on Advances in Computing and Communication, pp. 644–653. India (2011)
16. Grover, J., Gaur, M.S., Laxmi, V: Position forging attacks in vehicular ad hoc networks: implementation, impact and detection. In: International Conference on Wireless Communications and Mobile Computing Conference, pp. 701–706. IEEE Press, Turkey (2011)
17. Huang, D., Williams, S.A., Shere, S.: Cheater detection in vehicular network. In: 11th IEEE International Conference on Trust, Security and Privacy in Communications and Networking, pp. 193–200. IEEE Press, Liverpool (2012)
18. Khan, U., Agarwal, S., Silakari, S.:A detailed survey on misbehavior node detection in VANETS. In: 2nd International Conference on İnformtion System Design and İntelligent Applications, pp. 11–19. Springer, India (2015)

S. Rakhi completed her M.Tech. in VLSI and Embedded Systems in the year 2010 from Visvesvaraya Technological University (VTU), Belgaum, Karnataka, India securing first rank with gold medal. She is currently pursuing her Ph.D. under VTU. She is currently working as Assistant Professor in the Department of ECE, Atria Institute of Technology, Bengaluru. Her research interests include security in wireless networks and vehicular networks. She has presented research papers in the areas of VLSI, Vehicular Adhoc Networks and Wireless Networks.

Dr. K. R. Shobha. received her M.E. degree in Digital Communication Engineering from Bengaluru University, Karnataka, India and Ph.D. from Visveswaraya Technological University. She is currently working as an Associate Professor in the Department of Telecommunication Engineering, M.S. Ramaiah Institute of Technology, Bengaluru. Her research areas include Mobile Adhoc Networks, IoT and Cloud Computing. She has more than 25 Papers publications to her credit.

She is a Senior IEEE Member serving as Secretary of IEEE Sensor Council, Bengaluru Section. She is also an active member of IEEE Communication Society and Women in Engineering under IEEE Bangalore Section.

Maximal Frequent Itemset Mining Using Breadth-First Search with Efficient Pruning

K. Sumathi, S. Kannan and K. Nagarajan

Abstract Maximal Frequent Patterns can be mined using breadth-first search or depth-first search. The pure BFS algorithms work well when all maximal frequent itemsets are short. The pure DFS algorithms work well when all maximal frequent itemsets are long. Both the pure BFS and pure DFS techniques will not be efficient, when the dataset contains some of long maximal frequent itemsets and some of short maximal frequent itemsets. Efficient pruning techniques are required to mine MFI from these kinds of datasets. An algorithm (MFIMiner) using Breadth-First search with efficient pruning mechanism that competently mines both long and short maximal frequent itemsets is proposed in this paper. The performance of the algorithm is evaluated and compared with GenMax and Mafia algorithms for T40I10D100K, T10I4D100K, and Retail dataset. The result shows that the proposed algorithm has significant improvement than existing algorithms for sparse datasets.

Keywords Breadth-First search with pruning · Mining MFIs

K. Sumathi (✉)
Kalasalingam Academy of Research and Education, Krishnan Koil, Virudhu Nagar, India
e-mail: sumathirajkumar2006@gmail.com

S. Kannan
Madurai Kamaraj University, Madurai, India
e-mail: skannanmku@rediffmail.com

K. Nagarajan
Tata Consultancy Service, Chennai, Tamil Nadu, India
e-mail: nag12317@hotmail.com
URL: http://in.linkedin.com/in/nagarajank

S. Smys et al. (eds.), *International Conference on Computer Networks and Communication Technologies*, Lecture Notes on Data Engineering and Communications Technologies 15, https://doi.org/10.1007/978-981-10-8681-6_31

1 Introduction

Mining frequent patterns play an important role in data mining, and it is widely used in applications regarding association rules, customer behavior analysis etc. All frequent itemsets can be constructed from Maximal Frequent itemsets (MFI), since every subset of MFI is a frequent itemset. So many data mining applications need only MFI instead of the frequent itemset.

Bayardo presented MaxMiner algorithm [1] for mining maximal frequent itemset in 1998. MaxMiner used set-enumeration tree as the conceptual framework and followed breadth-first searching technique, as well as superset pruning strategy and dynamic recording strategy. Pincer Search [2] algorithm was presented by D. Lin and Z. M. Kedem. Pincer search used both top-down and bottom-up search for mining all MFI. The Depth Project algorithm adopted depth-first approach with superset pruning strategy and dynamic recording methods to mine all MFI [3]. Burdick proposed an algorithm called MAFIA [4] which used depth-first search and stored dataset in longitudinal bitmaps. MAFIA used three pruning strategies including Parent Equivalence Pruning (PEP), Frequent Head Union Tail (FHUT) and Head Union Tail MFI (HUTMFI). GenMax [5] combined the pruning and mining process, utilizing two approaches for mining MFI efficiently. The first approach is GenMax constructs Local Maximal Frequent Itemsets (LMFI) for fast superset checking. Another approach is GenMax that uses Diffset Propagation method for fast support calculation when the dataset is highly dense in nature.

2 Proposed Work

The MFIMiner algorithm that efficiently mines both long and short maximal frequent itemsets is proposed in this paper. The MFIMiner algorithm runs in bottom-up direction and employs pruning to efficiently mine all MFI from the dataset

In general, the structure of the dataset may be in two different ways—horizontal data format and vertical data format. This approach uses vertical data representation of dataset. In this format, data is represented in item-tidset format, where item is the name of the item and tidset is the set of transaction identifiers containing the item.

MFIMiner computes the frequency of an item by counting the number of tids in the tidset and computes the frequency of an itemset by intersecting the tidsets of corresponding items. The data structures used in the MFIMiner algorithm are FI which denotes set of frequent itemsets, CI denotes the set of candidate items, MFI denotes the set of Maximal Frequent Itemsets.

2.1 *MFIMiner Algorithm*

The first step of MFIMiner algorithm is to mine all frequent items in the dataset. The frequent items are reordered in increasing order of their support and added to FI. Candidates for each frequent item in FI are obtained and added to CI. For example candidates of frequent item FI_1 are added to CI_1, candidates of FI_2 are added to CI_2 and so on. Let FI_j ($FI_j \varepsilon$ FI) be the frequent item and CI_j ($CI_j \varepsilon$ CI) is the set of candidates for FI_j.

The frequent extension of FI_j is (FI_{j1}, FI_{j2} … FI_{jn}) where n is the number of candidates in CI_j. Frequent extension of FI_{j1} is constructed from CI_{j1} i.e. $CI_{j1} \subseteq$ (FI_{j2}, FI_{j3} … FI_{jn}). Initial candidates of FI_{j1} are constructed from (FI_{j2}, FI_{j3} … FI_{jn}) and candidates of FI_{j2} are constructed from (FI_{j3}, FI_{j4} … FI_{jn}) and so on. Exact candidates for each frequent extension are constructed using GenerateCandidate method. The exact candidates of a frequent itemset do not include the infrequent candidates of the frequent itemset. When a frequent itemset has no candidate and has no superset in MFI, then frequent itemset is added to MFI.

In the next step, the $FI_i \cup CI_i$ (i is varied from 1 to n) is removed from FI and CI, respectively, and checked, whether it has superset in MFI or not. If it has superset in MFI then ignored. Otherwise, the frequency of $FI_i \cup CI_i$ is computed. If $FI_i \cup CI_i$ is frequent then it is added to MFI. The next pair ($FI_{i+1} \cup CI_{i+1}$) is taken for the test. This step is repeated until an infrequent $FI_k \cup CI_k$ is obtained

If $FI_k \cup CI_k$ has no superset in MFI and it is not frequent, then the following steps are done.

1. Parent Equivalent Pruning
2. Frequent Extension to the next level

When an infrequent $FI_k \cup CI_k$ is found, parent equivalent pruning is done. This pruning which was introduced by Calimlim in Mafia algorithm [4] removes items from the candidate set (CI_k) and add those items to the frequent itemset(FI_k) if the tidset of any item in CI_k includes the tidset of FI_k.

The frequent itemset (FI_k) is extended to the next level using items of CI_k in a bottom-up direction. The candidates of each frequent extension are generated and infrequent candidates are pruned. The frequent extensions and candidates of each frequent extension are added to FI and CI respectively.

The data structures FI and CI are updated when new frequent extension and candidates of frequent extension are found. MFI is updated when the frequent extension has no candidate and no superset in MFI or $FI_i \cup CI_i$ has no superset in MFI and is frequent.

Table 1 Vertical data representation of the transactional database DB

Items	Tidsets
A	T1, T2, T3, T4, T5
B	T2, T5, T6
C	T1, T2, T4, T5, T6
D	T1, T3, T4, T5
E	T2, T3, T4, T5, T6
F	T2, T4, T6

Table 2 Frequent items and candidate items in database DB with support 3 transactions

Frequent Items	Candidates of frequent Item
B	C, E
F	C, E
D	A, C, E
A	C, E
C	E
E	–

MFIMiner algorithm is explained with the following example. Consider the transaction database DB (Table 1) which includes six different items, I = {A, B, C, D, E, F} and six transactions T = {T1, T2, T3, T4, T5, T6}. The vertical data format of the database DB is given below.

All frequent items are extracted and reordered in ascending order with respect to the support. The support of an item is directly given by the number of transactions in the tidsets. For example, the minimum support is considered to be of 3 transactions. In database DB (Table 1), all the items are having more than two tids in the tidset, and so all the items are frequent.

The items A, B, C, D, E, and F are frequent items and will be considered to the next level. These items are sorted in ascending order of their support.

The reordered frequent items are B, F, D, A, C, E and candidates of each frequent item are obtained. Frequent items and candidates of each frequent item of database DB (Table 1) is given in Table 2.

FI_1 and CI_1 are removed from FI and CI, respectively, and Mining MFI is started by checking the frequency of $FI_1 \cup CI_1$, continued until an infrequent $FI_n \cup CI_n$ is obtained. When an infrequent $FI_i \cup CI_i$ is obtained, the parent equivalent pruning is done and frequent extension of the FI_i is generated using CI_i and new candidates of frequent extension are also obtained. The frequent extensions of FI_i are generated from CI_i and each frequent extension and its candidates are added to FI and CI, respectively. Frequent extensions of a frequent itemset are generated as follows.

Table 3 Maximal frequent itemsets

Tid	Items	Maximal frequent itemsets Minimum support = 3	Maximal frequent itemsets Minimum support = 2
1	ACD		ACEF
2	ABCEF	ACD	ACDE
3	ADE	ACE	ABCE
4	ACDEF	ADE	BCEF
5	ABCDE	BCE	BCEA
6	BCEF	CEF	

frequentExtension (FI_i, CI_i)
For each x ε CI_i
$FI_{i+1} = FI_i \cup x$; *//new frequent extension*
CI_i.remove(x); *//initial candidates of new frequent extension*
CI_{i+1}=generateCandidate (FI_{i+1},CI_i); *//exact candidates of new frequent extension*

If the CI_{i+1} is empty and the FI_{i+1} has no superset in MFI, then FI_{i+1} is added to MFI. In database DB (Table 1), the first two frequent item and its candidates are {B}-{C,E} and {F}-{C,E}, has no superset in MFI, are frequent and added to MFI. The next pair {D}-{A,C,E} is not frequent and parent equivalent pruning is done. The item A is trimmed from the candidate set and added to the frequent item D. The frequent extension of {DA} is generated from {C, E} and candidates of each frequent extension are generated. Frequent extension and candidates are {D,A,C}-{} and {D,A,E}-{}. Both of these frequent itemsets, have no superset in MFI, are frequent, and are added to MFI. The next pair, is {A}-{C, E} has no superset in MFI, is frequent and added to MFI. The subsequent pairs, are {CE} {} and {E}{}, have superset in MFI and ignored. MFIMiner algorithm returns a subset of the MFI and would require post-pruning to eliminate non-maximal patterns. MFIs with minimum support 3 in database DB are {ACD}, {ACE}, {ADE}, {BCE}, and {CEF}.

Number of MFI generated depends on the support value which is specified by the user is shown in Table 3. MFIs with minimum support 2 in database DB are ACEF, ACDE, ABCE and BCEF.

Algorithm –MFIMiner

Input: dataset D, support S

Output: Maximal Frequent Itemsets MFI

MFIMiner (Dataset D, Support S)

BEGIN

1. Generate frequent items and reorder them in ascending order of their support.
2. Generate candidate items for each frequent item and reorder the candidates in increasing order of support.
3. For each x ε FIs // FI-> set of frequent items

```
If candidate(x)  is empty and x has no superset in MFI

        AddtoMFI(x); continue;
        c=cand(x);
        FI=FI/x  // remove the frequent item from the set of frequent item(FI)
        CI=CI/cand(x) // remove the candidates of FI from candidate set
    if x U c  has superset in MFI continue;
    if size of x U c  is 1 or 2 then add x U c  to MFI;
            continue;
    else if x U c  is frequent then add x U c  to MFI;
            continue;
    else
        for each y  ε c

            If(tidset(y).containsAll(tidset(x))
             x.add(y);
             c.remove(y);
        Bottomup_ExternFI(x,c,tidset(x))
```

END

Bottomup_ExternFI Method

Input: Frequent Itemset, Candidate set, tidset of Frequent Itemset

Output: Frequent extensions of frequent itemset and candidates of each frequent extension, MFIs if generated.

Bottomup_ExternFI(frequent,candidate, tidset(frequent))

BEGIN

For each x ε candidate

```
        Nfrequent=frequent ∪ x  // current frequent itemset
        candidate.remove(x) // candidates of current frequent itemset
         if candidate is empty && Nfrequent has no superset in MFI
        MFI.add(Nfrequent); continue;
    newcandidate=GenerateCandidates(Nfrequent,candidate, tidset(frequent))
        if newcandidate is empty
                if Nfrequent has no superset in MFI
                   MFI.add(Nfrequent);
                else
                   FI.add(Nfrequent);
// New frequent itemset is added to FI.
                   cand(Nfrequent)=newcandidate;
// candidates of new frequent item is added to the candidate set.
END
```

GenerateCandidates

Input: Frequent Itemset, Candidate set, tidset of Frequent Itemset

Output: Exact Candidates of Frequent Itemset.

```
GenerateCandidates(frequent,candidate, tid(frequent))
BEGIN
        cand=null; // cand will contain exact candidates of frequent item(frequent)
        for each x ε candidate
                If(tid(frequent) ∩ tid(x) ≥ support)
                cand.add(x); // candidates are stored in increasing order of support.
        return (cand);
END
```

The MFIMiner algorithm follows breadth-first search strategy and candidate generation method. When a candidate of a frequent itemset contains all tidsets of the frequent items, the candidate is removed from the candidate set and added to the frequent itemsets. The tidset of frequent itemset is sent to quickly generate exact candidates of the frequent extensions. MFIMiner generates maximal frequent itemsets before generating all frequent Itemsets. Frequent itemsets are added to the MFI directly, if it has no superset in MFI.

2.2 *Pruning*

MFIMiner algorithm using four pruning techniques to reduce the search space is given below.

1. Reordering of items with respect to its support in ascending order. This reordering technique is introduced by Roberto Bayardo in MaxMiner algorithm [1] for

mining maximal frequent itemsets. Once frequent items are generated, they are reordered with respect to its support in ascending order.
2. The $FI_i \cup CI_i$ can be directly added to MFI, if they have no superset in MFI. The elements of frequent $FI_i \cup CI_i$ are not used for finding frequent extension.
3. Infrequent candidates are pruned in bottom-up direction using GenerateCandidate method. Infrequent candidates of frequent itemset are removed from initial candidates to construct exact candidates.
4. Parent equivalent pruning is done to trim the item in candidate set of the frequent itemsets. This pruning was introduced by Calimlim in Mafia algorithm. Items in the candidate set are added to the frequent itemset without performing frequency computation if the tidset of item contains all tidset of frequent item.

3 Performance Evaluation

To evaluate the performance of the proposed MFIMiner algorithm, it is implemented in Java on a 2.93 GHz Intel core PC with 2.96 GB of main memory. The performance of the proposed MFIMiner is compared with Java version of GenMax [5] and Mafia [4]. Timings in the figures are based on total time including all preprocessing costs such as horizontal-to-vertical conversion.

The performance of MFIMiner algorithm is compared with two different algorithms. It is observed that the performance can vary significantly depending on the dataset characteristics. To evaluate the performance of MFIMiner algorithm, three different benchmark datasets are used. All these datasets can be downloaded from FIMI Repository [6]. Datasets taken for the experiment are

- T10I4D100K Dataset,
- T40I10D100K Dataset,
- Retail Dataset.

3.1 T10I4D100K Dataset

The first dataset is T10I4D100K [6], which contains 1000 attributes and 100,000 records. The average record length is 10. In T10I4D100K dataset, the number of frequent items is huge, the length of frequent itemset is short and frequent itemset will have a small number of candidates.

This dataset has very high concentrations of itemsets around two and three items long. The number of maximal frequent itemsets is not much smaller than the number of frequent itemsets. The mean pattern length is very small and it is around 2–6.

The performance of MFIMiner algorithm has been compared with GenMax and mafia algorithm for various support and results show that MFIMiner algorithm generates all MFI quickly than the existing algorithms. Performance of MFIMiner is

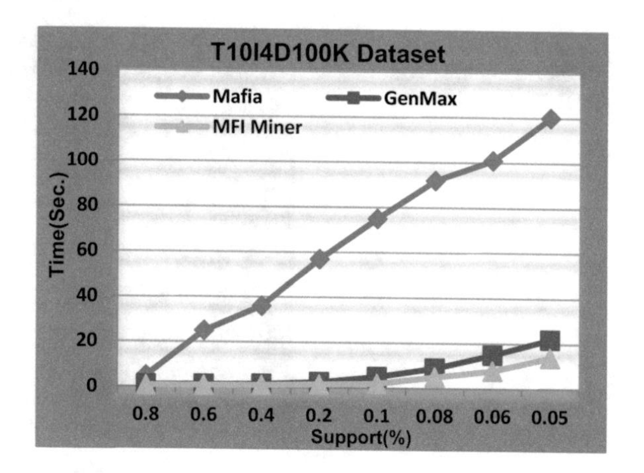

Fig. 1 Performance of MFIMiner algorithm on T10I4D100K dataset

significantly increased than GenMax, when support values are decreased as shown in Fig. 1. For T10I4D100K dataset, MFI Miner performance is about 8–17 times better than Mafia for the taken support. Figure 1 illustrates that the MFIMiner algorithm has better performance when compared to conventional GenMax and Mafia algorithms.

3.2 T40I10D100K Dataset

The second dataset is T40I10D100K [6] which contains 1000 attributes and 100,000 records. The average record length is 40. In T40I10D100K dataset, the number of frequent items is huge, the length of frequent itemsets is short and frequent itemsets have a small number of candidates. The T40I10D100K is characterized by very short maximal frequent itemsets, where the average length of maximal frequent itemsets is very small and it is around 2–6 items.

The maximum number of maximal patterns is of length one or two. The number of maximal patterns is not much smaller than the number of frequent itemsets.

The performance of MFI Miner algorithm has been compared with GenMax, Mafia algorithm for various support and results show that MFI Miner algorithm gives better performance than the existing algorithms. For T40I10D100K dataset, the performance of MFI Miner is slightly better than GenMax as shown in Fig. 2. For T40I10D100K dataset, MFI Miner performance is about 1.5–3 times better than Mafia for the taken support. Figure 2 illustrates that, the MFI Miner algorithm has better performance than GenMax and Mafia algorithms on T40I10D100K dataset.

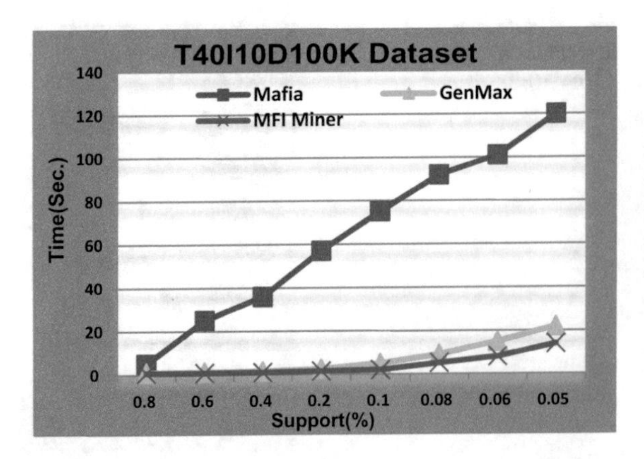

Fig. 2 Performance of MFIMiner algorithm on T40I10D100K dataset

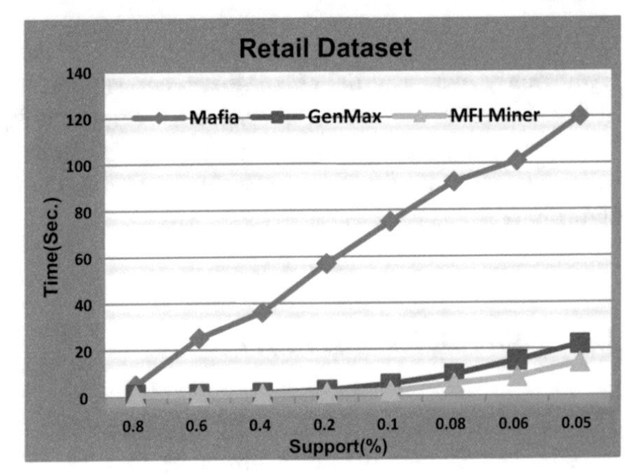

Fig. 3 Performance of MFIMiner algorithm on retail dataset

3.3 Retail Dataset

The third dataset is Retail which contains 16,470 items and 88,162 transactions. The average number of distinct items in each transaction is 13 and most transaction contains items between 7 and 11 items. Mean pattern length is long when compared to sparse datasets. In retail dataset, the number of frequent items is huge and frequent itemsets have large number of candidates.

The performance of MFI Miner algorithm has been compared with GenMax and Mafia algorithm for various supports and results show that MFI Miner algorithm gives better performance than the existing algorithms. On Retail dataset, we find that MFI Miner is the fastest, having a slight edge over GenMax. For Retail dataset, MFI Miner is about 7–38 times faster than Mafia and up to 2.5 times better than GenMax. Figure 3 illustrates that, the MFI Miner algorithm has better performance than GenMax and Mafia algorithms on Retail dataset.

Figure 3 illustrates that, the MFIMiner algorithm has better performance than GenMax and Mafia algorithms on Retail dataset.

4 Result and Discussion

T10I4D100K and retail dataset have more number of frequent items and each frequent item have less number of candidates. MFI contains small patterns and there is not much repetition of patterns in the dataset. For these dataset, the number of frequent extensions is small and MFIMiner generates all MFI quickly. T40I10D100K dataset has more number of frequent items and each frequent item has large number of candidates when it is compared to T10I4D100K and Retail dataset. The maximal pattern length is very low (around one or two). MFIMiner takes extra time to generate exact candidates. However the performance of MFIMiner is the same as the performance of GenMax for T40I10D100K dataset.

The performance of MFIMiner is fine, when teh number of frequent extension generations is small. The performance of MFIMiner is not up to the level, when frequent items in the dataset have a large number of candidate items and length of maximal frequent itemsets generated for these candidates is small.

5 Conclusion

MFIMiner algorithm for mining maximal frequent patterns is introduced in this paper. MFIMiner algorithm employs breadth-first search and pruning to remove the infrequent candidates. The MFIMiner algorithm is compared with GenMax [5] and Mafia [4] algorithms and the results show that MFIMiner algorithm performs well than existing algorithms for T40I10D100K, retail and T10I4D100K dataset and performance of MFIMiner is improved.

References

1. Bayardo, R.: Efficiently mining long patterns from databases. In ACM SIGMOD Conference (1998)
2. Lin, D., Kedem, Z.M.: Pincer-Search: a new algorithm for discovering the maximum frequent set. In: Proceedings of VI International Conference on Extending Database Technology (1998)
3. Agrawal, R., Aggarwal, C., Prasad, V.: Depth first generation of long patterns. In: 7th International Conference on Knowledge Discovery and Data Mining, pp. 108–118 (2000)
4. Burdick, D., Calimlim, M., Gehrke, J.: MAFIA: a maximal frequent itemset algorithm for transactional databases. In: International Conference on Data Engineering, pp: 443–452, April 2001, doi: 10.1.1.100.6805
5. Gouda, K., Zaki, M.J. (2005). GenMax: an efficient algorithm for mining maximal frequent itemsets. Proc. Data Min. Knowl. Discov, **11**, 1–20 (2005)
6. www.fimi.cs.helsinki.fi/fimi03/datasets.html
7. Agrawal, R., Imielienski, T., Swami, A.: Mining association rules between sets of items in largedatabases. In: Bunemann, P., Jajodia, S. (eds.) Proceedings of the 1993 ACM SIGMOD Conference on Management of Data, pp. 207–216. ACM Press, Newyork (1993)
8. https://arxiv.org/ftp/arxiv/papers/1109/1109.2427.pdf

9. http://www.aaai.org/Papers/KDD/1997/KDD97-060.pdf
10. http://sci-hub.tw/http://www.aaai.org/Papers/KDD/1997/KDD97-060.pdf
11. http://www.philippe-fournier-viger.com/spmf/LCM2.pdf
12. https://link.springer.com/chapter/10.1007/3-540-44957-4_65

Dr. K. Sumathi currently serves as an Assistant Professor in Computer Science and Information Technology Department, Kalasalingam University, India. Her overall teaching experience includes 15 years in various Engineering/Arts and Science/Polytechnic Colleges. Her notable publications include nine papers in international journals. She has presented papers in various national and international conferences. She is a recipient of Elite Certificate in Programming in C++ and Elite and Gold Medal in Problem solving using programming in C, NPTEL online course. She also serves as a reviewer in International Conference on Fuzzy Systems and Data Mining (FSDM 2017), Hualien, Taiwan. Her research interest includes mainly in big data and data mining. She is a life member in ISTE.

Dr. S. Kannan is an Associate Professor of Computer Application Department at Madurai Kamaraj University. He received his B.Sc. in Physics, M.Sc. in Physics, M.Sc.in Computer Science from Madurai Kamaraj university in 1985, 1987, and 1996, respectively. He received his Master of Philosophy and Ph.D. degrees in Computer Science from Madurai Kamaraj university in 1998 and 2010, respectively. His notable publications include seventeen papers in various international journals. His experience also includes presenting more than 20 papers in conference and seminars on Associative Classifications and Maximal Frequent Itemset Mining. His research interests focus on Data Mining and Image Processing.

Nagarajan Karuppiah Head-Machine Learning CoE (Retail), TCS, Chennai. He has been with Tata Consultancy Service, Chennai for over 15 years and has held leadership positions in various retail accounts across the globe. He joined TCS in Mumbai, India in 1997. An active member of IEEE and published various white papers on Enterprise Data warehouse, Business Intelligence, Decision Sciences and Data Mining. He has filed a patent on advanced search algorithms and guided analytics (pending). His research work focuses on advanced Big Data Architectures and databases. He is a TOGAF 8.0 Certified and Silver Certified Netezza Architect.

User Configurable and Portable Air Pollution Monitoring System for Smart Cities Using IoT

M. S. Binsy and Nalini Sampath

Abstract Pollution occurs at an unprecedented scale around the globe. Among various types of pollution, air pollution is a major threat to life and ecosystem as a whole. This paper presents the application of the Internet of Things to model a smart environment by developing a user-configurable air pollution monitoring device. This user configurable device monitors air pollution, collects location coordinates and send the collected data along with the location to an online Internet of Things platform called ThingSpeak. The Public can make use of the information in this platform. This device consists of sensors, Raspberry Pi 3, and GPS module. User configurability is achieved by developing a Bluetooth-based Android application thereby making the device more flexible.

Keywords Air pollution monitoring · Raspberry pi 3 · GPS · User configurable Internet of things

1 Introduction

Pollution can be defined as the introduction of harmful chemicals or other substances into earth's environment, which has a serious adverse effect on life and the ecosystem as a whole. There are different types of pollution, namely air pollution, water pollution, soil pollution, thermal pollution, radioactive pollution and noise pollution. Among these, air pollution causes a major threat to life and earth's environment. Air pollution can be defined as the introduction of harmful substances known as pollutants into the earth's atmosphere. The natural composition of air mainly consists of 78% of Nitrogen, 20.95% of Oxygen, 0.93% of Argon and 0.03% of Carbon diox-

M. S. Binsy · N. Sampath (✉)
Department of Computer Science and Engineering, Amrita School of Engineering, Amrita Vishwa Vidyapeetham, Bengaluru, India
e-mail: s_nalini@blr.amrita.edu

M. S. Binsy
e-mail: binsy.ms@gmail.com

S. Smys et al. (eds.), *International Conference on Computer Networks and Communication Technologies*, Lecture Notes on Data Engineering and Communications Technologies 15, https://doi.org/10.1007/978-981-10-8681-6_32

ide, along with water vapor and very minor traces of Methane, Hydrogen, Krypton, Xenon, Neon, Helium, Nitrous oxide and Nitrogen dioxide [1]. The Introduction of pollutants into the atmosphere causes air pollution by changing this natural composition and damaging the ecosystem thereby making it difficult for the living organisms to survive.

Air pollution is a serious problem in the developing world, which increases with increase in population [2]. As per United Nations estimates, by the year 2050, 70% of the world population will be residing in cities, leading to an increase in urban air pollution. The latest urban air quality database by World Health Organization (WHO) reports that WHO air quality guidelines are not met by 98% of cities in low and average income countries with more than 100,000 inhabitants. As urban air quality declines, the risk of various diseases such as stroke, cardiovascular diseases, various types of cancers namely lung cancer, and chronic and acute allergic respiratory diseases, including asthma, increases for its residents. Poor air quality causes premature deaths, increased disorders in newborn babies such as autism and low birth weight. Apart from health effects, other consequences of increased rate of an air pollution are global warming, greenhouse effect, change in seasonal patterns, the wide variation of temperatures, change in rainfall patterns and droughts [3, 4].

The three important steps required to maintain air quality are monitoring at regular intervals, recognizing the source of pollution and implementation of various preventive measures. Bengaluru known as the “IT capital of India”, is the second fastest growing metropolis in India, mainly due to the growth of IT industry. In past few decades, its urban area has grown in 46.6% and its population has grown by 51.39%. The expansion of city limits was possible by encroaching nearby villages and illegally destroying area under vegetation to accommodate the growing population. This rapid increase in population and its after effects affected the ecological services and thus resulting in increased air pollution [5].

Airpocalypse assessment of pollution in Indian cities reports that in Bengaluru city, the major contributors of PM_{10} are 42% from transport, 20% from road dust resuspension, 14% from construction, 14% from industries, 7% from diesel generator set, and 3% from domestic. The assessment also reports that air quality standards prescribed by the World Health Organization (WHO), are not met by any of the Indian cities and moreover Central Pollution Control Board (CPCB) standards are met by very few cities [6]. Polluted air has been linked to climate as cloud formation and rainfall pattern is affected by particulate matter which absorbs and reflect sunlight [5].

Today air pollution in Bengaluru city is monitored by government operated static pollution monitoring stations. These stations can measure various types of air pollutants and the measurements taken are highly reliable and very accurate. In spite of these advantages, fixed stations are large and expensive. Also, fixed monitoring stations require a significant amount of maintenance. And these factors limit the number of static stations in and around the city. To overcome these limitations, this paper aims to develop IoT based, user configurable, portable indoor and outdoor air pollution monitoring system, which can provide information about air pollution to people. Since the system is portable it can be used anywhere as required and the system is cheap compared to the static measurement stations.

This paper is organized as follows. In Sect. 2, related works till now are discussed under the heading "Literature survey". In Sect. 3, architecture and design of the system are described under the heading "System Design". In Sect. 4, implementation of the system is explained. Section 5 deals with experimental results and Sect. 6 give conclusion and future enhancements

2 Related Work

This section provides an overview of the related works done in air pollution monitoring.

In paper [3], Gupta et al. concentrates on the harmful effects of air pollution in Agra. It monitors air quality situation in Agra and conducts a survey to know the understanding of the public and their perception and attitudes toward air quality using Polltech fine dust sampler to monitor Particulate matter. The results show a very high concentration of PM_{10}, 186.8 $\mu g\ m^{-3}$ violating international standards (WHO, USEPA, EUPAQ) including that of NAAQS, India. In paper [7], Pokric et al. present IOT-based real-time monitoring of air pollution, namely CO, CO_2, NO, NO_2, O_3, $PM_{2.5}$, and PM_{10} along with temperature, air pressure and humidity. It is easy to deploy, use, and maintain the system with a large number of sensors. It can measure air pollution at the static location and as a mobile platform. The sensors used are Alphasense B series and are very costly. In paper [8], Marquez-Viloria et al. presents a portable device with a low-cost microprocessor with integrated Wi-Fi. Using Wi-Fi, the device can send the georeferenced data to the cloud in real-time using MQTT protocol and data can be published on an open access platform. The Internet of Things (IoT) is used for the acquisition and visualization of the data and ThingSpeak IOT platform is used to store data. For this system to work, Wi-Fi connection is a must and the system is not mobile.

In paper [9], Dam et al. introduced two devices Envirosensor1.0 and its next version Envirosensor2.0. Envirosensor1.0 uses temperature and humidity Sensor DHT 22, ozone sensor—MQ131, carbon monoxide sensor—MQ7, particulate matter sensor—Shinyei and Arduino Uno. In Envirosensor2.0, the MQ sensors and dust sensor are replaced by Spec sensors and Sharp dust sensor respectively resulting in a performance increase. Future work of this paper plans to develop a mobile application that connects with the sensor hardware to enable real time personal exposure analysis.

In paper [10], Yi et al. presents a system with plug-and-play sensor modules and multiple WSNs compatibility. This system is easy to use and maintain. It can be conveniently reconfigured and expanded for different monitoring scenarios. An Android application implemented via Bluetooth is also present in the system for receiving data from sensor nodes and latter transmitting the data to sink via the cellular network.

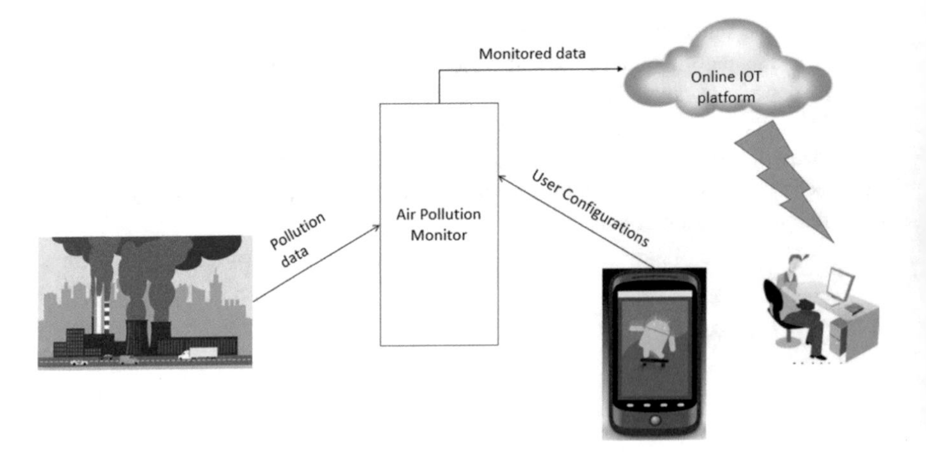

Fig. 1 System architecture

This paper introduces user configurability into air pollution monitoring device thereby reducing power consumption and improving flexibility. The device presented in this paper is an air pollution monitor which is user configurable and portable and can monitor indoor and outdoor air pollution, resulting in a smart environment [11]. This system includes gas sensors, particulate matter sensor, temperature sensor, Raspberry pi 3 and GPS module. Presence of CO, CO_2 and particulate matter ($PM_{2.5}$ and PM_{10}) are sensed along with temperature using a temperature sensor. The collected data along with location is sent to an online IoT platform. User configurability of the device is achieved by developing a Bluetooth based android application.

3 Proposed System

Figure 1 shows the system architecture and Fig. 2 shows the block diagram of the proposed system.

This system monitors carbon dioxide (CO) using MQ-7 sensor module, gases like carbon dioxide (CO_2) along with ammonia, alcohol, benzene and smoke using MQ-135 sensor module, particulate matter ($PM_{2.5}$ and PM_{10}) using Nova SDS011 PM sensor and temperature using DS18B20 sensor. GPS module Neo-6MV2 is used to track the location. Raspberry Pi 3 collects data from all the sensors and GPS module. The collected data along with location are sent to an online platform called ThingSpeak via Wi-Fi enabled on Raspberry Pi 3. User configurability is achieved by developing a Bluetooth based Android application. This application installed in the mobile phone acts as an interface between this system and user. Portability of the device accounts to use of the device at any place. It can be either fixed at a place or it can be attached to a vehicle.

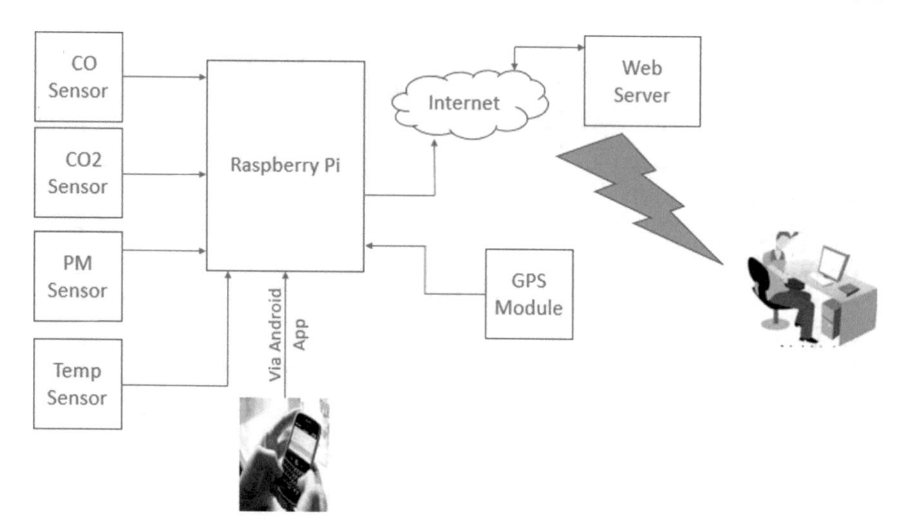

Fig. 2 System design

4 Implementation

4.1 Gas Sensors with Raspberry Pi 3

MQ-7 and MQ-135 used are analog sensors. Since Raspberry Pi can detect only digital signals, MQ-7 and MQ-135 are connected via Analog to digital converter PCF8591P as shown in Figs. 3 and 4 [12–15].

4.2 PM Sensor SDS011 Module with Raspberry Pi 3

PM sensor SDS011 is connected via USB2TTL interface. UART protocol is used for communication with a baud rate of 9600 bits/second. 10 bytes of data obtained from SDS011 has the format as mentioned in datasheet [16].

From the data received, $PM_{2.5}$ and PM_{10} are calculated as follows after validating the checksum.

Checksum: Checksum = DATA1 + DATA2 + ⋯ + DATA6
$PM_{2.5}$ value: $PM_{2.5}$ (μg/m^3) = (($PM_{2.5}$ High byte * 256) + $PM_{2.5}$ low byte)/10
PM_{10} value: PM_{10} (μg/m^3) = ((PM_{10} high byte * 256) + PM_{10} low byte)/10

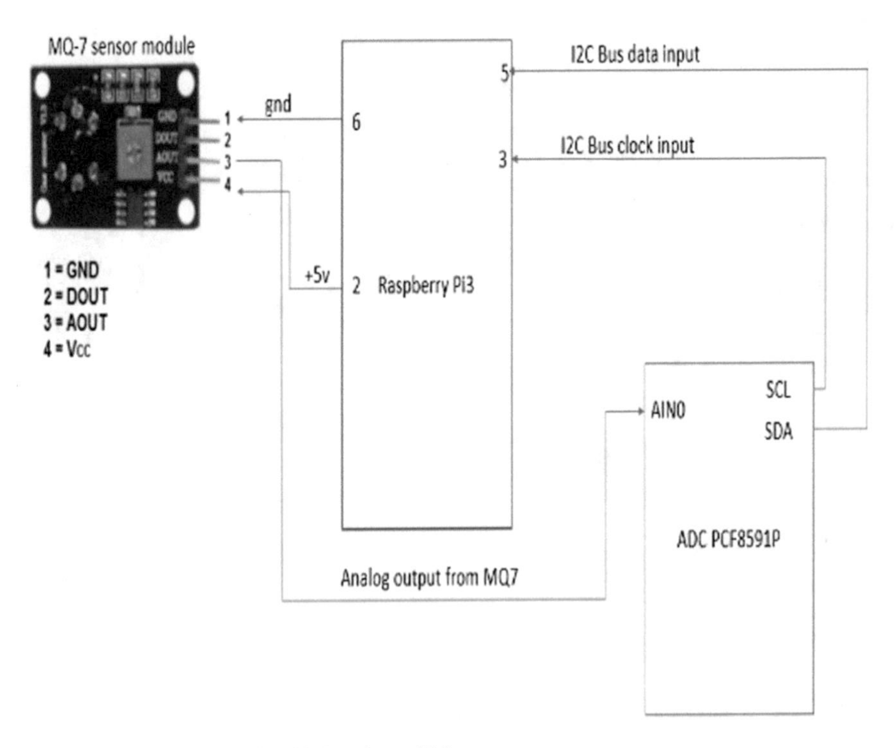

Fig. 3 MQ-7 sensor module with Raspberry Pi 3

4.3 Temperature Sensor DS18B20 Module with Raspberry Pi 3

Temperature module consists of Raspberry Pi 3 with microprocessor Broadcom. BCM 2837 and temperature sensor DS18B20. 1-Wire is the communication mechanism used to communicate between Raspberry Pi 3 and temperature sensor. Python is used for programming as shown in Fig. 5 [17].

4.4 GPS Module Neo-6MV2 Module

GPS module Neo-6MV2 gives the location data of the monitoring device. This is connected via a USB2TTL interface with Raspberry Pi. UART protocol is used with a baud rate of 9600 bits/second. Data from the GPS module is obtained as many NMEA sentences. Out of the many NMEA sentences, the one starting with "$GPRMC" is required to find location coordinates. The format of such a sentence- $GPRMC,204530.000,A,3091.5007,N,10075.9969,N,2.49,115.69,31051 4,,,A*74 The details of the fields are mentioned in the datasheet [18]. Fields 3 and 5 gives us

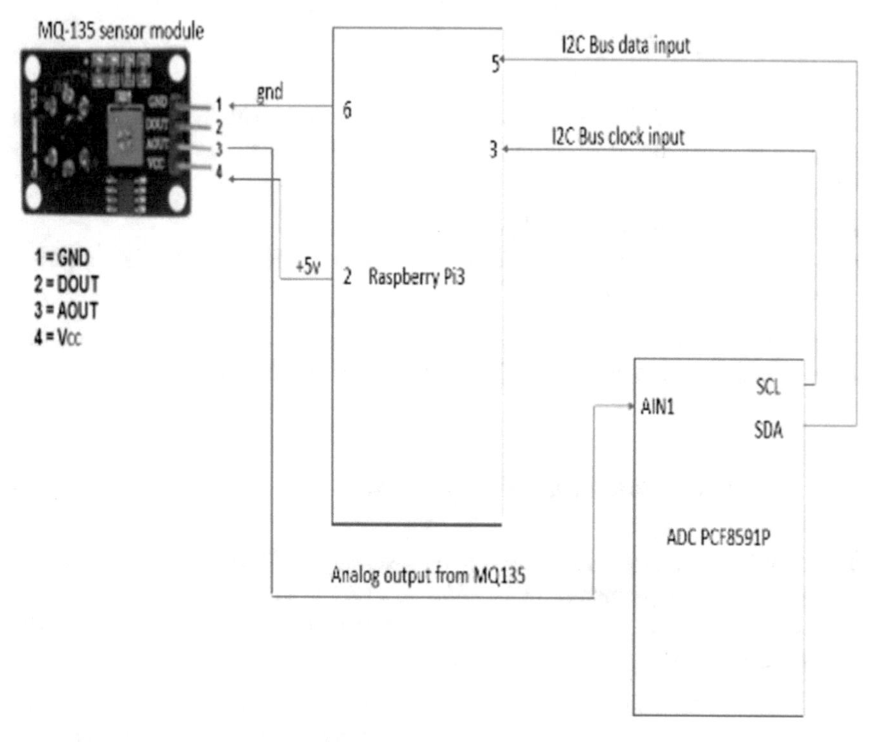

Fig. 4 MQ-135 sensor module with Raspberry Pi 3

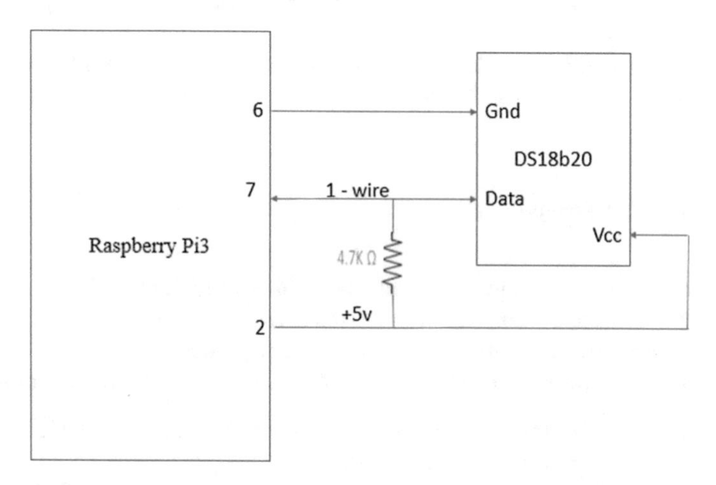

Fig. 5 Temperature module DS18B20 with Raspberry Pi

latitude and longitude. So, if sentence type is GPRMC and data is present and active, i.e., the second field is "A", data from third and fifth fields are extracted.

4.5 Sending Data to ThingSpeak

Collected data is sent to ThingSpeak IoT platform along with location coordinates using the Wi-Fi available on Raspberry Pi. The following steps are done to configure ThingSpeak:

- Sign into ThingSpeak using Mathworks Account.
- Create a new channel and create fields for MQ-7, MQ-135, DS18B20, $PM_{2.5}$, PM_{10}, latitude, and longitude.

Save the settings and select public view, so that data can be used by people. Using the provided API keys, we can read or write from or to the channel.

4.6 Android Application Development

Bluetooth is activated on Raspberry Pi by installing PyBluez. The feature of user configurability is implemented by developing an Android app using Android studio 3.0.1. Create a project in Android studio following the guidelines in webpage [19]. Following are the three main files used in application development.

MainActivity.java: This is the entry point of your application. When you build and run the app, the system launches an instance of this Activity and loads its layout.
activity_main.xml: Layout for the activity's user interface is defined by this XML file.
AndroidManifest.xml: This file is placed in the root directory. This file provides all the information regarding the developed Android application to the Android system.

4.7 Modes of Operation

The developed Android application acts as an interface between the air pollution monitoring system and user. The user can make the device work in any of the following modes based on the intensity of pollution at a particular place or based on the time at which monitoring is done. For example, Available modes of operation are

Aggressive Mode: Continuous monitoring of pollutants at an interval of 30 min.
Economy mode: Monitoring is done once in 3 h.
Auto Mode: Monitoring is done only during peak hours of traffic with an interval of 30 min in between. Peak hours of traffic are from 9 a.m. to 1 p.m. and between 6 p.m. to 9 p.m. At all other remaining hours measuring is done once in 3 h. If monitoring is done at hours of peak traffic, continuous monitoring is required. If monitoring of a less polluted tourist spot is to be carried out the monitoring need not be continuous. Similarly, if the location is a highly polluted area "aggressive mode"

Fig. 6 Home screen of android application

Fig. 7 Screen showing list of paired devices

can be chosen by the user. If the location is an area with very less pollution user can choose the economy mode. Figure 6 shows the home screen of the developed Android application.

When the user clicks on “PAIRED DEVICES”, paired devices with active Bluetooth in the vicinity of the mobile phone is displayed as shown in Fig. 7. When the user clicks on “Raspberry pi”, mobile phone and Raspberry Pi gets connected and screen to select the mode of operation is displayed as shown in Fig. 8. Three modes are displayed in the menu—namely Aggressive mode, Economy mode, and Auto mode.

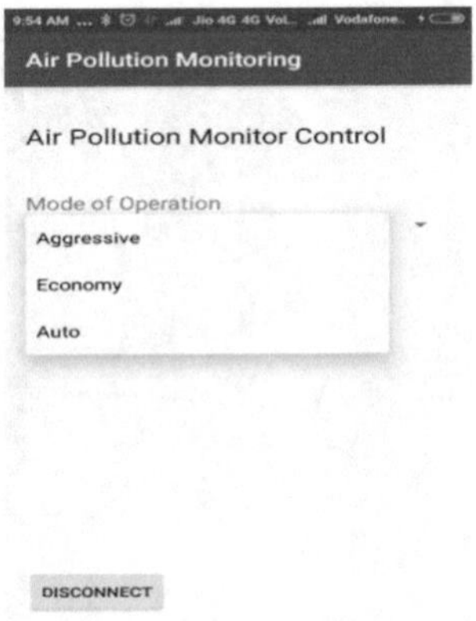

9:54 AM ... Jio 4G 4G VoL... Vodafone...

Air Pollution Monitoring

Air Pollution Monitor Control

Mode of Operation

Aggressive

Economy

Auto

DISCONNECT

Fig. 8 Screen to select the mode of operation

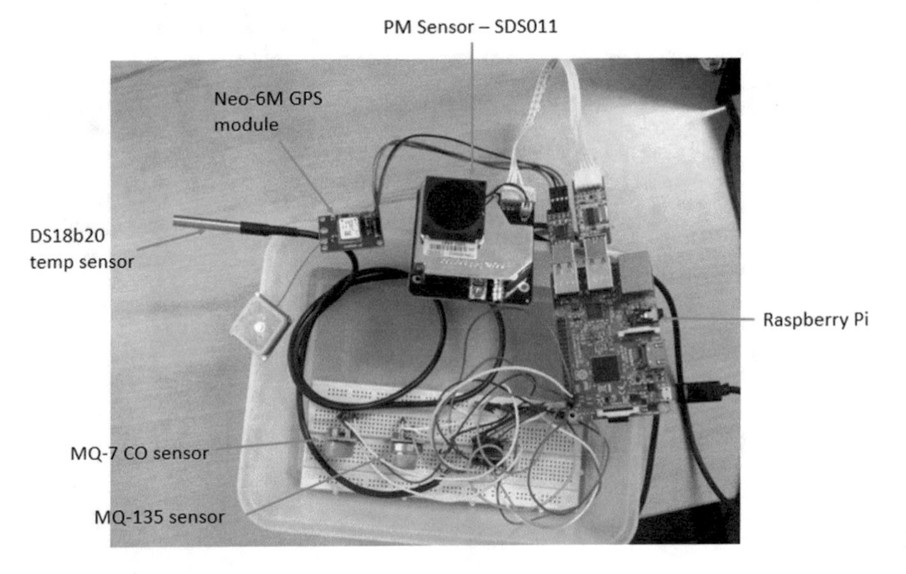

Fig. 9 Air pollution monitoring system

Using the data provided by the system, which is made available to the public via ThingSpeak, people can be aware of the pollution in their environment and people with acute respiratory diseases like asthma can reroute the path in transportation if possible. Figure 9 shows the implementation of the air pollution monitoring system.

5 Experimental Results

5.1 *Test Scenario 1*

The testing of this device is done indoors. Values of CO, gases like CO_2 along with ammonia, alcohol, benzene and smoke, $PM_{2.5}$ and PM_{10} are monitored and found that they are all within the healthy limits. Switching between various modes are tested using the developed Android app (Figs. 10 and 11).

Location: Indoor
Mode: Aggressive
Results: Monitored the parameters and are send to ThingSpeak platform
Average amount of CO = 20.75 ppm
Average amount of gases like CO_2 along with ammonia, alcohol, benzene and smoke = 128 ppm
Average amount of $PM_{2.5}$ = 26.15 $\mu g/m^3$
Average amount of PM_{10} = 42.46 $\mu g/m^3$

5.2 *Test Scenario 2*

The testing of this device is done outdoors. Values of CO, gases like CO_2, along with ammonia, alcohol, benzene and smoke, $PM_{2.5}$ and PM_{10} are monitored and found that values of all the monitored parameters show an increase in value compared to values obtained during indoor testing. PM_{10} value shows a great increase compared

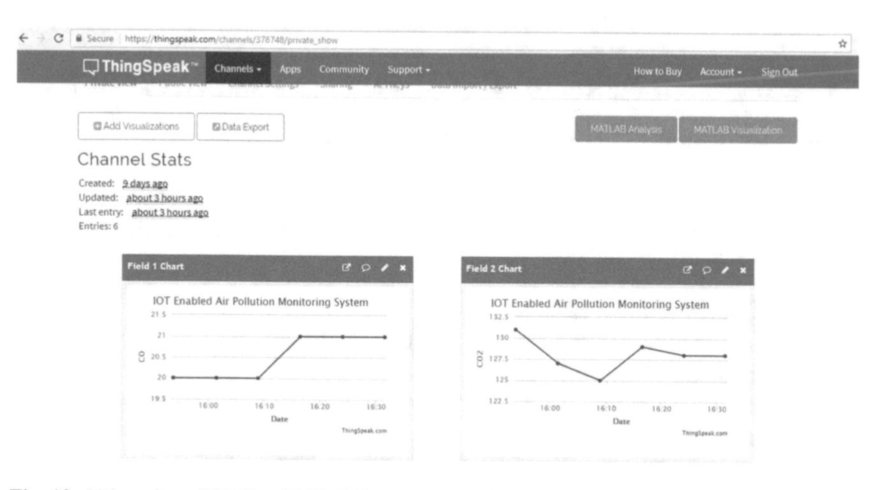

Fig. 10 Values from MQ-7 and MQ-135 sensors at ThingSpeak

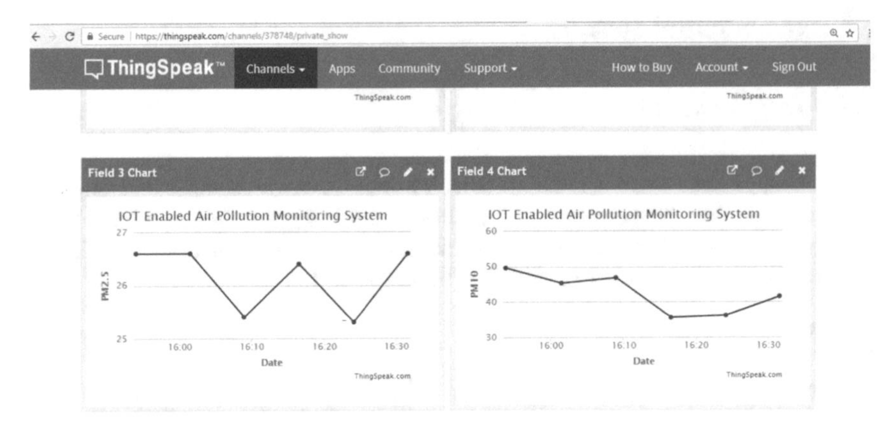

Fig. 11 Values from DS18B20 and GPS module

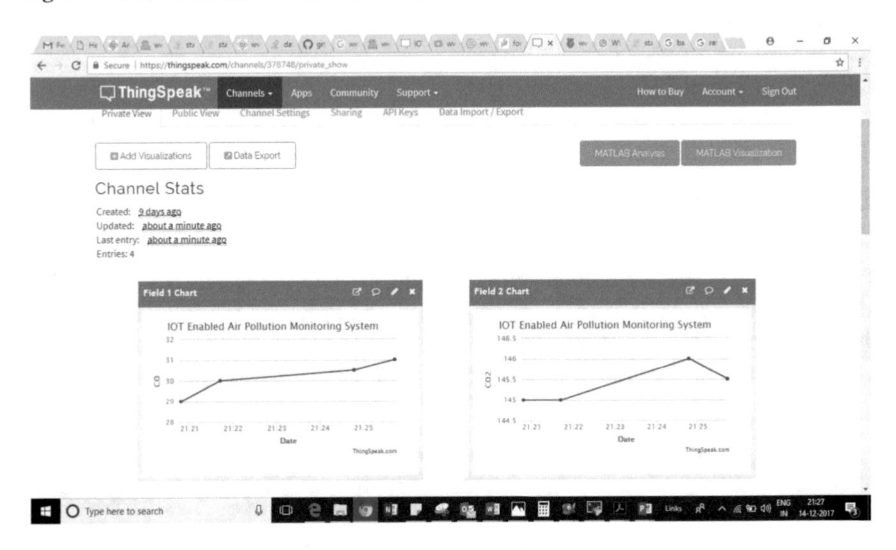

Fig. 12 Values from MQ-7 and MQ-135 sensors at ThingSpeak

to the indoor values. Switching between various modes are tested using the developed Android app (Figs. 12 and 13).

Location: Outdoor, Mode: Economy
Results: Monitored the parameters and are send to ThingSpeak
Average amount of CO = 30.1 ppm
Average amount of gases like CO_2 along with ammonia, alcohol, benzene and smoke = 145.5 ppm
Average amount of $PM_{2.5}$ = 42.8 $\mu g/m^3$ and PM_{10} = 93.8 $\mu g/m^3$

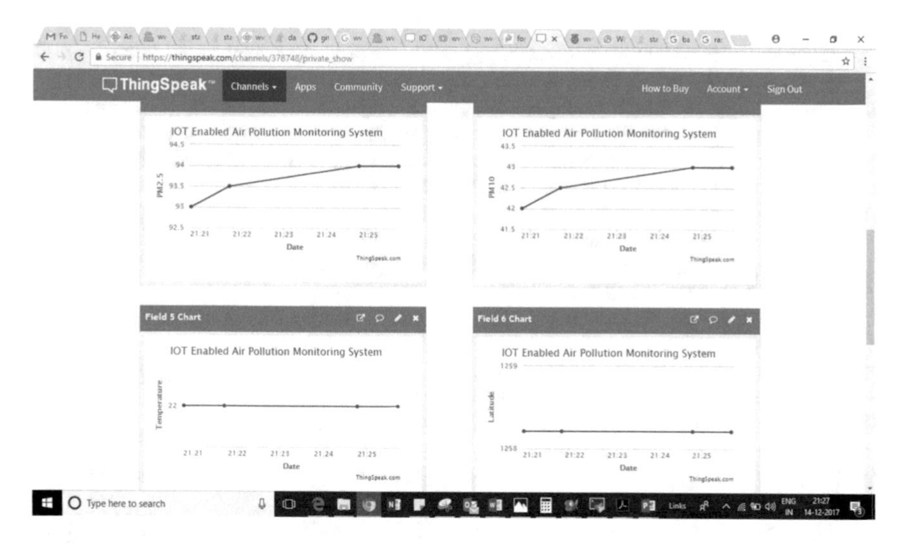

Fig. 13 Values from SDS011, DS18B20 and GPS

6 Conclusion and Future Enhancement

The proposed user configurable Air Pollution monitoring system collects data from sensors and GPS module and sends it to online IoT platform called ThingSpeak via Wi-Fi. This system is made user configurable with the help of a Bluetooth based Android application. This system can work in any of the three modes selected by the user thereby reducing the power consumption and improving the flexibility. Future enhancement of the system focuses on the analysis of data in ThingSpeak, updation of Google Maps based on pollution levels and developing a self-configurable air pollution monitoring system based on data analytics.

References

1. Natural composition of air available. https://en.wikipedia.org/wiki/Atmospheric_chemistry
2. George, J.E., Aravinth, J., Veni, S.: Detection of pollution content in an urban area using landsat 8 data. In: International Conference on Advances in Computing, Communications and Informatics (ICACCI), pp. 184–190. IEEE (2017)
3. Gupta, P., Kumar, R., Singh, S.P., Jangid, A.: A study on monitoring of air quality and modeling of pollution control. In: Humanitarian Technology Conference (R10—HTC), (pp. 1–4). IEEE Region 10 (2016)
4. Sidharth Ajith, Harivishnu, B., Vinesh, T.K., Sooraj, S., Geena Prasad: Automated gas pollution detection system. In: 2nd International Conference for Convergence in Technology (I2CT), pp. 483–486. IEEE (2017)
5. Thakur, A.: Study of ambient air quality trends and analysis of contributing factors in Bengaluru, India. Orient. J. Chem. **33**(2), 1051–1056 (2017)

6. Airpocalypse report. https://secured-staticgreenpeace.org/india/Global/india/Airpoclypse-Not-justDelhiAir-in-most-Indian-cities-hazardous—Greenpeacereport.pdf
7. Pokric, B., Kreo, S., Drajic, D., Pokric, M., Jokic, I., Stojanovic, M.J.: ekoNET—environmental monitoring using low-cost sensors for detecting gases, particulate matter, and meteorological parameters. In: 8th Eighth International Conference on Innovative Mobile and Internet Services in Ubiquitous Computing, pp. 421–426. IMIS, IEEE (2014)
8. Marquez-Viloria, D., Botero-Valencia, J.S., Villegas-Ceballos, J.: A low cost georeferenced air-pollution measurement system used as early warning tool. In: XXI Symposium on Signal Processing, Images and Artificial Vision (STSIVA), pp. 1–6. IEEE (2017)
9. Dam, N., Ricketts, A., Catlett, B., Henriques, J.: Wearable sensors for analyzing personal exposure to air pollution. In: Systems and Information Engineering Design Symposium (SIEDS), pp. 1–4. IEEE (2017)
10. Yi, W.Y., Leung, K.S., Leung, Y., Meng, M.L., Mak, T.: Modular sensor system (MSS) for urban air pollution monitoring. Sens. IEEE (2016)
11. Vijai, P., Sivakumar, P.B.: Design of IoT systems and analytics in the context of smart city initiatives in India. Procedia Comput. Sci. **92** (2016)
12. Datasheet of Raspberry Pi 3. http://docs-europe.Lectrocomponents.com/webdocs/4ba/0900766b814ba5fd.pdf
13. MQ-7 datasheet. http://edge.rit.edu/edge/R13401/public/FinalDocuments/Monitor/Appendix%20B%20B%20Sensors.pdf
14. MQ-135 datasheet. https://upverter.com/datasheet/05a4d494d8c28d681c71285ffeaa8c509a60537.pdf
15. ADC PCF8591P data sheet. https://www.aurel32.net/elec/pcf8591.pdf
16. Nova PM sensor datasheet. https://ecksteinimg.de/Datasheet/SDS018%20Laser%20PM2.5%20Product%20Spec%20V1.5.pdf
17. DS18b20 temperature sensor datasheet available. https://datasheets.maximintegrated.com/en/ds/DS18B20.pdf
18. GPS Neo6MV2 datasheet. https://www.ulox.com/sites/default/files/products/documents/NEO-6_DataSheet_%28GPS.G6-HW-09005%29.pdf
19. Android studio references available. https://developer.android.com/training/basics/firstapp/creating-project.html

M. S. Binsy studying PG: M.Tech. in Embedded Systems at Amrita School of Engineering, Bengaluru (2016–2018) and completed her UG: B.Tech. in Electronics and Communication from College of Engineering, Poonjar (CUSAT) (2006). Her research area includes wireless communication systems and embedded systems.

Nalini Sampath currently working as an Assistant Professor (Sr. Grade), Computer Science and Engineering Department, Amrita School of Engineering, Bengaluru. She completed her PG: M.Tech. in Computer Science and Engineering from RV College of Engineering, Bengaluru (2013) and completed her UG: B.Tech. in Computer Science and Engineering from C.M.R.I.T, Bengaluru. Her research area includes wireless communication systems and embedded systems.

Smart HealthCare System Using IoT with E-Commerce

Aswin Baskaran, A. Sriram, Saieesha Bonthala and Jahnavi Venkat Vatti

Abstract In this fast-paced environment, people often forget to take medicines on time, which is highly essential for one's health. The presence of this situation is more in senior citizens. They even have problems using e-commerce websites. Smart HealthCare System is a permanent and long-term solution for elders to organize their medicines and take care of their health. This product reminds them at right time and also maintains the summary of intake of pills. Both doctor and caretaker can monitor the medicine intake of the patient. It also aids the patient to order medicines online in an easier way which also provides the provision for separately programmed pillboxes. Customers can schedule an appointment with the doctor. No-stock feature is available in this system which reminds the user about placing an order. All the features are regulated by a portable CPU (Raspberry Pi) which is connected with the system and Android app. Orders are confirmed via Android app and pharmacist handles them by using the e-commerce website. The user can simply order medicines by a single click of a button and also differentiate all medicines.

Keywords Rapberry Pi · Android app · E-commerce website · Pillbox

1 Introduction

In this era of application, advancements in medical field and integration of electronics into the medical field are much needed. The population is increasing exponentially, as there is no time for children and other working professionals to take care of

---

A. Baskaran (✉) · A. Sriram · S. Bonthala · J. V. Vatti
Department of ECE, SRM Institute of Science and Technology, Kattankulathur, India
e-mail: aswin.baskar97@gmail.com

A. Sriram
e-mail: sriram.a@ktr.srmuniv.ac.in

S. Bonthala
e-mail: saieeshabonthala@gmail.com

J. V. Vatti
e-mail: jahnavivatti@gmail.com

S. Smys et al. (eds.), *International Conference on Computer Networks and Communication Technologies*, Lecture Notes on Data Engineering and Communications Technologies 15, https://doi.org/10.1007/978-981-10-8681-6_33

their parents. There has been a lot of advancement in technology in reporting live health status of the patient using the cloud. This Healthcare system offers medication indicator for patients at the right time and with LED indication of the pill to be consumed. It provides the user, a friendly interface with an option for doctor reviews and prescription management. The main advanced feature is to order the medicine, so that patients will not run out of stock.

It uses the Internet of Things technology for reporting alert in the form of message. IoT allows electronic components to communicate with the real-time world using web services. Android application is developed to maintain patient's profile, to view the description of medicine, details of caretaker and to update the prescription. Ordering the medicine is performed through the android app. This system comes as a whole package that consists of an E-commerce website for the pharmacist to respond with order acceptance and estimated delivery time. This system modernizes the lifestyle of a society as it can be improvised and implemented in many ways.

2 Literature Review

- Health care analysis is performed which is reported using the Internet of Things applications in [1]. The Raspberry Pi board is used which runs on a Raspbian OS, which has the GPIO pins for sensing the heartbeat rate and interfacing it to the database using MySQL server. This project does not provide a friendly interface to track their records and medicine organizer, for the pill to be consumed during emergency conditions.
- In paper [2], the Medicine pill box is organized and other health care analysis—ECG, Blood pressure is measured and communicated via GPIO pins in the Raspberry Pi and shared to the cloud, Relay board is used for the pill to be dispensed by driving DC motors. The drawback is that, there is no flexibility in ordering medicines, as the prescribed tablets are subject to change over time. Moreover, there is no feature for scheduling appointments with the doctor in the mobile application.
- In the research paper [3], medicine box is constructed and LED indicators are used for timely alerts. The action of the customer is intimated to the caretaker using email. There is no feature to order medicines, which is the main drawback of this system.

3 Design Methodology

This smart health care system consists of Raspberry Pi 3-Model B which has 40 GPIO pins for interfacing. Raspberry Pi has an in-built Wi-Fi module which is connected to internet clock using Python ver2.0 by importing "time" and "datetime".

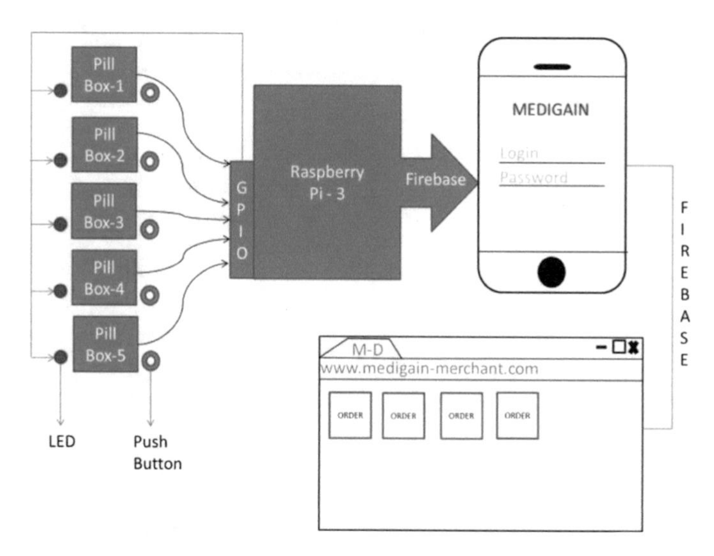

Fig. 1 System model

GPIO.setup(pin, IN/OUT)
Current-time = time.strftime("%HH:%MM")
Pill-intake-time = HH:MM
While True:
If(Current-time = Pill-intake-time)
//Turns ON LED and Buzzer
//Alarm in mobile application

The five GPIO pins are connected to five LEDs, where each medicine box consists of an LED. GPIO is connected with five push buttons which are utilized as order buttons. The pill intake time is set in the Raspberry Pi using mobile application. During the pill intake time, the corresponding LED glows for 10 min. A buzzer is also connected to GPIO which notifies the user using audio beep. The Mobile application throws an alarm at the pill intake time. Then, the pill intake must be acknowledged by the customer by pressing "OK" button in the mobile application. Push buttons are used to order the medicines. The initial conditions are made "true" for all the push buttons. If the particular button is pressed, it is set to "false". This "false" state is detected by the Raspberry Pi and updates the order in the Firebase database. Firebase is imported to python for communicating with database (Fig. 1).

From firebase import firebase
Firebase = firebase.FirebaseApplication("database Url")
//get method
get-data = firebase.get("/header", argument)
//post method
Post-data = firebase.post("/header",(JSON object))

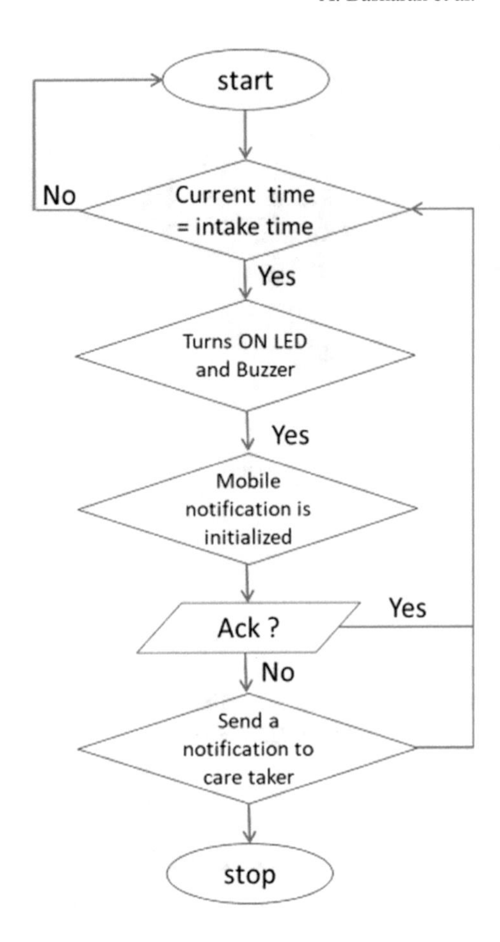

Fig. 2 Pill intake flowchart

The customer needs to sign up with his Email ID and password in the mobile application. When the customer logs in the application for the first time, it gets the customer's basic information, which can be updated later.

The pill intake procedure is represented in Fig. 2. When it is time for the customer to consume medicine, the corresponding GPIO pins of LEDs and buzzers are set "true". The LED glows for 10 min and the buzzer goes ON for 1 min which turns On the alarm in the mobile application. Then, the mobile application waits for the acknowledgement from the customer for about 30 min. A notification is sent to the caretaker, if the customer fails to acknowledge. After this process, the system waits for next pill intake time and repeats the process.

The procedure is followed as shown in Fig. 3 to order the medicines. When the order button is pressed, the corresponding GPIO pins is turned low, which is detected by the raspberry pi and corresponding information are communicated to the database by "post" and "get" method.

Fig. 3 Order flowchart

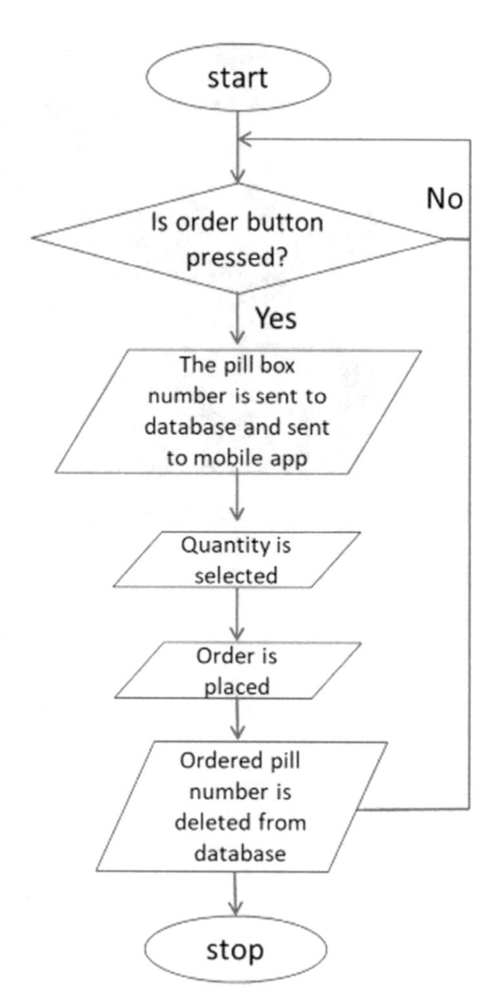

4 Hardware Setup

There are 40 pins in a Raspberry Pi 3 for expansion and 27 pins within the system are used for interfacing with real-time sensors. They can be used as both input and output pins. A total of five 5 mm LEDs are used which are connected to five boxes (Figs. 4 and 5).

> "time.exe" is imported to the Python script and LEDs are configured to the time by the user. When the pill intake time is reached, the buzzer goes ON. Firebase database is connected with Python and when order button is pressed, the particular identity of the button is recognized by the GPIO and the number is stored in the database, which is then retrieved by the mobile application and adds the certain medicine to the cart.

```
pi@raspberrypi: ~
File Edit Tabs Help
pi@raspberrypi:~ $ cd /home/pi
pi@raspberrypi:~ $ python Medigain.py
Waiting for pill intake time...
Time = 21:00
LED is ON !
Buzzer is ON !
Hold for 1 minute..
Buzzer is OFF !
Hold for 9 more minutes..
LED is OFF !
```

Fig. 4 Python script execution in Raspberry Pi

Fig. 5 Login page

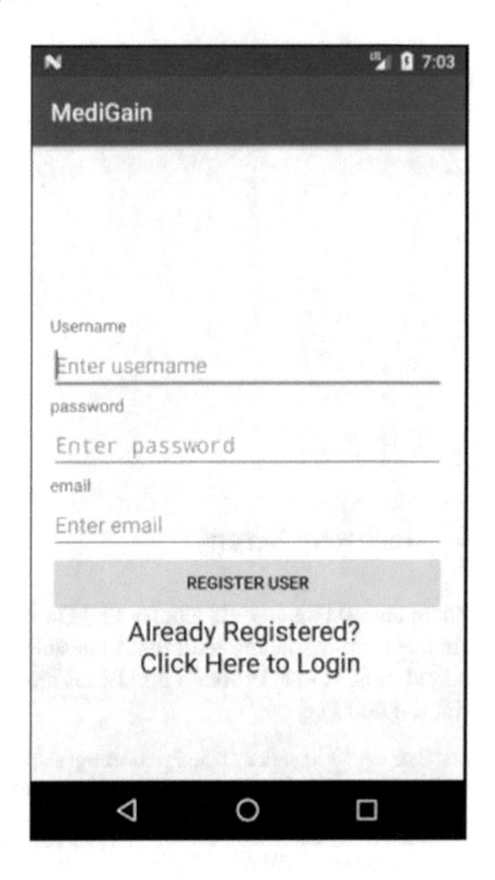

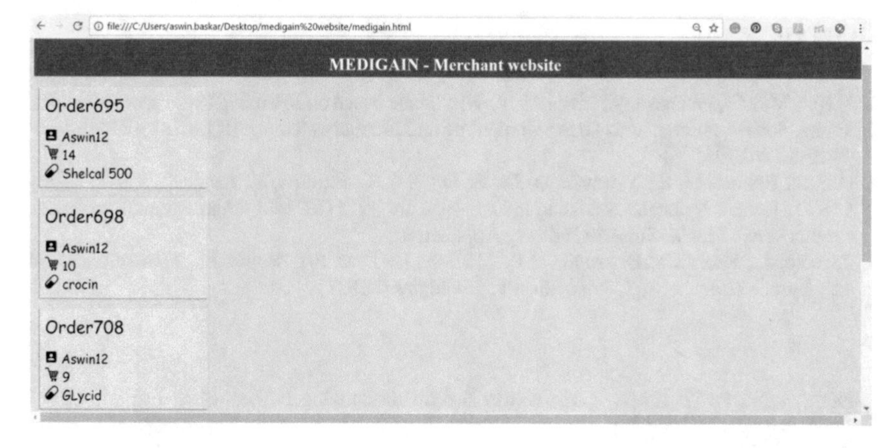

Fig. 6 Web application

5 Android Application

Android application is created using the android studio in which the user should authenticate using his email. In the startup, few questions are to be answered for registration and the medicine details are entered by the pharmacist. Then, Android application starts pushing notifications at the pill intake time. It retrieves order details from the database and waits for the user confirmation for the order to be placed. The quantity of pills can be changed while placing the order. It contains details of the patient, order summary and pill intake summary. Doctor's available time can be viewed an and appointment can be scheduled for regular a check-up of the patient (Fig. 6).

6 Conclusion

In this study, a new approach is presented to perform actions in an automated way, which will be a relief for many patients and old age people around the world. As the world is filled with inventions and smartphone plays a significant role in one's life, healthcare system inbuilt in a smartphone will create a more positive impact. Raspberry Pi is used for its multi-tasking capability and it can be extended for many tasks in the near future. This is just the prototype of such system and has been developed to facilitate the doctor, pharmacist and the patient.

References

1. Gupta, M.S.D., Patchava, V., Menezes, V.: Healthcare based on IoT using Raspberry pi. In: 2015 International Conference on Green Computing and Internet of Things (ICGCIoT). 978-1-4673-7910-6/15/IEEE
2. Aakash Bharadwaj, S., Yarravarapu, D., Reddy, S.C.K., Prudhvi, T., Sandeep, K.S.P., Reddy, O.S.D.: Enhancing healthcare using m-Care Box. In: 2017 (ICIMIA) International Conference on Innovative Mechanisms for Industry Applications
3. Jayanth, S., Poorvi, M.B., Sunil, M.P.: MED-ALERT: an IoT device. In: 2016 International Conference on Inventive Computation Technologies (ICICT)

Mr. Aswin Baskaran studying B.Tech.-SRM Institute of Science and Technology-2018 Under the Department of Electronics and Communication. He will continue MS in Computer Engineering (Electrical) at Arizona state University. His Current research—Portable smart health care devices.

Mr. A. Sriram currently working as an Assistant Professor in SRM Institute of Science and Technology in Department of Electronics and Communication. He has completed B.E.-Electronics and Communication Engineering in Anna University-2009. He had completed M.E.-Communication Systems in Anna University-2012. His research area includes wireless communication systems.

Ms. Saieesha Bonthala completed her B.Tech.-SRM Institute of Science and Technology-2018 Department of Electronics and Communication and she is currently Working—NTT Data (will start from August 2018).

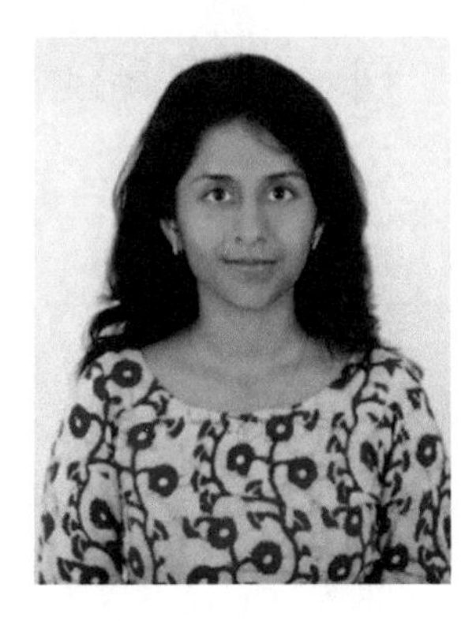

Ms. Jahnavi Venkat Vatti completed B.Tech.-SRM Institute of Science and Technology-2018 Department of Electronics and Communication.

Comparative Analysis of TCP Variants for Video Transmission Over Multi-hop Mobile Ad Hoc Networks

Sharada U. Shenoy, Sharmila Kumari M. and Udaya Kumar K. Shenoy

Abstract A comparative study with simulation has been carried out in this paper to choose a combination of routing protocol with the suitable TCP variant for video transmission over mobile ad hoc networks. Simulation is carried out to choose a routing protocol out of AODV, AOMDV, DSR and DSR protocols for ad hoc networks that suit the need of ad hoc scenario. AODV protocol is found to perform better than other routing protocols. Performance of TCP variants such as TCP Reno, TCP NewReno, TCP Tahoe, TCP Vegas, and TCP Sack is then compared for the video transmission in ad hoc networks by adapting AODV as routing protocol. The simulation results show that TCP Vegas with AODV gives good performance.

Keywords Ad hoc wireless networks · MANET · Video transmission
Routing protocols · TCP performance · TCP

1 Introduction

Mobile Ad hoc Networks (MANETs) are infrastructure-less networks that can be framed on the go. In multi-hop ad hoc wireless networks, the topology is unpredictable and varies rapidly. The mobile nodes may change position or the wireless channel condition may fluctuate. This requires a very robust and adaptive communication protocol that can handle the challenges of the multi-hop wireless networks very smoothly [1]. Transmission Control Protocol (TCP) is the widely used protocol in wireless networks. The characteristics of wireless networks such as mobility,

S. U. Shenoy (✉) · U. K. K. Shenoy
Department of Computer Science and Engineering, NMAM Institute of Technology, Nitte, India
e-mail: sharudivardhan@gmail.com

U. K. K. Shenoy
e-mail: ukshenoy@gmail.com

S. Kumari M.
Department of CSE, PA College of Engineering, Mangalore, India
e-mail: sharmilabp@gmail.com

S. Smys et al. (eds.), *International Conference on Computer Networks and Communication Technologies*, Lecture Notes on Data Engineering and Communications Technologies 15, https://doi.org/10.1007/978-981-10-8681-6_34

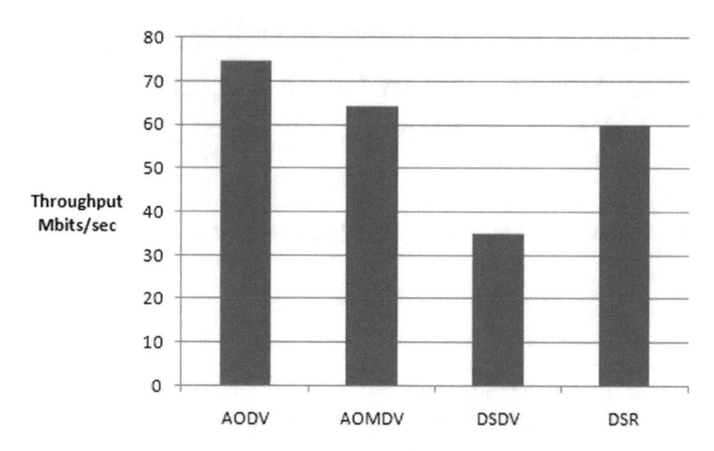

Fig. 1 Throughput in Mbits/s for DSR, DSDV, AOMDV and AODV protocols

link-layer contention, high bit error rate, asymmetric path, network partition, hidden exposed nodes and dynamic routing do not fit the requirements of TCP for a good reliable data delivery [2]. İt also gives an extensive survey on problems of TCP and the effects of mobility. With increased use in video data, there is a large attention paid for video transmission, and attempts are made by researchers to improve upon the existing TCP protocols to suit to the need of wireless networks. Since ad hoc wireless networks are infrastructure-less networks formed in battlefield or in case of disasters, the transmission becomes still worst. TCP has congestion control, generally, slow start and congestion avoidance. At the point when a connection starts, the slow start calculation is utilized to increment congestion window to achieve the threshold. At the point when the sender deduces that few packets are lost, it decreases the congestion window. After generally approximating the threshold limit, TCP changes to the congestion calculation, which estimates congestion window more slowly to test for extra transmission capacity [2].

Many efforts have been made by the researchers to compare the existing TCP protocols to choose for modification to suit the needs of ad hoc scenario. The four routing protocols such as Optimized Link State Routing (OLSR), Ad hoc On-demand Distance Vector (AODV), Dynamic Source Routing (DSR) and Temporally Ordered Routing Algorithm (TORA) are considered for analysis in ad hoc scenario and found that AODV outperforms [3]. Then, TCP Reno, TCP New Reno and TCP Selective Acknowledgement (SACK) are considered for analysis, which shows that out of the three, the Selective Acknowledgement (SACK) variant can adapt relatively well to the changing network sizes, while the Reno performs most robustly in different mobility scenarios [4]. The comparison carried out in our paper also gives AODV as the good performer for the current scenario. References [5, 6] also show the performance of TCP over DSDV, TORA, DSR and AODV. The experiments carried out in [7, 8] show the behaviour and unfairness problems of TCP, which gives the initial results for further improvements in TCP.

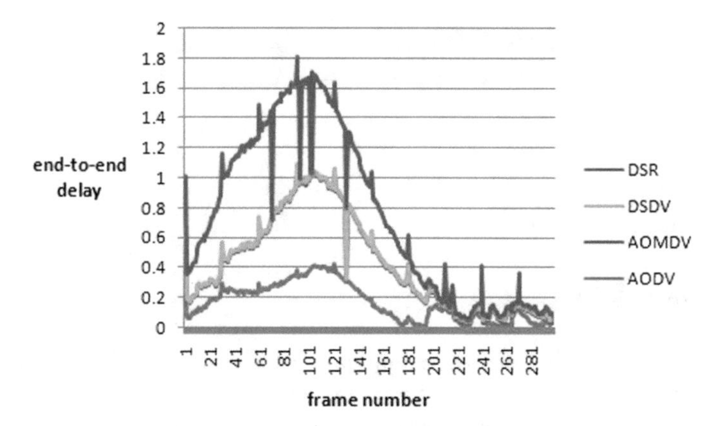

Fig. 2 End-to-end delay in seconds for DSR, DSDV, AOMDV and AODV protocols

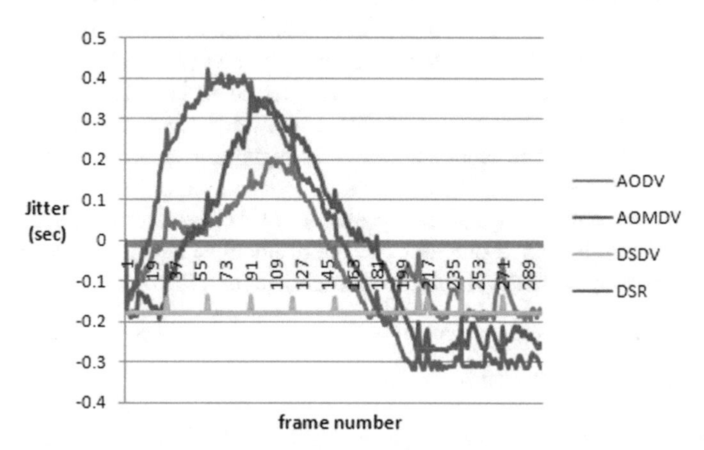

Fig. 3 Jitter in seconds for AODV, AOMDV, DSDV and DSR protocols

2 Results and Analysis

2.1 Simulation Set-Up

Simulation has been carried out using NS 2.25 simulator and Evalvid 2.7 tool. Experiments are conducted for different node speeds and many iterations are taken.

The parameters used for simulation

Payload size is 1000 Bytes/packet, channel type is Channel/WirelessChannel. Two-ray ground model has been chosen with network interface type as Phy/wirelessPhyExt. MAC type is Mac/802_11Ext, interface queue type is Queue/DropTail/PriQueue. With link-layer type as LL, antenna model chosen is

Antenna/OmniAntenna. Maximum packet in the queue is 50 with four nodes. It uses AODV, AOMDV, DSR and DSDV routing protocols under network topology size 1500 × 1500.

2.2 *Results and Analysis*

2.2.1 Simulation Scenario 1

Performance analysis is done for routing protocol to choose a right protocol that is further used for performance analysis of the TCP protocol. We have chosen four well-known ad hoc routing protocols [9] AODV, AOMDV, DSDV, and DSR [10].

The AODV protocol gives a maximum throughput of 74 Mbps (Fig. 1). The average time taken by the packet in transit is called end-to-end delay. It is clear that end-to-end delay for AODV is lesser compared to other protocols (Fig. 2). Jitter is defined as a variation in the delay between the received packets. The sender transmits packets in a continuous stream and spaces them evenly apart. Because of network congestion, link errors, improper queuing or configuration errors, the delay between packets will vary and are received with varying delays. Jitter (Fig. 3) for AODV is lesser compared to AOMDV and DSR protocols. DSDV has lower jitter than AODV, but it suffers with lowest throughput. So, the protocol AODV is suitable with high throughput and lesser delays for further comparison of TCP variants.

2.2.2 Simulation Scenario 2

Comparison of five TCP variants such as TCP Reno, TCP New Reno, TCP SACK, TCP Vegas and TCP Tahoe has been carried out with the AODV as routing protocol [11, 12]. AODV was selected from the simulation from scenario1. Figures 4 and 5 show that throughput and packet delivery ratio (PDR) are high for Vegas compared to other variants. Number of packet drop is also almost half of what has been lost by other packets (Fig. 6). That means more packets are received by the receiver using TCP Vegas compared to the other TCP variants. Congestion peak values for TCP New Reno is 42 (Fig. 7), TCP Reno is 45 (Fig. 8) and for Vegas, Tahoe and SACK is around 52 (Fig. 9). This implies that Vegas has good congestion management. Good throughput, less delay, PDR and good congestion growth makes Vegas a better combination with AODV protocol.

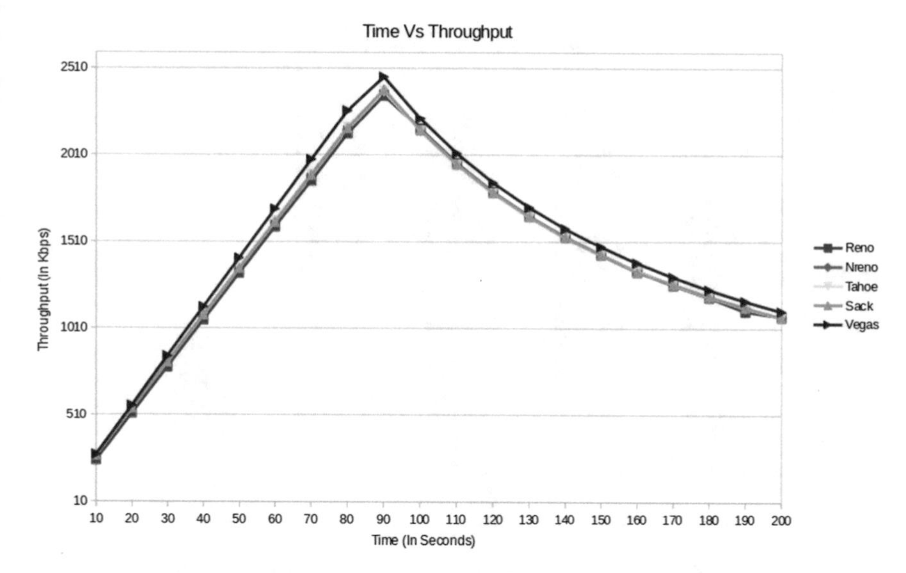

Fig. 4 Time versus throughput growth for TCP Reno, TCP New Reno, TCP SACK, TCP Vegas and TCP Tahoe using AODV protocol

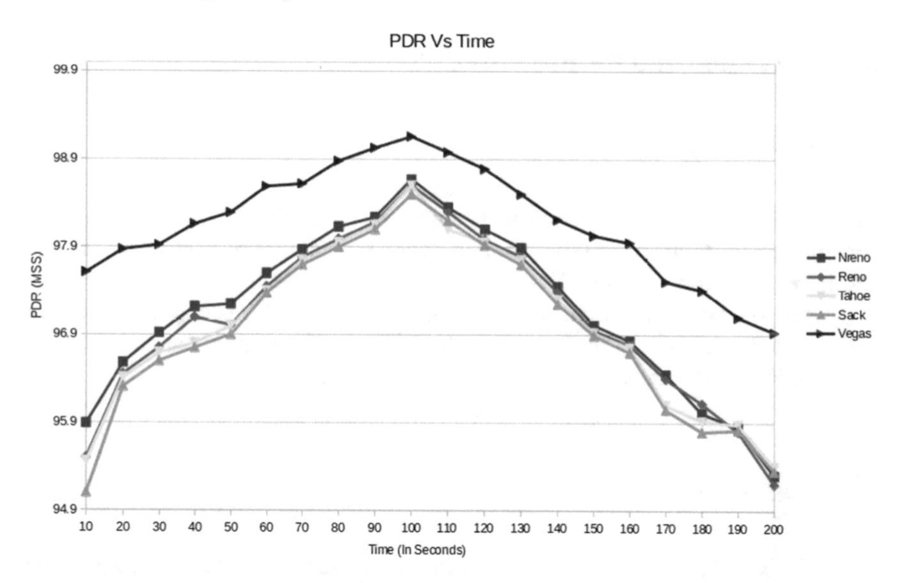

Fig. 5 Time versus packet delivery ratio growth for TCP Reno, TCP New Reno, TCP SACK, TCP Vegas and TCP Tahoe using AODV protocol

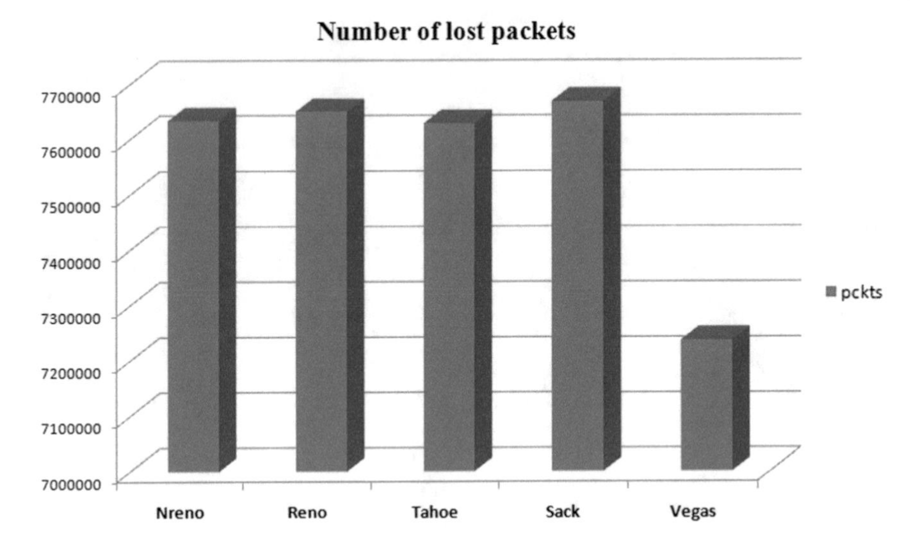

Fig. 6 Number of lost packets in transit for TCP Reno, TCP New Reno, TCP SACK, TCP Vegas and TCP Tahoe using AODV protocol

3 Conclusion and Future Work

3.1 Conclusion

The two simulation scenarios show that AODV with TCP Vegas together yields good performance in terms of throughput, and hence the number of packets lost is less and Packet Delivery Ratio (PDR) is more. The congestion window growth is high compared to other TCP variants that have been considered. The congestion window is stable in case of Tahoe, Vegas and SACK. TCP New Reno sends out more packets compared to other variants.

3.2 Future Work

So, TCP Vegas and AODV can be used for further simulation since it gives throughput and PDR higher than the other TCP protocols. Packet loss is half of the other variants and congestion window is large. We can also do modifications to TCP Vegas to improve the performance further.

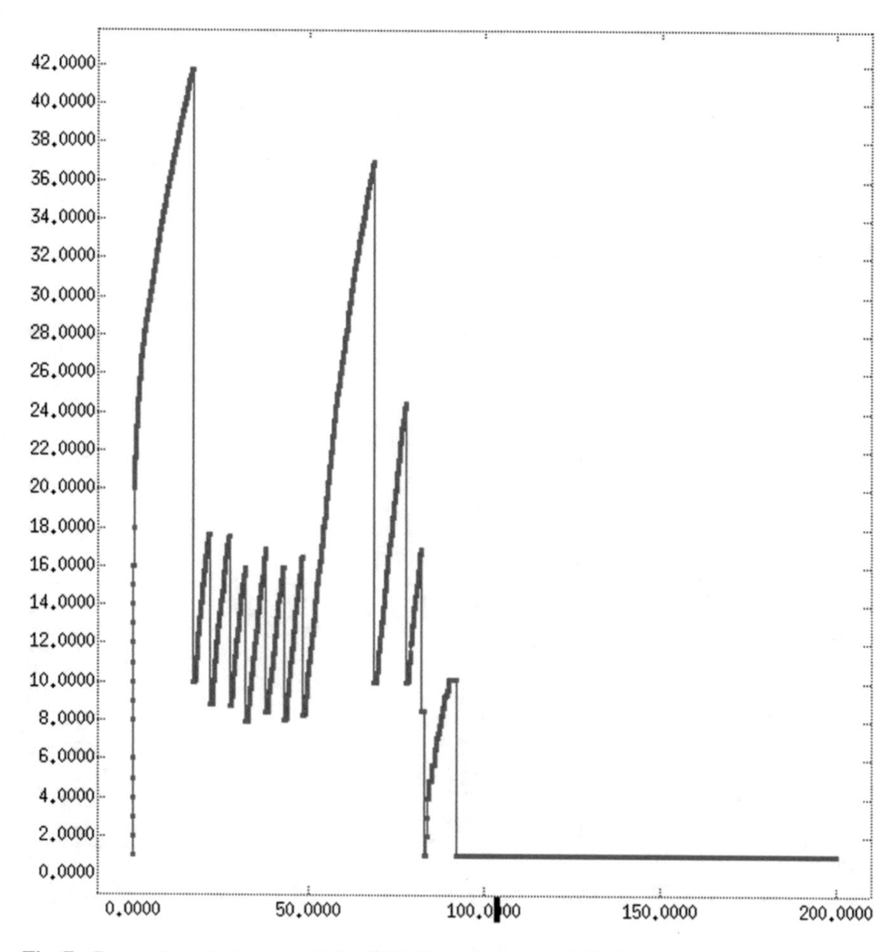

Fig. 7 Congestion window growth for TCP New Reno using AODV protocol

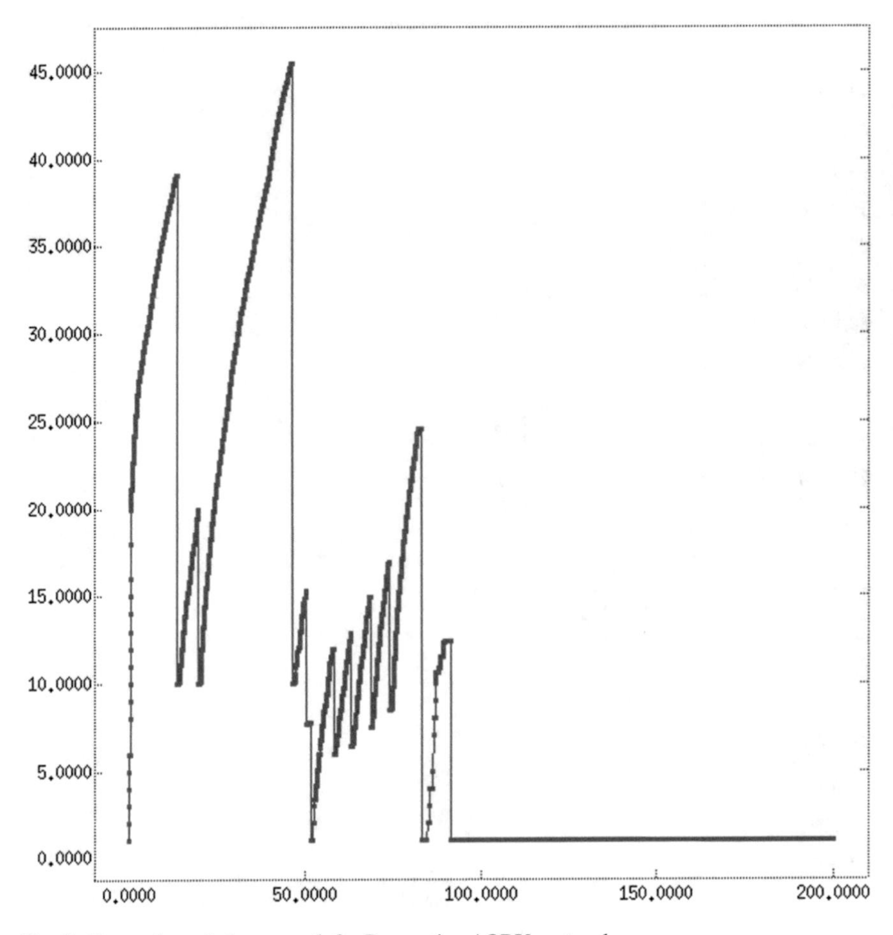

Fig. 8 Congestion window growth for Reno using AODV protocol

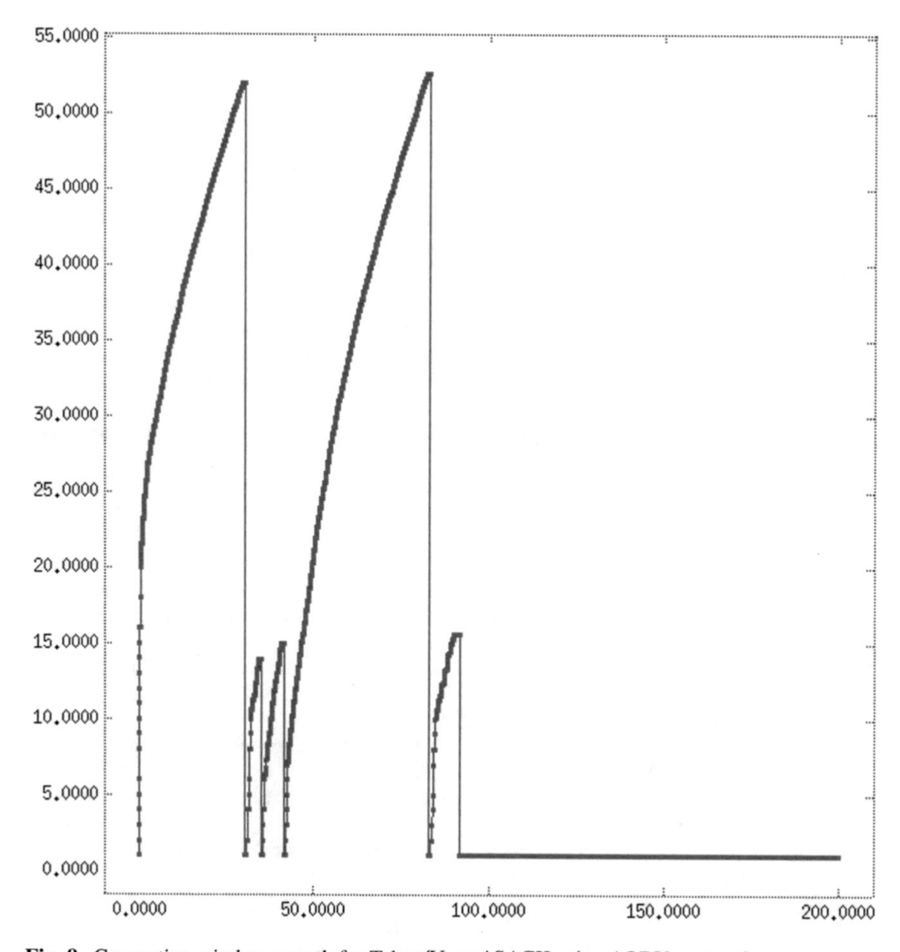

Fig. 9 Congestion window growth for Tahoe/Vegas/ SACK using AODV protocol

References

1. Thangam, S., Kirubakaran, E.: A survey on cross-layer based approach for improving TCP performance in multi hop mobile adhoc networks. In: International Conference on Education Technology and Computer, pp. 294–298. Singapore (2006). https://doi.org/10.1109/icetc.2009.63
2. Hanbali, A.A., Altman, E., Nain, P.: A survey of TCP over mobile ad hoc networks. In: IEEE Communications Surveys & Tutorials, vol. 7, no. 3, pp. 22–36, Third Quarter (2005). https://doi.org/10.1109/comst.2005.1610548(2005)
3. Ade, S.A., Tijare, P.A.: Performance comparison of AODV, DSDV, OLSR and DSR routing protocols in mobile ad hoc networks. Int. J. Inf. Technol. Knowl. Manag. **2**, 545–548 (2010)
4. Senthamilselvi, M., Muruganandam, A.: Performance evaluation of routing protocols by TCP variants in mobile ad-hoc networks. Int. J. Comput. Sci. Mob. Comput. **4**(2), 348–356 (2015)
5. Ahuja, A., Agarwal, S., Singh, J.P., Shorey, R.: Performance of TCP over different routing protocols in mobile ad-hoc networks. In: IEEE Vehicular Technology Conference 2000, vol. 3, pp. 2315–2319. Tokyo (2000)
6. Abdullah-Al-Mamun, M., Rahman, M.M., Tan, H.P.: Performance evaluation of TCP over routing protocols for mobile ad hoc networks. In: First International Conference on Communications and Networking in China, Beijing, pp. 1–3 (2006). https://doi.org/10.1109/chinacom.2006.344668
7. Al-Jubari, A.M. Othman, M., Ali, B.M., Hamid, N.A.W.A.: TCP performance in multi-hop wireless ad hoc networks: challenges and solution. EURASIP J. Wirel. Commun. Netw. (2011). https://doi.org/10.1186/1687-1499-2011-198
8. Hafslund, A., Landmark, L., Engelstad, P., Liz, F.Y.,: Testing and analyzing TCP performance in a wireless-wired mobile ad hoc test bed. In: Proceedings of IEEE International Workshop on Wireless Ad-Hoc Networks, pp. 115–119 (2004)
9. Abolhasan, M., Wysocki, T., Dutkiewicz, E.: A review of routing protocols for mobile ad hoc networks. J. Ad Hoc Netw. **2**(1), 1–22 (2004) (Elsevier)
10. Murthy, C.S.R., Manoj, B.S.: Adhoc Wireless Networks Architectures and protocols. Pearson Education Inc. (2004)
11. Ding, L., Zhang, W., Xie, W.: Modeling TCP throughput in IEEE 802.11 based wireless ad hoc networks. In: Communication Networks and Services Research Conference, 2008 (CNSR 2008), 6th Annual, Halifax, NS, pp. 552–558 (2006) https://doi.org/10.1109/cnsr.2008.66
12. Istikmal, Kurniawan, A., Hendrawan: Throughput performance of transmission control protocols on multipath fading environment in mobile ad-hoc network. In: 11th International Conference on Telecommunication Systems Services and Applications (TSSA 2017), Lombok, pp. 1–5 (2017) https://doi.org/10.1109/tssa.2017.8272939

Mrs. Sharada U. Shenoy Associate Professor in the Department of CSE, NMAM Institute of Technology, Nitte, Udupi District, Karnataka, India. Has 19 years of teaching experience. Received B.E. degree from Mangalore Unviversity, M.Tech. from VTU, Belgaum and pursuing Ph.D. in Computer Science from VTU Belgaum. Area of interest Multimedia Processing and Computer Networks.

Dr. Sharmila Kumari M. Head and Professor, Department of CSE, PA College of Engineering, Mangaluru, Karnataka, India. Received her B.E. degree from Mysore University, M.Tech. in Computer Science and Engineering from VTU, Belgaum and Ph.D. in Computer Science from Mangalore University. Has 20 years of experience in teaching. Her area of interest are Pattern recognition, Digital image processing, Computer networks.

Dr. Udaya Kumar K. Shenoy Received the bachelor's degree (1992) in Computer Science, master's degree (2000) in Computer Application from Mangalore University and Ph.D. (2009) from National Institute of Technology Karnataka, Surathkal. He is currently working as Professor in department of CSE, NMAM Institute of Technology, Nitte, Karnataka, India. His research interests include Wireless Networks, Optimization, Multimedia communication and Network Security.

Scrutinizing of Cloud Services Quality by Exerting Ultimatum Response Method, Miscegenation Cryptography, Proximate Approach and PFCM Clustering

N. V. Satya Naresh Kalluri, Tanimki Sujith Kumar, Korada Rajani Kumari and Divya Vani Yarlagadda

Abstract Cloud computing is forthwith educing like nevermore heretofore, with companies of all shapes and sizes mending to this neoteric technology. Industry experts accredit that this fantasy will only abide to grow and flourish even further in the forthcoming few years. Cloud services are bestowed by profuse clients. There are umpteen services in cloud, and clients without any erudition about the quality traits of cloud services modestly utilize the service and feel mournful, depressed if the essence of service quality is not congenial. Cognition of quality characteristics of cloud services is very important for clients to use cloud service and to feel ecstatic, satisfied by utilizing cloud service. In this paper, by exerting proximate technique and clustering technique, cognition of quality traits of services in cloud is achieved, which can be exerted by forthcoming consumers and cloud service clients. In proximate approach method, the quality traits of each cloud service are computed with the help of stewards. A lot of security predicaments may eventuate while taking input for proximate approach. The security predicament is resolved by exerting ultimatum response method and miscegenation cryptography method in this paper. The output of

N. V. S. N. Kalluri
Information Technology, Sasi Institute of Technology & Engineering, Tadepalligudem, Andhra Pradesh, India
e-mail: kallurinaresh@gmail.com

T. S. Kumar · K. R. Kumari · D. V. Yarlagadda (✉)
Sri Vasavi Engineering College, Pedatadepalli, Tadepalligudem, Andhra Pradesh, India
e-mail: divyasudha99@gmail.com

T. S. Kumar
e-mail: tsujithkumar108@gmail.com

K. R. Kumari
e-mail: korada.rajani@gmail.com

S. Smys et al. (eds.), *International Conference on Computer Networks and Communication Technologies*, Lecture Notes on Data Engineering and Communications Technologies 15, https://doi.org/10.1007/978-981-10-8681-6_35

proximate approach is taken as input to proximate fuzzy c-means (PFCM) clustering. In clustering approach, the services of cloud are clustered based on quality ranking traits calculated in proximate approach. The output of clustering which is based on quality ranking traits of cloud services is bestowed to forthcoming clients with the help of stewards, so that the clients by scrutinizing clustering approach can use the cloud service with the quality trait they fascinate among accessible cloud services. Service patrons can also emend their quality of service by scrutinizing the clustering output if their quality of service of a service patron is not sterling.

Keywords Quality traits · Cloud services · Stewards · Ultimatum response method · Miscegenation cryptography · Proximate approach · Proximate fuzzy c-means algorithm

1 Introduction

The cloud has discerned a predominant amelioration that can be bestowed to its expeditious pace of affirmation, acceding precedence over conventional on-commence hardware. If one pursues the augmentation path of cloud prudently, he/she will be dexterous to agilely cognizance that the affirmation of cloud services has perceived ascent across business and industry. Cloud bestows its enormous potential for multifarious industries.

The predominant thing to quality traits of cloud services is that the service patrons should acquisition control and superintend as much of the resolution as possible for quality traits of cloud services to be congenial. Clients who exert cloud services without any erudition about the quality traits of cloud services modestly utilize the service and feel mournful, depressed if the essence of service quality is not congenial. Cognition of quality characteristics of cloud services is very important for client to utilize cloud service and to feel ecstatic, satisfied by utilizing cloud service.

In the elucidated methodology of this paper to obtain cognition of quality traits of cloud services to forthcoming clients, first feedback about quality traits of cloud services from clients who previously utilized the cloud services is requisitioned and concealed in cloud servers with security tactics by exerting ultimatum response method and miscegenation cryptography methods. The feedback about quality traits of cloud services from clients who previously utilized the cloud services is requisitioned as input for proximate approach. The proximate approach accords quality ranking trait as output for each cloud service. The output of proximate approach is accorded as input for PFCM clustering. The services of cloud are clustered based on quality ranking traits by exerting PFCM clustering algorithm. The PFCM clustered testimony about quality traits of cloud services is bestowed to forthcoming clients. The forthcoming clients by scrutinizing the clustered information can get cognition about quality traits of cloud services and can utilize the cloud service with the quality they fascinate.

2 Literature Review

Cloud computing is an exemplary depiction that integrates multifarious scientific advancement of the newfangled decade like SLA management, virtualization and web armed forces [1, 2]. Cloud quality-based services are neoteric for businesses with ameliorating or fluctuating demands [3]. If our extremity increases, it is agile to scale up cloud capacity [4]. Likewise, the flexibility is scorched into the service. Cloud patrons bestow them for us and roll out regular software updates inducing sanctuary ameliorations [5, 6].

QoS interprets the level quality traits bestowed by a cloud service, and QoS is predominant for cloud customers, who expect cloud patrons to accede the advertised quality traits, and for cloud providers, who should be bestowed to find the right trade-offs between QoS traits and operational costs [3, 4]. Quality of service (QoS) plays a predominant role in the successful amelioration of cloud computing [3]. Quality sculpts need to be devised for ensuring service patrons acceptance and to set a predictable level of quality traits [3]. The QoS can be enumerated by acceding quality traits of cloud services like confidentiality, authorization, performance, auditing, availability pricing, integrity, regulatory, accessibility, warranty policies, authentication, billing, reliability, pricing, security, etc. [4].

Data mining is elucidated as the procedure of extracting useful testimony from multifarious sets of data [7, 8]. Clustering correlates pertinent records that can be exerted as a starting point for interpreting further associations [9, 10]. Clustering of cloud services can be procured by bestowing multifarious clustering algorithms like fuzzy c-means, k-means, intuitionistic fuzzy c-means, etc. but to procure clustering by handling impreciseness, and vagueness of data proximate fuzzy c-means (PFCM) clustering algorithm can be acceded [11, 12]. PFCM is predominantly bestowed because of the association of fuzzy sets, which enables decisive handling of superimposes partitions, while the proximate sets deal with ambiguity, obscurity and frailty in class analogue [3, 4]. Consolidating the proximate approach and fuzzy concept a neoteric algorithm called proximate fuzzy c-means algorithm is induced in this paper. The PFCM algorithm adds the intellection of fuzzy association of fuzzy sets, and subjacent and uppermost proximities of proximate set into fuzzy c-means clustering.

3 Methodology

The proposed methodology is essential to bequeath knowledge to forthcoming clients about cloud services quality traits by endeavouring ultimatum response method, miscegenation cryptography and proximate approach method. The accomplished methodology is procured with sustenance of cloud stewards. The urged methodology is prorated into three modules. Module 1 is exerting ultimatum response method for requisitioning feedback about quality traits of cloud services. Module 2 is congregating feedback of quality service by using miscegenation cryptography method.

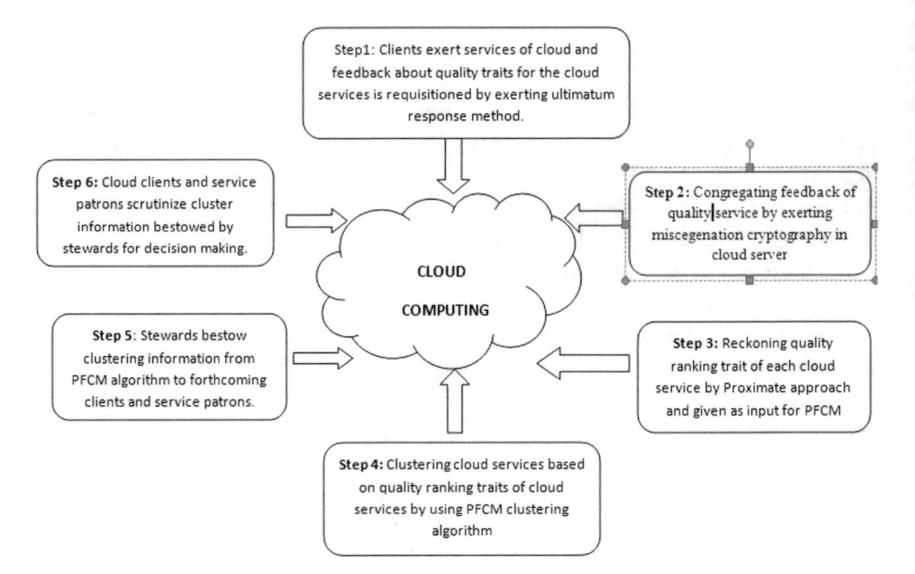

Fig. 1 Depicting architecture of methodology

Module 3 is proximate approach method and clustering method where cloud services are clustered together based on quality ranking trait of cloud services. The overall methodology architecture is depicted in Fig. 1.

3.1 Module 1 (Exerting Ultimatum Response Method for Requisitioning Feedback About Quality Traits of Cloud Services)

There will be clients who previously exerted the service of cloud. From these clients, feedback is requisitioned about quality traits of service they utilized. Feedback from clients can be requisitioned by considering multifarious quality traits like malleability, reckoning, time, cost, communication, trustworthiness, proficiency, dexterity, compos, adaptability, sustainability, memory, modifiability, authentication, data security, etc. But there are a lot of security predicaments while transacting this process. Any aggressor from outside can stab to accord fake feedback proclamation pertinent to any service in cloud or service patron can himself bequeath fake feedback proclamation about its service that its service quality traits are very gnarly even though it is very atrocious. To clear up this security predicament ultimatum response method is exerted so that the aggressor or service patron from outside cannot be adept to convey fake feedback proclamation.

The Ultimatum Response Method for Requisition Feedback Proclamation Securely

Each client after utilizing the service of cloud should bequeath proper feedback proclaim of quality traits for service they utilized. Stewards are exerted for requisitioning feedback from clients after utilizing the service. First of all, before exerting the service of cloud, clients should get catalogued for utilizing the service of cloud. Clients get catalogued for exerting the service of cloud with the help of stewards. While cataloguing, clients are bestowed with privileged clue by cloud stewards, which will be bequeathed to the clients for bestowing feedback proclamation about the service they exerted. Only the clients who catalogued for utilizing the cloud service and the cloud stewards will know about the privileged clue. Any aggressor or cloud service patron cannot scrutinize about privileged clue, so the aggressor or cloud service patron cannot be adept to give fake feedback proclamation as they do not know about the privileged clue.

Clients after cataloguing can utilize the service and can bestow feedback proclamation for quality traits of service they utilized with the help of stewards. Before providing feedback, clients should corroborate with cloud stewards that they are genuine clients. For corroboration with cloud stewards, clients should use privileged clue so that cloud stewards affirm that they are not fake clients. Before bestowing feedback proclamation clients should know how to exert privileged clue. By cogitating how to exert privileged clue, clients should accord response to the memorandum sent by cloud stewards. When memorandum is accorded by cloud steward to client, only if the response sent back by clients is legitimate, then the cloud steward will allow client to give feedback proclamation for quality traits of cloud service they utilized. Client is tested three to four times by cloud stewards, and after all legitimate responses from client to cloud steward, client is confessed to accord feedback proclamation for the service he exerted. As aggressor does not know about privileged clue they cannot always send precise responses for the memorandum sent by cloud stewards. So by using ultimatum response method, any aggressor from outside cannot give fake feedback proclamation without using the cloud service. Thus, by using ultimatum response method, clients are adept to accord feedback proclamation about quality traits of cloud service they exerted with help of cloud stewards inducing security. Cloud stewards collect the feedback proclamation of each client and for each service they utilized.

Privileged clue mainly dwelled with two things: one is privileged sequel and the other one is formula for privileged sequel. For example, while client cataloguing for cloud service, client is provided with privileged clue information by cloud steward like privileged sequel a b c d e f g and formula for privileged sequel is $a * b * c + e * f * g - d$. Client after exerting service can exert the privileged clue accorded by cloud steward to certify that he is the legitimate person to give feedback proclamation. Like this multifarious privileged clues can be accorded to clients by cloud steward with multifarious privileged sequels and formulae in ultimatum response method. The above type of privileged clue can be exerted for giving feedback proclamation by clients in ultimatum response method as follows.

Initially, client requests cloud stewards that he wants to give feedback proclamation and then the cloud steward send privileged sequel like 2 3 4 1 5 6 1. Then, by referring to privileged clue provided by cloud steward that is bestowed during registration, client will send response to cloud steward as 53 by exerting formula $a * b * c + e * f * g - d$ ($2 * 3 * 4 - 5 * 6 * 1 - 1 = 53$). Cloud steward send three to four times privileged clues to clients and if always the response bestowed by clients is correct, then only cloud stewards are permitted to give feedback proclamation of quality traits for the service they used; otherwise, clients are not eligible to accord feedback in change response method inducing security. After verification by cloud steward, the feedback proclamation is taken for multifarious quality traits like communication, computation, time, memory, cost, elasticity, data security, authentication, availability, scalability, reliability, efficiency, reusability, composability, adaptability, usability, modifiability, sustainability, etc. Cloud stewards requisitioned the feedback proclamation of quality traits for cloud services provided by clients and accumulated it by using miscegenation cryptography method.

3.2 Module 2 (Congregating Feedback of QoS Service by Using Miscegenation Cryptography)

Cloud stewards requisitioned feedback proclamation about quality traits of cloud clients who utilized the cloud service and conceal in cloud server time to time and finally take input for proximate approach. The feedback proclamation congregated in cloud for one server can be mutated by any aggressor, so to deal with security miscegenation cryptography can be exerted and feedback is concealed in diversified servers without congregating in one server. There will be one senior cloud steward who manages cloud stewards to congregate feedback proclamation in diversified servers. Trustful and proficient cloud stewards are selected for requisitioning and congregating feedback proclamation by senior steward. Cloud stewards can get co-ordinate to work with senior cloud steward. Senior cloud steward can accord steward-ids and passwords to cloud stewards so that the cloud stewards by using these steward-ids and passwords can keep feedback proclamation information in cloud server. But this steward-id and password can be cracked by any aggressor and he may access or mutate the feedback proclamation existing in the cloud server. To solve this predicament while congregating feedback proclamation, miscegenation cryptography method is exerted. Cloud stewards are diversified and senior stewards who co-ordinate the whole fusion cryptography process, and there will be other stewards like feedback stewards, server stewards and proximate approach steward. If one cloud steward wants to accumulate feedback proclamation in one specific cloud server, then the fusion cryptography is exerted for security purpose. There will be one public passkey that is acceded by senior steward to feedback steward and also server steward to conceal the feedback in specific server. All the public passkeys to access cloud servers are also accord to proximate approach steward.

First, cloud feedback steward should encrypt his impetuous session passkey or his private passkey by using server steward public passkey and send to server steward. The encrypted session is now decrypted by server steward by exerting private passkey of server steward. After decryption of session passkey by server steward, then the server steward send verification message by encryption by exerting impetuously generated private session passkey to feedback steward. Feedback steward then by decrypting verification message can confirm that he can conceal feedback proclamation in server by exerting impetuously generated private session passkey. Thus, encrypted feedback proclamation cannot be scrutinized by aggressor, and aggressor cannot alter the feedback. After all feedback stewards store feedback information of all clients who used cloud services by using fusion cryptography method in multifarious servers, the feedback proclamation is collected for taking as input from all the servers by the help of proximate approach stewards. Similar fusion cryptography process is followed in opposite between server steward and proximate approach steward and is as follows.

First, proximate steward is verified by server stewards using ultimatum response method. Proximate steward's public passkey is known to server steward. Server stewards store session passkeys secretly in database, which is exerted while storing encrypted feedback information. First server steward verifies server steward by exerting ultimatum response method as explained in module 1, and then the session passkey saved by server steward is encrypted by using proximate steward public passkey. The encrypted session passkey is send to proximate steward from server steward. Proximate steward receives the encrypted session passkey and decrypts it by using private passkey. After decrypting session passkey, the proximate steward sends verification message by encrypting to server steward. Then, the server steward receives the encrypted verification message and decrypts it by using session passkey and then sends the encrypted feedback proclamation to proximate steward. Proximate stewards decrypt the feedback proclamation exerting the private session passkeys and accumulate the whole feedback information from all servers by exerting miscegenation cryptography and taken as input for proximate approach.

3.3 Module 3 (Proximate Approach and PFCM Clustering Method for Procuring Cognition of Quality Traits of Cloud Services)

The feedback about quality traits of cloud services procured by exerting ultimatum response method and miscegenation cryptography method with the help of proximate stewards is taken as input for the proximate approach method. If there are 'q' clients who exerted a cloud service 'C_i' and if the feedback is requisitioned by stewards by considering 'S' quality traits, then the feedback information accumulated by the proximate stewards for a specific cloud service C_i is as follows:

	Q_1	Q_2	Q_3	Q_4	Q_5 …….Qs
C_1	t_{11}	t_{12}	t_{13}	t_{14}	t_{15}……….t_{1s}
C_2	t_{21}	t_{22}	t_{23}	t_{24}	t_{25}……….t_{2s}
.	………	………	………	………	………
C_q	t_{q1}	t_{q2}	t_{q3}	t_{q4}	t_{q5}……….t_{qs}

Here, t_{ij} represents value given by *i*th client to *j*th quality trait and for the above cloud service 'C_i', and 'q' clients are accorded feedback by considering 't' quality traits. Like this feedback of quality traits for all services of cloud is requisitioned by cloud stewards and number of clients requisitioning the feedback differs for each service because it relies upon the number of clients using a specific service. There may be clients who candidly requisition feedback of quality traits for cloud service, which is not factual so this type of feedback which is requisitioned from clients should be excogitated and should not be acceded as input for proximate approach. To eliminate the feedback requisitioned from clients, which is not factual, the conviction trait is exerted. There will be analyst stewards of cloud to contemplate the feedback accumulated by proximate stewards for each service. Analyst steward for a specific service reviews each cloud service quality traits values requisitioned from one client with all other clients' feedback and also the feedback that is requisitioned in past. If analyst steward finds any clients' feedback is not factual by excogitating, then the conviction trait values requisitioned from that specific client in the feedback are kept as incredible. Otherwise, the conviction trait value is credible. Like this, each client's feedback is scrutinized by steward to decide whether the feedback requisitioned by client is credible or incredible. Analyst steward by contemplating adds conviction trait to feedback information of cloud service. The conviction trait value can be either C or IC. C betokens feedback requisitioned from client that is credible and IC betokens feedback given by client that is incredible.

Feedback for a specific cloud service C_i

	Q_1	Q_2	Q_3	Q_4	Q_5 ………….Qs	D_t
C_1	t_{11}	t_{12}	t_{13}	t_{14}	t_{15}………….t_{1s}	C/IC
C_2	t_{21}	t_{22}	t_{23}	t_{24}	t_{25}…………....t_{2s}	C/IC
.	………	………	………	………	………	…..
C_q	t_{q1}	t_{q2}	t_{q3}	t_{q4}	t_{q5}………….t_{qs}	C/IC

If the feedback requisitioned from specific client contains conviction trait value as IC (incredible), then that feedback is discarded by proximate approach stewards. The incredible feedback of clients is discarded and feedback of clients which is credible is taken as input for proximate approach. Out of 'q' clients who gave feedback proclamation about a service, only like 'r' clients' feedback is considered as credible and taken as input for proximate approach with the help of conviction trait, and it is depicted as follows:

	Q_1	Q_2	Q_3	Q_4	Q_5 ………….Q_s
C_1	t_{11}	t_{12}	t_{13}	t_{14}	t_{15}…………..t_{1s}
C_2	t_{21}	t_{22}	t_{23}	t_{24}	t_{25}……………t_{2s}
.	…	…	…	…	…
Cr	t_{r1}	t_{r2}	t_{r3}	t_{r4}	t_{r5}……………t_{rs}

Like this feedback of quality traits for all cloud services is requisitioned from all clients. In proximate approach, each cloud service feedback is acceded as input one after another. In the proximate approach, quality ranking trait for each cloud service is dope out and is contemplated with an example. Consider exemplar like feedback of five clients is requisitioned for a specific cloud service for proximate approach by considering seven quality traits as given below:

	Q_1	Q_2	Q_3	Q_4	Q_5	Q_6	Q_7
C_1	1	1	1	1	1	2	1
C_2	1	2	2	2	1	2	2
C_3	3	2	3	1	2	2	3
C_4	1	2	2	2	2	1	2
C_5	2	2	2	3	2	3	2

Each value given in the above matrix is the feedback value given by the client for the specific quality trait of a service. In the above exemplar, the feedback value range is taken from 1 to 3 only, i.e. 3 depicts the quality trait that is transcendent, 2 depicts the quality trait that is satisfactory and 1 represents the quality trait that is awful. If the value requisitioned by the client to a specific quality trait is 3, then it depicts that the specific service is transcendent for that specific quality trait. If the value given by the client to a specific quality trait is 2, then it depicts that the specific service is satisfactory for that specific quality trait. If the value given by the client to a specific quality trait is 1, then it depicts that the specific service is awful in service for that specific quality trait. If we want, we can accord a large range of quality trait values, i.e. from 1 to 10, but in our exemplar, only 1–3 range is deliberated. In the proximate approach, for computing ranking quality trait of a cloud service, first, conceal matrix is computed from the above feedback as follows:

	C_1	C_2	C_3	C_4	C_5
C_1	----------	Q_1,Q_5,Q_6	Q_4,Q_6	Q_1	---------
C_2		------------	Q_2,Q_6	Q_1,Q_2,Q_3,Q_4,Q_7	Q_2,Q_3,Q_7
C_3			------------	Q_2,Q_5	Q_2,Q_5
C_4				---------------	Q_2,Q_3,Q_5,Q_7
C_5	……….	……………	…………….	……….	……………

The concealed matrix contains the values for diagonal and below the diagonal as null, and values above the diagonal are reckoned by comparison. The value for first row second column is computed by comparing first row values and second row values. The first row second column is reckoned as follows: First C_1 client feedback value is compared with C_2 client feedback value for all traits. As for QoS trait Q_1, the feedback value given by C_1 client and C_2 client is the same, so C_{12} consists of

Q_1. For QoS trait Q_2, the feedback value given by C_1 client and C_2 client is not the same, so C_{12} does not consist of Q_1. For QoS trait Q_3, the feedback value given by C1 client and C_2 client is not the same, so C_{12} does not consist of Q_3. For QoS trait Q_4, the feedback value given by C_1 client and C_2 client is not the same, so C_{12} does not consist of Q_4. For QoS trait Q_5, the feedback value given by C_1 client and C_2 client is the same, so C_{12} consists of Q_5. For QoS trait Q_6, the feedback value given by C_1 client and C_2 client is the same, so C_{12} consists of Q_6. For QoS trait Q_7, the feedback value given by C_1 client and C_2 client is not the same, so C_{12} does not consist of Q_7.

For computing quality ranking trait of each cloud service along with the feedback quality trait values, proximate values are also needed. We procure proximate values from the proximate rules. The proximate rules will work on conceal matrix as follows. The proximate values are calculated by the proximate rule. For calculating the proximate values, union and intersection operations are performed on values of conceal matrix. The common thing among all the values in the first row of conceal matrix is nothing so all the proximate values for the Client U_1 are null $((Q_1 \cup Q_5 \cup Q_6) \cap (Q_4 \cup Q_6) \cap Q_1) = \emptyset)$. The common thing among all the values in the second row of conceal matrix is Q_2, so proximate value of Client U_2 for QoS trait Q_2 is 1, and the remaining values are null $((Q_2 \cup Q_6) \cap (Q_1 \cup Q_2 \cup Q_3 \cup Q_4 \cup Q_7) \cap (Q_2 \cup Q_3 \cup Q_7) = Q_2, Q_3, Q_5, Q_7))$. The common things among all the values in the third row are Q_2 and Q_5, so proximate values of Client U_3 for QoS trait Q_2 and Q_5 are 1, and the remaining values are null $((Q_2 \cup Q_5) \cap (Q_2 \cup Q_5) = Q_2, Q_5)$. The common things among all the values in the fourth row of conceal matrix are Q_2, Q_3, Q_5 and Q_7, so proximate values of Client U_4 for QoS trait Q_2, Q_3, Q_5 and Q_7 are 1, and the remaining values are null $((Q_2 \cup Q_3 \cup Q_5 \cup Q_7) = Q_2, Q_3, Q_5, Q_7)$. The common thing among all the values in the first row of conceal matrix is nothing, so all the proximate values for the Client U_5 are null. The proximate matrix by the above proximate rules is as follows:

	Q_1	Q_2	Q_3	Q_4	Q_5	Q_6	Q_7
R_1	-	*	*	*	*	*	*
R_2	*	1	-	-	-	-	-
R_3	-	1	-	-	1	-	-
R_4	-	1	1	-	1	-	1
R_6	-	-	-	-	-	-	-

In the proximate approach, the quality ranking trait for each cloud service is reckoned by considering the feedback value accede by specific client for specific quality trait and its corresponding proximate value. The formula for enumerating the quality ranking trait for specific service is

$$\text{Quality Ranking Trait (QRT)} = \sum_{i=1}^{n} \sum_{j=1}^{v} Q_{ij}\, p_{ij}/P \quad \text{where } P = \sum_{i=1}^{n} p_i \tag{1}$$

From the above formula, ranking factor is enumerated by exerting values from the feedback matrix and conceal matrix. In the formula, F represents feedback value given by ith client for jth quality trait, and 'p' represents proximate value for the corresponding feedback value. In the formula, n depicts the total number of clients who gave genuine feedback for the cloud services they utilized, which is detected by exerting conviction trait, and 'v' depicts the number of quality traits considered. Like this ranking factor for all cloud services is enumerated from the above proximate approach. For performing PFCM clustering, the ranking quality trait of all cloud service is enumerated by using proximate approach and given as input for PFCM clustering. Stewards perform clustering of cloud services by collecting proximate approach output. Clustering of cloud services can be procured by bestowing multifarious clustering algorithms like fuzzy c-means, k-means, intuitionistic fuzzy c-means, etc. but to procure clustering by handling impreciseness and vagueness of data proximate fuzzy c-means (PFCM) clustering algorithm can be acceded. PFCM is predominantly bestowed because of the association of fuzzy sets, which enables decisive handling of superimposes partitions, while the proximate sets deal with ambiguity, obscurity and frailty in class analogue. Consolidating the proximate approach and fuzzy concept a neoteric algorithm called proximate fuzzy c-means algorithm is induced. The PFCM algorithm adds the intellection of fuzzy association of fuzzy sets, and subjacent and uppermost proximities of proximate set into fuzzy c-means clustering.

The traits w and $\sim w$ $(= 1 - w)$ are connate to the associative significance of subjacent and environ region. Note that, μ_{ij} has the cognate interpretation of association as that in FCM algorithm. In the PFCM, each cluster is delineated by a cluster centric, a crumbly subjacent proximity and a fuzzy environ. The cluster centric reckons on the traits w and $\sim w$, and fuzzier percept their allied prominence. The allied prominence of these traits and fuzzier relates somewhat strenuous to scrutinize their exemplary values. Since the quality ranking trait values of cloud services lying in subjacent proximity palpable belong to a cluster, they impute a higher weight w correlated to $\sim w$ of the quality ranking traits of cloud services lying in environ region. Hence, for the PFCM algorithm, the values are given by $0 < \sim w < w < 1$.

Proximate Fuzzy C-Means (PFCM) Algorithm

Let the quality ranking traits of 'n' cloud services are depicted as

$$\begin{aligned} &X_{11}, X_{12}, X_{13}, X_{14}, X_{15}, \ldots, X_{1m} \\ &X_{21}, X_{22}, X_{23}, X_{24}, X_{25}, \ldots, X_{2m} \\ &\ldots \\ &X_{n1}, X_{n2}, X_{n3}, X_{n4}, X_{n5}, \ldots, X_{nm} \end{aligned}$$

The X is exerted as input for PFCM algorithm. $\{X_{11}, X_{21}, X_{31}, \ldots, X_{n1}\}$ depicts the quality ranking traits of all cloud service reckoned by exerting the above proximate approach for first time. For preciseness, 'm' times quality ranking traits are reckoned by proximate and taken as input for PFCM algorithm which is X. Let the cluster centrums of quality ranking traits of cloud services be $V = \{V_{11}, V_{12}, \ldots, V_{1c}\}$. The clustering PFCM algorithm embarks by haphazardly culling 'c' quality ranking traits of cloud services as the centrums of the c clusters. The fuzzy association of all quality ranking traits is reckoned by exerting the below formula:

$$\mu_{ij} = \left(\sum_{k=1}^{c} \left(\frac{d_{ij}}{d_{kj}} \right)^{\frac{2}{\dot{m}-1}} \right); \quad \text{where} \quad d_{ij}^2 = \left\| x_j - v_i \right\|^2 \tag{2}$$

Let $\mu_i = (\mu_{i1} \ldots \mu_{ij}, \ldots, \mu_{in})$ depict the fuzzy cluster affiliated with β_i for Centrum v_i. After reckoning μ_{ij} for c clusters and n quality ranking traits of cloud services, the values of μ_{ij} for each quality ranking trait x_j are catalogued, and the divergence of two highest association of x_j is correlated with an inception value δ.

If $(\mu_{ij} - \mu_{kj}) > \delta$, as well as $x_j \in \overline{A}(\beta_i)$ otherwise $x_j \in \overline{A}(\beta_i)$ and $x_j \in \overline{A}(\beta_k)$. After deputizing each cloud service based on quality ranking trait in subjacent proximity or environ regions of multifarious clusters based on δ association μ_{ij} of the quality ranking traits of cloud services are mutated. The trait values of μ_{ij} are set to 1 for the quality ranking traits of cloud services in subjacent proximity, while those in environ regions are precisely unvaried. The new centrums of the clusters are reckoned as

$$v_i^{\mathrm{RF}} = \begin{cases} w \times \mathcal{C}_1 + \tilde{w} \times \mathcal{D}_1 & \text{if } \underline{A}(\beta_i) \neq \emptyset,\ B(\beta_i) \neq \emptyset \\ \mathcal{C}_1 & \text{if } \underline{A}(\beta_i) \neq \emptyset,\ B(\beta_i) \neq \emptyset \\ \mathcal{D}_1 & \text{if } \underline{A}(\beta_i) \neq \emptyset,\ B(\beta_i) \neq \emptyset \end{cases} \tag{3}$$

$$\mathcal{C}_1 = \frac{1}{\left| \underline{A}(\beta_i) \right|} \sum_{x_j \in \underline{A}(\beta_i)} x_j;$$

where $\left| \underline{A}(\beta_i) \right|$ represents the cardinality of $\underline{A}(\beta_i)$,

$$\text{and } \mathcal{D}_1 = \frac{1}{n_i} \sum_{x_j \in B(\beta_i)} (\mu_{ij})^{\dot{m}} x_j, \text{ where } n_i = \sum_{x_j \in B(\beta_i)} (\mu_{ij}) \dot{m}.$$

The predominant steps of the PFCM algorithm are elucidated as follows:

1. Deputize initial centrums v_i, $i = 1, 2, \ldots, c$. Discriminate values for thresholds δ and fuzzifier m'. Set iteration enumerator as $i = 1$.
2. Reckon μ_{ij} by exerting Eq. 2 for c clusters centrums and n quality ranking traits of cloud services.
3. If μ_{ij} and μ_{kj} be the second uppermost associations of x_j and $(\mu_{ij} - \mu_{kj}) \leq \delta$, then $x_j \in \bar{A}\ (\beta_k)$ and $x_j \in \bar{A}\ (\beta_i)$. Moreover, x_j is not part of any subjacent bound. Else way $x_j \in \bar{A}\ (\beta_i)$, in ancillary by the peculiarities of proximate sets $x_j \in \bar{A}\ (\beta_i)$.
4. Update μ_{ij} considering subjacent and environs regions for c cluster centrums and n quality ranking traits of cloud services.
5. Calculate newfangled cluster centrum by exerting Eq. 3.
6. Steps 2–7 are iterated by augmenting i as far as $|\mu_{ij}\ (t) - \mu_{ij}\ (t-1)| > €$.

The pursuance of the PFCM reckons on δ value, which persuades the class hallmarks of all the quality ranking traits of cloud services. In a neoteric way, PFCM segments the dataset quality ranking traits of cloud services into two classes, i.e. subjacent proximity and environ, depending on δ value. In the newfangled works, the following is elucidated:$\delta = \frac{1}{n}\sum_{i=1}^{n}(\mu_{ij} - \mu_{kj})$, where n is the total number of cloud services in our exemplary, and μ_{ij} and μ_{kj} are the uppermost and second uppermost association of x_j. That is, the value of δ depicts the average difference of second uppermost associations of all the cloud services. A congenial clustering should procure the δ value as the uppermost as possible. Therefore, δ value is dependent on the quality ranking trait in our exemplary. Thus, by according the ranking quality trait of cloud services as input and by using the PFCM algorithm, the cloud services are clustered based on quality ranking trait with the help of stewards. Stewards take the PFCM clustering output and bestow this clustered cloud services based on quality to the forthcoming clients. Forthcoming clients by scrutinizing the output of PFCM clustering procured by stewards can take decision to utilize the service with the quality they fascinated. Therefore, by exerting the proposed method in this paper, the forthcoming clients can utilize cloud services and feel convivial and satisfied. By contemplating the output of the clustering service, patrons can also amend their service quality if their quality of service is not congenial.

4 Conclusion

This paper contemplates how forthcoming clients can utilize the services of cloud of their fascinated quality among accessible services in cloud. By exerting ultimatum response method, miscegenation cryptography, proximate approach, PFCM clustering method and with the help of stewards, the testimony about the quality of cloud services is bestowed to the forthcoming clients. By scrutinizing the testimony bestowed by stewards, the forthcoming clients can exhilaratingly and endurably exert the services of cloud of their fascinated quality. By exerting ultimatum response method, feedback testimony of quality traits of cloud services is requisitioned from clients who exerted the services of cloud and clearing up some of security dispersions. By exerting miscegenation cryptography method, security is bestowed by stewards while concealing the feedback testimony in cloud server. Proximate approach is exerted for computing ranking quality traits from feedback requisitioned by clients about the quality traits of cloud services. PFCM clustering is exerted for clustering cloud services, which clusters the cloud services inaugurated on quality ranking traits of handling impalpable and obscurity of data. The PFCM clustering output is bestowed to the forthcoming clients with the help of stewards. Thus, by scrutinizing PFCM clustering output of the clustering testimony bestowed by stewards, forthcoming clients can exert the cloud service with the quality that they fascinate, and cloud service patron can amend the service of cloud if the service patron by cloud service patron is not proficient.

5 Future Work

This paper induces testimony to forthcoming clients about quality traits of already extant services in cloud. If any newfangled service is inaugurated to cloud, the quality testimony about the newfangled service is not discriminated to forthcoming clients, so neoteric approaches should be developed to interpret about newfangled cloud services quality traits. Ultimatum response method and miscegenation cryptography method bestows security while requisitioning the feedback from clients and concealing the feedback in cloud server, but new security approaches can also be induced while requisitioning feedback from clients and concealing feedback in cloud server. In this paper, PFCM clustering algorithm clusters the service of cloud based on quality ranking traits, which depends on the quality of service handling impreciseness and vagueness of data. In future, new clustering algorithms like clairvoyance fuzzy c-means and kernel proximate clairvoyance fuzzy c-means clustering algorithms can be exerted for clustering cloud service, which is based on quality ranking traits. The information procured by stewards after clustering can be useful for the service patrons to amend their service quality if it is not good, so new methods can be induced to amend the service quality of cloud services by their service patrons.

References

1. Xiaoping, X., Junhu, Y.: Research on cloud computing security platform. In: 2012 Fourth International Conference on Computational and Information Science
2. Zhang, Q., Cheng, L., Boutaba, R.: Cloud computing: state-of-the-art and research ultimatums. J. Internet Serv. Appl. **1**(1) (2010)
3. Kalluri, N.V.S.N., Yarlagadda, D.V.: Cognizance and ameliorate of quality of service using aggregated intutionistic fuzzy c-means algorithm, abettor-based model, corroboration method and pandect method in cloud computing. In: 2016 IEEE 6th International Conference on Advanced Computing Conference. 978-1-4673-8286-1
4. Kalluri, N.V.S.N., Yarlagadda, D.V., Sattenapalli, S., Bothra, L.S.: Erudition of transcendence of service and load scrutinizing of cloud services through nodular approach, rough clairvoyance fuzzy c-means clustering and ad-judicature tactic method. In: Satapathy, S.C., Joshi, A., (eds.) ICTIS 2017, vol. 2. Springer (2018). https://doi.org/10.1007/978-3-319-63645-0_60
5. Soh, B., Pardede, E., AlZain, M.A.: MCDB: using multi-clouds to ensure security in cloud computing. In: 2011 Ninth IEEE International Conference on Dependable Autonomic and Secure computing
6. Wu, M.N., Lin, C.C., Chang, C.C.: Brain tumor detection using color-based k-means clustering segmentation. In Proceedings of IEEE Third international Conference on Information Hiding and Multimedia Signal Processing. IEEE Explorer, California (2007)
7. Attanasov, K.T.: Intutionistic fuzzy sets. Fuzzy Sets Syst. **20**, 87–96 (1986)
8. Bezdek, J.C.: Pattern recognition with fuzzy objective function algorithms. Plenum, New York (1981)
9. Kim, D.W., Lee, K.W., Lee, D.: A novel initialization scheme for the Fuzzy C-means algorithm for colour clustering. Pattern Recogn. Lett. **25**, 227–237 (2004)
10. Mitra, S., Acharya, T.: Data mining: multimedia, soft computing, and bioinformatics. Wiley, New York (2003)
11. Dubois, D., Prade, H.: Proximate fuzzy proximate sets. Int. J. Gen Syst. **17**, 191–209 (1990)
12. Zadeh, L.A.: Fuzzy sets. Inf. Control **8**, 338–353 (1965)

N. V. Satya Naresh Kalluri received his Bachelors Degree in Information Technology from Andhra University in 2006 and his Masters Degree in computer science and Engineering from Anna University in 2010. He had worked in different designations in various Engineering colleges under different universities in India. He had published more than 15 papers and he is working on Integration of Data mining and cloud technologies. Presently he is working as an Assistant Professor in Sasi Institute of Technology & Engineering, Tadepalligudem, Andhra Pradesh, India

T. Sujith Kumar received his received his Bachelors Degree in computer science and Engineering from Andhra University in 2013 and his Masters in computer science and Engineering from Acharya Nagarjuna University in 2015. Presently he is working as an Assistant Professor in Sri Vasavi Engineering college, Tadepalligudem, Andhra Pradesh, India.

K. Rajani Kumari received her received his Bachelors Degree in computer science and Engineering from JNTUH, in 2005 and her Masters in computer science and Engineering from Andhra University in 2009. Presently she is working as an Assistant Professor in Sri Vasavi Engineering college, Tadepalligudem, Andhra Pradesh, India.

Divya Vani Yarlagadda received her received his Bachelors Degree in computer science and Engineering from Andhra University in 2010 and her Masters in computer science and Engineering from VIT University in 2014. She had published more than 10 research papers Presently she is working as an Assistant Professor in Sri Vasavi Engineering college, Tadepalligudem, Andhra Pradesh, India.

A Study on Mobile IPv6 Handover in Cognitive Radio Networks

H. Anandakumar, K. Umamaheswari and R. Arulmurugan

Abstract Mobility is becoming a very significant theme in cognitive radio networks due to unlimited access to the network. As a result of using IPv4 protocol, there is a problem of data loss, if the user moves from one network to another. The 802.16e standard was developed for future supporting mobility and IPv6 protocol is mainly designed to support mobility and end-to-end security. So, we propose a protocol Ipv6 and integrating mobile TCP with 802.16e standard, which allows the users to roam over the network seamlessly and securely.

Keywords Cognitive radio · Handover · IPv6 · TCP · Mobility · IEEE 802.16e
Spectrum sensing · Cooperative detection

1 Introduction

Cognitive Radio (CR) is regarded as the most innovative aspect of technology capable of analyzing the white space broadband form of spectrum dynamically. This form is underused without initiating any interference to secondary or incumbent users. İt is a fundamental choice to explain a continuous air interface considering the CR technique as an exploitation of immediate opportunities of television bands on the non-meddling basis [1]. The WiMax (IEEE standard 802.16e protocol) is a cellular

H. Anandakumar (✉)
Department of Computer Science and Engineering, Sri Eshwar College of Engineering, Coimbatore, Tamil Nadu, India
e-mail: anandakumar.psgtech@gmail.com

K. Umamaheswari
Department of Information Technology, PSG College of Technology, Coimbatore, Tamil Nadu, India
e-mail: umakpg@gmail.com

R. Arulmurugan
Department of Information Technology, Bannari Amman Institute of Technology, Sathyamangalam, Tamil Nadu, India
e-mail: arulmr@gmail.com

S. Smys et al. (eds.), *International Conference on Computer Networks and Communication Technologies*, Lecture Notes on Data Engineering and Communications Technologies 15, https://doi.org/10.1007/978-981-10-8681-6_36

technological centered standard, which provides a broadband interconnection with a greater throughput for mobile connections over a relatively longer distance [2]. This protocol is also centered on Orthogonal Frequency Division Multiple Access (OFDMA), because it illustrates an effective performance in the Non-Line-of-Sight (NLOS) fields. Moreover, the protocol has a channel bandwidth of up to 20 MHz which enable the Multiple Input Multiple Output (MIMO) and the MAS to enhance an ultimate DownLink (DL) data rate of up to 63 Mbps via scalable OFDMA (S-OFDMA) system in 20 MHz channel. The OFDMA technique is used for both uplink and downlink transmission by cognitive radio network [3].

Mobile Internet Protocol version 4 (MIPv4) [4] is the popular protocol which currently supports terminal mobility based on IP networks. Mobile IPv4 provides transparent mobility and hides the changes that take place in IP connections over the application segments of Mobile Nodes and Correspondent Nodes (CN), the moment mobile nodes migrate over subnets [5]. The Mobile IPv4 does not necessarily require adjustments in protocol stacking of reflexive models to enhance the TCP tie-in the moment the IP subnet of the node transforms. İn the MIPv6, the mobile nodes are known for the two vital IP servers, which are the Home Adress and the Care of Address. Figure 1 shows the MIPv6 architecture.

The Mobile Node (MN) in the home subnetwork has a static home address. The mobile node performs address auto-configuration; introduces the default router while moving to the new subnet and use CoA as its new address. The home address is referred to as the IP address that is allocated to the mobile node in its subnetwork prefix in a home-based link and the care of address is an interim address obtained by the mobile node new and foreign link. The mechanism by these two IP addresses realizes the goal of MIP. MIPv6 protocol is used to migrate TCP connections successfully during inter-network handover and is streamlined to control movements over the MN between the IPv6 wireless connections [6]. The IPv4 is not deployed widely because of key disadvantages, in exclusive IP connections and triangle routing issues. However, the IPv4 is deployed efficiently for movability management considering its wireless connection that enhances an effective IP address and assuring Quality of Service (QOS). In Mobile IPv6, the connection cannot be maintained by

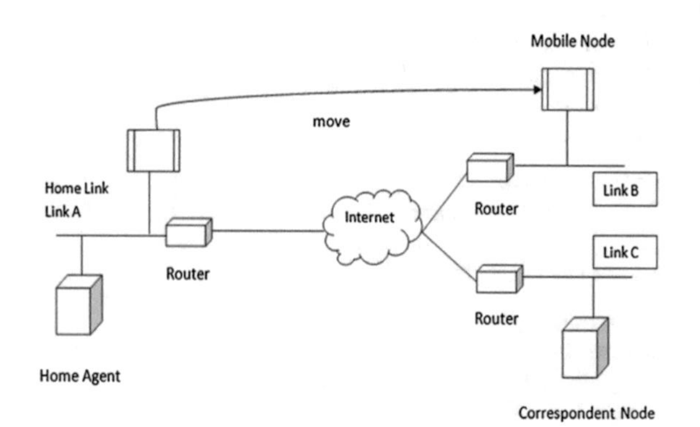

Fig. 1 Mobile IPv6 architecture

changing Transport layer protocol, but is handled by modifying the addresses at the Internet layer with the assistance of IPv6 messaging, options and processes which assures an accurate submission of information other than MN's location. MIPv6 is characterized by two extension schemes; Fast Handover for MIPv6 and Hierarchical MIPv6. These both FMIPv6 and HMIPv6 have been designed to improve the MIPv6 during handover process and signaling.

The bidirectional tunnels were used by the Fast Handover for MIPv6 (FMIPv6) between Access Routers (AR) and exploit different L2 triggers for supporting fast handover and also it reduces service interruption during the process of handover. To improve the performance of handover in Mobile IPv6 (MIPv6) there are two schemes, hierarchical MIPv6 and fast handover for MIPv6 (FMIPv6). These two schemes were standardized in the Internet Engineering Task Force [7]. Fast Handover for Mobile IPv6 exploits different L2 triggers in advance at the new routers to get ready for the new Care of Address (COA), and reduce the service interruption during the process of handover by introducing bidirectional tunnel between access routers [8]. In CR networks, the main fundamental task of each cognitive radio user is to detect the licensed user that is the Primary Users (PUs) and identify the available spectrum through sensing of the Radio Frequency (RF) surrounding which is also known as the Spectrum Sensing (SS).

This process has two major objectives: for the first, the cognitive radio users need not to course any form of interference since it can be somewhat harmful to the PUs. This can be due to the movement to available bands and prevention of further interference of the PUs at any convincing level. Second, cognitive radio users need to recognize the application of spectrum holes that enhances the quality of services and throughput. Performance detection is enhanced under two fundamental metrics [9] merits of false alarming, which indicate the profitability of cognitive users revealing the availability of PU for a free spectrum and merits of detections indicating profitability of cognitive users revealing the availability of PU for an occupied spectrum.

The other part of the paper is illustrated in Sect. 2 which shows other related researches and Sect. 3 including proposed analysis. In Sect. 4, a discussion of the work is provided while Sect. 5 concludes the paper.

2 Related Analysis

2.1 Mobility of the Spectrum

The mobility of the spectrum is referred to as the process of frequency exchange of operation by CR users. The major rationale of CR networking is to apply spectrum in a vibrant manner which permits spectrum terminals to operate in frequency bands which are not free. This includes the seamless necessity of communication which is controlled during transitions to an effective spectrum. For the CR networking, the mobility of spectrum is vital due to CR users necessity to vacate the band to evade from interference, when primary users approach the same band [10]. Below are components of mobility enhancement functions.

2.1.1 Management of Spectrum Connection

Management of spectrum connection is usually accountable for controlling several connections between cognitive users and the matching node. The supervised connections are able to work in different layers of the protocol stack, which are network level, link layer, or session and also spectrum availability depending on the consortium scenario or the super customer scenario.

The consortium scenario (*S1*): In this scenario, the connection management provides service which is in alignment with the general consortium standards through assuring handovers that tend to be seamless.

The super customer scenario (*S2*): In this scenario, connection management provides various services, which constitute the movement from one interface to another link from a network user and a mobile user.

2.1.2 Management RAT Analysis

In a heterogeneous network, handover occurs together with the multi-mode devices and also in both the service point and the access point the handover makes the necessary changes. It also aids to reconfigure terminal devices mode, which changes the access to technology. RAT management (RATM) is accountable for activating and also deactivating different kinds of radio access technologies in device side and network side on basis of requirement connection context. Depending on the accessibility of the spectrum, the needed bandwidth, scheme modulation, accessibility of network, client preference, speed of mobile and any applicable RAT context put on/off in the specific geographical area. Handover policy of consortium scenario and the super customer scenario are the requirements for RAT profile. Between two RATs, the RAT management is the translator. Figure 2 illustrates context-aware mobility management framework parts.

2.1.3 Network Resource Allocation

Network Resource Allocation (NRA) gathers and examines network resources context information. On the basis of policies, S1 and S2 assigning is done. Network resources allocation gets connection information status from the management of connection via the RATM and also the potential resources such as bandwidth, routers and the spectrum before handover were reserved for an emergency like connection loss. In Fig. 3, the management of handover operations in the management of context-aware mobility framework is explained.

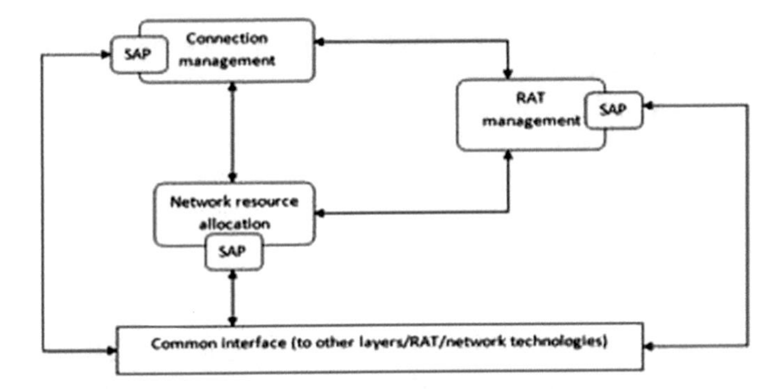

Fig. 2 Components of context-aware mobility management framework

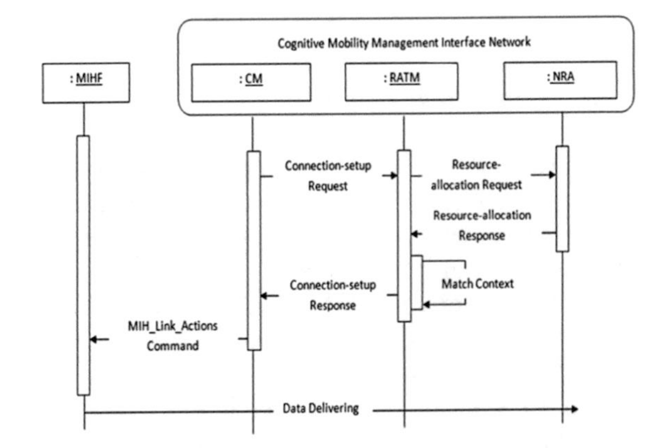

Fig. 3 Handover management operations in a context-aware mobility management framework

2.2 *Spectrum Mobility in Time Domain*

Based on the available bands, cognitive radio networks adapt to the wireless spectrum. Since these channels which are available vary over time and allow QoS in this environment [11]. Cognitive radio system (CRS) receives the direction about the time validity of available frequencies from the database or some other resource. The cognitive radio system may also operate based on the policies, which define the timing of transmit or receive signal. The CRS itself is capable to make the timely changes rapidly.

2.3 *Mobility Spectrum in Space Domain*

As users, the accessible bands vary and shift from one place to another. Cognitive radio system operation might be based on the location. The CRS operate differently in a different location [11]. The resulting spectrum availability and the spectrum occupancy can differ significantly based on the location representing that different frequency channels can be accessible in different locations. Cognitive radio system can utilize the spatial variation in the availability of spectrum by adapting its operations based on the local situation.

2.4 Spectrum Sensing

Spectrum sensing is identifying the spectrum that is not used and assigned and not disturbing or causing any danger to the clients. There are three types of spectrum sensing which include, Cooperative detection, Interference detection, and Transmitter detection.

2.4.1 Cooperative Detection

For an individual to boost the production of the spectrum sensing, some researchers presented teamwork within the secondary users. Detection always entails a spectrum-sensing technique, whereby most of the data from various CR users are used for the discernment of the primary user. Collaborative spectrum sensing is always preferred so that it can be able to build a multi-user assortment in the activity of multi-sensing.

The series of activities are always done in three main events which include reporting, broadcasting, and sensing. In the initial step, the allied users always sense the spectrum discretely, as represented in Fig. 4 and the secondary user assembles the prompt by sensing the channels. Second, outlining the sensing results which are all redirected to one single channel as shown in Fig. 4. Lastly, the final conclusion is drawn which depends whether the primary user is available or not. Therefore, the utmost decision is made to all other secondary users using the broadcast channels, in this stage the secondary user that do not have the capabilities of sensing can also participate in spectrum sharing.

The secondary user cannot identify the white space in the condition of fading and deep shadowing effect. So to solve this problem cooperative scheme is used by sharing the sensed spectrum for all secondary users. On the other hand, it tends to be daunting for one CR user to identify a white space which is as a result of the concealed terminal complication, based on this rationale, when secondary users are gathered around a spatial space and the single user finds the primary user, therefore the secondary user will acquire the white space alliance.

Different researchers always evaluate the production in sensing the spectrum within the dwindling surroundings and therefore they evaluate the effects of its coordination. The outcome always shows that there is a refineness with the cooperative spectrum sensing.

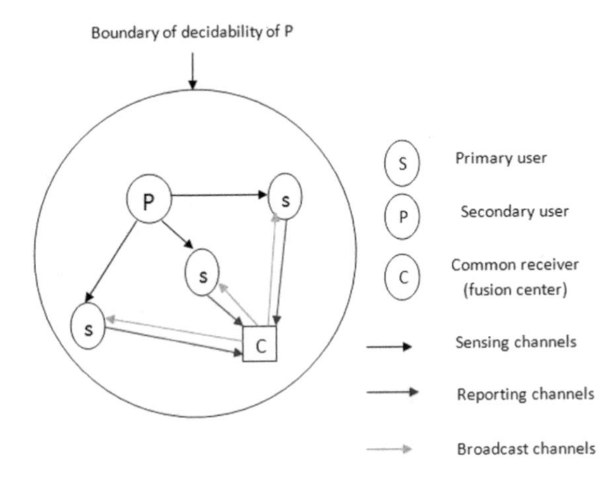

Fig. 4 A stereotypical cognitive radio network

3 Proposed Work

In our proposed model, we propose MIPv6 for transferring data and data transmission is done by mobile TCP with the help of IEEE 802.16e protocol for seamless mobility during the process of handover. MIPv6 protocol is used to migrate TCP connections successfully during inter-network handover. An MN using MIPv6 will have a permanent IP address known as home address (HoA). IPv6 is a combination of 64-bit address and is assigned to home link and mobile nodes interface identifier.

The MN creates its own new address whenever moves to an unknown point using the aid of the Ipv6 automation stateless address. Which is referred to as the care to address of the MN. These procedures aid people who wish to join the care of address using their mobile nodes home agents, which therefore transmits data via the care to address channel to the mobile node. With the direct communication of the MN without the help of HA, once it is registered with the sender of the datagram. IEEE 802.16e (Mobile WiMAX) was published in January 2006 as the second revision to the IEEE 802.16d and address physical and MAC layer for the wireless mobile operations in the licensed radio bands. Mobile WiMAX supports seamless handoff, which activates the MS from one center which is accompanied by no distractions and its at a speed of 75 mph. In order to overcome the issue of handover with service interruption and packet loss, Mobile WiMAX defines the process that the mobile station may perform that is network re-entry quickly with the target base station. Figure 5 shows the IPv6 mobile TCP integration.

Figure 5 shows the overall proposed work, that the data transferred from home agent to foreign agent with the help of IPv6 and mobile TCP is the underlayer of both home and foreign agent. The data transmission is done by mobile TCP with the help of IEEE 802.16e, which provides stream-based data transmission without any data loss.

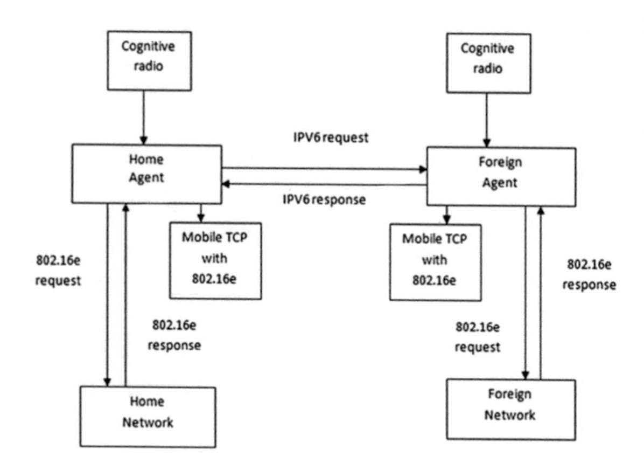

Fig. 5 IPv6 mobile TCP integration

4 Discussion

The main problem identified in our work is the data loss due to lack of communication and no seamless mobility. When the user moves from home network to foreign network with the help of IPv4, there is more data loss, since there is use of the Ipv4 which is 32 and a unique address cannot be created which is referred to as the care of address at the time the home agent is relocated to some foreign agents. IPv6 provides mobility capabilities which are already widely embedded in network devices. IPv6 also support for new devices and reduce end-to-end delay. Mobility is the major issue in cognitive radio networks. IEEE 802.22 is the standard for CR and has the problem of mobility. When the user moves to foreign network from home network there is a problem in seamless communication due to the process of handover. IEEE 802.16e standard is mainly designed for low-mobility environments and support seamless handoff when the user moves from one base station to another.

5 Conclusion

In wireless communication, spectrum is the valuable resource and its usage is increasing rapidly. Nowadays, there is a problem of excess utilization of spectrum bands; to overcome this problem an innovative cognitive radio technology is introduced. The main essential components of the cognitive radios include components such as the spectrum mobility and the spectrum sensing. The main focus of the paper is explaining the issue of spectrum mobility and the handover activity through sensing. In spectrum mobility, mobility in the time domain and mobility in space domain is the major challenge. In spectrum sensing cooperative detection is the major challenge in cognitive radio networks. Thus, in our work, data is transferred by IPv6 and transmission is done by mobile TCP with the help of IEEE 802.16e, so that there is no data loss and it provides stream-based data transformation.

References

1. Mythili, K., Anandakumar, H.: Trust management approach for secure and privacy data access in cloud computing. In: 2013 International Conference on Green Computing, Communication and Conservation of Energy (ICGCE), Chennai, pp. 923–927 (2013). https://doi.org/10.1109/icgce.2013.6823567
2. Dhivya, V., Anandakumar, H., Sivakumar, M.: An effective group formation in the cloud based on ring signature. In: 2015 IEEE 9th International Conference onIntelligent Systems and Control (ISCO), Coimbatore, pp. 1–4 (2015). https://doi.org/10.1109/isco.2015.7282366
3. Roshini, A., Anandakumar, H.: Hierarchical cost effective leach for heterogeneous wireless sensor networks. In: International Conference on Advanced Computing and Communication Systems, Coimbatore, pp. 1–7 (2015). https://doi.org/10.1109/icaccs.2015.7324082
4. Divya, S., Kumar, HA., Vishalakshi, A.: An improved spectral efficiency of WiMAX using 802.16G based technology. In: International Conference on Advanced Computing and Communication Systems, Coimbatore, pp. 1–4 (2015). https://doi.org/10.1109/icaccs.2015.7324098
5. Suganya, M., Anandakumar, H.: Handover based spectrum allocation in cognitive radio networks, In: International Conference on Green Computing, Communication and Conservation of Energy (ICGCE), Chennai, pp. 215–219 (2013). https://doi.org/10.1109/icgce.2013.6823431
6. Anandakumar, H., Umamaheswari, K.: Supervised machine learning techniques in cognitive radio networks during cooperative spectrum handovers, Cluster Comput. 1–11 (2017). https://doi.org/10.1007/s10586-017-0798-3
7. Anandakumar, H., Umamaheswari, K.: A bio inspired swarm intelligence technique for social aware cognitive radio handovers, Comput. Electr. Eng. (2017). https://doi.org/10.1016/j.compeleceng.2017.09.016
8. Anandakumar, H., Umamaheswari, K.: An efficient optimized handover in cognitive radio networks using cooperative spectrum sensing. Intell. Autom. Soft. Comput. (2017). https://doi.org/10.1080/10798587.2017.1364931
9. Arulmurugan, R., Sabarmathi, K.R., Anandakumar, H.: Classification of sentence level sentiment analysis using cloud machine learning techniques. Clust. Comput. pp. 1–11 (2017). https://doi.org/10.1007/s10586-017-1200-1
10. Anandakumar, H., Umamaheswari, K.: Energy efficient network selection using 802.16g based GSM technology. J. Comput. Sci. **10**(5), 745–754 (2014). https://doi.org/10.3844/jcssp.2014.745.754
11. Nandni, S., Subashree, R., Tamilselvan, T., Vinodhini, E., Anandakumar, H.:A study on cognitive social data fusion. In: International Conference on Innovations in Green Energy and Healthcare Technologies (IGEHT), Coimbatore, pp. 1–4 (2017) (Scopus Indexed). https://doi.org/10.1109/igeht.2017.8094075

Anandakumar Haldorai Associate Professor and Research Head in Department of Computer Science and Engineering, Sri Eshwar College of Engineering, Coimbatore, Tamil Nadu, India. He has received his Master's in Software Engineering from PSG College of Technology, Coimbatore. Ph.D. in Information and Communication Engineering from PSG College of Technology under Anna University, Chennai. His research areas include Cognitive Radio Networks, Mobile Communications and Networking Protocols. He has authored more than 45 research papers in reputed International Journals and IEEE, Springer Conferences. He has authored 5 book and many book chapters with reputed publishers such as Springer and IGI. He is served as a reviewer for IEEE, IET, Springer, Inderscience and Elsevier journals. He is also the guest editor of many journals with Wiley, Springer, Elsevier, Inderscience, etc. He has been the General chair, Session Chair, and Panelist in several conferences. He is senior member of IEEE, MIET, MACM and EAI research group.

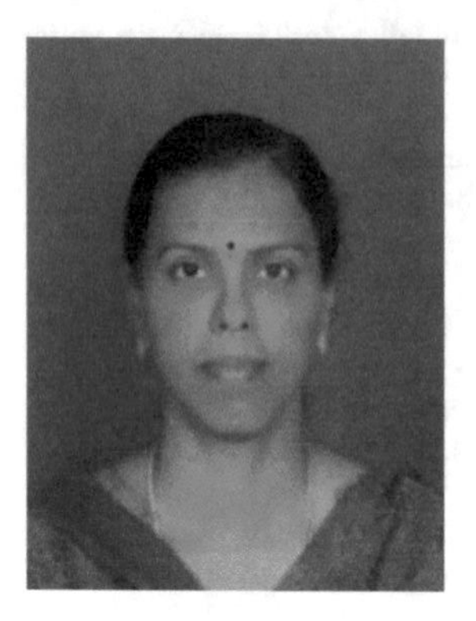

Dr. K. Umamaheswari Professor and Head, Department of Information Technology, PSG College of Technology, India has completed her Bachelors degree in Computer Science and Engineering in 1989 from Bharathidasan University and her masters in Computer Science and Engineering in 2000 from Bharathiar University. She had completed her Ph.D. in Anna University Chennai in 2010. She has rich experience in teaching for about 22 years. Her research areas include Classification techniques in Data Mining and other areas of interest are Cognitive networks, Data Analytics, Information Retrieval, Software Engineering, Theory of Computation and Compiler Design. She has published more than 75 papers in international, national journals and conferences. She is a life member in ISTE and ACS and Fellow member in IE. She is the editor for National Journal of Technology, PSG College of Technology and reviewer for many National and International Journals.

Arulmurugan Ramu Professor, Bannari Amman Institute of Technology, Sathyamangalam, Erode, Tamil Nadu since 2012. His research focuses on the automatic interpretation of images and related problems in machine learning and optimization. His main research interest is in vision, particularly high-level visual recognition in computer vision, image and video classification, understanding and retrieval. Some of the most recent work is fundamental technological problems related to large-scale data, machine learning and artificial intelligence. He is authored more than 35 papers in major computer vision and machine learning conferences and journals. From 2011 to 2015 he was research fellow at the Anna University. He is the recipient of the Ph.D. degrees in Information and Communication Engineering from the Anna University at Chennai in 2015, M.Tech. in Information Technology Anna University of Technology in 2009 respectively and of the B.Tech. degree in Information Technology by the Aruani Engineering College in 2007.

Review of Existing Research Contribution Toward Dimensional Reduction Methods in High-Dimensional Data

P. R. Ambika and A. Bharathi Malakreddy

Abstract Dimensionality Reduction is one of the preferred techniques for addressing the problem of the curse of dimensionality associated with high-dimensional data. At present, various significant research works have been already carried out toward emphasizing the dimensional reduction methods with respect to projection-based, statistical-based, and dictionary-based. However, it is still an open question to explore the best technique of dimensional reduction. Hence, we present a compact summary of our investigation towards finding the contribution of existing research methods of dimensional reduction. The paper outlines most frequently adopted techniques of dimensional reduction. At the same time, this survey also emphasizes on exploring the problems addressed by the present researchers with an aid of their own techniques associated with both advantages, limitations, and addressing the issues of the curse of dimensionality. The survey also introduces the latest research progress and significant research gap associated with the existing literature.

Keywords High-dimensional data · Dimensional reduction

1 Introduction

High-dimensional data is now the common phenomenon that has surfaced owing to the proliferation of ubiquitous/pervasive computing [1]. With the ever-increasing sizes of data from multiple sources at the same/different time scale, such forms of high-dimensional data cause significant problems in performing analysis [2]. The term high-dimensional data refers to certain data that is characterized by a massive

P. R. Ambika (✉)
Department of CSE, City Engineering College, Bengaluru, India
e-mail: ambikatanaji@gmail.com

A. Bharathi Malakreddy
Department of CSE, B M S Institute of Technology and Management, Bengaluru, Karnataka, India
e-mail: bharathi_m@bmsit.in

S. Smys et al. (eds.), *International Conference on Computer Networks and Communication Technologies*, Lecture Notes on Data Engineering and Communications Technologies 15, https://doi.org/10.1007/978-981-10-8681-6_37

number of features, where some are known and some are unknown [3]. Some of the examples of high-dimensional data are a large quantity of Magnetic Resonance Images (MRI), microarray data, satellite imagery, text mining, protein classification, intrusion detection, etc. The usage of such forms of data are normally evident in machine learning area and there is a technique called as Dimensionality Reduction (DR) meant exclusively for addressing the problem of the curse of dimensionality problem associated with high-dimensional data that usually evolves during the process of either structuring or analyzing the data [4]. The process of DR is mainly concerned about minimizing the arbitrary variables involved in the given set of High-Dimensional Data (HDD), where the key procedure is to apply the selection as well as extraction of feature [4]. Basically, the selection process aims for minimizing dimensionality with the elimination of unwanted data (or noise). It also aims for enhancing knowledge discovery process with better predictive accuracy. On the other hand, extraction of the feature is mainly associated with computation of representation of lower dimensional attributes using either supervised or unsupervised approaches. Similarly, the mechanism of feature reduction pertains to utilize all the source features where the processed data (or feature) exhibit linear characteristics, whereas the process of feature selection only performs a selection of a subset of source feature.

The operation involved in DR is associated with the conversion of the dataset characterized by a larger data dimension into smaller dimension by confirming that the reduced set of the data do bear the similar knowledge. Implication of DR methods offers potential advantage toward processing high-dimensional data, e.g., (i) assisting in compression data, minimizing the storage complexity, (ii) minimizing the size of dimension also directly lower any form of computational complexity associated with the data, (iii) ensuring faster response time for analyzing massive data, (iv) addresses the problems of multicollinearity in order to enhance the system performance, and (v) elimination of any form of redundant data, etc. [5]. The usual technique to carry out DR methods is exploring missing values, lower variance, decision tree, random forest, high correlation, backward feature elimination, factor analysis, Principal Component Analysis (PCA) [6]. It has been also seen that majority of the existing DR methods are mainly governed by unsupervised learning techniques. However, it should be known that the problems and research complexity does not only reside within high-dimensional data but also its associated process. With the increasing adoption of cloud, the storage complexity is reducing but the problem associated with the massive processing of data is increasing day by day. In order to ensure smoother processing of high-dimensional data, it is essential that data processing should become not only easier but also faster (reduced algorithm processing time). There are vast number of research work carried out toward addressing the problem of dimensional reduction, whereas it is still an open challenge to understand the best research contribution till date. Hence, we present the review work where less repeated theoretical discussion on DR methods have been carried out and has emphasized on highlighting the true picture of existing techniques of DR methods. The manuscript discusses the frequently adopted techniques of DR along with exploring the limitation, research problems, research techniques, open research issues, etc. This survey paper is organized as follows. In Sect. 2, we review the DR techniques and investigate the existing research contributions toward Dimensionality Reduction. In Sect. 3, we discuss research trend and research gap. Finally, the conclusive remarks are provided in Sect. 4.

2 Related Works on Dimensionality Reduction

2.1 Background

In the existing system, there are various research attempts toward approaches in DR. There are certain review works being carried out towards discussing the existing approaches of it. The work carried out by Engel et al. [7] have presented a review of DR methods in highly elaborated manner; however, the discussion carried out is more focused on theoretical aspect and the study does not offer much visualization towards the limitations of the existing system. A specific form of review work is presented by Blum et al. [8] with respect to regularization-based techniques, projection-based techniques, and selection of subset techniques. The paper has also discussed about Bayesian-based approach. The review introduced by Sorzano et al. [9] have offered comprehensive visualization of the base techniques used in DR methods while Pindah et al. [10] have discussed about DR methods explicitly from the viewpoint of biological data of high dimension.

At present, there has been a considerable amount of research work being carried out towards addressing the problems of dimensional reduction associated with the high-dimensional data. This section will briefly discuss about the standard techniques or most frequently adopted techniques for solving dimensionality reduction problem as well as some of the recent work being carried out.

2.2 Standard Techniques

Basically, the conventional process of dimensionality reduction algorithm normally consists of four stages, i.e., (i) taking the input of high-dimensional data, (ii) applying dimensional reduction techniques, (iii) obtain low dimensional data, and (iv) data for processing. However, every standard Dimensionality Reduction (DR) techniques adopted by the existing researchers are found to emphasize on feature selection operations. The significant extraction of feature offers exploring an optimal set of solution from a given set of problems. It also assists in minimizing the unnecessary data as well as a contribution to dimensionality reduction. The process of feature selection consists of the generation of subset, the evaluation process of subset, criteria of stopping, and validation [11]. At present, the various approaches of feature selection are (Fig. 1):

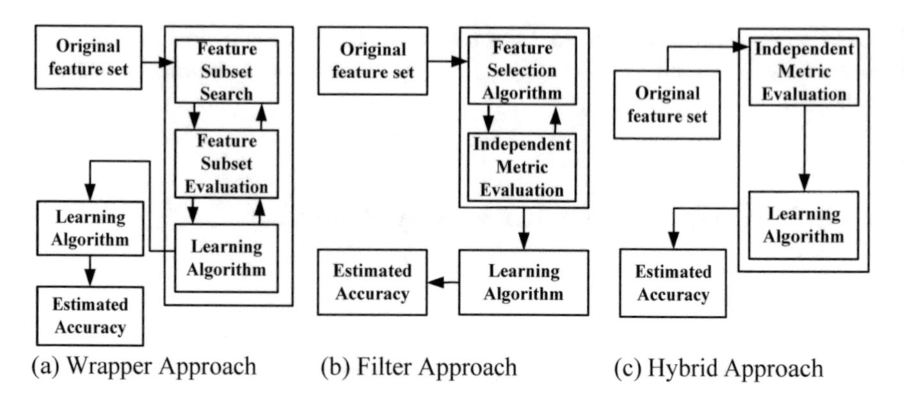

Fig. 1 Approaches of feature selection

- **Wrapper Approach**: Such approach normally uses learning mechanism in order to extract the features and solve the dimensionality reduction problem. As shown in Fig. 1a, the feature subset search algorithm explores a good subset using the induction algorithm. The final subset is selected based on the highest evaluation on which the learning algorithm runs.
- **Filter Approach**: These approaches use statistical procedures for identifying the necessary feature. The filter approach shown in Fig. 1b selects the features during the preprocessing step based on feature selection algorithm, some metrics can be used for efficient evaluation of features. The main drawback is that the features are filtered by ignoring the induction algorithm.
- **Hybrid Approach**: Such approaches are meant for managing massive dataset. It provides a mixture of multiple feature selection criteria to find and evaluate a small subset of features using some metrics to predict the result with an adequate level of accuracy as shown in Fig. 1c.

The various mechanism used for exploring the process of feature selection are (a) Symmetric uncertainty, (b) Information theory, (c) Gain Ratio, (d) Mutual Information, (e) Chi-square, (f) Correlation based. At the same time, standard methods used for extraction of features are Fischer criterion, principal feature analysis, Principal Component Analysis (PCA), and linear discriminant analysis. At present, there are three standard methods of dimensionality reduction techniques, i.e., techniques based on dictionaries (Fig. 2), techniques based on Statistics and Information Theory (Fig. 3), and techniques based on projections (Fig. 4). Figures 2–4 also showcases various subsets of techniques introduced by the three different methods. A closer look into the taxonomies of the standard techniques of the existing system shows that Principal Component Analysis (PCA) and Singular Value Decomposition (SVD) are among the most dominant techniques used in dimensionality reduction. Hence, methods based on statistics and information theory are much dominantly considered as the solution towards DR problems in high-dimensional data.

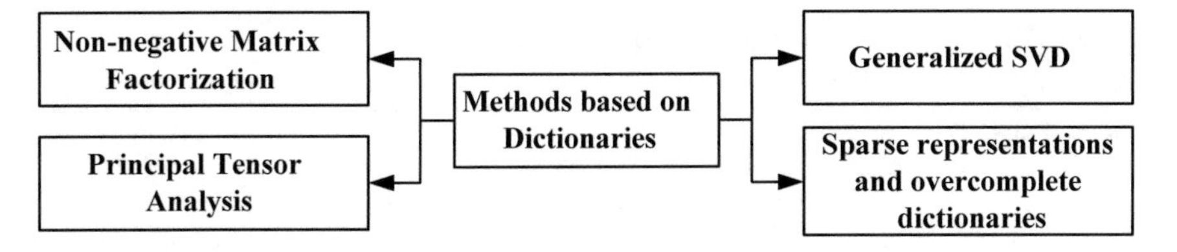

Fig. 2 Methods based on dictionaries

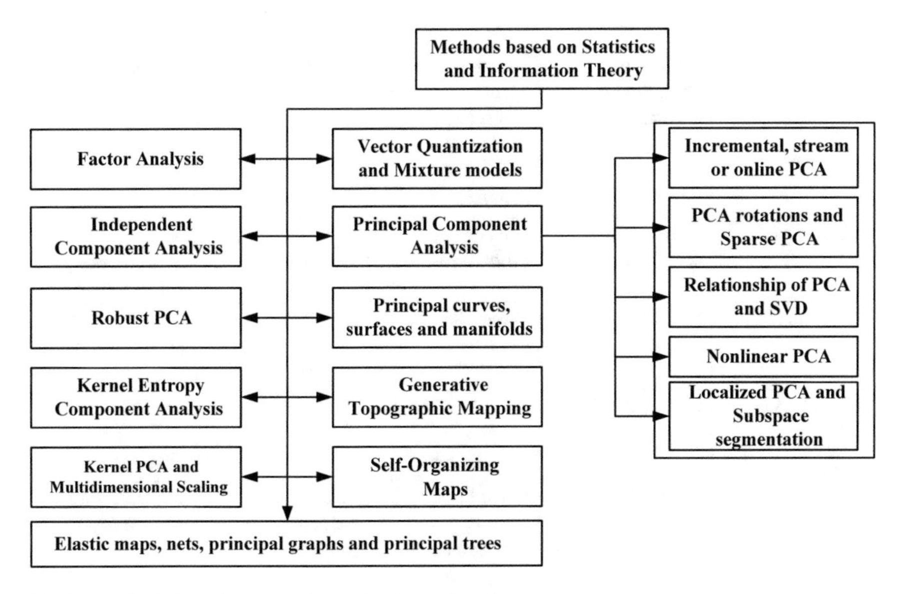

Fig. 3 Methods based on statistics and information theory

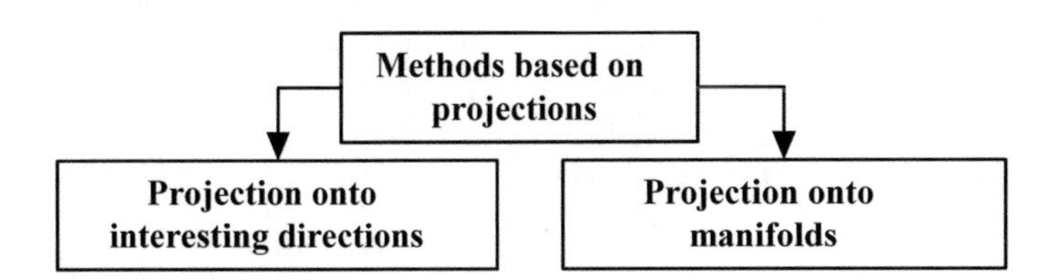

Fig. 4 Method based on projections

2.3 *Existing Research Contribution*

Apart from the standard methods of DR techniques, most recently there has been an evolution of various techniques where the problems of DR has been further broken down to anomaly detection, cross visualization, increasing accuracy, complexity reduction, etc. Table 1. Highlights some of the significant research contribution during the past few (4) years.

Table 1 Existing approaches of dimensionality reduction

Authors	Problems	Techniques	Advantage	Limitation
Hong et al. [12]	Variability of complex spectral, multicollinearity	Local manifold representation	Better classification accuracy	Does not consider SC approach
Wang et al. [13]	Non-orthogonality of locality projection	Incorporate orthogonal	Locality minimization, globality maximization	Specific to imagery data only
Wang et al. [14]	Enhancing classification accuracy	Non-negative sparse divergence, constraint discriminant analysis	Minimizes computational load	Doesn't consider SC approach
Yairi et al. [15]	Anomaly detection	Clustering, probability	Highly scalable and robust	No benchmarking of outcomes
Wan et al. [16]	Large margin DR	Target detection, stochastic approach	Excellent detection performance	Doesn't consider SC approach
Liu et al. [17]	Cross view DR	Sparse learning, automatic, nonparametric	Better stability	Variable performance on different datasets
Yuan et al. [18]	Detection accuracy	Local binary pattern, regression, component analysis	Enhanced accuracy performance	Tradeoff with complexity.
Zhao et al. [19]	Unsupervised 2D DR	Adaptive structural learning, k-means	Enhanced clustering outcomes	More processing time
Yu et al. [20]	Sensitivity of locality preserving projection	Kernel-based mapping	Better convergence performance	No much focus on subspaces
Wu et al. [21]	Redundancies in data	Principal component analysis, Hadoop, Apache Spark (AR)	Better accuracy and computational performance	Does not focus on classification much
Niu and Ma [22]	Investigation on manifold learning	Manifold learning (local), Taylor expansion, Eigenvalue	Simpler convergence performance	No SC approach, No benchmarking

(continued)

Table 1 (continued)

Authors	Problems	Techniques	Advantage	Limitation
He et al. [23]	Problems in sparse graph embedding	Weighted sparse graph	Concatenates both sparsity and locality	Includes more processing time
Chen et al. [24]	Enhancing accuracy	Sparse graph learning, unsupervised	Enhanced accuracy performance	Higher iteration offers more complexity
Cortés et al. [25]	Feature extraction, reduced complexity	Classifier design, feature selection	minimize model overtraining	High computational cost
Dong et al. [26]	DR of hyperspectral image	Metric learning (adaptive)	Reduced utilization of trained samples, efficient data handling	Higher processing time, and less computational efficient
Griparis et al. [27]	Identification and removal of redundancies	Visual descriptors	Better distinction of classes	Outcome associated with ambiguities
Huang et al. [28]	Anomaly detection	Distance-based approach, covariance, subspace	Supports both centralized and non-centralized approach	Doesn't include SC approach
Gao et al. [29]	Data redundancy	Graph-based, tensor property, patch alignment	Improved alignment performance	More scope of optimization exists
Jiang et al. [29]	Enhancing DR algorithm performance	Latent variable model (shift and rotation)	Supports properties of shift	Does not include SC approach
Lai et al. [30]	Orthogonality problems during DR	Sparsity, singular value decomposition	Geometric structure is better preserved	Iterative process and its possible impact on computational time are not discussed
Laparra et al. [31]	Minimization of reconstruction errors	Unsupervised DR, invertible transforms, Principle Component Analysis, regression	Improved data compression, enhanced prediction performance	No typical clustering is supported
Liu et al. [32]	Poor discrimination of dictionary in DR approach	Dictionary learning	Efficient storage, robust algorithm	No analysis on subspace data

(continued)

Table 1 (continued)

Authors	Problems	Techniques	Advantage	Limitation
Liang et al. [33], An et al. [32]	Determining projection vector, the constraint of orthogonality	Lagrange multiplier, decentralized DR, low-rank subspace	Supportability for distributed tensor data	Does not emphasize on algorithm complexity
Cai et al. [34]	Low-rank representation	Integrated system of DR approach, conventional classification	Better accuracy performance, reduced detection time	Still posses computational complexity

3 Research Gap

After a thorough study of the statistics of the existing studies toward high-dimensional data, we find that there are large numbers of research manuscript publication (approximately 1225 Journals, 4394 Conference papers, 75 Early Access Articles, and 7 Books in IEEE Xplore itself). A typical trend in this regard is projection-based techniques, gaining more attention with respect to the manifolds as well as dictionaries. Majority of them are claimed to offer supportability on real-time applications. On the other hand, studies using vector quantization is losing its grip on manuscript publication as well as technical adoption. Adoption of PCA-based approach with a slight improvement on it is another research trend to address DR problems. However, the present ongoing research activity is found not definite to stick on to any one specific problem but the problems have become more modular. Research gaps that could be addressed based on the survey made as listed in Table 1:

- *More inclined to hyperspectral image*: It is known that hyperspectral image is bigger in size and dimension with more complexity. However, there are various other forms of data apart from it that can be also said to be high-dimensional data that is not explored much in the existing system.
- *Ignorance to Subspace Clustering* (*SC*): A closer look into Table 1 will show that majority of the existing approaches for addressing DR problems has not used subspace clustering approach, which is one of the effective and smart mechanism of exploring information from high-dimensional data.
- *Lack of Benchmarked Models*: The area of research in high-dimensional data is quite old from a theoretical viewpoint and very novel in practical implementation. Hence, there are very few benchmarked models to be found.
- *Lesser Optimization*: Need of optimization algorithm is very high in solving DR problems in order to minimize the computational complexity and obtaining better accuracy.

4 Conclusion

This paper discusses the study and reviews of the existing approaches for solving DR problems in high-dimensional data, where it is explored that there are a large number of scattered work being carried at present. The usage of standard techniques, e.g., dictionary-based, statistics-based, projection-based, etc., is more dominant at present. However, an essential logic of solving DR problems using subspace clustering and also the work carried out using subspace clustering toward DR problem is very rare to find in the existing system. The present subspace clustering is more focused on improving clustering performance and not much research has been reported to assist in solving DR problems. Hence, our research work will be in the direction of using subspace clustering for solving DR problems associated with high-dimensional data from both computational efficiency and scalable clustering viewpoint.

References

1. Kumar, P., Tiwari, A.: Ubiquitous Machine Learning and Its Applications. IGI Global (2017)
2. Michaelis, S., Piatkowski, N., Stolpe, M.: Solving Large Scale Learning Tasks. Challenges and Algorithms. Springer (2016)
3. Frigessi, A., Buhlmann, P., Glad, I., Langaas, M., Richardson, S., Vannucci, M., Statistical Analysis for High-Dimensional Data. Springer (2016)
4. Bolón-Canedo, V., Sánchez-Maroño, N., Alonso-Betanzos, A.: Feature Selection for High-Dimensional Data. Springer (2015)
5. Song, Y., Li, Q., Huang, H., Feng, D., Chen, M., Cai, W.: Low dimensional representation of fisher vectors for microscopy image classification. IEEE Trans. Med. Imagin. **36**(8), 1636–1649 (2017)
6. Ray, S.: Beginners guide to learn dimension reduction techniques. https://www.analyticsvidhya.com/blog/2015/07/dimension-reduction-methods/. Accessed 7 Aug 2017
7. Engel, D., Hüttenberger, L., Hamann, B.: A survey of dimension reduction methods for high-dimensional data analysis and visualization. OpenAccess Series in Informatics (2012)
8. Blum, M.G.B., Nunes, M.A., Prangle, D., Sisson, S.A.: A comparative review of dimension reduction methods in approximate bayesian computation. Stat. Sci. **28**(2) (2013)
9. Sorzanol, C.O.S., Vargas J., Pascual Montanol, A.: A survey of dimensionality reduction techniques. arXiv, vol. 28, no. 2 (2014)
10. Pindah, W., Nordin, S., Seman, A., Said, M.S.M.: Review of dimensionality reduction techniques using clustering algorithm in reconstruction of gene regulatory networks. In: 2015 International Conference on Computer, Communications, and Control Technology (I4CT), Kuching, pp. 172–176 (2015)
11. Liu, C., Wang, W., Zhao, Q.: A New Feature Selection Method Based on a Validity Index of Feature Subset, vol. 92. Elsevier (2017)
12. Hong, D., Yokoya, N., Zhu, X.X.: Learning a robust local manifold representation for hyperspectral dimensionality reduction. IEEE J. Sel. Topics Appl. Earth Obs. Remote Sens. **10**(6), 2960–2975 (2017)
13. Wang, R., Nie, F., Hong, R., Chang, X., Yang, X., Yu, W.: Fast and orthogonal locality preserving projections for dimensionality reduction. IEEE Trans. Image Process. **26**(10), 5019–5030 (2017)
14. Wang, X., Kong, Y., Gao, Y., Cheng, Y.: Dimensionality reduction for hyperspectral data based on pairwise constraint discriminative analysis and nonnegative sparse divergence. IEEE J. Sel. Topics Appl. Earth Obs. Remote Sens. **10**(4), 1552–1562 (2017)

15. Yairi, T., Takeishi, N., Oda, T., Nakajima, Y., Nishimura, N., Takata, N.: A data-driven health monitoring method for satellite housekeeping data based on probabilistic clustering and dimensionality reduction. IEEE Trans. Aerosp. Electron. Syst. **53**(3), 1384–1401 (2017)
16. Wan, L., Zheng, L., Huo, H., Fang, T.: Affine invariant description and large-margin dimensionality reduction for target detection in optical remote sensing images. IEEE Geosci. Remote Sens. Lett. **14**(7), 1116–1120 (2017)
17. Liu, H., Liu, L., Le, T.D., Lee, I., Sun, S., Li, J.: Nonparametric sparse matrix decomposition for cross-view dimensionality reduction. IEEE Trans. Multimed. **19**(8), 1848–1859 (2017)
18. Yuan, F., Xia, X., Shi, J., Li, H., Li, G.: Non-linear dimensionality reduction and gaussian process based classification method for smoke detection. IEEE Access **5**, 6833–6841 (2017)
19. Zhao, X., Nie, F., Wang, S., Guo, J., Xu, P., Chen, X.: Unsupervised 2D dimensionality reduction with adaptive structure learning. Neural Comput. **29**(5), 1352–1374 (2017)
20. Yu, Q., Wang, R., Li, B.N., Yang, X., Yao, M.: Robust locality preserving projections with cosine-based dissimilarity for linear dimensionality reduction. IEEE Access **5**, 2676–2684 (2017)
21. Wu, Z., Li, Y., Plaza, A., Li, J., Xiao, F., Wei, Z.: Parallel and distributed dimensionality reduction of hyperspectral data on cloud computing architectures. IEEE J. Sel. Topics Appl. Earth Obs. Remote Sens. **9**(6), 2270–2278 (2016)
22. Niu, G., Ma, Z.: Local non-linear alignment for non-linear dimensionality reduction. In IET Comput. Vis. **11**(5), 331–341, 8 (2017)
23. He, W., Zhang, H., Zhang, L., Philips, W., Liao, W.: Weighted sparse graph based dimensionality reduction for hyperspectral images. IEEE Geosci. Remote Sens. Lett. **13**(5), 686–690 (2016)
24. Chen, P., Jiao, L., Liu, F., Gou, S., Zhao, J., Zhao, Z.: Dimensionality reduction of hyperspectral imagery using sparse graph learning. IEEE J. Sel. Topics Appl. Earth Obs. Remote Sens. **10**(3), 1165–1181 (2017)
25. Cortés, G., Benítez, M.C., García, L., Álvarez, I., Ibanez, J.M.: A comparative study of dimensionality reduction algorithms applied to volcano-seismic signals. IEEE J. Sel. Topics Appl. Earth Obs. Remote Sens. **9**(1), 253–263 (2016)
26. Dong, Y., Du, B., Zhang, L., Zhang, L.: Exploring locally adaptive dimensionality reduction for hyperspectral image classification: a maximum margin metric learning aspect. IEEE J. Sel. Topics Appl. Earth Obs. Remote Sens. **10**(3), 1136–1150 (2017)
27. Griparis, A., Faur, D., Datcu, M.: Dimensionality reduction for visual data mining of earth observation archives. IEEE Geosci. Remote Sens. Lett. **13**(11), 1701–1705 (2016)
28. Huang, T., Sethu, H., Kandasamy, N.: A new approach to dimensionality reduction for anomaly detection in data traffic. IEEE Trans. Netw. Serv. Manage. **13**(3), 651–665 (2016)
29. Gao, Y., Wang, X., Cheng, Y., Wang, Z.J.: Dimensionality reduction for hyperspectral data based on class-aware tensor neighborhood graph and patch alignment. IEEE Trans. Neural Netw. Learn. Syst. **26**(8), 1582–1593 (2015)
30. Lai, Z., Wong, W.K., Xu, Y., Yang, J., Zhang, D.: Approximate orthogonal sparse embedding for dimensionality reduction. IEEE Trans. Neural Netw. Learn. Syst. **27**(4), 723–735 (2016)
31. Laparra, V., Malo, J., Camps-Valls, G.: Dimensionality Reduction via Regression in Hyperspectral Imagery. IEEE J. Sel Topics Signal Process. **9**(6), 1026–1036 (2015)
32. Liu, L., Ma, S., Rui, L., Lu, J.: A novel locality constrained dictionary learning for nonlinear dimensionality reduction and classification. IET Comput. Vis. (2017)
33. Liang, J., Yu, G., Chen, B., Zhao, M.: Decentralized dimensionality reduction for distributed tensor data across sensor networks. IEEE Trans. Neural Netw. Learn. Syst. **27**(11), 2174–2186 (2016)
34. Cai, Z., Shen, C., Guan, X.: Mitigating behavioral variability for mouse dynamics: a dimensionality-reduction-based approach. IEEE Trans. Human-Mach. Syst. **44**(2), 244–255 (2014)

P. R. Ambika received her B.E. degree and M.Tech. degree in Computer Science and Engineering from Visvesvaraya Technological University (VTU), Belgaum, Karnataka. She is a Research Scholar in the Department of CSE at BMSIT&M, Bengaluru under Visvesvaraya Technological University. Currently, she is working as an Assistant Professor in the Department of CSE, City Engineering College, Bengaluru, Karnataka. Her areas of interest includes data mining, data science, big data analytics and Machine learning.

Dr. A. Bharathi Malakreddy has B.E. degree in Computer Science and Engineering, M.Tech. in CSE and Ph.D. in Computer Science and Engineering, University of JNTU Hyderabad. Presently working as Professor Department of CSE at BMSIT&M, Bengaluru. Worked as visiting faculty for IGNOU and Infosys campus connect program. Areas of interest are Wireless Sensor Network, Medical Imaging, IoT, Big Data and Cloud Computing. Published papersin IEEE, Springer, Elsevier and LNSE.

IoT-Based Patient Remote Health Monitoring in Ambulance Services

C. M. Lolita, R. Roopalakshmi, Sharan Lional Pais, S. Ashmitha, Mashitha Banu and Akhila

Abstract Ambulatory health care is a type of remote patient monitoring that allows a medical caretaker to use the medical device in the ambulance to perform a routine test and send the test data to a healthcare professional in real time. Even though there are various methods to observe the health condition of the patient at home or in the hospital, the necessity of the quick measures to treat the person in case of emergencies are not yet fulfilled. If the person suddenly falls ill and is being carried to the hospital, the doctor will get to know the condition or the cause of the illness only after diagnosing the patient which will consume more time. There is a need for monitoring technology in ambulances since in case of emergency lots of time is wasted in carrying patient to the hospital and diagnosing. To overcome this problem, online system for remoting health parameters of a patient in the ambulance is proposed in this paper . The experiment is conducted to compare the system values with the values obtained by the standard devices and the results are in a good format and the system is efficient.

Keywords Health care · Ambulatory services · Patient monitoring

1 Introduction

The act of taking preventative or necessary medical procedures to improve a person's well-being is health care. Health care is one of the major challenges for the mankind. Presently, various wearable devices are available to remotely monitor the health

C. M. Lolita (✉) · R. Roopalakshmi · S. L. Pais · S. Ashmitha · M. Banu · Akhila
Alvas Institute of Engineering and Technology, Shobhavana Campus, Mijar, Moodbidri, Mangalore 574225, India
e-mail: lolitacmenezes@gmail.com

R. Roopalakshmi
e-mail: drroopalakshmir@gmail.com

S. L. Pais
e-mail: lionalpais@gmail.com

S. Smys et al. (eds.), *International Conference on Computer Networks and Communication Technologies*, Lecture Notes on Data Engineering and Communications Technologies 15, https://doi.org/10.1007/978-981-10-8681-6_38

parameters of the patients. Precisely, ambulatory health care is a type of remote patient monitoring that allows a medical caretaker to use the medical device in the ambulance to perform a routine test and send the test data to a healthcare professional in real time. However, due to poor infrastructure and undisciplined drivers, ambulances were not able to reach the place of emergency on time. For example, as per All India Institute of Medical Sciences in 2012, 28% of emergency calls received between March 2009 and May 2010 in the city were refused because of the shortage of time for ambulances to reach to the place of emergency. Further, WHO suggests that there should be at least 1 ambulance per 100,000 people [1]. But in Indian cities, there is 1 ambulance for every 144,736 people, which is of the ratio 1:1.44. Even though there are various methods to observe the health condition of the patient at home or in the hospital, the necessity of the quick measures to treat the person in case of emergencies are not yet fulfilled. For example, if the person suddenly falls ill and being carried to the hospital, the doctor will get to know the condition or the cause of the illness only after diagnosing the patient which will consume more time. To avoid the time delay incurred in carrying and diagnosing is minimized by providing a primary diagnosis in the ambulance. The doctor can even suggest the quick measures to be taken in case of the emergencies. To reduce the delay in primary diagnosis, the concept of IoT can be used. Using IoT, a system can be developed where primary diagnosis is done in the ambulance and save many lives.

2 Related Work

In 2012, Gundale and Kamble [2] proposed Wireless Data logger for Health Monitoring System. A low-cost Environmental Parameter Monitoring System is developed to give clearer and more detailed view of indoor air quality and will be beneficial in many low-cost applications. However, the drawback is the initial cost of purchasing the equipment. In 2015, Katake and Kute [3] proposed Health Monitoring Management Using Internet of Things (IoT). Patients diagnose can be monitored in the monitor screen of a computer using Raspberry Pi. However, in the proposed system Raspberry Pi is useful but its memory is limited, so large data cannot be stored. In 2015, Phyo et al. [4] proposed Wireless Patient Monitoring System Using Point to Multi-point ZigBee. ZigBee sensor network for data acquisition and monitoring is presented in this System. Data collected by sensor devices are sent to the base station PC, which is set as Wireless Sensor Network. Sorte et al. [5] proposed ZigBee and GSM-based Health Monitoring System. This system will help in monitoring the patients centrally over a large scale. However, ZigBee technology is not secure like Wi-Fi-based secure system and replacement of ZigBee component costs more. In 2015, Hassan Alieragh et al. [6] proposed Health Monitoring and Management Using Internet-of-Things (IoT). Sensing with Cloud-based Processing reviewed the current state and projected future directions for integration of remote health monitoring technologies into the clinical practice of medicine. However, collecting health parameters remotely and storing in the cloud lacks in security and privacy. In 2015, Zanjala et al. [7] proposed Medicine Reminder and Monitoring System for Secure Health, which implements to improve the efficiency of prescribed drug and reduce economic factor. However, if the system fails then it will not alarm at the right time and elderly persons may forget to take medicine. In 2017, Sathya and Kumar [8] proposed a secured remote health monitoring system implements the medical sensor senses the patient's physiological data and transmits them over the wireless channels which are more susceptible than wired networks. However, remote parameters of patients were not accurate and efficiency was less. To summarize, parameter accuracy in the existing systems were not sufficient due to noise, some results showed deflections and were appeared many times. There is a need for monitoring technology in ambulances since in case of emergency lots of time is wasted in carrying a patient to the hospital and diagnosing. To overcome this problem online system for remoting health parameters of a patient in the ambulance is proposed in this paper.

3 Proposed Framework

The online system to remote health parameter using IoT is a combination of web as well as Android application. The proposed system will be reliable, and is an energy-efficient patient monitoring system, it enables the doctors to monitor patient's health parameters in real time. In the proposed system, the patient health is constantly monitored in ambulance using different sensors which are connected to the Arduino board and the acquired data is sent to the server using the Ethernet shield attached to the Arduino board and it also sent to an Android application installed in the doctor's smartphone. The patient will be carrying hardware having sensors and the sensors will sense the parameter value of patient and these data are transferred to Android smartphone via Wi-Fi. The system has the database which stores all information about patient's health and the doctors will prescribe medicine using this information stored.

Figure 1 describes different input components to be used. Pulse rate acquisition of a body can be counted by a change in blood flow in blood vessels. Body temperature is depending on the activities performed by the human body, which measures temperature. Blood pressure acquisition is the pressure of circulating blood on the walls of blood vessels. Sugar-level acquisition is to measure glucose levels in tissue fluid. Collected data is processed and sent to the monitor placed in ambulance parallel data is sent to the Hospital Database as well as Android application used by the doctor.

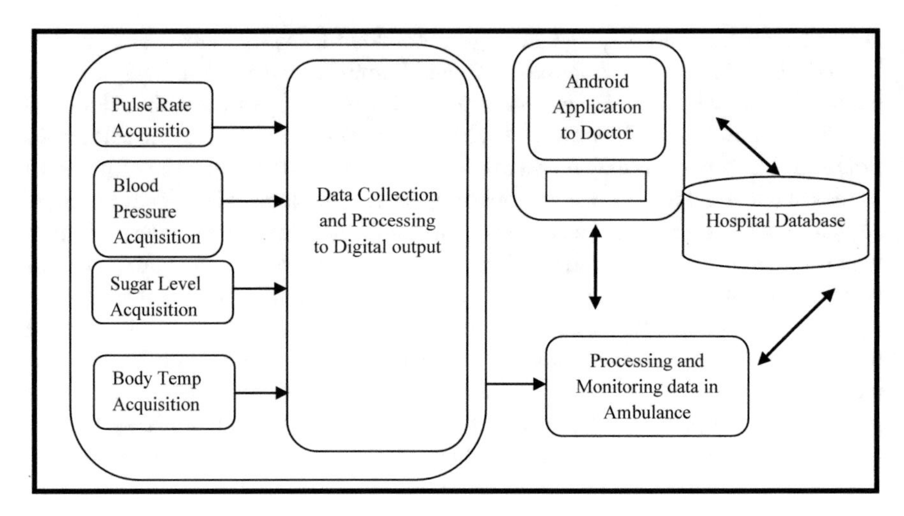

Fig. 1 Block diagram

Sending Patient data from Ambulance to Doctor is represented in the following steps.

Algorithm

Step 1: Admin login.
Step 2: Ambulance and doctor registration by admin.
Step 3: Ambulance login and add patient's details.
Step 4: Monitor the parameters using the device.
Step 5: Update the parameters.
Step 6: Collected data sent to the server.
Step 7: Data received in the designed Android application.
Step 8: Send a message to the doctor and doctor can give a quick response.

The module is used to send the information about the patient's health to the priorly registered hospital. Using various sensors, the readings will be recorded by the system which will be available inside the ambulance. The data stored in the server will be available to the hospital management and doctor, so that they can make necessary prerequisites till patient reaches the hospital. Taking patient from his place to the hospital in case of heart-related diseases is the most difficult task. The doctor can guide which actions should be taken until the patient reaches hospital.

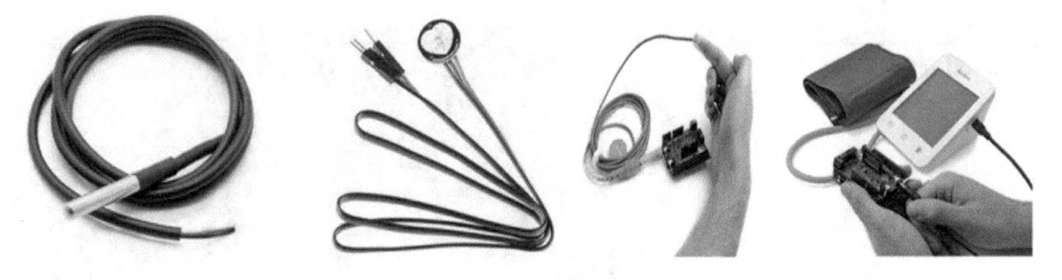

Fig. 2 Different sensors

4 Experimental Setup

Figure 2 represents different sensors including body temperature, pulse rate, glucose level, and blood pressure sensors. The change in body temperature depends on the activities performed by the human. The output of temperature sensors is proportional to the temperature in Celsius. The pulse rate of a body can be counted by a change in blood flow in blood vessels. Heartbeat sensor is designed to give a digital output of heat beat when a finger is placed. This digital output can be connected to Arduino directly to measure the beats per minute rate. Glucose-level sensor is a way to measure glucose levels in real time. A tiny electrode is inserted under the skin to measure glucose levels. It is the way of testing the concentration of glucose in the blood. Blood pressure (BP) is the pressure of circulating blood on the walls of blood vessels. Blood pressure is usually expressed in terms of the systolic pressure, i.e., maximum during one heartbeat and diastolic pressure minimum in between two heartbeats and is measured in millimeters of mercury (mmHg), above the surrounding atmospheric pressure.

Figure 3 represents the snapshot of the proposed framework. The digital outputs can be connected to Arduino. Body temperature sensor DS18B20 communicates over a 1-Wire bus that by definition requires only one data line for communication with a central microprocessor. The processed data are then transmitted to the server by wireless transmission using Ethernet module. The data is accessed through the server using a mobile phone. Using Android application, the doctor can view his patient's medical history date wise, event wise, etc. The collected data is communicated to the doctor through Android application.

5 Results and Discussion

Figure 4 illustrates how the patient's data is sent to the server. A well-functioning system prototype was built composed of the following hardware components: pulse rate, temperature , blood glucose, blood pressure sensor, and Arduino integrated

Fig. 3 Snapshot of the proposed framework

Fig. 4 Sending the patient data to the server

together to monitor the vital parameters of the patient. When the patient is being carried to the hospital the details of the patient is collected and stored in the database. Then, the device consisting of various sensors is connected to the patient's body to monitor the vital parameters. A connection is established between the device and the server. After successful establishment of the connection, the values are captured.

Figure 5 represents the patient details displayed in the Android application. The vital parameters of the patient can be accessed through the application installed in the doctor's Android phone. The doctor can also suggest quick measures to be taken in case of emergencies. Figures 6 and 7 represent the patient details displayed in the monitor page of the hospital and ambulance, respectively. The hospital management can access the parameters of the patients using the web page. The web page displays the id of the ambulance in which the patient is carried, vehicle number, and the

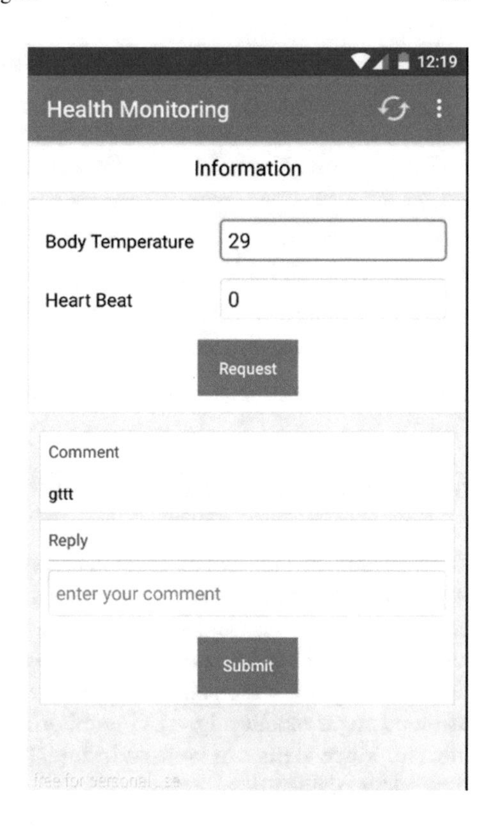

Fig. 5 Patient details displayed in the android application

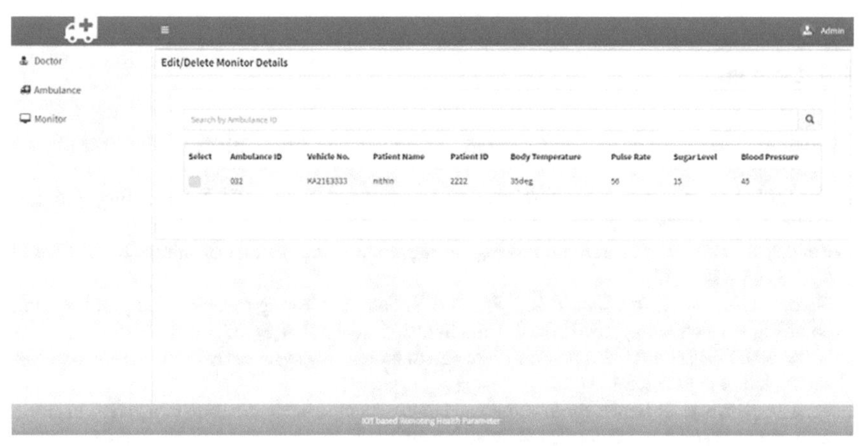

Fig. 6 The patient details displayed in the monitor page of the hospital

parameters of the patient. Using this information the required measures can be taken by the hospital to treat the patient as quickly as possible.

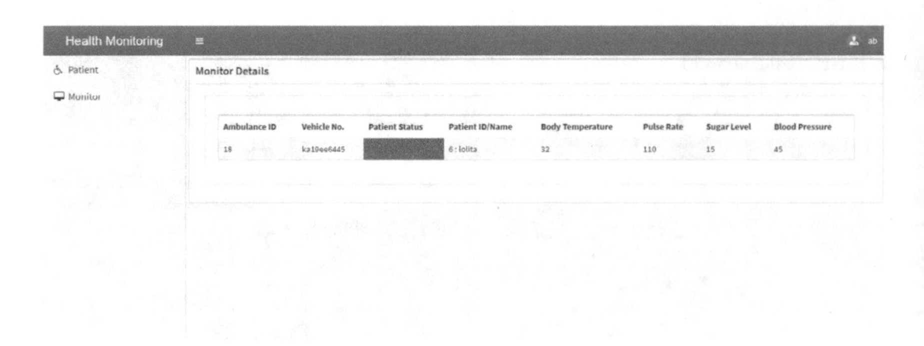

Fig. 7 The patient details displayed in the monitor page of ambulance

6 Conclusion and Future Work

To overcome the problem of delay in carrying a patient to the hospital and diagnosing, online system for remoting health parameters of a patient in the ambulance is proposed. It is an efficient system and easy to handle and thus provides flexibility and serves as an improvement over other conventional monitoring systems. The system able to monitor the patient health in real time. In future, other parameters such as blood sugar monitoring, ECG monitoring, blood group detection can be implemented. Voice alerts can be used to initiate the various controlling of devices and their status of operation.

References

1. Indian ambulance emergency. https://blogs.wsj.com/indiarealtime/2014/10/16/indias-ambulance-emergency/
2. Gundale, A.S., Kamble, P.A.: Wireless data logger using ZigBee. IEEE Int. J. Eng. Sci. Res. Technol. **3**(8), 1799–1802 (2014)
3. Katake, R., Kute, B.: Health monitoring management using internet of things. IEEE Sens. J. **5**(5), 1144–1147 (2015)
4. Phyo, S., Tunand, Z.M., Tun, H.M.: Wireless patient monitoring system using point to multi point Zigbee technology. IEEE Int. J. Eng. Sci. Res. Technol. **4**(6), 271–274 (2015)
5. Sorte, S.: ZIGBEE and GSM based patient health monitoring system. IEEE Int. J. Eng. Sci. Res. Technol. **3**, 445–451 (2016)
6. Page, A., Soyata, T.: Health monitoring and management using internet-of-things (IoT) sensing with cloud-based processing. In: Proceeding of IEEE International Conference on Services Computing, vol. 47, no. 4, 285–292 (2015)
7. Zanjala, S.V., Talmaleb, R.: Medicine reminder and monitoring system for secure health using IOT. In: Proceeding of Elsevier International Conference on Information Security and Privacy, vol. 7, no. 3, pp. 471–476 (2015)
8. Sathya, D., Kumar, P.G.: Secured remote health monitoring system. Proc. Int. Comput. Conf. **4**(5), 228–232 (2017)

C. M. Lolita Studying in the Alvas Institute of Engineering and Technology. Her research area includes IoT and health care applications.

Dr. R. Roopalakshmi is currently working as Professor in Information Science and Engineering (ISE) Department of Alva's Institute of Engineering and Technology affiliated under VTU, Belgavi. She has started her Engineering career, Bachelor of Engineering (B.E.) from A.M.S. College of Engineering, Periyar University and completed her M.Tech. (CSE) from P.E.S. Institute of Technology (PESIT)—Bangalore. Followed by her post graduation, she has pursued her Full-time Ph.D., in Information Technology Department of National Institute of Technology Karnataka (NITK)—Surathkal, which is one of the most prestigious and oldest NITs of India. She has published 5 journal articles in very high quality journals such as Elsevier Signal Processing, Springer Signal, Image and Video Processing and so on. She is technically sound enough for having her Paper presentations in more than 10 Tier-I International Conferences like SPCOM-2012 (which was held at IISc, Bangalore), IMSAA-2011 (held at IIIT-Bangalore), ANT-2011 (conducted at Niagara Falls, Canada) and SIRS-2014 (held at IIIT-Kerala). Henceforth, her research articles are widely available in digital libraries such as IEEE Xplore, ACM, Elsevier science direct and Springer. She proved to be a Best Researcher, by involving herself as Principal Investigator in Funded research projects such as DST-Women Scientist Research Grant of worth Rs. 18.20 lakhs from MHRD of India, New Delhi. Further, she is also honoured with various Awards such as SSLC—Salem District First Student Award, Rajya Puraskar Award from President of India for Bharat Scouts and Guides and Best Paper Award in International Conference and so on. She is also recipient of various state and national level scholarships such as National Merit scholarship, GATE scholarship and Rural Development scholarship. Currently, she is guiding 4 doctoral candidates at Visvesvaraya Technological University (VTU), Karnataka and also received guideship from Anna University, Chennai.

Sharan Lional Pais working in the Alvas Institute of Engineering and Technology. Her research area includes IoT and health care applications.

S. Ashmitha Studying in the Alvas Institute of Engineering and Technology. Her research area includes IoT and health care applications.

Mashitha Banu Studying in the Alvas Institute of Engineering and Technology. Her research area includes IoT and health care applications.

Akhila Studying in the Alvas Institute of Engineering and Technology. Her research area includes IoT and health care applications.

Design of CR-OFDM in 900 MHz Band

S. B. Mule and G. C. Manna

Abstract Research shows that there are few GSM channels available for the assignment of control channels and remaining channels dynamically are available for service as secondary channels users of CR network. For the design of CR-OFDM in 900 MHz GSM band, it is necessary that the channel shall be considered as frequency selective and time variant within the scope of practical limits. The OFDM parameters need to be calculated within the availability of vacant GSM bands, that is, 200 kHz which is very narrow in comparison with the existing bands deploying OFDM technology that is above and in multiples of 1.25 MHz. The results have been obtained and compared using three methods: (a) Measurement of OFDM efficiency at 2600 MHz band and GSM spectral efficiency at 900 MHz band both at Chakan, Pune to ascertain the geographical location. This will help to map 2600 MHz data to 900 MHz band for OFDM. The Atoll planning tool is used with the same geographical location parameters at 2600 MHz OFDM technology and is validated with the obtained results. The GSM 900 MHz RF parameters were used for simulation of Atoll parameters using OFDM technology, (b) Parameters pertaining to frequency selective and time-variant channel, viz., coherence bandwidth and Doppler delay were calculated based on the practical data available for Indian Geographical locations and matched with the ITU Table for same parameters. The OFDM parameters for GSM 900 MHz band with 200 kHz bandwidth was calculated subject to the limitation imposed by OFDM technology. (c) The obtained result was verified using Atoll planning tool and finally, the spectral efficiency was calculated. Although OFDM subcarriers are very low in size to combat coherence bandwidth, not much research work has been done to use OFDM technology for narrow bandwidth. The present work shows the design of cognitive OFDM system for frequency selective and time-variant channel in 900 MHz GSM band. The suggestions are based on the calculation done and the design has been verified using data collected for live network and results are sum-

S. B. Mule (✉)
Department of Electronics Engineering, G. H. Raisoni College of Engineering, Nagpur, India
e-mail: mulesb1@gmail.com

G. C. Manna
Indian Telecommunication Service, BSNL, Jabalpur, India
e-mail: gcmanna@gmail.com

S. Smys et al. (eds.), *International Conference on Computer Networks and Communication Technologies*, Lecture Notes on Data Engineering and Communications Technologies 15, https://doi.org/10.1007/978-981-10-8681-6_39

marized on the basis of adaptive CP length, finally, the result has been calculated using three methods, i.e., from live network, from mathematical calculations, and from Atoll tool has been summarized.

Keywords OFDM · GSM · Cyclic prefix (CP) · CR

1 Introduction

Since the introduction of mobile technologies, the demand increased data rate compels the designer to shift design from sub-GHz band to higher frequency bands. Higher carrier frequency levels provide higher bandwidths which can be allocated to a user in demand. Higher bandwidth provides higher data capacity. Accordingly, ITU published IMT 200 specifications based on WCDMA technologies, which cover bandwidth in the range of 2.5, 5, 10 MHz, etc., in FDD mode in 2000 MHz band. This technology is popularly known as third-generation (3G Mobile Communication). The original specification was meant to provide data rate of 2 Mbps. WCDMA technologies were further improved to provide asymmetric data rate during download and upload resulting in peak download throughput of 14.4 Mbps. However, the design suffered from the radio impairments of intersymbol interference when the attempt was made to improve data rate.

2 Literature Survey

Data rate improvement process makes that the Cyclic Prefix size may double channel Root Mean Square delay spread [1]. In [2], it was suggested to change sampling rate to match the CP length according to the radio condition. The problem with this method is it required a very large bandwidth. In older papers, the Power Delay Profile (PDP) is in negative exponential decaying for the channels fitted into a Negative Exponentially Decaying Profile (NEDP). For variable CP [3–7] length, a mathematical formula is derived by the help of NEDP for different radio conditions and environment. The calculation for the CP length needed for GSM 900 band cellular phone condition difficulty is studied and it was tried to remove in this study. In this research, changing Cyclic Prefix size introduced on the thought that of the end sample (tap) in Power Delay Profile (PDP) of the radio channel, where the PDP is not taken as the NEDP. It suggests Cyclic Prefix size is governed by Maximum Excess Delay Spread of the channel.

A way by round-robin arrangement known as the Cyclic Prefix is appended to the start Orthogonal Frequency Division Multiplexing symbols. In view to remove intersymbol interference and inter-carrier interference, the Cyclic Prefix size should be larger. In comparison to the delay spread in a number of channels [8], this method of the Cyclic Prefix contributes to lower bandwidth efficiency, which reduces systems data rate. As CP contains unuseful data, which makes transmitter to consume more power. Also, the Signal-to-Noise Ratio (SNR) reduces by addition of a cyclic prefix, the bandwidth efficiency is low. Hence, the choice of the Cyclic Prefix needs optimization. Typically, it is chosen based on the given operating radio conditions and the multipath channel duration [9]. Earlier OFDM WiMAX system in around 2600 MHz band uses big and fixed Cyclic Prefix length to manage the poorest type of channel and radio environment. Static CP length does not consider instant radio channel conditions, which makes channel less efficient and also wastes transmitter energy. If the design of receiver is dependent on the in-built delay spread, cellular moving phone will get multiple ISIs, as it remains in moving condition. The design of the receiver consumes more power and hence reduces battery life. Hence, it is desired to alter the CP length according to the radio conditions. References [9–11] OFDM deploys Cyclic Prefix such that it is chosen multiple times larger in comparison to Root Mean Square delay spread in the existing radio condition. Under such conditions, valuable radio spectrum is not used properly and from a device perspective, the battery is misused. Thus, it is required that present design should use optimum Cyclic Prefix size to GSM 900 band of 200 kHz bandwidth, that can achieve adequate spectral radio efficiency. "In 2004, Zhang proposed a method to alter the CP length based on the channel in the channel delay spread". In [12] OFDM for cognitive radio was studied for 802.11a/g, 802.16 and LTE. In [13], cognitive radio for next-generation technology was studied for 802.22 in 614–698 MHz. So, it is concluded that CR-OFDM was studied and implemented for several bands in 700 MHz, ISM 2.4 GHz band, 2.6 GHz bands, 3.3 GHz band, and 5.4 GHz ISM band. There is ample scope for design and implementation of cognitive radio in 900 MHz band using the OFDM technology [14].

From the above discussion, it is clear that the no attempt has been made to study for CR-OFDM at 900 MHz for data throughput, channel efficiency, and spectral efficiency.

3 Design Parameters for OFDM System in GSM 900 MHz

In a mobile communication system, there are three objects, viz., the transmitter, receiver, and objects of the environment, where any one or all can move. Such a heterogeneous situation gives rise to delay spread of the transmitted signal as perceived by receiver and Doppler spread.

Lemma *Determination of symbol time and subcarrier frequency bandwidth is the prime issues for which the two impairment factors and available bandwidth of the spectrum are required to be taken into consideration. Delay spread introduces Inter-symbol Interference (ISI), whereas Doppler spread introduces Inter-Carrier Interference (ICI). The Symbol Time (T_s) must be such that*

$$T_s \gg Delay\,spread$$
$$and \quad T_s \ll Coherence\,time\,due\,to\,Doppler\,spread$$

(1) ***Delay Spread***

Delay spread is dependent on exponential power delay profile, which is measured under different environmental conditions. Statistical path loss model can be written as

$$L = L_0 + 10\gamma \log_{10}(d/d_0) + \sigma \tag{1}$$

where L_0 =path loss in dB at a distance of d km from base station provided a minimum of d_0 km has already been crossed and σ is the RMS deviation. Here, γ is the path loss exponent and had a value of 2 in free space.

Let us consider a simple geometry, where d is the line of sight distance from a transmitter at a location where the receiver is located. An impulse signal transmitted by the transmitter is received by the receiver at time t and suffers a path loss L_1. The same signal is received at the same location by the receiver when $d + r$ is the distance traversed including reflected path at time $t + \tau$. Let us consider a signal level loss of 10 db for 90% reduction.

The additional distance r covered can then be written as

$$10 = 10\gamma \log_{10}(d + r/d)$$

i.e.,

$$\text{i.e.,} \quad r = d * \left(10^{1/\gamma} - 1\right) \tag{2}$$

and

$$\begin{aligned} \tau = \tau\text{max} &= r/C \\ &= \left\{d * \left(10^{1/\gamma} - 1\right)\right\}/C \end{aligned} \tag{3}$$

where

τ_{max} Maximum value of delay spread and
C Velocity of light km = $3 * 10^5$ km/s

Let us consider three different types of environments, e.g., Urban General, Rural Nomadic, and Rural Vehicular for practical calculation delay spread and Coherent time. The results for maximum delay spread τ using Eq. (3) have been shown in Table 1. The delay may be considered as maximum delay spread and represented as τ_{max} for all practical purposes. These values are at par with different values of τ_{rms} under different environments shown in Table 2, when we consider $\tau_{max} = 5 * \tau_{rms}$ as considered for different calculations of OFDM design. In Table 1 of present design, τ_{max} rural vehicular environment is 21.71 μs and for vehicular environment τ_{max} =

Table 1 Maximum delay spread in different environments

Environment	d in km	C	Γ	r in km	τ in secs
Urban general	3	300000	4.5	2.004302	6.68101E−06
Rural nomadic	5	300000	3.5	4.653489	1.55116E−05
Rural vehicular	7	300000	3.5	6.514884	2.17163E−05

Table 2 Power delay profile of ITU-A channel

Tap No.	Indoor A		Pedestrian A		Vehicular A	
	Delay (ns)	Power (dB)	Delay (ns)	Power (dB)	Delay (ns)	Power (dB)
1	0	0	0	0	0	0
2	50	−3.0	110	−9.7	310	−1.0
3	110	−10.0	190	−19.2	710	−9.0
4	170	−10.0	410	−22.8	1090	−10.0
5	290	−26.0	NA	NA	1730	−15.0
6	310	−32.0	NA	NA	2510	−20.0
	τRMS = 35 ns		τRMS = 45 ns		τRMS = 370 ns	

$5 * \tau_{rms} = 5 * 4000$ ns = 20 μs. This shows that maximum delay spread as per Eq. (3) is valid.

(2) **Coherence Time Calculation**

For calculation of coherence time due to Doppler shift, let us consider that a vehicle is moving at a speed v when a transmitter is stationary and let C is the velocity of the electromagnetic wave.

The Doppler shift in frequency f_{ed} can be written as

$$f_{\mathrm{d}} = [\{c/(v+v)\} - 1] * f_0 \tag{4}$$

where

f_0 frequency of the electromagnetic wave 900×10^6 Hz where GSM operates.

$$\text{Coherence time } t_{\mathrm{coh}} = 0.16/f_{\mathrm{d}} \tag{5}$$

The variation of Doppler Coherent time under different environments are calculated and shown in Table 4. The negative sign is due to a reduction in frequency which was evident. The coherence is found to be 3840 μs considering the condition and maintaining parity with slandered design, maximum delay spread is 21.4 μs and minimum Doppler coherence time is 3840 μs. Hence, the symbol time should lie within these two values.

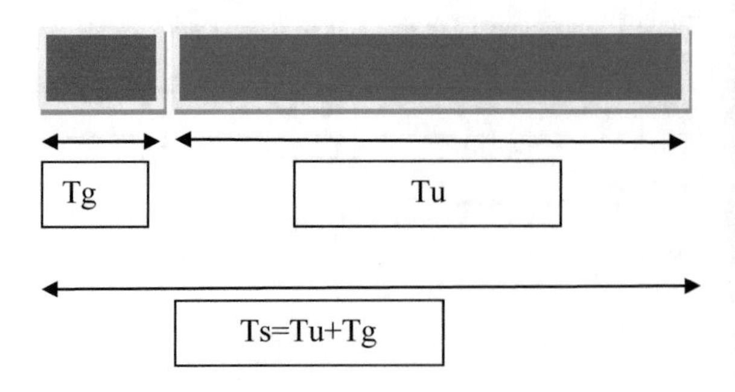

Fig. 1 Symbol length calculation

(3) **CP Length Calculation for GSM 900 MHz Band**

Orthogonal Frequency Division Multiplexing symbol contains—(a) Distinguished sub-channels contain information subcarriers. (b) Pilot subcarriers used for channel identity, information and synchronization, (c) Direct current subcarrier. (d) Guard subcarriers (That are used to provide high inter-channel interference margin).

$$
\begin{aligned}
T_{\mathrm{u}} &= 1/\Delta f \\
T_{\mathrm{u}} &= 1/10.9375\,\mathrm{KHz} = 91.43\,\mu\mathrm{s} \\
T_{\mathrm{g}} &= G * T_{\mathrm{u}} \left(\text{where } G \text{ is } T_{\mathrm{g}}/T_{\mathrm{u}}\right) \\
T_{\mathrm{s}} &= T_{\mathrm{u}} + T_{\mathrm{g}} \\
T_{\mathrm{s}} &= T_{\mathrm{u}} + G * T_{\mathrm{u}}
\end{aligned}
$$

For all practical purposes, $G = 1/8$ is considered as optimum value as shown in Fig. 1.

$$
\begin{aligned}
T_{\mathrm{s}} &= 91.43\,\mu\mathrm{s} + 1/8 * 91.43\,\mu\mathrm{s} \\
T_{\mathrm{s}} &= 102.85\,\mu\mathrm{s}
\end{aligned}
$$

The following condition is satisfied from above Table

Delay spread $\ll$ Symbol Time T_{s} $\ll$ Doppler Coherent Time and maintaining parity with other standard designs

$$
20\,\mu\mathrm{s} \ll 102.85\,\mu\mathrm{s} \ll 3840\,\mu\mathrm{s}
$$

Thus, the Lemma is satisfied.

(4) **OFDM Design Parameters for 900 MHz Band**

The OFDM design parameters are for GSM 900 MHz band for a bandwidth of 200 kHz cognitive radio working in FDD mode is compared with the existing OFDM standards shown in Table 3.

(5) **OFDM Spectral Efficiency for GSM 900 MHz Band**

Table 3 Comparison table for OFDM parameters

System parameter	Existing OFDM standards	OFDM design parameters for GSM 900 MHz band
Bandwidth (B)	10 MHz	200 kHz
Number of data subcarriers (N_{mod})	560	12
Number of pilot subcarriers	280	2
Guard subcarrier	184	1
DC subcarrier	1	1
FFT size (N)	1024	16
Subcarrier spacing (Δf)	10.9375 kHz	10.9375 kHz
Useful symbol time	($T_{\text{u}} = 1/$_f) 91.43 μs	91.43 μs
Ratio of CP time to useful time (G)	Variable	Variable
Cyclic prefix (T_{g})	$T_{\text{g}} = G * T_{\text{u}}$	$T_{\text{g}} = G * T_{\text{u}}$
Modulation scheme	QPSK/16-QAM/64-QAM	QPSK/16-QAM/64-QAM
Total symbol time $T_{\text{s}} = T_{\text{u}} + T_{\text{g}}$	102.85 μs	102.85 μs

Spectral efficiency can be given as

$$\eta = \frac{N_{\text{mod}}}{N} \cdot \frac{\log_2 M}{BT}$$

N_{mod} Number of useful subcarriers for data
N Number of subcarriers
M Constellation level
B Bandwidth = 200 kHz

$$T = T_{\text{u}}/N$$

In a digital communication system, the modulation constellation is chosen based on the available signal-to-noise ratio. Every SNR and the selected constellation are subject to error as shown in Fig. 2.

The packet error rate in terms of SNR and selected constellation can be written as

$$P_{\text{s}}, \text{MQAM} = 2\left(1 - \frac{1}{\sqrt{M}}\right)\text{erfc}\left(\sqrt{\frac{3}{2(M-1)}\frac{E_{\text{S}}}{N_{\text{O}}}}\right) - \left(1 - \frac{2}{\sqrt{M}} + \frac{1}{M}\right)\text{erfc}^2\left(\sqrt{\frac{3}{2(M-1)}\frac{E_{\text{S}}}{N_{\text{O}}}}\right)$$

where

M = Constellation size and E_s/N_o is a signal-to-noise ratio. Drive Test was conducted at Chakan, Maharashtra. Both voice and data were recorded separately. How-

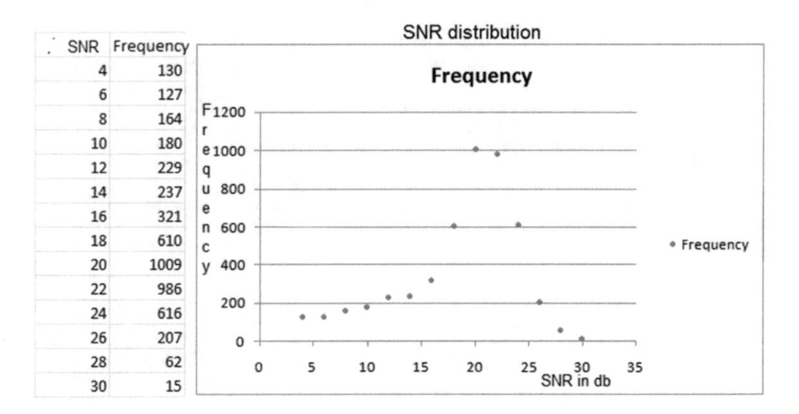

SNR	Frequency
4	130
6	127
8	164
10	180
12	229
14	237
16	321
18	610
20	1009
22	986
24	616
26	207
28	62
30	15

Fig. 2 SNR distribution of Chakan

ever voice data records provided SNR, signal strength, and interference details, data drive test provided throughput and related modulation scheme.

In GSM/EDGE technology, the maximum structural efficiency achievable is 270/200 = 1.35 bits/Hz as shown in Table 4.

So maximum efficiency is 4.1 bit/Hz

At 2600 MHz, CINR = 15.8 (as higher the frequencies have more losses as compared to the lower band of EM wave communication frequencies) is not sustainable, hence we cannot go with 2600 MHz band and, therefore, 900 MHz band is the only solution as it sustains higher CINR values as thus higher modulation types to deliver more data rate. By **2020, GSM 900** band will be vacant as GSM is not delivering the throughputs as compared to other radio technologies.

Reason

In OFDM 2600 MHz, 64 QAM is rarely shifted, mostly it is 16-QAM. In OFDM 900 measurement, C/I allowed 64 QAM in more than 85% cases, which allows higher throughput in OFDM 900 band and hence spectral efficiency in OFDM 900 MHz band is four times higher as compared to OFDM 2600 MHz band as shown in Fig. 3.

4 Verification of Design Results

Atoll is the planning tool used to simulate the radio condition as per choice with different radio conditions, such as frequency, modulation, access type, and radio type. The tool cover frequency band ranges for 450, 800, 900, 1800, 1900, 2100, 2300, 2600, and 3300 MHz used by LTE and WiMAX.

(a) Validation of Atoll using the measurement of OFDM WiMAX

Table 4 SER versus efficiency

BER	$M = 2$	SER	Efficiency	$M = 4$	SER	Efficiency	$M = 6$	SER	Efficiency
19		0	1.36		<0.2	2.72		<0.2	4.1

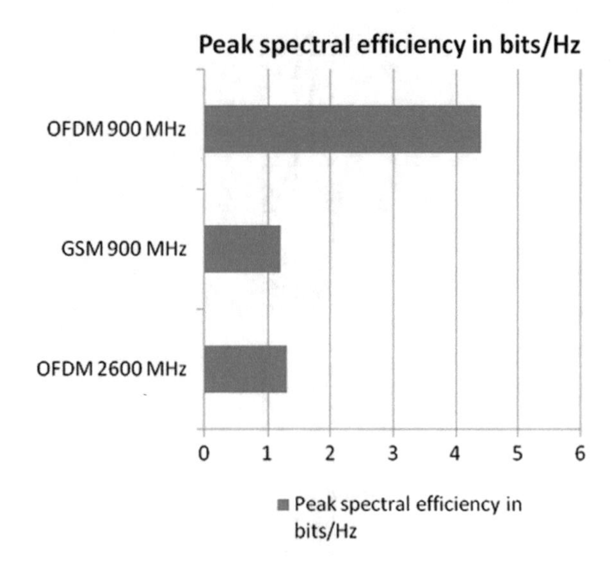

Fig. 3 Comparison of spectral efficiency as per calculation

In order to establish the applicability of the tool, drive test for existing WiMAX Coverage at Chakan was conducted. The base station of Chakan is Multi-Input Multi-Output featured.

Figure 5 is taken from Atoll 3.1.1 planning and optimization tool on real-time basis by having both the 2600 and 900 MHz transmitters in OFDM access type. The snapshots so taken are used to collect data points for comparing the spectral efficiency of both the systems.

(b) Simulation results of OFDM 900 MHz using GSM 900 C/I result

The graph is indicating spectral efficiency and CINR ranges from 4 to 8 is obtained, which is the same as the range obtained from GSM data at Chakan. The result is matching with mathematical and with the results of Chakan collected practically by real-time data traffic drive given below. Validation of OFDM design results through simulation result is shown in Fig. 4.

Now the spectral efficiency is calculated as:

38000000/10000000 = 3.8 bit/Hz

For the below RSSI plots, Fig. 5 is the snapshot showing the throughput in kbps for OFDM 900 MHz. The various legends taken have been shown in the diagram, it clear that as RSSI decreases the throughput values decrease significantly. The range of RSSI is taken from −70 to −110 dBm, which covers an almost entire range of RSSI and throughput ranges from 0 to 50,000 kbps. The various data points have been taken at various RSSI and the result is converted to the graph as shown in Fig. 5.

The design result given in equation provides a peak spectral efficiency of 4.1 bits/Hz for 83% of data.

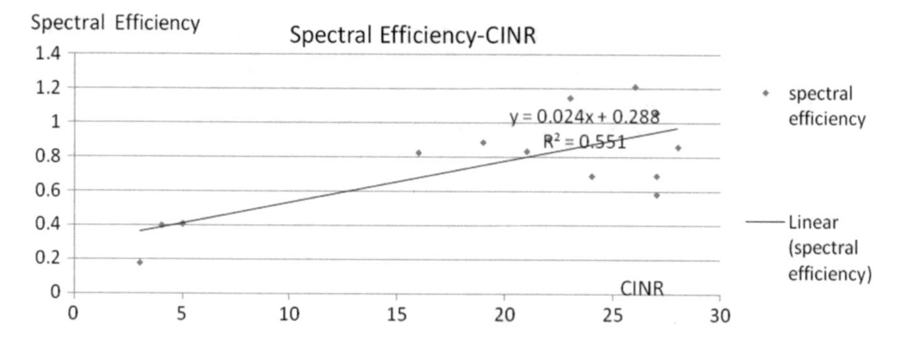

Fig. 4 Spectral efficiency—CINR

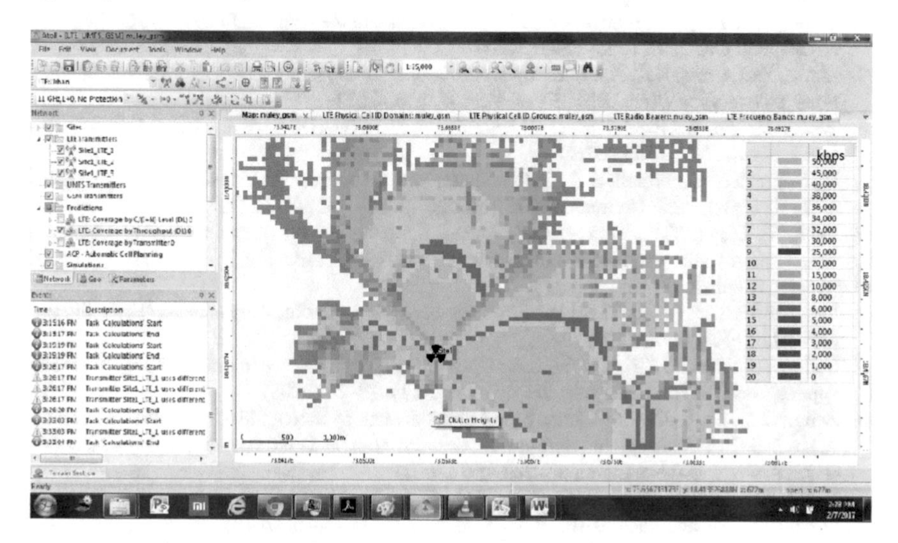

Fig. 5 Map of throughput by transmitter at 900 MHz

5 Conclusion

In the experimentations done, the results are independent of a carrier frequency. If the cognitive radio is to be deployed at 900 MHz frequency band for better coverage advantage, the OFDM technology shall be a better choice due to (a) High SINR operation, i.e., noise immunity nearly linear throughput gain with MIMO configuration, (b) Small operational bandwidth, i.e., better scalability. In the above discussion, we have designed and tabulated OFDM parameters for 200 kHz bandwidth in GSM 900 MHz band. The design parameters were ported in planning tool and the spectral efficiency was calculated and verified with the practical data obtained from experimentation and applied to the tool. The matched result establishes the correctness of the design parameters.

The use of CR-OFDM technology employing unused channels of the same band shall add up to produce appreciably higher bit rate than that obtainable with existing technologies and, hence a subject of interest for future researchers for implementation as an optimal technology solution for cognitive radio.

References

1. Al-Jzari, A., Iviva, K.: Cyclic prefix length determination for orthogonal frequency division multiplexing system over different wireless channel models based on the maximum excess delay spread. Am. J. Eng. Appl. Sci. (2015)
2. Zhang, Z.Y., Lai, L.F.: A novel OFDM transmission scheme with length-adaptive cyclic prefix. J. Zheijang. Univ. Sci. 1336–1342 (2004)
3. Liang, Y.C., Chen, K.C., Li, G.L. Mahonen, P.: Coginitive radio networking and communications: an overview. Proc. IEEE Trans. Veh. Technol. 2011
4. Wang, B., Liu, K.J.R.: Advances in cognitive radio networks: a survey. IEEE J. Sel. Topics. Sig. Process. 1932–4553 (2011)
5. De Domenico, A., Strinati, E.C., Di Benedetto, M.G.: A survey on MAC strategies for cognitive radio networks. IEEE Commun. Surv. Tutorials (2012). 1553-877X112
6. Zeng, Y., Liang, Y.-C., Hoang, A.T., Zhang, R.: A review on spectrum sensing techniques for cognitive radio: challenges and solutions. EURASIP J. Adv. Sig. Process. **2010**(1) (2010) (Article Number: 381465)
7. Gardner, W.A.: Exploitation of spectral redundancy in cyclostationary signals. IEEE Sig. Process. Mag. **8**(2), 14–36 (1991)
8. Sutton, P.D., Nolan, K.E., Doyle, L.E.: Cyclostationary signatures in practical cognitive radio applications. IEEE J. Sel. Areas in Commun. **26**(1), 13–24 (2008)
9. Zeng, Y., Liang, Y.-C.: Covariance based signal detections for cognitive radio. In: Proceedings of the 2nd IEEE International Symposium New Front in Dynamic Spectrum Access Networks (DySPAN'07), Dublin, Irland, pp. 202–207 (2007)
10. Zhang, R., Lim, T.J., Liang, Y.-C., Zeng, Y.: Multi-antenna based spectrum sensing for cognitive radios: a GLRT approach. IEEE Trans. Commun. **58**(1), 84–88 (2010)
11. Mule, S.B., Manna, G.C., Nathani, N.: Comparison of spectral efficiency of mobile OFDM-WiMAX technology with GSM and CDMA for cognitive radio usage. Procedia Comput. Sci. J. (Elsevier, Scopus Indexing). ISSN No. 1877–0509
12. Weiss, T.A., Jondral, F.K.: Spectrum pooling: an innovative strategy for the enhancement of spectrum efficiency. IEEE Commun. Mag. **42**(3), 8–14 (2004)
13. Musavian, L., Aissa, S.: Capacity and power allocation for spectrumsharing communications in fading channels. IEEE Trans. Wireless Commun. **8**(1), 148–156 (2009)
14. Mule, S.B., Manna, G.C., Nathani, N.: Assessment of spectral efficiency about 900 MHz using GSM and CDMA technologies for mobile cognitive radio. In: International Conference on Convergence of Technology—2014. IEEE Explore (2014)

S. B. Mule studying as a Research Scholar in Department of Electronics Engineering, G. H. Raisoni College of Engineering, Nagpur, India. His research area includes OFDM.

Dr. G. C. Manna working as an Ex-Chief General Manager, Indian Telecommunication Service, BSNL, Jabalpur, India. His research area includes OFDM.

New Routing Protocol in Ad Hoc Networks

Bilal Saoud and Abdelouahab Moussaoui

Abstract Ad Hoc network is a collection of wireless devices (nodes) forming a temporary network which has any established infrastructure or centralized administration. These devices have the ability to move which means that the topology of Ad Hoc network changes according to the mobility of nodes. In order to maintain the connectivity between nodes in Ad Hoc network it is necessary that mobile nodes work together. Many routing protocols have been proposed to keep the connectivity between nodes, including the AODV protocol (Ad Hoc On-Demand Distance Vector), which is maintained as a standard by the IETF. This protocol has limitations in terms of stability of links between nodes. To overcome these limitations, we proposed an improvement of AODV protocol to find stable links, which endure between source and destination. Our idea is based on the quality of the received signal strength between the different nodes. In this article we proposed a new strategy of finding routes and predicting ruptured links. The results of simulation and comparison with the standard version of AODV show the effectiveness of our propositions.

Keywords Ad hoc networks · Routing protocols · AODV · Link stability
Ruptured link

1 Introduction

A wireless Ad Hoc network (MANet Mobile Ad Hoc Network) is a set of mobile nodes that communicate together without an existing infrastructure to ensure the communication. In this type of network, the nodes can be cell phones, laptops, handheld digital devices, personal digital assistants, or wearable computers [1]. Nodes

B. Saoud (✉)
Electric Enginnering Department, Bouira University, 10000 Bouira, Algeria
e-mail: Bilal340@sgmail.com

A. Moussaoui
Intelligent Systems Laboratory, Setif University, 19000 Setif, Algeria
e-mail: Moussaoui.abdel@gmail.com

S. Smys et al. (eds.), *International Conference on Computer Networks and Communication Technologies*, Lecture Notes on Data Engineering and Communications Technologies 15, https://doi.org/10.1007/978-981-10-8681-6_40

can leave or join the network at any time. In fact, the connections between nodes are created and destroyed at any moment which leads to change the network topology. Moreover, each mobile node uses radio waves with a limited range as a medium of communication. Also, nodes behave like routers to relay nodes in order to transfer data from a source node to a destination node [2].

Routing in Ad Hoc networks has always been a difficult and tough task due to the dynamic topology and error rate of wireless channel. There are a number of issues that has to be considered while routing a data packet from the source to the destination in the Ad Hoc network like lack of centralized control, constantly moving nodes, limited energy and bandwidth, etc. Routing of data packets becomes much more challenged with the increasing of nodes' mobility [3]. Routing protocols [1, 4, 5] are classified into three classes: proactive, reactive, hybrid. Proactive routing protocols try at each node of the network to maintain the best existing routes to all possible destinations. The routes are saved even if they are not used. The permanent saving of routes is done by a continuous exchange of messages. For example, OLSR [6], DSDV [7] are proactive protocol. Reactive routing protocols (on demand) create routes according to requirements, as soon as a node needs a given route, a discovery procedure is activated. The found route is maintained by a route maintenance procedure until the destination is inaccessible from the source node or the source node no longer needs that route. Among these protocols we can find AODV [8, 9], DSR [10]. Some protocols combine proactive and reactive strategies to find and maintain routes. This is the case for ZRP [11] protocol.

All the nodes have the ability to move in Ad Hoc network. They are free to join or leave the network at any moment. According to the mobility of nodes links will be dynamically change. Links are formed when two nodes moves into transmission range of each other and are broken down when they move out of transmission range. State of links in Ad Hoc network is based on the proximity of one node to another node. In addition, topology of Ad Hoc network undergoes unpredictable changes and routing of data packets from the source to the destination becomes a difficult task. For these reasons we propose a new routing protocol based on AODV protocol to establish stable links and predict links which may be broken down between nodes. The paper is organized as follow. Our approach is detailed in Sect. 2. Experimental results are shown in Sect. 3. Finally, we conclude in Sect. 4.

2 Proposed Routıng Protocol

In this section we are going to describe our contribution to improve the protocol AODV. Our work is based on received signal strength between nodes (Fig. 1). We have used this information about the quality of received signal to establish routes between source nodes and destination nodes and to predict ruptured link in Ad Hoc network topology.

A stable route consists of a set of stable links between source node and destination node. The link stability between two nodes is measured by the received signal strength

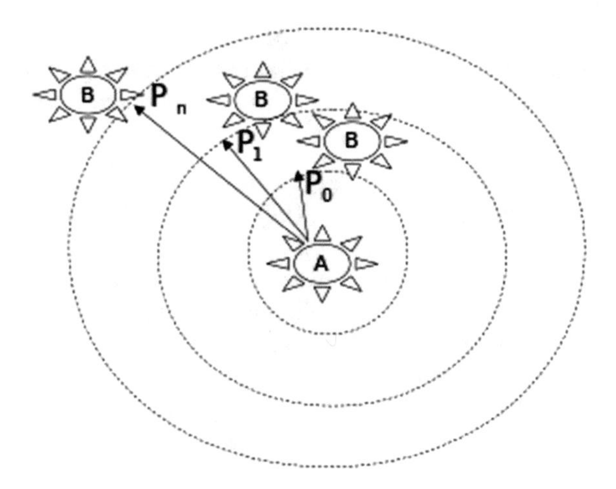

Fig. 1 The received signal strength according to the mobility of nodes

between them. We propose a new mechanism based on this information to find routes between source nodes and destination nodes. Starting from the AODV protocol, we combined the information about received signal strength and the Hello messages, which are sent periodically, to determine the quality of links (good or not) according to the mobility of nodes. This allows us not only to find the most stable links, but also to anticipate links which have a high probability to be broken down in the future, in order to improve the performance of this protocol.

2.1 Route Establishment

To choose a stable route, which lasts for long time, it will be necessary to change the AODV selection mechanisms. For each node transmits a signal with a certain power, and receives a signal with a certain power that varies according to the distance between nodes. In fact, links between nodes are chosen in our proposition with a better reception quality of signal that means the distance separating the nodes is minimal. We try to establish a path between a source node and a destination node with intermediate nodes very close (separating distances between them are minimal). These paths have a high degree of stability.

A communication is between a node source and a destination node to send data in Ad Hoc network. When the source node does not have the destination address in its routing table, then a RREQ broadcast is started. We defined the RREQ package as follows:

- Address of the source node (@source)
- Source sequence number
- Broadcast ID (Broadcast ID)
- Address of the destination node (@destination)

- Destination sequence number
- Signal-power-List (list containing received signal strength of route)
- Intermediate-Node-List (nodes addresses between source node and destination node)

The pair (ID, @source) identifies RREQ. The ID is incremented each time the source broadcasts a new RREQ packet. Each node receives a RREQ packet, it will start by checking whether it has already received this RREQ with the same ID and packet sender address, or whether its address belongs to the Intermediate-Node List, if it is the case so the node must ignore the packet. Otherwise, it processes the RREQ packet. Then the pair (@source, ID) will be saved in the RREQ history table to reject future duplicates. If the intermediate node finds a route to the destination node and its destination sequence number is greater than or equal to the sequence number in the RREQ packet, it sends a route response packet (RREP) to the source node. In the case where the intermediate node does not find a route to the destination, then the value of the received signal of RREQ packet and its address (@node) are saved respectively in Signal-Power-List and Intermediate-Node-List. The updated RREQ packet re-broadcast to all its neighbors. If the RREQ packet arrives at the destination node, it extracts the Signal-Power-List, and finds the minimum value of the list (*Ps*). Destination node inserts the value *Ps* into the RREP packet and sends it to the source node. Note that the route response packet (RREP) is unicast.

The modified RREP package then contains the following fields:

- Source address (@source)
- Destination address (@destination)
- Destination sequence number
- Lifetime (lifetime of the packet)
- Val-signal (minimum value of the received signal of RREQ between nodes of the route)

Each intermediate node receives the RREP response, and adds an entry to its routing table if

- No route to the destination is known.
- The sequence number for the destination in the response packet is greater than the value present in the routing table.
- The sequence numbers are equal and the value Val-signal of RREP is greater than the old value.

For the same RREQ request, the source node can receive many RREP responses. The route that has the highest Val-signal value it will be selected by the source node. The source node *A* (Fig. 2) received two RREP packets, the first form the route *DCBA* with Val-signal = $2e^{-12}$ and the second with the value Val-signal = $3e^{-12}$ from the route *DEFGA*. The source node takes the route *AGFED* to send data packets to the destination node *D*.

Figures 3 and 4 give a general form of our RREQ and RREP treatment.

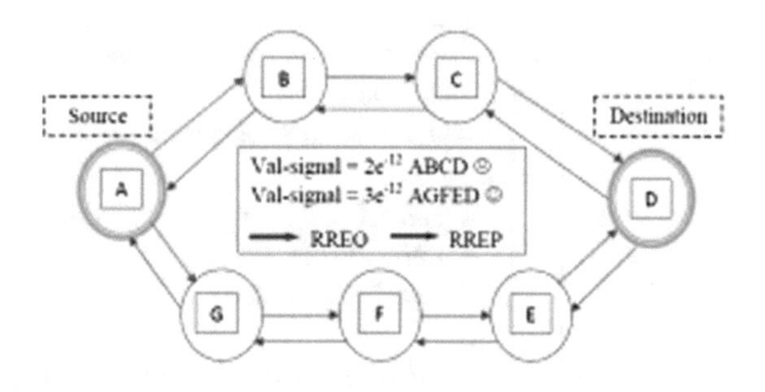

Fig. 2 Route choice based on received signal strength

1 **if** *RREQ.ID and @ source* $\in$ *Routing table* **then**
2     $destroy(RREQ)$;
3 **else**
4   **if** @ *node* $\in$ *RREQ.Intermediate-Node-List* **then**
5         $destroy(RREQ)$;
6   **else**
7   **if** @ *node* == @ *Destination node* **then**
8             $P_s = min(RREQ.Signal - power - List)$;
9             $creat(RREP)$;
10             $RREP.Val - signal = P_s$;
11             $send(RREP)$;
12   **else**
13             $P_r = getSignalPower(RREQ)$;
14             $add(RREQ.Signal - power - List, P_r)$;
15             $add(RREQ.Intermediate - Node - List, @node)$;
16             $send(RREQ)$;

Fig. 3 Algorithm of RREQ treatment

2.2 *Routes Maintenance*

According to the strategy of the AODV protocol, if a node E (Fig. 2) does not receive a Hello message from its neighbor F, it means that F has moved out of range of the E node. In this case, the node E checks its routing table to find if F is a next hop. If F is a next hop that means F is an essential node in a route between source and destination nodes. Node E activates a route maintenance procedure. It may take a long time before another route will be established. For this reason that, we prefer to predict the ruptured routes rather than to detect them. Hence the proposal algorithm of prediction ruptured routes is based on the received signal strength.

We can say that a node B is in the ruptured zone of node A, if node A received Hello message form B with a weak signal strength. That means the node B has a

```
1  if @ node == @ Source node then
2      if A new route then
3          add(routing table, new route);
4      else
5          if RREP.Sequence − number >
           routing table.Sequence − number then
6              update(routing table);
7          else
8              if (RREP.Sequence − number ==
               routing table.Sequence − number) and
               (RREP.Val − signal >
               routing table.Val − signal) then
9                  update(routing table);
10 else
11     if A new route then
12         add(routing table, new route);
13         send(RREP);
14     else
15         if RREP.Sequence − number >
           routing table.Sequence − number then
16             add(routing table, new route);
17             send(RREP);
18         else
19             if (RREP.Sequence − number ==
               routing table.Sequence − number) and
               (RREP.Val − signal >
               raouting table.Val − signal) then
20                 add(routing table, new route);
21                 send(RREP);
```

Fig. 4 Algorithm of RREP treatment

high probability to get out of the range of A. So, the node B is in the ruptured zone of node A. This information is very important for our proposition to maintain routes between source and destination nodes.

Hello messages are sent periodically between nodes. We had based on these messages to measure the quality of links between nodes. For instance, if a node F is located in the ruptured zone of node E (Fig. 2) that means the node F has a high probability to get out of the range of E. So, the node E activates the route maintenance procedure. In our proposition, once the nodes, which are located in the ruptured zone and are parts of routes between source and destination nodes, are found the route maintenance procedure will be activated.

3 Experiment and Results

In this section, we evaluate the effectiveness and efficiency of our proposition by comparison with AODV protocol. We run a simulation on OPNET simulator [12]. Table 1 presents the parameters of the simulation.

The results of the performance evaluation of the improved algorithm are presented in the form of graphs (Fig. 5). A comparative study of the performances of our proposition and AODV protocol are formulated.

Figure 5a shows the total RREP response packets delivered by all destination nodes in the network. We can observe clearly in this graph that our proposition generates more RREP packets than AODV protocol. Because, in our proposition we try to find among several routes the one that has a high degree of stability. In other words, destination node sends RREP for all received RREQ. However, in AODV, each destination responds only to the first received RREQ.

Figure 5b illustrates the total RERR error packets sent in the network. We can observe that the total delivery of RERR packets in our proposition is lower than AODV protocol. This result can be explained by the fact that the routes in the AODV protocol are less stable than the routes of our proposition. AODV selects routes based on hop count, but in our proposition we select routes that are more stable with good received signal strength between intermediate nodes of routes. In addition, our proposition predicts ruptured links but AODV protocol detects only broken links. This result shows that our proposition is more powerful than AODV because data packet reached destination node.

Figure 5c illustrates the total RREQ packets sent in the network. We can observe that the total delivery of RREQ packets in our proposition is lower than AODV protocol. In AODV protocol routes are less stable than the routes of our proposition. With our proposition routes can still exist for long time and there is a mechanism to predict ruptured routes, for this reasons the number of RREQ is fewer than AODV.

Figure 5d illustrates the total number of packets lost in the network. In this graph, we can clearly see that the ratio of packets lost in AODV is greater than our proposi-

Table 1 Simulation parapeters

Parameter	Value
Number of nodes	35
Area size of simulation	2000 × 2000 meters
Mobility model	Random Way-Point
Speed of nodes Time of pause Time of simulation	0–25 m/s 4 s 30 min

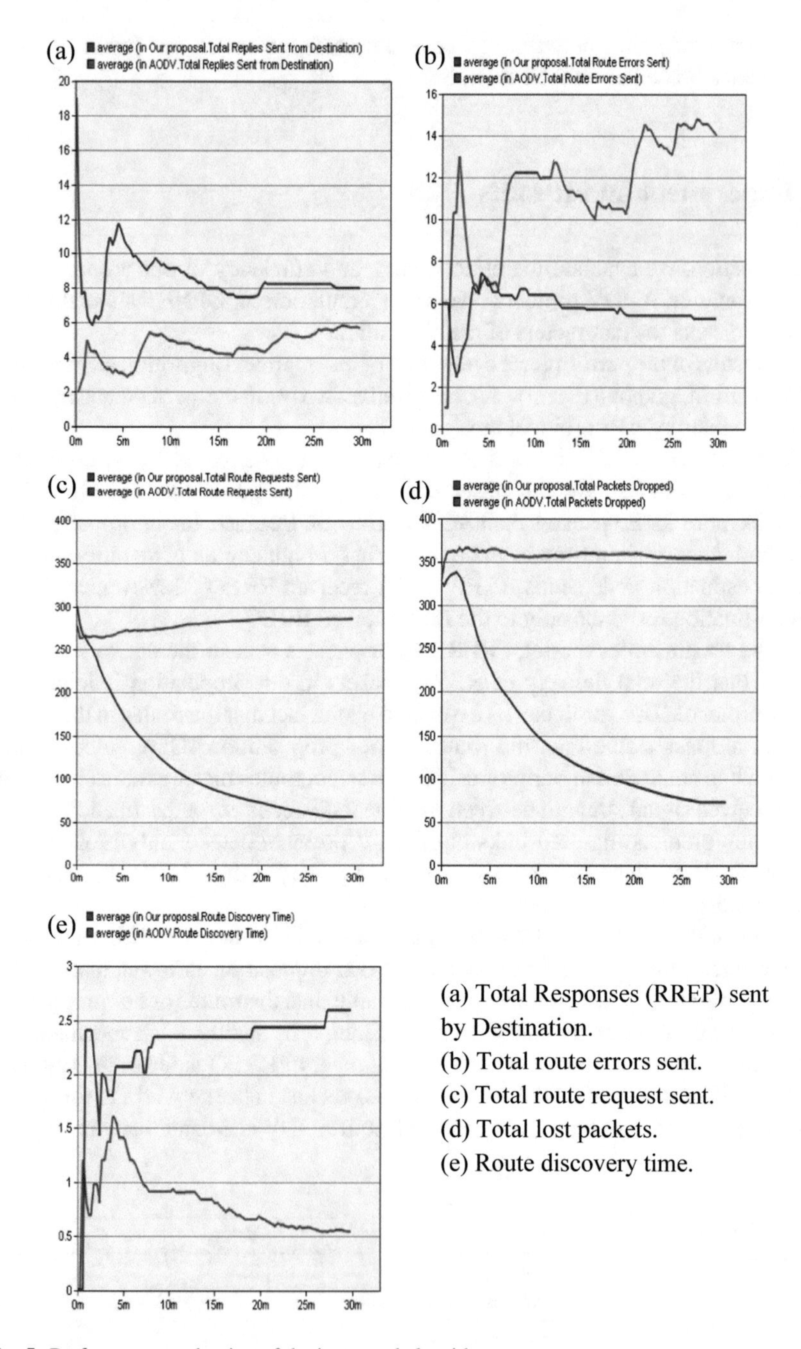

Fig. 5 Performance evaluation of the improved algorithm

tion. Because the intermediate nodes predict the ruptured links instead of detecting them with our proposition and the routes are more stable than AODV protocol. In our proposition, we install stable routes. These routes can still exist between source and destination nodes for long time. This feature of our proposition allows data packets to reach their destinations and minimize the ratio of packets lost. Furthermore, our proposition predicts ruptured routes, with this strategy our proposition gives more extra time to routes.

Figure 5e illustrates the time to discover routes in network. AODV is faster than our proposition to install a route between source and destination nodes. AODV can find routes very fast. However, the routes found by AODV are less stable than our proposition. In our proposition, the time to install routes between source and destination nodes is greater than AODV, because we select the best route (in term of stability of links) among many routes. Our proposition is slower than AODV to install routes but it gives stable routes, which means routes that last for long time between source and destination nodes.

4 Conclusion

To guarantee stable communication in an Ad Hoc network between nodes we have improved the AODV protocol. Our proposition is based on the metric of stable links in order to establish routes between the communicating nodes (Source and Destination). We have introduced a new mechanism to predict ruptured links in order to maintain them before they disconnect the network locally or from end-to-end (from the source to the destination). The evaluation of the effectiveness of our proposition which was compared with the AODV protocol showed perfectly the advantages of our proposition. Other simulations could be done with different scenarios to compare our improved protocol with AODV protocol or other protocols like DSR, SSA.

References

1. Chlamtac, I., Conti, M., Liu, J.: Mobile Ad Hoc networking: imperatives and challenges. Ad Hoc Netw. **1**(1), 13–64 (2003)
2. Conti, M., Giordano, S.: Mobile ad hoc networking: milestones, challenges, and new research directions. IEEE Commun. Mag. **52**(1), 85–96 (2014)
3. Menon, V.G., Prathap, P.M.J.: Routing in highly dynamic ad hoc networks: issues and challenges. Int. J. Comput. Sci. Eng. **8**(4), 112–116 (2016)
4. Boukerchea, A., Turgutb, B., Aydinc, N., Ahmadd, M.Z., Blnid, L., Turgutd, D.: Routing protocols in ad hoc networks: a survey. Comput. Netw. **55**(13), 3032–3080 (2011)
5. Cadger, F., Curran, K., Santos, J., Moffett, S.: A survey of geographical routing in wireless Ad-Hoc networks. IEEE Commun. Surv. Tutorials **15**(2), 621–653 (2013)
6. Jacquet, P., Muhlethaler, P., Qayyum, A., Laouiti, A., Viennot, L., Clausen, T.: Optimized link state routing protocol (OLSR). IETF Draft (2001)

7. Perkins, C., Bhagwat, P.: Highly dynamic destination sequenced distance-vector routing (DSDV) for mobile computers. ACM SIGCOMM. pp. 234–244 (1994)
8. Perkins, C.E., Royer, E.M.: Ad-Hoc On-Demand distance vector routing. In: Proceedings of Second IEEE Workshop on Mobile Computing Systems and Applications, pp. 90–100. New Orleans, USA (1999)
9. Perkins, C., Belding-Royer, E., Das, S.: Ad hoc on-demand distance vector (AODV) routing, IETF Request for Comments, pp. 3561 (2003)
10. Johnson, D.B., Maltz, D.A., Broch, J.: DSR: The dynamic source routing protocol for multihop wireless Ad Hoc networks. In: Perkins, C.E. (eds.) Ad Hoc Networking, pp. 139–172. Addison-Wesley (2001)
11. Samar, P., Pearlman, M., Haas, S.: Independent zone routing: an adaptive hybrid routing framework for Ad Hoc Wireless networks. IEEE/ACM Trans. Networking **12**(4), 595–608 (2004)
12. www.opnet.com (2018)

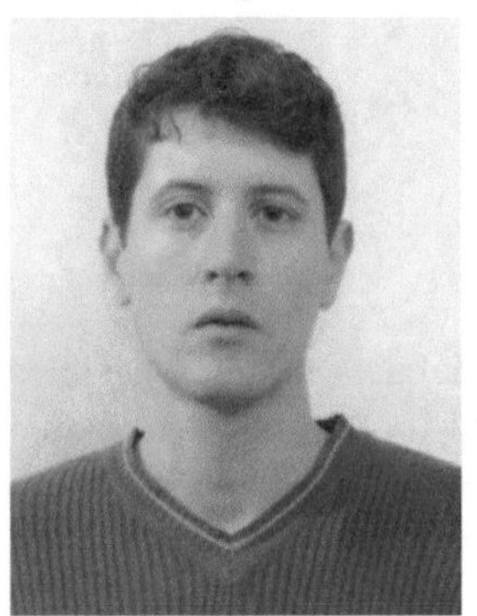

Bilal Saoud holds a Ph.D. degree in computer science from the University of Bejaia, Algeria (2017), Magister degree (post-graduation) degree in computer science from the University of Bejaia, Algeria (2012) and Engineer degree in computer science from the University of BBA, Algeria (2009). He was lecturer at the University of M'sila, Algeria (2012–2013). He was military officer during his military service. He is a teacher-researcher at the University of Bouira, Algeria form 2015. His research areas include Ad Hoc network, wireless sensor network, social network and community detection in network, statistical models and copula, computational intelligence.

Abdelouahab Moussaoui received the doctorate in computer science from the Setif University, Algeria (2005). He received his Magister degree from Sidi Bel-Abbes University, Algeria (1995) and his Engineer degree from Bab-Ezzouar University, Algeria (1991). He is a professor at the department of computer science at Setif University His research interests include clustering, image processing, and machine learning.

Mobile Communication Based Security for ATM PIN Entry

G. Nandhini and S. Jayanthy

Abstract This paper proposes a prototype for multi-level secured PIN (Personal Identification Number) entry authentication. This system is based on ARM microcontroller in which Stegano PIN method is implemented. If in Emergency situation to withdraw amount, owner of an ATM card gives it to a third party, there is possibility to withdraw extra amount. Moreover even if a user is directly accessing the ATM machine there is a possibility of shoulder surfing to track PIN number. To achieve security and usability, a practical indirect PIN entry method Stegano PIN is designed and implemented. A random key is generated using the Stegano pin method in the keypad matrix each time the card is inserted. This prevents the shoulder surfing attack. If the card is given to third party, the card owner will send a message to the server placed in the ATM machine specifying the amount limit for that card number. If the third party tries to withdraw extra amount the ARM processor will process only the amount specified by the owner through the message from the GSM. The extra amount entered by the third party is not processed by the ATM machine. Thus a highly secured and user-friendly system is developed and tested.

Keywords Authentication · Shoulder surfing · Personal identification number (PIN) · ARM · GSM · Keypad

1 Introduction

A personal identification number (PIN) is a numerical code used in many electronic financial transactions. Personal identification numbers are usually issued in association with payment cards and may be required to complete a transaction. Personal

---

G. Nandhini (✉) · S. Jayanthy
Department of Electronics and Communication Engineering, Sri Ramakrishna Engineering College, Coimbatore, India
e-mail: nandhu.suren92@gmail.com

S. Jayanthy
e-mail: jayanthy.s@srec.ac.in

S. Smys et al. (eds.), *International Conference on Computer Networks and Communication Technologies*, Lecture Notes on Data Engineering and Communications Technologies 15, https://doi.org/10.1007/978-981-10-8681-6_41

Identification Number (PIN) is a unique number used for personal authentication when, for example, unlocking the phone, making payment with a payment card, withdrawing cash, and so on [1].

A personal identification number (PIN) is a security code for verifying your identity, like a password, your PIN should be kept secret because it allows access to important services like the ability to withdraw cash, change personal information, and more. Unlike most passwords, a PIN is numeric only and there are no letters or special characters in a PIN. However, the password submission process is inclined to direct observational attacks such as shoulder-surfing [2] (Shoulder surfing happens when a crook is looking over your shoulder while you are carrying out a transaction at a cash dispenser. By doing this, he hopes to get to know your secret code. Once he has seen it, he will try to divert your attention in order to get hold of your bank card), is a source of security concerns. There is another critical issue that a third party can withdraw an extra amount. To overcome the above-mentioned problem, a "**Mobile Communication based Security for ATM PIN Entry**" is proposed to secure the personal identification number (PIN) and shoulder surfing attack.

2 Literature Review

Kwon and Na [3] has proposed a new PIN entry method for shoulder surfing attack. Shoulder-surfing attack poses a great threat for the users performing the banking transactions in the modern world. These attacks are more common in the densely crowded areas especially in the ATM and POS systems. Often when a user withdraws money from an ATM, card holder will enter the identical four-digit PIN number. A new approach proposed in this project is that it enables the end user to have a secured money withdrawal from ATM. This system introduces a security mechanism to the ATM system through GSM, RFID reader. The ARM processor processes the signal and gives the appropriate output from the predefined code loaded in it. Cost and time is saved during the operation which can be achieved by minimizing the unnecessary procedure of authentication system. The user can be benefited by using handheld devices like mobile number communicated with the ATM system through GSM. Machine access control mechanism, can be implemented to allow only authorized persons to use the ATM cards. By using this system the user can prevent their ATM cards from the attackers. The major drawback is attackers can easily access the card holder's ATM card without the permission to the user.

Mali and Mohanpurkar [4] has proposed a new method called Grid-Based Authentication System and rules for secure PIN-entry method by examining the current routines under the new structure. Grid authentication is a method of securing user logins by requiring the user to enter values from specific cells in a grid whose content should be only accessible to him and the service provider. Because the grid consists of letters and numbers in rows and columns, the method is sometimes referred to as bingo card authentication. Typically, the grid is provided to the user on a wallet-size card that contains randomly generated characters in rows and columns. Grid authentication protects against replay attacks because the same characters selected for one login cannot be reused. However, there is no mechanism in place to prevent copying of the entire grid.

Kwon and Hong [5] proposed another method which defines the personal identification number (PIN), consisting of four decimal digits. The problem in their method lies with the susceptibility of the PIN entry process to direct observational attacks, such as human shoulder-surfing and camera-based recording because the 10-digit keypad is very simple. Their system consists of an elegant adaptive black-and-white coloring of the 10-digit keypad, where half the keys are colored in white and half the keys are colored in black. Even though the method required uncomfortably many user inputs, it had the merit of being easy to understand and use. This system combines the information obtained through the four-color choices to single PIN digit. Hence by having the four-round procedure the attackers standing back of the user can easily get confused when user types the same number four times. It also mixes with the two-color combination by using black and white. This technique acts in a zigzag manner. The major drawback in the system, using ATM card and password cannot verify the client's identity exactly.

Dinesh Kumar et al. [6] have proposed a keypad-based authentication system having several possibilities of password guessing by means of shoulder movements. The PIN entry can be observed by human or device attackers. To overcome this problem, he has proposed a shuffled Automated Teller Machine keypad which shows the shuffled numbers in the Liquid Crystal Display keypad which confuses person who is standing near to guess the password. A Linear Feedback Shift Register is used to generate the random numbers. The major drawback is user must give the PIN numbers within the pre-defined timeframe, if it exceeds, then the login page will be terminated automatically.

The above-mentioned papers discuss the concepts of PIN entry authentication system using various methods like Biometric recognition, Cognitive method, Circular Hough Transform, Grid based high-performance security and usability [7–9]. Compared to those papers, the proposed system aims to enhance the process by secured PIN entry authentication system.

The proposed system uses ARM microcontroller in which stegano pin method is implemented. To secure the personal identification number (PIN) ARM microcontroller acts as the main server. The card owner sends a message to the server through phone, which is received by the GSM, which is interfaced with the ARM microcontroller. This message sets the amount limit for the ATM card.

To achieve security and usability, a practical indirect PIN entry method Stegano PIN is implemented to physically block shoulder surfing attacks. A random key is generated using the Stegano pin method in the keypad matrix, each time whenever the card is inserted.

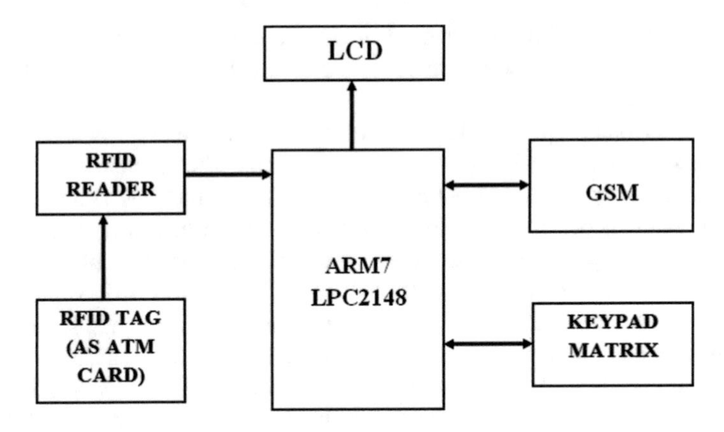

Fig. 1 Block diagram

3 Block Diagram

Figure 1 shows the block diagram of Mobile Communication based PIN Entry Authentication System. This system involves LCD, RFID tag, RFID reader, GSM module, Keypad matrix, and ARM microcontroller. ARM microcontroller acts as the main server. The message from the card owner is sent to the server through phone and is received in the server by the GSM is connected with the microcontroller. This sets the amount limit for the card. RFID tag number is read by the RFID reader and sent to the ARM microcontroller. When the card number matches, the LCD will ask for the PIN. During this time a random key is generated in the Stegano pin format in the keypad matrix. PIN is entered in the randomly generated Keypad matrix. This method avoids the shoulder surfing attack. After the PIN is entered, LCD will ask for the Amount. If the third party enters more amount ATM machine will provide only the specified amount predetermined by the owner.

4 Schematic Diagram

This paper is to design a development of PIN entry authentication system where the users can directly access their PIN by sending a message to their server by blocking their card to take only limited amount of money by the third party.

The proposed system is developed in the ARM microcontroller, which act as a main server. The message from the card owner is sent to the server through phone, and message is received by the GSM placed along with the microcontroller. This message sets the amount limit for the card. RFID tag number is read by the RFID reader and is sent to the ARM 7 microcontroller. When the card number matches, the LCD will ask for the PIN.

During this time, a random key is generated in Stegano pin format in the keypad matrix. PIN is entered in the randomly generated Keypad matrix. This method avoids the shoulder surfing attack. After the PIN is entered LCD will ask for the Amount. If the third party enters more amount ATM machine will provide only the specified amount predetermined by the owner.

4.1 Implementation

The system is developed as mentioned in the schematic diagram shown in Fig. 2. The PIN entry authentication system has been designed with ARM microcontroller. It reads the value of RFID reader and GSM module simultaneously.

The proposed system has different modules. The ARM microcontroller acts as a main server. The RFID reader transmitter (TXD) pin is connected to the P0.1 (RXD) receiver of ARM microcontroller, VCC to pin 43 and GND to pin 42. The LCD whose data D4 is connected to P0.20, D5 is connected to P0.21, D6 is connected to P0.22, D7 is connected to P0.23, Enable pin is connected to P0.13, RS is connected to P0.10, VSS to GND and VDD to 5v. RFID tag number is read by the RFID reader and is sent to the ARM 7 microcontroller. When the card number matches, the LCD will ask for the PIN.

The switch pins are connected to the ARM microcontroller which P0 is connected to P1.19, P1 is connected to P1.20, P2 is connected to P1.21, P3 is connected to P1.22, P5 is connected to P1.18, P6 is connected to P1.17, and P7 is connected to P1.16. During this time, a random key is generated in Stegano pin format in the keypad

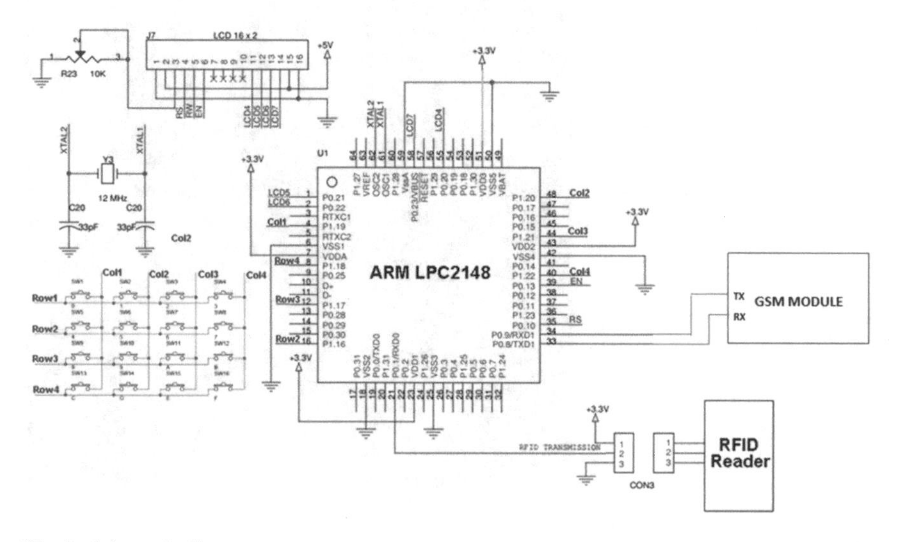

Fig. 2 Schematic diagram

matrix. PIN is entered in the randomly generated Keypad matrix. After the PIN is entered LCD will ask for the Amount. Simultaneously the message is sent to the server through phone and is received in the server by the GSM placed along with the controller. GSM (TX) pin is connected to P0.9/RX receiver pin of the ARM microcontroller and RX pin is connected to P0.8/TXD1 transmitter pin of ARM microcontroller. This message sets the amount limit for the card.

5 Experimental Results

5.1 Experimental Setup

Figure 3 shows the Experimental setup of this paper. Embedded C program are coded using keilμvision4 software and ported into the ARM microcontroller using Flash magic. There are two sections in this project. First section deals with ARM microcontroller, RFID reader and Tag, GSM. In this section the card owner sends the message to server through phone and received by the GSM which is connected to the ARM microcontroller. RFID reader reads the RFID tag and sent it to the ARM Microcontroller. If the third party tries to withdraw extra amount the ATM machine will pop up the unauthorized message. Second section deals with the ARM microcontroller, Keypad matrix, RFID reader and Tag. Stegano pin method provides for each and every RFID tag a random key is generated from the keypad matrix.

Fig. 3 PIN entry authentication system

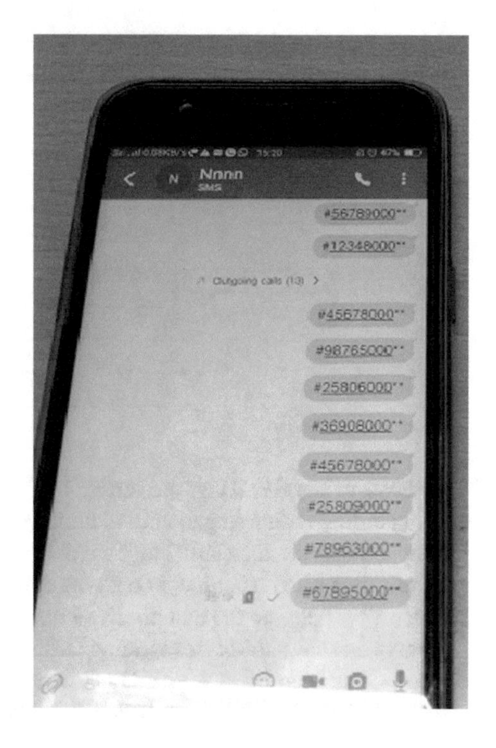

Fig. 4 Message sent to the server

5.2 *Result Analysis*

The results are analyzed from the ten RFID tags are used to secure the personal identification number using Keypad matrix, GSM module, ARM microcontroller, RFID Tag and reader module. μVision, the popular IDE from Keil Software is used to simulate the code and the code is ported into the ARM microcontroller using Flash magic.

The card owner sends the message to the server through phone and received by the GSM is connected with the ARM microcontroller. Figure 4 shows the four digit password (7438) and required amount (5000) are sent to the server through phone and received by GSM. If third party tries to enter more amounts is not processed by the ATM machine.

5.3 *Customer Tag Number*

10 RFID tags are used. For each and every RFID tag unique number is present. This unique numbers are assigned as customer numbers as shown in Table 1.

Table 1 10 RFID tag and customer numbers

S. No.	RFID number	Customer numbers
1.	9721614932352	Customer no 1
2.	9486614930002	Customer no 2
3.	6600942812	Customer no 3
4.	0E008E845C	Customer no 4
5.	0E008E86E6	Customer no 5
6.	080067ACB6	Customer no 6
7.	64F88C00451	Customer no 7
8.	85E525756112	Customer no 8
9.	73314CA4500	Customer no 9
10.	08721B56587	Customer no 10

Whenever any RFID tags are sensed by the RFID reader a random keys are generated according to this stegano pin method and corresponding keys are pressed from the keypad matrix. Each RFID tags are assigned different password. If the passwords are same then ATM machine process the correct password message otherwise ARM processor will pop up the unauthorized message.

Figure 5a shows the corresponding customer number of the RFID tag is displayed on the LCD screen. Here 10 RFID tag numbers are displayed.

Figure 5b shows the randomly generated keypad in the LCD screen. According to this numbers are displayed in the LCD screen, the corresponding keys in the keypad matrix are assigned respectively.

During withdrawal of the amount, by the card holder default Personal identification number is used. Figure 6a shows the cash withdrawal by entering the default PIN number (1234) by the card holder.

The third party entered the four digit password and sent to the GSM module rather than default password as shown in Fig. 6b.

If the RFID reader sense the RFID tag and the corresponding password are pressed using stegano PIN method. If the passwords are same then ATM machine process the correct password message as shown in Fig. 7.

If the extra amount (5050) entered by the third party as shown in Fig. 8a, the unauthorized message will pop as shown in Fig. 8b and ATM machine will not process further.

When the third party enters correct amount as shown in Fig. 9a and the ATM machine will process the amount to take the cash message will pop up as shown in Fig. 9b.

(a)

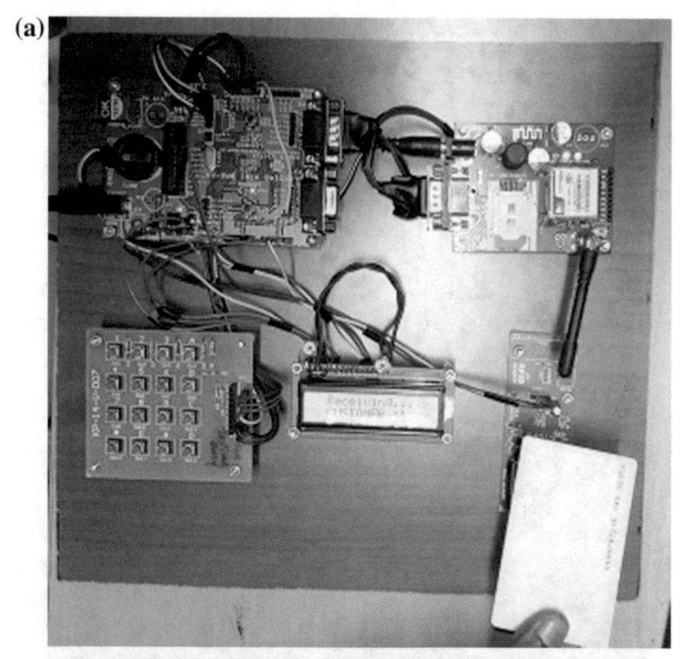

(b)

Fig. 5 **a** Displaying the corresponding customer number 1 of the RFID tag. **b** Random keypad generation 1

(a)

(b)

Fig. 6 **a** Card holder entered the default PIN number. **b** Four digit passwords entered by third party

Fig. 7 Process the correct password

(a)

(b)

Fig. 8 **a** Extra amounts entered by third party. **b** ATM machine pop up the unauthorized message

(a)

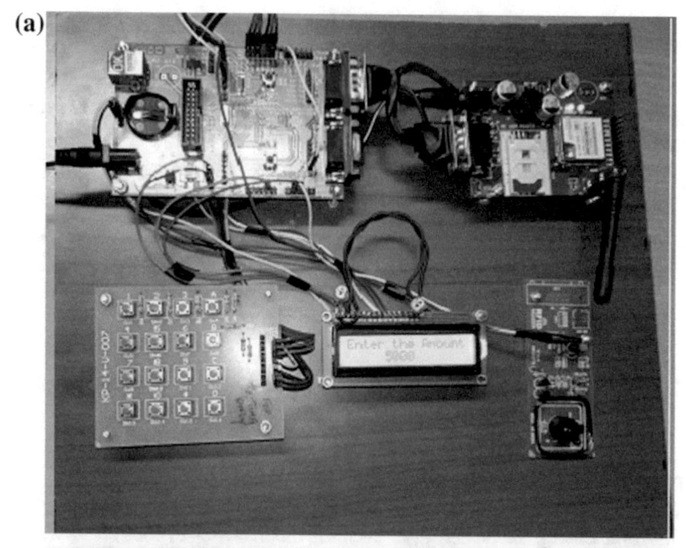

(b)

Fig. 9 **a** Enter the correct amount. **b** Delivered the cash

6 Conclusion

In this paper, design and implementation of ATM-based secured PIN entry authentication system is proposed. The proposed system secures the personal identification number (PIN) by preventing more amount by the third party than specified from the ATM card owner. To avoid the shoulder surfing attack, a random key is generated using the Stegano pin method in the keypad matrix, each time whenever the card is inserted. Stegano PIN system provides more security in public places.

References

1. Lee, M.K.: Security notions and advanced method for human shoulder-surfing resistant PIN-Entry. IEEE Trans. Inf. Forensics Secur. **9**(4), 144–169 (2014)
2. Shi, P., Zhu, B., Youssef, A.: A PIN Entry scheme resistant to recording-based shoulder-surfing. Third Int. Conf. Emerg. Secur. Inf. Syst. Technol. **5**(6), 237–241 (2009)
3. Kwon, T., Na, S.: SteganoPIN two-faced human-machine interface for practical enforcement of PIN entry security. IEEE Trans. Hum. Mach. Syst. **46**(1), 143–150 (2016)
4. Mali, Y.K., Mohanpurkar, A.: Advanced pin entry method by resisting shoulder surfing attacks. IEEE Int. Conf. Inf. Process. **8**(14), 37–42 (2015)
5. Kwon, T., Hong, J.: Analysis and improvement of a PIN-entry method resilient to shoulder-surfing and recording attacks. IEEE Trans. Inf. Forensics Secur. **10**(2), 278–292 (2015)
6. Dinesh Kumar, G., Kumaresan, S., Radhika, S.: Design of secured ATM by wireless password transfer and shuffling keypad, International Conference on Innovations in Information Embedded and Communication Systems ICIIECS15, (2015)
7. Soares, J., Gaikwad, A.N.: Fingerprint and Iris biometric controlled smart banking machine embedded with GSM technology for OTP. Int. Conf. Autom. Control Dyn. Optim. Tech. (ICACDOT) **9**(10), 409–414 (2016)
8. Yang, Y.: ATM terminal design is based on fingerprint recognition. 2nd Int. Conf. Comput. Eng. Technol. **1**, 92–95 (2010)
9. Rajesh, V., Vishnupriya, S.: IBIO-A new approach/or ATM banking system. In: International Conference on Electronics and Communication Systems, pp. 1–5 (2014)

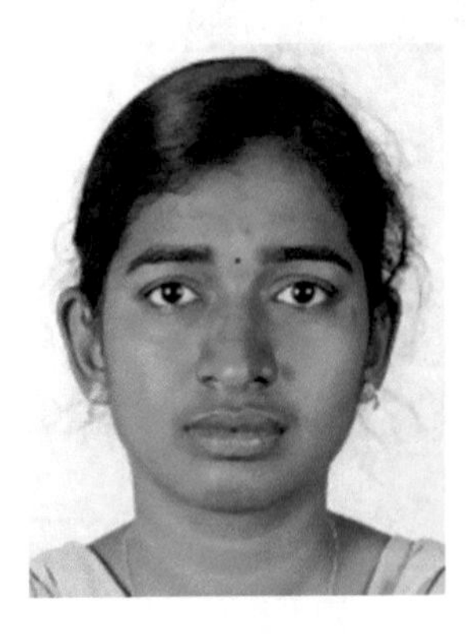

G. Nandhini studying M.E., in Department of Electronics and Communication Engineering at Sri Ramakrishna Engineering College and located in Coimbatore. Her research area is Embedded systems and mobile Communication.

Dr. S. Jayanthy working as a Professor of Department of Electronics and Communication Engineering in Sri Ramakrishna Engineering College and located in Coimbatore. Her research interest includes VLSI Testing, Computer networks, Embedded systems and IOT.

Design of Direct Digital Frequency Synthesizer with the Technique of Segmenting in Quarter Wave Symmetry

S. Karpagavalli, K. Hariharan, G. Dheivanai and M. Gurupriya

Abstract In recent communication systems, Direct Digital Frequency Synthesizer (DDFS) plays dominant role in signal generation. DDFS is highly stable and highly controllable circuitry. DDFS requires read only memory (ROM) for signal generation, but usage of high ROM size leads to high power consumption and more hardware requirements. In this paper, on utilizing a quarter wave symmetry technique and storing the difference value between consecutive segments in ROM Look-up table (LUT) is proposed. The proposed architecture has reduced ROM size with beneficial effects in terms of speed and power. For 8-bit resolution SNR is of 44.92 dB and SFDR is of 50.64 dBc.

Keywords DDFS architecture · Read only memory · Spectral purity
ROM compression · Signal-to-noise ratio · Spurious free dynamic range
Spectrum

1 Introduction

Digital Frequency Synthesizer (DDFS) [1] is capable of generating sinusoidal signal with fine frequency tuning, fast frequency switching, high stability, phase resolution, and low phase noise. A DDFS generates the signal based on stored version of waveform in read only memory (ROM). By advancing the phase in fixed increment determines the frequency of the signal which is generated.

S. Karpagavalli (✉) · K. Hariharan · G. Dheivanai · M. Gurupriya
ECE Department, Thiagarajar College of Engineering, Madurai, India
e-mail: karpagavallistce@gmail.com

K. Hariharan
e-mail: khh@tce.edu

G. Dheivanai
e-mail: dheivagd@gmail.com

M. Gurupriya
e-mail: gurupriyamaniece@gmail.com

S. Smys et al. (eds.), *International Conference on Computer Networks and Communication Technologies*, Lecture Notes on Data Engineering and Communications Technologies 15, https://doi.org/10.1007/978-981-10-8681-6_42

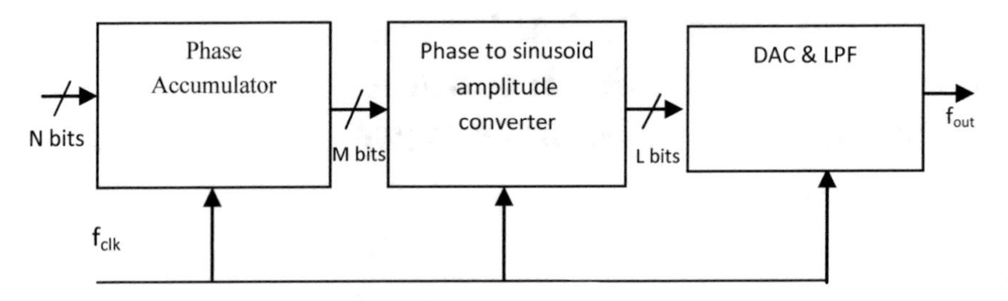

Fig. 1 Conventional DDFS architecture

A conventional DDFS architecture [2] is shown in Fig. 1 consists of Phase Accumulator, Phase to Sinusoid Amplitude converter (PSAC), Digital to Analog Converter (DAC) and filter. Frequency Control Word (FCW) corresponds to phase increment at regular interval for sinusoidal waveform. FCW added for each $1/f_{clk}$ period to produce linearly increasing digital value [3]. PSAC [4] consist of ROM which stores the number corresponding to the voltage required for each value on phase.

Spectral purity of the conventional DDFS is also determined by the resolution of the values stored in ROM LUT [3]. The sequence of the digital value is passed to DAC and filter to construct a sinusoidal waveform. In conventional DDFS architecture, PSAC implementation is based on basic look-up table (LUT). The LUT stores the pre-estimated amplitude value for the sinusoidal generation. The amplitude sequence of the sinusoidal signal is based on the output of the phase accumulator. Phase accumulator generates address which steps through each memory location, the corresponding digital amplitude of a signal which drives DAC to generate analog output signal. Each address in the look-up table corresponds to phase point in the sine wave from 0° to 360°. The output frequency depends on the clock frequency, frequency control word, and total number of bits used in phase accumulator. The output frequency is given by

$$f_{\text{out}} = \frac{\text{FCW}^* f_{\text{clk}}}{2^j}, \tag{1}$$

where f_{out} is clock frequency, j indicates total number of bits in phase accumulator. 2^j represents the maximum size of the ROM LUT.

A ROM compression technique such as Quarter wave symmetry technique, sinusoid amplitude compression technique is used. In quarter wave symmetry method [5], ROM only stores $\pi/2$ radians of sine points and generate full range of 2π radians by means of computation. The first two MSB bits among M-bit input to phase accumulator is used to estimate the quadrant of the signal. The remaining $(M - 2)$ bits are used to address one quadrant of sine points. To achieve high spectral purity we are in need of more sample points. In sinusoidal amplitude compression technique [6], the approximate sinusoidal value is estimated by using approximation circuitry, error value $e(x)$ is calculated from ideal sinusoidal wave and approximated value. ROM compression is not done effectively in the above techniques.

ROM1 provides a lower resolution phase and additional phase resolution is provided by ROM2. In piecewise linear approximation technique [7, 11], sine quadrant is divided as $S = 2^N$ linear segments. The slope coefficient C_i and initial amplitude M_i were stored in separate ROM LUT. Both the coefficients were accessed at single clock cycle. This technique is called time sharing. Based on segmentation, size of LUT is compressed. There are two types of segmentation: Uniform and non-uniform segmentation. In uniform segmentation, full cycle of sine points is divided into 2^S equal segments. Lower degree polynomial [8] is evaluated for each sub interval. The polynomial coefficient for entire segment is stored in ROM LUT. In Non-uniform segmentation [9, 10] is based on threshold value δ_i. Initially entire cycle of 2π radians is segmented to n macro segments. When $\delta_i = 0$, j and $j + 1$ macro segments were joined together and computed as a single segment. When $\delta_i = 1$, j and $j + 1$ were considered as two separate segments. DDFS characteristics is evaluated by Spurious free dynamic range (SFDR) which is the difference between amplitude of wanted sinusoid signal and amplitude of largest undesired frequency component. SFDR in non-uniform segmentation is high compared to uniform segmentation. Similar to the uniform segmentation, polynomial coefficient for each segment is stored in ROM LUT. Compared to uniform segmentation memory location to hold polynomial coefficients is less compared for non-uniform segmentation. The limitation in segment approach is deciding the segment boundaries.

2 Proposed DDFS Architecture

The proposed DDFS design provides effective reduction in ROM size compared to other techniques based on quarter wave symmetry technique is represented in Fig. 2. The output of the phase accumulator is of M bits. The input to ROM1 and ROM2 is of $(M - d)$ bits. Initially one quarter of the sine points is extracted, based on the number of samples and the difference between the consecutive segments, the value which is responsible for the growth of the segment is stored in ROM2 look-up table. Due to the gradual increase in the slope for pure sine wave, the sine points at ith instant are comparable to adjacent one. The starting point of each segment in stored in ROM1. It is expressed by equation,

$$x(i) = \mathrm{ROM1}(i) + \left(\mathrm{ROM2}(i)^{*}d\right) \tag{2}$$

i varies from 1 to n, where n represents the total segments for 0 to $\pi/2$ radians, i.e., one quarter wave sine points.

Figure 3 describes the proposed DDFS architecture. In this architecture, difference value is repeated for some interval of time. Therefore, unique difference value is stored in ROM2 and the starting point of each segment is stored in ROM1. In this architecture, frequency control word (FCW) corresponds to the phase increment of a waveform. FCW is incremented for each clock cycle. For each clock, phase register adds a FCW value which is stored in frequency register. Phase register generates

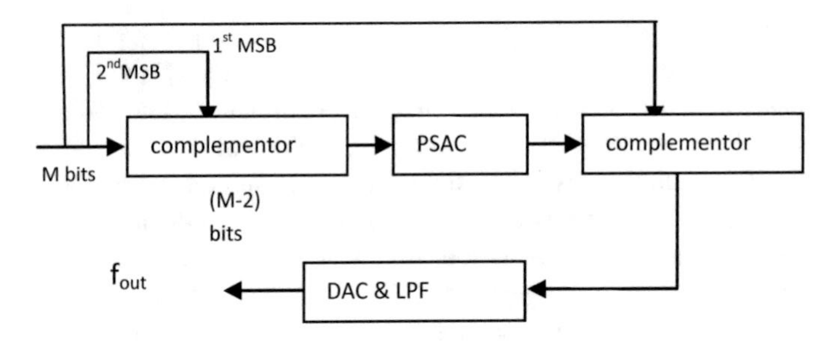

Fig. 2 Quarter wave symmetry

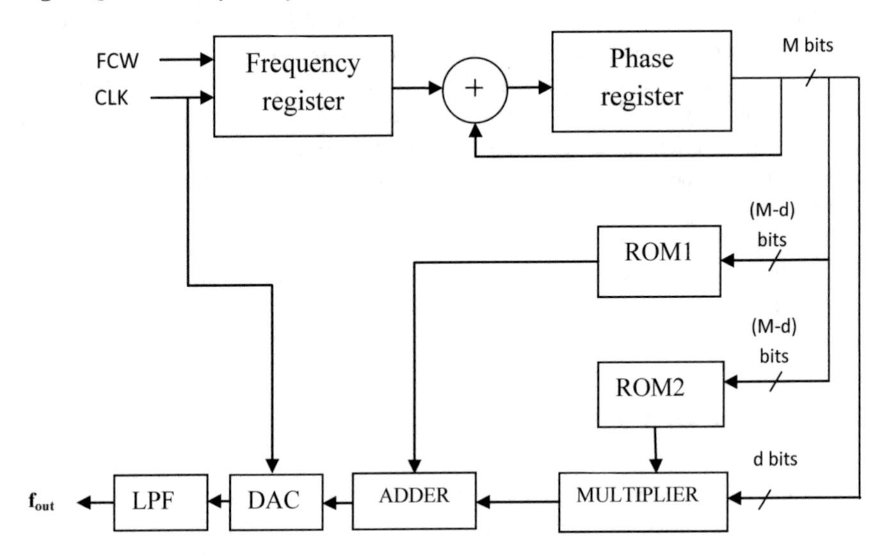

Fig. 3 Proposed DDFS architecture

M-bit address for each clock cycle. Among those M-bit address, $(M - d)$ bits are given as input to two ROMs (ROM1 and ROM2), d bits serves as input to multiplier. Only one quarter wave of the sinusoidal signal is implemented (0° to 90°), the full wave is reconstructed from quarter wave based on quarter wave symmetry technique [7]. İn this proposed technique, instant sinepoints can also generated. For example if phase accumulator generates address 10100001(8 bit). $M = 8$ bit, $d = 6$ bit, $(M - d) =$ 2 bits. The 2 bit MSB (10) addresses ROM1 and ROM2. 6 bitLSB(100001) addresses the multiplier. Figure 4 describes the sine wave generation from 2 ROM. Both ROM memory is accessed at a single clock. The sine amplitude in volt is extracted from specified ROM address and further computation is done by using adder and multiplier. The quarter wave of sinusoidal signal is divided into four segments. The sinusoidal signal generation is illustrated in Fig. 4 and its corresponding value is illustrated in Table 1.

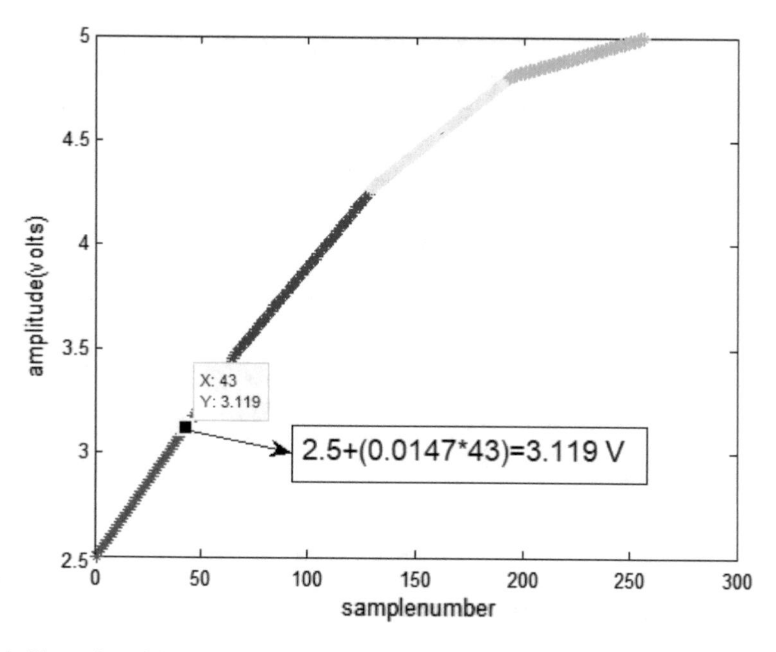

Fig. 4 Illustration of sinusoidal signal generation for quarter wave

Table 1 Computation of proposed design

Segment	ROM1	ROM2	Multiplier output at end of each segment	Computed output
1 (1:64)	2.5000	0.0147	0.0147 * 64	2.5000 + (0.0147 * 64)
2 (65:128)	3.4567	0.0125	0.0125 * 128	3.4567 + (0.0125 * 128)
3 (129:19)	4.2678	0.0084	0.0084 * 192	4.2678 + (0.0084 * 192)
4 (193:25)	4.8097	0.0030	0.0030 * 256	4.8097 + (0.0030 * 256)

3 Results and Discussion

The proposed architecture is simulated using MATLAB software. 2^j represents the maximum size of the ROM LUT. For 8 bit resolution 1024 sampling points, maximum size of ROM for conventional LUT is 1024. In quarter wave symmetry technique, 256 sine values needs to store in LUT whereas in proposed technique only 8 sine values were stored in LUT. From the above provided details, it is observed that the size of the LUT is drastically reduced. Figure 5 describes the sine generation by accessing ROM1(starting point at each segment value) and ROM2(difference value)

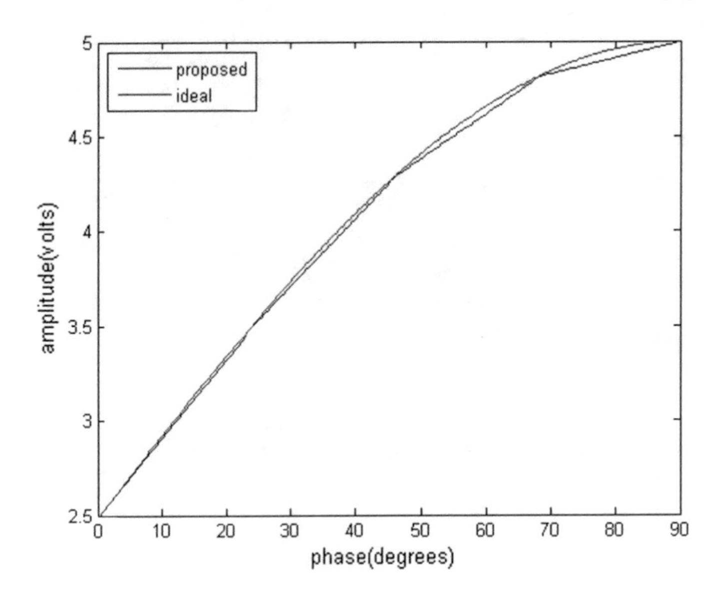

Fig. 5 Sine wave generation(quarter wave- 0 to π/2 radians)

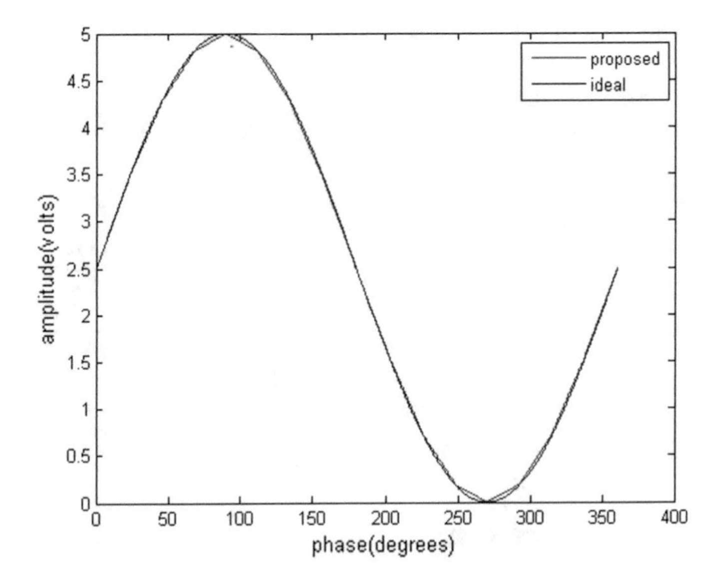

Fig. 6 Full wave reconstruction

range from 0 to π/2 radians. Figure 6 represents the reconstruction of sine wave from 0 to 2π radians based on quarter wave symmetry technique. The spectrum of proposed signal is illustrated in Fig. 7.

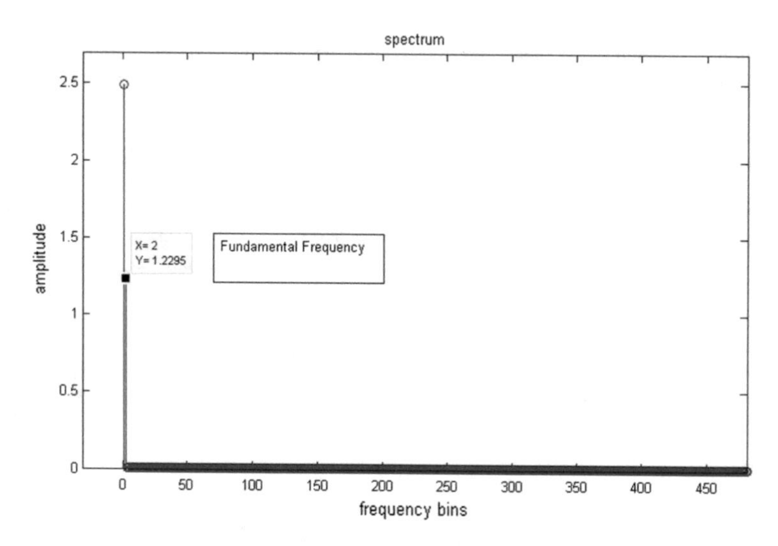

Fig. 7 Spectrum of proposed signal

4 Conclusion

Thus the Direct Digital Frequency synthesizer with compressed ROM is implemented with the SNR of 44.92 dB and SFDR is of 50.64 dBc based on segmentation. In conventional LUT For 1024 samples, maximum size of LUT needed is 1024, whereas for quarter wave symmetry technique maximum size is 256, for this proposed work, maximum size requirement is only 8. Hardware requirement for 1024 samples in conventional LUT 100% of the total size of LUT is used. In quarter wave symmetry technique, only 25% of the LUT is used for storing value compared to conventional LUT. In proposed technique, only 5% of the LUT is stored comparable to conventional LUT. Hardware overhead is reduced compared to other technique.

References

1. Kester, W.: Fundamental of direct digital synthesis, Analog Devices (2009)
2. Kester, W.: Analog-Digital conversion. Analog Devices, ISBN 0-926550-27-3 (2004)
3. Vankka, J., Halonen, K.A.I.: Direct digital synthesizers: theory, design and applications. In: Springer International series in Engineering and Computer Science (2001)
4. Langlois, J.M.P., Al-Khalili, D.: Phase to sinusoid amplitude conversion techniques for direct digital frequency synthesis. IEEE Proc. Circuits Devices Syst. **151**(6), 519–528 (2004)
5. Jafari, H., Ayatollahi, A., Mirzakuchaki, S.: A low power, high SFDR, ROM -less direct digital frequency synthesizer. In: IEEE Proceedings on Electron Devices Solid State Circuits (2005)
6. Chimakurthy, L.S.J., Ghosh, M., Dai, F.F., Jaeger, R.C.: A novel DDS using nonlinear ROM addressing with improved compression ratio and quantization noise. In: IEEE Transaction on Ultrasonics Ferroelectric Frequency Control (2006)

7. Ashrafi, A., Adhami, R.: Theoretical upper bound of the spurious free dynamic range in direct digital frequency synthesizers realized by polynomial interpolation methods. In: IEEE Transaction on Circuit System (2007)
8. Chen, Y.H., Chau, Y.A.: A direct digital frequency synthesizer based on a new form of polynomial approximations. IEEE Trans. Consum. Electron. **56**(2), 436–440 (2010)
9. De Caro, D., Petra, N., Strollo, A.G.M.: Direct digital frequency synthesizer using nonuniform piecewise-linear approximation. IEEE Trans. Circ. Syst. **18**, 1308–1312 (2011)
10. Langlois, M.P., Al-Khalili, D.: Novel approach to the designof direct digital frequency synthesizers based on linear interpolation. IEEE Trans. Circuits Syst. II, Analog Digit. Signal Process **50**(9), 567–578 (2003)
11. Omran, Q.K., Islam, M.T.: An efficient ROM compression technique for linear-interpolated direct digital frequency synthesizer. IEEE Conf. Semicond. Electron. **48**, 2409–2418 (2014)
12. Nicholas, H.T., Samueli, H., Kim, B.: The optimization of direct digital frequency synthesizer performance in the presence of finite word length effects. In: IEEE Transaction on Analog Digital Signal Process (2008)

S. Karpagavalli is pursuing Master degree in Wireless technologies at Thiagarajar College of Engineering, Madurai, Tamil Nadu. She received his B.E. degree in Electronics and Communication Engineering from Mepco Schlenk Engineering College. Her areas of interest includes digital circuits, embedded system design and communication.

Dr. K. Hariharan is an Associate Professor in Electronics and communication Engineering department at Thiagarajar College of Engineering, Madurai, Tamil Nadu, He received his doctoral degree in VLSI Testing from Anna University Chennai and M.E. degree in communication systems from Thiagarajar College of Engineering, Madurai and B.E. degree in Electronics and Communication Engineering from Madras University. His research interest includes digital Computer Architecture, VLSI system design, Mixed signal IC testing and embedded system design. He has registered two Indian patents in these relevant areas.

G. Dheivanai is pursuing her B.E. degree in Electronics and Communication Engineering at Thiagarajar College of Engineering, Madurai. Her areas of interest includes embedded system design, microprocessor and microcontroller.

M. Gurupriya is pursuing her B.E. degree in Electronics and Communication Engineering at Thiagarajar College of Engineering, Madurai. Her areas of interest includes embedded system design, microprocessor and microcontroller.

Link Layer Traffic Connectivity Protocol Application and Mechanism in Optical Layer Survivability Approach

K. V. S. S. S. S. Sairam and Chandra Singh

Abstract In today's prospective and perspective nature optical communication and networks have been emerged by means of Fiber Distribution Connectivity (FDC), Optical Relocation Mechanism (ORM), Optical Link Path Assortment Approach (OLPAA) and System Survivability Network Integrated Approach (SSNIA). Further it can be enhanced in the direction of Link Layer Traffic Connectivity Protocol Applications and Mechanism (LLTCPAM) is enlightened in our paper. In future course demand queuing is a versatile factor, it will be used as a resource facility for the users hence the different mechanism such as Parcel List Algorithm (PLA) have been analyzed in order to compare the Network Connectivity Parameter (NCP).

Keywords Fiber distribution connectivity (FDC) · Optical relocation mechanism (ORM) · Optical link path assortment approach (OLPAA) · System survivability network integrated approach (SSNIA) · Link layer traffic connectivity protocol applications & mechanism (LLTCPAM) · Parcel list algorithm (PLA)
Network connectivity parameter (NCP) · Cost connectivity matrix (CCT)
Link connectivity matrix (LCM) · Traffic connectivity matrix (TCM)

1 Introduction

Survivability is a computation of network recovery capabilities when the failure occurs in the networks. It consists of Span, Hub, COs, Switches, and Gateways across many network segments [1]. It represents the layered architecture in order to represent the components with methods such as Fiber Distribution Connectivity (FDC), Optical Relocation Mechanism (ORM), Optical Link Path Assortment Approach (OLPAA) and System Survivability Network Integrated Approach (SSNIA) [2]. It

K. V. S. S. S. S. Sairam · C. Singh (✉)
Department of E&CE, NMAM Institute of Technology, Nitte, Karkala, Karnataka, India
e-mail: chandrasingh146@gmail.com

K. V. S. S. S. S. Sairam
e-mail: drsairam@nitte.edu.in

S. Smys et al. (eds.), *International Conference on Computer Networks and Communication Technologies*, Lecture Notes on Data Engineering and Communications Technologies 15, https://doi.org/10.1007/978-981-10-8681-6_43

also describes the network routing with network elements. Fiber optic restoration is vital scenario where the point to point, span Architecture is used to overcome multilevel link failures. The design tool here used such as c-programming and ns2 simulators. It provides optimum design in distinguish with different network topologies with parameters.

2 Problem Definition

2.1 *Fiber Distribution Connectivity (FDC)*

In FDC Point to Point Architecture is in the form of non-centralized is used as shown in Fig. 1. In this central office and hubs are connected through the clusters and the categories of clusters are served by a connectivity which is known as gateway. In turn these gateways are connected by fiber systems [3]. These fiber systems are designed in the form of fiber joints and these fiber joints are used in point to point as compared to hub hierarchy. The Point-to-Point Fiber Joint Architecture (PPFJA) depicts hubbing architecture by itself. It determines the traffic in each link to obtain the fiber network system and throughput in the form of path matrix and data matrix [4].

In our work three different topologies viz 6 × 6, 8 × 8, and 12 × 12 non-centralized PPFJA is considered to estimate Fiber distribution Connectivity (FDC) which is depicted in Table 1. The parameter are path which determines the connectivity between user to user and also to provide end to end connection whereas path is well associated in the form of data which is processed in the form of traffic in each link [5].

Graph which determines the connectivity in terms of traffic in each link is depicted by three different topologies as shown in (Fig. 2).

Table 1 Fiber distribution connectivity (FDC)

Parameter									
Network topology	6 × 6			8 × 8			12 × 12		
Path	(1–2)	(2, 3)	(4, 5)	(2, 8)	(7, 8)	(4, 8)	(12, 11)	(7, 12)	(5, 7)
Cost	7	9	11	6	8	10	9	10	4
Traffic Connectivity Matrix	44	50	65	37	45	70	78	86	36

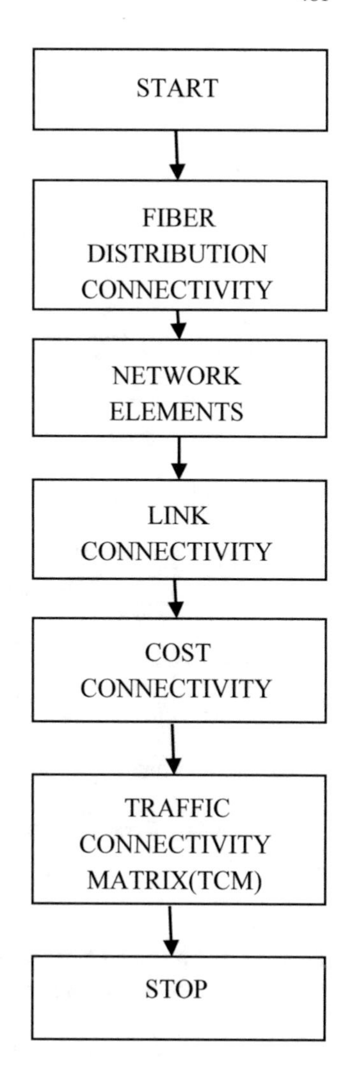

Fig. 1 Flowchart of fiber distribution connectivity (FDC)

2.2 *Optical Relocation Mechanism (ORM)*

In ORM the FDC is further extended in order to protect and restore the traffic in each link for given network topology. This architecture ensures the protection at different digital signal levels which are assigned randomly between a common node (server) to different users to vice versa. The implementation provides the estimation in the form of distinction between Associated Non-Associated Path Quasi Modes (ANQM) [6] further the relocation mechanism supports huge no demands through PPJFIA which facilitates and also extends the mechanism but an IRS (Integration Restoration Scheme). In this ORM three different topologies, i.e., 6×6, 8×8, and

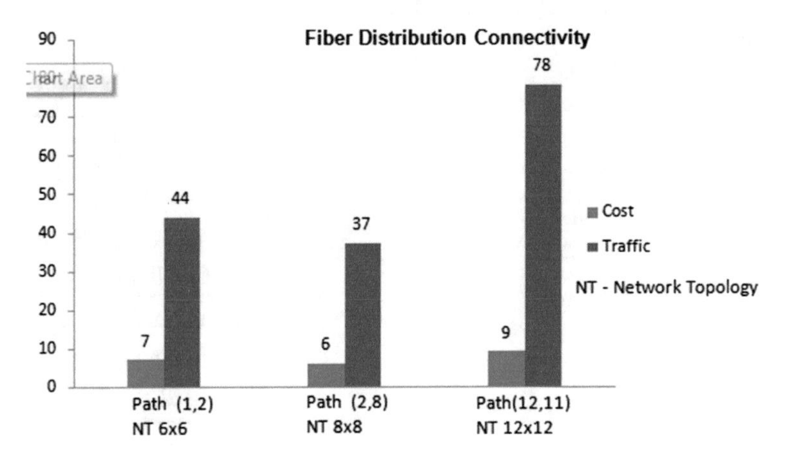

Fig. 2 FDC output

Table 2 User relocation factor

Parameter	Network topology								
	6 × 6					8 × 8			12 × 12
Total demand	120	128	140	130	152	160	150	172	195
Demand after failure	105	108	115	115	138	140	135	142	168
Protection scheme	1:2DP	1:2DP	1:2DP	1:2DP	1:2DP	1:2DP	1:2DP	1:2DP	1:2DP
User relocation factor (URF)	87.5%	84.37%	82.14	88.46%	90.7%	87.5%	90%	82.55%	86.15%

12 × 12 and also can be measured towards NxN where the demands are assigned via central node to end to end system (Fig. 3).

In this work for example main user (Mu) is connected according to topology configuration. Further demands are assigned through different Digital Signal Link Level Design (DSLLD). Where it is considered w.r.t total no of demands.

Protection Mechanism is assigned for Individual link Failure (ILF) and Multi-Link Failure (MLF). Protection Scheme such as 1:2DP, SHR, APS are used. Failed path is taken into consideration and alternate path (Backup Path) is employed. Finally relocation mechanism is estimated, i.e., failed demand to the total no demand through fiber network restoration mechanism (Table 2).

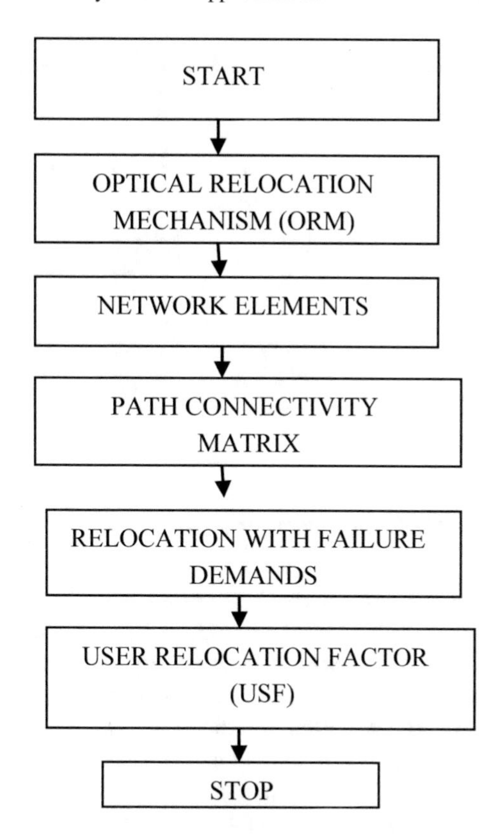

Fig. 3 Flowchart of Optical Relocation Mechanism (ORM)

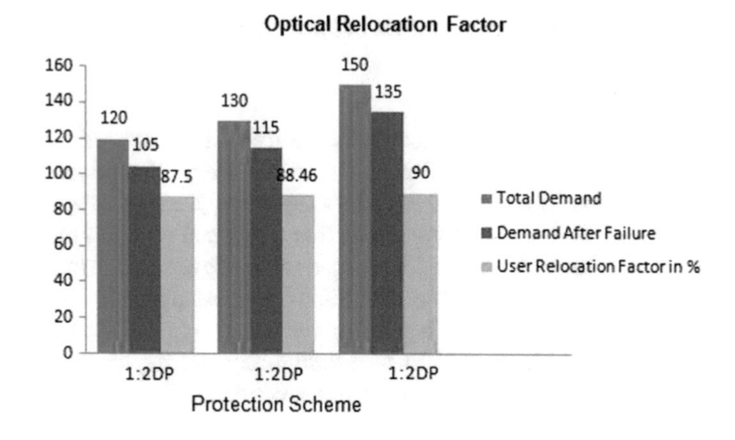

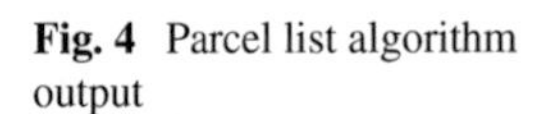

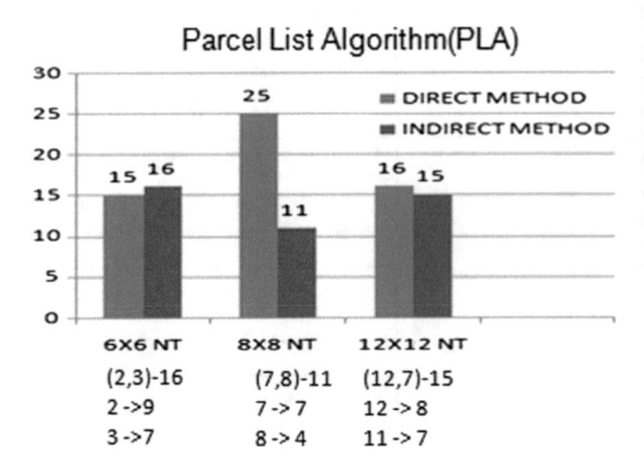

Fig. 4 Parcel list algorithm output

2.3 *Optical Link Path Assortment Approach*

In optical link layer survivability approach, the extension of first and second problem and is further evaluated in the form of Optical Link Path Assortment Approach (OLPAA). It describes the network cross connectivity in terms of demands. That is, routed through direct and indirect path. In direct path the signal transmits the data directly between source to destination in indirect path routing in terms of hubs via parcel list. It results in multilevel period optimality on each routing path for a given network topology. In this approach it increases multilevel demands connectivity's for multiperiod where the max line rate is taken into consideration (Table 3; Fig. 4).

2.4 *System Survivability Network Integrated Approach (SSNIA)*

These three approaches are further extended System Survivability Network Integrated Approach (SSNIA). In these it facilitates fiber distribution, fiber relocation mechanism, OLPAA Approach; finally the capacity is measured through integrated approach. It also facilities digital signal transformation levels and also end-to-end connectivity for multiyear demands and it calculates the automatic restoration facility regarding working and protection fiber. These all SSNIA in these planning models determines the growth strategy and from beginning year to nth year is calculated. This total network depicts multiplex performance design for selecting appropriate fiber optic network survivable architecture and planning period [7, 8].

Table 3 Parcel list algorithm (PLA)

Parameter	Network topology								
	6 × 6			8 × 8			12 × 12		
Digital signal levels	DS1			DS1			DS1		
Circuit configurations	28	28	28	28	28	28	28	28	28
Direct path assortment	(1,2)-	(2,3)-	(4-5)-	(2,8)-	(7,8)-	(4-8)-	(12,11)-	(7,12)-	(5,7)-
	20 → 20	>15	>26	25 → 25	>18	>26	16 → 16	>15	>16
Parcel list algorithm (PLA)	(1-2)-20	(2,3)-	(4,5)-	(2-8)-25	(7,8)-	(4,8)-	(12,11)-17	(7,12)-	(5,7)-
	1 → 11	>16	>17	2 → 13	>11	>19	12 → 10	>15	>27
	2 → 9	2->9	4->8	8 → 12	7->7	4-	11 → 7	7->8	5-
		3->7	5->9		8->4	>11		12->7	>11
						8->8			7-
									>16

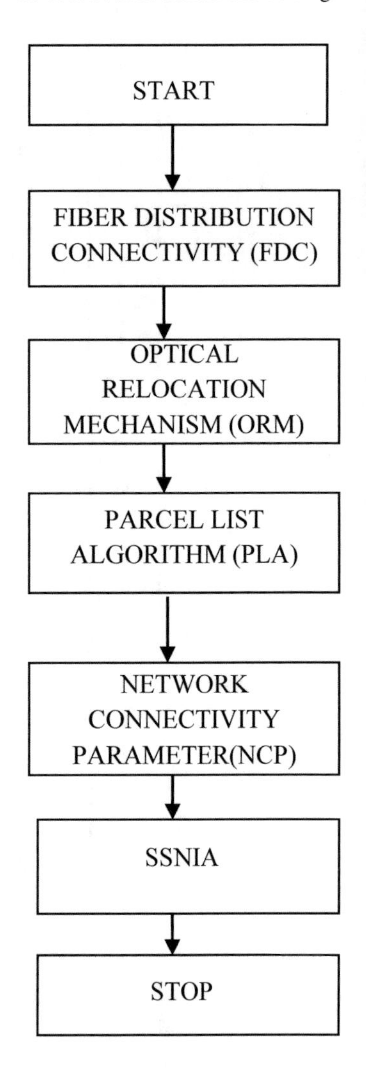

Fig. 5 Flowchart of system survivability network integrated approach (SSNIA)

In this model the topology is in the form of ring architecture whereas incremental demand should be less than maximum line rate. In planning period, an incremental demand on any ring should remain less than maximum line rate. For high-speed architecture is well as considered by using hubbing/diverse/diverse architecture (Table 4; Fig. 5).

Table 4 System survivability network integrated approach (SSNIA)

Parameter	Network topology								
	6 × 6			8 × 8			12 × 12		
Traffic connectivity matrix	44	50	65	37	45	70	78	86	36
User relocation factor (URF)	87.50%	84.37%	82.14	88.46%	90.70%	87.50%	90%	82.55%	86.15%
Parcel list algorithm (PLA)	(1-2)-20	(2,3)-	(4,5)-	(2-8)-25	(7,8)-	(4,8)-	(12,11)-	(7,12)-	(5,7)-
	1→11	>15	>17	2→13	>11	>19	16	>15	>27
	2→9	2->8	4->8	8→12	7->7	4->11	12→9	7->8	5->11
		3->7	5->9		8->4	8->8	11→7	12->7	7->16
DS-STS-STM-OC	115	118	124	126	129	138	150	165	220

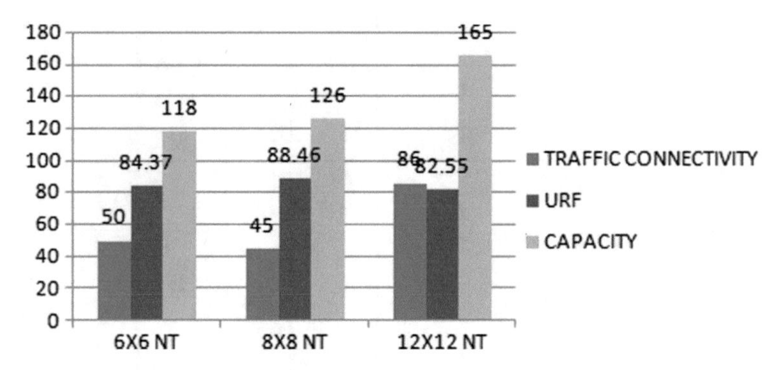

Fig. 6 SSNIA output

Graph
(See Fig. 6).

3 Conclusion

The reconfigurable optical network provides users equipment and optical fiber equipment with each other. The PPJFA approach enhances the performance of survivable fiber network and correlation of these results collaborates this algorithm validation. These architectures are well used in the context of incremental changes towards the exit network in order to improve survivability throughout in optical network. Further this work can be extended by convergence of distributed algorithm by using synchronous competition and scholastic approximation and also transition from ring-to-ring-mesh-hybrid approach.

References

1. Skoog, R., Von Lehmen, A.: Metro network design methodologies and demands. IEEE J. Light Wave Technol. **22**(11), 2680–2692 (2015)
2. Kavian, Y.S., Rashvand, H.F., Leeson, M.S., Ren, W., Hines, E.L., Naderi, M.: Network topology effect on QoS delivering in survivable DWDM optical networks. J. Telecommun. Inform. Technol. 68–71 (2016)
3. Zhou, D., Subramaniam, S.: Survivability in optical networks. IEEE Netw. **14**, 16–23 (2015)
4. Strand, J., Chiu, A., Tkach, R.: Issues for routing in the optical layer. IEEE Commun. Mag. **39**, 81–87 (2013)
5. Zang, H., Ou, C., Mukherjee, B.: Path-protection routing and wavelength assignment (RWA) in WDM mesh networks under ductlayer constraints. IEEE-ACM Trans. Networking **11**(2), 248–258 (2015)

6. Shao, X., Zhou, L., Cheng, X., Zheng, W., Wang, Y.: Best Ef-fort shared risk link group (SRLG) failure protection in WDM networks. In: Proceeding IEEE International Conference on Communication, pp. 5150–5154 (2015)
7. Katiyar, P., Dwivedi, R.K., Jain, A.K.: Network survivability in WDM. In: Proceeding IEEE International Conference on Communication, pp. 150–154 (2015)
8. Strand, J., Chiu, A., Tkach, R.: Issues for routing in the optical layer. IEEE Commun. Mag. **39**, 81–87 (2011)

Dr. K. V. S. S. S. S. Sairam B.E., M.Tech., Ph.D. working as a Professor, E&CE Department and also IEEE student branch Counselor in NMAMIT, NITTE . He obtained his B.E. (ECE) from Karnataka University Dharwad in 1996, M.Tech (Industrial Electronics) from SJCE, Mysore University Mysore,1998 and Ph.D. (ECE) (OPTICAL COMMUNICATIONS) JNTUH, Hyderabad in 2013. He is having 20 years experience in Teaching and Research too. He published 45 papers in International Journals, National Journals, Conferences and Workshops. He is Springer Reviewer, IEEE Reviewer and Editorial Member and reviewed 32 International papers. He authored 2 Books, Optical Communications, Laxmi Publications, New Delhi and Basic Electrical and Electronics Engineering, Krishnan Publications, Chennai. He is guiding five PhD students, guided 40 M.Tech Projects and B.E. Projects too. His research areas are optical communications, optical Networks and Wireless Communication. He is associated with IEEE Mangalore Subsection and conducted several events like I2 CONNECT, The International Year of the Light-2015, Light beyond the bulb in JNNCE, shivamoga, Technical talks and etc. He gave a talk on "Optcial Communication", organized under IEEE Student Branch, Canara Engineering College. He achieved the First Prize for addressing the IEEE Student branch contest organized by IEEE Bangalore Section in April 2015. He was awarded as IEEE Student Branch Counselor Award organized by IEEE Hyderabad Section in 2004. He was awarded as the BEST FACULTY in Dr. M.G.R. Deemed University, Chennai for his best performance towards 100% RESULT in the University Examinations in 2000. He received the YUVA ENGINEER AWARD from MHRD in MAY 2014 for his outstanding contribution in the field of OPTICAL COMMUNICATIONS. He organized 100 events regarding IEEE student Branch Activities in FDP, Guest Lectures, Seminar Talks, Workshops, Globally Student Exchange Program, Research Methodologies and Paper Publication etc., Special Lectures were given on. Recent Trends in Optical Communications, JAN 19, 2005, IEEE Students Branch, ECE Department, NIT WARANGAL, A.P. State, INDIA, Internetworking Design by Using FOST, August 6, 2005 IEEE—CS Chapter, CSE Department, Osmania University, Hyderabad, A.P. State, INDIA. Recent Trends in Optical Communications, January 20, 2006, SBIT, Khammam, Andhra Pradesh, India. Future Technology—An Overview, March 10, 2006, S.R.K Engg College, Chennai, Tamilnadu State, INDIA. Recent Trends in OPTICAL COMMUNICATIONS, Canara Engineering College, IEEE subsection on March 2015. Research Opportunities in Engineering, PA College of Engineering in May 2016. He organized ISRO SPACE EXHIBITION, as a part of the event the demonstration water rocket launch experiment was carried out on both sessions on 27th and 28th September 2017. There were expert lectures from Sri. Guruprasad B R on both days of the event. All schools and colleges of Udupi and South Canara districts were invited for the event. More than 2000 students and faculty from various institutions have actively participated in the event. The event was held at two places parallely. There were not only exhibits and panels displayed during the exhibition, but also rare video capsules from ISRO were played continuously in the indoor auditorium.

Mr. Chandra Singh he has obtained M.tech (DEC) from NMAM Institute of technology Nitte in the year 2018. He has obtained B.E (E&C) at Srinivas School of Engineering, Mukka (2015), Mangaluru. He has published about 10 papers in M.tech. It contains IEEE papers and even Springer lecture note series too. I have received best paper award for the paper entitled "Fiber Optic Network Integration by using Survivability Approach" at *Malviya National Institute of Technology, Jaipur*. Even Received Second Prize at Project exhibition held at NMAMIT nitte in the year 2018 and also received 3rd Prize for my B.E project entitled "Novel Iris Recognition System" at SSE Mukka in the year 2015. My area of interest are Optical Networking and communication, wireless communication and Image processing. I am a active participant of IEEE Mangaluru sub-section. I have served as NMAMIT NITTE student branch president for the academic year 2016–17. As execom member of IEEE Nitte Student Branch we have conducted many technical talks and workshops. One of the main event which we conducted in my tenure was I2 CONECCT-2017 which was held in NMAMIT Nitte.

DORBRI: An Architecture for the DoD Security Breaches Through Quantum IoT

Beatrice A. Dorothy and Britto S. Ramesh Kumar

Abstract Internet of Things is the one based on data analytics and the competence to gain an accurate and deep understanding about the sensors. Security is the most crucial and vital need for developing IoT Communication wireless network. Classical Cryptography has provide a way to ensure the fundamental needs of communication networks but the security is not to the optimal level and the downside may be dealt by Quantum technology. To constrain these insights, IoT applications depend on neural networks and machine learning to extract Knowledge data based decision support. The DORBRI architecture proposed in this paper addresses the issues that are faced by the cryptosystem through the Post-Quantum cryptography based Quantum IoT. The block chain allows us to ensure the soundness and permanence of the information since you cannot modify or capture data without that particular action get recorded. Communication over the proposed secure channel will be faster and considerably reduce the loss of data.

Keywords Quantum IoT · Internet of things · Block chain
Quantum cryptography

1 Introduction

The Internet of Things is the one that has prime source connected to sensors which senses the environment and control the data over the wireless network communication. Sensors engage in recreation of a essential role in the perception of IoT to explore the status of Things. One of the resources of access or control that provides the energy to set and keeps IoT in motion is data from Wireless Sensor Networks.

B. A. Dorothy (✉) · B. S. Ramesh Kumar
Department of Computer Science, St. Joseph's College, Tiruchirappalli, India
e-mail: adorothybrice@gmail.com

B. S. Ramesh Kumar
e-mail: brittork@gmail.com

S. Smys et al. (eds.), *International Conference on Computer Networks and Communication Technologies*, Lecture Notes on Data Engineering and Communications Technologies 15, https://doi.org/10.1007/978-981-10-8681-6_44

What sets the success of the connecting devices to make them more competent is dependent relative upon access, storage, retrieval, and processing of data.

The Internet of Things clutches with it a bunch of inconceivable latent like Intelligence of sensors, connectivity networks, convolution, Volume and Space. Regrettably, security and privacy aspects have been left unsubstantiated. One of the security facet is data encryption and decryption. The massive amount of data that spring from sensors is the Edge computing; where sensors will be sending out data to a gateway device, where the preprocessing of data in the cloud or data centers is reduced. Edge computing is superlative to set out IoT applications, for the reason that it consent data analytics and reduces complex system network travel.

Quantum computing is a paradigm shift. Quantum cryptography is the branch of Cryptology which consists of quantum mechanical properties as its key aspect. Quantum cryptography applies Heisenberg's uncertainty principle and the no-cloning theorem for its tremendous performance.

2 Literature Review

The Internet of Things will be a promising future wireless network which will have the ability to sense the environment, collect the information, and preprocess the data which is facilitated through Artificial Intelligence and neural network.

Ariel Amster has discussed that cryptography is most important task in quantum computing. Normal encryption did not allow making the file security. This cryptography with the help of quantum computing, can access the secure file with crypts key [1].

Scott Amyx has talked about that in the Internet of Things; there are some pervasive and difficult-to-solve challenges. Security is one of the biggest challenges hampering the growth of IoT. There are other issues that prevent IoT from truly taking off on a large-scale, e.g., interoperability, network latency, privacy, etc. Quantum computing has the potential to not only break all current Cryptography also creates new bullet-proof security that cannot be hacked by classical computers or even quantum computers. With IoT, the volume of data and different data types are increasing. Predictive data analytics will be able to support machine learning on the edge on a cloud-like scale without ever going to the Internet [2].

In the blog Telefonica it is said that Quantum encryption is a discipline that is currently booming since it allows information to be secured beyond what we could have imagined. As we know, in a world in which cyber attacks are growing, it is extremely important to safeguard our information. To solve problems of the neural networks used in machine learning, you have to be able to mathematically optimize certain functions with a huge amount of data [3].

Martín Serrano has verbalized that in the network, all real and virtual items have specific identification and physical sensory data in order to achieve the goal of information sharing through seamless connection of intelligent interface [4].

Mark Gerasimas in his paper has conferred that Radio retransmission is a common military technique used to provide long-range, secure, line-of-sight (LoS) voice communications in remote and austere environments. Units establish retrans stations by slaving together two amplified radios which function to receive any signals on one frequency within range and subsequently retransmit them on a separate frequency. Due to the lower signal strength of tactical VHF radios, the retransmitted signal is typically amplified to ensure the signal is received at all nodes operating in the area. The station location is often on a prominent terrain feature to ensure maximum reach, especially in areas prone to breaks in communication or "dead space" such as mountainous terrain [5].

3 Proposed System

3.1 Problem Definition

The Internet of Things requires enormous scalability in the network space to lever the heave of devices and consign to dock them. The sensors in the network will patch big data to the data centers, where there arise Data processing and management complexity [6], Heterogeneity between the devices and Security problem [7]. Artificial intelligence, Neural network and Big Data Preprocessing are issues inherent to the nature of the IoT.

By analyzing the IoT security vulnerabilities, there is a security breaches in the secure channel that comes between the Router to Cloud/Data center and Cloud/Data center to the End users. The entire cryptography will be based on Quantum properties, combining quantum computing with Edge computing, cloud technology and IoT.

IoT sensor devices are especially at high risk. They are easily exposed to attack by malicious node or susceptible activity. RSA, ECC, and Digital signatures are the finest cryptosystem which are used to secure the data during transmission [8–10]. It is proven that a classical encrypted data can be break in a feasible amount of time by supercomputing.

3.2 Methodology

Quantum computing is a change of assumption for using the cryptography in an effective way. The key Quantum computing principles are that includes Heisenberg's uncertainty principles, Quantization, Quantum superposition, entanglement and Quantum decoherence. Quantum Key Distribution (QKD) is the most significant facet of Quantum Cryptography and for Quantum Secure channel Communication. Quantum Communication is the one which consists of Quantum information processing, Quantum teleportation, Quantum network and Quantum Channel.

In this proposed architecture for Department of Defence (DoD), the sensor nodes that are deployed in the war field and various fields of monitoring areas collect data from the environment. The collected data from the sensors will be sent to the sink node or the Gateway. A gateway allocates a place to preprocess the data at the edge in prior to transferring the information to the data center. When data is cumulative, concise and strategically analyzed at the edge, it minimizes the volume of data that needs to be forwarded on to the data center. By refining the data from the noisy data it can reduce response times and network transmission travel.

A base station is a radio receiver or transmitter that serves as the hub of the local wired or wireless network. It consists of a low power transmitter which broadcast the data to the router. From the router, the data are sent to the data center through a secure channel. Secure channel is a way of transferring data that is resistant to fabrication and modification. In this secure quantum channel, we can deploy both the quantum and classical information. Quantum Channel is a linear map between spaces of operators or matrices. The photon's time of onset is used to lay down quantum information for the quantum cryptography. The Quantum channel is proficient of transmitting base states as well as superposition's of these states. The coherence of the state is sustained during transmission through the Quantum channel.

Quantum Key Distribution distributes random number keys securely by make use of the Heisenberg uncertainty principles. In these uncertainty principles, measurements of certain scheme cannot be done without affecting the method until the operations include extremely low-noise technology. The quantum information coded by quantum states allows tasks to perform much more efficiently in contrast of using classical information. Quantum Key Distribution offers an unconditional mechanism and secure communication. The key distribution can be used with any existing encryption algorithm to encrypt and decrypt a data. These encrypted or decrypted data can then be transmitted over a secured communication channel based on discrete variable entangled states (Fig. 1).

Quantum encryption and decryption associated computational complexity for such data makes this decryption impossible to break. Quantum teleportation uses block chain, which contains the information of a cryptographic hash of the previous block, a timestamp and transaction data. A block chain is intrinsically resistant to fabrication and modification of the data. Block chain allows us to ensure the security and stability of the information since you cannot modify or capture data without that particular action get recorded in the block. Communication over the secure channel will be faster and considerably reduce the loss of data. We will be able to achieve more efficient energy management.

The end user, Department of Defence (DoD) consists of Central Intelligence agency (CIA), Technological and logistics Directorate, Army Headquarters, Navy, Intelligence and Air force. These departments receive the preprocessed data and use them for various tasks to control and maintain the position of the troop, to deploy orders based on the Intelligence report which has been already preprocessed in the data center with the backing of machine learning support. Once the outlier or noisy data is reduced we can deliver Quality of Service (QoS) in a reduced time and network travel.

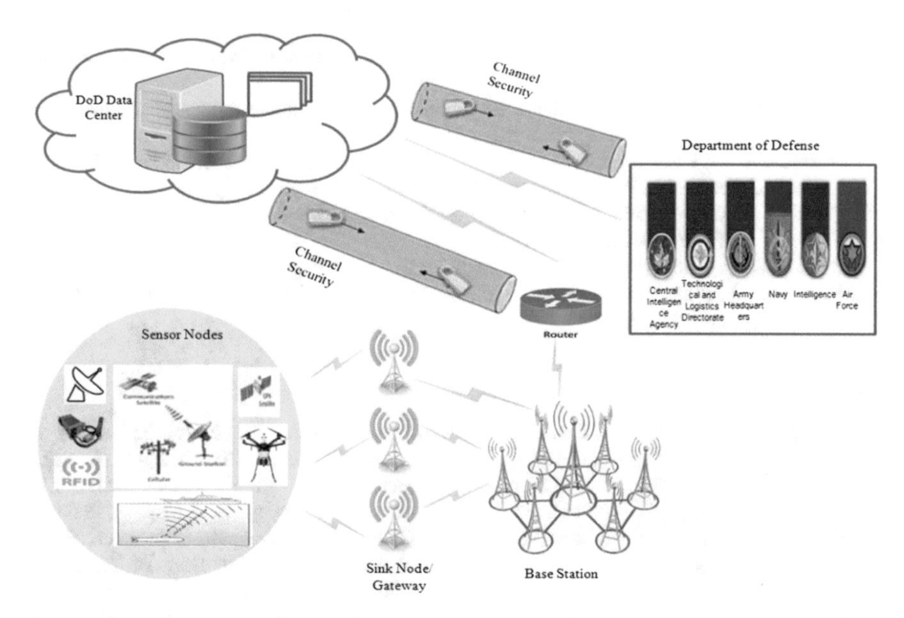

Fig. 1 DORBRI architecture for secure DoD communication

4 Conclusion

The existing security issues with the sensors by changing it into malicious nodes can be secured by the block chain. To avoid this integer and discrete mathematical issues, this proposed algorithm has been used. Even though, experimental quantum computers lack processing power to break the existing cryptographic algorithm. To secure against quantum threat we can rely on McEliece, lattice-based schemes and most symmetric-key algorithms. Quantum encryption is the one that allows information to be secured. Cyber attacks are growing; it is tremendously significant to uphold the sensitive information. The main problem for quantum computing is decoherence. It is the course of action of using a shared key to establish a secure quantum communication. The End user, Department of Defence (DoD) will also be able to access the data that are stored in the Data center more securely and effectively.

References

1. Amster, A.: Internet of Things: Big Data and Data Security Problems (2016)
2. Amyx, S.: The Potential Impact of Quantum Computing on the Internet of Things (2018)
3. https://iot.telefonica.com/blog/quantum-computing-in-the-future-of-the-iot
4. Serrano, M., Barnaghi, P., et al.: IoT Semantic Interoperability. Eur. Res. Clusters **4**, 450–500 (2015)

5. Gerasimas, M.: A Big Mesh—Military Tactics and the Future of Mobile Communications (2015)
6. Beatrice Dorothy, A., Britto Ramesh Kumar, S.: Internet of things—data management and security. Int. J. Control Theor. Appl. **9**(26), 115–120 (2016)
7. Beatrice Dorothy, A., Britto Ramesh Kumar, S., Jerlin Sharmila, J.: IoT Based Home Security Through Digital Image Processing Algorithms. IEEE, pp. 20–23 (2017)
8. Cheng, C., Lu, R., Petzoldt, A., Takagi, T.: Securing the Internet of Things in a Quantum World, pp. 116–120. IEEE (2017)
9. Routray, S.K., et al.: Quantum Cryptography for IoT—A Perspective. IEEE (2017)
10. Buchmann, J., et al.: Post-Quantum Cryptography—State of the Art, The New Codebreakers. Springer, pp. 88–108 (2016)

Beatrice A. Dorothy is pursuing Doctor of Philosophy in Department of Computer Science, St. Joseph's College, (Autonomous), Tiruchirappalli, Tamil Nadu, India. She received her Master of Philosophy in Computer Science from St. Joseph's College, Tiruchirappalli. She received her Master of Computer Application degree from Bishop Heber College, Tiruchirappalli. She has published research articles in the International Conferences and Journals. Her area of interest is Internet of Things and Quantum Computing.

Britto S. Ramesh Kumar is working as Assistant Professor in the Department of Computer Science, St. Joseph's College (Autonomous), Tiruchirappalli, Tamil Nadu, India. He has published many research articles in the National/International Conferences and Journals. His research interest includes Internet of Things, Neural Networks, Data Mining, Web Mining, and Mobile Networks.

An Innovative Way of Increasing the Efficiency of Identity-Based Cryptosystem by Parallelization of Threads

Sandeep Kumar Pandey, Tapobrata Dhar, Ryan Saptarshi Ray, Aritro Sengupta and Utpal Kumar Ray

Abstract In today's world, the most popular method of utilizing the public-key cryptosystem for secure communication between users is through a public-key infrastructure (PKI). But this infrastructure is time consuming, error prone, not user friendly as it requires authentication of public key using a valid Certifying Authority (CA). The most suitable alternative to PKI is ID-Based Cryptosystem (IBC). It is faster, secure, and user friendly as the public keys are constructed using the unique IDs of the registered users in the system. This infrastructure does not require a CA for ensuring public-key validity, thus speeding up the cryptographic procedure. However, the overall time consumed by the IBC algorithm can be significantly reduced by parallelizing. In this article, we have implemented Boneh–Franklin's IBC scheme using bilinear pairings. We have parallelized the decryption phase, thereby reducing the total communication time. The performance result of the parallelization is recorded in tabulated form as well as graphical form.

S. K. Pandey
Department of Information Technology,
Jalpaiguri Government Engineering College, Jalpaiguri, West Bengal, India
e-mail: pandey.1996.sandeep@gmail.com

T. Dhar (✉)
Department of Information Technology, Indian Institute of Engineering Science and Technology, Shibpur, West Bengal, India
e-mail: tapobrata.dhar91@gmail.com

R. S. Ray · U. K. Ray
Department of Information Technology, Jadavpur University, Kolkata, West Bengal, India
e-mail: ryan.ray@rediffmail.com

U. K. Ray
e-mail: utpal_ray@yahoo.com

A. Sengupta
Ministry of Electronics and Information Technology, Government of India,
Electronics Niketan, New Delhi, India
e-mail: aritro.sengupta@meity.gov.in

S. Smys et al. (eds.), *International Conference on Computer Networks and Communication Technologies*, Lecture Notes on Data Engineering and Communications Technologies 15, https://doi.org/10.1007/978-981-10-8681-6_45

Keywords Boneh–Franklin IBC · Threads · Elliptic curve cryptography
Bilinear pairing

1 Introduction

In a public-key cryptosystem, a pair of keys, a public key and a private key is used to ensure secure transmission of data over an insecure channel. The sender encrypts a plaintext message using the public key and creates the ciphertext. The receiver decrypts the ciphertext using the private key to get the plaintext [1, 2]. One of the most significant limitations of public-key cryptography is that it is heavily dependent on the public-key infrastructure (PKI). Most of the public-key systems use a third party to certify the validity of the public keys [3]. The sender and the receiver generate their own private- and public-key pairs and submit their own certificate request to a valid Certificate Authority (CA) along with the proof of identity. After successful authentication, the CA sends signed certificates, which is used by the entities to authenticate each other and communicate securely. This process is time consuming as there is an overhead of sending and receiving certificates from CA. Also, this method is error prone and not user friendly, as there is a chance that individuals who can receive encrypted messages cannot send secure messages to others due to lack of a proper certificate from CA [4].

Considering the complexity and drawbacks of PKI-based cryptography, several message exchange protocols are proposed as an alternative to PKI. Identity-based cryptography (IBC) helps to overcome these drawbacks by allowing the message recipient to receive the message without any certificate from CA. The only problem faced by IBC scheme is numerous XOR operations are needed while encrypting and decrypting.

Since XOR operations consume a lot of CPU cycles, the overall time taken for communication would substantially increase. In this article, we have successfully reduced the overall time consumed by the IBC scheme by parallelizing the decryption part of the scheme. The parallelization fastens the process and decreases the time consumption, which is very much desirable.

2 Identity-Based Cryptography

In identity-based cryptography, a publicly known unique string belonging to an individual or organization is used as a public key. The public string can be an email address, domain name, or a physical IP address [5]. Unlike PKI, identity-based cryptography requires the presence of a trusted third party, known as private key generator (PKG). The PKG is responsible for generating the private keys for its entities. In order to operate, the PKG first generates a master public key (pk_{PKG}) using random generators and retains the corresponding master private key (sk_{PKG}). Using

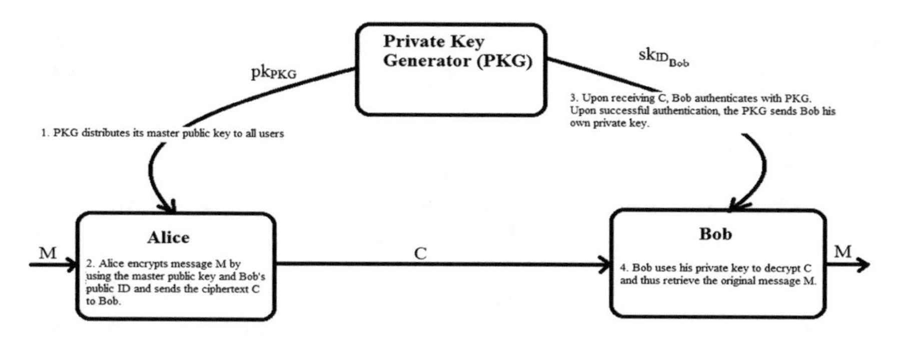

Fig. 1 ID-based cryptosystem

the master public key, any entity (say Alice) who wants to communicate with an intended receiver (say Bob) computes the public key by combining the master public key along with the unique identity value ID of Bob. To obtain the corresponding private key, the intended receiver having the corresponding unique identity ID contacts the PKG, which uses the master private key to generate the private key sk_{IDBob} [6].

2.1 *ID-Based Scheme*

The process of encryption and decryption [7, 8] proceeds as follows.

Alice prepares plaintext message M for Bob. She uses Bob's identity ID_{Bob} and the PKG's public key pk_{PKG} to encrypt M, obtaining ciphertext message C. Alice then sends C to Bob. Note that ID_{Bob} and pk_{PKG} were both already known to Alice before beginning the encryption process, so she requires no prior coordination or preparation on Bob's part to encrypt a message for him.

Bob receives C from Alice. It is assumed that the ciphertext C comes with instructions for contacting the PKG to get the private key required to decrypt it. Bob authenticates with the PKG, essentially sending it sufficient proof that ID_{Bob} belongs to him, upon which the PKG transmits Bob's private key sk_{IDBob} to him over a secure channel. Bob decrypts C using his own private key sk_{IDBob} and gets back the plaintext message M [4]. The details of the working of ID-Based Cryptography are shown in Fig. 1.

2.2 *Boneh–Franklin ID-Based Cryptosystem*

The Boneh–Franklin IBC uses bilinear pairing over elliptic curves and finite fields. It uses a public identity of an entity as the public key and hence does not require authentication from any CA. A trusted authority is still required as a Private Key

Generator (PKG). The PKG is involved only once, i.e., while the registration of an entity. As the scheme is based on pairings, all computations are performed in two groups, $\mathbf{G}_1$ and $\mathbf{G}_2$:

For $\mathbf{G}_1$, let $\mathbf{p}$ be prime, $\mathbf{p} \equiv \mathbf{2} \bmod \mathbf{3}$ and consider the elliptic curve $\mathbf{E}$: $y^2 = x^3 + 1$ over $\mathbb{Z}/\mathbf{p}\mathbb{Z}$. Since $4a^3 + 27b^2 \neq 0$, the curve is not singular. It only equals $\mathbf{0}$ for the case $\mathbf{p} = \mathbf{3}$, which is excluded by the additional constraint. Let $\mathbf{q} > \mathbf{3}$ be a prime factor $\mathbf{p} + \mathbf{1}$ (which is the order of $\mathbf{E}$) and find a point $\mathbf{P} \in E$ of order $\mathbf{q}$. $\mathbf{G}_1$ is the set of points generated by $\mathbf{P}$: $\{\mathbf{nP} | \mathbf{n} \in \{\mathbf{0}, \ldots, \mathbf{q} - \mathbf{1}\}$.

$\mathbf{G}_2$ is the subgroup of order $\mathbf{q}$ of $\mathbf{GF}(\mathbf{p}^2)^*$. We do not need to construct this group explicitly (this is done by the pairing) and thus do not have to find a generator [9].

(i) *Key Generation*

The PKG is required for generating private keys for an entity Bob. Let Bob's unique ID be his unique email id 'bob@mymail.com'. Therefore, Bob's public key is given by $P_{BOB} = H_1$(bob@mymail.com), where H_1 is a public hash function $\mathbf{H}_1$: $\{\mathbf{0}, \mathbf{1}\}^* \rightarrow \mathbf{G}_1^*$. Bob's private key is $D_{BOB} = sP_{BOB}$. The PKG transfers D_{BOB} to Bob securely through a secure channel. In order to encrypt data intended for Bob, anyone can compute P_{BOB}. But, Bob himself cannot compute 's' from D_{BOB} (DLP Assumption) [10, 11].

Also, the public key of PKG, P_{PKG} is calculated and made public.

(ii) *Encryption*

Alice plans to send an n-bit message M to Bob. She then computes the public key of Bob by $P_{BOB} = H_1$(bob@mymail.com) $\in G_1$ and $g = e(P_{Bob}, P_{PKG}) \in G'$. The hash function H_1 is publicly available. Alice chooses a random element 'r' $\in Z_a^*$. Next, Alice computes the ciphertext $C = (rP, M \oplus H_2(g^r)) \in G \times \{0, 1\}^n$. Here, 'r' is the session secret. H_2 is another hash function and $H_2(g^r)$ is used as a mask to hide the message. Now, anybody can send messages to Bob and no certificates are required to do so.

(iii) *Decryption*

Bob decrypts the received ciphertext received as $(U, V) \in G \times \{0, 1\}^n$. He has his private key D_{Bob} and computes the element $g' = e(D_{Bob}, U) \in G'$ and the mask $H_2(g')$. He retrieves the message $M = V \oplus H_2(g')$. The correctness for this setup is defined as

$$\begin{aligned} g' &= e(D_{Bob}, U) = e(D_{Bob}, rP) = e(sP_{Bob}, rP) \\ &= e(P_{Bob}, P)^{sr} = e(P_{Bob}, sP)^r = e(P_{Bob}, P_{PKG})^r = g^r \end{aligned}$$

Using the notations in Table 1, the total process is described below:

a. Key Generation

The private key generator (PKG) chooses the following:

1. The public group $\mathbf{G}_1$ (with generator $\mathbf{P}$) and $\mathbf{G}_2$ as stated above, with the size of $\mathbf{q}$ depending on security parameter.

Table 1 Notations used in the scheme

Notations	Description
$\mathbf{P_{Bob}}$	Public key of bob
$\mathbf{D_{Bob}}$	Private key of bob
s	Master secret key of PKG
$\mathbf{P_{PKG}}$	Public key of PKG
$\mathbf{H_1, H_2}$	Hash functions
P	Generator point
e	Pairing function
r	Session random number
$\mathbf{G_1, G_2}$	Public groups

2. The corresponding pairing function **e**.
3. A random private master key $\mathbf{s} \in \mathbb{Z}_q^*$.
4. A public key $\mathbf{P_{PKG}} = \mathbf{sP}$.
5. A public hash function $\mathbf{H}_1$: $\{\mathbf{0}, \mathbf{1}\}^* \to \mathbf{G}_1^*$.
6. A public hash function $\mathbf{H}_2$: $\mathbf{G}_2 \to \{\mathbf{0}, \mathbf{1}\}^{\mathbf{n}}$ for some fixed **n**.
7. The message space and the cipher space $\mathbf{m} = \{\mathbf{0}, \mathbf{1}\}^{\mathbf{n}}$, $\mathbf{C} = \mathbf{G}_1^* \times \{\mathbf{0}, \mathbf{1}\}^{\mathbf{n}}$.
8. The private key of Bob $\mathbf{D_{Bob}} = s\mathbf{P_{Bob}}$, which is sent to Bob through secure channel.

b. Encryption

To create the public key for $\mathbf{ID} \in \{\mathbf{0}, \mathbf{1}\}^*$, the sender Alice

1. Computes $\mathbf{P_{Bob}} = H_1(ID) \in \mathbf{G}_1^*$
2. Chooses random $\mathbf{r} \in \mathbb{Z}_q^*$
3. Computes $\mathbf{g}_{ID} = \mathbf{e}(\mathbf{P}_{Bob}, \mathbf{P_{PKG}}) \in \mathbf{G}_2$ and
4. Set $\mathbf{c} = (\mathbf{rP}, \mathbf{m} \oplus \mathbf{H}_2(g_{ID}^r))$
5. Send **c** to Bob.

c. Decryption

Given $\mathbf{c} = (\mathbf{u}, \mathbf{v})$, the plaintext can be retrieved using the private key $\mathbf{D_{Bob}}$. $\mathbf{H}_2$ is available with Bob.

$$\mathbf{m} = \mathbf{v} \oplus \mathbf{H}_2(e(\mathbf{D}_{Bob}, u)).$$

3 Parallel Processing

A computation in which many executions or calculations of processes are carried out simultaneously is known as parallel processing [1]. Often large problems are divided into smaller or modular problems and these modular problems are solved simultaneously. This method fastens the overall computational time. Parallel computing can be categorized into three types: bit-level, instruction-level, data, and task parallelism. In high-performance computing, parallelism has been deployed for many years, but interest in it has grown lately due to the physical constraints preventing frequency scaling. As power consumption (and consequently heat generation) and completion time taken by computers has become a concern in recent years, parallel computing has become the dominant paradigm in computer architecture, mainly in the form of multi-core processors [12].

Parallel programming is a way of implementing parallel processing. There are quite a few approaches towards parallel programming, namely, C*, threads, MPI, OpenMP. In this article, the parallelization is done using threads and the programming is done in C. The thread programming is implemented in C using pthread library. The pthread library is used in the program by including the pthread.h header file [13].

The basic routines of the pthread library are mentioned below.

- **pthread_create()**—This function is used for creation of threads.
- **pthread_join()**—This function is used to block the calling thread until the thread specified by its first argument terminates.
- **pthread _exit()**—This function terminates the calling thread and returns a value that is available to another thread in the same process that calls pthread_join().

4 Parallelization of Boneh–Franklin IBC

4.1 Problem Description

In this article, Boneh–Franklin Identity-Based Encryption is implemented using C language. PBC library is included for performing various cryptographic operations and pthread library is included for thread programming.

The time consumed by the application increases as the size of the input file increases and after a certain file size, the time consumption is far from desirable. Thus, the time consumption should be reduced and it can be done by parallelization of the scheme, as the operation is divided into smaller parts and then performed simultaneously. In this way, the time consumption can be reduced to a great extent.

Here, the decryption phase of the scheme is parallelized using thread programming and the performance result is recorded.

4.2 Implementation

The application is implemented in C language using the PBC and Type A elliptic curve [14]. The decryption part of the application is parallelized using pthread library. The application is implemented in three phases, namely, Key Generation, Encryption, and Decryption.

(a) *Key Generation*

The key generation phase uses Type A curve operations to generate public and private keys. It randomly selects the secret key 's' and a generator P and computes the public key $P_{PKG} = sP$. It also chooses two hash functions H_1 and H_2 and makes them public. P_{BOB} is generated by using hash function over Bob's public ID. D_{BOB} is generated by using s.P_{BOB}. Next, the values of P, P_{PKG}, $P_{BOB,}$ and D_{BOB} are written in files.

(b) *Encryption*

In encryption phase, the program accepts receiver's public ID as its runtime input. Then, P_{BOB} is generated by using hash function over Bob's public ID. It takes the value of P, P_{BOB}, P_{PKG}, the plaintext filename and the output filenames (C1 and C2) as runtime inputs. The session key is randomly selected and using the values of P, r, P_{BOB}, and P_{PKG} the plaintext file is encrypted.

Then, C1 is generated by using 'r' times P and C2 are generated by applying XOR operation on the plaintext file (in binary form) with the output of the pairing operation raised to the power of r; (g^r). Both the output files C1 and C2 serve as the ciphertext.

(c) *Decryption*

In decryption phase, the program takes D_{BOB}, C1, C2, and the output filename as runtime inputs. Using D_{BOB} and C1, the program will decrypt the contents of the C2 and recover the plaintext.

First, g1 is computed by pairing of D_{BOB} and C1. Next, the threads are created using pthread_create() function for parallelizing of the XOR operation of C2 and g1. In the thread function XOR, C2 is divided into different threads and XOR operation is applied on C2 with g1 to the decrypted data. After successful XOR, the result is joined together using pthread_join() and written in the output file.

The C code snippet mentioned below is for parallelization of the XOR operation. The function void *XOR(int *num_ptr) is parallelized using threads.

```
/* Creation of threads*/

i = 0;
for(i =0;i<NUM_THREAD;i++)
{         th_id[i]= i;
pthread_create(&tid[i],NULL,XOR,&th_id[i]);
}
for(i =0;i<NUM_THREAD;i++)
{
          pthread_join(tid[i],NULL);
}

/* End of Creation of threads */

/* Thread Function*/

void *XOR( int  *num_ptr)
{
int j,i = 0;
int num,*number_ptr;
number_ptr=num_ptr;
num=*number_ptr;
unsigned long long z;
z = (((num*C2len)/NUM_THREAD)%g1len);
for((j=(((num*C2len)/NUM_THREAD)));j<(((num+1)*C2len)/NUM_   THREAD);j++)
     {
               if((C2data[j]=='1'&&g1bin[z]==1)||(C2data[j]=='0'&& g1bin[z]==0))

mesbin[j]=0;
                    else
                              mesbin[j]=1;

                    z++;
          if(z==g1len)  z=0;

          }

          pthread_exit(0);
}

/* End of Thread Function*/
```

The Key Generation, Encryption, and Decryption algorithms using threads are discussed below.

Algorithm

Key Generation

// Type A elliptic curves are used for generation of keys.

// P, P_{PKG}, P_{BOB}, D_{BOB}, and s are generator, public key of PKG, Bob's public key, Bob's private key and master secret key respectively.

// Bob's ID := bob@mymail.com.

// H_1 is a public hash function $\mathbf{H}_1 : \{\mathbf{0}, \mathbf{1}\}^* \rightarrow \mathbf{G}_1^*$

Step 1: The secret key's' and generator P are selected randomly.

Step 2: P_{PKG} = sP, is computed.

Step 3: P_{BOB} := H_1(bob@mymail.com).

Step 4: $D_{BOB} := sP_{BOB}$, is given to the user.
Step 5: The values of P, P_{PKG}, P_{BOB} and D_{BOB} are stored in files.

Encryption

// The values of the program takes the value of P, P_{BOB}, P_{PKG}, the plaintext filename and the output filenames (C1 and C2) as runtime inputs.
// 'r' is the secret session key.

Step 1: Session key r is randomly selected.
Step 2: $C1 = r.P_{BOB}$.
Step 3: $g' = e(P_{BOB}, P_{PKG})$ i.e. g' is computed by pairing of P_{BOB} and P_{PKG}.
Step 4: C2 is generated by applying XOR operation on the plaintext file (in binary form) with the output of the pairing operation raised to power of 'r' (g^r).
Step 5: The ciphertext (C1, C2) are stored in files.

Decryption

// The program takes D_{BOB}, C1, C2, and the output filename as runtime inputs.

Step 1: $g1 = e(D_{BOB}, C1)$ i.e. g1 is computed by pairing of D_{BOB} and C1.
Step 2: Threads $\{T_1, T_2, \ldots T_n\}$ are created for parallel processing and C2 is divided among different threads.
Step 3: Within every thread T_i, XOR operation is performed between g1 and respective segment of C2 to get the decrypted data.
Step 4: The decrypted data is stored in memory.

5 Test Scenarios and Performance Results

5.1 *Purpose*

This section describes the test results of the parallelization of the decryption phase of the Boneh–Franklin IBC done over text files. These documents also lay down the hardware configuration in which these activities have been carried out.

5.2 *Scope*

The scope for testing the application involves testing for effective parallelization of the decryption phase of the above scheme and recording the speedup and efficiency achieved due to parallelization.

5.3 *Test Objectives*

The main objective of the testing was to determine that effective parallelization using threads has been achieved. It was also to record the speedup (if any) and subsequent efficiency achieved due to parallelization.

5.4 *System Specification*

- Hardware Configuration
    1. Processor: Intel® Xeon® CPU
    2. Frequency: 2.40 GHz
    3. No. of CPU Cores: 6
    4. No. of Threads: 12
    5. Memory (RAM): 16 GB
- Operating System

    Fedora 2.6.32-353 (64 bits)
- Software Configuration

1. Programming Language used: C
2. Compiler: GCC compiler (version 4.4.7)

Mr. Jones, of the Manor Farm, had locked the hen-houses for the night, but was too drunk to remember to shut the pop-holes. With the ring of light from his lantern dancing from side to side, he lurched across the yard, kicked off his boots at the back door, drew himself a last glass of beer from the barrel in the scullery, and made his way up to bed, where Mrs. Jones was already snoring.

As soon as the light in the bedroom went out there was a stirring and a fluttering all through the farm buildings. Word had gone round during the day that old Major, the prize Middle White boar, had had a strange dream on the previous night and wished to communicate it to the other animals. It had been agreed that they should all meet in the big barn as soon as Mr. Jones was safely out of the way. Old Major (so he was always called, though the name under which he had been exhibited was Willingdon Beauty) was so highly regarded on the farm that everyone was quite ready to lose an hour's sleep in order to hear what he had to say.

At one end of the big barn, on a sort of raised platform, Major was already ensconced on his bed of straw, under a lantern which hung from a beam. He was twelve years old and had lately grown rather stout, but he was still a majestic-looking pig, with a wise and benevolent appearance in spite of the fact that his tushes had never been cut. Before long the other animals began to arrive and make themselves comfortable after their different fashions. First came the three dogs, Bluebell, Jessie, and Pincher, and then the pigs, who settled down in the straw immediately in front of the platform. The hens perched themselves on the window-sills, the pigeons fluttered up to the rafters, the sheep and cows lay down behind the pigs and began to chew the cud. The two cart-horses, Boxer and Clover, came in together, walking very slowly and setting down their vast hairy hoofs with great care lest there should be some small animal concealed in the straw. Clover was a stout motherly mare approaching middle life, who had never quite got her figure back after her fourth foal. Boxer was an enormous beast, nearly eighteen hands high, and as strong as any two ordinary horses put together. A white stripe down his nose gave him a somewhat stupid appearance, and in fact he was not of first-rate intelligence, but he was universally respected for his steadiness of character and tremendous powers of work. After the horses came Muriel, the white goat, and Benjamin, the donkey. Benjamin was the oldest animal on the farm, and the worst tempered. He seldom talked, and when he did, it was usually to make some cynical remark--for instance, he would say that God had given him a tail to keep the flies off, but that he would sooner have had no tail and no flies. Alone among the animals on the farm he never laughed. If asked why, he would say that he saw nothing to laugh at. Nevertheless, without openly admitting it, he was devoted to Boxer; the

Fig. 2 Cryptographic and parallelization operations had been carried over this input file

5.5 *Experimental Results*

The experiment has been conducted in the machine, whose specifications are given below. The application was created using C language with GCC compiler version 4.4.7. Here, two different scenarios have been explored:

A. Encryption and Decryption using Boneh–Franklin's IBC of text file of size 170 MB
B. Encryption and Decryption using Boneh Franklin's IBC of text file of size 250 MB

The decryption part of this cryptographic operation is parallelized using threads and the decryption time using different numbers of threads are recorded.

From Figs. 2 and 4, it can be seen that both input and output files are the same. Hence, it can be concluded that the cryptographic operation was carried out perfectly. The results of the experiments conducted are given below (Fig. 3).

A. **Text File of Size 170 MB**

The performance result of the parallelization of decryption phase (for file size of 170 MB) is shown in Table 2.

From the above-mentioned table, it can be easily said that execution time decreases as the number of threads increases.

The performance result of the parallelization of decryption (for file size of 170 MB) is shown in graphical form in Fig. 5.

The graph clearly shows that as the number of threads increases the execution time decreases. Thus, parallelization decreases execution time.

The graph in Fig. 6 clearly shows that as the number of threads increases the speedup also increases (almost linearly).

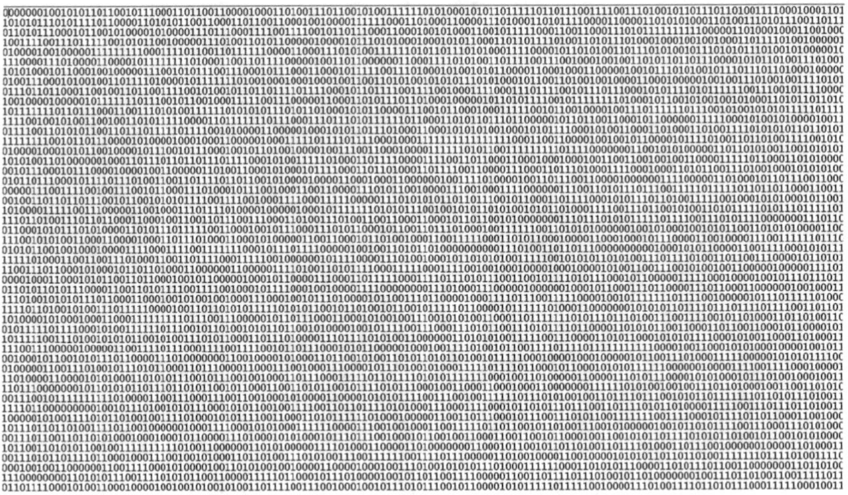

Fig. 3 Ciphertext (C2) after the encryption operation

Mr. Jones, of the Manor Farm, had locked the hen-houses for the night, but was too drunk to remember to shut the pop-holes. With the ring of light from his lantern dancing from side to side, he lurched across the yard, kicked off his boots at the back door, drew himself a last glass of beer from the barrel in the scullery, and made his way up to bed, where Mrs. Jones was already snoring.

As soon as the light in the bedroom went out there was a stirring and a fluttering all through the farm buildings. Word had gone round during the day that old Major, the prize Middle White boar, had had a strange dream on the previous night and wished to communicate it to the other animals. It had been agreed that they should all meet in the big barn as soon as Mr. Jones was safely out of the way. Old Major (so he was always called, though the name under which he had been exhibited was Willingdon Beauty) was so highly regarded on the farm that everyone was quite ready to lose an hour's sleep in order to hear what he had to say.

At one end of the big barn, on a sort of raised platform, Major was already ensconced on his bed of straw, under a lantern which hung from a beam. He was twelve years old and had lately grown rather stout, but he was still a majestic-looking pig, with a wise and benevolent appearance in spite of the fact that his tushes had never been cut. Before long the other animals began to arrive and make themselves comfortable after their different fashions. First came the three dogs, Bluebell, Jessie, and Pincher, and then the pigs, who settled down in the straw immediately in front of the platform. The hens perched themselves on the window-sills, the pigeons fluttered up to the rafters, the sheep and cows lay down behind the pigs and began to chew the cud. The two cart-horses, Boxer and Clover, came in together, walking very slowly and setting down their vast hairy hoofs with great care lest there should be some small animal concealed in the straw. Clover was a stout motherly mare approaching middle life, who had never quite got her figure back after her fourth foal. Boxer was an enormous beast, nearly eighteen hands high, and as strong as any two ordinary horses put together. A white stripe down his nose gave him a somewhat stupid appearance, and in fact he was not of first-rate intelligence, but he was universally respected for his steadiness of character and tremendous powers of work. After the horses came Muriel, the white goat, and Benjamin, the donkey. Benjamin was the oldest animal on the farm, and the worst tempered. He seldom talked, and when he did, it was usually to make some cynical remark--for instance, he would say that God had given him a tail to keep the flies off, but that he would sooner have had no tail and no flies. Alone among the animals on the farm he never laughed. If asked why, he would say that he saw nothing to laugh at. Nevertheless, without openly admitting it, he was devoted to Boxer; the

Fig. 4 The output file after decryption operation

The graph in Fig. 7 clearly shows that the efficiency of the parallelization process revolves between 0.87 and 1, which is considered to be good in case of parallel processing.

B. Text File of Size 250 MB

The performance result of the parallelization of decryption (for file size of 250 MB) is shown in Table 3 in tabulated form.

From the above-mentioned table, it can be easily said that execution time decreases as the number of threads increases.

Table 2 Performance result of decryption phase of file size 170 MB

No. of threads (*P*)	Execution time (*s*)	Speedup (s)	Efficiency
1	26	1	1
2	13	2	1
3	9	2.89	0.96
4	7	3.71	0.93
5	6	4.33	0.87
6	5	5.2	0.87

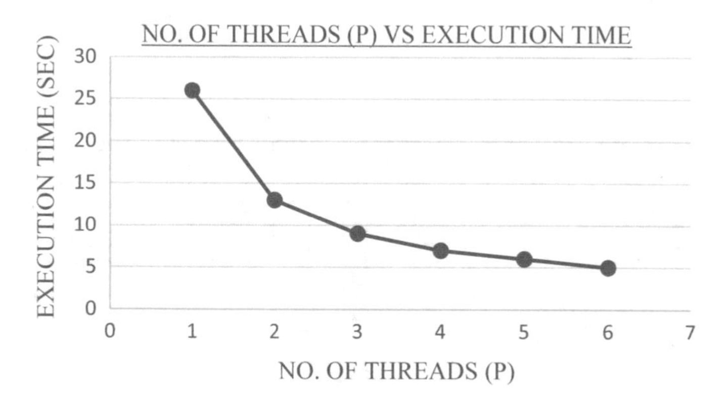

Fig. 5 Graphical analysis of performance result of decryption phase of file size 170 MB

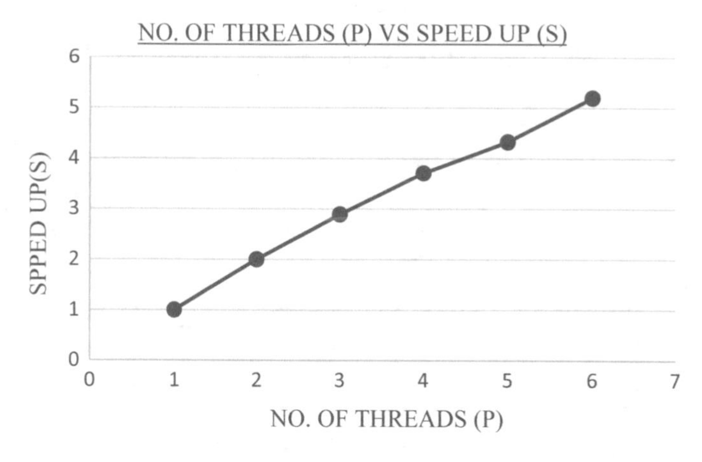

Fig. 6 Graphical analysis of performance result of decryption phase of file size 170 MB

The performance result of the parallelization of decryption (for file size of 250 MB) is shown in graphical form in Fig. 8.

The above-mentioned graph clearly shows that as the number of threads increases the execution time decreases. Thus, parallelization decreases execution time.

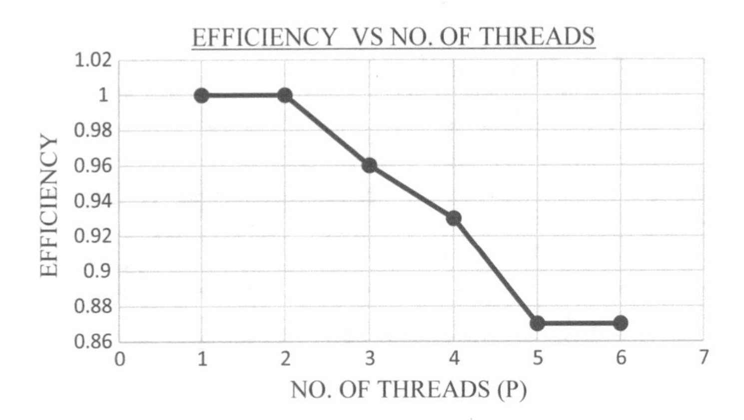

Fig. 7 Graphical analysis of performance result of decryption phase of file size 170 MB

Table 3 Performance result of decryption phase of file size 250 MB

No. of threads (*P*)	Execution time (*s*)	Speedup (*s*)	Efficiency
1	38	1	1
2	19	2	1
3	14	2.71	0.9
4	10	3.8	0.95
5	8	4.75	0.95
6	7	5.43	0.91

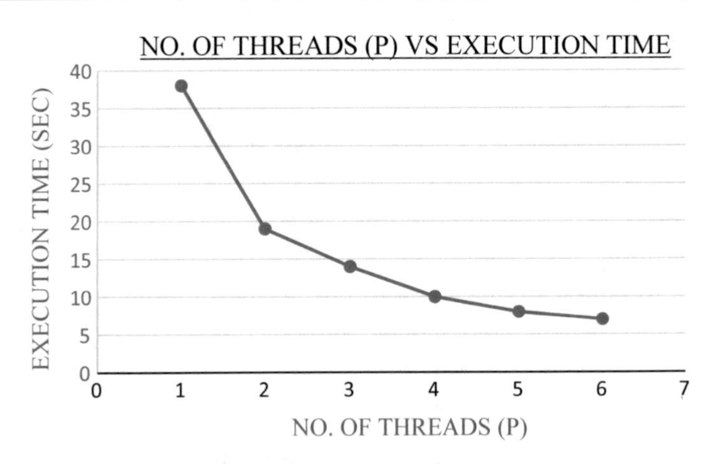

Fig. 8 Graphical analysis of performance result of decryption phase of file size 250 MB

The graph in Fig. 9 clearly shows that as the number of threads increases the speedup also increases (almost linearly).

The graph in Fig. 10 clearly shows that the efficiency of the parallelization process revolves between 0.9 and 1, which is very good in case of parallel processing.

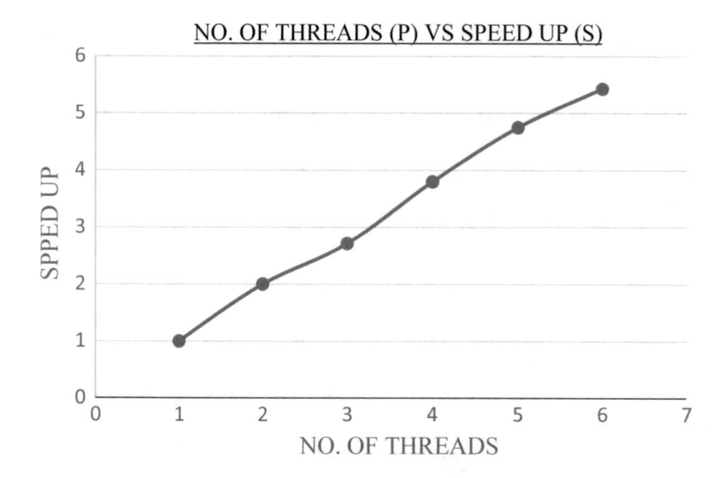

Fig. 9 Graphical analysis of performance result of decryption phase of file size 250 MB

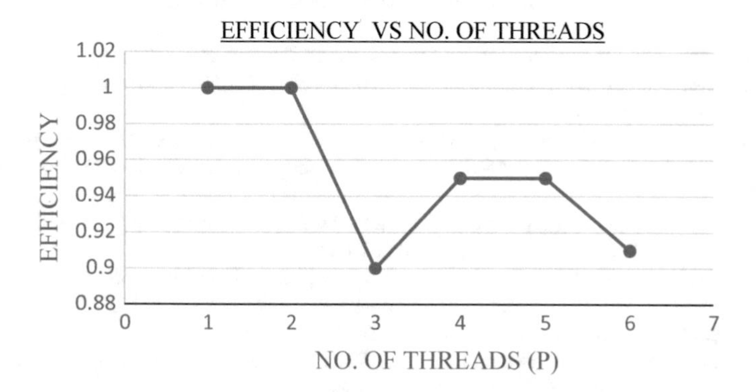

Fig. 10 Graphical analysis of performance result of decryption phase of file size 250 MB

6 Conclusion and Future Work

6.1 Conclusion

In this article, the ID-Based Cryptosystem is implemented and the decryption part is parallelized using C language. Parallelization is done through thread programming. Here, the execution time decreases as the number of threads used for parallelization increases. Thus, a nice time improvement is attained.

6.2 Future Work

- The application can be implemented using sockets which will reduce the disk I/O time significantly.
- The application can be implemented on machines with larger memory and more number of threads for performing the cryptographic operation on much larger files.

References

1. Stallings, W.: Cryptography and Network Security Principles and Practice. Pearson, Boston (2017)
2. InstituteSANS™.: Information Security Resources. Retrieved 10 Nov 2016, from https://www.sans.org/security-resources/ (2000)
3. Lander, S.: Disadvantages of Public Key Encryption. Retrieved 4 Jan 2017, from Small Business Chron. http://smallbusiness.chron.com/disadvantages-public-key-encryption-68149.html (2017)
4. Youngblood, C.: An Introduction to Identity-Based Cryptography. Retrieved 10 Feb 2016, from https://courses.cs.washington.edu/courses/csep590/06wi/finalprojects/youngblood_csep590tu_final_paper.pdf (2005)
5. Shamir, A.: Identity-Based Cryptosystems and Signature Schemes. Advances in Cryptology Lecture Notes in Computer Science, pp. 47–53. https://doi.org/10.1007/3-540-39568-7_5 (1985)
6. Al-Riyami, S.S., Paterson, K.G.: Certificateless Public Key Cryptography Advances in Cryptology—Proceedings of ASIACRYPT 2003 (2003)
7. Boneh, D., Boyen, X.: Efficient selective-ID secure identity-based encryption without random oracles. LNCS. Advances in Cryptography—EUROCRYPT 2004, vol. **3027**, pp. 223–238. Springer. https://doi.org/10.1007/978-3-540-24676-3_14 (2004)
8. Das, A.: Elliptic-Curve Cryptography (ECC). Retrieved 2 Nov 2016, from http://cse.iitkgp.ac.in/~abhij/download/doc/ECC.pdf (2016, March 20)
9. Boneh, D., Franklin, M.K.: Identity-Based Encryption from the Weil Pairing Advances in Cryptology—Proceedings of CRYPTO 2001 (2001)
10. Richard, C., Carl, P.: Prime Numbers: A Computational Perspective (Chapter 5), 2nd edn. Springer
11. Stinson, D.R.: Cryptography: Theory and Practice, 3rd edn. CRC Press, London. ISBN 978-1-58488-508-5 (2006)
12. https://en.wikipedia.org/wiki/Parallel_computing
13. http://www.llnl.gov/computing/tutorials/parallel_comp
14. On the Implementation of Pairing-Based Cryptosystems. Ben Lynn. https://crypto.stanford.edu/pbc/thesis.pdf

Sandeep Kumar Pandey is a final year student of Jalpaiguri Government Engineering College of Information Technology Department. His area of interests includes Cryptography, Parallel Processing and Programming, Machine Learning and Deep Learning.

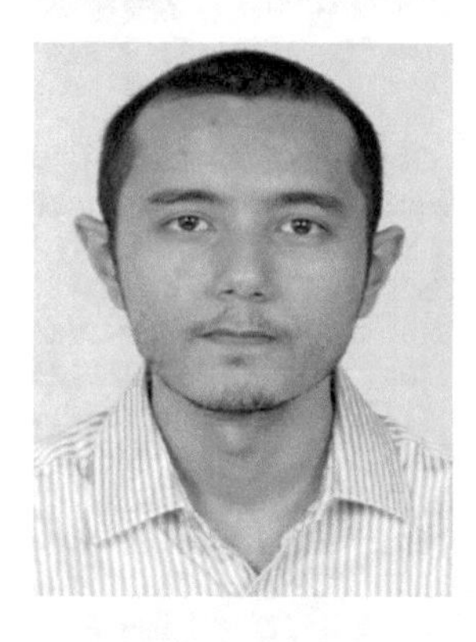

Tapobrata Dhar received his M.E. degree in Software Engineering from Jadavpur University, Kolkata, West Bengal, India in 2017. He is currently pursuing Ph.D. Degree in Information Technology from Indian Institute of Engineering Science and Technology, Shibpur, India. His research interests include identity-based cryptography and hardware security.

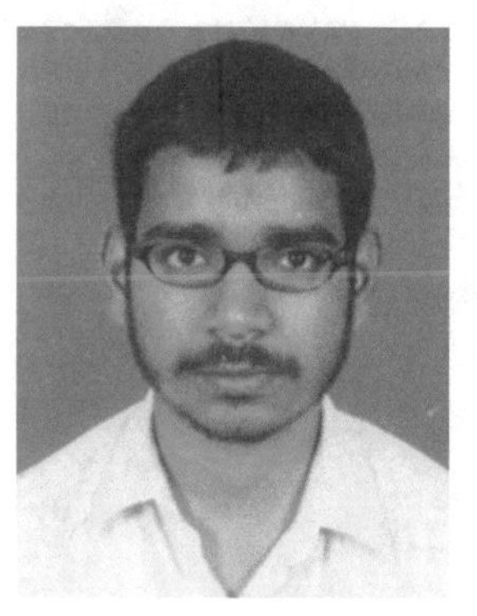

Ryan Saptarshi Ray received the degree of B.E. in I.T. from School of Information Technology, West Bengal University of Technology, India in 2007. He received the degree of M.E. in Software Engineering from Jadavpur University, India in 2012. Currently he is Ph.D. Scholar in the Department of Information Technology, Jadavpur University, India.

He was employed as Programmer Analyst from 2007 to 2009 in Cognizant Technology Solutions. He has published 2 papers in International Conferences, 9 papers in International Journals and also a book titled "Software Transactional Memory: An Alternative to Locks" by LAP LAMBERT ACADEMIC PUBLISHING, GERMANY in 2012 co-authored with Utpal Kumar Ray.

Aritro Sengupta received the degree of B.Tech. in Information Technology from West Bengal University of Technology, Kolkata, India in the year 2012. He received the degree of Master of Engineering in Software Engineering from Jadavpur University, Kolkata, India in the year 2014. Currently, he is working as a Scientist in Ministry of Electronics and Information Technology under the Government of India. He had worked as a research scholar in Jadavpur University from 2014 to 2017 in the area of Cryptography. He has published one article in SCI listed Journal and one article in International Conference. His area of research interest includes Cryptography and Network Security, Elliptic Curve Cryptography and Identity Based Cryptography.

Utpal Kumar Ray received the degree of B.E. in Electronics and Telecommunication Engineering in 1984 from Jadavpur University, India and the degree of M.Tech. in Electrical Engineering from Indian Institute of Technology, Kanpur in 1986.

He was employed in different capacities in WIPRO INFOTECH LTD., Bengaluru, India; WIPRO INFOTECH LTD., Bengaluru, India, Client: TANDEM COMPUTERS, Austin, Texas, USA; HCL America, Sunnyvale, California, USA, Clent: HEWLETT PACKARD, Cupertino, California, USA; HCL Consulting, Gurgaon, India; HCL America, Sunnyvale, California, USA; RAVEL SOFTWARE INC., San Jose, California, USA; STRATUS COMPUTERS, San Jose, California, USA; AUSPEX SYSTEMS, Santa Clara, California, USA and Sun Micro System, Menlo Park, California, USA for varying periods of duration from 1986 to 2002. From 2003 he is working as Assistant Professor in the Department of Information Technology, Jadavpur University, India. He has published 22 papers in different conferences and journals. He has also published a book titled "Software Transactional Memory: An Alternative to Locks" by LAP LAMBERT ACADEMIC PUBLISHING, GERMANY in 2012 co-authored with Ryan Saptarshi Ray.

SEER—An Intelligent Double Tier Fuzzy Framework for the Selection of Cluster Heads Based on Spiritual Energies of Sensor Nodes

Maddali M. V. M. Kumar and Aparna Chaparala

Abstract The selection of cluster heads in a wireless sensor network (WSN) is a very challenging area of research. Many algorithms have been proposed for the selection of cluster heads to prolong network lifetime, but maintaining an energy efficient network remains a problem. To overcome this, we propose a new algorithm for cluster head selection based on the Spiritual Energy of the whole WSN, which is known as the SEER (simple energy-efficient reliable) protocol. In addition, the implementation of the double tier fuzzy algorithms on the SEER protocol makes the network more energy efficient. The proposed algorithm has been simulated with MATLAB and compared with the other energy efficient algorithms such as CLERK and LEACH. The results proved to be very promising in terms of the reduction of energy consumption.

Keywords Spiritual Energy · FEER · LEACH · DEEDA · Fuzzy
Double tier fuzzy logics · Cluster heads

1 Introduction

The selection of a cluster head is major research challenge and many protocols have been proposed aiming for the most energy efficient solution to the problem. Several protocols have been designed and implemented but an efficient mechanism has not been found. The SEER protocol, which has been proposed depends on the Spiritual Energy of the network along with other parameters, such as distance, relative received signal strength (RSSI), proximity, and centrality. The proposed protocol works on a new intelligent framework called double tier fuzzy implementation I in which

M. M. V. M. Kumar (✉)
Acharya Nagarjuna University, Guntur, Andhra Pradesh, India
e-mail: mvmkumar.m@gmail.com

A. Chaparala
Department of CSE, RVR & JC College of Engineering, Guntur, Andhra Pradesh, India
e-mail: chaparala_aparna@yahoo.com

S. Smys et al. (eds.), *International Conference on Computer Networks and Communication Technologies*, Lecture Notes on Data Engineering and Communications Technologies 15, https://doi.org/10.1007/978-981-10-8681-6_46

qualifying rounds are employed for cluster head selection and a backup cluster head is chosen. This chapter is organized with the related work in Sect. 2, the proposed algorithm's working principles employing fuzzy rule sets are discussed in Sects. 3 and 4 addresses performance evaluation, and Sect. 5 presents the conclusions.

2 Related Work

The work by Bidaki et al. [1] introduced a fuzzy clustering technique which changed the probabilistic cluster head choice in the LEACH protocol. The fuzzy framework utilized in the inference engine plan is the Mamdani fuzzy framework, which is a straightforward rule-based strategy. This plan makes symmetric groups and the total hub to the cluster head separate. This in turn lessens the energy utilization of the sensors and enhances the lifetime of the WSN more than the LEACH protocol. Sharath et al. [2] employed a backoff-based distributed clustering protocol (LEACH) and Clustering-Fuzzy Logic has proposed the extension of time to network steadiness before the passing of the primary hub and the diminishment of shaky time before the death of the last hub. This protocol depends on the race of bunch head by the adjust of the probabilities of the rest of the energy for every hub. In this chapter, the authors propose enhancing clustering procedures through the use of clustering fuzzy logic (Clustering-FL). A fuzzy-based reproduction framework for WSNs is proposed, keeping in mind the end goal of computing the lifetime of a sensor by considering the rest of battery control, the rest time rate, and transmission time rate. We assessed the framework by NetBeans IDE reproductions and demonstrated that it performs the measurement of sensor lifetime efficiently. In this system, a couple of hubs progress toward becoming the bunch head and this results in the energetic heterogeneity of the system, and thus the conduct of the sensor organization turns out to be exceptionally precarious when the life of the primary hub is passed. Lee and Jeong et al. [3] proposed a fuzzy relevance-based cluster head determination algorithm (FRCA) to take care of issues found in existing remote portable specially appointed sensor systems, for example, the hub conveyance found in unique properties as a result of portability and level structures, and aggravation of the group arrangement. The proposed mechanism utilizes fluffy importance to choose the bunch set out toward grouping in the remote portable specially appointed sensor systems. In the simulation performed using the NS-2 test system, the proposed FRCA was contrasted and calculations performed, for example, the cluster-based routing protocol (CBRP), the weighted-based adaptive clustering algorithm (WACA), and the scenario-based clustering algorithm for mobile specially appointed systems (SCAM). The simulation results demonstrated that the proposed FRCA accomplishes the preferred execution more efficiently than the other existing mechanisms. Subramaniam et al. [4] studied a protocol supporting an energy proficient clustering, cluster head choice, and information routing technique to broaden the lifetime of sensor organization. Simulation results show that the proposed technique delays organization lifetime because of the utilization of proficient grouping, cluster head determination, and information directing. The results also demonstrate that toward the end of some specific piece of running the energy efficient clustering schemes (EECS) and fuzzy-based clustering algorithm increase the quantity of alive hubs in contrast to the LEACH and HEED techniques, and this can prompt an increase in the lifetime of sensor organization. By using the EECS technique, the aggregate number of messages received at the base station is increased compared to the LEACH and HEED strategies. The fuzzy-based grouping strategy contrasted and the k-means clustering by methods for emphasis tally and time taken to kick the bucket first hub in remote sensor organize, and the outcome shows that the fuzzy-based clustering technique performs well with k-means clustering strategies. Gajjar and Sarkar et al. [5] proposed a cluster head choice protocol utilizing fuzzy logic (CHUFL). This protocol uses hub parameters, namely remaining energy, reach and ability from its neighborhood, nature of communication connects with its neighborhood, and separation from the base station as fuzzy info factors for use in cluster head determination. A relative investigation of CHUFL with cluster head choice component utilizing fuzzy logic; Cluster Head Election instrument utilizing Fuzzy logic (CHEF) and group head chose the technique for remote sensor systems in light of fuzzy logic demonstrating that CHUFL is up to 20% more energy productive and sends 72% more bundles to base stations in contrast with protocol one of the energy efficient clustering protocol. Mishra et al. [6] concentrated on progressive protocols. In such conventions, energy proficient clusters are formed with a pecking order of group heads. Each cluster has its delegate cluster head in charge of gathering and collecting information from its separate cluster and transmitting this information to the base station, either straightforwardly or through the chain of importance in other cluster heads. Fuzzy logic has been effectively connected in different areas, including communication and has demonstrated promising outcomes. In any case, the possibilities of fuzzy logic in remote sensor organization should still be investigated. The streamlining of remote sensor systems includes different tradeoffs, for instance, reducing transmission control versus longer transmission span, multi-bounce versus coordinate correspondence, calculation versus correspondence, and so forth. Fuzzy logic is appropriate for applications having clashing requirements. In addition, in WSNs, as the vitality measurements shift generally with the type of sensor hub execution stage, fuzzy logic has the benefit of being effortlessly adaptable to such changes. Varghese et al. [7] most importantly considered hubs having a threshold value. A hub over this value was termed as qualifying for the cluster head. From these qualified hubs, the hub having the most extreme vitality, accessibility, and the hub with the least separation from the sink and the greatest throughput is chosen as the cluster head hub. Utilizing fuzzy rules, the potential value for every hub can be calculated. In the wake of arranging hubs by potential value, they should be checked again for malevolent conduct, so that there is no way for a malevolent hub to end up as the bunch head. This shows that improved cluster head determination is possible and can improve security of information exchange through WSNs. Secured clustering improves the lifetime of remote sensor arrangement, identifies pernicious hubs and prevents them from becoming the bunch head. Hub replication attack detection protocol is used to recognize pernicious hubs. Clones are produced by some pernicious hubs. Hub replication attack detection protocol recognizes clones and replay assaults can be distinguished by use of the replay assault discovery protocols, which depend on timestamps. Kumar et al. [8] proposed the use of a dynamic energy efficient distance aware (DEEDA) technique in energy efficient cluster selection mechanisms in the WSNs. The primary principle is that the selection of the cluster head is based on the principle of residual energy and distance (RED) algorithms. These algorithms focus on the selection of the cluster head in the network based on the distance, RSSI, and a new term called rank of the nodes. Low energy consumption has been achieved based on distance and signal strength. Cluster head selection is based on RED principles. Nithya et al. [9] proposed a Sugeno-type fuzzy inference-based clustering (SFIC) algorithm for improving system lifetime through more efficient use of energy. The dynamic cluster arrangement is created and relapse-based edge estimation is used to choose the cluster head. At that point, a super cluster head is chosen in light of Sugeno fuzzy inference rules and a coordinate information transmission technique is incorporated to further reduce energy loss from the system. Experiments were conducted and the results demonstrated that SFIC calculation increases system lifetime, packet conveyance, and end to end defer more than the fuzzy logic control (FLC) algorithm.

3 Proposed Algorithm

The proposed SEER protocol uses the Spiritual Energy calculation of the networks and works in two different phases.

3.1 Selection Phase (I Tier Fuzzy Operation)

In this phase, selection of nodes for the cluster head is calculated based on internal energy, neighboring nodes, traffic, and data gathering capability and is forwarded to the sink. In this case, energy and neighboring nodes are taken as the input for fuzzification and the Tier-I fuzzy rules are given in Table 1.

3.2 Optimization Phase (II Tier Fuzzy Operation)

This is the most important phase of the protocol. Here, different parameters, such as Spiritual Energy, distance, and RSSI are calculated for each node and sent to the sink. By applying the fuzzy rule sets, the cluster head is selected with the following the condition of threshold for the above mentioned parameter. This phase selects the cluster head and also the backup cluster head (BCH) in order to maintain energy consumption and network lifetime. The final selection of the cluster head is made by the maximum qualification measurements, which are as follows: distance of the

Table 1 Fuzzification rule engine for the Tier-I qualification

Energy	Neighboring nodes	Traffic	Tier qualification
Low	Low	Low	Very small
Low	Low	Medium	Small
Low	Low	High	Slightly small
Low	Medium	Low	Medium
Low	High	Medium	High
Low	High	High	Slightly high
Medium	High	High	High
High	High	High	Very high

Table 2 Fuzzification rule sets for the Tier-II cluster head qualification

Spiritual Energy	Proximity parameter	Centrality parameters	RSSI	Tier-II qualification
Low	Low	Low	High	Very high
Low	Low	Low	Medium	High
Low	Low	Low	Low	Slightly high
Low	Low	Medium	Low	High
Low	Low	High	Medium	High
Low	Low	High	High	Rather high
Low	High	High	High	Medium
Low	Medium	High	Medium	Average
Low	High	Medium	High	Medium
Low	High	High	Low	Low/little
Medium	High	High	Medium	Low
High	High	High	High	Very low
High	Low	Low	Low	Low
Medium	Medium	Medium	Low	Medium

cluster head to the other nodes is classified as proximity method, the location of the cluster head is evaluated based on parameters such as centrality and RSSI. Fuzzy rules are applied and the final qualification of the cluster head is given in Table 2.

The fuzzy rule engine was designed based on a Madami/Sugeno engine for the implementation of cluster head selection in network systems. This method of double tier fuzzy-based cluster head selection is efficient and enables the backup cluster head to be chosen based on Spiritual Energy calculations.

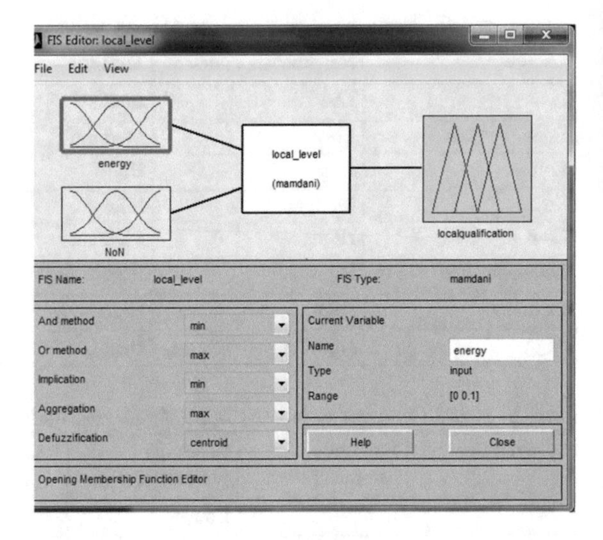

Fig. 1 Shows the fuzzification of the Tier-I selection of the cluster head

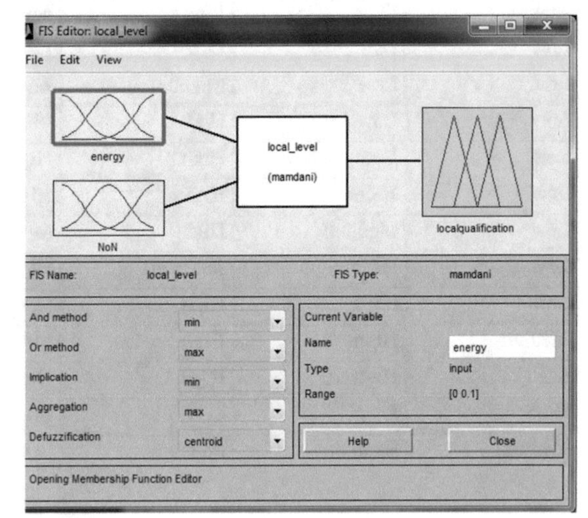

Fig. 2 Shows the fuzzification of the Tier-II selection of the cluster head

4 Performance Evaluation

The SEER protocol was simulated in MATLAB version R2014 in four different scenarios with a distribution of 30–90 nodes in 50×50 and 100×100 m^2 areas. Simulations were carried out for 100–200 rounds and different parameters of the network were calculated in order to evaluate the performance of the proposed protocols. The different stages of implementation are as follows (Figs. 1, 2, 3, 4, 5, 6 and 7).

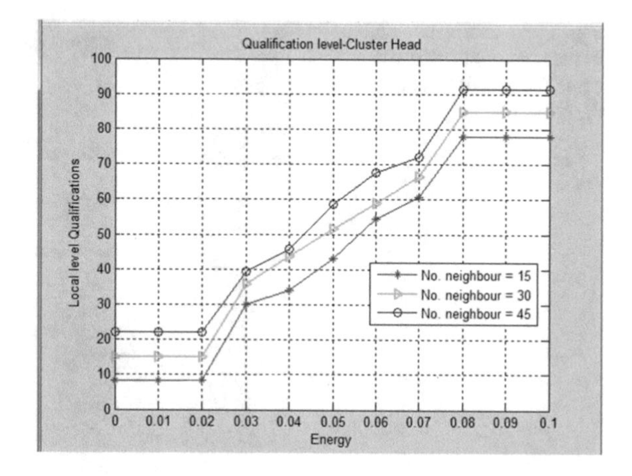

Fig. 3 Energy calculation in the optimization phase using the Tier-II selection of the cluster head

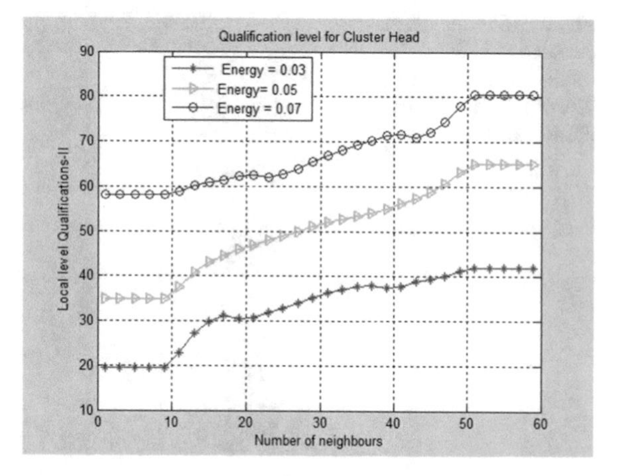

Fig. 4 Energy calculation in the optimization phase using the Tier-II selection of the cluster head

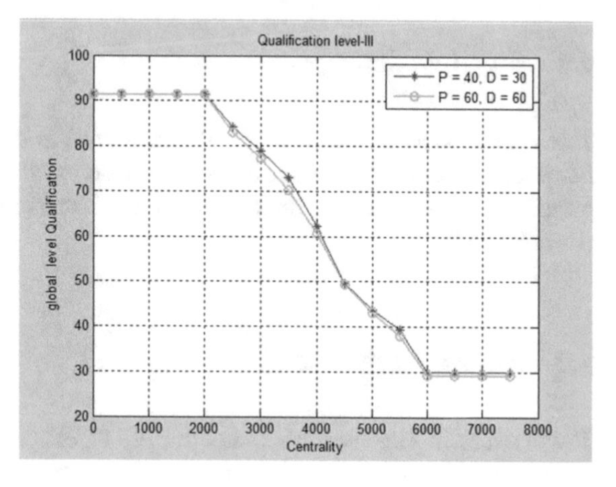

Fig. 5 Centrality calculation in the optimization phase using the Tier-II selection of the cluster head

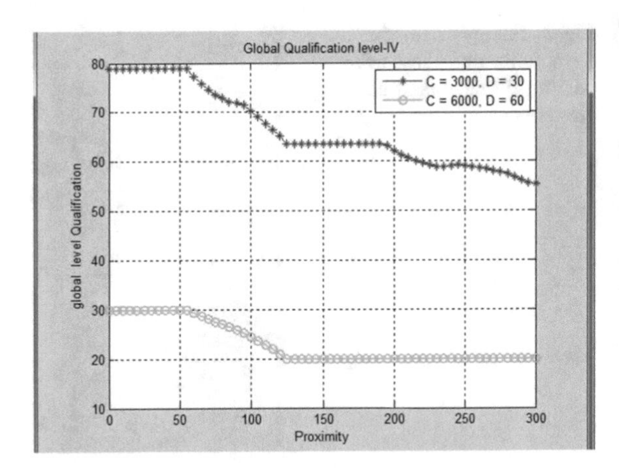

Fig. 6 Proximity calculation in the optimization phase using the Tier-II selection of the cluster head

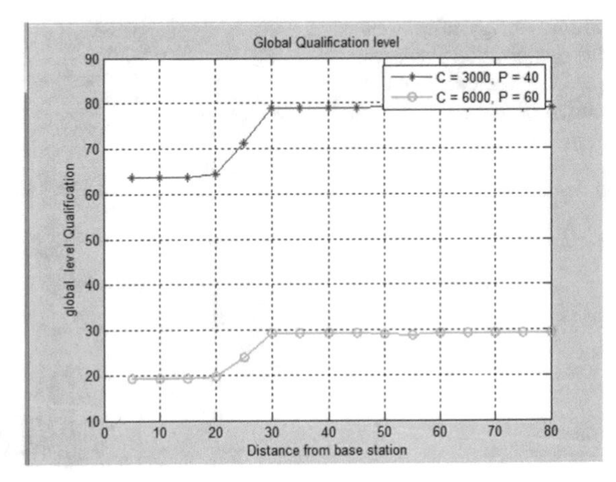

Fig. 7 RSSI and distance calculation in the optimization phase using the Tier-II selection of the cluster head

4.1 Spiritual Energy

This is defined as the average energy maintained in the network when the first and last node dies out. The average energy was calculated for both LEECH and CLERK protocols. In comparison, the SEER protocol consumed only 50% of the energy (Figs. 8 and 9).

4.2 Network Lifetime

The network lifetime was calculated for the SEER protocol, in which the lifetimes of the last dead nodes are calculated. An increase of 50% was observed compared with the other algorithms.

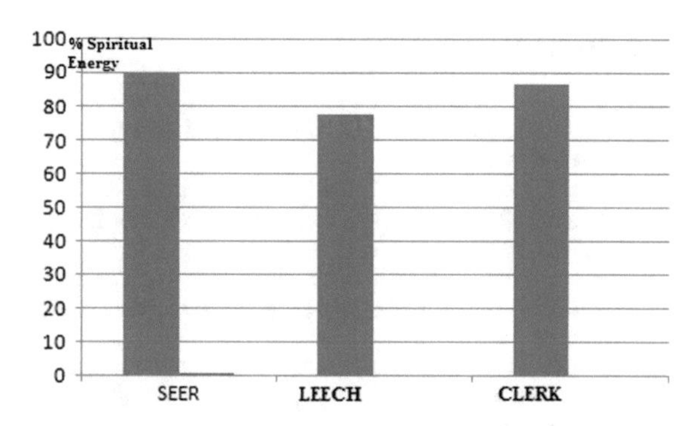

Fig. 8 Comparative analysis of Spiritual Energy of the proposed protocols after 100 rounds of simulation

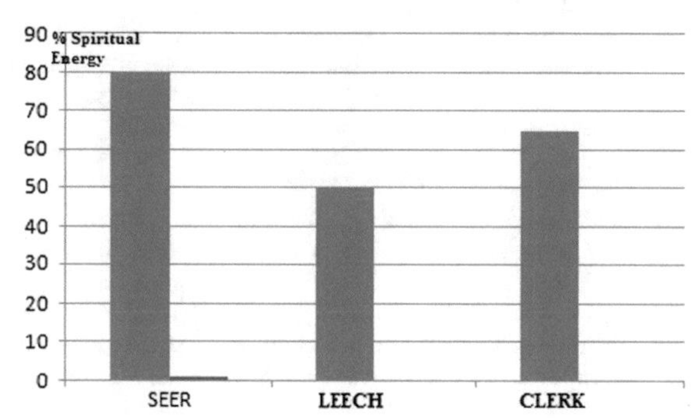

Fig. 9 Comparative analysis of Spiritual Energy of the proposed protocols after 200 rounds of simulation

5 Conclusion

The SEER protocol is more efficient than other energy efficient protocols used in the selection of the cluster head. Double tier fuzzy algorithms are still needed to improve the evaluation of additional parameters, such as the packet delivery ratio (PDR) and throughput etc.

References

1. Bidaki, M., Tabbakh, S.R.K.: Fuzzy based dynamic cluster head selection based on wireless sensor networks
2. Sharath Kumara, Y., Geetha, N.B., Mohamed, R.: Prediction of sensor lifetime by using clustering-fuzzy logic in wireless sensor networks. International Journal of Computer Science and Mobile Computing **4**(4), 835–841 (2015)
3. Lee, C., Jeong, T.: FRCA: A fuzzy relevance-based cluster head selection algorithm for wireless mobile Ad-Hoc sensor networks. Sensors **11**, 5383–5401 (2011). https://doi.org/10.3390/s110505383
4. Nirmala, A.M., Subramaniam, P., Anusha Priya, A., Ravi, M.: Enriched performance on wireless sensor network using fuzzy based clustering technique. Int. J. Adv. Stud. Comput. Sci. Eng. **2**(3), 11–18 (2013)
5. Gajjar, S., Sarkar, M., Dasgupta, K.: Cluster head selection protocol using fuzzy logic for wireless sensor networks. Int. J. Comput. Appl. **97**(7), 38–43 (2014)
6. Mishra, N.K.: Sensor energy optimization using fuzzy logic in wireless sensor networking. Int. J. Res. Rev. Eng. Sci. Technol. **2**(1), 11–14 (2013)
7. Varghese, B., Soumya, P.: Cluster head selection in wireless sensor networks. Int. J. Latest Res. Eng. Technol. **2**(4), 60–66 (2016)
8. Kumar, M. M., Chaparala, A.: Dynamic energy efficient distance aware protocol for the cluster head selection in the wireless sensor networks. In: 2nd IEEE International Conference on 2017, Recent Trends in Electronics Information and Communication Technology (RTEICT), pp. 147–150, (2017). https://doi.org/10.1109/rteict.2017.8256575
9. Nithyadevi, K., Radhapriya, S.: Improved fuzzy logic based clustering algorithm for enhancing the lifetime of wireless sensor network. Indian J. Educ. Inf. Manag. **5**(5) (2016)
10. Zhu, R., Qin, Y., Wang, J.: Energy-aware distributed intelligent data gathering algorithm in wireless sensor networks, Hindawi Publishing Corporation. Int. J. Distrib. Sens. Netw. **235724**, 1–13 (2011)
11. Chen, W.P., Hou, J.C., Sha, L.: Dynamic clustering for acoustic target tracking in wireless sensor networks. In: Proceedings of the 11th IEEE International Conference on Network Protocols (ICNP'03), pp. 1092–1648/03 (2003)
12. Yang, Y., Fonoage, M., Cardei, M.: Improving network lifetime with mobile wireless sensor networks, Elsevier. Comput. Comm. **33**, 409–419
13. Dong, Q., Dargie, W.: A survey on mobility and mobility-aware MAC protocols in wireless sensor networks, IEEE Commun. Surv. Tutorials First Q. **15**(1), 81–100 (2013)
14. Gupta, I., Riordan, D., Sampalli, S.: Cluster-head election using fuzzy logic for wireless sensor networks. In: Proceedings of the 3rd Annual Communication Networks and Services Research Conference, pp. 255–260 (2005)

Mr. Maddali M. V. M. Kumar received his Master of Technology in CSE from JNTUK and currently pursuing his Ph.D. in Computer Science and Engineering from Acharya Nagarjuna University, Guntur. He has 5 years of Teaching Experience and published papers in International/National Journals and Conferences. He is a Life Member in CSI and ISTE. His research focuses on the Computer Networks.

Dr. Aparna Chaparala had been awarded Ph.D. by JNTU, Hyderabad for her thesis titled "Efficient classification techniques for Brain Computer Interface (BCI)". She obtained her master's degree from the department of CSSE, Andhra University in 2004. She is currently working as an Associate Professor in the department of CSE at RVR & JC College of Engineering. She has over Ten research publications. Her research interests include data mining, databases, information security and computer networks.

Design of Slotted Pentagonal Structure Patch Antenna for RF Energy Harvesting in Mobile Communication Band

S. Venkata Suriyan and K. J. Jegadish Kumar

Abstract Printed antennas are most suited for aerospace and mobile application due to their inherent characteristics namely low profile, reduced weight, less difficulty in fabrication and integration. Regardless of these advantages, they suffer from narrow bandwidth, less gain, and poor directivity. In this project, to design a microstrip planar antenna for 2G, 3G, 4G, and 5G mobile communication applications. The multiband antenna design should be highly directive and should have high gain. The antenna should resonate at 0.924, 1.88, 2.112, 3.4, and 3.844 GHz. The multiband antenna so designed can be adaptively used for mobile communication application, as the antenna so designed satisfies the gain and VSWR requirements along with radiation pattern and can also be used for RF Energy harvesting, which is being studied and to be designed in the future.

Keywords Multiband antenna · Slotted antenna · Mobile communication band
Microstrip patch antenna

1 Introduction

Antenna is a necessary device for transmitting and receiving the radio waves in wireless communication systems. Microstrip patch antenna design is one main focus of today's technologies and it provides reduction in size, cost, and power consumption [1]. Microstrip antenna provides the benefit of reduced weight, less cost, and conformity to a surface that is shaped and also affinity to integrated circuits antenna. The fundamental design of the microstrip patch antenna consists of substrate with a conducting patch printed on it, with an impendence bandwidth of 1–2% [2]. Miniaturized microstrip patch antennas are required in many of the current trending applications

S. Venkata Suriyan (✉) · K. J. Jegadish Kumar
SSN College of Engineering, Chennai, India
e-mail: ernestosuriya@gmail.com

K. J. Jegadish Kumar
e-mail: jegadish.kumar@gmail.com

S. Smys et al. (eds.), *International Conference on Computer Networks and Communication Technologies*, Lecture Notes on Data Engineering and Communications Technologies 15, https://doi.org/10.1007/978-981-10-8681-6_47

like that of wireless local area networks (WLANs), mobile cellular handsets, Bluetooth, HiperLAN, ZigBee, and other short-range wireless communication application. Lately, the requirement for miniaturized antennas which will be able to operate at various frequencies with adequate bandwidth has been developed [3].

To develop an effective communication system with increased antenna performance, antennas with reduced size are fixed in handheld devices. Physical size of the antenna is one major factor for the performance of the antenna based on the application, the radiation characteristics of return loss of the antenna varies, hence in the communication system various antennas are used [4]. Antennas with less frequency are used made use in cellular mobile communication, military operation, and WiMax, whereas high-frequency antennas are used for satellite communication, weather monitoring system, etc [5]. There are a huge number of techniques to achieve compact size. Every compact antenna design either modifies the regular structure or adopts a particular miniaturization technique, but each has its own pros and cons. Compact size can be achieved by a simple antenna design together with inset feed [6]. Due to the inset structure, the impendence of the feed matches with the impedance of the radiating structure and, therefore, the desired result can be achieved with minimum dimension when compared to the traditional design structure. The current dispersion changes if there is any defects in the etches in the ground plane, which, in turn, increases the storage of effective capacitance and inductance [7]. This accordingly decreases the refection coefficient and the required output is met together with compact size. Other techniques such as meander lines are also adopted. Meander lines are nothing but the random fold of lines etched in the patch, which greatly affects the current distribution. These randomly folded increases the distance traveled by the current, which, in turn, is equal to the dimension that is required for the antenna to resonate in frequency bands such as L-band, C-band, etc., [8] complementary split ring resonator is one of the miniaturized techniques that is followed in order to get a compact size. They are either incorporated in the patch or ground [9]. A very common technique is either to etch slots in the patch in the appropriate positions or designing the patches in the form of slots which is adopted for almost all designs. This greatly helps in achieving a compact size [10].

2 Antenna Design

A multiband antenna consisting of rectangular patch with various slot lines to operate at 0.924, 1.88, 2.112, 3.4, and 3.844 GHz with a minimum bandwidth requirement of 50 MHz and directional radiation pattern. This is achieved by using pentagonal structure and symmetric slots. A satisfactory outcome for return loss, bandwidth, radiation pattern, and gain has been achieved such that it is suited for 2G, 3G, 4G, and 5G mobile communication system and RF energy harvesting. In order to achieve multiband operation, antennas with strips and slots are designed, such antennas can be used for wireless communication. The entire antenna design occupies a size of 100 mm × 80 mm × 1.6 mm. The antenna is fed by a microstrip line of width 3 mm

Table 1 Design specification

Components	Values
Substrate material	Fr4_Epoxy
Relative dielectric constant	4.3
Thickness of the substrate	1.6 mm
Design frequency	0.924, 1.88, 2.112, 3.4, and 3.844 GHz

Fig. 1 Patch

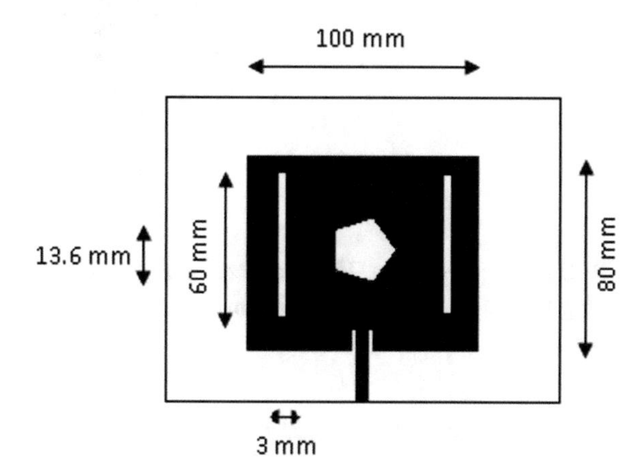

with an input impedance of 50 Ω [11]. The antenna that is proposed has a simple design and provides efficient control over three operating bands by changing the dimension of the pentagonal and symmetric slots present in the patch (Table 1).

For the lower frequency, a patch with dimension of 100 mm × 80 mm × 1.6 mm is used, i.e., for 0.924, 1.88, 2.112, 3.4, and 3.844 GHz bands. Figure 1 shows the lower resonant mode and the return loss simulated plots for the designed structure. In order to determine the resonant modes, the length and width of the patch can be used, using the slots on the patch the desired. To study the resonant modes that are desired, the slots in the patch are made use of. At high frequencies, in order to achieve noval resonant modes slots dimension and position was studied [12]. The pentagonal slot is responsible for 2.112 GHz band. To enhance the desired multiband frequency, slots of various lengths are used. Symmetric slot is responsible for 5G GHz, i.e., 3.4 GHz. The ground plane and the feed length are varied to get the perfect resonance and to optimize the results at the desired frequencies [13].

Pentagonal antenna operates at 5 bands. Reflection coefficient of various frequencies have been plotted and measured using CST (Fig 2).

Multiband antenna consisting of a rectangular patch with pentagonal and symmetric slots to operate at 0.924, 1.88, 2.112, 3.4, and 3.844 GHz with minimum bandwidth requirement of 50 MHz and directional radiation pattern. This is achieved by using slots [14]. The satisfactory results for return loss, bandwidth, radiation pattern, and gain have been achieved such that it is suitable for mobile communication system and

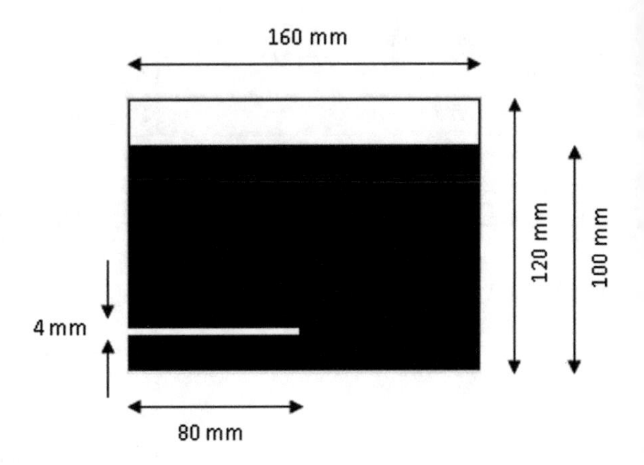

Fig. 2 Ground plane

RF Energy harvesting. The proposed multiband antenna is designed and analyzed using CST Microwave Studio 2016.

3 Simulation and Results

3.1 S-Parameter Performance

The operating frequency of the design, i.e., at which frequencies the antenna resonance is mentioned by looking at the S_{11} parameter, which is also known as return loss of an antenna.

Figure 3 illustrates the simulation results for the antenna using the CST software, the designed multiband antenna satisfies the frequencies at 0.924, 1.88, 2.112, 3.4, and 3.844 GHz. All the five frequencies have the reflection coefficient of the antenna. The radiation pattern will also change by changing the position and dimension of the slots in the patch. The values that are obtained in graph (CST) are extracted. The values are exported and plotted in sigma plot.

Figure 3 illustrators the S11 parameter for the designed antenna resonating at 0.924 GHz with return loss of −21.115 dB, 1.88 GHz with return loss of −22.106 dB, 2.112 GHz with return loss of −26.647 dB, 3.4 GHz with return loss of −26.462 dB and 3.844 GHz with return loss of −21.049 dB.

3.2 Radiation Pattern

The far-field radiation pattern of each resonating frequency is being plotted and studied at frequency 0.924, 1.88, 2.112, 3.4, and 3.844 GHz. CST software are used

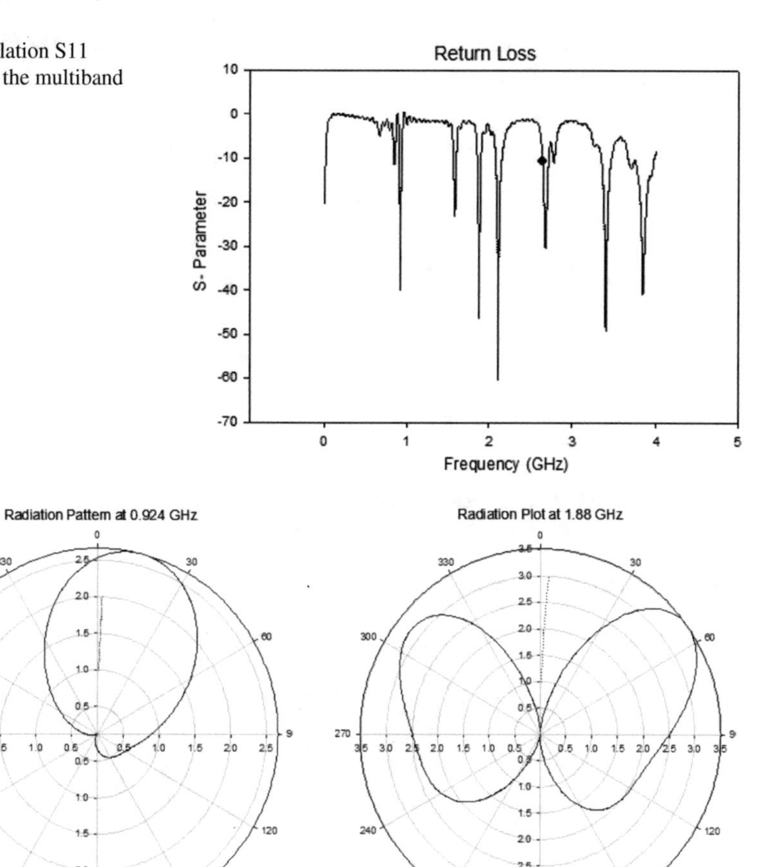

Fig. 3 Simulation S11 parameter of the multiband antenna

(A) Simulated far field Pattern at 0.924 GHz

(B) Simulated far field Pattern at 1.88 GHz

Fig. 4 Simulated far-field pattern

for plotting the patterns. The pattern shows that for all the 5 frequencies the radiation pattern required is a directional pattern which can communicate in a particular direction.

Figure 4 represents the radiation pattern at 0.924 and 1.88 GHz with beamwidth of 109.8°, gain of 2.66° and 119.9°, gain of 3.5, respectively.

Figure 5 represents the radiation pattern at 2.112 GHz and 3.4 GHz with beamwidth of 75.5°, gain of 2.4 along with some loss of −3.7 dB and 48.7°, gain of 2.47 along with some loss of −1.4 dB, respectively.

Figure 6 represents the radiation pattern at 3.844 GHz with beamwidth of 46.7°, gain of 2.55 along with some loss of −6.0 dB.

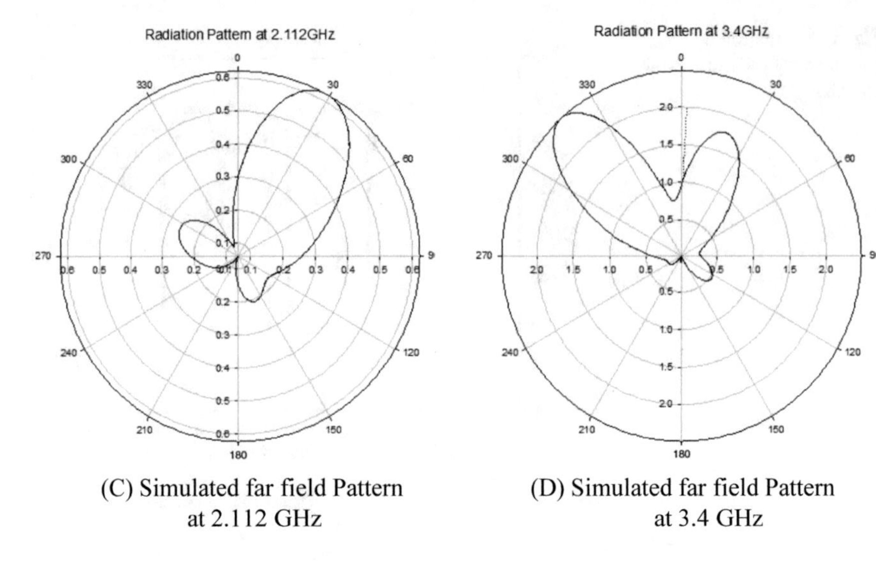

(C) Simulated far field Pattern at 2.112 GHz

(D) Simulated far field Pattern at 3.4 GHz

Fig. 5 Simulated far-field pattern

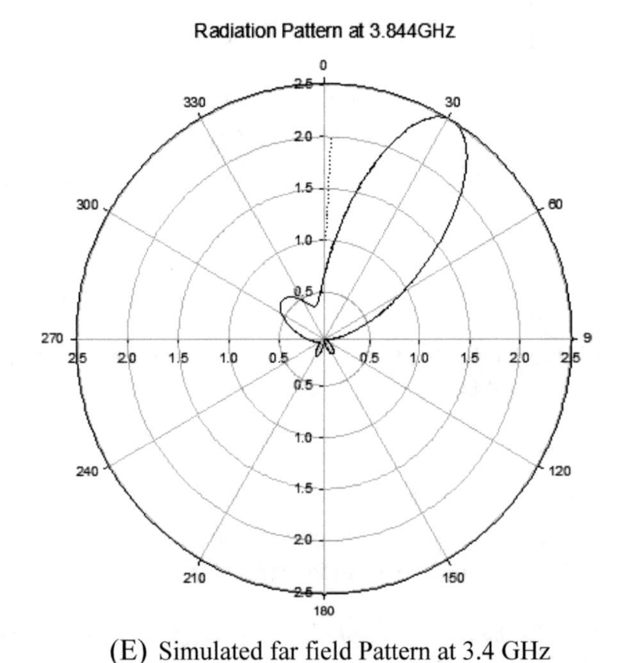

(E) Simulated far field Pattern at 3.4 GHz

Fig. 6 Simulated far-field pattern

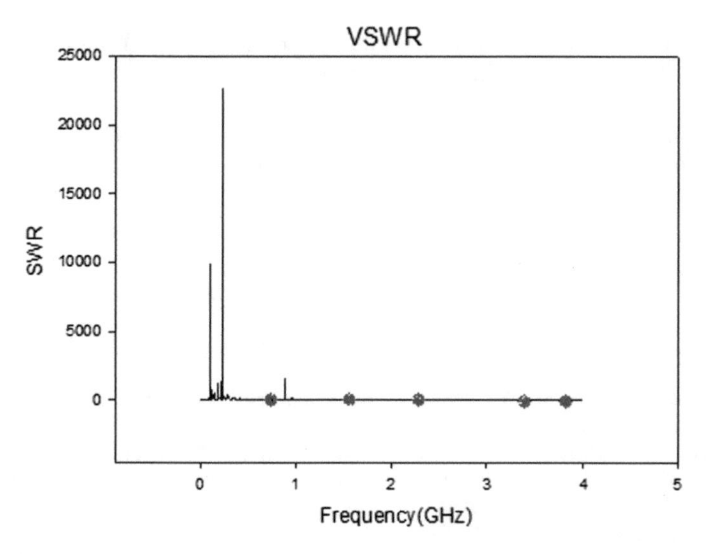

Fig. 7 VSWR

Table 2 Performance measure of the designed antenna

Parameters	Simulation results					Calculated results
	0.924 GHz	1.88 GHz	2.112 GHz	3.4 GHz	3.884 GHz	
Return loss	−21.115	−22.106	−26.647	−26.462	−21.049	<−10 dB
Gain	2.66	3.5	2.4	2.47	2.55	<1
Directivity (beamwidth)	109.8	119.9	75.5	48.7	46.7	(0.5–120)
VSWR	1.19	1.17	1.09	1.09	1.86	(1–2)

3.3 VSWR

Figure 7 represents the VSWR value of the rectangular patch antenna for 0.924, 1.88, 2.112, 3.4, and 3.884 GHz with values 1.1929, 1.1703, 1.0976, 1.0998, and 1.8638 respectively (Table 2).

4 Conclusion

Mobile communication is trending technology widely used in wireless communication system. 4G and 5G are the emerging spectrums required for today's mobile communication. A multiband antenna such design that it covers 2G, 3G, 4G, and 5G spectrum. The design antenna satisfies the directivity, gain, VSWR, and bandwidth requirement for above-mentioned applications.

Acknowledgements The authors would like to thank Dr. K. J. Jegadish Kumar for guiding me, Anthony Xavier Research Assistant in Simulations and support.

References

1. Balanis, C.A.: Antenna Theory: Analysis and Design, 2nd edn, pp. 542–544. Wiley, New York (1997)
2. Ying, Z.: Antennas in cellular phones for mobile communications. Proc. IEEE **100**(7), 2286–2296 (2013)
3. Naser-Moghadasi, M., Sadeghzadeh, R.A., Fakheri, M., Aribi, T., Sedghi, T., Virdee, B.S.: Miniature hook-shaped multiband antenna for mobile applications. IEEE Antennas Wirel. Propag. Lett. **11**, 1096–1099 (2012)
4. Bayatmaku, N., Lotfi, P., Azarmanesh, M., Soltani, S.: Design of simple multiband patch antenna for mobile communication applications using new E-shape fractal. IEEE Antennas Wirel. Propag. Lett. **10**, 873–875 (2011)
5. Garba, M.S.: Design of tri-band z-shaped patch antenna for WLAN and WiMAX applications. Int. J. Res. Electron. Commun. Technol. **2**(4), 670–673 (2015)
6. Yildirim, B.: Multiband and compact WCDMA/WLAN antenna for mobile equipment. IEEE Antennas Wirel. Propag. Lett. **10**, 14–16 (2014)
7. Huang, H., Liu, Y., Gong, S.: Broadband dual-polarized omnidirectional antenna for 2G/3G/LTE/WiFi applications. IEEE Antennas Wirel. Propag. Lett. **15**, 576–579 (2015)
8. AbuTarboush, H.F., Nilavalan, R., Peter, T., Cheung, S.W.: Multiband ınverted-F antenna with independent bands for small and slim cellular mobile handsets. IEEE Trans. Antennas Propag. **59**(7), 2636–2645 (2014)
9. Chou, Y.J., Lin, G.S., Chen, J.F., Chen, L.S., Houng, M.P.: Design of GSM/LTE multiband application for mobile phone antennas. Electron. Lett. **51**(17), 1304–1306 (2015)
10. Dhande, P.: Antennas and its Applications: DRDO Science Spectrum, pp. 66–78 (2012)
11. Huang, H., Liu, Y., Gong, S.: A dual-broadband, dual-polarized base station antenna for 2G/3G/4G applications. IEEE Antennas Wirel. Propag. Lett. **16**, 1111–1114 (2016)
12. Hsu, C.K., Chung, S.J.: Compact antenna with U-shaped open-end slot structure for multi-band handset applications. IEEE Trans. Antennas Propag. **62**(2), 929–932 (2013)
13. Bait-Suwailam, M.M., Siddiqui, O.F., Ramahi, O.M.: Mutual coupling reduction between microstrip patch antennas using slotted-complementary split-ring resonators. IEEE Antennas Wirel. Propag. Lett. **9**, 876–878 (2010)
14. Habashi, A., Nourinia, J., Ghobadi, C.: A rectangular defected ground structure (DGS) for reduction of mutual coupling between closely-spaced microstrip antennas. In: 20th Iranian Conference on Electrical Engineering (ICEE), pp. 1347–1350 (2015)

Venkata Suriyan has completed his post graduation in Communication Systems in Sri Sivasubramaniya Nadar College of Engineering. He received his Bachelor's degree in Electronics and Communication Engineering from Annamalai University in 2016. His research interests include Antenna, RF, Image Processing and Digital Electronics.

Dr. K. J. Jegadish Kumar has 14 years of teaching and research experience, including 6 years of research experience in the field of Cryptographic techniques for various wireless communication applications. He received his B.E. (ECE) degree with first class from Manonmaniam Sundaranar University, M.E. Communication systems first class with distinction from Madurai Kamaraj University. He completed Ph.D. in JNTU, Hyderabad. He has published over 26 research publications in refereed international journals and in proceedings of international and National conferences. His current research work includes Low power and area efficient cryptographic techniques for wireless communication applications, RF energy scavenging systems, Antenna designs and Wireless Embedded systems etc.

Design of Fractal-Based Dual-Mode Microstrip Bandpass Filter for Wireless Communication Systems

G. Abirami and S. Karthie

Abstract In this paper, a dual-mode microstrip bandpass filter is designed employing Peano fractal geometry. The filter design uses the square patch with a perturbation element added at the corner of the patch for dual-mode configuration. The proposed bandpass filter structure has been designed based on the second iteration Peano fractal curve at 2.45 GHz using two substrates (RT/Duroid 6010 and FR-4) and their performance is simulated for return loss and insertion loss. The design process is continued up to second iteration of Peano fractal in order to achieve the center frequency at 2.4 GHz, which can be used for WLAN applications. The comparison is made on the simulation results of the proposed fractal filter between the two substrates.

Keywords Peano fractal geometry · Bandpass filter · Size reduction · Return loss

1 Introduction

Fractal geometry has been used in almost all the fields of science and art. In electromagnetics, fractal geometries have been applied widely in the fields of passive microwave circuit design and application of modern wireless communication systems [1]. In modern wireless and mobile communication systems, the narrowband microstrip bandpass filters are always playing essential roles and transmit the signal between the transmitter and receiver. These geometries have two common properties; space-filling and self-similarity. Space-filling property of fractal shapes has been successfully applied to the design of dual-mode microstrip bandpass filter, while the self-similarity property has been utilized to reduce the size. This property can be exploited for the miniaturization of microstrip antennas, resonators, and filters [2].

G. Abirami (✉)
SSN College of Engineering, Chennai, India
e-mail: abhe1604@gmail.com

S. Karthie
Department of ECE, SSN College of Engineering, Chennai, India
e-mail: karthies@ssn.edu.in

S. Smys et al. (eds.), *International Conference on Computer Networks and Communication Technologies*, Lecture Notes on Data Engineering and Communications Technologies 15, https://doi.org/10.1007/978-981-10-8681-6_48

In this paper, a dual-mode microstrip bandpass filter design has been presented for use in the modern compact communication systems. The iterations of filter structure are based on Peano fractal curve geometry and it has been designed at a frequency of ISM bands. The resulting filter is expected to possess a considerable miniaturization owing to its remarkable space-filling property together with good transmission and return loss responses [3]. Peano fractal geometries are used in the design of square ring resonators in order to make high-performance miniaturized dual-mode microstrip bandpass filters and also to suppress the noise [4, 5]. The proposed Peano filter structure can be used to design dual-mode coupler. The high space-filling property of this fractal geometry is used to design bandpass filters with high size reduction levels [6, 7]. Peano fractal curve has been also used as a defected ground structure in the design of a microstrip bandpass filter operating at the S-band microwave frequency [8]. The same design has been carried out for different substrates and the performance of the filter is compared with the simulation results of insertion loss and return loss. Compared to RT/Duroid 6010, FR-4 substrate is easily available at low cost to fabricate [9].

Microstrip filters, in general, are used in the design of RF and microwave filters. Fractal filters are used in communication systems for suppressing the harmonics, to isolate a communication of a signal from various channels to improve the unique message signal from a modulated signal [10]. The filters are based on $\lambda/4$ resonators, which are twice as short as conventional $\lambda/2$ resonators, so they represent good candidates for the realization of compact filtering circuits. At the same time, $\lambda/4$ resonators have better potential for minimizing interferences at higher resonances [11].

2 The Peano Fractal Filter Structure

Microstrip bandpass filter with Peano fractal geometry has been designed on two different substrates such as RT/Duroid 6010 and FR-4. The frequency is shifted from 2.65 to 2.4 GHz by increasing the iteration and the design continues up to 2nd iteration [5] (Fig. 1).

2.1 Design Using FR4

At first, a microstrip resonator based on the second iteration Peano fractal geometry, has been designed at a frequency of 2.4 GHz. It has been supposed that the modeled filter structures have been etched using a substrate with a relative dielectric constant of 4.4 and a substrate thickness of 1.6 mm. The resulting resonator dimensions have been found to be 26 mm × 26 mm. In addition, a perturbation element is added at the corner of the patch. The guided wavelength λ_g at the design frequency with the stated substrate parameters is calculated as

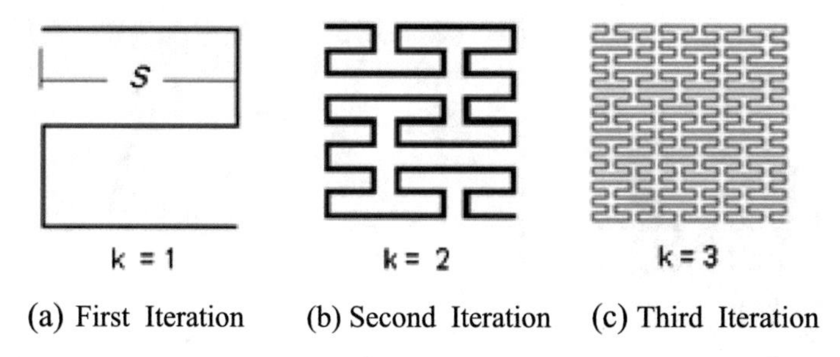

Fig. 1 Iterations of Peano Fractal. *Source* PIERS, p. 889

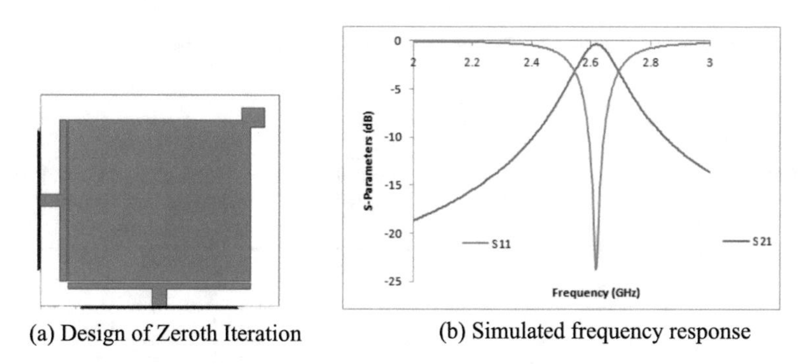

Fig. 2 The return loss S_{11} and the insertion loss S_{21} of the zeroth iteration

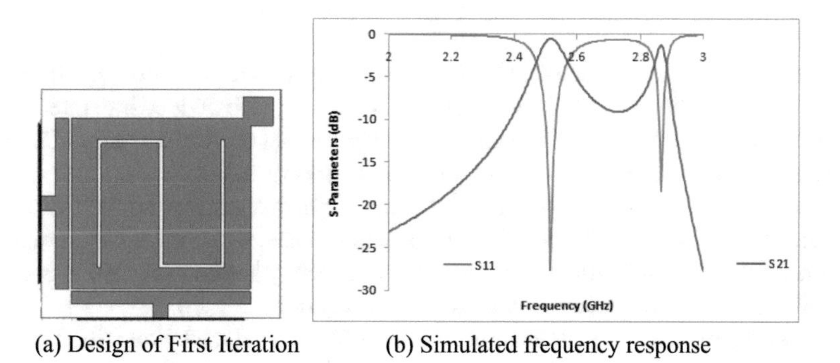

Fig. 3 The return loss S_{11} and the insertion loss S_{21} of the first iteration

$$\lambda_g = \frac{c}{f\sqrt{\varepsilon_{\text{eff}}}} \tag{1}$$

where $\varepsilon_{\text{eff}} = \frac{\varepsilon_r + 1}{2}$ is the effective dielectric constant and $c = 3 \times 10^8$ m/s. The filter layout and simulated frequency response of Peano fractal filter from zero to second iteration using FR-4 substrate are shown in Figs. 2, 3, and 4. Here, the input port and output port are placed orthogonal to each other in the filter structure.

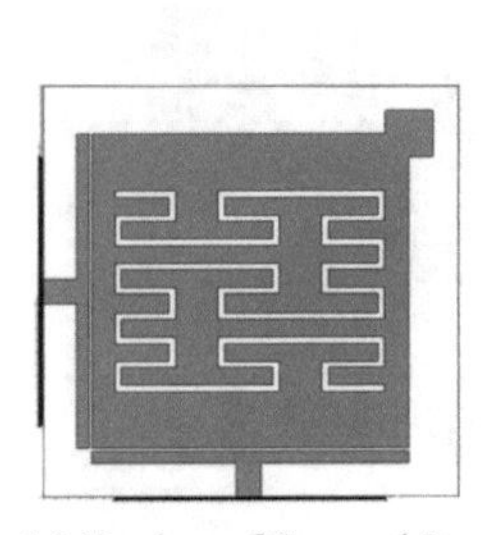

(a) Design of Second Iteration

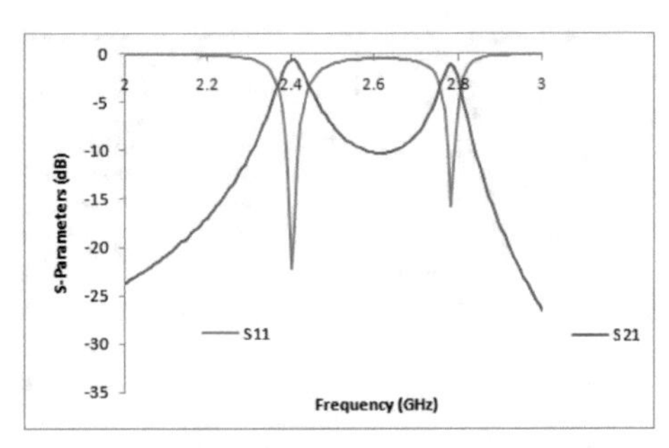

(b) Simulated frequency response

Fig. 4 The return loss S_{11} and the insertion loss S_{21} of the second iteration

Table 1 Fiter performance on FR-4 substrate

FR-4 substrate					
Iterations	Center frequency (GHz)	Return loss S_{11} (dB)	Insertion loss S_{21} (dB)	Bandwidth (GHz)	Fractional bandwidth (%)
Zeroth iteration	2.64	24.78	2.85	0.19 **FF**	7.19
First iteration	2.52	27.86	2.38	0.17	6.7
Second iteration	2.40	24.06	2.50	0.12	5

Peano fractal is constructed by a sequence of steps, where the first iteration step constructs a set with the length and width of the square patch divided into three equal loops or sequences and the one-element sequence consist of its center point. Connect those equal loops such that to form a structure as shown in Fig. 3.

The width of the Peano fractal filter is 0.2 mm, with the relative permittivity of 4.4. The size of the filter is reduced while the design utilizes fractal geometry compared to non-fractalized filter. As seen in Fig. 4, the frequency has been shifted in order to achieve the center frequency with increase in iteration.

The performance of the filter has been simulated and analyzed. From this designed structure and with the help of center frequency, return loss, insertion loss, the bandwidth, and the fractional bandwidth are tabulated in Table 1.

2.2 *Design Using RT/Duroid 6010*

The modeled filter structures have been etched using a substrate with a relative dielectric constant of 10.2 and a substrate thickness of 1.27 mm The resulting resonator patch dimensions have been found to be 17 mm × 17 mm. The design will be continued until it achieves the center frequency. The size of the patch has been calculated

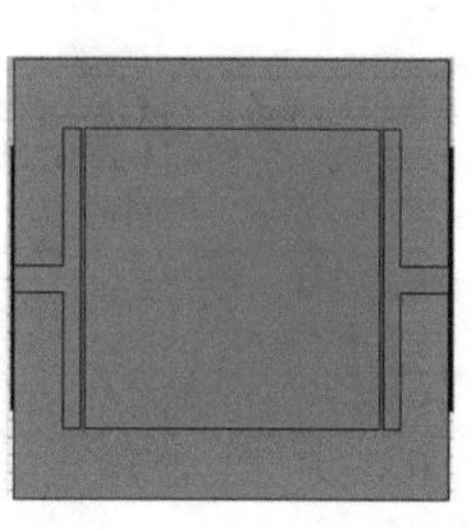

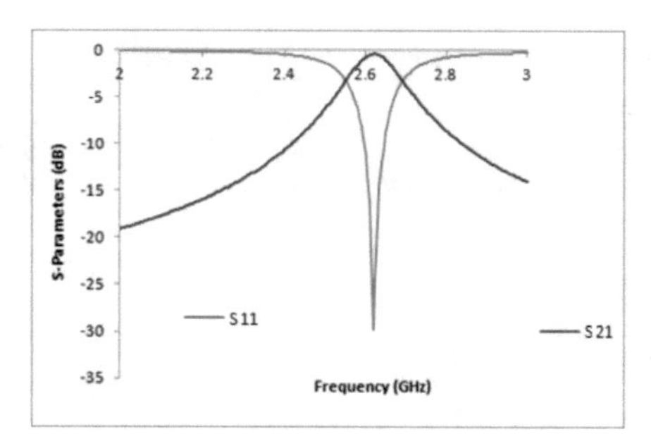

(a) Design of Zeroth Iteration (b) Simulated frequency response

Fig. 5 The return loss S_{11} and the insertion loss S_{21} of the zeroth iteration

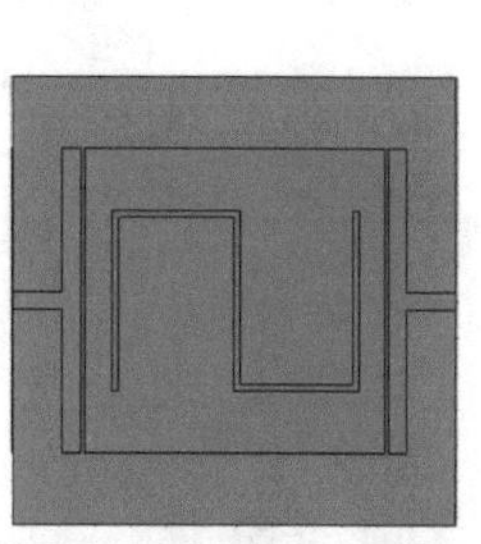

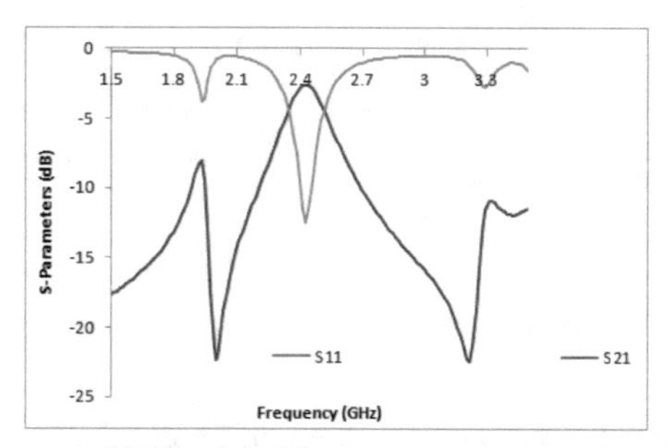

(a) Design of First Iteration (b) Simulated frequency response

Fig. 6 The return loss S_{11} and the insertion loss S_{21} of the first iteration

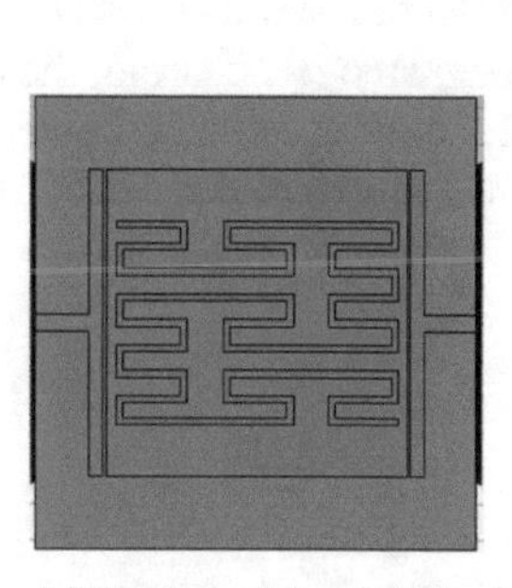

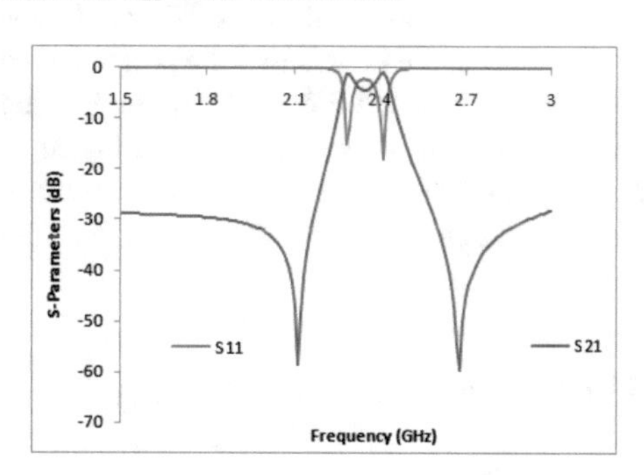

(a) Design of Second Iteration (b) Simulated frequency response

Fig. 7 The return loss S_{11} and the insertion loss S_{21} of the second iteration

from Eq. (1). The layout and simulated frequency response of Peano fractal filter from zero to second iteration using RT/Duroid substrate are shown in Figs. 5, 6, and 7.

From this designed structure and with the help of center frequency, return loss, insertion loss, the bandwidth, and the fractional bandwidth are tabulated in Table 2.

Table 2 Fiter performance on RT/Duroid substrate

RT/Duroid substrate					
Iterations	Center frequency (GHz)	Return loss S_{11} (dB)	Insertion loss S_{21} (dB)	Bandwidth (GHz)	Fractional bandwidth (%)
Zeroth iteration	2.62	25.78	0.39	0.22 **FF**	8.3
First iteration	2.49	22.06	0.30	0.16	6.4
Second iteration	2.40	14.94	0.22	0.14	5.8

Table 3 Comparision of bandpass filter simulation results on FR4 and RT/Duroid substrate

Substrates	Center frequency (GHz)	Return loss S_{11} (dB)	Insertion loss S_{21} (dB)	Bandwidth (GHz)	Fractional bandwidth (%)
FR-4	2.40	24.06	2.50	0.12	5
RT/Duroid	2.40	14.94	0.22	0.14	5.8

3 Comparison of Filter Performance

The Peano fractal designed on different substrates is compared with simulated frequency response. These fractal structures are designed on two substrates, FR-4 and RT/Duroid with their relative dielectric constant and thickness. The comparative study is performed on FR-4 and RT/Duroid substrates with same resonating frequency 2.4 GHz and it is seen that the performance of the fractal filter on RT/Duroid is better than the filter on FR-4. Due to the low loss tangent of RT/Duroid, the filter provides better performance. The performance of the filter is evaluated using the insertion loss and return loss [4, 9, 12]. The comparison results of the Peano fractal BPF on both the substrates for the second iteration are shown in Table 3.

4 The Proposed Filter Modeling

In this paper, the proposed filter structure based on Peano fractal geometry is designed to obtain a dual-mode microstrip bandpass filter by employing the fractal on the conventional square ring resonator using RT/Duroid 6010 substrate. At first, the square patch resonator has been divided into four quarters, then replacing each of the four parts with the second iteration Peano structure and the fractal structure is to be etched out. Meanwhile, a perturbation is added at the midway of the resonator. The resonator is coupled to the S_{11}/S_{21} ports via two couplers. This shape of perturbation is the most convenient to satisfy the required coupling. In the dual-mode configuration, the signal branches into two-way paths, which have different electrical lengths and in

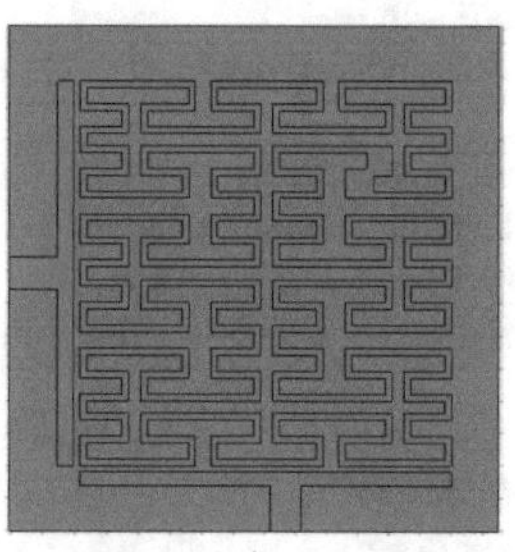

(a) Design of Dual-Mode BPF

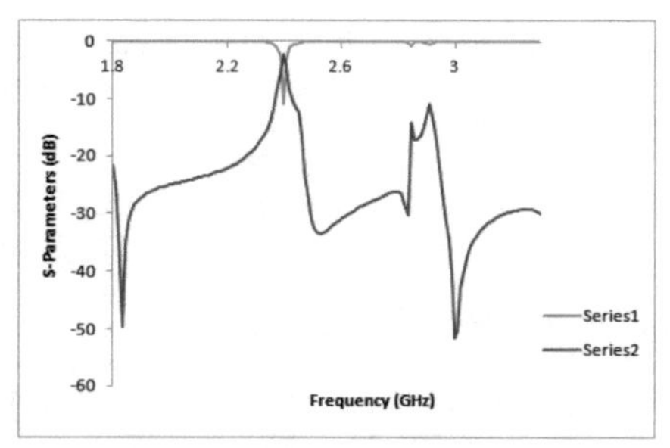

(b) Simulated frequency response

Fig. 8 The return loss S_{11} and the insertion loss S_{21} of dual-mode BPF

that way, two passbands are formed at the output [12]. The performance evaluation of the resulting dual-mode microstrip bandpass filter, with the layout shown in Fig. 8, has been modeled at 2.45 GHz.

5 Conclusion

The proposed dual-mode microstrip bandpass filter structure has been designed on RT/Duroid 6010 substrate in the form of second iteration Peano fractal curve modeled at 2.4 GHz. The resulting filter has reasonable passband performance besides the size reduction gained making it suitable for a wide variety of wireless communication applications. The use of different substrates for fractal-shaped resonators can vary the frequency ratio of the passband, which determines the selectivity, insertion loss, and return loss.

References

1. Pozar, D. M.: Microwave Engineering, 4th edn. Wiley (2011)
2. Ali, J.K., Alsaedi, H., Hasan, M.F., Hammas, H.A.: A Peano Fractal-Based Dual-Mode Microstrip Bandpass Filters for Wireless Communication Systems. University of Technology, Baghdad, Iraq: Progress. In: Electromagnetics Research Symposium Proceedings, Moscow, Russia (2012)
3. Ali, J. K., Mezaal, Y. S.: A New Miniature Narrowband Microstrip Bandpass Filter Design Based on Peano Fractal Geometry. Department of Electrical and Electronic Engineering, University of Technology, Baghdad, IEEE Conference Publications, pp. 27–29 (2009)
4. Mezaal, Y. S., Jehad, W. A.: Miniaturized bandpass filter based on Peano fractal geometry higher harmonic suppression. Al-Qadisiya J. Eng. Sci. **4**, 57–60 (2011)
5. Jarry, P., Beneat, J.: Design and Realizations of Miniaturized Fractal RF and Microwave Filters. John, Hoboken, New Jersey (2009)

6. Ahmed, H.S., Salim, A.J., Ali, JK.: A compact dual-band bandstop filter based on fractal microstrip resonators. In: Progress In Electromagnetics Research Symposium PIERS Proceedings, Prague, Czech Republic (2015)
7. Sakotic, Z., Crnojevic-Bengin, V., Jankovic, N.: Tri-band Coupler Based on Patch Resonator with Peano Fractal Curve Elements, Vol. 15, pp. 276–279. Infoteh-Jahorina (2016)
8. Ahmed, H.S., Salim, A.J., Ali, J.K.: A compact dual-band bandstop filter based on fractal microstrip resonators: Progress. In: Electromagnetics Research Symposium PIERS Proceedings, Prague, Czech Republic (2015)
9. Chanu, T.R., Kumar, A., Kumar, A.: Comparative study between FR4 and RT Duroid CPA. Int. J. Eng. Technol. Manag. Appl. Sci. **5**(4) (2017). ISSN 2349–4476
10. Hong. J.S., Lancaster, M.J.: Microstrip Filters for RF/Microwave Applications. John, New York (2007)
11. Crnojevic-Bengin, V.: Advances in Multi Band Micro Strip Filters. Cambridge University (2005)
12. Ali, K.: A new miniaturized fractal bandpass filter based on dual-mode microstrip square ring resonator. In: Proceedings of the 5th International Multi-Conference on Signals, Systems and Devices, IEEE SSD'08, Amman, Jordan (2008)

G. Abirami studying M.E. (Communication Systems), SSN College of Engineering, Chennai. Her research area includes Wireless Communication Systems.

S. Karthie working as an Assistant Professor, Department of ECE, SSN College of Engineering, Chennai. His research area includes Wireless Communication Systems.

A Sixth Sense Door using Internet of Things

John Britto, Viplav Chaudhari, Deep Mehta, Akshay Kale and Jyoti Ramteke

Abstract This paper provides a detailed explanation of a system that aims at detecting and capturing the movements of people in the proximity of a door, using Internet of Things (IOT). The door will be able to sense the movements of people within a predefined range. Once the movement is sensed, an attached camera will click a high quality image of the person. The image will be uploaded to the cloud in real time and transmitted to the owner's phone which will appear as a push notification on a custom application. The owner can view the incoming image and decide the future course of action, which may include sending back a prerecorded message, talking to the visitor over the phone, or setting off an alarm in case of suspicious activity. It can also interact with other appliances inside the home to pretend that someone is there. This system can be a great tool in crime detection and prevention.

Keywords IOT · Door · PIR sensor · Camera · Android application · Raspberry Pi

1 Introduction

The city of Mumbai is a busy one. People here end up working for days together. They do not get a break for long stretches of time, let alone a small period of relaxation. In such a scenario, it becomes very difficult to balance home with work. Since they are out most of the time, they don't come to know who visited their home and are not able to send a message in case of emergency. Also in question here is the ever increasing rate of crime. People get scared to even leave the city for one day. Thus there is the need for a home automation system. Consider an employee on a normal weekday, gone to work. Once he is safely in the comfort of his office, a thief sneaks

J. Britto (✉) · V. Chaudhari · D. Mehta · A. Kale · J. Ramteke
Computer Engineering Department, Sardar Patel Institute of Technology,
Mumbai 400055, India
e-mail: johnbritto.nadar@spit.ac.in

J. Ramteke
e-mail: jyoti_ramteke@spit.ac.in

S. Smys et al. (eds.), *International Conference on Computer Networks and Communication Technologies*, Lecture Notes on Data Engineering and Communications Technologies 15, https://doi.org/10.1007/978-981-10-8681-6_49

into the house by either breaking a window or a door. The proposed system would enable the client to monitor his home when a door or a window sensor triggers the alarm. The app will recognize the visitor if he/she has been marked as safe by the client. The clients can monitor their home with camera and could immediately inform local authorities or the police. Some of the major points and gaps that will be covered in this system are IOT, data in cloud for reliability and safety issues, Android programming, increase in efficiency, image compression algorithms, face recognition, sensor network control, optimization and maximum use of resources, safety and training of unskilled manpower by GSM based systems, etc.

2 Literature Review

The advent of home automation during the 1970s hardly did anything to make life easy for users. First, to pinpoint the exact economic benefits of such technologies is extremely difficult. Second, the costs of implementing such technologies must be accompanied along with a justification of their value. Home automation technologies must be cost efficient, easy to install and compatible with any network infrastructure and appliance. In 2003, Housing Learning and Improvement network presented a smart home definition offered by Interetec, stating that a smart home is "a dwelling incorporating a communications network that connects the key electrical appliances and services, and allows them to be remotely controlled, monitored or accessed". In the year 1995, Welfare TechnoHouses were set up in Japan. These experiments were performed to provide health monitoring for the elderly, sick and disabled people at home by using fully automated measurements to support daily health care and increase the quality of life [1]. Over the past 7 years, The University of Texas at Arlington has worked on the Managing an Adaptive Versatile Home (MavHome) project. It is a home environment to perceive environment states through sensors and intelligently perform actions on the environment though controllers. These sensors connect to form an ad hoc network and make appropriate decisions.

The most significant part of any home security system is the accurate face detection and recognition of incoming visitors [2]. An entrance security guard can be managed remotely since detecting visitors while they are at the door and alerting the user on mobile phone is the most straightforward way to ensure security [3, 4]. Many systems have extra features like viewing video stream on the mobile phone [5]. Additionally, there is also voice alert or siren for alerting the neighbors when intruder is detected. The system detects the visitor's presence, captures and sends the image through email instantly to the user for recognizing the visitors. The system can generate a voice output when a person tries to enter the house [6, 7]. The owner can directly login and interact with the system in real time without needing to maintain an additional server. Additional features include energy efficiency, intelligence, low cost, portability and high performance. In future, more attention must be given to the research for IOT authenticity, confidentiality, and integrity of data [8].

3 Implementation Steps

3.1 Important Components

- Raspberry Pi 3 Model B
- 8 MP Pi Camera
- PIR Sensor and Speakers
- Ethernet Cable, USB cord, 8 GB + SD Card and Jumper Wires

3.2 Flowchart of the System

See Fig. 1

3.3 Setting up the Raspberry Pi

Download the Raspbian Operating System (OS) image from www.raspberrypi.org. Using an image writing tool such as Etcher, write and install the downloaded image to an SD card of a capacity of at least 8 GB.

The norm is to use a screen, a keyboard and a mouse for programming with the Raspberry Pi, but for ease of use, a laptop can be used. Raspberry Pi can be connected with a laptop through a Secure Shell (SSH) client. As of 2016, Raspberry Pi has disabled SSH for security reasons, hence it needs to be enabled manually. For this purpose, load the Raspbian installed SD card on to a computer using a card reader. Access the boot partition and create a file named "ssh", without any extension. When the Pi boots, it will find the "ssh" file, enable SSH and delete the file. The "ssh" file can contain only text or be kept blank.

3.4 Accessing the Raspberry Pi GUI

In order to access the Raspberry Pi Graphical User Interface (GUI) via an SSH client, an Internet Protocol (IP) address needs to be assigned to the Raspberry Pi. This is achieved by connecting the Raspberry Pi with the laptop through an Ethernet cable. Once this is done, allow other network users to connect through the laptop's Wi-Fi (This can be done in the Wi-Fi settings). This provides the ethernet port and devices connected via the Ethernet, with a dynamic IP address. To find this IP address, check the IPv4 properties of the Ethernet adapter. The IP address displayed is the starting IP address assigned by the laptop to the Ethernet port.

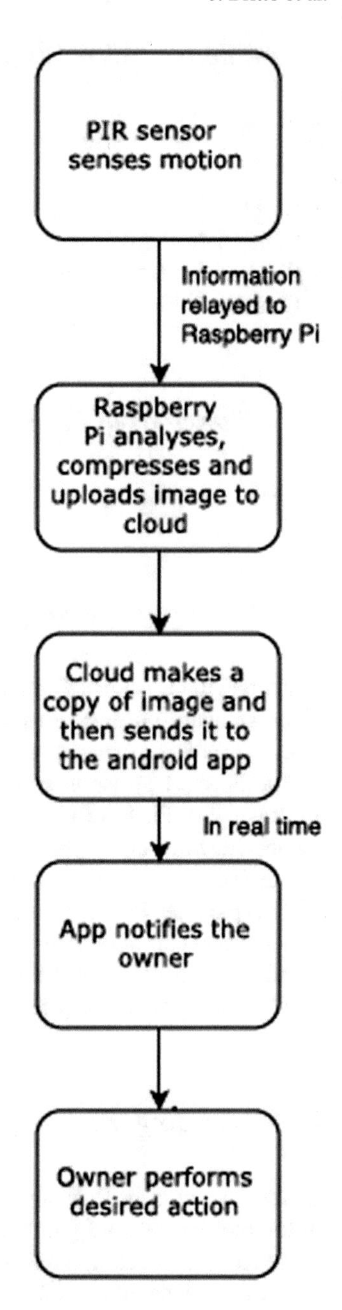

Fig. 1 Flow of the system

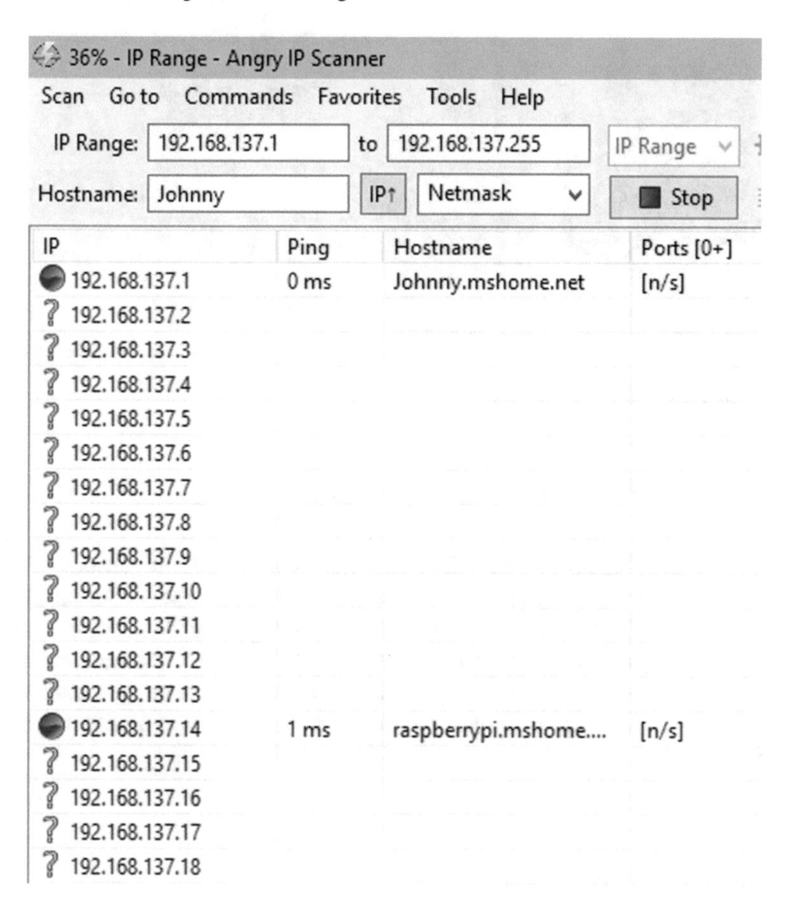

Fig. 2 The IP address assigned to the Raspberry Pi

To find the Raspberry Pi's IP address, a tool named Angry IP Scanner can be used. The Raspberry Pi's IP address can be found by entering the IP address of the ethernet port and scanning within that block of 255 IP addresses. The Raspberry Pi's IP address will show up on the screen. Use this IP address to access the Raspberry Pi via SSH. Download and install MobaXterm and open an SSH session. Enter the Raspberry Pi's IP address and start the session. Log in using the default username "pi" and the default password "raspberry". This will grant access to the Raspberry Pi's terminal. Type in the command "startlxde" and after a few seconds, the Raspberry Pi's GUI will open (Fig. 2).

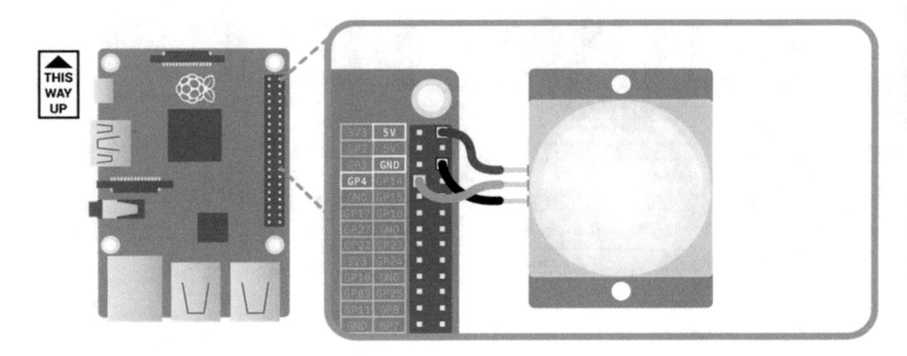

Fig. 3 Connecting the PIR sensor [9]

3.5 *Sensing the Motion and Capturing the Image*

Using Python 2.7.13, a code was written to sense the motion of the sensor. A Passive Infrared (PIR) sensor was used to detect the motion. When the output on the PIR sensor went high, the output "Motion Detected" was displayed. At the same time, the code for capturing the image runs simultaneously and the 8 MP Pi Camera, connected for clicking images, captures the image and stores a copy of the image locally on the Raspberry Pi as a backup, in case of extreme conditions such as an internet failure (Fig. 3).

3.6 *Image Compression*

In order to speed up the transmission of images and optimize the system, the transmitted images have to be compressed. This increases the overall efficiency of the system. For this, visually lossless image compression was performed using Discrete Cosine Transform (DCT). DCT expresses a finite sequence of data points in terms of a sum of cosine functions oscillating at different frequencies [10].

To perform the compression, the image is first divided into pixel blocks of 8×8. Treating the image as a matrix of 8×8, DCT is applied to each block. Once every block has been transformed, the blocks are compressed through quantization. This quantization is very precise, thus reducing the space without compromising on the quality of the image. The array of compressed blocks that constitute the image can be stored in a drastically reduced amount of space. The original image is then reconstructed using Inverse Discrete Cosine Transform (IDCT). To put this into perspective, the average original size of the images clicked through the Pi Camera was roughly 5 MB, which was brought down to around 3 MB by performing DCT without any visible distortion of the image.

3.7 Uploading to and Storing the Image on the Cloud

There are several cloud servers available for this purpose, such as Amazon Web Services, Firebase and Google Cloud, of which the latter has been used to implement the system. Google Cloud has several free features, and very convenient to use, since the database provided by Google Cloud for storage is Google Cloud's own version of MySQL, which is one of the more convenient databases. To connect the Raspberry Pi with Google Cloud, a Google Cloud SQL Instance must be created. After that, the Google Cloud SQL Instance access must be configured. This can be done by adding the IP address of the MySQL client in the authorized networks table. Once the cloud has been set up, the code for uploading the clicked image to the cloud must be run. For this purpose, a connection needs to be established with the database using a MySQL connector. The SQL Instance ID, IP address of the host machine, username and password of the database must be provided to the MySQL connector for establishing a successful connection.

For security reasons, the image needs to be encoded and then encrypted as a string before uploading on to the cloud. Once the image has been encrypted, an SQL query must be written which will insert the image into the database as an argument while also initializing the flag variable to 0. This flag variable will be used to indicate whether the image has been sent to the mobile application or not. The security of the transmitted images is also beefed up by establishing an encrypted link between the Raspberry Pi and Google Cloud, through Secure Sockets Layer (SSL).

3.8 Sending the Image to the Android App

This is the trickiest phase. For security reasons, Google Cloud does not allow direct transmission of images to an Android app. This, however, permits to add an additional layer of security and reliability to the system by exporting the whole back end to a web app which will not only act as a fail-safe but also make it extremely difficult for anyone to prevent the eventual transmission of the image since the potential hacker would have to delete the image at two places.

To connect with the web app made for storing the images, a PHP App Engine must be used. The web app needs to be deployed to the Google Cloud Platform. For this, a Cloud SQL Instance must be generated. This instance will be used to grant permission to the App Engine to connect with the database. The Cloud SQL Instance Connection Name, database, username and password must be provided to the App Engine to establish a connection. Once this has been done, an SQL query must be written to fetch the image from the database and send it to the web app. From within the directory in which the app's app.yaml configuration file is stored, start the local development server. Type the command "gcloud app deploy" from within the root directory to complete the successful deployment of the web app.

3.9 Face Recognition and Counter-Measures

An Android skin needs to be coded to fetch the image from the web app. The image appears as a push notification on the Android phone. On clicking the notification, the image opens up and the visitor is identified from the database of trusted visitors predefined by the user. Again, for this purpose, DCT has been used. The feature vector of the face is obtained by computing the DCT of the image. Feature extraction is basically extracting information which is most relevant in distinguishing a person. To perform feature extraction, the lowest DCT frequency coefficients have been selected. This is because most of the energy of the image is centered on the low-frequency coefficients. This approach is simplistic yet efficient because it neither evaluates the image's DCT coefficient, nor performs calculations or comparison [11, 12]. Furthermore, a Euclidean classifier has been used to find the best match between the received image and the images from the database.

If the person is not recognized, the app merely displays the image. In both the cases, the user must select the action to be performed next. The user can call any of the two emergency contacts he/she has saved on the app. He/she can also set off another IOT device to pretend as if someone is there at home, in case the visitor is suspicious. He/she can also interact with the visitor using an audio transceiver attached to the Raspberry Pi. In extreme cases, he/she is also allowed to sound off an alarm (Figs. 4 and 5).

4 Experimental Results and Conclusion

The total time taken from the clicking of the image to the transmission on the phone, on an average, was around 8 seconds. This is a very positive result since a lot of the existing systems take much more time for the same. The whole system costs around 250$ which is a far cry from certain professional systems priced around 2000$. The DCT face recognition algorithm had an accuracy of 87.23%.

Thus, the sixth sense door is an excellent home automation cum security system. In an age where the world is fast accepting IOT as the game-changer in the near future, the sixth sense door has the potential to stand out as a simple yet effective IOT device.

Fig. 4 The notification that appears when the app receives the image

Fig. 5 Face recognition

References

1. Bhaskar Rao, P., Uma, S.K.: Raspberry Pi home automation with wireless sensors using smartphone. Int. J. Comput. Sci. Mob. Comput. **4**(5) (2015)
2. Anwar, S., Kishore, D.: IOT based smart home security system with alert and door access control using smart phone. Int. J. Eng. Res. Technol. (IJERT) **5**(12) (2016)
3. Jeong, J.: A study on the IoT based smart door lock system. Lecture Notes in Electrical Engineering, vol. 376, Springer, Singapore (2016)
4. Gupta, R.K., Bala Murugan, S., Aroul, K., Marimuthu, R.: IoT based door entry system. Indian J. Sci. Technol. **9**(37) (2016)
5. Han, J., Choi, C.S., Lee, I.: More efficient home energy management system based on ZigBee communication and infrared remote controls. IEEE Trans. Consum. Electron. **57**(1), 85–89 (2011)
6. Ambika, Gadgey, B., Pujari, V., Pallavi, B.: Smart Door using IOT, Int. J. Res. Appl. Sci. Eng. Technol. **5**(VI) (2017)
7. Kodali, R., Jain, V., Bose, S., Boppana, L.: IoT based smart security and home automation system. International Conference on Computing, Communication and Automation, Noida, India, April 2016
8. Yinghui, H., Guanyu, L.: Descriptive models for internet of things. In: IEEE International Conference on Intelligent Control and Information Processing, pp 483–486, Dalian, China (2010)
9. Connecting the PIR motion sensor. The parent detector. https://projects.raspberrypi.org/en/projects/parent-detector/3. Accessed 18 July 2017
10. Nagaria, B., Hashmi, M.F., Patidar, V., Jain, N.: An optimized fast discrete cosine transform approach with various iterations and optimum numerical factors for image quality evaluation. In: 2011 International Conference on Computational Intelligence and Communication Networks, Gwalior, India, Oct 2011
11. Chadha, A.R., Vaidya, P., Roja, M.M.: Face recognition using DCT for global and local features. In: International Conference on Recent Advancements in Electrical, Electronics and Control Engineering, Sivakasi, India, Dec 2011
12. Vishwakarma, V.P., Pandey, S., Gupta, M.N.: A novel approach for face recognition using DCT coefficients re-scaling for illumination normalization. In: International Conference on Advanced Computing and Communications, India, Dec 2007

John Britto studying in Computer Engineering Department at Sardar Patel Institute of Technology, Mumbai—400058. His research areas include IoT and Machine Learning.

Viplav Chaudhari studying in Computer Engineering Department at Sardar Patel Institute of Technology, Mumbai—400058. His research areas include IoT and Network Security.

Deep Mehta studying in Computer Engineering Department at Sardar Patel Institute of Technology, Mumbai—400058. His research areas include IoT and Network Security.

Akshay Kale studying in Computer Engineering Department at Sardar Patel Institute of Technology, Mumbai—400058. His research areas include IoT and Network Security.

Jyoti Ramteke working as Assistant Professor of Computer Engineering Department at Sardar Patel Institute of Technology, Mumbai—400058. Her research areas include Data Warehousing and Mining and Operating Systems.

Secured IoT Based on e-Bulletin Board for a Smart Campus

V. Anuradha, A. Bharathi Malakreddy and H. N. Harinath

Abstract Notice board is an important thing in any institution. But sticking various information on notice board is a tedious process. A separate manpower is required to take care of that traditional notice board. In this paper, we are using an electronic LED or LCD display to display the notices on our campus. In this electronic bulletin board, we are displaying both the text and multimedia messages entered by an authenticated user. The authenticated users can also delete the messages whatever they entered in the bulletin board. The messages entered by an authenticated user is displayed on e-bulletin board by using the HTTP protocol.

Keywords LED display · Bulletin board · Multimedia messages

1 Introduction

The Internet of Things (IoT) is the network of connected devices where each device has a unique identification to communicate with other devices with or without human intervention. In today's world of connectedness, people are becoming accustomed to ease of information [1]. Whether it is through the Internet or television but people want to know up-to-date information with the latest information happening around the world. Traditional notice board as many limitations like using papers to display the notices and someone should be taken care of the notices what all should be present and which all should be removed from the notice board.

V. Anuradha (✉) · A. Bharathi Malakreddy · H. N. Harinath
B M S Institute of Technology and Management, Bengaluru, Karnataka, India
e-mail: anuradhashetty369@gmail.com
URL: https://bmsit.ac.in/

A. Bharathi Malakreddy
URL: https://bmsit.ac.in/

H. N. Harinath
URL: https://bmsit.ac.in/

S. Smys et al. (eds.), *International Conference on Computer Networks and Communication Technologies*, Lecture Notes on Data Engineering and Communications Technologies 15, https://doi.org/10.1007/978-981-10-8681-6_50

Electronic or moving boards are being used in a variety of applications for communicating the information to people quicker and in the most cost-effective manner when compared to a traditional notice board. While an e-mail is a way to converse privately with one or more people over the Internet, electronic bulletin boards are totally public [2]. Any messages that may be text or multimedia images that are posted by an authenticated user can be viewed by anyone else in the organization.

To take up this project is to make our campus into a "**Smart Campus**" and thereby it can be easily notified for students as well as teachers in a more précise manner. The organization of the paper is as follows: In this paper, in Sect. 1 the Introduction gives a brief introduction of an IoT and electronic notice board. In Sect. 2 related work is described, which shows about the already existing systems, Sect. 3 explains about the proposed methodology, and also we have discussed how our system is organized, in Sect. 4 we discuss the design of an e-bulletin board, in Sect. 5 we explained about how the electronic notice board is implemented, in Sect. 6 we discuss about the results and discussions of the bulletin board and finally we concluded the paper and future scope of this paper.

2 Related Work

Alase and Chinchur proposed a system which accepts the digital advertisements of real-time data like news, live traffic, weather updates, etc., that can be remotely monitored. They used the web interface for their project. They used the control panel to enter and delete the data [3].

Khera, Shukla, and Ashavthi developed a simple, low cost and Android-based device that remotely sends the information on the LCD screen. For the communication of data between the devices, they used WiFi, Bluetooth, and Arduino board. All Bluetooth devices have a unique MAC address and this address is assigned to a network interface. The microcontroller is programmed in such a way that it displays the messages entered by a user's [4].

In [5], they have discussed the GSM-based digital notice board. The microcontroller receives the messages when a user enters the data. Here, they used a SIM in GSM to receive a data from users as text messages. This GSM, in turn, uses ARM processor to display the data on 16 * 2 LCD [5].

3 Proposed Methodology

The traditional method uses a manual method of displaying the required information on the notice board. But this method is difficult for maintaining in real-time system. So, to overcome this problem, in our paper, we are discussing about displaying data sent from an authorized user over a network within the organization that data are displayed on e-bulletin board and students can access this e-bulletin page on their mobile itself by using an authenticated IP address.

3.1 Key Objectives of the Proposed System

- The authorized users can enter or delete the data.
- Then data entered by an authorized user wirelessly communicate within the network of an organization and the data is displayed on an e-bulletin page.
- The e-bulletin page also displays the temperature and humidity of our campus.
- It also displays the pH value of drinking water.
- It automatically refreshes for every 5 min.
- So, the students can get up-to-date information.

4 System Design

Figure 1 depicts the flow of an e-bulletin board. The various components in the flow diagram are discussed as follows:
Web user: It is a user interface of the system [6]. An authenticated user can access the system and enter the messages that should be displayed on an e-bulletin board.

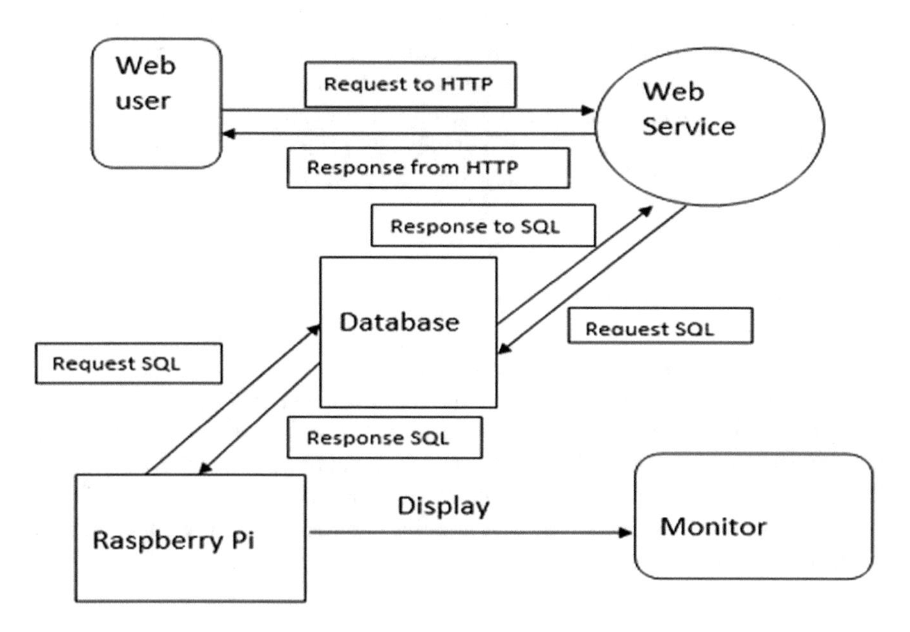

Fig. 1 Flow diagram of an e-bulletin

Web service: A web service is any piece of software that makes itself available over the Internet and uses a standardized XML messaging system. XML is used to encode all communications and its role is to register the commands from the remote user in the database, retrieve the current state of the notice, and return a failure/success message under the HTTP response on the commands performed by the user.
Database: A database management system is used to store a data entered by an authorized user and avoids storing of a duplication data.
Monitor: The monitor is used to display an information.
Raspberry Pi: It is a small single-board computer. The use of Raspberry Pi is intentional since these devices are widely accessible, inexpensive and available from multiple vendors [7, 8]. Raspberry Pi has an HDMI port provides both video and audio output.

5 Implementation

First, we have created a login page for the users. The authorized users in our system are principal, head of the department (HOD), placement, and an admin. We have provided the username and password for them. Once authentication is successful for the authorized users, it will forward them to enter both the text and multimedia messages. If there is any duplicated data that they are trying, it will notify them that this data is already present on the e-bulletin board. Once they enter the data remotely, it communicates within the organization IP address via HTTP protocol and displays the data on e-bulletin page. The authorized users can also delete the data. The admin of the system has full control over an e-bulletin board. He can maintain what are the necessary things that should be displayed on an e-board.

It also displays a temperature and humidity of our campus and pH value of the drinking water. The temperature and humidity and pH sensors send the data into a temporary storage of college IP address via WiFi module.

6 Results and Discussion

Figure 2 shows an output of an e-bulletin page.

The e-bulletin board page is divided into four main quadrants and one rectangular space. In the first three quadrants, it displays the text messages entered by the authenticated users in respective quadrants as shown in Fig. 2 and in the last quadrant it displays the images entered by all authenticated users.

In rectangular space, it displays the humidity, temperature of the campus, and pH value of the drinking water.

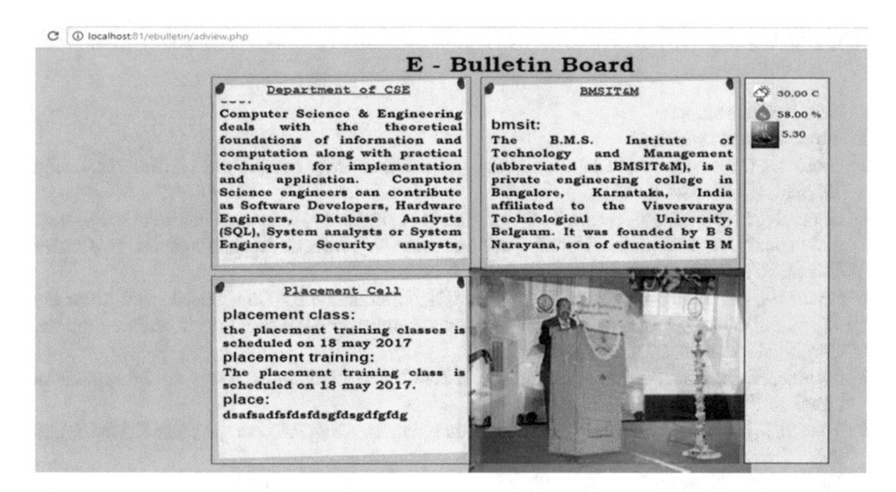

Fig. 2 Screenshot of an e-bulletin page

7 Conclusion and Future Scope

Now, the world is moving toward automation, so in this world, if we want to do some changes in the previously used system we have to use the new techniques. The wireless operation provides fast transmission over long-range communication. It saves resources and time. Data can be sent from a remote location. User authentication is provided to enter both text and digital images that can be remotely accessed by the students within the organization. In future, this application of e-notice board can be further extended to include the following features:

- **Categorization of Notice**: Notices can be categorized into different categories, so that it is possible for the user to easily manage the notices [9]. Categorization can also be done by making groups. Defining the notice to be circulated in a particular group can make it more secure.
- **Documents and PDF files**: The attachments can be further improved to include PDF files or Doc files. Then, there will not be much need to send images with the notices. A single file would serve all the purposes.
- **Feedback**: Feedback on the notices can also be taken. It can increase communication among connected members and any issue can be easily sorted out on the spot.

References

1. http://www.ermt.net/
2. http://www.answer.com/
3. Alase, S., Chinchur, V.: IoT based digital signage board using Raspberry Pi 3. Int. Res. J. Eng. Technol. (IRJET) **04** (2017)
4. Khera, N., Shukla, D., Ashavthi, S.: Development of simple and low cost Android based wireless notice board. In: International Conference on Reliability, Infocom Technologies and Optimization (ICRITO) (Trends and Future Directions) (2016)
5. Pramanik, A., Rishikesh, Nagar, V., Dwivedi, S., Choudhary, B.: GSM based smart home and digital notice board. In: IEEE Conference on Computational Techniques in Information and Communication (ICCTICT) (2016)
6. Lamine, H., Hafedh, A.: Remote control of a domestic equipment from an Android application based on Raspberry pi card (STA) (2014)
7. Tupe, S.J., Salunke, A.R.: Multi-functional smart display using Raspberry-Pi. Int. J. Adv. Found. Res. Comput. (IJAFRC) **2** (2015)
8. Jadhav, V.B., Nagwanshi, T.S., Patil, Y.P.: Digital notice board using Raspberry Pi. Int. Res. J. Eng. Technol. (IRJET) **03** (2016)
9. http://www.slideshare.net/

V. Anuradha studying Master Student in BMS Institute of Technology and Management, Bengaluru-560064. Her research area includes IoT and smart automation.

A. Bharathi Malakreddy working as Professor in BMS Institute of Technology and Management, Bengaluru-560064. Her research area includes IoT and smart automation.

H. N. Harinath working as an Assistant Lab Instructor, IoT lab in BMS Institute of Technology and Management, Bengaluru-560064. His research area includes IoT and smart automation.

An Augmented Line Segment-Based Algorithm with Delayed Overlap Consideration and Reconfigurable Logic for RSMT Problem

V. Vani and G. R. Prasad

Abstract An Augmented Line Segment-Based (ALSB) algorithm with delayed overlap consideration is proposed for the construction of Rectilinear Steiner Minimum Tree (RSMT). The algorithm incrementally augments the four line segments drawn from each point in four directions. When the line segments of two points interconnect, the edge connecting those two points is included to RSMT. The selection of the L-shaped layout is done after all the edges are identified. After the RSMT edges are selected, the two layouts of each edge are compared based on overlap count, and the layout with maximum overlap is added to RSMT, thereby reducing the total length of RSMT. RSMT has wide application in the field of VLSI during global routing.

Keywords Global routing · VLSI design · Rectilinear Steiner minimum tree
Computer networks

1 Introduction

Routing plays an important role in VLSI design. Routing identifies the connection of the nets that are placed using placement algorithm. RSMT algorithm is used for this purpose since the wires are allowed to run in horizontal and vertical directions. RSMT connects the given collection of points $P_n = \{P_1, P_2, P_3, \ldots, P_n\}$ using only horizontal and vertical lines with shortest length. Few additional points called Steiner points are included to the point set to connect in rectilinear manner.

Hanan et al. [1] were the first to work on the rectilinear version of Steiner tree, and exact solutions for $n < 5$ were provided. They also proved that all the possible

V. Vani (✉)
Department of I.S.E, Bangalore Institute of Technology, K. R. Road, Bengaluru, India
e-mail: vanisrin@gmail.com

G. R. Prasad
Department of C.S.E, B.M.S College of Engineering, Basavanagudi, Bengaluru, India
e-mail: gr_prasad@yahoo.com

S. Smys et al. (eds.), *International Conference on Computer Networks and Communication Technologies*, Lecture Notes on Data Engineering and Communications Technologies 15, https://doi.org/10.1007/978-981-10-8681-6_51

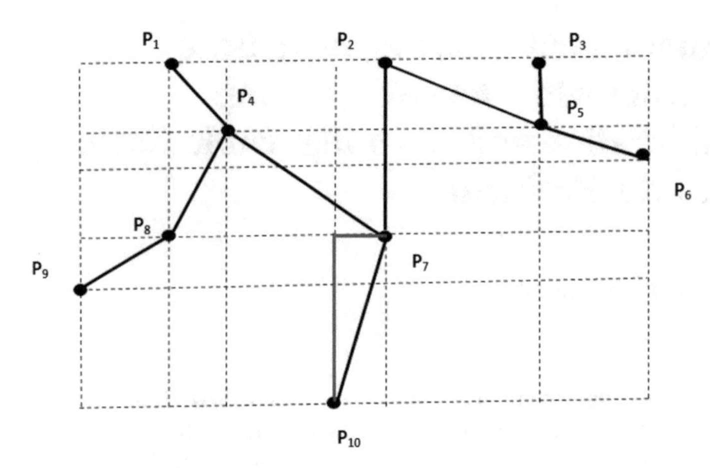

Fig. 1 Minimum spanning tree and Hanan grid

candidate Steiner points always lie on the Hanan grid. Hanan grid is the grid obtained by drawing horizontal and vertical lines through every point in the point set. The points obtained at junction of this grid are known as Hanan points.

1.1 Definitions

- Degenerate edge: An edge connecting two points with same value of x or y co-ordinates is termed as degenerate edge, and the edge is either a horizontal or vertical line, e.g. edge (P_3, P_5) in Fig. 1.
- Non-degenerate edge: An edge connecting two points which do not have same x or y co-ordinates is termed as degenerate edge and the edge is either an upper L-shaped layout or lower L-shaped layout.
- Overlap count: It is the length of a layout of an edge that overlaps with the layout of another edge. It indicates the amount of length reduction that can be attained by selecting a layout.
- L-shaped layout: For every non-degenerate edge, two L-shaped layouts can be identified (upper L-shaped layout and lower L-shaped layout) as shown by red lines in Fig. 1.

2 Related Work

Significant contributions for the construction of RSMT were that the problem of generating RSMT is NP-complete by Garey et al. [2], and that the ratio of the cost of Rectilinear Minimum Spanning Tree (RMST) to the cost of RSMT is $\leq 3/2$ by Hwang et al. [3]. Ho et al. [4] proposed an MST-based algorithm that converts the MST into RSMT by replacing each MST edge with an L or Z-shaped layouts that provided maximum overlap. Kahng et al. [5] presented an iterated Steiner tree algorithm which incrementally adds a Steiner points and constructs an RSMT unless no more cost reduction can be obtained. A batched variation that inserts a batch or group of Steiner points at a time was also proposed.

Chu et al. [6] suggested an algorithm FLUTE that generates the RSMT using a pre-computed table if $n \leq 9$ else applies the net breaking algorithm until the data is small enough to apply the table. Chen et al. [7] presented a refined single trunk algorithm which first draws a horizontal or vertical line (trunk) that goes through the computed median of the given points and then all the points are either interconnected to the trunk or the stem (a horizontal or vertical line that connects a point to the trunk) depending on the shortest distance. A comparative study of existing RSMT algorithm [8] and a clustering-based RSMT algorithm [9] was presented that distributes the given set of points into identified collection of clusters, generates the RMST for each cluster and lastly interconnects the clusters to obtain the final tree was proposed.

Augmented line segment-based algorithm [10, 11] that augments the length of line segments from all the points in four directions was developed and edges are added to the RSMT when the line segments of two points intersect. Further, an updated version of the algorithm which reverses the layout previously selected was also proposed [12].

3 Proposed Methodology

Augmented Line Segment-Based (ALSB) with delayed overlap consideration is an efficient algorithm as compared to ALSB algorithm [10], as it constructs the RSMT with reduced total length. The layout selection for each of the edges is delayed until all the edges are identified. Once all the edges are identified, the layouts are intelligently selected one by one based on the overlap count so that each edge is replaced by the layout that provides maximum overlap and thereby promotes length reduction.

The steps for construction of RSMT are as follows:

1. Augment the length of line segments: Four line segments drawn in four different directions from each point are incrementally augmented. If the number of points is n, then the line segments are augmented until $n - 1$ edges are added to the RSMT.
2. Addition of edges to the RSMT: When the line segments of two points interconnect, the edge connecting those two points is added to the RSMT along with both the corresponding layouts. The selection of the layout for each edge is delayed until all the edges are identified and added to the RSMT.
3. Computation of overlap count: Once all the edges are identified, each edge is examined to compute the overlap count for the two layouts of an edge.

4. Construction of final RSMT based on overlap count: The final RSMT is constructed by incrementally adding the layout of an edge with maximum overlap count and the corresponding layouts of the other edges overlapping with this edge.

3.1 Algorithm

```
n- Number of points
num_edges=0  actual_edges=0
while num_edges!=n-1 -------do in parallel
do
          Increment the length of line segments drawn in four directions from all
          the points
          If two line segments intersect and do not form a loop
                    Add the edge i.e, the both layouts of the edge to temporary
RSMT
                    num_edges=num_edges+1
          End if
done

for each of the edges identified ----do in parallel
          Compute overlap count for both the layouts of an edge
End for

while actual_edges!=num_edges
do
          Select the edge or layout with maximum overlap count
          If maximum overlap count !=0
                Add the layout and the 'm' overlapping layouts to the RSMT
                actual_edges=actual_edges + (m+1)
                Update the overlap count of rejected layout and its overlapping lay-
outs
          else
                Add all the remaining edges by randomly selecting any one layout
          End if
done
```

3.2 Demonstration

The following section shows the illustration of the working of ALSB with delayed overlap consideration algorithm on a sample set of nine points p_1 to p_8. Figure 2a shows the augmentation of the line segments in four directions from all the points.

Figure 2b illustrates the selection of edges when the line segments of two points intersect. An edge between two points is represented using either upper or lower L-shaped layouts. The selection of the layout is delayed until all the edges are identified.

After all the edges or layouts are identified as illustrated in Fig. 3a, the overlap count for each and every edge is computed. The layout of an edge with maximum overlap count is selected. While including that layout to RSMT, the corresponding overlapping layout of other edges are also included. For example, the overlap count of upper L-shaped layout of edge (p_1, p_4) is 0 as this layout does not overlap with any other layout and that of lower L-shaped layout is more than 1 as it overlaps with lower L-shaped layout of edge (p_0, p_1). Therefore, the lower L-shaped layouts of edge (p_1, p_4) and edge (p_0, p_1) are added to RSMT. This step is repeated unless the layouts of all the edges are selected and the final RSMT is constructed.

Figure 3b displays the final RSMT constructed for given set of nine points p_1 to p_8 using the ALSB with delayed overlap consideration algorithm.

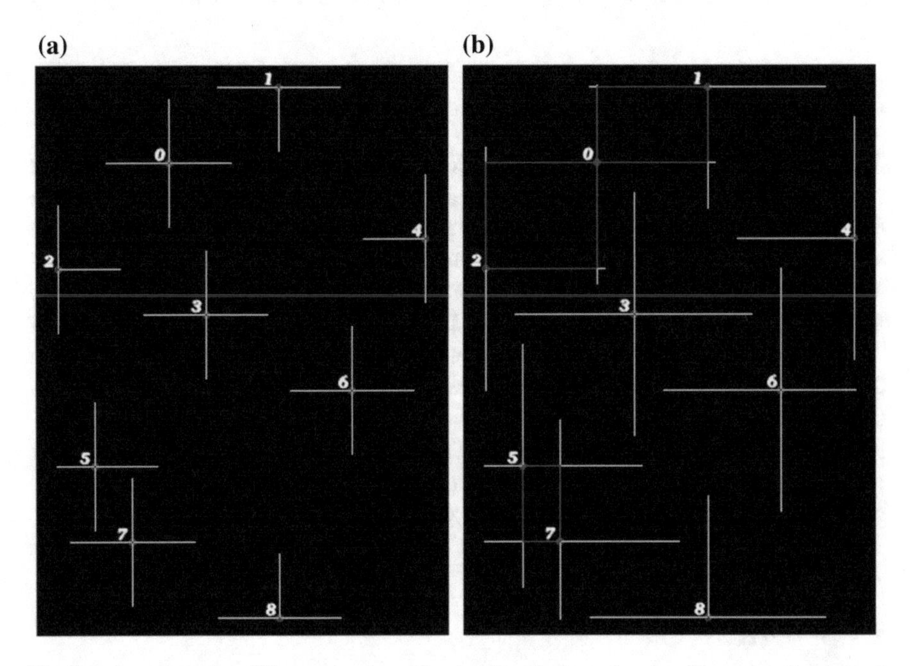

Fig. 2 **a** Augmentation of line segments and **b** selection of edges when two line segments intersect

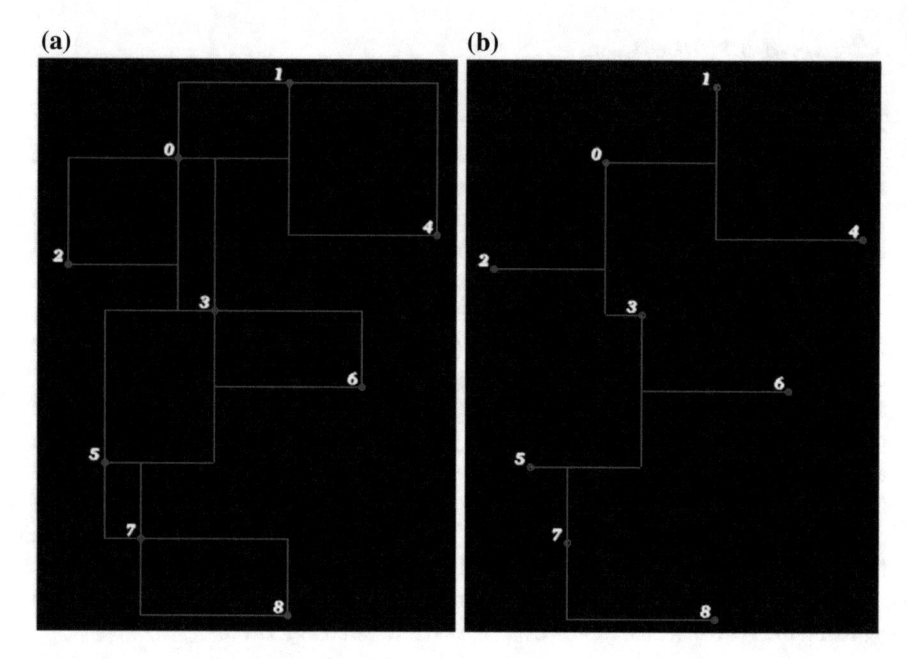

Fig. 3 **a** Selection of edges when two line segments intersect and **b** final RSMT

3.3 Hardware Design

The hardware design of the above-proposed algorithm is as shown in Fig. 4. The input and output are retrieved using software programme and the actual RSMT construction is carried on FPGA using SRAM to store the pixel information. Each pixel value stores two types of information: NODE and overlap count. When the line segments are augmented, the node or the point information is stored in NODE to indicate the node visiting that pixel if NODE! $= -1$ and the value of overlap count is incremented. Thus the line segments get the information of the nodes that have already visited the pixel and intersection of line segments can be identified. Layout for each edge is selected based on the value of overlap count.

Figure 5 shows the architecture of RSMT generator illustrating the various steps of the algorithm. The relationship between the node and its four line segments identified by top, bottom, left and right values is depicted in Fig. 6.

4 Results

The proposed algorithm was implemented using 'C' programming language and has been verified on various sets of data. The algorithm computes the RSMT efficiently and produces near-optimal solutions. Table 1 provides the comparative analysis of

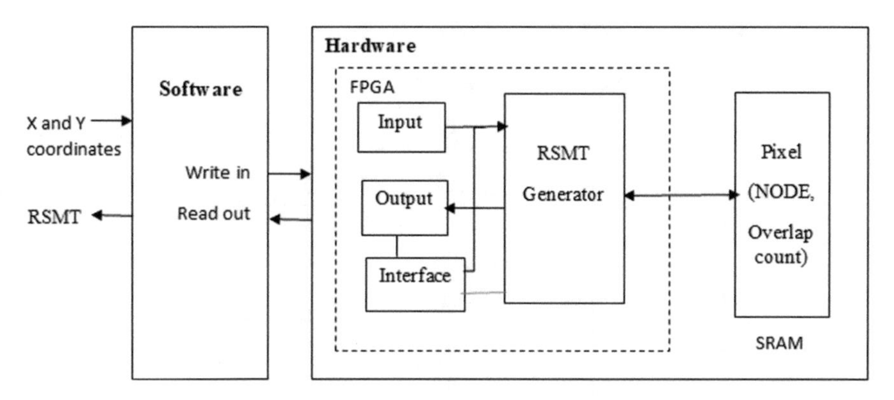

Fig. 4 Hardware design of ALSB with delayed overlap consideration algorithm

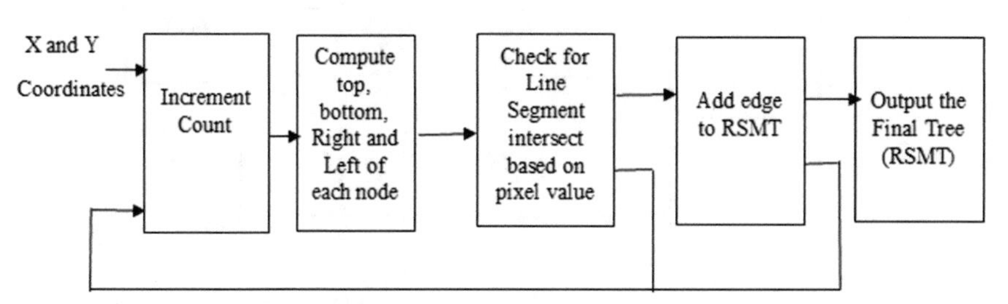

Fig. 5 Architecture of RSMT generator

the proposed ALSB with delayed overlap consideration algorithm with RMST and ALSB algorithm. Figure 7 shows the snapshot of the RSMT obtained for 250 set of points.

5 Conclusion

The proposed ALSB with delayed overlap consideration algorithm effectively computes the RSMT for the given input data. A considerable cost or length reduction is also attained. The proposed algorithm generates near-optimal solution irrespective of the size of input data. Hardware design of the algorithm is also presented. Future efforts would be focused on hardware implementation of the proposed design.

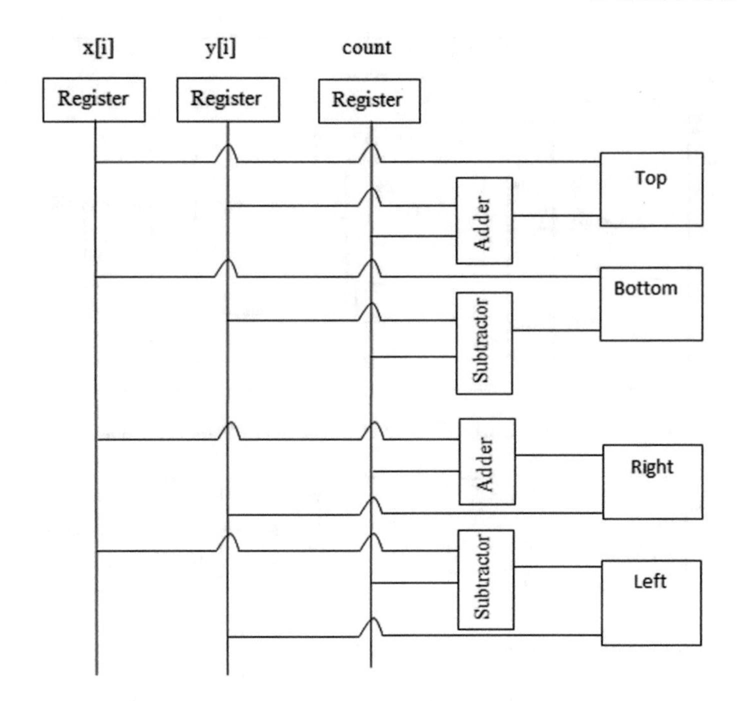

Fig. 6 Proposed representation of a node and its relationship with four line segments

Table 1 Comparison of ALSB with delayed overlap consideration algorithm with other algorithms

# of input points	RMST length	Total length of RSMT using ALSB algorithm	Total length of RSMT using ALSB with delayed overlap consideration algorithm
50	2225	1871	1810
100	3007	2655	2595
150	3576	3161	3107
200	4290	3754	3674
250	4976	4366	4256
300	5402	4755	4587
350	5763	5031	4827
400	6228	5425	5137
450	6634	5808	5480
500	6961	6142	5851

Fig. 7 Output for 250 set of points

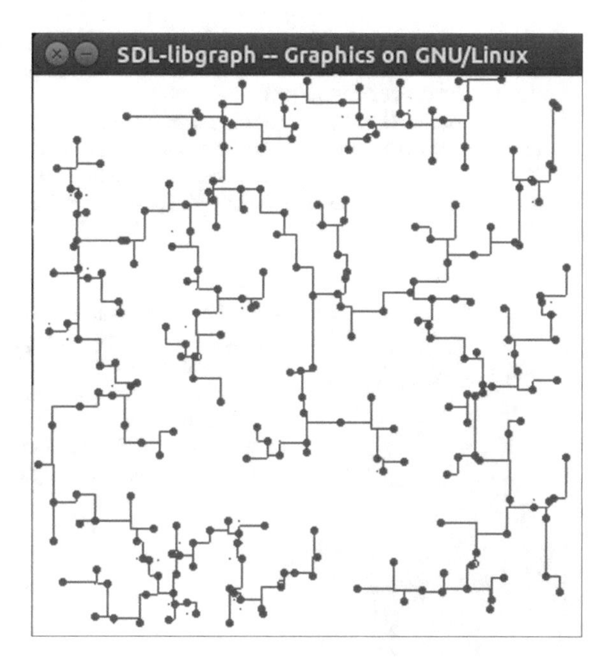

References

1. Hanan, M.: On Steiner's problem with rectilinear distance. SIAM J. Appl. Math. **14**(2), 255–265 (1966)
2. Garey, M.R., Johnson, D.S.: The rectilinear Steiner tree problem is NP-complete. SIAM J. Appl. Math. **32**(4), 826–834 (1977)
3. Hwang, F.K.: On Steiner minimal trees with rectilinear distance. SIAM J. Appl. Math. **30**(1), 104–114 (1976)
4. Ho, J.-M., Vijayan, G., Wong, C.K.: New algorithms for the rectilinear Steiner tree problem. IEEE Trans. Comput. Aided Des. Integr. Circuits Syst. **9**(2), 185–193 (1990)
5. Kahng, A.B., Robins, G.: On the performance bounds for a class of rectilinear Steiner tree heuristics in arbitrary dimension. IEEE Trans. Comput Aided Des. Integr. Circuits Syst. **11**(11), 1462–1465 (1992)
6. Chu, C., Wong, Y.-C.: FLUTE: fast lookup table based rectilinear Steiner minimal tree algorithm for VLSI design. IEEE Trans. Comput Aided Des. Integr. Circuits Syst. **27**(1), 70–83 (2008)
7. Chen, H., et al.: Refined single trunk tree: a rectilinear Steiner tree generator for interconnect prediction. In: Proceedings of the 2002 International Workshop on System-Level Interconnect Prediction. ACM (2002)
8. Vani, V., Prasad, G.R.: Performance analysis of the algorithms for the construction of rectilinear Steiner minimum tree. IOSR J. VLSI Sig. Process. **3**(3), 61–68
9. Vani, V., Prasad, G.R.: Algorithm for the construction of rectilinear Steiner minimum tree by identifying the clusters of points. In: 2014 International Conference on Information Communication and Embedded Systems (ICICES). IEEE (2014)

10. Vani, V., Prasad, G.R.: Augmented line segment based algorithm for constructing rectilinear Steiner minimum tree. In: International Conference on Communication and Electronics Systems (ICCES). IEEE (2016)
11. Vani, V., Prasad, G.R.: An improved augmented line segment based algorithm for the generation of rectilinear Steiner minimum tree. Int. J. Electr. Comput. Eng. (IJECE) **7**(3) (2017)
12. Vani, V., Prasad, G.R.: Rectilinear Steiner minimum tree formation using an improved augmented line segment based algorithm with edge reversal. IJETTCS **7**(1) (2018)

V. Vani is an Assistant Professor in the Department of Information Science and Engineering, Bangalore Institute of Technology, Bangalore. She received B.E. degree in Computer Science and Engineering from Visveshwaraiah Technological University in 2003 and M.Tech. degree in Computer Network Engineering from VTU in 2007. She is currently pursuing Ph.D. degree in VTU in the area of Reconfigurable Computing.

G. R. Prasad is an Associate Professor in Department of Computer Science & Engineering, BMSCE, Bangalore. He holds a Ph.D. from National Institute of Technology, Karnataka, Surathkal, India. He received his M.Tech. degree in Computer Science & Engineering from Bangalore University in 1999 and B.E. Degree in Computer Science & Engineering from Bangalore University in 1995. His research interests include Reconfigurable computing.

Automatic Accident Rescue System Using IoT

Karanam Niranjan Kumar, C. H. Rama Narasimha Dattu, S. Vishnu and S. R. Jino Ramson

Abstract Nowadays, the road accidents in modern urban areas are increased to uncertain level. In highly populated countries like India, more than 410 people get swooped up every day. A leading cause of the global burden of public health and fatalities is road accidents. The loss of human life due to accident is to be avoided. There is no technology in current times for detection of accident. Reaching of ambulance to the accident location is mostly delayed due to the conjested traffic that increases the chance of victim death. To reduce loss of life or save life of person due to accidents and reduce the time taken by ambulance to reach the hospital, there is a need of system which needs to come into force in our daily lives. To bar loss of human life due to accidents, we introduce a scheme called automatic accident rescue system (AARS). There is an automatic detection of accident by crash sensors in the vehicle. A GPS module in the vehicle will send the location coordinates of the accident using IoT platform to central unit which will notify and send an ambulance from the nearest hospital in the vicinity to the accident spot. This scheme is fully automated; thus, it finds the accident spot, helping to reach the hospital in time. This system can help in reducing the loss of lives of human which happen by the accident.

Keywords IoT · GPS · Accident · AARS

1 Introduction

Promptly, as the population is increasing rapidly the usage of vehicles is also increased and this has raised the inclusive accident rate. In most cases, the loss

K. Niranjan Kumar · C. H. Rama Narasimha Dattu · S. Vishnu (✉) · S. R. Jino Ramson
Department of Electronics and Communication Engineering, Vignans Foundation for Science Technology and Research, Guntur, Andhra Pradesh, India
e-mail: vishnuvazhamkuzhiyil@gmail.com

S. Smys et al. (eds.), *International Conference on Computer Networks and Communication Technologies*, Lecture Notes on Data Engineering and Communications Technologies 15, https://doi.org/10.1007/978-981-10-8681-6_52

of victim's life is due to the hindrance in reaching of the ambulance to the area where the accident has occurred. This in turn has an adverse effect on the economy of the country as well as the loss of lives. So problem given above will become worst in the future.

The main objective behind this AARS is to diminish the time gap between the occurrence of accident and time required for the ambulance to reach the location of the accident for treating the victim. When an accident takes place, a lot of time is wasted to search the location of accident; such a time our system work faster and avoid the loss of life due to the time delay.

In the previously proposed system, an accident of the vehicle is detected automatically with the help of vibration and fire sensors. The location coordinates of accident vehicle are detected using Global Positioning System (GPS) and then transfer the message to the central unit using GSM [1]. A GSM module is used for transfer of the message to the central unit which indeed causes delay and interference.

In our proposing system, we are replacing GSM module with the IoT platform due to delay in its working [2]. The victim vehicle coordinates are sent to the nearby ambulance, thereby reaching the location of the accident without delay. As this system can be manually turned off, it is not used in the minor accident cases, but for emergencies where an ambulance is needed by the victim. Such a system is helpful for providing very fast medical treatment to a victim of vehicle.

2 Related Works

This idea of automatic ambulance rescue system (AARS) is all about procuring the life of mishap individual. In previously proposed system if a vehicle has met with an accident, vibration sensor or fire sensor gives the electric signal to the microcontroller through signal conditioner. The location coordinates are identified using GPS and sent to the control centre using GSM modem. In control section, the GSM modem receives a message about the accident and sends it to PC. PC identifies the nearest ambulance and instructs to pick up the patient [3] (Fig. 1).

In this system, communication gets delayed due to the usage of GSM modem which is a slow process of transfer of a message and communication using GSM modem causes interference.

3 Proposed System

To overcome the existing problem we will implement a new system in which there is automatic detection of the accident. A crash sensor is fitted in every vehicle and when an accident occurs, signals from the crash sensor are sent to the microcontroller [4]. The signal is transferred from microcontroller to the central unit using IoT platform. The GPS module provides the latitude and longitude coordinates of victim vehicle

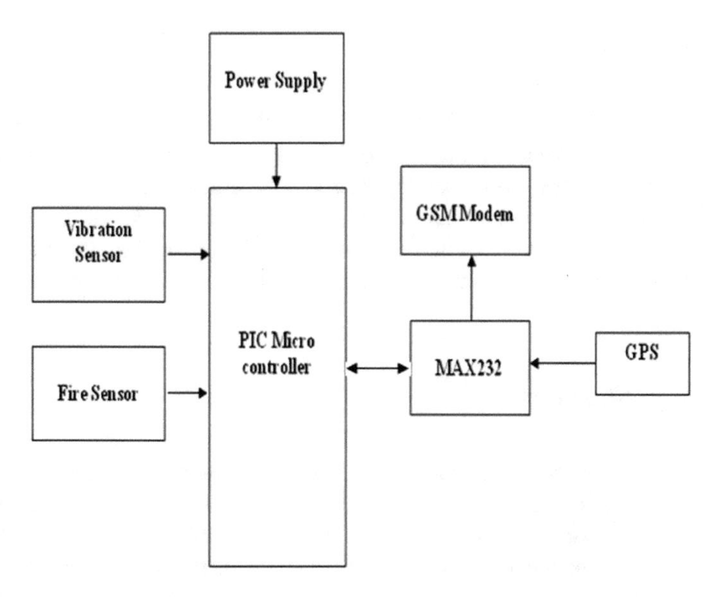

Fig. 1 Block diagram of vehicle unit

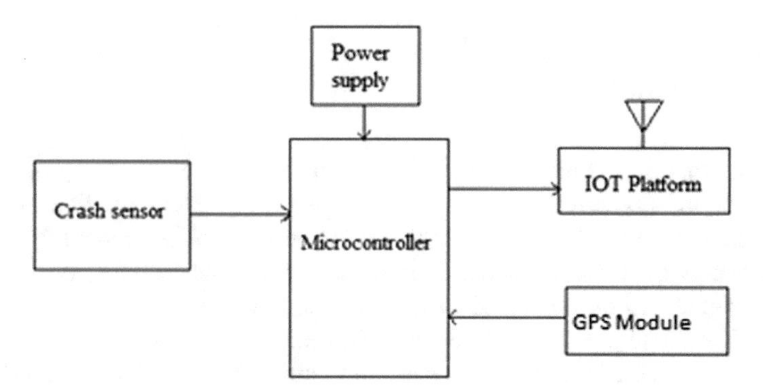

Fig. 2 Block diagram of vehicle unit

which are sent to the control using IoT platform. The central unit sends the location coordinates to the nearest ambulance and is instructed to pick up the victim. The central unit will be placed in a police station or a hospital that receives the signals from vehicle unit. It sends an alert message to the ambulance that is nearer to the location of the accident. The ambulance is also equipped with a GPS receiver for tracking of the accident location. This helps ambulance to reach the location in time and save the victim (Fig. 2).

3.1 Crash Sensor

The sensor employed in this vehicle unit is M510 Crash sensor. It is responsible for detecting sudden deceleration in a collision and converts it into corresponding signals within a matter of milliseconds. The vehicle unit is installed with the crash sensor. It turns mechanical quantity into electricity. When this sensor collided with an object, it will cause circuit switch close. As a result signals from the crash sensor are sent to the microcontroller, thereby indicating that an accident has occurred.

3.2 Microcontroller

The signals from the crash sensor and GPS module is sent to the microcontroller, for further processing. The microcontroller used in this system is Arduino Uno R3. As Arduino can be connected using USB, it is most preferred. The Clock speed of the Arduino is 16 MHz so it can perform a particular task faster than the other processor or controller.

3.3 GPS Module

GPS stands for Global Positioning System by which anyone can unchangingly obtain the position information. Initially, the time signal is sent from a GPS satellite at a given point to the vehicle equipped with GPS receiver [5, 6]. Subsequently, the time difference between GPS time and the point of time clock which GPS receiver receives the time signal will be calculated to generate the loftiness from the receiver to the satellite. The same process will be washed-up with three other satellites. When a wrecking occurs, the location coordinates of victim vehicle are identified using this receiver with an accuracy of 20 m and sent to the microcontroller.

3.4 IoT Platform

The IoT platform is used in this system to diminish the gap between the device sensors and data networks. The location coordinates of the victim vehicle are sent to the central unit using IoT platform. This IoT platform can be 4G/LTE networks. As the communication in 4G is 10 times faster than 3G or GSM, there would not be any delay in transfer of the message to the central unit [7].

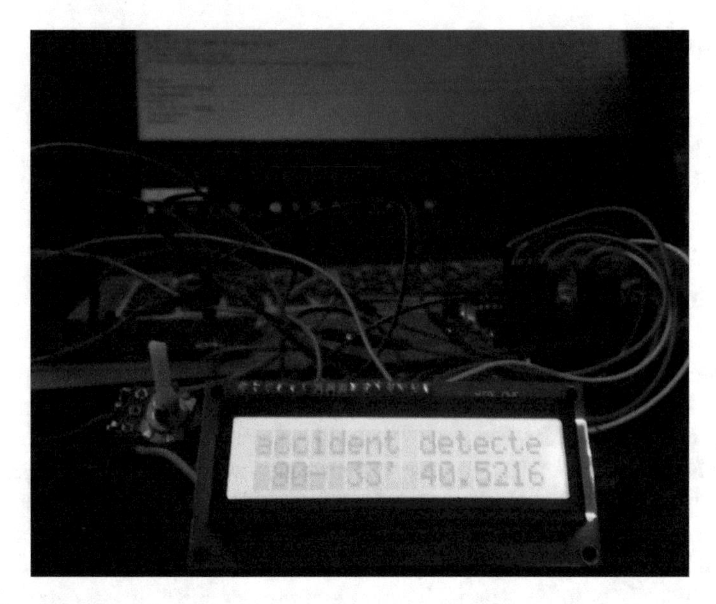

Fig. 3 Practical model of vehicle unit

4 Results

We have implemented the hardware section of vehicle unit. The circuit is made that if an accident occurs to the vehicle, the crash sensor in the vehicle unit sends the signal to the Arduino, and location coordinates from GPS module are collected. The location is sent to the central unit using Wi-Fi module ESP8266. The central unit can be a cloud storage in a hospital or a police station. The location coordinates of the vehicle unit are displayed on the above LED screen (Fig. 3).

5 Advantages

1. A totally advanced version of ambulance system [6].
2. With the help of GPS, we get the latitude and longitude of the detected position more precisely.
3. When we get the exact location of the vehicle, the ambulance will reach there in a few minutes.
4. This scheme is fully automated; thus, it finds the accident spot, helping to reach the hospital in time.

6 Applications

1. Biomedical applications.
2. Crash recorders.
3. Dead reckoning.

7 Future Improvements

In the proposing system, we are further extending to replace GPS with NAVIC chip as it has standard positioning service with an accuracy of 5 m which is much better than GPS. It is Indian Regional Navigation Satellite System (IRNSS) with its operational name as NAVIC [8]. The system was developed partly because access to foreign government-controlled global navigation satellite systems is not guaranteed in hostile situations. This system can be further extended for ambulance control system with intelligent traffic light system. This system can also be extended in facilitating connectivity to the nearest hospital and provide medical assistance through live streaming from the ambulance and also through video conferencing.

8 Conclusion

In this paper, a novel idea is proposed to save the lives of victims in an accident. This system can detect the location of accident spot automatically and accurately, and realize the automation of information transmission. Thus, AARS if implemented in countries with a large population like INDIA can produce better results. The AARS is more accurate with no loss of time. Such functions can be useful for 'help' and 'safety' of humans and society.

References

1. Fating, P., Bisen, P., Khursange, P., Krade, M.: Automated accident rescue system. Int. J. Res. (IJR) **2**(3), ISSN:2348795X (2015)
2. Wei, W., Hanbo, F.: Traffic Accident Automatic Detection and Remote Alarm Device (2011)
3. Iyyappan, S., Nandagopal, V.: Automatic Accident Detection and Ambulance Rescue with Intelligent Traffic Light System (IJAREEIE). P-ISSN(Print):2320–3765, ISSN (Online): 2278–8875 (2013)
4. Li, X., Shu, W., Li, M., Huang, H.-Y., Luo, P.-E., Wu, M.-Y.: Performance evaluation of vehicle-based mobile sensor networks for traffic monitoring. IEEE Trans. Veh. Technol. **58**(4), 1647–1653 (2009)
5. Chang, B.-J., Huang, B-J., Liang, Y.-H.: Wireless sensor network-based adaptive vehicle navigation in multihop-relay WiMAX networks. In: Proceedings of the 22nd International Conference on Advanced Information Networking and Applications (AINA), pp. 56–63 (2008)

6. Bhagya Lakshmi, V., Savitha, H., Mhamane, S.: FPGA based vehicle tracking and accident warning using GPS. Int. J. Sci. Eng. Res **5**(2), (2014)
7. Lu, X.: Develop web GIS based Intelligent transportation application systems with web service technology. In: Proceedings of International Conference on its Telecommunications (2006)
8. Department of Space, Indian Space Research Organization. https://www.isro.gov.in/irnss-programme

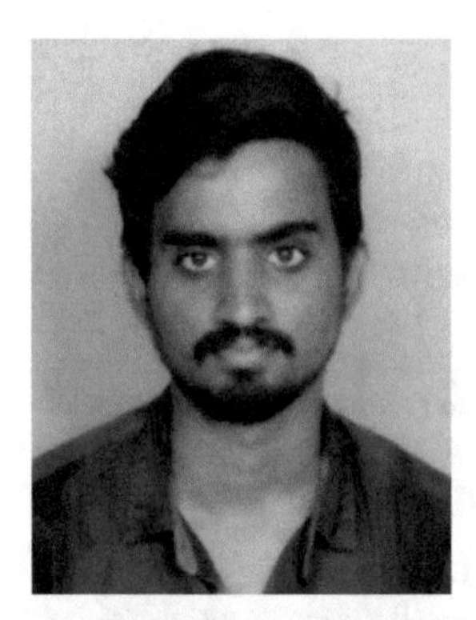

Karanam Niranjan Kumar is pursuing B.Tech. 3rd year (Electronics and Communication) in the year 2018 at Vignan's Foundation for Science, Technology and Research, Guntur, Andhra Pradesh, India. His main research interest includes Big data analysis and Internet of Things.

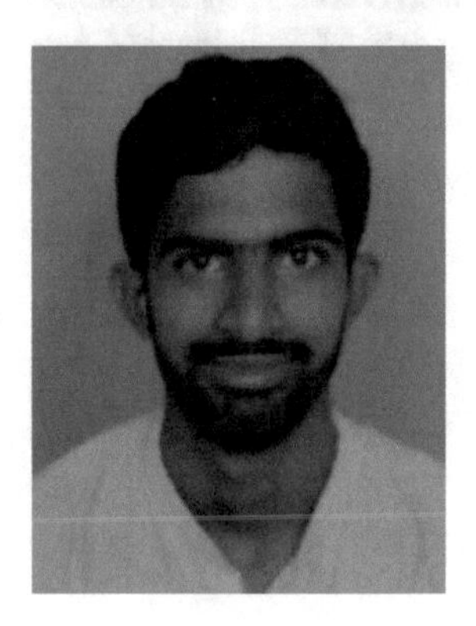

C. H. Rama Narasimha Dattu is a student of B.Tech. 3rd year (Electronics and Communication) at Vignan's Foundation for Science, Technology and Research, Guntur, Andhra Pradesh, India.

S. Vishnu received the B.Tech. degree in Electronics and Communication Engineering from Mahatma Gandhi University, India and M.E. degree in Applied Electronics from Coimbatore Institute of Technology, India. Currently he is working as an Assistant Professor in Vignan's Foundation for Science, Technology and Research, India. His main research interest includes remote health monitoring systems and applications in the field of Things of Internet.

S. R. Jino Ramson received the B.E. degree in Electronics and Communication Engineering from Anna University, India and M.Tech. degree in Networks and Internet Engineering and Doctoral degree in Electronics and Communication Engineering from Karunya University, India. Currently he is working as an Associate Professor in Vignan's Foundation for Science, Technology and Research, India. Also he received the Master of Business Administration in Manonmaniam Sundaranar University, India. His main research interest is development and evaluation of real time remote monitoring systems using Wireless Sensor Networks for various applications.

Enhanced Possibilistic C-Means Clustering on Big Data While Ensuring Security

Shriya R. Paladhi, R. Mohan Kumar, A. G. Deepshika Reddy, C. Y. Vinayak and T. P. Pusphavathi

Abstract Data clustering is the most important technique in knowledge discovery and data engineering. Recently, the possibilistic C-means algorithm (PCM) was proposed to address the drawbacks associated with the constrained memberships used in algorithms such as the fuzzy C-means (FCM). Among different variations of clustering, possibilistic C-means (PCM) uses constrained membership functions. But the drawback of PCM converges to coincident clusters. The purpose of this paper is to overcome the drawback of PCM while preserving the privacy of sensitive data. The proposed algorithm in this paper is tensor PCM (TPCM), which makes use of tensor relational data model. TPCM algorithm is further modified as privacy-preserving TPCM algorithm to preserve the privacy by using Brakerski–Gentry–Vaikuntanathan (BGV) technique.

Keywords Cloud computing · C-means · Homomorphic encryption
Tensor relational data model · MapReduce

S. R. Paladhi (✉) · R. Mohan Kumar · A. G. Deepshika Reddy
C. Y. Vinayak · T. P. Pusphavathi
Department of CSE, RUAS, Bengaluru, India
e-mail: shriyapaladhi@gmail.com

R. Mohan Kumar
e-mail: mohanmanagowda96@gmail.com

A. G. Deepshika Reddy
e-mail: shika.reddy17@gmail.com

C. Y. Vinayak
e-mail: vinayakcy123@gmail.com

T. P. Pusphavathi
e-mail: acepushpa@yahoo.com

S. Smys et al. (eds.), *International Conference on Computer Networks and Communication Technologies*, Lecture Notes on Data Engineering and Communications Technologies 15, https://doi.org/10.1007/978-981-10-8681-6_53

1 Background

The rapid growth of boosting technologies is producing huge amounts of the data, and there is a need to process them. Big data refers to large volumes of datasets that are collected from various domains like the social media interactions (Facebook, Twitter), Banking sector (customer daily transactions), Search Engines and so on. There is a need to process the big data to extract useful information from them that are useful to make the domain brand famous, better customer services, increased sales of the products and better understanding of the market. By definition, big data are huge volumes of information that require new techniques of processing to enable enhanced decision making, insight discovery and process optimization (Gartner Research) [1]. The processing of big data at once is very tedious job, so they should be broken into fragments so that each fragment can be processed separately and then integrated together [2]. Clustering algorithm will be used to process the data. A formal definition of Clustering is that a *cluster* is a collection of objects which are 'similar' between them and are 'dissimilar' to the objects belonging to other clusters [3]. This aids in working with similar clusters at once and processing them easily, which was a major challenge. The reason of choosing clustering and not any other data mining techniques like classification because big data is generally unstructured in nature (meaning they do not have a specific format for representing the data), unlabelled and heterogeneous in nature [3]. The major goal of clustering is to determine the intrinsic grouping in a set of unlabelled data. Dealing with high dimensions and large number of data items can be problematic; because of time complexity, there is a need to select right clustering algorithm to address this issue.

In the following section of this paper, we will discuss the various clustering algorithms along with the advantages and disadvantages of each algorithm. And also, we will discuss the current clustering technique adopted in this paper.

Big data security is another major issue these days because the users need privacy for their data [4]. It has been observed that privacy and security affect the wholesome big data storage and processing because they are usually accessed by third-party services. Security should be provided in the form of processing the big data without exposing sensitive information of the customers [5]. Current technologies are on static dataset, but generally data change dynamically; this challenge will be addressed in this paper.

The key factor of this paper is to find better clustering algorithm which handles a large amount of data while preserving the privacy.

The rest of the paper is organized as follows. Section 2 reviews the related work on the big data clustering methods and encryption techniques. Section 3 explains proposed implementation of algorithm. The whole paper is concluded in Sect. 4.

2 Related Work

This section explains clustering methods followed by encryption techniques. Section 2.1 explains various types of clustering on big data, and Sect. 2.2 explains the encryption techniques.

2.1 Big Data Clustering

The main issues of big data clustering are as follows: it is heterogeneous in nature and its large volume. There are various types of clustering algorithm such as graph theory, matrix factorization and fuzzy logic. The fuzzy C-means algorithm has the 60 clusters and that will be decided prior to execution of the algorithm based on Euclidian distance. The membership function is defined as the degree to which the data point belongs to the cluster [6]. The main drawback associated with this technique is that single data point can be part of several other clusters; thus, the name fuzzy is mentioned in this algorithm [7]. There is presence of ambiguity to decide which data point lies in which cluster. To overcome this drawback, the paper talks about possibilistic C-means (PCM) algorithm, which is improved version of fuzzy C-means [8]. In PCM algorithm, each data point is part of single cluster only [9]. It evaluates how compatible the input vectors are with respect to already formed clusters and thus mentioned by the name possibilistic. The drawback associated with this algorithm addresses is to be applied only for structured data (but big data are generally unstructured in nature), and it is capable of clustering only high-quality data (big data generally are not of high quality and have missing values in them). To overcome this drawback, this paper proposes an algorithm which is possibilistic C-means based on tensor space. This algorithm requires high-performance servers, large-scale memory and powerful computing unit to cluster big samples unlike the IOT systems, as we are dealing with the tensor space which supports high dimensional data [10, 11].

Tensor PCM algorithm can be broken in three steps, where the first step addresses the unsupervised feature learning from the big data. In this step, the algorithm learns the features from incomplete data by analysing the existing model part by part. The algorithm basically calculates the vector outer product and then joins the vectors together. The second step does feature selection where the learned features are used to model the non-linear correlations over different modalities. The final step creates the high dimensional clusters, by implementing the algorithm on dataset in tensor space. This algorithm was proposed by this paper because it is capable of clustering high-quality and incomplete data efficiently. Another use of this algorithm is that it is able to capture the high dimensional distribution associated with big data. Current distance measuring algorithms work on vector space and it is quite simple, but measuring in tensor space is quite complicated and needs to be taken care. Another challenge associated with proposed algorithm is that individual features learnt in step 1 are tough to make joint representation of them [12].

2.2 *Encryption Techniques*

The growing popularity and development of big data technologies bring a serious threat to the security of individual's sensitive information. Sometimes data needs to store on cloud tend to expose when executing tensor PCM. Companies' financial data or some person's medical data they are all typical big data. It consists of large amounts of private data which may contain data like mail ID, organizations' turn over with different companies and diagnostic data which is present in that records. Expose of that private information may have an adverse effect on company or person's personal life. Hence, it is must for safeguarding data which will be stored on cloud; we propose a privacy-preserving tensor PCM algorithm which uses BGV encryption technique that provides proficiency [13].

BGV encryption is a homomorphic encryption technique that permits computation on encrypted data and it produces an encrypted result. The operation of decryption will provide the same results which were given as plaintext [6, 14], which is not possible in other encryption techniques. The main advantage of homomorphic is as follows: it allows computation on encrypted data. Without having access to unencrypted data cloud computing platform allows complex computations on homomorphically encrypted data [15]. Homomorphic encryption permits to securely concatenate several services without exposing sensitive data [16]. Fully homomorphic encryption allows performing operations on encrypted data while data remains unchanged, without any secret key for manipulation of data [17].

During execution of tensor PCM on cloud tend to expose the data. Hence, BGV encryption technique is performed to encrypt, so that data do not get exposed to data engineer. BGV is basically homomorphic encryption allowing computation on encrypted data and performing operations on encrypted data will give the same result as operations performed on plain text [18].

BGV basically performs four operations namely, encryption, decryption, secure addition and secure multiplication. BGV doesn't support few operations like division and exponential functions which will be in required for few computations. To manage this issue, Taylor's theorem is being used. This theorem transforms polynomial functions into linear functions. The purpose of performing BGV encryption is to safeguard the sensitive data by executing tensor PCM on the cloud. In this paper, we are proposing privacy-preserving tensor PCM algorithm, which is modified version of TPCM.

3 Proposed Implementation

The objective is to provide security to big data while executing tensor PCM on it. Privacy-preserving tensor PCM will be executed on distributed system to improve the efficiency of algorithm. In distributed system, the Hadoop approach executes parallel operation [19]. The most important steps of tensor PCM is to calculate the membership matrix and the clustering centres. Therefore, the use of Map function to calculate the membership matrix and use the Reduce function to calculate the clustering centres. Hence the efficiency will be improved. Hadoop file system will be used as distributed system to implement the algorithm based on MapReduce [20].

For the experimental purpose, hospital sector data will be used. The experiment will be conducted in three steps:

Step 1: Tensor PCM will be executed on dataset.
Step 2: Tensor PCM will be executed on distributed system.
Step 3: Privacy-preserving tensor PCM will be executed on distributed system.

The algorithms will be compared based on execution time taken by each algorithm while the size of the dataset is increased. Another comparison is based on clusters formed by each algorithm. It is expected that tensor PCM takes more time to execute compared to tensor PCM on distributed system and privacy-preserving tensor PCM on distributed system. Because distributed system supports parallel execution. It is also expected that the clusters formed by all the three algorithms will be similar.

4 Conclusion

In this paper, we propose tensor possibilistic C-means for heterogeneous big data. Before clustering the data, it will be encrypted using BGV homomorphic technique to ensure the privacy. The algorithm will be implemented in distributed system using MapReduce. The performance of algorithms will be analysed based on execution time and the clusters formed. It is expected that performance of tensor PCM in distributed system will be better than other two algorithms.

References

1. Quizlet: Chapter 8—UNDERSTANDING BIG DATA & IT'S IMPACT Flashcards [online]. Available at: https://quizlet.com/228094043/chapter-8-understanding-big-data-its-impact-flash-cards/
2. Ji, C., Li, Y., Qiu, W.: Big data processing in cloud computing environments. In: IEEE Conference Publication [online]. Ieeexplore.ieee.org. Available at: http://ieeexplore.ieee.org/abstract/document/6428800/ (2018)
3. Anon [online]. Available at: https://home.deib.polimi.it/matteucc/Clustering/tutorial_html/ (2018)
4. Wang, C., Wang, Q., Ren, K.: Privacy-preserving public auditing for data storage security in cloud computing. In: IEEE Conference Publication [online]. Ieeexplore.ieee.org. Available at: http://ieeexplore.ieee.org/abstract/document/5462173/ (2018)
5. Big Data Security Solutions—Big Data Encryption for Hadoop & NoSQL [online]. Available at: https://safenet.gemalto.com/data-encryption/big-data-security-solutions/
6. Yu, J., Cheng, Q., Huang, H.: Analysis of the weighting exponent in the FCM. IEEE J. Mag. [online]. Ieeexplore.ieee.org. Available at: http://ieeexplore.ieee.org/document/1262532/ (2018)

7. Fuzzy c-Means Clustering Algorithm—Data Clustering … [online]. Available at: https://sites.google.com/site/dataclusteringalgorithms/fuzzy-c-means-clustering-algorithm
8. Mohamed Jafar, O.A., Sivakumar, R.: A study on possibilistic and fuzzy possibilistic C-means clustering algorithms for data clustering. In: IEEE Conference Publication [online]. Ieeexplore.ieee.org. Available at: http://ieeexplore.ieee.org/document/6513887/ (2018)
9. Zhang, J.-S., Leung, Y.-W.: Improved possibilistic C-means clustering algorithms. IEEE J Mag. [online]. Ieeexplore.ieee.org. Available at: http://ieeexplore.ieee.org/abstract/document/1284323/ (2018)
10. Ermiş, B., Acar, E., Cemgil, A.: Link Prediction in Heterogeneous Data Via Generalized Coupled Tensor Factorization (2018)
11. Kuang, L., Hao, F., Yang, L.T.: A tensor-based approach for big data representation and dimensionality reduction. IEEE J. Mag. [online]. Ieeexplore.ieee.org. Available at: http://ieeexplore.ieee.org/document/6832490/ (2018)
12. High-Order Possibilistic c-Means Algorithms Based on … [online]. Available at: https://www.sciencedirect.com/science/article/pii/S1566253517302245
13. Zhang, Q., Yang, L.T., Chen, Z.: Privacy preserving deep computation model on cloud for big data feature learning. IEEE Trans. Comput. **65**(5), 1351–1362 (2016)
14. En.wikipedia.org. Homomorphic Encryption [online]. Available at: https://en.wikipedia.org/wiki/Homomorphic_encryption (2018)
15. Atayero, A.A., Feyisetan, O.: Security Issues in Cloud Computing: The Potentials of Homomorphic Encryption—Covenant University Repository (2018)
16. Fontaine, C., Galand, F.: A Survey of Homomorphic Encryption for Nonspecialists (2018)
17. Networking116.wikispaces.com. Networking116—Encryption Advantages and Disadvantages [online]. Available at: https://networking116.wikispaces.com/Encryption+Advantages+and+Disadvantages?responseToken=3d8899636829f505ffce5897150d8352 (2018)
18. Hariss, K., Chamoun, M., Samhat, A.E.: On DGHV and BGV fully homomorphic encryption schemes. In: IEEE Conference Publication [online]. Ieeexplore.ieee.org. Available at: http://ieeexplore.ieee.org/document/8242007/ (2018)
19. Shvachko, K., Kuang, H., Radia, S.: The hadoop distributed file system. In: IEEE Conference Publication [online]. Ieeexplore.ieee.org. Available at: http://ieeexplore.ieee.org/abstract/document/5496972/ (2018)
20. Pandey, S., Tokekar, V.: Prominence of MapReduce in big data processing. In: IEEE Conference Publication [online]. Ieeexplore.ieee.org. Available at: http://ieeexplore.ieee.org/document/6821458/ (2018)

Shriya R. Paladhi working in Department of CSE, RUAS, Bangalore, India. Her research area includes clustering technology and big data analytics.

R. Mohan Kumar working in Department of CSE, RUAS, Bangalore, India. Her research area includes clustering technology and big data analytics.

A. G. Deepshika Reddy working in Department of CSE, RUAS, Bangalore, India. Her research area includes clustering technology and big data analytics.

C. Y. Vinayak working in Department of CSE, RUAS, Bangalore, India. Her research area includes clustering technology and big data analytics.

T. P. Pusphavathi working in Department of CSE, RUAS, Bangalore, India. Her research area includes clustering technology and big data analytics.

IoT-Based Smart Login Using Biometrics

C. G. Sarika, A. Bharathi Malakreddy and H. N. Harinath

Abstract The main idea of smart login system is to provide security to our beloved things, places, libraries, institutions, etc. To keep our data safe or to avoid them being stolen by third party, we use smart login. Nowadays, it is essential to protect or secure our things or valuable data and even money. This system is based on biometrics that is fingerprint template to unlock the door which provides security to many banks, institutes, and various organizations. The door lock system using biometrics interfaces with a biometric reader that is fingerprint scanner and a door lock that will secure a room or environment. This system is a replacement for keys, locks, and cards. This system consists of fingerprint scanner, Arduino board, magnetic solenoid lock, and LCD to unlock the door. Kiel compiler is used in the programming to interact with the fingerprint reader for triggering purpose. With this system, securing access will be provided while providing convenience and efficiency. If the fingerprint template which is authorized matches, then the door lock opens, otherwise it remains in closed state.

Keywords Arduino · LCD · Fingerprint sensor

C. G. Sarika (✉) · A. Bharathi Malakreddy · H. N. Harinath
B.M.S. Institute of Technology and Management, Bengaluru, Karnataka, India
e-mail: cgsarika@gmail.com
URL: https://bmsit.ac.in

A. Bharathi Malakreddy
URL: https://bmsit.ac.in

H. N. Harinath
URL: https://bmsit.ac.in

S. Smys et al. (eds.), *International Conference on Computer Networks and Communication Technologies*, Lecture Notes on Data Engineering and Communications Technologies 15, https://doi.org/10.1007/978-981-10-8681-6_54

1 Introduction

Internet of things connects the various embedded systems to the Internet [1]. Devices can be controlled from anywhere when the object or device represents digitally. So, this connectivity helps us to get more information from other places by increasing its efficiency, safety, and security. Hence, IoT is force which helps many companies to improve their performance through analytics and provides security to deliver or for better results.

The Internet of things (IoT) generally defines a connectivity that means it connects various devices or objects to Internet to exchange the information over Internet so that we can have less or minimal human intervention; hence, it transmits an information over the network without human or computer interaction.

Smart locks using biometrics are available and allow the user to integrate to unlock the door. Even one can connect the locks to cloud to have remote access, so that we can share the access to other people at the doorstep. The main idea of smart login is to open and close the door automatically by biometrics without manual locks to prevent the access from unauthorized access to enter the room in order to provide security and privacy to our confidential to our data and property.

Section 1 describes the domain of the project and the project description, Sect. 2 elaborates on the survey of the project, Sect. 3 describes the proposed system and its objectives, Sect. 4 describes the system design with the components description, Sect. 5 describes the implementation and dataflow diagram of the project, Sect. 6 describes the experimental outcomes or the results, and finally Sect. 7 gives the conclusion of this project.

2 Related Work

Dilum Bandara [2] discussed that biometrics as a set of technologies to measure some unique characteristics of an individual to identify or verify an individual which cannot be lost, forgotten, borrowed, and stolen. Biometrics measures the person's unique characteristics to authenticate the person's identity. Characteristics include keystroke analysis, lip movement, signature, gestures, retina, iris, facial characteristics, and fingerprint templates.

Prabhakar et al. [3] discussed that biometrics is based on pattern recognition as it recognizes an individual based on a feature derived from specific characteristics that individual possesses. And also, they discussed that it operates in two modes that are verification and identification; verification is done by enrolling individual's template, and identification is done by searching the entire database to get a match for that if it exists.

Aditya Shankar et al. [4] discussed how to protect the things or objects which can be stolen or lost, or to provide security from the unauthorized person by providing the high accuracy. And they discussed that skin on our bodies have some pattern say, for example, pal has ridges called friction ridges, and it is unique for every individual's even in case of twins also. So, they told that it makes a unique identification for every individual. Fingerprint reader or scanner scans the fingerprint template of individuals by authenticating hence they told that it has more accurate and duplication is highly impossible.

3 Proposed System

The motivation to implement a smart login by using biometric fingerprint to make our campus into a "**Smart Campus**" and thereby solving the security issues related to opening and closing of doors which are accessed by unauthorized users.

3.1 Objectives Are as Follows

- Easy to implement and use.
- It cannot be duplicated.
- Small in size since the components are integrated into single device.
- Accuracy is high.
- Errors are less or negligible.
- There will not be much intrusions with this system.

4 System Analysis and Design

The below diagram gives the overview description of the components of smart login. The process starts with providing power supply to the Arduino board where the fingerprint sensor, i.e., RS03 is embedded to Arduino board and displays or pop-ups the corresponding messages on LCD (Fig. 1).

4.1 Description of the Components

- **Fingerprint sensor**: Fingerprint sensor is used to take user fingerprint template and compare that with previously stored templates. The fingerprint scanner used in this system is capable to store 120 fingerprints in its flash memory. The comparison can be done or made based on 1:1 or 1:N to identify the person's fingerprint template. This is interfaced with Arduino board (microcontroller).
- **Arduino board**: Arduino board is used to dump our all code in order to perform the actions so that it works or functions correctly, so by doing this it controls

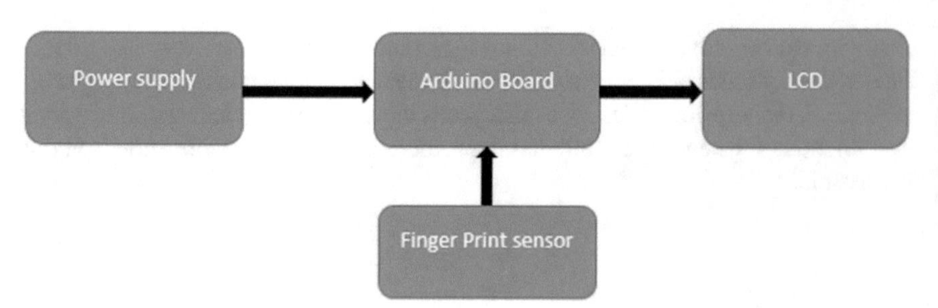

Fig. 1 Block diagram of smart login using biometrics

other devices. Arduino board has a capacity to control fingerprint sensor and other outputs. In this project, we have used Arduino Uno (8-bit) to control the fingerprint scanner. The program is written in C language, and it is uploaded into the Arduino microcontroller. So, every time when the power is supplied to the Arduino board automatically the program gets executed. The main advantage of using Arduino microcontroller is that it is user-friendly, it consumes of 12 V maximum, and it is low cost.
- **Liquid crystal display (LCD)**: It is used to display the corresponding messages when the user tries to login for better enhancement.

5 Implementation

Arduino language has been selected as it is a simplified language which is made up of C and C++ languages. Arduino microcontroller controls all the devices connected to it. The program has fed into Arduino board and connected to the power supply.

The devices are triggered according to the flow of the code. First, the fingerprints of the authorized users have to be stored inside the fingerprint sensor. Upon entering the fingerprint, the door either locked or unlocked. For opening, the users' fingerprint template has to match with the previously stored template.

- **For authorized user**: If the user is authorized and the fingerprint is validated successfully by the sensor, the Arduino will start and tell magnetic lock to unlock the door.
- **For unauthorized user**: If the user who tries to enter the door is found as unauthorized and if the validation fails, the Arduino will start and tell the magnetic lock to be in unlock state.

Figure 2 is used to depict the step-by-step process of the complete project flow.

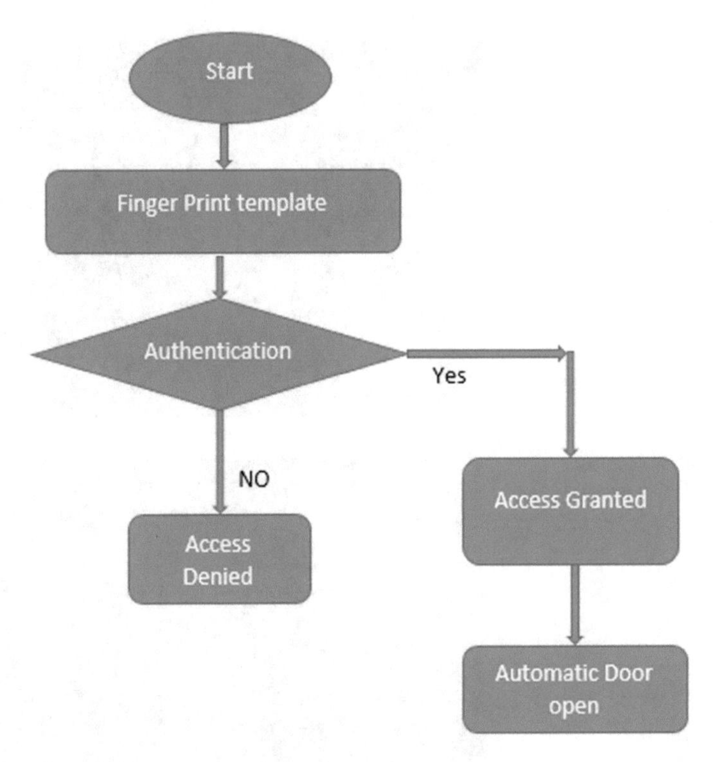

Fig. 2 Dataflow of implementation phase

Fig. 3 Admin login to enroll the users

6 Experimental Results

After the implementation stage, we observed the results will be displayed onto the LCD (Figs. 3, 4, 5, 6, 7 and 8).

Fig. 4 Select the ID to enroll

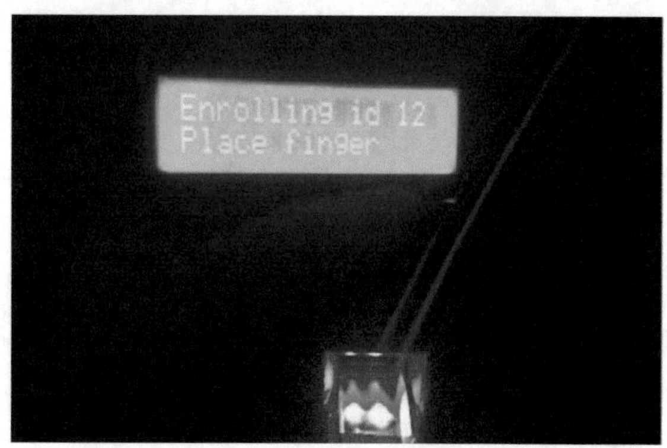

Fig. 5 Asks the user to place finger

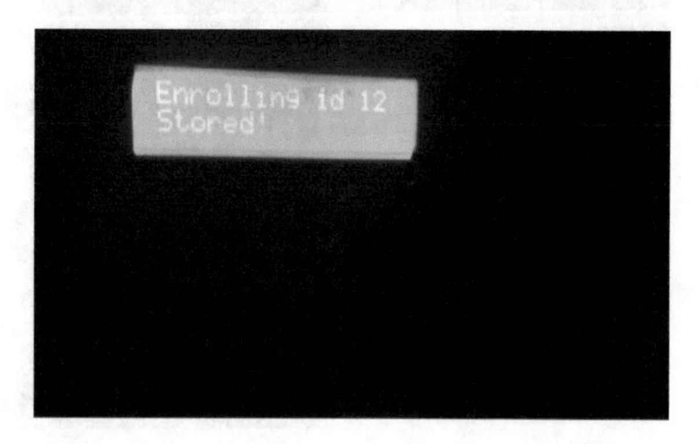

Fig. 6 Template is stored for particular ID

Similarly, we can delete the particular user ID by selecting the ID, i.e., final snapshot for deleting particular ID (Figs. 9 and 10).

Fig. 7 User login

Fig. 8 Lock is opened

Fig. 9 Select an ID to delete

Fig. 10 ID is deleted

7 Conclusion

There are many door locks using biometric fingerprint technology and most of them integrate the fingerprint device into the door lock itself. This proves that the system can accurately identify and compare fingerprint templates at a high rate whether it is to enroll a new fingerprint template or just verify if the captured template is in the memory or already enrolled. Through the use of this design, people will have an easier way of having a convenient, secured, and authorized entrance in a certain room or establishment as there would be no keys, passwords or cards will be used. Owners would just register trusted fingerprints that could enter its premises. With this system, we can automate door locks and help people especially security guards and utility men, administrators, and owners to secure its premises.

References

1. https://www.happiestminds.com/Insights/internet-of-things
2. "Biometrics", written by Dilum Bandara, a PhD Candidate in the Computer Networking Research Laboratory, Department of Electrical and Computer Engineering, Colorado State University, USA
3. Prabhakar, S., Pankanti, S., Jain, A.K.: Biometric Recognition: Security and Privacy Concerns
4. Aditya Shankar, A., Sastry, P.R.K., Vishnu Ram, A.L., Vamsidhar, A.: Students of IV B.Tech ECE, Student member - IEEE, IET, IETE, Assoc. Prof. and Head of the Department, ECE, Dadi Institute of Engineering and Technology, Visakhapatnam-531002, Andhra Pradesh, India
5. Thakur, M.K., Kumar, R.S., Kumar, M., Kumar, R.: Wireless fingerprint based security system using Zigbee. Int. J. Inventive Eng. Sci. (IJIES) **1**(5) (2013). ISSN: 2319–9598
6. Lourde, M., Khosla, D.: Fingerprint identification in biometric security systems. Int. J. Comput. Electr. Eng. **2**(5) (2010)
7. Fingerprint Recognition. http://en.wikipedia.org/wiki/Fingerprint_recognition
8. www.bayometric.com/how-do-fingerprint-door-locks-work
9. https://www.youtube.com/watch?v=KA7ZqTPvInw

C. G. Sarika studying Master Student in B.M.S. Institute of Technology and Management, Bengaluru-560064. Her research area includes IoT and smart automation.

A. Bharathi Malakreddy working as a Professor in B.M.S. Institute of Technology and Management, Bengaluru-560064. Her research area includes IoT and smart automation.

H. N. Harinath working as an Assistant Lab Instructor, IoT lab in B.M.S. Institute of Technology and Management, Bengaluru-560064. His research area includes IoT and smart automation.

Big Genome Data Classification with Random Forests Using VariantSpark

A. Shobana Devi and G. Maragatham

Abstract We are in the middle stage era of the digital revolution where all consumers and businesses demand decisions to be based on evidence collected from data. The challenge of big data is mainly well defined in the health and bioinformatics space where, for example, whole genome sequencing (WGS) technology leads researchers to cross-examine nearly all three billion base pairs of the human genome. When compared to traditional big data disciplines, such as astronomy, Facebook, YouTube and Twitter, data acquisition in this genomic space is predicted to outpace more, and nearly 50% of the world's population would be aimed for a medical decision by 2030. As of now, the analysis of using medical genomics data is having the front position of this growing need to apply complicated machine learning techniques to large high-dimensional datasets. Biology is even more complicated than this genomic space. In this paper, we explored the parallel algorithm for random forests with large genomic data, which is classified to find the ethnicity nature of a human being. Apache Spark's MLlib is mainly designed for the common use cases where the genomics data have nearly thousands of features and that need to be scaled up for the millions of genomic features. The study is conducted using VariantSpark based on Spark core, and it has the parallelization computing to coordinate with genomic massively distributed machine learning job.

Keywords SPARK MLlib · WGS · Genomic data · VariantSpark
Random forests

A. Shobana Devi (✉) · G. Maragatham
Department of Information and Technology, SRM University, Chennai, India
e-mail: shobanak07@gmail.com

G. Maragatham
e-mail: maragathamhaarish@gmail.com

S. Smys et al. (eds.), *International Conference on Computer Networks and Communication Technologies*, Lecture Notes on Data Engineering and Communications Technologies 15, https://doi.org/10.1007/978-981-10-8681-6_55

1 Introduction

In this paper, we provide how machine learning to be used to solve the main problems in genomic medicine. Genomics means information structure and function processes that are encoded using the DNA sequences of living cells. Precision medicine is the process of tailoring treatment based on the patient's all relevant information and also including the patient's genome. In terms of data, these disciplines are undergoing a sudden increase in growth [1–4]. The problem domains play a major role to develop the machine learning algorithms and thereby it improving the long time and quality of life for the millions of persons suffering from a genetic disorder at present and in future years to come up [5]. A genome is a building block of an organism. In ancient days, the DNA molecules are used as the physical medium of genetic information storage [6] and now the basic raw information of a distinctive human genome are used for human genome projects [7, 8]. The greater challenge in genomic projects is how to interpret the function, structure or meaning of the gene information. It is very hard to know how the genetic information is structured into distinct genes and each gene is a description of how to build an individual family of molecules. Protein-coding genes are to describe the build process of large molecules that are made from amino acid chains (proteins) and the non-coding genes are to describe the build process of small molecules that are made from ribonucleic acid chains (RNA) [9, 10]. Each human genome has approximately 20,000 protein-coding genes [11] and 25,000 non-coding genes [12] and in that some are very crucial for life and health and some can be removed without any harm from its whole family. A gene structure has the presence of introns and exons as alternating regions. The patterns in the nucleotide sequence are used to identify the boundaries between those regions and by disrupting the patterns, which will lead to disease-causing mutations act. The major genetic cause of infant mortality is the spinal muscular atrophy (SMA), [13] which causes the missing genome in the SMN1 gene or it has the damaged version resulting in the deficient production of the survival monitor neuron (SMN) protein. The SMN2 acts as an alternate version of a gene for the production of the SMN protein [14, 15]. The major example of a heterogeneous disease is the Cancer. Cancer is a disease of all having similar symptoms with multiple casual pathways but requires different treatments [16]. Genomic data have become more prominent for giving more detailed treatments and diagnoses for cancer [17].

For over a century, doctors are using blood type for blood transfusions and this precision medicine concept is not new [18]. But today, the fast growth of genomic data collected from the patient and wider community to be processed more quickly and cheaply to gain the potential insights by sharing the data. When compared to traditional use of genomic data in laboratory sets, the scale and complexity of data increase to 20–50 measurements [19]. Here, the main focus is on how machine learning applications are used in genomic medicine, to find out the targeted therapies using genomic characteristics and to discover the disease risks for possible preventive measures. The vision is to make the genomic medicine a reality; we should develop computer systems in such a way that can correctly interpret the text of the genome like as the machine does inside the cell. When compared to laboratory experiments and model organisms, it is a very difficult challenge to enable the effects of genomic variation and possible therapies that should be discovered quickly, cheaply and more accurately [20]. Protein-coding exons are the current state of the art in genomic medicine, and it is the most understood regions in the genome. Nonsense mutation means, if a mutation introduces a stop codon in its sequence structure, then the protein will be deleted as a general rule. Prediction of whether the mutation will affect the stability of final protein molecule is a long-lasting open research problem [21].

The most popular cognitive tasks like visual object detection, image processing and speech recognition, the human beings are not capable to perceive and interpret genomic sequences they need to understand all the methods, techniques, pathways and interactions that go inside a living cell. To make a moment, a system with a superhuman analytical ability should be required. Some of the few research groups which are emerging use the genome data using their machine learning techniques and genome biology expertise knowledge. There are numerous ongoing opportunities for machine learning researchers in the area of genomics and biology. It is now being recognized that resources would be better spent on developing new computational techniques rather than on pure data collection. The systems that have the ability to read the text of the genome should be used in a variety of ways to hold up genomic medicine [22].

Recently, gene editing technique is used by scientists to alter the genomes of already existing living cells with more efficiently. Gene therapies are used to add targeted modifications like removing harmful mutations or including new sequences at determined locations in a genome. Gene editing technology gives a way to extraordinary opportunities in genomic medicine field [23, 24] (Fig. 1).

2 Using Machine Learning for Genome Data Classification

2.1 Overview of Classification Techniques

In principle, predicting phenotypes like traits and disease risks from genome data is a type of supervised machine learning problem. The stretch of a DNA sequence relevant to biology is the input and the phenotypes are outputs. The sheer complexity of the relationship between a genotype and its phenotype should be found first. In a single cell, the state of the cell is directed by the genome through many layers of interconnected and complicated biophysical processes and control techniques that have been formed ad hoc by evolution [9, 25]. When we attempt to infer the outcomes of this complex regulatory process by using only genomes and phenotypes, it looks like a computer playing chess by investigating binary code and wins and losses and ignoring which move was taken. Second, if we find some model for predictive of disease risks, the hidden layers of that model will not have the biological mechanisms of the genome. In order to develop targeted therapies and also to provide

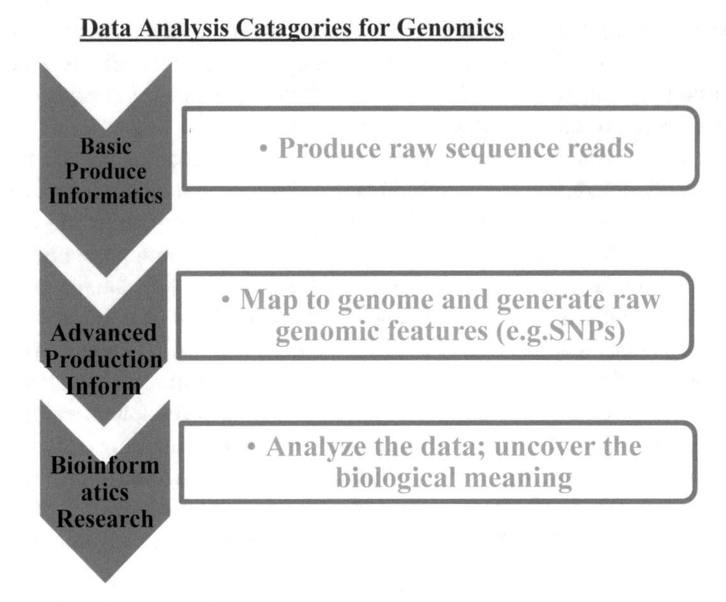

Fig. 1 Different categories of genomic data analysis

complementary information for phenotype screens, it is very important to insight into disease mechanisms. Genomic information is most widely used in medical practice. A most preferably performed task in such application domains is grouping individuals based on their genomic profile to identify association in population or to expose involvement in disease vulnerability [26, 27].

2.2 *Existing Techniques Issues*

The frequently used tool is ADMIXTURE, which is used to find maximum likelihood estimation ancestries from multilocus SNP genotype datasets [28]. It has now become an economical to study with sample sizes of genomic data due to the decreasing sequencing cost. By increasing the feature space by orders of magnitude, the whole genome sequencing enables the even somatic mutations in the analysis [29]. A massively parallel approach is required for data processing due to the severe increase in both sample numbers and features of datasets [30]. The old parallelisation techniques like MPI/OPENMP are not able to scale with variable data sizes during runtime.

To address that problem, Apache Hadoop MapReduce technique is introduced to transform data into key-value pairs and that will be split and distributed among the multiple nodes of a cluster commodity hardware based on the size of the problem [31]. MapReduce is used more increasingly in bioinformatics field (for Refs. [5,

32, 33]). The mainly used fields are read mapping Schatz [34], sequence analysis tasks [35], variant calling [36, 37] and duplicate removal and also used for genome-wide analysis study tasks [38, 39]. Apache has also introduced a machine learning library tool named Mahout [40], which is used efficiently in clinical applications like medical health records. However, the MapReduce paradigm does not always play a best and optimal solution for bioinformatics applications, since it requires an iterative in-memory computation. Hadoop is basically relying on hard disk for all input–output operations, and this will lead to a bottleneck problem in processing speed [41].

The more recent parallelized computing engine that overcomes many of drawbacks of Hadoop is Apache Spark [42]. The key benefit of this framework is that it allows the data to be processed in memory and thereby reducing the bottleneck problem of disk IO. It utilizes caching and can speed up to 100 times faster than Hadoop MapReduce. SPARK doesn't require programs to be coded exactly like MapReduce paradigm, which in turn gives more comfortable software design. By recognizing the capability of SPARK, recently SPARKSEQ [43] is developed for sequence data analysis with high throughput and the Big Data Genomics (BDG) group recently illustrated the strength of SPARK using ADAM in a genomic clustering application, ADAM contains a set of formats and APIs as well as processing stage implementations for processing the genomic data [44]. ADAM is accepted to be one of the cornerstones for Genomics and Health of the Precision Medicine [45]. The speedup and performance of ADAM was extraordinary nearly 50 fold speedup while comparing with traditional methods. In this paper, we hence applied random forest for a genomic data classification approach in SPARK, to classify the data based on HipsterIndex phenotype in order to find the ethnicity of an individual. We used the SPARK's MLlib machine learning library, and the input data are interface to the standard data format named as variant call format (VCF), which gives way to a wide range of genomic data analysis using different machine learning algorithms.

3 Proposed Methodology

3.1 Dataset Description

We apply Apache Spark to the freely available public genomic data and demonstrated the classification of the gene data based on the variable importance score of each individual genome. The dataset with a subset of samples and variants from the 1000 genomic project in VCF format http://www.1000genomes.org/data. The dataset contains 17,000 individuals' genomic data with 1,130,245 variants dimensional dataset. The metadata for this dataset has the labels and features associate for each individual. Implementation is done using VariantSpark in scala, with Apache Spark and MLlib. The random forests algorithm from MLlib is used to classify the data.

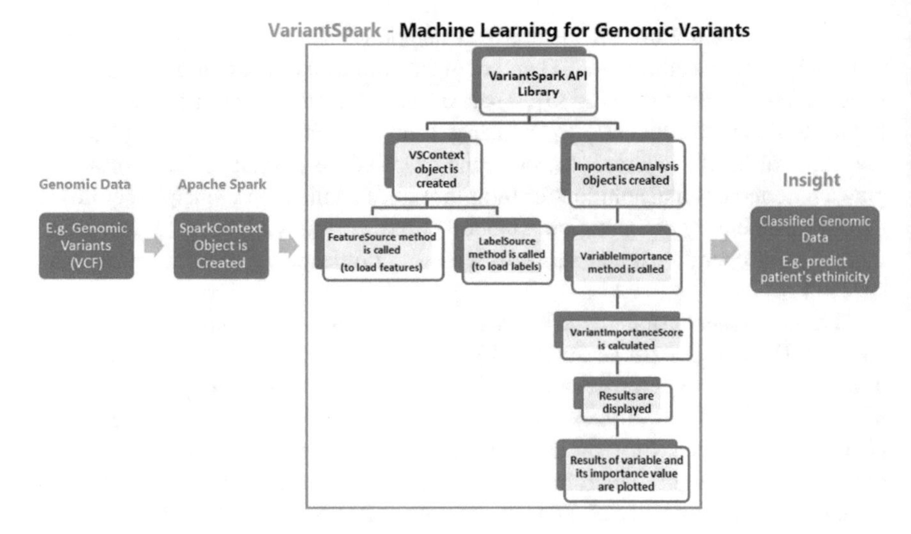

Fig. 2 Architectural view of VariantSpark

3.2 *VariantSpark*

VariantSpark is recently introduced as a machine learning library that is built on top of Apache Spark and i written in Scala. It is mainly used for the real-time genomic data analysis with thousands and millions of sample variants. It is developed and authored by the team of CSIRO Bioinformatics which is held up in Australia. In this paper, we mainly used a custom random forests algorithm implementation to find out the most important variants corresponds to the phenotype of interest. It also uses a synthetic phenotype HipsterIndex loaded in the CSV format and it has main factoring various real phenotypes of individuals like monobrow, beard, coffee consumption, etc. [46].

VariantSpark is mainly developed for genomic data application area to process many samples 'big' termed as 'n' and with large feature vector per sample 'wide' termed as 'p'. It was initially tested on datasets with samples $n = 3000$ where as each one containing about $p = 80$ million feature vectors. It is implemented using either with unsupervised clustering approaches like K-means or with the supervised learning approaches with output values as categorical means classification or continuous means regression. Even though VariantSpark is basically developed for genomic variant data, it can also be used for other application feature-based datasets like methylation, transcription and non-biological applications [47] (Fig. 2).

Table 1 Sample synthetic HipsterPhenotype index

ID	SNP ID	Chromosome	Position	Phenotype	Reference
B6	rs2218065	chr2	2.2E+08	Monobrow	Adhikari et al. [50] Nat Commun.
R1	rs1363387	chr5	1.3E+08	Retina horizontal cells (checks)	Peterson et al. [51] Nature
B2	rs4864809	chr4	5.5E+07	Beard	Adhikari et al. [50] Nat Commun.
C2	rs4410790	chr7	1.7E+07	Coffee consumption	Yang et al. [52] PLoS Genet.

3.3 HipsterIndex

HipsterIndex is one of the main illustrative examples of VariantSpark, where it uses the association testing feature to find out the genes that are associated with a 'Hipster' phenotype like coffee consumption, facial hair, etc. The formula to calculate the synthetic phenotype HipsterIndex for gene data is as follows:

$$\begin{aligned}\text{HipsterIndex} = {} & ((2 + \text{GT[B6]}) * (1.5 + \text{GT[R1]})) \\ & + ((0.5 + \text{GT[C2]}) * (1 + \text{GT[B2]}))\end{aligned}$$

In the above formula, the GT stands for the genotype of an individual. The location of genotype is referred with the homozygote encoded as 0, heterozygote encoded as 1 and homozygote alternative is encoded as 2. The individuals data with a HipsterIndex score above 10 are termed as hipsters, and the rest of the data are termed as non-hipsters. This HipsterIndex score calculation creates a binary annotation for the individuals' genomic data in the 1000 Genome Project dataset.

Table 1 gives the four sample synthetic HipsterIndex that was created using genotypes from dataset and with the formula. The table has ID of the genotype, SNP ID, the chromosome number with its position, the phenotype of the corresponding SNP ID and the reference paper for this HipsterIndex is listed. Only the four sample phenotypes are taken for this implementation such as monobrow, Retina horizontal cells, beard and coffee consumption. The given genomic data are classified using random forests approach only for the below listed phenotypes. Figure 3 gives the clear representation of these HipsterIndex phenotypes with their corresponding chromosome values.

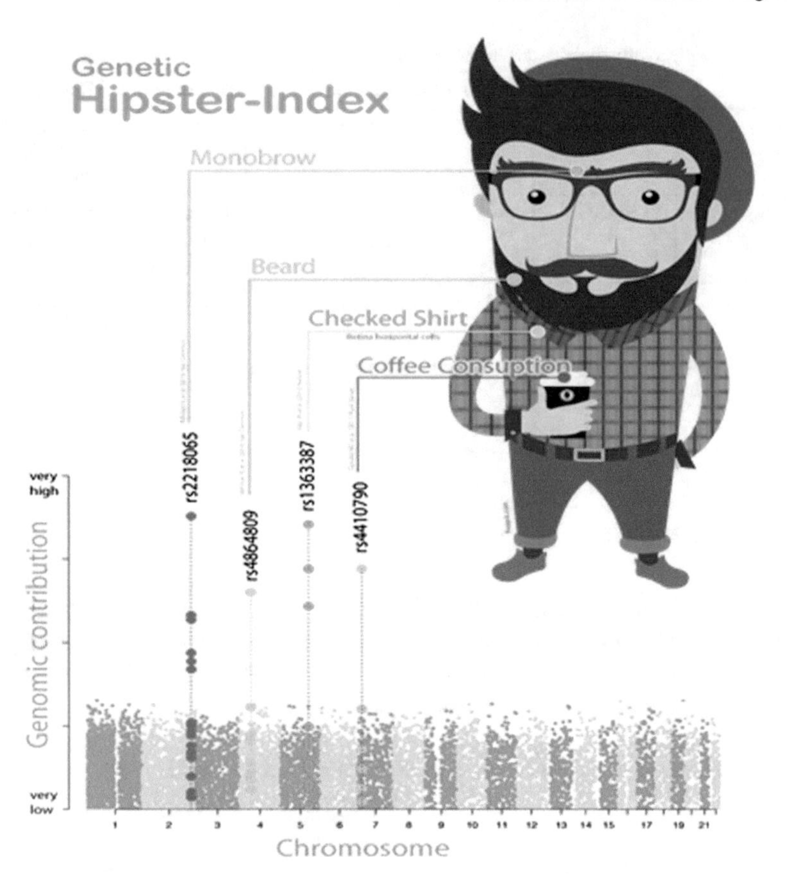

Fig. 3 Genetic HipsterIndex of an individual

4 Experimentation and Evaluation

In this paper, the experiment is conducted for genomic data classification on a virtual machine (VM) hosted on a local computer system. The VM is an Ubuntu 16.04.03 OS with four cores, and each having 8 GB memory and 20 GB hard disk space. For the whole genome data classification, we used our inbuilt Hadoop cluster with Hadoop 2.7.0 and Spark 2.2.0 and the scala of version 2.11.11. This is implemented using a single node cluster framework.

The main objective of this paper is to demonstrate the usage of VariantSpark for genomic data to find the association and classification of selected four SNPs phenotype HipsterIndex to the phenotype of interest that is the data being a Hipster or not. The following steps give the flow of the implementation and how it is classified.

- First, the downloaded datasets are loaded into the distributed file system (HDFS) for processing.

Table 2 Top 10 results of variable and its importance scores

Variable	Importance
2_109511398	8.451471757744014e−8
2_109511454	6.933887239515818e−8
2_109511463	0.0000015260923207317157
2_109511467	0.000004555482402038446
2_109511478	1.3271838818669408e−8
2_109511497	3.13557047416921e−8
2_109511525	2.384271359562883e−7
2_109511527	0
2_109511532	0

- The Spark Context object is created as **spark**.
- Scala is used to import the **VSContext** and **ImportanceAnalysis** objects from the VariantSpark API library.
- Instance for **VSContext** object is created and the Spark Context object **spark** is passed to it.
- The **featureSource** method is called on the created **vsContext** object instance to pass the target feature file path and to load the variants data from VCF file.
- The following is a list of 10 sample names data: List (HG00096, HG00097, HG00099, HG00100, HG00101, HG00102, HG00103, HG00105, HG00106 and HG00107).
- The **labelSource** method is called on the **vsContext** object instance to pass the target label file path and to load the hipster labels.
- The following is a list of 10 sample phenotype labels: List $(0, 1, 0, 0, 0, 1, 0, 1, 1, 1)$.
- Scala is used to create the **importanceAnalysis** object instance.
- Next, the **featureSource** (target feature file), **labelSource** (target label file) and number of trees (here, $n = 1000$) are passed to the above-created instance.

When compared to other supervised statistical approaches, the random forests algorithm has the main advantage of the data need not be extensively pre-processed. Hence, the analysis is started on the loaded raw data directly from HDFS. The full analysis process has taken around 4–5 min on a single node cluster.

- In order to calculate the variant importance attributing to the phenotype, the **variableImportance** method is called on the **importanceAnalysis** object instance.
- The results are cached and stored in the **SparkSQL** table.
- The results are displayed in a tabular format.
- From **SparkSQL** table, the results are queried to display top 30 values in descending order.
- The results are plotted using the visualization feature of Python (Table 2).

4.1 Variant Importance Score

VariantSpark uses a computing metric named as 'importance' score to each tested variant data which in turn reflects its associated phenotype of interest. This importance score is computed using the decision trees Gini impurity scores in the random forests built on VariantSpark. In the building process of decision tree, the VariantSpark assigns variants from the loaded VCF dataset at each node of the tree. The sample data at node is split to calculate the Gini impurity scores before and after each split.

For a given node N, the Gini impurity score decrease by splitting the feature p is illustrated as [48, 49]

$$\Delta I = I(N) - (I(L) + I(R))$$

where

$I(N)$ is the Gini impurity at node N,
$I(L)$ is the Gini impurity scores of the left child node, and
$I(R)$ is the Gini impurity scores of the right child node.

The importance score of variant data p is calculated as the mean value of the decrease in Gini impurity score which is defined by splitting the variant p across all the nodes and trees in random forests. A variant data which has higher importance score results as more strongly associated with the given sample synthetic phenotype of interest.

4.2 Gini Impurity

Gini impurity score is mostly used in the decision tree algorithms. It is a measure of how frequently a randomly selected element from the dataset is incorrectly labelled, if the element was randomly labelled based on the distribution of labels in the dataset. The Gini impurity score is computed using the summation of probability of an item p_i with its label i, being chosen wrongly times is, it results in zero, if all nodes belong to a single target class [48, 49].

$$\sum_{k \neq i} p_k = 1 - p_i$$

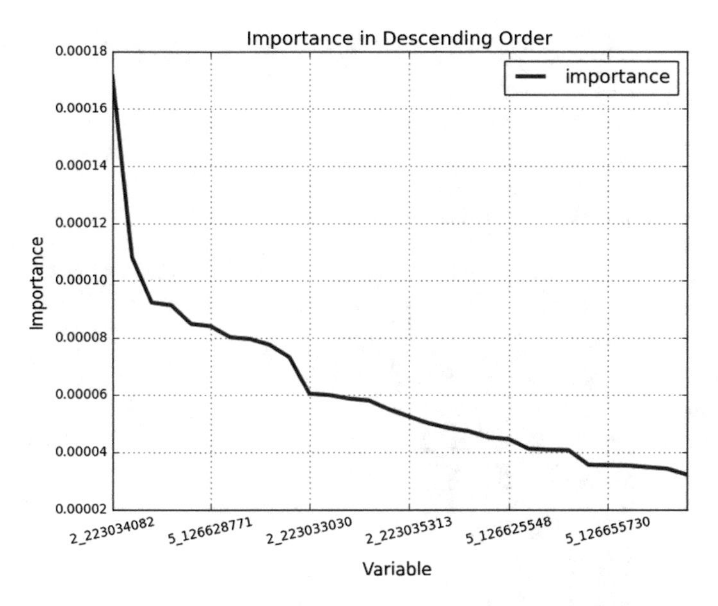

Fig. 4 Graphical representation of 30 variable importance score values

5 Results and Discussion

The following graph shows the top 30 results of the variable and importance scores of the genotype in descending order. The importance score for each variant is calculated using the formula of variant importance score of VariantSpark (Fig. 4).

The plot graph below shows that from the given genomic data, the VariantSpark has classified the correct genotypes of this multivariate phenotype with its associated interacting features of having multiplicative and additive effects.

- The encoding for monobrow phenotype is **chr2_223034082** (rs2218065) shows the most important feature.
- The SNPs encoding of in Retina horizontal cell formation **chr5_126626044** shows the second most important feature, explaining why hipsters prefer checked shirts.
- The encoding marker for increased coffee consumption **chr7_17284577** (rs4410790) is ranked as third most feature.
- The encoding marker for beards **chr4_54511913** (rs4864809) is the fourth feature.

The last two values when compared to the formula of the HipsterIndex are in swapped order with weight values as 0.5 and 1, and they may be very difficult to differentiate it (Fig. 5).

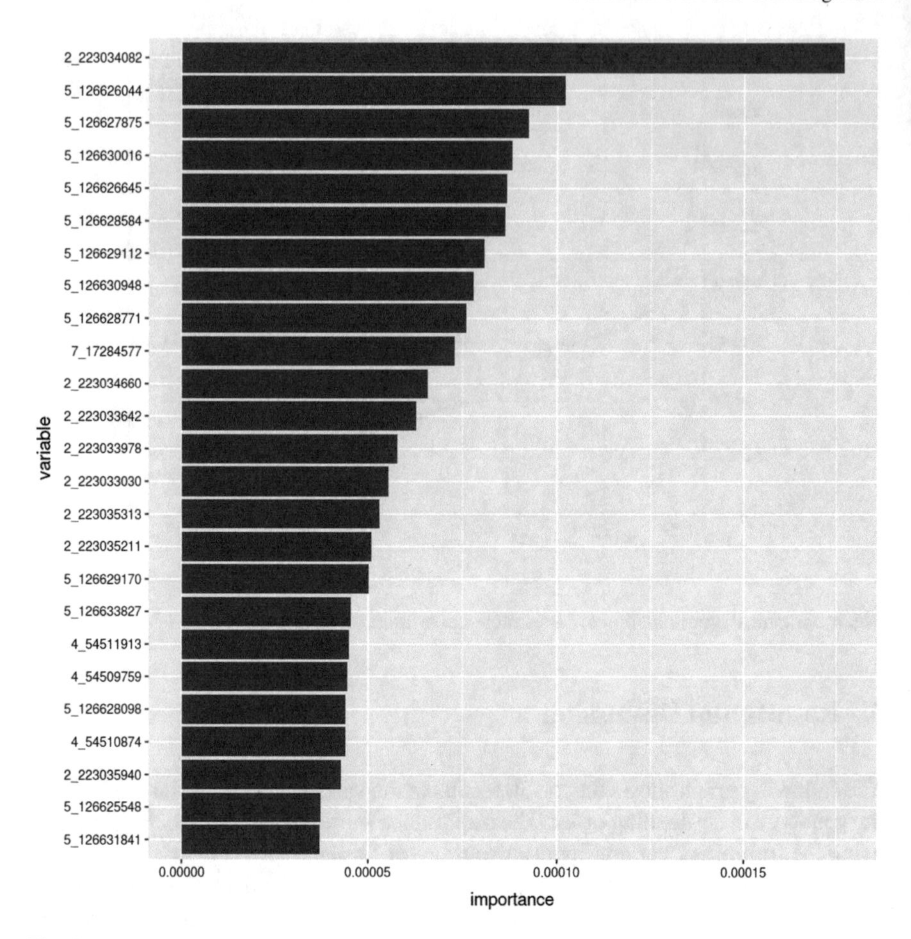

Fig. 5 The plot graph of classified genotypes

6 Conclusion and Future Work

In this paper, we have implemented the need to a parallel algorithm for random forests, and using that algorithm, the large genomic data are classified to find the ethnicity nature of a human being. Apache Spark's MLlib is mainly designed for the common use cases where the genomics data have nearly thousands of features and that need to be scaled up for the millions of genomic features. The study is conducted and done using VariantSpark based on Spark core and it has the parallelization computing to coordinate this genomic massively distributed machine learning task. First, the variant importance score of the data is computed and based on the score value, the higher importance score is strongly associated with the phenotype. The data are classified based on sample phenotypes, and the results are interpreted as a graph. As an extension of this work, the data can be still experimented for different synthetic phenotype hipsters to know more about the ethnicity of an individual using their gene data. It can be implemented as a multi-nodes cluster for better scalability and performance of the approach and to gain high throughput.

Acknowledgements This article draws on research currently being conducted in Department of Information Technology, SRM Institute of Science and Technology, Chennai, India.
Declarations:
Ethics approval and consent to participate
Not applicable.
Consent for publication
Authors prove consent of publication for this research.
Availability of data and material
Not applicable.
Competing interests
The authors declare that they have no competing interests.
Funding
Not applicable.
Authors' contributions
AS worked on the preparation of the manuscript. She also worked on the implementation presented in the paper. GM is responsible for the overall organization and the content of this survey article. Both authors read and approved the final manuscript.
Authors' information
Not applicable.
Author details
[1]Research Scholar (Full Time), Department of Information and Technology, SRM University, Chennai, India, shobanak07@gmail.com.
[2]Research Supervisor, Department of Information and Technology, SRM University, Chennai, India, maragathamhaarish@gmail.com.

References

1. Ashley, E.A.: The precision medicine initiative: a new national effort. JAMA **313**(21), 2119–2120 (2015)
2. Marx, V.: Biology: the big challenges of big data. Nature **498**(7453), 255–260 (2013)
3. Schatz, M.C., Langmead, B.: The DNA data deluge. IEEE Spectr. **50**(7), 28–33 (2013)
4. Stephens, Z.D., Lee, S.Y., Faghri, F., Campbell, R.H., Zhai, C., Efron, M.J., Iyer, R., Schatz, M.C., Sinha, S., Robinson, G.E.: Big data: astronomical or genomical? PLoS Biol. **13**(7), e1002195 (2015)
5. Taylor, R.C.: An overview of the Hadoop/MapReduce/HBase framework and its current applications in bioinformatics. BMC Bioinform. **11**(12), S1 (2010)
6. Wagstaff, K.: Machine learning that matters. arXiv preprint arXiv:1206.4656 (2012 June 18)
7. Lander, E.S.: Initial impact of the sequencing of the human genome. Nature **470**(7333), 187–197 (2011)
8. Lander, E.S., Linton, L.M., Birren, B., Nusbaum, C., Zody, M.C., Baldwin, J., Devon, K., Dewar, K., Doyle, M., FitzHugh, W., Funke, R.: Initial sequencing and analysis of the human genome. Nature **409**(6822), 860–921 (2001)
9. Genomes Project Consortium: An integrated map of genetic variation from 1,092 human genomes. Nature **491**(7422), 56–65 (2012). Alberts, B.: Molecular Biology of the Cell. Garland Science (2017 Aug 7)
10. Strachan, T., Read, A.: Human Molecular Genetics. Garland Science, New York (2010)

11. De Klerk, E., AC't Hoen, P.: Alternative mRNA transcription, processing, and translation: insights from RNA sequencing. Trends Genet. **31**(3), 128–139 (2015)
12. Harrow, J., Frankish, A., Gonzalez, J.M., Tapanari, E., Diekhans, M., Kokocinski, F., Aken, B.L., Barrell, D., Zadissa, A., Searle, S., Barnes, I.: GENCODE: the reference human genome annotation for The ENCODE Project. Genome Res. **22**(9), 1760–1774 (2012)
13. Cartegni, L., Krainer, A.R.: Disruption of an SF2/ASF-dependent exonic splicing enhancer in SMN2 causes spinal muscular atrophy in the absence of SMN1. Nat. Genet. **30**(4), 377–384 (2002)
14. Hua, Y., Sahashi, K., Rigo, F., Hung, G., Horev, G., Bennett, C.F., Krainer, A.R.: Peripheral SMN restoration is essential for long-term rescue of a severe spinal muscular atrophy mouse model. Nature **478**(7367), 123–126 (2011)
15. Naryshkin, N.A., Weetall, M., Dakka, A., Narasimhan, J., Zhao, X., Feng, Z., Ling, K.K., Karp, G.M., Qi, H., Woll, M.G., Chen, G.: SMN2 splicing modifiers improve motor function and longevity in mice with spinal muscular atrophy. Science **345**(6197), 688–693 (2014)
16. Hanahan, D., Weinberg, R.A.: Hallmarks of cancer: the next generation. Cell. **144**(5), 646–674 (2011)
17. Rubin, M.A.: Make precision medicine work for cancer care: to get targeted treatments to more cancer patients pair genomic data with clinical data, and make the information widely accessible. Nature **520**(7547), 290–292 (2015)
18. Collins, F.S., Varmus, H.: A new initiative on precision medicine. N. Engl. J. Med. **372**(9), 793–795 (2015)
19. Crick, F.H., Barnett, L., Brenner, S., Watts-Tobin, R.J.: General nature of the genetic code for proteins. Nature **192**(4809), 1227–1232 (1961)
20. Moult, J., Hubbard, T., Fidelis, K., Pedersen, J.T.: Critical assessment of methods of protein structure prediction (CASP): round III. Proteins: Struct., Funct., Bioinf. **37**(S3), 2–6 (1999)
21. Lindblad-Toh, K., Garber, M., Zuk, O., Lin, M.F., Parker, B.J., Washietl, S., Kheradpour, P., Ernst, J., Jordan, G., Mauceli, E., Ward, L.D.: A high-resolution map of human evolutionary constraint using 29 mammals. Nature **478**(7370), 476 (2011)
22. Hindorff, L.A., Sethupathy, P., Junkins, H.A., Ramos, E.M., Mehta, J.P., Collins, F.S., Manolio, T.A.: Potential etiologic and functional implications of genome-wide association loci for human diseases and traits. Proc. Natl. Acad. Sci. **106**(23), 9362–9367 (2009)
23. Lunshof, J.E., Ball, M.P.: Our genomes today: time to be clear. Genome Med. **5**(6), 52 (2013)
24. Watson, J.D., Crick, F.H.: Molecular structure of nucleic acids. Nature **171**(4356), 737–738 (1953)
25. Weinstein, J.N., Collisson, E.A., Mills, G.B., Shaw, K.R., Ozenberger, B.A., Ellrott, K., Shmulevich, I., Sander, C., Stuart, J.M.: Cancer genome atlas research network. The cancer genome atlas pan-cancer analysis project. Nat. Genet. **45**(10), 1113–1120 (2013)
26. Gao, X., Starmer, J.: Human population structure detection via multilocus genotype clustering. BMC Genet. **8**(1), 34 (2007)
27. Laitman, Y., Feng, B.J., Zamir, I.M., Weitzel, J.N., Duncan, P., Port, D., Thirthagiri, E., Teo, S.H., Evans, G., Latif, A., Newman, W.G.: Haplotype analysis of the 185delAG BRCA1 mutation in ethnically diverse populations. Eur. J. Hum. Genet. **21**(2), 212–216 (2013)
28. Alexander, D.H., Novembre, J., Lange, K.: Fast model-based estimation of ancestry in unrelated individuals. Genome Res. **19**(9), 1655–1664 (2009)
29. Stein, L.D.: The case for cloud computing in genome informatics. Genome Biol. **11**(5), 207 (2010)
30. Reyes-Ortiz, J.L., Oneto, L., Anguita, D.: Big data analytics in the cloud: Spark on hadoop vs mpi/openmp on beowulf. Proc. Comput. Sci. **1**(53), 121–130 (2015)
31. Borthakur, D.: The hadoop distributed file system: architecture and design. Hadoop Proj. Website **11**(2007), 21 (2007)
32. Qiu, J., Ekanayake, J., Gunarathne, T., Choi, J.Y., Bae, S.H., Li, H., Zhang, B., Wu, T.L., Ruan, Y., Ekanayake, S., Hughes, A.: Hybrid cloud and cluster computing paradigms for life science applications. BMC Bioinform. **11**(12), S3 (2010)

33. Zou, Q., Li, X.B., Jiang, W.R., Lin, Z.Y., Li, G.L., Chen, K.: Survey of MapReduce frame operation in bioinformatics. Brief. Bioinform. **15**(4), 637–647 (2013)
34. Schatz, M.C.: CloudBurst: highly sensitive read mapping with MapReduce. Bioinformatics **25**(11), 1363–1369 (2009)
35. Jourdren, L., Bernard, M., Dillies, M.A., Le Crom, S.: Eoulsan: a cloud computing-based framework facilitating high throughput sequencing analyses. Bioinformatics **28**(11), 1542–1543 (2012)
36. Langmead, B., Schatz, M.C., Lin, J., Pop, M., Salzberg, S.L.: Searching for SNPs with cloud computing. Genome Biol. **10**(11), R134 (2009)
37. McKenna, A., Hanna, M., Banks, E., Sivachenko, A., Cibulskis, K., Kernytsky, A., Garimella, K., Altshuler, D., Gabriel, S., Daly, M., DePristo, M.A.: The Genome Analysis Toolkit: a MapReduce framework for analyzing next-generation DNA sequencing data. Genome Res. **20**(9), 1297–1303 (2010)
38. Guo, X., Meng, Y., Yu, N., Pan, Y.: Cloud computing for detecting high-order genome-wide epistatic interaction via dynamic clustering. BMC Bioinform. **15**(1), 102 (2014)
39. Huang, H., Tata, S., Prill, R.J.: BlueSNP: R package for highly scalable genome-wide association studies using Hadoop clusters. Bioinformatics **29**(1), 135–136 (2012)
40. Owen, S., Anil, R., Dunning, T., Friedman, E.: Mahout in Action. Greenwich, CT (2011)
41. Ko, K.D,, Kim, D., El-ghazawi, T., Morizono, H.: Predicting the severity of motor neuron disease progression using electronic health record data with a cloud computing Big Data approach. In: 2014 IEEE Conference on Computational Intelligence in Bioinformatics and Computational Biology, 2014 May 21, pp. 1–6. IEEE
42. Zaharia, M., Chowdhury, M., Das, T., Dave, A., Ma, J., McCauley, M., Franklin, M.J., Shenker, S., Stoica, I.: Resilient distributed datasets: a fault-tolerant abstraction for in-memory cluster computing. In: Proceedings of the 9th USENIX Conference on Networked Systems Design and Implementation, 2012 Apr 25, pp. 2–2. USENIX Association
43. Wiewiórka, M.S., Messina, A., Pacholewska, A., Maffioletti, S., Gawrysiak, P., Okoniewski, M.J.: SparkSeq: fast, scalable and cloud-ready tool for the interactive genomic data analysis with nucleotide precision. Bioinformatics **30**(18), 2652–2653 (2014)
44. Massie, M., Nothaft, F., Hartl, C., Kozanitis, C., Schumacher, A., Joseph, A.D., Patterson, D.A.: Adam: genomics formats and processing patterns for cloud scale computing. EECS Department, University of California, Berkeley, Tech. Rep. UCB/EECS-2013-207 (2013 Dec 15)
45. Paten, B., Diekhans, M., Druker, B.J., Friend, S., Guinney, J., Gassner, N., Guttman, M., James Kent, W., Mantey, P., Margolin, A.A., Massie, M.: The NIH BD2 K center for big data in translational genomics. J. Am. Med. Inform. Assoc. **22**(6), 1143–1147 (2015)
46. Leung, M.K., Delong, A., Alipanahi, B., Frey, B.J.: Machine learning in genomic medicine: a review of computational problems and data sets. Proc. IEEE **104**(1), 176–197 (2016)
47. O'Brien, A.R., Saunders, N.F., Guo, Y., Buske, F.A., Scott, R.J., Bauer, D.C.: VariantSpark: population scale clustering of genotype information. BMC Genom. **16**(1), 1052 (2015)
48. https://databricks.com/
49. https://spark.apache.org/
50. Adhikari, K., Fuentes-Guajardo, M., Quinto-Sánchez, M., Mendoza-Revilla, J., Chacón-Duque, J.C., Acuña-Alonzo, V., Jaramillo, C., Arias, W., Lozano, R.B., Pérez, G.M., Gómez-Valdés, J.: A genome-wide association scan implicates DCHS2, RUNX2, GLI3, PAX1 and EDAR in human facial variation. Nat. Commun. **7**, 11616 (2016)
51. Peterson, B.K., Weber, J.N., Kay, E.H., Fisher, H.S., Hoekstra, H.E.: Double digest RADseq: an inexpensive method for de novo SNP discovery and genotyping in model and non-model species. PLoS ONE **7**(5), e37135
52. Yang, J., Manolio, T.A., Pasquale, L.R., Boerwinkle, E., Caporaso, N., Cunningham, J.M., De Andrade, M., Feenstra, B., Feingold, E., Hayes, M.G., Hill, W.G.: Genome partitioning of genetic variation for complex traits using common SNPs. Nat. Genet. **43**(6), 519 (2011)

A. Shobana Devi is a research scholar in the Department of Information Technology, SRM Institute of Technology, Kattankulathur, Chennai, India under the supervision of Dr. G. Maragatham. Her primary research interests are Data Mining, Big Data especially with the application in the area of Medical Healthcare. She has published 5 research papers in reputed international Journal and Conferences.

G. Maragatham is working as an Associate Professor in the Department of Information Technology, SRM Institute of Technology, Kattankulathur, Chennai, India. Her primary research interests are in the areas of Data Mining, Big Data Analytics, Machine Learning and Bio-inspired Computing. Her current focus in the last few years is on the research issues in Data Mining application especially in Social Impacts related areas, E Governance and Healthcare. She has published more than 20 research papers in reputed international Journal and Conferences.

IoT-Based Framework for Automobile Theft Detection and Driver Identification

P. Chandra Shreyas, R. Roopalakshmi, Kaveri B. Kari, R. Pavan, P. Kirthy and P. N. Spoorthi

Abstract Recently, almost everyone in the world owns a vehicle. On the other hand, there is an effective increase in the automobile theft, which is becoming a major problem in the present traffic scenario. However, in the current scenario, there is a lack of integrated systems which can effectively track and monitor the driver using Global Positioning System (GPS), GSM and camera. To overcome these issues, an effective anti-theft tracking system is introduced in this paper, which makes use of GPS to collects the latitude and longitude location of the vehicle and also the camera to take the picture of the intruder for further analysis. The resultant information is sent to the server, and the server sends message about intruder of the vehicle to the owner using GSM module. The evaluated results of the experimental setup illustrate the better performance of the proposed framework in terms of accurate identification of intruder and the location of the vehicle, and thereby, this framework can be employed in real time to prevent automobiles thefts.

Keywords Intelligent transportation systems · RFID technology · Anti-theft tracking · GPS

1 Introduction

Nowadays though most of the drivers and passengers use smartphones, they are unaware of new technologies; hence, they are not making use of their smartphones for effective communication. On the other hand, there is an effective increase in the

---

P. Chandra Shreyas (✉) · R. Roopalakshmi · K. B. Kari · R. Pavan · P. Kirthy · P. N. Spoorthi
Alvas Institute of Engineering and Technology, Shobhavana Campus, Mijar, Moodbidri, Mangalore 574225, India
e-mail: chandrashreyasp@gmail.com

R. Roopalakshmi
e-mail: drroopalakshmir@gmail.com

K. B. Kari
e-mail: kaverib.kari@gmail.com

S. Smys et al. (eds.), *International Conference on Computer Networks and Communication Technologies*, Lecture Notes on Data Engineering and Communications Technologies 15, https://doi.org/10.1007/978-981-10-8681-6_56

automobile theft, which is found to be a major problem in the present traffic scenario. For example, as per the recent National Insurance Crime Bureau (NICB) report in 2015, more than 7.5 lakhs of vehicles are being stolen in India [1]. The main reasons for unawareness of owner about the vehicle theft are as follows: (1) due to longer distance (range), siren is not heard; (2) similar siren sounds of cars; (3) physically, alarms can be disabled on theft attempts and also (4) alarm sound can be mitigated in crowded areas. Further, in the current scenario, there is a lack of an integrated system, which can track and monitor the driver using Global Positioning System (GPS) and camera. Precisely, the Global System for Mobile communication (GSM) is the most popular and accepted standard for mobile phones in the world. Also in wireless data transmission GSM and SMS technology is a common feature with all mobile network service providers. Only an Integrated vehicle tracking system with GPS, GSM and camera can provide an effective as well as secured transportation system, which can the effectively track the intruder and the location of the vehicle.

2 Related Work

In the literature survey, recently Tian and Lauer [2] proposed a new tracking approach which utilizes the part-based trackers, based on correlation filters for tracking an automobile. In this method, both the number and the size of part filters are adapted to the current appearance of the object, to eliminate the influence of occluded parts. Due to a sophisticated design, this approach fails to detect automobile in case of shape change of an automobile. Khader [3] defined remote monitoring system, which includes vehicle diagnostics and geographical position, by using on-board microcomputer system for remote monitoring of vehicles. The on-board smart box is equipped with an integrated Global Positioning System (GPS) receiver, which manages the processes of local data acquisition and transmissions. However, this system fails to track the automobile, if it changes its acceleration. In 2012, Manh La et al. [4] proposed a system in which the Signal received from Xbee RF module can be used to control the fronts and rear vehicle wheels through motors. In [5], Siddiqui et al. proposed the framework that guides the driver to move in a specific direction to avoid accidents. However, it provides control over only in a lateral direction. Singh and Mhalan proposed an engine control unit (ECU) that is connected to the information security circuit board and sensors inside the vehicle bus. The bus communicates with other vehicles, road side transportation and mobile phones with wireless interfaces. The performance of this system is less, due to the poor timeliness and network delays which are needed for secure car communications. On the other hand, recently, radio frequency identification (RFID) [6] is used in intelligent computerized anti-theft system [ICAT]. However, keyless RFID cards can be easily stolen, which results in the loss of an automobile [7]. To summarize, in the existing literature, less focus is given towards acceleration change of automobiles, which are essential for their tracking. Further, the literature lacks the strategies for proper specification of the latitude and longitude

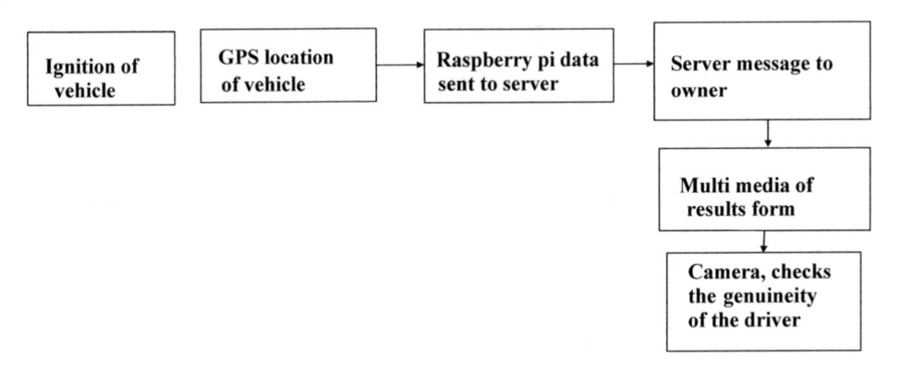

Fig. 1 Block diagram of the proposed automobile theft detection framework

area measurements. To solve these issues, an Android-based system is designed in this paper to track an automobile theft.

3 Proposed Framework

Figure 1 introduces the block diagram of the proposed framework, which can track and monitor the theft of the vehicle, so that and the owner can identify, the intruder of the vehicle. When the intruder starts the vehicle, GPS collects the changing latitude and longitude area of the vehicle and also the camera which is embedded in the steering of the car, captures the intruder's picture and thereby sends it to the Raspberry Pi device. Later on, the information is sent to the server, which collects all the data and by using the GSM module and it sends it to the owner of the vehicle. The owner evaluates the picture and in case of an intruder, the owner can turn off the vehicle.

4 Experimental Setup and Results

Figure 2 shows the experimental setup, which includes the components such as Battery, Raspberry Pi, current sensor, GSM, GPS and camera. Figure 3 shows the implementation of proposed system's experimental setup, including various components and their respective connections. Initially, the system turns on when the vehicle is ignited. After this step, the battery drives the Raspberry Pi device, and then all the other components incorporated into this Raspberry Pi device get activated and perform their respective functions. The Raspberry Pi device is programmed in such a way that all the components connected to it perform their task and send back the result to system. GPS collects the exact latitude and longitude location of the vehicle and sends it to Raspberry Pi then the information is sent from Raspberry Pi to the

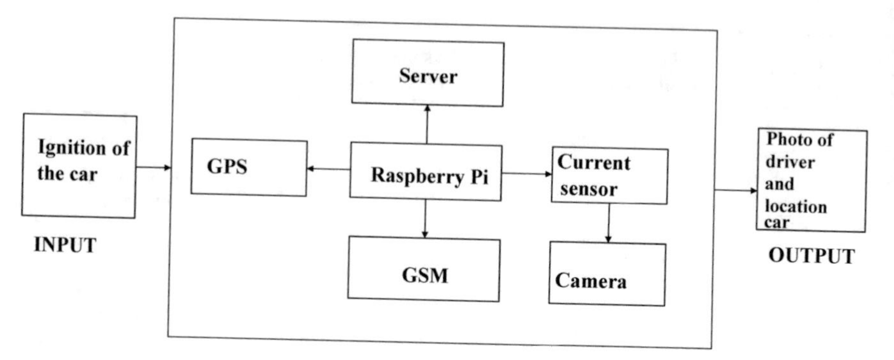

Fig. 2 Components used in the frameworks

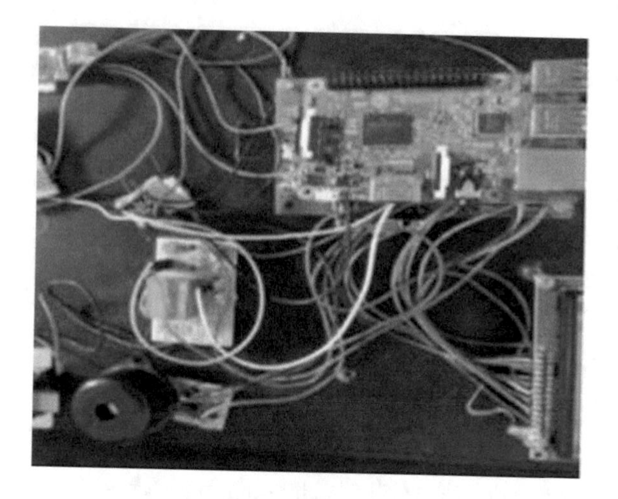

Fig. 3 Snapshot of experimental setup of the proposed framework

server. Later, by using this information, the GSM sends the message to the owner of the vehicle. The camera takes photos of the person who is driving the car and sends to Raspberry Pi; again, it gets stored in the server and sent to the Android app. All the transactions between the devices are stored in the server.

The owner can monitor the device through the Android application using the information stored in server. When the owner sends command to turn off the device, the current sensor holds the responsibility to turn the system off. Figure 4 shows the result of the proposed framework. In Fig. 4, the top portion shows the alert message sent to the owner of an automobile by using the GSM module. The middle portion represents the picture of a pilfer, and the bottom portion indicates the location of the automobile captured by the GPS. The various parameters used to locate the automobile are latitude and longitude. With this information, the owner will get the information of the pilfer as well as the location of the automobile through Google Earth.

Fig. 4 Snapshot of results of the proposed framework

5 Conclusion and Future Work

The proposed system precisely shows the location of the stolen car and the person who is driving the car. The proposed system is low cost, efficient, reliable and secure, which can be implemented for anti-theft of the vehicle. The information represented in the Android app may vary in accordance with the mobile phone of the owner and the owner can turn off the vehicle at any instance of time. The model can be implemented in bikes with adjustments made to sparkplug, battery and key. The system can be further improved with speed control mechanism, to stop the engine if the speed exceeds certain limits. Also, the system can be further being enhanced for providing parental guidelines such as to stop the vehicle, if it crosses a certain range of distance.

References

1. National Insurance Crime Bureau (NICB) report. https://www.nicb.org/ (2015)
2. Tian, W., Lauer, M.: Tracking vulnerable road users with severe occlusion by adaptive part filter modelling. In: Proceedings of 12th International Conference on Vehicular Electronics and Safety (ICVES), pp. 152–261, Vienna, Austria (2017)
3. Khader, O.B., Chandrasekhar, H., Nilayangode, K.: Password protected vehicle access system. Int. J. Innovative Sci. Modern Eng. (IJISME) **2**, 319–386 (2014)

4. Manh La, H., Lim, R.S., Du, J., Zhang, S., Yan, G., Sheng, W.: Real time vehicle theft identity and control system based on ARM 9. Int. J. Latest Trends Eng. Technol. (IJLTET) **2**(1), 240–245 (2013)
5. Siddiqui, N.A., Qayuml, M.A.: Control of autonomous cars for intelligent transportation system. In: Proceeding of International Conference on Informatics Electronics & Vision, pp. 7574–7580 (2014)
6. Al-Khedher, M.: Hybrid GPS-GSM localization of automobile tracking system. Int. J. Comput. Sci. Inf. Technol. (IJCSIT) **9** (2011)
7. Supriya, S., Bombale, U.L., Patil, T.B.: An intelligent vehicle control and monitoring using arm. Am. J. Appl. Sci. 709–716 (2012)

Mr. P. Chandra Shreyas is currently pursuing final year in Information Science & Technology in Alva's Institute of Engineering & Technology affiliated under VTU, Belagavi. He completed his schooling from Nisarga Vidyanikethana school, Kollegal and 2nd year PUC from Nisarga Independent PU College, Kollegal. He went to Japan under "Sakura Exchange Program" Organized by Japan Science and Technology, Japan, where he learnt about new technologies like Explosive Welding, Creation of metal alloy(Mg) & Shock waves technology.

Dr. R. Roopalakshmi is currently working as Professor in Information Science & Engineering (CSE) department of Alva's Institute of Engineering and Technology affiliated under VTU, Belgavi. She has started her Engineering career, Bachelor of Engineering (B.E.,) from A.M.S. College of Engineering, Periyar University and completed her M.Tech. (CSE) from P.E.S. Institute of Technology (PESIT)—Bangalore. Followed by her post graduation, she has pursued her Full-time Ph.D., in Information Technology Department of National Institute of Technology Karnataka (NITK)—Surathkal, which is one of the most prestigious and oldest NITs of India. She has published 5 journal articles in very high quality journals such as Elsevier Signal Processing, Springer Signal, Image and Video Processing and so on. She is technically sound enough for having her Paper presentations in more than 10 Tier-I International Conferences like SPCOM-2012 (which was held at IISc, Bangalore), IMSAA-2011 (held at IIIT-Bangalore), ANT-2011 (conducted at Niagara Falls, Canada) and SIRS-2014 (held at IIIT-Kerala). Henceforth, her research articles are widely available in digital libraries such as IEEE Xplore, ACM, Elsevier science direct and Springer. She proved to be a Best Researcher, by involving herself as Principal Investigator in Funded research projects such as DST-Women Scientist Research Grant of worth **Rs. 18.20 lakhs** from MHRD of India, New Delhi. Further, she is also honoured with various Awards such as SSLC—Salem District First Student Award, Rajya Puraskar Award from President of India for Bharat Scouts & Guides and Best Paper Award in International Conference and so on. She is also recipient of various state and national level scholarships such as National Merit scholarship, GATE scholarship and Rural Development scholarship. Currently, she is guiding 4 doctoral candidates at Visvesvaraya Technological University (VTU), Karnataka and also received guideship from Anna University, Chennai.

Ms. Kaveri B. Kari is currently working as Assistant Professor in Information Science & Engineering (ISE) department of Alva's Institute of Engineering and Technology affiliated under VTU, Belagavi. She has started her Engineering career, Bachelor of Engineering (B.E.,) from S G Balekundari Institute of Technology, Belgaum, VTU and completed her MTech. (CSE) from Alva's Institute of Technology, Moodbidri. She has published the paper on Detection of diabetic retinopathy using SVM in very high-quality journal publisher like Springer. Henceforth, her research article is available in digital library such as Springer.

Mr. R. Pavan perceived engineering from Alva's Institute of Engineering and Technology Moodbidre affiliated to VTU Belagavi. He Completed his Schooling from Ganodaya Convent & High School, Banglore. He Completed his 2nd year PUC from Soundarya Composite College.

Mr. P. Kirthy perceived engineering from Alva's Institute of Engineering and Technology Moodbidre affiliated to VTU Belagavi. He Completed his Schooling from Pt. Jawaharlal Nehru memorial high school. He Completed his 2nd year PUC from Vijay PU College.

P. N. Spoorthi perceived engineering from Alva's Institute of Engineering and Technology Moodbidre affiliated to VTU Belagavi. She was awarded Best Student during school and had participated in many inter-college debate competitions. She had won many prizes in sports.

IoT-Based Smart Food Storage Monitoring and Safety System

Saleem Ulla Shariff, M. G. Gurubasavanna and C. R. Byrareddy

Abstract It does not take much to see and experience the rapid advancement of technology in the world. We can observe personal technologies are getting smaller, smarter and overall more efficient but that is not to say that technological advancement is stopping at consumer level. People of this generation are busy in their day-to-day life as such that they do not get enough time to maintain the food storage facilities at home. Due to busy life schedule, they have to shop for the food commodities such as grains, etc. in bulk quantity to be maintained and used for over a period. We are proposing a system to monitor food grains and to maintain storage system at home. The proposed system is equipped with auto SMS and email alert system to alert the owner regarding the food storage level and the information related to the food spoilage. We come across local unscientific food storage systems in home-based kitchens. There is a chance for food grains getting spoilt due to moisture, humidity, temperature and various other factors. Hence, it becomes very important for us to monitor the food storage level and maintain it to lead a tension-free healthy life. For this purpose, we are going to use various sensors. For our design, an advanced board such as Renesas GR Peach has been used as a central processing unit with different sensors embedded to create a smart home food and grains storage maintenance and monitoring system.

S. U. Shariff (✉)
Department of Electronics and Communication, University Visvesvaraya College of Engineering (UVCE) (Under Bangalore University), KR Circle, Bengaluru 560001, India
e-mail: saleem_shariff@yahoo.co.in

M. G. Gurubasavanna
Department of Electronics and Communication Engineering, Government Sri Krishnarajendra Silver Jubilee Technological Institute, KR Circle, Bengaluru 560001, India
e-mail: gurumg2005@gmail.com

C. R. Byrareddy
Department of Electronics and Communication Engineering, Bangalore Institute of Technology, Bengaluru 560004, India
e-mail: byrareddycr@yahoo.co.in

S. Smys et al. (eds.), *International Conference on Computer Networks and Communication Technologies*, Lecture Notes on Data Engineering and Communications Technologies 15, https://doi.org/10.1007/978-981-10-8681-6_57

Keywords Smart home · Food monitor · Storage · GR peach
Personal technologies · SMS · Email

1 Introduction

Smart containers are in use to monitor and maintain the food grain storage levels from earlier days. It is often observed that homemakers keep different types of utensils for storing various food commodities by printing their names or using transparent glass-based utensils. The same concepts have been used here, by connecting various sensors inside the utensils which are used for storing the food commodities such as grains, etc. We are going to set the threshold. If it is observed that the food storage level goes below that threshold level then an auto alert system has to be activated. The owner or homemaker or both should be alerted with SMS as well as email alert. An advanced board such as The GR Peach board by Renesas can be utilized as a central processing unit by connecting the GR Peach board with the available various low-cost sensors to create a smart food and grains storage monitoring and maintenance system. In the proposed system, we are going to monitor various different parameters with the help of the moisture level, humidity and temperature level conditions inside the kitchen and storage area and also the same will be tracked. An alert should be triggered if unnecessary changes in the conditions are detected by the sensors which can damage the food grains. Implementation of the alert system is done via SMS; a GSM module can be interfaced with the GR Peach board for the same. As we know that the GR Peach is an advanced board; we can add extra features such as safety system as well into the design. We can use various low-cost safety sensors such as fire detection sensor, LPG gas sensor, etc. which are available easily at an affordable cost in the market. All these sensors should be interfaced by properly programming the GR Peach using the correct pin details to create a safety alert system.

For designing the proposed system, we can choose three utensils, in which we can use different sensors such as using LDR (light dependent resistor) in Utensil 1, IR infrared (obstacle sensor) sensor in Utensil 2, Ultrasonic sensor in Utensil 4 and moisture sensor in Utensil 3. The concept behind using these sensors is very simple and straightforward; to explain it, let us take up the case of LDR. The functionality of LDR is that LDR usually detects light and its resistance will vary depending on the intensity of the light. For instance, if we keep the LDR at a fixed height in Utensil 1, and the grains are filled inside the utensil and once if the grains are full or above the threshold value then LDR is not fed with enough light hence its resistance will be quite high, similarly if the grains are filled below the threshold value inside the utensil then LDR will get sufficient light automatically and hence its resistance will gradually decrease depending on the amount of light intensity received. We can design an alert system using this approach such that whenever the light with suitable intensity falls on LDR, it means the grains are below the threshold level otherwise it is considered as above the threshold level.

Similar and simple idea behind using the IR sensor is that IR sensors are used to detect the obstacle coming in front of them; hence if the IR sensors are placed at a certain threshold height, then they will sense the obstacles presence easily. Hence here in Utensil 2, if the food grain, for instance rice is filled above the IR sensor level, then it will detect it as an obstacle; hence, we can interpret it as rice is above the threshold level; the utensil is filled full with rice, and similarly, the IR sensor will not be detecting any obstacle if the rice is below the threshold level. The idea behind using the ultrasonic sensor is same as IR. Ultrasonic sensor can sense the obstacle and the distance can be calculated. Hence by using the ultrasonic sensor, the threshold level height can be adjusted easily by programming. The concept of using the moisture sensor is to design a system which can detect the liquid and moisture content. The sensor can be placed at a certain height and when the sensor detects the moisture then it means the liquid is above the threshold level else it should be interpreted as the liquid is below the threshold level. As discussed earlier, the ultrasonic sensor can measure the distance and hence exact height up to which the utensil is filled can be known and tracked easily. Compared to other sensors, the ultrasonic sensor is more effective but comes with an increase in cost. With these simple, low-cost sensors, we can design easily an effective smart food storage maintenance and monitoring system.

2 Literature Survey

A. *Motivation for the Project*

India has many cosmopolitan cities with IT hub Bangalore one among the metropolitan cities having the high density of population. We can observe that people of today's generation are lazy enough due to technical advancements and are very much dependent on the technology. We can blame this dependency on the busy lifestyle. It is often observed that people shop and store the food grains in bulk quantity to be maintained and used over a particular period. In this paper, we are proposing a system which is smart and low cost to maintain and monitor food grains storage at home with less human intervention. The proposed design which has an auto SMS and email alert system equipped to alert the owner/homemaker regarding the storage level and also on the status if the food is on verge of decay or getting spoilt. If we are to conduct a survey, we can come across many local food storage systems in home-based kitchens. Usually, they are designed in unscientific way with fewer applications. Hence, there will be a chance for food grains getting spoilt due to moisture, humidity, temperature and various other factors. The time-saving nature of the people by storing the food in bulk quantity in unscientific manner sometimes leads to uncalled problems such as sudden shortage of the food grains due to improper tracking of food storage level, negligence, spoilage, etc. It is very important for us to monitor the food quality by properly storing and maintaining it to lead a tension free healthy life. Hence to maintain and monitor the food grain storage levels, we are designing a system by choosing the low cost but effective sensors with GR Peach board. Renesas GR Peach board will be used as a central processing unit along with the various other easily available sensors to create a smart home food and grains storage maintenance and monitoring system.

B. *Survey for the Literature*

The GR Peach board is a new board which has been developed by Renesas Company as part of their Gadget Renesas series. The details have been collected from the various websites and the same has been used and mentioned in the references section.

The Internet of Things: A Survey authored by Atzori et al. [1] has discussed how the Internet of things concepts is evolving. The inspiration for the proposed project is the concept of IPv6 which powers the idea to have a unique identification to be kept or used for every device. IPv6 routing protocol for low-power and lossy networks [2] has been explained in the paper [2] by Winter et al. Convergence of MANET and WSN in IoT Urban Scenarios [3] by Bellavista et al. gives an idea how the concepts can be utilized effectively in IoT. Smart Cities and the Future Internet: Towards Cooperation Frameworks for Open Innovation [4] by Schaffers explains how the smart cities can be designed using the concepts evolving from IoT and how future of innovation will be in IoT field. The concept of IoT has changed how the design of our future smart cities would be. The paper explaining the Internet of things for smart cities [5] by Zanella et al. shows case how the technology can make human civilization smarter. With the concept of smart cities, the idea to have the better and effective healthcare system arises; an effective approach has been explained for healthcare applications by giving a solution based on the Internet of things [6] by Bui and Zorzi.

Secure communication is very important for achieving smart IoT concept; the paper on Secure Communication for Smart IoT Objects: Protocol Stacks, Use Cases and Practical [7] by Bonetto et al. explains how the secure communication can be established. The concept of smart food monitoring has been designed using the proposed approach whose details are available in the video uploaded in YouTube [8] and the videos on smart room automation [9] by Shariff.

The details about the GR Peach are obtained from Renesas website [10]. The website designed and developed for the students to create awareness about project ideas [11] for helping and sharing ideas such as helpshareideas.com and other websites are the basis for the idea for this project. The project details uploaded in toolcloud.renesas.com [12] explains the details regarding how a smart food monitoring device can be developed using GR Peach. The idea has been taken from the [13] GR Peach India Design contest. The ideas for a smart container has been provided in [14] and explained briefly regarding smart food tracking [15, 16] by Srivastava et al. Next sections will explain briefly about the components which are used in the project.

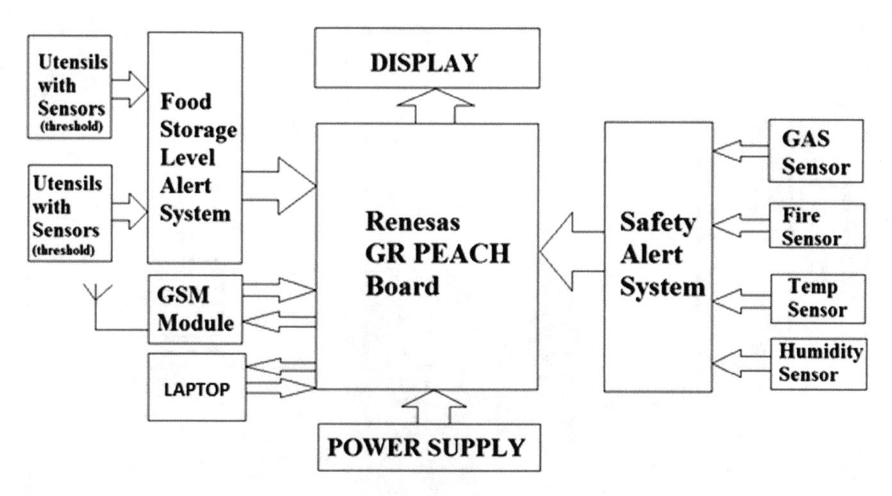

Fig. 1 Block diagram of the proposed system

3 Block Diagram/Schematic of the Proposed System

The proposed block diagram of smart food monitoring and safety alert system is given in Fig. 1. We have built three sections like food storage and monitoring, safety and temperature monitoring system with alert system. The central processing units operations will be handled by the GR Peach. We have tried to use low-cost sensors such as LDR, IR and ultrasonic sensors. The safety system is operated by monitoring different parameters such as gas, fire, smoke, temperature and humidity. Alert SMS can be triggered via GSM module or through any bulk SMS operator. The schematic diagram of the proposed system is shown below in Fig. 2. All the components are interfaced with GR Peach using the pin details mentioned below.

4 Algorithm of the Proposed System

The proposed algorithm of smart food monitoring and safety alert system is given in Table 1; GR Peach will be continuously monitoring for the changes in the surrounding and trigger the alarm whenever required.

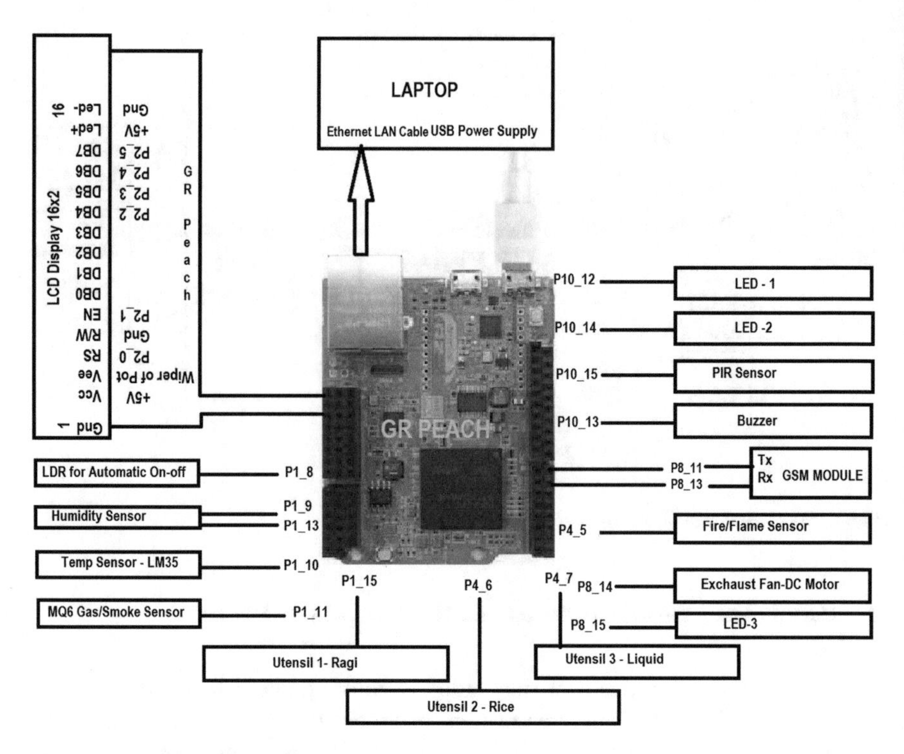

Fig. 2 Schematic of the proposed system

Table 1 Algorithm of the proposed system

Input: Level of Food Storage (grains in containers) and temperature humidity conditions, Smoke or Fire alert
Output: Alert SMS to the owner.
1. GR Peach Continuously receives the data related to food storage level, temperature, humidity and smoke & fire related activities.
2. Comparison of the received data with the stored data from the database.
3. An Alert will be issued depending on the necessities.
4. Alert SMS is triggered to the owner
5. Buzzer will trigger the alarm initiated by GR Peach.
6. The Process repeats and GR Peach monitors continuously.

5 Result and Performance Analysis

After the components have been connected, the results have been studied, which will be presented in this section.

A. *Utensils with Food Grains*

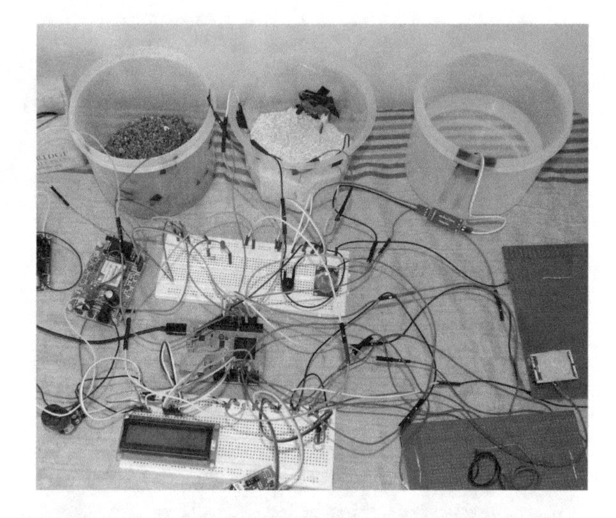

Fig. 3 Sample image of the devices connected with GR Peach

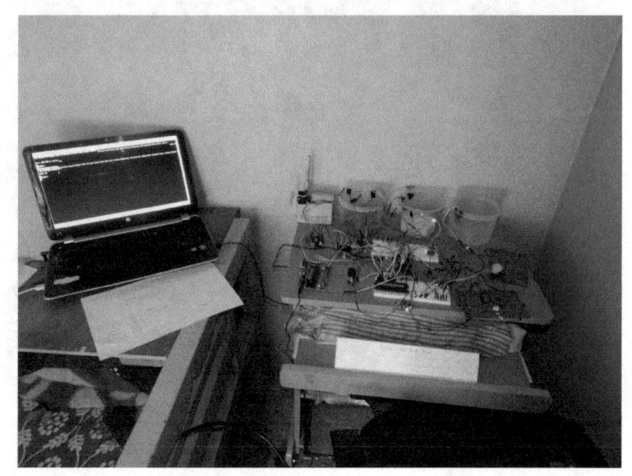

Fig. 4 Full setup along with components turned on

The utensils are filled with the grains such as rice rage, etc. Below picture in Fig. 3 depicts the same (Fig. 4 shows the full setup along with all components turned on).

B. *Results*

The prototype of the proposed system has been implemented and the results have been discussed in this section.

Terminal Application Screenshots: We have used Tera Term [17] terminal application software. The screenshots taken during the demonstration of the model explains the system behaviour. Screenshots shows how GR Peach was triggering the messages to the PC via terminal applications software [17]. Figures 5, 6, 7 and 8 depict the same.

Figure 6 shows how LDR sensor is monitoring the food storage level. When the GR Peach detects the ragi (millet) level below the threshold in Utensil 1 then the

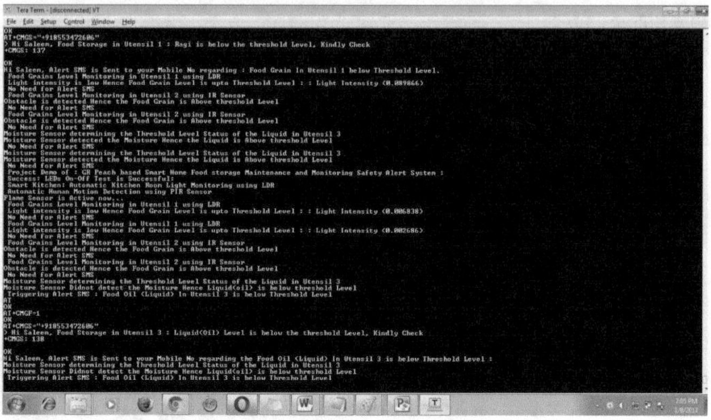

Fig. 5 Terminal application results

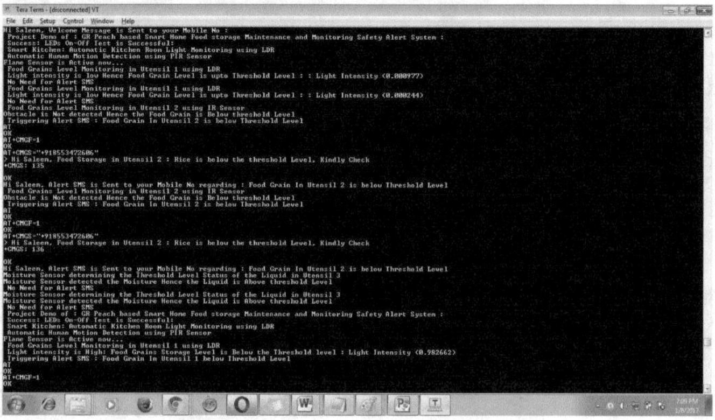

Fig. 6 Terminal application results—LDR sensor

alert gets triggered. Similarly, Fig. 7 shows how IR Sensor or ultrasonic sensor is monitoring the food storage level in Utensil 2. When the GR Peach detects the rice level below the threshold in Utensil 2, then the alert gets triggered.

Figure 8 depicts how PIR Sensor is monitoring the Human motion for room automation. Similarly, Fig. 9 shows how smoke sensors like MQ6 Sensor are monitoring the LPG or other gas leakage with smoke detection. When the GR Peach detects the gases, then the alert gets triggered.

Figure 10 shows how GSM is used by GR Peach for triggering the SMS by using AT commands for monitoring the food storage level. Similarly, Fig. 11 explains how alert was raised while monitoring the food storage level in Utensil 1. When the GR Peach detected the ragi millet level below the threshold in Utensil 1, then the alert got triggered.

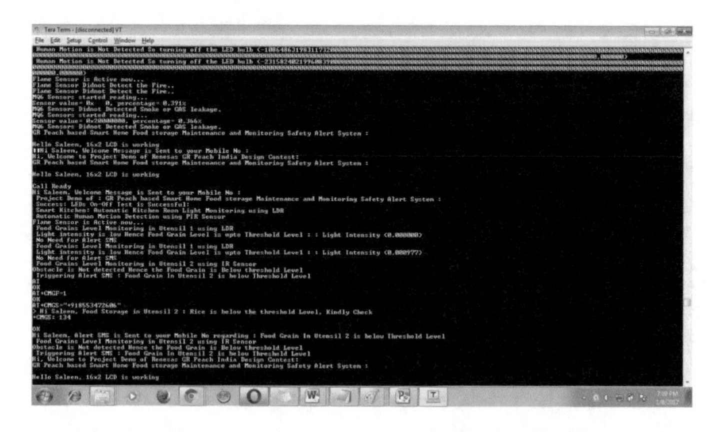

Fig. 7 Terminal application results—IR sensor

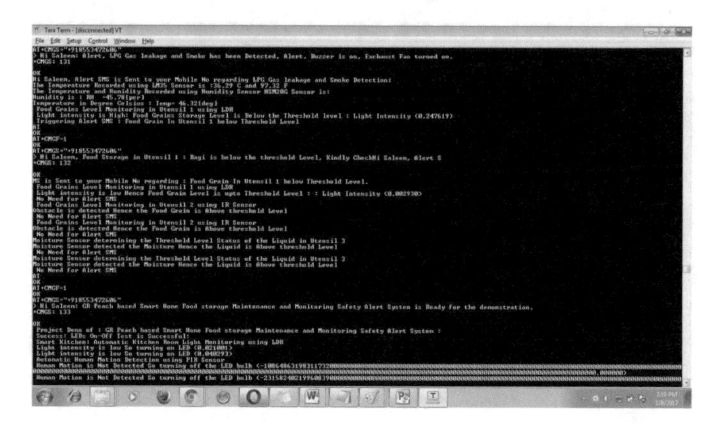

Fig. 8 Terminal application results—PIR sensor

Figure 12 shows how alert was raised while monitoring the food storage level in Utensil 2. When the GR Peach detected the Rice level below the threshold in Utensil 2 then the alert got triggered. Figure 13 explains room automation.

Figure 14 shows how GSM is being used by GR Peach for triggering the SMS by using AT commands for monitoring the safety in the surrounding. Also, Fig. 14 explains how alert was raised while monitoring the smoke and gas leakage.

Alert [18, 19] SMS screenshots taken from the mobile when alert triggered SMS's each time explains how GR Peach has been interfaced with GSM module for triggering the alert to the registered mobile number. The SMS received screenshots [18] taken from the mobile sent by GSM module, which were triggered by the GR Peach for alerting the owner are shown in Figs. 15, 16 and 17.

```
Tera Term - [disconnected] VT
File Edit Setup Control Window Help
 Success: LEDs On-Off Test is Successful:
 Smart Kitchen: Automatic Kitchen Room Light Monitoring using LDR
 Light intensity is low So turning on LED (0.001954)
Hi, Welcome to Project Demo of Renesas GR Peach India Design Contest:
GR Peach based Smart Home Food storage Maintenance and Monitoring Safety Alert System :

Hello Saleem, 16x2 LCD is working
AT
OK
AT+CMGF=1
OK
AT+CMGS="+918553472606"
> Hi, Welcome to GR Peach Design Contest Project Demo: GR Peach based Smart Home Food storage Maintenance and Monitoring Safety Alert SystemHi Saleem, Welcome Message i
s Sent to your Mobile No :

+CMGS: 128

OK
AT+CMGS="+918553472606"
> Hi Saleem: GR Peach based Smart Home Food storage Maintenance and Monitoring Safety Alert System is Ready for the demonstration. Project Demo of : GR Peach based Smar
t Home Food storage Maintenance and Monitoring Safety Alert System :
 Success: LEDs On-Off Test is Successful:
 Smart Kitchen: Automatic Kitchen Room Light Monitoring using LDR

+CMGS: 129

OK
 Light intensity is low So turning on LED (0.002686)
 Light intensity is low So turning on LED (0.021978)
 Automatic Human Motion Detection using PIR Sensor
Human Motion is Detected So turning on LED  bulb(-5306.046875)
Human Motion is Detected So turning on LED  bulb(-62665822848446419000[illegible].000000)
Flame Sensor is Active now...
Flame Sensor Didnot Detect the Fire..
Flame Sensor Didnot Detect the Fire..
MQ6 Sensors started reading...
Sensor value= 0x20000000, percentage= 16.996%
MQ6 Sensors Detected Smoke or GAS leakage, Buzzer is on...
AT
OK
AT+CMGF=1
OK
AT+CMGS="+918553472606"
> Hi Saleem: Alert, LPG Gas leakage and Smoke has been Detected, Alert. Buzzer is on. Exhaust Fan turned on.
+CMGS: 130

OK
Hi Saleem, Alert SMS is Sent to your Mobile No regarding LPG Gas leakage and Smoke Detection:
MQ6 Sensors started reading...
Sensor value= 0x20000000, percentage= 17.436%
MQ6 Sensors Detected Smoke or GAS leakage, Buzzer is on...
AT
OK
AT+CMGF=1
OK
```

Fig. 9 Terminal application results—MQ6 sensor

```
Tera Term - [disconnected] VT
File Edit Setup Control Window Help
OK
Hi Saleem, Alert SMS is Sent to your Mobile No regarding : Food Grain In Utensil 1 below Threshold Level.
 Food Grains Level Monitoring in Utensil 2 using IR Sensor
Obstacle is Not detected Hence the Food Grain is Below threshold Level
 Triggering Alert SMS : Food Grain In Utensil 2 is below Threshold Level
AT
OK
AT+CMGF=1
OK
AT+CMGS="+918553472606"
> Hi Saleem, Food Storage in Utensil 2 : Rice is below the threshold Level, Kindly CheckHi Saleem, Alert SMS is Sent to your Mobile No regarding : Food Grain In Utensil
 2 is below Threshold Level
 Food Grains Level Monitoring in Utensil 2 using IR S
+CMGS: 124

OK
ensor
Obstacle is Not detected Hence the Food Grain is Below threshold Level
 Triggering Alert SMS : Food Grain In Utensil 2 is below Threshold Level
AT
OK
AT+CMGF=1
OK
AT+CMGS="+918553472606"
> Hi Saleem, Food Storage in Utensil 2 : Rice is below the threshold Level, Kindly CheckHi Saleem, Alert SMS is Sent to your Mobile No regarding : Food Grain In Utensil
 2 is below Threshold Level

+CMGS: 125

OK
Moisture Sensor determining the Threshold Level Status of the Liquid in Utensil 3
Moisture Sensor Didnot detect the Moisture Hence Liquid(oil) is below threshold Level
 Triggering Alert SMS : Food Oil (Liquid) In Utensil 3 is below Threshold Level
AT
OK
AT+CMGF=1
OK
AT+CMGS="+918553472606"
> Hi Saleem, Food Storage in Utensil 3 : Liquid(Oil) Level is below the threshold Level, Kindly Check
+CMGS: 126

OK
Hi Saleem, Alert SMS is Sent to your Mobile No regarding the Food Oil (Liquid) In Utensil 3 is below Threshold Level :
Moisture Sensor determining the Threshold Level Status of the Liquid in Utensil 3
Moisture Sensor detected the Moisture Hence the Liquid is Above threshold Level
 No Need for Alert SMS
AT
OK
AT+CMGF=1
OK
AT+CMGS="+918553472606"
> Hi Saleem: GR Peach based Smart Home Food storage Maintenance and Monitoring Safety Alert System is Ready for the demonstration.
+CMGS: 127

OK
```

Fig. 10 Terminal application results—GSM message triggering

6 Enhancement to the Proposed Project

1. The interfacing of the GR Peach with the laptop with Ubuntu operating system can be easily done. Audio voice acknowledgement can be developed.
2. Whenever an alert SMS gets triggered or storage level goes below the threshold value, then an audio alert can be issued.
3. A text file can store all of the alert messages coming from the GR Peach which can be accessed.
4. LAMP server can be installed on the laptop, and a simple website can be designed, which can display the smart containers' status for food storage maintenance and monitoring from remote areas using Internet.

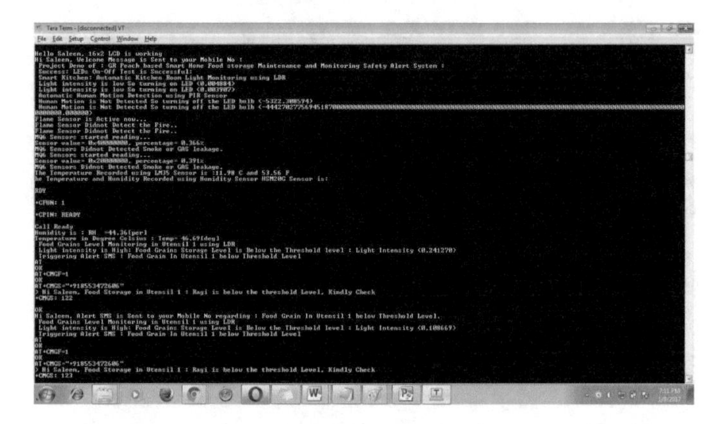

Fig. 11 Terminal application results—GSM Utensil 1 Alert

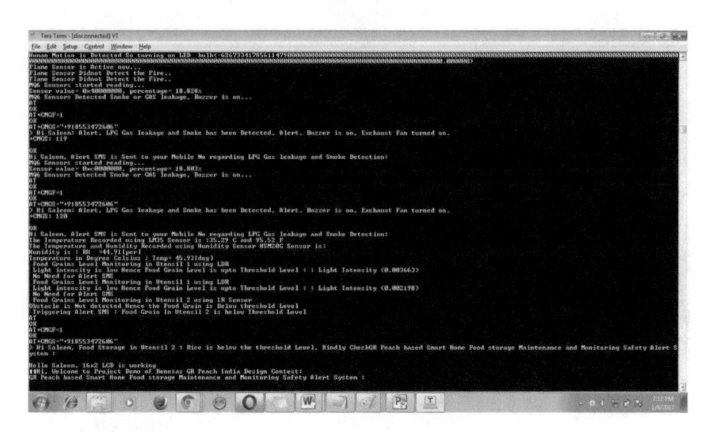

Fig. 12 Terminal application results—GSM Utensil 2

7 Applications of Proposed Project

1. Project can be used in food storage godowns as smart containers.
2. It can be used in restaurants and hotel kitchen rooms.
3. In homes to create a smart kitchen.
4. In rural areas, it can be used to store the food crops and monitor them.
5. As the project is low cost, it is used for the purpose of food storage maintenance and monitoring as well as for safety purpose.
6. The system can be used in hotels or restaurants food storage facilities.

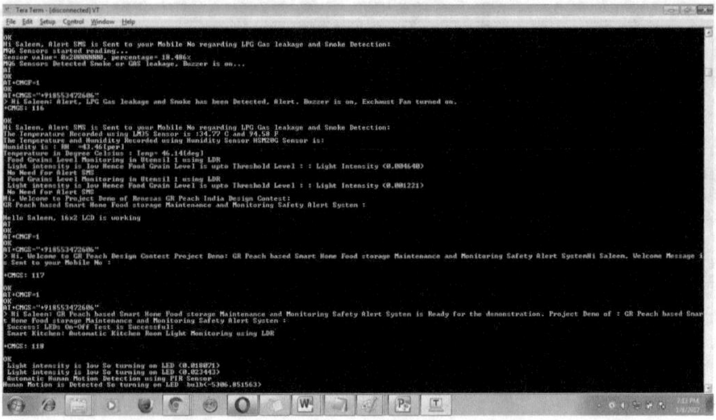

Fig. 13 Terminal application results—room automation

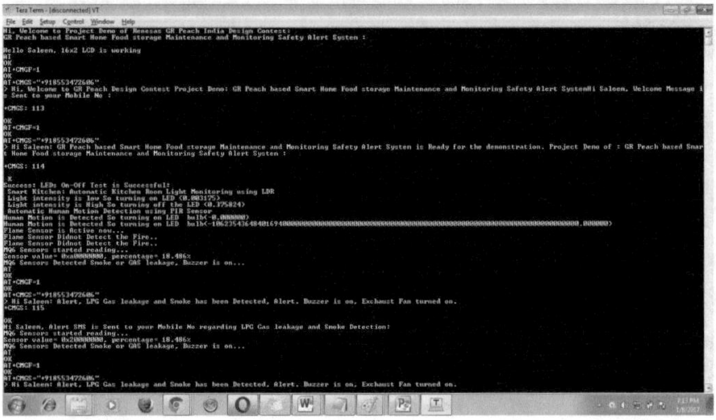

Fig. 14 Terminal application results—smoke alert

8 Conclusion

As GR Peach is an advanced board, the Internet of things (IoT) is allowing for a veritable network created not from computers as we know them, but of things, like containers and equipment. These things are embedded with an array of sensors, and tracking technology such as GPS along with RFID chips allows manufacturers, carriers and shippers to easily monitor temperature, vibration, losses and location of their cargo more accurately.

The proposed GR Peach-based smart home food storage monitoring maintenance and safety alert system has been designed with the current latest trends in mind. The proposed system will be of utmost importance for day to day life of the busy working-class families who depend on the accurate data for their monthly home budget preparation. The proposed system has been designed based on the principle of multilayer security. We have demonstrated successfully how smoke and fire sensor

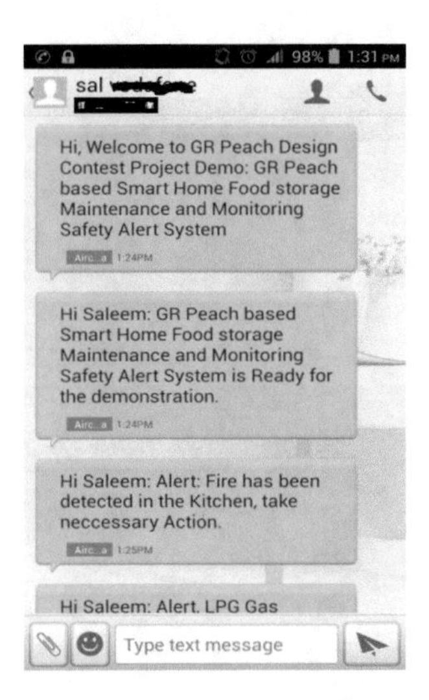

Fig. 15 SMS alert delivered to mobile

Fig. 16 SMS alert delivered to mobile

can be used integrated with temperature and humidity sensor for monitoring and developing safety alert feature, and also it is a low-cost easily affordable device. The proposed system not only works well as a food level monitoring system but also as a safety device in kitchens and food storage godowns.

Acknowledgements The authors are thankful to their friends and family for their support, and also thankful to Renesas India GR Peach Design contest for providing us with a free GR Peach Board for developing the system.

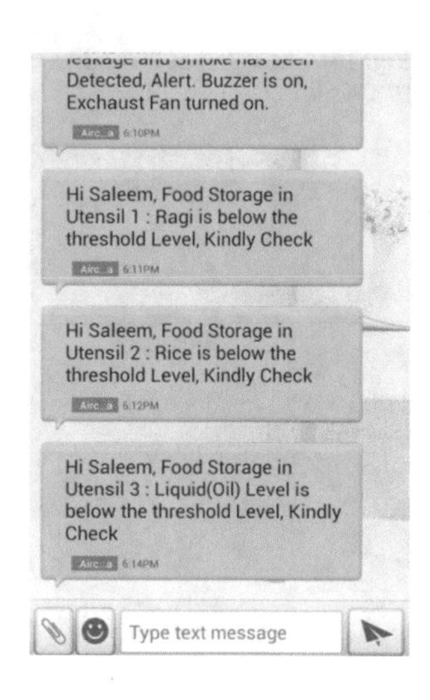

Fig. 17 SMS alert delivered to mobile

References

1. Atzori, L., Iera, A., Morabito, G.: The internet of things: a survey. Comput. Netw. **54**(15), 2787–2805 (2010)
2. Winter, T., Thubert, P., Brandt, A., Hui, J., Kelsey, R., Pister, K., Struik, R., Vasseur, J.P., Alexander, R.: RPL: IPv6 routing protocol for low-power and lossy networks, RFC6550, s.l.: IETF Mar. 2012 [online]. Available: http://tools.ietf.org/html/rfc6
3. Bellavista, P., Cardone, G., Corradi, A., Foschini, L.: Convergence of MANET and WSN in IoT urban scenarios. IEEE Sens. J. **13**(10), 3558–3567 (2013)
4. Schaffers, H., Komninos, N., Pallot, N., Trousse, B., Nilsson, M., Oliveira, A.: Smart cities and the future internet: towards cooperation frameworks for open innovation. In: The Future Internet. Lecture Notes Computer Science, vol. 6656, pp. 431–446 (2011)
5. Zanella, A., Bui, N., Castellani, A., Vangelista, L., Zorzi, M.: Internet of things for smart cities. IEEE Internet Things J. 22–32
6. Bui, N., Zorzi, M.: Health care applications: a solution based on the Internet of Things. In: Proceedings of ISABEL, pp. 1–5, Barcelona, Spain, Oct 2011
7. Bonetto, R., Bui, N., Lakkundi, V., Olivereau, A., Serbanati, A., Rossi, M.: Secure communication for smart IoT objects: protocol stacks, use cases and practical examples. In: Proceedings of IEEE IoT-SoS, pp. 1–7, San Francisco, CA, USA (2012)
8. GR Peach based Smart Home Food Storage Maintenance & Monitoring Safety Alert System Demo. Link: https://www.youtube.com/watch?v=zvPenKl_8-g
9. Renesas GR Peach based Smart Automatic Room light Monitoring using LDR. Link: https://www.youtube.com/watch?v=9dBs1jD8MGk
10. GR Peach details from Renesas website http://gadget.renesas.com/en/event/2016/jul/GR_Peach_Contest.html
11. Help share ideas website http://helpshareideas.com
12. Project details uploaded on to the Renesas cloud https://tool-cloud.renesas.com/en/atelier/detail.php?id=95
13. GR Peach design contest web details. http://gadget.renesas.com/en/event/2016/jul/GR_Peach_Contest.html
14. https://blog.orbcomm.com/smart-containers-are-just-the-start-how-connected-assets-will-drive-the-digital-supply-chain-revolution/, https://blog.orbcomm.com/smart-containers-are-just-the-start-how-connected-assets-will-drive-the-digital-supply-chain-revolution/

15. Srivastava, A., Paliwal, K., Reddy, S.R.N.: Smart food tracker—smart monitoring system for food safety. In: National Conference on Product Design (NCPD 2016), July 2016
16. Srivastava, A., Gulati, A.: iTrack: IoT framework for smart food monitoring system. Int. J. Comput. Appl. **148**(12), 0975–8887 (2016)
17. Tera-Term Terminal Application Software, https://ttssh2.osdn.jp/index.html.en
18. Shariff, S.U., Hussain, M., Shariff, M.F.: Smart unusual event detection using low resolution camera for enhanced security, 2017 International Conference on Innovations in Information, Embedded and Communication Systems (ICIIECS), (2017)
19. Shariff, S.U., Swamy, J.N., Seshachalam, D.: Beaglebone black based e-system and advertisement revenue hike scheme for Bangalore city public transportation system, 2016 2nd International Conference on Applied and Theoretical Computing and Communication Technology (iCATccT), (2016)
20. How to Mechatronics web details. www.howtomechatronics.com
21. tool-cloud.renesas.com

Mr. Saleem Ulla Shariff was graduated from Bangalore Institute of Technology Bangalore (under Visvesvaraya Technological University Belgaum Karnataka India in 2011) with B.E. in E&C. He has worked for IBM India Pvt. Ltd. as a Software Engineer prior to joining for Master of Engineering (ME) in Electronics and Communication in University Visvesvaraya College of Engineering (UVCE) Bangalore (under Bangalore University). His area of research interest includes microwave communication, antennas and wireless communication and embedded system design with a keen interest in processors and microcontrollers like 8051, ARM microcontrollers, Arduino, Beagle Bone Black and Raspberry Pi.

Mr. M. G. Gurubasavanna has completed his B.E. in Electronics and Communication from Gulbarga University and M. Tech in Applied Electronics from Dr. M.G.R Deemed University and Research Centre, Chennai. He is a research scholar in Electronics Engineering at Visvesvaraya Technological University, Belgaum. He has more than 15 years of teaching experience and is presently working as an Assistant Professor in Government Sri Krishnarajendra Silver Jubilee Technological Institute, Bangalore, Karnataka, India. His areas of research interest include wireless communication, sensor networks and antenna system design with a keen interest in processors like Arduino, Beagle Bone Black and Raspberry Pi.

Dr. C. R. Byrareddy graduated from Bangalore University with B.E. in Instrumentation Technology and the M.E. in Electronics in 1990 and 1999, respectively. He is currently Professor in Department of Electronics and Communication Engineering, Bangalore Institute of Technology, Bangalore. He got his Ph.D. from SV University of Engineering College, Tirupati. His area of research interest is microwave communication, antennas and wireless communication, with a keen interest includes analysis and design of patch antenna for wireless communication. He has published more than 20 papers in national/international journals. He has also presented paper at IconSpace2011 in Malaysia.

Implementation of Haar Cascade Classifier for Vehicle Security System Based on Face Authentication Using Wireless Networks

P. B. Pankajavalli, V. Vignesh and G. S. Karthick

Abstract This research work presents a vehicle security system for safeguarding the vehicle from theft issues under the architectural design of capturing and comparing the vehicle user's face. Since the face of human beings are unique and has different biometric characteristics which are complex to make fraudulent activities. Authenticating the vehicle users with face recognition mechanism is highly secured than token-based and knowledge-based security mechanisms. This research ultimately models and classifies the vehicle users into authorized and unauthorized users. Initially, an experimental prototype for vehicle security system is developed, and the application of image processing algorithms is incorporated into the model. The system uses Haar feature-based cascade classifier and AdaBoost method which is a machine learning algorithm used for detecting the authorized user's face effectively. The algorithm is trained initially with appropriate amount of positive and negative images, and the feature gets extracted. When the person tries to access the vehicle, the experimental system captures the image of the person and makes comparison with extracted features to identify the authorized user. Finally, the results obtained from the prototype system are satisfied and beneficial against the issue of vehicle theft.

Keywords Haar cascade · Wireless network · Face detection · GSM · OpenCV

P. B. Pankajavalli (✉) · V. Vignesh · G. S. Karthick
Department of Computer Science, Bharathiar University, Coimbatore, Tamil Nadu, India
e-mail: pankajavalli@buc.edu.in

V. Vignesh
e-mail: vigneshveeraperumal@gmail.com

G. S. Karthick
e-mail: karthickgs@outlook.com

S. Smys et al. (eds.), *International Conference on Computer Networks and Communication Technologies*, Lecture Notes on Data Engineering and Communications Technologies 15, https://doi.org/10.1007/978-981-10-8681-6_58

1 Introduction

Nowadays, vehicle thefts are increasing rapidly across the globe due to less attraction of researchers toward finding an anti-theft system with high security constraints. According to the investigation reports provided by various crime bureaus, in the United States, 721,053 of vehicles had stolen since 2012 [1] which caused loss of eight billion US dollars and this acquired the importance of designing an anti-theft system. Some of the widely used basic vehicle security methods are locking system, beepers, fire alarm, and password security systems which are expensive. These methods have some limitations: distance coverage, disabling of alarms on theft, and alleviation of alarm sound in crowded areas.

At present, researchers are intended to focus on developing a smart anti-theft system with high security constraints. In recent years, the person identification techniques are used to avoid vehicle thefts which include knowledge-based, token-based, and biometric-based technique. The knowledge-based technique uses the password or Personal Identification Number (PIN) for authenticating the vehicle users whereas token-based technique uses evidences for authenticating users. But the biometric-based technique uses biometric modalities for authenticating the vehicle users. The biometric-based technique is more efficient when compared to other two techniques. The biometric-based security system acquired importance due to identity theft is low, imitating biometric modalities is impossible and recollecting authenticating information is not required. The designing of anti-theft systems are growing parallel to the security breaches, in order to compete with vehicle theft issue, it is essential to design a mechanism with high security constraints. This research work presents a mechanism for capturing and analyzing the face of the person who is trying to access the vehicle in real time. The face of human beings has different characteristics, and it is unique which has motivated to develop this system. In this proposed system, Haar feature-based cascade classifier has been applied in the embedded system for authenticating the authorized users to gain access on vehicle. This proposed system captures the face image whenever a person tries to access the vehicle and set of features are extracted from the captured image. The extracted features are then compared with the pattern to identify the authorized user.

2 Related Work

Suganya and Kashwan [2] provided an anti-theft intelligence traffic management system for controlling the traffic and theft issues in overcrowded locations. Song et al. [3] created a sensor network-based vehicle anti-theft system (SVATS) which is used for identifying the unauthorized users and the system can able to deal with vehicle intruders but cannot be applied to any other attacks. Agustine proposed a vehicle security system which has Global Positioning System (GPS) for tracking of vehicle and Global System for Mobile Communication (GSM) for transmitting the geographical location of vehicle via a wireless communication technology. This system triggers an alert message when the vehicle crosses the predefined locations [4]. Jia et al. [5] presented a system which uses an input from single monocular camera for vehicle detection and tracking which is best suitable for street-like environments. It uses Continuously Adaptive MeanShift (CamShift) Algorithm for segmenting the street environment videos into multiple regions and the vehicle is tracked based on moving region by using Sobel operator. Zhong et al. [6] designed a system using fusion technology for reducing the car theft issue and false alarm rate. It uses improved D-S evidence theory for handling the conflict evidence.

Meng et al. developed an intelligent framework for handling the vehicle thefts which uses the behavioral patterns for authenticating the users. This system contains high security constraints than other static security mechanisms, because it uses the dynamic biometric characteristics which are harder to imitate [7]. Schmidt and Kasiński [8] analyzed the performance of Haar cascade classifier by comparing the results with the Lienhart's for face detection and Castrillon-Santana's for eyes detection. Sharma et al. presented a real-time car key detection in an image using Haar cascade classifier. This work consists of training and detection phases where training phase uses several object images and the detection phase tests the new images [9]. Kasinski and Schmidt [10] presented the three-stage architecture for face and eye detection based on the Haar cascade classifier which has been tested with 1000 images, and it accurately detected the 94% of eyes.

2.1 Main Attributes of Our System

This research work consists of various aspects. First, face-oriented biometric feature of vehicle users is applied to this proposed experimental system. Therefore, utilizing this biometric authentication provides high-level security because it cannot be shared, and it is hard to estimate the biometric modalities. Next, the proposed system is very strong and effective for real-time implementation since it captures the image of the person trying to access the vehicle. It does not use any additional biometric variables such as fingerprint, iris, and retina. This system has chosen the Haar feature-based cascade classifier for detecting the face which owes a high degree of accuracy. It also reduces the false positive rate of algorithms and reduces the intensive computational requirements.

This proposed intelligent security system can be installed easily on any kind of vehicles with respect to functional modifications. During the installation of the system, positioning the camera plays a vital role, which must be placed based on the constraints such as placement of camera; image captured must be face-centric; environment lighting system should not create any impacts on camera; and vehicle vibration should not affect the camera's performance. Finally, a methodology for capturing and analyzing the users face into computational patterns. Section 3 of this paper presents the experimental model of face authenticating-based vehicle security system. Section 4 introduces the application of Haar classifiers based on Haar-like

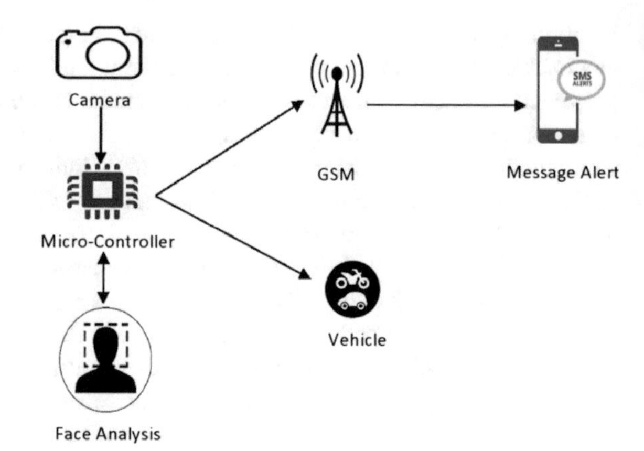

Fig. 1 Proposed system architecture

features for face detection. Section 5 shows the experimental results with respect to various performance metrics and the last section presents the conclusion.

3 Experimental Design

Figure 1 shows the architecture of the vehicle security system based on face authentication. The proposed system consists of three components, which includes face capturing subsystem, face analysis subsystem, and actuator subsystem. In face capturing subsystem, a microcontroller is used to receive the image of the vehicle user from the simulation environment. The image captured is transferred to the face analysis subsystem for further processing; the methodology described in the following section is applied to the image retrieved for differentiating the authorized and unauthorized users. For differentiating the user's model, libraries are created from the input images by machine learning algorithm during the training phase. After creating models, it can be used as a classifier in the system for authenticating users. When a person tries to access the vehicle, the person face is captured, and pattern-matching operation is performed using the created model. This system provides three chances to users for proving their identity. If all the three attempts are failed, actuator subsystem sends an alert message to the owners mobile regarding the unauthorized access of the vehicle.

4 Face Detection Using Haar Classifier Algorithm

In this proposed system, Haar feature-based cascade classifier has been used for face detection. Figure 2 shows the commonly used rectangular filters used on Haar-like features to be applied to the image area. Every individual Haar feature includes two or

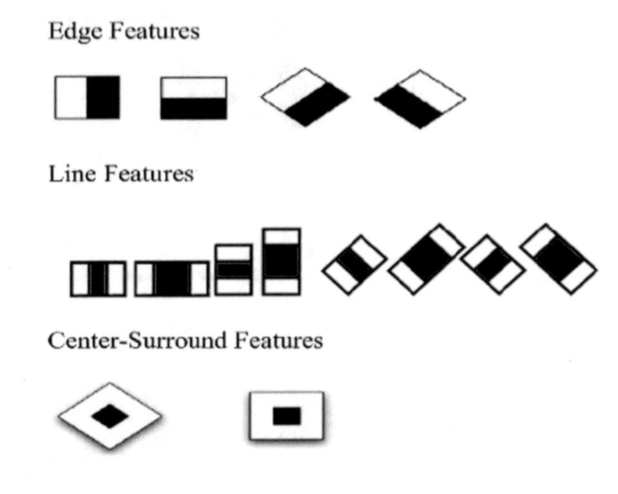

Fig. 2 Common rectangular filters used on Haar feature-based cascade classifier

three combined black and white rectangles. Initially, a rectangular Haar-like feature is identified by calculating the difference between the sums of pixels in the black and white rectangle [13–16].

$$f(x) = \sum \text{Sum of Pixel Gray Level (Black Rectangle)} \\ - \sum \text{Sum of Pixel Gray Level (White Rectangle)}$$

The Haar-like features are calculated using the integral image, which is the summation of pixels in the top-left portion from the point (x, y). The black area in Fig. 3 depicts the integral image computation. Likewise, all the features are identified as a cascading process by Open Source Computer Vision (OpenCV) software. Every successive stage of cascading process removes the weak classifiers (negative sub-windows) passed by the initial stage and at the end of cascading process, the maximum of negative sub-windows are discarded. The training is done by AdaBoost algorithm [11, 13–16], which predicts only the promising area in the image for face detection. The model is created by training the classifier with "n" number of positive images (authorized user face image) and negative image (non-face image) in accordance to achieve a cascaded classifier. At the end of the training phase, an XML is generated which can be used by classification algorithm for detecting the authenticated user.

The Haar cascade classifier is implemented in OpenCV, in which a threshold value is provided to identify the percentage of face matching with the pattern stored in the database. The vehicle can be accessed by the user, if the captured image matches the pattern and attains the threshold. In case of Haar cascade classifier fails to identify the face or detects any unmatched face images, then the system discards the access rights and sends an alert message via actuator subsystem. Figure 4 presents the overall system workflow.

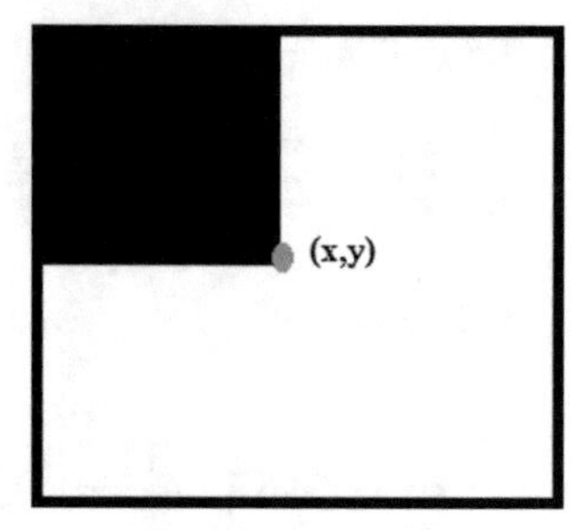

Fig. 3 Integral image computation at point (x, y)

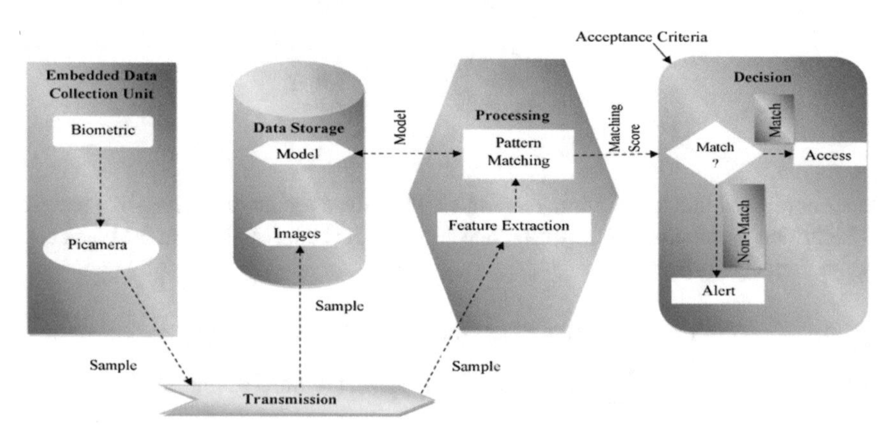

Fig. 4 System workflow

5 Experimental Results

In this proposed experimental system, real-time images are used and evaluated using OpenCV to prove the accuracy of the Haar cascade classifier. The main intention of the system is to identify the authenticated vehicle users by using the face image. The images are collected and then classified through Haar cascade classifier into two categories: two people as authorized user and other eight people as the unauthorized users. Figure 5 shows the sample result of this system, which represents a classification of authorized user and unauthorized user.

5.1 Evaluation Metrics

The performance of this experimental system is represented in terms of error rates such as False Acceptance Rate (FAR) and False Rejection Rate (FRR) [12]. The FAR is considered as a severe security error, which is the measure of unauthorized

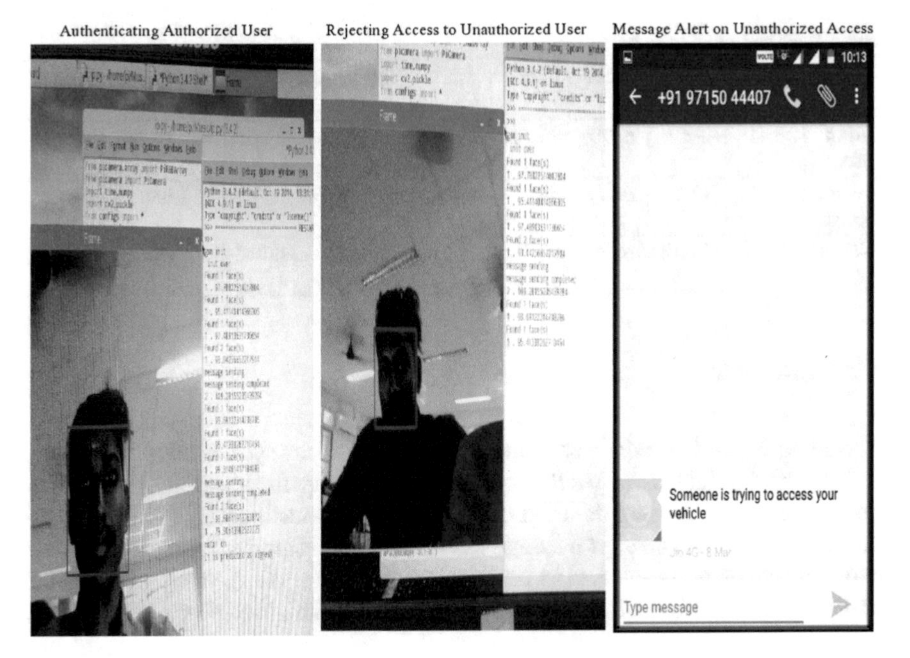

Fig. 5 Sample result

users identified as authorized person and provided access to the vehicle. It can be calculated using the formula:

$$\mathrm{FAR} = \left(\frac{\text{Number of false users authenticated to access vehicle}}{\text{Total number of attempts made}}\right)$$

The FRR is a biometric performance measure, which calculates the probability of the system which fails to authenticate the authorized users and calculated using the formula:

$$\mathrm{FRR} = \left(\frac{\text{Number of true users rejected to access vehicle}}{\text{Total number of attempts made}}\right)$$

The minimum FAR and FRR values is an indication of highly secured system. Among ten attempts made, this proposed experimental is able to classify the authorized and unauthorized users accurately with zero FAR and zero FRR error rate. Table 1 presents the experimental results of classifying users with the error rate threshold of 90 by our proposed security system.

Table 1 Experimental results

	User 1	User 2	User 3	User 4	User 5	User 6	User 7	User 8	User 9	User 10
Error rate	95.41	93.69	79.90	75.34	94.12	92.67	97.45	91.14	93.03	96.76
Vehicle access	No	No	Yes	Yes	No	No	No	No	No	No
Correctly identified authorized users: 02						Correctly identified unauthorized users: 08				

6 Conclusion

This research work developed a high-level vehicle security system using face detection mechanism and addresses the solution for vehicle thefts. By capturing and analyzing the user face using the Haar cascade classifier, unauthorized users are accurately identified and an alert message delivered to the authorized user. Finally, this experimental system evaluated to verify the security level and the results proved the system is highly affordable against the vehicle thefts with a high degree of accuracy.

References

1. Nored, L.S., Carlan, P., Downey, R.A.: A Brief Introduction to Criminal Law. Jones & Bartlett Publishers (2015)
2. Suganya, R., Kashwan, K.R.: FPGA implementation of anti-theft intelligent traffic management system. In: Reddy, M.S., Viswanath, K., Shiva Prasad, K.M. (eds.) International Proceedings on Advances in Soft Computing, Intelligent Systems and Applications. Advances in Intelligent Systems and Computing, vol. 628. Springer, Singapore (2018)
3. Song, H., Zhu, S., Cao, G.: Svats: a sensor-network-based vehicle anti-theft system. Networking and Security Research Center, Department of Computer Science and Engineering, Pennsylvania State University, Technical Report NAS-TR-0076-2007, August 2007
4. Agustine, L., Pangaliela, E., Pranjoto, H.: Vehicle security and management system on GPS assisted vehicle using geofence and Google map. In: Pasila, F., Tanoto, Y., Lim, R., Santoso, M., Pah, N. (eds.) Proceedings of Second International Conference on Electrical Systems, Technology and Information 2015 (ICESTI 2015). Lecture Notes in Electrical Engineering, vol. 365. Springer, Singapore (2016)
5. Jia, L., Wu, D., Mei, L., Zhao, R., Wang, W., Yu, C.: Real-time vehicle detection and tracking system in street scenarios. In: Zhao, M., Sha, J. (eds.) Communications and Information Processing. Communications in Computer and Information Science, vol. 289. Springer, Berlin, Heidelberg (2012)
6. Zhong, W., Chen, G., Qi, H., Qi, X.H., Cheng, Z.: The design of vehicle anti-theft system based on the improved D-S evidence theory. In: 2014 8th International Conference on Future Generation Communication and Networking, Haikou, pp. 84–87 (2014)
7. Meng, X., Ou, Y., Lee, K.K., Xu, Y.: An intelligent vehicle security system based on modeling human driving behaviors. In: Wang, J., Yi, Z., Zurada, J.M., Lu, B.L., Yin, H. (eds.) Advances in Neural Networks—ISNN 2006. Lecture Notes in Computer Science, vol. 3973. Springer, Berlin, Heidelberg (2006)

8. Schmidt, A., Kasiński, A.: The performance of the haar cascade classifiers applied to the face and eyes detection. In: Kurzynski, M., Puchala, E., Wozniak, M., Zolnierek, A. (eds.) Computer Recognition Systems 2. Advances in Soft Computing, vol. 45. Springer, Berlin, Heidelberg (2007)
9. Sharma, P., Gupta, M.K., Mondal, A.K., Kaundal, V.: HAAR like feature-based car key detection using cascade classifier. In: Singh, R., Choudhury, S. (eds.) Proceeding of International Conference on Intelligent Communication, Control and Devices. Advances in Intelligent Systems and Computing, vol. 479. Springer, Singapore (2017)
10. Kasinski, A., Schmidt, A.: The architecture of the face and eyes detection system based on cascade classifiers. In: Kurzynski, M., Puchala, E., Wozniak, M., Zolnierek, A. (eds.) Computer Recognition Systems 2. Advances in Soft Computing, vol. 45. Springer, Berlin, Heidelberg (2007)
11. Viola, P., Jones, M.: Rapid object detection using a boosted cascade of simple features. In: Conference on Computer Vision and Pattern Recognition (2001)
12. Maltoni, D., Maio, D., Jain, A.K., Prabhakar, S.: Handbook of Finger-Print Recognition. Springer Science & Business Media, Berlin (2009)
13. Pavani, S.-K., Delgado, D., Frangi, A.F.: Haar-like features with optimally weighted rectangles for rapid object detection. Pattern Recogn. **43**(1), 160–172 (2010)
14. Wilson, P.I., Fernandez, J.: Facial feature detection using Haar classifiers. ACM J. Comput. Sci. Coll. **21**(4), 127–133 (2006)
15. Viola, P., Jones, M.J.: Robust real-time face detection. ACM Int. J. Comput. Vis. **57**(2), 137–154 (2004)
16. Tian, H., Duan, Z., Abraham, A., Liu, H.: A novel multiplex cascade classifier for pedestrian detection. ACM Pattern Recogn. Lett. **34**(14), 1687–1693 (2013)

P. B. Pankajavalli is an Assistant Professor in the Department of Computer Science, School of Computer Science and Engineering, Bharathiar University, Coimbatore, Tamil Nadu, India. She obtained her Post Graduation MCA under Bharathiar University in 2003 and M.E., degree under Anna University in 2011and Ph.D., degree in Computer Science from Mother Teresa Women's University in the year 2013. She has published 30 papers in journals and in Conferences both at National and International level. She has received Best Teacher Award from Lions Club, Erode. Her areas of interest include Ad-hoc Networks, Wireless Sensor Networks and Internet of Things.

V. Vignesh is a Post Graduate Student in the Department of Computer Science, School of Computer Science and Engineering, Bharathiar University, Coimbatore, Tamil Nadu. He has received Under Graduation B.Sc. Computer Science in 2016 from Dr. SNS Rajalakshmi College of Arts and Science Affiliated to Bharathiar University, Coimbatore, Tamil Nadu.

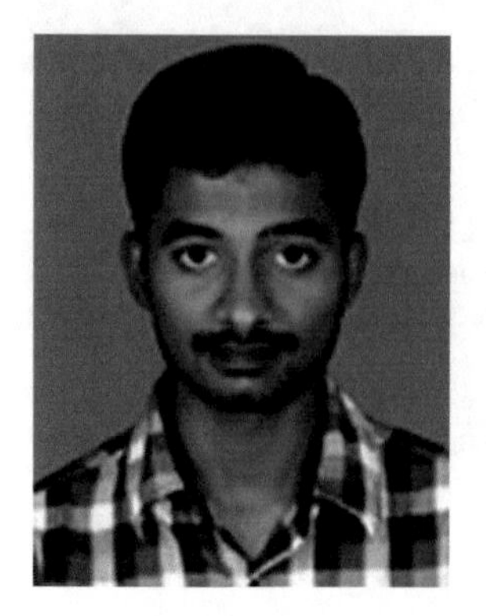

G. S. Karthick is a Ph.D Research Scholar in the Department of Computer Science, School of Computer Science and Engineering, Bharathiar University, Coimbatore, Tamil Nadu. He has received Post Graduation M.Sc. Computer Science in 2017 from Bharathiar University (UD), Coimbatore, Tamil Nadu. He has published research papers in International Journals and presented papers in International and National Conferences. His area of specializations and research interests are Wireless Sensor Networks, Internet of Things and Analysis of Algorithms.

Performance Comparison of Asynchronous NoC Router Architectures

Rose George Kunthara and Rekha K. James

Abstract Network-on-Chip (NoC) approach emerged as a promising alternative to overcome the bottleneck and scalability problems of traditional bus-based and point-to-point communication architectures. NoC implementation can be either fully synchronous, Globally Asynchronous and Locally Synchronous (GALS) or fully asynchronous. Asynchronous system is a potential solution to overcome the synchronous NoC design issues in deep submicron regime. GALS NoCs have reduced data throughput due to latency overheads caused by the synchronization interfaces whereas in asynchronous NoCs, each packet needs to traverse only two synchronization interfaces. This paper makes a comparative study of the state-of-the-art asynchronous NoC routers.

Keywords Network-on-Chip · Asynchronous circuits · GALS
Asynchronous NoC router

1 Introduction

Advancements in IC technology have led to miniaturization where the transistor densities have skyrocketed due to fall in transistor feature sizes to ultra-deep submicron levels, leading to the creation of complex System-on-Chip (SoC). SoC typically consists of predesigned functional blocks or IP (Intellectual Property) cores, with an interconnection architecture and various interfaces to the peripheral devices. The traditional SoC interconnect architecture was based on classical shared bus and point-to-point intercommunication using dedicated wires. The shrinking technology amplified the imbalance between gate delay and on-chip wire delay, which resulted in

R. G. Kunthara (✉) · R. K. James
Division of Electronics, School of Engineering, Cochin University of Science and Technology, Kochi, Kerala, India
e-mail: rosekunthara87@gmail.com

R. K. James
e-mail: rekhajames@cusat.ac.in

S. Smys et al. (eds.), *International Conference on Computer Networks and Communication Technologies*, Lecture Notes on Data Engineering and Communications Technologies 15, https://doi.org/10.1007/978-981-10-8681-6_59

increased power consumption, unpredictable delays, on-chip synchronization errors, etc. The packet-switched Network-on-Chip (NoC) concept was evolved to overcome the predictability, scalability and bottleneck issues faced by the traditional SoC interconnection architecture by improved parallelism and much better scalability [1, 2].

The problems like clock distribution, clock skew and power consumption associated with the synchronous design become more and more complex to deal with in the deep submicron levels. The International Technology Roadmap for Semiconductors (ITRS) [3] details about the asynchronous communication protocols requirements for the control and synchronization in Integrated Circuits (IC). Due to the scaling of CMOS technologies, they predict that by 2022, almost 45% of the chips will be using local handshaking circuits such as clockless or GALS ICs. GALS NoC has lower power consumption compared to a single frequency synchronous NoC as GALS NoC can allow controlling the operating clock or frequency of each router associated with individual IP cores. But on the downside, synchronization interfaces must be added to each of the routers and cores, operating in different clock frequencies.

Asynchronous NoC is an approach to overcome this issue of too many synchronization interfaces in the GALS NoC. Thus, there is only less latency overhead in an asynchronous NoC architecture as each packet needs to traverse only two synchronization interfaces, one at the output port of the sender side and another one at the input port of the receiver. Since most of the commercial EDA tools are tailored for purely synchronous designs, efficient implementations of clockless and GALS designs are difficult to obtain due to the absence of a global clock. But research on asynchronous NoC router design and implementation is gaining momentum due to advantages such as zero dynamic power consumption when idle, higher speed of operation, no complex clock related design issues, etc.

The remainder of the paper is organized as follows: Sect. 2 gives an overview of the fundamentals of NoC. Section 3 discusses the various possible timing techniques employed in NoC. Section 4 presents the related work in the field of asynchronous NoC routers, and finally, Sect. 5 concludes the paper.

2 Background

A brief overview of the fundamentals of NoC is described in this section.

2.1 Network-on-Chip

A typical NoC consists of routers, Network Interfaces (NI) and physical links forming an interconnect system which provides an efficient communication between the different IP cores, as shown in Fig. 1. The routers are connected physically to each other over a topology forming regular topologies such as mesh, tree, torus, etc. and

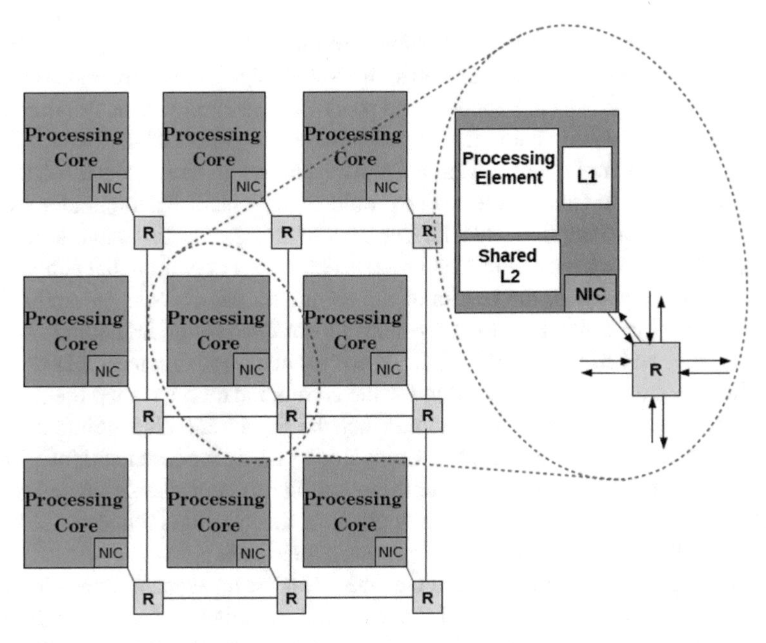

Fig. 1 A 3×3 mesh-type NoC

irregular topologies like butterfly fat tree which is application specific. Communication between the cores is by packet transmission in the form of flits.

The basic building blocks of NoC are topology, routing, switching, flow control and router microarchitecture. Topology is the arrangement or physical layout and connection pattern of the routers and channels in the interconnection network. Routing algorithm determines the path taken by a packet to reach the destination. Switching defines when and how the data flows through the routers and also determines the granularity of data transfer. Flow control manages the allocation of resources to packets as they propagate along their route. The main functions of a NoC router are buffering, route computation, arbitration of flits (prioritization) when contention occurs and switching from input port to output port [1, 2].

3 Possible Timing Methodologies of NoC

Several timing organizations are possible for the NoC-based SoC implementation.

Most of the works related to NoC implementations assume a globally synchronous operation of the entire system. A fully synchronous NoC has to operate at high frequency so as to meet the varying bandwidth requirements of the most demanding core, resulting in wasteful power consumption on the routers leading to bottleneck at the interconnection. But in reality, SoC-based NoCs contain different IP blocks operating with independent clock domains. Several methods exist to pass the signals between the different clock domains. In some NoC designs, the entire NoC is implemented as one single clocked module and so synchronization will be required only on the interfaces between the independently clocked IP cores and the NoC [4, 5].

It is difficult to implement NoC as a simple synchronous module as it has a chip-wide structure due to which there can be problems of clock distribution, clock skew, dynamic power consumption when idle, etc. So, other schemes that can tolerate timing uncertainty have been devised such as mesochronous operation that can tolerate phase difference between the clocks of the communicating blocks [6] and GALS operation [7, 8] in which the clock domains are confined to the individual IP cores and the individual routers of the NoC, where asynchronous request-acknowledge based handshaking protocols are used for the communication between them. Thus, some form of synchronization is required when the signals enter the individual clock domains in both the above approaches. Compared to a fully synchronous NoC, GALS NoC has minimal dynamic power consumption due to controllable operating frequency of each router. But the data throughput of the system is reduced due to the latency penalties caused by the synchronization interfaces.

Another approach is to implement the entire NoC as an asynchronous circuit [9, 10]. This is an extension of the GALS approach as no clock domain extends along the entire chip and so, synchronization will be needed when the data enters the clock domain of an individual IP core. The timing of handshaking signals is very much needed for an error-free transmission and reception of the data between the IP cores in a NoC. So, timing analysis has to be carried out. But due to the absence of a global clock signal and a well-defined asynchronous CAD flow, it is difficult to perform this timing analysis to find the effect of parasitic resistance and capacitance of the connections on the design.

Asynchronous NoC is more advantageous due to its ability to overcome the problems of clock skew, higher speed due to the average case performance, inherent flow control and zero dynamic power consumption when idle. The asynchronous NoC implementation is in such a way that it becomes insensitive to the gate and wire delays. This results in the design of higher speed circuits with increased bandwidth [11].

4 State of the Art

Most of the recent literature on asynchronous NoC router is based upon five port architectures. It is difficult to make a fair comparison of the various asynchronous NoC routers since different authors use different delay models and process technologies. In this section, a fair comparison of some of the asynchronous NoC router implementations is made.

An asynchronous NoC architecture targeting GALS architecture is proposed in [12], which provides low latency transfers with the support of two virtual channels, one for best-effort traffic and the other one for low latency real-time traffic. It employs four-phase handshaking protocol using a quasi-delay insensitive asynchronous NoC topology featuring wormhole flow control mechanism and odd–even turn model adaptive routing algorithm. Their proposed router has 5 GB/s throughput for 0.13 μm CMOS technology.

MANGO [9], which stands for Message-passing Asynchronous Network-on-Chip providing Guaranteed services through OCP interfaces, used virtual channels for providing connection-oriented Guaranteed Services (GS) and connectionless Best-Effort (BE) routing. The system level integrity was provided by guaranteed services. This flexible architecture had inherent support for a modular GALS system-based design flow since the implementation finds its basis on asynchronous circuit techniques. It is basically a wormhole NoC router that has inherent end-to-end flow control implemented using 0.12 μm CMOS standard cell approach. The 32-bit flit sized, 5×5 MANGO router employed four-phase bundled-data protocol between routers and was able to support connectionless BE routing and 32 independent connection-oriented GS simultaneously. It has comparable features, area and performance to that of similar clocked or synchronous NoC router.

In [13], the authors proposed a low latency wormhole asynchronous NoC router which uses sliced sub-channels and look-ahead asynchronous pipeline techniques. The C-element tree in the completion detector circuit of quasi-delay insensitive pipeline is removed by channel slicing, and thereby, a channel is converted into multiple independent sub-channels so as to minimize the cycle period. Look-ahead pipeline [14], which uses an early evaluation protocol, is used on the critical cycle to improve the throughput and their results showed reduction in cycle period and area overhead.

Hermes-A [15] gives the architecture and implementation details of an asynchronous NoC router employing wormhole flow control, distributed XY routing and scheduling scheme for a 2D mesh network topology. Independent arbitration was employed at each of the router ports to achieve the power consumption constraints of NoC by assigning specific voltage levels to specific paths of NoC with dynamic voltage level schemes. The router used delay insensitive four-phase handshaking protocol and dual-rail encoding and had a maximum throughput of 3.6 Gbits/s for 180 nm technology ASIC implementation. In [16], the architecture and implementation of Hermes-AA are detailed. Similar to [15], Hermes-AA employs independent arbitration at each of the port, wormhole flow control, distributed routing and scheduling mechanism for a 2D mesh topology. This fully flexible asynchronous router uses DI, four-phase dual-rail encoding scheme also. West-first turn model algorithm, which is an adaptive routing algorithm, is used in Hermes-AA.

An asynchronous NoC router using two-phase bundled-data handshake latch and employing Mousetrap latch controller [14] was designed and extended to a full five-port router architecture to obtain increased performance in [17]. This design is then compared to xpipesLite [18], which is a lightweight synchronous switch architecture. The authors also came up with two efficient and concurrent asynchronous elements, a transition signalling circular FIFO and a four-way arbiter. A semi-automated design flow exploiting commercial synchronous CAD tools, but specific for four-phase bundled-data design was used to create partially reconfigurable macro-blocks. The post-layout results exhibited asynchronous NoC switch had reduced power consumption, area and reduced energy per flit.

BAT-Hermes [19] describes a low power asynchronous NoC router using Mousetrap template or transition signalling bundled protocol [14]. It had features similar to the design described in [17] but employed a flit counter mechanism-based packet control technique. BAT-Hermes had minimal energy consumption but at an increased area cost compared with a synchronous counterpart (YeAH! Router [20]) that also used fully distributed control logic and functionality analogous to Hermes router [21].

In [22], the authors proposed Argo NoC, which is specifically developed to be used in multicore platforms intended for hard real-time applications. The combination of Time Division Multiplexing (TDM) and the usage of asynchronous NoC routers to have a hardware efficient NoC is the main feature of Argo. Argo supports message passing across end-to-end virtual channels and these virtual channels are implemented by a static TDM schedule of the resources so as to offer real-time guarantees. Argo is based on GALS architecture and consists of asynchronous routers, mesochronous network interfaces and IP cores which are all independently clocked. The timing analysis of the asynchronous router network and safely bringing the NoC out of reset to start the normal operation are two important issues this paper focus on. Kasapaki et al. [23] provided extensive details on the architecture, design, implementation and layout of 4×4 bi-torus Argo NoC along with speed, power and area results using 65 nm CMOS technology. Argo employs a network interface design combining TDM schedule and DMA controllers, and using asynchronous NoC routers result in end-to-end message transfer eliminating flow control, synchronization and buffering.

The real traffic may be unevenly distributed across IP cores imposing the need for heterogeneous NoC in practical applications. The authors of [24] made a comparison of synchronous and asynchronous heterogeneous NoC under uniform and exponential traffic conditions based on parameters such as number of cores and number of Virtual Channels (VC) in a router. Their simulation results showed that for a given number of cores and VCs, lower end-to-latency, higher throughput and bandwidth were offered by asynchronous NoC compared to their synchronous counterpart.

Representative routers of synchronous and asynchronous design styles were designed separately under the same functional specifications and compared precisely in terms of area, power, latency and fairness to show the areas of specialty of the routers in [25]. From their results, the authors conclude that asynchronous NoC routers are suitable for real-time control systems where short packets have to be quickly passed with minimum power whereas synchronous routers can be used in multimedia data processing systems due to their ability to efficiently handle large continuous data packets.

The design and implementation of a transition signalling bundled-data asynchronous NoC router suitable for Vision-System-on-Chip (VSoC) context and using Mousetrap asynchronous pipelining template are proposed in [26]. The basic architecture and optimization of the designed five-port asynchronous router are based on the router proposed in [17]. The authors make a comparison of synchronous and asynchronous GALS NoC router implementation based on common synthesis tools.

Their results showed that asynchronous implementation has better resource reduction in terms of area, latency and energy per flit.

A comparison of some of the related works discussed above is summarized in Table 1, including the architecture details, asynchronous style used and certain features used in them.

Table 1 Comparison of asynchronous NoC routers

Proposed system	Topology/Routing/Flow control	Asynchronous style used	Highlighting features
Bjerregaard et al. [9]	5×5 2D mesh, wormhole router	Four-phase, bundled-data protocol	32-bit flits, virtual channels provide both BE and GS routing, worst-case port speed of 515 MHz for 0.12 μm CMOS standard cell implementation
Renaudin et al. [12]	Odd–even turn model routing, wormhole	Four-phase, QDI	Provides low latency transport with 2 virtual channels support, 5 GB/s throughput for 0.13 μm CMOS technology
Sheibanyrad et al. [8]	2D mesh, XY routing, wormhole	Four-phase bundled data protocol, double rail encoding	32-bit flit size, maximum throughput 1131 Mflits/s for 90 nm technology
Soug et al. [13]	Wormhole	Look-ahead pipeline [14]	2.35 GB/s throughput per port for 0.13 μm CMOS standard cell library approach
Pontes et al. [15]	2D mesh, XY routing, wormhole	DI, four-phase dual-rail encoding	8-bit flits, maximum throughput of 3.6 Gbits/s for 180 nm technology, ASIC implementation
Pontes et al. [16]	2D mesh, west-first turn model algorithm, wormhole	DI, four-phase dual-rail encoding	8-bit flits, maximum throughput of 7.75 Gbits/s for 65 nm technology, ASIC implementation

(continued)

Table 1 (continued)

Proposed system	Topology/Routing/Flow control	Asynchronous style used	Highlighting features
Ghiribaldi et al. [17]	5 × 5 2D mesh, XY routing, wormhole	Two-phase bundled data, Mousetrap latch controller [14]	32-bit flits, two efficient and concurrent asynchronous elements—a transition signalling circular FIFO and a four-way arbiter used, 900 Mflit/s throughput in 40 nm standard cell library-based implementation
Kasapaki et al. [23]	4 × 4 bi-torus, source routing, wormhole	Two-phase, Mousetrap latch	32-bit flits, employs a network interface design combining TDM schedule and DMA controllers to have a hardware efficient NoC intended for hard real-time applications, 65 nm CMOS technology implementation
Swain et al. [24]	2D mesh, XY routing, wormhole	Four-phase handshaking	32-bit packet size, link data rate of 16 GB/s for 500 MHz frequency of operation
Imai et al. [25]	4 × 4 2D mesh, XY routing, wormhole	Two-phase, bundled-data path	Representative routers of synchronous and asynchronous design styles are designed and compared for packet sizes of four and eight flits in 130 nm CMOS technology
Russell et al. [26]	5 × 5 2D mesh, XY routing, wormhole	Two-phase bundled data, Mousetrap latch controller	Design and implementation of asynchronous NoC router suitable for vision-system-on-chip context, with better performance in terms of area, latency and energy per flit

5 Conclusion

In this paper, a comparative study of the existing asynchronous NoC router designs is done. A complete evaluation of the various asynchronous NoC routers is difficult to make since different authors have employed different delay models and process technologies for comparison. Much of the work on asynchronous NoC routers have used a bundled-data path scheme and transition signalling-based protocol to reduce the latency and power dissipation in order to have higher efficiency and performance. Due to the robustness under high process variation conditions, less area overhead, reduced latency and power consumption, asynchronous NoC is suitable for real-time control applications where short packets have to be passed very quickly.

References

1. Dally, W.J., Towles, B.: Principles and Practices of Interconnection Networks. Elsevier Science Publishers (2003)
2. Dally, W.: Route packets, not wires: on-chip interconnection networks. In: Proceedings of the 2001 Design Automation Conference, pp. 684–689. ACM Press, New York (2001)
3. Semiconductor Industry Association: The International Technology Roadmap for Semiconductors, ITRS 2008 Edition
4. Dallosso, M., Biccari, G., Giovannini, L., Bertozzi, D., Benini, L.: Xpipes: a latency insensitive parameterized network-on-chip architecture for multi-processor SoCs. In: Proceedings of International Conference on Computer Design (ICCD), pp. 536–539. IEEE Computer Society Press (2003)
5. Goossens, K., Dielissen, J., Radulescu, A.: The ethereal network on chip: concepts, architectures, and implementations. IEEE Des. Test Comput. **22**(5), 414–421 (2005)
6. Mesgarzadeh, B., Svensson, C., Alvandpour, A.: A new mesochronous clocking scheme for synchronization in SoC. In: Proceedings of International Symposium on Circuits and Systems, pp. 605–608 (2004)
7. Chapiro, D.M.: Globally-Asynchronous Locally-Synchronous Systems. PhD thesis, Stanford University (1984)
8. Sheibanyrad, A., Greiner, A., Miro-Panades, I.: Multisynchronous and fully asynchronous NoCs for GALS architectures. IEEE Des. Test Comput. **25**(6), 572–580 (2008)
9. Bjerregaard, T., Sparso, J.: A router architecture for connection-oriented service guarantees in the MANGO clockless network-on-chip. In: Proceedings of Design Automation and Test in Europe (DATE), pp. 1226–1231. IEEE Computer Society Press (2005)
10. Dobkin, R., Vishnyakov, V., Friedman, E., Ginosar, R.: An asynchronous router for multiple service levels networks on chip. In: Proceedings of International Symposium on Advanced Research in Asynchronous Circuits and Systems, pp. 44–53. IEEE Computer Society Press (2005)
11. Sparso, J.: Asynchronous design of networks-on-chip. In: Norchip Conference (2007)
12. Beigne, E., Clermidy, F., Vivet, P., Clouard, A., Renaudin, M.: An asynchronous NOC architecture providing low latency service and its multi-level design framework. In: Proceedings of the 11th IEEE International Symposium on Asynchronous Circuits and Systems (ASYNC'05), pp. 54–63 (2005)
13. Song, W., Edwards, D.: A low latency wormhole router for asynchronous on-chip networks. In: 15th Asia and South Pacific Design Automation Conference (ASP-DAC) (2010)
14. Nowick, S., Singh, M.: High-performance asynchronous pipelines: an overview. IEEE Des. Test Comput. **28**(5), 8–22 (2011)

15. Pontes, J.J.H., Moreira, M.T., Moraes, F.G., Calazans, N.L.V.: Hermes-A—an asynchronous NoC router with distributed routing. In: International Workshop on Power and Timing Modeling, Optimization and Simulation (PATMOS'10), LNCS, pp. 150–159 (2010)
16. Pontes, J.J.H., Moreira, M.T., Moraes, F.G., Calazans, N.L.V.: Hermes-AA—a 65 nm asynchronous NoC router with adaptive routing. In: 23rd IEEE International SoC Conference (SOCC'10), pp. 493–498. IEEE Computer Society, Las Vegas (2010)
17. Ghiribaldi, A., Bertozzi, D., Nowick, S.: A transition-signaling bundled data NoC switch architecture for cost-effective GALS multicore systems. In: Proceedings of Design Automation and Test in Europe (DATE) Conference, pp. 332–337 (2013)
18. Stergiou, S.: xpipesLite: a synthesis-oriented design flow for networks on chip. In: DATE 2005, pp. 1188–1193 (2005)
19. Gibiluka, M., Moreira, M.T., Moraes F.G., Calazans, N.L.V.: BAT-Hermes: a transition signaling bundled-data NoC router. In: IEEE 6th Latin American Symposium on Circuits & Systems (LASCAS) (2015)
20. Moreira, M.T., Heck, L.S., Heck, G., Gibiluka, M., Calazans, N.L.V., Moraes, F.G.: The YeAH! NoC router. Technical Report, Faculty of Informatics, PUCRS (2014)
21. Moraes, F., Calazans, N., Mello, A., Moller, L., Ost, L.: HERMES: an infrastructure for low area overhead packet-switching networks on chip. Integr. VLSI J. **38**(1), 69–93 (2004)
22. Kasapaki, E., Spars, J.: The Argo NoC: combining TDM and GALS. In: European Conference on Circuit Theory and Design (ECCTD) (2015)
23. Kasapaki, E., Schoeberl, M., Srensen, R.B., Mller, C., Goossens, K., Spars, J.: Argo: a real-time NoC architecture with an efficient GALS implementation. IEEE Trans. Very Large Scale Integr. VLSI Syst. **24**(2), 479–492 (2016)
24. Swain, A.K., Rajput A.K., Mahapatra, K.: Parametric performance analysis of synchronous and asynchronous heterogeneous network on chip. In: IEEE International Symposium on Nanoelectronic and Information Systems (iNIS) (2016)
25. Imai, M., Chu, T.V., Kise K., Yoneda, T.: The synchronous vs. asynchronous NoC routers: an apple-to-apple comparison between synchronous and transition signaling asynchronous designs. In: Tenth IEEE/ACM International Symposium on Networks-on-Chip (NOCS) (2016)
26. Russell, P., Dge, J., Hoppe, C., Preuer, T.B., Reichel, P., Schneider, P.: Implementation of an asynchronous bundled-data router for a GALS NoC in the context of a VSoC. In: IEEE 20th International Symposium on Design and Diagnostics of Electronic Circuits & Systems (DDECS) (2017)

Rose George Kunthara is a Research Scholar at Cochin University of Science & Technology (CUSAT), Kerala, India. She acquired her B.Tech degree in Electronics and Communication Engineering in 2009 from School of Engineering, CUSAT, followed by M.E. in Embedded Systems from Birla Institute of Technology & Science (BITS) Pilani, India. Her areas of interest include Network on Chip architectures and low-power circuits. She is a member ISTE and student member IEEE.

Rekha K. James is a Professor at Cochin University of Science & Technology, Kerala, India. She got her B.Tech degree in Electronics and Communication Engineering in 1989 from College of Engineering, Trivandrum (CET), University of Kerala, followed by M.Tech. in Digital Electronics and Ph.D. in Computer Engineering from Cochin University of Science and Technology (CUSAT), Kochi, India. Her research interests include the design of Multicore architectures, Decimal arithmetic, Reversible logic and Low-power circuits. She is a Member IEEE, ISTE (L) (India), and IETE (India).

Smart Omnichannel Architecture for Air Travel Applications Using Big Data Techniques

Hari Bhaskar Sankaranarayanan and Jayprakash Lalchandani

Abstract The goal of the paper is to present an omnichannel architecture for air travel applications that enables seamless access and up-to-date travel information across channels. Air travel applications are characterized by diversity and various actors participating in the travel lifecycle. The life cycle is fulfilled by different systems starting from travel inspiration to post-trip activities. The traveler as a consumer of travel applications has multiple options like websites, call center, and travel agent offices to search, book, track, and manage the travel activities. The presence of multiple channels makes the traveler experience easy in terms of time-saving and convenience. The challenge with multiple channels is to ensure the consistency of travel data and keeping them up-to-date. The multichannel does not mean an omnichannel experience. A truly omnichannel experience means that traveler gets travel information flowing seamlessly from one channel to the other. It can be achieved by the architecture building blocks which are powered by big data paradigms including personalization, context management, scaling, and sharing the travel data states. In this paper, we will present the approach, architecture in detail and also discuss the survey results that motivated the research for omnichannel interactions. The architecture includes an in-memory layer that can serve across the channels with traveler data like preferences, interaction information, and history in a scalable way.

Keywords Omnichannel · Big data · Architecture · Airlines · Travel

H. B. Sankaranarayanan (✉) · J. Lalchandani
International Institute of Information Technology, Bengaluru, India
e-mail: s.haribhaskar@iiitb.org

J. Lalchandani
e-mail: jtl@iiitb.ac.in

S. Smys et al. (eds.), *International Conference on Computer Networks and Communication Technologies*, Lecture Notes on Data Engineering and Communications Technologies 15, https://doi.org/10.1007/978-981-10-8681-6_60

1 Introduction

Airlines of today are aspiring to provide best-in-class shopping and service experience to travelers through innovative approaches across various channels. The paradigms like New Distribution Capability (NDC) are providing the ability to the airline to implement offers, orders, and shopping cart type of services where users can pick and choose travel recommendations like price, flight choices, and ancillary services like food, seat selection, and wireless Internet access [1]. Also, Online Travel Agencies (OTA) are innovating to drive travelers to book from their sites with offers and discounts. The common denominator is to delight the traveler and improve the travel satisfaction by providing various touch points through a multichannel strategy. The traveler has multiple channels to access travel information and travel providers to plan a trip. Travelers resort to multiple avenues for searching, booking, modifying services like seat selection, food, checking in, logging issues, complaints and posting queries during the life of cycle of travel including inspiration, booking, during trip, and post trip. The channels available for them include the following: call centers, helpdesk or visiting a travel agent, websites and Chatbots, mobile applications, and emails.

Call centers—Airlines and travel agents invest significantly on this channel with manpower, processes, and training. This is one of the primary channels for traveler touch point. Helpdesk or travel agent—Helpdesk refers to a person is available to listen and resolve issues, queries. It can be a check-in agent, airport helpdesk agents, etc. Websites and Chatbots—Websites provide a way to log issues through a form which consists of fields related to the issues and the passengers. Chatbot provides a more modern way to have an interactive approach with a trained computer agent that can answer queries based on natural language processing and machine learning techniques. Emails—Passengers can send email to listed email ids with problem details and receive responses. The diversity of channels introduces a complexity of synchronizing travel data with most recent interactions data are available across channels. In this paper, we will discuss one such architecture approach where a true omnichannel experience can be realized using big data approaches related to personalization and context management using an analytical layer interaction which provides information to the omnichannel component. The paper is organized into following sections: Sect. 2 discusses the related work and motivation behind this research, Sect. 3 proposes the approach and discusses survey results, Sect. 4 discusses the proposed architecture, Sect. 5 highlights the limitations and future work, and Sect. 6 provides the conclusion.

2 Related Work

The omnichannel approach is prominent in retail industry and highlighted in many studies in the literature. Also, there is state-of-the-art survey literature that encourages researchers to explore more opportunities toward omnichannel approach.

eCommerce architecture patterns are highlighted in certain literature but not specific to omnichannel strategy [2–6]. There are examples and lessons learnt from customer experience in omnichannel retailing [7–10]. Facebook implement Memcached to achieve the scalable access to social data for high availability and performance [11]. The main motivation comes from the paper which mentions an approach with case studies on omnichannel for Customer Relationship Management (CRM) [12].

3 Our Approach

Architecture evolution is typically done by building multiple viewpoints like a business, technology, applications, and data. In this paper, we use modern approaches like enterprise architecture and big data principles. As a first step, we have followed the principles of stakeholder analysis to identify and define the architecture needs. In this case, the traveler is a key stakeholder using the omnichannel services, and it is important to get the pulse of the traveler through a survey. The survey section details the approach, questions, and outcomes of omnichannel experience. Subsequently, the practices of big data come into play. As the omnichannel experience involves personalization, context-driven, and relevance-based approach, it would involve data crunching and analysis of traveler interaction across the channel. The traveler in every interaction with a channel generates a good amount of transaction data which can be crunched to get crucial insights like habits, preferences, and history of the traveler. The big data section details the approach used for evolving the architecture. As a next step, we have studied the practices used in social media and retail websites to understand how we can cross-pollinate those practices into the architecture under consideration. The section on eCommerce and Social media architecture patterns details the same.

3.1 Survey

As part of the research goal, an air travel survey was done to understand the traveler experience during air travel and a subsection captures expectations out of the channels used by them. The survey was responded by 824 travelers. All the questions have one choice to respond except question 1 where they can choose one or many options. The survey questions are the following:

1. I use the following touch points during air travel for issues/queries especially last time you had faced an issue:
   a. Call Center.
   b. Travel Agent Phone.
   c. Website, Chatbots/Mobile apps.
   d. Helpdesk person at counters.
2. I am satisfied in general with the travel information/updates about my flight through various channels like mobile app, computer, emails, airport displays, etc.
   a. Strongly Agree.
   b. Agree.
   c. Neither Agree or Disagree.
   d. Disagree.
   e. Strongly Disagree.
3. When I raise a complaint/issue with airlines, I get relevant and consistent updates and resolution progress at any touch point (e.g., Call Center, Mobile, Helpdesk counter, etc.)
   a. Strongly Agree.
   b. Agree.
   c. Neither Agree or Disagree.
   d. Disagree.
   e. Strongly Disagree.
4. Despite the availability of technology, I still feel that I end up repeating my issues, queries across the touch points (e.g., Call Center, Internet, Helpdesk person, etc.)
   a. Strongly Agree.
   b. Agree.
   c. Neither Agree or Disagree.
   d. Disagree.
   e. Strongly Disagree.
5. I always find it cumbersome to track my entire trip like air, hotel, events and if they change I end up spending lot of time to manage them
   a. Strongly Agree.
   b. Agree.
   c. Neither Agree or Disagree.
   d. Disagree.
   e. Strongly Disagree.

The survey results show that:

1. 83.86% respondents are satisfied with the information and updates provided in the channel which shows that they seek the channel for information and updates for their satisfaction and comfort.
2. 64.81% respondents use call center, 46.48% respondents use websites/Chatbots/mobile apps and 47.94% respondents use help desk interactions. All the channels are used widely which makes a strong case for omnichannel approach.
3. 52% respondents feel that they get relevant and consistent updates across all channels and still 48% respondents would not endorse the same. This shows a mixed outcome and good opportunity to explore this gap.
4. 59.71% respondents feel that they end up repeating themselves about the queries and issues across the channel. This is a significant opportunity that can be addressed as well.
5. 56.43% respondents feel that they find the trip tracking and managing them as cumbersome. Multichannel experience is a key driver for achieving a better travel management.

Overall, the survey results show the traveler as a stakeholder and how the results are in favor of having a solution building block that addresses the concerns of information consistency, relevance, and context of the traveler in a multichannel environment.

3.2 Usage of Big Data

The big data paradigm here means the crunching of data on traveler preferences, history of interactions, and finding relevance across the channels. The data is mined, processed, and understood for insights. For instance, a traveler ordering a meal through the website check-in page is tracked and used as a data point to supply discount offers at the airport food court area based on proximity marketing when the user turns on the mobile. When the traveler logs into the voice recorder-based call center application the application default welcome the traveler with the status of the last open issue logged. The crunching involves machine learning and other personalization algorithms which providers the transaction details and relevance to the application layer. The application layer can then push that intelligence into the omnichannel layer thus reducing the direct interaction with the analytical layer. The evolution of machine learning algorithms that analyzes big data for customer segmentation like clustering techniques helps to capture insights about the travel preferences and segment them for the relevant updates.

3.3 eCommerce and Social Media Architecture Patterns

In the eCommerce and Social Media architectures, the pattern like a shopping cart, most recently used profile information data, activity data like last accessed posts are used as a base to display news feeds more quickly with relevance. Also, page refreshes are really limited to preserve the client and server communication. They use techniques on asynchronous and background refresh in mobile applications. Also, they have application storage and cookies in browsers that contains client-side data to reduce the refresh rates on a constant basis. The goal in such website patterns is to reduce the response time and also load content faster for the user to make a purchase or relevant news feeds in a highly responsive user interface. The practices in social media and eCommerce are very relevant for airline and travel agent websites especially for delivering the omnichannel experience. The caching layer in the modern architectures stores key and value of personal data that can be retrieved faster and optimized for background synchronization and replications through a cluster setup of high availability. We can use the similar practices and patterns to solve the omnichannel problem. In the next section, we will review the proposed architecture in detail evolved based on the tenets of the above approaches.

4 Proposed Architecture

The proposed architecture is built around the following guiding principles:

1. **Information relevance** means the access to a relevant channel related traveler information. The services across channel vary based on the underlying travel system services. For example, the business rules like changing ancillaries are not possible at OTA website but only possible in airline website.
2. **Information consistency** means the recent and updated traveler information reflected across channels showing all the changes made across channels. For example, a query logged in a call center is visible to the website or a chatbot application with the latest status of the query.
3. **Information context** means the contextual information like location services and state of the traveler in the journey. The state information includes the journey state like checked in, boarded, in transit, etc. The state information can be accessed by travel applications like call center applications or mobile where they are provided reaccommodation options automatically without explaining in detail if the flight is disrupted during a transit like cancelation. Currently, the traveler walks to help desk to get reaccommodation to a different flight, and it is not automatic.

Figure 1 depicts the proposed architecture of omnichannel design. The omnichannel layer has the following attributes in design:

1. **In-memory objects**—Top "x" Context objects stored and retrieved where "x" is an integer that can be configured.

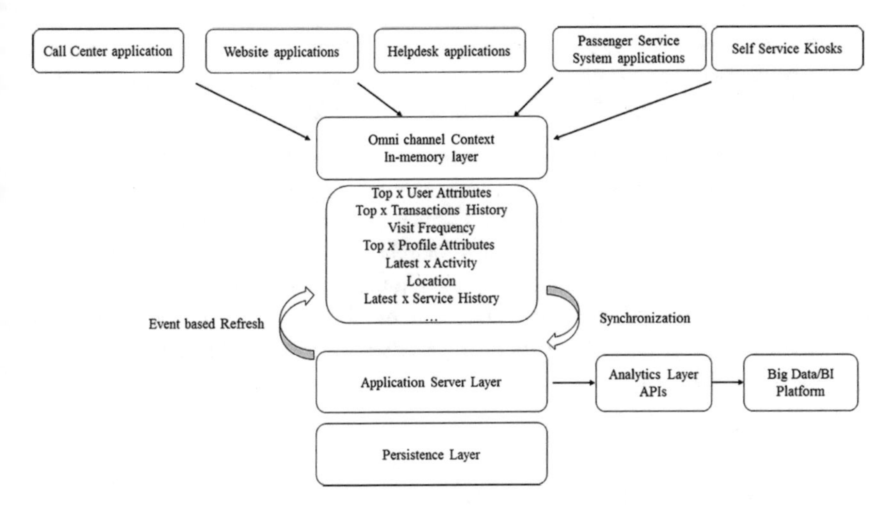

Fig. 1 Omnichannel architecture for air travel applications

Omnichannel is an in-memory key-value data store with various context objects like user, transaction history, visit frequency, profile, latest activity, location, and issue history that can be queried easily.

2. **Read optimized**
   The data structure used is key-value pair; hence, it is meant for read access. The background synchronization process through events and time-bound expiry can configure to ensure that the channel is refreshed frequently.
3. **Fast access**
   The channels can access the data fast with a simple lookup based on the key supplied by the channel layer like a user identification number or profile identity which can use hashing algorithms for addressing security needs.
4. **Synchronization and refresh in background with underlying layers**
   This is based on the time interval or when the user activity occurs the refresh job is triggered to ensure consistency of the omnichannel data store which resides in memory.
5. **High availability across channels**
   The omnichannel layer can be replicated across the clusters and also synchronized for consistency.
6. **Traveler identity key-value-based access**
   The traveler is identified in the omnichannel layer as a single entity principle without any possible duplication which means the profile information is stored in the channel once the traveler accesses the system. Anonymous access is not allowed in omnichannel context as it is difficult to model interactions.
7. **No persistence**

The proposal is not to have any persistence at omnichannel layer to avoid any write overheads involved. The persistence is addressed in persistence layer or the analytics layer where the intelligence is retrieved from.

The application layer connects with analytics layer through a set of interfaces that can provide the content for omnichannel intelligence. The persistence layer is part of the underlying applications that can interact for transaction access. The architecture implementations guidelines include thin clients—website cookies and implementation like Memcached on application side, thick clients—stored in local file system, mobile applications—managed in mobile storage, kiosk, and command line applications—preferably in application server in-memory through Memcached type of implementation.

5 Limitations and Future Work

The architecture approach has the following limitations:

1. The omnichannel interactions are meant to be stateful hence the underlying layer need to support some stateful parameters like context for ease of access and performance reasons.
2. The intuitiveness achieved in omnichannel to share interactions has some performance overheads and accuracy issues like preventing stale data.
3. In-memory implementations in the application server add to the overhead on the application server; hence, implementation and deployment choices to be reviewed holistically with benchmarks.

The architecture work can extend to functional areas like hyper-personalization, context-driven travel models and building ontology on the omnichannel layer for travel objects interactions and on the technical area like interoperability with underlying architecture like micro-services, service-oriented architecture, and client–server architectures.

6 Conclusion

The convenience of a traveler is increased with the presence of multiple channels but it poses a challenge of maintaining a good hand over between channels to provide a seamless experience for the traveler. As the airline industry is moving toward eCommerce and social media type of approaches to attract travelers, some of the practices and patterns are essentially applied as a building block to the newer architectures to achieve the retail-like experience. The architecture must address the concern of scale, traveler context, and store travel information that is powered by big data analytics and insights. A truly omnichannel experience can be built with such architectures and realization of the same can be achieved across various channels and different types of clients or applications using them.

References

1. IATA New Distribution Capability: http://www.iata.org/whatwedo/airline-distribution/ndc/Pages/default.aspx
2. Yuan, X., Fernandez, E.B.: Patterns for business-to-consumer e-commerce applications. arXiv preprint arXiv:1108.3342 (2011)
3. Lazaris, C., Vrechopoulos, A.: From multi-channel to "omnichannel" retailing: review of the literature and calls for research. In: 2nd International Conference on Contemporary Marketing Issues, (ICCMI), pp. 18–20 (2014)
4. Mirsch, T., Lehrer, C., Jung, R.: Channel integration towards omnichannel management: a literature review (2016)
5. Trautmann, H., Vossen, G., Homann, L., Carnein, M., Kraume, K.: Challenges of data management and analytics in omnichannel CRM (No. 28). Working Papers, ERCIS-European Research Center for Information Systems (2017)
6. van Delft, L.: Omni channel shopping behavior during the customer journey. Doctoral dissertation, Master thesis. Eindhoven University of Technology, Eindhoven, Netherlands (2013)
7. Verhoef, C., Kannan, P.K., Inman, J.J.: From multi-channel retailing to omni-channel retailing: introduction to the special issue on multi-channel retailing. J. Retail. **91**(2) (2015)
8. Cook, G.: Customer experience in the omni-channel world and the challenges and opportunities this presents. J. Direct, Data Digital Mark Pract. **15**(4) (2014)
9. Beck, N., Rygl, D.: Categorization of multiple channel retailing in multi-, cross-, and omni-channel retailing for retailers and retailing. J. Retail. Consum. Serv. **27** (2015)
10. Hansen, R., Sia, S.K.: Hummel's digital transformation toward omnichannel retailing: key lessons learned. MIS Q Executive **14**(2) (2015)
11. Nishtala, R., Fugal, H., Grimm, S., Kwiatkowski, M., Lee, H., Li, H.C., McElroy, R., Paleczny, M., Peek, D., Saab, P., Stafford, D.: Scaling Memcache at Facebook. In: nsdi, vol. 13, pp. 385–398 (2013)
12. Carnein, M., Heuchert, M., Homann, L., Trautmann, H., Vossen, G., Becker, J., Kraume, K.: Towards efficient and informative omni-channel customer relationship management. In: International Conference on Conceptual Modeling, pp. 69–78. Springer, Cham (2017)

Hari Bhaskar Sankaranarayanan working in International Institute of Information Technology, Bangalore. His research area includes channelization algorithms and big data analytics.

Jayprakash Lalchandani working in International Institute of Information Technology, Bangalore. His research area includes channelization algorithms and big data analytics.

An Architecture to Enable Secure Firmware Updates on a Distributed-Trust IoT Network Using Blockchain

George Gabriel Richard Roy and S. Britto Ramesh Kumar

Abstract The world has evolved to a point where it is impossible to imagine interaction without networks or without connections between devices and people. The advent of the fifth industrial revolution saw the rise of interconnected devices and 'things'. Internet of Things (IoT) has become the defacto standard for devices and people communicating with each other. The newer the devices that are released in the market so has to be the security be updated to counter security attacks. Every device has to be updated to the latest firmware to keep it secure and work optimally; this paper proposes an architecture to facilitate a secure firmware update in a distributed trust-based blockchain IoT network.

Keywords IoT · Blockchain · Trustless · Firmware · Update · Payload · Secure Distributed trust

1 Introduction

IoT has crossed its infancy and now it is in a stage where enormous growth. The number of IoT devices is expected to be 5 billion by 2020, and the number will continuously increase. The IoT devices are tiny and small, while those are mostly embedded devices designed for specific operations, e.g. sensing, automation, etc. [1]. Today's technology of interconnected devices allows us to be connected as well as being in control of them [2, 3]. The status of security given to those devices are very basic and its vulnerable to a plethora of attacks, this is the reason IoT devices are susceptible to many malware attacks. The malware attacks are growing exponentially and large-scale IoT implementations are the target of antisocial groups converting

G. G. R. Roy (✉)
Department of Information Technology, St. Joseph's College, Tiruchirappalli, India
e-mail: georgegrroy@gmail.com

S. Britto Ramesh Kumar
Department of Computer Science, St. Joseph's College, Tiruchirappalli, India
e-mail: brittork@gmail.com

S. Smys et al. (eds.), *International Conference on Computer Networks and Communication Technologies*, Lecture Notes on Data Engineering and Communications Technologies 15, https://doi.org/10.1007/978-981-10-8681-6_61

them into botnets which can lead to massive outages and attacks. This brings the current need to provide proper security to those IoT devices [4, 5].

With the reduction in cost of implementing IoT-based devices, nowadays mostly every manufacturer wants to implement Internet-enabled devices. A report suggests that there will be around 20 billion IoT devices online, with a majority of them will be consumer-based products which are still more vulnerable because of the always-on Internet connections, lack of knowledge to operate devices securely and infrequent updation of firmware [6]. This is heaven for the attackers because they can easily find loopholes and possible backdoors in the devices and exploit them to their own advantage [3, 7].

Security starts from the device itself and to keep the device up to date, its firmware needs to be updated frequently and securely. This will help delay the attackers from gaining control of the devices and thereby also patching any loopholes or backdoors [1, 5]. Centralized servers are like sitting ducks waiting to be as prey. The attackers know that all that flows from the centralized server can be either modified or stolen. This centralized focus of control is receptive to be corrupted or vulnerable to many attacks [8, 9].

Considering the above factors, this paper proposes an architecture to enable secure firmware updates on a distributed-trust IoT network using blockchain. In the proposed scheme, there are four separate entities: the Author Application Server (AuAS), Vendor Distribution Server (VeDS), the Payload Distribution Server (PDiS) and the blockchain of IoT devices.

This paper is organized as follows. Section 2 presents some ideas about blockchain and distributed trust. Section 3 presents the proposed high-level architecture flow of operation. Section 4 contains the conclusion.

2 Blockchain and Distributed Trust

2.1 Blockchain

Blockchains follows a protocol where two separate entities communicate with each other and establish a transaction over the Internet without the need for a centralized server. When you digitally transfer value from one account to another on the blockchain, you are trusting the underlying blockchain system to both enable that transfer and ensure sender authenticity and currency validity [10, 11].

When this digital transaction occurs a value is transferred from one account to the other on the chain, this means that the blockchain is trusted to enable the transfer. The information here is not copied but it is distributed. This information is not stored at a single location; the records of the blockchain database are kept public and easily verifiable. This is the reason why it cannot be corrupted easily by any hacker because it does not have a centralized storage but it is hosted on an array of computers worldwide.

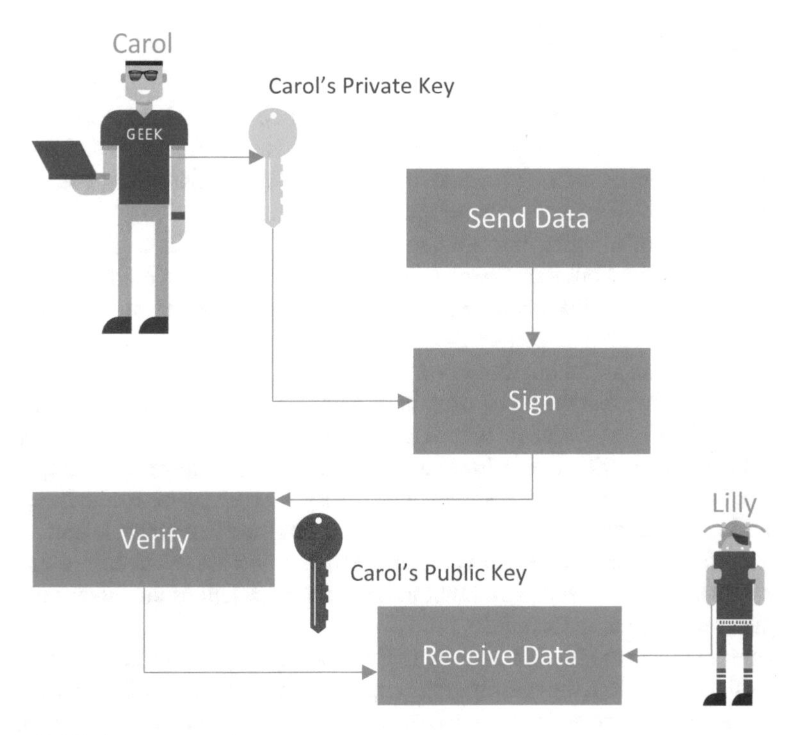

Fig. 1 Public key cryptography

This leads to the point that blockchains do not have a single owner, it cannot be owned by a single entity. Blockchains also do not have a single point of failure. Stating that blockchains are a trustless network is not exactly true. They minimize the trust that is needed from a single entity in the network, this trust is shared among various entities in the network [10] (Fig. 1).

A blockchain is a shared ledger or a distributed ledger where every transaction that takes place is done using public key cryptography. Each peer-to-peer transaction involves a Public Key (PuK) and Private Key (PiK) to authenticate the transaction. The PuK is visible to everyone but the PiK is visible only to the owner. This PiK issues a digital signature to ensure that the transaction has been originated from the authenticated user and also that the transaction is not susceptible to any attacks or modification, thereby leading to failure of the transaction if there is any modification if any.

2.2 Consensus Mechanism

Any Blockchain (BC) is in a state of consensus, that is, the blockchain is in general agreement with a set of nodes that holds the system together. The BC is designed in a way that it is driven by interactions and transactions. Being a distributed peer-to-peer distributed ledger, every entry is in the form of a transaction.

In a BC network, digital information be it in the form of transactions, files or tokens are nor copied but they are distributed. Thus, information on the blockchain exist as shared entity; this is where the Consensus Mechanism (CoMe) comes into play.

This CoMe is an algorithm that has one of the proofs (work, activity, stake, burn). This is to make sure that the block contains the only truth and the only version of the truth, and to block any attackers who intend to change the integrity of the BC [10].

The BC uses a Merkle tree (Mtree) to write blocks. The Mtree is a cryptographic hash function tree where every non-leaf node has a label with the hash of the names of child nodes. The Mtree is used to verify the data stored and that which is transmitted between different nodes in the network. This Mtree drives the BC to ensure that any duplication should be avoided and that none of the transactions are altered or have corrupted information in them [12].

Figure 2 depicts the Merkle tree where Carol wants to send some data to Lilly, he initiates the transaction, the information is passed as a transaction, this transaction is stored in the repository which is then connected to the repository protocol layer, the protocol specifies the defined rules and the transaction is applied with values. This information is sent to the overlay network, and it is later verified and stamped where at the final stage, it gets stored on the BC. This information is then encrypted into a non-repudiable, irreversible cryptographically secure block to the chain. The trust is distributed to other blocks in the chain, and Lilly gets the file from the BC [12].

3 Entities of the Firmware Update

To facilitate a secure transfer of the update payload over the network, certain entities are included along with IoT nodes (Fig 4). They are the author of the firmware payload, the author's application server, the vendor's distribution server and finally the firmware update server (Fig. 3).

3.1 Firmware Author (FA)

Every device when it leaves the production contains an embedded firmware as default. This version will be the first version that provides the functionality to the device and allows it to connect to other devices as well. If in case there is any vulnerability

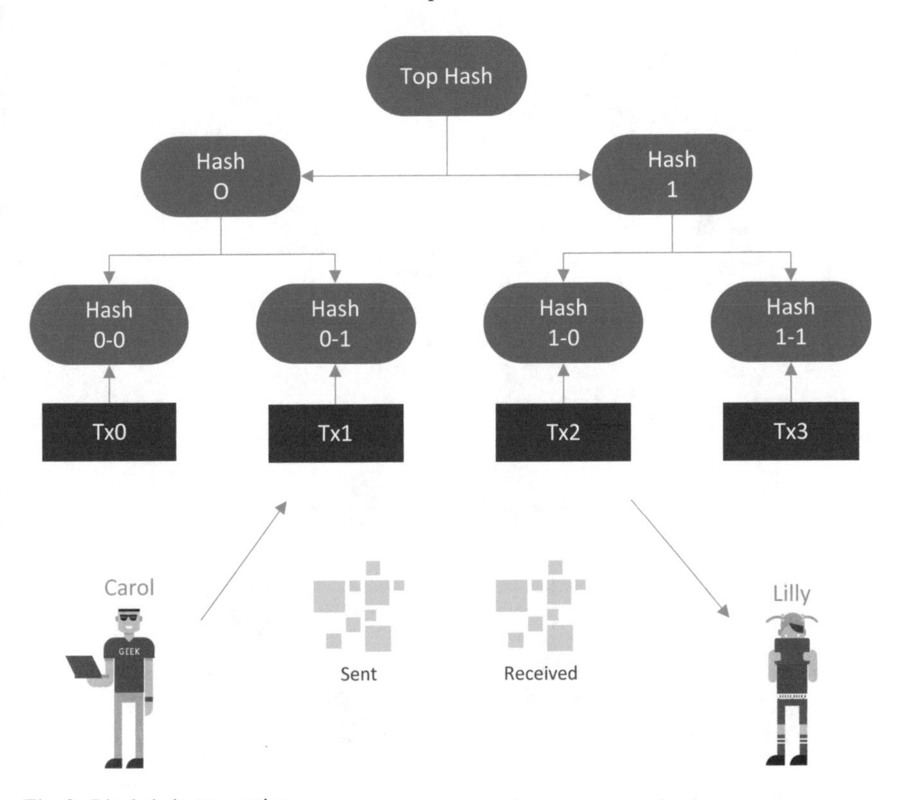

Fig. 2 Blockchain transaction

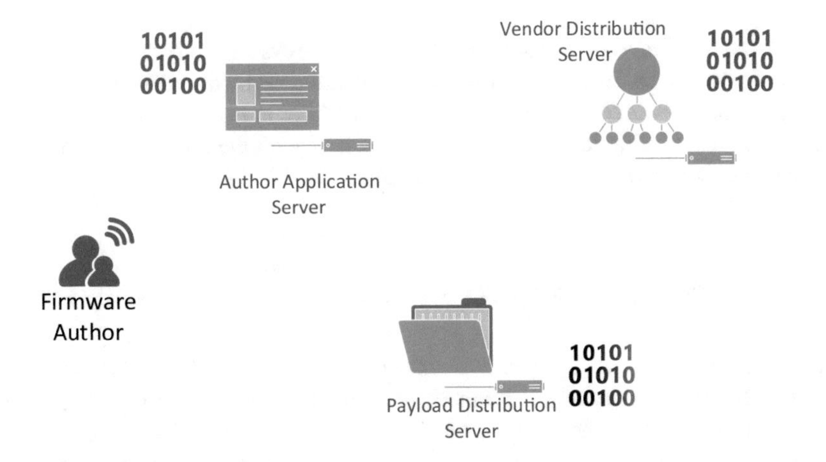

Fig. 3 Entities needed for firmware update

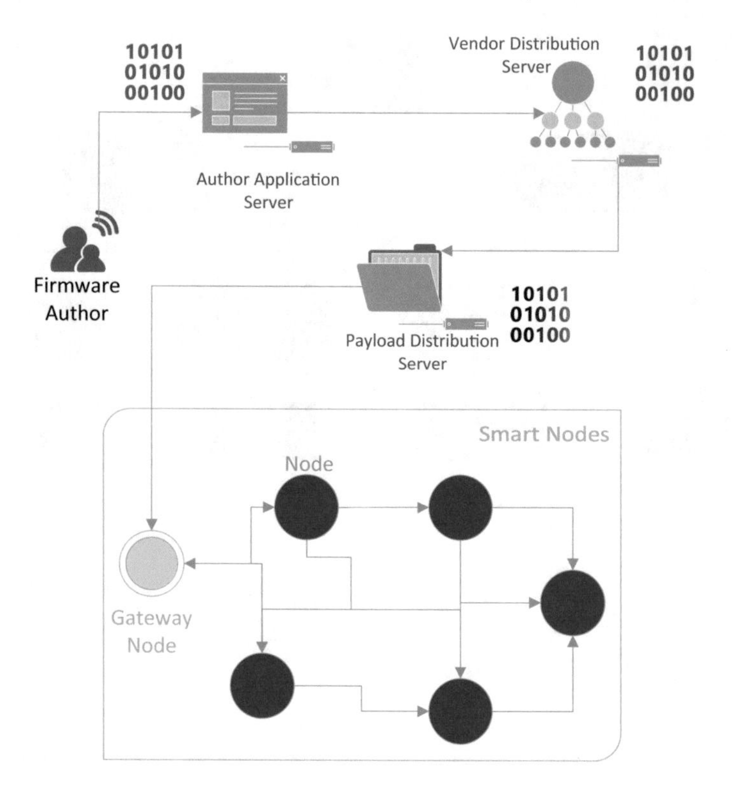

Fig. 4 Proposed architecture for secure firmware update

in that first version, then a requirement for a new firmware is needed to protect the device from attacks. Every firmware should be authored by an entity, this authoring entity can be outsourced or the device vendor can have in-house firmware authoring. The FA is responsible to rectify the error which was on the previous firmware and create a new firmware to be sent to the devices.

3.2 Author Application Server (AuAS)

After the FA creates the new firmware, the payload is sent to the AuAS where it is verified for its integrity and is secured by a manifest file containing a Hash. This Hash is used to check the integrity of the payload, whether the contents have been tampered with or not. The AuAS is responsible in sending the secured payload to the VeDS.

3.3 Vendor Distribution Server (VeDS)

The VeDS is the Original Equipment Manufacturer's (OEM) server; this server receives the payload checks for the integrity of the file and later checks that if the firmware satisfies the protocol and standards set by the OEM, the manifest is later then updated with the seal of approval in the form of a digital signature and sent to the payload distribution server.

3.4 Payload Distribution Server (PDiS)

The firmware update server is responsible to check the payload for the signature, the FA and the VeDS. If the integrity check goes well, then the payload is stored in the PDiS ready for distribution.

4 Proposed Architecture

The nodes on the network are on a blockchain, that is, they are on a peer-to-peer network. Every node is interconnected with each other on a shared-distributed trust. This incorporates security in the device level where the integrity of the node has very least chance of being compromised, if in any event a node is compromised; the node is removed from blockchain, this ensures that the attacker will find it very difficult to attack the network. In addition to this, the inclusion of the gateway node establishes a secure connection with all the nodes in the network and there is a ping to check the integrity of the node in the network.

In the event of a firmware upgrade, the PDiS sends a push notification to the gateway node to reduce traffic and bandwidth. The gateway node checks the nodes for the version, if the nodes have an outdated version, the gateway sends the request to the PDiS and PDiS sends the payload to the gateway node. The Gateway Node (GaN) checks the payload for the FA's digital signature and the VeDS's signature, and download the payload. The GaN then sends the payload to the other nodes through peer-to-peer torrent protocol.

The GaN acts as a seed and the other nodes act as leechers; this is only possible in the blockchain network. If any node is compromised, it is forced to leave the network.

5 Conclusion

This paper discusses the issue of an outdated IoT device and the threat of not securing devices. The IoT would benefit even more if it is built on a blockchain. The distributed trust networks work in a way that establishes trust by sharing the trust among all the nodes in a network. This paper concludes that a secure update will provide additional security and will keep the attacker at bay. The proposed architecture utilizes blockchain mechanism to establish a peer-to-peer data sharing in a secured manner.

Future work will contain the implementation of this architecture in a simulator along with suitable algorithms and a mathematical model.

References

1. Lee, B., Malik, S., Wi, S., Lee, J.-H.: Firmware verification of embedded devices based on a blockchain. In: Institute for Computer Sciences, Social Informatics and Telecommunications Engineering, pp. 52–61 (2017)
2. Prada-Delgado, M.A., Vázquez-Reyes, A., Baturone, I.: Trustworthy Firmware Update for Internet-of-Thing Devices Using Physical Unclonable Functions (2017)
3. Chang, C.-Y., Kuo, C.-H., Chen, J.-C., Wang, T.-C.: Design and implementation of an IoT access point for smart home. Appl. Sci. (2015)
4. De Donno, M., Dragoni, N., Giaretta, A., Mazzara, M.: AntibIoTic: Protecting IoT Devices Against DDoS Attacks (2017)
5. Arias, O., Wurm, J., Hoang, K., Jin, Y.: Privacy and Security in Internet of Things and Wearable Devices. IEEE (2015)
6. Barreto, L., Celesti, A., Villari, M., Fazio, M., Puliafito, A.: An Authentication Model for IoT Clouds (2015)
7. Barrera, D., Molly, I.: IDIoT: Securing the Internet of Things Like it's 1994 (2017)
8. Arias, O., Ly, K., Jin, Y.: Security and Privacy in IoT Era, Smart Sensors at the IoT Frontier. Springer, Berlin (2017)
9. Ling, Z., Luo, J., Xu, Y., Gao, C., Wu, K., Fu, X.: Security Vulnerabilities of Internet of Things: A Case Study of the Smart Plug System. IEEE (2018)
10. Kasi Reddy, P.: https://medium.com/@preethikasireddy/eli5-what-do-we-mean-by-blockchains-are-trustless-aa420635d5f6 (2016)
11. Constantinos, K., Devetsikiotis, M.: Blockchains and Smart Contacts for the Internet of Things. IEEE (2017)
12. Sharma, T.K.: https://www.blockchain-council.org/blockchain/what-is-merkel-tree-merkel-root-in-blockchain/ (2017)

George Gabriel Richard Roy is an Assistant Professor of Information Technology at St. Joseph's College (Autonomous), Tiruchirappalli. He received his MCA from Bishop Heber College, Tiruchirappalli. His research interests are Wireless Communication, Internet of Things and Blockchain.

Dr. S. Britto Ramesh Kumar is an Assistant Professor of Computer Science at St. Joseph's College (Autonomous), Tiruchirappalli. His research interests include software architecture, wireless and mobile technologies, information security and Web Services. He has published many journal articles and book chapters on the topics of Mobile payment and Data structure and algorithms. His work has been published in the International journals and conference proceedings, like JNIT, IJIPM, IEEE, ACM, Springer and Journal of Algorithms and Computational Technology, UK. He was awarded as the best researcher for the year 2008 in Bishop Heber College, Tiruchirappalli. He has completed a minor research project. He has visited countries like China, South Korea and Singapore.

A Survey of MTC Traffic Models in Cellular Network

T. N. Sunita and A. Bharathi Malakreddy

Abstract Machine-Type Communicating (MTC) Devices (MTCD) are usually wireless devices, for example, sensors, actuators and smart metres which can talk with each other through exchanging data and take decisions with little or no human intervention. Cellular networks are considered to be one of the best technologies for accommodating MTC. The characteristics generated by MTC traffic are completely different from Human-to-Human (H2H) communications. There will be no control on the increasing number of MTCDs and because of which volume of MTC traffic keeps on increasing. In this paper, we make a comprehensive survey on MTC traffic issues over cellular network and solutions to improve the MTC over cellular network. This paper provides a brief overview of 3GPP standards supporting MTC. We also survey MTC traffic issues in heterogeneous networks consisting of cellular networks, capillary networks. Finally, we review recent standard activities and discuss the open issues and research challenges.

Keywords M2M · Machine-to-machine · Machine-type communication
Traffic issues · Traffic models

1 Introduction

The web has experienced enormous changes since its introduction to the world and has turned into a critical communication medium anyplace whenever. The current network protocols, traffic profile and infrastructure are predominantly intended for Human-to-Human (H2H) communications. Of late, Machine-to-Machine (M2M) or Machine-Type Communication (MTC) has picked up prominence in the industry. Machine-Type Communicating (MTC) Devices (MTCD) are usually wireless

T. N. Sunita (✉) · A. Bharathi Malakreddy
Department of Computer Science, B.M.S. Institute of Technology, Bengaluru, Karnataka, India
e-mail: sunita.neelagiri@gmail.com

A. Bharathi Malakreddy
e-mail: bharathi_m@bmsit.in

S. Smys et al. (eds.), *International Conference on Computer Networks and Communication Technologies*, Lecture Notes on Data Engineering and Communications Technologies 15, https://doi.org/10.1007/978-981-10-8681-6_62

devices, for example, sensors, actuators and smart metres which can talk with each other through exchanging data and take decisions with little or no human intervention. It is assessed that by 2020, these gadgets will increment by 50 billion to connect different devices in the network.

Existing cellular system, with wide coverage and flexible data rates, primarily designed for H2H communications and expected to support MTC. The main characteristic required for MTC is to have MTCDs with low power consumption and with less deployment cost. To tackle some of the MTC issues, Third Generation Partnership Project (3GPP) Release 10 proposed some of the service requirements for supporting MTC in Long-Term Evolution-Advanced (LTE-A) cellular networks. One of the issues of this release was that it was not able to differentiate newly introduced MTCDs with User Equipments (UE). The characteristics of MTC were different than the normal H2H communication, because of which there was need of some design modification of cellular systems to support MTC. Niyato et al. [1] proposed some changes for LTE/LTE-A for supporting MTC.

MTC networks are mainly divided into two categories: M2M cellular networks and M2M capillary networks as shown in Fig. 1. In M2M cellular networks, communication is mainly through licensed spectrum band, such as LTE/LTE-A. Existing cellular network infrastructure would support MTC because of its flexible data rates and wide coverage. In M2M capillary network, the communication is through unlicensed spectrum band, and furthermore, it incorporates Bluetooth Low Energy (BLE), Halow, ZigBee and Smart Utility Networks (SUN). Low-power and long-range MTC procedures are expected to produce connections for wireless devices utilizing Low Power Wide Area Network (LPWAN) technologies. LPWAN technologies incorporate LoRa, SigForx, IEEE 802.11ah and LTE-MTC [2]. This paper mainly focuses on cellular M2M traffic issues and some of the traffic issues in capillary network as well.

In Heterogeneous Network (HetNet), M2M communication involves the combination of M2M cellular network and M2M capillary network. Here, MTCDs try to connect to M2M servers using either licensed or unlicensed band depending on the type of spectrum availability. Sometimes MTCDs form a group or cluster and one of the members of cluster will send packets to the M2M gateways or small cells.

In the current cellular network, MTCDs and H2H devices share the same network infrastructure. There will be no control over the increasing number of MTCDs and because of which volume of M2M traffic keeps on increasing. The volume of M2M traffic is directly dependent on the number of machines trying to access the network at a given instance of time. The MTCDs will grow exponentially and the demand for network resource also increases; in fact, MTC starts competing with existing HTC for network resource, which ultimately leads to huge M2M traffic.

Subsequently, it is essential to understand the MTC traffic patterns and how they are not quite the same as H2H traffic. With the understanding and knowledge of traffic patterns, which can prompt better administration of shared network resources and can guarantee better QoS for both types of devices. There have been several researches to address the M2M traffic issues. Shafiq et al. [3] have attempted to characterize MTC traffic in cellular networks and contrast MTC traffic and Smartphone traffic for

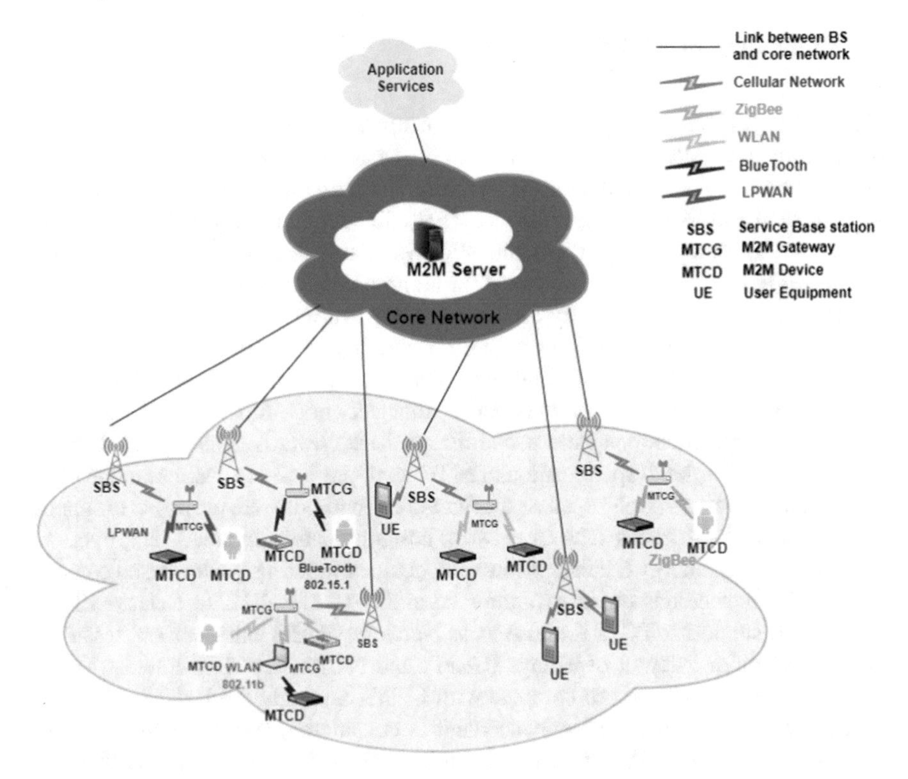

Fig. 1 M2M communication over cellular network LTE/LTE-A and capillary network/Bluetooth low energy M2M communication

various aspects like traffic patterns, device mobility, application usage and network performance. The most critical initial step for MTC is to distinguish MTCDs in the network. Laner et al. [4] have proposed certain feature or characteristics based on Internet Protocol layer measurements to detect MTC traffic patterns with accuracy up to 99% in a supervised case. Smiljkovic et al. [5] did some investigation on existing M2M traffic model and proposed how to incorporate M2M traffic into existing LTE cellular network utilizing contention-based access mechanism.

This paper first introduces a review of 3GPP models and talks about MTC traffic issues of M2M correspondence and recommended solutions for these issues. The remainder of the paper is sorted out as follows. In segment II, a concise overview of 3GPP standards is examined. In segment III, discussion on M2M characteristics, categories and M2M traffic models is given. In IV segment, some of the current advancements and discussion on the open issues and research challenges in M2M traffic are given, followed by the conclusions given in Sect. 5.

2 Overview of 3GPP Standards

The M2M applications have gigantic effect on numerous businesses, and thus, there is a requirement for standardization. The technical standards are being created by a few standard bodies, for example, Third Generation Partnership Project (3GPP), Institute of Electrical and Electronic Engineers (IEEE), Telecommunications Industry Association (TIA), European Telecommunication Standards Institute (ETSI) and so forth attempted to give a few standards to M2M communication and answers for some of M2M issues. This paper mostly focuses on M2M traffic issues in LTE/LTE-An of 3GPP, and hence, brief overviews of 3GPP standards are discussed.

In 3GPPs Rel-8, facility of M2M communication was first defined and did some assessment of architecture for MTC in cellular networks. In Rel-8 and Rel-9, initial essential specifications were added. From there onwards, considering different aspects of MTC a detail specification of MTC were presented to accommodate MTC in cellular network. In Rel-10, more specifications methods were proposed to support MTC. Other M2M issues such as network communication enhancement, overload control, security, energy efficiency, low cost, device triggering, group-based communications enhancement, etc. were addressed in Rel-10 to Rel-12. In Release 13 particularly to support MTC, a Radio Access Network (RAN) enhancement was done wherein Cellular Internet of Things (CIOT) and Narrow Band Internet of Things (NB-IOT) were suggested to compete with LPWA arena. In the Rel-14, which are still in progress, narrowband enhancement is considered, and expanding the LTE platform to support V2X application is also considered. More about this literature can be found in [6–13].

3 M2M Categories, Characteristics and Traffic Models

3.1 M2M Categories

According to Shafiq et al. [3], MTCDs are developed focusing on particular application and usage. Likewise, have critical contrasts among MTCDs. Henceforth, it was important to classify MTCDs, to better comprehend the MTC traffic pattern and their qualities. To categorize the MTCDs, the author started off by identifying the device model utilizing the Type Allocation Code (TAC), which is only a one of a kind identifier of MTCDs. Just by remarkably distinguishing every device was insufficient to decide their application categories and to recognize their M2M category, in light of the fact that a large number of the devices are being utilized as a part of multiple applications. Thus, device classification scheme utilized by cellular service provider was used as a reference template and around 150 devices were classified as MTCDs and divided into six categories based on their application usage as asset tracking, building security, fleet, metering, telehealth and miscellaneous.

3.2 M2M Characteristics

Shafiq et al. [3] additionally attempted to characterize MTC traffic in cellular networks and contrasted MTC traffic and Smartphone traffic for different viewpoints like traffic patterns, device mobility, application utilization and network performance. Compared M2M traffic and Smartphone traffic in the following aspects:

Aggregate Traffic Volume On examination of aggregated uplink and downlink traffic demonstrated that MTCDs have less downlink traffic when compared with smartphone. Thus, the network operators need to come up with a superior approach for allocating spectrum and spectrum administration to evade contention between the MTCDs low volume, substantial uplink traffic and smartphone's high volume, overwhelming downlink traffic.

Traffic Volume Time Series The investigation of traffic volume time series of MTCDs and Smartphone demonstrates that various MTCDs indicate different kind of diurnal practices than Smartphone. Network resources are normally provisioned relying upon the peak usage. At the point when the peak hours of both MTC traffic and HTC traffic coincide can cause contention problem. This requires a better administration of capacity planning for the shared network resources. One additional finding was that traffic produced by MTCDs is in a synchronized manner, accommodating such traffic within the limited spectrum resource was not possible and which can lead to congesting the radio network. Hence, network usage should be randomized.

Traffic Sessions The analysis of session level for active time, session length and session inter-arrival time shows that huge traffic volume is not dependent or not generally associated with more active time. MTCDs have distinctive session length and different inter-arrival time. These factors can be useful for device manufacturers so that they can enhance battery management system. These factors are additionally useful for network operators to improvise radio network parameter for MTCDs. Henceforth, better network resource allocation schemes are required so that contention can be avoided within the two types of traffic.

Application Usage MTCDs or HTCDs are designed or used for different application uses. The analysis of aggregate traffic volume of both devices with respect to application usage shows that M2M traffic usually uses custom protocols. And these protocols come with a baggage of some disadvantage or have some adverse effects. Consequently, it is troublesome for the network operators to grasp such protocols, which are not standard ones. And also, network operators might not be able to handle the adverse effects of these protocols. Thus, it is critical to standardize the M2M protocols so it is helpful for both M2M application service supplier and network operators.

Network Performance The analysis of network performance for the M2M traffic as compared to the HTC traffic shows that it is dependent on the devices radio technology (2G, 3G or 4G). Still, there are some devices which use old technologies because of which the Round Trip Time (RTT) of MTC is larger than HTC. Due to which it causes an overhead problem for the network operators to maintain some older services also. Even the packet loss ratio is higher for MTC because of MTCDs being placed in locations where signal reception is poor.

3.3 MTC Traffic Models

3GPP standard organizations Rel-11 and Rel-12 of evolved packet system and its class-based network-initiated QoS control scheme were released to help MTC traffic. Likewise, the network administrators moved from single to multi-service offering, because of which the quantity of connected devices expanded alongside their traffic volume began expanding quickly. Henceforth it is vital to have a traffic model to measure data traffic and depending on the statistical data obtained from the traffic model result can be used to optimize the network performance and QoS for MTC.

3GPP at first proposed a generic traffic model to replicate activity patterns around access plane. Simple poison process was modelled for different kind of network access (Synchronous or Asynchronous) coordinated or uncoordinated M2M traffic. Subsequently, the various arrival rate $\lambda(t)$ was dynamic in nature depending on the scenario.

This generic traffic model did not consider different kinds of applications use the MTCDs are involved. The behaviour of the MTCDs was not considered and treated all the devices as same. Due to which the complex behaviour (e.g. surveillance video) and their statistical properties could not be captured. Hence, this called for different kinds of traffic models requirement for analysing the source traffic. In source traffic model, individual devices were considered to capture the traffic patterns depending on their application usage.

Source Traffic Modelling Modelling the traffic of large number of individual devices simultaneously is called source traffic modelling. After the analysis of different applications, we came to know that they have different traffic pattern and associated states. Nikaein et al. [14] did some analysis of the MTC applications and has uncovered that MTC has three basic fundamental traffic patterns.

Periodic Update (PU) This kind of traffic pattern happens from those devices which transmit data in a timely manner to one of the core central unit. Additionally, this can be viewed as an event activated by the device at a regular interim. PU is not a real-time based, but a regular interval based along with constant data size. Also, these transmitting intervals are configurable by the server. E.g. for PU is smart metre reading (e.g. gas, electricity, water).

Event Driven (ED) This kind of traffic pattern is found in those devices where an event is activated and its corresponding information has to be transferred. The event trigger cause could be either by some of the parameters crossing the calculated threshold or may be because of server commands sent to the device to control it remotely. Most of the EDs are realistic traffic with variable time pattern, data size is different in both uplink and downlink. ED, e.g. for uplink would be notifications of the devices used in the healthcare systems and for downlink would be Tsunami alarm.

Payload Exchange (PE) This kind of traffic pattern is seen after an event, tailing one amongst the traffic types (PU or PE). Every one of the situations where extensive measure of data is exchanged within the device and the server can be considered as (PE) and the traffic is probably going to be more uplink predominant. The data size can be steady (e.g. telemetry) or variable size (e.g. image transmission).

When you consider the real-world applications, most of the time traffic is a combination of above-mentioned traffic types. Hence, modelling a traffic model based on the above-mentioned three types of elementary types would provide high degree complexity and accuracy.

Nikaein et al. [14] for convenient modelling of MTC traffic, proposed ON–OFF structure and tried to integrate the above-mentioned traffic pattern into Markov structure with the four totally different states: OFF, PU, ED and PE. The classification of the state into ON and OFF mode was to deal with the structure where there would be no data transmission or no packets being transmitted between the phases of activity. This helped to predict some parameters for each state, e.g. the attribute 'latency < 100 ms' was added to the ED state, with a specific end goal to guarantee quick sending off alarms.

To model the state transitions, the author used Semi Markov Models (SMM). The purpose behind utilizing SMM models for MTC was that (i) they permit capturing a wide range of spectrum traffic attributes, particularly the nearly vanishing data rate; (ii) enable augmented modelling if side information is available (e.g. the correct number of states is known); and (iii) fitting mechanisms to be established, which allow for good fitting quality, even if nothing but raw traffic measurements are given.

Aggregated Traffic Modelling Treating the collected data from all MTC devices as single stream is named as aggregated traffic modelling. As indicated by Smiljkovic et al. [5] for various access and capacity assessments wherein we need not bother with each and every node behaviour; in such case, aggregated traffic modelling is appropriate. This model is less precise when compared with source traffic model and will not have the capacity to catch genuine traffic characteristics. For a huge range of machines, this model is the best suited approach. 3GPP proposed few traffic models that are based on aggregated traffic [6, 7].

Model 1: 3GPP uncorrelated aggregated traffic model, which treats uncoordinated events triggering data traffic over a given time period.
Model 2: 3GPP correlated aggregated traffic model, which treats coordinated events triggering data traffic over a given period of time.

Simulated Traffic Models Laner et al. [15] utilized traffic modelling for more accurate traffic description when compared with the 3GPP aggregated traffic models. This model had two major issues: (i) huge quantity of devices to be modelled in parallel, and (ii) the strong spatial and temporal correlation between the machines. To beat these issues, Coupled Markov Modulated Poisson Processes (CMMPP) was proposed which has low complexity as compared to generic source traffic model and gives better accuracy. As indicated by this model, one master background process was used to establish correlation between the devices temporal behaviour and spatial behaviour. The state transition matrix for the machine in instance t time was calculated as

$$p_n[t] = \theta_n[t]p_c + (1 - \theta_n[t])p_u \quad (1)$$

p_c Transition matrix of coordinated machine.
p_u Transition matrix of uncoordinated machine.
$\theta_n[t]$ Correlation sample, where n refers to the nth machine, while t refers to an instance in time.
$p_n[t]$ The state transition matrix for the nth machine in instance t time.

For example, Markov chain for two state transition matrices

$$p_c = \begin{bmatrix} 0 & 1 \\ 1 & 0 \end{bmatrix} \quad p_u = \begin{bmatrix} 1 & 1 \\ 0 & 0 \end{bmatrix} \quad (2)$$

In Eq. (2), the focus for uncoordinated matrix is to never trigger any alerts, whereas for the coordinated matrix, trigger one alerts in one time slot and then return to regular operation.

Smiljkovic et al. [5] did some analysis on the existing M2M traffic model (above-mentioned models) and proposed how to integrate M2M traffic into existing LTE cellular network utilizing a contention-based access scheme. After the simulation investigation, it was discovered that source traffic model gives more exact outcomes and is appropriate for less range of machines, whereas aggregated traffic model outcome is less precise and suitable for a huge range of machines. The CMMPP model is trade-off between the two models, which has an accuracy of source traffic model and can be modelled for substantial number of devices. The outcomes additionally give some knowledge on the design parameters to serve diverse number of machines with expected rejection probability.

The model of Olav et al. [16] tried to cover a wide range of MTCDs (sources) having different traffic characteristics varying from regular to bursty of traffic sources. This model was called modulated renewal processes for packet level traffic modelling per MTCD.

Specific distribution of inter-arrival times and number of arrivals for a particular state was modelled by a random integer variable. Here, the sum of the inter-arrival time between the arrivals for a particular state is considered as total time spent on a particular state. Once the total time spent has expired for a particular state a new Renewal Process (RP) of arrival sequence with a specific distribution is triggered on another state as per modulated Markov chain. Hence the modulated RP is an extension of legacy RP wherein depending on a state variable, the distribution of arrival process changes. The variance of the packet count was derived by applying Laplace transform of the Z-transform for this variable. There was some asymptotic behaviour for large time intervals. A two-state model was also proposed wherein the inter-arrival time was exponentially distributed, and the number of arrivals was geometrically distributed in a particular state, for a total of four model parameters. This model also has the option of specifying the packet size distribution depending on the state variable. Hence, a wide range of M2M sources with different traffic can be captured using this model. There are some advantages and disadvantages of this model. If the source type and traffic pattern are known then modulated RP may not give accurate results. This model is more suitable for the source type which has random traffic generation pattern. This model cannot be considered as best model because it involves a large number of parameters. This model cannot be used for aggregated traffic.

Thomsen et al. [17] tried to assess the impact of traffic on the Base Station (BS) caused by spikes or bursts which have spatial characteristics. Stochastic geometry was used to consider the spatial aspects of MTCDs and events. Devices can be in either regular mode or alarm mode depending on whether the device location falls within the vicinity of event location. Poisson Point Processes (PPP) was used to capture the position of the device and event location and then derive an analytical expression for the total rate. This model was further extended to study temporal correlation using two-state Markov chain for each device. This model did not include various types of MTCDs and events.

Grigoreva et al. [18] came up with traffic model for MTC automotive applications which can be used for technology evaluations, network planning and techno-economic analysis. This model uses Coupled Markovian Arrival Process (CMAP) to model individual traffic sources and includes time and space correlation of sources. Time correlation was used to capture the burstiness and multimodality of the real traffic. Space correlation was captured by distributing the sources over an abstract road topology. This model was completely dependent on the characteristics measured on the sources and has higher computational complexity.

4 Open Issues and Challenges

There have been continuous efforts on improvement of M2M traffic issues for MTC; there are still some open issues and research challenges. In this section, we discuss some of the issues and propose future research directions: (i) a brief review on Narrow Band–Internet of Things (NB-IoT) for supporting IoT in cellular networks, (ii) discussion on Software Defined Networking (SDN)-enabled M2M communications and (iii) discussion on cloud-based M2M communications.

4.1 Narrow Band IoT

NB–IoT gives enhanced coverage to enormous number of low-throughput low-cost devices with low device power utilization in delay-tolerant applications. NB–IoT can be built on existing LTE network. NB–IoT supports three operation modes: (i) standalone, using standalone bearer; (ii) guard band, using the unused resource blocks inside an LTE carrier's guard band; (iii) in band, using resource blocks inside a typical LTE carrier. Ratasuk et al. [19] did examination on NB–IoT, which included battery life, scope, coverage, capacity and delay. The simulation results show that it meets all three operation modes design requirement.

The NB–IoT utilizes the traditional Orthogonal Frequency Division Multiple Access (OFDMA) for downlink and for uplink Single Carrier–Frequency Division Multiple Access (SC-FDMA). According to Lin et al. [20], the network bandwidth for NB-IoT is 180 kHz for both downlink and uplink. The downlink supports 15 kHz subcarrier spacing and uplink supports 3.75 kHz subcarrier spacing. The author utilized NB-IoT Physical Random Access Channel (NPRACH) design and three coverage classes to exhibit that NPRACH can satisfy the extended coverage demand and accomplish the uplink synchronization precision. More detailed literature about NB-IoT random access mechanisms and several other receiver algorithms for detecting NPRACH can be found in [20]. Ksairi et al. [21] proposed Multiservice Oriented Multiple Access (MOMA) scheme to support M2M in various classes of users and each class depends on data rate, service type and traffic pattern. The proposed MOMA scheme was implemented in both 1.4 MHz bandwidth (LTE-M) and 200 kHz bandwidth (NB-IoT) and the simulation results demonstrate that MOMA beats the other M2M related proposed techniques for LTE.

4.2 Software Defined Networking (SDN) Based M2M

SDN gives more flexibility in wired network management by abstracting the underlying infrastructure (control plane) with the data plane and reducing the communication overhead and delay in both Core Network (CN) and Random Access Network (RANs). Hesham et al. [22] proposed a Network Access Control (NAC) solution based on SDN for resource allocation and access control of MTC without any hardware and software complexity. Li et al. [23] used SDN controller to dynamically adjust the number of resource blocks that are used in the random access in virtual M2M network. M2M applications have different priorities and SDN controllers need this priority information of MTCDs for better resource allocation and path selections in routing. But the challenging issue is that how to provide a massive amount of priority information to SDN controllers.

4.3 Cloud-Based M2M

In M2M applications, depending on the traffic pattern MTCDs communicate with M2M servers. Deploying and maintaining an M2M server is both time and money consuming for service providers. Hence, cloud-based M2M can save the money for maintaining M2M server. And also, cloud-based service provides better security for M2M application. Gu et al. [24] directly connected MTCDs with data aggregators or access points to reduce the multi-hop ad hoc access delay and energy consumption in large cloud-based M2M HetNets. But for M2M application in a wide coverage area can have a long delay in the cloud-based approach. Hence to reduce the delays, edge nodes can be deployed near to MTCDs, where edge nodes can make local decisions and respond to MTCDs immediately.

5 Conclusion

M2M communication will play a vital role in future wireless communications. The main characteristic of requirement for MTC is to have MTCDs with low power consumption and with less deployment cost. In this paper, a comprehensive survey of M2M traffic issues over cellular networks, especially LTE/LTE-A, is presented. This paper presents a brief comparison of H2H and M2M communications and confirms that MTC characteristics are different from HTC. And with the increase in number of MTCDs, the traffic volume keeps on increasing. Hence a better understanding of MTC traffic pattern is required to provide better management of shared resources.

Next, a brief overview of 3GPP standardization and different releases is presented to be in sync with recent advancement in LTE/LTE-A networks. A brief discussion on MTC characteristics and categories is presented. Next, an extensive survey on M2M traffic models is presented for both source and aggregate models. The comparison among these traffic models was also discussed. At the end of this work, this paper identifies several open issues and challenges for future research.

References

1. Niyato, D., Wang, P., Kim, D.I.: Performance modeling and analysis of heterogeneous machine type communications. IEEE Trans. Wireless Commun. **13**(5), 2836–2849 (2014)
2. Xia, N., Chen, H.-H., Yang, C.-S.: Radio resource management in machine-to-machine communications—a survey. IEEE Commun. Surv. Tutor. (2017)
3. Shafiq, M.Z., Ji, L., Liu, A.X., Pang, J., Wang, J.: Large-scale measurement and characterization of cellular machine-to-machine traffic. IEEE/ACM Trans. Network. **21**(6), 1960–1973 (2013)
4. Laner, M., Svoboda, P., Rupp, M.: Detecting M2M traffic in mobile cellular networks. In: 2014 International Conference on Systems, Signals and Image Processing (IWSSIP), pp. 159–162. IEEE (2014)

5. Smiljkovic, K., Atanasovski, V., Gavrilovska, L.: Machine-to-machine traffic characterization: models and case study on integration in LTE. In: 2014 4th International Conference on Wireless Communications, Vehicular Technology, Information Theory and Aerospace & Electronic Systems (VITAE), pp. 1–5. IEEE (2014)
6. GPP. Study on RAN Improvements for Machine-Type Communications. Technical Report, TR 37.868, Release 10 (2012)
7. GPP Service Requirements for Machine-Type Communications. TR 23.368 v11.6.0. Release 11 (2012)
8. Ratasuk, R., Prasad, A., Li, Z., Ghosh, A., Uusitalo, M.A.: Recent advancements in M2M communications in 4G networks and evolution towards 5G. In: 2015 18th International Conference on Intelligence in Next Generation Networks (ICIN), pp. 52–57. IEEE (2015)
9. Ratasuk, R., Mangalvedhe, N., Ghosh, A.: Overview of LTE enhancements for cellular IoT. In: 2015 IEEE 26th Annual International Symposium on Personal, Indoor, and Mobile Radio Communications (PIMRC), pp. 2293–2297. IEEE (2015)
10. GPP Study on Architecture Enhancements for Cellular Internet of Things, TR 23.720 V13.0.0 Release 13 (2016-03)
11. Rico-Alvarino, A., Vajapeyam, M., Xu, H., Wang, X., Blankenship, Y., Bergman, J., Tirronen, T., Yavuz, E.: An overview of 3GPP enhancements on machine to machine communications. IEEE Commun. Mag. **54**(6), 14–21 (2016)
12. Ali, A., Hamouda, W., Uysal, M.: Next generation M2M cellular networks: challenges and practical considerations. IEEE Commun. Mag. **53**(9), 18–24 (2015)
13. GPP Study on Extended Architecture Support for Cellular Internet of Things (CIoT), TR 23.730, Release 14 (2016)
14. Nikaein, N., Laner, M., Zhou, K., Svoboda, P., Drajic, D., Popovic, M., Krco, S.: Simple traffic modeling framework for machine type communication. In: Proceedings of the Tenth International Symposium on Wireless Communication Systems (ISWCS 2013), pp. 1–5. VDE (2013)
15. Laner, M., Svoboda, P., Nikaein, N., Rupp, M.: Traffic models for machine type communications. In: Proceedings of the Tenth International Symposium on Wireless Communication Systems (ISWCS 2013), pp. 1–5. VDE (2013)
16. Østerbø, O.N., Zucchetto, D., Mahmood, K., Zanella, A., Grøndalen, O.: State modulated traffic models for machine type communications. In: Teletraffic Congress (ITC 29), 2017 29th International, vol. 1, pp. 90–98. IEEE (2017)
17. Thomsen, H., Manchón, C.N., Fleury, B.H.: A Traffic Model for Machine-Type Communications Using Spatial Point Processes. arXiv preprint arXiv:1709.00867 (2017)
18. Grigoreva, E., Laurer, M., Vilgelm, M., Gehrsitz, T., Kellerer, W.: Coupled Markovian Arrival Process for Automotive Machine Type Communication Traffic Modeling. In: ICC (2017)
19. Ratasuk, R., Vejlgaard, B., Mangalvedhe, N., Ghosh, A.: NB-IoT system for M2M communication. In: Wireless Communications and Networking Conference (WCNC), 2016 IEEE, pp. 1–5. IEEE (2016)
20. Lin, X., Adhikary, A., Eric Wang, Y.-P.: Random access preamble design and detection for 3GPP narrowband IoT systems. IEEE Wireless Commun. Lett. **5**(6), 640–643 (2016)
21. Ksairi, N., Tomasin, S., Debbah, M.: A multi-service oriented multiple access scheme for M2M support in future LTE. IEEE Commun. Mag. **55**(1), 218–224 (2017)
22. Hesham, A., Sardis, F., Wong, S., Mahmoodi, T., Tatipamula, M.: A simplified network access control design and implementation for M2M communication using SDN. In: Wireless Communications and Networking Conference Workshops (WCNCW), 2017 IEEE, pp. 1–5. IEEE (2017)
23. Li, M., Yu, F.R., Si, P., Sun, E., Zhang, Y., Yao, H.: Random access and virtual resource allocation in software-defined cellular networks with machine-to-machine communications. IEEE Trans. Veh. Technol. **66**(7), 6399–6414 (2017)
24. Gu, L., Lin, S.-C., Chen, K.-C.: Small-world networks empowered large machine-to-machine communications. In: Wireless Communications and Networking Conference (WCNC), 2013 IEEE, pp. 1558–1563. IEEE (2013)

25. GPP Network Architecture. TS 23.002. Release 11. v11.5.0 (2012)
26. Masek, P., Zeman, K., Uhlir, D., Masek, J., Bougiouklis, C., Hosek, J.: Multi-radio mobile device in role of hybrid node between WiFi and LTE networks. Int. J. Adv. Telecommun. Electrotech. Signals Syst **4**(2), 49–54 (2015)
27. Bendlin, R., Chandrasekhar, V., Chen, R., Ekpenyong, A., Onggosanusi, E.: From homogeneous to heterogeneous networks: a 3GPP long term evolution rel. 8/9 case study. In: 2011 45th Annual Conference on Information Sciences and Systems (CISS), pp. 1–5. IEEE (2011)
28. Yunzheng, T., Long, L., Shang, L., Zhi, Z.: A survey: several technologies of non-orthogonal transmission for 5G. China Commun. **12**(10), 1–15 (2015)
29. Wang, B., Wang, K., Lu, Z., Xie, T., Quan, J.: Comparison study of non-orthogonal multiple access schemes for 5G. In: 2015 IEEE International Symposium on Broadband Multimedia Systems and Broadcasting (BMSB), pp. 1–5. IEEE (2015)
30. Au, K., Zhang, L., Nikopour, H., Yi, E., Bayesteh, A., Vilaipornsawai, U., Ma, J., Zhu, P.: Uplink contention based SCMA for 5G radio access. In: Globecom Workshops (GC Wkshps), 2014, pp. 900–905. IEEE (2014)
31. Bockelmann, C., Pratas, N., Nikopour, H., Au, K., Svensson, T., Stefanovic, C., Popovski, P., Dekorsy, A.: Massive machine-type communications in 5G: physical and MAC-layer solutions. IEEE Commun. Mag. **54**(9), 59–65 (2016)
32. Lee, H.-K., Kim, D.M., Hwang, Y., Yu, S.M., Kim, S.-L.: Feasibility of cognitive machine-to-machine communication using cellular bands. IEEE Wireless Commun. **20**(2), 97–103 (2013)
33. Zhang, Y., Yu, R., Nekovee, M., Liu, Y., Xie, S., Gjessing, S.: Cognitive machine-to-machine communications: visions and potentials for the smart grid. IEEE Netw. **26**(3) (2012)

T. N. Sunita received her B.E degree and M.Tech degree in Computer Science and Engineering from Visvesvaraya Technological University (VTU), Belgaum, Karnataka. She is a Research Scholar in the Department of CSE at BMSIT & M, Bengaluru under Visvesvaraya Technological University. Currently, she is working as Assistant Professor in the Department of CSE, Rajiv Gandhi Institute of Technology College, Bengaluru, Karnataka. Her areas of interest include Mobile communication, IOT, Networking, Machine-To-Machine Communication, Data Structures and Machine Learning.

A. Bharathi Malakreddy has B.E degree in Computer Science and Engineering, M.Tech in CSE and Ph.D. in Computer Science and Engineering, University of JNTU Hyderabad. Presently working as Professor Department of CSE at BMSIT & M, Bengaluru. Worked as visiting faculty for IGNOU and Infosys campus connect program. Areas of interest are Wireless Sensor Network, Medical Imaging, IoT, Big Data and Cloud Computing. Published papersin IEEE, Springer, Elsevier and LNSE.

An Integrated Approach for Quality of Service (QoS) with Security in VANETs

C. Kalyana Chakravarthy and Y. Vinod Kumar

Abstract The capabilities of new generation wireless networks can be utilized in Vehicular Ad Hoc Networks (VANETs) to provide improved collision warnings, traffic information and infotainment applications. Information about one vehicle can be forwarded to another vehicle by configuring the IEEE 802.11p standard protocol with security until it reaches a Road Side Unit (RSU) providing autonomous vehicles with secure information such as the speed and direction of the approaching vehicle. This paper evaluates the performance of Ant Colony Optimization (ACO) based Routing with Improved QoS and Security for VANETs (RIQSV) by comparing the metrics delay and jitter with the AODV protocol for vehicular ad hoc networks with QoS without security. Packet level encryption/decryption of the data is carried out using substitution cyphers to provide additional security apart from that offered by the 1609.2. The QoS metrics were measured at three RSUs, averaged over a 2-min period with 0.07 s time interval. Results indicate substantial reduction in average delay and jitter of traffic, proving this suitable for real-time traffic with delay guarantees.

Keywords VANETs · Security · Ant colony optimization · QoS
Average latency · Jitter

1 Introduction

Vehicular ad hoc networks (VANETs) are defined as a category of Mobile Ad Hoc Networks (MANETs) with the main difference that the nodes present here are the vehicles. A node's motion is guided by several factors such as the road direction, traffic density and the traffic regulations. In view of these restrictions, VANET is

C. Kalyana Chakravarthy (✉) · Y. Vinod Kumar
Department of CSE, MVGR College of Engineering (Autonomous), Vizianagaram, India
e-mail: kalyan@mvgrce.edu.in

Y. Vinod Kumar
e-mail: Vinodkumar.yandamuri@gmail.com

S. Smys et al. (eds.), *International Conference on Computer Networks and Communication Technologies*, Lecture Notes on Data Engineering and Communications Technologies 15, https://doi.org/10.1007/978-981-10-8681-6_63

supported by some fixed infrastructure that assists with some services of the VANET and provides access to stationary networks.

VANETs are composed of On-Board Units (OBUs) installed on the participating vehicles and Road Side Units (RSUs) installed by the side of roads. These units of network provide hybrid communications, that is, both Vehicle to Vehicle (V2V) and Vehicle to Infrastructure (V2I) communications [1] using the Wireless Access in Vehicular Environments (WAVE) standards. The improvements in convenient interchanges allow differentiated usage for vehicular systems in the fundamental street map, metropolitan and applications with different QoS prerequisites [2]. VANETs provide several novel applications and opportunities including transportation safety and entertainment applications. Security is one of the most primary concerns in VANETs deployment as it is compulsory to assure public as well as transportation safety. The IEEE 1609 family encompasses the architecture, protocol set, services and interfaces that enable all WAVE stations to coordinate and operate within the VANET environment.

The IEEE proposes 802.11p protocol for wireless access in VANETs, while the IEEE Wireless Access in Vehicle Environment (WAVE) 1609.2 standard defines secured message structures necessary to protect the messages [3, 4]. The Secure Protocol Data Units (SPDUs) are signed using digital certificate chains. The hash algorithms used are SHA-256 and SHA-384 and the encryption algorithm used is Elliptic Curve Integrated Encryption Scheme (ECIES) defined by the IEEE 1363a standard (Fig. 1).

VANET supports various applications which require communication with minimum delay. Quality of Service (QoS) focuses on multiple parameters that include signal-to-noise ratio, handover delays, jitter, end-to-end delays, packet loss ratios, packet delivery ratio, connection lifetime, residual channel capacity, etc. Due to unique characteristics and dynamic topology, it is a difficult task to ensure QoS in VANETs [6, 7].

2 Routing with Improved QoS and Security for VANETS (RIQSV)

Different Quality of Service parameters in VANETS include throughput, end-to-end delay and jitter. These parameters have different priorities based on the traffic type. Throughput may be defined as the number of packets transmitted successfully over a given time.

$$T_i(t) = \sum (P_i(t)/t_i) \tag{3}$$

where $P_i(t)$ is the number of packets transmitted in time t_i.

The average end-to-end delay is the total time it takes for the packet to reach the destination from the source vehicle. It includes the total time for generation,

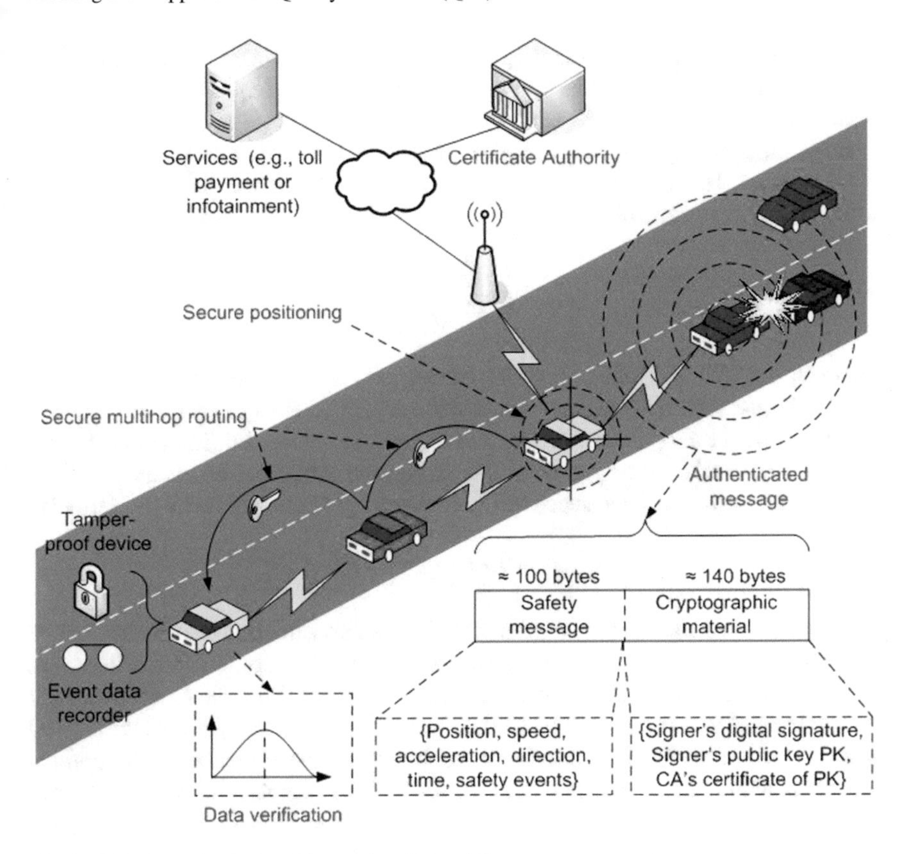

Fig. 1 VANET architecture with security. *Source* [5]

transmission and receiving the packet from the source (vehicle) to the destination (RSU).

$$D_i(t) = \sum (R_i(t)) - \sum (S_i(t)) \quad (4)$$

where $S_i(t)$ is the total delay for packet transmission at the source vehicle that includes the Queuing time, and $\sum R_i(t)$ is the time at the destination vehicle.

Jitter is the difference in packet delays at the receiver measured over a particular flow of packets

$$J_i(t) = \sum (D_{i+1}(t)) - \sum (D_i(t)) \quad (5)$$

Here, $T_i(t)$, $D_i(t)$, $J_i(t)$ are the throughput, delay and jitter at the ith RSU, respectively.

For example, real-time streaming video or urgent messages require low delay, while weather report and other non-critical messages may require only throughput guarantees.

In ACO, generally, the ants travel in the path that has the highest concentration of pheromone. The ants in the colony move randomly from one node 'x' to another node 'y' with the probability

$$P_{x,y} = \frac{\left(t^{a}_{x,y}\right).\left(c^{b}_{x,y}\right)}{\sum \left(t^{a}_{x,y}\right).\left(c^{b}_{x,y}\right)} \tag{1}$$

where '$t_{x,y}$' is the pheromone concentration which is being deposited at edge x, y. '$c_{x,y}$' is the user-defined parameter which defines the effective quality for the path x, y. 'a' and 'b' are the user-defined parameters for controlling the real-time proportionate values for $t_{x,y}$ and $c_{x,y}$, respectively. The pheromone trials can be updated on each iteration, after the ants have successfully finished a proper solution by the formula

$$t_{x,y}(t) = \left[e.t_{x,y(t-1)}\right] + \Delta t_{x,y} \tag{2}$$

where '$\Delta t_{x,y}$' is the sum of the moves of the ants for constructing the entire solution. 'e' is the evaporation coefficient of the pheromone and is user-defined whose values lie between 0 and 1. After the iteration, some portion of the pheromone can get evaporated and the process will be further iterated. The inherent parallelism in ACO makes it favourable for obtaining faster feedbacks and thereby choosing an optimal path for reaching the destination, not only for VANETs but also for other problems. There has been considerable work on Ant Colony Optimization for Quality of Service in VANETs. In [8], the authors apply ACO to VANETs and carry out simulations using Opportunistic Network Environment (ONE) simulator to observe an improved throughput and reduced delay and reduced packet dropping when compared with the existing 802.11p protocol. The routing protocol in [9] uses the quality of relaying of road segments to adaptively determine the most efficient routes from the source to the target intersection based on the delay and the delivery ratio. In [10], the authors suggest an ACO-based protocol with improved convergence, due to the pheromone concentration increasing exponentially with time and integration of local and global routes. The authors in [11] propose a multi-objective routing algorithm that updates the pheromone based on the number of nodes, hop count and the Euclidean distance between nodes.

Authors in [12] propose an ACO-based reliable clustering AODV protocol for VANETs that uses data aggregation at the RSUs to determine the demand and allocate time slots using the TDMA scheme. Authors claim that the method reduces the end-to-end delay and improves the packet delivery ratio compared with existing methods. The authors in [13] propose another clustering-based ACO routing protocol for VANETs, which builds over the existing QoS, OLSR protocol for MANETs. It uses cheating prevention and Multipoint Relay (MPR) recovery to improve the stability of the clusters. Authors conclude that the proposed protocol performs better in terms of many factors such as the reduced number of MPRs, lesser number of hops from source to destination and improved packet delivery ratio.

Authentication, integrity, confidentiality and non-repudiation form the security requirements for VANETs. Authentication allows a receiving node to identify whether the transmitter belongs to the network, the position of the transmitter and prevents impersonation attacks. Integrity prevents the message from being changed on its path from the source to the destination and uses signatures in some forms from RSUs or from known sources in the neighbourhood. Confidentiality ensures that information can only be accessible by the intended receivers and critical information such as driver profiles and vehicle IDs are kept confidential from intruders. It can be achieved using public or private key encryption schemes. Non-repudiation involves non-denial by the source of sending any messages and can be enforced by storing information related to the sending vehicle in a tamper-proof device (Fig. 2).

Our current work combines ACO-based AODV routing protocol with secure routing. The advantage of AODV is, it requires lesser memory and in the existing links, AODV does not generate additional traffic for communication. However, in AODV, to make all routing updates from other nodes invalidated, the attacker can advertise or promote routes having a smaller distance metric against the original distance metric

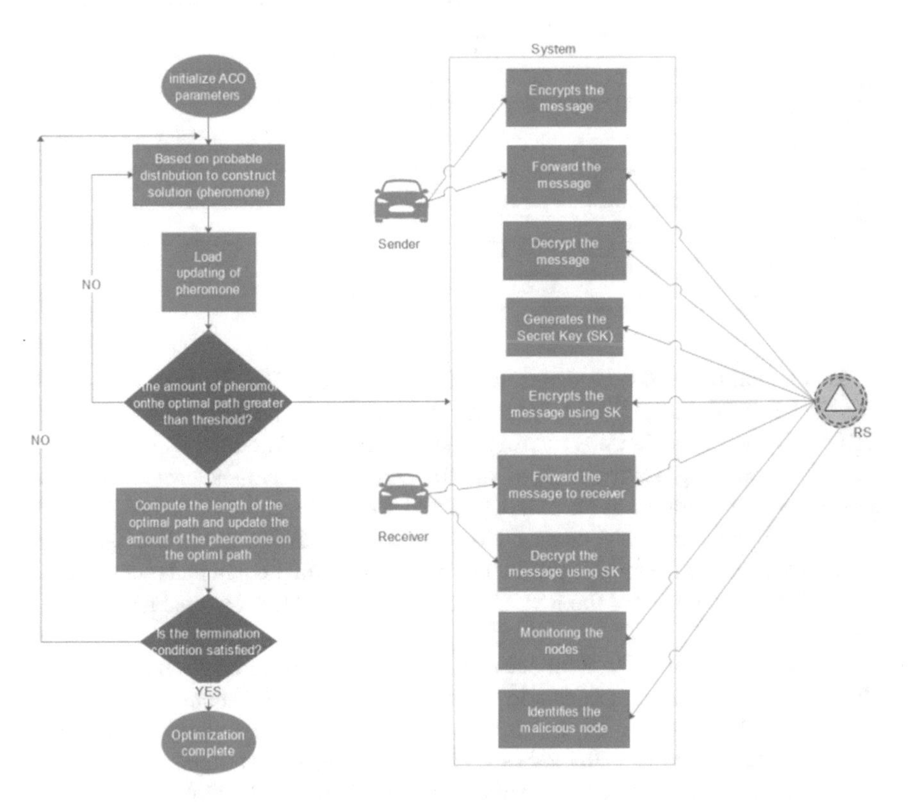

Fig. 2 ACO with encryption and decryption mechanism

or advertise route updates with large sequence number. Our proposed protocol would avert such kind of attacks.

The rationale is to devise a protocol that provides security using packet level encryption, while still managing to reduce the delay as compared to the original routing protocol, to be used for real-time message communication. This is essential, as most of the times, addition of security has adverse effects of increasing packet delays and sometimes reducing the throughputs due to reduced number of packets processed. Our proposed protocol using ACO reduces the delay and jitter considerably with increasing traffic, making it suitable for communication of emergency messages.

3 Results and Conclusion

The parameters considered for simulation are listed below:

Parameters	Values
Simulator	NS-2.35
Simulation area	800 m * 800 m
Number of nodes	20
Packet size	1000 bytes
Packet interval	0.07 s
Transmission range	50 m
MAC mechanism	IEEE 802.11p
Traffic source model	Constant bit rate
Mobility model	Random waypoint

Two scenarios were considered: (i) AODV for VANETS with QoS using ACO (VANET-AODV-ACO); (ii) Routing with Improved QoS and Security for VANETS (RIQSV). The simulation was run for 2 min with increased node mobility and exponentially increasing traffic rate, and various QoS parameters—latency, throughput and jitter—are averaged at the RSUs 0, 2 and 6 over the entire period. For AODV using ACO, the pheromone count $t_{x,y}$ (t) is updated as the weighted average of the QoS parameters which include the latency, jitter and throughput in that order. Results show high initial delays at the RSUs, which reduce and stabilize gradually (Figs. 3 and 4).

For AODV using ACO with security, encryption is applied on links with the pheromone count below a threshold value and the pheromone counts are updated. Results clearly indicate a reduction in delay compared to the above case. This is due to the fact that paths with increased latency become more reliable on improving confidentiality and are chosen more often, resulting in reduced initial discovery time. Also, the jitter at the RSUs is reduced as the delay between the end vehicles to the

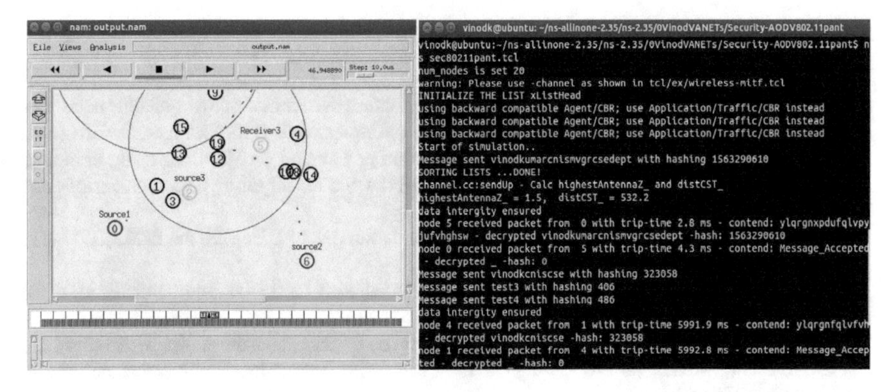

Fig. 3 Ns2 simulation in VANETs

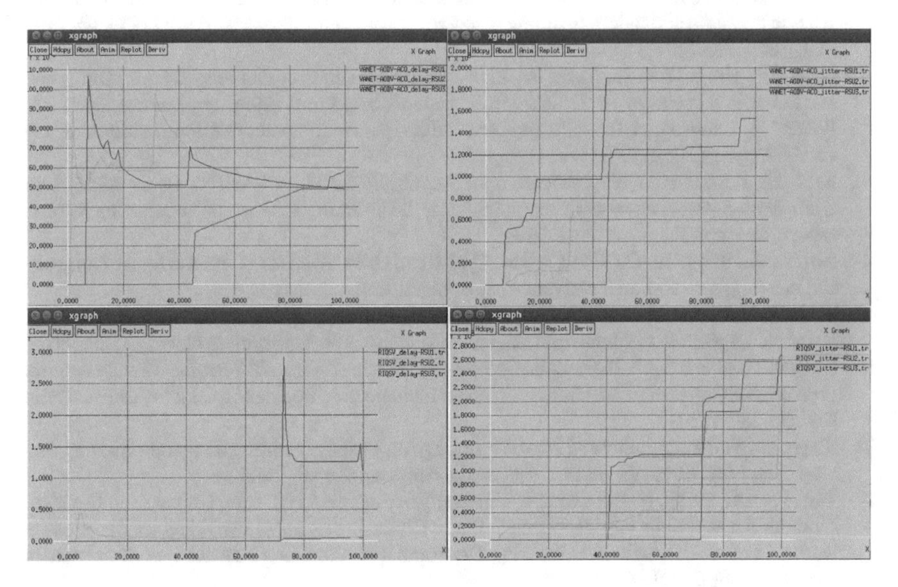

Fig. 4 Average delay and jitter graphs for **a** VANET-AODV-ACO and **b** RIQSV

RSU becomes more consistent. This makes the current protocol, the RIQSV desirable for real-time message communication.

Fog computing is an emerging concept and could be applied to VANETs to reduce end latency to the RSUs for the 5G wireless networks. Further, newer trust models could be built using ACO to improve link reliability and stability, in turn reducing the end delay for real-time traffic.

References

1. Kumar, R., Dave, M.: A novel approach for information dissemination in vehicular networks. In: High Performance Architecture and Grid Computing, pp. 548–556. Springer, Berlin (2011)
2. Mane, U., Kulkarni, S.A.: QoS realization for routing protocol on VANETs using combinatorial optimization. In: Fourth International Conference on Computing Communications and Networking Technologies (2013)
3. Bhoi, S.K., Khilar, P.M.: Vehicular communication: a survey. Inst. Eng. Technol. **3**(3), 204–217 (2013)
4. Raya, M., Papadimitratos, P., Hubaux, J.-P.: EPFI.: securing vehicular communications. IEEE Wirel. Commun. **13**(5), 8–13 (2006)
5. Laberteaux, K.P., Haas, J.J., Hu, Y.C.: Security certificate revocation list distribution for VANET. In: VANET'08 Proceedings of the Fifth ACM International Workshop on VehiculAr Inter-NETworking
6. Sadiq, A.S., Bakar, K.A., Ghafoor, K.Z., Lloret, J., Khokhar, R.: An intelligent vertical handover scheme for audio and video streaming in heterogeneous vehicular networks. Mob. Netw. Appl. **18**(6), 879–895 (2013)
7. Guan, W., He, J., Bai, L., Tang, Z., Zhou, Y.: QoS guarantee with adaptive transmit power and message rate control for DSRC vehicle network based road safety applications. In: U- and E-Service, Science and Technology, Communications in Computer and Information Science, vol. 264, pp. 130–135 (2011)
8. Kaur, H., Kumar, R.: QoS Realization for Routing Protocol on VANETs Using Ant Colony Optimization. Department of Computer Science, Bhai Gurdas Institute of Engineering & Technology, Sangrur, Punjab, India (2016)
9. Amjad, Z., Song, W.-C.: Road Aware QoS Routing in VANETs. Department of Computer Engineering, Jeju National University, Jeju, South Korea (2015)
10. Persis, D.J., Robert, T.P.: Ant Based Multi-objective Routing Optimization in Mobile AD-HOC Network. Department of Industrial Engineering, College of Engineering, Guindy, India (2015)
11. Ganguly, S., Das, S.: A Novel Ant Colony Optimization Algorithm for the Vehicle Routing Problem. Department of Electronics & Telecommunication Engineering, Jadavpur University, Kolkata, India (2013)
12. Singh, H., Singh, P.: Enhanced New Clustering Ant Colony Optimization based Routing Protocol AODV-R. Guru Nanak Dev University, Amritsar, Punjab (2017)
13. Wahab, O.A., Otrok, H. from Khalifa University of Science Abu Dhabi, United Arab Emirates, Mourad, A.: VANET QoS-OLSR: QoS-Based Clustering Protocol for Vehicular Ad hoc Networks. Lebanese American University, Department of Computer Science and Mathematics, Beirut, Lebanon (2013)

Dr. C. Kalyana Chakravarthy holds a Ph.D. from Andhra University in the area of Wireless Networks and is currently serving as Professor in Department of Computer Science and Engineering, MVGR College of Engineering (Autonomous). He has more than 18 years of academic, administrative and research experience and has several reputed journal and conference publications to his credit. His main research work focuses on Wireless Networks, Software Defined Networks and Internet of Things. Engineering Education Research is also his parallel line of interest.

Mr. Y. Vinod Kumar has Post graduated in Computer Networks and Information Security at MVGR College of Engineering (Autonomous). His areas of work include Wireless Networks and Vehicular networks. He holds industrial certifications in Networking.

Real-Time Network Data Capturing Using Attack Generators and Data Processing Method

Karuna S. Bhosale, Maria Nenova and George Illiev

Abstract Nowdays, use of the Internet for transaction of knowledge is very common. The end users who are accessing Internet or system are vulnerable to malicious user attacks which results in legitimate user being prevented from accessing the websites. Recently, there are several methods presented for application layer DDoS attacks by considering the different properties of attacks. However, most of the methods suffered from the poor accuracy performance of DDoS attack detection at the application layer. Hence, DDoS attacks have been low volume and act on its own as a legitimate transaction on layer seven means application layer hence such attacks are not detected easily by IDS (Intrusion Detection Systems) or firewall systems. We believe that the accuracy and efficiency of attacks detection are based on correctness of capture data traffic. In state-of-the-art methods, there is no provision to remove the noisy data from the capture logs and hence leads to incorrect detection results. In this paper, we presented the real-time computer networks data capturing for normal as well as attack infected traffics, then designed the preprocessing algorithm to remove irrelevant data to optimize the attack detection performance. In this architecture, we use LOIC DDoS attack generator tool to create attack at packet capturing time and also use KDD cup data set for other attacks trace such as DoS, R2L, L2R, Probe, etc., of the communication network. The experimental results show that proposed data preprocessing method is easy and efficient as compared to state-of-the-art solutions.

Keywords Irrelevant data · Real-time capturing · DDoS · IDS · LOIC · DoS R2L · L2R · Probe

---

K. S. Bhosale (✉) · M. Nenova · G. Illiev
Faculty of Telecommunication, Technical University of Sofia, Sofia, Bulgaria
e-mail: bhosale.karuna@gmail.com

M. Nenova
e-mail: mvn@tu-sofia.bg

G. Illiev
e-mail: gli@tu-sofia.bg

S. Smys et al. (eds.), *International Conference on Computer Networks and Communication Technologies*, Lecture Notes on Data Engineering and Communications Technologies 15, https://doi.org/10.1007/978-981-10-8681-6_64

1 Introduction

The DDoS takes advantage of the Internet protocols and fundamental benefit of delivering data packets from nearly any source to any destination in an open architecture network. Lot of the volume of packets devices of network overwhelms along with the servers, or packets are normally not completed for rapid decrease of server elements. Most of the attacks utilizing the IP addresses of spoofed source for elude the identification of source. DDoS is the multilayered attacks and activated in many layers of OSI. It is very critical to make detection, this type of attacks utilized spoofed IP addresses by amplifying attacking power into the big scale. Because of the big load created through packet flood destination routers, firewalls and servers will be rendered not available for the valid transaction. The existing bandwidth into the bottleneck link has been totally used through attack traffic which is dropping legitimate packets. Overview has been given the major effect of analysis for new DDoS attacks [1].

Noise is "meaningless or irrelevant data". For most of the present data cleaning techniques, the concentration is on the finding and removal of noise (low-level data errors) that is the output of an imperfect data collection process. The need to address this type of noise is clear as it is detrimental to almost any kind of data analysis. However, ordinary data objects that are not relevant or only relevant to a specific data analysis but not useful can also significantly hinder the data analysis, and thus these objects should also be considered as noise, at least in the context of a specific analysis. For instance, in document data sets that consist of news stories, there are many stories that are only weakly related to the other news stories. If the aims are to use clustering to find the strong topics in a set of documents, then the analysis will suffer unless irrelevant can be eliminated. Consequently, there is a need for data cleaning techniques that remove all kind of noise [7].

Noise into the data set has been generally small in several cases. For instance, for ignoring the data errors, it is claimed that for the business are generally 5% or minimum if the company normally taking count. The recent experimental data for the protein complexes has generally several false positive protein communication are available. As this is the instance of the sets of data which is big quantity of noise because of the errors of data collection and because the objects of unfavorable data quantity of noise will also be big. These examples involve the web data and recently mentioned document data sets. Therefore, data cleaning techniques for the enhancement of data analysis also need to be able to reduce the time duration as well as improve the accuracy [9].

For the communication networks and resource security domain, there are a number of different types of packets continuously generating and these flows are recorded for the analysis of security threats detection. However, the real-time data packet capturing is performed with a number of irrelevant texts in each flow. The presence of such irrelevant information in captured data may lead to inaccurate detection attacks. Therefore, before processing to feature extraction and attacks detection using data mining algorithms, we need to perform the data preprocessing to remove such irrelevant text information from capture data in order to improve the efficiency and accuracy of security analysis. In this paper, we presented an algorithm to capture the real-time network packets which consist of with attack flows. We used the attack generator tool to explicitly generate the different types of attacks such as R2L, U2R, DoS, and Probe. In short, the capture longs for each interval is a combination of normal traffic flows and attack infected traffic flows. After the log generated, we applied proposed preprocessing algorithm to remove the noisy data.

This research organized as follows. The method on the data processing and noise suppressing has been discussed in Sect. 2. Information about KDD cup & KDD cup datasets is explained in Sect. 3. Algorithms have been discussed in Sect. 4. Data analysis and practical results are presented in Sect. 5. Finally, conclusion and future work are presented in Sect. 6.

2 Related Work

This section presents the review of previous methods precisely.

Kiruthika Devi et al. [1]
In this paper, the author proposed that DDoS cyber weapon is highly promoted by some features involves, hactivitism, private vengeance, dissatisfy customer, anti-government force, political cause and theoretical, cyber intelligence, etc. The IP spoofing is a technique utilized by attackers to damage the available services in the Internet by mimic as a trusted source. Since the distortion function has identical assets from that of the trusted to integrity one detection and sifting turns out to be basic. The proposed system represents that the online monitoring system (OMS), parodied function detection module, and interface-based rate contrived (IBRL) algorithm. The OMS provides DDoS attack assessment regularly by notifying the deception in the host and system computation calculation. The distortion activity discovery module integrates with HCF to verify the legal framework assets and techniques using source IP address and it' is also differentiate hop to destination node. The disadvantage of this system is it does not work on any kind of noise that may be present in data packets.

Subbulakshmi et al. [2]
The DDoS dataset with different features is created to verify 14 features and ten types of the new DDoS attack. The benefit of creation DDoS dataset the EMCSVM (Enhanced Multi-class Support Vector Machines) is used to identification of the attacks in different classes. The computation of the EMCSVM is evaluated across SVM with various argument consider and segment capacities. The EMCSVM system performance is better for the DDoS dataset using 10 new DDoS attacks. It is compared with the KDDcup99 database which has 6 types of DDoS attacks.

Subbulakshmi et al. [3]
The goal of researchers is to monitor the system online which accordingly starts location tools if there is any suspicious activity and furthermore resistance the hosts

from being landed at the system. Both non-spoofed and spoofed IPs are detected in this system. None of the deride IP are prominent exploit ESVM (Enhanced Support Vector Machines) and ridicule IPs that are detected exploit HCF elements. The detected IPs are used to start the process. The attacking quality is feature calculated using Lanchester Law which initiates the obstacle system. In light of the discover attack any of the privacy plans, for instance, history based IP and rate based limitation partition are regularly initiated to delete the packets from the theorize IP. The IDS technology for detection and security of DDoS attack is used in a searching for supplies ground. The online model is analyzed to be obvious in the field of combined DDoS detection and protection.

Singh and De [4]
To protect and secure web server from the attacker, it is essential to know the behavior and nature of legitimate and illegitimate user. To providing the access for actual customer and also provide the security approach to legitimate user is of prime importance.
The DDoS attack is a fundamental difficulty in the Internet network. By exploiting its application layer protocols, DDoS can create a large destruction by quietly making an entry to the web server as it goes about as one of the true blue customers? The paper utilizes parameter of the system packet like HTTP GET, POST asks for and delta time to figure the exactness in discovering the conceivable assault. We utilize distinctive classifiers like Naive Bayes, Naive Bayes Multinomial, Multilayer Perception, RBF organize, and Random Forest and others to arrange the assault produced dataset.

Alias et al. [5]
The implementation of packet capturing technology has considerable impact in packet capturing whereby both programming and equipment arrangements are intended to streamline parcel catch execution. Packet capture is the way toward catching information parcels over a PC organizes. Packet capturing has turned out to be additional testing with the quick speed systems. The complex problem of capturing in high parcel rates and high data transfer capacity makes the requirement for more modern packet catch instrument progressively arrange movement. Existing arrangements can catch packets; however, at specific rates in fast system, parcels will be lost.

Jun et al. [6]
Author discusses the real-time data modeling, analysis and acquisition, examination and design of the real-time data warehouse system. In this system, the presentation of a multilevel constant information reserve encourages the continuous information stockpiling. The utilization of the Web administration to catch ongoing information influences the constant information to catch and load effectively. This paper likewise discusses the real-time data analyzed in the SOA based ongoing information distribution center. It joins the Web administrations and information stockroom, expands information distribution center's system capacity, explains information trade effectiveness with the server and enhances customer's execution proficiency. So it is helpful to choose significant data and after that Companies Corporation and individual can get data advantageously.

Xiong et al. [7]
Subsequently, there is a requirement for information cleaning strategies that evacuate the two sorts of noise. Since data collections can contain a lot of noise, these strategies additionally should have the capacity to dispose of a discard expansive portion of the information. Three of these techniques depend on customary exception location systems: separate based, bunching based, and an approach in light of the Local Outlier Factor (LOF) of a question. The other system, which is another strategy that we are proposing and is a hyperclique-based data cleaner (HCleaner). These procedures are assessed regarding their effect on the resulting information investigation, particularly, grouping, and affiliation examination. Our exploratory outcomes demonstrate that these techniques can give better bunching execution and higher quality affiliation designs as the measure of noise being evacuated increments, despite the fact that HCleaner by and large prompts better grouping execution and higher quality relationship than the other three strategies for paired information.

Prasad et al. [8]
Data quality change is regularly an iterative procedure which for the most part includes composing an arrangement of information quality tenets for institutionalization and end of copies that are available inside the information. Existing information purging apparatuses require a considerable measure of customization at whatever point moving starting with one client then onto the next and starting with one space then onto the next. In this paper, we display an information quality change instrument which helps the information quality expert by demonstrating the attributes of the substances introduce in the information. The device distinguishes the variations and equivalent words of a given element introduce in the information which is a vital undertaking for composing information quality principles for institutionalizing the information. Writer shows a ripple down control structure for keeping up information a quality tenet which helps in diminishing the administrations exertion for including new standards. Authors also exhibit a run of the millwork process of the information quality change process and demonstrate the helpfulness of the apparatus at each progression. Authors also display some exploratory outcomes and dialogs on the convenience of the apparatuses for lessening administrations exertion in an information quality change.

Uddin [9]
Researcher examines part one of the main work in the field of information science, analytics, and mining. Group of calculations is created to anticipate the instructive importance of people's gifts through the focal point of identity highlights (unstructured and semiorganized) and scholastic/profession information. The huge information (unstructured and semiorganized) contains loads of significant data that can be mined and broke down. Nonetheless, such handling and use of information present difficulties of managing noise (unimportant, superfluous and excess information). Notwithstanding the idea of information preparing and use for a given issue or an application, commotion includes superfluous time and cost. Creator quickly talks about the general research work and after that presents Noise Removal and Structured Data Detection (NR-and-SDD) calculation and related math builds. NR and

SDD identify the noise to minimize the handling cost and enhance organized information discovery in the importance of identity highlights.

Kayacık et al. [10]
KDD 99 interruption detection datasets, which depend on DARPA 98 dataset, give named information to specialists working in the field of interruption location and is the main named dataset openly accessible. Various scientists utilized the datasets in KDD 99 interruption detection datasets to ponder the use of machine learning for interruption identification and revealed location rates up to 91% with false positive rates under 1%. To substantiate the execution of machine learning constructs finders that are prepared with respect to KDD 99 preparing information; we research the importance of each component in KDD 99 interruption location datasets. To this end, information gain is employed to determine the most discriminating features for each class.

3 KDD Dataset

The KDD99 dataset are intrusion detection based, which is initiated in DARPA, 1998 for IDS creators with framework and different methodologies. If we consider the three target machines are running in different operating systems and services, the simulation system are created for the imitative military network. Also, for spoofing various IP addresses utilized three machines, this helps to generate the traffic into the various IP addresses. Lastly, with the use of TCP dump format, there is sniffer which records all traffic of network. The whole simulation time is 7 weeks. General connection has been made for profile which is necessary into the attacks fall and military network into the four categories: Denial of Service; User to Root; Probe and Remote to Local.

- Denial of Service (dos): With the use of service attackers try to ignore legitimate users.
- Remote to Local (r2l): On the victim machine attackers do not have the account, so he tries to increase access.
- User to Root (u2r): Attackers have the local access to victim machine and it is trying to increase superuser privileges.
- Probe: About the target, host attackers try to increase information about the target.

In 1999, the first TCP dump documents were preprocessed for usage in the Intrusion Detection System benchmark of the International Knowledge Discovery and Data Mining Tools Competition. To do as such, packet data in the TCP dump document is compressed into associations. In particular, "an association is a succession of TCP packets beginning and accomplishment at some very much characterized circumstances, between which information streams from a source IP deliver to an objective IP address under some all around characterized convention". This procedure is finished by utilizing the Bro IDS, bringing about 41 highlights for every

Table 1 F-measure, recall, and precision values

	Case-based [5]	Collaborative [6]	Proposed
F1-Measure	85.63	87.41	88.95
Recall rate	87.20	88.45	90.45
Precision rate	90.33	90.86	92.72
Accuracy	94.56	95.68	97.06

association, which are definite in Appendix 1. Highlights are gathered into four classes:

- Basic Features: Basic features can be derived from packet headers without inspecting the payload.
- Content Features: Domain knowledge is used to assess the payload of the original TCP packets. This includes features such as the number of failed login attempts;
- Time-based Traffic Features: These features are designed to capture properties that mature over a second temporal window. One example of such a feature would be the number of connections to the same host over the second interval;
- Host-based Traffic Features: Utilize a historical window estimated over the number of connections—in this case 100—instead of time. Host-based highlights are in this way intended to evaluate assaults, which traverse interims longer than two seconds. The KDD 99 interruption location benchmark comprises three segments, which are itemized in Table 1. In the International Knowledge Discovery and Data Mining Tools Competition, just "10% KDD" dataset is utilized to train [8]. This dataset contains 22 assault writes and is a more succinct rendition of the "Entire KDD" dataset. It contains a greater number of cases of assaults than the ordinary associations and the assault writes are not spoken to similarly. Due to their tendency, disavowals of administration assaults represent most of the dataset. Then again the "Redressed KDD" dataset furnishes a dataset with unexpected factual appropriations in comparison to either "10% KDD" or "Entire KDD" and contains 14 extra assaults [10].

4 Proposed Methodology

In this section, the proposed method architecture and algorithms are presented. The algorithm presented below helps to optimize the capture data flow from the network in order to improve the further processing of feature extraction and attack detection. In this technique first, we accept capture packets in original form in file D. Then we do next process for all lines in dataset. First, check line contains delimiters or not if it is present then remove it. A delimiter is one or more characters that separate text strings. Common delimiters are commas (,), semicolon (;), quotes (", '), braces ({ }), pipes (|), or slashes (/ \). Then we check for stop words if present then that also remove. Here, we consider stop words means "the", "that", "so", "such" like that. Then, we try to remove URL's present in data like "cdn.content.prod.cms.msn.com". At last, we also remove unwanted HTTP message. The aim of proposed data preprocessing algorithm is to improve the detection accuracy and speed as compared to the conventional methods. If we apply preprocessing algorithm then it helps to find accurate features so it increases accuracy and also data is reduced so the time required for further processes reduces.

Algorithm 1 Data Preprocessing

Input: Capture Packets Data file D
Result: Data file without noise
Step 1: Accept each line of file D as T[i]
Step 2: Check if (T[i] == Delimiters) then go to Step 3 else go to step 4
Step 3: Remove Delimiters
Step 4: Check if (T[i] == Stop Words) then go to Step 5 else go to steps 6
Step 5: Remove all stop words
Step 6: if (T[i] == URL) then go to Step 7 else go to Step 8
Step 7: Remove URL from T[i]
Step 8: if (T[i] == HTTP message) then go to Step 9 else go to Step 10
Step 9: Remove HTTP message
Step 10: If EOF D then go to Step 11 else go to Step 2
Step 11: Stop

In this algorithm, we eliminate all unwanted fields which are not required detecting attacks. Below diagram shows the flow of an algorithm.

5 Results and Discussion

The experimental analysis is conducted using Java tool with attack generator as mentioned in above section. The data preprocessing algorithm is designed and applied on each data captured by data capturing program. The evaluation of proposed algorithm is done by using three core performance metrics such as accuracy, precision, recall, and F1-measure. Figures 1, 2, 3, and 4 show the comparative results.

Figure 2 depicts the improved performance of proposed approach for F-measure as compared to other techniques of noise removal. There is a significant improvement in the proposed algorithm in performances. Similarly, the precision rate and recall rate is better compared to all existing techniques. The values of measured results are in range of [0, 100].

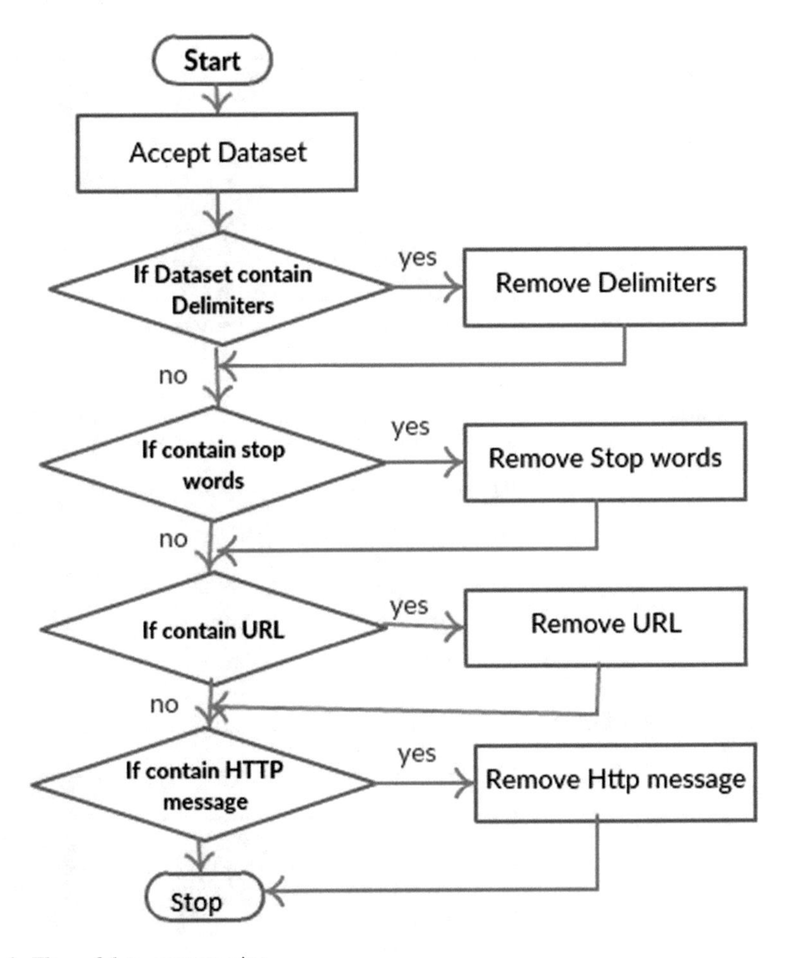

Fig. 1 Flow of data preprocessing

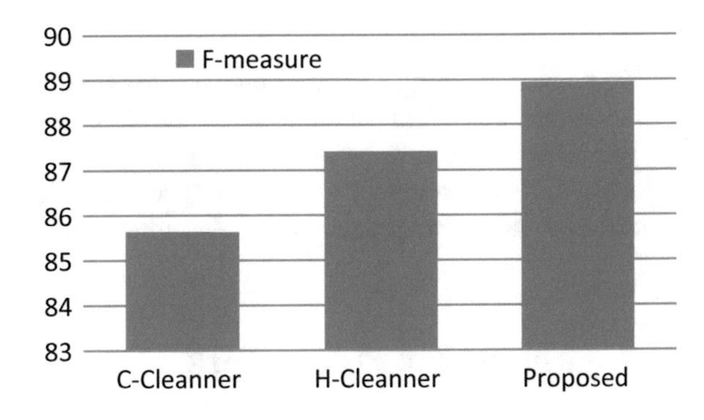

Fig. 2 Performance analysis of F-measure

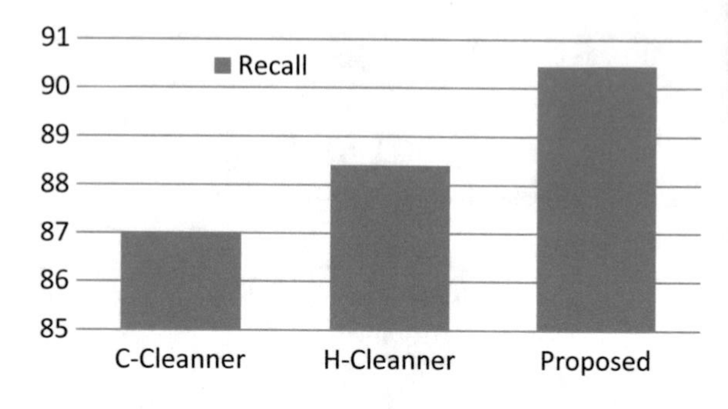

Fig. 3 Recall rate performance analysis

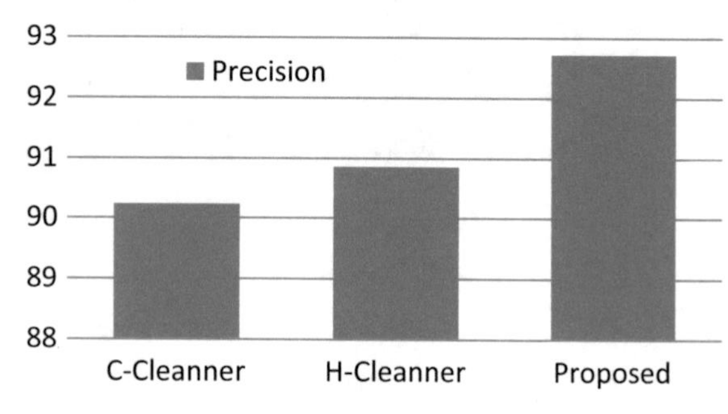

Fig. 4 Precision rate performance analysis

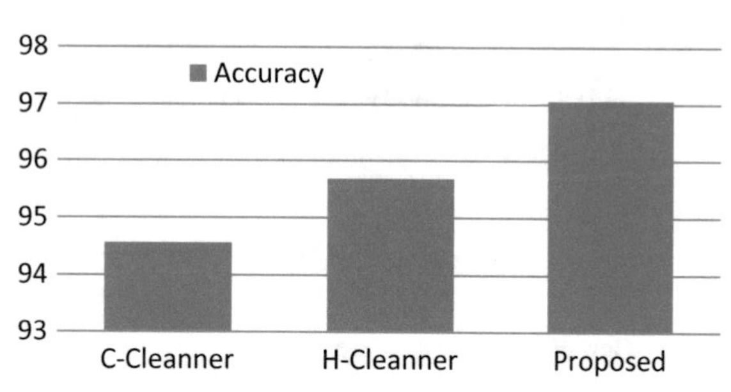

Fig. 5 Performance analysis of accuracy

From the above comparative results, the proposed simple and easy preprocessing method achieved better performance against the state-of-the-art methods.

The Table 1 is showing the actual outcomes for all three performance metrics F-measure, recall and precision rates respectively. These results are satisfying the objective of our research designs and algorithms (Fig. 5).

6 Conclusion and Future Work

This paper discusses the real-time data capturing with the DDOS attack with data preprocessing noise removal algorithm. There are a number of malicious activities to perform attacks on network, but overloading website with large unwanted traffic is common attack. The data collection errors and the amount of noise due to irrelevant data objects can also be large. Examples include the document data sets mentioned earlier on Web data. Therefore, data cleaning techniques for the enhancement of data analysis also needs to be able to reduce the time duration. In this paper, we presented the data cleaning method and evaluated its performance against the state-of-art solutions. For future work, we suggest to work on further part of communication networks threats detection using the novel data mining algorithms.

References

1. Devi, B.S.K., Preetha, G., Selvaram, G., Mercy Shalinie, S.: An impact analysis: real time DDoS attack detection and mitigation using machine learning. In: International Conference on Recent Trends in Information Technology (2014)
2. Subbulakshmi, T., Mercy Shalinie, S., GanapathiSubramanian, V., BalaKrishnan, K., Anand-Kumar, D., Kannathal, K.: Detection of DDOS attacks using enhanced support vector machines with real time generated dataset. In: International Conference on Advanced Computing (2011)
3. Subbulakshmi, T., Parameswaran, P., Parthiban, C., Mariselvi, M., Adlene Anusha, J., Maha-lakshmi, G.: A unified approach for detection and prevention DDOS attacks using enhanced support machine and filtering mechanism. ICTACT J. Commun. Technol. (2013)
4. Singh, K.J., De, T.: An approach of DDOS attack detection using classifiers. In: Emerging Research in Computing, Information, Communication and Applications. Springer, New Delhi (2015)
5. Alias, S.B., Manickam, S., Kadhum, M.M.: A study on packet capture mechanisms in real time network traffic. Adv. Comput. Sci. Appl. Technol. (ACSAT) (2013)
6. Jun, L., ChaoJu, H., HeJin, Y.: Application of web services on the real-time data warehouse technology. Adv. Energy Eng. (ICAEE) (2010)
7. Xiong, H., Pandey, G., Steinbach, M., Kumar, V.: Enhancing data analysis with noise removal. IEEE Trans. Knowl. Data Eng. **18**(3) (2006)
8. Prasad, K.H., Faruquie, T.A., Joshi, S., Chaturvedi, S., Subramaniam, L.V., Mohania, M.: Data cleansing for large enterprise datasets. In: Annual SRII Global Conference (2011)
9. Uddin, M.F.: Noise removal and structured data detection to improve search for personality features. Adv. Soc. Netw. Anal. Min. (ASONAM) (2016)
10. Kayacık, H.G., Zincir-Heywood, A.N., Heywood, M.I.: Selecting features for intrusion detection: a feature relevance analysis on KDD 99. In: Third Annual Conference on Privacy, Security and Trust (2005)

Karuna S. Bhosale working in Faculty of Telecommunication, Technical University of Sofia, Bulgaria. His research area includes data analytics and noise avoidance techniques.

Maria Nenova working as an associate professor in Faculty of Telecommunication, Technical University of Sofia, Bulgaria. Her research area includes data analytics and noise avoidance techniques.

George Illiev working as a professor in Faculty of Telecommunication, Technical University of Sofia, Bulgaria. His research area includes data analytics and noise avoidance techniques.

IoT Fog Cloud Model for Digital Reach in Rural India

Saniya Zahoor and Roohie Naaz Mir

Abstract Rural India faces a multitude of problems due to its digital divide as compared to the urban areas of the country. Internet of Things (IoT) may help to reduce the divide and lead to the digital development of rural India. Rural IoT can be designed to support smart village vision, which aims to provide automation in many sectors such as noise monitoring, energy consumption, smart lighting, environmental monitoring, health monitoring, education, agriculture, etc. and improve life quality of the people. This paper gives an insight of IoT, cloud computing, and fog computing concepts; and proposes a general model based on these computing paradigms with an objective to have digital inclusion of rural India in cyberspace. The paper presents the gram panchayat as the local grid for the data aggregation and transmission from IoT devices to cloud with nonconventional energy resources to minimize the effects of rural power supply inefficiencies that lead to the digital growth of rural India with urban India.

Keywords Digital inclusion · Internet of things · Cloud computing · Local grid
Fog computing · Gram panchayat

1 Introduction

Despite the rapid spread of digital technologies in India, there are around a billion people that still need to get online for pushing growth, creating jobs, and accessing public services. Rural India is considered as passive and static as compared to urban India in terms of mobility, technological and globalization processes [1]. According to recent Census Estimate report, urban India has close to 60% online penetration, reflecting a level of saturation and rural India has only 17% online penetration, therefore, there

S. Zahoor (✉) · R. N. Mir
NIT Srinagar, Srinagar, Jammu and Kashmir, India
e-mail: saniyazahoor@nitsri.net

R. N. Mir
e-mail: naaz310@nitsri.net

S. Smys et al. (eds.), *International Conference on Computer Networks and Communication Technologies*, Lecture Notes on Data Engineering and Communications Technologies 15, https://doi.org/10.1007/978-981-10-8681-6_65

is a need as well as the large scope of development in rural India. Connectivity and digital inclusion are the two major concerns for rural communities to compensate for their remoteness, but the problem is that they are least connected and included. Digital Rural India policies aim at combining both connectivity and inclusion issues for poorly connected and digitally excluded rural communities. These growth policies, therefore, focus on bridging the digital divide that exists between urban and rural India. The digital inclusion is transforming the world—helping offline users to get online, aiding information flow, tackling a problem of social isolation in terms of health, education, agriculture, etc. and thus creating huge opportunities for growth [2].

IoT is a revolutionary technology that can bring a real revolution in rural India by providing the much-needed bridge between urban and rural India. The IoT technology can be leveraged to offer benefits in many sectors across the rural India such as health, agriculture, education, local business, etc. Internet of Things is the biggest supplement to digital India [3]. Currently, the IoT is negligible in rural areas as compared to urban areas which imply there is a good scope of digital growth in rural areas. "Digital India" and "Make in India" are the revolutionary initiatives taken by Government of India that aim to make rural areas in India digital. IoT is considered as an important step in rural India where 70% of the population resides and it is going to transform the isolated rural areas into a network of objects talking to each other through data generation which can be analyzed remotely to derive valuable information.

The IoT is basically a large network of devices that have the capability of transferring data via Internet such as remote healthcare, monitoring of patient's biometrics, remote control of home appliances, real-time monitoring of crops to measure temperature and soil moisture, etc. [4]. Maintaining such large number of objects that transfer an enormous amount of data will need the cloud to work as a catalyst to provide right kind of IoT-cloud-based framework that can assist to make IoT work [5]. The integration of IoT and cloud computing is of great significance. Powerful storage, processing, and serviceability of cloud computing combined with the ability of information collection of IoT compose a real network between people and objects and the objects themselves [6]. There are problems of delay in transmission of data from IoT data to Cloud, hence a new paradigm of fog computing is introduced. Fog computing makes use of edge devices at the network edge to process the data generated by IoT devices quickly and if the application needs to maintain historic data, it can be transferred to cloud for storage and further analysis. Thus, rural India can transform itself by taking the technology leap and adopting real-life solutions based on IoT.

This paper discusses how IoT can lead to digital inclusion in rural India. Further, it proposes a model in IoT applications that allow rural people to enter into cyberspace with physical data from various sensors used for multi-applications. The organization of this work is as follows: Sect. 2 discusses Internet of things, Sect. 3 discusses cloud computing and Sect. 4 focuses on fog computing, Sect. 5 discussed proposed model, and conclusions are discussed in Sect. 6.

2 Internet of Things

The Internet of Things (IoT) is a computing paradigm that connects everyday physical objects to the internet which can identify themselves to other devices [4]. Initially, researchers could relate Internet of Things with RFID (Radio Frequency Identification) technology only. Later on, researchers did relate IoT with other technologies, such as sensors, actuators, gateways, etc. IoT components are the atomic components that connect real world to the digital world and these include RFID, sensors, actuators, object identifiers, gateways, and IP addresses.

Internet of Things has been identified as an emerging technology and will play an important role in many areas. There are a lot of application areas where IoT can lead to significant growth in rural India to transform villages into "smart villages". Table 1 discusses the literature survey done on IoT based applications.

3 Cloud Computing

The cloud computing has significantly changed over the last decade and has resulted in new computing architectures that affect areas such as connecting people and devices, data-intensive computing [28]. The number of IoT devices will reach 20 billion by 2020 and the world population will reach 8 billion in 2020 according to Gartner Report. That implies there will be 2.5 IoT devices per person, thus a number of IoT devices and data manageability are the major concerns of the society. The ability of cloud computing to handle a vast number of devices with huge amounts of data is responsible for the rapid expansion of IoT. Cloud computing resources are inexpensive in terms of availability, flexibility and user location is irrelevant to using with the cloud. IoT devices are more expensive in terms of development, deployment, and are not flexible and are generally stuck in one location. The cloud computing overcomes the limitations of IoT devices, and thus share a symbiotic relationship that leads to the success of IoT.

4 Fog Computing

Pushing the data from IoT devices to cloud introduces processing delay because the data and applications processed in a cloud are time consuming. Instead of simply pulling raw data and sending it off to the cloud for processing and analysis, a new push has begun for the IoT device to have an enhanced role in storing data and performing analytics on it as well. This is known as both fog or edge computing [29, 30]. Table 2 shows the comparative analysis of IoT devices, cloud computing, and fog computing in detail.

Table 1 IoT-based applications

IoT applications	Techniques
Smart agriculture	A prototype platform controls network information integration to study the actual situation of agricultural production while operating from remote location [11]
	Yang et al. [12] discusses problems of IoT applications in agriculture and summarizes applications in the modern methods of breeding, crop growth, quality, and safety of agricultural products using RFID enabled IOT devices
Smart healthcare	Wearable sensors devices take decision when to call remote doctors for assistance to patients in real time [13]
	Istepanian et al. [14] aims at health care in sports
	Hwang and Chang [15] focuses on diabetic patient assistance by monitoring glucose level and other physiological parameters in real time
Smart education	Heinemann and Uskov [16] focuses on creating optimal learning environment with a lot of resources for mobile learning.
	Lazarescu [17] proposes a framework for a smart campus with the help of cloud computing and internet of things in education
Environment monitoring	Glasgow [18] presents a platform based on WSN used for long-term environmental monitoring IoT applications.
	High-resolution cameras for real-time monitoring of plants and their environment; and sensors for monitoring temperature, humidity, rainfall, wind, and solar radiation is detailed in [19]
	Shrouf et al. [20] discusses sensorscope as a solution for environmental monitoring that works with wireless sensor network
Smart energy	Abedin [21] proposes an approach for energy management in smart factories based on the IoT
	Othman and Shazali [22] addresses the energy efficiency issues across diverse IoT-driven networks by proposing a system model for G-IoT and energy efficient scheme for the IoT devices to extend the life expectancy of the whole IoT network
Water quality and dam water monitoring	Real-time monitoring of water quality with WSN sensors can be done [23]
	Li et al. [24] proposes a dam monitoring and pre-alarm system based on the internet of things and cloud computing which provides real-time monitoring of the saturated line, water level, and the dam deformation
Disaster mitigation	Tadic et al. [25] focuses on the Internet of Things application that makes use of several sensors in greenhouse including air temperature, relative humidity, soil temperature, etc. in reducing the influence of low-temperature disaster on solar greenhouse
Road monitoring	Built-in IoT sensors in the vehicles help to monitor the road conditions by gathering information on the smartphones connected to these sensors [26]
Traffic monitoring	Lu and Zeng [27] focuses on managing road vehicles efficiently by IoT-based ecosystem

Table 2 IoT, cloud computing, and fog computing

Parameters	IoT devices	Cloud computing	Fog computing
Latency	Least	High	Less
Location	On the device	Within internet	Edge of network
Processing delay	Nil	High	Less
Communication capability	Least	High	Less
Storage capability	Least	High	Less
Processing capability	Least	High	Less
Bandwidth requirements	Data is stored locally	Bandwidth problem	Less demand of bandwidth
Distribution	Application specific	Centralized	Distributed
Response time	Quick response time	Slow response time	Avoid response time issues
Scalability	No scalability issues	Scalability problems arise	Avoid scalability issues

Fog computing overcomes the limitations of IoT devices and Cloud computing, thus seems desirable to use in various IoT applications.

5 Proposed Model

The proposed model focuses on bringing rural India close to urban India with the help of IoT and Cloud computing. There are a number of applications of Internet of Things in rural India as discussed in Sect. 4. General detailed design and functionality of our proposed layers or components for a particular cloud-based IoT application fog consist of five layers, viz., perception layer, network layer, edge layer, cloud layer, and application layer.

The perception layer is the physical layer whose function is to gather and transform data to readable digital signals with data acquisition hardware such as sensors, actuators, cameras, GPS, RFID tags, biometrics, barcodes, etc. This layer senses some physical parameters specific to a particular application and turns it into useful data. Only some part of processing can take place in this layer due to the resource-constrained nature of IoT devices. Network layer is the physical layer which collects data perceived by the perception layer and sends digital signals to corresponding platforms via network. The data from the sensors is needs to be aggregated first and then converted from analog form into digital form for further processing. The Data Acquisition System connects to the sensor network, aggregates data, and performs the analog-to-digital conversion. The data generated after aggregation and A/D conversion may require further processing in edge layer before it enters the data center. This is where edge devices come into play to perform more analysis and provide

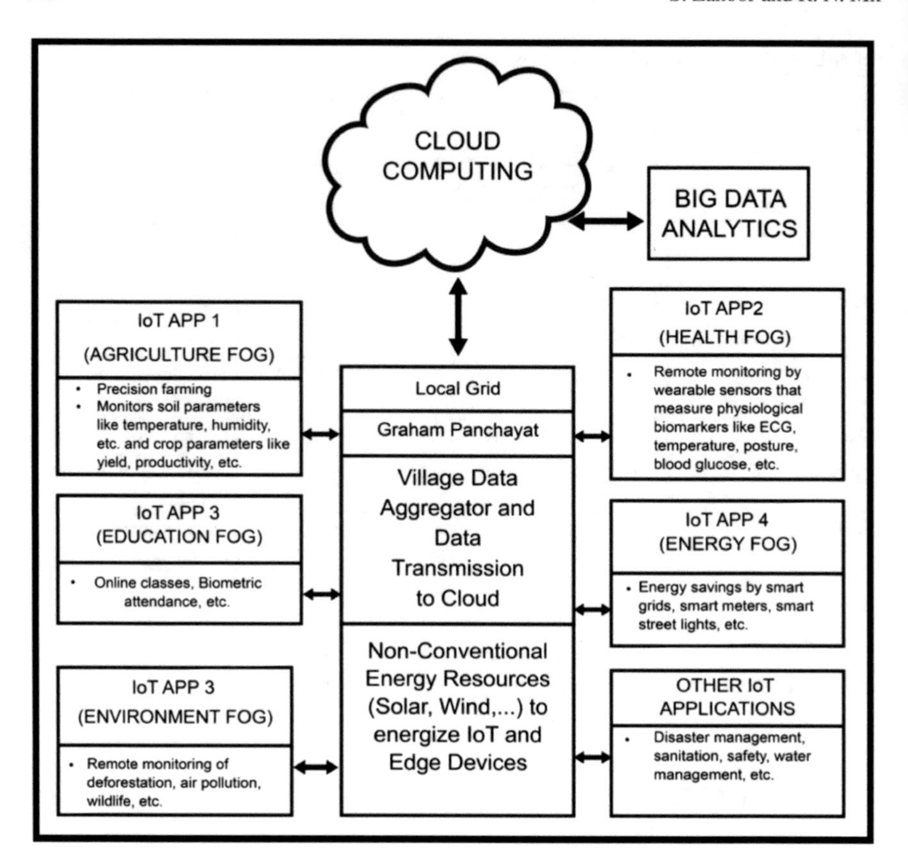

Fig. 1 Proposed model

quick feedbacks. Edge devices perform further analysis of data to lessen the burden on cloud in terms of Bandwidth usage, processing delays, etc.

Data that needs more processing gets forwarded to data center or cloud-based systems, which can analyze, manage, and securely store the data. This encounters processing delays, but processing and analysis are done more in-depth. In most cases, the virtual layer is deployed on cloud computing infrastructure eliminating the need of owning, housing, and maintaining computing resources. The last layer is related to the scale of the application that can vary from an individual to national scale. Here, the data collected at the sensor level is utilized and value is derived from the IoT. This layer provides a user interface to use IoT to execute commands, sends alerts, or draws conclusions from information.

These layers form the main components for the working of a particular application. Figure 1 shows the proposed model for various cloud-based IoT applications to ensure digital inclusion rural India.

The model consists of various IoT applications represented as:

$$
\begin{aligned}
A = \{&\text{IoT APP1, IoT APP2, IoT APP3}, \ldots, \\
&\text{IoT APPN} \mid \text{IoT APPi} \in \{\text{Health, Education, Agriculture}, \ldots\}\}
\end{aligned}
$$

For agriculture fog,

$$\text{IoT APP1} = \Sigma(\text{IoT agriculture1, IoT agriculture2}, \ldots, \text{IoT agriculture})$$

For health fog,

$$\text{IoT APP2} = \Sigma(\text{IoT health1, IoT health2}, \ldots, \text{IoT health } n)$$

Similarly, other IoT Applications can be represented in similar manner.

The proposed model is represented in the following main steps—data acquisition through IoT sensing, data aggregation and transmission to Edge devices, processing in edge devices, data transmission from edge devices to cloud, processing in cloud, and transmission of processed results from cloud to local grid (gram panchayat) for necessary action at the village level. The IoT applications make use of sensors, RFID, actuators, object identifiers, gateways, IP addresses, and layers to collect useful data that needs to be analyzed and converted into actions to remotely control these objects. The edge devices perform the required processing of the data and return results immediately if required or store the results onto the cloud for further storage, analytics, and visualization. Nonconventional energy resources are used to minimize the effects of rural power supply inefficiencies by encouraging solar, wind, etc. to support the much-needed power requirements for these devices at village level. Moving away from the traditional source of batteries which gave the sensor a very limited lifetime and had to be replaced manually, energy harvesting sources means power can be gathered from environmental sources such as solar, thermal, and vibration and wireless RF energy. These alternative sources need the availability of the corresponding energy source and are essential in extending the longevity of sensors.

6 Conclusion

Rural development has always been an important issue pertaining to the economic development of developing Countries like India as rural India comprises a substantial majority of the population. The proposed model will encourage in developing a strategy for digital inclusion of rural India through data acquisition, data transmission, and data processing. It will benefit the rural areas while providing them with the facilities to make quick and intelligent decisions pertaining to agriculture, health, education, skill development, sanitation, etc. The model will minimize the need for much hardware infrastructure at village level except few edge level devices at gram panchayat level. Also, the nonconventional resources used will help to include the villages having electricity problems.

References

1. Clasen, T., et al.: The effect of improved rural sanitation on diarrhoea and helminth infection: design of a cluster-randomized trial in Orissa, India. Emerg. Themes Epidemiol. **9**(1), 7 (2012)
2. Madon, S., et al.: Digital inclusion projects in developing countries: processes of institutionalization. Inf. Technol. Dev. **15**(2), 95–107 (2009)
3. Natarajan, K., Prasath, B., Kokila, P.: Smart health care system using internet of things. J. Netw. Commun. Emerg. Technol. (JNCET) **6**, 3 (2016). www.jncet.org
4. Sahana, M.N., et al.: Home energy management leveraging open IoT protocol stack. In: 2015 IEEE Recent Advances in Intelligent Computational Systems (RAICS). IEEE 2015
5. Patil, V.C., et al.: Internet of things (IoT) and cloud computing for agriculture: an overview. In: Agro Informatics and Precision Agriculture (AIPA 2012) (2012)
6. Botta, A., et al.: Integration of cloud computing and internet of things: a survey. Future Gener. Comput. Syst. **56**, 684–700 (2016)
7. Want, R.: An introduction to RFID technology. IEEE Pervasive Comput. **5**(1), 25–33 (2006)
8. Yick, J., Mukherjee, B., Ghosal, D.: Wireless sensor network survey. Comput. Netw. **52**(12), 2292–2330 (2008)
9. Gubbi, J., et al.: Internet of things (IoT): a vision, architectural elements, and future directions. Future Gener. Comput. Syst. **29**(7), 1645–1660 (2013)
10. Atzori, L., Iera, A., Morabito, G.: The internet of things: a survey. Comput. Netw. **54**(15), 2787–2805 (2010)
11. Zhao, J., et al.: The study and application of the IOT technology in agriculture. 2010 3rd IEEE International Conference on Computer Science and Information Technology (ICCSIT) vol. 2. IEEE (2010)
12. Shifeng, Y., et al.: Application of IOT in agriculture. J. Agric. Mech. Res. **7**, 190–193 (2011)
13. Yang, G., et al.: A health-Iot platform based on the integration of intelligent packaging, unobtrusive bio-sensor, and intelligent medicine box. IEEE Trans. Ind. Inf. **10**(4), 2180–2191 (2014)
14. Ray, P.P.: Generic Internet of Things architecture for smart sports. In: 2015 International Conference on Control, Instrumentation, Communication and Computational Technologies (ICCICCT). IEEE (2015)
15. Istepanian, R.S.H., et al.: The potential of Internet of m-health things m-IoT for non-invasive glucose level sensing. In: 2011 Annual International Conference of the IEEE Engineering in Medicine and Biology Society, EMBC. IEEE (2011)
16. Hwang, G.-J., Chang, H.-F.: A formative assessment-based mobile learning approach to improving the learning attitudes and achievements of students. Comput. Educ. **56**(4), 1023–1031 (2011)
17. Heinemann, C., L. Uskov, V.L.: Smart University: literature review and creative analysis. In: International Conference on Smart Education and Smart E-Learning. Springer, Cham (2017)
18. Lazarescu, M.T.: Design of a WSN platform for long-term environmental monitoring for IoT applications. IEEE J. Emerg. Sel. Topics Circ. Syst. **3**(1), 45–54 (2013)
19. Glasgow, H.B., et al.: Real-time remote monitoring of water quality: a review of current applications, and advancements in sensor, telemetry, and computing technologies. J. Exp. Mar. Biol. Ecol. **300**(1), 409–448 (2004)
20. Mitoi, M., et al.: Approaches for environmental monitoring sensor networks. In: 2015 International Symposium on Signals, Circuits and Systems (ISSCS). IEEE (2015)
21. Shrouf, F., Ordieres, J., Miragliotta, G.: Smart factories in Industry 4.0: a review of the concept and of energy management approached in production based on the Internet of Things paradigm. In: 2014 IEEE International Conference on Industrial Engineering and Engineering Management (IEEM). IEEE (2014)

22. Abedin, S.F., et al.: A system model for energy efficient green-IoT network. In: 2015 International Conference on Information Networking (ICOIN). IEEE (2015)
23. Othman, M.F., Shazali, K.: Wireless sensor network applications: a study in environment monitoring system. Procedia Eng. **41**, 1204–1210 (2012)
24. Sun, E., Zhang, X., Li, Z.: The internet of things (IOT) and cloud computing (CC) based tailings dam monitoring and pre-alarm system in mines. Saf. Sci. **50**(4), 811–815 (2012)
25. Li, Z., et al.: Forewarning technology and application for monitoring low temperature disaster in solar greenhouses based on internet of things. Trans. Chin. Soc. Agric. Eng. **29**(4), 229–236 (2013)
26. Tadic, S., et al.: GHOST: a novel approach to smart city infrastructures monitoring through GNSS precise positioning. In: 2016 IEEE International Smart Cities Conference (ISC2). IEEE (2016)
27. Xiao, L., Wang, Z.: Internet of things: a new application for intelligent traffic monitoring system. JNW **6**(6), 887–894 (2011)
28. Lu, G., Zeng, W.H.: Cloud computing survey. Appl. Mech. Mat. **530**, 650 (2014)
29. Yi, S., Li, C., Li, Q.: A survey of fog computing: concepts, applications and issues. In: Proceedings of the 2015 Workshop on Mobile Big Data. ACM (2015)
30. Bonomi, F., et al.: Fog computing: a platform for internet of things and analytics. In: Big data and internet of things: a roadmap for smart environments, pp. 169–186. Springer, New York (2014)

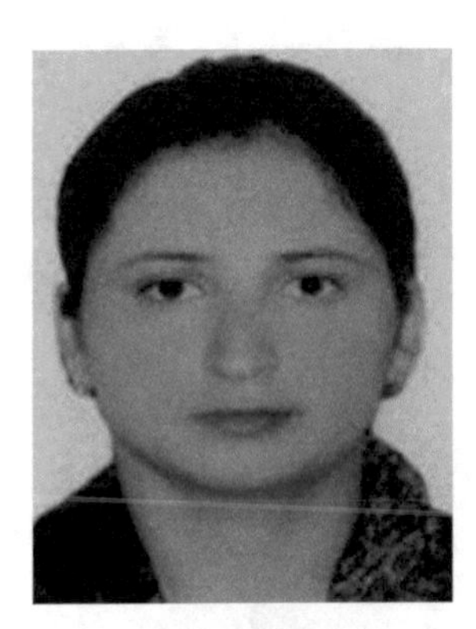

Saniya Zahoor is a research scholar in the Department of Computer Science and Engineering at NIT Srinagar, INDIA. She received her B.E. in Computer Science and Engineering from kashmir University, Kashmir (India) in 2011 and M.E. in Computer Engineering from Pune University, Maharashtra (India) in 2015. Her research interests include Internet of Things, Virtualization and Pervasive Computing.

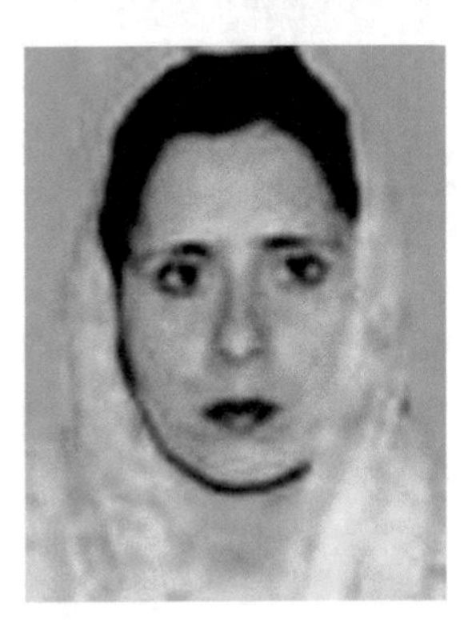

Roohie Naaz Mir is a professor in the Department of Computer Science and Engineering at NIT Srinagar, INDIA. She received B.E. (Hons) in Electrical Engineering from University of Kashmir (India) in 1985, M.E. in Computer Science and Engineering from IISc Bangalore (India) in 1990 and Ph.D. from University of Kashmir, (India) in 2005. She is a Fellow of IEI and IETE India, senior member of IEEE and a member of IACSIT and IAENG. She is the author of many scientific publications in international journals and conferences. Her current research interests include reconfigurable computing and architecture, mobile and pervasive computing, security and routing in wireless Ad hoc, sensor networks and Internet of Things.

Improved Vertical Handoff Decision Scheme in Heterogeneous Wireless Network Based on SCS

M. Naresh, D. Venkat Reddy and K. Ramalinga Reddy

Abstract Seamless continuity is the main objective and challenge in fourth era remote networks (FERNs). To help intelligent constancy in heterogeneous networks, the traditional vertical handover management (VHM) approaches are sufficiently bad. Thus, it is essential to manage those difficulties like a choice of a system and activating of contradictory handover. In heterogeneous remote systems, principle test is a consistent relationship of diverse systems like Wi-Fi, WI-Max, WLAN, and WPAN and so forth. This paper proposed a system called scatter cuckoo searches (SCS) algorithm for vertical handoff, the handover choice stage as well as to mitigate the handling delay and improve the fitness of nodes and quality of a network. We additionally contrasted along and GRA and TOPSIS techniques over SCS in view of the mobile terminal (MT) to get the availability with the best system it estimating by quality of service (QOS) parameters.

Keywords FERNs · VHM · TOPISS · GRA · SCS · QOS

1 Introduction

Mobility is a very important feature of a 4G wireless networks system. Mobile terminals ought to have the capacity to pick the best system among the accessible systems including WLAN, Wi-MAX, and satellite frameworks and after that make handover [1]. The conventional strategies where the handover is performed on the

M. Naresh (✉)
ECE Department, Matrusri Engineering College, Hyderabad, India
e-mail: nareshmuddamalla@gmail.com

D. Venkat Reddy
ECE Department, Mahatma Gandhi Institute of Technology, Hyderabad, India
e-mail: dasari_reddy@yahoo.com

K. Ramalinga Reddy
ETM Department, GNITSW, Hyderabad, India
e-mail: kattareddy2000@yahoo.com

S. Smys et al. (eds.), *International Conference on Computer Networks and Communication Technologies*, Lecture Notes on Data Engineering and Communications Technologies 15, https://doi.org/10.1007/978-981-10-8681-6_66

premise of the assessment of signal strength are insufficient. They do not consider different mobile user connection alternatives, for example, the present setting or the connection of the client choices. 4G advances must consider adaptive and keen approach for vertical handover and consider different factors, for example, financial cost, security, control utilization, and the portable terminal speed.

Handover technique has the four phases; Handover initiation, system revelation, handover decision, and handoff execution. In this paper, two vertical handover techniques named as gray relational analysis (GRA) and Technique for Order Preference by Similarity to Ideal Solution (TOPSIS). TOPSIS [2] is advanced than the gray relational analysis vertical handover decision scheme and SCS is the extended work of GRA and TOPSIS. In our work, the proposed efficient method is compared with GRA and TOPSIS methods. The bandwidth, delay, PDR, energy consumption, and throughput are the parameters took by the MT as the decision parameters for handover. Through performance analysis, we show that our extension method is successful not only in minimizing handover failures and unnecessary handover instances but also in giving the best QoS for MTs.

2 Related Work

At present, a significant number of the handoff choice calculations are proposed in the writing. In [3], comparison is done among SAW, technique for order preference by similarity to ideal solution(TOPSIS), gray relational analysis (GRA), and multiplicative exponent weighting (MEW) for vertical handoff decision. In [4], author talks about that the vertical handoff choice calculation for heterogeneous remote system, here the issue is planned as Markov choice process. In [5], the vertical handoff choice is planned as fluffy different trait basic leadership (MADM). Frequent mobility of users among different wireless networks requires strong authentication during vertical handoff (VHO). In [5], author presents a novel authentication algorithm anonymous authentication for vertical handoff (AAVHO).

In [6], they will likely decrease the over-burden and the handling delay in the portable terminal so they proposed novel vertical handoff choice plan to evade the preparing delay and power utilization. In [7], a vertical handoff choice plan DVHD utilizes the MADM strategy to stay away from the handling delay. In [5], Lahby Mohammed et al. gave an alternate approach by consolidating the ANP and TOPSIS strategies. Their approach is additionally given by considering and disposing of the differentiated weights of criterions and history attributes for handoff. The scheme proposed in [8] will reduce the consumption of power as well as processing delay.

In [5], a novel appropriated vertical handoff choice plan utilizing the SAW technique with a circulated way to keep away from the disadvantages. In [9], author describes the four stages of coordinated procedure for MADM [10] based system determination to take care of the issue. All these proposition works are primarily centered on the handoff choice and compute the handoff choice criteria on the mobile terminal side and the talked about plan are utilized to lessen the preparing delay by the computation procedure utilizing MADM in a disseminated way.

3 Proposed Method

In this paper, we proposed framework named Scatter Cuckoo searches [11] algorithm. We can describe the process of our network; the change of SS (scatter search) calculation was an expert by using Cuckoo Search it named as nature-inspired swarm intelligent algorithm. Finding the closest worldwide optimum solution demonstrated by Cuckoo search algorithm, it can be considered reasonable time and it shows good performance by taking care of some combinational issues. To enhance the SS algorithm through few spots since the algorithm is made out of several stages. However, to improving the SS algorithm by the effective methods are applied named as Subgroup propagation Method, Renovation Method, and Note Set Update Method.

When we endeavor to enhance the SS calculation, the improved method noticed that time was a big problem. To each new solution produced from sequence method where the Renovation Method is applying on all populations, so it will take a lot of time, locating the ideal solution in reasonable time is the fundamental objective of it in taking care of the problems and the SS calculation as one of the meta-heuristic algorithms [12] get influenced.

Algorithm

Step 1 Initialize the handoff problem in a number of nodes in network
Step 2 diversification generation method involves and uses to initialization process
Step 3 Number of nodes taken by applying renovation method
Step 4 Note set update process for reconfiguration
Step 5 While (itr < Maxitr)
 Itr → Iteration
 MaxItr → Maximum iteration
Step 6 While (Note set is changed) Go back to step 4
Step 7 Get a cuckoo search process randomly taken by levy flight
Step 8 Evaluate the function it as quality/fitness 'F_i'
Step 9 Choose randomly consider nodes among next one as 'n'
Step 10 If ($F_i > F_j$) then replace 'j' by new approach attribute
Step 11 New nodes are built after a fraction of malicious nodes are abandoned
Step 12 Our method set into network particular condition named as subgroup propagation
Step 13 while (counter <> 0)
 Sequence process applicable for network
Step 14 New approach starts up

Step 15 It is applied based on Note set methodology
Step 16 End while
Step 17 End while
Step 18 End while

However, the results were good and were obtained in reasonable time, when trying to improve the SS algorithm in note set update method in SS calculation. The best solutions can be taken explored and retrieved from the steps of CS to complete SS steps. The improved SS algorithm using CS is shown in below algorithm.

NoteSet1 of b1 of the best solutions and NoteSet2 of b2 of diversity of solutions will be chosen, in note set update method. While transferring from CS to SS, the new advances are added when NoteSet1 will enter. The NoteSet1 which is benefit from the neighborhood search in the cuckoo search steps can achieve more diversity from the new advances. As the substitution operator forms the cuckoo solutions, the updated note set will contain more enhanced solutions than the old.

To fabricate and keep up a note set, a reference update method is applied. While keeping astounding arrangements, the goal is to guarantee diversity. For example, the NoteSet1 arrangements with the best target capacity are chosen before including NoteSet2 arrangements with the ideal decent variety arrangements (Note Set = NoteSet1 + NoteSet2). A few subsets of its answers as a reason for making joined arrangements are delivered by A Subgroup propagation Method and to work on the note set. The subgroup propagation method into at least one consolidated arrangement vectors delivers a solution sequence method to change a given subset of arrangements.

Initially, the mobile node request for handoff, in this for checking which network is available. If the network is available then broadcasting for handoff parameters to analyze hop count, RSS, signal delay, and node movement direction. The request for the selected network. The process of proposed framework is shown in Fig. 1. Once the request is taken then handoff process completed. In this process, a network is not available for mobile nodes it terminates the communication.

4 Results and Discussions

Simulation analysis is carried out using the NS-2 simulator. Simulation parameters are shown in Table 1.

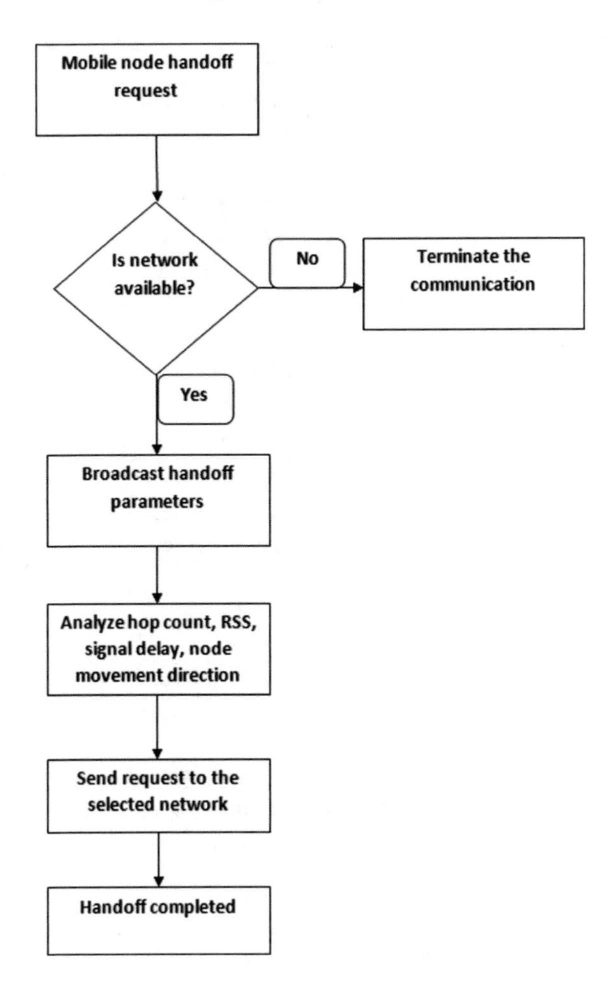

Fig. 1 Flowchart of the proposed method

4.1 Result Analysis

The simulation procedure starts with the objective of handoff process. Here, the framework shows node how to convey in the network of support of a routing convention. In this network, different networks taken so here vertical handoff process performed based on network scenarios. In this process, different types of nodes act as the router, access point, server, networks, and mobile nodes. Here, network process depends on all considering nodes. Here, the mobile node searches the network and collects data onto network1 after that mobile node move to some other location so it disconnects from network1 and goes for the next available network. For this time, the handoff process will be occurring. This process will continue for up to available of networks. In this, mainly check the time interval for a handoff process and reducing the time of handoff process. The results have verified that our mobile nodes are

Table 1 Simulation parameters

Parameter	Value
Application traffic	CBR (Constant bit rate)
Transmission rate	20 packets/s
Antenna model	Omni antenna
Noise model	AWGN
Packet interval	0.05 s
Frame duration	0.008
Advertisement interval	1.0
Iterations (min, max)	1,12
Radio range	250 m
Packet size	1000 bytes
Channel data rate	20 Mbps
Maximum speed	60 m/s
Simulation time	20 s
Number of nodes	8
Area	1000 × 1000
Network	802_11, 802_16
Method	GRA, TOPSIS and SCS
Routing protocol	NOAH, DSDV

capable of mitigating the handoff time and allowing the different networks normally. In addition, it has also shown that the protocol can reduce the time and be waiting time for next networks easily to its after handling time.

4.2 Performance Analysis

The performance of network process is being shown in Figs. 2, 3, 4, 5 and 6. Figures 2–6 demonstrates that the performance analysis of bandwidth, energy consumption, a packet delivery ratio, delay, and throughput with the time for conventional and proposed algorithms. These charts measure the proposed framework performance is more powerful and effective to look at GRA, TOPSIS [6] methods.

The simulation results have shown that our handoff time mitigating method does not bring significant routing overhead to the performance of network. Meanwhile, Fig. 2 shows that the bandwidth is higher when the handoff occurs. When a network checks the data to transmit, handoff process will be considered. The routing process should be depending on how much mobile node traveling distance. Figure 3 shows that individual node energy levels based on network routing process and routing levels are more efficient and it will be effective in the process of vertical handoff time intervals. The packets transmitted based on time intervals and maintain the routes for more data packets should be transmitted it represents in Fig. 4. Figure 5

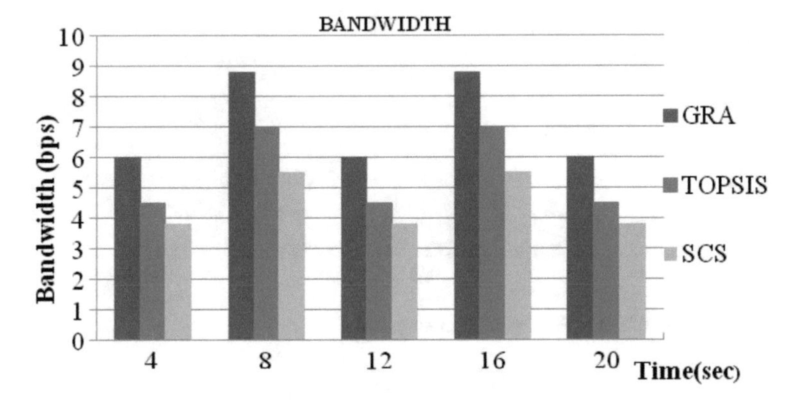

Fig. 2 Comparison of bandwidth with simulation time

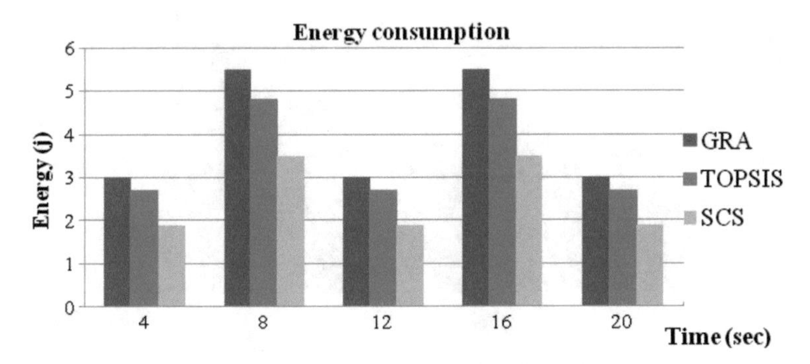

Fig. 3 Comparison of energy consumption with simulation time

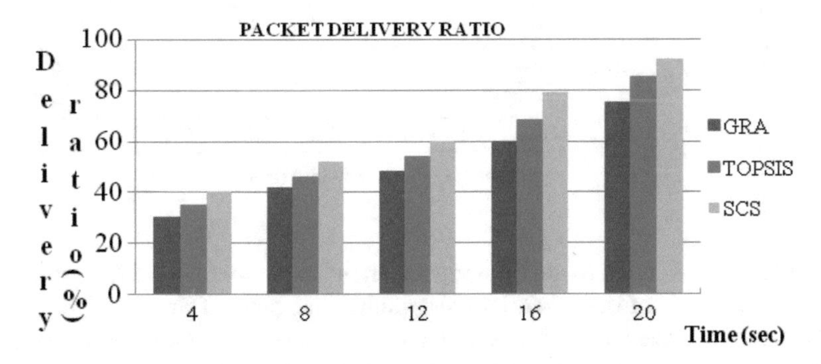

Fig. 4 Comparison of packet delivery ratio with simulation time

shows that delay is a little higher when the node considers as handoff time. When a normal node wants to communicate with a high level node, it will buffer some packets first and then start the route discovery process. As one of network has check the routes at certain time, the broadcasted RREQ may spend more time finding it

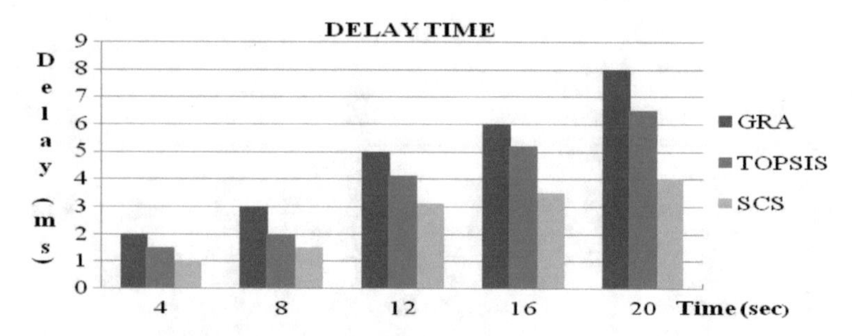

Fig. 5 Comparison of delay time with simulation time

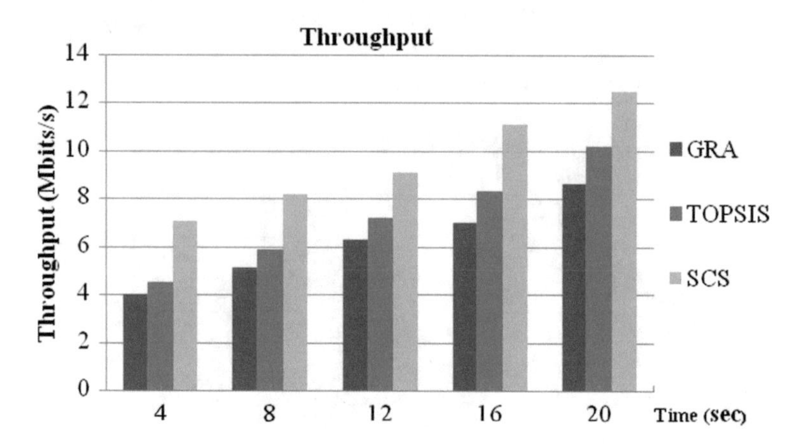

Fig. 6 Comparison of throughput with simulation time

Table 2 Time versus bandwidth

x/*y*	4	8	12	16	20
GRA	6	8.8	6	8.8	6
TOPSIS	4.5	7	4.5	7	4.5
SCS	*3.8*	5.5	3.8	5.5	3.8

$x \rightarrow$ time (s), $y \rightarrow$ bandwidth (bps)

and will include it in the route. For instance, in our experiment, the RREQ traversed around almost half of the network to discriminate the handoff time. As a result, the first buffered packets will be delayed for a longer time before getting to the receiver. With the increase of the number of high level nodes and network scale, the probability of this situation increases and its effects on average delay are more obvious (Tables 2, 3, 4, 5 and 6).

Table 3 Time versus energy

x/y	4	8	12	16	20
GRA	3	5.5	3	5.5	3
TOPSIS	2.7	4.8	2.7	4.8	2.7
SCS	1.9	3.5	1.9	3.5	1.9

$x \rightarrow$ time (s), $y \rightarrow$ energy (J)

Table 4 Simulation time versus packet delivery ratio

x/y	4	8	12	16	20
GRA	30	42	48	60	75
TOPSIS	35	46	54	68	85
SCS	42	52	60	79	92

$x \rightarrow$ time (s), $y \rightarrow$ PDR (%)

Table 5 Time versus delay time

x/y	4	8	12	16	20
GRA	2	3	5	6	8
TOPSIS	1.5	2	4.1	5.2	6.5
SCS	1.0	1.5	3.1	3.5	4

$x \rightarrow$ time (s), $y \rightarrow$ delay (m/s)

Table 6 Time versus throughput

x/y	4	8	12	16	20
GRA	4.0	5.1	6.3	7.0	8.6
TOPSIS	4.5	5.9	7.2	8.3	10.2
SCS	7.1	8.2	9.1	11.1	12.5

$x \rightarrow$ time (s), $y \rightarrow$ throughput (M Bits/s)

5 Conclusion and Future Work

In this paper, we proposed a vertical handoff decision algorithm for heterogeneous wireless networks and the results for the performance comparison between three different vertical handoff decision algorithms namely; GRA, TOPSIS, and SCS. The attributes that we considered in the simulation model included Throughput, delay time, packet delivery, energy consumption, and bandwidth. Results show that GRA and TOPSIS provide similar performance to all five traffic classes. SCS provides slightly higher throughput, packet delivery and lower delay and energy consumption.

Our future work will focus on optimization techniques for our proposed scheme.

References

1. IEEE 802.21 media independent handover IEEE 802.21 tutorial. July 2006
2. Mouâd, M., Cherkaoui, L.: A comparison between fuzzy TOPSIS and fuzzy GRA for the vertical handover decision making. Intell. Syst. Comput. Vision (ISCV). IEEE April 2017
3. Savitha, K., Chandrashekar, C.: Vertical handover decision schemes using SAW and WPM for network selection in heterogeneous wireless network. GJCST **11**(9) (2011)
4. Stevns-Navoam, E., Swong, V.W.S.: Comparison between vertical handoff decision algorithm for heterogenous wireless network. IEEE (2006)
5. Omheni, N., Zarai, F., Obaidat, M.S., Hsaio, K.F. (2014) A novel vertical handoff decision making algorithm across heterogeneous wireless network. IEEE (2014)
6. Naresh, M., Venkat Reddy, D., Ramalinga Reddy, K.: Discriminate vertical handoff time using GRA nad TOPISS algorithm in heterogenous networks by effect of metrics. IJERA **7**(7), 49–54 (2017)
7. Bhuwaneswari, A., Geroge, E., Prakash Raj, D.: An overview of vertical handoff decision making algorithm. IJCNIS **4**(9), 55–62 (2012)
8. Zarai, F., et al.: Seamless mobility in heterogenous networks. IJNGN 2(4)
9. Preethi, G.A., Chandrashekar, C., Priya, N.: Handoff between heterogenous networks based on MADM methods. IJSER **2**(12) (2013)
10. Bazrafkan, A., Pakravan, M.R.: An MADM network selection approach for next generation heterogeneous networks. Electrical Engineering (ICEE). IEEE May 2017
11. Xu, Y., Qu, R.: A hybrid scatter search meta-heuristic for delay constrained, multicast routing problems. Appl. Intell. **36**, 229–241 (2010)
12. Al-obaidi, A.T.S.: Improved scatter search using cuckoo search. IJARAI **2**(2), 61 (2013)

M. Naresh is working as Assistant Professor in ECE Department, in Matrusri Engineering College, Saidabad (MECS) Hyderabad. He is obtained B.Tech from JNTU in 2006, M.Tech from OSMANIA UNIVERSITY in 2007 and pursuing Ph.D. from JNTU Hyderabad. He has 9 years teaching Experience. He has 08 research International journals/National journals papers in his credit. And He has attended 10 FDPs/ workshops.

Dr. D. Venkat Reddy is working as Professor in ECE Department, in Mahathma Gandhi Institute of Technology (MGIT) Hyderabad. He is obtained B.Tech in 1989 from NAGARJUNA UNIVERSITY in Electronics and Communication Engg., M.Tech and PhD from JNTU Hyderabad. He has a rich teaching experience of 20 years and 4 years of research experience. He has 20 research International journals/National journals papers in his credit. He is a professor in ECE and Member of ISTE, IETE. He has an experience in NBA, ISO certifications and is In-charge of Training and Placement division. He has arranged many seminars/hands on workshops.

Dr. K. Rama Linga Reddy is working as Professor and Head of the Department in Electronics and Telematics (ETM) department in G. Narayanamma Institute of Technology and Science (GNITS). He completed his B.E. from Vasavi Engineering College in 1989 and M.Tech from SVU Engineering College in 1991, Ph.D. from JNTU Hyderabad. He won the academic excellence award in 2005 for his academic contribution in GNITS. Dr. K. Rama Linga Reddy has 24 years teaching experience (16 years in GNITS and 8 years in CBIT). He has been working as HOD for ETM department since 2002. He has 45 research papers in his credit. He has been invited as M.Tech and Ph.D. examiner by different universities. He is the BOS member and Research Review Committee member for JNTUH (ECE Department) and BOS member of ECE department for many Engineering Colleges. He Chaired Different technical sessions of national conferences & Student Symposiums. He participated over 80 T.V programs (Panel Discussions related to quality of engineering education and EAMCET counseling) in different T.V news channels. He worked as advisor for 15 engineering colleges in NBA accreditation process.

The Future of Cybersecurity: Major Role of Artificial Intelligence, Machine Learning, and Deep Learning in Cyberspace

B. Geluvaraj, P. M. Satwik and T. A. Ashok Kumar

Abstract Artificial intelligence (AI), and in particular machine learning(ML), deep learning (DL) has seen huge pace in recent years and is now set to really start influencing all aspects of community and occupations in which people are engaged. This growth has been charged the advancement in computing power, combined with headway in algorithms and cybersecurity is no exception. AI is "an area of computer science that deals with giving machines the ability to look like they have natural brilliance." Systems which are based on AI, sometimes called as cognitive systems, are helping us automate many jobs and gear up difficulties which are more complex than most humans are capable of solving. New generations malware and cyber-attacks can be difficult to detect with traditional cybersecurity procedures. They develop over time, so more vigorous approaches are necessary. Solutions for these problems in security rely on ML use data from earlier attacks to respond to newer. Another significant advantage of AI systems in cybersecurity is that they will free up a huge amount of time for IT employees. AI is most commonly used to detect threats and attacks. The systems are developed in a such a way that it must be able to act quickly to the situation on its own. And AI systems do not make errors in completing their tasks. So each threat is responded in the most effective and proper way. In future days, there will be a rapid increase of an international clash in cyberspace. This may include attacks on infrastructure and utilities, as well as damaging normal operations of government and financial bodies, traditional society institutions such as banks, press, law enforcement, and judicial. So taking the cybersecurity as an issue for this paper and let us see the challenges and what is the role of AI, ML, and DL in avoiding cybercrime in future.

B. Geluvaraj (✉) · P. M. Satwik · T. A. Ashok Kumar
Department of CS and IT, Garden City University, Bengaluru, India
e-mail: geluvaraj999@gmail.com

P. M. Satwik
e-mail: satwik.pm@gadrencitycollege.edu

T. A. Ashok Kumar
e-mail: ashok.kumar@gardencity.university

S. Smys et al. (eds.), *International Conference on Computer Networks and Communication Technologies*, Lecture Notes on Data Engineering and Communications Technologies 15, https://doi.org/10.1007/978-981-10-8681-6_67

Keywords Artificial intelligence (AI) · Machine learning (ML)
Deep learning (DL) · Cybersecurity (CS) · Supervised learning (SL)
Unsupervised learning (USL) · Neural networks (NN)

1 Introduction

The Security Infringement Survey 2016–17 revealed 47% of all global businesses identified at least one security breach or attack in the last few years. This includes medium-sized companies (67%) and large companies (69%). Securing the personal data of their client is now a business aspect. Using conventional data, such as Name, Locality, E-mail ID's, Bank ACC No, Phone nos, and Biometrics that are unique to the independent. This private data is present via public record or can be purchased on the darknet [1]. The lack of readiness on the business level and the cybersecurity team itself is also having a breathtaking tough time keeping up with a requirement. By 2021, there might be 3.4 million vacant cybersecurity jobs around the globe [2]. Modern big data analytics powered by ML, data science, and AI capabilities are emerging as the powerful solution. Building machine, powered with adaptive baseline behavior models, will be super effective in detecting new unknown attacks. Considering previous and current data with predictive analytics and machine intelligence for security including intelligence will boost the CS perspective tremendously [3]. The modern approach of these technologies seems to be arriving just in time to fill the gaps in the previously used rule-based data security system. So in this paper, we will be looking at how the AI or ML or DP can be the future of cybersecurity, recent attacks, and applications of the new technologies and challenges and remedies.

2 Traditional Methods

CS process fundamentally focused on preventing attacks from the external sources. As malware are advanced, so cybersecurity came up with requiring authentication to access respective departments and organizations, and supplementary firewalls continued to be attached for more confidentiality.

2.1 The Types of Traditional Security

(i) Access control (ii) Antivirus software (iii) Cryptographic software (iv) Intrusion detection system (v) Intrusion prevention system (vi) Sandbox (vii) SIEM (Security Information and Event management) etc. [4]. So in spite of all these approaches why the world wants to adopt new methodologies such as AI, ML, or DL below you can see the types new attacks happened recently around the globe and later let us look into those methods.

3 2016–17 Cyber-Attacks in the World

A group of Indian organizations was infected by the ransomware issue. WannaCry was the largest cyber-attack knocked system across 99 nations. Among the Indian firms damaged by the malware were 2 South Indian banks, two Delhi-based Indian production companies, and a Mumbai-based FMCG company. More than hundred systems of AP police have also been damaged [5]. And some Business tycoons like HBO, Verizon, Sonic, Uber, and all experienced massive customer data breaches.CIA documents and the records of California voters were all disclosed. Even CS guru and John McAfee had his Twitter account hacked [6].

4 AI-Based Attacks

With the profusion of free and statistical significance of AI executions, in a no more time, the bad lads jumped on AI trend for attacking the systems without revealing their original identities. They do this by entering with new techniques to closely related proxies and adopting new type of methodologies or resources to breach. And coming years will be a broader cyber-criminal adoption of AI. So let us see few AI related attacks in the upcoming paper.

4.1 Different Types of AI-Based Attacks

1. Fraudulent exercising of sending E-mails by utilizing Chat bots: Automation is the process in which confidential reports can be assembled for more advanced strikes. AI can be used for automated collection of relevant intelligence of an organization, its systems and identities, before attacks this can be done by implementing legal sources such as help desks, exterior code depositories, and other internet sources for relevant data that can make easy for the attacker's job.
2. The approach of making repetitive trials for estimating the passwords can be done through high level programming to translate information into a code, which is especially used to prevent unofficial access such as passwords this can be done with smarter attacks with AI or ML password guessing effort by using the new methodologies.
3. By the utilization of old encrypting procedures, the much more advanced process can be used for model identification that can be easily made by using the new

technologies of AI which deduces the problem of estimating the passwords by not wasting a quality time. Cyphertext attacks make a tremendous way to demonstrate the base for AI technologies [7].

5 How the AI or ML or DP Works

When we are trying to use ML to find threats. Primarily, we use it to identify malicious entities done by the bad actors so we name them as hackers, attackers, but to find threats, one of the biggest challenges is to interpret what is normal.

5.1 Unsupervised Learning

Let us reckon clustering, dimensionality reduction and association rule learning as the main approaches within USL. These methods make the greater data set easier to analyze [8].

5.2 Supervised Learning

This is where ML has made the biggest influence in CS. The two common use cases are malware stratification and spam detection. Recognizing whether a file is executable or not without worrying about the damages it may cause to the systems. Nowadays this area of research has greatly benefited from DL where it has helped to drop flawed positive rates to very manageable numbers while also reducing the flawed negative rates at the same time. Malware identification works so well because of the billions of free labeled samples. These samples allow us to train the deep neural networks extremely well. The identification of the spam is becoming very easy because there is a lot of available data which can be trained to teach our algorithm or program from wrong to the right path [8].

6 The Future Role of These Technologies

SL needs labeled training datasets which is less suited to CS. USL does not need labeled training data and is better suited for finding strange activity. CS is an intolerance area and one successful attack is a failure of the security of the organization. ML provides 24 * 7 monitoring and handles larger data loads than a human can deal with. But still, it requires human involvement who are experts. DL is also something to watch for the future. This form of AI is different from ML because it is based on algorithms that work how neurons in the brain behave. The aim of DL is to recognize known and unknown attack types. However, it deals with incomplete, filthy, and composite data better than the other algorithms [9].

6.1 The Top Use Cases of AI or ML or DL in Security in Real World

1. **To detect threats and Stop attacks**: NHS agency of London spotted the attack within a seconds using their algorithms and the threat was diminished without effecting the damage to the organization. They concluded that the recent ransomware issue infected more than 200 K victims across 150 countries. And stated that none of their customer were harmed by the WannaCry attack including those systems which were not updated.
2. **To analyze mobile end points**: Google started using ML threats against mobile end points. In October 2018, MobileIron and Zimperium companies would integrate each other using the technology of ML-based threat detection and security compliance which would address challenges like detecting device, network and application threats immediately take automated actions and Protect organization data.
3. **Enhance Human Analysis**: In 2017, MIT's Computer science and AI Lab succeeded in building a model called AI^2. It is an adaptive ML security platform that aids the analysts to find key things in a larger data base. Examining millions of logins each day, the system was capable to filter data and pass it to the human analyst. Decreasing watchful down to 100 a day. This was practically carried out by companies called CSAIL and PatterEx and found the attack detection rate rose to 85% [10].

7 How DL Boosts CS

DL, also known as NN's, is "influenced" by the brain's ability to learn and distinguish different kinds of objects. DL is more advanced and high intellectual, which results in greater perfection and quick actions on the job. Even the most advanced technologies which use dynamic analysis which a can respond to an issue vigorously and traditional ML have great difficulty in identifying new malware. This endangers the organizations to data breaches, data theft, and seizure for ransomware, data corruption, and destruction. We can solve this problem by applying DL to CS [11].

8 Threats and Challenges

1. **Ransomware Evolution**: Ransomware is the nightmare of CS. Which blocks consumer and business information and documents. The attack can be discarded only if the victim meets the cyber-criminal's outstanding demands.
2. **AI Expansion**: Automation may help safeguard against incoming cyber-attacks. This technology evolution comes with many advantages first is, you do not have to pay automated techniques which are made of powerful algorithms. They work for free once you have them. First and foremost is they can work 24 * 7.
3. **IoT Threats**: Recent generation is always plugged in. The problem is that having all constituent parts linked or connected makes consumers highly susceptible to cyber-attacks.
4. **Serverless Apps Vulnerability**: Serverless apps are a warm welcome for cyber-attacks. Consumer data is particularly at risk when users access your application off-server, i.e., from pattern own devices. If the data is on the server or if it is stored in the cloud, you have total control over the information you saved and the proper security will be implemented by the service providers [12].

9 Future Predictions of CS

CS is a technology-driven field understanding which new technologies may disrupt or change CS practice is vital in making accurate predictions.

Predicting cyber impacts resulting from the following emerging technology:

(i) Quantum Computing. (ii) Cloud Computing. (iii) Predictive Sematics. (iv) Behavioral Identity (v) Dynamic Networks

(i) **Quantum Computing**: Quantum computers are unbelievable high-powered systems which use new kind of procedure for processing information. By equipping real-time actions of the system, these systems can run modern types of algorithms to process information more Systematically [13].

(ii) **Cloud Computing**: Cloud computing introduces significant new security hurdles which have not been worked through. More importantly, cloud computing illustrates the need to apply new systematic lifecycle management and enterprise integration to CS.

(iii) **Predictive Semantics**: Semantic Tech is one of the least understood and most powerful emerging trends in IT. Few firms made the connection between the Semantic technology and CS. However, semantic tech gives new way to integrate and interpret data this will help the future generation in predictive and visual analytics. This will help to merge biometrics, identity management, and network behavior.

(iv) **Behavioral Identity**: One of the most important advances in the next decade will be the evolution in thinking about what identity represents. Identity is a credential or biometric marker or both. By 2021, identity will be evaluated by the dozens of variables as well as through the real-time behavior.

(v) **Dynamic Networks**: A dynamic networks takes us from being a fixed target to moving target this represents a vast change in how CS defends Critical assets. This represents next generation network management. Dynamic networks will support higher level of automation, self-repair, and performance.

10 Remedies

(i) Usage of strong passwords. (ii) Regular patching must be done. (iii) Using updated antivirus. (iv) Using latest operating systems. (v) Avoid usage of OS such as windows XP, cold fusion, etc.

11 Conclusion

Cybercriminals will generate jobs for security professionals over the next few years and this will be done extraordinarily by those professionals. Unfortunately, there seems to be no end to hackers who want to access your data and business information to use them for their own purpose. In coming days, AI will boost the CS, IoT, and serverless apps will increase the under the protection of the data which is saved on the personal devices. Further, there may be advanced global dispute in cyberspace. This includes attacks on organizational structure and utilities, as well as disrupting normal operations of government and financial systems.

References

1. https://www.raconteur.net/technology/the-role-ofai-and-machine-learnin-in-personal-data-security
2. https://www.infosecurity-magazine.com/next-gen-infosec/ai-future-cybersecurity
3. http://www.ibmbigdatahub.com/blog/cyber-security-powered-ai-and-machine-learning
4. https://en.wikipedia.org/wiki/Computer_security#Types_of_security_and_privacy
5. https://economictimes.indiantimes.com/tech/internet/indian-companies-and-government-institutions-hit-by-massive-lobal-cyber-attacks/articlesshow/58660902.cms
6. https://blog.loginradius.com/2018/01/2018-cybersecurity-trends/
7. https://www.csoonline.com/article/3244924/data-protection/2018-cybersecurity-trends-and-predictions.html
8. https://towardsdatascience.com/ai-and-machine-learning-in-cyber-security-d6fbee480af0
9. https://www.csoonline.com/article/3250850/security/artificial-intelligence-and-cybersecurity-the-real-deal.html
10. https://www.csoonline.com/article/3240925/machine-learning/5-top-machine-learning-use-cases-for-security.html

11. https://www.darkreading.com/analytics/introducing-deep-learning-boosting-cybersecurity-with-an-artificial-brain/a/d-id/1326824
12. https://www.globalsign.com/en/blog/cybersecurity-trends-and-challenges-2018/
13. https://www.research.ibm.com/ibm-q/learn/what-is-quantum-computing/

Mr. B. Geluvaraj Research Scholar in School of Computational Sciences and IT, Garden City University, Bangalore, My area of research is Data science. I completed B.E. in Information science and engineering from Visvesvaraya Technological University (VTU) and M.Tech in computer science and engineering from Visvesvaraya Technological University (VTU). And worked as Ad-hoc faculty in National institute of technology, Surathkal, Karnataka and in engineering colleges. My field of interest is Data science, Artificial intelligence, machine learning, neural networks.

Mr. P. M. Satwik Research Scholar in School of Computational Sciences and IT, Garden City University, Bangalore, His area of research is Data Science. He has done MCA from Visvesvaraya Technological University(VTU). His field of Interest is Data Science and Analytics.

Dr. T. A. Ashok Kumar working as Associate Professor, School of Computational Sciences, Garden City University, Bengaluru, Ph.D. in Computer Science MS University, Tirunelveli, Tamilnadu, MBA from Bhartiar University, M.Phil from Bharathidasan University.he has served previously for Christ University as associate professor, CMS College of Engineering as Associate Professor and Director, He is member and editor of various international Journals, He has contributed to research in Data Mining, Data Analytics.

Exploration of Various Cloud Segments Transference and Convenience for Cloud Computing

Srihari Bodapati and A. Akila

Abstract Cloud registering is rising standard to provisioning administrations over the web. It may be the zone about investigating the place a cloud may be used to get data, files, product, and so on. It permits offering from claiming networks servers, information applications, and capacity what's more benefits likewise registering assets. For cloud registering, there will be a collaboration around different heterogeneous free cloud platforms. This may be conceivable best through interoperability which assumes a way part over giving work to get to cloud registering in distinctive situations. While doing different straightforward expressions it implies that the client necessities interoperability should ship their possessions from you quit offering on that one cloud on different cloud [1]. Multi-tenancy approach knowledge and programs are living with knowledge and programs of alternative firms and that get entry to exclusive knowledge is imaginable through shared systems, shared garage, and shared networks. This section defines the critical consideration which should be addressed when designing for convenience and interoperability.

Keywords Cloud · Computing · Convenience · Interoperability · Tuning

1 Introduction to Interoperability

Interoperability is the requirement for the parts of a cloud eco-device to paintings in combination to succeed in their meant outcome. In a cloud computing eco-gadget, the parts would possibly smartly come from other resources, each cloud and conventional, private and non-private cloud implementations [2]. Interoperability mandates that the one's parts will have to be replaceable by way of new or other parts from other suppliers and proceed to paintings, as will have to the change of knowledge

S. Bodapati (✉) · A. Akila
Department of Computer Science, Vel's University, Chennai, India
e-mail: bodapati2709@gmail.com

A. Akila
e-mail: akila.scs@velsuniv.ac.in

S. Smys et al. (eds.), *International Conference on Computer Networks and Communication Technologies*, Lecture Notes on Data Engineering and Communications Technologies 15, https://doi.org/10.1007/978-981-10-8681-6_68

among techniques. Over the years, Companies made selections that result in the will to switch suppliers purposes for this preferred amendment come with:

(i) An unacceptable building up in value at agreement renewal time
(ii) The power to get the similar carrier at a less expensive worth
(iii) Supplier ceases industry operations
(iv) A supplier abruptly closes a number of products and services getting used without appropriate migration plans
(v) Unacceptable lower in carrier high quality, corresponding to a failure to satisfy key efficiency necessities or succeed in carrier degree agreements.

A dispute in the industry among the cloud consumer and supplier is interoperability. A loss of interoperability can result in being locked to a specific cloud carrier supplier. The level to which interoperability may also be accomplished or maintained while bearing in mind a cloud venture regularly depends on the level to which a cloud supplier makes use of open, or revealed, architectures and same old protocols. Despite the fact that many providers of "open" and "requirements primarily based" cloud provision supply propriety hooks and extensions and improvements that may obstruct each interoperability and convenience.

2 An Advent to Convenience

Convenience defines the convenience of skill to which software parts are moved and reused somewhere else without reference to supplier, platform, OS, infrastructure, region, garage, the layout of knowledge, or API's. Convenience and interoperability will have to be thought to be whether or not the cloud migration is to public, personal, or hybrid cloud deployment answers. They are necessary parts to believe for carrier type variety without reference to whether or not a migration technique is to Device as a Carrier, Platform as a Carrier, or Infrastructure as a Carrier [3].

Convenience is a key side to believe while settling on cloud suppliers due to the fact that it could each lend a hand save you supplier lock-in and ship industry advantages via permitting equivalent cloud deployments to happen in several cloud supplier answers, both for the needs of crisis restoration or for the worldwide deployment of an allotted unmarried answer. Attaining convenience for a cloud carrier is normally reliant at the products and services running in the similar architectural octant of the Cloud Dice, as outlined in Area One. The products place and services function in several octants, then porting a carrier on a regular basis method migrating the carrier again "in-space" ahead of re-outsourcing it to an alternate cloud carrier.

Failure to correctly cope with convenience and interoperability in a cloud migration would possibly lead to failure to succeed in the specified advantages of shifting to the cloud and may end up in pricey issues [4] or undertaking delays as a result of elements that are meant to have shied away from corresponding to:

(i) Software, supplier, or supplier lock-in—number of a specific cloud answer would possibly prohibit the power to transport to any other cloud providing or to any other dealer.
(ii) Processing incompatibility and conflicts inflicting disruption of carrier—supplier, platform, or software variations would possibly divulge incompatibilities that lead to programs to malfunction inside of a special cloud infrastructure.
(iii) Sudden software reengineering or industry procedure amendment—shifting to a brand new cloud supplier can introduce a want to transform how a procedure purposes or require coding adjustments to keep unique behaviors.
(iv) Pricey knowledge migration or knowledge conversion—loss of interoperable and transportable codecs would possibly result in unplanned knowledge adjustments while shifting to a brand new supplier.
(v) Retraining or retooling new software or control device.
(vi) Lack of knowledge or software safety—other safety coverage or regulate, key control or knowledge coverage among suppliers would possibly open undiscovered safety gaps while shifting to a brand new supplier or platform.

3 Proposals: Interoperability Suggestions

3.1 Hardware—Bodily Pc Hardware

The hardware will necessarily range or amendment over the years and from supplier to supplier leaving unavoidable interoperability gaps if direct hardware gets entry to is needed [5].

(i) On every occasion imaginable, use virtualization to take away many hardware degree considerations, remembering that virtualization does not essentially dispose of all hardware considerations, particularly on the present techniques.
(ii) If hardware will have to be instantly addressed, it is very important make sure that the similar or higher bodily and administrative safety controls exist while shifting from one supplier to some other.

3.2 Bodily Community Units

The community units together with safety units might be other from carrier suppliers to carrier suppliers at the side of its API and configuration procedure.

To care for interoperability [6], the community bodily hardware and community and safety abstraction will have to be in the digital area as far as imaginable APIs will have to have the similar functionally.

3.3 Virtualization

At the same time as virtualization can lend a hand to take away considerations approximately bodily hardware, particular variations exist among not unusual hypervisors (akin to ZEN, VMware, and others).

(i) The use of open virtualization codecs corresponding to OVF to lend a hand makes sure interoperability.
(ii) Record and remember which particular virtualization hooks are used regardless of the layout. It nonetheless would possibly not paintings on any other hypervisor.

3.4 Frameworks

Other platform suppliers are offering other cloud software frameworks and variations do exist among them that have an effect on interoperability.

(i) Check out the API's to decide the place variations lie and plan for any adjustments essential that can be required for software processing while shifting to a brand new supplier.
(ii) Use open and revealed APIs to make sure the broadest give a boost to for interoperability among parts and to facilitate migrating programs and information will have to convert a carrier supplier develop into important.
(iii) Programs within the cloud incessantly interoperate over the Web and outages can also be expected to happen. Deciding how failure in a single element (or a sluggish reaction) will have an effect on others and steer clear of stateful dependencies that can possibility device knowledge integrity while a far-flung element fails.

3.5 Garage

Garage necessities will range for several types of knowledge. Based knowledge will so much ceaselessly require a database gadget, or require software particular codecs. Unstructured knowledge will generally apply any of a variety of not unusual software codecs utilized by Phrase Processors, Spreadsheets, and Presentation Managers. Right here the worry will have to be to transport knowledge saved with one carrier to some other seamlessly [7].

Retailer unstructured knowledge in a longtime moveable layout.

(i) Check the will for encryption for the information in transit.
(ii) Test for suitable database techniques and determine conversion necessities if wanted.

3.6 Safety

Knowledge within the cloud depends on techniques and the consumer does not personal and most probably has most effective restricted regulate over. Quite a lot of essential pieces to believe for interoperable safety come with:

(i) Use SAML or WS-Safety for authentication so the controls can also be interoperable with different requirements—primarily based methods.
(ii) Knowledge ahead of it is positioned into the cloud will make certain that it can not be accessed inappropriately inside of cloud environments. See area eleven for extra element on encryption [8].
(iii) While encryption keys are in use, inspect how and the place keys are saved to make sure that getting entry to present encrypted knowledge is retained. See area eleven for extra element on key control.
(iv) Be mindful your obligations and liabilities will have to a compromise happen as a result of unanticipated "gaps" in coverage strategies presented via your carrier supplier.
(v) Log record knowledge will have to be treated with the similar degree of safety as all different knowledge shifting to the cloud. Make sure that log information are interoperable to make sure continuity of log research pre- and publish transfer in addition to compatibility with no matter what log control device is in use [9].
(vi) While finishing a transfer make certain that all knowledge, logs, and different knowledge are deleted from the unique gadget.

4 Conclusion

Interoperability and convenience standardization have an enormous effect on at the cloud adoption and utilization. Standardization will build up the adoption of cloud computing [10]. The average knowledge type helps and gives a mechanism for making improvements to the convenience and interoperability side of cloud computing. Customers could have a variety of possible choices in cloud and companies can use their very own present knowledge middle tools seamlessly. Standardization additionally supplies the customers to make a choice and use products and services supplied via a variety of cloud companies in line with more than a few standards [11]. It additionally give advantages to the seller via offering further upper degree products and services aside from commonplace cloud products and services which might be wanted by way of the consumer. Therefore, standardization will open the best way against understanding the real possible and advantages of cloud computing.

References

1. Sheth, A., Ranabahu, A.: Semantic modelling for cloud computing. Part I and II. IEEE Internet Comput. Mag. **14**, 81–83 (2010)
2. McKendrick, J.: Does platform as a service have interoperability issues, 24 May 2010 [Online]. Available: http://www.zdnet.com/article/does-platform-as-a-service-have-interoperability-issues/
3. Cattedu, D., Hogben, G.: Cloud Computing benefits, risks and recommendations for information security, Nov 2009. filefile:///E:/30–10-2014%20backup/Downloads/Cloud%20Computing%20Security%20Risk%20Assessment.pdf
4. Webster, J., Watson, R.T.: Analyzing the past to prepare for the future: writing a literature review. MIS Q **26**, 13–23 (2002)
5. Jamil, D., Zaki, H.: Security issues in cloud computing and countermeasures. Int. J. Eng. Sci. Technol. **3**(4), 2672–2676 (2011)
6. Peter Mell, T.G.: Sept 2011 [Online]. http://csrc.nist.gov/publications/nistpubs/800-145/SP800-145.pdf
7. Wang, C., Ren, K., Wang, Q.: Security challenges for the public cloud. IEEE Comput. Soc. **16**(01), 69–73 (2012)
8. European Journal of ePractice. Three dimensions of organizational interoperability. http://www.epractice.eu/files/6.1.pdf
9. Cloud Standards Customer Council: Practical guide to cloud computing (2011). www.cloud-council.org/2011_Practical_Guide_to_Cloud%20Computing.pdf
10. Cloud Standards Customer Council: Practical guide to cloud SLAs (2012). www.cloud-council.org/2012_Practical_Guide_to_Cloud_SLAs.pdf
11. Cloud Standards Customer Council: Public cloud service agreements: what to expect and what to negotiate (2013) http://www.cloudcouncil.org/PublicCloudServiceAgreements2.pdf

Srihari Bodapati working as a Research Scholar, Department of Computer Science, Vel's University, Chennai. His research area includes cloud computing and security analysis.

Dr. A. Akila working as an Assistant Professor, Department of Computer Science, Vel's University, Chennai. Her research area includes cloud computing and security analysis.

Smart Blood Bank System Using IOT

Vahini Siruvoru, Nampally Vijay Kumar and Yellanki Banduri Santhosh Kumar

Abstract Today, in the present scenario, many people were facing a lot of problems in getting blood for patients at right time. The needy does not know where to get the blood or how to get access to a required quantity of blood quickly in an emergency. Blood donation is a big issue and consumes a lot of time to the donor. The blood bank systems have no proper tool for managing the blood collected from the camps. The needy in an emergency could not find the compatible blood or platelets with the patient. There are many existing mobile apps for blood donation. Though smartphones are available it is not possible to find a right donor who has compatible blood group with the patient in right time. We propose an integrated solution which will connect blood banks, donors, and the needy. This solution provides efficiency and convenience for the needy to search for required blood group or platelets in your neighbourhood easy and simple. Get instant help form blood banks without signing up required. The solution works for the encouragement of blood donation and ensures the availability of blood or platelets accessible to the needy. It also assists quality management programs for blood platelets transfusion services. Finally, this solution is used to notify/communicate with the nearest blood bank organizations timely to ensure the availability of safe and quality blood accessible to the needy with minimum effort.

Keywords Blood bank management system · Blood banks · Needy · Donors Smart inventory management system

---

V. Siruvoru (✉) · N. V. Kumar
Computer Science and Engineering, S R Engineering College, Warangal, Telangana, India
e-mail: vahini_s@srecwarangal.ac.in

N. V. Kumar
e-mail: vijay_kumar_n@srecwarangal.ac.in

Y. B. Santhosh Kumar
Entrepreneur, Hyderabad, India
e-mail: yellankisantosh@gmail.com

S. Smys et al. (eds.), *International Conference on Computer Networks and Communication Technologies*, Lecture Notes on Data Engineering and Communications Technologies 15, https://doi.org/10.1007/978-981-10-8681-6_69

1 Introduction

In India, with its population of 1.27 billion people, someone needs blood for every two seconds. According to WHO survey, our country needs 12 million units of blood every year but collects only 9 million units [1]. Every year, need for blood is gradually increasing by at least 5%. Presently, there is blood deficit of 30–35%. One of the most pressing concerns in India today is the highest number of deaths is due to lack of donated blood at right time. So there is urgent need to create app for finding willing donor at the right time near to the needy [2].

People lack in blood for various reasons. They suffer from various kinds of diseases and hence the urge for blood is raised. There are many diseases, the person may suffer from anaemia or may meet with an accident. The possibility of lack of blood can also be seen in a pregnant lady. These are the popular emerging situations where the help given should be quick and meet the requirements. More than 20 million units of red blood cells, platelets, and plasma is transfused every year to treat diseases like anaemia, leukaemia, and sickle cell disease.

Blood donation is one of the most significant contributions that a normal human being can make towards the society. It is easy and does not cost the donor anything. Moreover, it benefits the donor by keeping healthy and literally lifesaving. Nowadays, awareness is noticed in public to donate blood. Many clubs, colleges, societies, offices, etc. organize blood donation camps on different occasions. It is a healthy gesture. Most of the blood banks are running short of required blood. Lack of awareness is the sole reason because people are busy performing their regular duties and could not concentrate on ongoing blood camps.

Day to day, there is rapid growth in the dynamics of Android devices in our country. We propose an emergency request service is an android free quick blood donor finder app which gives immediate response without signing up. This app is used to find blood at the right time near you. It also sends notifications to all the registered users about the blood camps that are being organized around them.

This application will provide for the needy to broadcast the message automatically to the blood bank emergency system so that the needy gets a response from the blood bank within a short span of time with a single click. The only thing the needy has to do is to enter required blood group and the quantity of blood needed. This app is also used to make requests for platelets to the nearest blood banks when the patient is in need.

This proposal builds a quick blood donor finder app for android mobile devices makes searching for blood donors in your neighbourhood easy and simple. Get instant help form blood banks, hospitals, and your community no signing up required.

2 Architecture of Smart Blood Bank Management System

The proposed blood bank management system provides efficient access to all stakeholders involved in the system. The core are needy, donors, and blood bank. The sensor senses the blood packets and the information is stored in the server. Using the application, the inventory in-charge gets the information on the future demands. He sets the minimum stock level of each blood type to be maintained for that month. Prior to the expiry of current stock, the system sends an automatic alert message to inventory in-charge so that he can do necessary action like transferring to other blood banks or hospitals to avoid wastage. The system also sends notifications to camp organizer when the stock of blood type falls below the set value. The camp organizer uses the prediction tool to find out the probable dates and location of camps and informs to on-site organizer. When the on-site organizer accepts the request and confirms the date, the system sends automatic messages to all regular donors and registered users.

When there is urgent need of blood, the needy clicks button in the app. It immediately sends an alert message to issuing system in-charge so that he responds as soon as possible. The in-charge immediately informs the needy if the required blood type is available using the prediction tool otherwise he communicates with other blood banks or donors and help the needy by finding the blood type from all the sources (Fig. 1).

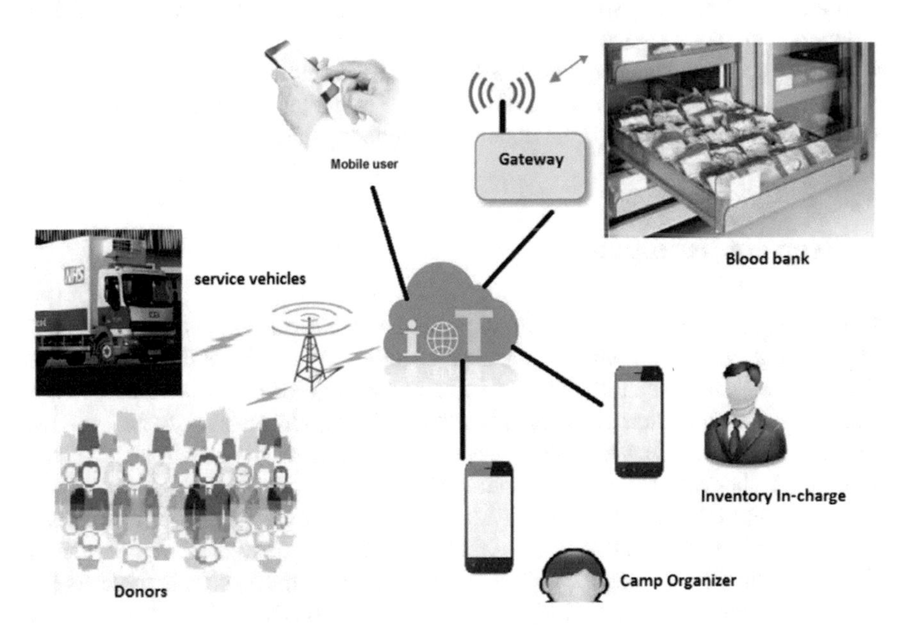

Fig. 1 Architecture of smart blood bank management system

3 Existing Mobile Apps

There are many research papers [3] and apps designed to find the blood required for the needy. The following are some of them:

- Indian blood donors App: This app is compatible with all devices; its simple, effective, and free. When the needy is in urge for some blood group, it can match with a blood donor near to him in minutes. The needy can view the details of all compatible donors in his city or area and can contact the donors seeing the details in the App.
- Simply Blood is another FREE blood donation app for Android mobiles. The needy can connect thousands of blood donors in three simple steps. This app ensures the privacy of a blood donor. This app also provides donor to select any specific date and location to donate blood and find a suitable person who needs blood. This app provides donors to earn badges for life-saving squad.
- D2D—Zindagi Wala: This a smartphone application to find a blood donor with the desired blood group type nearby (within the periphery of 50 km) the current location. This application is designed such that user-friendly, easily accessible and foremost, to connect people from each gender and age group (above 18 years) to voluntarily participate in saving the life via blood donation.
- Social Blood: This app connects donors and recipients. It solves through Facebook to find the blood.
- Blood Hero: This app also connects donors and recipients. It solves by geo-tracking donors with blood seekers.
- Haemovigilance: This app connects hospitals and recipients. It covers the whole transfusion chain, from the collection of blood to the follow-up of needy.
- Red Donor: This app helps the donors in scheduling donation.

4 Problem Definition

A hospital has its own system and limitations; the coordination between the hospital management system and blood banks is practically very less. Even though there are many blood banks available, when a patient is in urge of blood it becomes very difficult to find the compatible blood with a short period of time. In the present scenario, there are many android apps developed to meet the requirement of the needy [4, 5], but each has its own limitations.

According to the survey, we found some limitations from each stakeholder side. There are various stakeholders involved blood management system, we considered all of them into three categories: core, direct and indirect. The core is the ones for which the solution would be targeted and they are: **Needy, donors, blood bank** (Fig. 2).

After the study, we listed some of the limitations from each core stakeholder's perspective in Table 1.

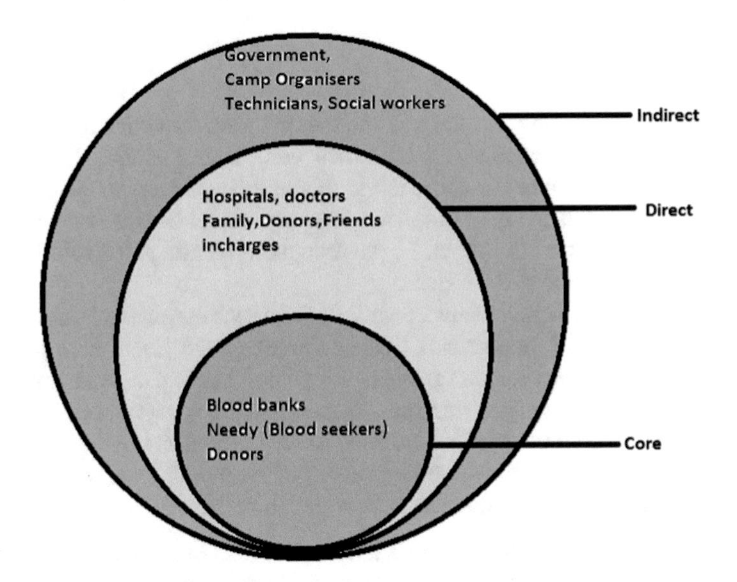

Fig. 2 Stakeholders involved in the blood bank management system

Table 1 Limitations observed from each of the core stakeholders

Blood bank	Donors	Needy
1. The blood bank is unable to maintain a supply of blood at the time of seasonal high demand for blood 2. Reducing the wastage of the blood is very difficult especially during the oversupply season 3. The human error leading to fatalities 4. No proper channel to contact other blood banks and hospitals for transferring blood 5. Securing donation campsite is a competition and no addition of new camps 6. No timely payment from the hospitals and no one to contact	1. General lack of knowledge about the process and demand 2. No prior information on upcoming donation camps 3. Fear of rejection at the campsite 4. Unable to keep a track on the donation dates 5. Distance and time are major constraints 6. No post-donation follow-up	1. Rely on the hospitals to find the blood bags 2. No unified source of information on the availability of blood and platelets 3. No real-time tracking of the donor 4. Lack of awareness in the process of issuing 5. Difficulty in finding donors leading to high anxiety

5 Proposed System

We studied all the existing solutions in the current scenario which addressed similar problems. It helped us to understand their benefits and failures. Analysing all the features, we found a better approach to this problem. Even though all of these provided good services but they failed in some way. It was because they were only catering to one section of the chain. So we thought of an integrated solution which will connect all the three stakeholders.

The proposed system gain many advantages from each stakeholder side. The very first is the urge of blood for a person, a sensitive case where you need to pay lots of attention. The person can search for the nearby blood banks and make a request to them. As the technology is growing rapidly, the app makes use of easiest ways that can be adopted to make the work of the receiver easy. The person who is in the urge has to click on search button where he is asked to mention the requirement details (phone number, blood group, quantity). Now the process is automatic.

(a) The request is sent in the form of a link to the blood banks nearby.
(b) The blood bank admin has to click on the link where he is asked to make confirmation regarding the donation.

The second is for making registrations. Only willing donors are requested to make registrations. This is a simple and one touch process where details of donors are recorded. Working with ease is focused in this app, hence the procedures, modules are made short and they are quick in responses.

(a) Click on 'Register'.
(b) Fill in the form and submit. Now the user is registered as donor.

We focus on blood donation camps conducted nearby in the user's location. All those who are interested in camps can click on this option where Google map is opened as a reference.

(a) With the help of user's location, the camps conducted in that particular location are traced and displayed on the screen.
(b) All those camps are spotted on the map.

The last module is where a person/organizer can seek help for advertisements from the app's admin.

These advertisements costs organizers and so we get benefitted. For the organizers to seek help for advertisements, they need to undergo a simple trusted payment process. Only after the payment is done, seek for help is approved and the advertisement is displayed on the advertisement page. This is the area where awareness can be brought among people. The need for camps, how it is to save life and all other matters can be focused here. The organizer can also mention the cause or the occasion for which they are doing so.

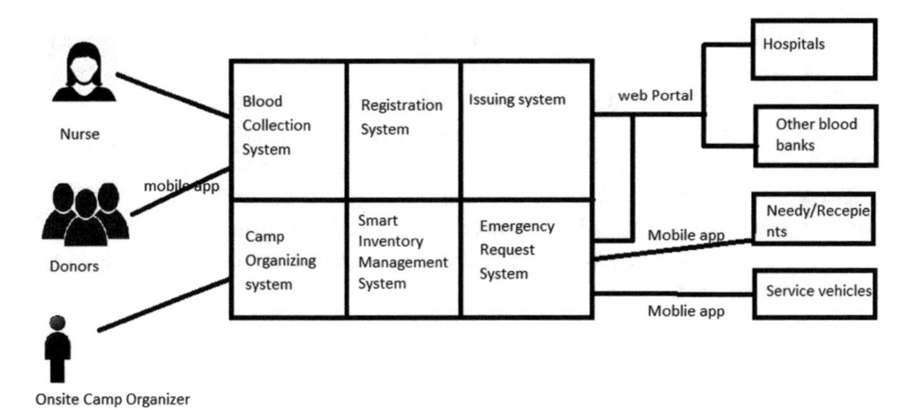

Fig. 3 Modules involved in smart blood bank system

6 Design Considerations

The solution would be like an enterprise product owned by blood bank who are connected to both recipients and donors. It focuses on three modules where each would have some goal to fulfil.

- **Application between donor and blood bank**—Effortless blood donation system with doorstep donation service.
- **Application within blood bank**—Smart Inventory Management System with effective distribution of donation camps using prediction tool
- **Application between needy and blood bank**—Emergency Request Service for the needy (Fig. 3).

6.1 Module 1: Effortless Blood Donation System with Doorstep Donation Service

Here we try to resolve all the limitations from the donor's side [3, 5]. When the on-site organizer decides the dates and confirms dates of camps, the dates and location are communicated to regular donors via message. This system provides doorstep donation service to the donors.

We provide service vehicles during the blood donation camps nearby the user's location. All those who are interested in camps can click on this option so that the service vehicle move to the user's required location.

(a) With the help of the app, the camps conducted in the particular location are traced and displayed on the screen.
(b) All those camps and service vehicles are spotted on the map.

The last module is where a person/organizer can seek help for advertisements from the app's admin.

These advertisements costs organizers and so we get benefitted. For the organizers to seek help for advertisements, they need to undergo a simple trusted payment process. Only after the payment is done, seek for help is approved and the advertisement is displayed on the advertisement page. This is the area where awareness can be brought among people. The need for camps, how it is to save life and all other matters can be focused here. The organizer can also mention the cause or the occasion for which they are doing so.

6.2 Module 2: Smart Inventory Management System with Effective Distribution of Donation Camps Using IOT

The blood bank system maintains the data like number of bags used per medical activity, historical data, average no of days spent in inventory by each blood group, historical data of collection from camps and blood group, blood group, and location of registered donors of the city [6]. Using the data, the Smart Inventory Management System can predict and provide the details like ideal dates to conduct camps, ideal no of bags to be collected blood group wise from each camp, future demand data, minimum stock level of each blood group to be maintained for any given period of time, and ideal location of camps.

The following are some of the solution features:

(a) Prediction tool [7] for future demand patterns based on the historical data of demands
(b) Set the minimum stock requirement of each blood group for that month
(c) Alert message sent to the blood bank in-charge about the shortage of any blood type
(d) Using IOT to generate dates for donation camps and number of bags of each blood group to be collected
(e) Decide the location of camps by visualizing the donation made in the past and blood group type on city map
(f) Facilitate the communication between the camp organizer and on-site organizer
(g) Pre-expiration (15 days prior) alert message sent to the in-charge
(h) Coordination between blood banks and hospitals to manage the excess stock in time
(i) Alert messages to the regular or registered donor and colleges/industries near campsite
(j) Multiple user login feature available

6.3 Module 3: Emergency Request Service for the Needy

This app provides on-demand service for the needy. This app helps needy to search for the blood in nearby blood banks and make a request to them. As the technology is growing rapidly, the app makes use of easiest ways that can be adopted to make the work of the receiver easy. The person who is in the urge has to click on search button where he is asked to mention the requirement details (phone number, blood group, quantity). Now the process is automatic.

(a) The request is sent in the form of a link to the blood banks nearby.
(b) The blood bank admin has to click on the link where he is asked to make confirmation regarding the donation.

This is a simple app that works on the fingertips of India in future with a great response, making life in India safe and secured. This will be an embodiment of care and responsibility. We all know the two sides of technology. Taking the positive side, the intention behind the app is to help India, grow India and make India a better place to live in.

This app aims for providing blood to needy people with minimum effort and it also provides donors convenience in donating blood. As technological development follows no physical or invisible boundaries and gradually expands its roots in all directions. Likewise, the Internet of Things has its applications in fields of home security, Industry, smart cities, and health care [8]. One of the main goals for this app is to involve the Internet of Things (IoT) to help blood bank authority, donors, and the needy by making better automation in blood donation and supply for the needy. With a wide range of technologies available across our daily life, we have found an appropriate way by producing smart system. Donors are able to use the application to enter their willingness and specify the location with the help of GPS coordinates. On the other hand, the application provides automation by enabling smart blood bank authority to manage the blood information and its availability. Finally, this app is used to notify/communicate with the nearest blood bank organizations timely to ensure the availability of safe and quality blood or platelets accessible to the needy with minimum effort.

7 Conclusion

In a medical emergency, blood is usually a very critical component. Crisis for blood was always there in the hospitals for a needy in an emergency situation. The needy could not know where they can go to get the blood or how to get access to a required quantity of blood quickly. A timely response in donating healthy blood can save many valuable lives. In this regard, we thought of providing a link between the donors, blood banks, and the needy using Internet of Things and a repository. The presented proposal will definitely make possible by the needy to find the required blood group or platelets from the nearby blood bank with minimum effort.

The essence of India lies in the way people feel closer and comfortable to move with. Our proposed solution provides efficiency and convenience to the core stakeholders: blood banks, donors, and the needy in doing their tasks.

References

1. Cheng, E., Chan C.W., Chau, M.: Data analysis for healthcare: a case study in blood donation center analysis. In: Americas Conference on Information Systems (AMICS) (2010)
2. Tushar, P., Niloor, S., Shinde, A.S.: A survey paper on E-blood bank and an idea to use on smartphone. Int. J. Comput. Appl **113**(6), 0975–8887 (2015)
3. Bhowmik, A., Nabila, N.A., Imran, M.A., Rahman, M.A.U., Karmaker, D.: An extended research on the blood donor community as a mobile application. Int. J. Wirel. Microw. Technol. **6**, 26 (2015)
4. Rahman, M.S., Akter, K.A.: Smart blood query: a novel mobile phone based privacy-aware blood donor recruitment and management system for developing regions. Department of Computer Science and Engineering Bangladesh University of Engineering and Technology (BUET)
5. Priya, P., Saranya, V., Shabana, S., Subramani, K.: The optimization of blood donor information and management system by Technopedia. Int. J. Innov. Res. Sci. Eng. Technol. **3**, 1–390 (2014)
6. Catassi, C.A., Petersen, E.L.: The blood inventory control system-helping blood bank management through computerized inventory control. Transfusion **7**, 1–60 (2009)
7. Sharma, A., Gupta, P.C.: Predicting the number of blood donors through their age and blood group by using data mining tool. Int. J. Commun. Comput. Technol. **1**(6), 6 (2012)
8. Vijay kumar, N., Vahini, S.: Efficient tracking for women safety and security using IOT. Int. J. Adv. Res. Comput. Sci. (2017)

Vahini Siruvoru was born in warangal, Telangana, India, in 1984. She received the B.Tech degree in computer science and engineering from the Vidya Bharathi Institute of Technology, JNTUH, India, in 2006, and the M.Tech in software engineering from the Jyothismathi Institute of Technology and science, JNTUH, India, in 2011.

In 2006, he joined the Department of Computer Science Engineering, Chirstu Jyothi Institute of technology, Jangaon, Telangana, as an Asst. Prof. She has 10 years of experience as an Assistant Professor in the department of Computer Science and Engineering. Her current research interests include cloud computing, algorithms and Internet of Things.

Nampally Vijay Kumar was born in Narsampet, Telangana, India, in 1983. He received the B.Tech degree in computer science and engineering from the Jayamukhi Institute of Technology, JNTUH, India, in 2005, and the M.Tech in software engineering from the Vatsalya Institute of Technology and science, JNTUH, India, in 2010.

In 2006, he joined the Department of Computer Science Engineering, Chirstu Jyothi Institute of technology, Jangaon, Telangana, as an Asst. Prof. He has 11 years of experience as an Assistant Professor in the department of Computer Science and Engineering. His current research interests include automata and compilation, algorithms computation and Internet of Things. He is a Life Member of the Indian Society for Technical Education (ISTE).

Santosh Kumar Yellanki Banduri was born in Hyderabad, Telangana, india in 1985. He received the B.Tech Degree in IT (Information Technology) from VVIT, JNTUH INDIA in 2008.

From then worked for Metro polliten water works Hyderabad as a Network administrator for 1 year as on contract base. Later worked as a Tester in Aviorr medical IT till 2011. And also worked as Sr. tester and consultant in Asset telematics pvt. ltd. till 2014. Now started a start-up organization in IT industry till now. Also working for various projects as “Entrepreneur”.

Implementation of Open Shortest Path First Version 3 (OSPFv3) with Encryption and Authentication in IPv6 Network

Rahul Sharma and Nishi Yadav

Abstract With the huge use of smart devices, consumption of IPv4 is increased So, in that case, there is a demand of IPv6 to fulfill the demand. However, in modern days, all smart devices have options of this and due to use of IPV6, all the limitations of IPv4 are crossed. In this paper, we have considered the Open Shortest Path First version 3 (OSPFv3) and presented a topology which is implemented in IPv6 Network. The whole network topology is implemented in GNS3 and the results are obtained. In this network topology, we are finding the shortest path, encryption, and authentication between all nodes of network topology.

Keywords OSPFv3 · IPv6 · ESP · SHA1 · 3DES · GNS3

1 Introduction

For the communication of computers (smart devices) which is allocated in different geographical location we must have the medium, that medium is called Internet for making it useful there are various method available but all methods need one thing common that is Internet Protocol (IP) for identification of device over the Internet [1]. The router is one of the devices which is used for networking and all the routing protocols are configured on the router. There are various routing protocols in IPv6 network some of them are RIPng, EIGRP, and OSPFv3 which are used in IPv6 network. In this paper, we have considered the Open Shortest Path First version 3 (OSPFv3) for IPv6 network as a routing protocol. For authentication, we are using sha1 and for encryption ESP 3DES encryption algorithm.

R. Sharma · N. Yadav (✉)
School of Studies in Engineering and Technology, Guru Ghasidas Vishwavidalaya, Bilaspur, Chhattisgarh, India
e-mail: nishidv@gmail.com

R. Sharma
e-mail: 1408rahulsharma@gmail.com

S. Smys et al. (eds.), *International Conference on Computer Networks and Communication Technologies*, Lecture Notes on Data Engineering and Communications Technologies 15, https://doi.org/10.1007/978-981-10-8681-6_70

2 IPv6

Due to a huge number of Internet users, all the valid IPs of IPv4 are almost consumed or will be consumed. To avoid this type of problem, the upgrade version of IPv4 comes as IPv6 to provide valid IPs. The address size of IPv6 is 128 bits, which is providing almost unlimited internet addresses. It provides almost 50 octillion internet addresses per person that is alive in the world [2]. In IPv6 address, 128 bits or 16 bytes and their bits or bytes are separated by a colon and one example is given below Y:Y:Y:Y:Y:Y:Y:Y.

3 OSPFv3 in IPv6

Open Shortest Path First version 3 (OSPFv3) in IPv6 uses the Dijkstra's algorithm. There are following types of LSAs which in OSPFv3 and some of them are not in OSPF: Router LSA (Type 1): Contains a list of all links local to the router Network LSA (Type 2): Number of network on multi-access segment means it contains a list of all routers, attached to the Designated Router contains Interarea Pre x LSA (Type 3): Interarea Pre x LSA: a list of all destination networks within an area. Interarea-router LSAs for ASBRs (Type 4) Interarea-router LSAs for ASBRs contains a route to any ASBRs in the OSPF system contain routes to the destination Autonomous system external LSA (Type 5) this LSA mainly redistribute the route from the various autonomous system into OSPFv3 Link LSAs (Type 8) it provide the linklocal to all attached router. Address Area-Pre x LSAs (Type 9) provide a unique linkstate ID. [3] OSPFv3 defines these routers: Internal router (IR) a router whose interfaces are contained in a single area. Area Border Router (ABR) is a router that has interfaces in multiple areas. Autonomous System Boundary Router (ASBR) is a router between internal router and area border router. The addresses FF02::5 and FF02::6. are used for multicasting between the DR and BDR routers is now OSPFv3.

4 Implementation of OSPFv3 in IPv6

In this paper, we have proposed a network topology, which contains the three routers respectively R1, R2, and R3 and these are connected to each other by a cable. The sample design of the network topology is drawn in Fig. 1. In the network topology, there are three Cisco routers of series C7200. All the routers are connected to each other by Cross-over Ethernet cable. The router R1 has two Fast Ethernet port f0/0 with IP address 2001:db:acad:13::1/64 and f1/0 with IP address 2001:DB:ACAD:12::1/64 and one loopback 0 with IP address 2001::1:1/64 and OSPFv3 with area 0 and encryption with ESP 3des with the key 1313 for f0/0 and 1212 for f1/0. The router R2 has two Fast Ethernet port f0/0 with IP address 2001:db:acad:32::1/64 and f1/0 with IP address 2001:db:acad:12::2/64 and one loopback 0 with IP address 2001::2:1/64 and OSPFv3 with area 0 and encryption with ESP 3des with the key 3232 for f0/0

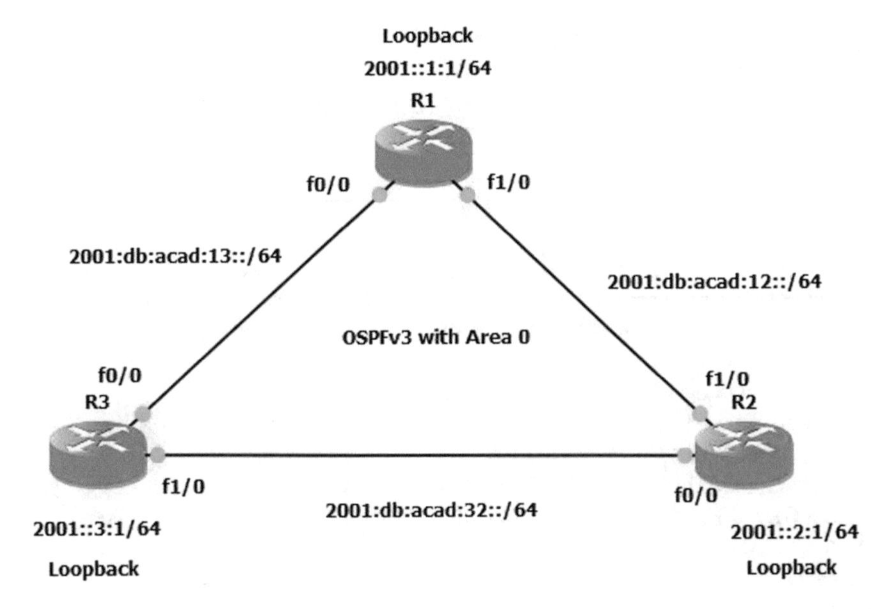

Fig. 1 Network topology

and 1212 for f1/0. The router R3 has two Fast Ethernet port f0/0 with IP address 2001:db:acad:13::2/64 and f1/0 with IP address 2001:db:acad:32::2/64 and one loopback 0 with IP address 2001::3:1/64 and OSPFv3 with area 0 and encryption with ESP 3des with the key 1313 for f0/0 and 3232 for f1/0.

4.1 Main Configuration Command

There are various commands which are given to the routers privilege mode R1, R2, and R3 and they are shown below:

[A] Router R1 Configuration
R1 sh run
Building configuration… Current configuration: 1982 bytes hostname R1
ipv6 unicast-routing ipv6 cef multilink bundle-name authenticated ip tcp synwait-time 5
interface Loopback0
ip address ipv6 address 2001::1:1/64 ipv6 ospf 10 area 0
interface FastEthernet0/0
ipv6 address 2001:DB8:ACAD:13::1/64
ipv6 ospf encryption ipsec spi 1313 esp 3des 123456789A123456789B123456789C 123456789DFFFFFFFF sha1 123456789A123456789B123456789C123456789D
ipv6 ospf 10 area 0
interface FastEthernet1/0

ipv6 address 2001:DB8:ACAD:12::1/64
ipv6 ospf encryption ipsec spi 1212 esp 3des 123456789A123456789B123456789C 123456789DFFFFFFFF sha1 123456789A123456789B123456789C123456789D
ipv6 ospf 10 area 0
ipv6 router ospf 10
router-id 1.1.1.1
R1
[B] Router R2 Configuration R2 sh run
Building configuration… Current configuration: 1982 bytes hostname R2
ipv6 unicast-routing ipv6 cef multilink bundle-name authenticated ip tcp synwait-time 5
interface Loopback0
ip address ipv6 address 2001::2:1/64 ipv6 ospf 10 area 0 interface FastEthernet0/0
ipv6 address 2001:DB8:ACAD:32::2/64
ipv6 ospf encryption ipsec spi 3232 esp 3des 123456789A123456789B123456789C 123456789DFFFFFFFF sha1 123456789A123456789B123456789C123456789D
ipv6 ospf 10 area 0
interface FastEthernet1/0
ipv6 address 2001:DB8:ACAD:12::2/64
ipv6 ospf encryption ipsec spi 1212 esp 3des 123456789A123456789B123456789C 123456789DFFFFFFFF sha1 123456789A123456789B123456789C123456789D
ipv6 ospf 10 area 0 ipv6
router ospf 10
router-id 2.2.2.2
end
R2
[C] Router R3 Configuration R3 sh run
Building configuration… Current configuration: 1982 bytes hostname R3
ipv6 unicast-routing ipv6 cef multilink bundle-name authenticated ip tcp synwait-time 5
interface Loopback0
ipv6 address 2001::3:1/64 ipv6 ospf 10 area 0 interface FastEthernet0/0
ipv6 address 2001:DB8:ACAD:13::2/64
ipv6 ospf encryption ipsec spi 1313 esp 3des 123456789A123456789B123456789C 123456789DFFFFFFFF sha1 123456789A123456789B123456789C123456789D
ipv6 ospf 10 area 0
interface FastEthernet1/0
ipv6 address 2001:DB8:ACAD:32::1/64
ipv6 ospf encryption ipsec spi 3232 esp 3des 123456789A123456789B123456789C 123456789DFFFFFFFF sha1 123456789A123456789B123456789C123456789 D
ipv6 ospf 10 area 0 ipv6
router ospf 10
router-id 3.3.3.3
end
R3

5 Experimental Results

The various results are given by the above network topology. These commands are applied in the privilege mode.

5.1 Routing Table

A. Command: R1 Show IPv6 Route: This command shows all the possible routes from R1 router to another router (Fig. 2).
B. Command: R2 Show IPv6 Route: This command shows all the possible routes from R2 router to another router (Fig. 3).
C. Command: R3 show IPv6 Route: This command shows all the possible routes from R3 router to another router (Fig. 4).

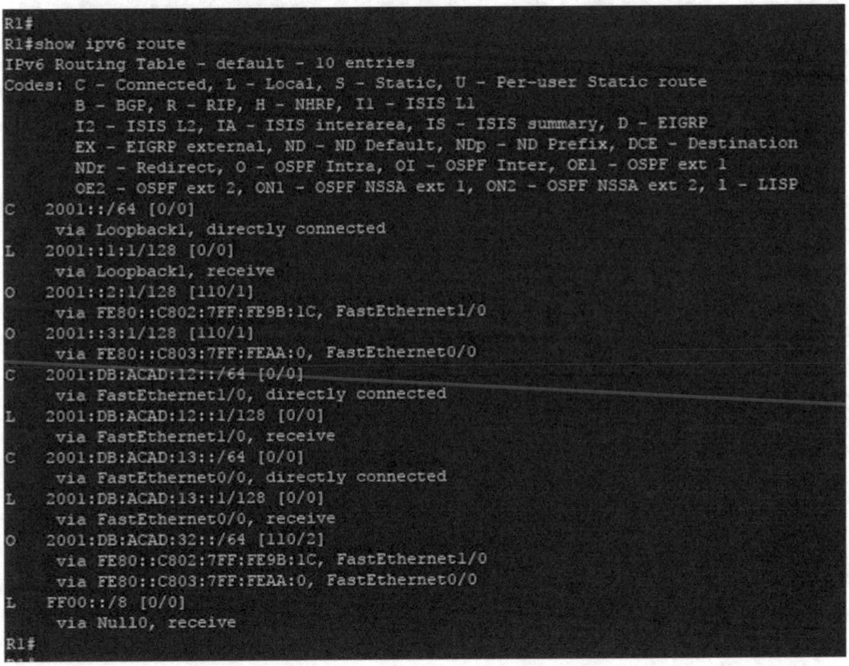

```
R1#
R1#show ipv6 route
IPv6 Routing Table - default - 10 entries
Codes: C - Connected, L - Local, S - Static, U - Per-user Static route
       B - BGP, R - RIP, H - NHRP, I1 - ISIS L1
       I2 - ISIS L2, IA - ISIS interarea, IS - ISIS summary, D - EIGRP
       EX - EIGRP external, ND - ND Default, NDp - ND Prefix, DCE - Destination
       NDr - Redirect, O - OSPF Intra, OI - OSPF Inter, OE1 - OSPF ext 1
       OE2 - OSPF ext 2, ON1 - OSPF NSSA ext 1, ON2 - OSPF NSSA ext 2, l - LISP
C   2001::/64 [0/0]
     via Loopback1, directly connected
L   2001::1:1/128 [0/0]
     via Loopback1, receive
O   2001::2:1/128 [110/1]
     via FE80::C802:7FF:FE9B:1C, FastEthernet1/0
O   2001::3:1/128 [110/1]
     via FE80::C803:7FF:FEAA:0, FastEthernet0/0
C   2001:DB:ACAD:12::/64 [0/0]
     via FastEthernet1/0, directly connected
L   2001:DB:ACAD:12::1/128 [0/0]
     via FastEthernet1/0, receive
C   2001:DB:ACAD:13::/64 [0/0]
     via FastEthernet0/0, directly connected
L   2001:DB:ACAD:13::1/128 [0/0]
     via FastEthernet0/0, receive
O   2001:DB:ACAD:32::/64 [110/2]
     via FE80::C802:7FF:FE9B:1C, FastEthernet1/0
     via FE80::C803:7FF:FEAA:0, FastEthernet0/0
L   FF00::/8 [0/0]
     via Null0, receive
R1#
```

Fig. 2 Result of show IP route for router R1

```
R2#
R2#show ipv6 route
IPv6 Routing Table - default - 10 entries
Codes: C - Connected, L - Local, S - Static, U - Per-user Static route
       B - BGP, R - RIP, H - NHRP, I1 - ISIS L1
       I2 - ISIS L2, IA - ISIS interarea, IS - ISIS summary, D - EIGRP
       EX - EIGRP external, ND - ND Default, NDp - ND Prefix, DCE - Destination
       NDr - Redirect, O - OSPF Intra, OI - OSPF Inter, OE1 - OSPF ext 1
       OE2 - OSPF ext 2, ON1 - OSPF NSSA ext 1, ON2 - OSPF NSSA ext 2, l - LISP
C   2001::/64 [0/0]
     via Loopback1, directly connected
O   2001::1:1/128 [110/1]
     via FE80::C801:7FF:FE8C:1C, FastEthernet1/0
L   2001::2:1/128 [0/0]
     via Loopback1, receive
O   2001::3:1/128 [110/1]
     via FE80::C803:7FF:FEAA:1C, FastEthernet0/0
C   2001:DB:ACAD:12::/64 [0/0]
     via FastEthernet1/0, directly connected
L   2001:DB:ACAD:12::2/128 [0/0]
     via FastEthernet1/0, receive
O   2001:DB:ACAD:13::/64 [110/2]
     via FE80::C803:7FF:FEAA:1C, FastEthernet0/0
     via FE80::C801:7FF:FE8C:1C, FastEthernet1/0
C   2001:DB:ACAD:32::/64 [0/0]
     via FastEthernet0/0, directly connected
L   2001:DB:ACAD:32::1/128 [0/0]
     via FastEthernet0/0, receive
L   FF00::/8 [0/0]
     via Null0, receive
R2#
```

Fig. 3 Result of show IP route for router R2

5.2 Show Crypto IPsec Policy

In this section, policy name, policy refcount, Inbound ESP SPI, Outbound ESP SPI, Inbound ESP Authentication Key, Outbound ESP Authentication Key, Inbound ESP Cipher Key, Outbound ESP Cipher Key, and transform set are shown by the applying the command Show crypto IPsec policy on router privilege mode.

A. Command: R1 how crypto IPsec policy: section policy name, policy refcount, Inbound ESP SPI, Outbound ESP SPI, Inbound ESP Authentication Key, Outbound ESP Authentication Key, Inbound ESP Cipher Key, Outbound ESP Cipher Key, and transform set on router R1 are shown (Fig. 5).
B. Command: R2 how crypto IPsec policy: section policy name, policy refcount, Inbound ESP SPI, Outbound ESP SPI, Inbound ESP Authentication Key, Outbound ESP Authentication Key, Inbound ESP Cipher Key, Outbound ESP Cipher Key, and transform set on router R2 are shown (Fig. 6).
C. Command: R3 how crypto IPsec policy: section policy name, policy refcount, Inbound ESP SPI, Outbound ESP SPI, Inbound ESP Auth Key, Outbound ESP Auth Key, Inbound ESP Cipher Key, Outbound ESP Cipher Key, and transform set on router R3 are shown (Fig. 7).

```
R3#sh ipv6 route
IPv6 Routing Table - default - 10 entries
Codes: C - Connected, L - Local, S - Static, U - Per-user Static route
       B - BGP, R - RIP, H - NHRP, I1 - ISIS L1
       I2 - ISIS L2, IA - ISIS interarea, IS - ISIS summary, D - EIGRP
       EX - EIGRP external, ND - ND Default, NDp - ND Prefix, DCE - Destination
       NDr - Redirect, O - OSPF Intra, OI - OSPF Inter, OE1 - OSPF ext 1
       OE2 - OSPF ext 2, ON1 - OSPF NSSA ext 1, ON2 - OSPF NSSA ext 2, l - LISP
C   2001::/64 [0/0]
     via Loopback1, directly connected
O   2001::1:1/128 [110/1]
     via FE80::C801:7FF:FE8C:0, FastEthernet0/0
O   2001::2:1/128 [110/1]
     via FE80::C802:7FF:FE9B:0, FastEthernet1/0
L   2001::3:1/128 [0/0]
     via Loopback1, receive
O   2001:DB:ACAD:12::/64 [110/2]
     via FE80::C802:7FF:FE9B:0, FastEthernet1/0
     via FE80::C801:7FF:FE8C:0, FastEthernet0/0
C   2001:DB:ACAD:13::/64 [0/0]
     via FastEthernet0/0, directly connected
L   2001:DB:ACAD:13::2/128 [0/0]
     via FastEthernet0/0, receive
C   2001:DB:ACAD:32::/64 [0/0]
     via FastEthernet1/0, directly connected
L   2001:DB:ACAD:32::2/128 [0/0]
     via FastEthernet1/0, receive
L   FF00::/8 [0/0]
     via Null0, receive
R3#
```

Fig. 4 Result of show IP route for router R3

```
R1#show crypto ipsec policy
Crypto IPsec client security policy data

Policy name:      OSPFv3-1212
Policy refcount:  1
Inbound  ESP SPI:         1212 (0x4BC)
Outbound ESP SPI:         1212 (0x4BC)
Inbound  ESP Auth Key:    123456789A123456789B123456789C123456789D
Outbound ESP Auth Key:    123456789A123456789B123456789C123456789D
Inbound  ESP Cipher Key: 123456789A123456789B123456789C123456789DFFFFFFFF
Outbound ESP Cipher Key: 123456789A123456789B123456789C123456789DFFFFFFFF
Transform set:    esp-3des esp-sha-hmac

Crypto IPsec client security policy data

Policy name:      OSPFv3-1313
Policy refcount:  1
Inbound  ESP SPI:         1313 (0x521)
Outbound ESP SPI:         1313 (0x521)
Inbound  ESP Auth Key:    123456789A123456789B123456789C123456789D
Outbound ESP Auth Key:    123456789A123456789B123456789C123456789D
Inbound  ESP Cipher Key: 123456789A123456789B123456789C123456789DFFFFFFFF
Outbound ESP Cipher Key: 123456789A123456789B123456789C123456789DFFFFFFFF
Transform set:    esp-3des esp-sha-hmac
```

Fig. 5 Result of Show crypto IPsec Policy for Router R1

```
R2#sh crypto ipsec policy
Crypto IPsec client security policy data

Policy name:      OSPFv3-3232
Policy refcount:  1
Inbound  ESP SPI:         3232 (0xCA0)
Outbound ESP SPI:         3232 (0xCA0)
Inbound  ESP Auth Key:    123456789A123456789B123456789C123456789D
Outbound ESP Auth Key:    123456789A123456789B123456789C123456789D
Inbound  ESP Cipher Key: 123456789A123456789B123456789C123456789DFFFFFFFF
Outbound ESP Cipher Key: 123456789A123456789B123456789C123456789DFFFFFFFF
Transform set:    esp-3des esp-sha-hmac

Crypto IPsec client security policy data

Policy name:      OSPFv3-1212
Policy refcount:  1
Inbound  ESP SPI:         1212 (0x4BC)
Outbound ESP SPI:         1212 (0x4BC)
Inbound  ESP Auth Key:    123456789A123456789B123456789C123456789D
Outbound ESP Auth Key:    123456789A123456789B123456789C123456789D
Inbound  ESP Cipher Key: 123456789A123456789B123456789C123456789DFFFFFFFF
Outbound ESP Cipher Key: 123456789A123456789B123456789C123456789DFFFFFFFF
Transform set:    esp-3des esp-sha-hmac
```

Fig. 6 Result of show crypto IPsec policy for router R2

```
R3#sh crypto ipsec policy
Crypto IPsec client security policy data

Policy name:      OSPFv3-3232
Policy refcount:  1
Inbound  ESP SPI:         3232 (0xCA0)
Outbound ESP SPI:         3232 (0xCA0)
Inbound  ESP Auth Key:    123456789A123456789B123456789C123456789D
Outbound ESP Auth Key:    123456789A123456789B123456789C123456789D
Inbound  ESP Cipher Key: 123456789A123456789B123456789C123456789DFFFFFFFF
Outbound ESP Cipher Key: 123456789A123456789B123456789C123456789DFFFFFFFF
Transform set:    esp-3des esp-sha-hmac

Crypto IPsec client security policy data

Policy name:      OSPFv3-1313
Policy refcount:  1
Inbound  ESP SPI:         1313 (0x521)
Outbound ESP SPI:         1313 (0x521)
Inbound  ESP Auth Key:    123456789A123456789B123456789C123456789D
Outbound ESP Auth Key:    123456789A123456789B123456789C123456789D
Inbound  ESP Cipher Key: 123456789A123456789B123456789C123456789DFFFFFFFF
Outbound ESP Cipher Key: 123456789A123456789B123456789C123456789DFFFFFFFF
Transform set:    esp-3des esp-sha-hmac
```

Fig. 7 Result of show crypto IPsec policy for router R3

```
R1#
R1#ping 2001::3:1
Type escape sequence to abort.
Sending 5, 100-byte ICMP Echos to 2001::3:1, timeout is 2 seconds:
!!!!!
Success rate is 100 percent (5/5), round-trip min/avg/max = 20/28/52 ms
R1#
R1#
```

Fig. 8 Result of ping for router R1

```
R1#traceroute 2001::3:1
Type escape sequence to abort.
Tracing the route to 2001::3:1

  1 2001:DB:ACAD:13::2 20 msec 12 msec 24 msec
R1#
```

Fig. 9 Result of traceroute for Router

5.3 *Ping and Traceroute*

In this section, Ping and Traceroute commands are applied on the privilege mode of the router. Ping command is for checking the end-to-end connectivity and Traceroute command to show the route which is followed by ICMP packets.

A. Command: R1Ping 2001::3:1: On the router R1, ping command is applied to check connectivity from loopback interface 2001::3:1 of router R3 (Fig. 8).
B. Command: R1traceroute 2001::3:1: On the router R1, Traceroute command is applied to know the path followed by the ICMP packet from router R1 to loopback interface 2001::3:1 of router R3 (Fig. 9).

6 Conclusion

Routing is the basically process in which packets go from one to another network. Generally, routing has activities such as find best and backup path and transmit the packet over on that routes. Various routing protocols have various technique to decide these activities. OSPFv3 is the best routing protocols because it has the fast convergence than EIGRP. OSPFv3 has less administrative value than EIGRP which is a very good thing. OSPFv3 has classless but EIGRP not. In authentication of OSPFv3 we are using sha1 which is better than MD5 and for encryption, we are using ESP 3DES encryption algorithm which is better than all encryption algorithm. OSPFv3 protocol has better routing capabilities than other routing protocols hence OSPFV3 in IPv6 network will be mainly used in the future.

References

1. Goyal, Vikas, Arora, Geeta: Implementation of enhanced interior gateway routing protocol (EIGRP) in IPv6 network. Int. Res. J. Adv. Eng. Sci. **2**(1), 90–95 (2017)
2. http://en.wikipedia.org/wiki/IPv6
3. https://www.cisco.com/c/en/us/td/docs/iosxml/ios/iprouteospf/configuration/15-sy/iro-15-sy-book/ip6-route-ospfv3.html
4. Arafat, M.Y., Sobhan, M.A., et al.: Study on migration from IPv4 to IPv6 of a large-scale network. Mod. Appl. Sci. **8**(3), 67 (2014)
5. Hinds, A., Atojoko, A., et al.: Evaluation of OSPF and EIGRP routing protocols for ipv6. Int. J. Future Comput. Commun. **2**(4), 287 (2013)
6. Coltun, R., Ferguson, D., Moy, J., Lindem, A.: RFC 5340—OSPF for IPv6, IETF, July 2008
7. Gupta, M., Melam, N.: RFC 4552—Authentication/Confidentiality for OSPFv3, IETF, June 2006

Rahul Sharma working in School of Studies in Engineering and Technology, Guru Ghasidas Vishwavidalaya Bilaspur, (C.G), India. His research area includes authentication and routing protocol.

Nishi Yadav working in School of Studies in Engineering and Technology, Guru Ghasidas Vishwavidalaya Bilaspur, (C.G), India. Her research area includes authentication and routing protocol.

Re-LEACH: An Energy-Efficient Secure Routing Protocol for Wireless Sensor Networks

Sonali Pandey and Rakesh Kumar

Abstract Wireless sensor network most knowing topic in the current time. The popularity of wireless sensor network growing very fast because it is low cost devices as well as low energy consumption with higher utilization. Routing protocol is an important part of sensor network. Routing protocols play an important role in the wireless sensor network. LEACH protocol is cluster-based routing protocol. LEACH protocol work on two steps, the first step is setup phase and the second step is a steady phase. In the setup phase, we select the one cluster head in the group of nodes after that come to the steady phase here we aggregate the data or compress the data then send the data to sink node or gateway. In this paper, we propose a new protocol Reappointment LEACH (Re-LEACH) which is more energy efficient than other protocols as well as lighter overhead. Re-LEACH protocol increases the life period of sensor network with low energy consumption.

Keywords LEACH · Energy efficient · Wireless sensor network
Routing protocol · Re-LEACH

1 Introduction

The concept of wireless sensor network [1, 2] that is sensor node operated with sensitive data or information. Wireless sensor network is more complicated than wired network because wireless networks have limited and open resources. In other words, we can say that sensor network is distributed autonomous network where sensor nodes control the physical and environmental situation that is temperature, vibration, sound, pressure etc. Wireless sensor network deployed the thousand or more number of nodes in the unattended environment. Wireless sensor network

S. Pandey (✉) · R. Kumar
Computer Science and Engineering, MMMUT, Gorakhpur, India
e-mail: sonali79109@gmail.com

R. Kumar
e-mail: rkcs@mmmut.com

S. Smys et al. (eds.), *International Conference on Computer Networks and Communication Technologies*, Lecture Notes on Data Engineering and Communications Technologies 15, https://doi.org/10.1007/978-981-10-8681-6_71

incorporated a huge number of sensor nodes. Every sensor nodes have four basic components such as sensing unit which are used for collecting the information from the environment, processing unit is microcontroller with memory it is used for control the sensor node, transmission unit is used for transmission the data one node to other node, and power supply unit is used for supply the power of each node. Wireless sensor network connected to thousands of nodes in wireless medium. Each sensor node gathers the information from environment after that collected information or data send to base station then with the help of internet user use the data through wired or wireless medium.

Routing protocol [3] also describes the category of wireless sensor network. Routing protocol is basically divided into three parts on network structure basis. First is data-centric protocol here each sensor node directly interacts with the base station. In this type of routing protocol, every node send their data directly the sink node not required any type of cluster head node. DD and SPIN, etc., are examples of flat based routing protocol. Second is cluster-based routing protocol also known as hierarchical routing protocol. Hierarchical routing protocol is the most efficient communication protocol. Cluster-based protocol is more energy efficient type of protocol. In the hierarchical routing protocol make the some group of node, one node is decided as cluster head. This cluster head transmits the information of all nodes after that cluster head transmits the data or information to the base station. Some examples of cluster-based routing protocol are LEACH, APTEEN, TEEN, etc. The third is network structure based routing protocol is location-based routing protocol. Location-based protocol is addressed by its location, here every sensor node needs some location information in wireless sensor network. According to these information, data is transmitted to desire node rather than overall network. This information gets through Global Positioning System (GPS) signal, then this node sends data to sink node. Some example of location-based routing protocol such as GAF and GEAR, etc. Remaining paper is arranged as follows. Section 2 describes the LEACH protocol, Sect. 3 describes the related work, Sect. 4 describes the proposed work, Sect. 5 describes the simulation point, and Sect. 6 describes the conclusion and future scope [4].

2 LEACH Protocol and Its Description

2.1 Protocol Introduction

Low Energy Adaptive Clustering Hierarchy (LEACH) [5, 6] protocol represents the cluster based routing protocol. LEACH protocol is the oldest protocol of hierarchical routing protocol. This protocol is self-organized as well as adaptive type of protocol. It also decreases the unnecessary and unwanted energy cost of overall network. This protocol works on round process. In every round, one node is select as cluster head. Round process scheduled with the help of TDMA (Time Division Multiple Access). Each round is divided into two parts: first is setup phase and second is steady phase.

In other words, LEACH protocol depends on two steps setup phase along with steady phase. Setup phase is important part of LEACH protocol. In setup, phase make some clusters or group of nodes. Each group elects one node as cluster head with in cluster. Each node in the cluster makes the cluster head based on TDMA. With the help of TDMA scheduling, every sensor node makes the cluster head in the group of nodes. Steady phase takes long time duration than setup phase.

2.2 *Phases of LEACH Protocol*

- Two basic part of LEACH protocol

Setup phase: This phase [7] is starting and important part of LEACH protocol. Here, each node of sensor network generates the random number. This random number is lower than the threshold value then that node is decided as cluster head.

$$T(x) = \begin{cases} \frac{p}{1-p\left(r \bmod \frac{1}{p}\right)}, & x \in G \\ 0, & \text{otherwise} \end{cases} \tag{1}$$

where

P is desired percentage value for cluster head,
R represents the current round and
G represents the group of node.

Steady phase: This phase [7] takes a long time as compared to setup phase. This phase depends on TDMA for scheduling process. According to TDMA, scheduling allow the time slots of each node. Here, cluster head aggregates and compresses the data after that send to the sink node.

- Flowchart of LEACH protocol

Here how cluster head [8, 9] is selected in group of node, when cluster head announce in cluster is explained, then TDMA scheduling process is applied on the node in network. Flowchart describes both phases with easy graphical representation in very clear view. Figure 1 describes cluster head in group of node.

3 Related Work

Zang et al. [10] explain security, performance as well as lighter overhead in LEACH protocol. Here, author describes the symmetric key and random pairwise key for increase in the security in cluster-based routing protocol in wireless sensor network. But this method does not resolve the problem of energy consumption and network

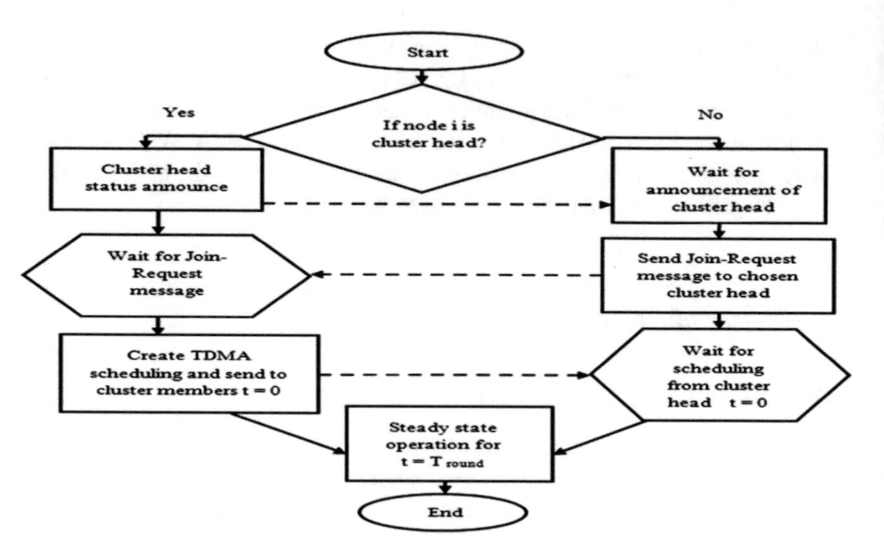

Fig. 1 Flow chart of LEACH protocol

lifetime. Here the proposed method increases the energy consumption and reduced the network lifetime. Kumar et al. [11] presented the taxonomy of routing protocol. Here, author describes the LEACH protocol and timeline as well as variants of LEACH protocol. But energy performance is not so much good. Gananambigai et al. [12] discuss many routing protocols related to LEACH protocol. Here, many drawback and issues of LEACH variants protocol are described. Author describes feature as well as the performance challenges of all related protocol of LEACH. So there is a need to improve the performance of LEACH variants protocol. Jan et al. [13] represent the survey-based hierarchical protocol. These protocols describe how the node is arranged in the cluster. Here, author explains the data transmission node to base station. But there are some disadvantages with these protocol. Manimala et al. [14] presented survey of LEACH with different cluster-based routing protocols. Here, author describe disadvantages and advantages of each protocol. So some improvement is required for each protocol in wireless sensor network. Braman et al. [15] presented the brief description of routing protocol in wireless sensor network. Here, author describes all design issues and challenges of routing protocol. According to author, LEACH protocol is more efficient protocol in routing protocol. But more enhancement of LEACH protocol is required. Usha et al. [16] describe the LEACH protocol with its descendant. Here, author explains mobility, hop count or reliability, etc. But the problem of energy consumption is not resolved.

4 Propose Protocol

Routing protocol concept is not only the cluster head selection process but also the involvement of transmission process. Many LEACH variants protocol are presented, all protocols describe the energy consumption and network lifetime. Some common problem are created for every protocols they are energy consumption and network lifetime. We propose the new protocol, which is reducing the energy consumption and increase the network lifetime. Here, a new protocol is proposed, which is known as Re-LEACH (Reappointment LEACH) protocol. This protocol is similar to LEACH protocol. So it accepts all feature of LEACH protocol. First, build clusters in available nodes in the network and selected the cluster head in network. Second data transmission phase, after transmission applies the reappointment algorithm [17] in next round on the cluster head, so selected cluster head will take for many rounds. The improved version known as reappointment algorithm. In this way, the energy consumption of proposed protocol is reduced. According to LEACH protocol two, steps are involved. But when we talk about Re-LEACH protocol the addition the some more steps and present a new protocol. Some steps are given below:

- Cluster setup phase,
- Reappointment phase,
- Data transmission phase.

Re-LEACH introduces a new concept that is reappointment, basically reappointment means one point or node assign for several time. This is very helpful for energy consumption and network lifetime.

4.1 Phases of Re-LEACH Protocol

Cluster setup phase: This phase is very popular in LEACH protocol. In this phase, cluster are formed after that cluster forming decides for the cluster head selection process. Cluster head selection is based on the threshold value. If the value of sensor node is less than the calculated threshold value then that node becomes a cluster head. Base station assigns a different sequence number of each cluster, with the help of sequence number cluster head assign TDMA scheduling for its own node. According to TDMA scheduling node transmit, the messages to cluster head in some gaps.

Reappointment phase: Reappointment concept basically works on the cluster head node in the cluster. According to this concept, if one node becomes a cluster head then it continues to become cluster head next several round. Here, cluster head does not hand over to other node next several times. In this way, we reduce the energy consumption. Here, we calculate the new threshold $T(x)'$ with the help of reappointment algorithm. If the value of the node is less than the calculated threshold then node becomes a cluster head.

$$T(x)' = \begin{cases} \frac{k}{\left(\left[N-K\left(\left(\text{floor}\left(\frac{r1}{T}\right) \bmod \frac{N}{K}\right)\right)\right]\right)} \\ 0 \end{cases} \quad (2)$$

where

N	denotes overall no of node in the network,
K	denotes the no of cluster head,
T	denotes the times for reappointment,
And $r1$	is considered as zero in the reappointment of every new round.

The reappointment concept applies until the end of T round (T_{round}). Cluster head reappointment is stopped at this point. In other word, we can say that reappointment process is stopped when cluster head totally loses their energy during the period of reappointment.

Transmission phase: Transmission phase is normal phase of any clusters. In this phase, cluster head sends the data or information from node to gateway or sink node. In this phase, data to be compressed and aggregated after that transmit to the base station or gateway.

4.2 Pseudo Code of Re- LEACH Protocol

Step 1: First, make the cluster in given node using many clustering techniques such as k-means clustering techniques etc. Otherwise, make cluster based on energy potential field.

Step 2: More than one cluster are formed than base station assign different sequence number of each cluster.

Step 3: Every cluster selects the cluster head based on phases of LEACH protocol.

Step 4: Every cluster calculates the threshold value $T(x)$

$$T(x) = \begin{cases} \frac{p}{1-p\left(r \bmod \frac{1}{p}\right)}, & x \in G \\ 0, & \text{otherwise} \end{cases} \quad (1)$$

Step 5: Each node of cluster calculates the random number $t(x)$, random number in between 0 and 1.

Step 6: If $T(x)>t(x)$, then node becomes cluster head, otherwise node is not cluster head.

Step 7: If node (i) becomes a cluster head then it generates TDMA gap for each node. So interference problem does not occur from one node to other node.

Step 8: With the help of new proposed protocol, reappointment algorithm, we calculate the new threshold value

$$T(x)' = \begin{cases} \frac{k}{\left(\left[N-K\left(\left(\text{floor}\left(\frac{r1}{T}\right) \bmod \frac{N}{K}\right)\right)\right]\right)} \\ 0 \end{cases} \quad (2)$$

Step 9: New calculated $T(x)'$ threshold value compared with cluster head number $t(x)$. If $t(x)$ is less than $T(x)'$ then it makes cluster head for several round or time.

Step 10: Cluster head collected the data or information from node in the cluster.

Step 11: This process continues until the cluster head node is totally exhausted or dead.

Step 12: Cluster head collects the information or data, after that transmission phase. In this phase, compress the data or aggregate the information.

Step 13: Performing above all steps at last transmit, the information or data to sink node or gateway.

Step 14: The process continues until the *step 11*.

Step 15: The process ends when one cluster head is totally dead.

4.3 Flowchart of the Proposed Protocol

See Fig. 2.

5 Simulation Scenario and Result

In this paper, we proposed the Re-LEACH protocol in wireless sensor network. Some assumption required for simulation of this protocol.

- Starting energy of nodes in the network is same.
- Base station is placed in center point of the network area.
- All nodes as well as clusters are static in nature.
- Cluster node directly transmits the data their cluster head node.

Many more new protocols are presented for increasing the energy efficiency. But some drawback also creates with these new protocols.

With the help of simulation of Re-LEACH protocol. We express the network lifetime and dead nodes in the network with single node and multi node routing protocol. This simulation is performed on the MATLAB.

Figure 3 shows that where the proposed protocol is in stable position, which protocol is more reliable. Figure 4 depicts that how much data packet is transferred from node to base station. Here, the proposed protocol transfers the maximum number data node to cluster head then base station (Table 1).

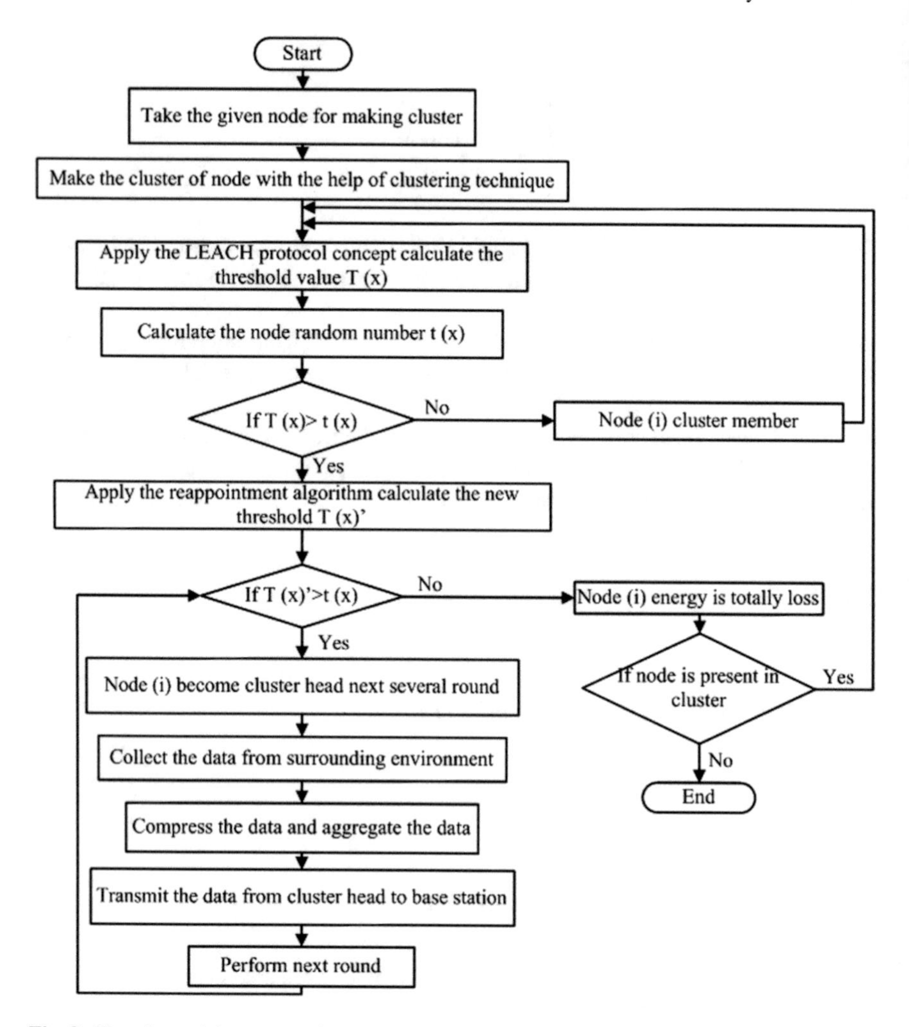

Fig. 2 Flowchart of the proposed protocol

6 Conclusion and Future Work

This paper basically focuses on the energy consumption of the node in network. Here, the LEACH protocol and its steps and setup phase and steady phase are explained. With the help of LEACH protocol, we present a new protocol that is known as Re-LEACH protocol. In this protocol, one cluster head selected for several rounds, this concept is known as reappointment LEACH protocol. The Reappointment process is continuously perform until node is totally exhausted or full loss of their energy. In next round, other node is select as cluster head. In this paper when we apply the

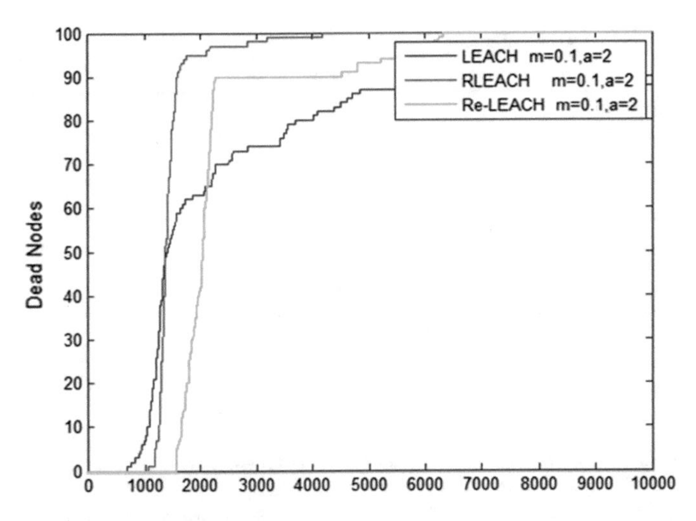

Fig. 3 Dead node in the network

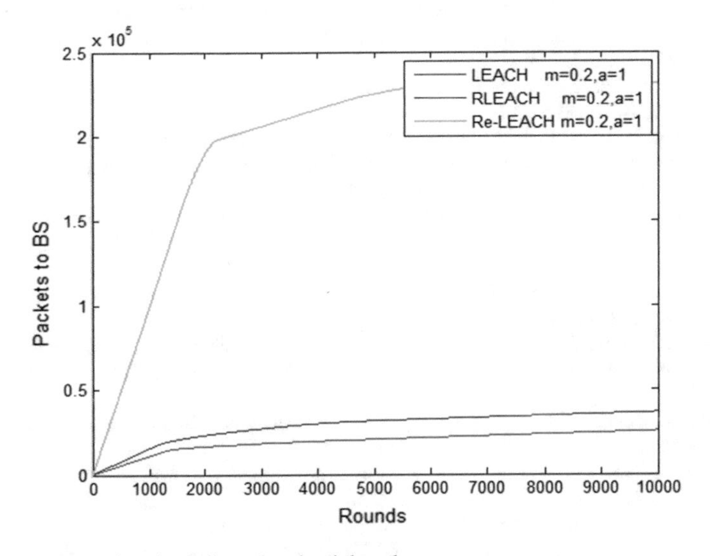

Fig. 4 Percentage of packet delivered to the sink node

reappointment concept at the node of the network then increase the network lifetime, less energy consumption as well as less overhead.

In future, we perform many new improvements in LEACH protocol on the way of energy consumption and network lifetime. In further, we will increase the stability factor of LEACH protocol with the help of new improvement.

Table 1 Values of network parameter

Parameter	Value
Number of nodes (n)	100
Number of round ($r_{\max}$)	10,000
Energy of free space (Efs)	1×10^{-11}
Energy of multipath (Ems)	1.3×10^{-15}
Probability of cluster node selection (p)	0.1
Field dimension ($xm \times ym$)	100×100
Transition power energy (ERX)	5×10^{-8}
Data aggregation (EDA)	5×10^{-9}

References

1. Walters, J.P., Liang, Z., Shi, W., Chaudhary, V.: Wireless sensor network security: a survey: security in distributed, grid, and pervasive computing, pp. 1–367 (2006)
2. Rajagopalan, K.S.S.D.S.P.: Security analysis with respect to wireless sensor network–review. Int. J. Eng. Comput. Sci. **6**, 21070–21080 (2017)
3. Anjali, S., Sharma, M.: Wireless sensor networks: routing protocols and security issues. In: Fifth International Conference on Computing, Communications and Networking Technologies (ICCCNT), pp. 1–5. Hefei (2014)
4. Tandel, R.I.: Leach protocol in wireless sensor network: a survey. Int. J. Comput. Sci. Inf. Technol. **7**, 1894–1896 (2016)
5. Deosarkar, B.P., Yadav, N.S., Yadav, R.P.: Clusterhead selection in clustering algorithms for wireless sensor networks: a survey. In: International Conference on Computing, Communication and Networking, pp. 1–8. St. Thomas (2008)
6. Palan, N.G., Barbadekar, B. V., Patil, S.: Low energy adaptive clustering hierarchy (LEACH) protocol: a retrospective analysis. In: International Conference on Inventive Systems and Control (ICISC), pp. 1–12. Coimbatore (2017)
7. Chauhan, M.S., Yadav, N.: LEACH-I algorithm for WSN. Int. J. Innov. Res. Comput. Commun. Eng. **4**, 3459–3465 (2016)
8. Hou, G., Tang, K.W., Noel, E.: Implementation and analysis of the LEACH protocol on the TinyOS platform. In: International Conference on ICT Convergence (ICTC), pp. 918–923. Jeju (2013)
9. Yadav, L., Sunitha, C.: Low energy adaptive clustering hierarchy in wireless sensor network (LEACH). Int. J. Comput. Sci. Inf. Technol. **5**, 4661–4664 (2014)
10. Zhang, K., Wang, C., Wang, C.: A secure routing protocol for cluster-based wireless sensor networks using group key management. In: 4th International Conference on Wireless Communications, Networking and Mobile Computing, pp. 1–5. Dalian (2008)
11. Kumar, V., Jain, S., Tiwari, S.: Energy efficient clustering algorithms in wireless sensor networks: a survey. Int. J. Comput. Sci. Issues (IJCSI) **8**(5), 259–268 (2011)
12. Gnanambigai, J., Rengarajan, N., Anbukkarasi, K.: Leach and its descendant protocols: a survey. Int. J. Commun. Comput. Technol. (IJCCT) **01**(2), 15–21 (2012)
13. Jan, M.A., Khan, M.: A survey of cluster-based hierarchical routing protocols. IRACST—Int. J. Comput. Netw. Wirel. Commun. (IJCNWC) **3**(2), 138–143 (2013)
14. Manimala, P., Selvi, R.S.: A survey on leach-energy based routing protocol. Int. J. Emerg. Technol. Adv. Eng. (IJETAE) **3**(12), 657–660 (2013)
15. Braman, A., Umapathi, G.R.: A comparative study on advances in LEACH routing protocol for wireless sensor networks: a survey. Int. J. Adv. Res. Comput. Commun. Eng. **3**(2), 5683–5690 (2014)
16. Usha, M., Sankarram, N.: A survey on energy efficient hierarchical (leach) clustering algorithms in wireless sensor network. Int. J. Innov. Res. Comput. Commun. Eng. (IJIRCCE) **2**(1), 601–609 (2014)

17. Li, Y.Z., Zhang, A.L., Liang, Y. Z.: Improvement of leach protocol for wireless sensor networks. In: Third International Conference on Instrumentation, Measurement, Computer, Communication and Control, pp. 322–326. Shenyang (2013)

Sonali Pandey working in Computer Science and Engineering, MMMUT, Gorakhpur. Her research area includes network security and wireless sensor networks.

Rakesh Kumar working in Computer Science and Engineering, MMMUT, Gorakhpur. His research area includes network security and wireless sensor networks.

Psi Slotted Fractal Antenna for LTE Utilities

Kirandeep Kaur, Chahat Jain and Narwant Singh Grewal

Abstract PSI (Ψ) shaped slotted fractal antenna (PSSFA) that operates in multiband behavior was designed for Long-Term Evolution (LTE) applications. The shape which is again repeated on base of PSI geometry is known as Modified PSI shaped slotted fractal antenna (MPSSFA). The simulated results were obtained by the use of Zeland IE3D v.14.1. The simulated values of reflection coefficient for PSSFA obtained are −23.68, −15.94, and −14.04 dB for the resonant frequencies 2.33, 4.60, and 5.08 GHz. As the numbers of repetitions were increased, the resonant frequencies were shifted in the left direction from the base geometry. The simulated values of resonant frequencies for MPSSFA are 2.32, 4.52, and 5.12 GHz at reflection coefficient −24.10, −10.18, and −24.28 dB. The dimensional analysis was also performed for PSSFA. The length of PSSFA was varied in comparison to the previous cases taken and it was exhibited a left shift in the resonant frequency.

Keywords PSI slot · Fractal · LTE · FR4 · Multiband

1 Introduction

Antenna is the device used to transmit RF signals. Small antennas are used in mobile devices because of low space. Nowadays, single antenna is needed which has properties of low-profile/cost, broadband, and several band operations. Word fractal can be defined as the insertion of same shape into base geometry but with reduced dimension [1]. The iteration of fractal geometry in the second stage was concluded by iteration factor. Fractal can be used in two different methods resulting in strengthening

K. Kaur (✉) · C. Jain · N. S. Grewal
Guru Nanak Dev Engineering College, Ludhiana, Punjab, India
e-mail: kirangill2194@gmail.com

C. Jain
e-mail: chahatjain26@gmail.com

N. S. Grewal
e-mail: narwant@gndec.ac.in

S. Smys et al. (eds.), *International Conference on Computer Networks and Communication Technologies*, Lecture Notes on Data Engineering and Communications Technologies 15, https://doi.org/10.1007/978-981-10-8681-6_72

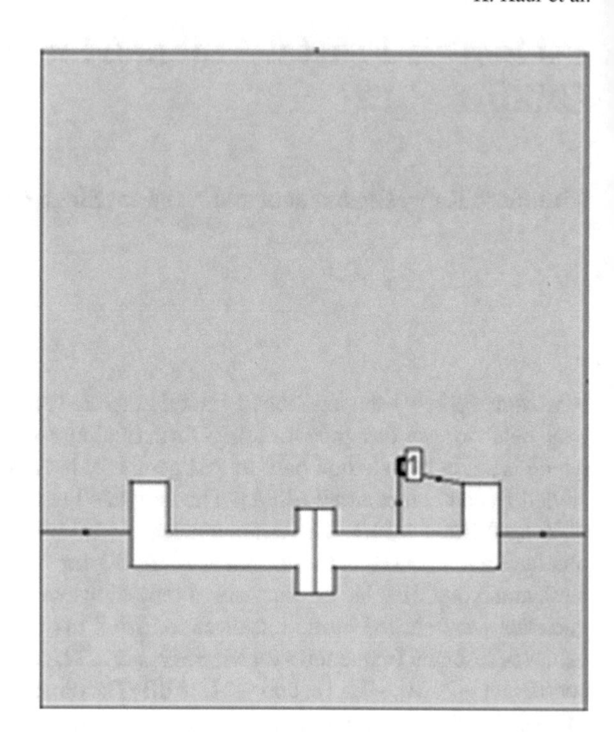

Fig. 1 PSSFA geometry

antenna design [2]. Using the first method to design miniaturized antenna element while the other is to use self-similarity in geometry [3–5]. Basic shapes of fractal antennas are Koch Curve/Snowflake, Hilbert Curve, Sierpinski Gasket/Carpet, Minkowski Island, etc. [6]. Fractal antenna has advantages of higher resonant frequency and lower reflection coefficient. Long-term evolution (LTE) technology finds its use in commercial, medical, and industrial fields [7]. To make an antenna work at LTE and WLAN standards, a single antenna which can operate at higher frequency band is needed [8]. PSI (Ψ) slotted fractal antenna operates at *S*, *L*, and *C* bands [9]. Microstrip Ψ-shape antenna with use of probe feed method provides better percentage of bandwidth [10]. Ψ-shaped antenna can be used for achieving wide bandwidth [11, 12].

Table 1 Important parameters PSSFA design

S. No.	Name of parameter	Value
1	Dielectric constant, ε_r	4.4
2	Substrate thickness, h	1.58 mm
3	tan δ	0.02
4	Patch length	29.83 mm
5	Patch width	38.51 mm

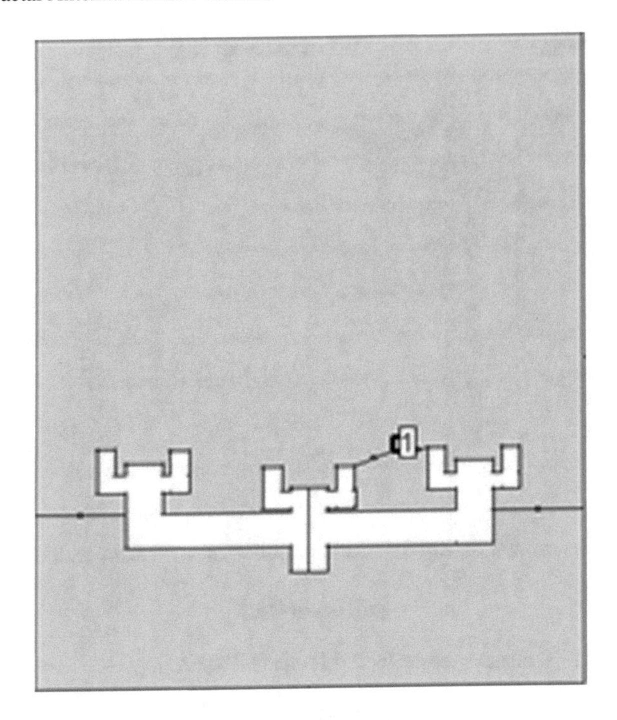

Fig. 2 MPSSFA geometry

Section 1 provides a brief overview of fractal antenna properties and LTE. Section 2 gives the description of antenna design. A discussion about simulated results is presented in Sect. 3 followed by the conclusion of the paper in Sect. 4.

2 Antenna Design

In present work, the geometries of PSSFA and MPSSFA were discussed. Modified MPSSFA was designed by iterating the previous PSSFA by factor 2. Thus, finally, an MPSSFA has been proposed. The geometry of PSSFA and MPSSFA was developed and simulations were performed in IE3D software. The vital parameters which are used for designing are given in Table 1.

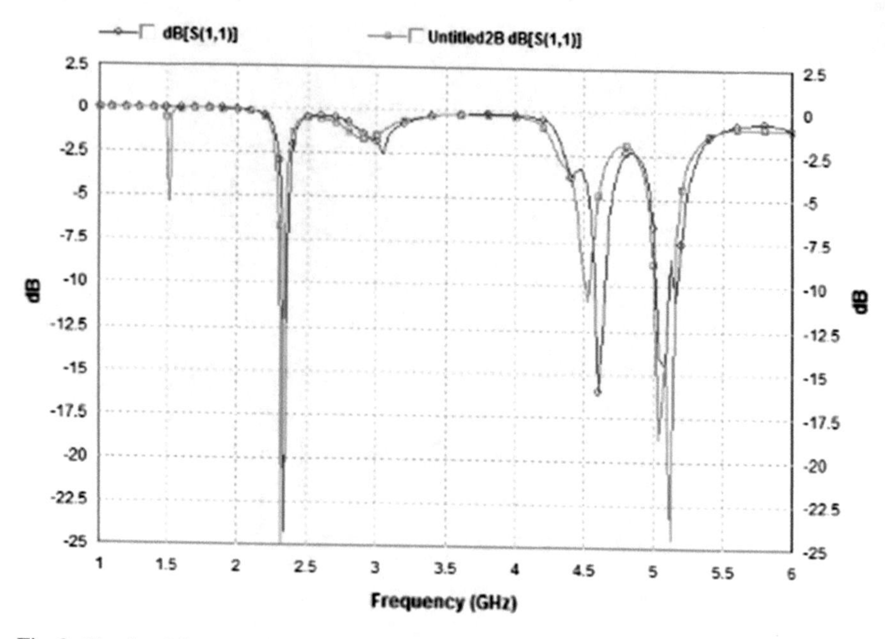

Fig. 3 Simulated S-parameters results for PSSFA and MPSSFA

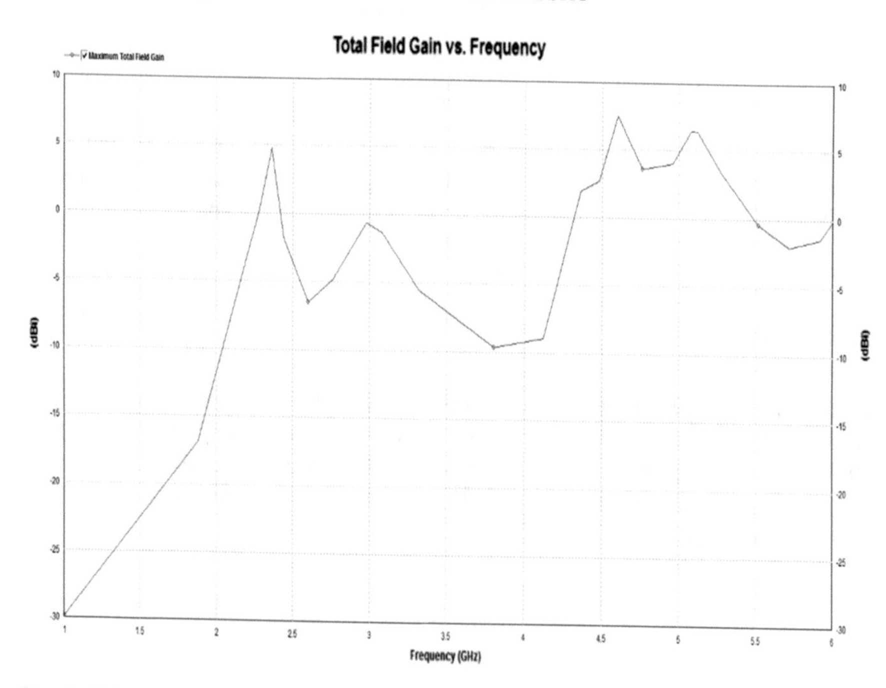

Fig. 4 Gain versus frequency plot for PSSFA

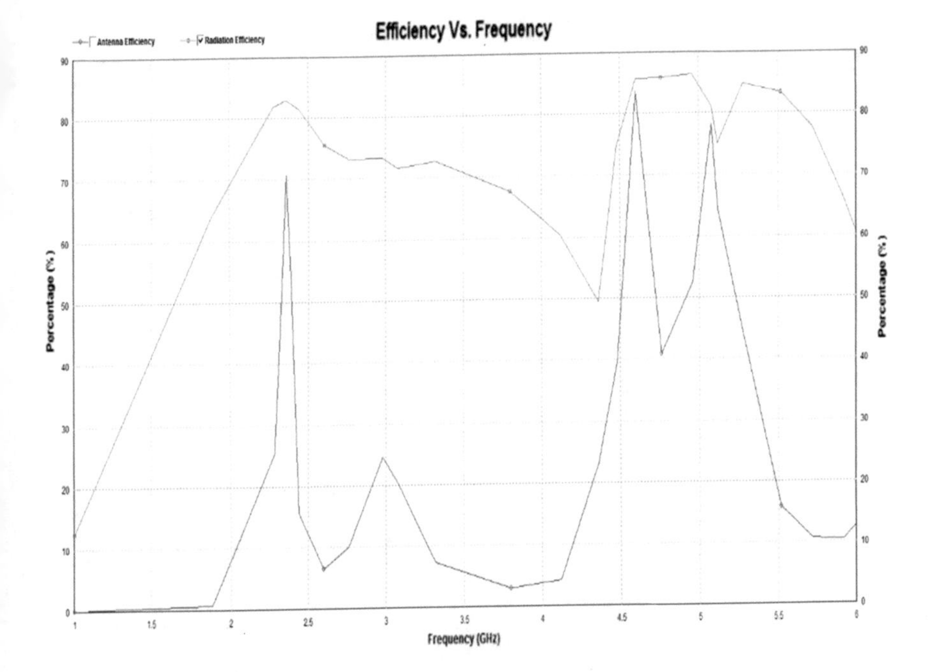

Fig. 5 Efficiency versus frequency plot for PSSFA

The PSSFA geometry is shown in Fig. 1 and MPSSFA geometry is shown in Fig. 2. *L* and *W* of the PSSFA are defined in Eqs. 1 and 5:

$$W = \frac{c}{2f_0}\sqrt{\frac{2}{\varepsilon_r + 1}} \tag{1}$$

$$\varepsilon_r = \frac{\varepsilon_r + 1}{2} + (\varepsilon_r - 1)\left[1 + 12\frac{W}{h}\right]^{-\frac{1}{2}} \tag{2}$$

$$\Delta L = 0.412h + \frac{(\varepsilon_{\text{reff+0.3}})\left(\frac{W}{h} + 0.264\right)}{(\varepsilon_{\text{reff+0.2580}})\left(\frac{W}{h} + 0.8\right)} \tag{3}$$

$$L_{\text{eff}} = \frac{c}{2f_0\sqrt{\varepsilon_{\text{reff}}}} \tag{4}$$

$$\mathrm{L} = L_{\text{eff}} - 2\Delta L \tag{5}$$

From Eqs. (1–5) [9], *L* and *W* of patch was observed as 29.83 and 38.51 mm, respectively. After that PSI shape slots were inserted into patch. The new width and length for repetition geometry were calculated from iteration factor (IF).

For next stage, new width of PSI slot $= a_1 x$ previous width of PSI slot

Iteration factor, a_1 is 1/2

For next stage length of PSI slot $= a_2 x$ previous length of PSI slot

Iteration factor, a_2 is 1/4

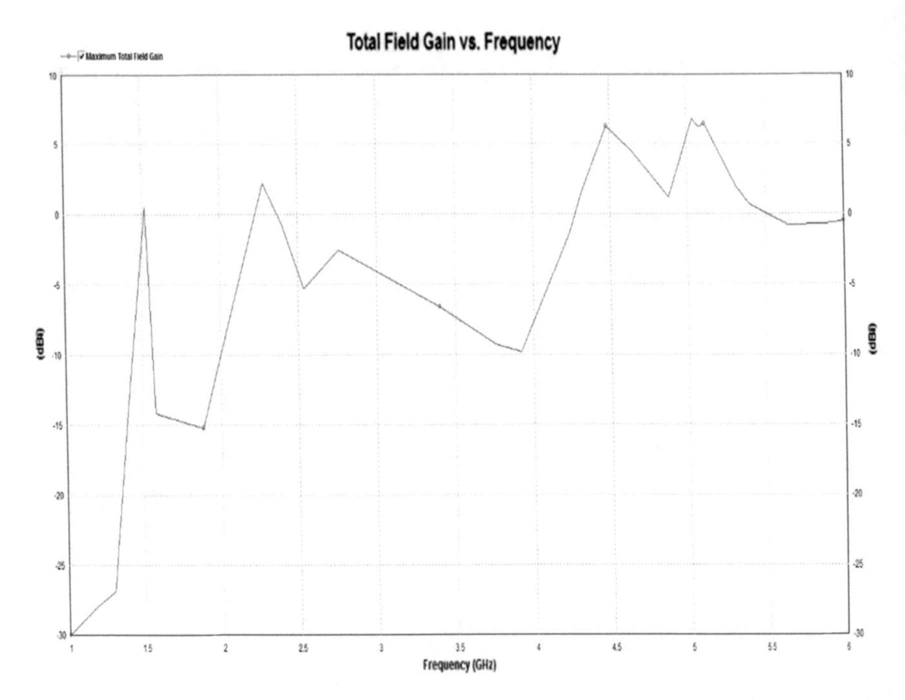

Fig. 6 Gain versus frequency plot for MPSSFA

3 Result and Discussion

S-Parameters obtained results for PSSFA are shown in Fig. 3 with resonant frequencies 2.33, 4.60, and 5.08 GHz with S-parameters −23.68, −15.94, and −14.04 dB and for MPSSFA were 2.32, 4.52, and 5.12 GHz with reflection coefficient −24.04, −10.56, and −24.68 dB.

Gain obtained values according to resonant frequencies are 3.33, 7.80 and 6.85 dB as shown in Fig. 4. **Efficiency** results were shown in Fig. 5 in which antenna efficiency was 56.66, 81.98, and 76.23% and radiation efficiency was 82.50, 85.65, and 80.23%.

For MPSSFA, **gain** obtained values were 1.22, 5.75, and 6.26 dB as shown in Fig. 6 and antenna efficiency values were 32.32, 63.49, and 79.73% and values of radiation efficiency were 81.12, 82.79, and 81.01% as shown in Fig. 7.

It has been observed that resonant frequency was decreased from the base PSI geometry having length 5 mm, as length of PSSFA was varied. The comparison was shown in Fig. 8. Table 2 describes the length variation results of PSSFA.

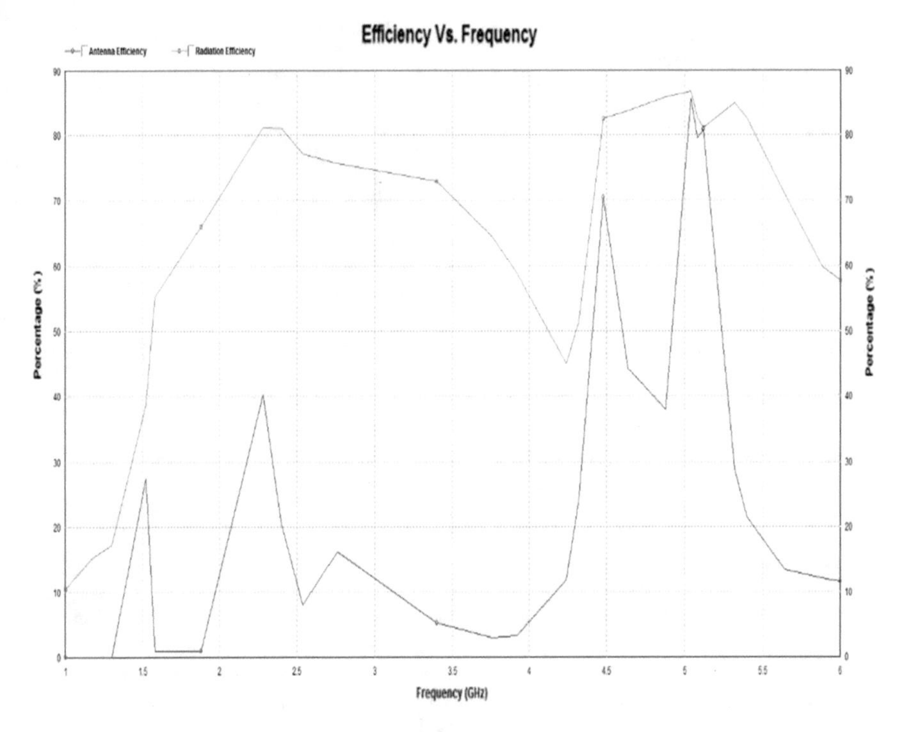

Fig. 7 Efficiency versus frequency plot for MPSSFA

Table 2 Length variation results of PSSFA

S. No.	Length	Resonant frequency (GHz)	Reflection coefficient (dB)	Gain (dB)
1	7	2.32	−26.18	3.27
2	9	1.49, 2.26	−10.77,−11.04	1.07,2.37
3	11	2.16	−13.02	4.35

4 Conclusion

From the simulated results, reflection coefficient was found to be −24.04, −10.56, and −24.68 dB at resonant frequencies 2.32, 4.52, and 5.12 GHz with Gain 1.22, 5.75, and 6.26 dB respectively. Microstrip antennas have disadvantages of more return loss, no multiband operation and narrow bandwidth. PSSFA were used to overcome these disadvantages. As iterations were increased, there is left shift in resonant frequencies. Also, with increase in length of PSSFA resonant frequency, gain, and return loss was decreased from PSSFA geometry results.

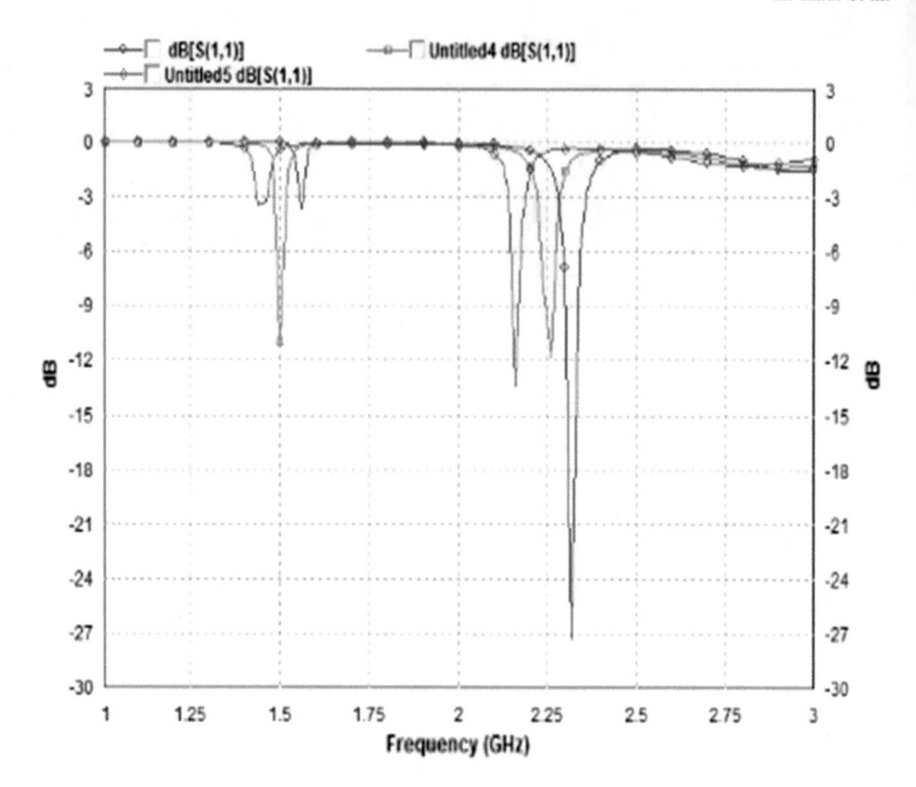

Fig. 8 Comparison of S-parameters for different lengths

References

1. Werner, D.H., Haup, R.L., Werner, P.L.: Fractal antenna engineering: the theory and design of fractal antenna arrays. IEEE Antennas Propag. Mag. **41**, 37–59 (1999)
2. Mandelbrot, B.B.: The fractal geometry of nature. Freeman, New York (1983)
3. Werner, D.H., Ganguly, S.: An overview of fractal antenna engineering research. IEEE Antennas Propag. Mag. **45**, 38–57 (2003)
4. Gianvittorio, J.P., Samii, Y.R.: Fractal antennas: a novel antenna miniaturization technique, and applications. IEEE Antennas Propag. Mag. **44**, 20–36 (2002)
5. Khanna, G., Sharma, N.: Fractal Antenna Geometries: A Review. Int. J. Comput. Applicat. **153**, 0975–8887 (2016)
6. Chowdary, P.S.R., Prasad, A.M., Rao, P.M., Anguera, J.: Design and performance study of Sierpinski fractal based patch antennas for multiband and miniaturization characteristics. Wirel. Pers. Commun. **83**, 1713–1730 (2015)
7. Mehdipour, A., Sebak, A.R., Trueman, C.W., Denidni, T.A.: Compact multiband planar antenna for 2.4/3.5/5.2/5.8-GHz wireless applications. IEEE Antennas Wirel. Propag. Lett. **11**, 144–147 (2012)
8. Al-Bawri, S.S., Jamlos, M.F., Soh, P.J., Junid, S.A.S., Jamlos, M.A., Narbudowicz, A.: Multiband slot-loaded dipole antenna for WLAN and LTE-A applications. IET Microw. Antennas Propag. **12**, 63–68 (2018)
9. Kaur, G., Rattan, M., Jain, C.: Design and optimization of PSI (W) slotted fractal antenna using ANN and GA for multiband applications. Wirel. Pers. Commun. **97**, 4573–4585 (2017)

10. Sharma, S.K., Shafai, L.: Investigations of a novel Ψ-shape microstrip patch antenna with wide impedance bandwidth. In: Antennas and Propagation Society International Symposium, pp. 881–884. IEEE Press, USA (2007)
11. Deshmukh, A.A., Mohadikar, P., Lele, K., Panchal, G., Parvez, A.: Psi-shaped ultra-wideband monopole antenna with a modified feeding structure. In: 6th International Conference on Advances in Computing and Communications, pp. 60–66. Cochin, India (2016)
12. Kaur, G., Jain, C., Rattan, M.: A novel multiband Psi (Ψ) slotted fractal antenna for S-band applications. In: International Conference on Computing, Communication and Automation, pp. 1563–1567. IEEE Press, Noida, UP (2016)

Kirandeep Kaur is currently pursuing M.Tech from Department of Electronics and Communication Engineering of Guru Nanak Dev Engineering College, Ludhiana. She received her B.Tech. Degree in Electronics and Communication Engineering from Gulzar Group of Institutes, Khanna in 2016. Her major research area is Antenna Design.

Chahat Jain is currently working as an Assistant Professor in the Department of ECE at Guru Nanak Dev Engineering College, Ludhiana. She did her B.Tech in ECE in 2009 from RIMT, Mandi Gobindgarh, and M.Tech in ECE from Guru Nanak Dev Engineering College, Ludhiana, in 2011. She has published several papers in international and national journals and conferences. Her current research interests are Electromagnetics, Antenna Design and Metamaterials Design.

Narwant Singh Grewal did his M.Tech in 2002. Currently, he is pursuing Ph.D. and working as Assistant Professor in Department of ECE at Guru Nanak Dev Engineering College, Ludhiana, Punjab. His research interests include Antenna Array failure System, Optimization Techniques and Digital System Design. He has guided more than 20 M.Tech students. He co-authored one book on Microwave and radar engineering and published several papers in international, national journals and conferences of repute.

An Enhanced Round-Robin-Based Job Scheduling Algorithm in Grid Computing

Turendar Sahu, Sandeep Kumar Verma, Mohit Shakya and Raksha Pandey

Abstract Nowadays, grid computing is a recognized name in the area of computation. Grid computing provides a strong platform for the jobs demanding high computational power. Resource scheduling and job scheduling are two broad categories of grid computing. A lot of work has been done in this area. In this paper, we present a new job scheduling algorithm which focuses on standard round-robin algorithm with the addition of new "Enhancement Factor". The experimental evaluation shows that the proposed algorithm reduces Average Waiting Time and Average Turnaround Time, and it takes very less time to process the jobs than other existing algorithms.

Keywords Job scheduling · Round robin · Grid computing · Waiting time

1 Introduction

GRID computing is a collection of large computational resources, services, or datasets. The term grid emerged from the analogy to power grid, where one can get the electricity or power without concerning about the place of power source or method of their arrival [1–2, 3]. There are basically two types of users [4, 5] in the grid. First, can be thought as those who demand for low processing cost of their job scheduling and second, who involves implementing the required demand. It is dynamic in nature, in which resources are combined or released by users or

---

T. Sahu · S. K. Verma · M. Shakya · R. Pandey (✉)
Department of Computer Science and Engineering. Sos in Engineering and Technology,
Guru Ghasidas Vishwavidyalaya, Bilaspur, Chhattisgarh, India
e-mail: rakshasharma10@gmail.com

T. Sahu
e-mail: turedarsahu@gmail.com

S. K. Verma
e-mail: skv007bond@gmail.com

M. Shakya
e-mail: mohitshakya797@gmail.com

S. Smys et al. (eds.), *International Conference on Computer Networks and Communication Technologies*, Lecture Notes on Data Engineering and Communications Technologies 15, https://doi.org/10.1007/978-981-10-8681-6_73

owners from grid environment [6]. Two broad categories of grid computing are job and resource scheduling. Many researches have been done in this area to optimize the performance of jobs. The grid is mainly divided into two categories: Data Grid and Computational Grid [7, 8], in which Grid is popular in the scientific area. Network Grid is another category of the grid. The overall aspect of the GRID computing is to get benefit whether by sharing resources or by reducing the processing cost of computation of jobs. GRID computing [9] is an approach which is basically sharing wide-area computational resources all over the world, evolving continuously, and becoming the backbone technology to form the large-scale virtual organization. It came [10] into existence in 1990s, which brings the revolutionary idea to form a complex computation environment that is capable of solving very huge problem in very short time which if solved by a normal system, will take too much time (in some cases may be months or years). The inspiration behind the GRID computing was the sharing of resources among many organizations to resolve large-scale problems such as weather forecasting, invention of medicines for new diseases, etc. The key element of GRID computing is the grid which refers to systems, services, or applications that integrate resources and services spread all across the multiple control domains. Computational grids bring the concept to share large-scale resources, such as database, personal computer, MPPs, clusters, and online instructions, which may be dynamic, cross-domain, or even heterogeneous. The continuous research in GRID computing [11] is going on in many fields and job scheduling is one of them. Job scheduling is very essential part which specifies how the user jobs are handled. The aspect of job scheduling mainly comes under operation system where scheduling is done by various schedulers.

2 Related Work

In the field of grid computing, the concept of operating system is required for the job scheduling on the distributed computing environment. In [12], researchers have done very much considerable work. There are many algorithms that have been proposed in recent years, each algorithm having their own characteristics and strength. Jobs submitted to a grid computing system need to be executed by the available resources. Best resources regarding processing speed, memory, and availability status are more likely to be selected for the submitted jobs during the scheduling process. In [11], the idea of scheduling is present at the different levels including cluster, OS, and global system. The operating system manages the implementation or execution of a process at the different stages of the life cycle of the processes. In the operating system, scheduling of process is handled by long-term, medium-term, and short-term schedulers. Long-term scheduler schedules the process from ready state to running state, mid-term scheduler schedules the process from running state to waiting state or waiting state to ready state, and short-term scheduler schedules the new process to the ready state in the operating system. In [6], job scheduling algorithms are a demand to make use of resources effectively in different fields of science and technology. Therefore, the execution performance increases and system load balancing is maintained. Grid computing is an environment where resources are provided by remote and ideal computers. In [13], job scheduling is done by batch mode algorithm which reduces total completion time, utilization of resources with load balancing. It shows better results than some of the existing scheduling algorithms. But if the jobs are lighter, then the load balancing is not so good.

3 Proposed Model

We are proposing a new job scheduling algorithm with new concept of Enhancement Factor in round-robin algorithm. Using this new enhanced round-robin algorithm works in a new way, and reduces waiting time, Average Turnaround Time and increases throughput of jobs.

The algorithm is divided into two broad sections:

(A) *Scheduling Strategy*:

Jobs come in FCFS order. These jobs are kept into the ready queue. At any particular time, we get a record of total number of jobs available in ready queue, and we calculate the enhancement factor with respect to all the available jobs in ready queue. The formula for calculation of Enhancement Factor is:

$$e = \text{ceil}(\text{sqrt}(\text{floor}(\text{Avg.ET}) * \max(\text{ET}))/\text{TQ})$$

After the calculation of e, we will check for each job in ready queue whether the job satisfies the following criteria:

(i) $\text{ET} <= e * j_{\text{tq}}$ and
(ii) $\text{ET} > 0$

If yes, then we will execute complete jobs otherwise, we will execute the job for our specified Time Quantum (TQ). We will continue the process for each number of jobs until the ready queue is empty.

(B) *Execution Strategy*:

The jobs are assigned to the resources by the grid scheduler (also called dispatcher). Jobs are assigned in FCFS order for the specific execution time. Grid scheduler gets the information such as MIPS, memory size, and other required characteristics of resources from the GIS (Grid Information Service).

The standard round robin decreases the average response time and increases the Average Waiting Time and Average Turnaround Time.

In our proposed algorithm, round-robin algorithm is enhanced by adding a new concept of enhancement factor.

Table 1 Process number or Job ID with their Execution time

Process no.	Execution time
P1	5
P2	25
P3	95
P4	40
P5	30
P6	80

Algorithm

1. Job comes in FCFS order.
2. Set Time Quantum for n no. of jobs = j_{tq}
3. Compute the average Execution Time of all jobs and set it as $ET_{avg.}$
4. Find the maximum execution time and set it as ET_{max}.
5. Compute the Enhancement Factor (e) as:
   - Enhancement Factor (e) = ceil(sqrt(floor(ET_{avg}) * (ET_{max}))/(j_{tq})
6. while (j<=joblist_size-1)
7. If($ET_j < e * j_{tq}$ && $ET_j > 0$)
   - Execute the jobs completely and terminate the current job after complete execution.
8. Else
   - Execute the job only for specified time quantum.
9. End of while loop.
10. End.

Let us take the example

In our purposed algorithm, Time Quantum is taken as 4 ns. And, it is increased by the Enhancement Factor (*e*) which is defined as (Fig. 1):

Enhancement Factor (*e*) is (Table 1):

$$e = \text{ceil}(\text{sqrt}(\text{floor}(\text{Avg.ET}) * \max(\text{ET}))/\text{TQ})$$

where

Avg. ET	Average of all Execution Time
ET	Execution Time
TQ	Time Quantum
sqrt	Square Root Function
max	Maximum Function
floor	Floor function to get integer part of any number

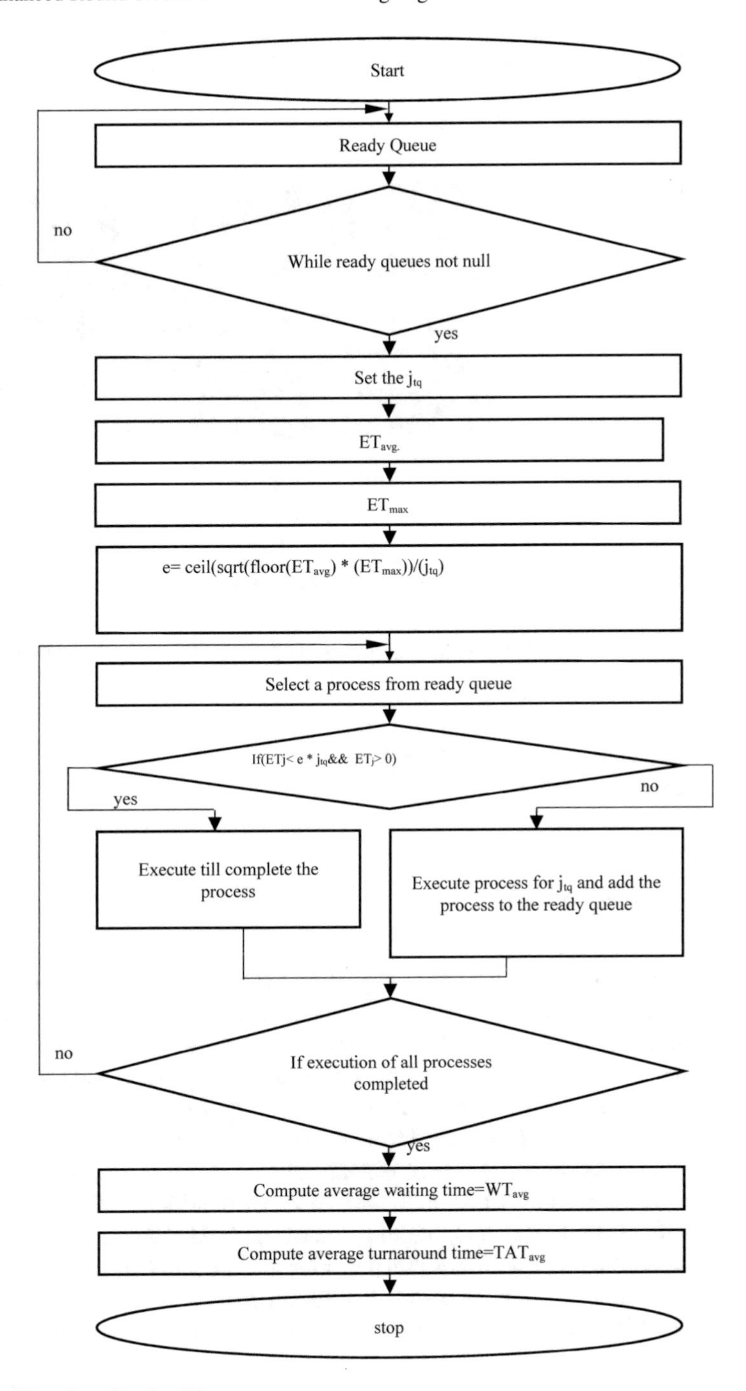

Fig. 1 Flowchart for algorithm

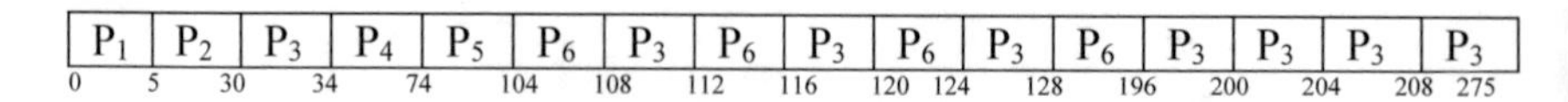

Fig. 2 Gantt chart

Chart 1 Comparison between FCFS, SRR, and ERR

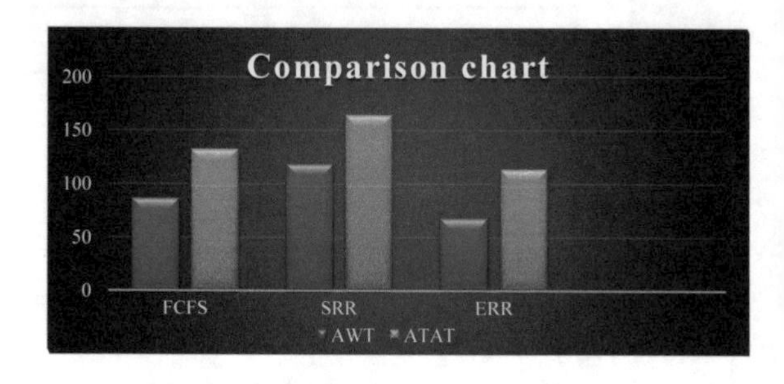

Therefore,

$$e = \text{ceil}(\text{sqrt}(\text{floor}(45.833) * 95)/4) = 17$$

Now, the Gantt chart according to the enhancement round-robin algorithm is (Fig. 2):

Average Waiting Time = 68.166667 ns
Average Turnaround Time = 114.000000 ns
Number of context switches = 16

4 Experimental Evaluation

In this paper, we are proposing a new algorithm to improve the performance and efficiency of job scheduling. The concept of round-robin algorithm is used in a new form. Algorithm performs well as the number of jobs increases. As the grid computing environment is used for a huge number of jobs so the proposed algorithm can work well in this area. Several algorithms have been designed on grid computing environment. The proposed Enhanced Round-Robin algorithm has been compared to some existing algorithms and experimental results shows proposed algorithm performs better. Average Waiting time and Average Turnaround Time of our algorithm is very less as compared to other algorithms, and we can see in Results as the number of jobs increases, the proposed algorithm performs much better (Charts 1, 2 and 3).

The following images show the comparison chart between First Come First Serve, Standard Round Robin, and Enhanced Round Robin in various number of jobs.

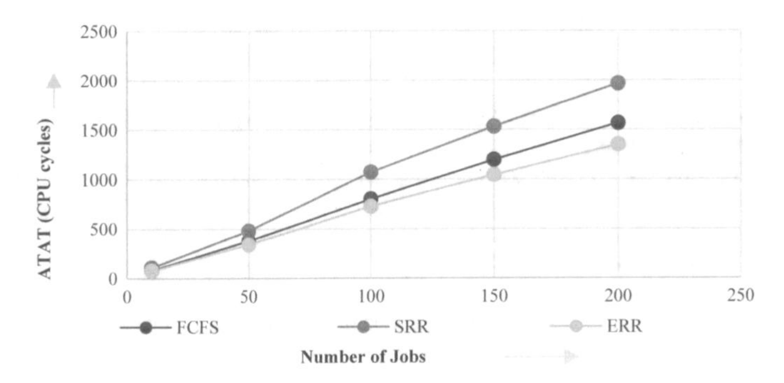

Chart 2 Comparison of ATAT between FCFS, SRR, and ERR

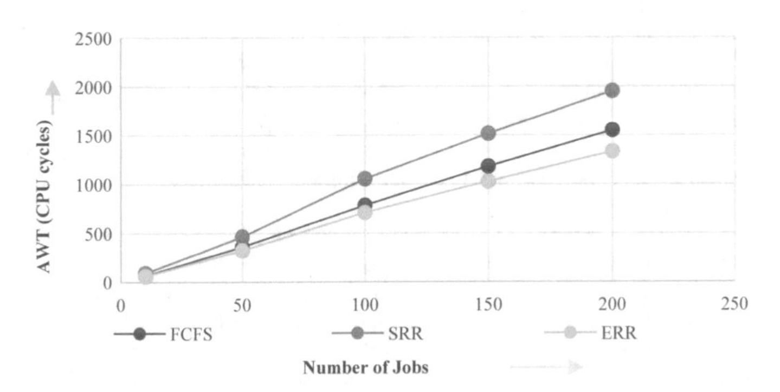

Chart 3 Comparison of AWT between FCFS, SRR, and ERR. Caption: FCFS—First Come First Serve; SRR—Standard Round Robin; ERR—Enhancement Round Robin; AWT—Average Waiting Time; ATAT—Average Turnaround Time

5 Conclusion and Future Work

When the number of job is large, the grid environment works well for job scheduling. Experimental evaluation shows that our algorithm performs better in terms of Average Waiting time, Average Turnaround Time, and Throughput as compared to FCFS and SRR. With the new term "Enhancement Factor", the proposed algorithm performs much better with large number of jobs and can work in the grid computing environment.

In future, we will try to implement this algorithm in real-time environment.

References

1. Jacob, B., Brown, M., Fukui, K., Trivedi, N.: Introduction to grid computing. In: IBM International Technical Support Organization, Dec 2005
2. Baker M., Buyya, R., Laforenza, D.: Grids and grid technologies for wide-area distributed computing. Softw. Pract. Exper. 2002 (in press) (2002). https://doi.org/10.1002/spe.488
3. Singh, M.: Incremental checkpoint based failure-aware scheduling algorithm in grid computing. In: International Conference on Computing, Communication and Automation (ICCCA2016) (2016)
4. Kaur, M.: FastPGA based scheduling of dependent tasks in grid computing to provide QoS to grid users. In: International Conference on Internet of Things and Applications (IOTA), Jan 2016
5. Yagoubi, Belabbas, Meddeber, Meriem: Distributed load balancing model for grid computing. Revue ARIMA **12**, 43–60 (2010)
6. Anitha, Avula: Job scheduling using coupling in grid. J. Comput. Commun. **3**, 1–12 (2015)
7. Jairam Naik, K., Jagran, A., Satya Narayana, N.: A novel algorithm for fault tolerant job scheduling and load balancing in grid computing environment. In: International Conference on Green Computing and Internet of Things (ICGCIoT) (2015)
8. Baria, A., Meena, D.S., Pandey, R.: An analysis of different scheduling approaches used in grid computing. Int. J. Sci. Res. Comput. Sci. Eng. Inf. Technol. (IJSRCSEIT) **2**, 2456–3307 (2017)
9. Soni, V.K., Sharma, R., Mishra, M.K.: Grouping-based job scheduling model in grid computing. Int. J. Comput. Electr. Autom. Control Inf. Eng. **4**(5), 334 (2010)
10. Kokilavani, T.: Enhanced round robin techdunique with variant time quantum for task scheduling in grid computing. Int. J. Emerg. Trends Sci. Technol. IJETST **04**, 6016–6021 (2017)
11. Prajapati, H.B., Shah, V.P.: Scheduling in grid computing environment. In: Fourth International Conference on Advanced Computing and Communication Technologies (2014)
12. Gomathi, S., Manimegalai, D.: An adaptive grouping based job scheduling in grid computing. In: International Conference on Signal Processing, Communication, Computing and Networking Technologies (ICSCCN 2011) (2011)
13. Maipan-uku, J.Y., Konjaang, J.K., Baba, A.: New batch mode scheduling strategy for grid computing system. Int. J. Eng. Technol. (IJET) **8**(2), 1314 (2016)

Turendar Shau studying 3rd year Bachelor of technology student, Department of Computer Science and Engineering, Sos in Engineering and Technology, Guru Ghasidas Vishwavidyalaya, Bilaspur (Chhattisgarh), India.

Sandeep Kumar Verma studying 3rd year Bachelor of technology student, Department of Computer Science and Engineering, Sos in Engineering and Technology, Guru Ghasidas Vishwavidyalaya, Bilaspur (Chhattisgarh), India.

Mohit Shakya studying 3rd year Bachelor of technology student, Department of Computer Science and Engineering, Sos in Engineering and Technology, Guru Ghasidas Vishwavidyalaya, Bilaspur (Chhattisgarh), India.

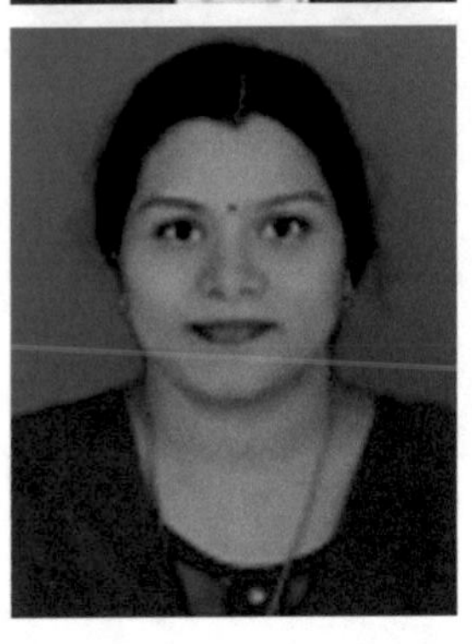

Raksha Pandey Assistant Professor, Department of Computer Science and Engineering, SoS in Engineering and Technology, Guru Ghasidas Vishwavidyalaya, Bilaspur (Chhattisgarh), India.

Wireless Sensor Network to Monitor River Water Impurity

Shweta Doshi and Sharad Dube

Abstract Advances in Wireless Sensor Network (WSN) has led to its utilization for development of broadscale and long-term natural checking. In the past, there have been difficulties experienced in meeting end-clients necessities for data gathering and real-time situation monitoring. The essential point is to provide ease of installation, real-time monitoring, and also to give more imperative reliability and productivity to affect choices of system hardware and programming. This paper suggests a method for better utilization of Wireless Sensor Network (WSN) for real-time water quality monitoring to prevent the delay in handling the situation. The method comprises of sensor nodes with a networking capability that can be deployed for continuous monitoring purpose. The temperature and turbidity parameters associated with the water quality assurance is estimated in real time by the sensors, and this information is updated in a remotely located server through GPRS.

Keywords Wireless sensor network · Water quality monitoring · GPRS

1 Introduction

Water is essential for agriculture, industry, and for creature's existence on earth including human beings.

Also, the human body is comprised of 60% of water. We utilize clean water to drink, develop crops for nourishment, work processing plants, and for swimming, surfing, angling, and cruising. Water is imperatively vital to each part of our lives and hence, monitoring the nature of surface water will help shield our conduits from contamination.

S. Doshi (✉) · S. Dube
Department of Electronics and Telecommunication, Cummins College of Engineering for Women, Pune, India
e-mail: ssdoshi22@gmail.com

S. Dube
e-mail: sharad.dube@cumminscollege.in

S. Smys et al. (eds.), *International Conference on Computer Networks and Communication Technologies*, Lecture Notes on Data Engineering and Communications Technologies 15, https://doi.org/10.1007/978-981-10-8681-6_74

It is observed that water is wasted by many uncontrolled ways. This problem is quietly related to poor water allocation, inefficient use, and lack of adequate and integrated water management. Therefore, efficient use and water monitoring are potential constraints for water management system as each living thing on earth needs water for survival.

The use of Wireless Sensor Network (WSN) for water quality checking consists of sensor nodes with networking capability that serves the purpose of real-time monitoring of data. The network gathers, transmits, and processes water quality parameters consequently, so generation productivity and economy advantage are enhanced incredibly. This framework setup emphasizes on the low cost, simplicity of establishment, handling, and maintenance.

The utilization of remote framework for monitoring will not only just diminish the general monitoring framework cost in terms of the setup of facilities and labor cost, but also gives flexibility as far as separation or area are concerned. Subsequently, the created stage is cost-effective and permits simple customization.

Farmers can utilize the data to manage better to deal with their property and harvests. Our nearby, state and national governments utilize monitored data to help control contamination levels. Water contamination checking can help with water contamination location, and also the release of poisonous chemicals and defilement in water.

The goal of this paper is to design and manage a Wireless Sensor Network (WSN) that helps to monitor the quality of water with the help of information sensed by the sensors immersed in water, so as to keep the water resource within a standard described for domestic usage and to be able to take necessary actions to restore the health of the degraded water body. Temperature and turbidity are the parameters collected in river/lake water pollution/quality monitoring systems, which are transmitted using GPRS and updated on remote server.

2 Literature Survey

Sharma [1], in this exploration work, an overview on Environmental Monitoring utilizing Wireless Sensor Networks and their advancements and gages was completed.

The absolute, most pertinent natural observing activities with genuine organizations were broken down what is more, the conclusions used to distinguish the difficulties that should be tended to. Remote sensor systems keep on emerging as an innovation that will change the way we measure, comprehend, and deal with the indigenous habitat. Out of the blue, information of various kinds and spots can be combined and gotten to from anyplace.

Some critical advance has been made in the course of the most recent couple of years keeping in mind, the end goal to cross over any barrier between hypothetical improvements and genuine organizations, albeit accessible plan strategies and arrangements are still moderately youthful. As an outcome, far-reaching utilization of WSNs for ecological proposes is not yet a reality.

Yadav et al. [2], in the present life condition is pivotal issue influencing life of individual and agribusiness part also. The requirement for checking rural field and its natural parameters to oversee legitimate water system and keep up soil quality has brought the consideration of technologist towards it.

The detonating populace, changes in temperature, and lessened water accessibility has made it important to oversee appropriate usage of different assets required for cultivating and to control condition parameters according to the necessity for every item also. Rising condition issues have an awesome effect on technology, science, social condition, and practical field.

This issue which is all-inclusive has raised numerous exchanges for discovering arrangement. Among different mechanical arrangements, some of the strategies are utilizing satellite framework, wireless system, sensor system, etc. This paper presents contemplate comes about utilizing Wireless sensor organize, Internet of things, sensors, and Raspberry Pi.

The framework is ease and low-control devouring framework. This framework is very adaptable as far as number of sensors and sort of sensors. With over a rot of innovative work in remote innovation and sensor advancement, Wireless Sensor Network innovation has been developing as proficient and monetary answer for different application territories.

Peralta et al. [3] the WISE-MUSE venture plans to the arrangement of a WSN for consequently and persistently observing and control the earth of galleries. The utilization of WSNs for natural checking of a gallery is, surely, a more dependable arrangement. It is additionally more affordable than manual information accumulation or than a wired focal checking framework.

This undertaking has proposed a few commitments, among them is the WISE-MUSE portable application. This application permits the ecological observing in light of remote sensor systems by means of a cell phone. In addition, it enables clients to continually check and break down information on the observed parameters, whenever and anyplace, empowering checking with greater adaptability and greater security, giving better perception of all information caught by various sensors through graphs, maps and authentic inclinations, progressively.

A standout among the most critical highlights of the proposed application is to give a warnings' framework to clients, in case any parameter displays a strange esteem. The tests did permit to infer that the WISE-MUSE application fits the most critical necessities of an exhibition hall, partner the productive show of precise estimation of natural factors, all caught by sensors, with the capacity of sending warnings, permitting preventive activity in protecting the history that these works speak to. B.

Manjurega et al. [4] In this manner according to the above undertaking the temperature, dampness and the gas show in the earth, that is, in the room has been distinguished by the sensors in the sensor hub and the comparing yield is shown utilizing a Liquid Crystal Display (LCD) with a solitary hub in the Wireless Mesh Network (WMN).

The proposed framework is exhibited to have the benefits of minimal effort joined with high unwavering quality and execution, and can be valuable in actualizing checking applications without the difficulties of complex remote systems administration issues. In the proposed framework, safety measures can be when required if the parameters surpass the ordinary esteem.

Corke et al. [5] this paper is worried about the use of remote sensor organize (WSN) innovation to long-span and extensive scale natural checking. The sacred chalice is a framework that can be sent and worked by space masters, not engineers, but rather this remaining parts some separation into the future. We show our perspectives regarding why this field has advanced less rapidly than numerous conceived it would over 10 years back.

We utilize genuine cases taken from our own particular work in this field to show the innovative troubles and difficulties that are involved in meeting end-client pre-requisites for data gathering frameworks. Unwavering quality and efficiency are key concerns and impact the plan decisions for framework equipment also, programming.

Pašalić et al. [6] incorporation of WSN in a domain for various applications and activities for condition information checking and handling has pulled in advance improvement of little and asset rich SNs.

Such WSN incorporation prompts better comprehension of the quality and conduct of observed condition and empowers to settle on better choices in different ecological applications and requirements. The proposed and portrayed down to earth WSN usage for condition information checking is extremely adaptable, reasonable and appropriate for various condition applications which require and utilize basic battery controlled SNs.

It is anything but difficult to build number of SNs to a required number, to program SNs for utilization of various sensors, and to program required information collection and handling, contingent upon solid use of the WSN. For such battery fueled SNs it is prescribed to utilize any conceivable vitality source in the earth to charge the battery to expand the battery substitution period and additionally periodical information transmission or on occasion event.

In this viable execution it was utilized sun oriented vitality produced by the sunlight based cells board for charging battery. Fitting techniques for lessening of vitality utilization in the sensor hubs and in the entire remote sensor arrange are additionally utilized as a part of the down to earth execution, what is essentially diminishing aggregate vitality utilization.

Khedo et al. [7] sensor systems are as of now a dynamic research territory predominantly because of the capability of their applications. In this paper we explore the utilization of Wireless Sensor Networks (WSN) for air contamination observing in Mauritius. With the quickly developing modern exercises on the island, the issue of air contamination is turning into a noteworthy worry for the strength of the populace.

We proposed an inventive framework named Wireless Sensor Network Air Pollution Monitoring System (WAPMS) to screen air contamination in Mauritius using remote sensors sent in enormous numbers around the island. The proposed framework influences utilization of an Air Quality to list (AQI) which is by and by not accessible in Mauritius. To enhance the productivity of WAPMS, we have planned and executed another information conglomeration calculation named Recursive Converging Quartiles (RCQ).

The calculation is utilized to combine information to wipe out copies, sift through invalid readings, and abridge them into a more straightforward frame which essentially diminishes the measure of information to be transmitted to the sink and along these lines sparing vitality. For better power administration, we utilized a various leveled directing convention in WAPMS and made the bits rest amid sit out of gear time.

Alhmiedat et al. [8] this paper targets minimal effort and low power execution for indoor air quality observing applications. The proposed framework in this paper offers ease plan, since XBee modules and ATTiny-85 microcontrollers has been conveyed, which thus offer ease. A rest state calculation and interface circuit have been planned and executed to accomplish least power utilization.

For future works, vitality productive calculations are required to be embraced keeping in mind the end goal to additionally limit the power utilization required for indoor air quality checking frameworks.

Sousa et al. [9] this paper reports the improvement and reconciliation of a remote sensor organize for ecological observing. The fundamental objectives of this framework incorporate seclusion, low power utilization, and simplicity of extension. The framework incorporates three primary components: sensor hubs, entryways, and a server. Every sensor hub can just associate with a passage, bringing about a star arrange format.

Information gathered from the diverse sensor hubs is put away in a database inside the server. An online UI for this framework was created and made accessible on the web.

Ye et al. [10] this paper proposes the WSN utilized as a part of natural checking. We receive IRIS bit equipment stage and plan an information procurement board, which predominantly assembles seven natural parameters, for example, barometric weight, encompassing temperature, climatic moistness, wind heading, wind speed, underground water level, and precipitation. The framework has made a few accomplishments in: (1) this framework is a viable natural observing application where sensor hubs occasionally sense their feeling and the data they got, which can be transmitted through multi-hop approach in the long run to the examination place for upper programming investigation. (2) Packets from the entryway can be transmitted to every one of the clients through the Internet which makes remote natural observing practical.

Ali et al. [11] and Saeed et al. [12] this paper proposes WSN application for monitoring issues associated with the oil and gas pipelines. This paper gives an in-depth explanation of an algorithm for setting up the network and controlling the sensed data in LWSN that implements the Zigbee protocol.

3 Proposed System

The system (Fig. 1) comprises of sensor nodes which will be placed within the river bed and which will gather real-time data and the data will be remotely conveyed

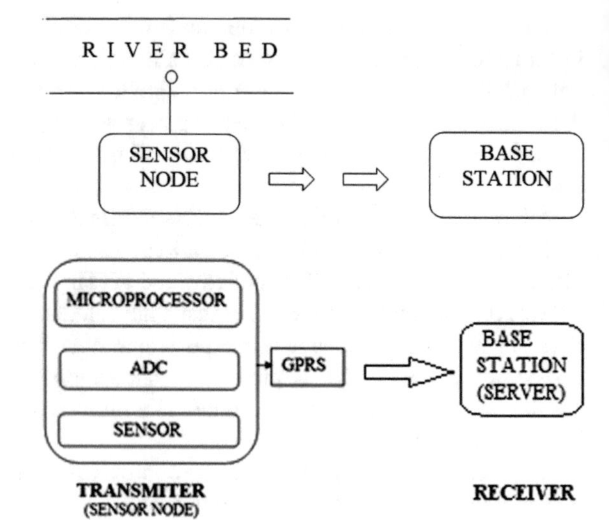

Fig. 1 General system overview

Fig. 2 Proposed system block diagram

through GPRS to keep the water quality inside a standard which is portrayed for local use and to have the capacity to take vital action to reestablish the nature of the degraded water body.

The proposed system (Fig. 2) for checking water quality uses two sensors, temperature and turbidity sensor. The turbidity sensor will monitor water purity and the temperature sensor will monitor the water temperature. The data collected through the sensors will initially process through a signal conditioning circuit.

Turbidity sensors reflect the turbidity level in terms of output voltage ranging from 0 to 4.5 V. The analog data of turbidity sensor is converted into digital signal using ADC MCP3204, so that it can be easily processed by Raspberry Pi. The ADC is interfaced with Raspberry Pi over SPI interface. The turbidity sensor identifies water quality by estimating the levels of turbidity. It utilizes the light to recognize suspended particles in water by estimating the light transmittance and scattering rate, which changes with the measure of Total Suspended Solids (TSS) in water. As the TTS builds, the fluid turbidity level increases. As per USEPA Method 180.1 for turbidity estimation, the Turbidity Sensors are a 90° dissipate nephelometer. The turbidity sensor coordinates a beam into the observed water. The light pillar reflects off particles in the water, and the resultant light force is estimated by the turbidity sensor's photodetector situated at 90° to the light shaft. The light power identified by the turbidity sensor is straightforwardly corresponding to the turbidity of the water .It utilizes the nephelometric optical system and provides direct measurements in units of NTU (Nephelometric Turbidity Unit).

Temperature sensor DS18B20 is used which reflects the temperature value ranging from −55 to 125 °C (−67–257 F) with 9-bit resolution. The sensor uses one-wire communication protocol which makes its use efficient.

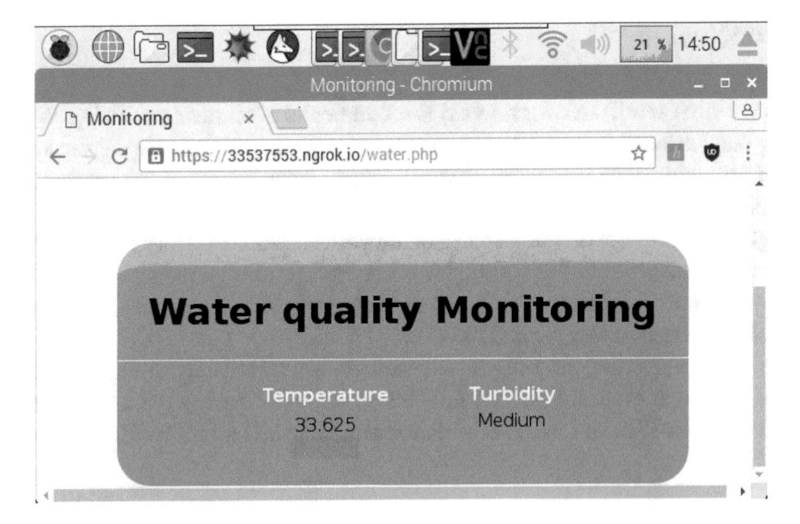

Fig. 3 Result of the proposed approach

GPRS usage practice has demonstrated GPRS innovation and can accomplish well inside the intricate condition of poor water quality which remains unmonitored.

The general structure and system of the framework, and the method for access to GPRS information for checking water quality are composed in view of GPRS technology. Water quality parameters gathered by sensors are transmitted to data monitoring unit through GPRS wireless communication. The framework automatically gathers, transmits, and processes the parameters, so as to gain the advantage of production efficiency as well as economy.

The information transmitted using GPRS is displayed on a remotely located server in real-time. The server is programmed using MySQL.

4 Result

The results are displayed on a server, programmed using MySQL database, and webpage is designed using html. It displays the following values:

(1) Temperature in degree Celsius, and
(2) Turbidity sensor is calibrated to display the water quality as:

   (a) Good (Turbidity value > 700),
   (b) Medium (200 < Turbidity value > 700), or
   (c) Poor (Turbidity value < 200).

The result of the proposed approach (Fig. 3).

5 Conclusion

In this research work, an overview on Environmental Monitoring utilizing Wireless Sensor Networks and their advancements was completed. Remote sensor systems keep on emerging as an innovation that will change the way we measure, comprehend, and deal with the common habitat.

The efficient design of wireless sensor network is considered based upon the study of various architectures. The information from temperature and turbidity sensor are collected by utilizing Raspberry Pi as a fundamental controlling unit and transmitted remotely to a base station over GPRS.

The performance of the proposed system shows promising results and suggest that the proposed system based on GPRS is twice more energy preserving, reliable as well as it also aids in ease of installation and real-time monitoring.

References

1. Sharma, P.: Wireless sensor networks for environmental monitoring. Int. J. Sci. Res. Eng. Technol. (IJSRET) (2014)
2. Yadav, R.B., Kanade, S.S. Padwa, S.C.: Environment monitoring using wireless sensor network for agricultural application. Int. J. Res. Appl. Sci. Eng. Technol. (IJRASET) 5(5) (2017)
3. Peralta, L.M.R.: Environmental monitoring based on wireless sensor network via mobile phone SENSORCOMM 2013. In: The Seventh International Conference on Sensor Technologies and Applications
4. Manjurega, B., Priyadharsini, M., Poovika, T., Sanjeevi, S., Rooban, P.: Environmental Monitoring system with wireless mesh network using Zigbee. Int. J. ChemTech Res. **10**(14), 249–252 (2017)
5. Corke, P., Wark, T., Jurdak, R., Hu, W., Valencia, P., Moore, D.: Environmental wireless sensor networks **0018**(11), 1903 (2010)
6. Pašalić, D., Bundalo, Z., Bundalo, D., Softić, F., Cvijić, B.: Environmental data monitoring using wireless sensor networks
7. Khedo, K.K., Perseedoss, R., Mungu A.: A wireless sensor network air pollution monitoring system. Int. J. Wirel. Mobile Netw. (IJWMN), 2(2) (2010)
8. Alhmiedat, T., Samara, G.: A low cost zigbee sensor network architecture for indoor air quality monitoring. Int. J. Comput. Sci. Inf. Secur (IJCSIS), 15(1) (2017)
9. Sousa, P.J., Tavares, R., Abreu, P., Restivo, M.T.: NSensor—wireless sensor network for environmental monitoring. IJIM **11**(5), 25 (2017)
10. Ye, D., Gong, D., Wang, W.: Application of wireless sensor networks in environmental monitoring. In: 2009 2nd International Conference on Power Electronics and Intelligent Transportation System
11. Ali, S., Ashraf, A., Qaisar, S.B., Afridi, M.K., Saeed, H., Rashid, S., Felemban, E.A., Sheikh, A.A.: SimpliMote: a wireless sensor network monitoring platform for oil and gas pipelines. IEEE Syst. J. (2016)
12. Saeed, H., Ali, S., Rashid, S., Qaisar, S., Felemban, E.: Reliable monitoring of oil and gas pipelines using wireless sensor network (WSN)—REMONG. In: System of Systems Engineering (SOSE), 2014 9th International Conference on, pp. 230–235. IEEE, 2014

Shweta S Doshi studied BE—ENTC: Modern Education Society's College of Engineering Pune. And studying Mtech—ENTC, Signal processing: MKSSS's Cummins College of Engineering Pune. Her research area includes Wireless Sensor and Water Impurity.

A Priority-Based Max-Min Scheduling Algorithm for Cloud Environment Using Fuzzy Approach

A. Sandana Karuppan, S. A. Meena Kumari and S. Sruthi

Abstract Cloud Computing has significantly changed the field of distributed computing systems today, with rapid advancements in web-based computing. As a promising utility service, cloud enables users to access distributed, scalable, virtualized hardware and/or software infrastructure over the internet. Cloud has a pay-per-use model which serves as a major advantage. Task scheduling is an essential and momentous part in a cloud environment. Scheduling mainly focuses to enhance resource utilization thereby reducing the total execution time. Several algorithms like Max-Min, Min-Min, ant colony optimization based, etc., have come into being to handle the complex issue of scheduling. Here, a Priority-Based Max-Min scheduling algorithm is proposed which aims to achieve lower makespan and maximized throughput. This algorithm schedules tasks based on the new priorities obtained from the designed fuzzy inference system, which takes user-set priority and Maximum Completion Time (MCT) of the tasks as inputs.

Keywords Cloud computing · Task scheduling · Priority · Max-min
Fuzzy logic · Mamdani

1 Introduction

Cloud computing is a buzzword that means different things to different people. However, in the view of an IT professional, cloud computing is all about providing services like hardware, platform, softwaren and much more, which can be accessed over the internet in a seamless way. It is predicted that, in 2018, at least half of IT

A. Sandana Karuppan · S. A. Meena Kumari · S. Sruthi (✉)
Department of Information Technology, SSN College of Engineering, Chennai, India
e-mail: sruthi.manian.ss@gmail.com

A. Sandana Karuppan
e-mail: sandanakaruppana@ssn.edu.in

S. A. Meena Kumari
e-mail: meena.meens.kumari36@gmail.com

S. Smys et al. (eds.), *International Conference on Computer Networks and Communication Technologies*, Lecture Notes on Data Engineering and Communications Technologies 15, https://doi.org/10.1007/978-981-10-8681-6_75

spending will be cloud based. Cloud has become the most promising technology in today's world, offering undefined storage limits for caching data of different formats. Cloud can be visualized as a parallel and distributed system consisting of a collection of interconnected and virtualized computers. The resources in cloud are stored and used dynamically based on service-level agreements (SLAs), established through negotiations between the service provider and consumer. Siri, Alexa, and Google Assistant—all are cloud-based natural language intelligent bots. These bots leverage the computing capabilities of the cloud to provide personalized context-relevant customer experiences. The next time you say, "Hey Siri!", remember that there is a cloud-based AI solution behind it. WhatsApp is also based on cloud infrastructure. All our messages and information are stored on the service provider's hardware, and this allows us to access the needed information from anywhere via the internet. Cloud computing has many advantages to offer. But few issues like limited customization, privacy issues, data leakage, lack of standards, and much more hamper the adoption of cloud technologies.

2 Literature Survey

Task scheduling is the process of allocating resources to many different tasks by an Operating System (OS). The OS handles prioritized job queues that are waiting for CPU time, and it should determine which job to be taken from which queue and the amount of time to be allocated for the job. Scheduling is one of the important research topics in distributed systems, especially in cloud computing. The heterogeneous nature of resources makes it difficult to develop an optimal solution for job scheduling. Another issue is the assigning of resources based on the user preferences and requirements.

In cloud environment, load balancing is required to distribute the workload between all nodes. It helps in allocation of resources in a promising way to achieve high user satisfaction. It helps to avoid bottlenecks. A study on max-min, min-min and improved max-min are conducted, and the results show that improved max-min produces better makespan with more reliable scheduling schema. Also, concurrent execution of tasks is ensured with high probability when compared with conventional max-min. The time complexity for improved max-min is same as that of max-min O (mn^2) [1].

Tasks scheduled using improved max-min sometimes increase the overall makespan, because the largest task is executed by the slowest resource while other tasks are concurrently executed on faster resources. A unique modification is done to the improved max-min, called the enhanced max-min, where the task with average execution time (instead of task with maximum completion time) is assigned to a resource with minimum completion time. The algorithm reduces overall makespan and properly balances load across resources [2].

Performance evaluation of min-min and max-min algorithms in federated cloud, conducted with 60 tasks assigned to a cloud with 4 resources with each resource having 1 CPU and 1 GB RAM, show that the scheduler which works on various policies is important in a cloud and it is on the basis of the scheduler that VMs are allocated to specific hosts. It is also observed that in small distributed systems, max-min scheduling algorithm achieves better makespan [3].

A single task can be divided into subtasks, and can be executed on virtual machines on different hosts or on multiple virtual machines in the same host. Handling such tasks requires the scheduling algorithms to consider the arrival time of the tasks and also the load conditions on each VM. The Improved Weighted Round Robin (IWRR) algorithm takes into account, the VM capabilities and also the length of each task, and then assigns the task to the appropriate VM. The algorithm first uses the static scheduler to allocate job requests to the participating VMs, and then by calculating the load on each of the configured VMs and their expected completion time using the dynamic scheduler, the tasks are scheduled. Response time is taken as the main Quality of Service (QoS) parameter. This algorithm is suitable for heterogeneous environments [4].

Scheduling of tasks in cloud is a NP-hard optimization problem. Min-min algorithm is applicable when small tasks are having high priority. But if both smaller and larger tasks are having high priority, it increases the makespan. To overcome this problem, the priority-based performance improved algorithm, considers user priorities of meta-tasks. The high prioritized tasks are scheduled using min-min algorithm and other tasks are scheduled using max-min algorithm. This algorithm outperforms the existing min-min algorithm in terms of makespan and resource utilization [5].

Every task is assigned with a weight based on the task length difference and considering the Quality of Service (QoS) parameters of the VM, the newer priorities were found out for the tasks. Based on the new priorities obtained, the tasks were assigned to virtual machines that were sorted based on Band Widths (BW). This proposed model outperforms the existing algorithms like FIFO and SJF in terms of processing time, processing costs, and average waiting time [6].

Apart from these, papers [7–16] were also referred for designing the proposed system.

3 Problem Description

Fuzzy logic deals with approximation instead of precise answers. In other words, it can be said that fuzzy logic deals with the degree of truth, wherein the truth value can range between completely true and completely false. The main advantage of fuzzy systems is that they deal with linguistic values instead of numbers. The usage of linguistic variables enhances human understandability of the problems. Due to this reason, fuzzy systems have found their application in many fields.

3.1 Membership Functions

Fuzziness is characterized by membership functions. Membership functions are the ones which denote the degree of truth in fuzzy logic. These are represented in graphical forms such as triangular, trapezoidal, Gaussian, bell-shaped, etc.

Mathematical notation of membership function:

A fuzzy set $\boldsymbol{A}$ in the universe of information U can be defined as a set of ordered pairs, and it can be represented mathematically as:

$$\boldsymbol{A} = \{(\boldsymbol{y}, \boldsymbol{\mu}_A(\boldsymbol{y}))|\boldsymbol{y} \in \boldsymbol{U}\} \quad (1)$$

Here, $\boldsymbol{\mu}_A$ $(\cdot)=$membership function of $\boldsymbol{A}$; this assumes values in the range from 0 to 1, i.e., $\boldsymbol{\mu}_A$ $(\cdot) \in [0, 1]$.

The membership function $\boldsymbol{\mu}_A$ $(\cdot)$ maps U to the membership space. The dot $(\cdot)$ in the membership function represents the element in a fuzzy set; whether it is discrete or continuous.

3.2 Fuzzy Inference Systems

- Fuzzy inference (reasoning) maps a given input to an output using fuzzy logic. Fuzzy Inference System (FIS) is an important unit of a fuzzy logic system which helps in decision-making. It uses the "IF-THEN" rules along with connectors "OR" or "AND" for drawing essential decision rules. The fuzzy inference systems are successfully applied in fields such as automatic control, data classification, decision analysis, expert systems, and computer vision.
- The Mamdani FIS is used and the architecture for the same is given in Fig. 1. This system was proposed in 1975 by Ebrahim Mamdani. Basically, it was anticipated to control a steam engine and boiler combination by synthesizing a set of fuzzy rules obtained from people working on the system.
- The steps of fuzzy reasoning (inference operations upon fuzzy IF-THEN rules) performed by FISs are:

For the proposed algorithm, INPUTS = MCT, User-set Priority (USP), OUTPUT = New Priority (NP).

1. Fuzzification: Comparing the input variables with the membership functions on the antecedent part to obtain the membership values of each linguistic label.
2. Combining the membership values on the premise part to get firing strength (degree of fulfillment) of each rule.
3. Generate the qualified consequents (either fuzzy or crisp) or each rule depending on the firing strength.
4. Defuzzification: Aggregate the qualified consequents to produce a crisp output.

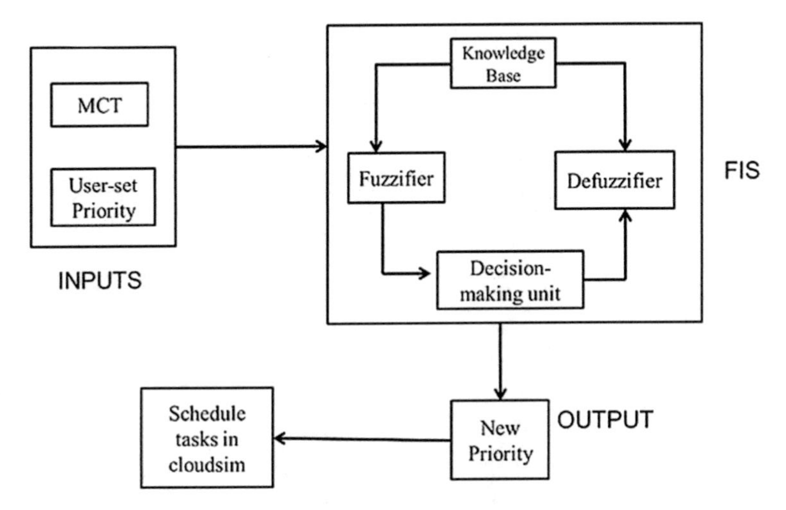

Fig. 1 Fuzzy inference system architecture for PBMM

– Centroid Defuzzification method is used. This method is also known as the center of area or the center of gravity method. Mathematically, the defuzzified output $\boldsymbol{x}^*$ will be represented as:

$$\boldsymbol{x}^* = \frac{\int \mu_A(\boldsymbol{x}).\boldsymbol{x}d\boldsymbol{x}}{\int \mu_A(\boldsymbol{x})d\boldsymbol{x}} \tag{2}$$

- The fuzzifier converts crisp input values crisp values of inputs MCT and USP to linguistic values according to their membership functions.
- The knowledge base contains IF-THEN rules which are needed for decision-making.
  For the proposed algorithm, the set of Fuzzy rules are given in Table 1.
- The decision-making unit operates with the rules from knowledge base.
- The defuzzifier converts fuzzy output value of new priority to crisp output value.

4 Proposed Algorithm

The proposed algorithm makes use of task priorities (user-set) and Maximum Completion Time (MCT) of all the tasks and computes new priority, for each of them, using a fuzzy inference system.

The max-min algorithm begins with a set of unscheduled tasks where the expected completion time of all tasks are calculated, and the task with overall maximum expected completion time is assigned to the resource with minimum overall execution time. The recently scheduled task is removed from the meta-tasks set and the steps

Table 1 Fuzzy rules for FIS in PBMM

Inputs		Output
USP	MCT	NP
Very low	Low	Low
Very low	Medium	Low
Very low	High	Very low
Low	Low	Medium
Low	Medium	Medium
Low	High	Very low
Medium	Low	High
Medium	Medium	Medium
Medium	High	Low
High	Low	Very high
High	Medium	High
High	High	Medium
Very high	Low	Very high
Very high	Medium	High
Very high	High	Medium

are repeated until the meta-tasks set become empty. This method does not guarantee utilization of the resources efficiently, and is suitable only when the number of shorter tasks is more than the larger ones.

In order to overcome the drawbacks of the max-min algorithm and to efficiently consume resources, user-set priority of the tasks is also considered.

4.1 Algorithm Workflow

1. Compute the expected completion time of all tasks in the job queue.
2. The User-Set Priorities (USP) of all the tasks is taken.
3. Taking the user-set priority and MCT of the tasks as inputs, the Mamdani fuzzy inference system calculates the new set of priorities, for the tasks, using fuzzy rules listed in Table 1.
4. The execution time of all available resources is found.
5. Schedule the tasks according to the new priorities.

Calculation of "New Priority" (NP) using Mamdani Fuzzy Inference System:

1. The inputs are measured using suitable linguistic variables. User-set priority is categorized as Very Low, Low, Medium, High, and Very High. The MCT is categorized as Low, Medium, and High.

2. Fuzzy rules are generated, using which the input fuzzy values are converted to output fuzzy value of the new priority. For PBMM, the fuzzy rules are listed in Table 1.
3. The crisp value of new priority is found using centroid defuzzification method of Mamdani inference system.

Thus, the task having the highest priority will be executed first using the resource having the Minimum Execution Time (MET). Concurrently, the other tasks are scheduled to the remaining resources thereby enhancing resource utilization and throughput.

4.2 Tools

Working and carrying out experiments on cloud is highly expensive. Due to this reason, simulation tools are developed. Simulation tools allow the user to view the simulation of the algorithms in real time by varying the parameters. Every simulator has its own pros and cons. It is essential to choose the appropriate simulator to meet user requirements. Some of them are listed below:

CloudSim, developed by GRIDS laboratory is a generalized framework, which helps to apply the experiments along with some design constraints and get the corresponding results. This tool provides classes for representing data centers, physical hosts, virtual machines, services to be executed in the data centers, users of cloud services, internal data center networks, and energy consumption of physical hosts and data centers elements. It supports dynamic insertion of simulation elements. All these factors make this tool a suitable one for building a cloud environment and testing the various scheduling and allocation policies.

CloudAnalyst, also developed at the GRIDS laboratory, is designed for some specific goals. The notable feature about this simulator is its graphical presentation of output, and the repetition of simulations to suit the user's requirements. CloudAnalyst is beneficial as it inherits the original features of CloudSim framework, extending it to simulate large-scale Internet applications. However, CloudAnalyst fails to handle prioritized traffic.

GreenCloud, though being an efficient tool, offers very less GUI support thus making it complex for the users to work with it.

Thus, after studying various simulation tools, it can be concluded that CloudSim is an ideal tool for simulation as it offers the following features:

- Support for modeling and simulation of large number of cloud computing resources.
- Flexibility to switch between space-shared and time-shared allocation.
- Provides platform for scheduling and load-balancing policies.
- Enables the user to customize the configurations.

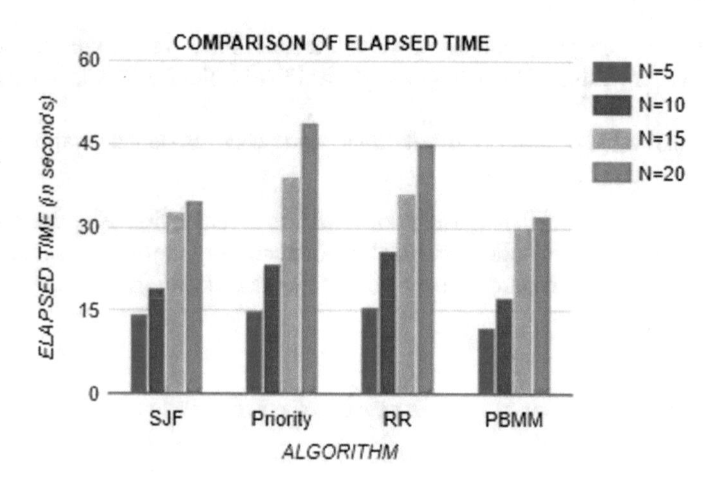

Fig. 2 Comparison graph for elapsed time

4.3 Experimental Results

In the above graph, X-axis shows the number of processes for three existing algorithms (SJF—Shortest Job First, RR—Round Robin) and also for the proposed algorithm, PBMM and the Y-axis shows the elapsed time (in seconds) for scheduling the tasks. It is proved that Priority-Based Max-Min (PBMM) takes lesser time for scheduling tasks compared to other algorithms (Fig. 2).

5 Conclusion

A Priority-Based Max-Min scheduling algorithm is proposed and the efficiency of the algorithm is compared with some of the existing scheduling algorithms. The experimental results of the proposed algorithm show that it outperforms the other scheduling algorithms in terms of execution time and usage of resources. The performance of the algorithm is to be tested with thousands of different randomly generated sets of tasks. This paper is concerned with static number of tasks and user priorities. The study can be further extended by considering the real-time dynamic tasks that are newly appearing in cloud environment.

References

1. Kaur, R., Luthra, P.: Load balancing in cloud system using max min and min min algorithm. In: National Conference on Emerging Trends in Computer Technology (2014)
2. Bhoi, U., Ramanuj, P.N.: Enhanced max-min task scheduling algorithm in cloud computing. Int. J. Appl. Innov. Eng. Manag. **2**(4), 259–264 (2013)
3. Priyadarsini, R.J.: Arockiam, L: Performance evaluation of min-min and max-min algorithms for job scheduling in federated cloud. Int. J. Comput. Appl. **99**(18), 0975 (2014)
4. Devi, D.C., Uthariaraj, V.R.: Load balancing in cloud computing environment using improved weighted round robin algorithm for non pre-emptive dependent tasks. Sci. World J. (2016)
5. Amalarethinam, D.I.G., Kavitha, S: Priority based performance improved algorithm for meta-task scheduling in cloud environment. In: Second International Conference on Computing and Communications Technologies (2017)
6. Razaque, A., Soni, N., Reddy Vennapusa: Task Scheduling in Cloud computing. In: Systems, Applications and Technology Conference (2016)
7. Kashyap, D., Viradiya, J.: A survey of various load balancing algorithms in cloud computing. Int. J. Sci. Technol. Res. **3**(11), 115 (2014)
8. Kansal, N.J., Chana, I.: Existing load balancing techniques in cloud computing: a systematic review. J. Inf. Syst. Commun. **3**(1), 87–91 (2012)
9. Dhinesh Babu, L.D., Krishna, P.V.: Honey bee behaviour inspired load balancing of tasks in cloud computing environments. Appli. Soft Comput. **13**, 2292–2303 (2013)
10. Ray, S., De Sarkar, A.: Execution analysis of load balancing algorithms in cloud computing environment. Int. J. Cloud Comput. Serv. Architect. **2**(5), 1–13 (2012)
11. Lin, C.T., Lee, C.S.G.: Neural-network-based fuzzy logic control and decision system. IEEE Trans. Comput. **40**(12), 1320 (1991)
12. Shojafar, M., Javanmardi, S., Abolfazli, S., Cordeschi, N.: A joint meta-heuristic approach to cloud job scheduling algorithm using fuzzy theory and a genetic method. Cluster Comput. **18**(2), 829–844 (2015)
13. Kushwaha, M., Gupta, S.: Various schemes of load balancing in distributed systems—a review. Int. J. Sci. Technol. Res. **4**(7), 741 (2015)
14. Kaur, D., Singh, S.: An efficient job scheduling algorithm using min-min and ant colony concept for grid computing. Int. J. Eng. Comput. Sci 3(7) (2014)
15. Kumari, R., Sharma, V.K. and Kumar, S.: Design and implementation of modified fuzzy based cpu scheduling algorithm. Int. J. Comput. Appl. 77(17) (2013)
16. Mathew, T., Sekaran, K.C., Jose, J.: Study and analysis of various task scheduling algorithms in the cloud computing environment. In: International Conference on Advances in Computing, Communications and Informatics (2014)

Mr. A. Sandana Karuppan is currently working as Assistant Professor in department of Information Technology SSN College of Engineering, Chennai. He completed his Bachelor of Engineering degree in Computer Science and Engineering from Kamaraj University, Madurai, in 2003. He completed his Master of Engineering in Computer Science and Engineering from Anna University, Chennai, India in 2006 and is pursuing his Ph.D. in Computer Science and Engineering, Anna University, Chennai, India. He has presented and published 5 technical papers in international and national conferences and 5 technical paper in international journal.

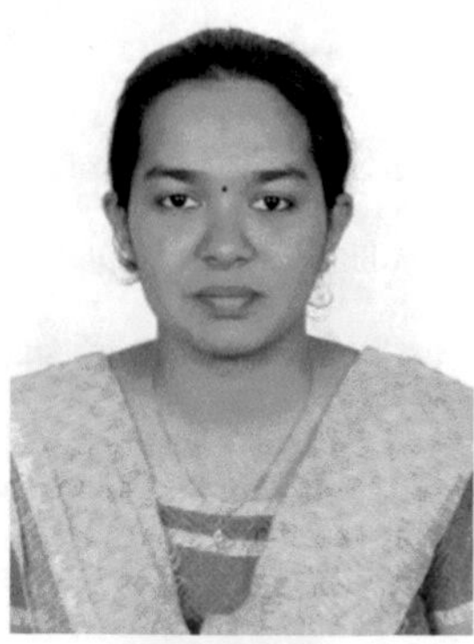

Ms. S. Sruthi has completed her Bachelor of Technology in Information Technology from SSN College of Engineering, Chennai. Her areas of interest include Cloud Computing and Analytics.

Ms. S. A. Meena Kumari has completed her Bachelor of Technology in Information Technology from SSN College of Engineering, Chennai. Her areas of interest include Cloud Computing and the Internet of Things (IoT).

Least Value Cloud Carrier Across a Handful of Cloud Vendors

U. M. Prakash, Dharun Srivathsav and Naveen Kumar

Abstract Diverse cloud advantage bearers outfit substances garage relationship with datacenters oversaw visit. These datacenters give particular get/organized latencies and unit charges for resource utilization and reservation. Consequently, then as settling on enormous CSPs' datacenters, cloud clients of essential administered programs (example—online social party) go up against two requesting occasions a technique to control bits of figuring out how to general datacenters to meet programming provider deal with objective (SLO) necessities, expansive of every datum recuperation torpidity and openness the ideal approach to managing disperse data and keep up resources in data centers having a place with boundless CSPs to compel the regard charge. To address those issues, we first frame the respect minimization trouble under SLO objectives using the whole grouping programming. Because of its NP-hardness, we by then present our heuristic alliance, sweeping at a shocking cost basically based estimations allocating the set of essentials and an impeccable asset reservation figuring. We what's more critical support three change techniques to decrease the segment charge and transporter torpidity coefficient-based completely genuinely bits of learning reallocation multicast-fundamentally chiefly based data moving ask for redirection-basically based blockage control. We, at last, adjust an establishment with permit the conduction of the estimations. Our sign driven trials on a supercomputing association and on certifiable fogs (i.e., Amazon S3, Windows Azure Storage, and Google Cloud Storage) demonstrate the sensibility of our counts for SLO guaranteed associations and client cost minimization.

U. M. Prakash (✉) · D. Srivathsav · N. Kumar
SRM Institute of Science and Technology, Chennai, Tamil Nadu, India
e-mail: Prakash.mr@srmuniv.edu.in

D. Srivathsav
e-mail: Dharun_srivathsav@srmuniv.edu.in

N. Kumar
e-mail: Naveen_balasubramanian@srmuniv.edu.in

S. Smys et al. (eds.), *International Conference on Computer Networks and Communication Technologies*, Lecture Notes on Data Engineering and Communications Technologies 15, https://doi.org/10.1007/978-981-10-8681-6_76

1 Introduction

Scattered amassing, for example, has transformed into a ceaseless undertaking advantage. A cloud ace focus (CSP) offers facts restrain get (checking Get and Put limits) utilizing its typical topographically spread datacenters. These days, a consistently growing wide variety of endeavors course their information workloads to the cloud capacity holding as a best need the quit target to spare capital impacts use of to make to and keep up the rigging foundations what is more unmistakable, hold a key partition from the multifaceted thought of adjusting to the datacenters.

2 Existing System

We name a datacenter that works a purchaser's application its customer datacenter. Likewise, concerning the exercises of a purchaser's clients, the purchaser server farm makes examine/make asking for to a capacity server farm securing the asked for sure nesses. A purchaser may additionally have several buyer server farms (implied by Dc). We make use of ci∈ Dc to exhibit the nth purchaser server farm of the customer. We make use of Ds to demonstrate all server farms which gave through all cloud suppliers and use pj to mean putting away server farm MA customer's Put/Get the request is dispatched from a buyer server ranch as far as possible server homestead of the requested data. One kind of SLO demonstrates the Get/Put restricted inaction and the level of offers fitting in with the due date [1]. Another sort of SLO guarantees the records straightforwardness as an affiliation likelihood [2] by strategies for guaranteeing a without question degree of duplicates in changing regions [3]. DAR considers the two sorts to outline its SLO and can change to both kinds decisively. This SLO chooses the due dates for the Get/Put asks for (Lg and Lp), the most outrageous genuine approve level of estimations Get/Put undertakings past the due dates (g and p), and the base extent of ages (appeared by methods for β) among restrain server farms [3]. Every single one of the estimations things from one customer has the same SLO fundamental. In the event that a purchaser has emerged strategies of substances with different SLOs, by then we treat each datum set openly in handling assurances task plan. A purchaser can in like way have a versatile due date require (e.g., arranged SLOs in different circumstances). We can without a mess of an extend oblige this need by utilizing component the entire time term to different spans in the meantime as deciding the limitations.

3 Problem Definition

A client can take part in the auction and enter the required space (e.g., specific SLOs in various periods). We can without a considerable measure of an expand oblige this fundamental by part the entire time to various periods while demonstrating the targets. For a client-server farm's Get ask for, any point of confinement server farm holding the asked for information (i.e., pantomime server farm) can serve this demand. An appropriated accumulating framework normally demonstrates the demand serving degree for each copy server farm of an information thing in the middle of charging period taken (e.g., 1 month).

4 Proposed System

We show slant task structure for getting provisioning in passed on hazes in which figuring property are made reachable fundamentally closer to customers, possibly inward switches themselves. It will be a vitality for cloud asset part estimations that work on asset obliged computational contraptions that serve constrained subsets of customers. In this blueprints with equipping database by utilizing HADOOP with SPARK, we can inquire about no repression of information and fundamental add number of machines to the social affair and we get happens with less time, high throughput and maintenance cost is less and we are utilizing joins, partitions and bucketing philosophy in Hadoop.

5 Literature Survey

The creators of [4] characterized in the distributed computing, sit out of gear assets can be incorporated and designated to clients as administration. An asset allotment component is in need to adequately dispense assets, rouse clients to join the asset pool, and maintain a strategic distance from misrepresentation among users. In this paper, we handle this issue by bringing microeconomic strategies into the asset administration and portion in the cloud condition. Blend of clump coordinating and turn around sell-off, an invert cluster coordinating sale instrument is proposed for asset allocation. Cloud transport Toolkit for Market-Oriented Cloud Computing. An Online Auction Framework for Dynamic Resource Provisioning in cloud computing. Cost-based Scheduling of Scientific Workflow Applications on Utility Grids. Double Combinatorial Auction-based Resource Allocation in Cloud Computing by Combinational utilizing of ICA and Genetic Algorithms, A Reverse Auction-Based Allocation Mechanism in the Cloud Computing Environment.

In his paper [5], Raj Kumar Buy a utilized IT ideal models promising to convey figuring as a utility. Characterizes the design for making market-situated Clouds and figuring climate by utilizing innovations, for example, virtual machines. Provide techniques that include both client driven administration and computational hazard administration to maintain SLA-arranged asset portion.

The creators of [6] have fabricated an Auction system that has as of late pulled in significant consideration as an effective way to deal with evaluating and asset designation in distributed computing. Speaks to the online combinatorial sale composed in the distributed computing worldview, which is general and sufficiently expressive to both advance framework effectiveness over the fleeting space rather than at a disconnected time point, model dynamic provisioning of heterogeneous.

The creators of [7] played out an investigation on utility figuring and have developed as another administration arrangement display. Clients expend the administrations when they have to and pay just for what they utilize. In the current past, giving utility registering administrations has been fortified by benefit situated grid figuring. It makes a foundation empowering client to devour utility administrations straightforwardly finished a protected, shared, adaptable, and standard overall system condition. Numerous grid applications, for example, bioinformatics and space science require work process handling in which assignments are executed in light of their control or information conditions. Rather, they want to utilize less expensive administrations with bringing down QoS that is adequate to meet their necessities.

The authors of [8], presented a novel pricing and resource allocation approach for batch jobs on cloud systems. The cloud supplier accordingly apportions a subset of these employments, taking into advantage the edibility of distributing assets to occupations in the cloud environment. Focusing on social welfare as the framework objective (particularly significant for private or in-house mists), we develop an asset portion calculation which gives a little estimate factor that methodologies 2 as the quantity of server's increments.

The creators of [9] led an examination that computational assets have turned out to be less expensive, more effective, and available than any other time in recent memory with the improvement of preparing and solid innovation and web success. One of the issues with distributed computing is identified with enhancing the designated assets. Asset allotment is finished with the point of limiting costs, times and because of the uniqueness of model.

6 System Architecture

The frameworks planner builds up the fundamental structure of the framework, in this we think about the whole procedure which will going ahead after customer window page. The information chose records which are split into parcels rely on the span of the document which we are transferred. In this way, from that point, the real procedure of secured encryption will go under picture, and afterward, the scrambled parcel subtle elements and customer points of interest will be put away into database for our reference purpose. The encoded information bundles will be

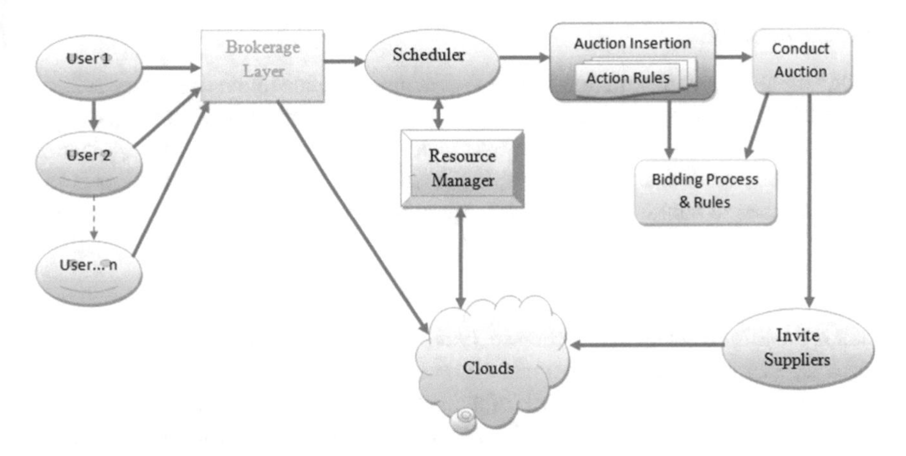

Fig. 1 System architecture

sent through system and after that, the decoding procedure will be performed and sent to collector. The collector checks the information got at customer side and stores the information forever (Fig. 1).

7 Module Description

i. User Interface Design.
ii. Quality of service.
iii. Auction-Based Scheduling.
iv. User Transaction.
v. Price Matching.
vi. Dynamic Workflow Scheduling.
vii. R-studio.

User Interface Design

In this UI plan, this is the underlying module of our venture. UI login page design is an essential part for the client to associate with login page to customer page or client page. This module has been made for client verification reason. In this login page, authorized clients can login with their legitimate certifications else they need to enlist with their points of interest like giving mail-id, address, telephone number, etc., subtle elements. Along these lines, from that point enrolled subtle elements will be put away into database and will be validated while logging time. It will check every last client data points of interest. On the off chance those points of interest does not match with database subtle elements, then it will give a blunder message and it will demonstrate the enrollment page consequently. Thus, here, we are skirting the illicit clients and giving more reconnaissance to our application.

Quality of Service

In this module, the information is given by client demands touch base at each front-end intermediary server. After the accepting the information it senses consequently to check the whether the server the aggregate number of Server. What's more, it in view of powerfully created DNS reactions, HTTP redirections, or utilizing tireless HTTP intermediaries to burrow demands. We expect that there exists an intermediary/DNS server gathered with each demand source.

Auction-Based Scheduling

This is the third module in our venture, here, the closeout-based planning procedure will happen while amid the client when a few sales have been proposed for dispersed frameworks. They displayed stack adjusting as a constrained minimization issue and exhibited a calculation that minimizes the normal culmination time of tasks. The proposed bidding calculation is depicted in view of nearsighted equilibrium strategies. They investigate discerning systems of users in a rehashed sell-off-based component in which users look for required assets by refreshing their offers. The effect of introverted specialists to cause misfortunes on the other agents taking an interest in an assignment booking instrument on related machines.

User Transaction

This is the fourth module in our undertaking, here symbolizes a unit of work performed inside a database administration framework (or comparative framework) against a database and treated in an intelligible and solid route autonomous of different exchanges. An exchange, for the most part, speaks to any adjustment in database and the client will exchange the sum to supplier.

Price Matching

This is the fifth module in our undertaking, the present cloud suppliers generally charge clients in light of a pay-as-you-go evaluating model. Regarding our multiple provider cloud demonstrate and the two considered objectives (makespan and money-related cost), CSP offers is the asset unit value, thus AI needs to get total soliciting cost from a CSP to the CSC demanded administration and match it with the CSCs offering price to locate the qualified exchange relationship among CSCs and CSPs. For VMS, CPS, DBS, and STS, the aggregate asking cost.

Dynamic Workflow Scheduling

This is the last module in our project; in this module, here, we will dispense the assets for clients which are handled after sale-based planning process. Here, we are actualizing the makespan and observing expense of the procedure which includes in powerful process. By utilizing the procedure profile of the client procedure, we will distribute the rank in light of the undertakings which are performed by the client. Here, we will be likewise presenting a period-based booking process which will include turn around sell-off instrument to offer the client for his decisions relies on proposed time and proposed cost of the sale procedure. After, the fulfillment of closeout processes they will gives us the points of interest of victor in that sale procedure which has been booked progressively.

R-Studio

Using R-Studio, we will collect the dataset from MySQL which later will be converted into Excel sheets. The dataset present in Excel sheet is taken and inserted in R-Studio for analysis. We can do various analysis according to the purpose in need. Pie Chart is used to check the percentage of winners and losers and represent them graphically. While, Bar chart is also used to find the users who opted for amount of storage for the auction.

8 Conclusion

We at first proposed a studying model and veritable structure for dynamic sorting out of a specific task in business multi-cloud condition. We used the part to book of insightful work shapes concerning change of two targets, influence cross and cash identified with cost, and demonstrated the legitimacy of the framework hypothetically. Since focus the cost of defiance in a multi-target issue is not sensible, we likely separated it and found that the made arrangements of our proposed instrument are around Pareto come full circle.

Acknowledgements This work was supported by Department of Computer Science and Engineering, School of Computing, Faculty of Engineering and Technology, SRM Institute of Science and Technology, Chennai, Tamil Nadu, India.

References

1. Hussam, A., Lonnie, P. Hakim, W.: RACS: a case for cloud storage diversity. In: Proceedings of the 1st ACM symposium on Cloud computing, pp. 229–240. Jun 2010
2. Service Level Agreements [Online]. Available: http://azure.microsoft.com/en-us/support/legal/sla/. Accessed on Jul 2015
3. Amazon S3 [Online]. Available: http://aws.amazon.com/s3/. Accessed Jul 2015
4. Wang, K., et al.: A reverse auction based allocation mechanism in the cloud computing environment. Feb 2013
5. Buyya, R.K. et al.: Cloud bus toolkit for market-oriented cloud computing. Nov 2010
6. Shi, W., et al.: An online auction framework for dynamic resource provisioning in cloud computing. Jan 2011
7. Yu, J. et al.: Cost-based scheduling of scientific workflow applications on utility grids. Accessed Jul 2005
8. Jain, N., et al.: A truthful mechanism for value-based scheduling in cloud computing, Sep 2007
9. Sabzevari, R.A., et al.: Double combinatorial auction based resource allocation in cloud computing by combinational using of ICA and genetic algorithms. Dec 2015

U. M. Prakash working in SRM Institute of Science and Technology, Chennai, Tamil Nadu, India. His research area includes cloud computing and data security.

Dharun Srivathsav working in SRM Institute of Science and Technology, Chennai, Tamil Nadu, India. His research area includes cloud computing and data security.

Naveen Kumar working in SRM Institute of Science and Technology, Chennai, Tamil Nadu, India. His research area includes cloud computing and data security.

A Hybrid Approach to Mitigate False Positive Alarms in Intrusion Detection System

Sachin and C. Rama Krishna

Abstract The aim of intrusion detection systems (IDSs) is to detect the malicious traffic and dynamic traffic which changes according to network characteristics, so intrusion detection system should be adaptive in nature. Many of IDS have been developed based on machine learning approaches. In proposed approach, experiment have been carried out on KDD-99 dataset with three classes DoS attack, other attacks and normal (without any attack). Paper checks the potential capability of optimization-based features with artificial neural network (ANN) classifier for the different types of intrusion attacks. A comparative analysis with ANN and other optimizer with ANN has been carried out. The experimental results show that the accuracy of intrusion detection using particle swarm optimization with genetic algorithm (PSO_GA) improves the results significantly by reducing false positive alarms and also improve individual class detection.

Keywords PSO · IDS · ANN · GA · KDD99 · IDPS

1 Introduction

Intrusion is an action executed to break the security of one's system and to misuse. Mainly there are two threats in any system first is malware and other is intruder. Intruder may be defined as a threat which always used to break the system and to mislead the system [1]. For the solution of intruder many researchers used to introduce a detection system termed as intrusion detection system. This paper gives a spotlight on the intrusion detection system and its working and how one can enhance its working by reducing false positive alarms. Researcher provides many techniques

Sachin (✉) · C. Rama Krishna
Department of Computer Science and Engineering, National
Institute of Technical Teachers Training and Research, Chandigarh, India
e-mail: sachin.cse@nitttrchd.ac.in

C. Rama Krishna
e-mail: rkc_97@yahoo.com

S. Smys et al. (eds.), *International Conference on Computer Networks and Communication Technologies*, Lecture Notes on Data Engineering and Communications Technologies 15, https://doi.org/10.1007/978-981-10-8681-6_77

like firewall, encryption [2] to protect the interior system from intruder or intrusion but because of its some drop out intruder make their way to affect the system without any harm to itself [2].

Now a day it is very difficult to make any system free from intrusion. The main function of IDS is to detect the unknown or abnormal activity in a particular system and to resolve this activity in very less interval of time [3]. IDS are used to protect or prevent various penetration and inner structure of computers. This system consist of several hardware and software which is used to determine unexpected events which is going to give an indication like attack is going to happen, attack is happening in your system or attack is happened in your system. These are such indication given by the IDS. It can be classified as its working type of the system like it warns before attack or it can warn while attack is in process or it warns after attack [3]. There are three components of IDS (a) sensor which is used to generate events and to sense the traffic of network or activity of the system (b) console which is used to control sensor, events and alerts and (c) detection engine which is used for the generation of alerts after receiving variation from security events. It is also used to maintain the data of sensors' events in any database.

This paper gives its contribution towards the enhancement of IDS system and its working on intrusions. KDD-99 is used in this paper as a data set having 41 features for analysis. The main goal of this paper is to attain accuracy in the working of IDS by mitigating the false positive alarms. This work represents the various processes processed for integrating IDS. Many machine learning approaches whether it is supervised or unsupervised techniques are utilized to enhance the efficiency of intrusion detection system which is explained clearly in Sect. 2. Organization of rest paper is as follows; Sect. 2 with related work, Sect. 3 gives the description about the proposed work, analysis and evaluation is explained in Sect. 4 and at the end summary of whole work is given in Sect. 5.

2 Related Work

Machine learning is a study that permits the computer or any system to learn without being programmed. There are two types of approaches which are used to enhance the working of IDS system. In supervised approach, they are capable to create the function from the given training data. There are several techniques given by several researchers. To apply any technique or approach it is important to have a study on the IDS system, its function appropriately so, [4] gives a brief study on the function of IDS which mainly focuses on four major classes of detection methods (a) classification (b) statistical (c) information theory and (d) clustering. Ahmed et al. [4] Also used to spotlight on the problems ascertain during its function. To develop the efficiency in the function of IDS [5] gives a fuzziness based semi-supervised learning approach which is termed as SSL. SSL is used to improve the performance of classifier for IDS where unlabeled samples are assisted with some other supervised learning approach. To improve the security of IDS [6] provides a deep neural network which is used to alert the system about attack and then sensor used to recognize the malicious attack. There is some other supervised approach to develop IDS like k-NN, fuzzy K-NN etc.

Ambusaidi et al. [7] Proposes a combination of IDS with LSSVM with dataset KD-99 which is also used in this paper for enhancing the intrusion detection system. Now if we talk about the un-supervised learning approach, it is a method of machine learning where a model is perfect for observations. There are number of approach like hybrid approach, domain approaches etc. Erin Liong et al. [8] Provides an un-supervised approach for learning purpose termed as DH approach which stands for deep hashing approach. Patel and Jhaveri [9] gives a study on the several machine learning techniques like ANN, SVM, Q-learning and Bayesian network to recognize the malware nodes and also recognize the misbehaving of nodes in the system. A xeromorphic cognitive approach is given by Alom and Taha [10] to detect intrusion in cyber security with the help of deep learning. These are some techniques given by researcher to enhance the IDS and make it efficiently.

3 Proposed Work

As discussed in Sect. 1 this paper works on KDD-99 data set. In this section, KDD-99 dataset with 41 features is processed for optimization, and learning to attain accuracy, precision and recall. Whole methodology is proposed into three phase, in first phase implementation of KDD-99 data set, in second phase optimization is processed on the feature and at last they allow for learning in third phase. KDD-99 dataset is a bench mark dataset and recognize by many users. For testing we use to select 10% of KDD-99 which contains 41 features from KDD-91. Where 10% KDD-dataset contains 494,021 connections and we set 311,030 connections for our work. These connections are labeled as normal or attack which classified into four categories: Denial of service which is termed as (DOS), Probe (port scanning) Unauthorized access to root person termed as U2R and last unauthorized remote login to machine [11]. There are three kind of features (1) basic feature (2) content based feature and (3) time based feature [11]. These features are labeled like $X_1, X_2 \ldots X_n$. After labeling these features are used for optimization in our work.

Optimization is done by two algorithm PSO and learning approach. Initially feature is applied on PSO for optimization to obtain fitness value. If any feature is not optimized by PSO then these un-optimized feature is allowed on learning algorithm for further optimization process. If all the features are optimized by these two algorithms, then it further moves to check convergence. If convergence is not accepted, then these feature again used for optimization by PSO and Learning algorithm. If convergence check is ok, then we move to next phase which is learning.

After labeling the dataset feature optimization is done by PSO which stands for particle swarm optimization where swarm denotes to collection of particles. In the process, PSO Particles floats through the hyper-dimensional search space. PSO is a population based search algorithm which is based on simulation on the social behavior of birds within a flock. Variation in the position of particle in a search space is depend upon the psychological tendency of each particle to imitate the development of other. In PSO swarm consists of a set of particles where each particle demonstrates a potential solution [12]. The position of particle is varying with respect to its own experience and of the neighbor particles. PSO is used to optimize the value of objective function. Every particle in space used to mobile to find the point where optimized function is obtained where 'z' is the position of particles in time 'h' having velocity 'u'. Every particle has its local and global best position in the space. Global best position is the position of a particle which is close to optimal value and all the particles move towards the global best position. The global position of particle will vary with the motion of particles. The changed position is obtained using equation [12].

$$z_r(h) = z_r(h-1) + u_r(\mathrm{h}) \tag{1}$$

$$u_r(h) = Iu_r(h-1) + L_1V_1\big(z_{pbest_r} - z_r(h)\big) + L_2V_2\big(z_{gbest} - z_r(h)\big) \tag{2}$$

where, 'I' is the weight of inertia and L_1 and L_2 are the learning factors and V_1 and V_2 are the random values. Using Eqs. 1 and 2 we can obtain fitness value of the feature after optimization. If all the features are optimized, then we proceed to further step that is convergence check otherwise we use genetic algorithm to optimize the feature which are left by PSO algorithm. In this work genetic algorithm is utilizes to obtain the optimal and near-optimal threshold for feature selection. Genetic algorithm is a technique which is used to evaluate true and approximate solution to optimization and search problem. In this paper GA is used for find the solution of optimization.

In GA, the main and initial step is to demonstrate the problem in such a manner that GA able to resolve the problem as it works on binary coding. In GA, chromosomes are used to gradually evolve with the help of biological operations. It might be possible to obtain greater feature from big data but it takes extra time and computational steps hence, we used GA for selected feature for random generation [13]. After initializing, every individual chromosome are computed by fitness function and according to this value of chromosome which is associated with fitness gives better result as compared to un-fit value. Crossover permits the search to determine the efficient way to obtain solution and optimization, and also allow chromosomal material from different parent to be combined in a single child. Crossover gives a way to introduce new information into the population [13, 14]. Here, GA uses Roulette selection for selecting best features from the feature sets. Roulette selection is similar to the game of roulette, every features get a slice of wheel but the features which is more fit gets larger slice as compare to other. In short, they are chosen in terms of their fitness value:

$$p_s = \frac{S_i}{\sum_k S_k} \tag{3}$$

After optimization with Genetic algorithm if, all the features are optimized they proceed to further step otherwise the optimization with GA is repeated. This procedure is repeated unless the entire features are optimized. After optimization every

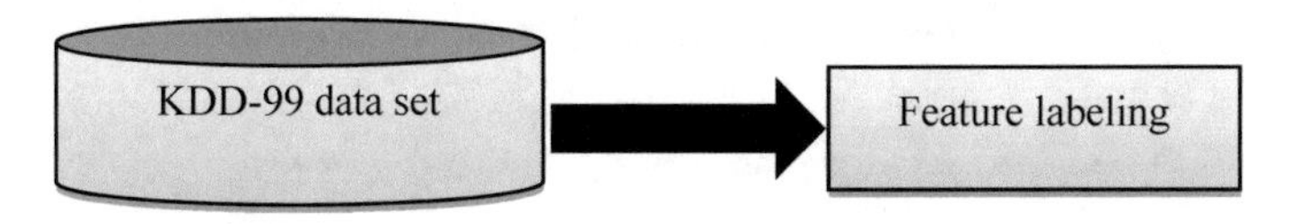

Fig. 1 Labeling of KDD-99 data set

feature convergence is tally if it is ok then our feature are allow for learning by neural networks [13, 14].

```
Genetic algorithm
Simple_Algorithm
{
Initialize the population;
Evaluate Fitness Function;
While ( The number of generation > maximum number)
        {
Selection ; // Natural selection, Survival of fittest
Crossover ; // Reproduction, Propagate favorable characteristics
Mutation; // Mutation
                Calculate fittest function;
                }
  }
```

Now, optimized feature are permitted to learn from neural networks which is our phase3. From the reference of bio-logical neurons, there is an introduction of artificial neural network which is a set of neurons similar to the neuron in human nervous systemin computer system. These neurons are used to learn patterns and relationship in the given data. In this work neuron network are used to learn the feature which is optimized from PSO and GA. NN do not require any explicit codes of any particular problems this is because of the learning rules in NN. These rules helps network to gain knowledge from the given or available dataset and implement this knowledge in problems according to requirement. NN is a network which contains number of nodes having node function related to it that is used to evaluate the output from the local parameters. This node function depends upon the variation in local parameters. Hence it is termed as information processing system where neurons are used to process the information and signals are transmitted through connection links. These links have some weight which is multiplied with respect to the incoming signals. Neural net may be of single layer or multiple layers [15] (Fig. 1).

Optimized features are converted into neurons and applied as input like N_1 and N_2 having weight M_1 and M_2. It may be single layer or multi-layer of neurons. In the input layer raw information i.e. optimized features are input in the networks. In hidden layer evaluation is done between input data and hidden layer in terms of their connecting weights. On the basis of activity of neurons in hidden layer output varies (Fig. 2).

Net weight is calculated as:

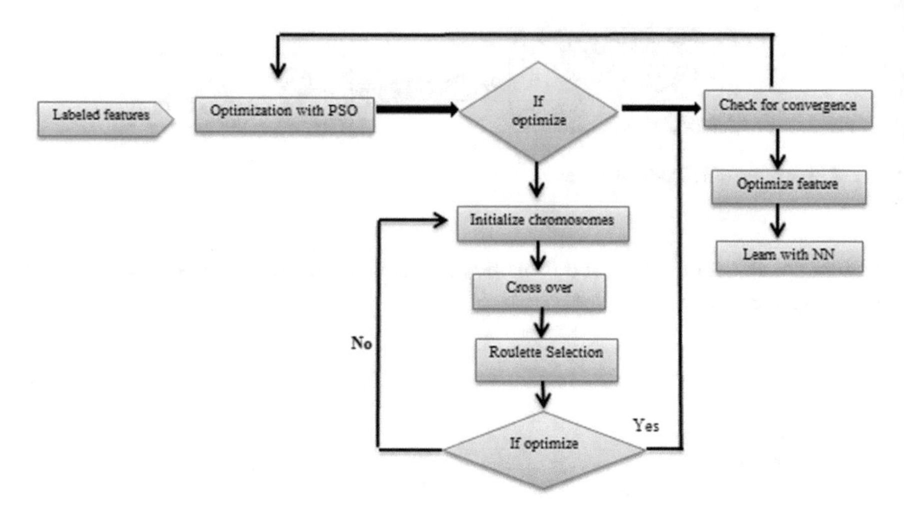

Fig. 2 Optimization process

$$\text{Net} = \text{N}_1\text{M}_1 + \text{N}_2\text{M}_2 \tag{4}$$

and can be written as

$$\text{Net input} = \sum_m N_m M_m \tag{5}$$

Algorithm

Step 1: Input the KDD-99 data set having 41 features and label all the features.
 	
Step 2: Optimize these features with PSO to obtain fitness value using equation1 and 2.
 	
Step 3: If we obtain optimized value by PSO then proceed further for convergence check otherwise apply genetic algorithm for optimization and uses unless we obtain optimized features. After obtaining optimized feature use to check convergence of whole feature set.
Rolette selection is used for feature selection in GA having equation 4.
Step 4: Check convergence of entire feature set.
If (convergence = yes)
{
Learn the feature with neural networks and evaluate accuracy, precision, F-measure and recall.
Else
{
Repeat step 3
}

Table 1 Evaluation of the parameters among ANN, ANN with GA, ANN with PSO and ANN with GA_PSO

	ANN	ANN with GA	ANN with PSO	ANN with GA_PSO
Accuracy	89	92.23	92.34	94.23
Precision	88	89.23	90.32	92.33
Recall	87	88.56	91.26	96.33
F-measure	85	88.25	86.23	91.13

Table 2 The attack type from KDD CUP 99 dataset

Normal	Dos	R2L	U2R	Probe
Normal	Smurf Processtable Pod Land	PHF Xlock Send-mail Guess_password	Root-kit Eject Perl Buffer overflow	Portsweep Satan Saint M-scan

4 Description of Dataset

As discussed above experiments are executed using KDD-99 having 41 feature sets. These features are used for optimization and then learning and now they are use to analyze in terms of attack. In this work, we use to evaluate the accuracy rate in an intrusion detection system. In the analysis we take data on the basis of number of intrusions. Attacks generally fall into four categories (1) Dos, (2) Probe, (3) R2L (4) U2R. In our analysis we uses three categories (1) other attack which consist of probe, R2L and U2R (2) DoS-attack, and (3) Normal attacks (non-attacks). In this work we evaluate the accuracy, precision, recall, and F-measure in various cases:

Case 1: Evaluation of accuracy, precision, f-measure and recall is given by ANN individually, ANN with GA, ANN with PSO and ANN with combined GA_PSO which is represented in Table 1. In this case we evaluate the efficiency of IDS by applying ANN individually or with GA and PSO or by hybrid of both algorithms with ANN.

Case 2: In this case, we evaluate the efficiency for single ANN on three attack condition (a) other attack consist of probe, R2L and U2R, (b) Dos attack, and (c) Normal or non-attacks condition. Similarly we evaluate the efficiency for ANN with PSO, ANN with GA and at last ANN with both GA_PSO. Here we evaluation efficiency of algorithms in terms of accuracy, precision, recall and F-measure. If we spot some light on attacks we are considered. Dos attack which stands for denial of service attack in Dos attack hidden attacks is done by user which is shown in the system. This type of attack may be done by single intruder or a group of intruders. It makes the system unavailable to its real user. Probe attack is a kind of attack where intruder used to break the security by trial methods. R2L attack stands for remote to user attack. And at last U2R attacks it is the type of attack where intruder starts on the system as a normal user and spoil all the activities of the systems [16] (Fig. 3).

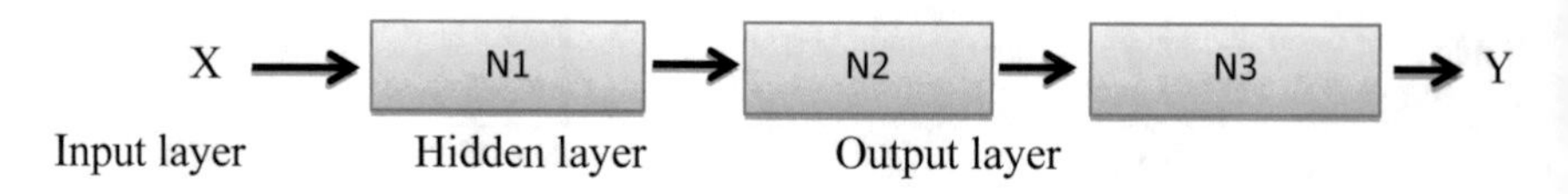

Fig. 3 Layer of MNN

Table 3 The static data to analyze the efficiency of the approaches for above discussed attack

Algorithm type	Types of attack	Accuracy	Precision	Recall	F-measure
(a) ANN	Other attack	86.23	87.23	86	83
	Dos attack	87.23	88.33	88	84
	Normal attack	91.23	90.13	87	86
(b) ANN with GA	Other attack	91.13	87.23	86.75	87.23
	Dos attack	90.23	90.13	84.23	86.23
	Normal attack	94.11	89.99	94.23	89.13
(c) ANN with PSO	Other attack	93.13	89.23	92.23	84.23
	Dos attack	92.13	88.13	89.13	83.23
	Normal attack	90.13	84.23	92.13	87.34
(d) ANN with GA and PSO	Other attack	95.62	90.23	92.33	89.23
	Dos attack	89.23	89.23	93.33	90.13
	Normal attack	96.23	93.13	99.23	90.23

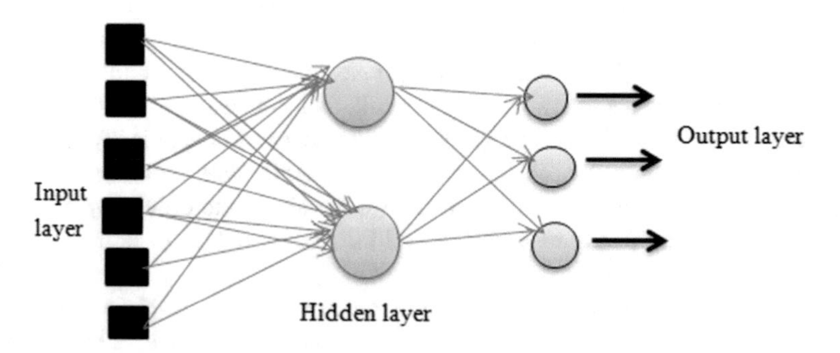

Fig. 4 Various layers in neural networks

4.1 Experimental Results

In this section we analyze the statistical data by simulation. Graphical result of Tables 1 and 3 is given by simulation process (Fig. 4; Table 2).

Figure 5 shows the simulated analysis of Table 1 in terms of accuracy, precision, recall, and F-measure. In this figure analysis on efficiency is demonstrated from all the

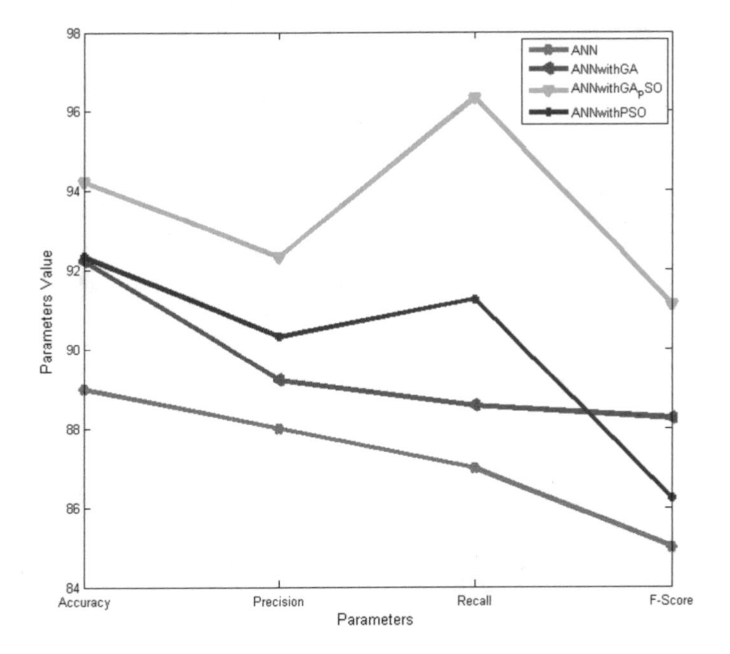

Fig. 5 Simulated graph of Table 1

four algorithm that are ANN represented by green line, ANN with PSO represented by purple line, ANN with GA represented by red line and ANN with both PSO and GA represented by blue line. Analysis demonstrates that ANN with both SPO and GA gives better result in terms of all the four parameters (accuracy, precision, recall, and F-measure).

Observation 1: In Fig. 5 Parameters analysis of different classifier and proposed approach has been shown. In analysis, parameters like precision, recall, accuracy and F-measure vary according to classifier but one analysis about proposed approach (PSO with GA in neural network) is very clear that it shows significant improvement in all parameters. If analysis have to be made only over proposed approach then recall parameter have shown significant improvement then other parameters. So it gives clear indication of reducing false negative rate and attacks identification is effective in proposed approach because of optimize weight given by PSO_GA approach.

Figures 6 and 7 gives the analytic result in terms of accuracy (represented by black line), recall (represented by red line), precision (represented by blue line), and F-measure (represented by green line) of two approaches that are ANN and ANN with GA for the parameters (Dos attack, other attack, and Non or normal attack).The entire four graphs demonstrate the better efficiency of the algorithm ANN with PSO_GA.

Observation 2: In Fig. 6, depth analysis of all three classes in ANN and ANN_GA has been shown. In this analysis, we have tried to show what is the significance of our approach. We continue this discussion in observation 3 also. So, at first point if analysis is done over normal class and on other class which not any attack

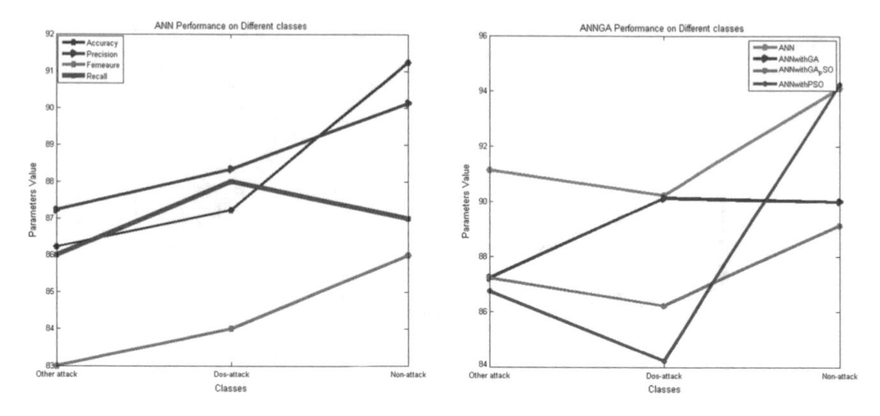

Fig. 6 Analysis with ANN and ANN _GA

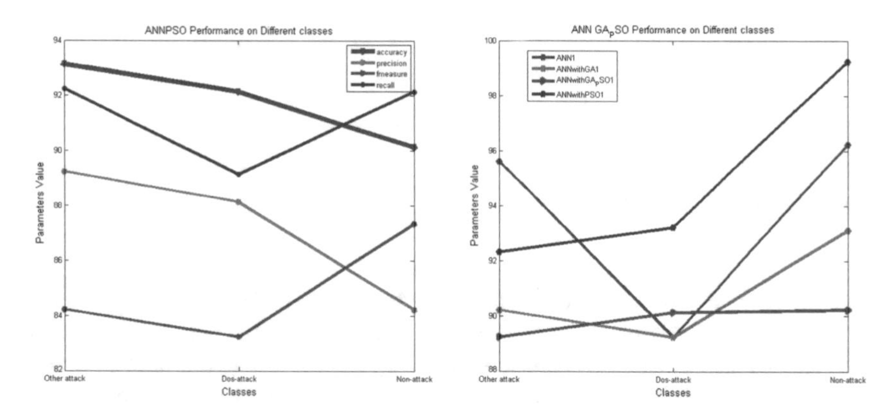

Fig. 7 Analysis with ANN _PSO and ANN with PSO_GA

in both cases ANN and ANN with GA performed well compared to other parameters like precision, recall, and F-measure but ANN_GA still has better accuracy than ANN, so feature weighted by optimization somehow performs well because of reducing overlapping information learning. If analysis is done through DOS attack only it also shows higher accuracy in ANN with GA, so we can conclude Feature optimize weight is better approach. So how it is going to improve optimization that have been discussed in next part.

At last from the whole analysis it can be concluded that algorithm ANN with PSO_GA gives better result for all the attacks we examined in our work.

Observation 3: In Fig. 7, analysis continues from observation 2 and tried to find out the significance of optimization and improvement in detection of different classes by classification. If analysis have to be made then both graph shows the effective recall but only for normal class, which reduces the false positive rate. This improvement can be seen with all classes like DOS attack and other attacks. But effective results have been seen in the in proposed approach where the improved detection have been done for other attacks too. So PSO optimization is effective but PSO with GA i.e. the proposed approach gives more improved results in other attack and normal class.

5 Conclusion

This paper investigates the optimization weight of feature and how to reduce the statistically overlapping between features and improving the optimization by hybridization is worthy or not. In the proposed approach optimization base features with artificial neural network has been used and experiment shows that this hybrid approach not only improves cumulative accuracy, precision, recall, and F-score but also improves all individual parameters among three classes (i.e. DOS attack, Normal and other attacks).

References

1. Ashfaq, R.A.R., et al.: Fuzziness based semi-supervised learning approach for intrusion detection system. Inf. Sci. **378**, 484–497 (2017)
2. Lin, W.C., Ke, S.W., Tsai, C.F.: CANN: An intrusion detection system based on combining cluster centers and nearest neighbors. Knowl. Based Syst. **78**, 13–21 (2015)
3. Harendra, V., Mirza, S., Mali, N.: Intrusion detection system. Int. J. Adv. Res. Eng. Sci. Technol. **3** (2016)
4. Ahmed, M., Mahmood, A.N., Jiankun, H.: A survey of network anomaly detection techniques. J. Netw. Comput. Appl. **60**, 19–31 (2016)
5. Ashfaq, R.A.R., et al.: Fuzziness based semi-supervised learning approach for intrusion detection system. Inf. Sci. **378**: 484–497 (2017)
6. Kang, M.J., Kang, J.W.: Intrusion detection system using deep neural network for in-vehicle network security. PLoS ONE **11**(6), e0155781 (2016)
7. Ambusaidi, M.A., et al.: Building an intrusion detection system using a filter-based feature selection algorithm. IEEE Trans. Comput. **65**(10): 2986–2998 (2016)
8. Erin Liong, V., et al.: Deep hashing for compact binary codes learning. In: Proceedings of the IEEE Conference on Computer Vision and Pattern Recognition, pp. 2475–2483 (2015)
9. Patel, N.J., Jhaveri, R.H.: Detecting packet dropping nodes using machine learning techniques in Mobile ad-hoc network: A survey. In: International Conference on 2015 Signal Processing and Communication Engineering Systems (SPACES), pp. 468–472. IEEE (2015)
10. Alom, MZ., Taha, T.M.: Network intrusion detection for cyber security on neuromorphic computing system. In: International Joint Conference on 2017 Neural Networks (IJCNN), IEEE (2017)
11. Tavallaee, M., et al.: A detailed analysis of the KDD CUP 99 data set. In: IEEE Symposium on 2009 Computational Intelligence for Security and Defense Applications, CISDA 2009. IEEE (2009)
12. Sierra, M.R., Coello, C.A.C.: Improving PSO-based multi-objective optimization using crowding, mutation and e-dominance. In: Evolutionary Multi-criterion Optimization, vol. 3410, Springer, Berlin, Germany (2005)
13. Kim, Kyoung-jae, Han, Ingoo: Genetic algorithms approach to feature discretization in artificial neural networks for the prediction of stock price index. Expert Syst. Appl. **19**(2), 125–132 (2000)

14. Lipowski, A., Lipowska, D.: Roulette-wheel selection via stochastic acceptance. Physica A: Stat. Mech. Appl. **391**(6), 2193–2196 (2012)
15. Vassiliadis, S., et al.: Artificial neural networks and their applications in the engineering of fabrics. In: Woven Fabric Engineering, InTech (2010)
16. Paliwal, S., Gupta, R.: Denial-of-service, probing and remote to user (R2L) attack detection using genetic algorithm. Int. J. Comput. Appl. **60**(19), 57–62 (2012)

Sachin studying in Department of Computer Science and Engineering, in National Institute of Technical Teachers Training and Research, Chandigarh, India. His research area includes denial of service attacks and network security.

C. Rama Krishna working as a professor in Department of Computer Science and Engineering, in National Institute of Technical Teachers Training and Research, Chandigarh, India. His research area includes denial of service attacks and network security.

Reliability Evaluation of Wireless Sensor Networks Using EERN Algorithm

M. S. Nidhya and R. Chinnaiyan

Abstract Wireless sensor network senses the environment where the human access is limited. Sensor nodes will collect the information and pass it to the sink node. When passing the information, malicious node or compromised node will corrupt it and forward it to the sink node. In the receiver side, a corrupted message will create a problem. The proper information will not be conveyed to the sink node. The presence of compromised node will generate incorrect report and it affects the event detection. In this paper, we proposed an EERN algorithm to provide reliability for the entire wireless sensor network against compromised node. To provide reliability and to save time, we introduced vigilant node and aggregator node. Aggregator node will detect the suspicious node. Vigilant node will identify the suspicious node as malicious, compromised or faulty node.

Keywords Wireless sensor network · Reliability · Vigilant · Malicious
Compromised node · Sink node

1 Introduction

Wireless sensor networks (WSNs) are the most essential technologies in material science and it has the ability to attain information in tough locations [1–4], and they are broadly used in military, farming, manufacturing and so on [5–7]. In WSNs, there is a one or limited sink nodes and many sensor nodes. Sensor nodes monitor the tough location and transmit messages, and sink nodes are responsible for gathering messages from all sensor nodes [1, 2]. A sink node will be installed in a specified

M. S. Nidhya
Department of Computer Science, VISTAS, Chennai, Tamil Nadu, India
e-mail: nidhyaphd@gmail.com

R. Chinnaiyan (✉)
Department of Master of Computer Applications, New Horizon College of Engineering, Bengaluru, Karnataka, India
e-mail: vijayachinns@gmail.com

S. Smys et al. (eds.), *International Conference on Computer Networks and Communication Technologies*, Lecture Notes on Data Engineering and Communications Technologies 15, https://doi.org/10.1007/978-981-10-8681-6_78

spot, and it has unlimited energy and prevailing calculation and transmission capabilities; sensor nodes are randomly positioned in the target area, and their energy and calculation capacities are limited by their volume.

In our proposed method, the entire wireless sensor network is divided into three groups. The first group is collecting the information from the environment. The second group will collect the information from primary sensors, and then it will forward it to the sink node. The third one is the sink node. Malicious node or compromised nodes may be present between the sensor nodes. Generally, malicious nodes are detected by two ways: (1) Signal strength and (2) Trust value, which is given by neighbouring node.

Several papers suggest that signal strength is sufficient to find out the malicious node. But the problem is normal sensor node, which will be compromised by the malicious node. So this type of the compromised node will be find out by signal strength.

Because signal strength of a compromised node is same as other normal nodes signal. Some situation will occur like, if the normal node is faulty. It may send wrong information that may create a suspicious among the neighbour. Generally, a suspicious message will be sent by compromised node, malicious node and faulty node. So, we have differentiated these nodes and provide reliability to the entire wireless sensor network. The trust value given by the neighbouring node is based on the behaviour of a node. If a node sends the information properly to the sink node and forward it to the packet of the next node that node will get a good trust among the other nodes in the network.

In some instances, a normal node will send a suspicious message due to fault. Malicious node will send the correct message to get good trust count from neighbours. The existing methods [8] use RSS to detect malicious node, and neighbour's trust to detect compromised node. Malicious node will attract the traffic from other nodes. It will create illusion among the nodes by advertising itself as a close node to the sink. Faulty node will send a suspicious message.

This paper will provide reliability to a wireless sensor network against malicious node, compromised node and remove the faulty node by introducing vigilant node and aggregator node. Finally, the results are compared with the existing approaches.

2 Related Work

Yang et al. [9] proposed to find the spoofing attack by received signal strength (RSS) which is derived from wireless nodes. It defines multiple attackers problem. To determine the number of attackers, cluster-based mechanism is developed [10]. Compromised node acted as a sinkhole which can be found out by the trust model evaluation [11]. Faulty node is detected and the shortest reliable path is created to forward a message from sensor node to sink node. Nidhya and Chinnaiyan [12] proposed energy factor and time delay factor calculation are used to eliminate dead node and time delay node in wireless sensor network. Chinnaiyan and Somasundaram [13] proposed task-based reliability modelling by defining reliability for a task but it a system reliability will not be defined. Chinnaiyan and Somasundaram [14] proposed reliability by a communication variable for object-oriented software systems.

3 Reliability Model

A reliability model for wireless sensor network systems represents a clear picture of the data functional interdependencies by providing a means to trade-off in WSN design alternatives, and to identify areas for WSN design improvement of databases. The reliability models for wireless sensor networks are also helpful in:

(i) Identification of critical items and single points of failure of data.
(ii) Allocating reliability goals to portions of the design of wireless sensor networks.
(iii) Providing a framework for comparing the estimated reliability of WSN methods.
(iv) Trading-off alternative fault tolerance approaches for wireless sensor network approaches.

Reliability of systems is presented by Chinnaiyan and Somasundaram [13–18]. The authors evaluated the reliability of systems with novel methods and ensured that the reliability models provides optimized reliability results. The same can be applied to all the wireless sensor network models for providing reliable and optimized results.

4 Aggregator Node (A-Node)

Aggregator node will collect the information from the sensor nodes in the network. In a network, number of aggregator nodes will be placed based on the population of the sensor node. Aggregator node (A-node) collects the information and passes it to the sink node. Before forwarding the information, it will check the output of the sensors. If the output value of the sensor node is different from the remaining sensor node, it will check the trust of the sensor. The trust value of a suspicious node is high, and node id will be passed to the vigilant node. If the trust value is low that particular node information will not be taken for a final decision. Low trust value for suspicious node is also investigated by the vigilant node. Aggregator node will not accept the message from the suspicious node until the vigilant node publishes the final report.

Suspicious information sent by the nodes will be found out by the aggregator. During aggregation, one or two nodes will give fake information, and it will be identified by aggregator node.

$$A = \sum_{s=1}^{n} T_n \times O_n \quad (1)$$

T_n Weight or trust of sensors.
O_n Output of the sensor.
A Aggregation.

where A is the aggregation result and T_n is the trust ranging from 0 to 1. An essential concern is about the definition of sensor node's output O_n. In practice, the output information O_n may be 'yes-1' or 'no-0' information or continues numbers such as temperature reading. Thus, the definition of output O_n is usually depending on the application where the sensor network is used. The following issue is to update the weight of each sensor node based on the correctness of information reported.

Updating the trust of each sensor node has two purposes. First, if a sensor node is compromised (becomes a malicious node) and frequently sends its report inconsistent with the final decision, its trust is likely to be decreased.

if $O_n \neq A$ then Check the trust value of the node. If the trust of the node is high, node id is sent to vigilant node. Or if the trust is low, then output given by that node will not be accepted for final decision. O_n-output of the node, later the vigilant node is free it will take a necessary step against this node.

Aggregator will check the information passed by the sensors. If one sensor sends a suspicious message, then that node id will be passed on to vigilant node. If more than one sensor sends a wrong message, then aggregator will send D_n to the vigilant node. V node will use the following formula to find out the ratio of compromised node.

$$D_n = c/s \quad (2)$$

D_n Number of sensors which is sends different message.
C Suspicious sensors.
s Total no. of sensor in the control of aggregator.

5 Vigilant Node

Vigilant node is placed in between sink node and aggregator node. It has limitless energy and powerful calculation and transmission capabilities. It will not participate in the transmission of message by primary sensors. It always listens to enquiry send by the aggregator node (Fig. 1).

Vigilant node got a node id from A-node, and it will check the last ten trust values of suspicious node. Decreasing order of trust value shows that the node is going to dead state or faulty. Varying order of trust value shows that the suspicious node is compromised node. Compromised node high trust value will be decremented by vigilant node. Then, other nodes in the network will omit the compromised node.

Fig. 1 Vigilant node and aggregator node in WSN

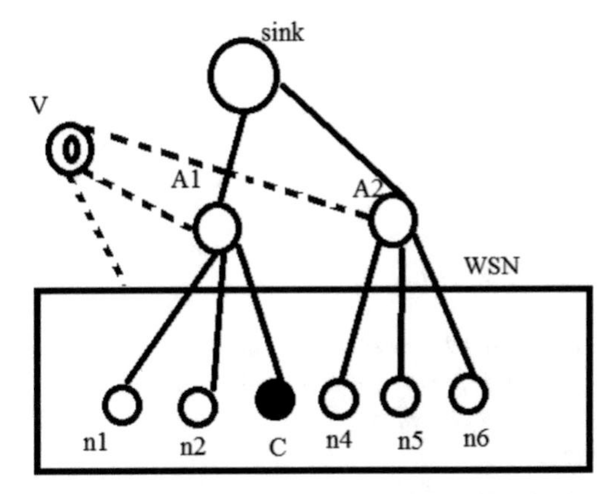

Vigilant node will differentiate the compromised node and faulty node. If the faulty node is found out by the vigilant, it will take necessary steps to repair it. It will disconnect or isolate the particular node from the network until the repair is over.

During detection of compromised or faulty, it should be checked that the weightage given for the particular sensor is high, normal or low. For the final decision, the particular sensor information is important or not. In that history, it retrieves the particular node information based on the enquiry sent by aggregator node. It will check whether the node maintained the good trust from all other nodes. That is, if the trust is greater than the threshold or equal, it will not be found out by the other node.

$$T_n = \begin{cases} T_n - \theta \times D_n & \text{If } O_n \neq A \\ T_n & \text{Otherwise} \end{cases} \tag{3}$$

where T_n is the trust of the node. Here, the aggregator node will send the node id of the suspicious node. The vigilant node once again will check the trust of the node with the aggregation value, if it is not equal, then trust of the suspicious node will be subtracted with the forfeit ratio θ.

6 Enhanced Way of Evaluating Reliable Nodes (EERN) Algorithm

Aggregator node (Algorithm 1)

For to send a message from Anode to sink node
If $O_n \neq$ *A.node then*
Compute D_n

For all node in D_n *do*
If $P(D_n)$*=high then*
Pass node.id to V.node
Else Buffer=node-id
End For.
End.

V Node (Algorithm 2)

Retrieve last 10 events trust of node.id[e]
If T(node.id[e]) == varying then
Node.id=compromised
Compute T_n
Else if T(node.id[e])=decreasing then
Node.id=Fault
else
Node.id=normal
End.
End.

6.1 Comparison of Existing Model with EERN

In the existing model, signal strength and trust model is used to calculate the reliability but it will not be identified, if a node is compromised. It is assumed that node n3 is a compromised node, but it maintains good trust. In Fig. 2, node n3 cannot be identified by the existing model [8]. The existing model can identify a malicious node because signal strength of node n3 will be high when compared with other nodes. If node n3 is compromised, it cannot be identified by the existing model.

Suspicious message send by node n3 is detected by aggregator node. A1 node will send the node id to the vigilant node. Vigilant node retrieves the history of trust. In that history, if the suspected node's trust level varies inconsistently means it is confirmed as a compromised node. Node n3 maintained good trust, and its signal strength is also normal. So, the existing model will not be able to find compromised node. Using an EERN algorithm, it is detected in the Fig. 3 that the varying trust level of node n3 is a compromised node.

7 Conclusion

In this paper, we proposed an EERN algorithm to provide reliability against compromised node. It will evaluate the reliability and identify suspicious node as a compromised node. A node with good trust and normal signal strength but acted

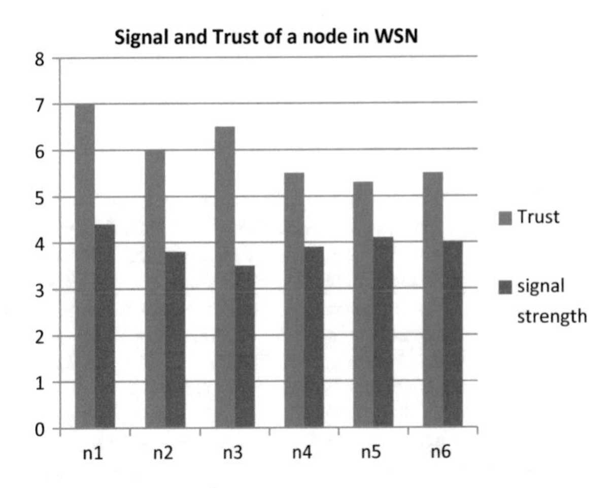

Fig. 2 Compromised node cannot be identified by signal strength and trust

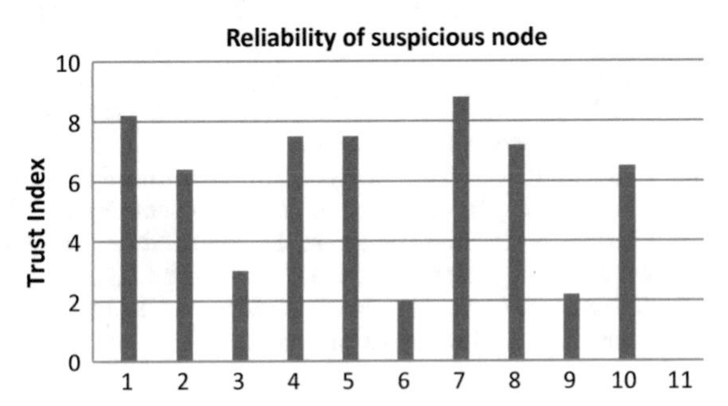

Fig. 3 Trust index of a suspicious node n3 maintaining good trust is identified as a compromised node

as a suspicious node will not be detected by Received Signal Strength (RSS) and localization methods. But our proposed algorithm and vigilant node will detect the suspicious node as compromised node or faulty node. Based on the trust history, *V* node will detect the highly trusted compromised node, and it decrements the trust value. It also found out that the faulty node in the network takes necessary steps to resolve it. In future, the EERN algorithm will work in flat Wireless Sensor Network.

References

1. Akyildiz, I.F., Su, W., Sankarasubramaniam, Y., et al.: Wireless sensor networks: a survey. Comput. Netw. **38**(4), 393–422 (2002)
2. Yick, J., Mukherjee, B., Ghosal, D.: Wireless sensor network survey. Comput. Netw. **52**(12), 2292–2330 (2008)
3. Van der Werff, T.J.: 10 emerging technologies that will change the world. MIT's Technol. Rev. **32**(2), 33–49 (2003)
4. Misra, S., Reisslein, M., Xue, G.: A survey of multimedia streaming in wireless sensor networks. IEEE Commun. Surv. Tutor. **10**(4), 18–39 (2008)

5. Rehman, A., Abbasi, A.Z., Islam, N., et al.: A review of wireless sensors and networks' applications in agriculture. Comput. Stand. Interfaces **36**(2), 263–270 (2014)
6. Oliveira, L.M., Rodrigues, J.J.: Wireless sensor networks: a survey on environmental monitoring. J. Commun. **6**(2), 143–155 (2011)
7. Alemdar, H., Ersoy, C.: Wireless sensor networks for healthcare: a survey. Comput. Netw. **54**(15), 2688–2710 (2010)
8. Yang, J., Chen, Y., Trappe, W., Cheng, J.: Detection and localization of multiple spoofing attackers in wireless networks. IEEE Trans. Parallel Distrib. Syst. **24**(1), 44–57 (2013)
9. Yang, J., Chen, Y., Trappe, W., Cheng, J.: Detection and localization of multiple spoofing attackers in wireless networks. IEEE Trans. Parallel Distrib. Syst. **24**(1), 1045–9219 (2013)
10. Nidhya, M.S., Chinnaiyan, R.: Improving the reliability of data transfer against sinkhole in wireless sensor networks—a review. Int. J. Appl. Eng. Res. (IJAER) **10**(82), 122–125 (2015)
11. Nidhya, M.S., Chinnaiyan, R.: Shortest reliable path for wireless sensor network. Int. J. Pure Appl. Math. (IJPAM) **117**(20), 105–108 (2017)
12. Nidhya, M.S., Chinnaiyan, R.: Conniving energy and time delay factor to model reliability for wireless sensor networks. IEEE xplore Digital Library (2017)
13. Chinnaiyan, R., Somasundaram, S.: Evaluating the reliability of component based software systems. Int. J. Qual. Reliab. Manag. **27**(1), 78–88 (2010)
14. Chinnaiyan, R., Somasundaram, S.: Reliability of object oriented software systems using communication variables—a review. Int. J. Softw. Eng. **2**(2), 87–96 (2009)
15. Chinnaiyan, R., Somasundaram, S.: Reliability assessment of component based software systems using test suite—a review. J. Comput. Appl. **1**(4), 34–37 (2008)
16. Chinnaiyan, R., Somasundaram, S.: An experimental study on reliability estimation of GNU compiler components—a review. Int. J. Comput. Appl. (0975 – 8887) **25**(3) (2011)
17. Chinnaiyan, R., Somasundaram, S.: Reliability of component based software with similar software components—a review. i-Manager's J Softw. Eng. **5**(2), 44 (2010)
18. Chinnaiyan, R., Somasundaram, S.: Monte carlo simulation for reliability assessment of component based software systems. i-Manager's J. Softw. Eng. **3**(4), 27 (2010)

M. S. Nidhya is working as Associate Professor in the School of Computing, Vels University, Chennai, Tamil Nadu, INDIA. She is having 16+ years of teaching experience. She is doing her research in Bharathiyar University-Coimbatore since 2014. Her research interest includes Security, Reliability of Wireless Sensor Networks, Machine Learning, Big Data, Cloud Computing nd Data Science.

Dr. R. Chinnaiyan is working as Professor in the Department of Computer Applications, New Horizon College of Engineering, Marathalli, Bangalore, Karnataka, INDIA. He is having 17+ years of teaching experience. He is a life member of ISTE and CSI of INDIA. He completed his research in Anna University- Chennai at Coimbatore Institute of Technology, Coimbatore in 2012. His research interest includes Object Oriented Analysis and Design, Qos and Software Reliability.

Secure Data Concealing Using Diamond Encoding Method

A. Christie Aiswarya and A. Sriram

Abstract In recent years, due to the fastest development in digital data exchange, the security of information becomes highly important for storing the data and its transmission process. The data size of text and image are different. So, for providing security to both text and image involves various methods. Choosing the best method to encrypt and decrypt the data also plays a crucial role. Traditional methods may require more time to encrypt and decrypt. Since the compilation time for one method differs from the other. The proposed model provides security for both text and image by means of embedding encrypted text into an encrypted image, and the same can be decrypted to view the original text and image. The algorithms used in the proposed model are chaotic algorithm for text and AES algorithm is used for image encryption and decryption process. Encrypted text is embedded in encrypted image using diamond encoding data hiding strategy and the same is decrypted to get the original text and image.

Keywords AES · Chaotic · DEMD · Decryption · Encryption · MATLAB

1 Introduction

In today's world with the usage of smart devices such as computers, mobiles, tablets etc., are used for data storage and transmission which in turn, increases the number of users. Whenever a data transmission occurs, the intruder tries to read or modify the information. The issue arises, here, with data security is due to unauthorized access to data. Since our surroundings are digitalized, there is serious demand for user's privacy and security in many applications. This problem can be solved by storing or

A. Christie Aiswarya (✉)
Department of EST, SRM Institute of Science and Technology, Kattankulathur, India
e-mail: christie.alury@gmail.com

A. Sriram
Department of ECE, SRM Institute of Science and Technology, Kattankulathur, India
e-mail: aramsri147@gmail.com

S. Smys et al. (eds.), *International Conference on Computer Networks and Communication Technologies*, Lecture Notes on Data Engineering and Communications Technologies 15, https://doi.org/10.1007/978-981-10-8681-6_79

transmitting the data in an encrypted format [1–3]. Which means before transmitting the data will be modified using some mathematical algorithms and a key and then the digital data is transformed into cipher code, and to get the original data at the receiving end, decryption has to be performed using mathematical algorithms, key, and cipher code [1, 4]. Though there are many traditional encryption algorithms, the same cannot be used for text and image. Since the size of the text and image are not same. The compilation time differs in both the cases. In the case of encrypting text, both the decrypted text and the original text should be one and the same. But for images, this condition may not be true because a decrypted image with some distortion may be acceptable.

In the proposed model, three different techniques are used at the encryption and decryption process. Image encryption is done by AES algorithm and then, text encryption is followed by chaotic algorithm. Encrypted text is inserted into the encrypted image using diamond encoding method [1, 2, 5].

2 Conventional Method

The five important elements in any cryptosystems is plaintext, encryption algorithm, secret key, ciphertext, and decryption [6] as shown in Fig. 1.

- Plaintext: The data which has to be modified is fed as input to the encryption algorithm.
- Encryption: Many iterations and transformations will be performed on the plaintext to get encrypted data.

Fig. 1 Block diagram

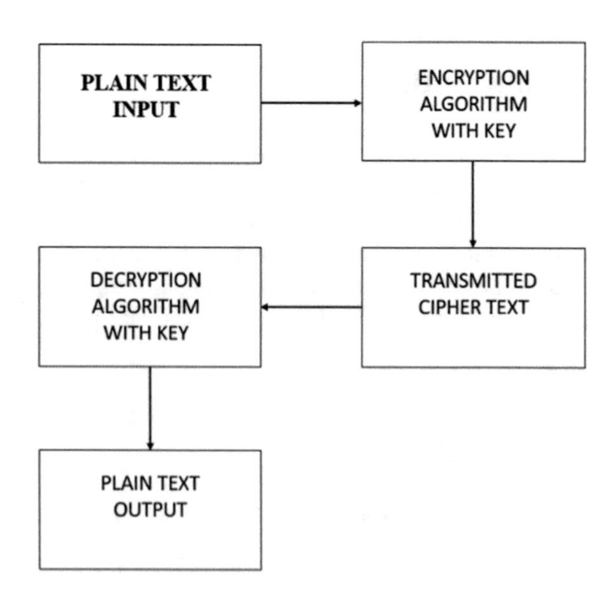

- Secret key: The key is different from the plain text. Depending on the key, output will be varied accordingly.
- Ciphertext: It depends on the secret key which is considered as output and seems to be scrambled.

Decryption: This process is similar to the encryption but the iteration takes place in reverse order. Ciphertext and secret key are given as input to decryption process for obtaining original plain text.

3 AES Algorithm

This algorithm uses a single key for encryption and decryption. Data is fixed which is of 128 bits but the length of the key can be 128,192, or 256 bits. Since it is an iterative algorithm, a number of iterations depends on the key. The 128-bit data is divided into 1 bytes, and is mapped to $4 * 4$ array. The suboperations of the algorithm will be performed on the array [1, 4].

AES decryption is not identical to encryption procedure. Since the steps to be followed to decrypt is the reverse of encryption. But we can define an equivalent inverse cipher with steps as for encryption using inverse of each step with a different key schedule [6].

Sub Bytes: A substitution box is created using multiplicative inverse and affine transformation which is a nonlinear byte substitution.

Shift Rows: It is a simple byte transformation. The first row is not changed but the bytes in the last three rows of the array are cyclically shifted. The first row cannot be altered, the second row is shifted by one byte, the third row is shifted by two bytes, and the fourth row is shifted by three bytes.

Mix Columns: In this transformation, bytes are considered as polynomials rather than numbers. Each column in the array is considered as vector, which is multiplied with a fixed matrix.

Add Round Key: It is simple XOR operation performed between the current array and the round key.

Expansion key: Both the sender and receiver know the key. But the key cannot be discovered by any unauthorized person since the algorithm is intended to employ any one of the key size AES-128, AES-196, or AES-256. Thus, the key is expanded by means of key expansion routine.

Figures 2 and 3 are the flowcharts of encryption and decryption of AES-128 in which input is an image, key is in hexadecimal format, and the output is similar to that of the original image.

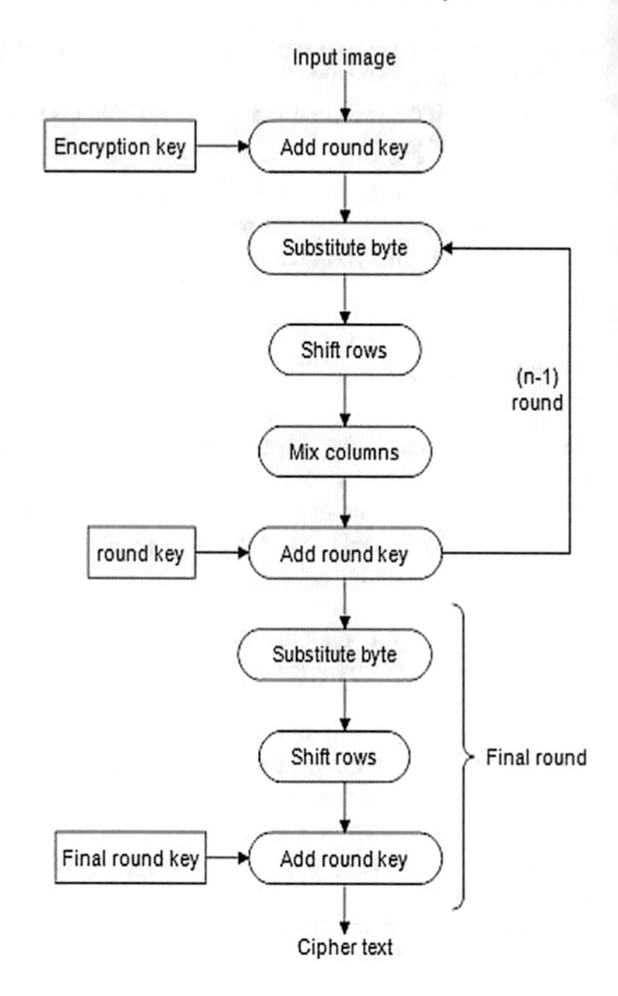

Fig. 2 Encryption process

4 Chaotic Algorithm

More consideration has been paid to data security with the fast advancement of innovative, for example, the cloud innovation. It has turned into a critical research field in software engineering of how to secure transmission [2, 5]. Shannon once stated: A good mixed transform field is obtained by two simple operations which are not commutative. Chaotic algorithm mainly deals with the systems which are very sensitive to initial conditions. These systems can be predictable for sometime, however, then shows up as arbitrary [7, 8].

Following equation is used to generate chaotic sequence-

$$W\text{new} = \mu^{*} W\text{old}^{*}(W\text{old} - 1) \quad (1)$$

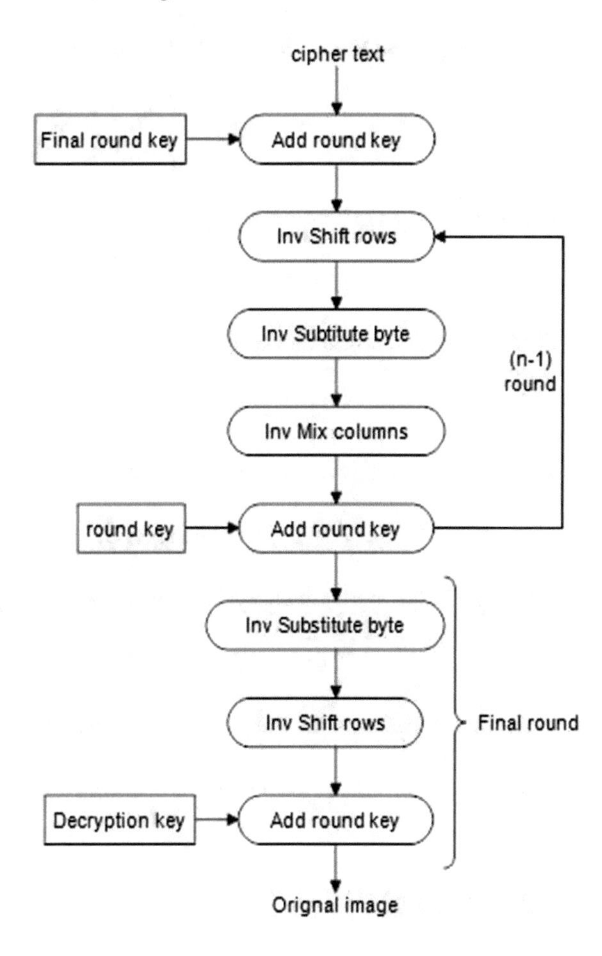

Fig. 3 Decryption process

where Wnew is new state and Wold is previous state.

Here, every value is depending on its previous state which in turn, proves that the system in a whole depends on initial conditions.
Algorithm:

Steps to be followed:

- Generate chaotic sequence using the following formula.

$$Wp+1 = \mu^* Wp^*(1-Wp),$$
$$Wp \in [0, 1], \mu \in [1, 4], p = 0, 1, 2, \ldots \quad (2)$$

- Obtain a random sequence as follows:

$$W = \{Wp|0 \leq Wp \leq 1, p = 1, 2, 3, \ldots \quad (3)$$

- Using Eq. (3) and binary function $T(W)$, we ge

$$T(W) = \begin{cases} 00 \le W \le 0.5 \\ 10.5 < W \le 1 \end{cases} \tag{4}$$

- From Eq. (4), we get the binary sequence

$$H = \{Hi|Hi = T(Wp), p = 1, 2, 3, \ldots \tag{5}$$

- Changing the data into binary sequence

$$S = \{s_1, s_2, s_3 \ldots s_s \tag{6}$$

- Using Eqs. (5) and (6), generate the key J

$$J = \{Jp|Jp = Hp \odot Sp, p = 1, 2, 3 \ldots \tag{7}$$

- Using the key J, perform encryption and obtain the ciphertext

$$Z = \{Zp|Zp = Jp \oplus Sp, p = 1, 2, 3 \ldots \tag{8}$$

- Perform decryption using the same key and plaintext is obtained.

$$D = \{Dp|Dp = Jp \oplus Zp, p = 1, 2, 3, \ldots \tag{9}$$

5 Proposed Model

Select an image of the file format .jpeg or .bmp or .png and resize it to the co-ordinates [256,256]. Encrypt the resized image using AES algorithm.

Consider the data as RA1612009010024. Encrypt the data by the following steps mentioned in the chaotic algorithm. The reason for selecting this method is due to randomness. Figure 4 shows the randomness in a system based on different μ values.

For different values of μ, output varies accordingly. If μ is 2.0, then it gives a constant value. If μ is 3.0, then it converges to a stable value. If μ is 3.5, then it bounces around. $W0$ in the Eq. (1) has to be in the range of [0,1]. If $W0$ is close to 1, a fastest initial decay will be observed. So, here in the proposed model, we are considering μ and $Y0$ as 3.9 and 0.4.

Figure 5 decribes how to conceal the data in the encrypted image using diamond encoding method [9].

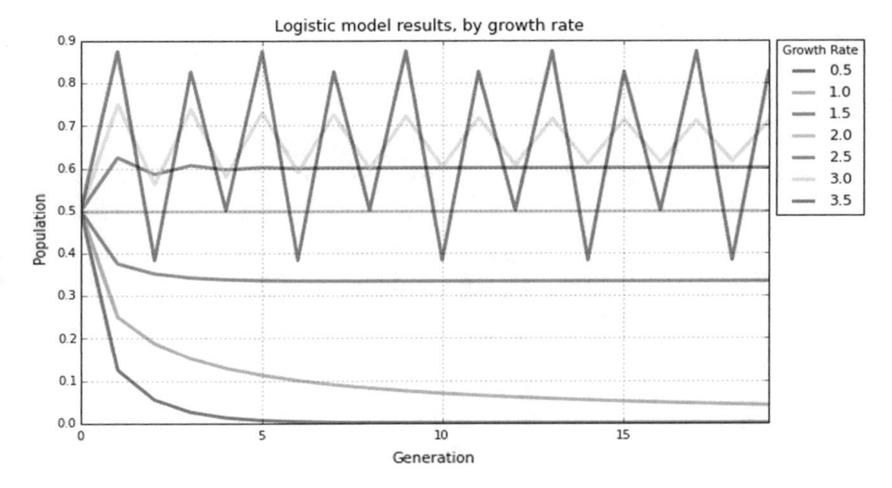

Fig. 4 System behavior based on different μ values

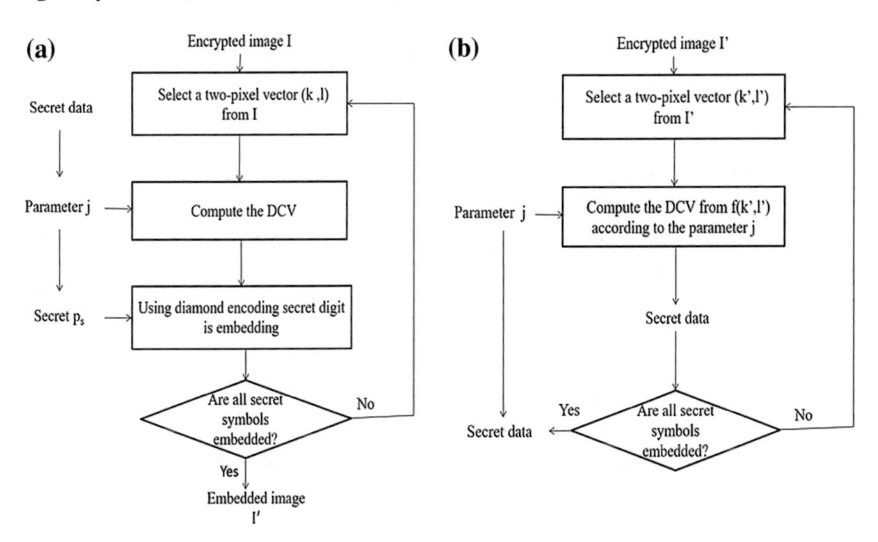

Fig. 5 Proposed model **a** data embedding process, **b** data extraction process

Diamond Encoding Method

Embedding Procedure:

Step 1: A parameter *j* is selected based on data size and data is now converted into diamond encoding digits [10].

Assume data size as *t*, and embedding parameter ***i*** is obtained by finding the minimal positive integer that satisfies the following inequality:

$$\left[((r^{*}s)/2)\boldsymbol{log}_{2}(2\boldsymbol{j}^{2}+2\,\boldsymbol{j}+1)\right]\geq p \tag{10}$$

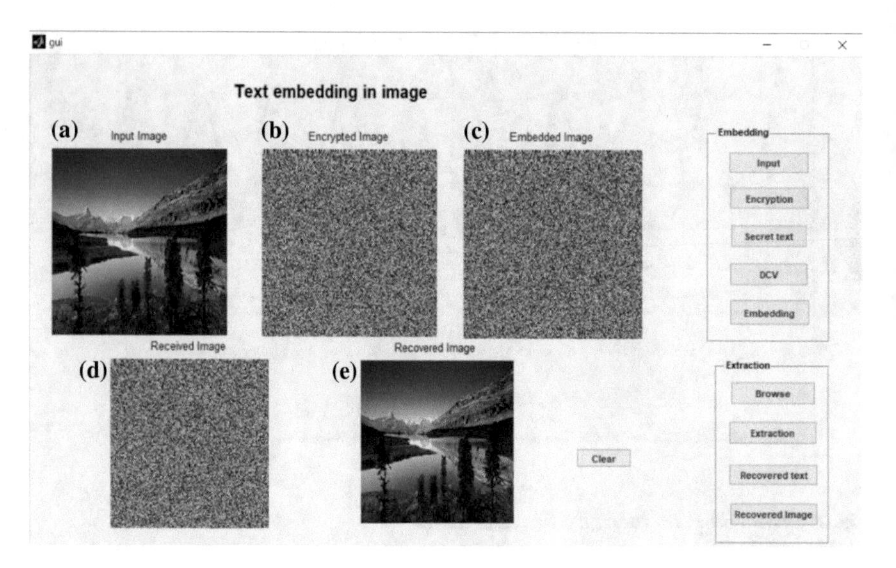

Fig. 6 Encryption and extraction of image and text **a** input image, **b** encrypted image, **c** embedded image, **d** received image, and **e** recovered image

This process hides $(2\boldsymbol{j}^2 + 2\boldsymbol{j} + 1)$-ary digits into an image pixel pair.
where j is the embedding parameter.

Step 2: In this procedure, the encrypted image I is divided into nonoverlapping two-pixel blocks. For text embedding, the selection of each block can be either from top-down or left-right.

Step 3: Compute the diamond correlation values

$$f(k, l) = ((2j + 1)^{*}k + l) \bmod j \tag{11}$$

Step 4: The embedded image pixel pair can be obtained by replacing $f(k, l)$ with $\boldsymbol{p}_s$ using the following equation.

$$z_p = (p_s - f(k, l)) \bmod j \tag{12}$$

Step 5: The symbol $\boldsymbol{z}_p$ gives the modulus between the $\boldsymbol{p}_s$ and $f(k, l)$.

Extraction Procedure:

Step 1: Embedded image I′ is divided into nonoverlapping two-pixel blocks. For data extracting, the selection of each block can be either from top-down or left-right.

Step 2: Compute the diamond correlation values for pixel values (k', l')

$$f(k', l') = ((2j + 1)^{*}k' + l') \bmod j \tag{13}$$

Step 3: Until all the secret digits have been extracted, steps 1 and 2 must be repeated for the corresponding pixel pairs of encrypted image.

Step 4: By transforming the secret symbols with base 2, the secret data can be obtained.

Once the secret data is obtained, then the original image can be obtained only after decryption process which is done using inverse AES algorithm.

After recovering the data, then perform decryption process for encrypted image using inverse AES algorithm, and the original image is obtained.

6 Results

From Fig. 6 we can observe that the image and secret data has been concealed and secured.

7 Conclusion

The proposed algorithm ensures that the data and the image are secured. It mainly concentrates on the data that needs to be protected such as a secret text or financial or business particulars, etc. The efficacy of the new algorithm is demonstrated by our experiments and the security of the new algorithm is improved and strengthened.

References

1. Saraf, K.R., Jagtap, V.P., Mishra, A.K.: Text and image encryption decryption using advanced encryption standard. Int. J. Emerg. Trends Technol. Comput. Sci. (IJETTCS) **3**(3), 118–126 (2014)
2. Zhang, Q.: Study on image encryption algorithm based on chaotic theory. In: International Conference on Information Science and Cloud Computing Companion, pp. 635–639 (2013)
3. Ramineni, V., Loganathan, A., Kalubandi, V.K.P., Vaddi, H.: A novel image encryption algorithm using AES and visual cryptography. In: 2nd International Conference on Next Generation Computing Technologies (NGCT), pp. 808–813 (2016)
4. Deshmukh, P.: An image encryption and decryption using AES algorithm. Int. J. Sci. Eng. Res. **7**(2), 210–213 (2016)
5. Wang, X.-Y., Gu, S.-X.: New chaotic encryption algorithm based on chaotic sequence and plain text. IET Inf. Secur. **8**(3), 213–216 (2013)
6. Radhadevi, P., Kalpana, P.: Secure image encryption using AES. IJRET Int. J. Res. Eng. Technol. **1**(2), 115–117 (2012)
7. Saleem, A., Trinadh Rao, A., Kumar, K.T.P.S., Chaitanaya, G., Keerthi, B.: An image encryption and decryption using chaos algorithm. IOSR J. Electron. Commun. Eng. (IOSR-JECE) (2015)
8. Abdalla, T., Al-Maadeed, S., Al-Ali, A.: A new chaos based image encryption and compression algorithm. IOSR J. Electron. Commun. Eng. (IOSR-JECE) 1–11 (2011)

9. Chao, R.-M., Wu, H.-C., Lee, C.-C., Chu, Y.-P.: A novel image data hiding scheme with diamond encoding. EURASIP J. Inf. Secur. 1–9 (2009)
10. Kuoa, W.-C., Laia, P.-Y., Wang, C.-C., Wuu, L.-C.: A formula diamond encoding data hiding scheme. J. Inf. Hiding Multimedia Signal Process. **6**(6), 1167–1175 (2015)

A. Christie Aiswarya Department of EST, SRM Institute of Science and Technology, Kattankulathur.

A. Sriram Department of ECE, SRM Institute of Science and Technology, Kattankulathur.

Enrichment of Security Using Hybrid Algorithm

Deeksha Ekka, Manisha Kumari and Nishi Yadav

Abstract In this paper, the concept of symmetric encryption method, AES (Advanced Encryption Standard) and asymmetric encryption method, RSA followed by an EX-OR operation are combined to introduce a new Hybrid Algorithm for the enrichment of data security over network. The proposed scheme adds up more complexity in data by increasing confusion and diffusion in ciphertext using AES for data encryption and RSA for key encryption. Thus, intruder will require more time to decrypt the text and it also resolves the brute-force attack, differential attack, and linear attack. The performance analysis of proposed scheme is done and compared with the AES and RSA on the basis of encryption and decryption time. The result shows that the proposed scheme takes less time for encryption process and more time for decryption process, hence it improves the security of the data.

Keywords Advanced encryption standard (AES) · RSA cryptosystem Encryption · Decryption · EX-OR operation

1 Introduction

On a public network, data is vulnerable to various attacks. Thus, security is a major concern during data handling, data transfer, and electronic communication. To achieve the confidentiality in data transfer and resistant to attacks, various cryptography techniques are introduced [1]. Cryptography is used for message randomization,

D. Ekka · M. Kumari
Computer Science and Engineering, School of Studies in Engineering and Technology, Guru Ghasidas University, Bilaspur, India
e-mail: deekshaekka88@gmail.com

M. Kumari
e-mail: manisha1921995@gmail.com

N. Yadav (✉)
School of Studies in Engineering and Technology, Guru Ghasidas University, Bilaspur, India
e-mail: nishidv@gmail.com

S. Smys et al. (eds.), *International Conference on Computer Networks and Communication Technologies*, Lecture Notes on Data Engineering and Communications Technologies 15, https://doi.org/10.1007/978-981-10-8681-6_80

confidentiality, and low modification rate to provide high security to data over network. Cryptography can be thought of as a process of converting the plaintext into a non-readable format and protects the message from unauthorized access. Thus, in the past years, various encryption techniques are introduced to provide a secure transmission of data over network [2–4].

Cryptography techniques are divided into two major categories such as symmetric key cryptography and asymmetric key cryptography. Symmetric key cryptography works in such a way that it uses the same cryptographic key for both encryption and decryption process. It could either be stream cipher or block cipher. In stream cipher, message digits are encrypted one at a time while in block cipher, some bits are taken and encrypted as a unit. Sometimes, padding in plaintext is required, so that it could make a multiple of block size. AES is an example of symmetric key cryptography [2, 5].

In asymmetric key cryptography, different keys are used for the encryption and decryption process. One key is kept public known as public key which is used for encryption process while another key is kept secret known as private key which is used for decryption process. RSA is an example of asymmetric key cryptography [2, 5].

AES is widely accepted in the past years because of its high security, fast speed, and high computational complexity. AES is one of the best symmetric block cipher algorithm introduced by NIST. AES is preferred extensively because of its multiple layer protection features which arises due to various rounds involved in the algorithm [3, 6].

2 Background Details and Related Work

2.1 Rivest—Shamir—Adleman (RSA) Algorithm

RSA is an asymmetric key cryptography, which uses different keys for encryption and decryption process such as public key and private key, respectively. The practical difficulty of factorization of two large prime numbers defines the security of RSA algorithm [7]. The RSA scheme is as follows:

Key generation:

1. Select two large random prime numbers, p and q of approximately equal size that their product $n = p * q$ is desired bit length.
2. Compute $n = p * q$ and $\text{Ø}(n) = (p - 1) * (q - 1)$.
3. Choose a positive integer e, such that $1 < e < \text{Ø}(n)$, such that GCD $(e, \text{Ø}(n)) = 1$.
4. Compute the value of secret Exponent d, $1 < d < \text{Ø}(n)$, such that $e * d = 1(\text{mod } \text{Ø}(n))$.

5. The pair (e, n) is a public key and (d, n) is a private key. The values d, p, q, and $Ø(n)$ should be kept as a secret.
   where
   - n is modulus,
   - e is the public exponent or encryption exponent or simply the exponent, and
   - d is the secret exponent or decryption exponent.

Encryption:
Suppose user A wants to send message "m" to user B.

1. Obtain the public key (e, n) of user B.
2. Represent the plaintext as positive integer m.
3. Compute the ciphertext $c = m^e \bmod n$, using user B's public key.
4. Send the ciphertext c to user B.

Decryption:
User B will retrieve the original message from cipher text.

1. Use private key (d, n) to compute $m = c^d \bmod n$.
2. Extract the plaintext m from c.

2.2 Advanced Encryption Scheme (AES)

AES is a symmetric encryption standard which is introduced by NIST in 2001 and originally, it was named as *Rijndael*. The main advantage of using AES is that it provides various key lengths for data encryption such as 128, 192, and 256 bits key, and also the rounds involved in the AES algorithm depends on the various key lengths such as 10, 12, and 14 rounds, respectively [8, 9].

AES consist of four main transformation functions except the last round is slightly different. $4*4$ matrix of bytes array are as state in these operation function. Four different rounds in AES are as follows [10]:

(a) *Byte substition*: Using substition matrix (S-Box), nonlinear byte-by-byte substition is applied.
(b) *Shift row*: It involves simple byte transformation where the left shifting of last three rows in circular way is done depending on the offset of row index.
(c) *Mix column*: It involves matrix multiplication. Each byte in the column is converted to a novel value which depends on the value of other four bytes in the same column.
(d) *Add round key*: It is a normal EX-OR operation between the recent state and the round key.

3 Proposed Algorithm

In this section, a brief discussion of the proposed algorithm is given. The proposed scheme is based on two standard cryptography techniques: AES (Advanced Encryption Standard) and RSA. These two techniques are combined to provide a high level of data security by increasing randomization of plaintext and cryptography key along with the employment of an EX-OR operation.

Here, AES increases the randomness in ciphertext and RSA encrypts the secret key and finally, AES and RSA are integrated using EX-OR operation, resulting in high computational complexity and low modification rate in ciphertext. The sole purpose of using AES for encryption is that it is the most complex and secure cryptography algorithm.

Modules in Proposed Work

The proposed scheme is comprised of three main modules such as:

I. *Encryption with AES and EX-OR operation*: The plaintext is first encrypted using AES algorithm with secret key SK_1 to generate the ciphertext C_1. The EX-OR operation is applied on ciphertext C_1 and key SK_1, and the result is stored in C_2.
II. *Encryption with RSA*: Now, the key SK_1 is encrypted using RSA algorithm to generate the encrypted text SK_1'.
III. *Integration of RSA and AES*: Now, apply the EX-OR operation on encrypted text SK_1' and C_2 and store the result in C_3. Send C_3 and SK_1' to the receiver for the decryption process.

Encryption:

Suppose user *A* wants to send a plaintext *m* to receiver *B*.

1. User *A* have to encrypt the plaintext *m* using AES scheme with secret key SK_1 to generate encrypted text C_1.
2. Now, do EX-OR operation with encrypted text C_1 and SK_1 to generate result C_2.
3. Encrypt the AES secret key SK_1 using RSA with its public key to generate result SK_1'.
4. Now, perform the EX-OR operation with SK_1' and C_2 to obtain final result C_3.
5. Now, send the pair of (C_3, SK_1') to the receiver *B*.

Decryption:

User B will receive the encrypted text as (C_3, SK_1'):

1. Perform EX-OR operation between C_3 and SK_1' to generate result C_2.
2. Decrypt the SK_1' using RSA scheme with its private key to obtain result SK_1.
3. Now, again perform the EX-OR operation between C_2 and SK_1 to generate result C_1.
4. Apply AES algorithm on C_1 and SK_1 to generate plaintext *m*.

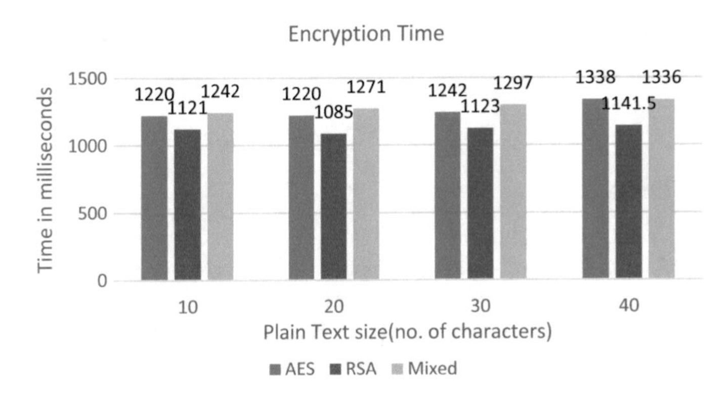

Fig. 1 Analysis of encryption time

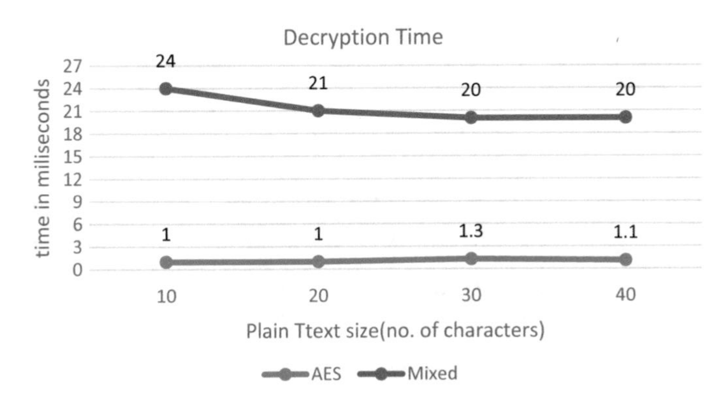

Fig. 2 Analysis of decryption time

3.1 Results and Discussion

This section represents the result of number experiments done on the proposed algorithm to deduce a fair comparison between proposed algorithm, AES and RSA on the basis of encryption and decryption time. The comparision with these two parameters are discussed below.

Simulation Parameters:

1. **Encryption time**: The new Hybrid Algorithm takes less time as compared with AES-256-bit algorithm for encryption of data. While the RSA algorithm is only used to encrypt the cryptography key, thus it takes less for encryption as compared with both AES-256-bit algorithm and the proposed algorithm which is shown in Fig. 1.
2. **Decryption time**: The proposed Hybrid Algorithm is more secured as compared with AES-256-bit as it takes more decryption time than AES-256-bit algorithm.

Figure 2 shows the fair comparison between proposed Hybrid Algorithm and AES-256-bit algorithm.

4 Conclusion

The proposed Hybrid Algorithm aims to provide high level of data security during the message transfer over the Internet. The proposed scheme provides a multilayer security to data as the data is first encrypted using AES-256-bit algorithm and then integrated with the encrypted key using an EX-OR operation. Thus, such computation on data increases the randomness of message in ciphertext and as a result, increases the decryption time, therefore the proposed algorithm enhances the security of data at higher level.

References

1. Sharma, A., Singh, A.: Hybrid improved technique for data security and authentication for RDIF tags. In: IEEE International Conference on Signal Processing, Computing and Control (ISPCC2017)
2. Shahara Banu, T., Shajeesh, K.U.: Secure reversible data hiding technique on textures using double encryption. In: International Conference on Innovations in Information, Embedded and Communication System (ICIIECS 2017)
3. Oyetola Oluwadamilola, K., Okubanjo Ayodeji, A., Osifeko Martins, O.: An improved authentication system using hybrid of biometrics and cryptography. In: International Conference on Electro-Technology for National Development (NIGRCON) (2017)
4. Harini, M., Pushpa Gowri, K., Pavithra, C., Pradhiba Selvarani, M.: A novel security mechanism using hybrid cryptography algorithm. (ICEICE2017)
5. Hazra, T.K., Mahato, A., Mandal, A., Chakraborty, A.K.: A hybrid cryptosystem of image and text files using blowfish and Diffie Hellman technique. (IEMECON2017) @IEEE
6. Rachmawanto, E.H., Amin, R.S., Setiadi, D.R.I.M., Sari, C.A.: A performance analysis Stego Crypt algorithm based on LSB-AES 128 bit in various image size. In: International Seminar on Application for Technology of Information and Communication (iSemantic2017)
7. Bhandari, A., Gupta, A., Das, D.: Secure algorithm for cloud computing and its applications. 2016 IEEE
8. D'souza, F.J., Panchal, D.: Advanced encryption standard (AES) security enhancement using hybrid approach. In: International Conference on Computing, Communication and Automation (ICCCA2017)
9. Vichare, A., Jose, T., Tiwari, J., Yadav, U.: Data security using authenticated encryption and decryption algorithm for Android phones. In: International Conference on Computing, Communication and Automation(ICCCA2017)
10. Ebrahim, M.A., El-Maddah, I.A.M., Mohamed, H.K.: Hybrid model for cloud data security using steganography. @2017 IEEE

Deeksha Ekka studying in Computer Science and Engineering, School of studies in Engineering and Technology, Guru ghasidas university, Bilaspur, India. Her research area includes network security and cryptography.

Manisha Kumari studying in Computer Science and Engineering, School of studies in Engineering and Technology, Guru ghasidas university, Bilaspur, India. Her research area includes network security and cryptography.

Nishi Yadav working as an Assistant professor of Computer Science and Engineering, School of studies in Engineering and Technology, Guru ghasidas university, Bilaspur, India. Her research area includes network security and cryptography.

Comparative Study of Transmitter-Side Spectrum Detection in Cognitive Radio Network

G. Shine Let, Songa Christeen, P. Lidiya Priya, B. Keerthi Reddy and P. Swetha

Abstract Cognitive radio network is an intelligent radio network that adapts its transmission parameters based on the sensed free spectrum. Sensing the vacant spectrum is a major issue in cognitive radio networks. The spectrum availability is dynamic. Before data transmission, secondary users sense the spectrum to check the occupancy of primary user. In this paper, the comparison of transmitter-side spectrum detection is carried out. The variation of probability of detection is analyzed with respect to probability of false alarm and SNR. Simulation results show that a signal transmitted with high SNR can be detected accurately by having fixed probability of false alarm. In this work, when the SNR is increased from −20 to −10 dB, the probability of detection is increased on an average of 38 and 57% in matched detection technique and energy detection technique, respectively. This paper also gives a comparative study of energy detection techniques, cyclostationary feature techniques, and matched-filter detection techniques.

Keywords Probability of detection · Probability of Fasle alarm · Cognitive radio
Energy detection · Matched-filter detection · Cyclostationary feature detection

1 Introduction

Nowadays, communication is observed as a basic necessity in day-to-day life. In olden days, people used to communicate through pigeons. Now, it has reached to a totally different extent. Wireless technology is being extensively used today. During the communication process, three terms are to be noted. They are transmitter, path or channel, and receiver. The main barrier in this communication is that there is scarcity of spectrum. Spectrum refers to a band of frequencies through which the signal or any data travels. To overcome the scarcity, the concept of cognitive radio was

G. Shine Let (✉) · S. Christeen · P. Lidiya Priya · B. Keerthi Reddy · P. Swetha
Department of Electrical Sciences, Karunya Institute of Technology and Sciences, Coimbatore, India
e-mail: shinelet@gmail.com

S. Smys et al. (eds.), *International Conference on Computer Networks and Communication Technologies*, Lecture Notes on Data Engineering and Communications Technologies 15, https://doi.org/10.1007/978-981-10-8681-6_81

introduced by J. Mitola in 1999 [1]. It is an intelligent radio that can be programmed and configured robustly. Cognitive radio helps us to solve spectrum scarcity problem. Cognitive radio comprises of spectrum sensing, spectrum management—selects the finest channel for communicating, spectrum sharing—coordinates the access with any other user and spectrum mobility—when primary user or licensed user is detected secondary user vacates the channel.

This paper focuses on different spectrum sensing techniques and comparison between them. Spectrum sensing is defined as the process of identifying the unused bands of frequencies. This helps us to send the data efficiently without any objection. Spectrum sensing involves in obtaining the custom distinctiveness across time, space, frequency, type of signal occupying the spectrum, etc. Also, there are few troubles in spectrum sensing like low channel capacity and SNR, more exploitation of energy, and unproductive usage of the spectrum. There are different spectrum sensing techniques, out of which the paper mainly focus on transmission-side detection technique in which it finds the primary users that are transmitting at any instant of time.

The problems in this technique are vagueness in receiver, also the power of the receiver signal varies due to the elements present between the transmitter and receiver. In little deeper concept, there are two types of detection techniques. They are coherent and noncoherent. This paper compares the performances of the three transmitter-based detection techniques called as energy detection, cyclostationary feature detection, and matched-filter detection. Energy detector is also known as radiometer. It uses squaring and integration methods. It is widely used due to its effortlessness [2, 3]. This technique is used to gain desired probability of detection which is based on the set threshold and false alarm [2]. Only when the threshold value exceeds, it shows that the presence of the signal [3]. The power of the received signal is calculated in this method and compares it with the fixed threshold value. The energy detector is noncoherent. The second method is matched-filter detection [4]. It is very accurate. A matched filter is a linear filter, which means it processes input signals that fluctuate with time to produce outputs. It has the property of coherence, which benefits in achieving high gain in less time. And, the third type is cyclostationary feature detection. From the name, cyclostationary signal refers to the one whose mean and autocorrelation keeps repeating periodically [5]. Also, feature detection is noticing the characteristics of the received signal and performing the processes accordingly. This method is very complex and expensive. Generally, modulated signals are delineated as cyclostationary. Both matched and cyclostationary techniques are coherent.

2 Preliminaries in Spectrum Sensing

The secondary user sense different spectrum bands to determine the activity of the primary user. If the primary user is occupying the channel [6], the received signal at the secondary user is given by (1):

$$y(t) = x(t) + w(t) \tag{1}$$

where $y(t)$ is the secondary user observed signal, $x(t)$ is the primary user signal, and $w(t)$ is the additive white Gaussian noise (AWGN). When the primary user is not present, the secondary user received signal is given by (2):

$$y(t) = w(t) \tag{2}$$

AWGN is an independent, identically distributed noise with zero mean and variance σ_w^2. The above secondary user sensing is related to two hypotheses

$$y(t) = \begin{cases} w(t), & H_0 \\ x(t) + w(t), & H_1 \end{cases} \tag{3}$$

H_0 represents the absence of primary user and H_1 represents the presence of primary user.

In cognitive radio network, the probability of detection (P_d) must be high for any spectrum sensing techniques. This represents protection in primary user communication and also, secondary user can perfectly detect the availability of primary user. If the probability of false alarm (P_f) is low from secondary user perspective, then more chances are available for the use of frequency band. The probability of missed detection (P_m) is given by $(1 - P_\mathrm{d})$. Next section brief about the different spectrum sensing techniques such as energy detector, matched-filter detection and cyclostationary detection.

2.1 *Energy Detection*

Energy detection is the simple, less complex spectrum sensing technique. The working of energy detection sensing technique is shown in Algorithm 1. In this technique, the secondary user sensed signal is passed through a bandpass filter and the output is integrated over the time interval. The output of the integration is compared with the threshold. The threshold value is set based on the power of the noise level in the environment [2]. At low SNRs, the energy detector performance is low and error estimated degrades the performance significantly [7]. The detection probability and false alarm probability depends on the threshold of a single user energy detector and the performance varies over a fading channel [8] is the main drawback.

Algorithm 1

1 **begin**
2 set the channel parameters
3 the received RF signal is added with the channel characteristics (y)

4 apply square root raised cosine function
5 compute the average power of the signal over the considered frequency band (P)
6 set a value for threshold (λ) according to the probability of false alarm
7 compare P and λ
8 **if** $P > \lambda$, **then** primary user is present
9 **else** primary user is absent
10 **end end**

2.2 *Matched-Filter Detection*

Matched-filter detection is a coherent detection model, so that the secondary user should have some information about the primary user signal [3]. The received signals' signal-to-noise ratio (SNR) is maximized first in this technique. After maximizing the received signal SNR, this technique correlates the signal with time modified version. The output is then compared with the predetermined threshold. This technique can be applied only if the secondary user is having some prior knowledge of the primary user [4]. Algorithm 2 shows the flow of matched-filter detection. The main operation of matched-filter detection is given by (4):

$$y(n) = \sum_{k=-\infty}^{\infty} h(n-k)x(k) \quad (4)$$

Algorithm 2

1 **begin**
2 set the channel parameters
3 the received RF signal is added with the channel characteristics (y)
4 generate the RF signal based on primary user signal characteristics (c)
5 cross-correlation function is applied between the received primary user signal (y) and the generated signal (c), the resultant output is (z)
6 set a value for threshold (λ) according to the probability of false alarm
7 compare z and λ
8 **if** $z > \lambda$, **then** primary user is present
9 **else** primary user is absent
10 **end end**

2.3 Cyclostationary Feature Detection

Cyclostationary feature detection has been applied to the signals which have periodic properties. Compared to energy detection and matched-filter detection, this technique is more immune to noise [5]. Based on spectrum correlation function, the primary user is detected. The detection depends on a number of samples considered in the received signal. Detection is high, if the number of samples "N" are high. As "N" increases, the complexity and the computation increases. This approach is more robust to noise and interference than other abovementioned techniques [9]. For noise, the peak of spectral correlation function will be at the zero cyclic frequency. Different unique cyclic frequencies are there for different modulated signals. Algorithm 3 gives the flow diagram of cyclostationary feature detection technique.

Algorithm 3

1 **begin**
2 set the channel parameters
3 the received RF signal is added with the channel characteristics ($r(t)$)
4 FFT of $r(t)$ is $R(f)$
5 correlation function is applied
6 averaging the correlated output
7 identifying the peak value (z) and comparing it with the threshold (λ)
8 **if** $z > \lambda$, **then** primary user is present
9 **else** primary user is absent
10 **end end**

3 Comparative Study

Different authors have proposed different techniques for detecting the dynamic appearance of primary users. Table 1 gives the information of different transmitter-side sensing techniques proposed by different authors.

4 Computer Simulations

To evaluate the performance of different transmitter-side spectrum sensing techniques, computer simulations are carried out in MATLAB. When the secondary user wants to transmit the data, it senses if the primary user is present. The channel considered is the AWGN (additive white Gaussian noise) channel. If the primary user is in attendance, the secondary user receives it along with the noise. Otherwise, only the noise is received by the secondary user.

Table 1 Literature review of different transmitter-side spectrum sensing techniques

References	Spectrum sensing technique	Parameters considered	Concluded analysis
[10]	Energy detection with eigenvalue detection	SNR = −15 dB The number of samples for each test, Ns = 100,000	The Probability of detection and false alarm vary for different samples When the noise is nonuniform, the probability of false alarm will be high and the efficiency will be low i.e., the probability of detecting the signal will be low
[11]	Energy detection technique with equal gain	Different P_d and P_f values are considered	As the SNR value increases, the probability of detection increases. In any condition, medium SNR is required to achieve good probability of detection.
[12]	Energy detection	For wireless microphone signal, sampling rate at receiver = 6 MHz $N = 10{,}000$ For independent and identically distributed signals, $K = 1$(PU), $M = 4$(receiver antennas)	As the SNR value increases, the probability of detection increases
[13]	Matched-filter detection with dynamic threshold	$N = 1000$ and threshold factor = 1 (SNR range = −20 dB to +20 dB	As SNR increases, P_d also increases This works at its best during low SNR values $P_d = 100\%$ when $N = 1000$ samples and SNR >−4 dB
[14]	Matched-filter detection	No. of simulations = 1000 SNR = 3 dB, $P_f = 0.001$	Improvement of P_d is proportional to SNR and Improvement of SNR is proportional to no. of sample pulses. Therefore, when the number of samples increases, the P_d improves
[5]	Cyclostationary feature detection with modulation schemes	Operating frequency = 1000 MHz Modulations: BPSK and QPSK No. of relays = 10	When the no. of relays used is greater, the detection will be better

(continued)

Table 1 (continued)

References	Spectrum sensing technique	Parameters considered	Concluded analysis
[15]	Cyclostationary feature detection for ultra-wideband communication	$P_f = 0.5$ dB For SNR = 5, 10, 15, and $P_d = 0.28$, 0.4, 0.48, respectively For relays = 10, 20, 30, and $P_d = 0.3$, 0.37, 0.5, respectively	P_d is directly proportional to SNR and no. of relays i.e., As SNR and No. of relays increase, P_d increases
[16]	Cyclostationary feature detection	For good $P_{d,}$ SNR = − 8 dB	At −8 dB, high P_d and low P_f are observed

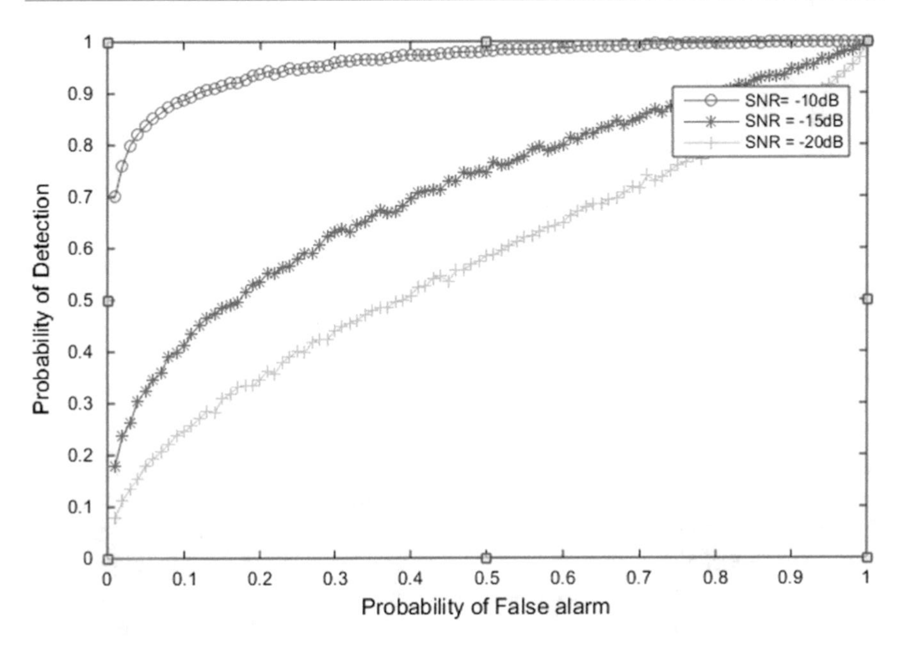

Fig. 1 Energy detection (P_f vs. P_d)

Figures 1 and 2 shows the probability of false alarm (P_f) versus the probability of detection (P_d) for energy detection and matched-filter detection techniques, respectively. For simulation, the channel is modeled as AWGN channel with zero mean and variance as one. Primary user signal is considered as a random signal which is modulated using BPSK. Secondary user while sensing receives primary user signal with noise or only noise (if primary user not available).

As the SNR varies, in energy detector, the variation of P_d is high, i.e., as noise increases the P_d is difficult. Matched-filter technique is highly resistant to variation of noise. For different SNR of the received signal, matched-filter technique is able to detect the received signal.

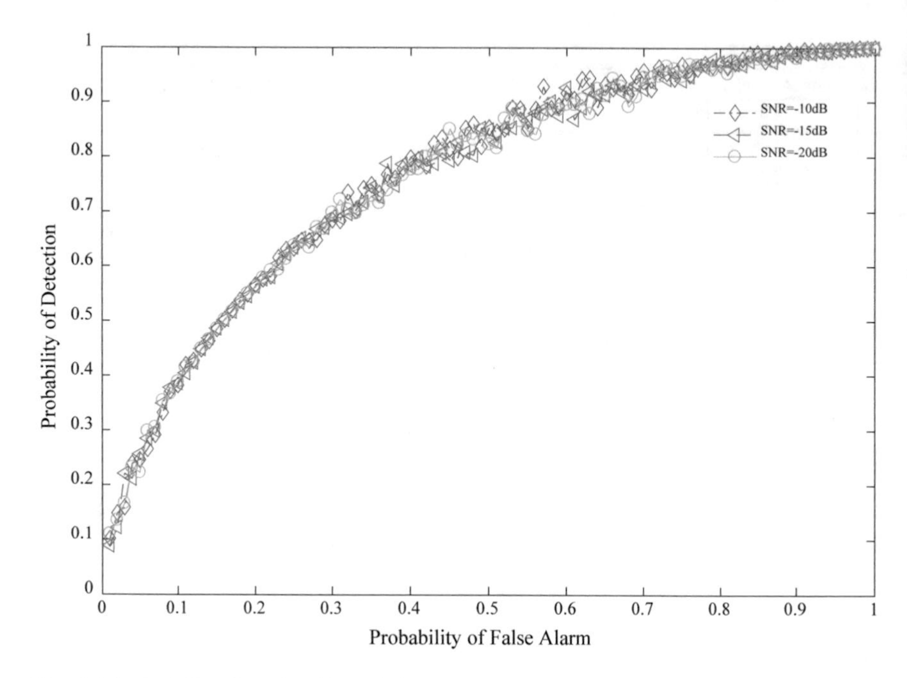

Fig. 2 Matched-filter detection (P_f vs. P_d)

5 Conclusion

In cognitive radio network, finding the activity of primary users is the challenging task. Various researchers have introduced different sensing techniques to detect the primary user' presence. In this paper, review on different transmitter-side spectrum sensing techniques is carried out. In transmitter-side sensing technique, optimal detection threshold is determined based on the probability of false alarm. The implementation of these techniques is simple, but the threshold value fixed will be susceptible to noise and interference. Due to dynamic nature of channel characteristics and primary user behavior, different optimization techniques can be applied for the detection of primary user within less computation time.

References

1. Mitola, J., Maguire, G.Q.: Cognitive radio: making software radios more personal. IEEE Pers. Commun. **6**(4), 13–18 (1999)
2. Abdulsattar, M.A., Hussein, Z.A.: Energy detection technique for spectrum sensing in cognitive radio: a survey. Int. J. Comput. Netw. Commun. **4**(5), 223–242 (2012)
3. Cabric, D., Mishra, S.M., Brodersen, R.W.: Implementation issues in spectrum sensing for cognitive radios. In: IEEE Conference Record of the Thirty-Eighth Asilomar Conference on

Signals, Systems and Computers, vol. 1, pp. 772–776 (2004)
4. Yawada, P.S., An, J.W.: Performance evaluation of matched filter detection based on non-cooperative spectrum sensing in cognitive radio network. Int. J. Comput. Netw. Commun. Secur. **3**(12), 442–446 (2015)
5. Aparna, P.S., Jayasheela, M.: Cyclostationary feature detection in cognitive radio using different modulation schemes. Int. J. Comput. Appl. **47**(21) (2012)
6. Liang, Y.C., Zeng, Y., Peh, E.C., Hoang, A.T.: Sensing-throughput tradeoff for cognitive radio networks. IEEE Trans. Wirel. Commun. **7**(4), 1326–1337 (2008)
7. Atapattu, S., Tellambura, C., Jiang, H.: Energy detection based cooperative spectrum sensing in cognitive radio networks. IEEE Trans. Wirel. Commun. **10**(4), 1232–1241 (2011)
8. Pandya, P., Durvesh, A., Parekh, N.: Energy detection based spectrum sensing for cognitive radio network. In: Fifth IEEE International Conference on Communication Systems and Network Technologies (CSNT), pp. 201–206 (2015)
9. Lin, Y., He, C.: Subsection-average cyclostationary feature detection in cognitive radio. In: IEEE International Conference on Neural Networks and Signal Processing, pp. 604–608 (2008)
10. Zeng, Y., Liang, Y.C.: Maximum-minimum eigenvalue detection for cognitive radio. In: IEEE 18th International Symposium on Personal, Indoor and Mobile Radio Communications, pp. 1–5 (2007)
11. Herath, S.P., Rajatheva, N.: Analysis of equal gain combining in energy detection for cognitive radio over Nakagami channels. In: IEEE GLOBECOM 2008 Global Telecommunications Conference, pp. 1–5 (2008)
12. Zeng, Y., Liang, Y.C., Zhang, R.: Blindly combined energy detection for spectrum sensing in cognitive radio. IEEE Sign. Process. Lett. **15**, 649–652 (2008)
13. Salahdine, F., El Ghazi, H., Kaabouch, N., Fihri, W.F.: Matched filter detection with dynamic threshold for cognitive radio networks. In: International Conference on Wireless Networks and Mobile Communications (WINCOM), pp. 1–6 (2015)
14. Odhavjibhai, B.A., Rana, S.: Analysis of matched filter based spectrum sensing in cognitive radio. Int. Res. J. Eng. Technol. **4**(4), 578–581 (2017)
15. Aparna, P.S., Jayasheela, M.: Cyclostationary feature detection in cognitive radio for ultra-wideband communication using cooperative spectrum sensing. Int. J. Future Comput. Commun. **2**(6), 668–672 (2013)
16. Verma, M.P.K., Dua, R.L.: A survey on cyclostationary feature spectrum sensing technique. Int. J. Adv. Res. Comput. Eng. Technol. (IJARCET) **1**(7), 300–303 (2012)

G. Shine Let received her M.E. degree in Communication Systems from Anna University, Chennai, India in 2007. Currently, she is an Assistant Professor at Karunya University, Coimbatore. Her interests are in Wireless Communication, Mobile Adhoc Network and Cognitive Radio Network.

Songa Christeen completed her B.Tech degree in Electronics and Communication Engineering from Karunya Institute of Technology and Sciences in 2018.

P. Lidiya Priya completed her B.Tech degree in Electronics and Communication Engineering from Karunya Institute of Technology and Sciences in 2018.

B. Keerthi Reddy completed her B.Tech degree in Electronics and Communication Engineering from Karunya Institute of Technology and Sciences in 2018.

P. Swetha completed her B.Tech degree in Electronics and Communication Engineering from Karunya Institute of Technology and Sciences in 2018.

Enabling Data Storage on Fog—An Attempt Towards IoT

A. Padmashree and N. Prasath

Abstract Internet of Things (IoT) devices is generating large volume and variety of data. These unconnected devices are generating more than two exabyte of data each day. Today's cloud models are not suited to manage these tremendous data. Moving them from these devices to the cloud for analysis would require huge bandwidth. Cloud servers communicate only through an Internet protocol (IP). But some of the IoT devices are machines that are connected to a controller using countless industrial protocols. Before sending the information retrieved from these devices to the cloud for analysis and storage, it must be translated to IP. The ideal place to analyze IoT data is near the devices that produce these data. This paper introduces fog computing or edge computing, a computational method that is suited for data generated by IoT devices and a few protocols to handle these data.

Keywords Internet of things · Fog computing · Cloud computing · IoT storage IoT protocols

1 Introduction

Today's electronic miniaturization and communication technologies have led to the development and deployment of large number of devices that are connected and execute many advanced functions. These have become smart devices and led to something called Internet of Things. Internet of things consists of networked objects connected over the internet. The devices are heterogeneous and large in number and so the data they provide. The data center operators are going to face problems in dealing with loads of new data generated from these connected devices. Also, these

A. Padmashree (✉) · N. Prasath
KPR Institute of Engineering and Technology, Coimbatore, India
e-mail: apadmashree.me@gmail.com

N. Prasath
e-mail: prasath283@gmail.com

S. Smys et al. (eds.), *International Conference on Computer Networks and Communication Technologies*, Lecture Notes on Data Engineering and Communications Technologies 15, https://doi.org/10.1007/978-981-10-8681-6_82

data must be accessed and managed in a unique way [1]. Now, this has become one of the biggest challenges with IoT.

Also, the bandwidth of WAN, the amount of data, and the amount of transactions provided by data centers also varies from one applications to another. We are in need of different models for prioritizing and classifying the data. Since IoT is used in a variety of applications like agriculture, health care, and smart home, we also need different models to handle these different data at a cheaper cost.

The life period of data is also a major concern. Consider a smart transport system which tracks the vehicle traffic and accidents, the healthcare system that provides data of patients. The data obtained from these system needs to be stored in the cloud for future reference. But consider a smart home system that controls the power consumption at a particular time [2]. Here, once the data is handled, the data need not be stored for future and is of least concern. These data need not go to the central place called cloud for analysis. Instead, these data can be stored and analyzed in localized centers called fog or edge.

2 Fog Computing

Fog computing or edge computing is a new model for analyzing and acting on IoT data. The processes and the resources are placed at the edge of the cloud, instead of establishing channels for cloud storage and utilization. It analysis the sensitive data on the network edge rather than in the cloud [3, 4]. Instead of sending the large amount of data to the cloud, it aggregates it on an access point, thereby reduces the bandwidth. Hence, it minimizes the cost and maximizes the efficiency. Data is collected at the extreme edges like vehicles, ships, roadways, railways, homes, children, rescue, and tactical operations [5], etc., analyzed and acted on within a few seconds [6].

2.1 Fog Computing Versus Cloud Computing

The time taken to make a decision on the data plays an important role in deciding where to place those data. The historical data that requires analytics are computed and stored in cloud [1, 7]. Whereas, the data required to make time-sensitive decisions are to be kept very close in the network edge [8, 9]. If we try to move these data from the network edge to the data center, the latency is increased [4, 8, 10] and the bandwidth capacity also outstrips. Also, the cloud server communicates only with the IP but not with other protocols (MQTT, XMPP, DDS, and AMQP) used by IoT devices.

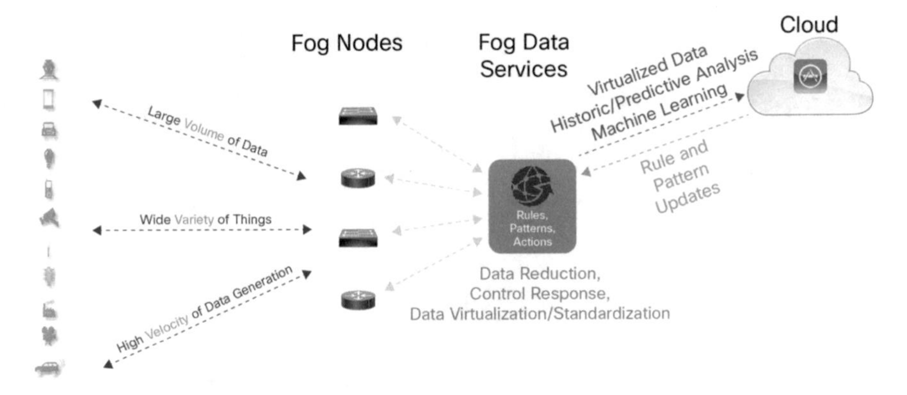

Fig. 1 Fog data services coordinate the movement of data from fog to cloud

3 Fog Computing Model

As IoT devices generate data constantly, analysis of those data must be very rapid. For example, when the temperature in a chemical factory is crossing the acceptable limit, corrective action must be taken immediately. In this case, if the temperature reading travels from fog to cloud and then analyzed in cloud, then there is an opportunity for any disaster. The fog extends the cloud to be closer to the things that produce and act on IoT data. These devices are called fog nodes that can be deployed anywhere with a network connection. Any device with computing, storage, and network connectivity can be a fog node. Examples include switches, routers, embedded servers, and video surveillance cameras. Analyzing IoT data close to where it is collected minimizes latency. It also minimizes the network traffic and keeps the sensitive data inside the network.

4 Working Principle

Developers write IoT applications for fog nodes at the network edge that is very close to the device. The network edge reads the data from IoT devices and directs different types of data to the optimal place for analysis (Fig. 1). Depending upon the time sensitivity, the data is handled. The most time-sensitive data is analyzed on the fog node and are stored for 1–2 h, the data that can wait for a few seconds or minutes are analyzed on the aggregate node and the data with less time sensitivity is placed on the cloud and then analyzed [11].

The IoT enables billions of devices to be connected and communicated to each other that lead to a better lifestyle. It enables creation of new business model, education system, transportation, and so on. To enable this, there is a need for different levels of communication. There are different levels of communication such as,

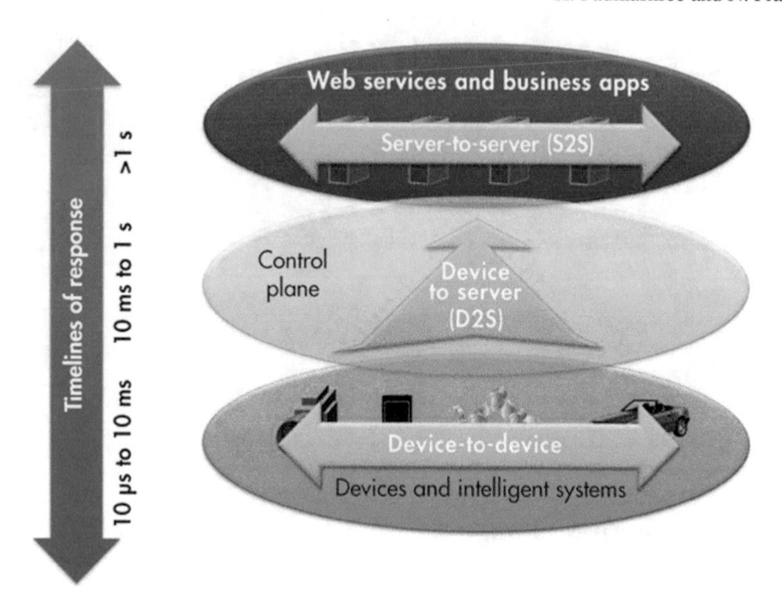

Fig. 2 Taxonomy for basic protocols [12]

a. Device to Device (D2D): The intelligent devices communicate with each other.
b. Device to Server (D2S): The data collected from the devices are sent to the server infrastructure.
c. Server to Server (S2S): The server infrastructure shares the data collected by the devices, provide it back to devices, and to analyze the data.

Today's Internet supports hundreds of protocols. The IoT will support hundreds more. It is important to understand the class of protocols to be used. MQTT and XMPP enable D2S communication. DDS enables D2D communication, whereas AMQP is designed to enable S2S communication.

Figure 2 shows the simple taxonomy for the basic protocols with the response time for each protocol [12].

4.1 Message Queue Telemetry Transport

MQTT, the Message Queue Telemetry Transport, is a simple and reliable protocol that enables device-to-device data communication in particular, in remote monitoring (Fig. 3). It collects data from many small devices and transports it to the IT infrastructure that is to be monitored or controlled from the cloud [12]. This protocol works on top of TCP, is comparatively slow and the time taken is measured in seconds. This

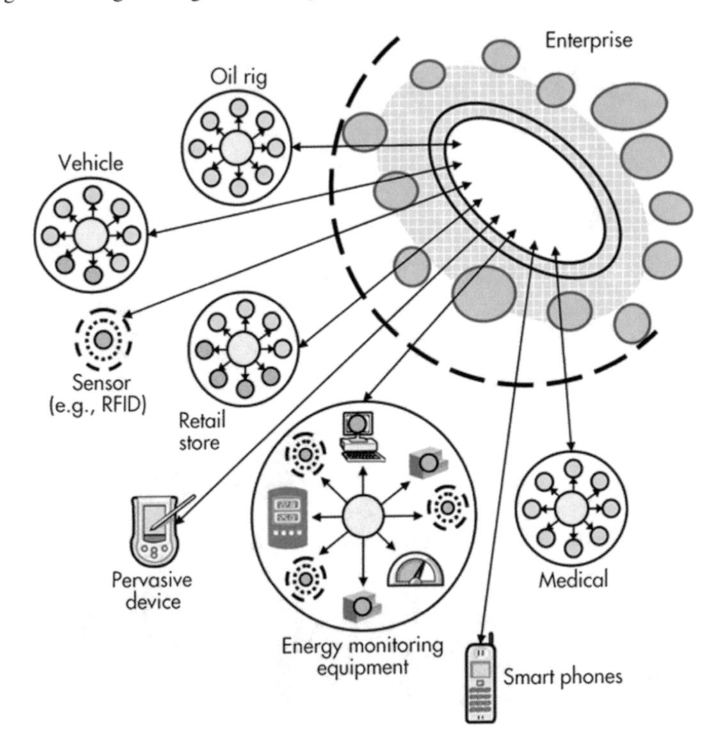

Fig. 3 MQTT implements a hub-and-spoke system [12]

protocol collects data from many resources, transport them into enterprise technologies, so that the IT infrastructure uses the data for analysis. Applications for MQTT include power usage monitoring, lighting control, intelligent gardening, and so on.

4.2 *Extensible Messaging and Presence Protocol*

XMPP—Extensible Messaging and Presence Protocol also called as Jabber, is a messaging protocol that connects people through text messages.

To make the communication natural, it uses XML text format. It runs over TCP, or over HTTP on top of TCP. In Fig. 4, it uses name@domain.com addressing scheme that helps connect people with huge Internet haystack [12]. This protocol is also not much faster and works in seconds but proposes an easy way to address a device. Most of the implementation uses polling, or checking for updates on demand [12]. Bidirectional streams over Synchronous HTTP or BOSH protocol lets the server to push messages. Many applications use XMPP, where the devices connected to a web server can be controlled by phone or tablets [12]. Since it has a good addressing, security, and scalability factors, it is well suited for consumer-oriented IoT applications.

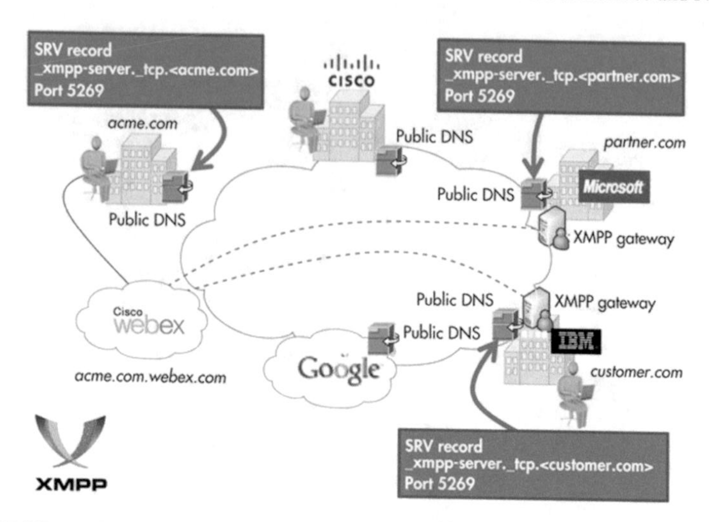

Fig. 4 XMPP provides text communication between points [12]

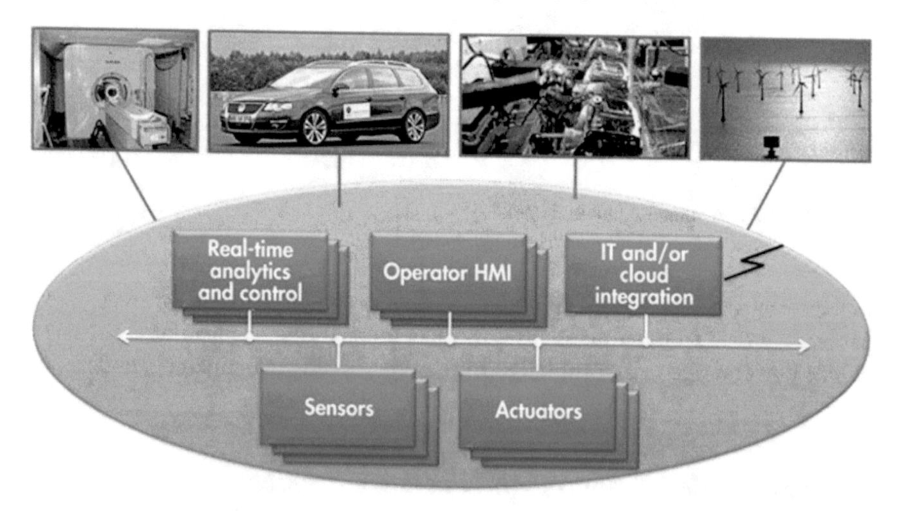

Fig. 5 Data Distribution Service (DDS) implements a publish/subscribe architecture [12]

4.3 *Data Distribution Service*

Sometimes, many high-performance devices work together as a single system. These devices communicate with each other, without the intervening of servers, i.e., the devices is connected to each other and the data is distributed between them. Unlike MQTT and XMPP, the Data Distribution Service (DDS) supports device-to-device communication (Fig. 5). It is a data-centric protocol and delivers messages at faster rate to many simultaneous receivers.

Normally, the devices demand data much faster than the IT infrastructure and the data is transferred in microseconds. Instead of TCP, DDS offers detailed Quality-of-Service (QoS) control, multicast, configurable reliability, and pervasive redundancy [12]. DDS implements "bus" communication with a relational data model. The data bus controls data access and updates by many simultaneous users.

DDS is used in many applications including military systems, wind farms, hospital integration, medical imaging, asset-tracking systems, and automotive test and safety.

4.4 Advanced Message Queuing Protocol

The Advanced Message Queuing Protocol (AMQP) is all about queues that send transactional messages between servers and can process thousands of queued transactions (Fig. 6).

AMQP is suitable for server-based analysis functions. It is mostly used in business messaging where the devices can be mobile handsets, communicating with back-office data centers. TCP is used for communications to enable reliable point-to-point connection. The endpoints must acknowledge the acceptance of each data focusing on no data loss. It provides a transaction mode with a multiphase commit sequence. The

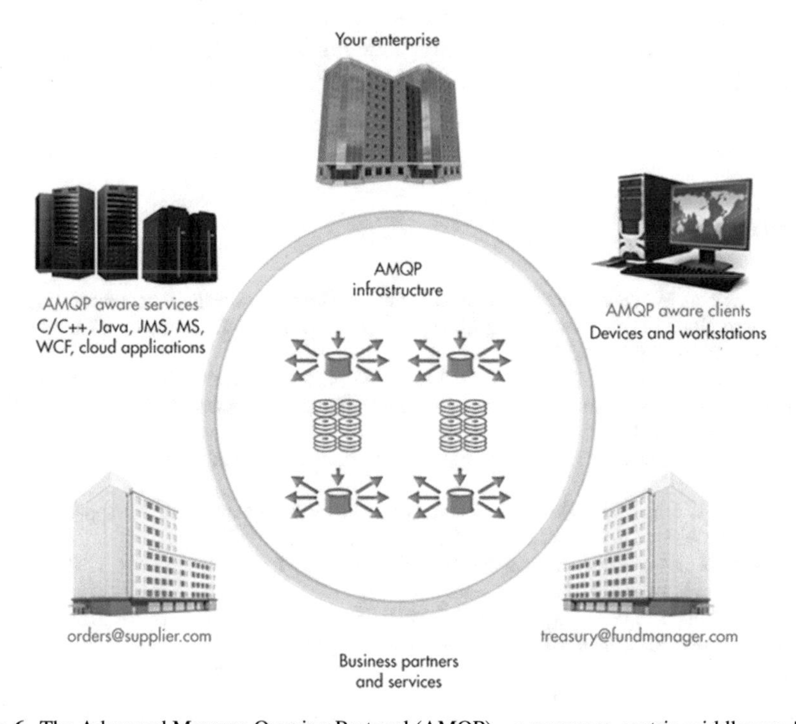

Fig. 6 The Advanced Message Queuing Protocol (AMQP)—a messages-centric middleware [12]

middleware focuses on ensuring each message are delivered as intended, regardless of failures or reboots [12].

Although we have discussed the four major protocols, the most critical distinction comes down to the intended applications. Inter-device data usage is totally different from device data collection. For example, to control the light switch use XMPP, to generate that power use DDS, to monitoring the transmission lines use MQTT and to analyze the power usage at the data center use AMQP [12].

Overlapping exists here. For instance, DDS can send and receive data from the cloud, and MQTT can send information back out to devices. The major difference between these protocols lies in their fundamental goals, the architectures, and the capabilities. All of these protocols are critical to the rapid evolution of the IoT.

5 Conclusion

Hence, we can say that the fog or edge computing is a good companion for cloud to handle the tremendous data generated daily from the IoT. It solves the issue of exploding data volume, variety, and velocity by processing data closer to where it is produced and needed. It reduces the latency by eliminating a round trip to the cloud for analysis. It reduces the cost of higher bandwidth by offloading gigabytes of network traffic from the core network. It also protects sensitive data by analyzing it locally. Hence, fog is best suited for organizations that look for increased business agility, higher service levels, lower price for storage, and improved safety to handle IoT data.

References

1. Sasitharagai, M., Padmashree, A., Dhanalakshmi, T., Gowri, S.: Dynamic resource management in large cloud environments using distributed gossip protocol. Int. J. Internet Comput. **1**(4) (2012). ISSN: 2231-6965
2. Zhou, J., Leppanen, T., Harjula, E., Ylianttila, M., Ojala, T., Yu, C., Jin, H.: Cloudthings: a common architecture for integrating the internet of things with cloud computing. In: CSCWD, 2013. IEEE (2013)
3. What is fog computing? Available: http://archive.thoughtsoncloud.com/2014/08/fog-computi ng/
4. Khan, S., Parkinson, S., Qin, Y.: Fog computing security: a review of current applications and security solutions. J. Cloud Comput. Adv. Syst. Appl. **6**(1), 19 (2017)
5. Prasath, N., Sengottuvelan, P.: Evaluation of link quality for routing in DSR. J. Theor. Appl. Inf. Technol. **59**(1) (2014). ISSN: 1992-8645
6. Thierry Coupaye: Fog Computing and Geo-Distributed Cloud, Nov 2014. Available: https://r echerche.orange.com/en/fog-computing-and-geo-distributed-cloud/
7. Botta, A., de Donato, W., Persico, V., Pescape, A.: Integration of cloud computing and internet of things: a survey. Future Gener. Comput. Syst. (2015)
8. Cisco Fog Data Services. Available: http://www.cisco.com/c/dam/en/us/solutions/collateral/tr ends/at-a-glance-c45-734964.pdf

9. Diaz, M., Martin, C., Rubio, B.: State-of-the–art, challenges, and open issues in the integration of internet of things and cloud computing. J. Netw. Comput. Appl. **67**, 99–117 (2016)
10. Fog Computing and the Internet of Things: Extend the Cloud to Where the Things Are, Cisco White paper, 2015. Available: http://www.cisco.com/c/dam/en_us/solutions/trends/iot/docs/computing-overview.pdf
11. Cisco Fog Computing Solutions: Unleash the Power of the Internet of Things, Cisco White Paper, 2015. Available: http://www.cisco.com/c/dam/en_us/solutions/trends/iot/docs/computing-solutions.pdf
12. Schneider, S.: Understanding the protocols behind the internet of things. Electron. Des. (2013)

Ms. A. Padmashree is a Senior Assistant Professor in the Department of Computer Science and Engineering, KPR Institute of Engineering and Technology, Coimbatore, Tamil Nadu, India. She received her B.E., in Computer Science & Engineering in 2005 and M.E. in Computer Science & Engineering under Anna University, Chennai, Tamil Nadu, India in 2011. She is having 10 years of teaching experience. Her areas of research and scientific interests include: Cloud computing, Image processing and IoT.

Dr. N. Prasath is an Associate Professor in the Department of Computer Science and Engineering, KPR Institute of Engineering and Technology, Coimbatore, Tamil Nadu, India. He completed his B.E. Electronics and Communication Engineering, under Anna University, Chennai, India in 2006, and M.Tech. Degree in Computer Science and Engineering from the SASTRA University, Thanjavur, India in 2009 and Ph.D. degree in Information and Communication Engineering from Anna University, Chennai, India in 2017. He is having 10 years of experience in teaching. He has published more than 10 research papers in refereed conferences, journals and served as a technical program committee in several national and international conferences. His research interests include MANET, Sensor Networks, Protocol Design, Medium Access Control (MAC), Routing Protocols and Qos.

Block Link Flooding Algorithm for TCP SYN Flooding Attack

C. M. Nalayini and Jeevaa Katiravan

Abstract In the recent years, SDN is an emerging architecture that is ideal for today's application. It is dynamic and also cost-effective in order to deal with the networking application. It consists of two planes, namely data plane and control plane that are separated from each other. Open flow is the default protocol that has been used in SDN, and it works on the basis of rules framed by the control plane. It may lead to a denial-of-service attack called control plane saturation attack. To overcome this attack, Avant-Guard technique [2] was introduced which overcomes the control plane saturation attack but it leads to another attack called Buffer saturation attack due to SYN flooding. Then, Line Switch technique was introduced with proxy to blacklist the spoofed IP address in order to block the node. But it was not a permanent solution to the problem. Therefore, we proposed a Block Link Flooding (BLF) algorithm which uses Legitimate User Table (LUT) to provide an effective solution. From the experimental results, our Block Link Flooding algorithm is able to identify and block the spoofed IPs and discards the link to the server. This technique is efficient for saving energy up to 33%.

Keywords SDN · Denial of service · Block link flooding · Buffer saturation Legitimate user table

1 Introduction

Nowadays, software automation beats the entire information technology world. A new paradigm came into the picture called SDN which provides directly programmable, virtual-centralized network control, and full automation with elasticity. It separates the network layer and data layer for effective flexibility

C. M. Nalayini (✉) · J. Katiravan
Department of Information Technology, VEC, Chennai, India
e-mail: nalayinicm13@gmail.com

J. Katiravan
e-mail: jeevaakatir@gmail.com

S. Smys et al. (eds.), *International Conference on Computer Networks and Communication Technologies*, Lecture Notes on Data Engineering and Communications Technologies 15, https://doi.org/10.1007/978-981-10-8681-6_83

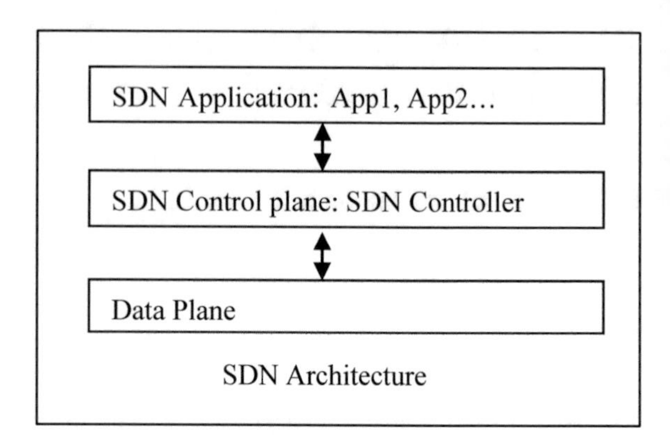

Fig. 1 SDN layers [2]

and automation of information. The network layer does controlling functionality also called as control plane. It acts as the middleware, which bundles the communication between two planes and the network infrastructure density to support independent direct programming for applications at the data plane. The data layer does the data forwarding functionality also called as data plane. The main challenge in SDN is the communication between the data and control plane. It becomes quite critical and leads to a severe bottleneck in terms of control plane saturation attack, which degrades the performance of the entire network.

Open flow Switch [1] is an existing methodology to support effective communication between the control and data plane through communication protocol. It has the centralized controller to maintain the forwarding rules in the respective flow tables. During routing, if there is no matching rule available in the flow table for the incoming packet, control plane is disturbed immediately to write new rules to serve the packet. Due to ineffective scalability and overloading, the control plane leads to control plane saturation attack [2, 3] which in turn, performs rigorous SYN flooding throughout the network. To handle such case, if the controller tries to manage n number switches may lead to severe network damage [4, 5] (Fig. 1).

Avant-Guard [2] methodology is an extension to Open Flow Switch paradigm to handle the control plane saturation attack. It has two features built inside the Open Flow Switch such as the Connection Migration (CM) and Actuating Trigger (AT). CM has the ability to control over control plane saturating attack by acting as the proxy for each and every new connection. But it is difficult to maintain those connections again and it becomes the overhead to the network. AT is another feature built to manage the control messages.

Line Switch is another paradigm [6] to overcome the problems of Avant-Guard. It coalesces both the blacklisting feature and the SYN proxy methodology. Proxy is maintained for every new connection, so that the number of SYN packets can imbibe a small probability of the proxy. The main theme of proxy is that it forces the attacker should use the original IP, therein the IP is immediately blacklisted and the respective link gets blocked.

2 SYN Flood Attack

SYN flooding attack is a type of denial-of-service attack. The aim of this attack is to send SYN request continuously such that the server could not process another request from the legitimate client. Denial-of-Service attack in SDN simply targets the control plane by bombarding it with n number of flows [7], this in turn severely affects the bandwidth and degrades the performance of the network. N flow entries can make the flow table full which leads to unnecessary traffic and packet loss [8].

2.1 Three-Way Handshake

A three-way handshake is a mechanism in which the client and the server communicate with each other. A client starts the handshake by sending SYN packets to the server. Then, the server replies with SYN + ACK packet to the client to continue with the further communication.

2.2 Dropped Connections

A proper three-way handshake is possible only with legitimate users. But when invader tries to enter the connection, it will lead to dropped connection resulting in half-opened state. The invader does not send ACK and hence makes use of server resources. It also prevents the server from further processing the request from another legitimate user, which in turn, consumes a large amount of time. This state of connection is called dropped connection (Fig. 2).

In Avant-Guard paradigm [2], the connection migration module handles the SYN flooding through its proxy method using the round trip mechanism. This, in turn, discards the spoofed connections. Its work is to delay the communication by sending the legitimate packet to the migration module also called delayed connection migration before sending it to the control plane. The impact of SYN flood at the data plane uses SYN cookie mechanism (Fig. 3).

3 Related Work

Saini et al. [9] proposed an SCTP (Stream Control Transmission Protocol) four-way handshake mechanism to defend the server from SYN flooding attack. It was built with the Cookie to deal with ECHO and ACK parameters. UPPAL model checker was used to validate the handshaking mechanism of SCTP.

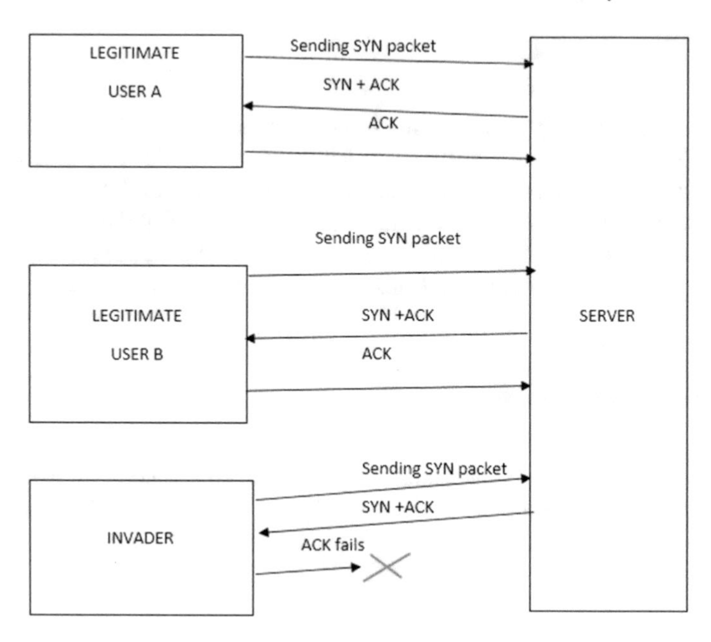

Fig. 2 TCP three-way handshake

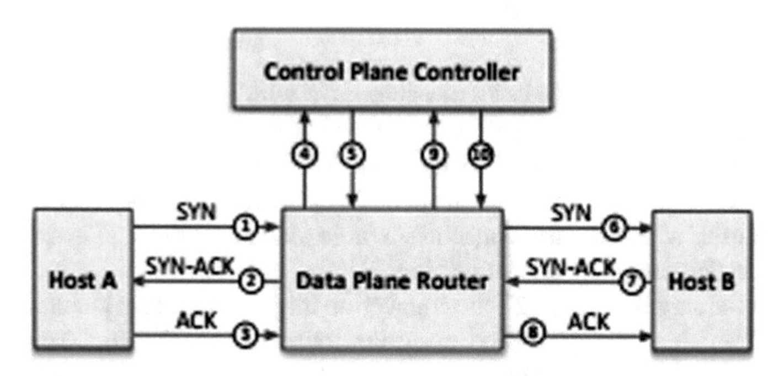

Fig. 3 Avant-Guards [1]—connection migration

Gkountis et al. [10] proposed a lightweight method with a set of complex rules to transfer the legitimate packets to the network. Experimental results are done by Testbed which includes DDoS mitigation methods to provide an effective solution that makes the ease for the SDN mobile clients.

Wang et al. [11] mainly focuses on the latency parameter. Minimization of latency becomes a big challenge between the controller and the network. Therefore, effective network partitioning should be done through the proposed optimized k-means algorithm in order to find the proper placement of controller to achieve the latency minimization. Their simulation results provide better solution for effective network partition.

Hu et al. [12] introduced the various new concepts of SDN scheme, its challenges, characteristics, and security to the new researchers. Their main aim is to explore SDN to every researcher so that many news schemes will come into reality to handle both wired and wireless network. Both pros and cons of SDN parameters are analyzed effectively including open flow technology to help the SDN users.

Hu et al. [13] found a new solution for the effective placement of controllers in the network path with full reliability. The number of flows in the path, number of backup paths to handle traffic, number of backup controllers, and latency should be decided by the proposed reliability-aware controller. Their simulation results proved the np hardness. Proper network partitioning should be done to achieve both latency and reliability

Shang et al. [14] proposed a flood defender to address the DOS attack at the open flow—SDN network. Three techniques such as table-miss, packet filtering, and flow management were introduced to analyze the traffic in the flow path, number of adjacent, delay, and packet loss. They measured the packets flows of the open switch scenario in both the hardware and software environment using flood guard and flood defender, and proved their methods controls 96% of the attack traffic in the network.

Ambrosin et al. [15] mainly addressed the control plane saturation produced by SYN flooding. They added many new features in the Line Switch technique based on the probability of proxy usage. They compared both open flow and Avant-Guard techniques and proved the advantages of using Line Switch wherein the overhead was reduced.

4 Proposed Approach: Block Linking Algorithm

There are several limitations in the existing system. Avant-Guard is the extension of open flow switches, but it produces control plane saturation attack. This, in turn, leads to another approach called Line Switch, though it provides a solution to control saturation attack, but blocking the link creates n chances of data transmission via the same affected link in future. We proposed an algorithm called Block Link Flooding algorithm. It easily identifies the spoofed IP by analyzing the data packets being received at the server. The main purpose of using BLF algorithm is to block the Link through which the attacker performs TCP SYN Flooding attack. The attacker sends SYN request to the respective server as a legitimate client. The server processes the request and sends SYN + ACK. If the server does not receive further ACK, then the user is identified as spoofed IP and the link with the server is blocked. Even though the attacker provides ACK, the size of data packets to be transmitted and the time taken to transmit data packets are analyzed. The server has fixed size of data packets and end time for transmission. If the size of data packets and time limit exceeds then Legitimate User Table (LUT) blacklist, the user as Spoofed IP and the Link of the user is blocked and discarded further (Fig. 4).

LUT table is very much useful in order to make a quick analysis of the data packets. BLF algorithm is of two-phase detection phase and Avoidance Phase. In

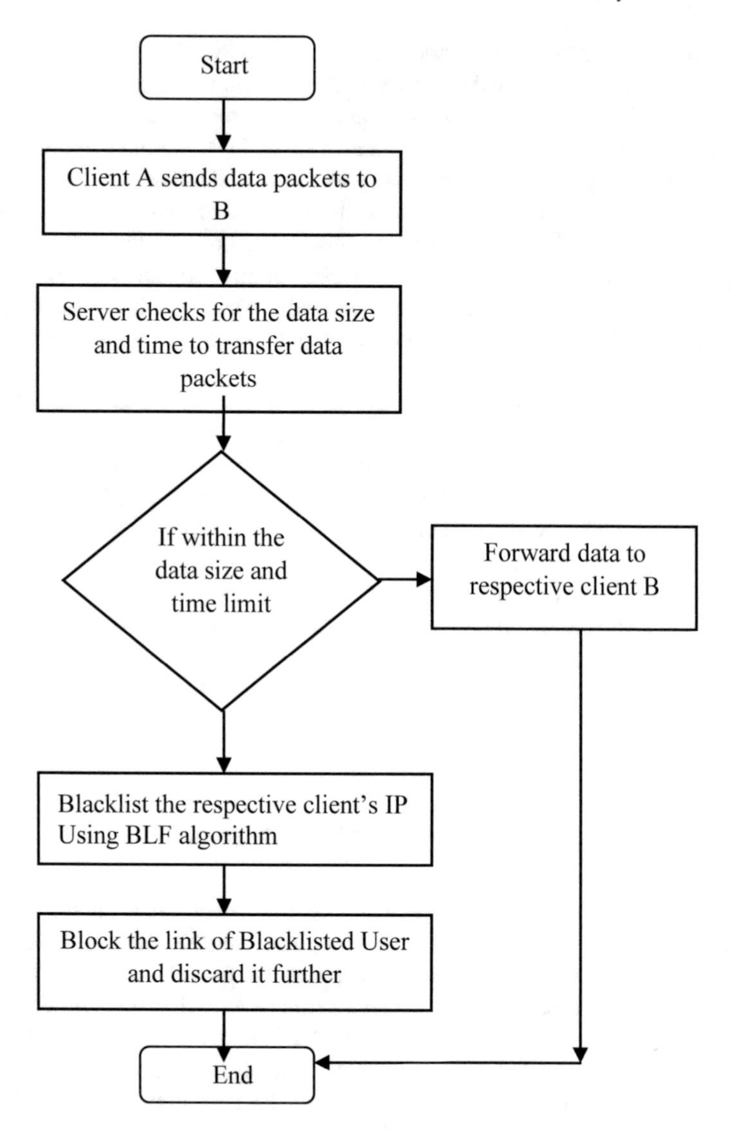

Fig. 4 An overview of the proposed algorithm

the Detection phase of BLF algorithm, the user with spoofed IP is identified and blacklist entry is made true in the LUT. In case of Avoidance phase, the blacklisted user is removed from the LUT table and also the Link of the user with the server is blocked. The use of BLF algorithm results in an energy saving up to 33% using this LUT table analysis.

4.1 BLF Detection Algorithm

Step: 1 Start.

Step: 2 The threshold value is assigned as "1" for successful delivery of data packet and "0" for failure in delivery of data packets. Data limit value is 3500 kb. The end time specified is 30 s.

Step: 3 if (LUT (table_size-1) ==legitimate user), it indicates that LUT is full.

Step: 4 If the user request for connection, then perform the following steps.

- Step: 4.1 if (LUT[k].SYN==Null)
  Then, set (LUT[k].SYN==SYN.1) sending SYN request to the server.
- Step: 4.2 Wait for SYN + ACK from the server. Where LUT[k].SYN + ACK==SYN.1 + ACK
- Step: 4.3 Server waits for ACK message from user. Where LUT[k].ACK==ACk.1
  If we do not receive ACK, then the user is blacklisted. Then, LUT [Blacklist Entry==True]
- Step: 4.4 We can check the user is legitimate or not by the amount of data being sent within the assigned time limit value.
  - Step: 4.4.1 When data size = 2000 kb time taken = 5.50 ms
    If (data size > data limit) and (time taken > time limit)
    If (2000 kb > 3500 kb) and (5.50 ms > 30 s)
    Under this analysis, we could identify the user is legitimate user. Threshold value is set to 1.
  - Step: 4.4.2 When data size = 2000 kb time taken = 35 s
    If(2000 kb > 3500 kb) and (35 s > 30 s).
    Then, the user is detected as spoofed IP. Threshold value is set to 0.
    Then we set, LUT [blacklist Entry == true].

Step: 5 Repeat from step 3 to step 4 until LUT overflows.

Step: 6 Stop.

In detection phase of BLF, the user makes an entry to the Legitimate User Table. The user first checks whether the LUT is full. Then, after making an entry, the user provides SYN request to server. Then, the SYN field of LUT is set to LUT[k]SYN = SYN.1 (it represents the first SYN request). The user waits for the ACK from the server as SYN + ACK. Then, the SYN +ACK field of the first user is entered in LUT as SYN + ACK.1, now the user must provide a perfect three-way handshake by sending ACK message to the server.

If we receive ACK message, then we undergo the process of comparing the size of data and time limit. If the condition fails, the user is spoofed Ip and it is marked in LUT [Blacklist Entry==True].

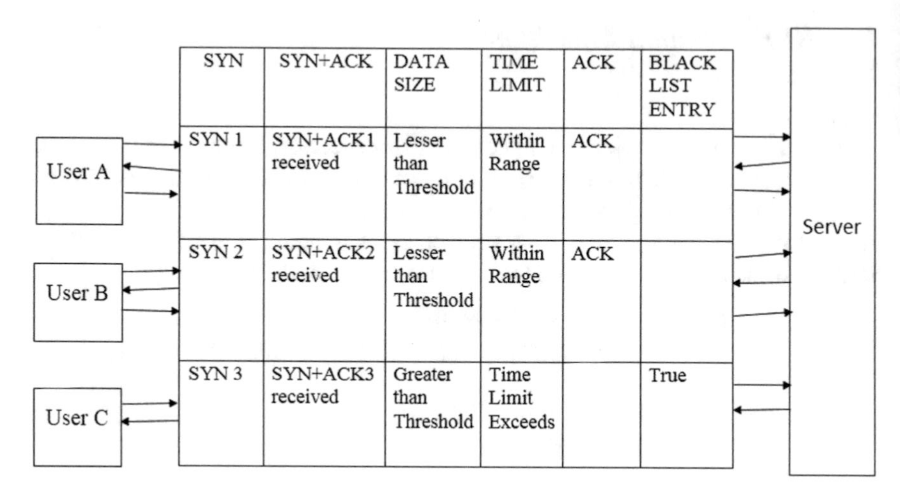

Fig. 5 Legitimate user table

If we do not receive ACK message, then eliminate the user from LUT and block the entire Link of the user from further communication.

4.2 *BLF Avoidance Algorithm*

Step: 1 Start.
Step: 2 Check if (Threshold value==0), then remove the user from LUT as data have been transferred successfully obtain free space for other users.
Step: 3 Also check if (Blacklist Entry==true) then block or remove the link of invader with spoofed IP address thereby avoiding the invader from connection establishment with the server.
Step: 4 Stop.

In avoidance phase, the need to check for threshold value which gives the success rate of the data packets. If the server receives threshold value as 0, then it represents the failure. Thus, the user must be removed from LUT table. Also, if the Blacklist Entry in LUT table is true, remove the User and block the link of the user with the server (Fig. 5).

5 Experimental Results

The below graph is based on the analysis of Packet Delivery Ratio. During this analysis, 2000 kb of data was taken and which was sent from sender to the receiver

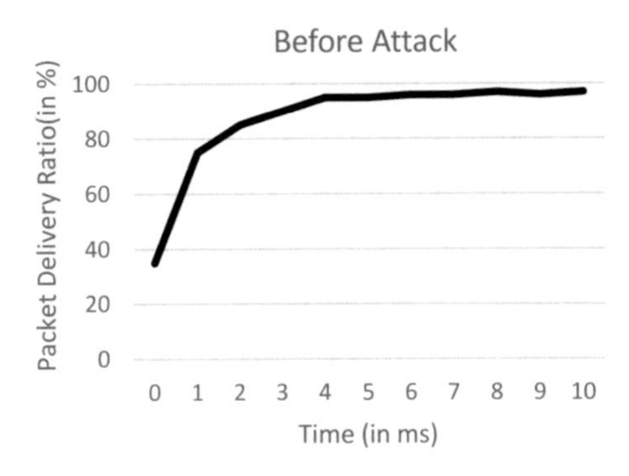

Fig. 6 Before attack

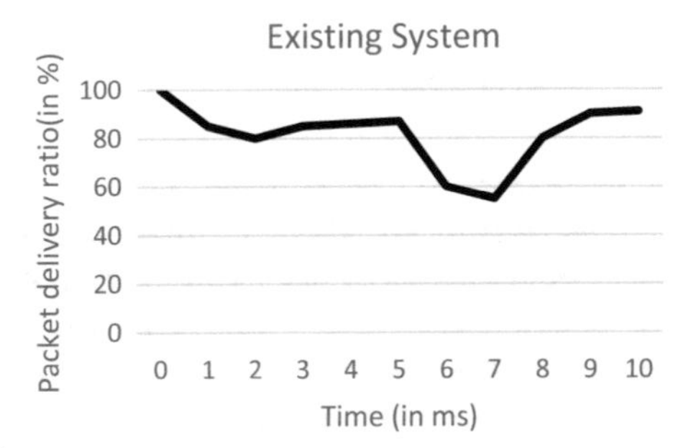

Fig. 7 Using existing system

within 5 ms. Thus, we were able to achieve 98% of Packet Delivery Ratio. The actual end time specified for packet transfer is 30 s and, we were able to achieve successful packet delivery within this time (Fig. 6).

Using the existing system, we infer that when the intruder arrives the packet delivery ratio is affected to a greater extent 2000 kb of data is being sent where we end up with Packet Delivery Ratio of 85%. There are many fluctuations where the existing system reaches minimum PDR as 55%. There is heavy data loss. The data being transmitted is received in 10 ms (Fig. 7).

From the graph analysis, we infer that using the BLF algorithm, we are able to analyze the data packets whether it is from spoofed IP or not and, then we perform transfer of 2000 kb data packets within the time limit of 5.67 ms, thus, we achieved packet delivery Ratio of 97.6%. Thus, our algorithm plays a vital role in saving energy up to 33% and it was possible only through LUT we have used. This table helps the server to analyze the data packets easily and blacklist spoofed IP (Fig. 8).

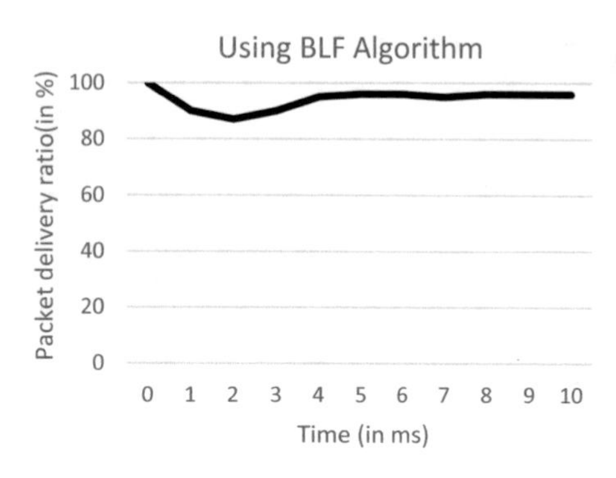

Fig. 8 Using BLF algorithm

6 Conclusion

In this paper, we have proposed BLF algorithm. Line Switch is one of the existing mitigation methods which identifies and blocks spoofed IP address, but there are many issues in this method. This method blocks the node of spoofed IP address but the link remains, through which data transfer may occur. Thus, our BLF algorithm overcomes this issue by blocking and discarding the entire link of the invader. It relies on the data size and time limit. The user is analyzed based on these constraints using Legitimate User Table. LUT table is an effective method for identifying invader. Our proposed algorithm proved that TCP SYN flooding attacks are identified and avoided using analysis from LUT table. In future, we will proceed with Expanded LUT tables.

References

1. Peng, T., Leckie, C., Ramamohanarao, K.: Survey of network-based defense mechanisms countering the DoS and DDoS problems. ACM Comput. Surv. **39**(1) (2007)
2. Shin, S., Yegneswaran, V., Porras, P., Gu, G.: Avant-Guard—Connection Migration. AVANT-GUARD: Scalable and Vigilant Switch Flow Management in Software-defined Networks. In: CCS'13, pp. 413–424 (2013)
3. Wang, H., Xu, L., Gu, G.: OF-GUARD: a DoS attack prevention extension in software-defined networks. USENIX Open Network Summit (2014)
4. Benton, K., Camp, L.J., Small, C.: Open flow vulnerability assessment. In: HotSDN, pp. 151–152 (2013)
5. Kloti, R., Kotronis, V., Smith, P.: Open flow: a security analysis. In: ICNP'13, pp. 1–6 (2013)
6. Ambrosin, M., Conti, M., De Gaspari, F.: University of Padua, Italy, Radha Poovendran University of Washington, USA. LineSwitch: efficiently managing switch flow in software-defined networking while effectively tackling DoS attacks. In: ASIA CCS'15, 14–17 Apr 2015, Singapore
7. Ahmad, I., Namal, S., Ylianttila, M., Gurtov, A.: Security in software defined networks: a survey. IEEE Commun. Surv. Tutor. **17**(4), 2317–2346 (2015)
8. Kandoi, R., Antikainen, M.: Denial-of-service attacks in open flow SDN. In: IFIP 2015

9. Saini, S., Fehnker, A.: Evaluating the stream control transmission protocol using Uppaal. In: Formal Analysis of Real Systems (MARS 2017) EPTCS 244, 2017, pp. 1–13. https://doi.org/10.4204/eptcs.244.1
10. Gkountis, C., Taha, M., Lloret, J., Kambourakis, G.: Lightweight algorithm for protecting SDN controller against DDoS attacks. In: 2017 10th IFIPValencia, Spain on Published in Wireless and Mobile Networking Conference (WMNC), 25–27 Sept 2017
11. Wang, G., Zhao, Y., Huang, J., Duan, Q., Li, J.: A K-means-based network partition algorithm for controller placement in software defined network. In: 2016 IEEE International Conference on Published in Communications (ICC), 22–27 May 2016 Kuala Lumpur, Malaysia
12. Hu, F., Hao, Q., Bao, K.: A survey on software-defined network and openflow: from concept to implementation. In: Published in IEEE Communications Surveys & Tutorials, vol. 16, Issue 4, Fourthquarter 2014
13. Hu, Y., Wang, W., Gong, X., Que, X., Cheng, S.: On reliability-optimized controller placement for software-defined networks. Published in: China Communications, vol. 11, Issue: 2, Feb 2014
14. Shang, G., Zhe, P., Bin, X., Aiqun, H., Kui, R.: FloodDefender: protecting data and control plane resources under SDN-aimed DoS attacks. www4.comp.polyu.edu.hk/~csbxiao/paper/…/2017_INFOCOM_FloodDefender.pdf
15. Ambrosin, M., Student Member, IEEE, Conti, M., Senior Member, IEEE, De Gaspari, F., Poovendran, R., Fellow, IEEE: LineSwitch: tackling control plane saturation attacks in software-defined networking. https://www2.ee.washington.edu/research/nsl/papers/TON.pdf

Dr. Jeevaa Katiravan is working as an Associate Professor and Head in the Department of Information Technology at Velammal Engineering College, Chennai. He has completed his UG degree on B.Tech-IT in Velammal Engineering College, Madras University and completed his M.Tech (IT) from Sathyabama University. He has obtained his Ph.D. from Anna University in Network Security. His areas of specialization are Network Security and Information Security. He has 15 years of teaching experience and handled more than 15 subjects including UG and PG. He is currently involved in New Gen IEDC research projects. He published many papers in International Journals, National and International Conferences.

Mrs. C. M. Nalayini is working as an Assistant Professor in the Department of Information Technology at Velammal Engineering College, Chennai. She has completed her UG degree on B.Tech-IT in Velammal Engineering College, Anna University and obtained her M.Tech (CSE) from Dr.M.G.R University and secured University First Rank. She is currently pursuing Ph.D. in Network Security at Anna University and her areas of Specialization are Network Security and Database Systems. She has 13 years of teaching experience and handled more than 15 subjects including UG and PG. She published many books and papers in National and International Conferences and International Journals.

Fuzzification of Context Parameters for Network Selection in Heterogeneous Wireless Environment

Shilpa Litake and Mukherji Prachi

Abstract In heterogeneous wireless networks, vertical handoff support provides seamless connectivity. There are multiple context parameters that influence the decision of vertical handover, hence Multiple Attribute Decision-Making (MADM) algorithms are leveraged in deciding the best network for handover. This paper proposes an MADM-based technique to select an appropriate radio access, out of available WLAN and WiMAX radio access technologies. The authors have proposed a novel HUETANS:Handoff Urgency Estimator and Target Access Network Selector module to facilitate vertical handoff. HUETANS ensures Quality of Service (QoS) by incorporating fuzzy logic controllers to handle the vagueness of wireless environment parameters. Usage of Grey Prediction technique for predicting Received Signal Strength (RSS) further restricts number of unnecessary handoffs at low level. Simulation results show that the proposed HUETANS module selects the best network.

Keywords Vertical handoff · MADM · Fuzzy logic · QoS · Grey prediction

1 Introduction

Users are offered with variety of services in Beyond 3rd Generation (B3G) wireless networks. Single Radio Access Technology (RAT) cannot support all types of services. The Next-Generation Wireless Networks (NGWN) consists of multiple overlapping wireless networks. Various RATs are integrated to provide Quality of Service (QoS) to users. Multiple radio interface Mobile Terminals (MTs) are used

S. Litake (✉)
Vishwakarma Institute of Information Technology, Pune, India
e-mail: shilpalitake@yahoo.co.in

S. Litake
PVG'S College of Engineering and Technology, Pune, Maharashtra, India

M. Prachi
M.K.S.S.S Cummins College of Engineering, Pune, Maharashtra, India
e-mail: prachi.mukherji@cumminscollege.in

S. Smys et al. (eds.), *International Conference on Computer Networks and Communication Technologies*, Lecture Notes on Data Engineering and Communications Technologies 15, https://doi.org/10.1007/978-981-10-8681-6_84

in NGWN to access different wireless networks. When MT switches from one wireless network to another, vertical handoff takes place. Vertical handoff consists of four phases, namely handover initiation, system discovery, handover decision and handover execution [1]. Vertical Handover Decision (VHD) algorithm evaluates the candidate networks to select the best network for handover.

The IEEE 802.21 specification [2] defines the Media Independent Handover Function (MIHF), to support vertical handovers. MIHF works as the layer between data link layer and network layer. It provides services to upper layers, which is independent of lower layer technologies. For efficient handoff, MIH framework provides timely information about the link quality. To provide network context information, IEEE 802.21 standard includes Information Server (IS). Response and query type of mechanism are implemented between IS and user terminal.

Co-existence of WiMAX and WLAN networks is considered for analysis in the proposed work. Generally, whenever WLAN network is available trend is to select WLAN network as serving network, but performance of WLAN degrades as the number of simultaneous users increases. In the traditional approaches, the handover decision is taken based on single parameters like RSS or bandwidth. This does not result into optimum performance of the system. Multiple context parameters like throughput, delay, jitter, service cost and security, etc., decide the QoS for the user. The authors have proposed a novel model HUETANS:Handoff Urgency Estimator and Target Access Network Selector. HUETANS evaluates the necessity of handoff by analysing the input parameters from serving wireless network. Handoff Urgency Parameter (HUP) is derived to trigger the network selection process. Degrading performance of serving network results in higher value of HUP. The MADM or Multiple Criterion Decision-Making (MCDM) algorithms have multiple objectives and multiple criteria which are conflicting that is some criteria are benefit type, whereas others are cost type. Alternatives are ranked according to the weights of the criteria. MADM algorithms, viz. Simple Additive Weighting (SAW), Technique for Order Preference by Similarity to Ideal Solution (TOPSIS) and VIKOR Serbian: VIseKriterijumsa Optimizacija I Kompromisno Resenje, meaning: multi-criteria optimization and compromise solution are used in the proposed approach. In wireless environment the criteria and consequences are not accurately known. Zadeh's work [3] provided mathematical formulation of fuzzy uncertainty. It is useful to use fuzzy set in MADM for modelling vertical handoffs. Input parameters in the form of crisp numbers are fed to the fuzzifiers, which transforms them into fuzzy set values of linguistic terms like low, medium and high. Fuzzy TOPSIS and Fuzzy VIKOR is utilized to implement network selection for dissimilar operating parameters.

The remainder of the paper is organized as follows: Sect. 2 gives insight on existing work of vertical handover. Detailed description of the proposed HUETANS module is presented in Sect. 3. Section 4 provides analysis of simulation results. Section 5 gives concluding remarks and future extensions.

2 Related Work

In this section, an overview of various vertical handover techniques is presented. Ahmad et al. [1] have provided a detailed discussion on various vertical handover schemes. They have classified the schemes into RSS based, QoS based, decision function based, network intelligence based and context based. In [4], L. Wang and G. S Kuo presented mathematical modelling techniques used for network selection. Comparison of various mathematical theories, viz. utility theory, multiple attribute decision-making, fuzzy logic, game theory, combinatorial optimization and Markov chain is provided. Fernandes and Karmouch [5] have discussed various wireless mobility scenarios along with their design challenges. The authors [5] have also proposed Context-Aware Mobility Management System (CAMMS) architecture for seamless handover. Fallon et al. [6] have proposed Fixed Route Adapted Media-Streaming Enhanced (FRAME) architecture for vehicular networks. FRAME enhances the performance of IEEE 802.21 MIH with the help of neural network triggers. Taleb and Ksentini [7] have proposed a VECOS, VEhicular COnnection Steering protocol framework for ensuring better QoS and Quality of Experience (QoE) of Long-Term Evolution (LTE) connected vehicles. Time-continuous Markov chain is used for network selection process. Mehbodniya et al. [8] have applied fuzzy logic to MADM VIKOR method for selecting the best network. Fuzzy VIKOR selects optimum network from simultaneously existing WLAN, WMAN and WWAN networks. Zhang [9] have proposed the algorithm that performs objective weight calculation using FAHP. Authors have also ranked the available networks using Grey Relational Analysis. In this access selection algorithm, call blocking probability is reduced with balanced network load. Kantubukta et al. [10] proposed Fuzzy TOPSIS-based Vertical Handover Decision (VHD) algorithm with application-specific rules to provide energy and QoS efficient network selection in heterogeneous wireless networks. Wilson et al. [11] proposed MADM algorithm based on Fuzzy logic controller. They have defined three phases preselection, discovery and decision-making. After defining simple rules a priori, the proposed system provides good approximation for levels of fitness.

3 Proposed System

The architecture of proposed HUETANS module is as shown in Fig. 1. It comprises of Handoff Urgency Estimator (HUE) module and Target Access Network Selector (TANS) module. The inputs required for the HUETANS Module are obtained from the User Equipment, Wi-Fi Access Point, WiMAX Base Station and IEEE 802.21 IS.

List of selected access networks is provided to the user equipment by the TANS module. HUE module is responsible for estimating necessity of handoff. It generates Handoff Urgency Parameter (HUP) to trigger the Target Access Network Selector

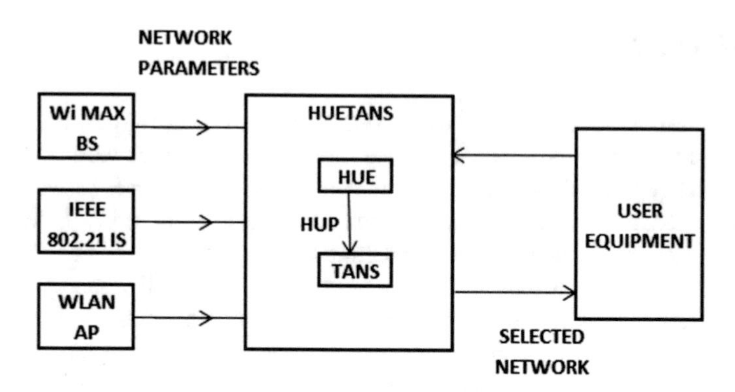

Fig. 1 System architecture

(TANS) block. If HUP is above threshold, it activates the TANS block for selecting the best network out of available candidate networks. The Handoff Urgency Estimator consists of Fuzzy Logic Controller. As depicted in Fig. 2, the obtained values of traffic class, predicted RSS, N/W load, user speed, delay and throughput of current Point of Attachment (PoA) are applied as an input to the fuzzifier. The RSS predictor block, predicts the RSS value after analysing the previous samples. This block helps in timely initiation of the handover process, by alerting the system regarding dropping signal strength before it actually goes below threshold. Grey prediction technique is used in the proposed approach for RSS prediction. Fuzzifier provides mapping of the crisp input values and fuzzy numbers. These fuzzy numbers are operated with inference rules to obtain HUP. Fuzzy inference rules are defined to ensure that requirements of traffic class are fulfilled. More value of HUP indicates greater necessity of handoff. As depicted in Fig. 3, Design of Target Access Network consists of fuzzifiers. It normalizes and brings the values of context parameters in the range of [0, 1]. Weight calculator block assigns priority to attributes by considering the requirements of traffic class. This block is designed using AHP and FAHP algorithms. These weights and normalized parameters are applied to MADM-based selector. After comparing with available networks, the selector block defines the best network out of available access networks.

The first step of the proposed module is to collect all attributes from the network nodes including the traffic class, viz. conversational, streaming, interactive and background. To decide the priority amongst these attributes, weights are assigned using Analytic Hierarchy Process (AHP) and Fuzzy Analytic Hierarchy Process (FAHP). The RSS predictor provides predicted value of RSS for HUP calculation. The HUP derived from fuzzy logic controllers is used to trigger selection of networks. MADM tools enable network selection step in association with IEEE 802.21-based IS, if obtained HUP is greater than the threshold.

For efficient handoff, attributes that are considered in the proposed system include Received Signal Strength (RSS), MS speed, distance between the MS and AP, net-

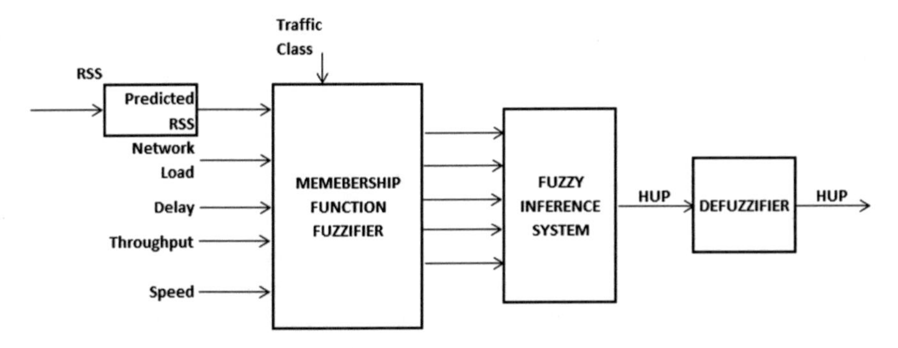

Fig. 2 Handoff urgency estimator module

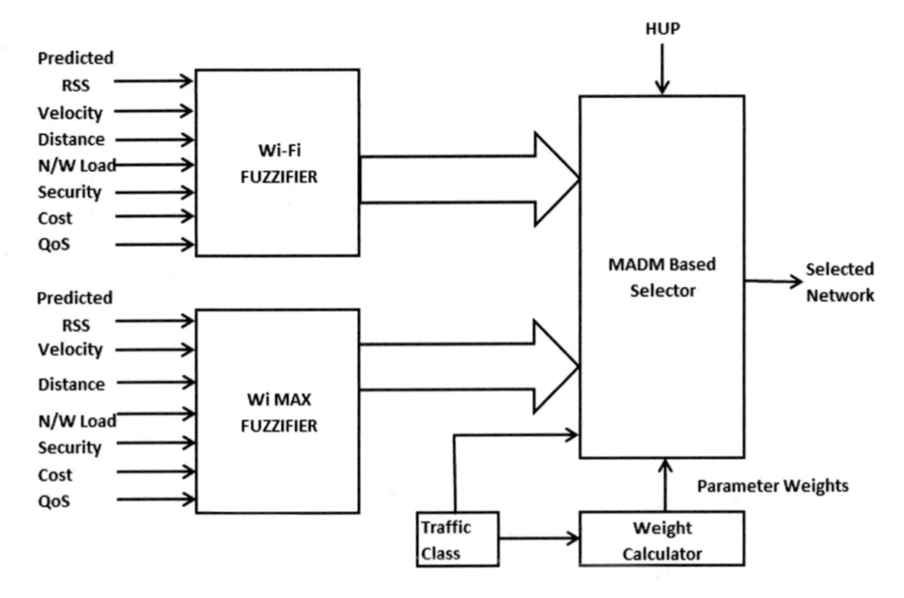

Fig. 3 Target access network selector module

work loading conditions, security of the network, monitory cost, QoS parameters like throughput, latency, jitter and Packet Loss Rate (PLR).

3.1 Weight Calculation Using Analytic Hierarchy Process (AHP)

The relative importance of the decision factors is defined by pair-wise comparisons between attributes after determining the objective and decision factors. To provide relative importance, AHP fundamental scale of importance is used [12].

While assigning the relative importance, QoS requirements of various traffic classes are taken into consideration.

3.2 Fuzzy Analytic Hierarchy Process (FAHP)

Uncertainty and vagueness associated with wireless network are handled efficiently using Fuzzy Analytic Hierarchy Process (FAHP) technique. AHP is extended with the use of Triangular Fuzzy Numbers (TFN) instead of crisp numbers while creating comparison matrix. TFN is represented as $x = (l; m; u)$ [13]; parameter l and u are lower and upper limits of each attribute and m is the threshold value.

3.3 RSS Prediction Using Grey Prediction Theory

Grey prediction technique was proposed by Deng [14]. The proposed scheme has utilized Grey differential Model GM (1,1) to predict RSS values of current Point of Attachment (PoA) and target network. Predicted values are used in fuzzy logic controllers that decide necessity of handoff. The input applied to GM(1,1) model [14] consists of n RSS samples.

3.4 Simple Additive Weighting (SAW)

Simple Additive Weighting [15] is the most simple MADM technique that is based on weighted average. The procedure to find evaluation score in SAW is as follows:

- Construct a pair-wise comparison matrix for criterion using Saaty's [12] scale.
- Construct Decision Matrix ($m \times n$) for attributes and criteria.
- Calculate normalized decision matrix for benefit criterion
- For every alternative Ai, evaluate the scores using weights of criteria

3.5 Technique for Order Preference by Similarity to Ideal Solution (TOPSIS)

TOPSIS [15] is an MADM ranking algorithm. After accepting preference weights of different criteria from decision makers, TOPSIS efficiently selects the best alternative. TOPSIS works on the principle of defining ideal solutions. The positive ideal solution which is defined based on best performance values for each alternative. Worst performance values of each alternative provide negative ideal solution. Distance between each alternative and these ideal solutions are computed. The alternative that is closest to the positive ideal solution is selected as the best alternative.

3.6 VIKOR

The concept of VIKOR was developed by Professor Serafim Opricovic in 1979. Evaluation of each alternative is done according to each criterion function. Measure of closeness to the ideal alternative decides the compromise ranking [16]. Determine the best and worst values of all criterion functions. For computing values of indices, we use the weights assigned to criteria.

3.7 Attribute Processing Using Fuzzy Logic

In the design of Fuzzy Logic Controller (FLC), membership functions are assigned to various attributes. Operating range of values for each attribute in WLAN and WIMAX network is utilized to define Universe of Discourse (UoD). Initially, all the measured parameters are normalized using FLCs. Parallel FLCs are implemented in the proposed approach. Fuzzy Inference Systems (FIS) of types Sugeno and Mamdani are utilized in these FLCs. The design of FIS includes rules that decide the working of FLC. The triangular and trapezoidal membership functions are used for input variables as they produce accurate results for real-time system. These Fuzzy techniques are utilized to enhance the performance of TOPSIS and VIKOR [17, 18].

4 Simulation Results

The network selection process is influenced by the operator as well as the user. User-defined criteria focus on maximizing the QoS with reduction in cost, whereas operator aims at optimum utilization of resources. MATLAB code is implemented for performing selection of best network using TOPSIS, FTOPSIS, VIKOR and FVIKOR. Inputs to the MADM method are taken from the fuzzy controllers after calculating its weights by AHP/FAHP method. The attributes that are considered for analysis are RSS, delay, jitter, PLR, throughput, velocity, network loading, security and cost.

Grey predicted values of RSS for WLAN and WiMAX network for few samples are indicated in Table 1. Weights obtained after AHP and FAHP technique for basic criteria are as shown in Fig. 4. As depicted in Fig. 4, more priority is given to RSS and QoS as compared to other parameters. The assignment of weight is subjective and can be changed for different applications. Weight assignments to QoS parameters for various traffic classes using AHP and FAHP technique are indicated in Table 2. It can

Table 1 Predicted RSS values

Network Type	RSSs sample (dbm)	Predicted RSS (dbm)
WLAN	[−110 −110 −112 −113]	−114.69
WiMAX	[−140 −150 −151 −155]	−157.08

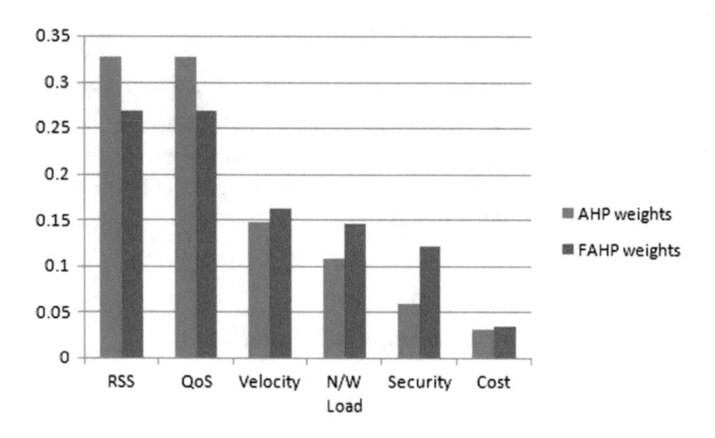

Fig. 4 Weights based on AHP and FAHP basic criteria

be observed from the readings that for conversational class, more weights is given to delay and jitter while in streaming class more weights to throughput. Similarly, interactive class gives more importance to throughput while in background class more weights for PLR and throughput. Context parameters range is different for various networks, depending upon its standard. It is necessary to normalize these context parameters in the range [0,1], so that they can be compared during the decision-making process. The representation of WLAN RSS, in the form of membership function is as depicted in Fig. 5. It defines the mapping of linguistic variables low, medium and high on operating range values of RSS. The universe of discourse is defined from −120 to −50 dBm. The fuzzy set linguistic variable 'low' is defined for the range −120 to −90 dBm, 'medium' for the range −100 to −60 dBm and 'High' over the range of −70 to −55 dBm. The input parameters for the test case of slow speed user are indicated in Table 3. These parameters are applied to FLC for normalization. The normalized values of attributes are as shown in Fig. 6.

Fuzzy Inference System (FIS) to calculate Handoff Urgency Parameter is shown in Fig. 7. The HUP value for current Point of Attachment is calculated from this. If the value of HUP is greater than the threshold, then the network selection module is activated. The threshold value of 0.45 is defined after doing experimentation trials.

Along with the membership functions, fuzzy inference system involves inference rules of IF-THEN format. The formulated fuzzy rule set gives higher value of HUP for degrading values of QoS parameters. The designed rules also ensure higher value of HUP for MS with higher speed in smaller coverage area networks. Evaluation of rule set required for working of FIS is as shown in Fig. 8. The defuzzification

Table 2 AHP and FAHP weights of QoS parameters for various traffic classes

	RSS	Delay	Jitter	PLR	Throughputt	SPEED	N/W load	Security	Cost
AHP									
Convs	0.327	0.107	0.122	0.030	0.066	0.146	0.108	0.059	0.031
Strmg	0.327	0.013	0.057	0.056	0.200	0.146	0.108	0.059	0.031
Intrac	0.327	0.085	0.042	0.098	0.100	0.146	0.108	0.059	0.031
Bckgd	0.327	0.016	0.034	0.116	0.160	0.146	0.108	0.059	0.031
Fuzzy AHP									
Convs	0.286	0.077	0.102	0.021	0.067	0.161	0.145	0.120	0.034
Strmg	0.286	0.032	0.077	0.024	0.133	0.161	0.145	0.120	0.034
Intrac	0.286	0.058	0.082	0.049	0.078	0.161	0.145	0.120	0.034
Bckgd	0.286	0.044	0.084	0.028	0.111	0.161	0.145	0.120	0.034

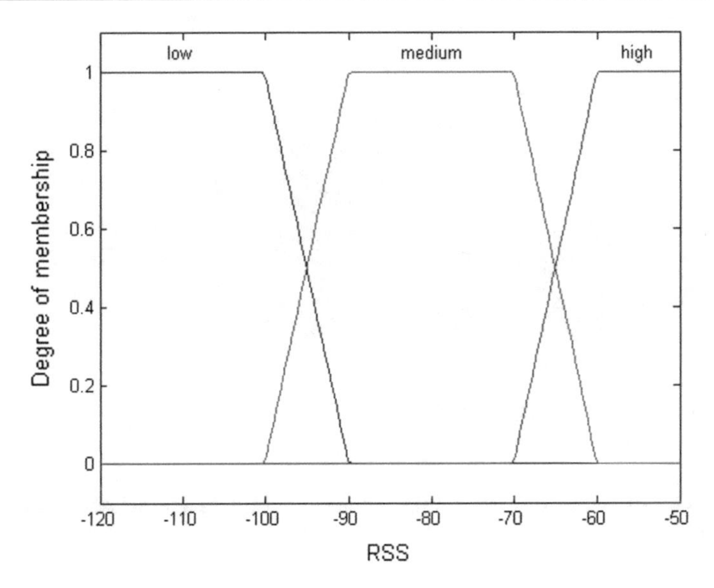

Fig. 5 Membership function for wireless RSS

process yields the crisp value of HUP. The variation of HUP w.r.t. speed of user is as depicted in Fig. 9. As speed increases HUP for WLAN increases sharply. The control surface shown in Fig. 10 indicate increase in HUP with an increase in MS speed and RSS-QoS factor. Network ratings obtained from TOPSIS using AHP and FAHP weights are depicted in Fig. 11. As shown, WLAN network gets highest rating for background class. Ratings for WLAN and WiMAX networks using TOPSIS and SAW techniques are shown in Fig. 12. As depicted, TOPSIS provides clear choice of network as compared to SAW. Selection of networks using VIKOR and FVIKOR

Table 3 Test case: parameter set of available networks for slow speed user

Parameters	WLAN	WIMAX
PRSS (dBm)	−114.05	137.40
Delay (ms)	130	20
Jitter (ms)	27	5
PLR (loss per 10^6 bytes)	3	4
Throughput (Mbps)	70	60
N/W load (%)	20	30
Security (1–10)	1	5
Cost (1–10)	3	4
MS speed (m/s)	2	2

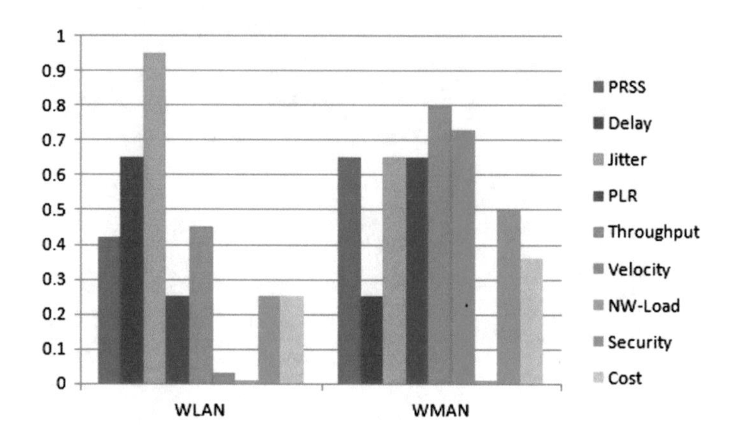

Fig. 6 Normalized network parameters

method is as shown in Fig. 13. Fuzzy technique helps in achieving clear choice of networks.

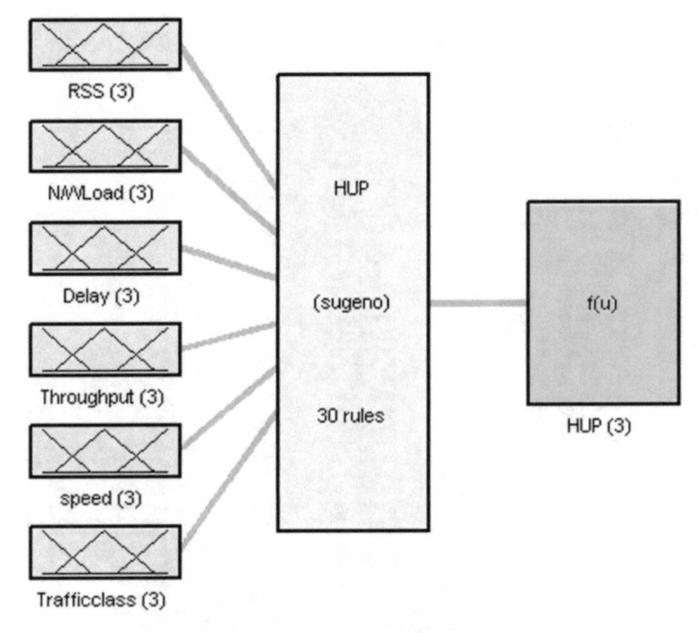

Fig. 7 Fuzzy inference system for HUP

5 Conclusion

For seamless network connectivity in heterogeneous wireless environment, optimal network selection is the key issue. This paper provides QoS-based scheme to select the best radio access network from the available alternatives using MADM, a popular decision-making tool. The proposed HUETANS module decides necessity of handover and also selects the best network by providing rankings to available alternative networks. The generation mechanism of HUP ensures timely initiation of handoff process. The usage of predicted RSS helps in avoiding ping-pong effect. The simulation results demonstrated better performance of fuzzy-based techniques over other selection schemes. The proposed scheme can be added as an extension to IEEE 802.21 Information Server (IS) services. In future work, integration of Markov Decision Process (MDP) with fuzzy MADM schemes can be implemented to achieve improved handover performance in heterogeneous network environment.

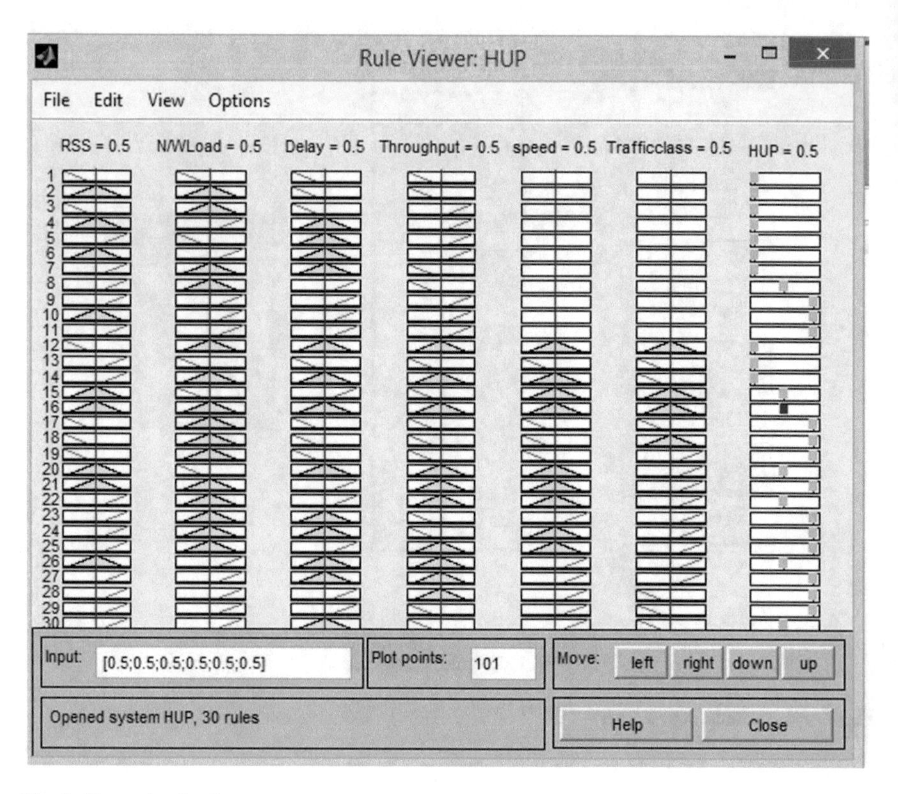

Fig. 8 Example of rule set

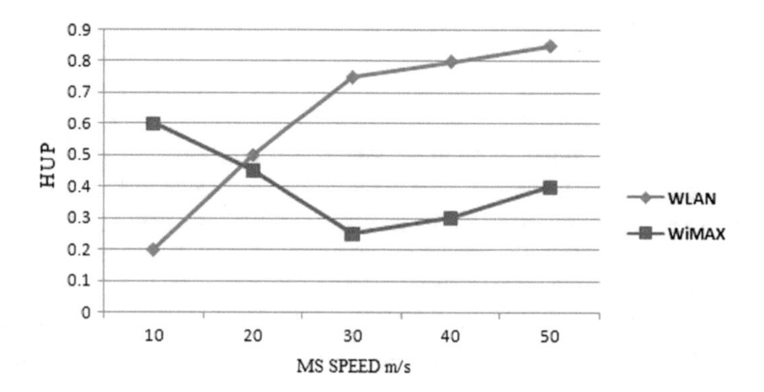

Fig. 9 Variation of HUP w.r.t. speed

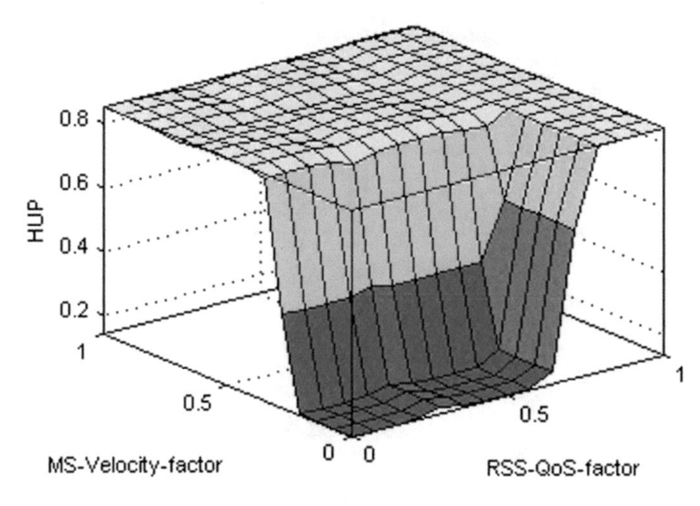

Fig. 10 Control surface for handoff urgency with AHP and FAHP weights parameter

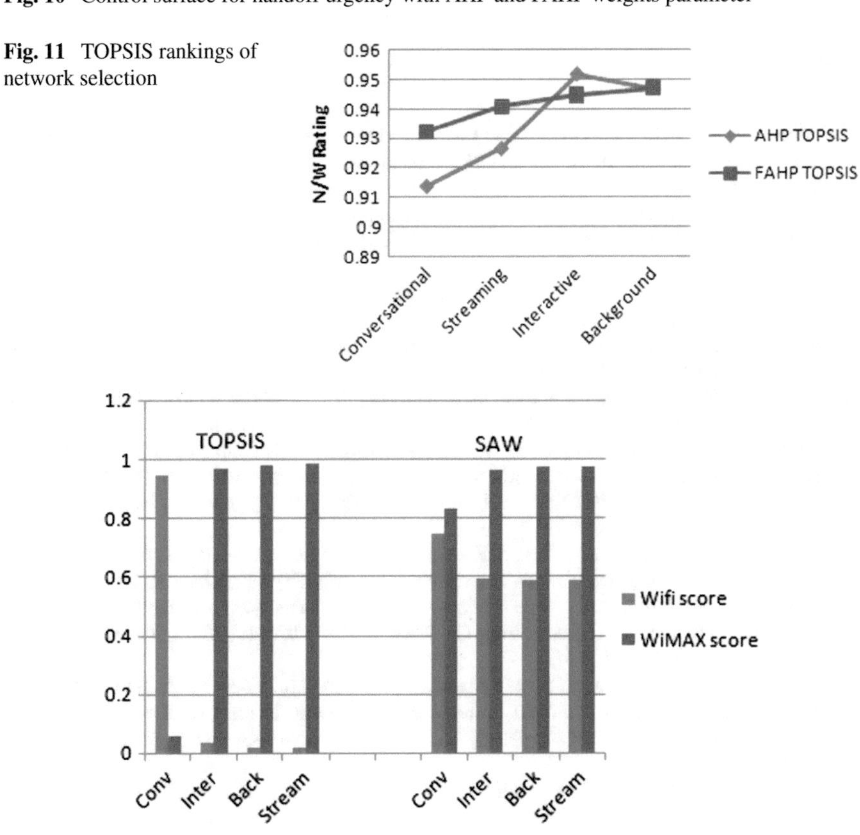

Fig. 11 TOPSIS rankings of network selection

Fig. 12 SAW and TOPSIS network selection rankings

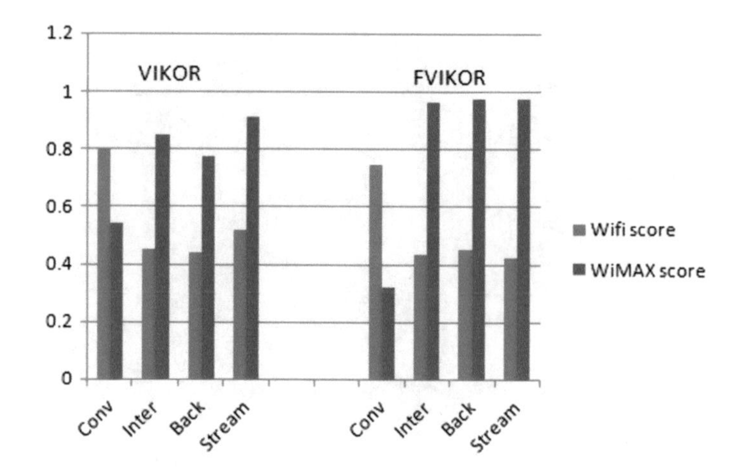

Fig. 13 VIKOR and FVIKOR network selection rankings

References

1. Ahmed, A., Boulahia, L.M., Gaiti, D.: Enabling vertical handover decisions in heterogeneous wireless networks: a state-of-the-art and a classification. IEEE Commun. Surv. Tutor. **16**(2), 776–811 (2014)
2. Ghahfarokhi, B.S., Movahhedinia, N.: A survey on applications of IEEE 802.21 Media Independent Handover framework in next generation wireless networks. Comput. Commun. **36**(10–11), 1101–1119 (2013)
3. Zadeh, L.A.: The concept of a linguistic variable and its application to approximate reasoning I. Inf. Sci. **8**(3), 199–249 (1975)
4. Wang, L., Kuo, G.S.G.: Mathematical modeling for network selection in heterogeneous wireless networks—a tutorial. IEEE Commun. Surv. Tutor. **15**(1), 271–292 (2013)
5. Fernandes, S., Karmouch, A.: Vertical mobility management architectures in wirelessnetworks: a comprehensive survey and future directions. IEEE Commun. Surv. Tutor. **14**(1), 45–63 (2012)
6. Fallon, E., Murphy, L., Murphy, J., Miro-Muntean, G.: FRAME Fixed route adapted media streaming enhanced handover algorithm. IEEE Trans. Broadcast. **59**(1), 96–115 (2013)
7. Taleb, T., Ksentini, A.: VECOS: a vehicular connection steering protocol. IEEE Trans. Veh. Technol. **64**(3), 1171–1187 (2015)
8. Mehbodniya, A., Kaleem, F., Yen, K.K., Adachi, F.: A fuzzy extension of VIKOR for target network selection in heterogeneous wireless environments. Phys. Commun. **7**, 145–155 (2013)
9. Zhang, W.: Handover decision using fuzzy MADM in heterogeneous networks. In: Wireless Communications and Networking Conference, 2004. WCNC, Mar 2004, IEEE, vol. 2, pp. 653–658. IEEE (2004)
10. Kantubukta, V., Maheshwari, S., Mahapatra, S., Kumar, C.S.: Energy and quality of service aware FUZZY-technique for order preference by similarity to ideal solution based vertical handover decision algorithm for heterogeneous wireless networks. IET Netw. **2**(3), 103–114 (2013)
11. Wilson, A., Lenaghan, A., Malyan, R.: Optimising wireless access network selection to maintain QoS in heterogeneous wireless environments. In: Wireless Personal Multimedia Communications, pp. 18–22 (2005)
12. Saaty, T.L.: Decision making with the analytic hierarchy process. Int. J. Serv. Sci. **1**(1), 83–98 (2008)
13. Pal, M.: Triangular fuzzy matrices. Iran. J. Fuzzy Syst. **4**(1), 75–87 (2007)

14. Deng, J.: Introduction to grey system theory. J. Grey Syst. **1**(1), 1–24 (1989)
15. Stevens-Navarro, E., Wong, V.W.: Comparison between vertical handoff decision algorithms for heterogeneous wireless networks. In: Vehicular Technology Conference, May 2006, VTC 2006-Spring, IEEE 63rd, vol. 2, pp. 947–951. IEEE (2006)
16. Opricovic, S., Tzeng, G.H.: Compromise solution by MCDM methods: a comparative analysis of VIKOR and TOPSIS. Eur. J. Oper. Res. **156**(2), 445–455 (2004)
17. Ndban, S., Dzitac, S., Dzitac, I.: Fuzzy topsis: a general view. Procedia Comput. Sci. **91**, 823–831 (2016)
18. Opricovic, S.: Fuzzy VIKOR with an application to water resources planning. Expert Syst. Appl. **38**(10), 12983–12990 (2011)

Shilpa Litake is a Ph.D. candidate of Vishwakarma Institute of Information Technology, Pune, India. She received the B.E. in Electronics and Telecommunication engineering from Savitribai Phule Pune University, India in June 2000. She obtained her Master's degree in Electronics and Telecommunication engineering from Mumbai University, India in June 2006. She is currently working as an assistant professor in PVG's COET, Pune, India. Her research interests are in wireless networks and communication system.

Dr. Prachi Mukherji obtained her Bachelor's degree in Electronics from Jiwaji University, Gwalior. She then obtained her Master's degree in Electronics and Telecommunication from Barkatullah University, Bhopal. She did her Ph.D. in Image Processing from University of Pune, focusing on Image Analysis using graph theory. Currently, she is Professor and Head of the Department of Electronics and Telecommunication, Cummins College of Engineering for Women, Pune. Her specializations include Image Processing, Signal Processing, and Communication.

Walking and Transition Irregularity Detection Using ANN Approach for Wireless Body Area Network

S. P. Shiva Prakash and Apurwa Agrawal

Abstract Human mobility and walking pattern interpret a lot about the activities and behavior of an individual. Tracing these patterns and detecting abnormalities in the traced patterns can be used for health and fitness monitoring. Human resident sensors like the pocket IMUs uses the Body Area Network (BAN) for motion tracking, measuring vitals and other human behavior. Several works have been carried out to determine the posture by researchers. From the survey, it is found that no attempt has been made to detect abnormality. Hence, in this work, a model is proposed to determine abnormalities in walking and transition pattern using artificial neural network. The experiments were conducted using Castalia Wireless Body Area Network (WBAN) simulator for standard UCI data set. The simulator result shows that the proposed model detects irregularity more accurately.

Keywords Artificial neural network (ANN) · Body area network (BAN)
Health care · IMU sensors · Inertial measurement unit (IMU) · Wearable sensors
Wireless body area network (WBAN) · Wireless sensors

1 Introduction

An increasing number of lightweight wearable sensors has entered the market and is rising exponentially with fitness and health care as dominant wearable applications. Wireless Body Area Network (WBAN), also referred as Body Sensor Network (BSN) is a wireless network of human resident sensors or devices which is con-

S. P. Shiva Prakash (✉) · A. Agrawal
Department of Information Science and Engineering,
JSS Science and Technology University, Mysuru, India
e-mail: shivasp@sjce.ac.in

A. Agrawal
e-mail: apurwaagr@gmail.com

S. P. Shiva Prakash · A. Agrawal
Sri Jayachamarajendra College of Engineering, Mysuru, India

S. Smys et al. (eds.), *International Conference on Computer Networks and Communication Technologies*, Lecture Notes on Data Engineering and Communications Technologies 15, https://doi.org/10.1007/978-981-10-8681-6_85

Table 1 IMU sensors

Sensors	Description
Accelerometer	Measures the acceleration of the body in its own rest frame
Gyroscope	Measures the angular acceleration along roll, pitch and yaw
Proximity sensor	Detects the presence of nearby object without any physical contact
Magnetometer	Measures the magnetic field around the body
EMG	Records the electrical activity produced by skeletal muscles

tributing towards the rapid growth of wearable technology. It uses various wearable or accompanied devices for communication, motion tracking, measuring the vitals or other human behaviors. Inertial Measurement Units (IMU) have been developed which integrates multi-axis combination of accelerometer, gyroscope, magnetometers, and pressure sensors to measure body's specific force, angular rate, displacement and sometimes the magnetic field surrounding the body. Pocket IMUs are used in smartphones motion sensing.

Figure 1 shows an image of a human body connected with body sensors such as accelerometer and gyroscope. An accelerometer is used to measure acceleration of a body about three axes, x-, y-, and z-axis. The raw data obtained is in terms of gravitational force. Accelerometer has multiple industrial uses and has been heavily used in the smart devices like smart bands, smartphones, etc., for motion tracking. Today, the accelerometer along with gyroscope and other piezoelectric sensors are used in health service area to track the human activity and alert them or provide with better diagnosis. Gyroscope is another pocket IMU sensor which measures angular velocity. It detects the roll, pitch and yaw motions of the body along the x-, y-, and z-axis respectively. Due to the law of conservation of energy the orientation of axis remains same while rotation. Gyroscope has wide application in compasses, pointing devices, aircrafts, etc. Other than these applications, since gyroscope measures orientation and rotation, it is embedded in many modern technologies. Using this integrated system in healthcare service, area would help keep a detailed track of human activities and help them in case health management and emergency alerts. Table 1 gives the description of different IMU sensors.

Artificial Neural Network is a vast network of artificial neurons. Each connection called a synapse transmits signal between one neuron to another, processes it, and sends signal to neuron connected to it. This method is used for cognitive problems. The system improves performance by learning from the examples or the training data sets provided to it. Once they have learned from the training examples, they are ready to make predictions on any sample based on their learning experience.

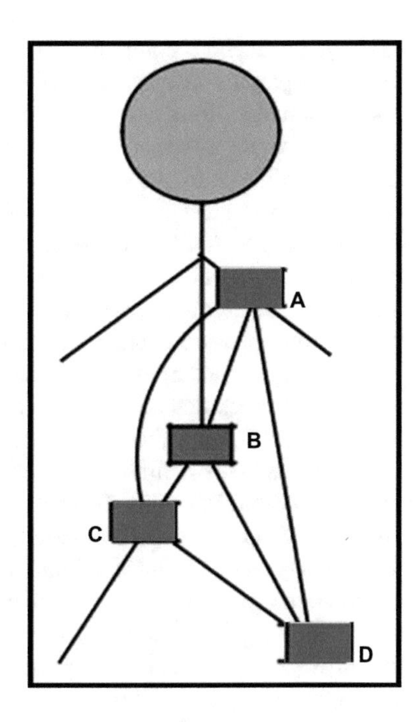

Fig. 1 Wireless body area network with sensors

The rest of the paper is divided into five sections. Section 2 discusses the related works carried out. Section 4 discusses proposed model. Algorithm is described in Sect. 5. The experimental details and results are in Sect. 6. Section 7 presents conclusion and future work.

2 Related Works

In this section, the related work carried out by several authors on the irregularity detection is discussed.

With the increasing awareness and self-consciousness about the health and fitness throughout the world, popularity of the miniature wearable sensors had been increasing exponentially in the last decade. A number of research works are performed and are also queued up to ease the life of an individual and provide them with all possible facilities in the minimal time. Alam et al. [1] presented an accurate and realistic simulation of body area network (BAN) and body-to-body networks using deterministic and semi-deterministic approaches. In deterministic approach, they came up with a biomechanical model by introducing dynamic distances which diversify the mobility pattern and scenarios applications. In the semi-deterministic approach, a real-time measurement campaign is performed which is further categorized through statistical analysis. The results proved semi-deterministic approach as the best option. Su et al.

[2] have presented a comprehensive survey of the advances in activity recognition with smartphones. They reviewed the core data mining techniques behind the activity recognition algorithm. Li et al. [3] have derived a method for accurate fast fall detection and posture information using gyroscopes and accelerometer. They divide the human activities into two categories: static posture and dynamic transitions between these postures: sitting, bending, standing, and lying. To determine if the transition between the postures is intentional or not both linear acceleration (acceleration reading) and angular velocity (gyroscope reading) are measured. Hence, a fall detection algorithm was produced. Morioka et al. [4] identified specific human with the accurate posture measurements by integrating the networked laser range finders in the intelligent space and a wearable acceleration sensor with the human. In the proposed model, only walk detection results were communicated among the sensors. Aqueveque et al. [5] talks about the Android Platform for real-time gait tracking using inertial measurement units. Their design is based on two elements: (a) IMU Sensors and (b) Android Platform. The data from IMU sensors are sent to the Android via Bluetooth links. Three IMU sensors are placed each at the hip, knee and ankle joints and, the angles of all the three joints were measured. The collected data was used to calculate the gait index and evaluate gait quality online. Veenendaal et al. [6] looked into abnormal gait and how drunken person walks. They used markers based tracking to train SVM classifiers for recognizing various abnormal actions while walking such as tripping, swaying, walking sideways, falling, dragging, and walking with support versus normal gait. Prado et al. [7] developed a novel approach for fall detection, posture analysis, mobility, and metabolic expenditure for elderly based on intelligent architecture, supported by WPAN. Similar kind of works can be found from [8–11] where authors have defined different mobility models for human walking patterns.

In all these works, the authors either classify different postures or uses IMU sensors for different activity recognitions but none of the works focus on finding abnormalities in these activities.

3 Problem Statement

Today smartphones have become a very important part of human's daily life. Modern technology has taken over a major part of humans life. With increasing trend in technology, health monitoring system is being embedded in our smartphones. From the literature, it can be found that very less work have been carried out on classifying different postures and no work is focused on determining abnormalities in human activities. Hence, there is a need to develop a model that focuses on wearable sensors that are used to observe the moving pattern of an individual and can conclude whether the moving pattern is normal or abnormal.

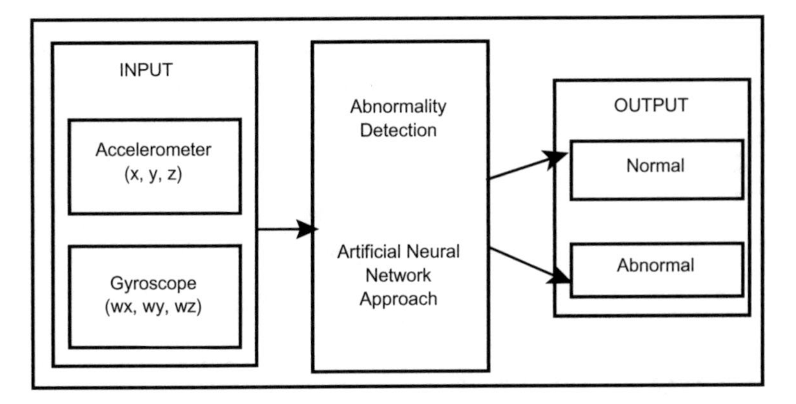

Fig. 2 Proposed WTID-ANN model design

4 Proposed Model

In this section, Walking and Transition Irregularity Detection using Artificial Neural Network (WTID-ANN) model design is presented. Figure 2 shows the proposed model design having three modules namely input, process, and output. Accelerometer and gyroscope are used to sense the coordinates x, y, z and wx, wy, and wz, respectively. The data generated by the input module are used by the process module consisting of abnormality detection using artificial neural network algorithmic approach. The existing ANN is modified to classify the obtained data to set the status of the walking pattern as normal or abnormal. Further, the detected walking and transition irregularity is classified and labeled as normal or abnormal in the output module.

5 Algorithm

In this section, a proposed WTID-ANN algorithm is presented. Algorithm 1 defines the steps to be followed to determine the postures. It takes training data set as input that consists of data generated by number of accelerometer and gyroscope. The training data set is loaded and a design matrix X and a target vector y are created. A learning model is instantiated for an appropriate value of k and then this classifier is fit in the model. The response is predicted and then the accuracy is evaluated. For the available data set KNN (k-nearest neighbor), classifier produces the most accurate result, hence KNN is fit into the model.

Table 2 WTID setup

Parameters	Values
Accelerometer	x, y, z
Gyroscope	wx, wy, wz
No. of people	4
Age	28, 31, 46 and 75.
Posture	Sitting, sitting down, standing, standing up, walking

Algorithm 1: WTID-ANN Algorithm

```
  input : Training dataset
  output: Classified labeled classes
1 begin
2     Load training dataset;
3     create design matrix X;
4     create target vector y;
5     classi f ier = kNeighborClassi f ier(n neighbors = k) # instantiate learning model
      (value of k)
6     classi f ier: f it(X train; y train) # fit the model
7     prediction = knn: predict(X test) # predict the response
8     evaluate accuracy
9 end
```

6 Experiments, Results, and Discussion

In this section, the experiments conducted and the results obtained are presented.

6.1 *Experimental Setup*

This section describes the experimental setup. Table 2 describes the parameters and its values for the setup.

Four sensors were used each mounted on waist, left thigh, right ankle, and right upper arm. The raw data obtained by these sensors are compared and the system is trained to identify whether the motion of the individual is normal or abnormal. The cause of abnormality may differ from person to person. The labels '*a*', '*b*', '*c*', and '*d*' in Fig. 1 represent the four nodes of the wireless body area network.

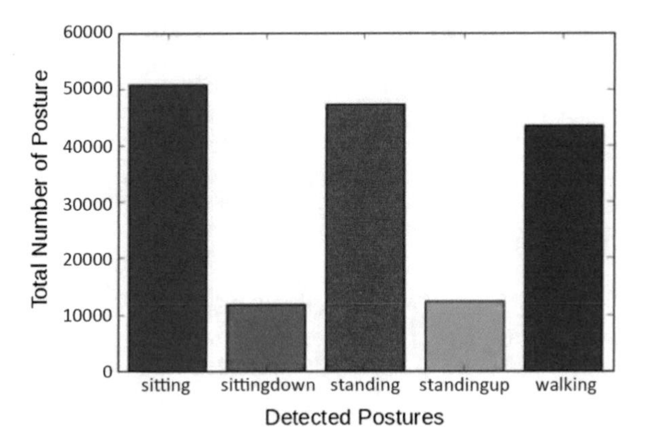

Fig. 3 Detection of posture versus total number of all postures

Table 3 Average accuracy for each posture

	Precision	Recall	$f1$-score	Support
Sitting	1.00	1.00	1.00	12,577
Sitting down	0.97	0.98	0.97	2910
Standing	0.99	1.00	0.99	11,887
Standing up	0.98	0.96	0.97	3116
Walking	1.00	0.99	0.99	10,919
Avg./total	0.99	0.99	0.99	41,409

Fig. 4 Confusion matrix to detect error

```
[[12569     1     0     6     1]
 [    1  2853    12    32    12]
 [    0     2 11853    13    19]
 [    8    58    47  2998     5]
 [    0    35    84    21 10779]]
```

6.2 *Results and Discussion*

In this section, the experiments conducted and the results obtained are presented. Different classifiers were fit in the data set to check for accurate solution and it was found that k-nearest neighbor (KNN) gives the most accurate classification, i.e., 99% accurate results. Figure 3 shows the graphical representation of the number of all postures and detected postures. It can be noticed that 50,000 sitting postures, 10,000 sitting down posture, 48,000 standing posture, 12,000 standing up posture, and 45,000 walking postures have been identified out of 165,000 data resulting in 100% accuracy in classification. This is due to the modification made in the existing ANN algorithm during instantiating, model fitting and predicting method. Figure 6a–o describes the histogram of each sensor data representing x, y, and z coordinates of various postures that are used in determining walking and transition pattern.

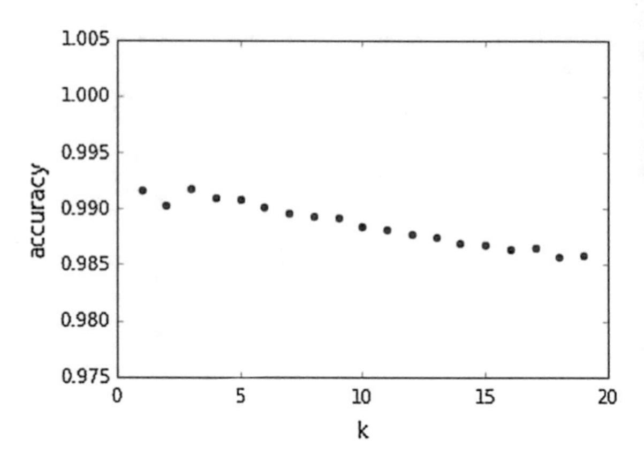

Fig. 5 *k* value for KNN

Several classifiers were implemented and tested to find the best classifier algorithm for the data set obtained by using accelerometer and gyroscope sensors. Table 3 shows the average accuracy for each posture. The confusion matrix in Fig. 4 provides an indication of 99% accuracy on test set.

Figure 5 shows the accuracy obtained by the variation of k value from 1 to 20. It can be noticed that the KNN results are more accurate for k value 3 (Fig. 6).

7 Conclusion and Future Work

In this work, a new model is proposed to determine the irregularities in postures. The proposed model design and algorithms are presented. The experiments were conducted using Castalia simulation tool. The experimental results show that the proposed WTID model accurately determines the irregularities in the postures. It is evident from the result that 100% accuracy of classification is achieved by modifying the existing ANN algorithm in the proposed model. The abnormalities can be classified further to detect different kinds of abnormalities and a pre-diagnostic system would suggest the user, solution to those abnormalities via their smartphones. Diseases like Alzheimer and dementia can be detected by observing their symptoms on a regular basis by collecting the data from the sensors and an alert can be sent in case of any abnormalities. The sensors' data can also be used to sense if a person is drunk by observing his walking pattern. The system can also be developed in such a way so as to provide help to people suffering from arthritis or any physical disabilities. A mobile app can be designed later to alert a person and his caretaker in case of any abnormalities in future.

Acknowledgements There is no human directly involved in the study.

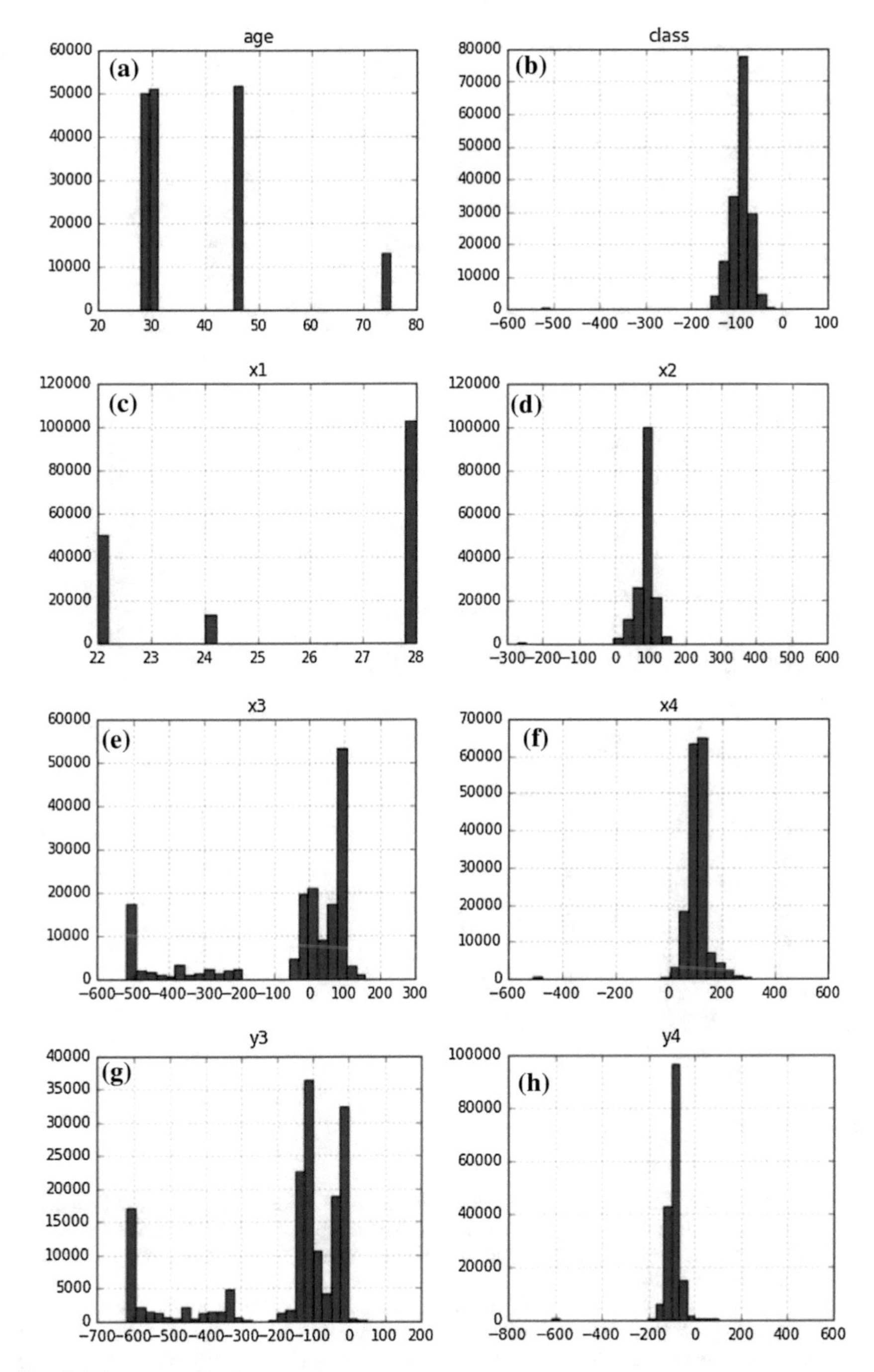

Fig. 6 Histogram of each sensor data

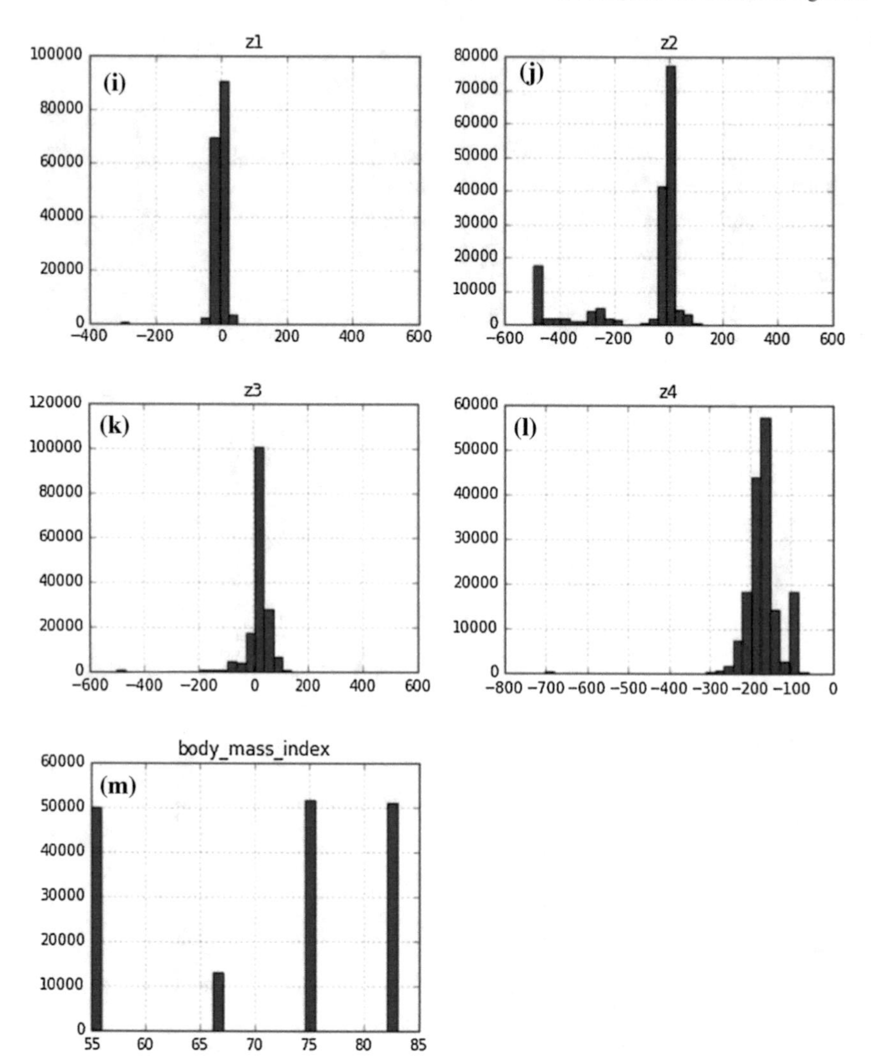

Fig. 6 (continued)

References

1. Alam, M.M., Ben Hamida, E., Ben Arbia, D., Manman, M., Mani, F., Denis, B., D'Errico, R.: Realistic simulation for body area and body-to-body networks. Sensors **16**(4), 561 (2016)
2. Su, X., Tong, H., Ji, P.: Activity Recognition with Smartphone Sensors. Tsinghua Sci. Technol. **19**(2), 235–249 (2014)
3. Li, Q., Stankovic, J.A., Hanson, M., Barth, A., Lach, J., Zhou, G.: Accurate, fast fall detection using gyroscopes and accelerometer—derived posture information. In: Wearable and Implantable Body Sensor Networks, BSN'09, Sixth International Workshop on Wearable and Implantable Body Sensor Networks, pp. 138–143. IEEE (2009)
4. Morioka, K., Hashikawa, F., Takigawa, T.: Human identification based on walking detection with accelerometer sensor and networked laser range sensors in intelligent space. Int. J. Smart

Sens. Intell. Syst. **6**(5), 2040–2054 (2013)
5. Aqueveque, P., Sobarzo, S., Saavedra, F., Maldonado, C., Gomez, B.: Android platform for realtime gait tracking using inertial measurement units. Eur. J. Transl. Myol. **26**(3), 262–267 (2016)
6. Veenendaal, A., Daly, E., Jones, E., Gang, Z., Vartak, S., Patward-han, R.: Drunken abnormal human gait detection using sensors. CSERJ (2013)
7. Prado, M., Reina-Tosina, J., Roa, L.: Distributed intelligent architecture for falling detection and physical activity analysis in the elderly. In: 2nd Joint EMBS/BMES conference, pp. 1910–1911. IEEE (2002)
8. Chan, C.K., Loh, W.P., Abd Rahim, I.: Human motion classification using 2D stick-model matching regression coefficients. Appl. Math. Comput. **283**, 70–89 (2016). ELSEVIER
9. Bourke, A.K., O'Brien, J.V., Lyons, G.M.: A threshold-based fall-detection algorithm using a biaxial gyroscope sensor. Med. Eng. Phys. **30**, 84–90 (2008)
10. Lee, K., Hong, S., Kim, S.J., Rhee, I., Chong, S.: SLAW: a mobility model for human walks. In: IEEE Communications Society IEEE INFOCOM (2009)
11. Pande, A., Zeng, Y., Das, A., Mohapatra, P., Miyamoto, S., Seto, E., Henricson, E.K., Han, J.J.: Demonstration Paper: Accurate Energy Expenditure Estimation Using Smartphone Sensors. ACM (2013)

Dr. S. P. Shiva Prakash have obtained Ph.D. degree in computer science from university of Mysore. Have secured BE degree in Information Science and Engineering and M.Tech in Software Engineering from Sri Jayachamarajendra College of Engineering, Mysuru. He is currently working as Assistant Professor in the department of Information Science and Engineering, J.S.S. Science and Technology University (Formerly SJCE), Mysore. He has published several articles in peer-reviewed reputed conferences and journal. He is a reviewer for various conference and journal papers. He has delivered invited talks at various platforms. His area of research interests are wireless mesh networks, energy aware routing, wireless body area networks and internet of things.

Ms. Apurwa Agrawal is pursuing her bachelors degree in the department of Information Science and Engineering, J.S.S. Science and Technology University (Formerly SJCE), Mysore. She is active member of CSI. She has developed few algorithm for health care applications. Her area of interest is deep learning, wireless body area networks and Internet of things.

Implementing MIMO-OFDM for the Improvement of the System Performance in WPAN

N. Rakesh, T. Anjali and B. Uma Maheswari

Abstract Developing cost-effective solutions to deliver broadband connection to undeserved or under-connected communities using fixed wireless access technology can be achieved by canopy networks. The former can be implemented using MIMO-OFDM technique. The implementations of MIMO-OFDM in systems have shown promising results in terms of system performance improvement. This paper presents high-speed FFT algorithms for wireless personal area network (WPAN) applications which uses high data rate. In WPAN, the FFT/IFFT block plays the major role. To eliminate the huge burden of the computational requirement of the FFT/IFFT process, the divide and conquer algorithm is employed here. This method not only improves the efficiency but is also easy to implement. In this work, the concept of Discrete Fourier Transform is discussed and the same is implemented using MATLAB for the MIMO-OFDM process.

Keywords OFDM · Fast Fourier Transform (FFT) · Discrete Fourier Transform (DFT)

1 Introduction

Forthcoming scenarios would witness consumers demanding both high speed and high quality. Consequently, it is significant that service providers come up with tech-

N. Rakesh · B. Uma Maheswari
Department of Computer Science & Engineering, Amrita School of Engineering, Amrita Vishwa Vidyapeetham, Bengaluru, India
e-mail: n_rakesh@blr.amrita.edu

B. Uma Maheswari
e-mail: b_uma@blr.amrita.edu

T. Anjali (✉)
Department of Computer Science & Engineering, Amrita School of Engineering, Amrita Vishwa Vidyapeetham, Amritapuri, India
e-mail: anjalit@am.amrita.edu

S. Smys et al. (eds.), *International Conference on Computer Networks and Communication Technologies*, Lecture Notes on Data Engineering and Communications Technologies 15, https://doi.org/10.1007/978-981-10-8681-6_86

nologies that could not only increase reliability but also the throughput. One such technology is the orthogonal frequency division multiplexing (OFDM). It escalates the data rate by separating the channel into a number of sub-channels placed orthogonal to each other and transmitting the data simultaneously. The other advantage is that it decreases the effect of path loss and data fading [1]. Therefore, OFDM technique is employed in many wireless communication systems such as wireless personal area networks (WPANs), wireless local area networks (WLANs), digital video broadcasting (DVB), third generation partnership project long-term evolution (3GPP LTE), mobile worldwide interoperability for microwave access (WiMAX), etc. Multiple-input multiple-output (MIMO) technique can be applied in order to achieve consistency without additional energy consumption and increased bandwidth. [2–4]. The fundamental concepts in digital signal processing are linear filtering and Fourier transform. According to the definition, the direct computation of $X(k)$ comprises of $N-1$ complex additions and N complex multiplications. On proceeding with the process, to calculate all N denominations of the DFT, it will require $N2$-N complex additions and $N2$ complex multiplications. In order to achieve supreme throughput with limited power consumption and area [5, 6], optimized use of FFT and IFFT is required.

2 Methodology

2.1 The Discrete Fourier Transform

The conversion of time domain sequences to their respective frequency domain helps to carry out frequency analysis on discrete time signals $x(n)$. It leads to the DFT computation which operates on an N point sequence of numbers, denoted to as $x(n)$ [3, 5–7]. The functions $X(k)$ and $x(n)$ are represented by

$$X(k) = \sum_{(n=0)}^{(N-1)} x(n)e^{(-j2\pi nk/N)}, \quad k = 0, 1, 2 \ldots N-1 \tag{1}$$

Here, $x(n)$ is the time domain depiction of input and N represents the count of inputs to the DFT. The value n represents the time index, and k is the frequency index. The conversion of frequency domain data to time domain is done using the inverse discrete Fourier transform. This results in setting back $X(k)$ to $x(n)$ [7, 8].

$$x(n) = \sum_{(n=0)}^{(N-1)} x(k)e^{(-j2\pi nk/N)}, \quad n = 0, 1, 2 \ldots N-1 \tag{2}$$

The IDFT and DFT apply the same type of computational algorithm. For simplicity, the variable WN is repeatedly identified to be the "Nth root of unity with its exponent evaluated modulo N" since

$$(\mathrm{WN})N = \mathrm{e}^{(-j2\pi)} = 1. \tag{3}$$

Here, the difference between Eqs. (1) and (2) is the type of coefficients applied to it. The computation of Fourier transform in DFT is inefficient as there is no productive use of the periodicity and symmetric properties of the phase factor WN. [9, 10].

Periodicity property:

$$\mathrm{WN}k + N = \mathrm{WN} \tag{4}$$

Symmetry property:

$$\mathrm{WN}k + N/2 = -\mathrm{WN}k \tag{5}$$

3 MIMO-OFDM

Multiple-input multiple-output Orthogonal frequency division multiplexing (MIMO-OFDM) is a sustainable solution for offering high data rate facilities in tough channel environments. Multi-rate, multi-carrier, and multi-symbol are the three dominant transmission principles. The data is distributed over a vast amount of carriers. These carriers are spaced apart at exact frequencies. The orthogonality is a result of this spacing. As a result, the demodulator is restricted from seeing other frequencies. The radio signals are transmitted and received using multiple antennas. MIMO-OFDM has two advantages; first, deployment of broadband wireless access systems that obligates non-line-of-sight (NLOS) operation. Second, the usage of the multipath properties of the environment using base station antennas that do not have LOS.

The transmission process in OFDM comprises of a single data stream through numerous low rate subcarriers. This enhances the strength against selective fading. With respect to a single carrier system, the entire link is weakened due to selective fading. In contrast, only a few subcarriers are affected for a multi-carrier system [11–13]. OFDM technique has two parts; a transmitter and a receiver, seen in Fig. 1. The transmitter consists of modulator and IFFT. The carrier is first divided into subcarriers; each carrier being orthogonal to one another. Data is then transmitted over the subcarriers, modulated, and IFFT is performed. Finally, guard intervals are added to avoid interference. This ends with the transmitter side. OFDM receiver portion first releases the guard interval, performs FFT and then demodulates the signal. Hence the data transmitted over the subcarriers can be retrieved individually without any errors.

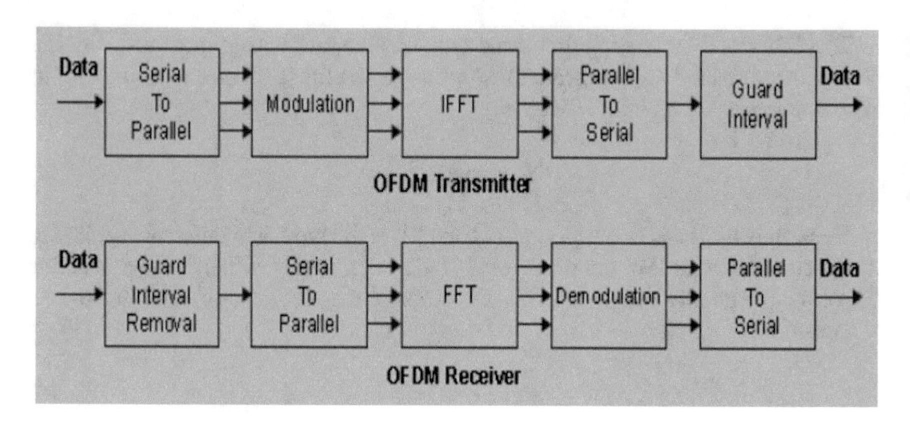

Fig. 1 Block diagram of OFDM

4 Results

The entire OFDM process is simulated in MATLAB. Figure 2 shows the discrete data (input) which undergoes the OFDM process. The input is a digital data which is plotted using MATLAB.

The input data is subjected to modulation. In this case, it is the quadrature phase shift keying which is shown in Fig. 3. The process involves modulating two bits at the same time. The carrier for phase shift can be one of the four, i.e., 0°, 90°, 180°, and 270°, respectively [4, 14, 15]. The QPSK permits to carry twice the information using the same bandwidth when compared to PSK.

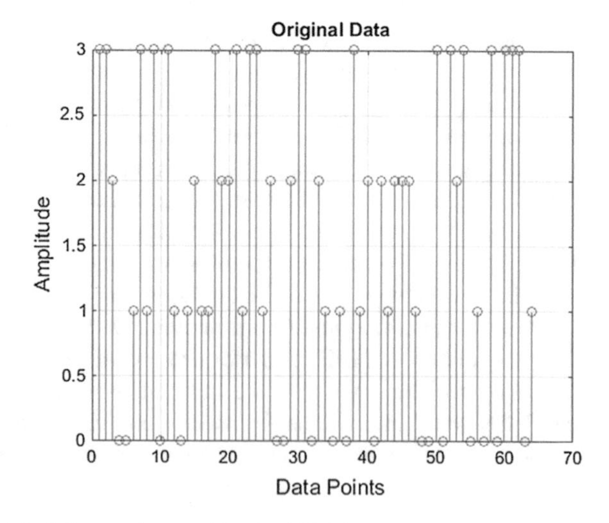

Fig. 2 The original discrete data used for the OFDM process

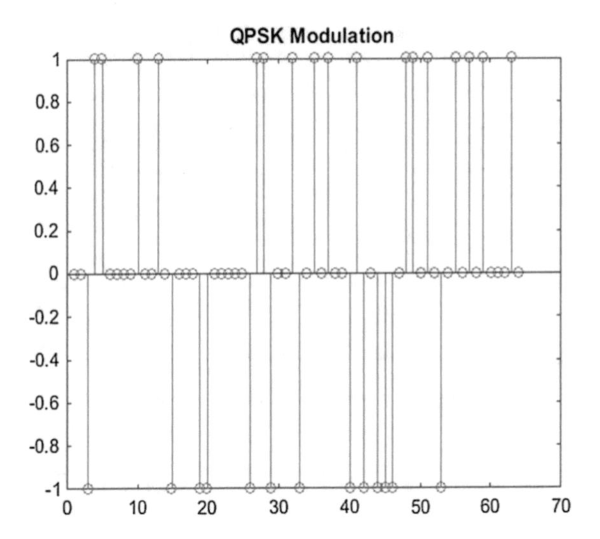

Fig. 3 Modulating the data

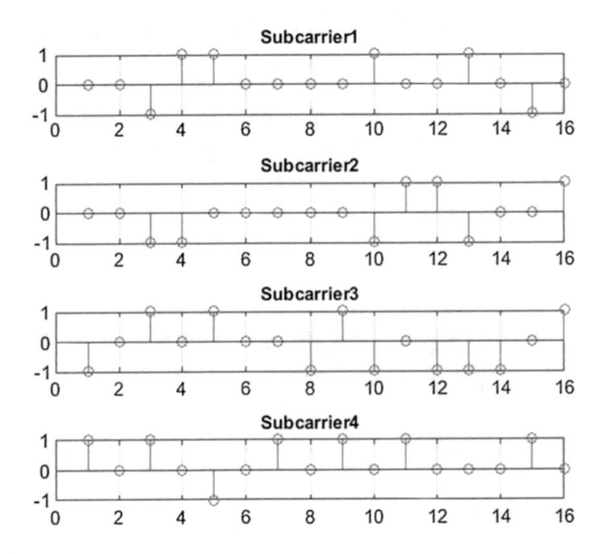

Fig. 4 Subcarriers

As seen in the concept of OFDM, the carrier has to be split into subcarriers. Figure 4 shows the original carrier being distributed into four subcarriers. After splitting into subcarriers, each carrier exclusively has the modulated data. At the transmitter side, the inverse fast Fourier transform is performed which is seen in Fig. 5. The IFFT is applied to all the subcarrier.

To avoid signal interference among the subcarriers, guard intervals are applied to each subcarrier, given in Fig. 6. This marks the end of the transmitter side.

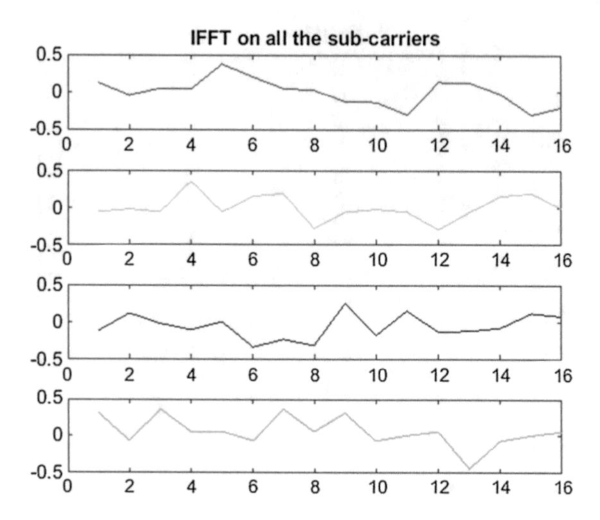

Fig. 5 Implementing the IFFT

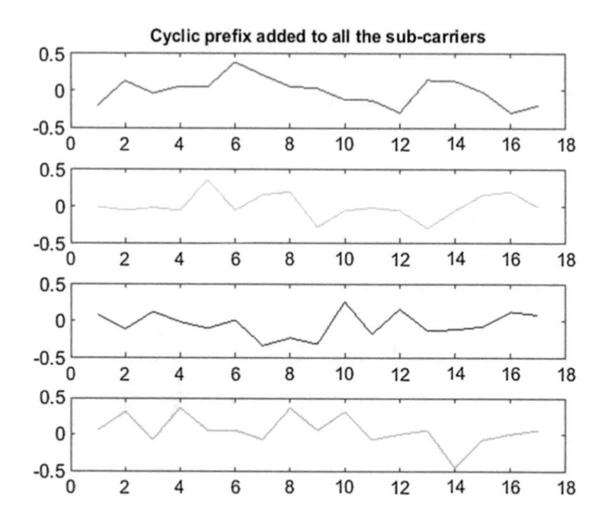

Fig. 6 Adding guard interval

The obtained OFDM signal (Fig. 7) is passed through a channel so that it is available at the receiver side. The channel acts as a bridge between the transmitter and the receiver terminals.

At the beginning of the receiver end, before retrieving the OFDM signal, the guard intervals are removed. Figure 8 depicts this process.

The receiver end generally sees the opposite of what occurred at the transmitter side. Therefore, FFT is performed. The result for this process is shown in Fig. 9.

The OFDM signal is recovered from the subcarriers (Fig. 10). This signal is seen with errors which are filtered.

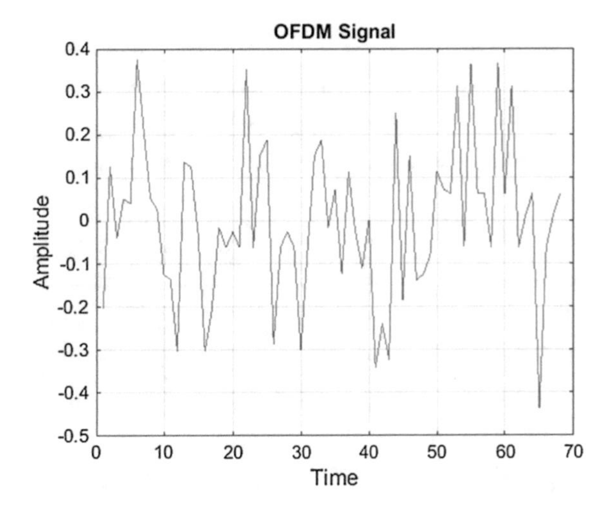

Fig. 7 Passing the OFDM signal through the channel

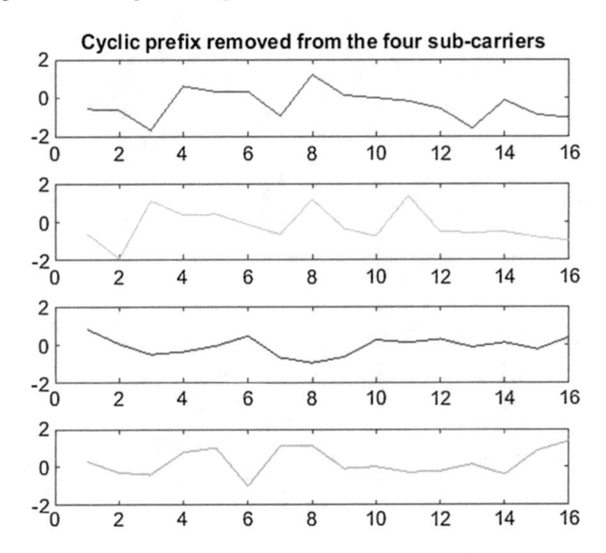

Fig. 8 Removing the guard interval before receiving the orthogonal signal

On removing the errors the final data signal carried by the four subcarriers is retrieved. Figure 11 shows each signal carrying the required information from individual subcarrier.

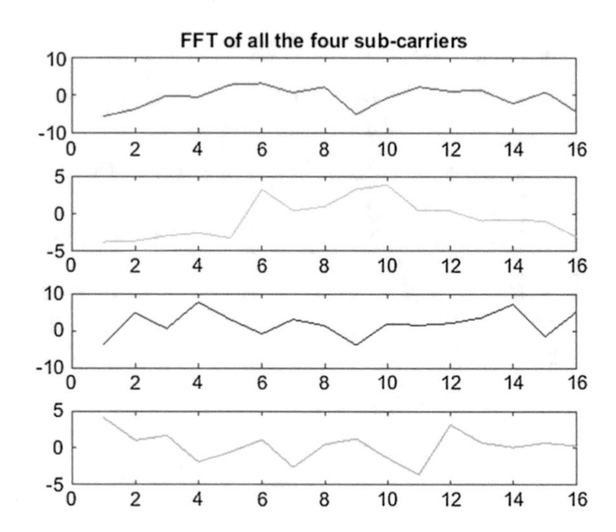

Fig. 9 Implementing FFT for all the subcarriers

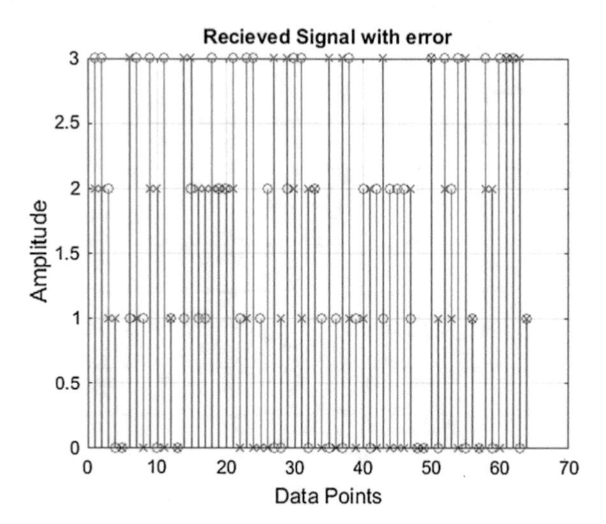

Fig. 10 The received signal with errors. Blue indicates the data and red indicates the error

5 Conclusion and Future Scope

Using FFT, there is an increase in computational efficiency thus an increase in overall throughput. The algorithm also reduces power and area consumption. This technology supports adaptive network management. Due to the limitations in size and cost, it calls for a new technique called the cooperative OFDM structure with single space diversity. This helps in transmitting both of their signals in a single time slot.

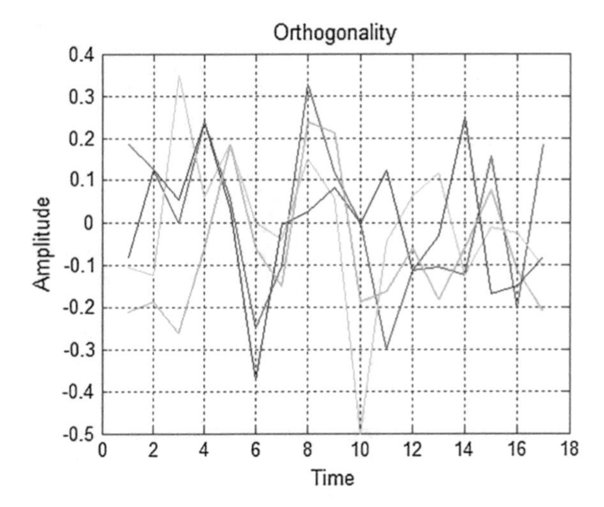

Fig. 11 The final data signals carried by the four subcarriers. Each color indicates individual carrier data

References

1. Bingham, J.A.C.: Multi-carrier modulation for data transmission: an idea whose time has come. IEEE Commun. Mag. **28**(5), 17–25 (1990)
2. Foschini, G.J., Gans, M.J.: On limits of wireless communications in a fading environment when using multiple antennas. Wirel. Pers. Commun. **6**, 311–335 (1998)
3. Tarokh, V., Seshadri, N., Calderbank, A.R.: Space-time codes for high data rate wireless communication: performance criterion and code construction. IEEE Trans. Inf. Theory **44**(2), 744–765 (1998)
4. Srivastava, J.R., Sudharshan, T.S.B.: Intelligent traffic management with wireless sensor networks. In: ACS International Conference on Computer Systems and Applications (AICCSA) (2013)
5. IEEE std 802.11a: Wireless lan medium access control (MAC) and Physical layer (PHY) specifications: high speed physical layers in the 5 GHz band (1999)
6. Wireless Medium Access Control (MAC) and Physical layer (PHY) specifications for High Rate Wireless Personal Area Networks (WPANs): Amendment 2: Millimeter-wave based Alternative physical layer extension. IEEE (2008)
7. Proakis, J.G., Manolakis, D.G.: Digital Signal Processing Principles, Algorithm and Applications, pp. 449–495. Prentice Hall (1996)
8. Burrus, C.S.: Efficient Fourier transform and convolution algorithms. In: Lim, J.S., Oppenhiem, A.V. (eds.) Advanced Topics in Digital Signal Processing. Prentice-Hall, Englewood Cliffs, NJ (1988)
9. Cooley, J.W., Tukey, J.W.: An algorithm for the machine calculation of Fourier series. Math. Comput. **19**, 297–301 (1965)
10. Duhamel, P., Vetterli, M.: Fast Fourier transforms: a tutorial review and a state of the art. Signal Process. **19**, 259–299 (1990)
11. Kumar, S., Toamr, P., Shukla, A.: Effectiveness of OFDM with antenna diversity. In: International Conference on Communication, Control and Intelligent Systems (CCIS), pp. 172–175 (2015)

12. Rakesh, N., Nalineswari, D.: Comprehensive performance analysis of path loss models on GSM 940 MHz and IEEE 802.16 Wimax frequency 3.5 GHz on different terrains. In: International Conference on Computer Communication and Informatics (2015)
13. Rakesh, N., Nalineswari, D.: Improvising the performance of shadowing effects in urban terrains using Empirical models. In: International Conference on Signal Processing, Communication and Networking (ICSCN) (2015). https://doi.org/10.1109/icscn.2015.75219836
14. Mukesh, M., Abhishek, L., Bhambare, R.R.: QPSK modulator and demodulator using FPGA for SDR. Int. J. Eng. Res. Appl. **4**(4), 394–397 (2014). ISSN: 2248-9622
15. Measel, R., Bucci, D.J., Lester, C.S., Wanuga, K., Primerano, R., Dandekar, K.R., Kam, M.: A MATLAB Platform for Characterizing MIMO-OFDM Communications with Software-Defined Radios, pp. 1–6. IEEE (2014)
16. Arun, C.A., Prakasam, P.: Design of high speed FFT algorithm for OFDM technique. In: Conference on Emerging Devices and Smart System (ICEDSS), pp. 66–71 (2016). https://doi.org/10.1109/icedss.2016.7587780
17. Ko, S.-W., Kim, J.-H., Yoon, J.-S., Song, H.-K.: Cooperative OFDM system for high throughput in wireless personal area networks. IEEE Trans. Consum. Electron. **56**(2), 458–461 (2010)
18. Alam, M.M., Amin, Z., Abrar, M.S.: Performance analysis of Reed Muller coded OFDM on Nakagami–m fading environment. In: Symposium on Communication and Control Theory, pp. 1–5 (2015)

Dr. N. Rakesh working has Vice chair in Department of Computer Science & Engineering, at Amrita University, Bengaluru Campus. He has almost 14 years of teaching along with research experience. His area of interests includes Computer Networks, Wireless Communication, Wireless sensor Networks, Internet of Things, Wireless Channel Modeling, Mobile Communication and topics related to Networks and Wireless Communication domains. He served has Session chair, Technical Program Chair, International Program Chair for various International Conference in India and abroad. Designated Reviewer for various International Journals and Conferences. Also currently leading "Computer Networks and Internet of Things" Tag.

T. Anjali currently serves as an Assistant Professor at the Department of Information Technology at Amrita School of Engineering, Amritapuri.

B. Uma Maheswari currently serves as Assistant Professor at the department of Computer Science, Amrita School of Engineering. She is currently pursuing her Ph.D.

A Reliable Network System for Railway Track Crack Detection

Pragati Jadhav, Shivani Kondlekar, Divyata Kotian, Navya Kotian and Preeti Hemnani

Abstract In the current railway system, it is important to have safety measures to avoid accidents and loss of human life and resources. The important issue that causes an accident is the obstacles on the track. This project deals with the efficient method to avoid accidents due to cracks on the track and obstacles. The main aim of this project is to detect the crack in the railway track and alert the nearby stations. A GPS system is being used to point the location of faults on tracks. The project presents a solution to provide an intelligent train tracking and management system to improve the existing railway transport service. The solution is based on a powerful combination of ultrasonic sensor, peripheral interface controller (PIC), global system for mobile communication (GSM), global positioning system (GPS) technologies, Bluetooth module, and Android application. Using Android application, we can send messages to nearest railway stations.

Keywords PIC · Bluetooth module · Ultrasonic sensor · Obstacle sensor Android handset · DC motor

P. Jadhav · S. Kondlekar (✉) · D. Kotian · N. Kotian · P. Hemnani
SIES Graduate School of Technology, Sector 5, Nerul, Navi Mumbai, India
e-mail: kondlekar.shivani@siesgst.ac.in

P. Jadhav
e-mail: jadhav.pragati@siesgst.ac.in

D. Kotian
e-mail: kotian.divyata@siesgst.ac.in

N. Kotian
e-mail: kotian.navya@siesgst.ac.in

P. Hemnani
e-mail: preeti.hemnani@siesgst.ac.in

S. Smys et al. (eds.), *International Conference on Computer Networks and Communication Technologies*, Lecture Notes on Data Engineering and Communications Technologies 15, https://doi.org/10.1007/978-981-10-8681-6_87

1 Introduction

The project is basically used to detect crack on the railway tracks. These are detected by ultrasonic sensors. It also consists of obstacle sensors, which is used for sensing any living object present in between the tracks. The PIC controller is used to stop the motor driver when the crack or living being is detected or. Then, the GPS and Bluetooth module are used for sending the location to the nearby station network. Also, off-line message will be provided through the Android app to the user.

2 Literature Survey

According to IEEE paper, Railway Security in Wireless Network printed on 2013 in the International Journal of Computer Science and Network, Vol. 2 and another paper, the Innovative Railway Track Surveying with Sensors and Controlled by Wireless Communication printed on 2013 in the International Journal of Advanced Electrical and Electronics Engineering, volume-2. Avinash Vanimireddy, D. Aruna Kumari et al. describe that the issue of crack detection in railways cannot be controlled by the existing approaches, which cost high amount of people's live. Ramavath Swetha inferred the ideas in the proposed crack detection in railway with the help of microcontroller. The infrared sensor is used to monitor the obstacles in the tracks, and also can trace the location by using the GPS module. It will intimate about the cracks in tracks via SMS using GSM application.

3 Proposed System

The prototype detects the crack on the track based on ultrasonic sensor, which is placed in such a position that it detects horizontal and vertical cracks on the track and also living being intervention in front of the rail cart is detected using obstacle sensor. The crack-detected location is made available on app, which is accessible to the concerned department head at each station, and also an SMS service is made available if off-line. The prototype is actually a rail cart with motors, sensors, Bluetooth module, and Android handset. Android handset is used to make GSM and GPRS services available when needed and Bluetooth module to connect with the handset (Refer Fig. 1).

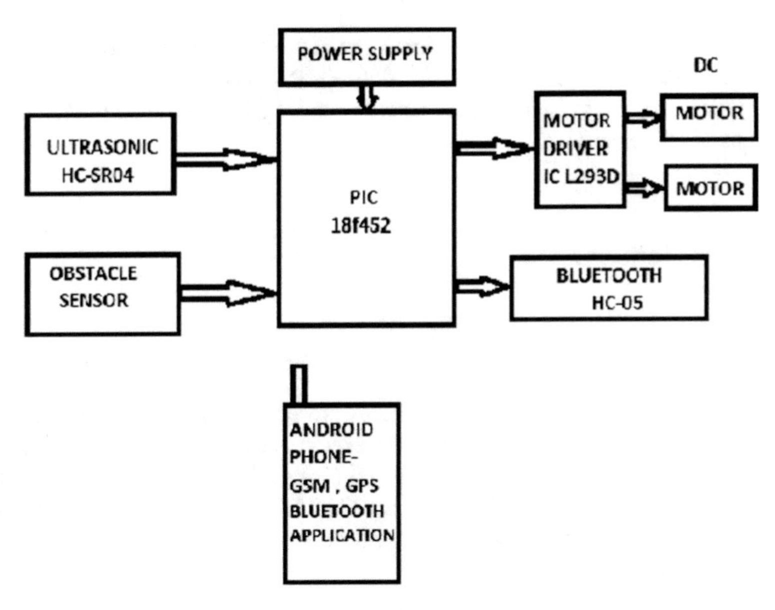

Fig. 1 PIC184F452 is the controller, which properly maintains the work of sensor, motor in this prototype. Controller is connected to ultrasonic sensor HC-SR04 for crack detection. Motor driver IC is connected to the controller, which would be controlling the motion of DC motor based on commands from PIC. Bluetooth module HC-05 is used to connect with Android mobile in rail cast, so as to upload the GPRS location on app as soon as the crack is detected

4 Block Diagram

4.1 Abstract

PIC184F452 controller is connected to ultrasonic sensor HC-SR04 for crack detection. Motor driver IC is connected to the controller, which would be controlling the motion of DC motor based on commands from PIC. Bluetooth module HC-05 is used to connect with android mobile in rail cast so as to upload the GPRS location on app as soon as the crack is detected.

4.1.1 Working of Project

PIC184F452 is the controller, which properly maintains the work of sensor and motor in this prototype. Controller is connected to ultrasonic sensor HC-SR04 for crack detection and an obstacle sensor, which consists of IR LED and photodiode used to detect motion on track. Ultrasonic sensor is mounted on rail cast facing the track below, which will detect horizontal and vertical cracks in the track. Motor driver IC is connected to the controller, which would be controlling the motion of DC motor based on commands from PIC. Bluetooth module HC-05 is used to connect with Android mobile in rail cast so as to upload the GPRS location on app as soon as the crack is detected. Android handset is used to send the location information on app as well as via SMS service.

5 Model Requirements

5.1 Abstract

The proposed system consists of both hardware and software requirements. The hardware system of the model consists of ultrasonic sensor HCSR04, obstacle sensor, motor driver IC L293D, four motors, PIC controller PCB, Bluetooth module HC-05, and Android phone user side and in the boot. Designing of PCB is done using MPLAB. The basic application of user and boot is programmed using basic for Android software.

5.1.1 Hardware Requirements

PIC Controller: The controller used here is the PIC18F452 and it can be compiled by using a C language. It consists of 356 bytes of EEPROM, 10-bit ADC, three-line serial port, and two-line asynchronous receiver transmitters.

Android Handset: Android smartphones can be connected to two types of cellular networks such as CDMA and GSM. When your Android is connected to a GSM network, it uses the general packet radio service (GPRS) to transfer data to and from the network.

DC Motor: Robot speed is 0.5 m/s. North polarization uses the red color indication and south polarization uses the green light indication.

Ultrasonic Sensor: Ultrasonic sensor emits an ultrasonic wave in one direction and started timing when its launched. Ultrasonics spread in air and return immediately when the obstacle is sensed. Ultrasonic sensor is used to measure the depth, distance, and thickness. Ultrasonic transmitter emitted an ultrasonic wave in one direction, and started timing when it launched.

Obstacle Sensor: The obstacle sensor used here is of two different types for transmitter and receiver, and its frequency range is between 32 and 42 kHz.

Bluetooth Module: Bluetooth used here is HC-05. The data of the crack location is transferred via Bluetooth in this application.

5.1.2 Software Requirements

MPLAB is used for PIC Programming. MPLAB is a proprietary freeware integrated development environment for the development of embedded applications on PIC and dsPIC microcontrollers, and is developed by Microchip Technology. Flash magic is used to dump the code to the microcontroller from PC. Another application is used for Windows and that is Flash Magic. It helps to do program for the microcontroller. This will be connected using the serial communication via RS232 link. PCB layout is a high-level engineering tool for board design featuring smart manual routing of high-speed and differential signals, shape-based auto-router, advanced verification, and wide import/export capabilities. For PCB layout, DipTrace is used. DipTrace features design process with real-time DRC, which reports errors on the fly before actually making them. B4A, i.e., basic for Android is used for Android programming so as to build App. Basic4Android is a rapid application development tool for native Android applications, developed and marketed by Anywhere Software Ltd.

6 Scope of Project

6.1 Abstract

The scope of the system explains the advantages of the system to the current railway system. It provides a better alternative to the current geodetic system used to detect cracks in railway system. The future developments of these projects include the use of CCTV and image processing system to capture the live images of crack, etc.

6.1.1 Advantages

The auto crack detection is more efficient than the traditional geodetic method of crack detection in technical field. Due to the use of ultrasonic sensor and obstacle sensors, quick response is achieved. The construction of structure is comparatively simple than the others systems and is cost-efficient. Maintenance and repair of the system are easy. No fire hazard problem is observed during overloading of the system. Operational cost of the system is less. Automatic alert and notification to station master is possible. The signal transmission is wireless.

6.1.2 Conclusion

This project is cost-effective. Monitoring and maintenance of railways using the proposed system are much convenient than the current system. This system will help us reduce the delay in train arrival due to mega blocks.

References

1. Rao, M., Jaswanth, B.R.: Crack sensing scheme in rail tracking system. Int. J. Eng. Res. Appl. **4**(1), 13–18 (2014). ISSN: 2248-9622
2. Parrilla, M., Nevado, P., Ibanez, A., Camacho, J., Brizuela, J., Fritsch, C.: Ultrasonic Imaging of Solid Railway Wheels. IEEE Xplore Conference: Ultrasonic Synopsium, (2008)
3. The Hindu Official Website: Ultrasonic railway crack detection system (2012). Available from: http://www.thehindu.com/todays-paper/tpnational/tp-kerala/students-develop-device-fordetecting-cracks-on-rail-tracks/article3916920.com
4. Jianhua, Q., Lin-sheng, L., Jing-gang, Z.: Design of rail surface crack-detecting system based on linear CCD sensor. In: IEEE conference on network sensing and control, vol 14, no 4, pp. 961–970, Apr 2008
5. Sharma, K., Maheshwari, S., Solanki, R., Khanna, V.: Railway Track Breakage Detection Method using Vibration Estimating Sensor Network. IEEE (2014)
6. Shekhar, R.S., Shekhar, P., Ganesan, P.: Automatic detection of squats in railway track. In: IEEE Sponsored 2nd International Conference on Innovations in Information Embedded and Communication Systems, vol. 3, iss. 6, pp. 413–413, Dec 2015
7. Greene, R.J., Yates, J.R., Patterson, E.A.: Rail crack detection: an infrared approach to in-service track monitoring. In: SEM Annual Conference and Exposition on Experimental and Applied Mechanics, vol. 112, nos. 23, pp. 291–301, May 2006
8. Somalraju, S., Murali, V., Saha, G., Vaidehi, V.: Robust railway crack detection scheme (RRCDS) using LDR assembly. In: IEEE International Conference on Networking, Sensing and Control, vol. 6, iss. 3, pp. 453–460, May 2012

Ms. Pragati Jadhav working in SIES Graduate School of Technology, Sector 5, Nerul, Navi Mumbai. Her research area includes sensors and android applications.

Ms. Shivani Kondlekar working in SIES Graduate School of Technology, Sector 5, Nerul, Navi Mumbai. Her research area includes sensors and android applications.

Ms. Divyata Kotian working in SIES Graduate School of Technology, Sector 5, Nerul, Navi Mumbai. Her research area includes sensors and android applications.

Ms. Navya Kotian working in SIES Graduate School of Technology, Sector 5, Nerul, Navi Mumbai. Her research area includes sensors and android applications.

Prof. Preeti Hemnani working as a professor in SIES Graduate School of Technology, Sector 5, Nerul, Navi Mumbai. Her research area includes sensors and android applications.

Research on a Novel B-HMM Model for Data Mining in Classification of Large Dimensional Data

C. Krubakaran and K. Venkatachalapathy

Abstract Data mining has been gaining widespread significance in recent times with the ever-increasing volume of data to be handled and processed in real time scenarios. This research paper investigates a hybrid scheme of data mining implemented for a classification problem on a repository consisting of large volumes of data. A hidden Markov model (HMM) has been utilized for modeling the high dimensional data pool and classification achieved using a naive bayesian classifier. A data repository with four different subjects and 50,000 instances has been mined using the proposed hybrid algorithm and compared against recent mining techniques like radial neural network and bayesian models and superior performance in terms of classification accuracy, computation time has been observed and reported.

Keywords Data mining · Classification · Big data · Hidden markov model
Naive bayesian classifier · Classification accuracy

1 Introduction

In recent times with improved state of the art data acquisition techniques and methodologies, the volume of data that needs to be stored, processed and transmitted has grown in enormous quantity. High definition data acquisition devices have enabled a detailed capture of events and descriptions of the subject under study giving the finest details and consequent clarity [1]. At the same time, the memory required for handling the data as well as the time taken to process them have also become

C. Krubakaran (✉)
Department of Information Technology, Bharathiyar College of Engineering & Technology, Karaikal, U.T. Puducherry, India
e-mail: Kirubabcet2018@gmail.com

K. Venkatachalapathy
Department of Computer and Information Science, Faculty of Science, Annamalai University, Annamalai Nagar, Chidambaram, Tamil Nadu, India
e-mail: omsumeetha@rediffmail.com

S. Smys et al. (eds.), *International Conference on Computer Networks and Communication Technologies*, Lecture Notes on Data Engineering and Communications Technologies 15, https://doi.org/10.1007/978-981-10-8681-6_88

complicated. This has led to the evolution of big data and cloud computing where distribution of services to clients is achieved on a global basis at any point of time as and when required. A cloud is normally a pool of data comprising of a composite mixture of data from various sources or sensors. Apart from cloud, there are also several instances where composite collection of data in a single point could be observed. One such instance could be quoted as the analysis of accidents occurring at a particular place from which a specific pattern could be extracted to predict the future occurrences [2]. Another instance could be the forecast of forex services or weather conditions based on collected inputs from several sources or sensors. This extraction mechanism which is concerned with information retrieval has paved the way for data mining concepts which obtain a pattern from the retrieved information.

1.1 Data Mining

Data mining is defined as the art or process of extracting meaningful information from a specific data set which may be homogenous in nature or a mixture of several data types like images, videos, audio signals, emails, digital signatures, etc. It is popularly known as KDD or knowledge discovery in data base as it deals with extracting useful and meaningful information from the pool of data and translating it into an understandable form. It is also to be observed that data mining is an intersection of several disciplines like engineering, mathematics and statistics, regression analysis, survey analysis, etc. Most of data mining problems are solved by learning and stochastic-based methods.

1.2 Big Data and Mining

As briefed in previous sections, recent advancements in data acquisition methods have led to evolution of huge volumes of data to be handled and stored constituting big data. A conceptual view of big data is depicted in Fig. 1.

Common and critical issues that poses a challenge to big data involves security of data, their processing methodologies and the method of storing the high dimensional data which occupies significant memory. In addition to it, the data which are high dimensional in nature may be of a heterogenous nature in terms of resources involving governmental agencies, private and public sectors which may be structured or scrambled. Research issues concerned with processing involves organization, handling and processing of data which may be sequential or in batches. In doing so, the data has to be kept secure from unauthorized access and tampering attacks as the data is distributed over different sources. All the above mentioned issues have been illustrated in Fig. 1 and indicated as the three versus namely velocity, variety and volume concerned with processing, nature of data and sources followed by memory capabilities of the system respectively.

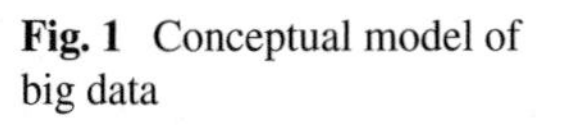
Fig. 1 Conceptual model of big data

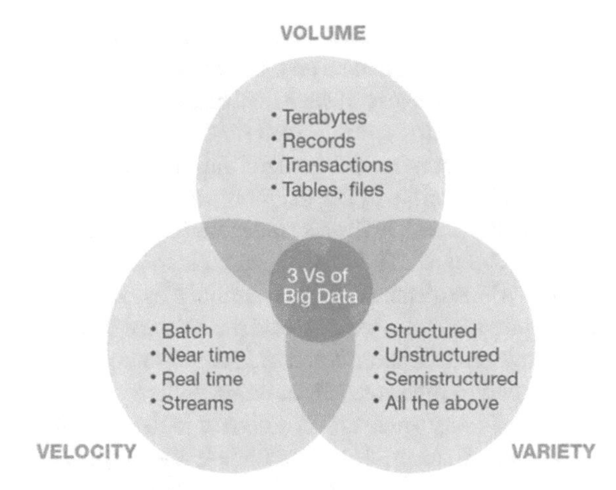

1.3 Problem Formulation

In general, data mining could be strategically categorized into a classification based problem, estimation-based approach, association rule mining approach, clustering-based approach. Classification problem is similar to clustering approaches excepting the fact that classification problems have a known target class. In other words, the given set is classified into one of the defined target classes. On the other hand, clustering-based methods have to work on unknown targets. For example given a data set of 1000 subjects, it is quite impossible to determine as to which target class these subjects would belong to and is greatly dependant upon the extraction pattern. This research paper has investigated the data mining using a data classification strategy using a naive bayesian classifier.

2 Related Works

A systematic survey of literature in the field of data mining presents various strategies which have been adopted based on the problem formulation like classification, clustering analysis, prediction, etc. [3–5]. It is found that the computation time for extraction of information is significantly reduced when a suitable platform is available for integrating the sources of information distributed over several servers or places [6]. Mining techniques have been investigated and experimented in the literature using statistical, learning or stochastic methods. Further data mining techniques have been found to have a wide utilization ranging from agriculture [7], social media [6], forensics [8] and health care [9]. Since the proposed paper is focussed towards implementing a classification based data mining, the survey of the literature has also been limited towards research contributions in the field of classification.

Literature [1] presents machine-based learning methods which could instance-based, bayesian classifiers and K-nearest neighbor methods. Instance based methods fall under the category of statistical data mining and normally termed as lazy learning algorithms due to their observed time delay in convergence until the final classification stage. Nearest neighbor algorithms are also a class of instance-based techniques in which the points in the neighborhood constitute the instances and the distances between the points in the are under study are computed using suitable distance computing techniques. The decision upon classification is arrived by studying the distance metrics where instances fall into similar class subsets if the distance metric is minimal or fall into various other classes if the distance metric is maximum. A slight variant of the above method is observed in K-nearest neighbor classifiers where given instance is assigned to the same subset based on its vote obtained among its neighbors. It is usually a binary value and is assigned to the same subset for a value of $k = 1$.

The other set of classifiers include the Bayesian networks which are a graphical based approach at establishing probability relationships between the variables. Bayesian classifiers [10, 11] are found to be probabilistic models which are capable of predicting the probability that a particular instance falls under a specific subset class. Bayesian networks are quite suitable for classification in real time which characterized by huge volumes of incoming data. From study in literature, it is observed that bayesian models exhibit superior performance over existing techniques such as CART, C4.5, Rule induction- and neural-based models.

3 Proposed Work

Data mining involves extraction of meaningful information and converting them into an understandable language or format from a pool of repository which may be homogenous or heterogeneous in nature. The proposed model is formulated keeping in mind the quantity of data to be handled. Such a big data to be processed in real time in the least minimum time as well as with precise accuracy involves complexity. The proposed approach utilizes a naive hybrid B-HMM approach in a hybrid combination for increasing the accuracy and simultaneously reducing the computation time for handling the big data. Naive Bayesian classifier is a derivative of Bayesian theorem and utilized in special cases where the volume of data to be handled is quite high. Maximum likelihood method is used in computation of parameters and many real world scenarios replicate Naive Bayesian formulation. Given a n dimensional vector with tuples denoted by D with the vectors specified as

$$V = \{v_1, v_2, v_3, \ldots v_n\} \tag{1}$$

and if the number of m classes $\{c_1, c_2, c_3, \ldots c_n\}$ the prediction of naive bayesian classifier for V belonging to C_n is successful if

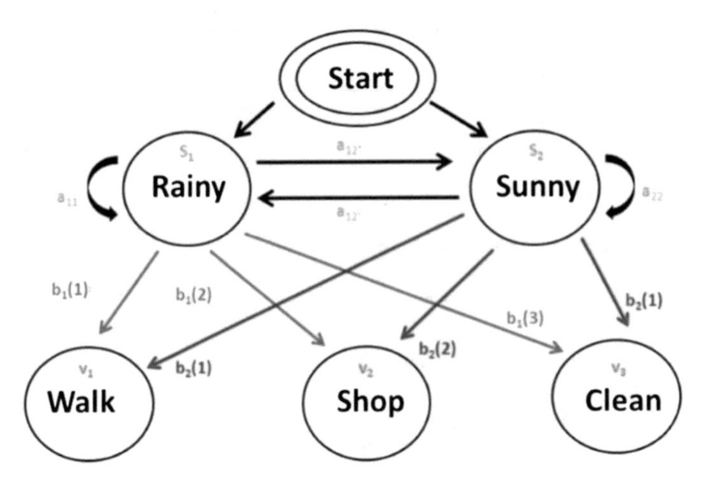

Fig. 2 Illustration of a hidden markov model for a sample case

$$P\left(\frac{C_n}{V}\right) > P\left(\frac{C_p}{V}\right) \text{with } 1 \le p \le n, n <> p \tag{2}$$

The above Naive Bayesian classifier is combined with a hidden Markov model which is capable of handling stochastic problem formulations in a big data environment. A hidden Markov model is characterised by unique states for each process associated with the events. A simple HMM model is depicted in Fig. 2 with a sample case taken with two input subsets namely sunny and rainy and the probability that a person will choose to walk, shop or stay indoors and do household cleaning.

Each chain in the HMM associated with probability of occurence of events with the total probability adding up to unity. The proposed algorithm reads the inputs from the repository taken for investigation and cumulatively adds them up in a feature count vector variable and is initialized with information regarding the frequency of repetition of the extracted pattern. The final phase involves the naive bayesian classifier used to map the given feature vector into their associated target subsets. A pseudocode is illustrated below with the initial number of levels of transition taken as I and the module blocks involved in the state transition taken as $\{M_1, M_2, M_3 \ldots M_n\}$ for the period initialized at T_t. In the proposed model the possible number of levels is taken as a feature vector set given by $\{S_1, S_2, S_3 \ldots S_n\}$ and the effects during transition of levels taken as U.

InitializeY(i) = $\beta(0_i)$, for $1 \le i \le J$
Choose the initialize point I^1.

1. *for* $i = 2$ *to J do*
2. *Select the sample* I^{cond}
3. *If* $S \ge 1$ *then* $I^1 = I^{cond}$
4. *else*
5. *Initialize a random integer between [0,1]*
6. *If* $S < I$
7. *then* $I^l = I^{cond}$
8. *else*
9. *Draw a new point from* $J(u)$
10. *for the given* u'
11. *Accept the change* $I^{t+1} = u'$,
12. *If rejected take new samples*
13. $I^l = I^{l-1}$
14. *end if*
15. *end if*
16. *end for*

4 Results and Discussion

A data mining model for classification problem using the hybrid Naive Bayesian—HMM model is investigated with a large data set obtained from accelerometer readings for observing the movements of four different classes of people or subjects. The number of instances measured for a two hour period and the subject details are depicted in Table 1.

The following metrics are used in the evaluation of the overall efficiency of the proposed mining hybrid algorithm. TP, TN, FP, and FN denote the number of true positives, true negatives, false positives and false negatives respectively.

$$\text{Accuracy} = \frac{\text{TP} + \text{TN}}{\text{TP} + \text{TN} + \text{FP} + \text{FN}} \tag{3}$$

$$\text{Error} = 100 - \text{accuracy} \tag{4}$$

Table 1 Details of data set used for experimentation

Subject	Genre	Age	Height (m)	Weight	Instances
A	Female	46	1.62	67	51,577
B	Female	28	1.58	53	49,797
C	Male	31	1.71	83	51,098
D	Male	75	1.67	67	13,161

$$\text{Precision} = \text{TP}/\text{TP} + \text{FP} \quad (5)$$

$$\text{Recall} = \text{TP}/\text{TP} + \text{FN} \quad (6)$$

Based on the above computations, the classwise classification accuracy has been depicted in Fig. 3 and it could be seen that precise accuracy has been achieved in the case of proposed B-HMM technique. In the first instance, it could be seen that ANN has extracted nearly 49,748 records pertaining to the class "standing" while Bayesian has produced 44,574 and the proposed B-HMM has presented a precise 29,984 records with an accuracy of nearly 98%. It was found that ANN had extracted records which matched both sitting and standing samples which has been corrected to precision by the proposed technique.

Table 2 presents the performance analysis of proposed method with respect to execution times and memory utilization in an Intel I5-6402P CPU running at 2.8 GHz with a total memory availability of 1 TB storage and a 8 GB RAM.

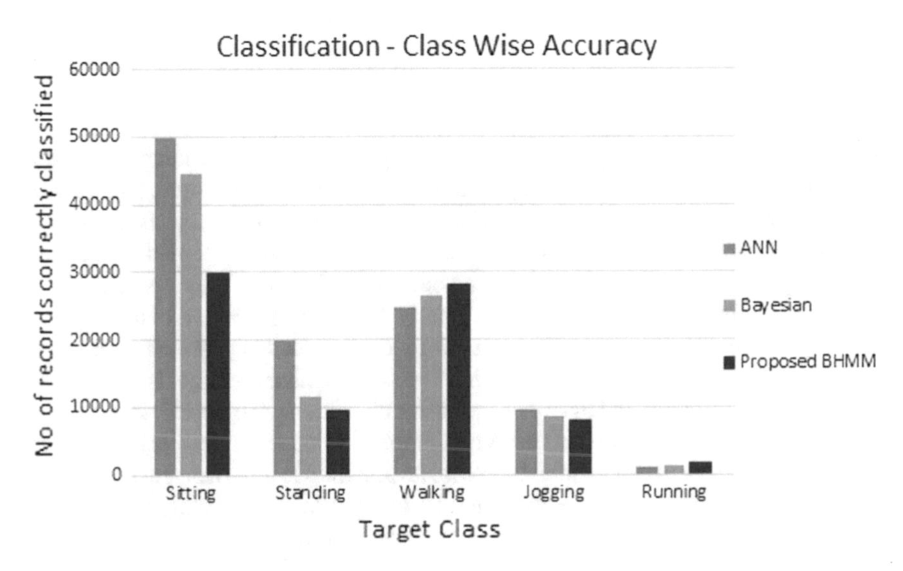

Fig. 3 Illustration of data utilization in cluster

Table 2 Comparative analysis of execution time

Instances	Execution time (ms)		
	ANN	Bayesian	B-HMM
10,000	9841	7540	2460
20,000	12,458	9841	2988
30,000	18,452	10,448	7514
40,000	24,225	11,585	8458
50,000	28,488	12,547	8500

Table 3 Comparative analysis of classification accuracy

S. No.	No. instances (5000)	Accuracy (%)		
		ANN	Bayesian (m)	B-HMM
1	Sitting	46	1.62	67
2	Standing	28	1.58	53
3	Walking	31	1.71	83

The observations from above table indicate superior performance of proposed B-HMM model in terms of execution time. No remarkable advantage is found in terms of memory utilization. Table 3 gives a comprehensive comparison of overall classification accuracy which further justifies the superior performance of proposed model.

5 Conclusion

The proposed research paper investigates the problem of data mining through a classification approach. A hybrid algorithm has been utilized to extract meaningful information from a data repository found in the literature. Four subjects have been studied and their movement patterns have been observed and recorded in the data repository for a period of observation amounting to 2 h. The proposed algorithm has been applied to a large data set of nearly 50,000 instances and efficiency of the proposed method evaluated in terms of its overall accuracy, accuracy in terms of individual target patterns, computation time and memory utility. The proposed method has been compared against benchmark algorithms like neural network and bayesian mining techniques and the output is found to be superior over the latter in terms of accuracy and computation time with no remarkable improvement in memory utility. A future scope of this paper could be thought of in improving the memory utility and computation time by utilizing the proposed algorithm on a cluster-based scenario.

References

1. Setiono, R., Baesens, B., Mues, C.: A note on knowledge discovery using neural networks and its application to credit screening. Eur. J. Oper. Res. **192**, 326–332 (2009)
2. Kiani, R., Mahdavi, S., Keshavarzi, A.: Analysis and prediction of crimes by clustering and classification. Int. J. Adv. Res. Artif. Intell. **4**(8), 11–17 (2015)
3. Thomassey, S., Fiordaliso, A.: A hybrid sales forecasting system based on clustering and decision trees. Decis. Support Syst. **42**, 408–421 (2006)

4. Kagdi, H., Collard, M.L., Maletic, J.I.: A survey and taxonomy of approaches for mining software repositories in the context of software evolution. J. Softw. Maint. Evol.: Res. Pract. **19**, 77–131 (2007)
5. Das, P., Das, A.K.: Behavioural analysis of crime against women using a graph based clustering approach. In: International Conference on Computer Communication and Informatics, pp. 1–6 (2017)
6. Kamruzzaman, S.M., Jehad Sarkar, A.M.: A new data mining scheme using artificial neural networks. Sens. J. **11**(5), 4622–4647 (2011)
7. Patel, S., Patel, H.: Survey on data mining techniques used in healthcare domain. Int. J. Inf. Sci. Tech. **6**(1), 53–60 (2016)
8. Cole, S., Tidke, B.: A survey of big data in social media using data mining techniques. In: Proceedings of International Conference on Advanced Computing and Communication Systems (2015)
9. Phyu, T.N.: Survey of classification techniques in data mining. In: Proceedings of the International Multi-Conference of Engineers and Computer Scientists, vol. 1, pp. 1–5 (2009)
10. Law, H.C.M., Figueiredo, A.T.M., Jain Anil, K.: Simultaneous feature selection and clustering using mixture models. IEEE Trans. Pattern Anal. Mach. Intell. **26**(9), 1154–1166 (2004)
11. Mucherino, A., Papajorgji, P., Pardalos, P.M.: A survey of data mining techniques applied to agriculture. Oper. Res. Int. J. **9**(2), 121–140 (2009)

C. Krubakaran working as an Assistant Professor, Department of Information Technology, Bharathiyar college of Engineering & Technology, Karaikal, U.T. Puducherry, India. His reasearch area includes image processing and network security.

Prof. Dr. K. Venkatachalapathy working as a Professor, Department of Computer and Information Science, Faculty of Science, Annamalai University, Annamalai Nagar, Chidambaram, Tamilnadu, India. His reasearch area includes image processing and network security.

Security Framework for Context Aware Mobile Web Services

P. Joseph Charles and S. Britto Ramesh Kumar

Abstract Service-Oriented Computing (SOC) could be served as a good platform for the growth of Context Aware Mobile applications with the multi-functionality smart mobile devices. The other key drivers for the evolution of these location aware systems are ease-of-use, convenience, social interactions and anytime-anywhere availability. Additionally, the Context Aware Mobile Applications facilitates day-to-day activities such as search services, business, entertainment, advertising, marketing, shopping, ticket purchasing, payment and mobile banking live and smarter. These applications utilize services as fundamental elements that support rapid, low-cost, loosely connected and dynamic services in heterogeneous environment. Service Computing (SC) promises the development of a world of services with cheaper and agile distributed applications. Underneath the hype and publicity over these new technologies, the design specifications and the physical properties of the involving devices define the capabilities and limitations for many distributed enterprise computing applications.

Keywords Service oriented computing · Context aware mobile applications
Service computing

1 Introduction

Context Aware Mobile Web Services is an emerging discipline that involves Context Aware Computing (CAC), Mobile Computing (MC) and Service Computing (SC). Context Aware Computing is considered as component of Ubiquitous Computing (UC) or Pervasive Computing (PC) [1]. In Context Aware Computing, the application adapts not only to changes in the availability of computing and commu-

P. Joseph Charles (✉)
Department of Information Technology, St. Joseph's College, Tiruchirappalli, India
e-mail: Charles_pjm74@yahoo.com

S. Britto Ramesh Kumar
Department of Computer Science, St. Joseph's College, Tiruchirappalli, India

S. Smys et al. (eds.), *International Conference on Computer Networks and Communication Technologies*, Lecture Notes on Data Engineering and Communications Technologies 15, https://doi.org/10.1007/978-981-10-8681-6_89

nication resources but also to the presence of situational information such as user context, environment context and computing context. The Context-sensitive applications require contextual information that must be obtained from various sources such as sensors that are embedded in the environment, devices that are carried by end users, repositories of historical data tracking use of the application, and information contained in user profiles.

There has been in these days a notable increase in consumer use of Context Aware Mobile Applications (CAMA). The consumer centric Context Aware Mobile Applications are realized as the next evolution of personal Mobile Computing. Context aware computing devices and applications respond to changes in the environment in an intelligent manner to enhance the computing environment for the user [2]. Context aware applications should be pre-emptive in obtaining contextual information and provide their response based on the collected information. Henceforth, context aware applications tend to be mobile applications. In this scenario, context-sensitive applications have been facilitated by mobile handheld devices, mobile networks and Internet [3].

1.1 Context Aware Mobile Web Services

Context aware computing is a mobile computing paradigm in which applications can discover and take advantage of contextual information such as user location, time of day, nearby people and devices, and user activity. In the recent years many researchers have studied and built several context aware applications to demonstrate the usefulness of this new technology [4].

1.2 Security Limitations in Context Aware Mobile Web Services

Security in Context Aware Mobile Web services is critical to their wide scale adoption and integration in Web-based enterprise systems and software. While shifting from the traditional client/server architecture to Web services technology is seen as an ratification of the Internet community's faith in the promise of the Web services paradigm. The goals of interoperability and ubiquity as envisioned by the Context Aware Web services technology can only reasonably be realized if the unique security challenges posed by this paradigm are appropriately addressed. The uniqueness comes from the fact that Web-based enterprise resources being exposed via Context Aware Web Services are typically dynamic and distributed in nature [5]. Also which requires some integrity among the data sharing between the client and server.

2 Literature Survey

Hayashi et al. [6] have proposed a probabilistic framework for Context Aware Scalable Authentication (CASA). The probabilistic framework is dynamically selecting an active authentication scheme that satisfies a specified security requirement given passive factors. The prototypes could select active authentication factors based on passive factors which balancing security and usability of user authentication. The basic idea of CASA is the passive multi-factor data can be used to modulate the strength of active authentication needed to achieve a given level of security. Kayes et al. [7] have presented a framework for Purpose Oriented Situation-Aware Access Control (PO-SAAC) software services. This framework specifies purpose oriented situations and it's related to situation-specific access control policies. To achieve context aware access control, the framework considers the states of the entities and the states of the relationships between entities.

Bandara et al. [8] have proposed many consumption aspects and techniques to manage context constraints. It also proposes an ontology-based context model for service provisioning. The important objective of this model is to provide contextualization of dynamically relevant aspects of Web service processes. Chakravorty et al. [9] have proposed a framework for data security and privacy. The framework provides security and privacy through sensor data from smart homes. Privacy is associated with collection, storage, use, processing, sharing or destruction of personally identifiable data. Storing the personally identifiable data as hashed values with holds identifiable information from any computing nodes.

Having reviewed the literature, it is evident that several architectures, frameworks, models, and security mechanisms exist. But each one has its own limitations. The literature review reveals that the existing architectures, models, and security mechanisms address only certain level of security issues and each one has its own limitations and security breaches. In addition, there is no end-to-end securing framework for context aware mobile web services adoptable by the healthcare industries to exchange their sensitive information.

Therefore, there is an urgent need to design a novel framework for Context Aware Mobile Web Services related to healthcare services. Thus, the scope of the proposed research gains national and international importance. The functionalities of the security framework are presented in detail in the forthcoming chapters.

3 Methodology for Conflict Identification and Resolution

SFCAMWS is responsible to provide well-defined, logical components, and artifacts of the proposed system. In addition, this framework provides the essential technical infrastructure like acquiring user information, connectivity, and secure communication to facilitate context aware mobile web services and acts as an intermediary between web users and the service providers. Figure 1 shows the security framework for the proposed model SFCAMWS.

The proposed Security Framework for CAMWS is divided into three major divisions, the Mobile Client is where the requests originate and where responses are sent to, various mobile devices can be adopted as mobile clients. The communication manager is responsible to send and receive the requests and responses respectively, these details are also stored in the Backup Manager. After authentication of the device the Connection Manager validates the request. The Service Manager is responsible to encrypt and decrypt the incoming and outgoing information. It also acts as a repository of the services. The Context Manager interprets the context of the IMU and performs various actions to interface with the Service Manager, it also has a log of the IMUs context for archival purposes.

The Service Provider Infrastructure is the depot for the registered services, it is responsible for the filtering of information and providing suitable services when necessary.

Figure 2 shows the multilevel security for the proposed architecture. The multilevel security is adopted for the proposed framework sing PKI. Additionally, CAMWS gateway facilitates secure communication among the service requestors and SPs, via HTTPS. It also offers security services with the help of SS using digital

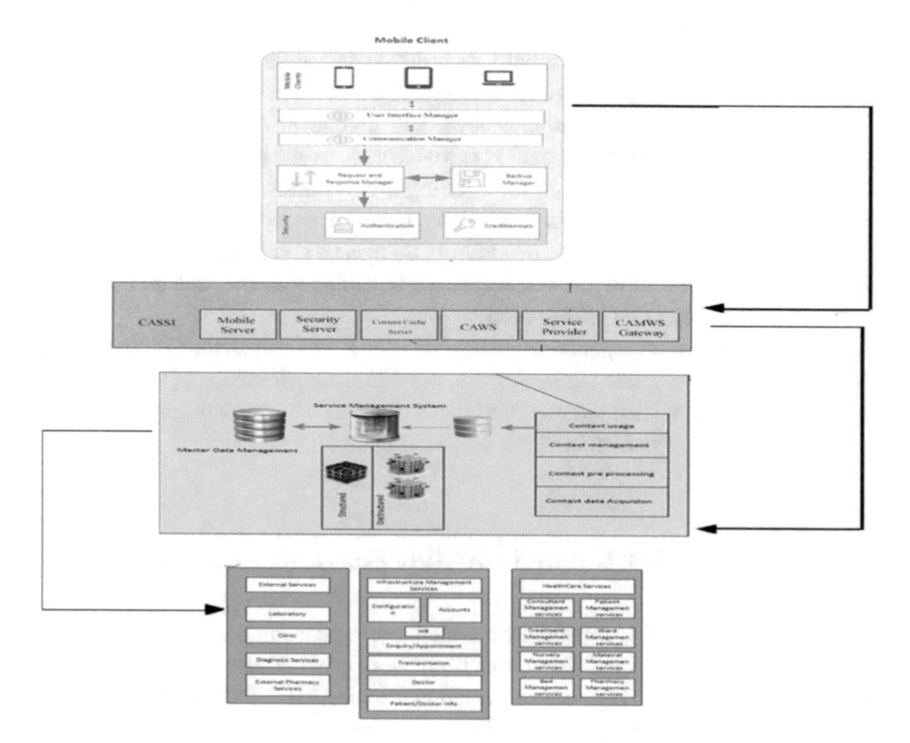

Fig. 1 Multilevel security for the proposed framework

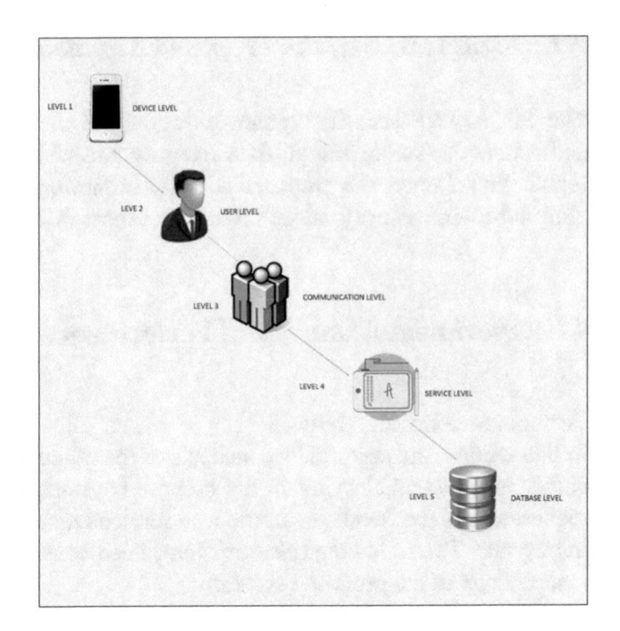

Fig. 2 Multilevel security for the proposed framework

signatures and standard encryption and decryption algorithms. The message confidentiality and IMU's authentication are accomplished by using RSA and digital signature. The digest algorithm known as SHA-1 ensures the integrity of the messages successfully. Another important security service, non-repudiation, is well supported by the digital signature. The security mechanisms are adopted in different levels and tested them successful. They are

1. Secure Communication between MC and CAMWS Gateway of CASSI
   - Secure IMU Registration
   - IMU Authentication
   - Mobile Client Device authentication
   - IMU and CAMWS Gateway Authentication
   - Service Authorization
   - Secure transmission of Service requests/responses between MC and CAMWS Gateway.
2. Secure Communication between CAMWS Gateway and SPI
   - CAMWS Gateway/SP Authentication
   - Secure Transmission of Service Requests/Responses
   - Secure Transmission of Data services.

3.1 Significance of the Proposed Architecture

The SFCAMWS security framework addresses the security issues such as user authentication, authorization, data integrity and confidentiality, privacy and non-repudiation. Hence, the proposed security framework for CAMWS is a novel one with end-to-end security using PKI to access the web services as well as data services.

4 Experimental Study and Performance Analysis

Performance Result Analysis
In this section, the performance analysis of the proposed system is described. Scalability is an essential factor in the proposed system where the number of service requesters and the frequency of the web service requests cannot be determined well in advance. The following measures have been considered to carry out the performance study of the proposed system:

- Throughput
- Hit Ratio
- Response Time

Such measures are supportive during the deployment of this web application as a commercial product. The performance tests have been conducted by simulating a heavy load on a CAMWS gateway.

4.1 Throughput of the System

The system throughput represents the amount of data in terms of bytes that the service requester receives from the Web server at any given second. The throughput of the system has been studied exclusively by the use of Hp LoadRunner tool [10]. Figure 3 illustrates the sample output screen shots to demonstrate the system throughput for two different loads on the CAMWS gateway with 1 and 100 service requesters. The clients and other involving entities are simultaneously authenticated as well. Proper and adjustable loads of the concurrent users are monitored. The purpose of this test is to check if the system's performance is changing when increasing the number of concurrent users.

Tests are done to find out the response times with 1, 5, 10, 20, 30, 40, 50, 60, 70, 80, 90, and 100 successive internal users generating request to the proposed system simultaneously. 100 users are chosen to represent the test scenario. It is observed that even when there are 110 simultaneous service requesters, the graph does not produce a straight line, thus it is observed that the throughput of the proposed system is nonlinear.

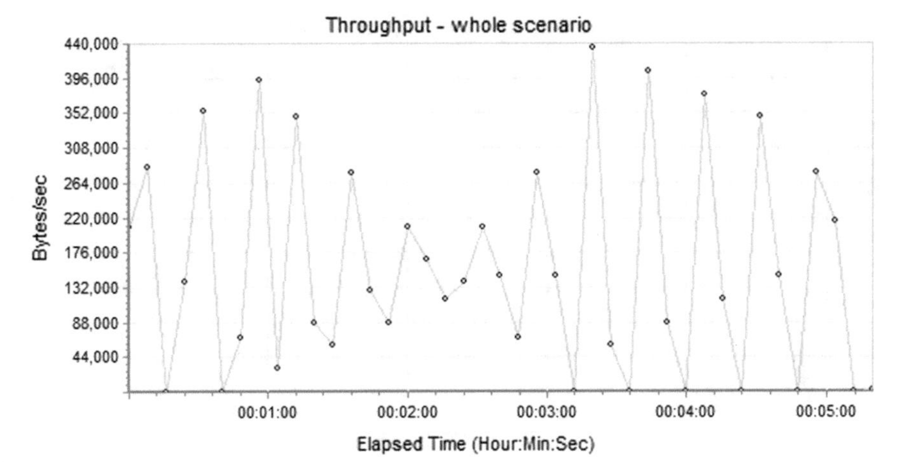

Fig. 3 Sample output screen shots to demonstrate the system throughput for two different loads

4.2 *Hit Ratio on the Server*

Figures 4 displays the screenshots for two different loads on the SFCAMWS by creating 100 and 1000 virtual service requesters. This graph is useful for determining the number of users at any given amount of time.

Figure 4 indicates that the service requesters are ramped up every 14 s at the arrival rate of 2 and this process continues until the number of service requesters reaches 1000. The service requester's load is continued up to 04:30 min. The sample

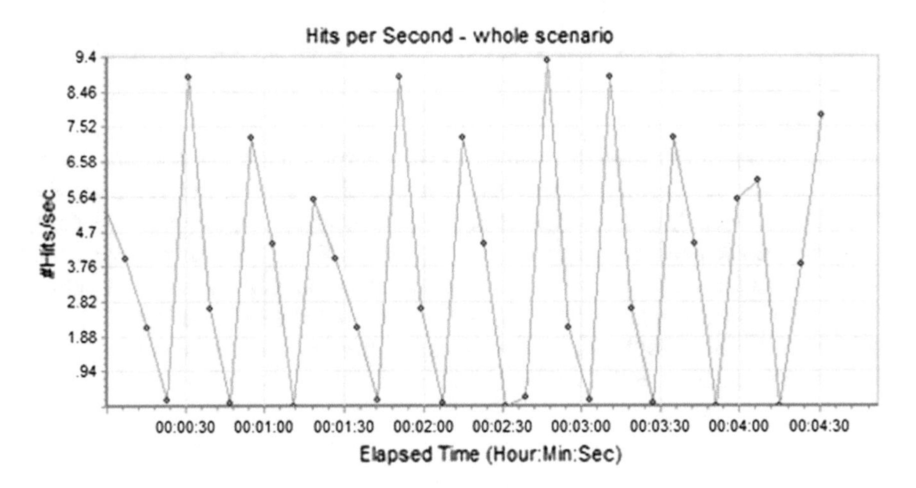

Fig. 4 Output screenshots for hit ratio per seconds

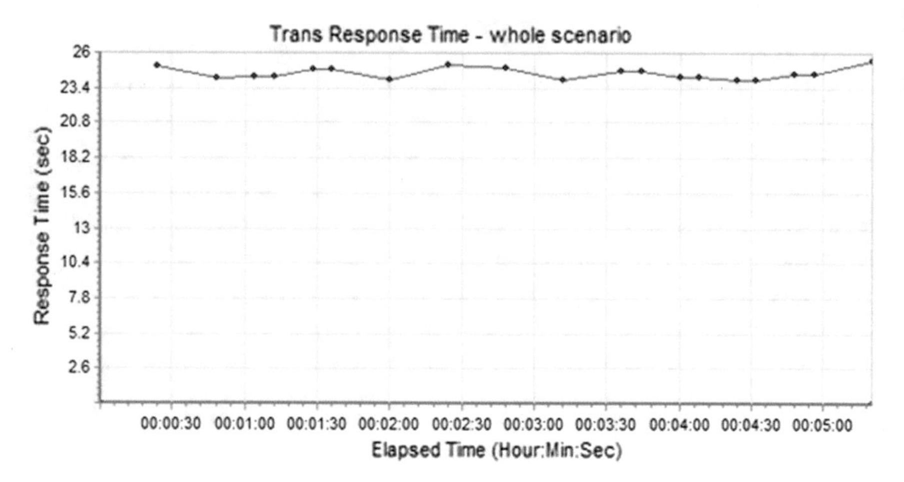

Fig. 5 Sample output screen shots to evaluate the transaction response time for two different loads

output screen shots to evaluate the amount of load generated by 100 and 1000 service requesters in terms of number of hits per second [11].

Figure 5 shows the sample output screen shots to evaluate the transaction response time for two different loads on the SFCAMWS by creating 10 and 20 virtual service requesters. This line graph is helpful to determine whether the web server performance is within the allowed minimum and maximum transaction performance time ranges which are defined for the proposed system. It also provides responses to the service requester requests with the reasonable response time and the web server load is found to be linear [12].

5 Conclusion

Since security plays a crucial role in the Context Aware Mobile Web services scenario, the design and development of security framework for complex web services, like healthcare services has become more promising research area. Furthermore, there has been considerable amount of research work carried out in the development and deployment of securing simple web services. In this paper, the author has put on efforts to propose a Security Framework for Context Aware Mobile Web Services (SFCAMWS) using Public Key Infrastructure to perform healthcare web services through Web-enabled mobile devices.

References

1. Saranya, C.M., Nitha, K.P.: Analysis of security methods in internet of things. Int. J. Recent Innov. Trends Comput. Commun. **3**(4) (2015)
2. Kaur, S., Singh, I.: A survey report on internet of things applications. Int. J. Comput. Sci. Trends Technol. **4**(2) (2016)
3. Nachouki, G., Chastang, M.: Multi-data source fusion approach. Int. J. Database Manag. Syst. (IJDMS) **2**(1), 25–32 (2010)
4. Siraj, S., Mikhailov, L., Keane, J.: Contribution of individual judgments toward inconsistency in pairwise comparisons. Eur. J. Oper. Res. **242**(2), 557–567 (2015). https://doi.org/10.1016/j.ejor.2014.10.024
5. Misra, S., et al.: Security Challenges and Approaches in Internet of Things. Springer Briefs in Electrical and Computer Engineering (2016)
6. Hayashi, E., Das, S., Amini, S., Hong, J., Oakley, I.: CASA: context-aware scalable authentication. In: Symposium on Usable Privacy and Security (SOUPS) 2013, Newcastle, UK
7. Kayes, A.S.M., Han, J., Colman, A.: OntCAAC: an ontology-based approach to context-aware access control for software services. Comput. Sci. Theory Methods Tools Comput. J. **58**(11) (2015)
8. Bandara, K.Y., Wang, M., Pahl, C.: An Extended Ontology-Based Context Model and Manipulation Calculus for Dynamic Web Service Processes. Springer, London (2013)
9. Chakravorty, A., Wlodarczyk, T., Rong, C.: Privacy preserving data analytics for smart homes. IEEE Security and Privacy Workshops (2013)
10. Vongsingthong, S., Smanchat, S.: A review of data management in internet of things. KKU Res. J. (2015)
11. Gubbia, J., Buyyab, R., Marusic, S., Palaniswami, M.: Internet of things (IoT): a vision, architectural elements, and future directions. Future Gener. Comput. Syst. **29**, 1645–1660 (2013)
12. http://cdn2.hubspot.net/hubfs/552232/Downloads/Partner_program/Smart_Environments_Flyer.pdf?t=1458917278396

P. Joseph Charles is an Assistant Professor of Computer Science and Information Technology at St. Joseph's College, Trichy. He received his degrees are M.Sc., M.Phil in computer science at St. Joseph's College, Trichy.

Dr. S. Britto Ramesh Kumar is an assistant professor of Computer Science at St. Joseph's College (Autonomous), Tiruchirappalli. His research interests include software architecture, wireless and mobile technologies, information security and Web Services. He has published many journal articles and book chapters on the topics of Mobile payment and Data structure and algorithms. His work has been published in the International journals and conference proceedings, like JNIT, IJIPM, IEEE, ACM, Springer and Journal of Algorithms and Computational Technology, UK. He awarded as a best researcher for the year 2008 at Bishop Heber College, Tiruchirappalli. He guides 8 Ph.D. research scholars and has completed a minor research project. He visited the countries like China, South Korea and Singapore.

Optimal SU Allocation to Multi-PU LCC CR Networks Consisting of Multiple SUs Using Cooperative Resource Sharing and Capacity Theory

Kaustuv Basak, Akhil Gangadharan and Wasim Arif

Abstract Resource Allocation is a significant issue in wireless communication and Cognitive Radio Networks in particular, in order to ensure effective utilization of spectrum and telecommunication resources. In this paper we discuss an optimal resource allocation policy for SU in a multi-PU Lost Calls Cleared (LCC) CR Network environment. Here, the primary user (PU) initiates a bargaining to optimally share its underutilized radio resources with the Secondary Users and optimize its profit subject to system constraints. In our proposed model, the Secondary User (SUs) also participate in the bargaining and jointly decides the optimum usage of the shared spectrum from the primary user (PU) satisfying its utility function. Since both the primary and the secondary users aim to maximize their profits by mutual cooperation, we model their interaction by a Nash Bargaining problem using Game Theoretic Approach. We introduce a two-step algorithm in order to obtain the optimal resource allocation for PU and SU. We propose a multi-PU LCC model for optimal resource allocation and obtain an optimal SU assignment policy satisfying the system constraints. The comprehensive simulation results of the analytical model are presented. The result shows encouraging values of average blocking probability and low degradation of system performance with enhanced interference levels.

Keywords Dynamic spectrum access · Cooperative resource sharing
Game theoretic model · Nash bargaining problem · Capacity theory
LCC networks

K. Basak (✉) · A. Gangadharan · W. Arif
Department of Electronics and Communication Engineering, National Institute of Technology Silchar, Silchar, Assam, India
e-mail: kbasak5190@gmail.com

A. Gangadharan
e-mail: akhilkgangadharan109@gmail.com

W. Arif
e-mail: arif.ece.nits@gmail.com

S. Smys et al. (eds.), *International Conference on Computer Networks and Communication Technologies*, Lecture Notes on Data Engineering and Communications Technologies 15, https://doi.org/10.1007/978-981-10-8681-6_90

1 Introduction

In the past few decades there has been a rapid growth in communication networks and services. With this boom there has been emergence of the issue of network congestions and in its response, the arrival of several new solutions in the form of either innovative concepts or technologies. One such technology is that of Software Defined Radios (SDRs) that emulate specialized communication hardware and functions (such as mixers, filters, amplifiers, modulators/demodulators, detectors, etc.) on a single hardware platform using software. A variant of such Software Defined Radio networks are the Cognitive Radio Networks (CRNs) which are used in spread spectrum communication. Possible functions of cognitive radio include the ability of a transceiver to determine its geographic location, identify and authorize its user, encrypt or decrypt signals, sense neighboring wireless devices in operation, and adjust output power and modulation characteristics. These networks with their advanced capabilities of sensing and cognition have been proposed as a solution to be able to use the spectrum efficiently and release the pressure on the network. Several models have been proposed to implement CRNs such as Dynamic Exclusive Usage Model, Hierarchical Access Model, Open Sharing Model. Among these the Dynamic Exclusive Usage Model ("Dynamic Licensing") is of interest to us. The Primary User (PU) has exclusive rights over the spectrum, which it might use as and when required. Sometimes this spectrum is underutilized. To improve the efficiency of Spectrum utilization, the primary user is assumed to be aware about the existence of secondary users and will share its underutilized radio resources (Spectrum, time slots, etc.) with the secondary users in order to obtain some remuneration (e.g. Improved performance [1–6], economic rewards [7–10], etc.).

Basically, the Dynamic Exclusive Usage Model may be broadly divided into two types of models:

1. Interference-Based Model.
2. Interference-Free Model.

Here, we use interference free model in System model-I [11] and introduce an interference term in System model-II.

2 System Model

Here, we consider the above Fig. 1a, b system model [11], where a single PU-pair (with a PU Tx and PU Rx) shares its licensed spectrum (time slots) with an infrastructure-based secondary network serving a set of SUs (denoted by the set $S = \{1, 2, \ldots, n\}$). The PU time slot is denoted by T. There is a fluctuation of the channel power gains for the PU and the SUs. Assuming, that there are C-possible states associated with the entire system, with the c-th state denoted by the vector $\{g_0^c, g_1^c, g_2^c, \ldots, g_s^c\}$. Here, g_0^c denotes the channel gain between the PU transmitter

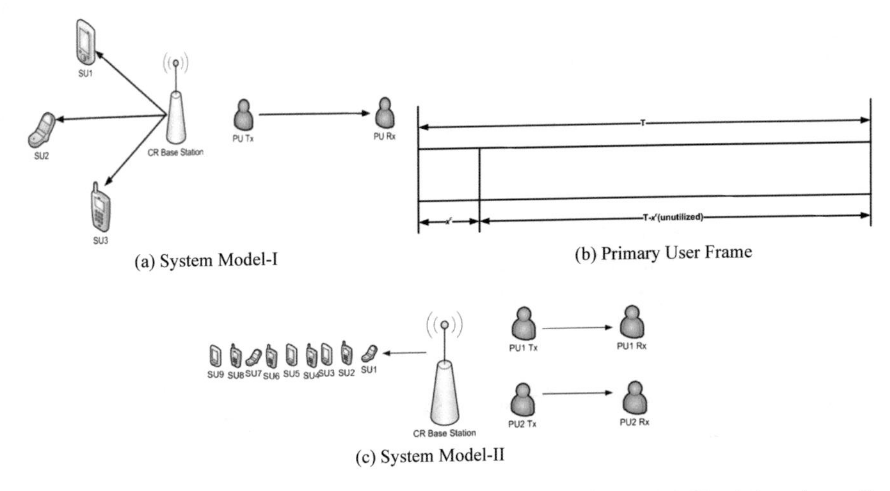

Fig. 1 **a** System model-I {one PU with Tx and Rx pair shares it's underutilized time slots with three SUs using an infrastructure CR base station}, **b** PU time frame {total frame length = T; time required for PU transmission = x^c; time shared with SUs = $T - x^c$}, **c** system model-II {consisting of two primary user channels and n-secondary users who are served on first come, first serve basis}

(PU Tx) and PU receiver (PU Rx) under channel state c and g_s^c, for all $s = 1, 2, \ldots, n$ denotes the channel gain from the CR base station to SU s under channel state c. The probability of occurrence of channel state c is denoted by Pr^c.

Further in System model-II, we utilize the results obtained from System model-I [11], to design an LCC (Lost Call Cleared) or *m*/*m*/queue system, where the n-SUs are assigned channel 1 or channel 2, corresponding to PU_1 and PU_2 respectively. The channel allocation is based on the availability of the time slots required by each secondary user with one/both of the PU time frames and a profit function computation, the details of which is provided in the subsequent sections.

3 Problem Formulation

The PU shares its time slots with the secondary users under each channel state c. Under each channel state c, the PU uses only a part of its slot denoted by x^c. The remaining $T - x^c$, part is shared with the SUs. Let t_s^c denote the time shared by the SU s under the channel state c and let p_s^c denote the power allocation to SU s under the channel state c. Hence, the SU's utility can be expressed in terms of it's average throughput as follows [11]:

$$U_s\left(t_s^c, p_s^c\right)_{\forall c} = \sum_c Pr^c t_s^c \log_2\left(1 + k\frac{p_s^c g_s^c}{n}\right), \tag{1}$$

where, n = noise power of background noise

Let d_s = payment of SU s. Then the net reward to the SU s is given by [11]:

$$R_s\left(t_s^c, p_s^c, d_s\right)_{\forall c} = U_s\left(t_s^c, p_s^c\right)_{\forall c} - d_s, \quad \forall s \tag{2}$$

If,

V primary user volume, to be delivered in the time slot T

x^c transmit time for the PU under channel state c

L additional energy cost to the PU, when sharing its spectrum with SU under channel state c

δ PU's marginal cost for it's energy consumption.

Then the interaction b/w the PU and the SUs can be modeled as a Nash Bargaining problem as follows [11]:

$$\text{NBP} = \max\left(v - \delta L(x^c)_{\forall c} - \Psi_0\right)^{\mu_0} \prod_{s \in S} \left(U_s\left(t_s^c, p_s^c\right)_{\forall c} - d_s - \Psi_s\right)^{\mu_s} \tag{3}$$

Subject to constraints

$$x^c + \sum_{s \in S} t_s^c \le T, \quad \forall c \tag{4}$$

$$T^{\min} \le x^c \le T^{\max}, \quad \forall c \tag{5}$$

$$0 \le t_s^c \le T, \quad \forall s, c \tag{6}$$

$$\sum_c \left(\sum_{s \in S} p_s^c t_s^c\right) Pr^c \le E_{\text{avg}} \tag{7}$$

$$p_s^c \ge 0, \quad \forall s, c \tag{8}$$

$$\sum_{s \in S} d_s \ge v \tag{9}$$

Here, the decision variables are:

$$\{x_c\}_{\forall c}, v, \left\{t_s^c, p_s^c\right\}_{\forall s, c}, \{d_s\}_{\forall s}$$

Ψ_0 Target reward that the PU aims to at least achieve

Ψ_s Target reward that the SU s aims to at least achieve

μ_0 Bargaining power of the PU

μ_s Bargaining power of the SU s

However, the difficulty that arises in the NBP objective function is its non-convexity, leading to no efficient algorithms to solve it. Hence, [11] identifies a property of the NBP, that correlates it with a social optimization problem, providing

Fig. 2 System model-II

maximum possible profit for all the parties involved in the problem based on the property [11] proposes a two-step algorithm to solve the SOP problem.

The two-step procedure in [11] addresses the following key points:

1. How the PU should share its time slots with the SUs and charge them for the same?
2. How the SUs share the allocated time slot from the PU among themselves efficiently and also what their preferred payment scheme is?

We then further use the above two-step algorithm [11] along with a cost function related to the maximization of the capacity/throughput of the SUs upon their assignment to any of the two PUs shown in the modified system model Fig. 2c to obtain an optimal PU channel assignment policy for the SUs. The model addresses the need to figure out an optimal channel allocation policy in a multi PU-SU environment. System model II consists of two PUs and n-SUs, who are served on first come first serve basis (LCC system) maximizing the overall gain of the system.

4 Bargaining Problem

Paper [11] sites the relationship between the NBP and an associated social optimization problem. They then look at a two-step procedure to solve the bargaining problem. They relate the bargaining problem with the social optimization problem (M), as follows [11]:

$$M = \max\left\{\sum_{s\in S} U_s\left(t_s^c, p_s^c\right)_{\forall c} - \delta L\left(x^c\right)_{\forall c}\right\} \tag{10}$$

M denotes the optimal (or maximum) value for the social optimization problem.

The SOP is however still difficult to solve due to its non-convexity, i.e., the non-convexity of $\sum_{s\in S} U_s(t_s^c, p_s^c)_{\forall c}$, to be exact.

The social optimization problem (M) can still be changed into the following problem (M_{eqv}) by changing the variable $\theta_s^c = p_s^c t_s^c$, denoting the allocated energy for the SU s under the channel state c, into the social optimization problem, as follows [11]:

$$M_{\text{eqv}} = \max \sum_{c} \left\{ \sum_{s \in S} Pr^c t_s^c \log_2 \left(1 + k \frac{\theta_s^c g_s^c}{t_s^c n} \right) - \delta L(x^c)_{\forall c} \right\} \tag{11}$$

The above M_{eqv} problem is a strictly convex optimization problem and it assumes a set of unique optimum solutions [11].

5 Two-Step Procedure

Two-step procedure to solve the NBP [11]:

Step 1: First, M_p is solved and the optimal values of resource allocations (optimal PU transmit time $\{x^{c*}\}_{\forall c}$, optimal transmit time of each SU s $\{t_s^{c*}\}_{\forall s,c}$ and its transmit power $\{p_s^{c*}\}_{\forall s,c}$) are obtained.

Step 2: By putting these optimal resource allocations into the NBP, we get the optimal PU's charge (v^*) and SU's payment ($\{d_s^*\}_{\forall s}$).

6 Algorithm 1 to Solve Step-1

Algorithm 1 tries to address the above optimality condition, by performing finite number of iterations to find the PU's optimal transmit time $\{x^{c*}\}_{\forall c}$ [11],

<u>**Steps**</u>:

1. Set, the upper bound for Ω as $\overline{\Omega} = \frac{T - T^{\min}}{T - T^{\min} \sum_c Pr^c \frac{n}{G^c} + E_{\text{avg}}}$.
2. Set, the lower bound for Ω as $\underline{\Omega} = \frac{T - T^{\max}}{T - T^{\max} \sum_c Pr^c \frac{n}{G^c} + E_{\text{avg}}}$.
3. Compute, $\Omega = \frac{1}{2}(\overline{\Omega} + \underline{\Omega})$.
4. Compute, $\{x^{c*}\}_{\forall c}$ under each channel state c using

$$x^c = \begin{cases} T^{\min}, & K_L^c(\Omega) \le K_R^c(T^{\min}) \\ \frac{V \ln 2}{1 + W\left[-\frac{1}{e} \left\{ K_L^c(\Omega) \frac{g_0^c}{\delta n} + 1 \right\} \right]}, & K_R^c(T^{\min}) < K_L^c(\Omega) < K_R^c(T) \\ T^{\max}, & K_L^c(\Omega) \ge K_R^c(T^{\max}) \end{cases}.$$

5. Compute, $C = \sum_c (Pr^c(T - x^c))$ and $D = \sum_c \left\{ Pr^c \frac{n}{G^c}(T - x^c) \right\}$ using the calculated x^c.
6. Calculate, $\Delta = C - \Omega \,(E_{\text{avg}} + D)$, using C and D.

7. Check for the convergence condition,
   $|\Delta| \leq \text{tol}$, where tol = tolerance error for convergence.
8. If $|\Delta| \leq \text{tol}$, then the algorithm converges, and we output the set of $\{x^c\}$, as the optimal set of $\{x^c\}$. Else, we go to step 9.
9. If, $\Delta \rangle \text{tol}$ we update the lower bound as $\underline{\Omega} = \Omega$. Else, if $\Delta \leq -\text{tol}$, we update upper bound as $\underline{\Omega} = \Omega$.
10. Go, back to step 3.

Here,

$$K_L^c(\Omega) = \frac{1}{\ln 2}\left\{1 + \ln\left(\Omega\frac{n}{G^c}\right) - \Omega\frac{n}{G^c}\right\}$$

And,

$$K_R^c(T^{\min}) = \delta\left\{2^{\frac{V}{x^c}}\left(1 - \frac{V}{x^c}\ln 2\right) - 1\right\}\frac{n}{g_0^c}$$

Now, the unique value of x^c for each value of c and Ω as is given by [11]:

$$x^c = \frac{V \ln 2}{1 + W\left[-\frac{1}{e}\left\{\mathrm{K}_L^c(\Omega)\frac{g_0^c}{\delta n} + 1\right\}\right]},$$

where, $W(.)$ is the Lambert W-function.

Hence, Algorithm 1 is used by the PU to determine it's optimal transmit time $\{x^{c^*}\}$ under all channel states. After knowing $\{x^{c^*}\}_{\forall c}$ each of the SUs s can calculate its optimal utility $U_s^* = U(t_s^{c*}, p_s^{c*})_{\forall c}$ and the PU can calculate its energy consumption cost $L^* = L(x^{c*})_{\forall c}$.Then, the optimal value for the social optimization can be obtained.

7 Algorithm A2 to Solve Step-2

Algorithm 2 is used to find the PU's charge and each SU's payments [11] Knowing $\{U_c^*\}_{\forall c}$ and δL^* from the social optimization problem, the objective NBP problem can be written as the following, *Charge Payment Problem* (C_p) [11]:

$$C_p\text{: } \max_{v,\{d_s\}}(v - \delta L - \Psi_0)^{\mu_0}\prod_{s\in S}\left(U_s^* - d_s - \Psi_s\right)^{\mu_s}$$

Since, $\{U_c^*\}_{\forall c}$ and δL are known, therefore, the above problem can be re-written in the log form, called $C_p - E$ (C_p-Equivalent) problem as follows [11]:

$$C_p - E = \max_{v,\{d_s\}} \mu_0 \ln(v - \delta L - \Psi_0) + \sum_{s\in S}\mu_s \ln\left(U_s^* - d_s - \Psi_s\right)$$

<u>**Steps**</u>:

1. Set, upper bound for χ as $\bar{\chi} = \frac{\min\left(\min_{s\in S}\left\{\frac{H_T^*}{\mu_s}\right\}, \frac{H_T^*}{\mu_0}\right)}{\text{tol}}$.
2. Set, lower bound of χ as $\chi = 0$.
3. The PU sets, $\chi = \frac{1}{2}\left(\overline{\chi} + \underline{\chi}\right)$.
4. The PU evaluates the expected charge $v = \mu_0 \chi + \alpha L^* + \Psi_0$.
5. The PU broadcasts the signaling χ to all the SUs.
6. On receiving χ, each SU calculates its preferred payment $d_s = U_s^* - \Psi_s - \mu_s \chi$, and relays it to the PU.
7. After receiving all the d_s values from the SUs, the PU compares the aggregate preferred payment with its expected charge v.
8. If $v > \sum_s d_s + \text{tol}$, then set upper bound, $\overline{\chi} = \chi$. Go back to step 3.
9. Else if, $v < \sum_s d_s + \text{tol}$, then set upper bound, $\underline{\chi} = \chi$. Go back to step 3.
10. Else, $v = \sum_s d_s + \text{tol}$ and the algorithm converges. In this case output the current values of v and $\{d_s\}$ as the optimal solution of $C_p - E$.

Here, $H_T^* = M^* - \sum_s \Psi_s - \Psi_0$ [11], represents the net benefit above target levels that the PU and the Sus get from participation in this mutual cooperation.

8 Algorithm A3 to Obtain Optimal SU Allocation with Two PU System

Here, we consider the system model of Fig. 2c.

Here, we assume an LCC system with n-SUs coming up to the CR Base Station on a random and sequential basis. Each SU attempt to access the PU network is assumed to be a fresh attempt and independent of any previous attempts or SU allocations. The simplified model consists of two primary channels, each having its own associated profit/cost of access. A SU arriving at any instant of time is allocated to one of the two available PU channels based on time slot availability and a cost function value, which is computed using the former two algorithms Algorithm 1 and Algorithm 2.

The parameters to be considered while computing the cost function value as:

$$\left\{t_s^c\right\}_{\forall s,c}, T_p, \left\{g_s^c\right\}_{\forall s,c}, \left\{p_s^c\right\}_{\forall s,c}, B_p, \left\{I_s^c\right\}_{\forall s,c},$$

where,

$\left\{t_s^c\right\}_{\forall s,c}$	Optimal SU transmit time obtained from Algorithm
T_p	Total time slot for PU_p.
$\left\{g_s^c\right\}_{\forall s,c}$	Information channel gain of SU s under channel state c.
$\left\{p_s^c\right\}_{\forall s,c}$	Power allocation for SU s under channel state c.
B_p	Bandwidth of the PU channel p
$\left\{\text{I}_s^c\right\}_{\forall s,c}$	Interference channel gain of SU s under channel state c

Using capacity theory, to maximize the profit to the SU in terms of improved transmission capacity/throughput, the profit function for the SU s can be represented as follows:

$$\mathrm{Pft}_p = \frac{t_s^c}{T_p} B_p \log_2\left(1 + \frac{g_s^c p_s^c}{I_s^c}\right) \tag{12}$$

Here, Pft_p represents the effective capacity of each channel, which varies for each PU and is a function of the decision variables $\{t_s^c\}_{\forall s,c}, T_p, \{g_s^c\}_{\forall s,c}, \{p_s^c\}_{\forall s,c}, B_p, \{I_s^c\}_{\forall s,c}$

Steps:

1. Initialize the optimal PU transmit times for PU_1 and PU_2 as, x_1^c for PU_1 and x_2^c for PU_2 respectively using the current CSI.
2. Obtain the time slots left in the frames corresponding to PU_1 and PU_2 as: $T_1 - x_1^c$ and $T_2 - x_2^c$, respectively.
3. Obtain t_s^c for the SU s arriving at the given instant for the current CSI using algorithm 1.
4. If $t_s^c > \left(T_1 - x_1^c\right)$ and $\left(T_2 - x_2^c\right)$, then block the SU and go back to step 1.
5. If $t_s^c > (T_1 - x_1^c)$ and $t_s^c \leq (T_2 - x_2^c)$, assign SU s to PU_2 and exit.
6. If $t_s^c \leq \left(T_1 - x_1^c\right)$ and $t_s^c > \left(T_2 - x_2^c\right)$, assign SU s to PU_1 and exit.
7. If $t_s^c \leq (T_1 - x_1^c)$ and $(T_2 - x_2^c)$, calculate Pft_1, Pft_2 corresponding to PU_1 and PU_2 and go to step 8.
8. If $\mathrm{Pft}_1 > \mathrm{Pft}_2$, assign SU s to PU_1 and exit.
9. If $\mathrm{Pft}_1 < \mathrm{Pft}_2$, assign SU s to PU_2 and exit.
10. If $\mathrm{Pft}_1 = \mathrm{Pft}_2$, assign SU s to PU_1 or PU_2 randomly and exit.

The algorithm is repetitive and assigns a SU to PU_1 or PU_2 at every instant a SU arrives to use the PU network.

9 Generalized Model for Algorithm A3

The above algorithm may be extended for m-PU, n-SU, LCC, CRN networks, as is quite intuitive.

Steps:

1. Initialize the optimal PU transmit times for PU_1, PU_2, PU_3, …, PU_n as, x_1^c for PU_1, x_2^c for PU_2, …, x_n^c for PU_n respectively using the current CSI.
2. Obtain the time slots left in the frames corresponding to PU_1, PU_2, …, PU_n as: $T_1 - x_1^c$, $\mathrm{T}_2 - x_2^c$, …, $T_n - x_n^c$ respectively.
3. Obtain t_s^c for the SU s arriving at the given instant for the current CSI using algorithm 1.
4. If $t_s^c > (T_i - x^c)_{\forall i}, i = 1, 2, \ldots, n$, then block the SU and go back to step 1.

5. If $t_i^c > (T_i - x_i^c)\forall i \neq j$ and $t_s^c \leq (T_j - x_j^c)$, assign SU s to PU_j and exit.
6. If $t_s^c \leq (T_i - x_i^c)$, for more than one i, calculate $\{\mathrm{Pft}_i\}$, corresponding to the set of i PUs and go to step 7.
7. If $\mathrm{Pft}_i > \mathrm{Pft}_j, \forall j, j \neq i$, assign SU s to PU_i and exit.
8. If $\mathrm{Pft}_i = \mathrm{Pft}_j, \forall i, j, i \neq j$, assign SU s to one of $\{\mathrm{PU}_i\}$ randomly and exit.

10 Results and Discussion

Assuming, each channel to be slow-fading and having a Normal distribution of probability of occurrence.

Let the state of a channel be modeled by the following states, with the associated relative gains (w.r.t mild/normal state) of each state assumed to be fixed as follows (Table 1).

These, states are assumed to be the result of channel scattering effects and slow-fading effect with weather conditions of the channel. Assuming the original, system model that we assumed in Fig. 1a, where we have one single PU and 3 SUs, we have (Fig. 3).

Here, we have considered the assumptions and model parameters from the results of System model-I [11].

Apart from that, we have considered two PUs and n-SUs forming an LCC system. The SUs are served on a first come first serve basis and any dropped attempt is cleared from the system. We have used a Poisson distribution to model the secondary transmit times requested by the SUs.

Table 1 Channel states with associated relative gains

Channel state	Associated relative gains
Good	1.2
Mild	1
Bad	0.8

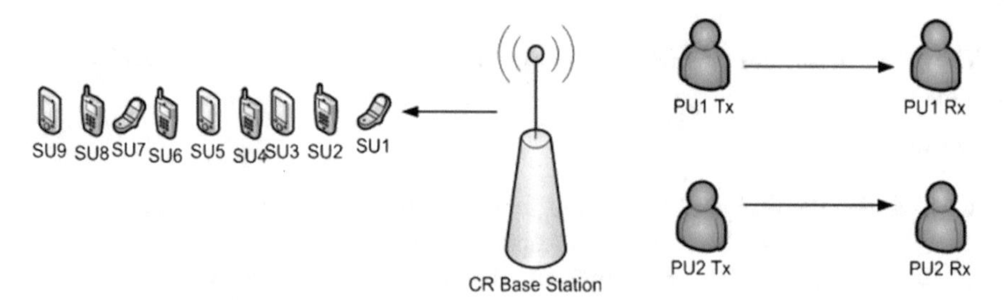

Fig. 3 LCC CRN system consisting of n-SUs and two PUs

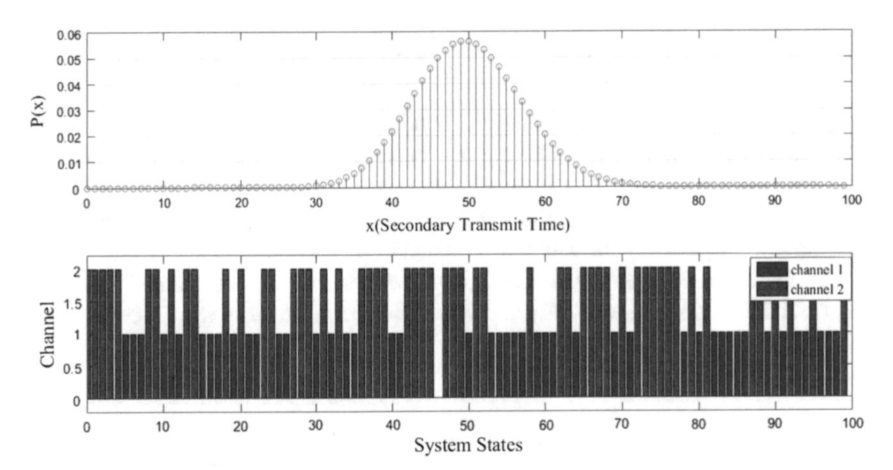

Fig. 4 Channel assignment for system model-II

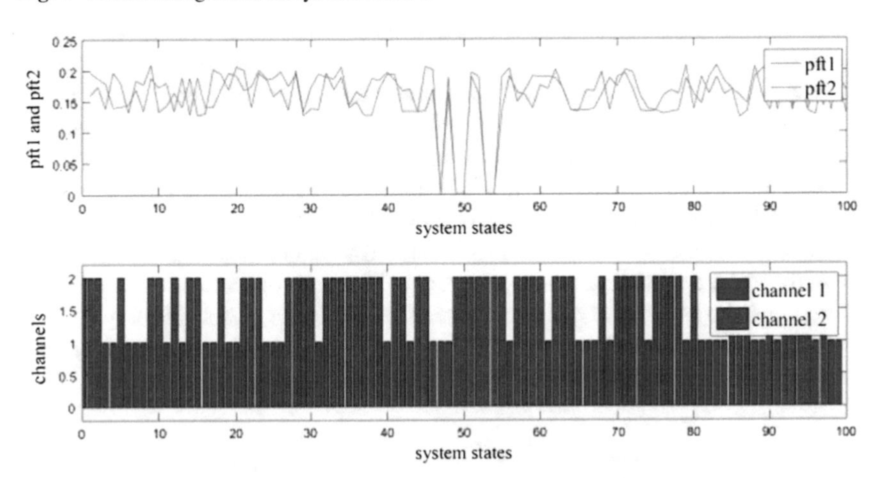

Fig. 5 Profit calculation for both the channels and corresponding channel assignment for system model-II

$$\text{i.e., } P(k) = e^{-\lambda}\frac{\lambda^k}{k!}, \quad k = 0, 1, 2, \ldots,$$

where, $\lambda =$average call holding time, $P(k)$ represents the probability of k units of secondary transmit time requested by the SU s with a mean secondary time duration of λ (Fig. 4).

The above plot assumes identical PU bandwidth, power and interference conditions. Only the instantaneous CSI and the corresponding calculations for Pft_1 and Pft_2 based on algorithm 3, for channel 1 and channel 2, corresponding to PU_1 and PU_2 varies. Here, $\lambda = 50$ and we find that 55 out of 100 calls are assigned to channel 2 and 44 to channel 1 with a GOS of 0.01 (Fig. 5).

10.1 Profit Values and Channel Assignment

The adjoined plot assumes identical PU bandwidth, power and interference conditions. Only the instantaneous CSI and the corresponding calculations for Pft_1 and Pft_2 based on algorithm 3, for channel 1 and channel 2, corresponding to PU_1 and PU_2 varies. The upper figure shows variations in Pft_1 and Pft_2 with channel states and the lower figure the corresponding channel assignments.

At any instant of time a SU *s* is assigned to either channel 1 or channel 2 depending on the relative values of Pft_1 and Pft_2. The above run shows a GOS/Blocking probability of 0.

10.2 Transmission Power and Channel Assignment

Figure 6 on the right shows a clear preference for channel 2 over channel 1, when the power allocated for transmission over channel 2 is much greater than that for channel 1. Here, the mean SU secondary time is assumed to be 5 units of time.

Figure 7 on the left shows a preference for both channel 2 and channel 1, when the power allocated for transmission over channel 2 is almost equal to that for channel 1. Here, the mean SU secondary time is assumed to be 5 units of time.

Figure 8 shows a clear preference for channel 1 over channel 2, when the power allocated for transmission over channel 1 is much greater than that for channel 2. Here, the mean SU holding time is assumed to be 5 units of time.

Figure 9 shows a nearly constant value of GOS/Blocking Probability of 0.07 up to a transmission power difference of 0.2 beyond which, blocking decreases linearly with the difference in transmission power and becomes nearly 0 at around a differ-

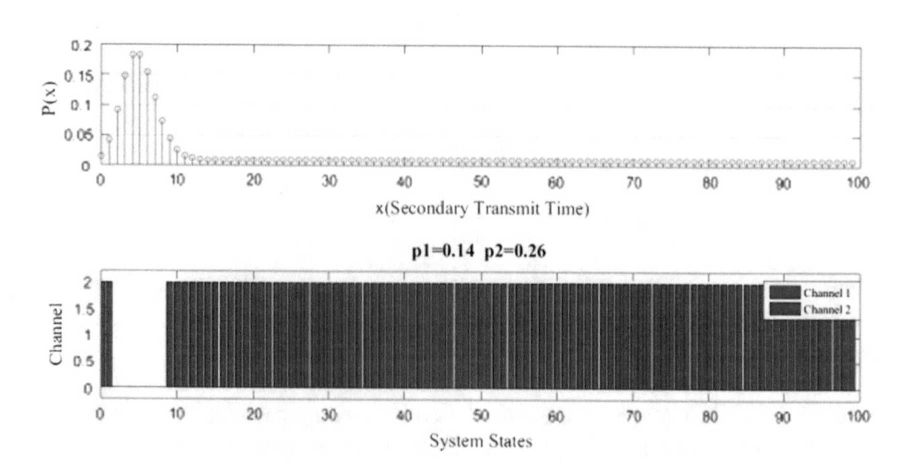

Fig. 6 Channel assignment policy when power allocation for channel 2 ≫ power allocation for channel 1

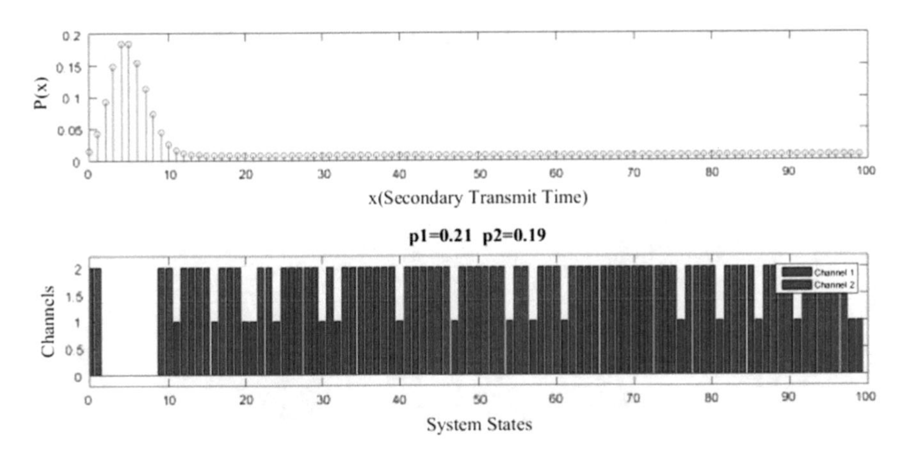

Fig. 7 Channel assignment policy when power allocation for channel 2 = power allocation for channel 1

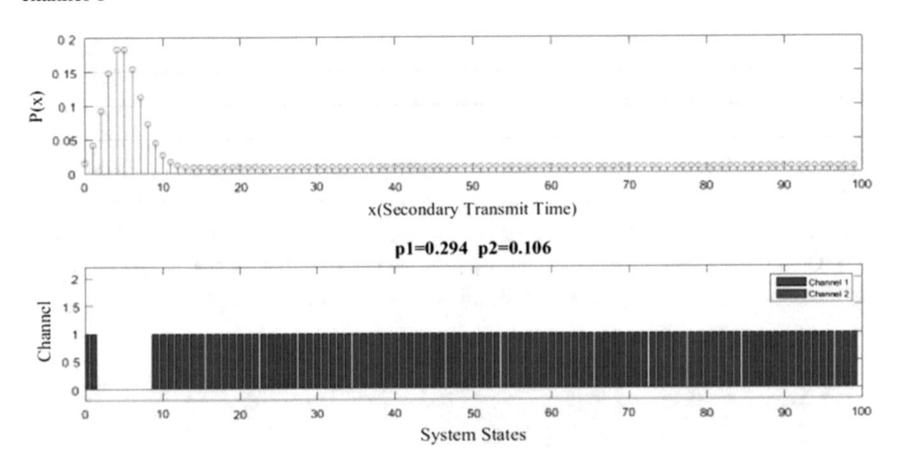

Fig. 8 Channel assignment policy when power allocation for channel 2 $\ll$ power allocation for channel 1

ence of 0.37. Thus, the blocking probability for the proposed system decreases as the difference between the transmission powers corresponding to the two channels becomes large. However, a very large difference may lead to one channel being preferred heavily over the other. Hence, a balanced power allocation above a difference of 0.2 is required.

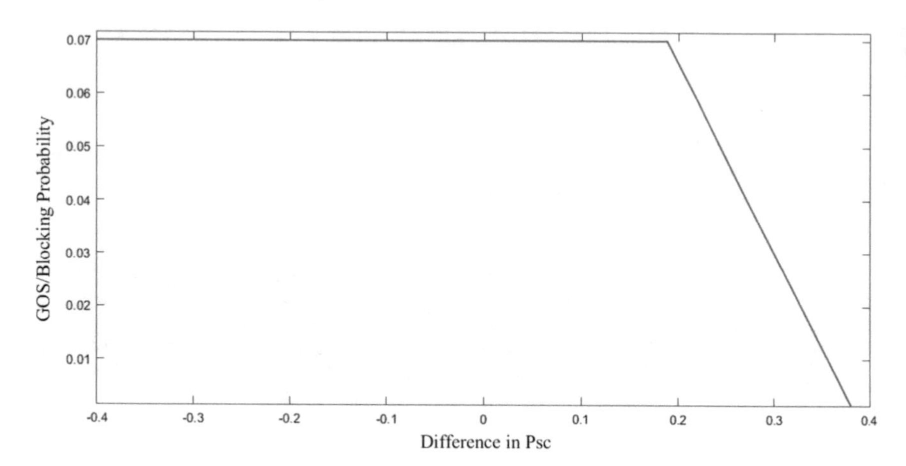

Fig. 9 GOS/blocking probability versus difference in transmission power b/w channel 1 and channel 2

10.3 *Interference Power and Channel Assignment*

Similar, to the last case, here too (Fig. 10), the blocking probability seems to decrease as the difference in interference powers between the two channels increases beyond 0.2. This shows that the proposed model is quite robust to the effects of interference as the GOS improves as the interference difference between the two channels increases beyond 0.2, however large the individual values of interference powers may be.

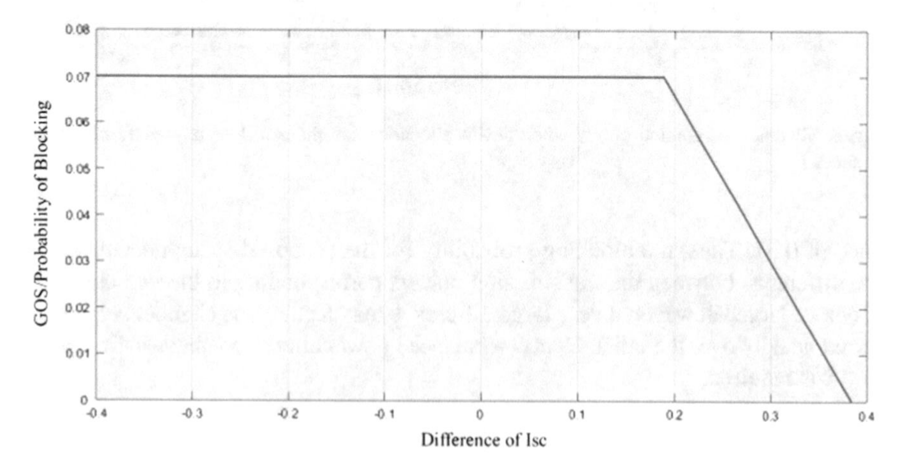

Fig. 10 GOS/blocking probability versus difference in interference power b/w channel 1 and channel 2

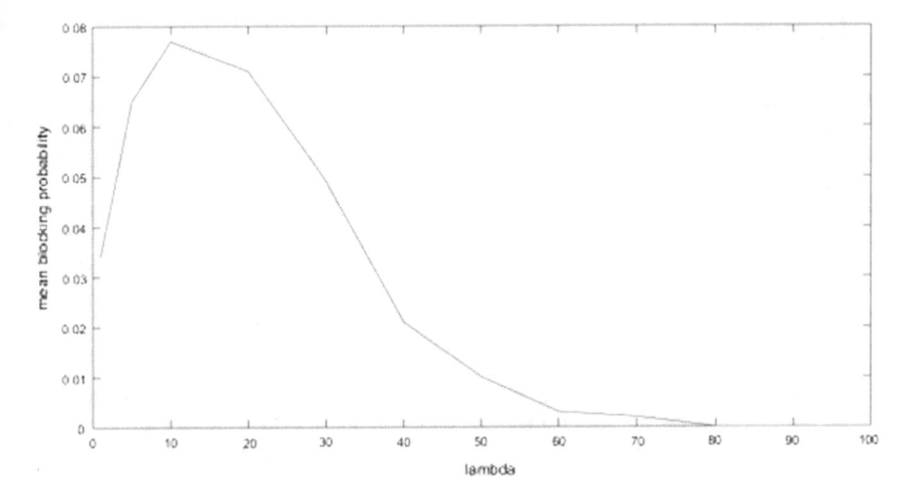

Fig. 11 Average GOS versus mean secondary time

10.4 Variation of Average GOS with Average Secondary User Transmit Time

Figure 11 shows that the mean GOS for the system increases for an increase in mean secondary time from 0 to 10 and then shows a gradual roll-off to zero as the mean secondary time increases from 10 to 100.The average GOS over the entire range of lambda from 1 to 100 is about 0.0033.

11 Conclusion

In this paper, we attempted to use the developments made in [11], to design an LCC system, where the optimal SU assignment is made to the available PU channels based on algorithm 3. Algorithm 3 optimizes both the PU transmit times and profit, while at the same time maximizing the throughput of the assigned SU. In most of the random runs we saw a blocking probability of zero. However, in some runs, the probability of blocking went up to as large as 0.07. However, over the 200 runs of the algorithm made with varying values of average SU transmit times (lambda), the overall average blocking probability is found to about 0.0033, which makes the system quite robust.

References

1. Zhang, J., Zhang, Q.: Stackelberg game for utility-based cooperative cognitive radio networks. In: Proceedings of the ACM MobiHoc, pp. 23–32
2. Wang, X., Ma, K., Han, Q., Liu, Z., Guan, X.: Pricing-based spectrum leasing in cognitive radio networks. IET Netw. **1**(3), 116–125 (2012)
3. Tao, M.X., Liu, Y.: Spectrum leasing and cooperative resource allocation in cognitive OFDMA networks. J. Commun. Netw. **15**(1), 102–110 (2013)
4. Gao, L., Xu, Y., Wang, X.B.: MAP: multi-auctioneer progressive auction in dynamic spectrum access. IEEE Trans. Mob. Comput. **10**(8), 1144–1161 (2011)
5. Jayaweera, S.K., Bkassiny, M., Avery, K.A.: Asymmetric cooperative communications based spectrum leasing via auctions in cognitive radio networks. IEEE Trans. Wirel. Commun. **10**(8), 2716–2724 (2011)
6. Liu, J., Chen, W., Cao, Z.G., Zhang, Y.J.A.: Cooperative beamforming for cognitive radio networks: a cross-layer design. IEEE Trans. Commun. **60**(5), 1420–1431 (2012)
7. Niyato, D., Hossain, E.: Spectrum trading: an economics of radio resource sharing in cognitive radio. IEEE Wirel. Commun. **15**(6), 71–80 (2008)
8. Niyato, D., Hossain, E.: Competitive pricing for spectrum sharing in cognitive radio networks: dynamic game, inefficiency of Nash equilibrium, and collusion. IEEE J. Sel. Areas Commun. **26**(1), 192–202 (2008)
9. Wysocki, T., Jamalipour, A.: An economic welfare preserving framework for spot pricing and hedging of spectrum rights for cognitive radio. IEEE Trans. Netw. Serv. Manage. **9**(1), 87–99 (2012)
10. Yu, H., Gao, L., Wang, X., Hossain, E.: Pricing for uplink power control in cognitive radio networks. IEEE Trans. Veh. Technol. **59**(4), 1769–1778 (2010)
11. Wu, Y., Song, W.-Z.: Cooperative Resource Sharing and Pricing for Proactive Dynamic Spectrum Access via Nash Bargaining Solution

Kaustuv Basak born at Barpeta Road, Barpeta, Assam, India to a family of teachers on 26th July, 1995 and brought up at Guwahati, Kamrup Metro, Assam, Kaustuv Basak pursued his Secondary and Higher Secondary Educations at Maharishi Vidya Mandir Barsajai, Guwahati, Assam, in science where he received A+ grades in both. Later on he took on a Bachelor of Technology course in Electronics and Communication Engineering from National Institute of Technology Silchar, where too he performed extraordinarily being in the top four of his batch. His primary research interests include Cognitive Radio Networks, Game Theory, Optimization, Software Defined Networks, Embedded Design, Data Networks, Neural Networks and Artificial Intelligence.

Akhil Gangadharan received B.Tech. degree in Electronics and Communication from National Institute of Technology, Silchar, India. Current research interest includes cognitive radio, spectrum sensing, spectrum utilization, cooperative communications, machine learning and artificial intelligence.

Wasim Arif B. Tech (University of Burdwan), M. E. (Jadavpur University), Ph.D. (NIT Silchar) Mr. Wasim Arif received the B.E degree Electronics and Communication Engineering from University of Burdwan (first class first in ECE) M. E. from Jadavpur University and pursuing Ph.D. degree in Electronics and Telecommunication Engineering from NIT Silchar. He worked as Lecturer in ECE in BIET Suri from 2003 to 2008 and as Sr. Lecturer in the same institute from 2008 to 2010. He joined NIT Silchar as an Assistant Professor.

Development of Advanced Driver Assistance System Using Intelligent Surveillance

G. Sasikala and V. Ramesh Kumar

Abstract Day-to-day vehicle usage count is gradually increasing. To provide safety and security to the driver the proposed Advanced Driver Assistance Systems (ADASs) is developed for vehicle/driver safety and better driving. An ADAS is a vehicle intelligent system that uses environmental vision data to improve traffic safety and driving comfort by helping the driver by recognizing and reacting to potentially hazardous in traffic environment. Since an ADAS performs autonomously with passive safety systems (Anti-Lock Braking System, Electronic Stability Control System, Active Steering System, and so on). While driver drives the vehicle, the ADAS-developed electronic system shall provide sufficient information like sleep warning and automatic braking, lane departure warning, blind spot detection, driver monitoring information, pedestrian collision warning information, and speed alert. A driver assistance system keeps on monitoring driving actions and if the driver out-of-at any time overridden by the driver.

Keywords OpenCV · Face detection · Eye blinking detection · Lane detection Traffic light detection

1 Introduction

Every hour in India more than 17 deaths on roads every hour are caused by vehicle-related accidents. Advanced Driver Assistance Systems (ADAS) were built by the motive to increase vehicle safety and to decrease the number of road casualties. ADAS is made of the different sensors like RADAR, LIDAR, cameras, ultrasonic sensors and night vision sensors which allows the vehicle to monitor the surround-

G. Sasikala (✉) · V. Ramesh Kumar
Department of ECE, Vel Tech Rangarajan Dr. Sagunthala R&D Institute of Science and Technology, Chennai, India
e-mail: sasikalaeverest369@gmail.com

V. Ramesh Kumar
e-mail: rameshkumar.viswanathan2888@gmail.com

S. Smys et al. (eds.), *International Conference on Computer Networks and Communication Technologies*, Lecture Notes on Data Engineering and Communications Technologies 15, https://doi.org/10.1007/978-981-10-8681-6_91

ing areas around it and to enhance vehicle, driver, and pedestrian's safety depending on factors such as weather, traffic, and dangerous road conditions. With upcoming technology, ADAS plays an important role in real time by warning the driver or by operating the control systems which help in the safety of vehicles and road accidents Passive safety technologies like seat-belts, airbags cannot prevent accidents. Traditional active safety technologies such as Anti-lock Braking System, Electronic Stability Control are intended to improve driving stability. Currently ADAS like Lane Departure Warning, Blind Spot Detection, Adaptive Headlight, Parking Assistance, Speed Bump Detection, Pedestrian Detection and etc. help to avoid accidents and provides safety for the driver.

The objective is

1. To enhance active safety of a vehicle
2. To minimize the number of accidents
3. To make driving safer
4. To alert the driver to potential problems
5. To improve vehicle stability

1.1 Adaptive Cruise Control

In Adaptive cruise control in the vehicle consists of cameras, scans the area in front road condition of the vehicle. In Fig. 1, the adaptive Cruise Control system automatically controls the speed of the vehicle with respect to speed of the nearest vehicle and applies automatic braking and reduces acceleration to follow the vehicle. Such system is good for long drives on highways and less traffic areas and also helps to avoid over speeding. It reduces the driver stresses for constant acceleration and braking.

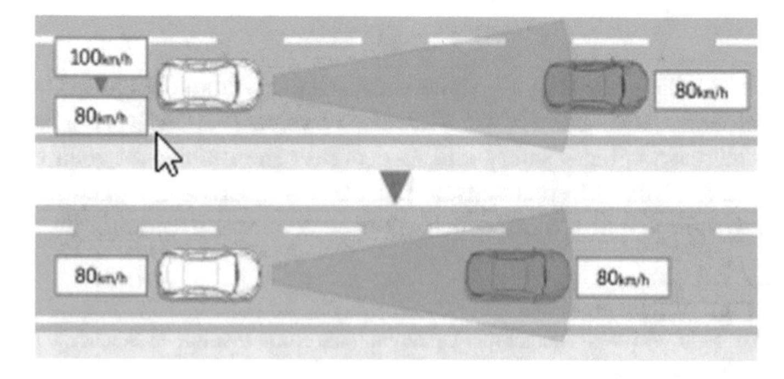

Fig. 1 Adaptive cruise control

Fig. 2 Lane departure warning

1.2 *Lane Departure Warning System*

In Lane departure warning system as shown in Fig. 2 helps to alert the driver if person is changing the road lanes without applying the signal by indicators. Image-based Hough Transform algorithm is applied for lane detection in which left lane and right lane markings are detected. By this results of this module will control the steering of the vehicle in case no action is taken by the driver. This module will ensure that the vehicles are in the driving lane. If the lane markings are not visible or faded then the working of the system may not be efficient. This module development is applicable only for lane marking on the road.

1.3 *Sleep Detection*

OpenCV-based automatic drowsy driver monitoring will provide information for sleep detection of the driver. In an Automatic drowsy driver monitoring is an accident control system, which is monitoring the changes in the eye-blink duration as shown in Fig. 3. In this proposed method detects visual changes in eye locations using the cascade classifier and hough circle transform algorithm. The face of the driver is identified in the image as an initialization stage of the image processing. After this first detection stage, eyes tracking image normalization, Edge detection, HOG description and SVM classification. This information is used to get the blinking frequency and also its duration.

By using the above method one who can detect eyes blinks via a front web-cam in real time. Stored experimental trained results in the Eye-Blink database showed that the proposed system will detects eye blinks with an appropriate accuracy.

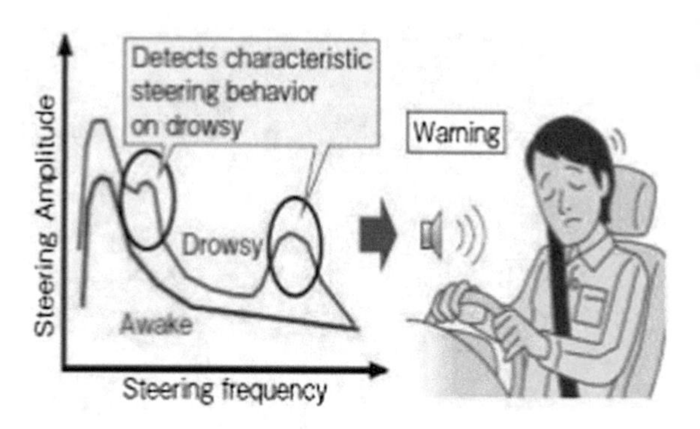

Fig. 3 Sleep detection

Digital Camera Equipped with devices will identify Sleep Detection of the driver and it provides alarm to the driver and also controls ESC (Electronic Stability Control) of the Vehicle.

1.4 Traffic Sign Detection

The traffic sign detection is an important part of Advanced Driver Assistance System (ADAS). Traffic symbols will provide the information about the traffic rules, route directions, and road conditions and such informative details shall provide to driver for better and safe driving. Traffic sign detection and recognition system has two main stages: The first stage which involves the traffic sign/symbol identification and the second stage classify the detected traffic signs into a particular class.

Digital camera equipped with devices is used to detect the Traffic symbols and it sends command to the driver for the actions. In traffic environments, Traffic Sign Recognition is used to intimate/warn drivers, traffic symbols, and traffic command or necessary actions need for ADAS System. Fast real time and automatic traffic sign detection and recognition will provide safety of vehicle and information for the driver. Generally, traffic signs provide the driver with a variety of traffic recognized information for better safe.

Each Traffic sign information is collected and it is stored in database; thus it becomes offline data and will not suit for the real-time scenario. This method is applicable for the standalone cars with predefined traffic sign references (Fig. 4).

2 Literature Survey

Luo and Hu [1] portray the ADAS (Advanced Driver Assistance System) which goes about as a help for the driver while driving has been a prevalent research focal point of numerous auto producers, colleges, look into foundations and organizations. Numer-

Fig. 4 Traffic sign identification by using OpenCV

ous sorts of ADAS frameworks, for example, Autonomous journey control, Lane takeoff cautioning, Adaptive Cruise Control, forward crash cautioning, Blind Spot Information System, Adaptive Front-lighting System, Pedestrian Collision Warning, Start/Stop framework have been exhibited. With investigation of these new frameworks, both within and outside driving condition would be moved forward. As of now a considerable lot of them are now embraced by vehicles and some are being tried in the genuine street conditions.

Mody et al. [2] disclose to empower different security parts of ADAS, camera-based frameworks are introduced in front, back, and encompass see. The front camera frameworks are utilized for applications, for example, Auto Emergency Braking and Forward Collision Warning. The back view and encompass see frameworks are utilized for stop help and cross activity ready applications. Front camera frameworks can utilize mono- or stereo camera setup. Stereo camera is utilized to get 3-D data by creating difference. Be that as it may, stereo camera frameworks are more costly contrasted with mono camera frameworks. Structure from Motion innovation empowers a solitary moving camera to acquire profundity by and large generally inquired about for its capacities in ADAS applications. Encompass see frameworks contains different cameras (four to six) set around the car.

Mandlik and Deshmuk [3] give path departure cautioning framework in view of the Hough change and Euclidean separation calculation. At first, histogram evening out is utilized for improving the differentiation level of info picture. So as to enhance the speed and precision of the framework and chosen ROI into two sub-regions. At that point Hough change calculation is utilized for path location in which left and right path checking are recognized free of each other. At long last, path flight distinguishing proof is completed utilizing path-related parameters, evaluated based on Euclidean separation.

Kim et al. [4] clarify the BSD (Blind Spot Detection) is one of the main parts of ADAS (Advanced Driver Assistance System). The underlying advance of Blind Spot location is to determine blobs in pre-characterized two ROI (Region of Interest). Each information Images/Videos/outlines are changed over to paired picture by utilization of edge with a specific end goal to isolate the splendid fog light region from foundation. After the brilliant blobs are recognized, the subsequent stage is to gauge the movement vectors of them. At the point when a vehicle is drawing nearer to inner self-vehicle in blind side, the blobs are getting higher esteem and have the movement vectors with descending position. On the off chance that two blobs are situated at that tallness of the position, at that point these two blobs are viewed as front light of same vehicle. At the point when the fog light gathering is identified in ROI, the caution flag is ON when the fog light gathering shows up in ROI and keeps while the gathering keeps up. The alert goes to OFF when the moving bearing of front light is not drawing nearer amid a couple of edges or the fog light moves out of ROI.

Simić et al. [5] clarify an ADAS gives new answers for non-minor true issues like mechanized expressway driving, which is caught by path location calculation, person on foot discovery, computerized stopping, braking, street signs identification/acknowledgment, driver rest recognition and some more. Driver checking has an extraordinary place among these calculations, as it is tending to the driver's tiredness issue, which is in charge of genuine number of street mischances. Driver's face and eyes position are utilized for distinguishing if driver is active or dormant.

Nayana et al. [6] clarify the Object identification innovation is to distinguish the particular area and size of a specific protest in a database of picture or a video scene. With the rising necessity of identification-based security and mechanical applications chiefly in the car field, the question recognition in a quick and solid way has been pulling in much extension and concern. Consequently, more dependable and high precise close ongoing articles identification application running on an implanted stage is urgent and basic, because of the rising security worries in the different fields. Planning an entire dependable framework is ready to all the while identify numerous articles on a scene of picture or video. The framework shows an outline of the course question recognition calculation and additionally Local Binary Pattern highlight determination utilized by course classifier. At that point, it proposes an OpenCV-based answer for numerous protest identification, lastly, presents the aftereffects of the correlation of exhibitions in a standard stage and an implanted gadget.

Shah et al. [7] clarifies the actualizing facial checking framework by implanting face identification and face following calculation found in MATLAB with the GPIO pins of Raspberry pi B by utilizing RasPi summon to such an extent that the variety of LEDS takes after the facial development by recognizing the face utilizing Haar classifier.

The Computer vision field centers on encouraging a PC to comprehend or translate visual data from pictures or video arrangements and it started in the late 1950s and mid-1960s (Buch et al. [8]). A normal vehicle observation framework is made out of four phases: the closer view for division organize, the depiction arrange, the order arrange and the following stage. The frontal area division stage could be refined by computing the contrast between a foundation display and the present edge, as observed in (Gupte et al. [9]). It could likewise be accomplished by registering the distinction between two progressive edges and thresholding the distinction to make the frontal area veil (Nguyen and Le [10]). Another strategy for frontal area estimation includes portioning the closer view autonomous of the information of the foundation. This could be acknowledged by looking at the angle picture of the present edge to the slope picture of a 3D demonstrate wireframe. The frontal area will then be processed by thinking about the match between the two pictures (Sullivan et al. [11]).

In the characterization phase of an activity framework a question with removed highlights is mapped to a protest class by an officially prepared classifier. A classifier is a prepared model acquired as a yield of a learning calculation sustained with a substantial arrangement of preparing information or highlights, a case of such a learning calculation is Adaptive Boosting (Ada-Boost) Some outstanding classifiers are the Haar course classifier (Viola and Jones [12]) and the Local Binary Patterns (LBP) course classifier.

The Haar course classifier was a quick protest locator. It accomplished its speed by presenting the figuring of highlight esteems with a basic picture and by utilizing fell Ada-Boost classifiers. By utilizing an Ada-Boost classifier the calculation is capable select segregate highlights from a pool of Haar highlights (Porwik and Lisowska [13]) and by falling discovery speed is improved since non-coordinating windows of information pictures are immediately disposed of in a beginning period and location would then be able to be centered around less windows with less figuring time. The calculation has connected to confront identification and it was speedier than any past technique for confront discovery close to 15 times better. It was viewed as a leap forward continuously confront discovery. The Haar course calculation has additionally been connected to the identification of walkers in a rush hour gridlock framework (Jones and Snow [14]).

Fan et al. [15] depicts the Traffic signs which incorporate numerous valuable condition which enable drivers to find out about the change/troubles of the street ahead and the driving necessities. In this strategy a novel activity signs acknowledgment calculation, which in light of machine vision and machine learning strategies

3 System Development

3.1 Adaptive Cruise Control System—Methodology

Opposite of the car vehicle is detected by using Haar Cascade classifier. This algorithm has applied by using OpenCV and it was trained with the 526 cars samples from the rear camera (360 × 240 pixels, no scale). Finally CarsSample.xml is generated and this data have used for car detection and distance calculation finally adaptive

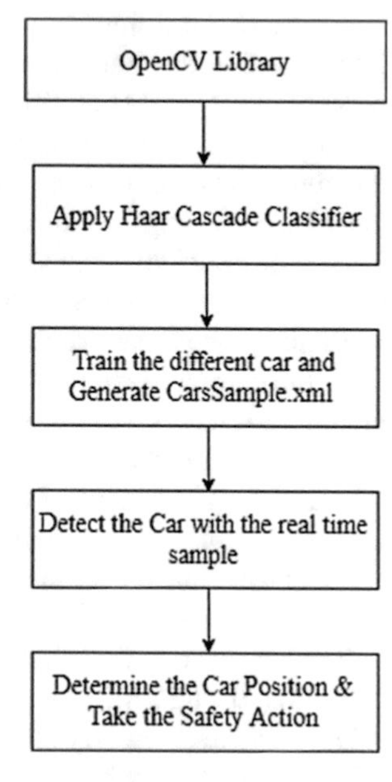

Fig. 5 Adaptive cruise control block diagram

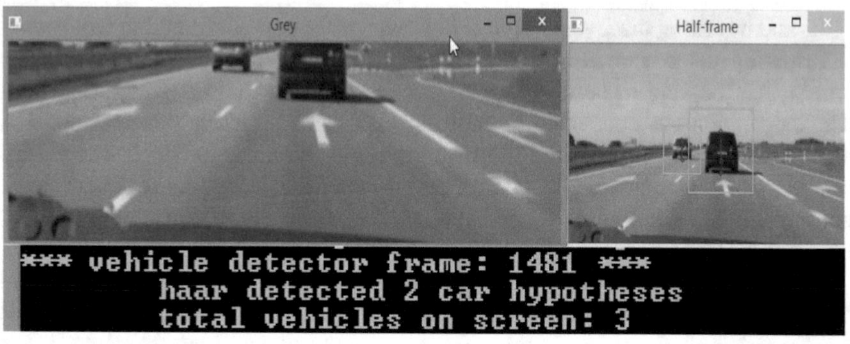

Fig. 6 Vehicle car detection

cruise control recognize module will take safety action of the vehicle as shown in Figs. 5 and 6.

3.2 *Lane Departure Warning System—Methodology*

Lane Departure of the vehicle is detected by Gaussian blur and edges detection and performs Hough transform. This algorithm has been applied by using OpenCV. Once different lane is detected, Lane Departure recognizes module will take safety action of the vehicle (Figs. 7, 8).

Fig. 7 Lane-departure warning system block diagram

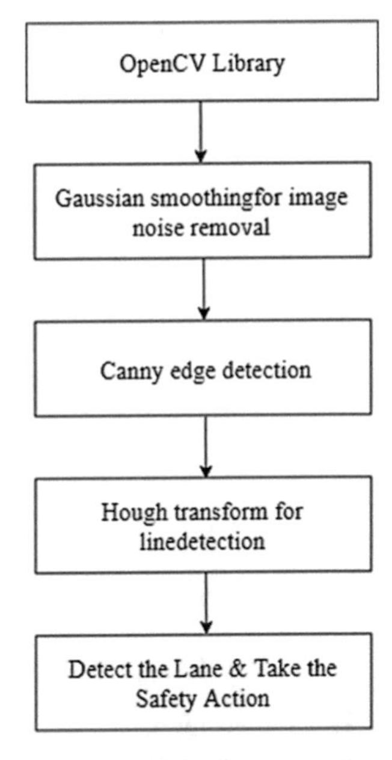

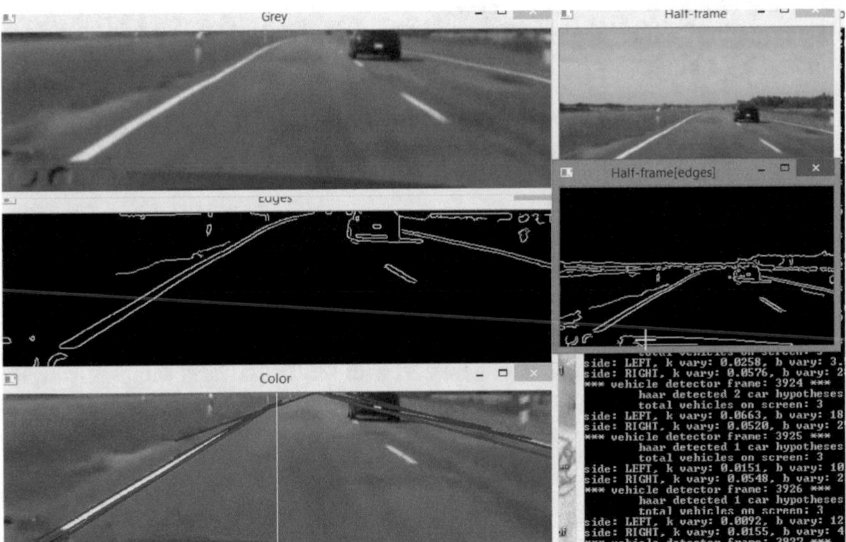

Fig. 8 Lane departure detection

3.3 Sleep Detection System—Methodology

Haar-like Ada-boost classifiers, which is the most common technique in computer vision for face and eye detection.

Fig. 9 Face & eye blink detection

It involves following steps for Sleep Detection

- Eye localization.
- Finding threshold whites of the eyes.
- Determining "white" region of the eyes disappears for a period of time which is indicating a blink.

By using above technique drowsiness shall be detected and its safety action will be taken by Sleep Detection recognizer module (Fig. 9).

3.4 Traffic Light Detection—Methodology

Traffic light is detected by using Haar Cascade classifier. This algorithm has been applied by using OpenCV and it was trained with the Traffic light samples from the rear camera and finally TrafficLightSample.xml is generated and this data have used for Traffic Light detection. Traffic Light Detection recognize module will take safety action of the vehicle (Fig. 10).

4 Future Enhancement of ADAS Application

Future Enhancement of Adaptive Cruise Control

- In existing module distance of the opposite CAR is not computed, this feature to be analyze
- Fastest CAR Detection mechanisms to be analyze

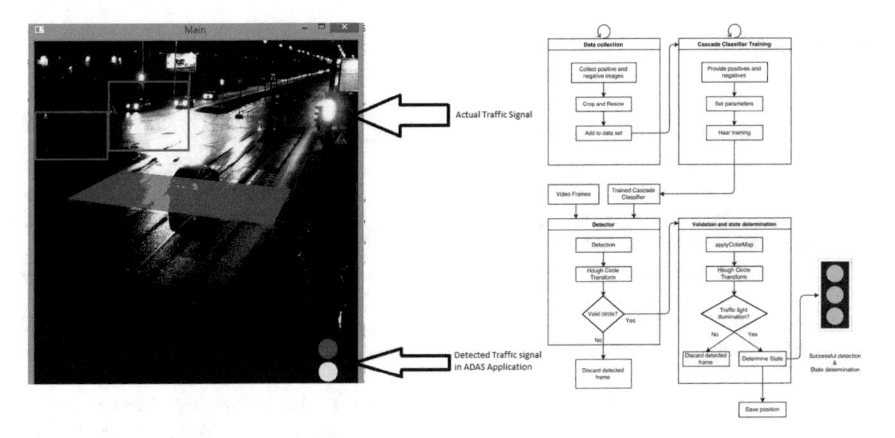

Fig. 10 Example of traffic light detection and it's block diagram

- Analyze the Night vision detection and prepare the different dataset in night vision system

Future Enhancement of Lane Departure Warning System

- Road area extraction and detection for roads without lanes and include road shadow removal
- Better lane tracking (probability methods), stability and accuracy
- Pedestrian Detection mechanisms to be analyze

Future Enhancement of Sleep Detection Mechanism

- Fastest Eye Blinking detection algorithm to be analyze
- Analyze the Night vision eye blink detection and prepare the different eye blinking dataset in night vision
- Pedestrian Detection mechanism shall be analyze

Future Enhancement of Traffic Light System

- Different kind of Indian Environmental Traffic sign and Road Condition dataset shall be analyze

5 Conclusion

Advanced Driver Assistance Systems (ADAS) will help to improve the driver safety and will make driving safer. It is different than the passive safety systems and other traditional safety features like Electronic Stability Control (ESC), Automatic Gear Control, Anti-lock Braking System (ABS).

References

1. Feng, L.U.O., Fengjian, H.U.: A comprehensive survey of vision based vehicle intelligent front light system. Int. J. Smart Sens. Intell. Syst. **7**(2) (2014)
2. Mody, M., Swami, P., Chitnis, K., Jagannathan, S., De-sappan, K., Jain, A., Poddar, D., Nikolic, Z., Viswanath, P., Mathew, M., Nagori, S., Garud, H.: High-performance front camera ADAS applications on TI's TDA3x platform. In: IEEE 22nd International Conference on High-Performance Computing (HiPC), pp. 456–463 (2015)
3. Mandlik, P.T., Deshmuk, A.B.: Image processing based lane departure warning system using hough transform and Euclidean distance. Int. J. Res. Sci. Innov. (IJRSI) **III**(X) (2016)
4. Kim, S.-G., Kim, J.-E., Yi, K., Jung, K.-H..: Detection and tracking of overtaking vehicle in blind spot area at night time. In: IEEE International Conference on Consumer Electronics (ICCE) (2017)
5. Simić, A., Kocić, O., Bjelica, M.Z.: Driver monitoring algorithm for advanced driver assistance systems. In: IEEE International Conference: 2016 24th Telecommunications Forum (TELFOR) (2016)
6. Nayana, H.C., Dr. Neelgar, B., Hiware, R.: Object detection and classification for vehicle ADAS. Int. J. Eng. Sci. Comput. (2017)
7. Shah, A.A., Zaidi, Z.A., Dr. Chowdhry, B.S., Dr. Daudpoto, J.: Real time face detection/monitor using Raspberry pi and MATLAB. In: IEEE 10th International Conference on Application of Information and Communication Technologies (AICT), 12–14 Oct 2016
8. Buch, N., Velastin, S., Orwell, J.: A review of computer vision techniques for the analysis of urban traffic. IEEE Trans. Intell. Transport. Syst. **12**(3), 920–939 (2011)
9. Gupte, S., Masoud, O., Martin, R., Papanikolopoulos, N.: Detection and classification of vehicles. IEEE Trans. Intell. Transport. Syst. **3**(1), 37–47 (2002)
10. Nguyen, P.V., Le, H.B.: A multi-modal particle filter based motorcycle tracking system. In: Pacific Rim International Conference on Artificial Intelligence, pp. 819–828. Springer, Berlin (2008)
11. Sullivan, G., Baker, K., Worrall, A., Attwood, C., Remagnino, P.: Model-based vehicle detection and classification using orthographic approximations. Image Vis. Comput. **15**(8), 649–654 (1997)
12. Viola, P., Jones, M.: Robust real-time face detection. Int. J. Comput. Vision **57**(2), 137–154 (2004)
13. Porwik, P., Lisowska, A.: The Haar-wavelet transform in digital image processing: its status and achievements. Mach. Graph. Vis. **13**(1/2), 79–98 (2004)
14. Jones, M., Snow, D.: Pedestrian detection using boosted features over many frames. In: 19th International Conference on Pattern Recognition, 2008. ICPR 2008, pp. 1–4. IEEE (2008)
15. Fan, Y., Zhang, W.: Traffic sign detection and classification for advanced driver assistant systems. In: 12th International Conference on Fuzzy Systems and Knowledge Discovery (FSKD) (2015)

Dr. G. Sasikala is born on 8th Jan 1974 at Harur, Dharmapuri District, Tamil Nadu. She has obtained her B.E. degree in Electronics and Communication Engineering from Government College of Technology, Coimbatore affiliated to Bharathiar University, Tamil Nadu and Master Degree in M.E. Embedded System Technology, VEL TECH Engineering College affiliated to Anna University, Chennai and finished her Doctor of Philosophy in Vel Tech Rangarajan Dr. Sagunthala R&D Institute of Science and Technology, Chennai, Tamil Nadu, India. She has a teaching experience of around 17.3 yrs. and industry experience of 2 yrs. She has presented more than 14 papers in various national and international conferences and published more than 12 papers in various reputed international journals. She has organized various workshops and seminars. She is currently working as an Associate Professor in the Department of Electronics and Communication Engineering in Vel Tech Rangarajan Dr. Sagunthala R&D Institute of Science and Technology, Chennai, Tamil Nadu, India. Her research interest includes Embedded System, Image Processing, Antennas and VLSI design.

Mr. V. Ramesh Kumar obtained his B.E. (ECE) in P.S.R. Engineering College, Sivakasi, Tamil Nadu and completed his M.Tech. in Embedded System & Technologies in Vel Tech Rangarajan Dr. Sagunthala R&D Institute of Science and Technology, Chennai, Tamil Nadu, India. He having domain Knowledge in Avionics for past 3 years and having knowledge in Automotive for past 3 years.

Design and Implementation of Hamming Encoder and Decoder Over FPGA

A. H. M. Shahariar Parvez, Md. Mizanur Rahman, Prajoy Podder, Mohammad Hossain and Muhammad Ashiqul Islam

Abstract In the digital world, the field of communication has developed many applications. In most sectors of communication, the input message or data is encoded at the transmitter and transferred through the communication channel. Data is received at the receiver and after the decoding of the received data, original data is recovered successfully. Noise can be added to the communication channel. Noise affects the input data and data might get corrupted during the transmission. So it is essential for the receiver to have applied some functions, which are able to localize the error at the receiver end data. There are many types of forward error-correcting code. Hamming code is one of them, which has got numerous applications. Hamming error detection and correction code are used in many common applications, including optical storage devices, random access memory, wireless communication system, high-speed modems (ADSL, xDSL) and digital television system. This paper tries to elucidate the design process of hamming code using VLSI because FPGA is more cost-effective than the other system. In this paper, the algorithm for hamming code encoding and decoding has been discussed and after hamming error detection and correction code has been implemented in Verilogger pro 6.5 to get the results. Hamming code is an improved version compared to the parity check method and it is implemented in Verilog in which n-bit of information data is transmitted with ($n -$ 1) redundancy bits. In order to find the value of these redundant bits, a hamming

A. H. M. Shahariar Parvez · Md. Mizanur Rahman
Department of CSE, Ranada Prasad Shaha University, Narayanganj, Bangladesh
e-mail: sha0131@gmail.com

Md. Mizanur Rahman
e-mail: mizan173@gmail.com

P. Podder (✉) · M. Hossain
Department of EEE, Ranada Prasad Shaha University, Narayanganj, Bangladesh
e-mail: prajoypodder@gmail.com

M. Hossain
e-mail: mohammadandhossain@gmail.com

P. Podder · M. A. Islam
Department of ECE, Khulna University of Engineering & Technology, Khulna, Bangladesh
e-mail: kuet0909018@hotmail.com

S. Smys et al. (eds.), *International Conference on Computer Networks and Communication Technologies*, Lecture Notes on Data Engineering and Communications Technologies 15, https://doi.org/10.1007/978-981-10-8681-6_92

code is written in Verilog, which will be simulated in Xilinx software. The result of simulation and test bench waveforms is also explained.

Keywords Hamming encoder · Bit error · Hamming decoder · Even parity Odd parity · Redundant bits · Most significant bit · Verilogger pro 6.5

1 Introduction

Secure and error-free data transmission from the transmitter to the receiver is the major issue in the field of telecommunication. At present, there are a number of technologies for error-free data transmission. One of technology Hamming code is used for error detection and correction to get error-free data at the desired destination. In case of Hamming code data, information is encrypted according to even and odd parity method before transmission of information at the source end [1, 2] to the receiver end. Hamming codes can correct one error per word. So, it widely uses to protect memories or registers from soft errors. When a radiation particle strikes the device and modifies the logical value of a memory cell or register then a soft error occurs. ECC may add block code that is associated with a block of store data sector and stream codes. Examples of ECC are Hamming codes, Reed Solomon codes, Bose-Chaudhuri-Hocquenghem code (BCH) and cyclic redundancy check codes. Hamming codes are simple to build for any word length as a result the encoding and decoding can be performed with low delay. It is SEC code and can be enlarged with a parity bit that covers all bits to execute an SEC-DED code [3]. Therefore, they are not only suitable for RAM and ROM applications but also to protect registers in digital circuits.

At present, digital communication system has been used in different types of error correction codes based on the type of channel noise. Few of them are Bose-Chaudhuri Hocquenghem code (BCH) [4], Reed Solomon code [5], Low-Density Parity Check code (LDPC) [6], Turbo code [7] and Hamming code [8]. Hamming codes are based on the work of R. W. Hamming at bell laboratories in 1950. Hamming discovered that by taking a data word and adding bits to it according to an algorithm, errors that cause bits to be changed can be detected and even corrected to obtain the original data. This is varying powerful concept. Parity, which is the most common form of error detection, can only report that an error occurred. Hamming codes word algorithm is a little involved, but it turns out that the circuitry needed to implement the algorithm is very simple combinatorial logic. There are two C programs; one is for producing Verilog code to generate Hamming codes and one is for producing Verilog code to decode Hamming codes. Both programs take the word size an input parameter and generate the appropriate Verilog function.

This paper has been organized as follows. Section 2 gives the overview of data encoding and decoding. Under this section, matrix properties and syndromes rules are explained clearly. Proposed system architecture of Hamming code is explained in Sect. 3. Hamming code error detection and correction has been narrated on Sect. 4.

Under this section, encoding, channel coding, redundancy, and decoding of bits are explained briefly. The experimental results are discussed in Sect. 5, where the schematic, pin diagram, and the timing diagram of a simple Hamming encoding and decoding process are explained and last of all, and Sect. 6 concludes the paper.

2 Data Encoding and Decoding

The basic concept of Hamming codes is generally based on the concept of Hamming distance (H_d) which can be defined as the difference between two binary numbers, i.e., the total number of bit position in which they differ from one another. For example, the hamming distance between two binary numbers 1000010 and 1010100 is 3 because there are three positions where the bits are different.

The general formula for hamming code can be stated by the following equation which can correct x errors and detect additional y errors.

$$2x + y + 1 \leq H_d \tag{1}$$

In order to transmit data and detect and correct errors, it is required to map every data word to set of Hamming codes. This can do by appending bits of each data word to increase their hamming distance. At first, it needs to create the smallest matrix H that has the following properties:

Matrix property 1: If the number of code bit is n and the number of data bits is m then the number of columns matrix H is $n+m$, and the smallest n can be found such that, $n + m \leq 2^n - 1$.
Matrix property 2: The number of rows n of matrix H is equal to the number of code bits to be appended.
Matrix property 3: The left side of matrix H is an $n \times n$ identity matrix.
Matrix property 4: The remaining columns of matrix H are a unique binary number.
Matrix property 5: Hamming code bits must be used for every data word in order to create matrix H, satisfying the following equation, $H \times p^{\mathrm{T}} = 0$, where p is the data word with the Hamming code bits added to the most significant bits. T is the transposed form of p vector. The 0 of right-hand side is an n-bit vector of all zeros.

The following 3×7 matrix can be created for a 4-bit data word,

$$H = \begin{bmatrix} 1\,0\,0\,1\,1\,0\,0 \\ 0\,1\,0\,1\,0\,1\,1 \\ 0\,0\,1\,0\,1\,1\,1 \end{bmatrix}$$

The currently unknown code bits can be appended onto the data word to create 7-bit vector,

$$p = x_2 x_1 x_0 y_3 y_2 y_1 y_0 \tag{2}$$

$$x_2 = y_3 \wedge y_2 \wedge y_0 \tag{3}$$

$$x_1 = y_3 \wedge y_1 \wedge y_0 \tag{4}$$

$$x_0 = y_2 \wedge y_1 \wedge y_0 \tag{5}$$

where x_j are the code bits and y_i are the data bits. In order to correct the errors to produce the original data, it requires creating a syndrome according to the following rules,

$$H \times p'^{t} = S \tag{6}$$

Syndrome rule 1: Use the following equation to produce a syndrome S.

$$H \times p'^{\mathrm{T}} = S \tag{7}$$

where p'^{T} is the data pulse code bits that were received.
Syndrome rule 2: No single-bit error has occurred when the syndrome is all zero. In this case, $p' = p$.
Syndrome rule 3: If the syndrome is not all zeros, the column in the matrix that matches the transposed syndrome corresponds to the bit that is an error. The bit in the receiver data must be changed and then the code bits were stripped off to obtain the correct data.

3 Proposed System Architecture

Hamming code is a linear error-correcting code. It can detect up to two coinciding bit errors and correct single-bit errors. So, the reliable communication will be possible if the hamming distance between the transmitted and received bit patterns is ≤ 1. Simple parity code can only detect an odd number of errors cannot correct the errors. For generating redundancy bit, Hamming code process works only two methods (even and odd parity). The amount of redundancy depends on the quantity of information data bits. So, the process of hamming encoding and decoding are given below,

---

Algorithm: Hamming Encoding and Decoding

Step 1: Start

Step 2: Declare variables: clock to output delay (DEL), Number of bits in a data word, single bit error(bit_err),error

Step 3: Assign new data possible corrupted and New EDC bits possibly corrupted

Step 4: Define the necessary module (Hamming encoder and decoder for 8 bit data) for simulation

Step 5: Initialize variables

clock$\leftarrow$ 1,bit error$\leftarrow$0,data_in$\leftarrow$0,start$\leftarrow$0

Step 5: Look at the rising edge of the clock

Step 6: On the first cycle, corrupt no bits

On each subsequent cycle, corrupt each bit from LSB to MSBthen repeat the process

6.1: If (data_in=0 & start $\neq$0) then

No errors and go to step 6.

Else

Start<=DEL

6.2: If (error$\neq$1) & (bit_err $\neq$0)

Then error at time is calculated and error signal is not asserted and go to step 6

6.3: If (error$\neq$0) & (bit_err = 0)

Then error at time is calculated and error signal is asserted and go to step 6

Step 7: Stop (Use stop for debugging the module)

---

4 Hamming Code Error Detection and Correction

4.1 Encoding

Forward error correction coding is normally employed in the field of communication system where the binary information source transmits a data or bit sequence to the encoding device. Then redundant (or parity) bits are inserted in the encoder inserts, thereby resulting in a longer sequence of code bits, called a ‘code word.’ These code words may be transmitted to the receiver via communication channel. A proper

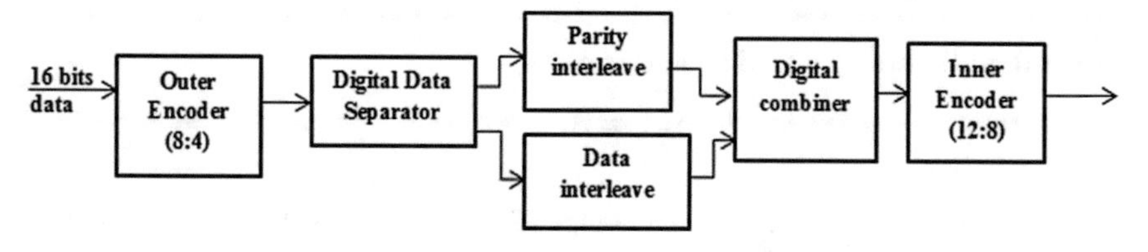

Fig. 1 Block diagram of Hamming encoder circuit

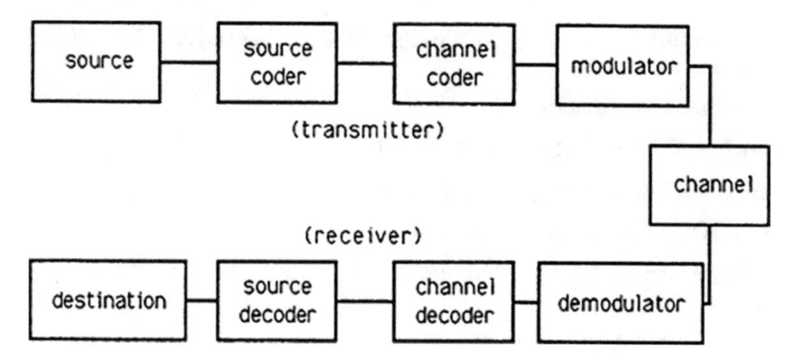

Fig. 2 Basic block diagram of communication system

decoding system must be used in the receiver section to extract the original data sequence.

The (x, d, t) code mentions a block of 'd' data bits that is mapped into an x bit code word (where $d < x$) and 'b_r' ($=x - d$) error control bits with the code whose have the ability of correcting 't' bits in the error. If the total number of bits in the transmitter section (i.e., code word) is 'x' ($=d + b_r$), 'b_r' must be able to specify at least '$x + 1$' ($=d + b_r + 1$) different states. Of these, one state means no error, and 'x' states indicate the location of an error in each of the 'x' positions. So '$n + 1$' states must be discoverable by 'b_r' bits; and '$2b_r$' different states can be indicated by 'b_r' bits. Therefore, $2b_r$ must be equal to or greater than '$n + 1$'. The value of 'b_r' can be determined by substituting the value of 'd' (where d is the original length of the transmitted data). For example, if the value of 'd' is '7,' the smallest 'r' value that can satisfy this constraint is '4' (Fig. 1).

4.2 Channel Coding

Digital communication systems, particularly used in the defense systems, need to perform accurately and reliably even in the presence of noise and interference occurred normally in the channel (Fig. 2).

Forward error correction coding is very agile and economical due to reducing transmission errors. It (also called 'channel coding') is used in the transmitter end of a digital system enabling the receivers to correct the errors. FEC is typically implemented using redundancy bits for error detection and correction. In the field of radio communication, FEC is one of the techniques unique to digital systems, used filtering, diversity reception, and equalization methods. This coding technique enables the decoder to correct errors without requesting retransmission of the original information. Hamming code is such type of example of FEC.

4.3 Redundancy

Redundant bits are those types of additional bits that can be used to detect or correct the error bits. They are appended to the information data at the transmitter end and remove at the destination end. The existence of the redundancy bit concedes the receiver to detect or correct corrupted bits. The ratio of redundant bits to the data bits and the robustness of the process are important factors in any coding scheme. Correct sync can therefore be obtained even if the correctly framed code word contains errors in the redundant bits.

4.4 Decoding

The Hamming decoder circuitry of Fig. 3 can be specified as $k{:}n$ where n and k is the number of encoded and message bits, respectively. The 48 bits encoded (input) data is divided into 4 sets of 12 bits and then each set is sent to inner decoder section. Each 12 bits to 8 bits with 1- or 2-bit error detection is decoded by the Inner decoder. If a single-bit error is detected, then the error is corrected by the correcting mechanism [9]. An error flag is generated when double bit error is detected. 4-bit data and 4-bit parity are included in each 8-bit data. These data and parities can be separated and given to the data and parity de-interleaver block to achieve the original form. Each 8-bit data are given to the outer decoder section. It decodes the 8-bit code to the 4-bit data stream with 1-bit error correction and 2-bit error detection. Therefore, the output of the outer decoder gives the original data.

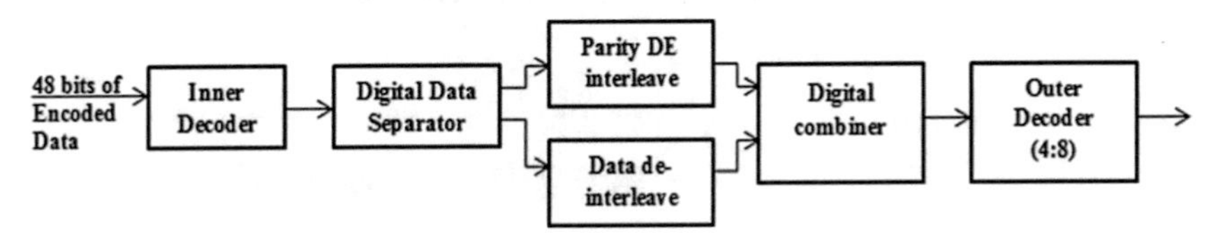

Fig. 3 Block diagram of Hamming decoder circuit

5 System Synthesize and Implementation

5.1 An (11, 7, 1) Hamming Encoder and Decoder

Figure 4 shows the encoder pin diagram and Fig. 5 displays RTL schematic of a simple (11, 7, 1) hamming encoder. Figures 6 and 7 indicate the i/o pin and RTL schematic of a simple (11, 7, 1) hamming decoder, respectively. The encoder module generates a (11, 7, 1) Hamming code encoder where $n = 11$ and $k = 7$. A 7-bit binary number is then converted into an 11-bit code data. On the other hand, decoder module generates same size Hamming code decoder that converts the encoded 11-bit code word back into a 7-bit binary code after fixing the single-bit error [10, 11]. These have been simulated in Xilinx.

5.2 Circuit Schematic

Figure 8 shows the top-level diagram of the Hamming encoder. The encoder module defines a generator of codes for detecting and correcting single-bit data errors.

Figure 9 and 10 show the i/o pin and top-level schematic of the Hamming decoder, respectively. The decoder module defines an error detector and corrector of single-bit errors using Hamming codes. The edc_in signal computes the parity bit. The error signal is 0 when the syndrome is zero. The syndrome is the connection between hardware elements.

5.3 Simulation Results

A simple counter generates a sequence of data words. The data word is put through a Hamming code generator and the error detecting and correcting bits are added to the word. Then for the first word, no bits are corrupted. For the second word, bit 1 is corrupted. For the third word, bit 2 is corrupted. For the fourth word, bit 3 is corrupted. When the most significant bit (MSB) is corrupted, the sequence begins again with no bits corrupted. Each time, the data is sent through the Hamming code

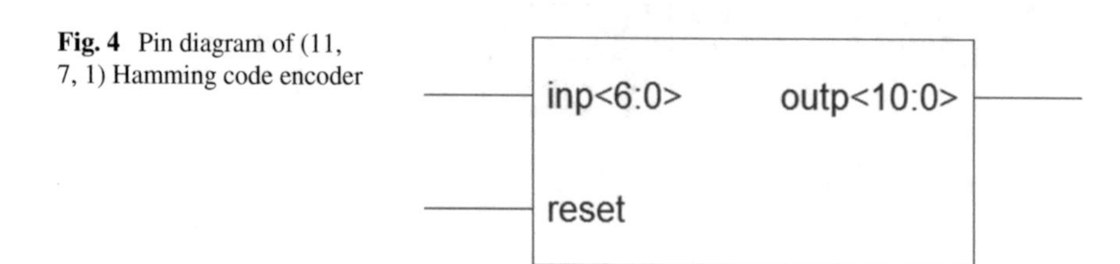

Fig. 4 Pin diagram of (11, 7, 1) Hamming code encoder

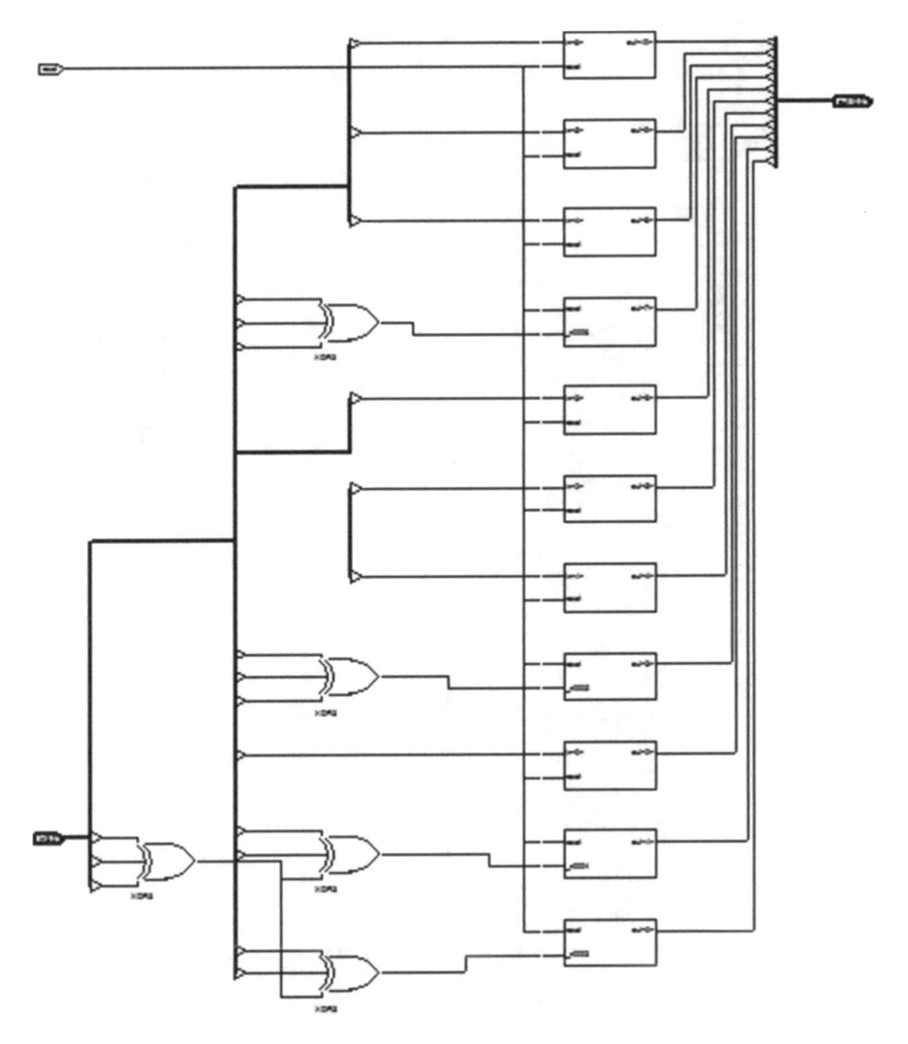

Fig. 5 RTL Schematic of (11, 7, 1) Hamming code encoder

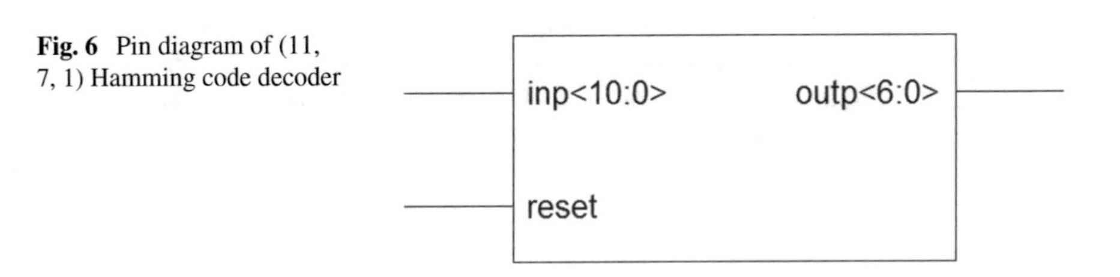

Fig. 6 Pin diagram of (11, 7, 1) Hamming code decoder

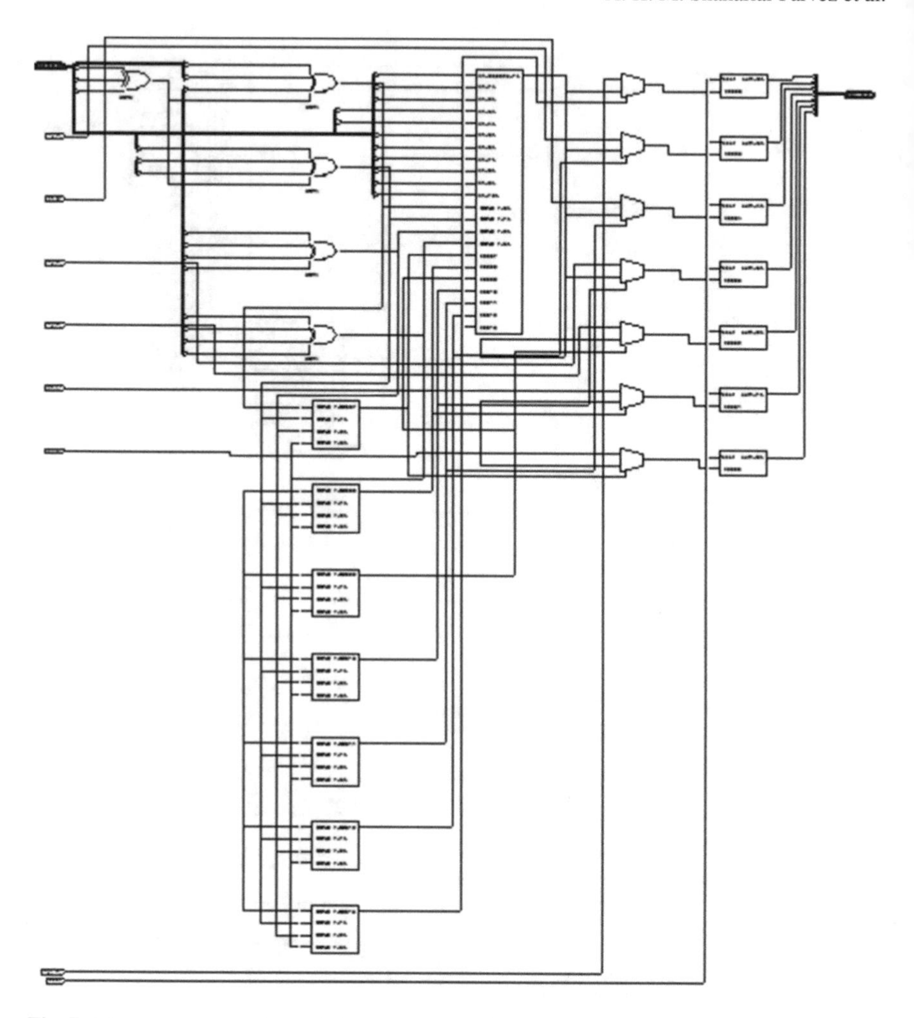

Fig. 7 RTL schematic of (11, 7, 1) Hamming code decoder

decoder to obtain the correct data. This data is compared to the original data, which should match perfectly or an error message is displayed and the simulation stops. The data word is normally incremented in the simulation until it wraps back around to zero. If this happens and no errors were found, the simulation ends successfully. The simulation diagram is obtained in Verilogger pro 6.5 software (Fig. 11).

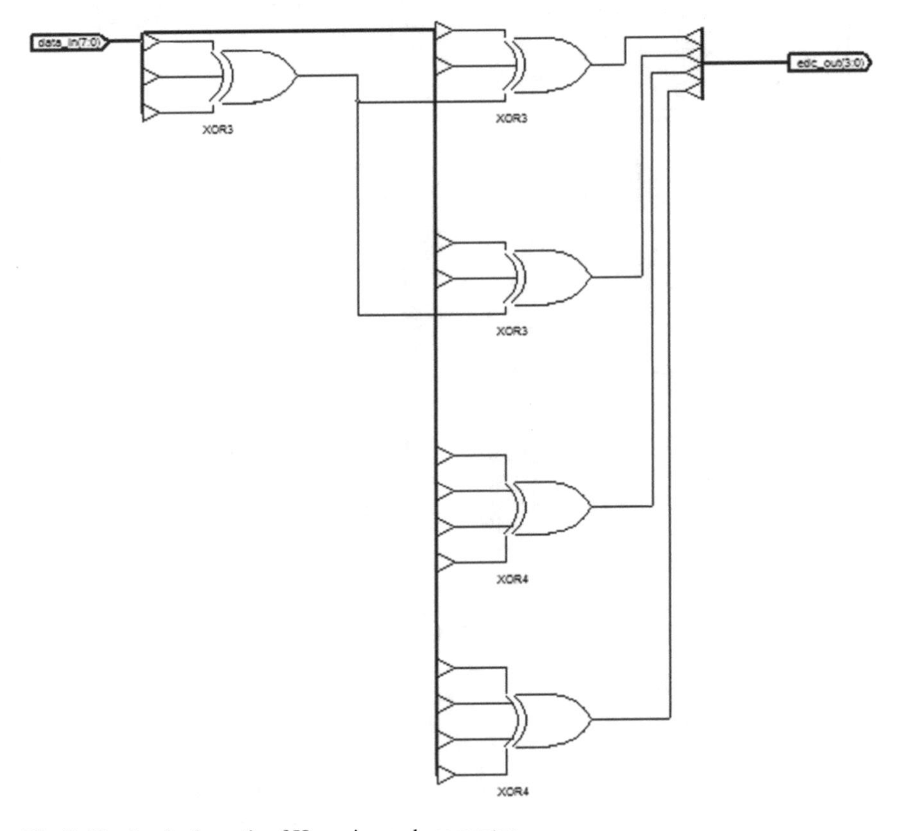

Fig. 8 Top level schematic of Hamming code generator

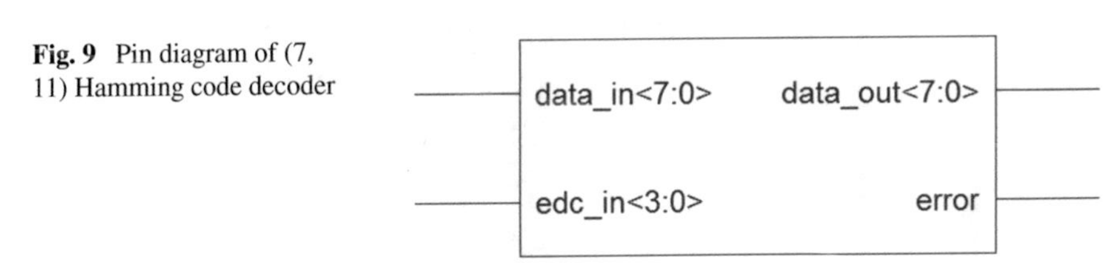

Fig. 9 Pin diagram of (7, 11) Hamming code decoder

5.4 *FPGA Implementation*

Field Programmable Gate Arrays can be used to effectuate the analytical function with small budget. The recommended design is fabricated in HDL, using Xilinx 6.3i with SPARTAN 2 family, FG320 platform logically verified and then synthesized using XST synthesis tool [12, 13]. The timing outline, device utilization analysis, and HDL synthesis report of hamming code decoder (from Fig. 10) are focused on

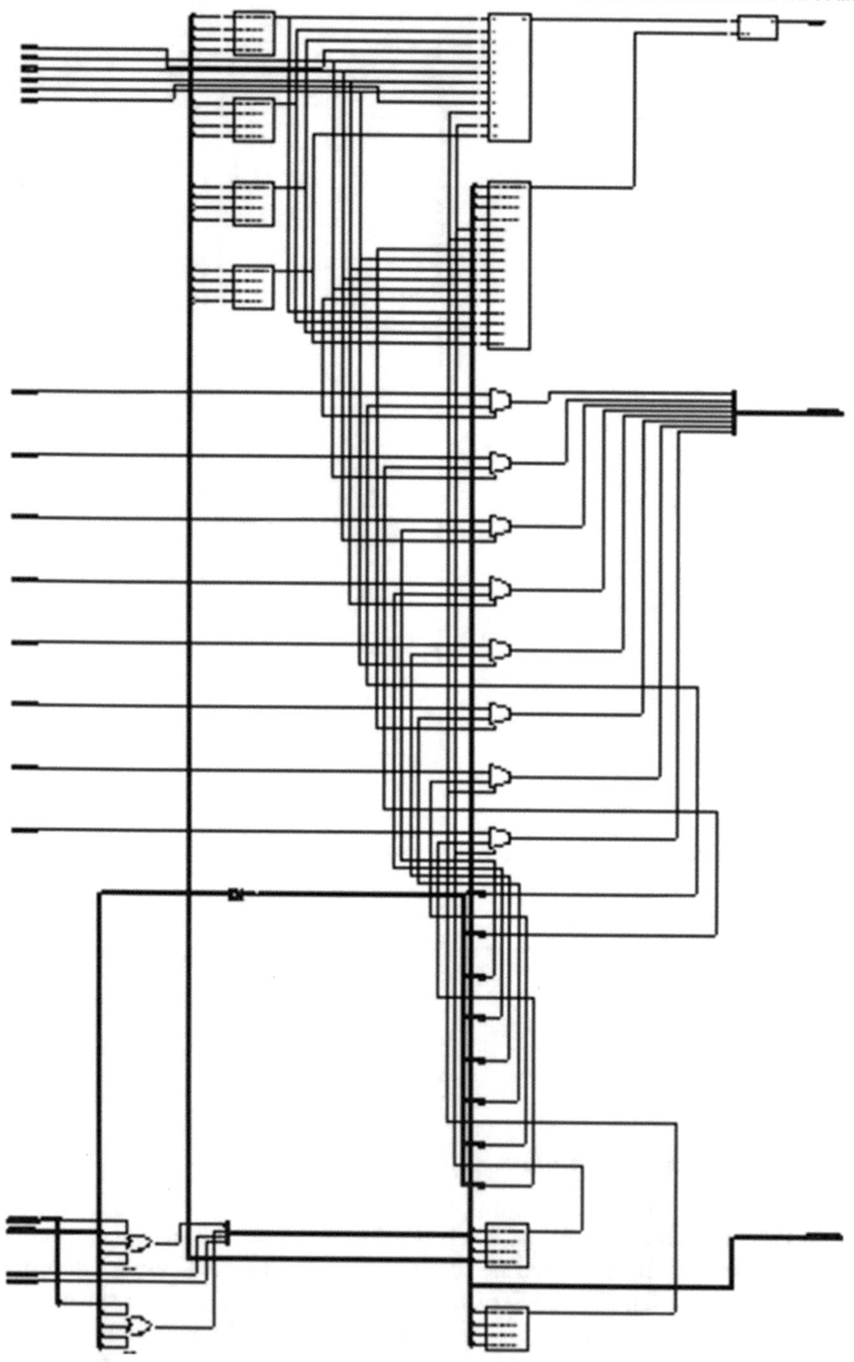

Fig. 10 RTL schematic of Hamming code decoder

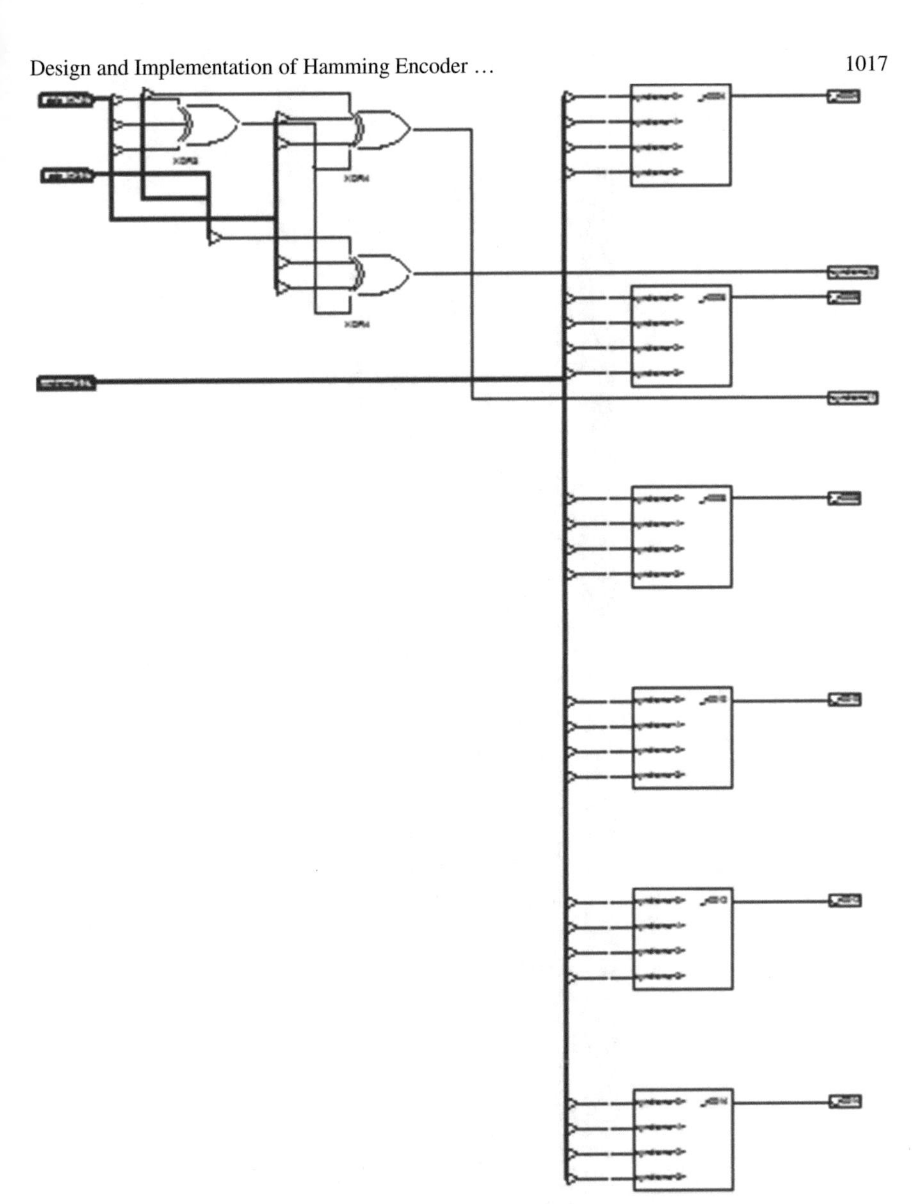

Fig. 10 (continued)

Tables 1, 2, and 3, respectively. The obtained maximum time delay is 14.507 ns. From Table 2, it is seen that single flip-flop is used out of 384. Table 3 illustrates the HDL synthesis report including the name of components and its quantity.

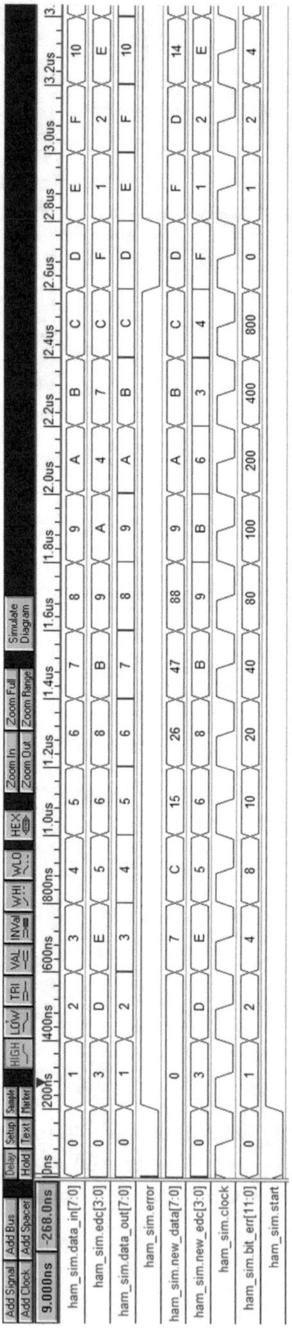

Fig. 11 Timing diagram of Hamming code simulation

Table 1 Timing outline of hamming code decoder

Parameters	Seconds (ns)
Minimum input arrival time before clock	7.758
Maximum output required time after clock	6.897
Maximum time delay	14.507

Table 2 Device utilization analysis of Hamming code decoder

Name	Used blocks	Percentages (%)
Number of slices	14 out of 192	7
Quantity of slice flip flops	1 out of 384(LD:6)	0
Amount of 4 input static RAM lookup table (LUT)s	24 out of 384	6
Amount of bonded input/output buffers	21 out of 90 IBUF (input buffer): 12 OBUF (output buffer): 9	23

Table 3 HDL synthesis report of Hamming code decoder

Name	Contents	Quantity
#Latches	1-bit latch	1
#Multiplexers	1-bit 2 to 1 multiplexer	8
#Exclusive ORs (Xor)	1-bit Xor3	1
	1-bit Xor5	2
	1-bit Xor4	2

6 Conclusion

Hamming code is an error correction code. It is useful for error-free communication. It can be used not only for detecting single-bit and burst errors but also for correcting single-bit errors that can occur at the time of transmitting binary data or information from one device into another device. In this paper, Hamming encoder and decoder have been pictorialized in FPGA by means of Xilinx software. FPGA has low costs compared to an application-specific IC for moderate volume applications and it is more flexible than the alternative approaches. So, proficient use of FPGA has been always covetable for the complicated circuit designing purpose due to requiring less hardware for a given system. The simulation diagram has been observed in the Verilogger pro 6.5. Timing outline and device utilization summary have also been found out in this paper. It has been observed that how many latches, multipliers, and digital logic circuits have been used to design the Hamming decoder.

References

1. Behrouz, A.F., Sophia, C.F.: Data Communication and Networking, 4th edn. Tata McGrawHill Publication (2007)
2. Gao, C.: Data Communications and Computer Networks, 2th edn, pp. 75–82. Publishing House of High Education, Beijing (1999)
3. Pedroni, V.A.: Digital Electronics and Design with VHDL. Morgan Kaufman Publishers (2008)
4. Berrou, C., Glavieeux, A.: Near optimum error correcting coding and decoding Turbo codes. IEEE Trans. Commun. **44**(10) (1996)
5. Lin, S.: Error Control Coding, 2nd edn. Pearson Education, India (1983)
6. Gross, W.J., Kschischang, F.R.: Applications of algebraic soft-decision decoding of Reed—Solomon codes. IEEE Trans. Commun. **54**(7) (2006)
7. Ashenden, P.J.: Digital Design: An Embedded Systems Approach Using Verilog. Elsevier (2007)
8. Panda, A.K.: FPGA implementation of encoder for (15, k) binary BCH code using VHDL and performance comparison for multiple error correction control. In: International Conference on Communication Systems and Network Technologies (2012)
9. Fan, C., Zhang, P., Wu, C.: Communication Principle, pp. 280–292. Publishing House of Electronics Industry, Beijing (2005)
10. Pedroni, V.A.: Digital Electronics and Design with VHDL. Morgan Kaufmann Publisher (2008)
11. Cohen, B.: Real Chip Design and Verification Using Verilog and VHDL. VhdlCohen Publishing (2002)
12. Podder, P., Md. Hasan, M., Khan, T.Z.: FPGA implementation of high performance Fast Page Mode dynamic random access memory, 8th International Conference on Software, Knowledge, Information Management and Applications (SKIMA 2014), Dhaka, 18–20 Dec (2014)
13. Paul, A., Khan, T.Z., Podder, P., Md. Hasan, M., Ahmed, T.: Reconfigurable architecture design of FIR and IIR in FPGA, 2nd International Conference on Signal Processing and Integrated Networks (SPIN), Noida, India (2015)

A. H. M. Shahariar Parvez is an Assistant Professor and Head in the Department of Computer Science and Engineering in Ranada Prasad Shaha University at Narayanganj. He is persuing Ph.D. in Computer Engineering from Dhaka University of Engineering & Technology (DUET). He has completed his B.Sc. and M.Sc. in Computer Science and Engineering from Kiev, Ukraine. He also got another M.Sc in E-Commerce from UK. He had successfully completed Microsoft Certified IT Professional (MCITP) on Server 2008 platform. He is the Co-author of a book "The Essentials of Computer Science" for the B.Sc (hons) students. His Research interest areas include Big Data, Bio informative, Brain engineering, Image Processing.

Md. Mizanur Rahman is currently working as a Lecturer in the Department of Computer Science and Engineering in Ranada Prasad Shaha University, Narayanganj, Bangladesh. He obtained his B.Sc. (Engg.) and M.S. degree in Computer Science and Engineering from Jatiya Kabi Kazi Nazrul Islam University, Trishal, Mymensingh, Bangladesh. His research interest includes VLSI design, FPGA system design, Computer networking, Steganography, Digital signal processing, Embedded system design, Data analytics.

Prajoy Podder is currently working as a Lecturer in the Department of Electrical and Electronic Engineering in Ranada Prasad Shaha University, Narayanganj, Bangladesh. He completed B.Sc. (Engg.) degree in Electronics and Communication Engineering from Khulna University of Engineering & Technology, Khulna-9203, Bangladesh. His research interest includes machine learning, pattern recognition, neural networks, computer networking, VLSI system design, image processing, embedded system design, data analytics. He published several IEEE conference papers and journals.

Mohammad Hossain is currently working as an Assistant Professor in the Department of Electrical and Electronic Engineering in Ranada Prasad Shaha University, Narayanganj, Bangladesh. He obtained his B.Sc. (Engg.) in Electronics and Communication Engineering from Khulna University, Bangladesh. He obtained his M.Sc. in Electrical and Electronics Engineering from Blekinge Institute of Technology, Sweden. His research interest includes wireless sensor network, digital communication, neural network.

Muhammad Ashiqul Islam is currently working as Specialist, Policy & Research under Regulatory Affairs division in Robi Axiata Limited, Bangladesh. He received B.Sc. (Engg.) in Electronics and Communication Engineering from Khulna University of Engineering & Technology, Khulna-9203, Bangladesh. His research interest includes wireless communication, OFDM, wireless optical communication, computer system networks, VLSI system design and Telecom market regulatory tools. He published an IEEE conference paper on analysing DCO-OFDM and Flip-OFDM for IM/DD Optical-wireless System with others.

Denial of Service (DoS) Detection in Wireless Sensor Networks Applying Geometrically Varying Clusters

S. S. Nagamuthu Krishnan

Abstract Wireless Sensor Networks (WSN) have great benefits of reduced costs, lesser scalability factor, and can be employed upon complex and dangerous locations for the purpose of control/automation of tasks and for sensing, processing, sharing/forwarding data. Denials of service (DoS) attacks hinder the regular functioning of such networks leading to compromise of the objectives of them. In this paper a hierarchical clustering approach is proposed to detect the compromise of nodes in WSN due to DoS attacks. This approach outweighs other approaches in the aspect of elimination of outliers and faster response time in detecting the attacks.

Keywords Service denial · Clustering · Centroid · Heuristics · Partition Propagation

1 Introduction

In WSNs the network is embedded with the environment and the nodes that form the network can effectively perform sensing and actuation to have a measure or impart its influence on the environment [1]. The processed information in each node is communicated in wireless fashion. The communication happens through radio signals among the nodes. The target applications could be industry-oriented, Science related, transportation enhancing, maintaining civil infrastructure, security related, etc. The processor within the sensor nodes could be in one of the three modes of sleep, idle and active. The power source and memory capacity of the nodes are very limited.

The nodes in the network may be participating as sources that involves in measuring data, sink listening to receive data from the network and act as actuators for control of devices based on the transmitted data. The interaction patterns between the various sources and sinks will be for detection of events, periodical measurement

S. S. Nagamuthu Krishnan (✉)
Department of MCA, R V College of Engineering, Bengaluru 560060, India
e-mail: ssnkrishnan@gmail.com

S. Smys et al. (eds.), *International Conference on Computer Networks and Communication Technologies*, Lecture Notes on Data Engineering and Communications Technologies 15,
https://doi.org/10.1007/978-981-10-8681-6_93

and reporting besides some secondary needs. The primary characteristics of WSNs include good scalability, verifying number of nodes for a given area, reprogrammable capability, and conveniently maintainable. The most important limitation of such network is the limited energy, which has to be efficiently utilized for communication, sensing, computation and actuation [2, 3]. The nodes very importantly collaborate among themselves to achieve a common goal primarily through preprocessing.

The energy of sensor nodes could be wasted due to their exposure to Denial of service attacks. One common attack at the physical layer is node tampering that leads to destroy of keys related to encryption and decryption [2]. The link layer attacks could be continuous interrogation by sending RTS message for handshaking that makes the nodes completely busy with responding requests, denying the sensor nodes to move to the sleep mode, thus reducing the battery life of sensors [4]. Network layer attacks could be of IP spoofing effected by masking the source address as bogus address, or the address of a victim, homing attacks targeting key managers for blocking and neglecting, and greed attacks aiming on neglecting routing of some messages and greedy of sending own messages [5, 6]. The sole aim of the work presented here is to devise a new method to nominate control elements/nodes in WSN and find out the distance between other nodes of a subset of nodes. The beginning of the process is to employ a clustering algorithm termed as CURE clustering [7] and reapplying the clustering technique to every other cluster determined. The cluster heads are determined as control nodes to detect harmful traffic if any. The next section discusses on some related works, followed by usage of CURE clustering for control node selection, followed by discussion of simulation results. Here the main contribution is highlighted and future developments are also indicated.

2 Related Works

It is very common that dynamics exist in topology as well as traffic. In order to accommodate that, clustering technique among nodes is adopted. Clustering groups a set of nodes such that the nodes in a cluster have similar properties. The similarity among the nodes (points) is normally defined using distance measures of Euclidian, Cosine, Jaccard, etc. Here all the nodes compete to become head and increase the chance of detecting denial of service attack [8] through the entry points of the sensor network.

Several clustering algorithms following the principles of hierarchical, heuristics based and partition based exist. Hierarchy-based clustering works on the principle that each point is cluster by itself. The key operation here is to merge two nearest clusters into the merged cluster to elect the head upon the nodes that are accessible within the range of radio signals [3, 9].

One popular clustering algorithm is K-Means that assumes Euclidian space for distances. BFR clustering algorithm for handling very large data sets considers clusters to be normally distributed around the centroid of the cluster [10].

Partitioning-based clustering algorithms [11, 12] involve movement of instances from one cluster to another. The number of clusters to be arrived is preset by the users and all possible partitions are exhaustively enumerated. Error minimizing algorithm with the basic idea of finding a structure minimizing certain error criterion, say sum of squared error for giving an approximate solution for minimizing errors. This is also employed by K-means algorithm which is the simplest and most common algorithm that also comes under this category [13].

Heuristics-based clustering algorithm works on the basis of definite heuristic. MaxMin D clustering [14] is a technique where there cannot be any node more than D hops distance from the head of the cluster. In this algorithm the cluster head is selected by performing $2d$ flooding rounds. In the first "d" round the nodes propagate largest node ID and the second "d" round is used by the nodes to propagate smallest node IDs. The rules adopted here are, if any node has received its id, it declares itself as cluster head, nodes look for pairs and select minimum node pair to become cluster head [13]. Linked cluster algorithm [14] is another example for this category where the cluster head is elected by choosing the lowest ID among non cluster head nodes or nodes that are at one hop distance from the cluster heads.

Low Energy based Adaptive Clustering Hierarchy (LEACH) and energy efficient hybrid clustering are examples for this. LEACH is a time division multiple access based protocol. Four phases involved here are, advertisement by cluster head for nodes to become member, setting up phase where the nodes answer to the heads, creation of schedule by the cluster head based on time division multiplexing and sending to cluster member during the time of data transmission [15].

Here we consider a hierarchical clustering algorithm CURE [7] with special capabilities of recognizing arbitrarily shaped clusters, not affected by outliers and linearly increasing storage requirements detection of Denial of Service attacks in a large network. The steps that are adopted here aim for inserting the node information of all the nodes into a tree and treating each node in the tree as a cluster for computing closest for each cluster, to be inserted into a heap. The closest clusters collect information on varied behavior indicating DoS activities and report to the elements of the cluster through the cluster head. This reduces the distance through which the information has to traverse.

3 Proposed Method

The proposed method employs DoS detection in a sensor network among the scattered nodes applying a hierarchical clustering algorithm proposed by [7]. Here the number of sensor nodes to be chosen is determined as a constant "C". The number of scattered nodes decides on the shape and extent of the cluster. The totally scattered points are shrunk towards the centroid decided by a fraction Cr. These points act as representatives of the cluster. The clusters of sensor nodes with the closest pair of representative nodes are merged at each step of the clustering algorithm thus alleviating the disadvantage of all points and centroid based clustering algorithm.

Table 1 Simulation parameters

	No. of nodes	Shape of clusters	No. of cluster
Data set 1	900	Small circles and ellipsoids	4
Data set 2	900	Small rings	6
Data set 3	1200	Circles	10

This process enables to correctly identify the clusters, and the approach is relatively less sensitive to the outliers as the number of nodes scattered are shrunk towards their mean thus dampening the adverse effects due to outlier nodes. This method of multiple scattered points enable discovering elongated clusters where the space within the vicinity of clusters is obviously non-spherical.

The algorithm begins by drawing a random sample of nodes representing the geometry of clusters, reasonably accurate for correctly clustering the input set of clusters and effective DoS detection. The minimum size of the sample is chosen to be in exponentially decreasing form and contain a fraction of nodes for every cluster. Here partitioning of the sample is carried out, and further clustering is done with very partition. Elimination of non-considered values (outliers) that contribute least for denial of service detection is carried out next, and partitions finally clustered during the final pass. The merging of clusters is repeated until arbitrary number of clusters is arrived, such that transfer of attack information is done in minimum time.

The algorithm builds clusters and inserts into a heap. At any point in time the cluster at the top of the heap is the closest to the next immediate cluster. Every iteration in the algorithm deletes the top element "Q" and then merges closest clusters at the next level. The representative points of the merged cluster are the union of representative nodes of the clusters that are merged. After merging the distance between them is recomputed.

For larger number of nodes the combination of random sampling and partitioning could be effective in identifying the clusters. Outlier handing technique is combined with random sampling to eliminate outliers. The partitioning constant is set to three here and clustering of partitions is continued, until the no. of nodes remaining is one-third of the number of nodes initially. Clusters that contain one node are eliminated as outliers. Then as the total clusters reach a value k, here too the outliers are removed as clusters that contain as many as five nodes.

The testing experiment is carried out through simulation with three data sets having nodes on two dimensions (Table 1).

The algorithm was run with the data sets 1–3 mentioned above and it moderately shrunk to a mean factor of 0.4 enabling it to be less sensitive to outliers.

The simulation experiment also prove that for a node representative value of greater than 10 right number of clusters, for effective DoS detection information propagation were found. The splitting of clusters was also carried out fairly during the instances of large distance between the representative nodes, enabling speedier communication among them during anomalous traffic. The study also proved that

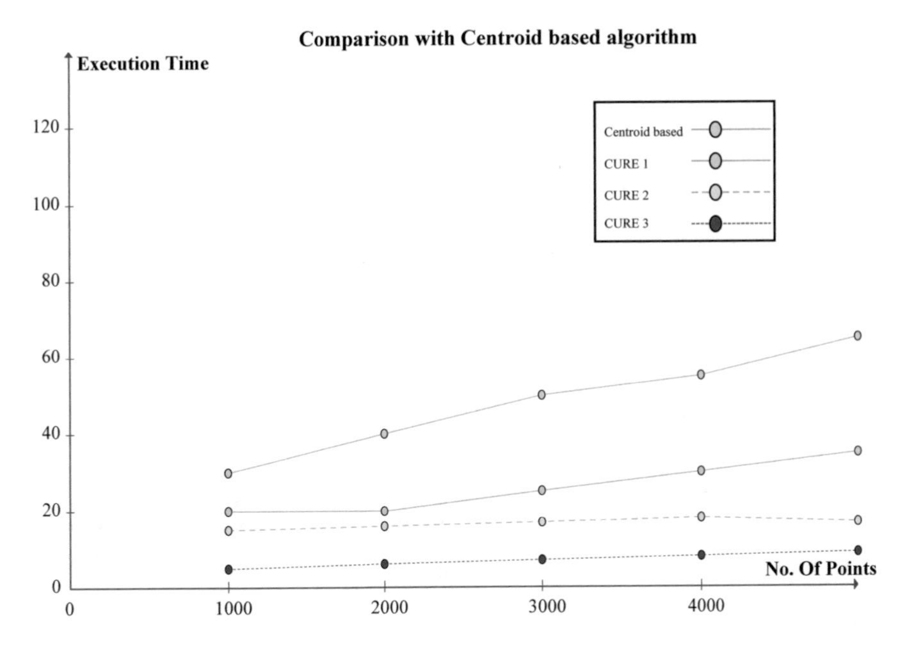

Fig. 1 Execution time comparison with centroid based algorithm

for a value close to 50 partitions the desired number of clusters was discovered for effective DoS detection and information propagation. When the number of partitions was successively increased the quality suffered due to reduction in number of representative points.

The execution time of the partitioning algorithm is also relatively lesser as compared to the other counterparts as it combines the techniques of random sampling and partitioning in a way to bring down the input size of the nodes. This is particularly seen as the number of nodes is increased. The execution time increases very little as the sample size remains the same (Fig. 1).

The graphical representation indicates the comparison of execution times of the clustering algorithm for three different shapes of clusters viz. small circles and ellipsoids, small rings, and circles. The execution times tend to improve in all the three cases when compared to that of Centroid based algorithm (Fig. 2).

The graphical representation given below shows the relationship between the no. of partitions taken for the sample run with the sample sizes of 1000, 2000 and 3000 nodes respectively. It clearly indicates the decrease in time of execution of the algorithm as the no. of partitions increase resulting in faster communication among the node representatives on DoS detection.

The third aspect analysis is on representative points versus execution time for the algorithm. It proves that the execution time tends to increase as the number of representative points increase for the three sample sets taken for the study. This recommends the choice of lesser representative points to achieve the execution time.

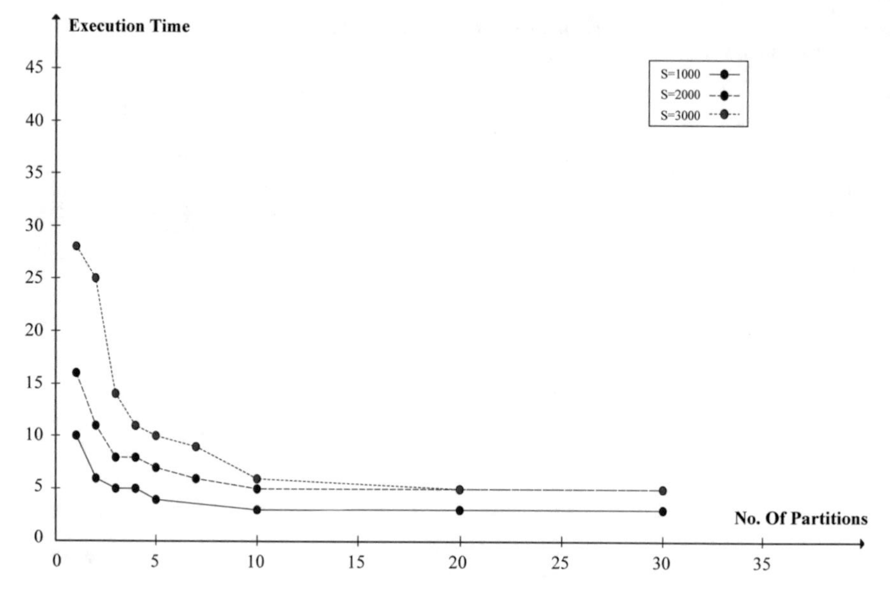

Fig. 2 No. of partitions versus execution time

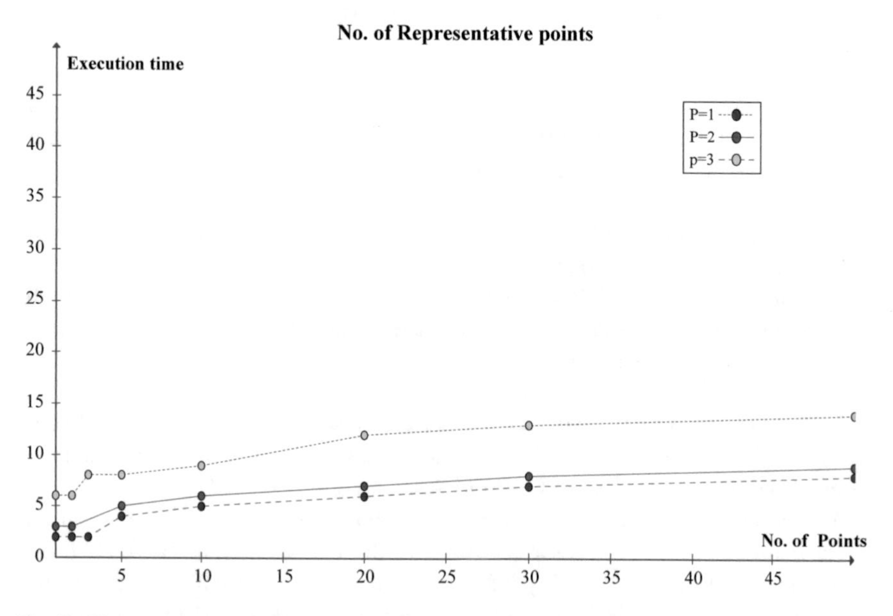

Fig. 3 Cluster representative points versus execution time

But, considering the speeder propagation of DoS attack information among the sensor nodes, a slight increase in execution time is inevitable to prevent further damage to the sensor network, on identification of anomalous behavior in a node (Fig. 3).

4 Conclusion

The algorithm uses multiple representative points (nodes) by selecting well-scattered nodes and then shrinking them to the center by a fractional number enabling effective identification and propagation of DoS attack information. This results in wide-varying geometrical shapes to the clusters. A Combination of random sampling with partitioning is used for larger number of nodes for faster identification of attacks. Filtering of outliers is also done effectively through random sampling. The time is estimated to the order of $O(S^2)$ for a sample of size S and space complexity varies linearly with "S". The primary benefit of the identification procedure applying the clustering algorithm is scalability for large number of nodes without compromising cluster quality. The rate of identification of attacks and the time period could be further improved by considering an alternative clustering procedure that improvises the distribution/scattering of nodes in a network.

References

1. www.vonbi.ac.ke/conferences/WSN/day1/introduction.pdf
2. Lai, G.H., Chen, C.-M.: Detecting denial of service attacks in sensor networks. J. Comput. **18**(4) (2008)
3. Jain, A.K., Dubes, R.C.: Algorithms for Clustering Data. Prentice Hall, Englewood Cliffs, New Jersey (1988)
4. Heizelman, W.R.: Energy efficient communication protocol for wireless microsensor networks. In: Proceedings of IEEE Hawaii international Conference in System Sciences (2000)
5. Wood, A.D., Stanklovic, J.A.: Denial of service in sensor networks. IEEE/Computer 49–56 (2002)
6. Dhara, B., Devesh, J.: Denial of service attacks in wireless sensor networks. In: International Conference on Current trends in Technology (2010)
7. Guha, S., Rastogi, R., Shims, K.: CURE: an efficient clustering algorithm for large databases. Inf. Syst. **26**(1), 35–58 (2001)
8. Guechari, M., Mokdad, L., Tan, S.: Dynamic solution for detecting denial of service attacks in wireless sensor networks. In: Proceedings of IEEE ICC 2012—Ad-hoc and Sensor Networking Symposium, pp. 173–177 (2012)
9. Olson, C.F.: Parallel algorithms for hierarchical clustering. Technical report, University of California at Berkeley (1993)
10. Springer: https://link.springer.com/chapter/10.1007/978-1-4939-2468-4-1
11. www.csie.ntpu.edu.tw/~tschen/course/96-1/wn-ch10.pdf
12. Raymond, D.R., Midkiff, S.F.: Denial-of-service in wireless sensor networks: attacks and defenses. IEEE CS Pervasive Comput. 74–79 (2008)
13. Amis, A., Prakash, R., Vuong, T., Hymnh, D.: Max-min D-cluster formation in wireless Ad-Hoc networks
14. https://web.stanford.edu/class/cs345a/slides/12-clustering.pdf
15. Meng, T., Volkan, R.: Distributed network protocols for wireless communication. In: Proceedings of IEEE ISCAS (1998)

Dr. S. S. Nagamuthu Krishnan obtained his Bachelor's degree in Physics from Madurai Kamaraj University during 1995 and Masters Degree in Computer Applications from Bharathiar University during 1998. He has completed his Ph.D. in Computer Science in the Department of Computer Science and Engineering, Bharathiar University during 2015. He specialized on security in Networks. His research area includes detection and prevention of Distributed denial of service attacks in networks. He has presented many research papers in National, International conferences and Journals. He has totally 19 years experience in teaching. He is a life member of Computer Society of India. He has served in various premier educational institutions in Tamil Nadu, India and at present, he is working as Assistant Professor in the Department of Computer Applications of R V College of Engineering Bangalore.

Performance Analysis of Fragmentation and Replicating Data Over Multi-clouds with Security

R. Sugumar, A. Rajesh and R. Manivannan

Abstract In cloud computing, outsourcing a data to third-party cloud providers is not trusted. Ensuring the security is major concern in clouds, our method proposed the security with minimal cryptographic system by fragmenting a data and replicates over Multi-Clouds. We implement RRNS encoding method for fragmenting the data and CTNA which has the centrality measure as a point of distributing and recovering the data. Even in the case of successful attack no information is revealed to anyone. We compute different fragmentation process analysis on several multi-cloud systems, impact in increasing number of nodes, fragments and also the analysis based on different dissemination and retrieval process efficiency with multi-cloud systems. The higher level of security and performance was observed.

Keywords RRNS (redundant residue number System) · CTNA (cloud tree node Assignment) · Centrality

1 Introduction

In order to provide security, with multi-cloud our basic idea is fragmenting data using the Redundant Residue Number system (RRNS) [1]. Suppose we have a set $\{M_1, M_2, \ldots, M_n\}$ of N positive and pair wise relatively prime modulo. Let M be

R. Sugumar (✉)
Department of CSE, Sri Chandrasekharendra Saraswathi Viswa Maha Vidyala (SCSVMV) University, Kanchipuram, Tamil Nadu, India
e-mail: sugumar_prof@rediffmail.com

A. Rajesh
C. Abdul Hakeem College of Engineering and Technology, Melvisharam, Vellore District, Tamil Nadu, India
e-mail: amrajesh73@gmail.com

R. Manivannan
Department of Computer Science and Engineering, Stanley College of Engineering and Technology for Women, Hyderabad, Telangana State, India
e-mail: drmanivannan@stanley.edu.in

S. Smys et al. (eds.), *International Conference on Computer Networks and Communication Technologies*, Lecture Notes on Data Engineering and Communications Technologies 15, https://doi.org/10.1007/978-981-10-8681-6_94

the product of modulo, then every number $X < M$ has unique representation in the residue number system which is the set of residues $\{|X|m_i;\ 1 < i < N\}$. A partial proof of this as follows, supposes X_1 and X_2 are two different numbers with the same residue set. The $|X_1|m_i - |X_2|m_i$, and so $|X_1 - X_2|m_i - 0$. The residues of integers 0 through 15 relative to the modulo: 2, 3, 5. Using the RRNS [1], we split a file into chunks where "p" represent number of chunks to construct the original file and "r" is the degree of redundancy. For example we consider $p = 5$ and $r = 3$ for a total of 8 fragment of files and 4 different cloud storage providers like A, B, C, D. so that we can store 22 chunks on each provider assume D is not available any more, the end user can retrieve long term availability each provider will have a partial content of information. In order to provide security that no cloud provider can have full access to stored files, in this paper we ensures that even in the case of successful attack no data will be revealed to the attacker. Our scheme provides minimal cryptographic techniques for data security as to avoid computational methods in generating keys to make a faster system, and ensures the controlled replication of file fragmentation to improve the purpose of security. In order to prevent unauthorized access we are providing each user with key to store their data in the dispersed manner. Each of the cloud nodes we represent them with different system like computing system, storage, physical and virtual machines to provide security, to improve the data retrieval time we used centrality measure to coordinate the cloud nodes.

2 Related Work

Antonio et al. [1] introduced a redundancy factor, using the RRNS, they split a file into $p + r$ chunks allows them to achieve long term availability in clouds, also enables data obfuscation to provide partial view of each file in order to improve security they encrypted each chunk using a symmetric algorithm. Authors [2] provided a community cloud computing concerns with security, compliance and jurisdiction a computation model for task oriented multi-cloud collaboration namely TOMC, the heterogeneous social network H can be extracted $\pi = (V, E)$, V represents a community cloud holding a set of cost (access cost and monetary cost). A file fragmentation system for multi-provider cloud [3] provided a new approach to store data into the cloud by means of splitting phase, dissemination phase, retrieval phase, reconstruction phase. In splitting phase, a file is fragmented with redundancy degree (r) and this process generates $p + r$ residue segments and made the XML residue segments in to several cloud storage providers; each provider is not able to know the content of whole file. In retrieval phase it downloads the XML chunks, in reconstruction phase process the XML de-capsulation, so the end users access the original file.

Author [4] represents automated reassembling of files, using Greedy algorithm where digital images are fragmented and there is no table information to reconstruct, they used a forensic analyst to reassemble the correct fragment order to reconstruct each image. Here a Greedy Heuristic algorithm provides four steps:

Greedy SUP: a Sequential Unique Path creates vertex disjoint paths reconstruction path ($p_1 = p_i \| t$) and the complexity of algorithm is $O(n) + O(n^2) + O(n^2 \log n) = O(n^2 \log n)$.

Greedy NUP: a Non Unique Path, find its best match until the images had reconstructed and repeat the process until all K images had reassembled and complexity of the algorithm is $O(n^2 \log n)$.

Greedy SPF UP: a Shortest Path UP algorithm a reconstruction path having the lowest average cost, it takes n calculation $n - n$, and it takes $\sum_{i=1}^{K} (n - n_i)$ of complexity $O(n^2 \log n)$.

Greedy PUP: A Parallel Unique Path, it is a variation of Dijkstra's single source shortest path to reassemble the images simultaneous best match takes constant time and complexity of algorithm is $O(n^2 \log n)$.

3 System Architecture

The usage of cloud storage providers is formalized by the customer to subscribe many cloud storage providers, sometimes it is also free (copy, drive, dropbox etc.,) they can easily manage upload and download a file. RRNS encodes the data as $(n + r)$ tuples of residues using $n + r$ keys, these residues are distributed over cloud nodes in the cluster, recovering the original data requires the fact of at least residues and that of corresponding moduli. Data can be reconstructed from the cloud provider with the help of CTNA process in the presence of $S \leq r$ residue losses, combined with up to $[(r - S)/2]$ corrupted residues. The size of the fragments affects the performance of the system as large records increase the information efficiency of encoding and implies a smaller number of messages to be exchanged when creating and accessing a file.

Our system gets connected with CTNA for distributing data over multi-clouds. CTNA organizes clusters of public and private clouds and has an instance of connection with head_of_cluster (hcl), which maintains a connected path of Active nodes (A_n), whose nodes are actively participant in the clusters. Once the configure active node starts working with the placement of fragments over multi-cloud, the fragmented copies gets replicated over cloud nodes via FRD system through CTNA the placement of replication takes place with different nodes that do not have the original primary copy by this way it replicates all of the fragmented files to other cloud nodes. Each of the k_i represent ($l_{i,j}$ +max $l_{(l,j)}$ leads to $K_i \geq K + 1$). CTNA will maintains path of all the visited cloud nodes and it holds the structure and act as a centrality point of distributing and recovering of data over multi-clouds. CTNA's structure performs and make way for reforming the pieces of files to the original file and available at the time of retrieval by users (Fig. 1).

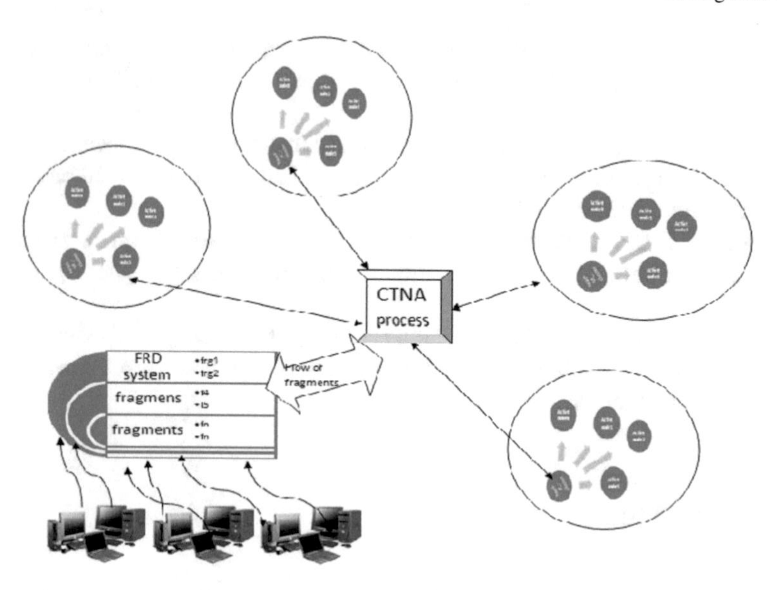

Fig. 1 FRD system with CTNA process

3.1 FRD with CTNA

- Initialize each fragment; initialize the counter value equals to zero. Set $\mathbf{l_{i,j}} = 0$, where $i = 1, 2, \ldots, m, j = 1, 2, \ldots, n$.
- Process the active node configuration. Set $\mathbf{l_{i,j}} \geq \max(k)$, set hcl $= i$ and cover node 1 with head of clusters.
- Proceed to fragment placement over cloud nodes, set hcl = active, follow all $K_{n(s)}$ to k_i and set path to cover to hcl(max(k)).
- Replicate the fragment over other clouds, set $k_i = \mathbf{l_{i,j}} + \max(\mathbf{l_{i,j}})$.
- Increase the access counter of active node and replicating nodes one by one, loop until all the nodes are fragmented and replicated.
- If counter of locale node is greater than the max($\mathbf{l_{i,j}}$), then goto step2.
- Process until all the fragments get placed and replicated over the cloud clusters (Table 1).

4 System Implementation and Experimental Setup

We implemented our method in a Lab with Intel corei3 processor with distributed system setup and we found comparisons of efficiency, performance with a 1-node, 2-node, 4-node and 6-node clusters. First we compute the ratio of error free fragments with the accuracy as a parameter with different fragmentation process. In our analysis passmark fragger a tool which splits a file, for examples 180 kb file is splitted

Table 1 Symbols and their meanings

Symbols	Meanings
A_n	Active node
Init	Initial node
l	1st initial node
$l_{i,j}$	Initialization of cloud node l in cluster
K	Set of participants
k	Participant in active node
Hcl	Head of clusters

in seven fragments and aggregates the fragmentation process accuracy from 38.44 to 49.94%, for RRNS accuracy calculation, we work on Gsplit tool, a fragmentation tool we had one file size of 1.82 Mb that was fragmented into six pieces with MD5 encryption those fragment were encode by RRNS method and were distributed to cloud nodes, each fragment were observed with average byte frequency and mean of each byte frequency are accounted and we calculated on how originally the file was restored, when we distribute over cloud nodes, the CTNA process coordinates and acts as a centrality to disperse and gather the fragments over cloud instance (nodes), the CTNA maintains a file descriptor content holds dispersed and gathered fragments details of active nodes (A_n) to reconstruct the original file. In the error free ratio calculation, we considered the network failure occurs between the cloud nodes and our method produces error free fragments when it was united/reversed from the distributed fragments. Our RRNS implementation with CTNA process were made at the developer interfaces of multicloud, scalr, rightscale and splunk with some parameters. We observed the utilization of our system is measured as 11.44 Mb on a single session of our CTNA process. In scalr 8.5 Mb utilization was observed, In rightscale we observed 10.66 Mb utilization, and in splunk we found that 9.88 Mb utilization. These performances were analyzed by Wireshark which captures all packets transferred and counts are calculated and we present the result as cloud bandwidth efficiently with increase in number of nodes and also we showed the efficiency of sample cloud clusters with different execution time. In the chart we showed the X-axis by number of counts minimum value and maximum value, rate of access and burst rate as cloud cluster with parameter dissemination and retrieval process.

When we implement the efficiency of cloud cluster based on 1-, 2-, 4-, 6-nodes on the cloud management tool a Scalr a open source tool for multi-cloud management, we computed results that as in our case we consider five packets were transferred for 1520 bytes with memory utilization and execution of accessing the cloud providers with the job_id and packet_id by RRNS encoding and fragmentation part it takes only 0.01 s only small number of cycles and minimal cryptographic computation are needed for all the fragmented packet though CTNA process and in the case of dissemination process construction of cloud nodes (i.e., participation of active node (A_n) it consumes only 0.06 s. when user wants to retrieve the original file with our system CTNA coordinates the node here, no process is carried over for retrieve from

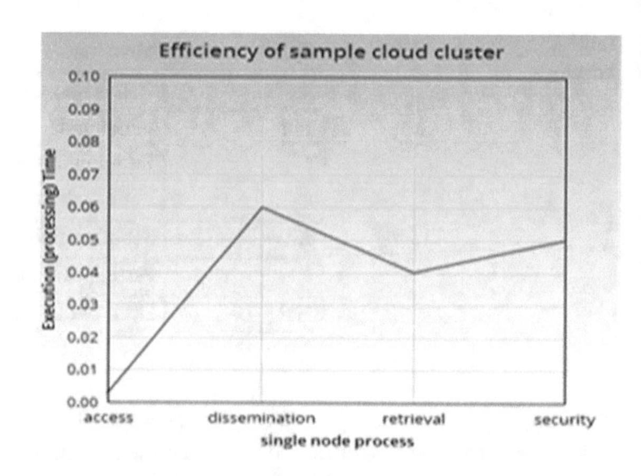

Fig. 2 Efficiency of cloud cluster (1-node)

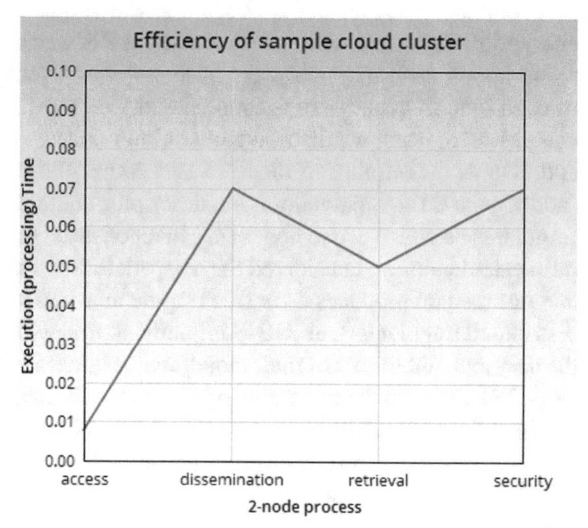

Fig. 3 Efficiency of cloud cluster (2-node)

several nodes but it waits for turnout time to complete the process of retrieving file (Figs. 2 and 3).

Next, we analyzed the efficiency of sample cloud clusters with 2-nodes accessing the cloud provider takes only 0.01 s, in the case of dissemination the fragments are distributed to the 2-nodes with the active participant in the cloud cluster and it assumes any one as the head_of_cloud cluster (hcl) and memory cycles and cpu cycles for dissemination process takes execution time as 0.07 s, in the case of retrieval hcl coordinates the fragments on the two nodes and gathers the content into original file to user, that takes time around 0.05 s and in the case of security from the user side and the provider side make the execution time around 0.07 s. Next we analyzed on 4-nodes in a cluster the access time takes around 0.05 s, since it needs to get accessed to 4 cloud provider and dissemination process takes 0.07 s, here fragmented file has to be distributed on 4-nodes so it follows the routing table provided by CTNA process

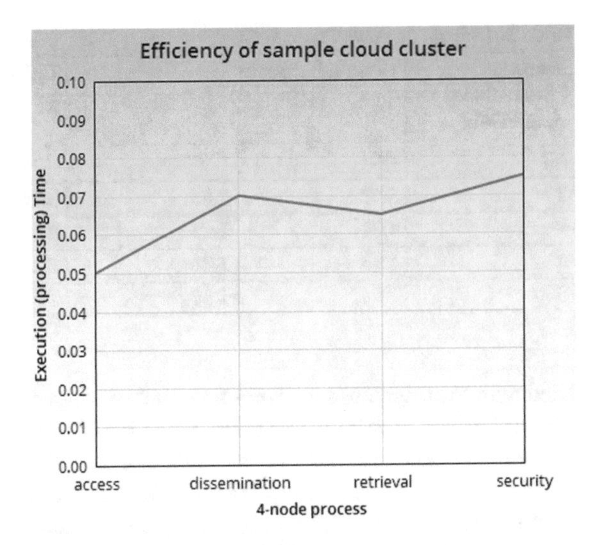

Fig. 4 Efficiency of cloud cluster (4-node)

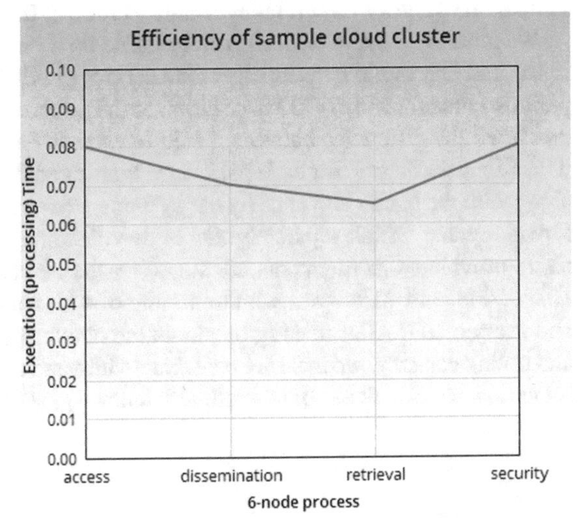

Fig. 5 Efficiency of cloud cluster (6-node)

and retrieval execution time around 0.06 s hence CTNA coordinates through hcl and give back the original file, and security mechanisms were accounted as strong from both the user side and provider side it takes execution time around 0.08 s (Figs. 4 and 5).

Next we analyzed on 6-nodes, here access time for accessing six nodes takes 0.08 s, dissemination process like we discussed earlier for (4-nodes) takes around 0.07 s, retrieval of original file takes 0.065 s and security mechanisms implemented in our system takes 0.08 s, all our analysis are shown clearly and difference in access, dissemination, retrieval and security are analyzed with different cycles, execution speed varies for different parameter used in our system. The access time was calcu-

Table 2 Uplink and downlink scenarios

Scenario (different file sizes in Mb)	Downlink capacity		Uplink capacity	
	Peak	Average	Peak	Average
2	2.52	1.63	3.75	0.86
4	4.81	1.45	2.56	1.2
6	5.63	1.21	2.24	0.56
8	7.54	1.87	1.92	0.74

lated with the fragment sent every time that was based on percentage of packets and five packets for 1520 bytes and transfer rate is 412 bits/s, considering all the data sent through IPV4 our analysis shows access time is 100% and dissemination execution process takes 0.06 s and retrieval process from 1-node cluster takes around 0.04 s and security aspects, we focus on monitoring from unauthorized access through firewall setup at the security level it takes execution time of 0.05 s (Table 2).

We monitored our system with daily traffic for 5 days which was calculated on In traffic and Out traffic between the cloud service providers, we accounted for certain period of time for 2–5, 5–10 Mb of file sizes in Dropbox, DriveBox and OneDrive and measured the difference between 5 min interval in each and every cloud providers with Cloudshark tool, for each packet we monitored the bit rate, i.e., data rate (bps) along with the bandwidth used by our systems, here we take a scenario for downlink separately and uplink separately that is downloading a file and uploading a file to cloud providers our fragments are stored by the blocks in cloud storage with 512, 1024, 2048, and 4096 bytes with minimum of 4 kb chunks as a piece of file in each and every cloud. After making blocks in the cloud storage with different downlink and uplink capacity, we measure peak bandwidth and average bandwidth in different scenarios, we calculated Bandwidth efficiency $B_{\text{eff}} =$ Data rate/bandwidth.

5 Conclusion

We proposed a system that fragments and replicates over the cloud nodes, by CTNA we assigns the nodes and also checks the participant and identify which are all active participant and sleep nodes and also maintains a table to identify where the data stored, this table helps in dissemination and retrieval process by the user stored in multi-clouds. In order to change the users mostly reliable in single clouds, we implement to store data in multi-clouds even in the minimal cryptographic methods we provide a higher level of security with even in the case of successful attack, users maintains integrity of their data over multi-clouds. Some factors driven to distributing the fragments over cloud nodes, it attains lower cost, increased scalability and increased availability.

References

1. Celesti, A., Fazio, M.: Adding long-term availability, obfuscation and encryption to multi-cloud storage systems. J. Netw. Comput. Appl. (2014) (Elsevier)
2. Hao, F., Min, G.: An optimized computational model for task oriented multi-community-cloud social collaboration. IEEE Trans. Serv. Comput. (2013)
3. Volari, M., Celesti, A., Fazio, M., Puliafito, A.: Evaluating a file fragmentation system for multi-provider cloud storage. In: 2013 SCPE, vol. 14, no. 4, pp. 265–277 (2013). https://doi.org/10.12694/scpe.vl4i4.932. ISSN: 1895-1767
4. Menon, N., Pal, A.: Automated reassembly of file fragmented images using Greedy Algorithms. IEEE Trans. Image Process. **15**(2), 385–393 (2006)
5. Ali, M., Bilal, K.: DROPS: division and replication of data in cloud for optimal performance and security. IEEE Trans. Cloud Comput. (2015). https://doi.org/10.1109/TCC.2015.2400460
6. Dasygenis, M., Ioannis, P.: A generic moduli selection algorithm for the residue number system. In: 10th International Conference on Proceedings on IEEE, 21–23 Apr 2015, pp. 1–2
7. Aruna, C., Siva Ram Prasad, R.: Resource Grid Architecture for Multi Cloud Resource Management in Cloud Computing. Springer International, Switzerland (2015)
8. Fadulilhahi, I.R.: Efficient algorithm for RNS implementation of RSA. Int. J. Comput. Appl. **127**(5) (2015)
9. Fazio, M., Celesti, A.: A message oriented middleware for cloud computing to improve efficiency in risk management systems. In: 2013 SCPE, vol. 14
10. Hudic, A.: Data confidentiality using fragmentation in cloud computing. Int. J. Commun. Netw. Distrib. Syst. **1**(¾) (2012)
11. Hasan, H.: Secure data portioning in multi cloud environment. In: 2014 Fourth World Congress on Information and Communication Technologies (WICT), 8–11 Dec 2014
12. Singhal, M., Chandrasekhar, S.: Collaboration in Multi Cloud Computing Environment: Framework and Security Issues. Research Feature. IEEE Computer Society (2013)
13. Chan, J.W.-T.: Online Tree Node Assignment with Resource Augmentation. Springer Science + Business Media LLC (2010)
14. Chiasserini, C.F.: An Energy Efficient Method for Nodes Assignment in Cluster Based Adhoc Networks. Kluwer Academic Publishers (2004)
15. Volari, M., Celesti, A., Fazio, M., Puliafito, A.: Evaluating a file fragmentation system for multi-provider cloud storage. In: 2013 SCPE, vol. 14, no. 4, pp. 265–277. https://doi.org/10.12694/scpe.vl4i4.932. ISSN: 1895-1767
16. Chang, R.-S., Chen, P.-H.: Complete and fragmented replica selection and retrieval in data grids. Future Gener. Comput. Syst. **23**, 536–546 (2007)
17. Zheng, Z., et al.: STAR: strategy-proof double auctions for multi-cloud, multi-tenant bandwidth reservation. IEEE Trans. Comput. **64**(7) (2015)

R. Sugumar received his B.E. and M.Tech. Degree in computer science and Engineering from Anna University and Dr. M. G. R. Educational and Research Institute University, Chennai, India. He is currently a Research Scholar in Department of Computer Science and Engineering at Sri Chandrasekharendra Saraswathi Viswa Maha Vidyala (SCSVMV) University, Kanchipuram, India. His area of interests includes Cloud Computing, Grid Computing, Computer Networks, Artificial Intelligence, Distributed Systems, and Distributed Databases.

Dr. A. Rajesh Professor & Principal at C. Abdul Hakeem College of Engineering & Technology, Tamilnadu, India, and he received his Ph.D. Degree from Dr. M. G. R. Educational and Research Institute University, Chennai, India, in March 2011 and his M.E. Degree from Sathyabama University, Chennai, India, in April 2005 in the field of Computer Science and Engineering. His area of interests includes Cloud Computing, Data Mining, Natural Language Processing, Semantic Web and Intelligence Systems.

Dr. R. Manivannan is a Professor in the department of Computer Science and Engineering, Stanley College of Engineering & Technology for Women, Hyderabad, India, and he received his Ph.D. Degree from Dr. M. G. R. Educational and Research Institute University, Chennai, India, in August 2014 and his M.E. Degree from Sathyabama University, Chennai, India, in April 2005 in the field of Computer Science and Engineering. His area of interests includes Data mining, Information Retrieval, and Cloud Computing. He is having 18 years of Experience in Teaching, and Administration.

Efficient Energy Re-organized Clustering Based Routing for Splitting and Merging Packets in Wireless Sensor Network

V. Naga Gopi Raju and Kolasani Ramchand H. Rao

Abstract The main objective of work is to develop an efficient energy clustering algorithm with merging and splitting. Energy efficiency is the main issues in WSNs, because of battery that is available in WSN cannot be replaced or recharged. The proposed method is an efficient energy re-organized clustered model which uses merging and splitting (EECSM). This algorithm performs merging and splitting for efficient energy in clusters. Previous works show that balancing of load is considered with distributed organized self manner. To eliminate the existing work drawback, we develop a system of proposed for use. EECSM will use energy state information related to sensor nodes for reducing consumption of energy and load maintenance and balancing. EECSM will prolong the network life by merging and splitting the clusters using cluster heads in a sensor network. The experimental results shows that EECSM will performance better in terms of life of the network, scalability, energy residual, and robustness.

Keywords Clustering · Lifecycle · Wireless sensor network · Self-organization Distributed · Monitoring

1 Introduction

Wireless Sensor Networks is defined as collection of nodes, the attractive research attention is based on its extensive range of applications in major areas like tracking of objects, detection of intrusion, monitoring of environment, of management inventory system in a company and also the applications related to health sciences and so on. In WSNs, the node energy in a network is limited.

V. Naga Gopi Raju (✉)
Department of Computer Science, Acharya Nagarjuna University, Guntur 522006, India
e-mail: bkrishnaklu36@gmail.com

K. R. H. Rao
Department of Computer Science, ASN Degree College, Tenali, India

S. Smys et al. (eds.), *International Conference on Computer Networks and Communication Technologies*, Lecture Notes on Data Engineering and Communications Technologies 15, https://doi.org/10.1007/978-981-10-8681-6_95

So, there is a need for efficient energy use and control in network topology. So manage efficient in energy, clustering is the suitable and mostly used un-supervised learning process in solving problems; likewise similar problems of its kind is dealt in building the structure by collection of data node which are unlabeled. WSN has a unique characteristic with the following constraints. At first it is a collection of numerous sensor nodes form a network based on certain topology. Therefore every WSN node posses several type of devices, which are based on constraints for processing and data storage capacity.

Second, each senor node of wireless network posses a battery which cannot be replaced or recharge, so all the batteries should be predominately managed effectively in order to increase the network life and also reduce consumption of energy in a WSNs.

Loose definition of clustering states that objects are re-organized into groups to form members of similar clusters in an appropriate way or some way. A cluster is defined as collection of similar objected grouped less than one and objects grouped under dissimilar belongs or grouped to another cluster.

The aim of clustering will determine the grouping intrinsic in a unlabeled set of data. Then we can decide about the good clustering. It has been noted that there is a absolute best criterion which will independently final the clustering aim in the network. As and then at the same time, usually there are numerous number of nodes in a WSN, the nodes only get a part of information in a topological network. Which results in need of clustering algorithms and also a need of appropriate cluster-sub in a local network topology based on partial information?

In WSN topology will be changing often based on changes which are done dynamically due to various reasons, including failure of battery, environmental destruction, node failure, location of change in sensor nodes, WSN nodes will communicate via wireless link alternation signal though a power node control or by the factors of environment, this can also addition of new node into the network which will enhance the accuracy of monitoring, etc.

WS networks should have the capability to adapt the variations and reconfiguration of the user task satisfactorily and dynamically. Likewise concerns in sensor networks will acquisition the data information regionally rather than node specific information, and thus the control mechanism topology of the existing will frustrate.

Based on energy constraints, node sensor will communicate directly with other nodes with a distance limit. In order to communicate with sensors and other communication range devices to form a sensor multi-hop network communication. Moreover nodes of groups will not only get energy saved but will reduce the network contention along with the nodes to be communication with a range of shortest path distance within a range of clusters and its cluster head.

Moreover, The nodes in the network are grouped under a cluster group, by grouping in the form of cluster, it will reduce energy consumer by saving the energy and also will reduce the contention in a network by which nodes are communicate between cluster based on the shortest distance based on the cluster heads.

Routing based on cluster in a network, which consists of various clusters, each of the cluster possess a cluster head (CH) and various type of members in the cluster (CMs).

Based on various limitations that occur in WS nodes, we build a node centralized cluster and also a method routing which will improve the energy efficiency in a WSN. Due to the limitation occurred a distributed process of cluster and its routing function will be normally used in wireless sensor networks. In case of routing in distributed clustering network, sensor nodes of each can reduce the consumption of energy.

Centralized structure in wireless sensor network is difficult to maintain as well as maintain the clustering was major issue. Distributed architecture is suitable for wireless sensor network due to decentralized network. With the help of distributed architecture model, can reduce the energy consumption even in high data transmission.

This has emerged from a range of bottom-up network of interactions, which will appear to be size of limitless, whereas there are top-down networks hierarchy which do not belong to self-organized. But they are typical network or organization related type, which is based on the limit size of nodes and the number of cluster required to the network is fixed. Here in this case, the cluster size should be properly adjusted, which can minimize the energy consumption in a network. The available cluster-based routing does not guarantee the size of cluster appropriately, because they do not have a localized sensor neighbor node to share the information.

As that CMs will transfer packets to the nearest CH, within the specific size of the cluster as if the cluster may be too long enough, as the consumption of energy may be higher which may consume more in a network. In addition the transmission experience of CH may not be delayed due to the causes of bottlenecks. Invariantly the size of the cluster may be too smaller or small, then its energy consumption may be decrease. Likewise, the consumption of energy in a network of cluster may be increase, if the size of cluster and number of clusters are increased. The energy consumption is measured with the range value of CH.

2 Related Works

In this part of work, we have identified certain pros and cons related to the routing protocols performance measure and are compared with the existing cluster based techniques.

2.1 LEACH

It is a micro sensor node network which uses a process of energy efficiency protocol for communication [10]. The major merits of this protocol is configuration is Ease, quality of network/improve the life of network and dissipation of energy.in a network.

But this type of protocol will consume very less energy and distribute the energy among clusters in a network.

This is also called as low consumption of energy cluster adaptive protocol and is called as self-organized cluster based method. This protocol [1] will decide the Cluster head randomly based on rotation, such as to distribute the energy required and consumed to the sensors of a network. Each of the node sensor will compared with the threshold value $T(n)$ which has numbers of random in electing a cluster head CH. The cluster heads will change periodically with time, such as to balance the node sensors energy consumption. The sensor nodes in a LEACH [2] will organize the cluster of its own, based on local communication and decision of local.

2.2 LEACH-ED

ED-LEACH is also a self-organized protocol of cluster based. The ED-LEACH [3] will decide the cluster heads on 2 types of thresholds. Initially, it gains the ratio value of the sensor nodes residual value of energy with that of sum of the energy consumed by whole of the network. Finally, or in the second case [8], it is the distance between the threshold value of some node sensor which should be less than the threshold distance, based on this only one node will act as a cluster head. The ED-LEACH [4] will improve the network life and balance on load in a network which is better compared to LEACH [11].

2.3 HEED

HEED [7] is a Hybrid based approach uses clustering method in balancing the energy efficiency. This method will decide the cluster head based [5] on two parameters. At first, sum of the energy of residual used by sensor nodes, then the second parameter calculates the cost of communication between intra clusters in solving the tie-breaks. The cost of intra cluster communication is as Cost of minimum degree (ii) cost of maximum degree and (iii) power ability of average and minimum cost. The HEED protocol will improve the network life longer than the LEACH protocol [12].

2.4 Clustering Based Data Collection algorithm

The clustering-based data collection algorithm focuses on the energy efficiency problem of clustering and prediction. In clustering, this algorithm uses dynamic splitting and merging clusters, in order to reduce the communication cost. However, these clustering [6], splitting, and merging methods are operated not for energy-efficient routing [9], but for using AR model-based similarity of features of CH and CMs.

To overcome the drawback of the existing system the proposed system is used. The proposed systems are intended to generate energy-efficient clustering of WSN. To reduce the energy consumption of CHs and CMs, EECSM adjusts the size of the clusters. When the size of a cluster is too large, EECSM splits the cluster into two clusters. On the other hand, when the size of a cluster is too small, EECSM merges the cluster into other clusters. Through splitting and merging clusters, EECSM prolongs the network lifetime our work has been organized into sections as follows. In Sect. 4 methodology of study has been proposed, the results are explained in Sect. 5, followed by conclusion in Sect. 6.

2.5 *Proposed System Used*

Efficiency energy self-organized clustering model with splitting and merging (EECSM).

3 Preliminaries

3.1 *Objectives*

The objective is for developing a self-organized efficient energy analytical clustering model which splits and merges the packet for reducing consumption of energy in CMs and CHs based on auto adjustment of cluster size. As the cluster size is larger, EECSM will split the cluster into two or more clusters. On the other, if the size of the cluster is too small, then EECSM will merge the cluster into another cluster. This processing of splitting and merging in EECSM will increase the life of the network.

3.2 *Model of Network*

The node sensors consume energy, while receiving and transmitting packets data in a WSN. In transmission of wireless data consumption of energy is correlated to the packet data size and the distance between the nodes of two sensors.

Extensive work of research has be done in the area of low-energy radios. With varying assumptions based on the characteristics of radio, including dissipation energy in receive and transmission nodes, which will change the merits of various protocols.

Assuming initially those sensor nodes can identify the energy of its own.

- Transmission of data packets through transmitter and amplifier
- Received data packets through lesser energy consumption
- Distance can be calculated

- Sensor nodes will receive the packets as well as send.

4 Energy-Efficient Self-organized Clustering with Splitting and Merging

EECSM consists of five different considerations which uses local data information. Initially first, node sensors which the most energy remaining from the neighbors will become the candidate node of CHs, since a lot of energy is consumed by CH. Second, the each node sensor, except CHs, will select the nearest CH, for reducing energy consumption in transmission data phase. Third, the clusters which has number of CMs and is less than the threshold of merging will merges into two or more clusters, this is due to minimizing the energy transmission of consumption that the data packets from the CHs to BS. Fourth part, the clusters which have number of CMs and which is larger than the threshold of splitting is splitting into clusters of two, this is done for reducing the CH energy overheads. Fifth, this will prevent breakage of CHs, a Backup CH mechanism will select a CH new which maximum the cluster energy in CH.

The mechanism process of EECSM will focus on clustering, splitting, and merging. To evaluate the performance of EECSM uses dynamic routing method between the clusters heads CHs. This means that CHs will directly transmit packet data and receive from its particular CMs to its BS.EECSM consist of three phases and a Backup CH Mechanism. The phase of clustering regroups the cluster in WSN. The phase of cluster merging will decide to merge clusters or not. The phase of data transmission will receives/sends data packets and the backup CH mechanism will elect the new CH as and when CH is broken down.

4.1 Clustering Phase

This phase begins as and when the sensor node fields are scattered around the network and is completely after the phase of data transmission. This is the phase which will decide the new CHs to be formed from new clusters in WSN. To gain efficiency in energy, the selection criterion is based on energy retained and energy remaining CH to that of CMs. To decrease the overhead of load in CHs, the EECSM will regulate the size of clusters (Fig. 1).

4.1.1 Step of Broadcasting

Initially the sensor nodes field is set to zero and are scattered in the network, CH is not available. Then selection of CHs is done among the scattered nodes automatically.

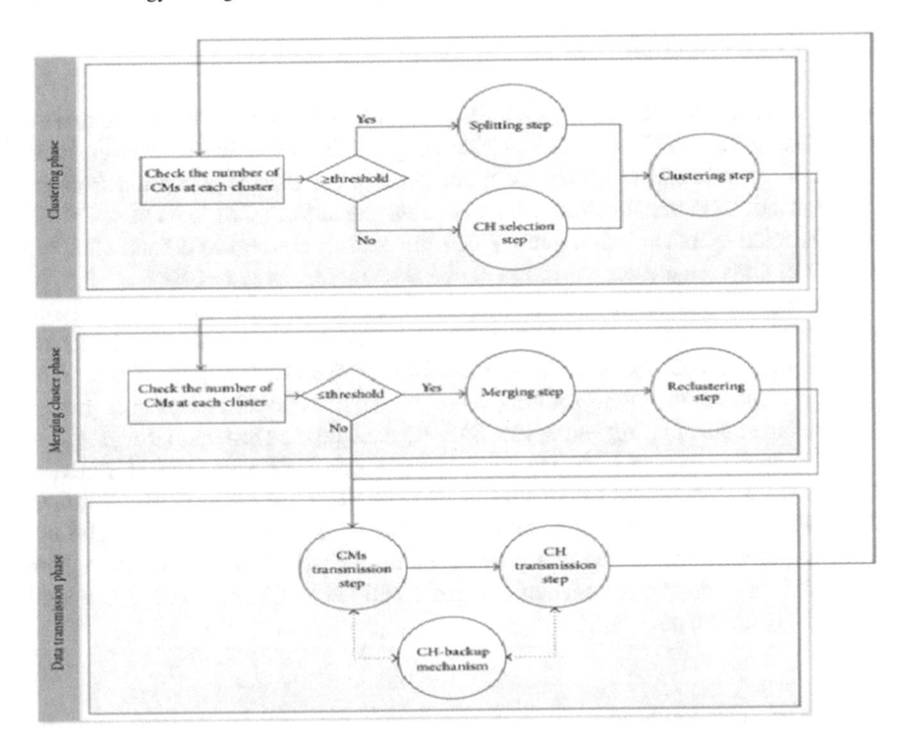

Fig. 1 Flow chart for EECSM

The EECSM then will use the energy to decide the next CHs and initial it as zero. Because all the CHs will be assigned with initial energy in the sensor nodes identically i.e., CHs values in all the sensor nodes will be initialized as period zero.

In order to avoid the situation of wastage, EECSM will select another CH method assuming zero period. Next we will decide the range of broadcast before explanation of selection of other CH methods. The range of broadcast is a rechargeable range of transmission of packets from the sensor nodes.

The selection of CH is initialized to zero for a period sense and also uses the neighboring nodes within a range by broadcasting the signal, in order to identify the correct CHs in the first round. The selection is based on the number of neighboring sensor nodes within a range of broad cast of CH, there should be a reduction of distance between CMs and CHs.

In order to identify and justify CHs, we broadcast range of energy signal to the neighboring nodes for just a period of zero.

4.1.2 Cluster Splitting Step

As and when clusters are configured, EECSM will consider and identify the related more clusters. Initially EECSM will divide a huge CH into small two or more clusters. The main issue is bigger cluster is, it consumes more energy utilization from its Cluster head. Data transmission can be communicated from CM to its near CH via from its relative transmission path. When the packets are received from CM, the number of CMs the cluster has in it will consume more energy in CH.

A bottleneck process can occur at CH in a big cluster network, in a transmission phase of the previous round of cluster, this is due to identify the number of clusters occur in the next round of clustering, EECSM will select next step process accordingly based on the threshold splitting value. As the CMs count of cluster is greater than the threshold splitting value, EECSM will execute the cluster splitting step.

Based on the count of CMs in a clustering network, if the value of CH is greater than the value of threshold split, then EECSM will execute the step cluster splitting. Where the cluster will be divided into clusters of two at the step of cluster splitting. A round process of next step will provide a cluster heads to the two cluster which are split from a cluster are recognized and identified as CH first and CH second respectively and so on.

4.1.3 CH Selection Step

The count of CMs in a cluster should be very much less than the threshold value of splitting, if the value is greater, EECSM will execute a process of splitting in a cluster step, in the other words, the step of cluster split step, will select only one CH based on the selection of one CH, the next round of cluster head will be decided based on the iteration step of above. Then a similar process of CH will decide the process of cluster splitting step.

4.1.4 Clustering Step

A cluster head will broadcast the packet signal to the cluster throughout the entire sensor nodes of clusters. The node sensors, of which its CHs node is expected, or the nodes which possess the states of discharge, or store, receive and the CH packet signal list.

Based on the reality that the information which is used can be available at the reclustering phase or step at the cluster merging phase, the stored EECSM will list out the temporally information, which will reduce the reclustering overheads. To minimize the consumption of energy of CMs from the phase of data transmission, the node sensor which is selected with the nearest cluster header CH of which assigning it as CH.

4.2 *Cluster Merging Phase*

As the current phase of clustering round, the EECSM will check the size of clusters, as the cluster size are larger in the previous phase, EECSM will split the clusters of larger size into smaller cluster for efficient consumption of energy, particularly as the packet data is transmitted from BS to the CH, as the consumption of energy is larger or much more from BS to CH than the energy consumption from CM to BS for larger transmission. For such a case smaller clusters is not much energy efficient as compared to larger cluster in energy efficient as so smaller cluster should be merged into larger clusters. The cluster merge phase on EECSM will be comprised into two steps as: cluster merging step and step reclustering as follows.

4.2.1 Step of Cluster Merging Cluster

As the cluster of CMs count is very less of its threshold merging value, the EECSM will execute the cluster merging step. Processes of combining small cluster in larger clusters are executed.

4.2.2 Step of Reclustering

This stage will execute the information stored in clustering step in the cluster phase; after this step is processed, the stored information in the clustering stage will be deleted.

4.3 *Data Transmission Phase*

After completion of cluster merging phase. EECSM will now enter into the transmission of data phase, it will start to intimate the sensor field node the situation that may occur to the BS external, by the process of data gathering. The phase of data-transmission is divided into two categories.

4.3.1 Transmission CM Step and Transmission CH Step

CM will create a data packet which contains the environment neighboring information for every period of time. The packets will be transmitted to CH. Each of the CH will aggregates in receiving packets data into packet data and will transmit the packet data directly to BS. This type of process is iterated for every clustering round.

4.4 Backup CH Mechanism

The WSNs are very useful in dangerous areas like disaster regions, military targets and environment hazardous, these networks protect human across danger. By using sensor nodes in these areas will cause breakdown of some individual nodes in a network. This breakdown will reduce the life of the network. Furthermore the breakdown of network will also affect the CH network lifetime and loss of information in a network. If there is a breakdown of CHs or fail of CH due to certain failures network loss and information received and transmitted from its CMs to the BS nodes will be lost.

In order to decrease the downfall of the network life, we have added in EECSM a backup CH mechanism. This mechanism will first, check the failure of CH break, the nearest CMs close to the CH will recognize the failure of their CHs.

The main reason why CMs can identify the failure of CH is that, CMs can identify whether the packet data is transmitted to BS from CH or not. This advantage will provide additional information to CMs in identifying the failure and overheads in the state of CH.

The backup CH mechanism in EECSM will decide the new CH, when there is a failure or breakdown of CH. This mechanism consists of two steps as follows: reelection CH step and Recovery of cluster Step.

4.4.1 Reelection CH Step

The reelection of CH step will be done immediately as there is a notification breakdown done by CM related to CH during the phase of data transmission. CM identifies a breakdown of CH of the nearest CM from CH. A broadcasting of energy signal is done by CMs twice in its range, for electing a CH new node. The main reason in broadcasting twice in the range of cluster is to reduce the consumption of energy of CMs. The CM which is having energy residual maximum will become the CH new in that range of WSN.

4.4.2 Recovery of Cluster Step

The recovered CH will broadcasts the signal to the whole of the network based on sensor field, because the new CH is not able to understand the cluster size exactly. The Cluster Method will decide the CH area not only for the existing clusters but also for the new CH cluster, based on the distance accordingly.

This will automatically minimize the consumption of energy related to CMs during the phase of data transmission. The backup mechanism in EECSM will restore a stable state in a network from unstable state. This is the basic feature of robustness in WSN.

5 Experimental Results

We have demonstrated experiments using simulation tool, our proposed algorithm performs effectively and superiorly when compared to the existing algorithms. Advantages of EECSM algorithms can be seen more and more as described below.

In this part, we show the experimental results of EECSM

(a) assuming the location of BS and sensor nodes as fixed
(b) Sensor nodes are deployed using random distribution
(c) The BS location is known before as 50 m × 50 m related to field sensor. As the field of sensor is set to 10 m for a broadcasting range.
(d) The packet signal size is 50 bits and packet data size is 1000 bits
(e) The initial energy for all the sensor nodes is 0.5 J
(f) Assuming that WSN nodes cannot be operated if the 30% of the nodes gets discharged.
(g) Assuming the count related to node sensor are 100 in our experiment.

Figure 2 shows the lifetime of network in EECSM is 167.9% (as and when first sensor node gets discharged) and 23.5% (when 30% of nodes of sensor get discharged in WSN), longer will be the life of the network in HEED. This may be assumed that five considerations in EECSM is used for energy information and operation properly to increase the life of the network.

Wireless Sensor Network which consist of numerous sensor nodes, routing protocols and clustering in WSNs has to posses' very high scalability. As EECSM is a self-organized modeling of clustering, it has very good scalability. The experiment shows that even increasing the sensor nodes will node decrease the scalability. The graph below shows a range of increase of sensor nodes from 100 to 500. Here the lifetime of network related to WSN.

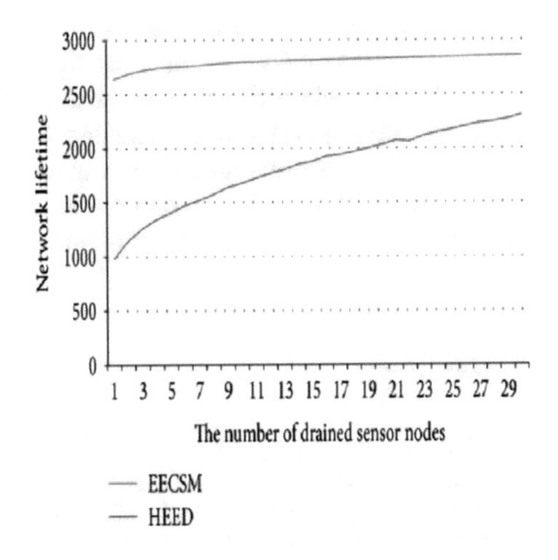

Fig. 2 Performance analysis of network lifetime

Fig. 3 Analysis of Performance in scalability based on lifetime of network

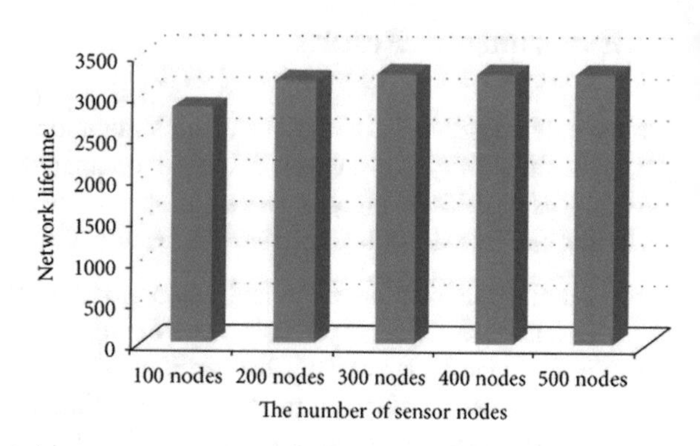

Figure 3, shows above the lifetime of network in EECSM, when the sensor nodes number is greater than 200, it s life will be longer than the lifetime in EECSM, As the sensor nodes number is equal to 100, from a minimum state of 11.67% and for a maximum state of 14.68%. Based on the increase of sensor node number, EECSM will tend to increase the life of the network without degrading the network life.

Based on the gained output, EECSM has high rate of scalability, based on merging and splitting method of EECSM will decrease the range of transmission and gain balance fair maintenance in consumption of energy.

6 Conclusion

Our work proposes an energy efficient self-organized cluster model in WSNs. Since EECSM uses self-organized method, this has good characteristics such as adaptability, distributed control, scalability, and robustness. In variantly EECSM will fix dynamically the cluster head CHs which is based on energy efficiency, which can resize the cluster and maintain its appropriate size, and further which can restore and damage the cluster of an own, and is based on its local information. Clustering with merging and splitting method has major interest in research. The basic research examples are auto decision methods of threshold merging and threshold splitting with information throughput, with neighbor localized information, ex: the no of neighbors, the actual state of neighbor and go on.

References

1. Al-Karaki, J.N., Kamal, A.E.: Routing techniques in wireless sensor networks: a survey. IEEE Wirel. Commun. **11**(6), 6–28 (2014)
2. Wang, Y.C., Peng, W.C., Tseng, Y.C.: Energy-balanced dispatch of mobile sensors in a hybrid wireless sensor network. IEEE Trans. Parallel Distrib. Syst. **21**(12), 1836–1850 (2015)

3. Jiang, H., Jin, S., Wang, C.: Prediction or not? An energy-efficient framework for clustering-based data collection in wireless sensor networks. IEEE Trans. Parallel Distrib. Syst. **22**(6), 1064–1071 (2014)
4. Dressler, F.: Self-organization in Sensor and Actor Networks. Wiley, New York (2015)
5. Prehofer, C., Bettstetter, C.: Self-organization in communication networks: principles and design paradigms. IEEE Commun. Mag. **43**(7), 78–85 (2015)
6. Heinzelman, W. R., Chandrakasan, A., Balakrishnan, H.: Energy-efficient communication protocol for wireless microsensor networks. In: Proceedings of the 33rd Annual Hawaii International Conference on System Siences (HICSS-33'00), January 2013
7. Sun, Y., Gu, X.: Clustering routing based ximizing lifetime for wireless sensor networks. Int. J. Distrib. Sens. Netw. **5**(1), 88–88 (2014)
8. Younis, O., Fahmy, S.: HEED: a hybrid, energy-efficient, distributed clustering approach for ad hoc sensor networks. IEEE Trans. Mob. Comput. **3**(4), 366–379 (2014)
9. Akkaya, K., Younis, M.: A survey on routing protocols for wireless sensor networks. Ad Hoc Netw. **3**(3), 325–349 (2015)
10. Younis, M., Youssef, M., Arisha, K.: Energy-aware management for cluster- based sensornetworks. Comput. Netw. **43**(5), 649–668 (2013)
11. Yick, J., Mukherjee, B., Ghosal, D.: Wireless sensor network survey. Comput. Netw. **52**(12), 2292–2330 (2013)
12. Mahfoudh, S., Minet, P.: Survey of energy efficient strategies in wireless ad hoc and sensor networks. In: Proceedings of the 7th International Conference on Networking (ICN '08), pp. 1–7, April 2014

V. Naga Gopi Raju pursuing Ph.D. in Department of Computer Science, Acharya Nagarjuna University, Guntur, India. His research area includes Sensor and computer networks.

Kolasani Ramchand H. Rao working as a Professor in Department of Computer Science, ASN Degree College, Tenali. His research area includes Sensor and computer networks.

Classification of Seamless Handoff Process in Wifi Network Based on Radios

Abhishek Majumder and Samir Nath

Abstract IEEE 802.11 wireless network is the widely used network which offers a good connectivity at low cost. A mobile node connects to the Access Point (AP) from which it receives the best signal. In present days, every user wants best service from the network domain. In IEEE wireless 802.11 network user change their position. This is the main causes of handoff. Real time applications should not experience any interruption when user travels to different network coverage. Handoff is the critical issue for continue network services in IEEE 802.11 networks. Different handoff process schemes are introduced for offering seamless handoff to the clients. In this paper, a classification of handoff schemes in 802.11 networks has been presented. The schemes have been classified based on number of radios used. The handoff schemes have also been discussed in brief. A comparative study of the schemes has been presented in this paper.

Keywords Handoff · Multi-radio based · Single-radio-based · Client

1 Introduction

Nowadays IEEE 802.11 wireless [1] networks is a popular platform for wireless communication and day by day its application is increasing in different places like airports, cities, shopping malls, etc. Because of mobility of the clients, the problem of handoff comes into the picture. Handoff is a critical issue in wireless networks. For offering QoS [2] in the IEEE 802.11 based networks seamless handoff is very important. Wireless networks are consists of several devices known as wireless Access Points (AP) or Base Station. Each of the AP is capable of providing

---

A. Majumder (✉) · S. Nath
Mobile Computing Laboratory, Department of Computer Science & Engineering, Tripura University, Suryamaninagar, Tripura, India
e-mail: abhi2012@gmail.com

S. Nath
e-mail: nathsamcse@gmail.com

S. Smys et al. (eds.), *International Conference on Computer Networks and Communication Technologies*, Lecture Notes on Data Engineering and Communications Technologies 15, https://doi.org/10.1007/978-981-10-8681-6_96

communications services to clients or mobile users in local or roaming area. Each AP can only serve up to a certain area and number of mobiles. So, when a mobile reaches out of the coverage area of any AP and enters other AP's coverage area, handoff [3, 4] is initiated. A handoff may also be triggered when signal strength of the serving AP becomes weak. In 802.11 wireless networks, there are two types of handoff process namely hard handoff and soft handoff. Hard handoff [5, 6] means "break before make" that means when a mobile moves from one coverage area to another, first it disconnects from current base station then connects to the new base station. Soft Handoff [7] means "Make before Break". When signal strength of the current base station is poor, the mobile client joins another base station which is serving good signal strength. In real scenario, soft handoff is better than hard handoff. Soft handoff is a critical issue in IEEE 802.11 based wireless network for real time applications such as audio, video streaming [8], etc. In this paper a classification of handoff schemes in wifi based network has been carried out. A comparison of the schemes has also been performed.

Section 2 presents a classification of handoff schemes in IEEE 802.11 networks considering number of radios used. It also discusses the handoff schemes in brief. In Sect. 3 a comparison of the schemes has been presented. Finally the paper is concluded in Sect. 4.

2 Classification

In IEEE 802.11 wireless network handoff process has been classified into two categories based number of radios used. The classification is shown in Fig. 1.

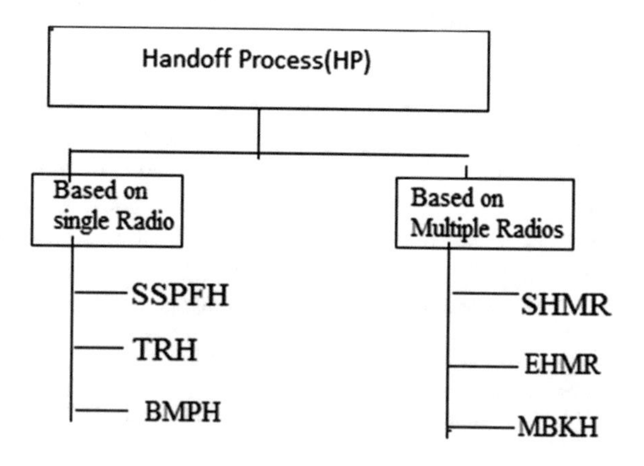

Fig. 1 Classification of handoff process

2.1 Multi-radio Based Handoff Process

In this type of handoff process, multiple radios have been used to minimize handoff latency. Some of the handoff schemes are, Seamless Handoff with multiple Radios (SHMR) scheme [9], Eliminating handoff latencies in 802.11 WLANs using Multiple Radios (EHMR) [10] and Make-Before-Break MAC Layer Handoff (MBKH) [11].

Seamless Handoff with Multiple Radios (SHMR) Scheme

Jin and Choi [9] proposed SHMR handoff scheme. Each radio stores number of reserve and operating channels information. One of the multiple radios is reserved for scanning purpose. Therefore, STA can scan the APs through the reserved channels. Each Mobile Station (STA) performs the Handoff operation in two ways:

Handoff via *reserve channel*

The process has the following steps, (a) STA sends NULL frame, i.e., a data frame to serving AP along with power saving mode (PM) bit set to 1, (b) After getting the data frame from STA, serving AP starts to buffer the data frames destined to STA, (c) STA then switches non-reserved channel to reserved channel for scanning the neighbor APs and send a null frame to serving AP for forwarding the buffered data frames with PM bit set to 0. (d) STA now broadcasts the probe request frames in the reserved channel. If no probe response found, it determines that there is no AP in the channel. (e) After receiving probe response frames from neighbor APs, STA selects the best AP. (f) Authentication and re-association is performed for the newly chosen AP, (g) After handoff is over, null frames with PM bit set to 0 and 1 are transmitted in the non-reserve and reserve channels respectively. (h) Now STA communicates with new AP for exchanging data frames.

Handoff perform stay in non-reserve channel

In this process STA stays in non-reserve channel and scans the APs for better signal compared to serving AP. The procedure has the following steps, (a) STA first transmits null frame to serving AP along with PM bit set to 1 and then serving AP buffers the data frames destined for the SAT, (b) STA switches to reserved channel and sends a null frame to serving AP. It sets PM bit to 0 that means it requests the serving AP to forward the buffered data frames, (c) STA broadcasts the probe request frame in the reserve channel for scanning the APs better than serving AP, (d) If the STA receives the probe response frames from APs through reserved channel but they are not better than serving AP, it returns to the non-reserve channel and continues data transmission.

Eliminating Handoff Latencies in 802.11 WLANs Using Multiple Radios (EHMR)

Brik et al. [10] proposed EHMR scheme which is also known as Multi-scan. This scheme uses two network interface cards namely primary interface and secondary interface cards on the same device. Multi-scan scheme is implemented on the client side.

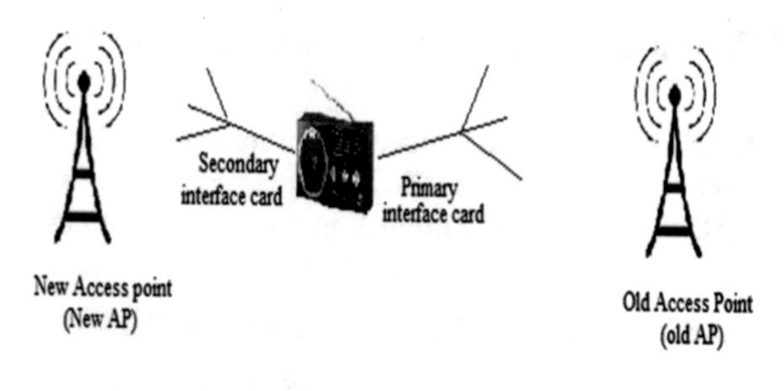

Fig. 2 Multiple radio handoff process in multi-scan

Fig. 3 Two card static algorithm

Step 1: Activity of control card
1.1 After every T_1 time interval
1.1.1 Probe all available channels;
1.1.2 Store information of neighbor APs;
Step 2: Activity of data card
2.1 Every T_2 time interval
2.1.1 Monitor channel quality;
2.1.2 If (channel quality degrades below threshold level)
2.1.2.1 Select best AP;
2.1.2.2 Disassociate with current AP;
2.1.2.3 Change channels;
2.1.2.4 Authenticate with new AP;
2.1.2.5 Associate with new AP;
Step 3: End

Multi-scan scheme has the following steps:

- Normal Operation: Primary and secondary interfaces are associated with old AP and new AP respectively. One is used for communication and other to perform scanning operation.
- Re-association: When signal strength degrades below threshold, secondary interface is used for association with new AP. Primary card will be still working for data communication with existing old AP.
- Interface Switch: When secondary interface gets associated with the new AP, all the data traffic destined to the node is sent through the secondary interface.
- Completion: Each interface card switches their job role after completion of handoff.

An example scenario is shown in Fig. 2.

Make-Before-Break MAC Layer Handoff (MBKH)

Ramachandran et al. proposed [11] MBKH. This scheme uses two algorithms namely: two card static algorithm and two card dynamic algorithm. Two card static algorithm and two card dynamic algorithm is shown in Figs. 3 and 4 respectively. Each algorithm uses two radio cards. One radio card is used for control operation and the other for data communication. For reducing authentication delay context cache [12] algorithm is used.

Fig. 4 Two card dynamic algorithm

```
Step 1: Activity of current control card
        1.1 Every T1 time interval
        1.1.1 Probe all 802.11 channels;
        1.2.2 Store information on neighbor APs;
Step 2: Activity of current data card
        2.1 Every T2 time interval
        2.1.1 Monitor channel quality;
        2.1.2 If (channel quality degrades)
          2.1.2.1 Select best AP;
          2.1.2.2 On current control channel
          2.1.2.2.1 Change channel;
          2.1.2.2.2 Authenticate with new AP;
          2.1.2.2.3 Associate with new AP;
          2.1.2.2.4 Assign current IP;
          2.1.2.2.5 Change routing table;
          2.1.2.3. Disassociate with current AP;
          2.1.2.4. Swap (current control channel, current data
        channel);
Step 3: End
```

2.2 *Single-Radio Based Handoff Process*

Seamless handoff problem can be solve or reduced using different schemes using single radio. Some schemes are: Sync Scan: Practical Fast Handoff (SSPFH) [13], Techniques to Reduce the IEEE 802.11b handoff time (TRH) [14], Behavior-Based Mobility Prediction for seamless handoffs in mobile wireless network (BMPH) [15].

Sync Scan: Practical Fast Handoff (SSPFH)

Ramani and Savage [13] proposed SSPFH scheme which manages the scanning operation in handoff process by maintaining the time synchronization. Number of APs working on channel 1 will broadcast beacon message at the same time period t. After d period of time, i.e., $(t+d)$ time APs working on channel 2 broadcast the beacon message same as channel 1 and APs working on channel 3 broadcast at time $(t+2d)$, and so on. So a client associated with an AP working on channel C can gather the knowledge of APs working on next channel, i.e., C + 1 channel by switching to C + 1 channel after d period of time when beacon message received from its own AP. In this way STA can get the information of all the APs working on different channels in its neighborhood. Security is maintain among new AP, old AP and STA applying inter access point protocol (IAPP) [16].

For getting the beacon messages from neighbor APs STA wait for some time. So, total delay can be calculated using Eq. 1.

$$SS_d = 2.ST + WT \tag{1}$$

Here, SS_d is the synchronous scanning delay. ST is the switching time, when STA t switches from current channel to next neighbor channel. WT is the waiting time. STA waits for beacon messages from APs working on that channel.

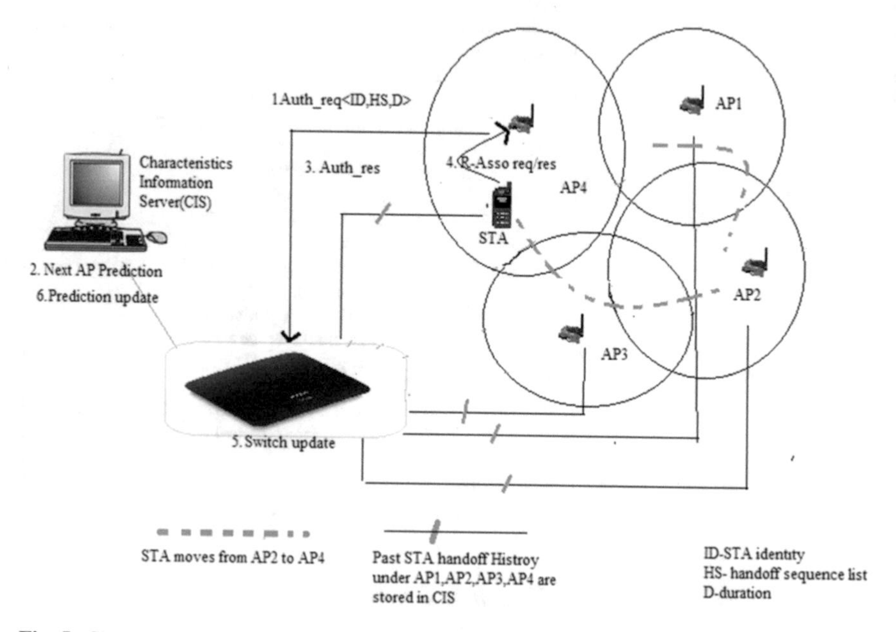

Fig. 5 Characteristics based mobility prediction for handoff

Techniques to Reduce the IEEE 802.11b Handoff Time (TRH)

Velayos and Karlsson proposed [14] TRH scheme focused on link layer handoff. The aim of this scheme is to reduce the duration of different stages faced during handoff process. Different stages are: detection, search and execution stage.

- Detection phase: Station can monitor the signal strength or any frame transmission error generated before handoff. When signal strength gets weak from current AP, transmission error also increases and neighbor APs' signal strength increase, station moves to search stage.
- Search phase: In this stage, station searches the available APs situated in the neighbor. Those APs can operate in any of the channel in the channel set available in IEEE 802.11 standard. In standard two scanning method are used namely passive scanning and active scanning. In TRH scheme only active scanning is applied. In active scanning two waiting time are focused, i.e., Min Channel Time (MCT) and Maximum Channel Time (MCTx). IEEE 802.11 standard does not specify MCT and MCTx value. Min Channel Time (MCT) is the waiting time of a station, if the channel is idle, i.e., no traffic available in the channel. Max Channel Time (MCTx) is the waiting time of a station, if traffics are available in the channel. Total search time is the sum of MCT value and MCTx value. This search time can be calculated using Eq. 2.

$$\text{Search time } (S_t) = C_u T_u + C_e T_e, \tag{2}$$

where C_u is the number of used channels and T_u is the scanning time of used channel. C_e is the number of empty channels and T_e is the scanning time of empty channel.
For discovering the neighbour APs operating in different channels the station broadcasts a probe request to the channel. For avoiding the collision, two consecutive probe requests are broadcasted. If the transmission delay is denoted by T_d, the scanning time of used channel and empty channel can be calculated as,

$$T_u = 2T_d + \text{MCT} \tag{3}$$

$$T_e = 2.T_d + \text{MCTx} \tag{4}$$

Therefore, total search time can be calculated using Eqs. 2 and 3 in Eq. 1. Search time S_t will increase if number of used channel is increased.

- Execution phase: In execution phase total handoff process will be completed. After completion of detection and search stage, station sends association message to the new AP. The AP replies with the confirmation message to the station.

Behavior-Based Mobility Prediction for Seamless Handoffs in Mobile Wireless Network (BMPH)

Wanalertlak et al. proposed [15] BMPH scheme. It is mainly focused on four characteristics of mobile user namely, location, group, time-of-day and duration. Past information of these characteristics is maintained by Characteristics Information Server (CIS).

- Location-based [17] mobility prediction: Mobility pattern of mobile users is either static or dynamic. Static mobility pattern means movement of users is fixed such as building, roads, etc. Dynamic pattern means users change their movement frequently based on their operating behavior. CIS stores handoff history of all the Mobile Stations (MSs) located in the network.
- Group-based mobility prediction: Group-based mobility prediction means collection of MSs having same characteristics behavior make a group. In a single group, all the MSs' mobility patter will be same.
- Time-of-Day-based mobility prediction: Time of Day (ToD) indicates the user's behavior changes as function of time. Different user performs different functions in different time. Let us consider in a particular session of time MSs are doing same function. Those MSs will be in a one ToD1. In this way the scheme creates a sector wise ToD based on MSs functioning behavior in different time of day.
- Duration-based mobility prediction: Duration-based mobility prediction makes the predicted group short-, long-, or medium based on their working duration. Duration represents how long a MS is present under particular AP. Short duration means MS stay in an AP for short time and then moves to another AP. Medium duration is the default duration time of MS doing a job in a particular AP. Long duration represents a MS performing some activity for long time under single AP.

Sequence of message transfer is shown in Fig. 5.

Table 1 Multi-radio based handoff process

Scheme	Properties	Advantages	Disadvantages
SHMR [9]	1. One radio of each AP works in the reserved channel. 2. STA scans the nearby channels in two modes i.e. scan at reserved channel another stay at non-reserved channel.	1. STA scans only the reserved channel. There is no need to scan other channels. 2. Significantly reduces packet loss. 3. Proposed scheme is feasible to be used in practical life.	1. The scheme uses two radios which is more costly. 2. The scheme takes long time for scanning process.
EHMR [10]	1. EHMR scheme is entirely implemented on client side. 2. Two radios are used for scanning and operating data communication.	1. In this scheme there is no need to change existing network. Change is required in the wireless client. 2. Two radio interface cards are used in a single device that work simultaneously. 3. Based on functionality two radios change their work behavior between them.	1. Cost increases because of using two radios.
MBKH [11]	1. MBKH scheme follows "Make Before Break" property. 2. The scheme focuses on layer-2 handoff in IEEE 802.11 based wireless network. 3. Two radio cards are used. They change their working role dynamically based on situation	1. Completely eliminates the handoff latency in layer-2 of IEEE 802.11 based wireless network. 2. For dynamic behavior no delay is incurred when a mobile moves from one place to another. 3. Frame or packet loss is also reduced drastically.	1. This scheme is difficult to implement.

3 Comparison

In this section the handoff processes have been compared. Tables 1 and 2 shows the comparison of multi-radio and single-radio-based handoff processes.

Table 2 Single-radio-based handoff process

Sl. no	Properties	Advantages	Disadvantages
1. SSPFH [13]	1. Time synchronization is greatly focused in this scheme. 2. All the APs belonging to a channel broadcast beacon message at a time. 3. Mobile station communicates with APs on a channel on every predefined interval.	1. It reduces handoff latency. 2. On every interval of time channels are scanned. Therefore, more time remains in hand before handoff. 3. Node can capture signal strength degradation by continuously monitoring the signal strength.	1. Maintaining a clock time is a big problem. 2. Interference may also occur during beacon broadcast. 3. An hidden cost is introduced by synscan. 4. This process is not dynamic.
2. TRH [14]	1. In this scheme total handoff process is divided into three phases. 2. Among three phases, the scheme focuses only on scanning process. 3. In this scheme MaxChannel and MinChannel waiting time is calculated.	1. Unnecessary waiting time has been minimized. 2. Reduces the search and detection time.	1. Search time increases if number of AP increases. 2. Null authentication algorithm is used.
3. BMPH [15]	1. Behavior of the mobile is used to predict the next handoff zone.	1. No need to perform full scan at the time of handoff. 2. Full scan can be performed at first time only. 3. Duration and season are also considered for making the behavior list.	1. If all the entries are failed in server, then the scheme again performs full scan and it is very time-consuming process.

4 Conclusion

In this paper different handoff processes for reducing handoff latency have been discussed. Most of the handoff time is spent for scanning the available channels in IEEE 802.11 based wireless network. The schemes for minimization of handoff delay have been classified based on number of radios used. Some of the schemes use multiple radios whereas some of them use single radio. In Multi-radio based schemes, handoff latency is minimized to zero, because two radio cards are used simultaneously. One card is continuously used for scanning operation and other card for data communication purpose. But commercial implementation of these schemes is very costly.

Single-radio-based schemes also minimized handoff latency. But these schemes are slow and take much longer scanning time. Finally the advantages and disadvantages of the schemes have been presented. Classification of schemes considering some more aspects remains as future work.

Acknowledgements The work is funded by Department of Electronics and Information Technology (DeitY), Ministry of Communications and Information Technology, Electronics niketan, 6 CGO Complex, Lodhi Road, New Delhi-110003, Government of India, Vide no. 14(8)/2014-CC&BT, Dated: 03.09.2015.

References

1. Crow, B.P., Widjaja, I., Kim, J.G., Sakai, P.T.: IEEE 802.11 wireless local area networks. IEEE Commun. Mag. **35**(9), 116–126 (1997)
2. Akyildiz, I., Altunbasak, Y., Fekri, F., Sivakumar, R.: AdaptNet: an adaptive protocol suite for the next-generation wireless internet. IEEE Commun. Mag. **42**(3), 128–136 (2004)
3. Zeng, Q.A., Agrawal, D.P.: Handoff in wireless mobile networks. In: Handbook of Wireless Networks and Mobile Computing, pp. 1–25. Wiley (2002)
4. Pack, S., Choi, J., Kwon, T., Choi, Y.: Fast-handoff support in IEEE 802.11 wireless networks. IEEE Commun. Surv. Tutorials **9**(1), 2–12 (2007)
5. Khan, R., Aissa, S., Despins, C.: MAC layer handoff algorithm for IEEE 802.11 wireless networks. In: IEEE Symposium on Computers and Communications, pp. 687–692 (2009)
6. Prakash, R., Veeravalli, V.V.: Adaptive hard handoff algorithms. IEEE J. Sel. Areas Commun. **18**(11), 2456–2464 (2000)
7. Kim, D.K., Sung, D.K.: Characterization of soft handoff in CDMA systems. IEEE Trans. Veh. Technol. **48**(4), 1195–1202 (1999)
8. Singh, G., Atwal, A.P.S., Sohis, B.S.: Mobility management technique for real time traffic in 802.11 networks. J. Comput. Sci. **3**(6), 390–398 (2007)
9. Jin, S., Choi, S.: A seamless handoff with multiple radios in IEEE 802.11 WLANs. IEEE Trans. Veh. Technol. **63**(3), 1408–1418 (2014)
10. Brik, V., Mishra, A., Banerjee, S.: Eliminating handoff latencies in 802.11 WLANs using multiple radios: Applications, experience, and evaluation. In: Proceedings of 5th ACM SIGCOMM Conference on Internet Measurement, pp. 27–27 (2014)
11. Ramachandran, K., Rangarajan, S., Lin, J.C.: Make-before-break mac layer handoff in 802.11 wireless networks. In: Proceedings of IEEE International Conference on Communications, vol. 10, pp. 4818–4823 (2006)
12. Mishra, A., Shin, M., Arbaush, W.A.: Context caching using neighbor graphs for fast handoffs in a wireless network. In: Proceedings of Twenty-third IEEE Joint Conference on Computer and Communications Societies, vol. 1 (2004)
13. Ramani, I., Savage, S.: SyncScan: Practical fast handoff for 802.11 infrastructure networks. In: Proceedings of 24th Annual Joint Conference on IEEE Computer and Communications Societies, vol. 1, pp. 675–684 (2005)
14. Velayos, H., Karlsson, G.: Techniques to reduce the IEEE 802.11 b handoff time. In: Proceedings of IEEE International Conference Communications. vol. 7, pp. 3844–3848 (2004)
15. Wanalertlak, W., Lee, B., Yu, C., Kim, Park, S.M., Kim, W.T.: Behavior-based mobility prediction for seamless handoffs in mobile wireless networks. Wirel. Netw. **17**(3), 645–658 (2011)
16. IEEE Computer Society LAN MAN standards Committee: IEEE. Recommended practice for multi-vendor access point interoperability via an inter-access point protocol across distribution systems supporting IEEE 802.11 operation. Jan 2002, IEEE Draft 802.1f/D3

17. Wanalertlak, W., Lee, B.: Global path-cache technique for fast handoffs in WLANs. In: Proceedings of 16th IEEE International Conference on Computer Communications and Networks, pp. 45–50 (2007)

Abhishek Majumer received his B.E. degree in Computer Science and Engineering from National Institute of Technology, Agartala and M.Tech. degree in Information Technology from Tezpur University, Assam in 2006 and 2008, respectively. He has received Ph.D. from the Department of Computer Science and Engineering, Assam University. He is currently working as an Assistant Professor in the Department of Computer Science and Engineering, Tripura University (A Central University), Suryamaninagar, India. His areas of interest are Wireless Network and Cloud Computing.

Samir Nath received his B.E. degree in Computer Science and Engineering from Tripura Institute of Technology, Agartala and M.Tech. degree in Computer Science and Engineering from Tripura University, Suryamaninagar. He is currently working as Junior Research Fellow in a project funded by Ministry of Electronics and Communication Technology, Government of India.

Author Index

S. Smys et al. (eds.), *International Conference on Computer Networks and Communication Technologies*, Lecture Notes on Data Engineering and Communications Technologies 15, https://doi.org/10.1007/978-981-10-8681-6

Printed in the United States
By Bookmasters